AF689160

Tissue Engineering

Second Edition

Tissue Engineering

Second Edition

Editors

Clemens A. van Blitterswijk and Jan de Boer

Associate Editors

Ali Khademhosseini, Matthias Lutolf,
Arnaud Scherberich, Jan Schrooten
and Molly Stevens

Editorial Assistant

Laura Zendman

ELSEVIER

AMSTERDAM • BOSTON • HEIDELBERG • LONDON
NEW YORK • OXFORD • PARIS • SAN DIEGO • SAN FRANCISCO
SINGAPORE • SYDNEY • TOKYO
Academic Press is an imprint of Elsevier

AP

Academic Press is an imprint of Elsevier
32 Jamestown Road, London NW1 7BY, UK
525 B Street, Suite 1800, San Diego, CA 92101-4495, USA
225 Wyman Street, Waltham, MA 02451, USA
The Boulevard, Langford Lane, Kidlington, Oxford OX5 1GB, UK

Copyright © 2015 Elsevier Inc. All rights reserved.

No part of this publication may be reproduced or transmitted in any form or by any means, electronic or mechanical, including photocopying, recording, or any information storage and retrieval system, without permission in writing from the publisher. Details on how to seek permission, further information about the Publisher's permissions policies and our arrangements with organizations such as the Copyright Clearance Center and the Copyright Licensing Agency, can be found at our website: www.elsevier.com/permissions.

This book and the individual contributions contained in it are protected under copyright by the Publisher (other than as may be noted herein).

Notices
Knowledge and best practice in this field are constantly changing. As new research and experience broaden our understanding, changes in research methods, professional practices, or medical treatment may become necessary.

Practitioners and researchers must always rely on their own experience and knowledge in evaluating and using any information, methods, compounds, or experiments described herein. In using such information or methods they should be mindful of their own safety and the safety of others, including parties for whom they have a professional responsibility.

To the fullest extent of the law, neither the Publisher nor the authors, contributors, or editors, assume any liability for any injury and/or damage to persons or property as a matter of products liability, negligence or otherwise, or from any use or operation of any methods, products, instructions, or ideas contained in the material herein.

ISBN: 978-0-12-420145-3

British Library Cataloguing-in-Publication Data
A catalogue record for this book is available from the British Library

Library of Congress Cataloging-in-Publication Data
A catalog record for this book is available from the Library of Congress

For information on all Academic Press publications
visit our website at http://store.elsevier.com/

Typeset by TNQ Books and Journals
www.tnq.co.in

Working together
to grow libraries in
developing countries

www.elsevier.com • www.bookaid.org

Contents

Contributors

M. Adelaide Asnaghi Departments of Surgery and of Biomedicine, University Hospital Basel, Basel, Switzerland

Jean-Marie Aerts Prometheus, Division of Skeletal Tissue Engineering, KU Leuven, Leuven, Belgium; Division M3-BIORES: Measure, Model & Manage Bioresponses, KU Leuven, Heverlee, Belgium

Enateri V. Alakpa Centre for Cell Engineering, Institute of Molecular, Cell and Systems Biology, College of Medical, Veterinary and Life Sciences, Joseph Black Building, University of Glasgow, Glasgow, Scotland, UK

Morgan R. Alexander Laboratory of Biophysics and Surface Analysis, University of Nottingham, Nottingham, UK

Mauro Alini Musculoskeletal Regeneration, AO Research Institute Davos, Davos, Switzerland

Jessica J. Alm Center for Hematology and Regenerative Medicine and Division of Clinical Immunology and Transfusion Medicine, Department of Laboratory Medicine, Karolinska Institutet, Stockholm, Sweden

Anthony Atala Department of Urology, Wake Forest University School of Medicine, Winston–Salem, NC, USA

Stephen F. Badylak McGowan Institute for Regenerative Medicine, University of Pittsburgh, Pittsburgh, PA, USA; Department of Surgery, University of Pittsburgh, Pittsburgh, PA, USA; Department of Bioengineering, University of Pittsburgh, Pittsburgh, PA, USA

Florence Barrère-de Groot Xpand Biotechnology BV, 3723 MB Bilthoven, The Netherlands

Cameron Black Bone and Joint Research Group, Centre for Human Development, Stem Cells and Regeneration, Institute of Developmental Sciences, University of Southampton, Southampton, United Kingdom

Mats Brittberg Department of Orthopaedics, Institute of Clinical Sciences, The Sahlgrenska Academy University of Gothenburg, Sweden

Chaenyung Cha School of Materials Science and Engineering, Ulsan National Institute of Science and Technology, Ulsan, Korea

Smadar Cohen The Avram and Stella Goldstein-Goren Department of Biotechnology Engineering, Ben-Gurion University of the Negev, Beer-Sheva, Israel; The Center for Regenerative Medicine and Stem Cell (RMSC) Research, Ben-Gurion University of the Negev, Beer-Sheva, Israel; The Ilse Katz Institute for Nanoscale Science and Technology, Ben-Gurion University of the Negev, Beer-Sheva, Israel

Matthew J. Dalby Centre for Cell Engineering, Institute of Molecular, Cell and Systems Biology, College of Medical, Veterinary and Life Sciences, Joseph Black Building, University of Glasgow, Glasgow, Scotland, UK

Paul D. Dalton Department of Functional Materials in Medicine and Dentistry, University of Würzburg, Germany

Noel L. Davison MIRA Institute for Biomedical Technology and Technical Medicine, and Department of Biomaterials Science and Technology, University of Twente, Enschede, The Netherlands; Xpand Biotechnology BV, 3723 MB Bilthoven, The Netherlands

Jonathan I. Dawson Bone and Joint Research Group, Centre for Human Development, Stem Cells and Regeneration, Institute of Developmental Sciences, University of Southampton, Southampton, United Kingdom

Jan de Boer Department for Cell Biology-inspired Tissue Engineering, Merlin Institute, Maastricht University, Maastricht, The Netherlands

Marco C. DeRuiter Department of Clinical and Experimental Anatomy, Leiden University Medical Centre, Leiden, The Netherlands

Adam J. Engler Department of Bioengineering, University of California, San Diego, USA

Liesbet Geris Prometheus, Division of Skeletal Tissue Engineering, KU Leuven, Leuven, Belgium; Biomechanics Research Unit, Universite de Liège, Liège, Belgium

Susan Gibbs Department of Dermatology, VU Medical Centre, Amsterdam, The Netherlands

David Gibbs Bone and Joint Research Group, Centre for Human Development, Stem Cells and Regeneration, Institute of Developmental Sciences, University of Southampton, Southampton, United Kingdom

Adriana C. Gittenberger-de Groot Department of Anatomy and Embryology, Leiden University Medical Centre, Leiden, The Netherlands

Dirk W. Grijpma MIRA Institute for Biomedical Technology and Technical Medicine, and Department of Biomaterials Science and Technology, University of Twente, Enschede, The Netherlands; Department of Biomedical Engineering, University of Groningen, University Medical Center Groningen, Groningen, The Netherlands

Chungmin Han Department of Mechanical Engineering, Pohang University of Science and Technology, Pohang, Gyeongbuk, Republic of Korea

Alan R. Harvey School of Anatomy, Physiology and Human Biology, The University of Western Australia, Perth, WA, Australia

Marietta Herrmann Musculoskeletal Regeneration, AO Research Institute Davos, Davos, Switzerland

Beerend P. Hierck Department of Anatomy and Embryology, Leiden University Medical Centre, Leiden, The Netherlands

Andrew L. Hook Laboratory of Biophysics and Surface Analysis, University of Nottingham, Nottingham, UK

Jeffrey A. Hubbell Institute of Bioengineering, Ecole Polytechnique Fédérale de Lausanne, CH-1015 Lausanne, Switzerland

Dietmar W. Hutmacher Institute of Health and Biomedical Innovation, Queensland University of Technology, Brisbane, Australia

Johan Joly Division of Rheumatology, UZ Leuven, Leuven, Belgium

Janos Kanczler Bone and Joint Research Group, Centre for Human Development, Stem Cells and Regeneration, Institute of Developmental Sciences, University of Southampton, Southampton, United Kingdom

Marcel Karperien Department of Developmental BioEngineering, University of Twente, Enschede, The Netherlands

Candace Kerr Department of Biochemistry and Molecular Biology, University of Maryland School of Medicine, Baltimore, MD, USA

Ali Khademhosseini Harvard-MIT Division of Health Sciences and Technology, Brigham and Women's Hospital, Harvard Medical School, Cambridge, MA, USA

Joseph M. Labuz Department of Biomedical Engineering, College of Engineering, University of Michigan, Ann Arbor, MI, USA; Biointerfaces Institute, University of Michigan, NCRC, MI, USA

Toon Lambrechts Prometheus, Division of Skeletal Tissue Engineering, KU Leuven, Leuven, Belgium; Division M3-BIORES: Measure, Model & Manage Bioresponses, KU Leuven, Heverlee, Belgium

Vanessa L.S. LaPointe Department for Cell Biology-inspired Tissue Engineering, Merlin Institute, Maastricht University, Maastricht, The Netherlands

Matthias W. Laschke Institute for Clinical & Experimental Surgery, University of Saarland, Homburg, Saarland, Germany

Katarina Le Blanc Center for Hematology and Regenerative Medicine and Division of Clinical Immunology and Transfusion Medicine, Department of Laboratory Medicine, Karolinska Institutet, Stockholm, Sweden

Shulamit Levenberg Department of Biomedical Engineering, Technion - Israel Institute of Technology, Haifa, Israel

Anders Lindahl Department of Clinical Chemistry and Transfusion Medicine Institute of Biomedicine, The Sahlgrenska Academy University of Gothenburg, Sweden

Ricardo Londono McGowan Institute for Regenerative Medicine, University of Pittsburgh, Pittsburgh, PA, USA; School of Medicine, University of Pittsburgh, Pittsburgh, PA, USA

Frank P. Luyten Division of Rheumatology, UZ Leuven, Leuven, Belgium; Skeletal Biology & Engineering Research Center, KU Leuven, Leuven, Belgium

Athanasios Mantalaris Biological Systems Engineering Laboratory, Centre for Process Systems Engineering, Department of Chemical Engineering, Imperial College London, London, UK

Ivan Martin Departments of Surgery and of Biomedicine, University Hospital Basel, Basel, Switzerland

Mikaël M. Martino World Premier International Immunology Frontier Research Center, Osaka University, Suita, Osaka 565-0871, Japan

Laura E. McNamara Centre for Cell Engineering, Institute of Molecular, Cell and Systems Biology, College of Medical, Veterinary and Life Sciences, Joseph Black Building, University of Glasgow, Glasgow, Scotland, UK

Dies Meijer Centre for Neuroregeneration, University of Edinburgh, Edinburgh, UK

Lorenzo Moroni Department of Complex Tissue Regeneration, MERLN Institute for Technology-Inspired Regenerative Medicine, Maastricht University, Maastricht, The Netherlands

Simon Myers Centre for Cutaneous Research, Blizard Institute, Barts and The London School of Medicine and Dentistry, Queen Mary, University of London, United Kingdom

Harshad Navsaria Centre for Cutaneous Research, Blizard Institute, Barts and The London School of Medicine and Dentistry, Queen Mary, University of London, United Kingdom

Nkemcho Ojeh Faculty of Medical Sciences, The University of the West Indies, Cave Hill, Bridgetown, Barbados

Richard OC. Oreffo Bone and Joint Research Group, Centre for Human Development, Stem Cells and Regeneration, Institute of Developmental Sciences, University of Southampton, Southampton, United Kingdom

Martin Oudega Departments of Physical Medicine and Rehabilitation, Bioengineering, and Neurobiology, University of Pittsburgh, Pittsburgh, PA, USA

Ioannis Papantoniou Prometheus, Division of Skeletal Tissue Engineering, KU Leuven, Leuven, Belgium; Skeletal Biology and Engineering Research Center, KU Leuven, Leuven, Belgium

Jaesung Park Department of Mechanical Engineering, Pohang University of Science and Technology, Pohang, Gyeongbuk, Republic of Korea; School of Interdisciplinary Bioscience and Bioengineering, Pohang University of Science and Technology, Pohang, Gyeongbuk, Republic of Korea

Francesco Piraino Institute of Bioengineering, École Polytechnique Fédérale de Lausanne, Lausanne, Switzerland

Mark F. Pittenger Department of Surgery, University of Maryland School of Medicine, Baltimore, MD, USA

Giles W. Plant Department of Neurosurgery, Stanford University, Stanford, CA, USA

Rob E. Poelmann Department of Anatomy and Embryology, Leiden University Medical Centre, Leiden, The Netherlands

Hong Qian Center for Hematology and Regenerative Medicine, Department of Medicine, Karolinska Institute, Stockholm, Sweden

Ilona Reischl Division clinical trials, Institute Surveillance, Federal Office for Safety in Health Care/Austrian Agency for Health and Food Safety, Vienna, Austria

Jeffrey J. Rice Department of Chemical Engineering, Tennessee Technological University, Cookeville, Tennessee, USA

Bernard A.J. Roelen Department of Farm Animal Health, Faculty of Veterinary Medicine, Utrecht University, Utrecht, The Netherlands

Jeroen Rouwkema Laboratory of Biomechanical Engineering, MIRA Institute for Biomedical Technology and Technical Medicine, University of Twente, Enschede, The Netherlands

Emil Ruvinov The Avram and Stella Goldstein-Goren Department of Biotechnology Engineering, Ben-Gurion University of the Negev, Beer-Sheva, Israel

Jason M. Sandberg Department of Urology, Wake Forest University School of Medicine, Winston–Salem, NC, USA

Arnaud Scherberich Laboratory of Tissue Engineering, Department of Biomedicine, University Hospital of Basel, Basel, Switzerland

Jan Schrooten Prometheus, Division of Skeletal Tissue Engineering, KU Leuven, Leuven, Belgium; Department of Metallurgy and Materials Engineering, KU Leuven, Leuven, Belgium

Evan A. Scott Department of Biomedical Engineering, Northwestern University, Evanston, Illinois, USA

Yulia Shandalov Department of Biomedical Engineering, Technion - Israel Institute of Technology, Haifa, Israel

Brian M. Sicari McGowan Institute for Regenerative Medicine, University of Pittsburgh, Pittsburgh, PA, USA; Department of Surgery, University of Pittsburgh, Pittsburgh, PA, USA

Timothy Smith Octane Medical Group, Kingston, Ontario, Canada

Jérôme Sohier Laboratory of Tissular Biology and Therapeutic Engineering - UMR 5305, Institute of Biology and Chemistry of Proteins, Centre National de la Recherche Scientifique, Lyon, France

Maarten Sonnaert Prometheus, Division of Skeletal Tissue Engineering, KU Leuven, Leuven, Belgium; Department of Metallurgy and Materials Engineering, KU Leuven, Leuven, Belgium

Shuichi Takayama Department of Biomedical Engineering, College of Engineering, University of Michigan, Ann Arbor, MI, USA; Biointerfaces Institute, University of Michigan, NCRC, MI, USA; Macromolecular Science and Engineering Center, College of Engineering, University of Michigan, Ann Arbor, MI, USA

Rahul Tare Bone and Joint Research Group, Centre for Human Development, Stem Cells and Regeneration, Institute of Developmental Sciences, University of Southampton, Southampton, United Kingdom

Roman Truckenmüller Department of Complex Tissue Regeneration, MERLN Institute for Technology-Inspired Regenerative Medicine, Maastricht University, Maastricht, The Netherlands

Penelope M. Tsimbouri Centre for Cell Engineering, Institute of Molecular, Cell and Systems Biology, College of Medical, Veterinary and Life Sciences, Joseph Black Building, University of Glasgow, Glasgow, Scotland, UK

Lesley-Anne Turner Centre for Cell Engineering, Institute of Molecular, Cell and Systems Biology, College of Medical, Veterinary and Life Sciences, Joseph Black Building, University of Glasgow, Glasgow, Scotland, UK

Dieter Van Assche Division of Rheumatology, UZ Leuven, Leuven, Belgium; Department of Rehabilitation Sciences, Faculty of Kinesiology and Rehabilitation Sciences, KU Leuven, Leuven, Belgium

Clemens A. van Blitterswijk Department of Complex Tissue Regeneration, MERLN Institute for Technology-Inspired Regenerative Medicine, Maastricht University, Maastricht, The Netherlands

Johan Vanlauwe Faculty of Physical Education and Physical Therapy, Faculty of Medicine and Pharmacy, VU Brussels and Department of Orthopedics & Traumatology, UZ Brussel, Brussels, Belgium

Sophie Verrier Musculoskeletal Regeneration, AO Research Institute Davos, Davos, Switzerland

Stellan Welin Department of Culture and Communication, Linköping University, Linköping, Sweden

David Wendt Departments of Surgery and of Biomedicine, University Hospital Basel, Basel, Switzerland

Dave A. Winkler CSIRO Material Science & Engineering, Parkville, Australia

Tim B.F. Woodfield Department of Orthopaedic Surgery and Centre for Bioengineering & Nanomedicine, University of Otago, Christchurch, Christchurch, New Zealand

Preface

Lectori salutem!

Based on the success of the first edition of *Tissue Engineering*, we, editors and authors, are proud to present a revised book which will be the basis of education for a new generation of tissue engineers, edited by a new generation of co-editors but in the same *meme* presented by the editors of the first edition. This book brings science into engineering and engineering into science because *building* tissues is something else than dissecting them, and building *tissues* is different from building a bridge. That is why tissue engineers should familiarize themselves with both sides of the coin, not only to apply both disciplines themselves (as biomedical engineers do) but also to be able to collaborate. A stem cell biologist should know the basics of material sciences when an organ-on-a-chip device is produced for him or her by a microfabrication expert. Clinicians should know about rapid prototyping when introducing three-dimensional printing into the clinic.

To structure the book and its content, we have devised a grid with two axes: from small to large and from basic sciences to applied. Each chapter has its own unique place on this grid and teachers can use it as a tool to explain the actors in the field. Those familiar with the first edition will recognize some changes. As editors, we believe that technology gains an increasing role in tissue engineering and many tissue engineering projects have reached maturity in the past years. Therefore, we decided to put more emphasis on the technology behind tissue engineering and more chapters on translational aspects of tissue engineering. Nevertheless, many chapters remained the same, although updated.

We hope that the students will enjoy reading and learning from this book, as we did in writing. Remember that behind every chapter, there are real scientists who are performing new experiments every day to generate the *Tissue Engineering* content of future editions. We all invite you to participate shaping the future!

Jan de Boer
Weiswampach

CHAPTER 1

Tissue Engineering: An Introduction

Lorenzo Moroni[1], Jan Schrooten[2], Roman Truckenmüller[1], Jeroen Rouwkema[3], Jérôme Sohier[4], Clemens A. van Blitterswijk[1]

[1]Department of Complex Tissue Regeneration, MERLN Institute for Technology-Inspired Regenerative Medicine, Maastricht University, Maastricht, The Netherlands; [2]Division of Skeletal Tissue Engineering, KU Leuven, Leuven, Belgium; [3]Laboratory of Biomechanical Engineering, MIRA Institute for Biomedical Technology and Technical Medicine, University of Twente, Enschede, The Netherlands; [4]Laboratory of Tissular Biology and Therapeutic Engineering _UMR 5305, Institute of Biology and Chemistry of Proteins, Centre National de la Recherche Scientifique, Lyon, France

In 1997, media all over the world were aroused by a BBC documentary, *Tomorrow's World*, showing what is now known as the Vacanti mouse (Cao et al., 1997). The term "tissue engineering (TE)" was no longer seen as an expression familiar only to a limited number of scientists working in the field—it had become well-known to millions of individuals worldwide. Although the Vacanti experiment (Box 1.1) is truly exemplary for the discipline of tissue engineering, it is fair to say that the media upheaval was not so much caused by the actual experiment but even more by the spectacular sight of a nude mouse that had apparently grown a human ear on its back. This was not the first media hype on tissue or organ repair and most certainly will not be the last. It would not be difficult to dedicate an entire chapter in this textbook to the promising aspects of our discipline that made it to the media and had a major public impact. Whereas in the first edition of this book, it would have been difficult to fill an entire chapter with actual clinical successes, in this second edition we are witness to an increasing number of tissue engineering and regenerative medicine strategies successfully translated to patients. Yet, the eagerness of the media to report on the advances in the field of tissue engineering may give still the opposite impression. This is not so much caused by publicity of eager scientists; the actual cause is the enormous demand in our society for technologies that are able to repair, or even better regenerate, damaged or worn out tissues and/or organs.

On an annual basis, many millions of patients undergo surgery for tissue reconstruction. A fair proportion are treated satisfactorily, another portion less effectively, and millions still await treatments that would help them at least in an acceptable way. A now almost classical analysis on the need and

Tissue Engineering.
Copyright © 2015 Elsevier Inc. All rights reserved.

commercial opportunity was published in *Science* in 1993 by Langer and Vacanti, the abstract of which is given below.

> The loss or failure of an organ or tissue is one of the most frequent, devastating, and costly problems in human health care. A new field, tissue engineering, applies the principles of biology and engineering to the development of functional substitutes for damaged tissue. This article discusses the foundations and challenges of this interdisciplinary field and its attempts to provide solutions to tissue creation and repair.
>
> **Langer and Vacanti, 1993 (p. 920)**

If the need is indeed so high, and the field has so extensively grown over the last decades, then why do we still have to improve upon the frequency of clinical successes? There are two major reasons for this phenomenon. First, bringing a relatively simple medical device from initial idea to a widespread clinical reality frequently takes a minimum of 10 years. As the underlying technology for developing tissue engineering products is less mature, and possibly more complex, it is to be expected that clinical progress in this field will be measured in decades rather than years. Second, tissue engineering is truly an interdisciplinary field where acquired knowledge from individual classical disciplines (e.g., quantum physics, polymer chemistry, molecular biology, anatomy) no longer suffices to make substantial leaps. Individuals active in this field will have to acquire multidisciplinary skills and be willing to look over the borders of their home discipline. The relatively young age of the field does not make it easy to acquire those skills as dedicated textbooks are still scarce and frequently do not address the appropriate audience by either offering a collection of research papers or by only dealing with a selected part of the entire discipline. Without the widespread availability of such textbooks, it is to be feared that the rapidly increasing number of graduate courses on tissue engineering may not be as effective as required, which may hamper the development of the field of tissue engineering into a mature scientific discipline.

With this in mind, in 2004 it was decided that it was time to bring together a representative group of internationally active scientists who would be willing to contribute to a textbook, which would address both the multidisciplinary nature of tissue engineering and the underlying base disciplines. From that time, the field has considerably matured. Tissue engineering witnessed not only more clinical successes, but also new discoveries in the basic sciences and new technologies and methods in the applied sciences converging into it. With respect to the first edition, the structure of the book has slightly changed. We still move from fundamental to translational research, but we have also introduced a scale level where topics will be framed from the small (e.g., small molecules) to the large scale (e.g., tissues and organs). This is framed in each chapter by a matrix identifying at which research and scale level the topic is located (Figure 1.1).

The rationale for this matrix originates from the intrinsic, interdisciplinary complexity of tissue engineering, the dynamics of the field, and also from the

	CHAPTER MATRIX	Scales of life			
		Molecules / elements	Cells / materials	Tissue / organ	Clinic/industry
From understanding to process	Biology				
	Chemistry - physics - mechanics				
	Tools from in-silico to in-vivo				
	Engineer - process				
	Integrate - scale up				

FIGURE 1.1
Diagram sketching the layout of the book.

belief of the authors that this matrix can serve as a map for students to enter the world of tissue engineering.

Tissue engineering is more than the sum of its ingredients. Their combinations create novel dimensions, insights, and questions that need to be addressed. Hence, a matrix-like chapter build-up represents better the interdisciplinary and translational road that tissue engineering has to travel, by showing logical links between all building blocks. The authors envision that this approach could turn this book into a handbook that can guide the next decade of research, bringing tissue engineering into the clinic and educating real tissue engineers.

The word "matrix" itself is already a good example of the new dimension that tissue engineering is creating. Matrix is an established term that is recognized both from the biology and from the engineering sides. For the former it represents a scale of tissue formation, an in vivo environment, and for the latter it is a tool that enables to develop a research strategy that builds on a multitude of variables and parameters. When these two definitions can be merged through the "matrix map" we have introduced, it would be a first sign that we have created a common language and that we are on the right track to engineer cell biology.

By using this "matrix map," the reader will first be confronted with the fundamentals of the cell type that makes tissue engineering truly a part of regenerative medicine: the stem cell. **Stem Cells** (Chapter 2) gives insight into the different aspects of this cell and will show that the stem cell is not a single cell type but, in reality, encompasses different categories of cells ranging from the multipotent embryonic stem cell to the apparently less potent adult stem cell. With this knowledge, the reader will subsequently be guided to **Tissue Formation during Embryogenesis** (Chapter 3), where the formation of tissues in the embryo is discussed. Although most tissue engineers do not venture into the realms of developmental biology and research with a strong focus on developmental biology is just emerging in tissue engineering, we feel this should be strengthened in the future and hope that this chapter will urge more and more young scientists to enter into such an essential area, as it will allow them to better understand the mechanisms behind tissue and organ development, and implement them in new tissue engineering strategies. Obviously most tissue engineering constructs will not be implanted into embryos but into human beings after birth. Tissues in both the embryo and in an individual after birth usually, although not always, contain multiple cell types. Understanding how these cells interact is recognized as pivotal for the success of tissue engineering. **Cellular Signaling** (Chapter 4) provides such insights. Furthermore, in spite of all media attention that befalls stem cells, in reality cells represent only a small part of the dry weight of living tissue. All, or at least most, cells interact with an extracellular matrix (ECM), which, in contrast to the errant opinion of some engineers and even biologists, presents much more than mechanical support and adds substantially to the biological interactions in our body. As tissue engineering typically combines scaffolds with biologically active components such as cells or growth factors, we felt that a chapter on the biological equivalent the **Extracellular Matrix as Bioscaffold in Tissue Engineering** (Chapter 5) could not be missed. As many components of the ECM are considered as biomaterial sources, in this chapter we also introduce how the ECM itself can be used as a scaffold for tissue engineering.

At this point in the book a shift is made to the more fundamental engineering aspects. Most tissue engineers actively involved in the design of both synthetic and natural scaffolds or matrices prefer to have these degrade after implantation. This is with good cause as the prolonged presence of foreign material in the body may induce a variety of unwanted effects such as implant-associated infection or mutagenesis. The chapter on **Degradation of Biomaterials** (Chapter 6) discusses aspects of implant degradation, providing a deeper insight into the interactions between materials and a living system; ceramic and polymer degradation is introduced, particularly in relation to their function and the release of degradation products. It is obvious that by changing the chemical, physical, and mechanical bulk or surface properties of biomaterials, cell–material interactions would be consequently affected. The influence of biomaterial properties at the interface between cells and biomaterials is discussed in Chapter 7, entitled **Cell–Material Interactions**. These changes could be infinite, thereby posing the challenge to develop new tools and assays to explore more efficiently how to engineer the cell–material interface. The chapter entitled **Materiomics: A Toolkit for Developing New Biomaterials** (Chapter 8) describes technologies and methods recently developed to implement high-throughput screening of cell–material interactions and the relevance of this strategy for tissue engineering. Although some applications would involve direct injection of cells into the tissues that one intends to regenerate, most researchers active in the field of tissue engineering, and working with cultured cells, strive for implanting combinations of cells and scaffolds. These so-called hybrid constructs will usually have three dimensions and need to give access to both cells and nutrients or have to allow an out-flux of active ingredients and waste products. The chapters **Microfabrication Technology in Tissue Engineering** (Chapter 9) and **Scaffold Design and Fabrication** (Chapter 10) present the reader with different technologies and ways to fabricate scaffolds and explain the criteria these scaffolds have to fulfill. The design of scaffolds may also involve incorporating biologically active ingredients that can then either be made available to the cells in the construct or be released to initiate effects in the tissues of the organism into which the device was implanted. The various aspects of such release systems are presented in the chapter on **Controlled Release Strategies in Tissue Engineering** (Chapter 11). Ideally the researcher working on a hybrid construct and following the chapters, as they have been presented so far, now faces one of the main obstacles of tissue engineering: how to initiate and monitor cell growth and ECM formation in three-dimensional (3D) constructs of clinically relevant dimensions (usually centimeters versus millimeters in many in vitro and animal studies). These challenges, fundamentals, and some solutions are discussed in the chapter entitled **Bioreactors: Enabling Technologies for Research and Manufacturing** (Chapter 12).

After considering the above, more fundamental, aspects, the text now gradually moves to knowledge that bears direct practical relevance for the actual process of tissue engineering. In addition, chapters start to increasingly show the multidisciplinary nature of the field, moving more and more from the fundamental to the translational level mentioned earlier. As already explained in the chapter on Stem Cells (Chapter 2), there are different cells that can be used for generating tissues. Chapter 13, entitled **Clinical Grade Production of Mesenchymal Stromal cells,** elaborates further on this subject, taking human bone marrow mesenchymal stromal cells as an example, and touches on the methods used to isolate, expand, characterize, and preserve these cells. Particular attention is placed when performing these steps to obtain clinical-grade cells. A further challenge for the success of tissue-engineered constructs in the clinics is the strategy applied to vascularize these constructs. Typically, this can be obtained in two ways, either by promoting in situ vascular ingrowth from the surrounding tissues or by engineering a vascular network in vitro in the tissue constructs. These strategies are further discussed in the context of improving the integration of tissue engineering constructs with the host tissue in the chapter entitled **Vascularization, Survival, and Functionality of Tissue-Engineered Constructs** (Chapter 14).

Although not complete or covering all of the subjects for which one could argue would be essential, the editors feel that these first 14 chapters provide a solid basis for students who wish to acquire an understanding for tissue engineering.

Much of the international research effort based on tissue engineering is along the lines presented in these chapters. But it is also fair to say that some of the major challenges are still to be found into making tissue engineering a widespread clinical reality. These clinical aspects are treated in the remaining nine chapters.

In making a choice on which clinical applications to discuss, the editors faced a difficult task. Researchers worldwide are currently focusing their attention on the repair or regeneration of many tissues and organs. Almost certainly most, if not all, of them deserve to be presented in this book, but this would lead to an unacceptably thick volume or a lack of sufficient depth per chapter. In selecting the subjects, we felt that those tissues that historically moved first into the tissue engineering arena, and have actually found relatively widespread application, certainly deserve to be dealt with here. We feel, therefore, that most would agree that **Skin Engineering and Keratinocyte Stem Cell Therapy** (Chapter 15) and **Cartilage and Bone Regeneration** (Chapter 16) chapters warrant addition to this book on the basis of such historical observations. Clinical relevance could be another selection criterion. Neural and cardiovascular diseases would be an example of this approach. Not only are these relevant research areas, but the prospect that one day tissue engineering might contribute to relieving the problems of those suffering from neural or cardiovascular lesions is truly inspiring—hence a chapter on **Tissue Engineering of the Nervous System** (Chapter 17) and a chapter on **Principles of Cardiovascular Tissue Engineering** (Chapter 18). Any scientist, of some reputation, working in the field of tissue engineering will be frequently confronted with a question like: *When will you grow a complete kidney or liver?* Substituting whole organs is a challenge, if only since it forces us to combine multiple tissues into one system. In a sense one might say that tissue engineering can be compared to say a house where the elements provided (i.e., windows, doors, etc.) need to fit together to create a home. This convinced us to contribute a chapter on **Tissue Engineering of Organ Systems** (Chapter 19). In the exquisite essence of the multidisciplinary character of tissue engineering, a chapter on the emerging topic of **Organs-on-a-Chip** (Chapter 20) could not be missed. Here, the combination of different technological platforms with biological and medical fundamentals results in the creation of in vitro 3D tissue equivalents able to recapitulate key functions of targeted organs. Despite the fact that the resulting tissues are not necessarily used in the classical conception of tissue engineering, they have become a reality where tissue engineering principles serve the scope of creating a better mimicking 3D model of a targeted organ functionality than conventional two-dimensional cultures to understand pathological mechanisms and find possible treatments. Furthermore, this emerging technology holds the potential to drastically reduce animal experimentation, thus potentially offering a solution to the ethical concerns associated to it.

There are some areas that may not directly involve actually generating a tissue or organ substitute, but are still pivotal to the overall development and acceptance of the field. Therefore, the editors decided to add three more chapters on quality controls, guidelines, and regulations needed to bring tissue engineering products to the clinics. Ultimately, tissue engineering constructs have to be placed into a human recipient. This puts major responsibility on the researchers and clinicians who are actively involved in this process. Naturally, a minimum requirement would be that the construct is safe and may not harm the patient; preferably it would be more effective than existing treatments. Furthermore, constructs should be produced in a reproducible and traceable manner, so that along the process every step can be monitored and controlled to ensure the same high quality. Principles behind these processes are described in the chapter entitled **Product and Process Design: Toward Industrial TE Manufacturing** (Chapter 21). Practical and regulatory aspects that are needed for the final clinical translation of tissue engineering products are then discussed in **Clinical Translation** (Chapter 22). Apart from this, an important issue for any new technology, which is frequently underestimated by scientists, concerns the ethical aspects of such technologies. It requires little imagination to see that tissue engineering would certainly rank among those disciplines most prone to ethical scrutiny. Not only should we be sincerely interested in the ethical aspects of our field, but we should thoroughly realize that public acceptance, and the ethical debate related to that, will be an important determinant for the widespread application of the technology we develop. In view of this, a final chapter on **Ethical Issues in Tissue Engineering** (Chapter 23) closes the book. A careful reader may have noticed that this chapter does not start by giving a definition on tissue engineering, but a schematic overview. This was a deliberate choice. Very often students are confronted with definitions which they are expected to accept as the ultimate truth. In reality, any definition has both supporters and opponents, and this is certainly true in the field of tissue engineering. Therefore, instead of expecting everybody to agree to the same set of definitions, we have endeavored to encourage the reader to use the definitions in which he or she believes.

It is the hope of the editors that the chapters that have been included in this book will provide a solid basis in tissue engineering. Apart from explaining what the purpose of this textbook is, and how we tried to achieve this goal, this introduction chapter might further be useful in that it should exceed the purpose of the other individual chapters. While each of the remaining chapters focuses on a particular element of tissue engineering, they are not specifically dedicated to giving an overall view of the field. More particularly, all those areas where developments take place are both exciting and crucial for the further development of our discipline.

In treating some of these aspects it may become clear as to where the field is moving toward. Not only will it reveal that, indeed, in the years to come tissue engineering may likely contribute to the well-being of millions, which gives justification to the field being there at all, but it will also provide something else: *Tissue Engineering is fun!* This is a subject frequently forgotten but not justifiably so. It is relevant that researchers not only find justification in their efforts leading to the benefits of society, but have fun on the way as well. After all, a subject like tissue engineering is quite adventurous and usually involves combining knowledge of different disciplines into one research project. The complexity of such combinations makes the field intellectually challenging and the possibilities are such that there is ample space for individual efforts still to be recognized. The fact that a successful experiment may bring the technology closer to clinical reality subsequently provides further rewards. That this opinion is also shared by those not professionally active in our field was illustrated by a *Time* Magazine article, around the turn of the century, ranking the "Tissue Engineer" among the hottest future professions of the twenty-first century (*Time* Magazine online, http://www.time.com/time/magazine/article/0,9171,997028,00.html).

Let us look back at some of the early statements at the beginning of this chapter. Many people active in the field feel we are progressing too slowly and wonder if we have not been overpromising. We should realize that comments on the slow progress of the field can only concern the apparent lack of clinical success. Science is progressing and at a rapid rate. In this perspective, it might be useful to look again at the Vacanti mouse (Cao et al., 1997; Box 1.1).

BOX 1.1 THE MOUSE AND THE EAR: TISSUE-ENGINEERED CARTILAGE IN THE SHAPE OF A HUMAN EAR

Although there have been several studies since the 1980s where the concepts of tissue engineering were first applied, a formal formulation of the discipline is traced back to the paper of Langer and Vacanti (1993) in *Science*. Since then, the number of studies in tissue engineering has grown rapidly. A landmark study published in 1997 in *Plastic and Reconstructive Surgery* by Cao et al. (1997) attracted the interest of a large audience, thanks also to a BBC service on the subject. In this paper, it is shown how to successfully regenerate the cartilaginous part of a 3-year-old child's ear. The work is useful also from an educational point of view, as most of the "ingredients" to perform a classic tissue engineering experiment are also included.

To highlight the multidisciplinarity of tissue engineering, the study can be divided into three parts:

- **Material Science**: First, a plaster mold of the ear of a 3-year-old child was cast from an alginate impression of the ear. Then, a 100 μm thick nonwoven mesh of poly(glycolic acid) (PGA) was immersed in a 1% weight/volume solution of poly(lactic acid) (PLA) in methylene chloride for 2 seconds and subsequently shaped into the plaster mold.
- **Biology**: Calf chondrocytes (cartilage cells) were harvested from the articular surfaces of a calf, isolated by collagenase digestion, filtered, washed in cell medium, labeled with BrdU to control their viability, and seeded onto the polymeric scaffolds (1.5×10^8 cells in total). The constructs were cultured in vitro for 1 week in an incubator under physiological conditions ($T = 37\,°C$; $p_{CO_2} = 5\%$) and then implanted into a subcutaneous pocket on the back of athymic mice. Three groups of scaffolds were considered: (1) scaffolds seeded with cells; (2) scaffolds reinforced with an external stent and seeded with cells; and (3) externally stented scaffolds with no cells.
- **Biochemistry**: After 12 weeks, the constructs were explanted, sectioned, and histologically stained with specific markers for typical cartilage ECM components (hematoxylin and eosin for general tissue formation; Alcian blue for glycosaminoglycan deposition; Masson's trichrome for collagen formation). Immunohistochemistry was also performed to confirm that present collagen was specific for cartilage (type II).

The results showed extensive cartilage formation in the scaffolds that were seeded with cells, while no cartilage was present in the unseeded scaffolds. Furthermore, scaffolds reinforced with an external stent for the first 4 weeks of implantation maintained

BOX 1.1 THE MOUSE AND THE EAR: TISSUE-ENGINEERED CARTILAGE IN THE SHAPE OF A HUMAN EAR *Continued*

the anatomical shape of the ear. In contrast, the other scaffolds lost their integrity partially and appeared of reduced size and distorted shape. From these findings the scientists concluded that cartilage formation is not mature enough in the first 4 weeks to counteract the contraction forces in the healing process.

This experiment was surely a success for those years and definitely contributed to boost the interest in the field. If we consider the state of the art nowadays, however, a number of drawbacks still characterize this study, some of which are mentioned here:

- Skin coverage is missing and is a critical element of any ear reconstruction.
- Bovine immature (young) chondrocytes were used, while clinical application thereof is, of course, highly unlikely and these cells are now known to be partially unrepresentative for the use of human cartilage cell sources. Such human sources tend to lose differentiation capacity quite fast and are furthermore frequently characterized by necrosis in the center of scaffolds with a clinically relevant size.
- An athymic or immunodeficient mouse model is used here. Obviously, this is nothing more than a useful screening model and large animal models are required to test clinical relevance in the presence of a functional immune system.
- Scaffolds need to provide an adequate mechanical stability to the construct at the time of implantation.
- Implications on the growth rate of the artificial ear compared to the growth rate of a 3-year-old child should be addressed before the final implantation in the patient.

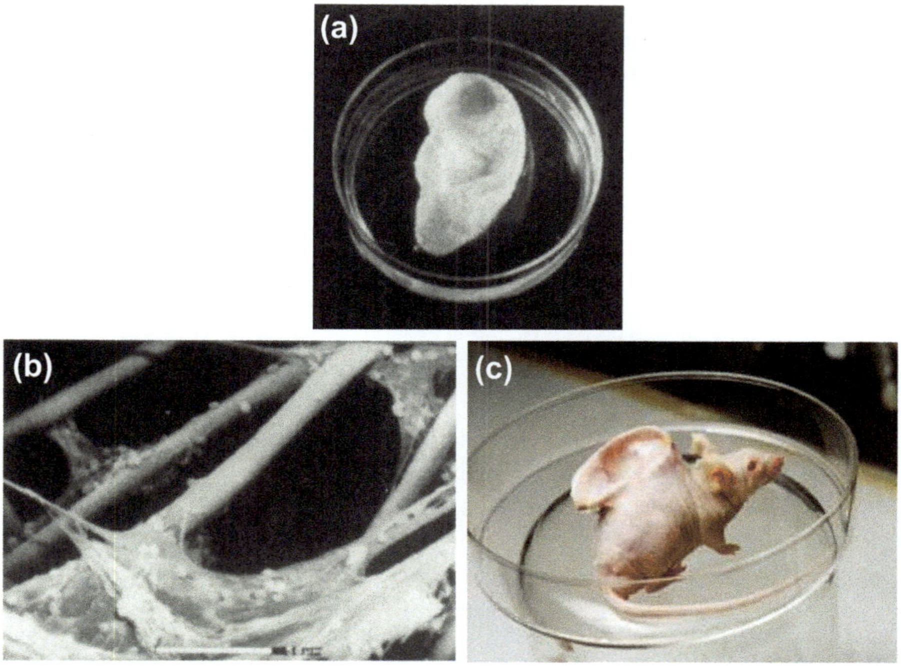

Ear anatomically shaped scaffold. (a) Scaffold structure before seeding; (b) scanning electron microscopy micrograph showing cells and ECM formed in the scaffold; (c) scaffolds implanted subcutaneously on the back of an immunodeficient mouse. *Reproduced with permission from Cao et al. (1997).*

At the time this was a state-of-the-art experiment. A cast of a human ear formed the fundamental for a porous nonwoven poly(glycolic acid) (PGA)/poly(lactic acid) (PLA) scaffold which was subsequently filled with cultured cartilage cells. Then the hybrid construct was placed under the skin of an immunodeficient mouse. Irrespective of the outcome of the specific Vacanti experiment, we now know that such an approach would, in general, present major problems. One reason being that implants of a clinically relevant size in general do not allow optimal nutrient transport to the cells in the center of such scaffolds, as illustrated in Figure 1.2. After implantation this lack of nutrients is directly related to poor and late ingrowth of blood vessels into the scaffold. At the time, this was a fundamental obstacle and even today it still presents a significant challenge. However, we are progressing. To illustrate this we should introduce the Levenberg mouse (Levenberg et al., 2005) (Box 1.2)—less well-known than that of Vacanti, the Levenberg mouse is perhaps even more interesting. The concept of this study was that a carefully generated hybrid construct of which the pores were filled with cultured tissue still faces a major risk after implantation. During culture nutrients can be forced through the scaffold by processes such as agitation or perfusion, but this usually is no longer the case after implantation. In most cases the scaffold will be dependent on nutrients coming from blood vessels. In an average natural tissue, cells will typically find such a source of nutrients (a capillary) within roughly 200 microns. As an implant of clinically relevant size may

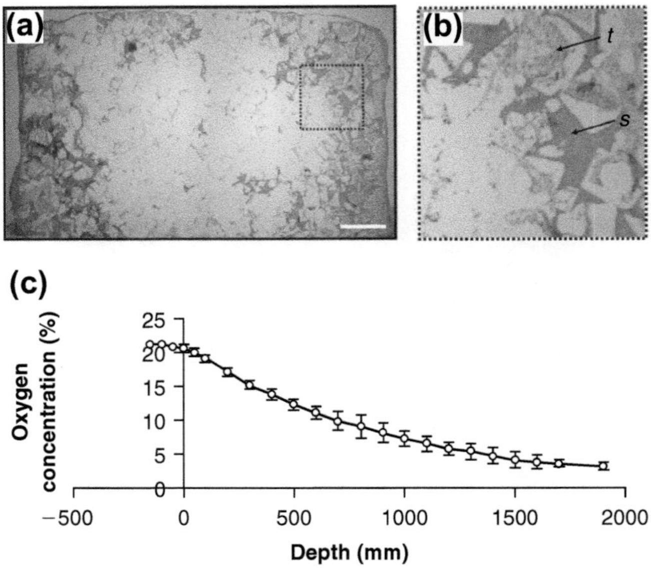

FIGURE 1.2
Insufficient nutrition transport is one of the main problems that tissue engineers have to face (scale bar = 1 mm).

BOX 1.2 CELLS NUTRITION: ENGINEERING VESSELS

One of the major limitations of tissue engineering is the inability to provide a sufficient blood supply in the initial phase after implantation. As long as a proper vascular network has not been established, the cells in the implanted tissue have to rely on diffusion for the supply of nutrients and the removal of waste products. This can lead to nutrient limitations, which can result in improper integration or even death of the implant. Therefore, in order to grow tissues on a clinically relevant scale, this problem has to be dealt with.

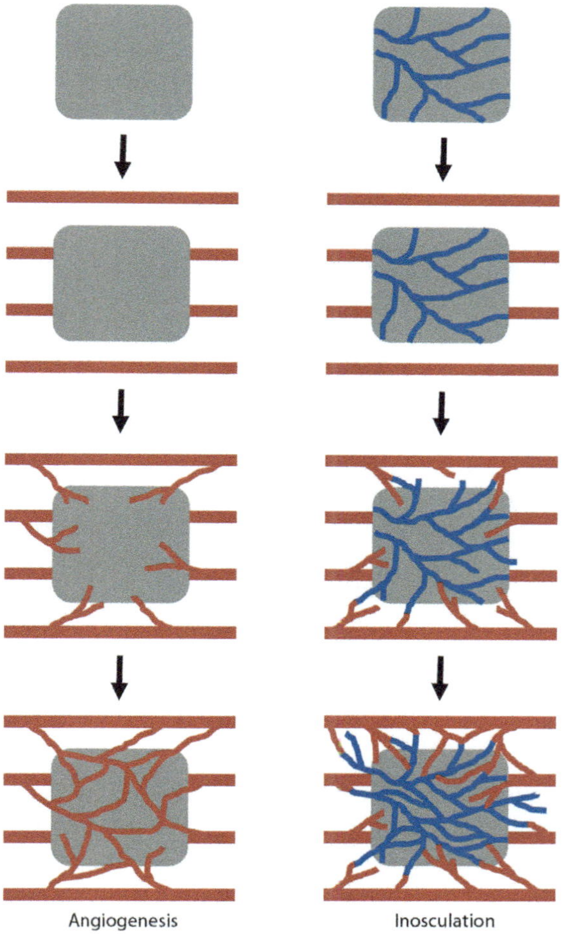

Schematic representation of the two principal vascularization strategies in tissue engineering. The angiogenesis approach is characterized by the ingrowth of vascular structures from the host vasculature (red vessels) into an implanted tissue construct, which finally form a new microvascular network. In case of the inosculation approach, a preformed microvascular network (blue vessels) is generated within a tissue construct prior to its implantation. After implantation, the preformed microvessels have to develop interconnections (inosculation) to the host vasculature to become functional and perfused with blood. *Reproduced with permission from Laschke and Menger (2012).*

Continued...

BOX 1.2 CELLS NUTRITION: ENGINEERING VESSELS *Continued*

Enhanced vascularization of an implanted tissue can be achieved by the stimulation of angiogenesis, or by the inosculation of preformed microvascular networks within the implants to the host microvasculature. Even though the stimulation of angiogenesis, for instance by the inclusion of angiogenic growth factors or an open scaffold design that allows for the unobstructed ingrowth of vascular structures, can shorten the time needed to vascularize an implanted tissue, it is clear that strategies based on inosculation possess the best potential to fully perfuse an implanted tissue with blood shortly after implantation.

An example of an approach that is based on inosculation is offered by a study where endothelial cells (HUVEC) were added to engineered muscle tissue, which lead to the formation of a vascular network inside the muscle tissue in vitro. After implantation, the muscle tissue integrated properly with the surrounding tissue, indicating that the added vascular structures did not have a negative effect on the differentiation of the muscle tissue. Moreover, the prevascular network inosculated with the host vasculature after implantation and became functional vessels transporting blood. This resulted in enhanced vascularization and survival of the implant.

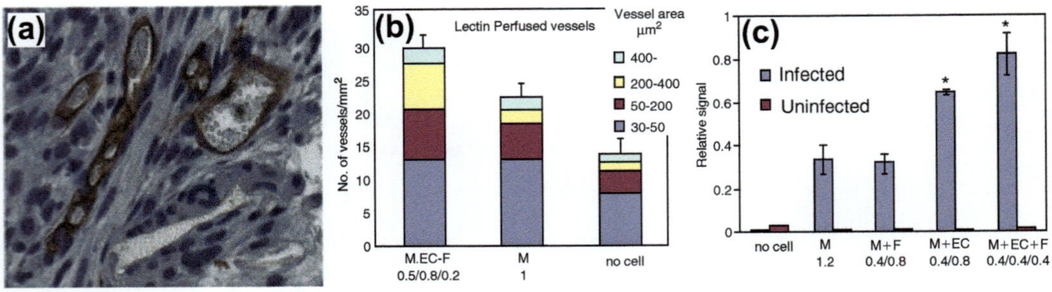

In vivo results after implanting a prevascularized muscle construct for 2 weeks. (a) Human vascular structures (brown, human-specific CD31 labeling) have connected to the host vasculature and carry blood. (b) Quantification of the amount of perfused vessels after 2 weeks of implantation. Note that prevascularized samples (M + EC + F) contain more perfused vessels. (c) Survival of cells in the implants determined using a luciferase assay. Note that prevascularized samples (M + EC and M + EC + F) display a better survival of the cells. *Figures (b) and (c) reproduced with permission from Levenberg et al. (2005).*

measure in centimeters, it becomes clear that most cells in such a scaffold may quickly starve to death by lack of vascular supply. After all, it may take a week before blood vessels have sufficiently penetrated a hybrid construct after implantation. A week is really a long time, as surgeons who have to operate on patients whose bowel was constricted for only a day can tell us, as much of the tissue may have become necrotic by then. Levenberg had an appealing idea: rather than waiting for spontaneous vascular ingrowth after implantation, she would provide a cultured vascular network prior to implantation. If Vacanti, with his mouse, provided the walls, taps, and sanitary equipment in a house, Levenberg would now provide the plumbing to make it all work and connect it to the water supply and sewer system. In an editorial this was described as a landmark paper (Jain et al., 2005), from which several other strategies have been originated and resulted in advancements in vascularization strategies of engineered tissues. Of these, a few have now progressed toward clinical assessment. This all fits in with the earlier remark that it easily takes a decade to get from early concept to clinical reality. The next example will show that it may take even longer than a single decade before a concept is translated into a clinical and commercial success.

Urist (1965) pioneered in the field of bone morphogenetic proteins (BMPs) in the 1960s and 1970s. Urist had discovered that the demineralized fraction of bone contains a protein fraction that is able to induce bone formation in a non-species specific way, even in nonbony sites. He first hypothesized that the bone formation was induced by a noncollagenous protein acting as bone morphogen. Obviously, this was a finding of great importance and researchers worldwide started to further investigate the phenomenon of bone induction. It was not until 15 years later that the BMPs were isolated and their existence definitely proven by Sampath and Reddi (1981, 1983). Nevertheless, their role in osteoinduction mechanisms has been debated—and is still today—although it has been clearly demonstrated over the last 25 years that BMPs alone could induce bone formation in nonbony sites of many different mammals. The lack of consensus around the BMPs action further delayed the development of BMPs as a potential tool for bone regeneration and it was not until the beginning of the 1990s that industries and researchers really saw the interest of the protein and went on the way to develop and commercialize it for clinical use in fracture, nonunion, and spinal fusion treatment. During the 10 years necessary for the protein to be approved by the regulatory bodies, it became the most extensively studied orthobiologic product in history. Researchers would focus on how to translate their findings into clinical practice by optimizing, standardizing, and industrializing the processes. In parallel, quality and regulatory protocols had to be developed as well. Finally, in 2001, BMP-7 was approved for spinal fusions and in 2002, BMP-2 followed. The protein is mixed with collagen sponges that allow its delivery to the site of surgery, from which it will induce bone formation and fusion of two vertebras. The treatment of tibial fractures by BMP-2 was further approved in 2004. Since then, the number of patients successfully treated with BMPs has not ceased to increase, making it one of the most successful commercial and healthcare successes in the field of orthopedics and in the field of tissue engineering. Overall, it took 30 years from the concept of BMPs expressed by Urist to the successful commercialization of a beneficial product. Interestingly, the availability and subsequent sales of the product now reveals that the market size for bone replacement might well be larger than ever expected, stimulating researchers in both industry and academia to try to do a better and, hopefully, faster job than those who went before them.

The example above clearly illustrates the difficulties that tissue engineers have to face. As much fundamental knowledge is still lacking, they have to conjointly investigate and unravel the basic mechanisms taking place during tissue homeostasis and regeneration, and at the same time develop the tools to exploit and enhance these natural mechanisms. This dichotomy becomes evident when one considers the range of matrices used for tissue engineering applications.

One of the first tissue engineering paradigms was the use of supportive matrices that provide associated cells with a substrate appropriate for attachment, multiplication, and differentiation toward the desired tissue. Further than the necessary degradability of such a substrate and eventual remodeling, the exact requirements of the matrix were not known with regard to cell interaction and response. Therefore, as a first step, many different matrices were evaluated to determine the most prone to replace the cell environment, from natural or synthetic origin. Collagen sponges made from purified bovine collagen fibers and scaffolds from poly(α-hydroxy) acids such as poly(lactic-*co*-glycolic) acids (PLGA) appeared suitable at first. The former is still widely in use and provides the cells with a nature-derived matrix familiar to them. However, the non-optimal control and reproducibility of the properties of this material combined with the potential transmission of diseases raises concerns for clinical applications. Polymeric scaffolds such as from PLGA, on the contrary, can be produced in a variety of reproducible shapes and properties such as degradation and mechanical properties while being free of any pathogenic substances. However, the scaffolds produced with such polymers are foreign to the cells, thus creating important drawbacks. The lack of interaction with the polymer results in uncontrolled response of the cells, among which is insufficient cell attachment (which cause tremendous difficulties during the seeding of the cells), or undesired changes in cell morphology. These nonphysiological responses are the result of the material noncompliance with the necessary cell requirements. But what are these requirements and what do the cells find in their natural surroundings that they do not in a polymer scaffold? While we are starting to understand more and more the complexity of native ECM and strategies to provide mimicking cues to synthetic scaffolds have been developed (Box 1.3), we are still at the tip of the iceberg when recapitulating in synthetic scaffolds the dynamic complexity and multiplexing character of cell–ECM communication. We have learned to provide both hydrogel and solid synthetic polymers with biological cues, in the form of either peptides or larger proteins, which can be directly remodeled by the cells via enzymatically degradable linking sites. As our knowledge on the full spectrum of the ECM—the matrisome—is progressing, it remains a challenge to solve for the future years to replicate such an exquisite complexity and dynamics in a synthetic ECM where interactions with cells are controlled in space and time in the same manner.

Aside from the importance of understanding and creating the most optimal environment for the cells, another open question of possibly higher

BOX 1.3 MIMICKING THE EXTRACELLULAR MATRIX

From the various attempts to create an optimal substrate for cell guidance and tissue formation, one crucial realization was the importance of the natural cell environment. In all tissues, cells exist within a 3D viscoelastic and fibrillar composite milieu, the ECM, whose roles are multiple. Being synthesized by cells, it provides the structural and mechanical properties of tissues while concomitantly serving as an important regulator of cell behavior through signaling moieties presented in time- and space-controlled fashions.

Accordingly, extensive efforts have been displayed in designing systems inspired from the natural cell situation,

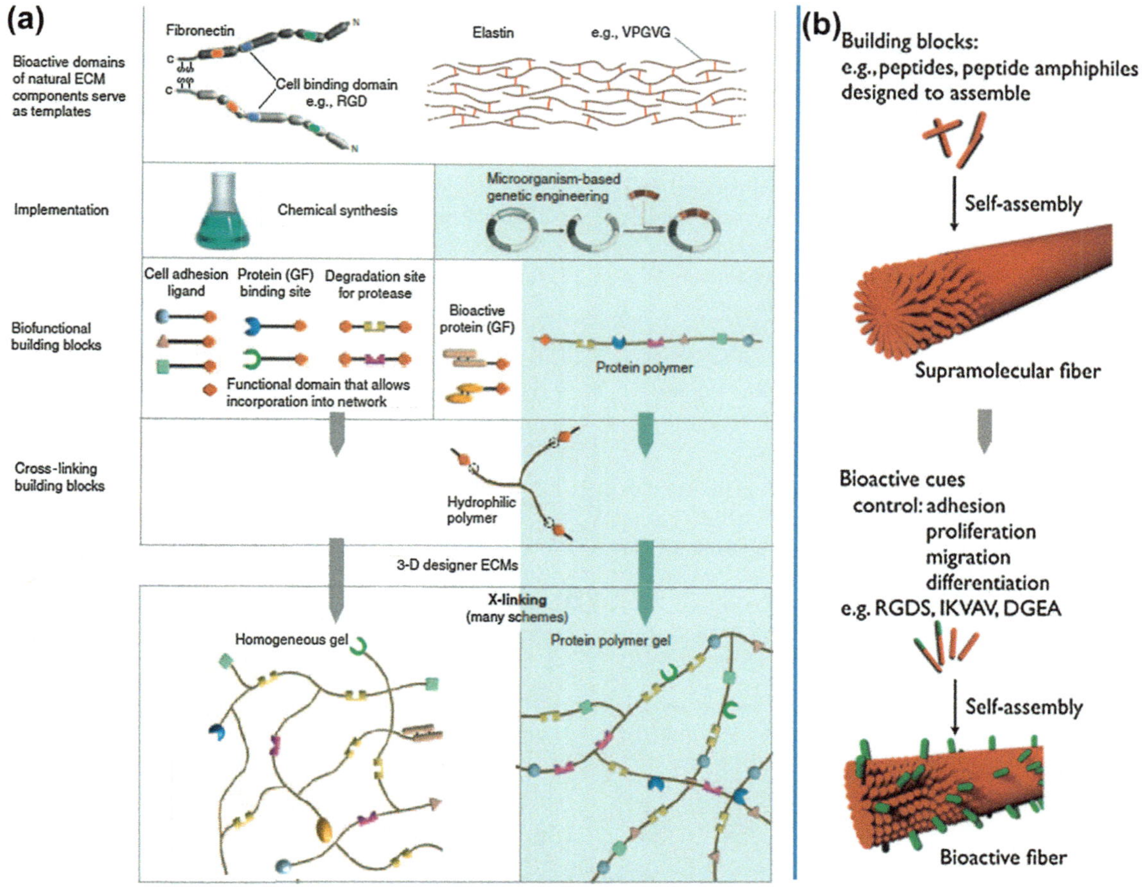

Design strategies of materials mimicking ECM key functions. (a) Bioactive domains of proteins are identified and synthesized by chemical strategies or by recombinant technologies. They include cell-adhesive ligands, domains for protease degradation, and growth factor binding sites with cleavage sequences. Synthetic hydrogels can then be obtained by chemical cross-linking. (b) Supramolecular materials that instruct cells can be designed to self-assemble into nanostructures similar to the structural components of ECM. *Figure (a) is adapted from Lutolf and Hubbell (2005) and (b) is adapted from Boekhoven and Stupp (2014).*

Continued...

BOX 1.3 MIMICKING THE EXTRACELLULAR MATRIX *Continued*

in view of replicating cells–ECM interactions and to provide tools to better control and guide tissue regeneration.

Two main approaches have been pursued in the last decades to mimic and integrate ECM key functions in biomaterial design. The first one, leaded by Hubbell and collaborators (Hubbell, 1995; Lutolf and Hubbell, 2005), focuses on designing modular synthetic hydrogels that:

- incorporate bioactive domains of natural ECM components recognized as interacting with the cells or providing structural function. The most employed recognition motif being the RGD tripeptide derived from fibronectin to provide cell attachment.
- provide cell-controlled local degradation to permit migration and remodeling of the synthetic hydrogel substrate. This can be achieved by the integration of sites that are sensitive for cell-synthesized proteases, such as serine proteases and matrix metalloproteinases (MMPs).
- possess growth factors binding sites that contain a protease cleavage site (Lutolf et al., 2003) to permit a cell-hydrogel dialogue in which growth factors are released locally by cells through cell-secreted or cell-activated proteases.

The other approach finds its inspiration in the aptitude of certain ECM proteins, such as collagen, to self-assemble into supramolecular architectures. As a result, it adds to the aforementioned features the possibility to mimic the versatile and dynamic nanofibrillar components of the ECM. Stupp and collaborators (Boekhoven and Stupp, 2014) have been pioneers in such a strategy by designing peptide amphiphile molecules that self-assemble into nanofibers through noncovalent bonds triggered by the ionic strength of the environment.

Numerous studies have shown the interest of these approaches for tissue engineering and regeneration over the past decade. We hereafter provide a small number of recent and representative examples.

The capacity of mimicking hydrogels to provide cells with localized growth factor release and hence improve tissue formation and repair has been demonstrated in a recent study from Martino et al. (2013). Through the reproduction of the heparin-binding domain of fibrin (Fg β15–66$_{(2)}$), it was possible to use its unique binding abilities to consistently deliver placenta growth factor (PlGF-2) and fibroblast growth factor (FGF-2) into a chronic wound mouse model. The sequence was incorporated into fibrin-mimetic hydrogels (poly ethylene glycol-based) containing additionally integrin-binding sequences (α_2PI$_{1-8}$-RGDSPG) and plasmin/MMPs degradation sequences. The cell-induced delivery of the growth factors led to faster wound closure and significantly more granulation tissue after

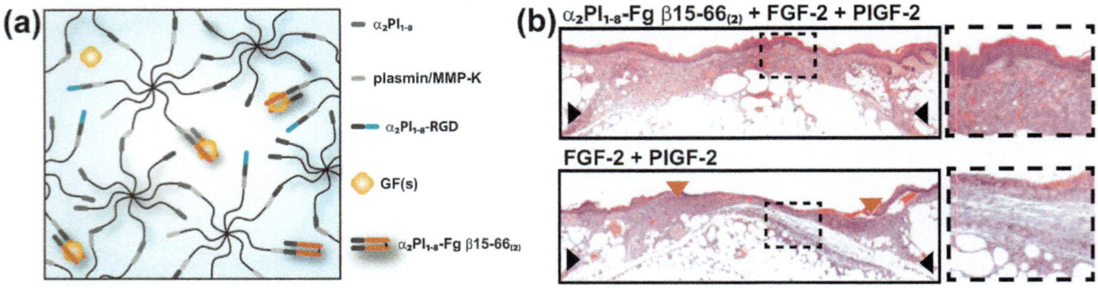

Strategy based on synthetic functionalized hydrogels. (a) Two eight-arm poly ethylene glycol (PEG)-peptide conjugates (PEG-plasmin/MMP-K and PEG-α_2PI$_{1-8}$), a cell-adhesion peptide (α_2PI$_{1-8}$-RGD), the heparin-binding domain of fibrin (α_2PI$_{1-8}$-Fg β15–66$_{(2)}$), and growth factors (GF(s)) form a fibrin-mimetic hydrogel through the transglutaminase factor XIIIa cross-linking. (b) Wound closure and granulation tissue are significantly enhanced when the heparin-binding sequence is present and allows a cell-controlled delivery of FGF-2 and PlGF-2 after 10 days. Red arrows indicate wound closure (tips of epithelium tongue) while granulation tissue is stained in pink-violet. *Adapted from Martino et al. (2013).*

BOX 1.3 MIMICKING THE EXTRACELLULAR MATRIX *Continued*

10 days than matrices lacking the heparin-binding domain. Moreover, angiogenesis within the granulation tissues was much more pronounced, demonstrating further the advantage of an on-demand delivery of growth factors by the cell-induced degradation of the mimetic hydrogels.

In a study by Tysseling-Mattiace et al. (2008), the ability of peptide amphiphile (PA) molecules to self-assemble and present bioactive cues when injected into tissues has been used to improve the restoration of motor function after spinal cord injury in mice. PA molecules have been

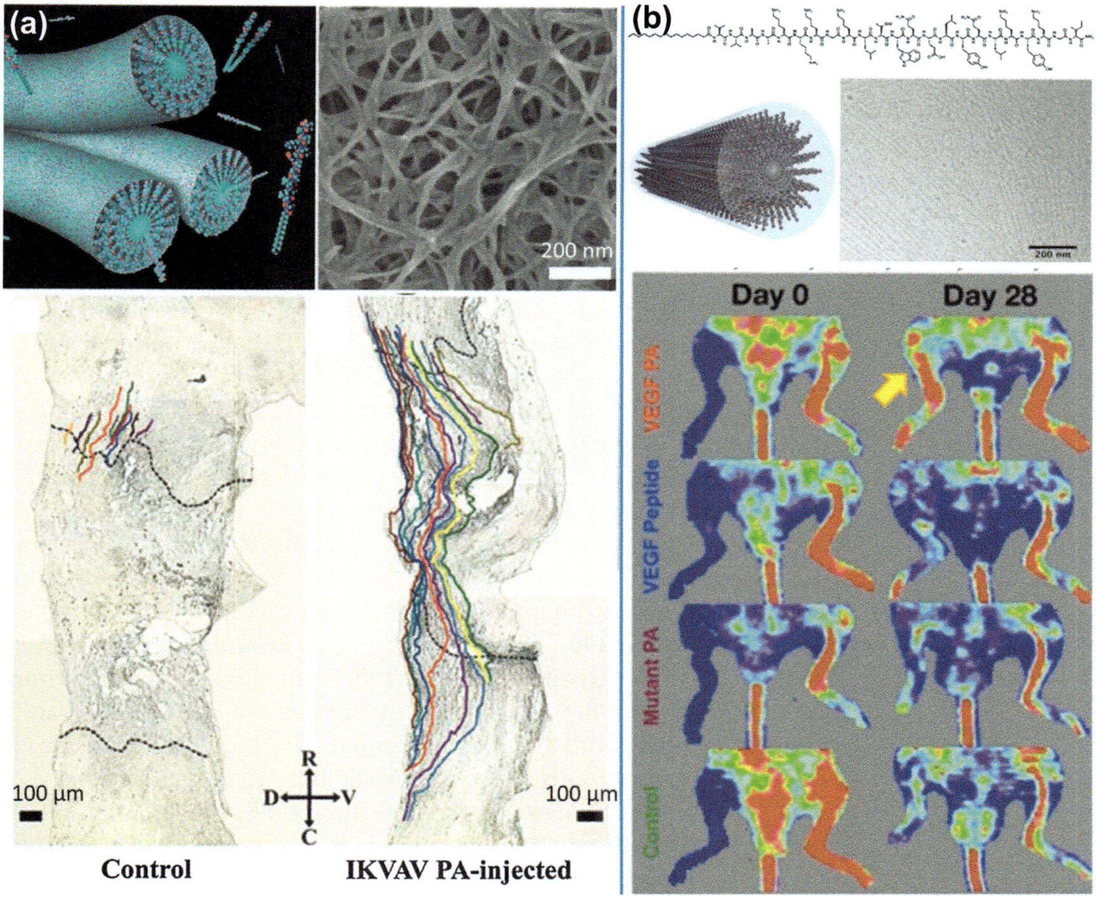

Strategies based on self-assembling molecules. (a) IKVAV-expressing PAs assemble into nanofibers once injected in tissues and present a high sequence density at their surface. Eleven weeks after spinal cord lesion (dotted lines), tracings of descending motor fibers illustrate the induced improvement. (b) A VEGF-mimetic PA forming nanofiber significantly enhances perfusion in a rat hind limb ischemia model (yellow arrow). *Adapted from Tysseling-Mattiace et al. (2008) and Webber et al. (2011).*

Continued...

BOX 1.3 MIMICKING THE EXTRACELLULAR MATRIX *Continued*

designed to form 6–8 nm nanofibers that assemble into bundles and present at their surface the laminin epitope IKVAV. Once presented at very high density on the injected nanostructures in vivo, this sequence, which was known to promote cell adhesion and induce neurite outgrowth of neural progenitor cells in vitro, reduced astrocyte differentiation as well and therefore scar formation. Moreover, the injected IKVAV-displaying PAs promoted the regeneration of descending motor fibers and ascending sensory fibers, thereby significantly improving motor function of the injured animals.

In another study by Webber et al. (2011), a vascular endothelial growth factor (VEGF)-mimetic peptide was attached to PA in view of activating VEGF receptors of endothelial cells and inducing angiogenesis. In vitro, the nanoscale filaments displaying a high density of the mimetic peptide enhanced the proliferation, survival, and migration of HUVECs while eliciting an angiogenic response in chicken embryo assays. More interestingly, using a mouse hind limb ischemia model, the injection of VEGF-mimetic PA significantly enhanced blood perfusion and motor function in previously ischemic limbs over 28 days through an improved microcirculatory angiogenesis.

Overall, these examples clearly highlight the importance of mimicking structure and key functions of the ECM, to selectively and modularly provide tools that can in turn guide and orchestrate tissue formation and regeneration.

importance concerns the cells themselves: What are the most suitable cells to use for any given tissue? Pivotal knowledge regarding this fundamental question is also lacking. One could logically think that the most suitable cell source for cartilage or bone, for instance, is cartilage and bone, respectively. This is logical indeed and was the first approach to be followed. However, many hurdles render this strategy not as straightforward as expected. The first difficulty was seen in the number of cells necessary to provide a basis for tissue growth. The extraction of cells from native tissue is generally done in humans by biopsy and usually results in an insufficient cell number (Figure 1.3). The cells so collected therefore have to be multiplied. This step is generally done by culturing them on polystyrene culture flasks that allow cell attachment. Although almost any cell types can be expanded, provided the right and specific conditions are supplied, their multiplication immutably induces the loss of their particular phenotype; in other words, they dedifferentiate. For instance, a cartilage cell of round morphology will turn to a fibroblastic-like cell (of stretched and elongated shape) after some divisions (von der Mark et al., 1977). As a result, once a sufficient number of cells have been reached, they have to be redifferentiated to the desired phenotype, which is not easily attainable.

Another approach, which has gained a constant increase in interest, consists of using undifferentiated cells as starting material. After all, a complete organism consisting of billions of highly specialized cells originates only from a single undifferentiated one. Even in adults, there are pools of undifferentiated cells which allow self-renewal of the organism over a lifespan; in other words, cells that form the stem of all the others. The first of these stem cells to be identified after the Second World War were the hematopoietic cells that allow

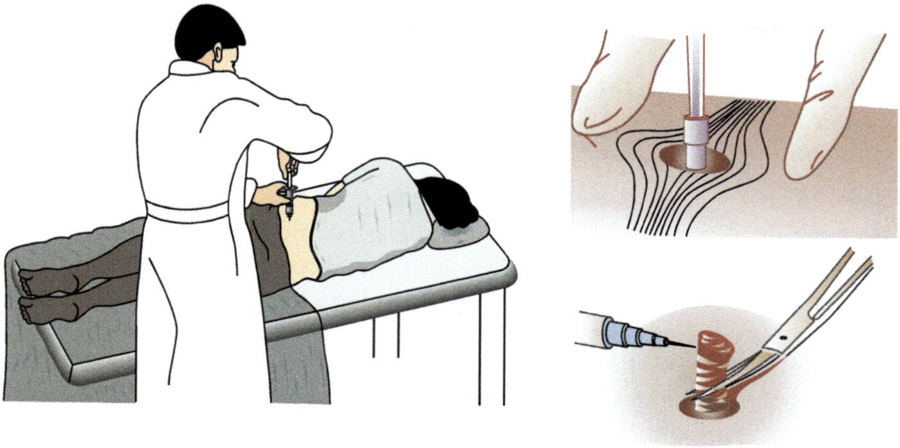

FIGURE 1.3
Although human biopsies are the main cell source, they do not provide a sufficient number of cells.

the renewal of blood (Till and McCulloch, 1961). Since then, other stem cells have been discovered in the bone marrow, peripheral blood, brain, spinal cord, dental pulp, blood vessels, skeletal muscle, heart, epidermis, mucosa of the digestive system, cornea, liver, and pancreas. These cells, once isolated, can be differentiated in highly specialized tissues (multipotency) if provided the right conditions, which are not entirely established. Furthermore, dedifferentiation also occurs with stem cells. More and more studies are elucidating the effect of the surface properties of culturing substrates on their activity. Although in the case of stem cells it is easier to redifferentiate them into the targeted specialized cell type, we should make sure that we have obtained full control on the final phenotype before implantation. One may even question if we should keep continuing to culture our cells in polystyrene flasks, as recent studies highlighted that stem cells have a memory of the stiffness of the substrate onto which they have been cultured and that their phenotype may be permanently modified (see Chapter 7). These are important issues that tissue engineering has to face in order to collect the fruits promised by the use of stem cells.

Even though the apparent slow progress of the field can be understood, it should not hide the fact that successful applications already exist. With current techniques, which may be qualified as crude by some, the effectiveness of tissue engineering for some applications can already be demonstrated clinically. Successful stories do not only comprise skin grafts for burn victims, instructive ceramics for bone regeneration, (matrix-assisted) autologous chondrocyte implantation for small articular cartilage defects, or the luminal study by Atala et al. (2006) on bladder regeneration. More recent successful attempts have seen the first cases of vascular regeneration (L'Heureux et al., 2007; Dahl et al., 2011), tissue-engineered trachea and esophagus implants

(Gonfiotti et al., 2014), reproductive organ regeneration (Raya-Rivera et al., 2014), among others. These examples should serve to convince those who do not see tissue engineering evolving fast enough that, although its progress is slower than all of us would wish, it has already reached the stage of treating numerous patients and adding valuable quality of life.

These final considerations will undeniably persuade the reader of the long way that tissue engineers have already progressed and of the long way that still remains. A journey where unexpected hurdles will certainly appear, always making room for renewed excitement and fun. Indeed, tissue engineers will have to become as multiple as the tissues they wish to regenerate and as pluripotent as the cells they use. We hope this textbook will contribute to this and to a faster progress of the field toward a clinical reality by enlightening young researchers to the state of the art of tissue engineering and to the challenges still lying ahead.

REFERENCES

Atala, A., Bauer, S.B., Soker, S., Yoo, J.J., Retik, A.B., 2006. Tissue-engineered autologous bladders for patients needing cystoplasty. Lancet 367 (9518), 1241–1246.

Boekhoven, J., Stupp, S.I., 2014. 25th anniversary article: supramolecular materials for regenerative medicine. Adv. Mater. 26 (11), 1642–1659.

Cao, Y., Vacanti, J.P., Paige, K.T., Upton, J., Vacanti, C.A., 1997. Transplantation of chondrocytes utilizing a polymer-cell construct to produce tissue-engineered cartilage in the shape of a human ear. Plast. Reconstr. Surg. 100 (2), 297–302, discussion 303–304.

Dahl, S.L.M., Kypson, A.P., Lawson, J.H., Blum, J.L., Strader, J.T., Li, Y., Manson, R.J., Tente, W.E., DiBernanrdo, L., Hensley, M.T., Carter, R., Williams, T.P., Prichard, H.L., Dey, M.S., Begelman, K.G., Nikalson, L.E., 2011. Readily available tissue-engineered vascular grafts. Sci. Transl. Med. 3 (68), 68ra9.

Gonfiotti, A., Jasus, M.O., Barale, D., Baiguera, S., Comin, C., Lavorini, F., Fontana, G., Sibila, O., Rombolà, G., Jungebluth, P., Macchiarini, P., 2014. The first tissue-engineered airway transplantation: 5-year follow-up results. Lancet 383 (9913), 238–244.

Hubbell, J.A., 1995. Biomaterials in tissue engineering. Biotechnology (N.Y.) 13 (6), 565–576.

Jain, R.K., Au, P., Tam, J., Duda, D.G., Fukumura, D., 2005. Engineering vascularized tissue. Nat. Biotechnol. 23 (7), 821–823.

Langer, R., Vacanti, J.P., 1993. Tissue engineering. Science 260 (5110), 920–926.

Laschke, M.W., Menger, M.D., 2012. Vascularization in tissue engineering: angiogenesis versus inosculation. Eur. Surg. Res. 48 (2), 85–92.

L'Heureux, N., McAllister, T.N., de la Fuente, L.M., 2007. Tissue-engineered blood vessel for adult arterial revascularization. N. Engl. J. Med. 357 (14), 1451–1453.

Levenberg, S., Rouwkema, J., Macdonald, M., Garfein, E.S., Kohane, D.S., Darland, D.C., Marini, R., van Blitterswijk, C.A., Mulligan, R.C., D'Amore, P.A., Langer, R., 2005. Engineering vascularized skeletal muscle tissue. Nat. Biotechnol. 23 (7), 879–884.

Lutolf, M.P., Hubbell, J.A., 2005. Synthetic biomaterials as instructive extracellular microenvironments for morphogenesis in tissue engineering. Nat. Biotechnol. 23 (1), 47–55.

Lutolf, M.P., Raeber, G.P., Zisch, A.H., Tirelli, N., Hubbell, J.A., 2003. Cell-responsive synthetic hydrogels. Adv. Mater. 15 (11), 888–892.

Martino, M.M., Briquez, P.S., et al., 2013. Heparin-binding domain of fibrin(ogen) binds growth factors and promotes tissue repair when incorporated within a synthetic matrix. Proc. Natl. Acad. Sci. U.S.A. 110 (12), 4563–4568.

Raya-Rivera, A.M., Esquiliano, D., Fierro-Pastrana, R., Lopez-Bayghen, E., Valencia, P., Ordorica-Flores, R., Soker, S., Yoo, J.J., Atala, A., 2014. Tissue-engineered autologous vaginal organs in patients: a pilot cohort study. Lancet. 384 (9940), 329–336.

Sampath, T.K., Reddi, A.H., 1981. Dissociative extraction and reconstitution of extracellular matrix components involved in local bone differentiation. Proc. Natl. Acad. Sci. U.S.A. 78 (12), 7599–7603.

Sampath, T.K., Reddi, A.H., 1983. Homology of bone-inductive proteins from human, monkey, bovine, and rat extracellular matrix. Proc. Natl. Acad. Sci. U.S.A. 80 (21), 6591–6595.

Till, J.E., McCulloch, C.E., 1961. A direct measurement of the radiation sensitivity of normal mouse bone marrow cells. Radiat. Res. 14, 213–222.

Tysseling-Mattiace, V.M., Sahni, V., et al., 2008. Self-assembling nanofibers inhibit glial scar formation and promote axon elongation after spinal cord injury. J. Neurosci. 28 (14), 3814–3823.

Urist, M.R., 1965. Bone: formation by autoinduction. Science 150 (698), 893–899.

von der Mark, K., Gauss, V., von der Mark, H., Muller, P., 1977. Relationship between cell shape and type of collagen synthesised as chondrocytes lose their cartilage phenotype in culture. Nature 267 (5611), 531–532.

Webber, M.J., Tongers, J., et al., 2011. Supramolecular nanostructures that mimic VEGF as a strategy for ischemic tissue repair. Proc. Natl. Acad. Sci. U.S.A. 108 (33), 13438–13443.

CHAPTER 2

Stem Cells

Mark F. Pittenger[1], Candace Kerr[2]

[1]Department of Surgery, University of Maryland School of Medicine, Baltimore, MD, USA;
[2]Department of Biochemistry and Molecular Biology, University of Maryland School of Medicine, Baltimore, MD, USA

LEARNING OBJECTIVES

- To recognize the defining properties of stem cells
- To understand that a variety of stem cells can be found in the body with different potentials
- To understand that stem and progenitor cells exist in metastable states
- Phenotype readout: how assays are instrumental in facilitating successful study of stem cells
- To understand the opportunities and challenges for tissue engineers when using stem cells

It is a riddle, wrapped in a mystery, inside an enigma: but perhaps there is a key.

Winston Churchill

Our questions about stem cells must be disassembled into manageable investigations that match our skill sets.

Mark Pittenger

2.1 INTRODUCTION

2.1.1 What Are Stem Cells?

Stem cells are cells that proliferate and also can differentiate to produce cells with new phenotypes and functions. As such, they provide a useful cell type for many tissue engineering applications. They provide certain challenges as well. The field can be divided into adult, embryonic, and induced stem cells, but may also be further subdivided in additional ways. For the purposes of this book, we are interested in those stem cells that can be isolated, manipulated, combined with other materials, analyzed, further engineered, and placed back in the in vivo environment to perform a specific function. We are still discovering many aspects

Tissue Engineering.
Copyright © 2015 Elsevier Inc. All rights reserved.

of stem cells, but stem cells and tissue-engineered constructs containing them are currently undergoing preclinical testing and clinical evaluation in many studies.

The cells, tissues, and organs of the body are under constant use and many cells die and are replaced by new cells on a daily, weekly, and monthly basis. Homeostasis requires the loss of cells to be accurately balanced by cell replacement. Injury and repair/regeneration can accelerate this cell replacement process, but aging seems to slow down this process. Understanding the process of normal cell replacement and its relationship to injury repair is a goal of modern medicine and the tissue engineer. Every day, millions of red blood cells die and millions are created to replace them. All of your red blood cells are replaced every 6 weeks by hematopoietic stem cells (HSCs) in the bone marrow. Skin cells are replaced frequently from epithelial stem cells in the basement layers and migrate to the surface as the outer differentiated cornified epithelial layer sloughs off (Chapter 15). Your whole skeleton is replaced every 6–8 years by the combined actions of the osteoclasts, which remove and remodel the calcified bone matrix, and the osteoblasts, which divide and then differentiate into new osteocytes and produce abundant extracellular matrix which becomes cross-linked and calcified over time (Chapter 16). The central nervous system—brain and spinal cord—were thought not to regenerate but work over the last 10–15 years has shown that the cells of these tissues may be replaced by neural stem cells (Chapter 17). The heart was thought to contain the same cells from birth to death and only recently has evidence emerged for new cardiomyocytes over time (Chapter 18). However, cardiomyocyte turnover is very slow and a 50-year-old person has replaced only about half of his/her cardiomyocytes (Bergmann et al., 2012). These examples demonstrate the potential of stem cells to produce new healthy cells as needed by the body but there are limitations. In this chapter, we will examine further the properties and characteristics of stem cells that make them useful for tissue engineering. Understanding the signals that promote proliferation of stem cells and their subsequent differentiation to functional cell types has become a rapidly advancing field that stretches across many disciplines—cell biology, molecular biology, bioengineering, material science, and many fields of medicine. The study of stem cells is not new, but recent advances have brought greater interest and the promise of using stem cells for tissue repair and regeneration, and spawned the rapidly advancing field of regenerative medicine.

2.1.2 What Defines a Stem Cell? Self-renewal and Differentiation

Stem cells are not easy to identify. They look like many other cells and they may be present among cells that are not stem cells. The daughter cells from the division of a stem cell may behave as stem cells or one or both may have already begun the process of differentiation. The "choice" is not always evident. In general, stem cells may be slow to divide and are mostly thought to be

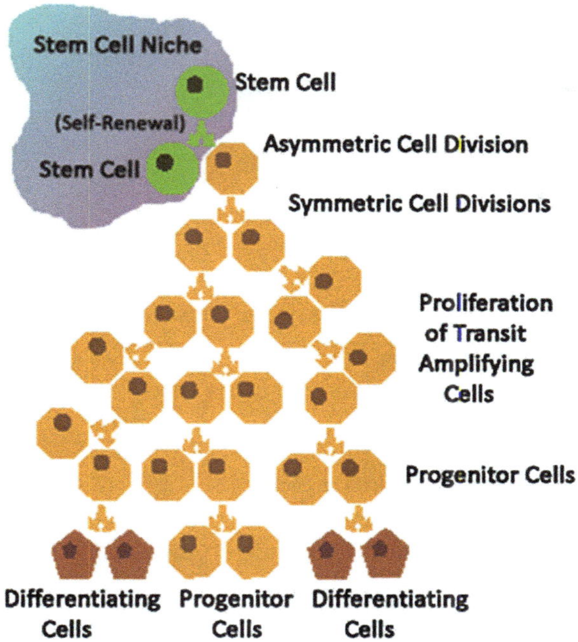

FIGURE 2.1

Stem cell replication can be symmetrical or asymmetrical. **Symmetrical division** creates two identical daughter cells. **Asymmetrical division** creates two nonidentical daughter cells. The asymmetry can be caused when cell division occurs to divide the cell or its organelles unequally, or the environment provides different signals to the two daughter cells. Proliferation of daughter cells as they move away from the **stem cell niche** can quickly amplify a population of progenitor cells to create a large number of cells, hence the name **transit amplifying cells**. From the transit amplifying stage, the cells begin differentiation and express genes and proteins characteristic of the differentiated adult cell type and produce largely one type of progenitor cell, and its differentiation may be dependent on its environment. If the progenitor cells move into environments that differ, this can produce completely different phenotypes.

quiescent in the body, preserving their potential until needed. Once a stem cell begins to divide, it may produce daughter cells that are subsequently given names, such as progenitor cells, transit amplifying cells, early differentiated cells, and fully differentiated cells, which reflect the loss of proliferative potential and the gain of functional attributes (Figure 2.1). Scientists continue to develop methods to identify stem cells but the two properties that are most closely associated with stem cells are self-renewal and differentiation.

2.1.3 Self-renewal

Self-renewal is a special form of cell proliferation. Self-renewal refers to the ability of a stem cell to divide and make identical copies of itself to assure that the stem cell population is not depleted, and exists throughout development, and some may remain for the life of the organism. Self-renewal in vivo requires a specialized location—the stem cell niche—where stem cells are maintained through the action of nearby supporting cells. In vitro conditions that approximate conditions found in the stem cell niche have been developed but these conditions are usually imperfect and may not be as favorable as those found in vivo. This remains an area of active study. Self-renewal allows proliferation of stem cells or it can produce one daughter cell that is a stem cell and one daughter cell that then proceeds along a differentiation pathway (Morrison and Kimble, 2006). This asymmetric division may be caused by internal cell fate signals accumulating in one daughter cell or can be caused by signals outside the cell such as another cell or a gradient of morphogens, growth factors, and so forth (Figure 2.2).

2.1.4 Proliferation

Proliferation of stem cells allows a single-celled fertilized egg—the ultimate stem cell—to grow into an adult organism. Along the way, tissues and organ systems develop and become functional. A subset of stem cells may self-renew and be stored in each tissue for later use in repair and regeneration. When potential stem cells are isolated from an embryo or an adult tissue, the first evaluation is usually whether the cells can proliferate and if so, how fast and for how long. The conditions for in vitro growth are found by experience and trial and error as cells are adapted to their new environment. The doubling time is measured and is usually in the 20–30 h range when conditions are favorable.

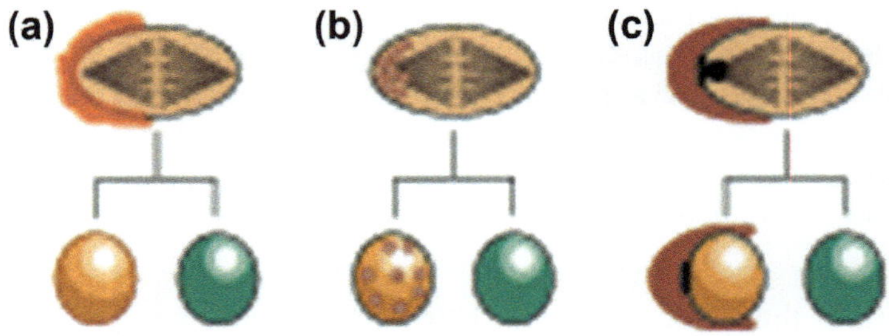

FIGURE 2.2

Controls of asymmetric stem cell division. Three simple mechanisms are shown. Stem cells are orange, and differentiated cells are green. (a) Asymmetric localization of cell polarity regulators (red) initiates the asymmetric division; (b) cell fate determinants (red) can be segregated to the cytoplasm of one daughter cell, as shown here, or they can be associated with the membrane, centrosome, or another cellular constituent that is differentially distributed to the daughters; (c) regulated orientation of the mitotic spindle retains only one daughter in the stem cell niche (red) such that only that daughter cell has access to extrinsic signals necessary for maintaining stem-cell identity. This mechanism achieves an asymmetric outcome, even though the division itself is intrinsically symmetric. In an alternative but similar model, the daughter cell placed away from the niche is exposed to signals that induce differentiation. *Reproduced with permission from Morrison and Kimble (2006).*

Eventually the dividing stem cells may use all the space available to them on the culture dish or metabolize all of a limiting, important nutrient, and their rate of cell division will decrease. Experimentally, at this point the cells in the culture will be subcultured or "split" into 2–4 new culture vessels with fresh new medium. This step is referred to as "passaging" the stem cells, and each passage usually represents 3–4 doublings of the stem cells. Stem cells are often cultured this way to keep them in log phase growth to produce millions of cells in a predictable time period. If stem cells are passaged too sparsely, say into 20 culture vessels, the cells will not remain in log phase growth and although they may eventually fill the 20 vessels, the timing and outcome is less predictable. Keep in mind that a single cell dividing every 24 h can produce 1 million cells in 21 days and 1 billion (10^9) in 31 days (Figure 2.3). As a point of reference, the average adult human is composed of about 3.7×10^{13} cells (Bianconi et al., 2013). Ideally, and by definition, stem cells can grow forever. In practice, it is difficult to keep stem cells growing indefinitely, yet enough cells can be produced for a particular purpose and the remainder may be cryopreserved (frozen) for later use, or discarded. Early studies on aging of cells in culture using human fibroblasts found that after about 50–60 divisions the cells slowed and stopped growing (Hayflick, 1965; reviewed in Hayflick, 1998). Prior to this work it was assumed that cells in culture could divide an infinite number of times if proper conditions could be found. Hayflick's work identified that normal cells were mortal and could divide a finite number of times, and would not form tumors when injected into animals. The concept of cellular aging accompanying cell division is still formidable and important but stem cells may have additional properties that allow them to surpass this limit as discussed next.

Telomeres are the name for the repetitive DNA sequence motifs on the ends of chromosomes. It requires a special DNA polymerase activity to maintain the chromosome ends and ensure that they are replicated with fidelity during each cell

Assuming all cells survive and divide again:

1 stem cell divides into 2
2 stem cells divide into 4
4 stem cells divide...
10 divisions yield 1024 stem cells
20 divisions yield 104,857 stem cells or ~1 million
30 divisions yield 1,073,741,824 stem cells or ~1 billion

Figure 2.3 shows 100 million MSCs = ~300 μl after collection by centrifugation or about ~300 μg or 0.3 mg.

Therefore, 1 mg of packed cells = ~333 million MSCs and for a "standard" 70 kg human contains 2.2×10^{13} cells.

Perhaps a more accurate calculation of the number of cells in the body comes from the study of Bianconi et al. (2013) who set the number at 3.37×10^{13} cells. The authors estimated the number of cells in organs of the body:

Liver = 3.6×10^{11} cells
Skin = 2.2×10^{12} cells
Blood vessel endothelial cells = 2.5×10^{12} cells
Pancreas = 3×10^9 cells
Heart = 2×10^9 cardiomyocytes + 2×10^9 fibroblasts

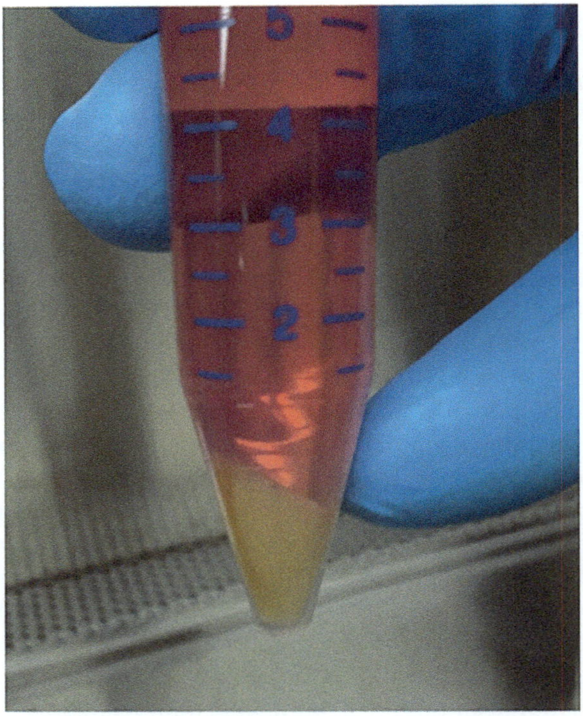

FIGURE 2.3

Cultured human mesenchymal stem cells can be expanded a millionfold and retain their multipotential properties. Figure shows that 100 million MSCs after harvest and gentle centrifugation is about 300 μl volume.

division and the enzyme is named telomerase. Telomerase is a ribonucleoprotein enzyme which carries a small RNA sequence that is complimentary to the telomere sequence of all the chromosome ends—TTAGGG—and which is conserved in all vertebrates, but differs in other organisms (Greider and Blackburn, 1985). When DNA polymerase has finished replicating virtually all of the chromosome, telomerase finishes off the process by adding ~200 base pair repetitive "cap" sequence to the free DNA end and thus protects the essential gene coding DNA sequences. Drs. Carol Greider and Elizabeth Blackburn were awarded the 2009 Nobel Prize in Physiology or Medicine for being the first to isolate telomerase and understand how it solved the replication problem of chromosome ends. The polymerase chain reaction (PCR)-based telomeric repeat amplification protocol (TRAP) assay is used to test for telomerase activity and produces a "ladder of repeats" representative of a particular species (Figure 2.4). Most adult somatic cells have limited telomerase activity but stem and progenitor cells including mesenchymal stem cells (MSCs), as discussed later, have telomerase activity (pittenger et al., 1999; Otte et al., 2013). Mutations in the telomerase gene are rare but are associated with several medical disorders. The association of telomerase activity with aging or cell senescence has been broadly studied but remains controversial. Cancer cell lines, such

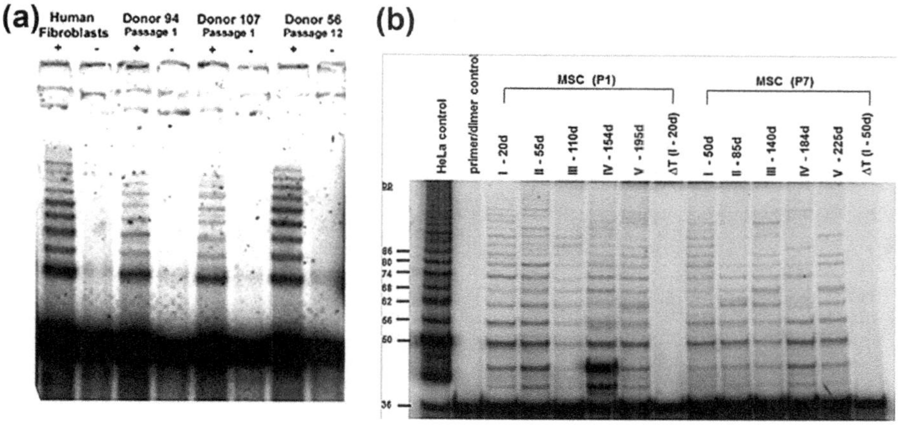

FIGURE 2.4

Telomerase adds a repeating cap sequence to protect the ends of chromosomes. The 6 base pair repeat—TTAGGG—forms a repeat "ladder" by telomerase activity from adult bone marrow MSCs (a) or umbilical cord MSCs (b), but activity is much less than transformed HELA cells. *Reprinted with permission from (a) Pittenger et al. (1999) web figures and (b) Otte et al. (2013).*

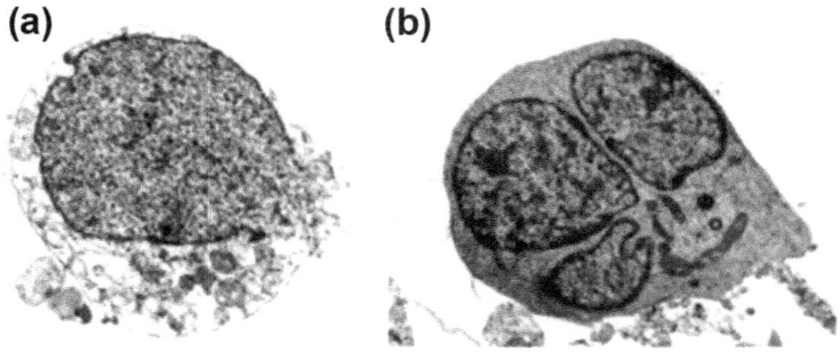

FIGURE 2.5

The natural process of programmed cell death or apoptosis can eliminate unwanted cells in a particular environment and this process can also eliminate old cells that lose their function. In necrosis (a) the cell membrane may become damaged and the nuclear membrane breaks down only later, but in apoptosis (b) the nuclear membrane and cellular organelles begin to break down while the cell membrane is intact. *EM photomicrograph reprinted with permission from Ueda et al. (2007).*

as HELA cells, are aneuploid with many chromosomes and usually have abundant telomerase activity.

Apoptosis is the natural process of cell death and is also called programmed cell death. It differs from necrosis wherein the cell is starved for nutrients by lack of blood supply or cell trauma wherein the cell becomes damaged and dies. In programmed cell death, the cell is said to die from within and the nuclear membrane will breakdown and the nucleus and DNA vesiculate within the cell while the cell membrane is still intact (Figure 2.5). Eventually, the cell plasma membrane breaks down and the cell contents are released. One of the natural reasons for apoptosis is a cell "recognizes" it is not in the right spot. For example, during development, apoptosis is part of the process of morphological change and many cells, such as the cells between the developing fingers, may die a natural and important death from this process. The suicidal enzymes involved in apoptosis are the caspases and the triggering of caspase activity causes a cascade that is hard to reverse once initiated. The signals that cause apoptosis are not understood completely, but the tissue engineer should recognize the occurrence of this natural process. It counterbalances proliferation that might otherwise proceed unchecked.

2.2 DIFFERENTIATION

2.2.1 Can a Stem Cell Become Everything It Wants to Be?

The second defining property of a stem cell is its ability to differentiate into a more specialized cell. The number of cell types a stem cell can differentiate into is determined by its potency:

- Totipotent stem cells have the ability to form an entire organism. The fertilized oocyte and the cells after the first cleavage divisions are considered to be totipotent.
- Pluripotent stem cells are able to form all three germ layers including germ cells, but not the extraembryonic tissue, the placenta, and the umbilical cord. Cells of the inner cell mass (ICM) of the blastocyst are pluripotent. When these cells are brought into culture, they are called embryonic stem cells.
- Multipotency means the ability to form multiple cell types. MSCs can differentiate into cells that form bone, cartilage, and fat.
- Oligopotent stem cells can differentiate into two or more lineages, for example, neural stem cells that can form a subset of neurons in the brain.
- Unipotency is the ability to form cells of a single lineage, for example, spermatogonial stem cells.
- The term omnipotence is not used for stem cells, but is used in religions as one of the God's characteristics.

Differentiation is the process whereby stem cells transform into more specialized cell types and can perform new functions through the expression of new genes, mRNA, and proteins. Differentiation involves the deactivation of some genes and the activation of a new set of genes. The signals that cause stem cells to differentiate are many, and often it is difficult to keep many pluripotent stem cells from differentiating during in vitro proliferation. Differentiation can be caused by a change in basal nutrients, a change in the cell's environment, stimulation—or a lack of stimulation—by a signaling molecule, a new cell-to-cell interaction,

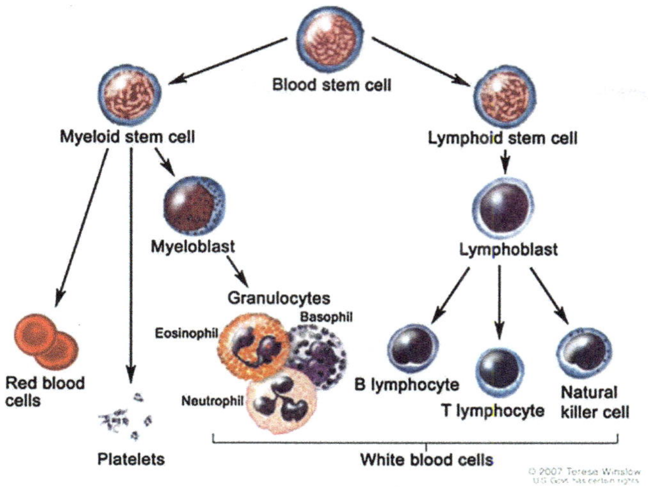

FIGURE 2.6

Lineages derived from the blood stem cell also known as the hematopoietic stem cells (HSCs) include the white blood cells as well as the red cells and platelets released from megakaryocytes. This type of diagram could only be drawn after 20 years of research. Such studies were an important part of establishing bone marrow transplant as a treatment option for leukemia patients. *(NCI website image.)*

and so forth. Some programs of differentiation are more common than others and some types of stem cells proceed more easily along certain lineage pathways to a "preferred" differentiated phenotype. Stem cells are poised for change, and it can be subtle or dramatic. A large effort over a number of years has identified differentiation conditions that can be applied in vitro to the study of stem cells and their potential for differentiation. Although these conditions may work very well to initiate differentiation, the fully differentiated phenotype found in adult tissue is not often evident in vitro. That is, additional signals may be necessary for maturation of the differentiated phenotype. It is generally assumed that the signals that cause in vitro differentiation are related to, but perhaps not the same as those that cause differentiation in vivo during development or in the adult. Nevertheless, numerous assays have been developed that reproducibly induce stem cell differentiation to desired phenotypes, allowing further experimentation and understanding of the signals involved in the differentiation process (Figure 2.6).

2.3 CHARACTERIZATION OF STEM CELLS: SURFACE PROTEIN EXPRESSION

We said that stem cells proliferate and differentiate into functional cell types. These two characteristics are usually mutually exclusive because the cell DNA must be organized into tightly packed chromosomes prior to cell division and this compact state of DNA is not compatible with the expression of the many genes needed for the differentiated phenotype. These two stem cell characteristics—proliferation and differentiation—are active functional attributes occurring over time; there is no assured a priori way to identify stem cells. However, trial and error studies have identified common characteristics of each stem cell type that allows purification, classification, and identity criteria. A useful technique that depends less on analyzing the stem cells over time is antibody-based flow cytometry which gives a snapshot of the expressed proteins and therefore the genes actively transcribed in the cell. Flow cytometry and the related technique fluorescence activated cell sorting (FACS) examine the surface phenotype of the cell by using antibodies that bind to the proteins on the cells of interest. In sorting mode, particular subsets of a cell population can be isolated for further study. The flow cytometry characterization of stem cells is usually performed during log phase growth with the majority of the cells in the G_1 state of the cell cycle although there may be some cells at all cell cycle stages in the sample. One of the powerful advantages of flow cytometry is that the machines are sophisticated enough that once the cells are bound with antibody, it is possible to analyze 10,000 cells in minutes and the large number of data points allows for strong statistical confidence (Figure 2.7). About 300 cell surface markers identified using monoclonal antibodies are designated as cluster of differentiation antibodies or CD antibodies. Many CD markers are used to characterize stem cells and multipotential cells, although positive reaction with the antibodies is not conclusive evidence of "stemness" but it is a start. Stem cells express a spectrum of receptors and surface proteins that allow them to interact and respond to information from their environment.

2.4 CHARACTERIZATION OF STEM CELLS: GENE EXPRESSION

Another technique that is universally employed in the study of stem cells and is complimentary to flow cytometry is gene expression analysis. Stem cells are maintained by a unique set of transcription factors that are characteristic of the different types of stem cells. Epigenetics refers to the stable alterations in gene expression that occurs during cell proliferation or differentiation. These alterations in gene expression are dependent on the degree of accessibility of transcription factors to genes and are related to the compaction of DNA into nucleosomes and controlled by methylation and demethylation (Figure 2.8).

While flow cytometry is most often employed to examine the proteins expressed on the surface of the cell, gene expression analysis can examine messenger RNA

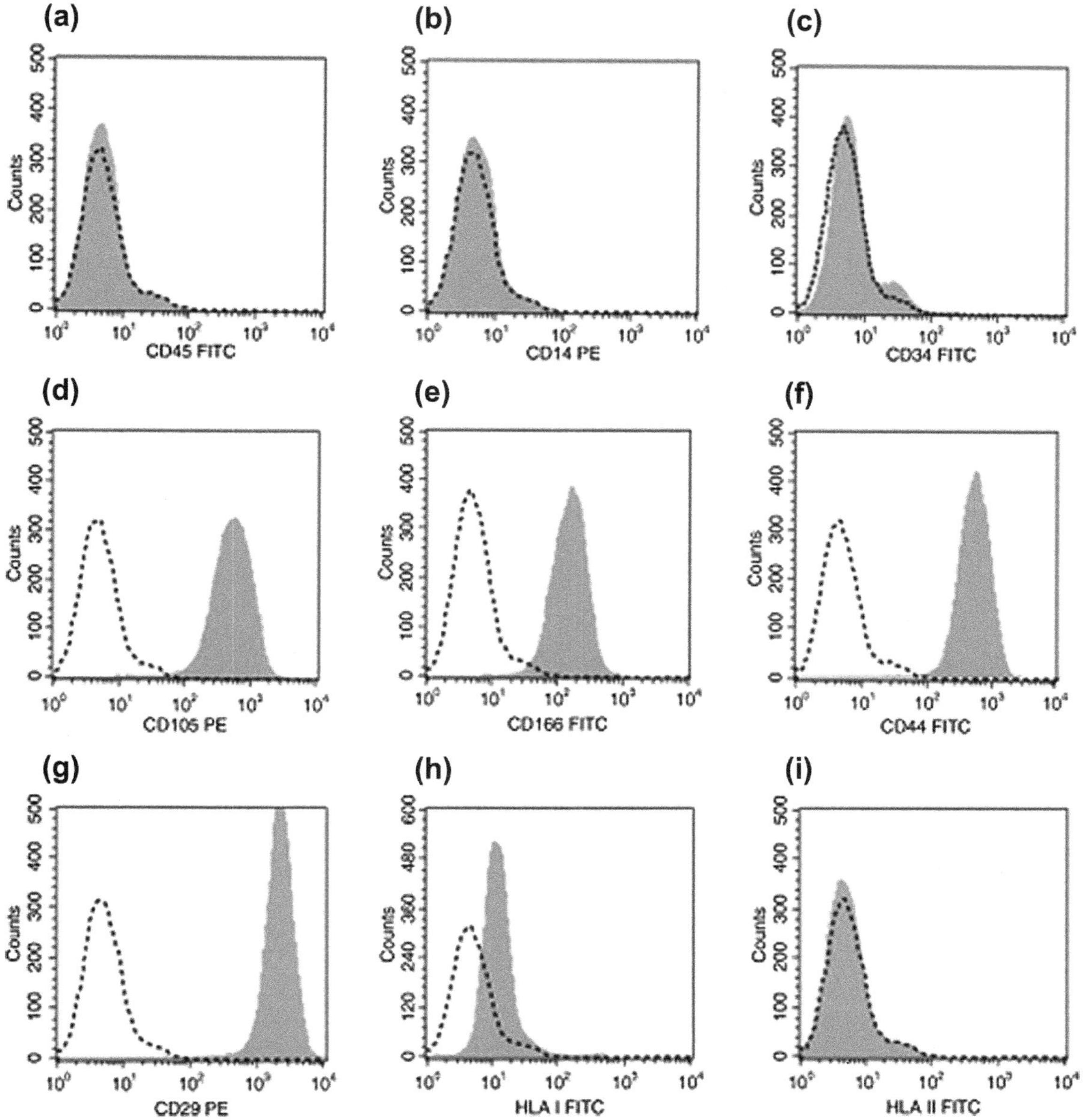

FIGURE 2.7

Flow cytometry, also referred to as fluorescence activated cell (FAC) analysis or FACS when used in the sorting mode, identifies cells that react with particular antibodies against surface molecules on cells. Dotted line is the signal for nonspecific antibody isotype control in each binding experiment. Mesenchymal stem cells bind positively with antibodies CD29, CD44, CD73, CD105, CD166, and HLA I, but negatively with antibodies CD14, CD34, and CD45 which bind to hematopoietic cells, as well as HLA II. *Reprinted with permission from Le Blanc et al. (2003b).*

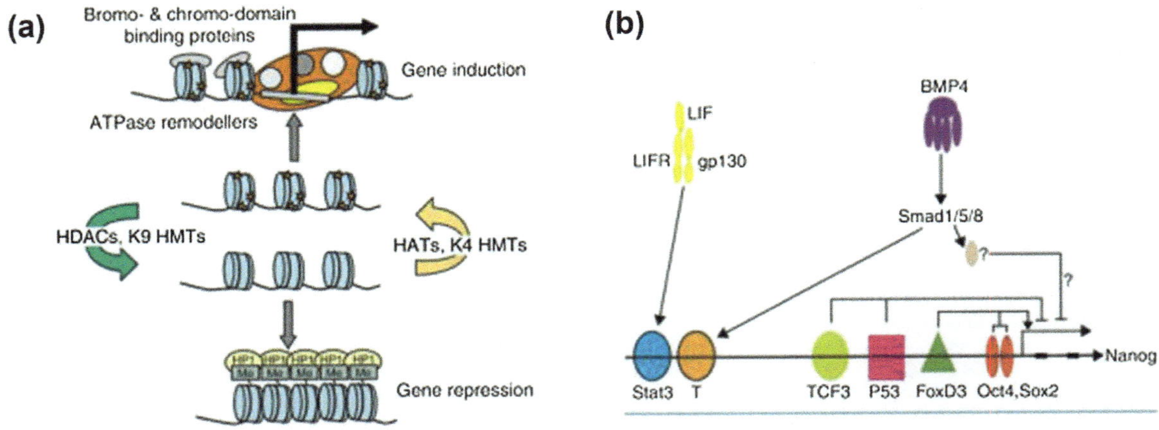

FIGURE 2.8

(Figures 1.7 and 1.13 from 1st edition as A and B) (a) The histone switch. Targeted modifications under the control of histone methylases (HMTs), histone acetyltransferases (HATs), and histone deacetylases (HDACs) alter the histone code at gene regulatory regions. This establishes a structure that contains bromo- and chromo-domains that permit recruitment of ATP-dependent chromatin remodeling factors to open promoters and allow further recruitment of the basal transcription machinery. Deacetylation, frequently followed by histone methylation, establishes a base for highly repressive structures, such as heterochromatin. Acetylated histone tails are shown as yellow stars. Methylation (Me) is shown to recruit heterochromatin protein 1 (HP-1). (b) Regulation of Nanog expression. FoxD3 and Oct4/Sox2 bind to the proximal region of Nanog promoter and support its expression. TCF3 and p53 also bind to the promoter and negatively regulate. *Reproduced with permission from Adcock et al. (2006).*

that is expressed, including those that do not result in cell surface products such as transcription factors and internal enzymes. Gene expression analysis can also examine RNAs such as regulatory micro-RNAs that are never translated into proteins. The two methods most employed for gene analysis of stem cells are reverse transcriptase-polymerase chain reaction (RT-PCR) and micro-array analysis. In both methods, the mRNA expressed by the stem cells is first isolated to eliminate any contamination from DNA. The mRNA is reverse transcribed to its more stable complementary DNA counterpart where it can be amplified manyfold by PCR so that even minor RNAs found in the cells become abundant and can be analyzed and compared to other cell types. Micro-array analysis starts with the same mRNA isolation and reverse transcription, but then hybridizes the mixture of complementary DNA to known samples of DNA representing genes of interest, anywhere from 40 to 200 or more, that then can be quantitatively analyzed. In general, these comparisons are sufficient for most studies, but there are methods that seek to compare the global expression of 2000 or all 23,000 genes at a time to understand different stem cells and/or their differentiated cells. These global comparisons of gene expression are usually given in relative transcript levels as red/green "heat maps" or as percentages of mRNA copies and results may be presented as numbers or in Venn diagrams or branching dendrograms of the relatedness of various stem cells (Figure 2.9).

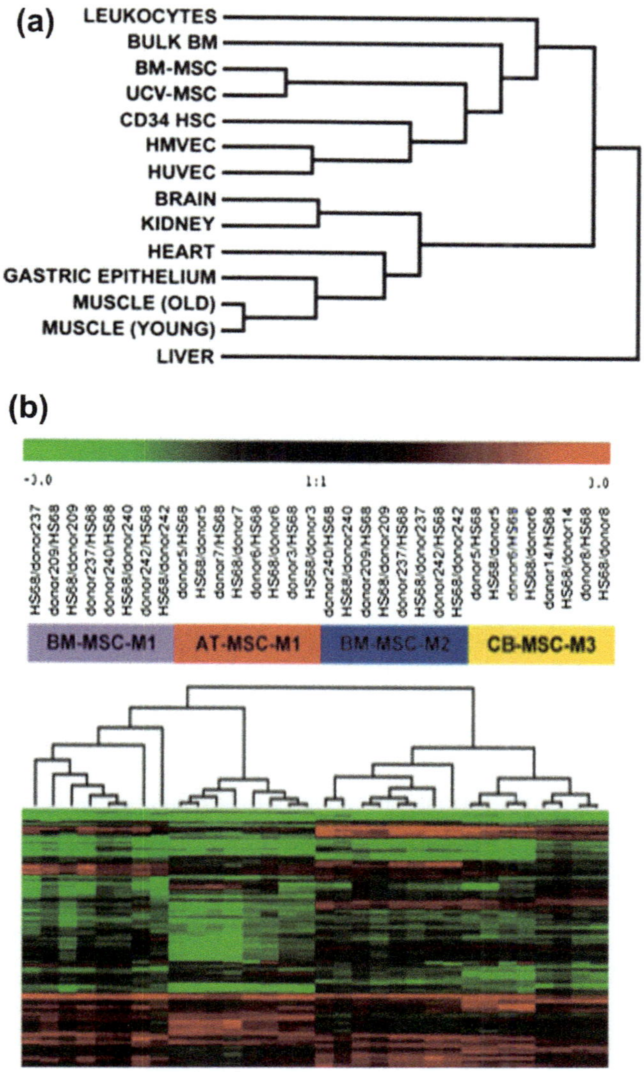

FIGURE 2.9

The "relatedness" of cells can be compared by analyzing the genes they express under particular conditions. For this, RNA is isolated from the different cells or the different conditions and converted to the more stable complimentary DNA. These "expressed sequence tags" (ESTs) are compared and those common to different cells reflect how similar the different cells might be. (a) The 500 most abundant transcripts across 14 types of cells were compared. *Reprinted with permission from Panepucci et al. (2004).* (b) A green to red heat map of the similarity of genes expressed by MSCs isolated from (BM), adipose (AT), and cord blood (CB); similar color represents similar level of expression for genes of interest, all compared to a skin fibroblast control cell line. *Reprinted with permission from Figure 3C in Wagner et al. (2005).*

2.5 METASTABLE STATES OF STEM CELLS

Similar cell types can be found in different locations in the body. That is, there are macrophages that scavenge dead and dying cells in each tissue and when isolated the macrophages from lungs may be evaluated to be somewhat different from those of the intestine. Similarly, stem cells isolated from bone marrow may have some properties that are measurably different from similar stem cells isolated from adipose tissue although many properties may be very similar (more on this later.) The stem cell niche provides a particular environment but as the daughter cells move out of the niche they encounter new signals and may change their phenotype as they adapt to their new environment. Another simple situation is the stem cell at the confluent center of a colony of stem cells which is surrounding and contacting many like cells, and compare this to the stem cell at the very edge of a colony where it contacts no other cells or has only a very limited cell–cell interaction. These two closely related but positionally different stem cells can respond very differently to differentiation signals such as those that cause differentiation to adipose (fat) cells. The confluent cells at the center will readily become adipocytes whereas those cells at the periphery of the colony will not, although only a day or two before they were part of the same parent cell! This ability of stem cells to respond to their surroundings is a key feature for tissue repair, as well as development. A metastable state best describes the different states, position, or stages stem cells find themselves as they move into different environments for extended periods. This metastable existence helps to explain stem cell plasticity as the cells adapt to new conditions, and a depiction of this for a stem cell is shown in Figure 2.10. The developmental potential of a stem cell can be changed by the influence of local factors and time spent in a particular environment. If the cell moves to a new location, it may retain some influence of its previous environment for a period of time. Embryonic stem cells also display metastable states (see Figure 2.13).

2.6 PLURIPOTENT STEM CELLS

Pluripotent stem cells can be isolated from embryos, germ cell progenitors of sperm, and fertilized egg or they can be derived from cells in culture through genetic engineering. These in vitro propagated cells include embryonic stem (ES) cells, embryonal carcinoma (EC) cells, embryonic germ (EG) cells, and induced pluripotent stem (iPS) cells. Each of these cell types have been derived in a multitude of species including rodents and humans.

2.6.1 Embryonic Stem Cells

Embryonic stem cells are generally derived from the ICM of a normal embryo at the blastocyst stage. The blastocyst consists of the outer layer of trophectoderm,

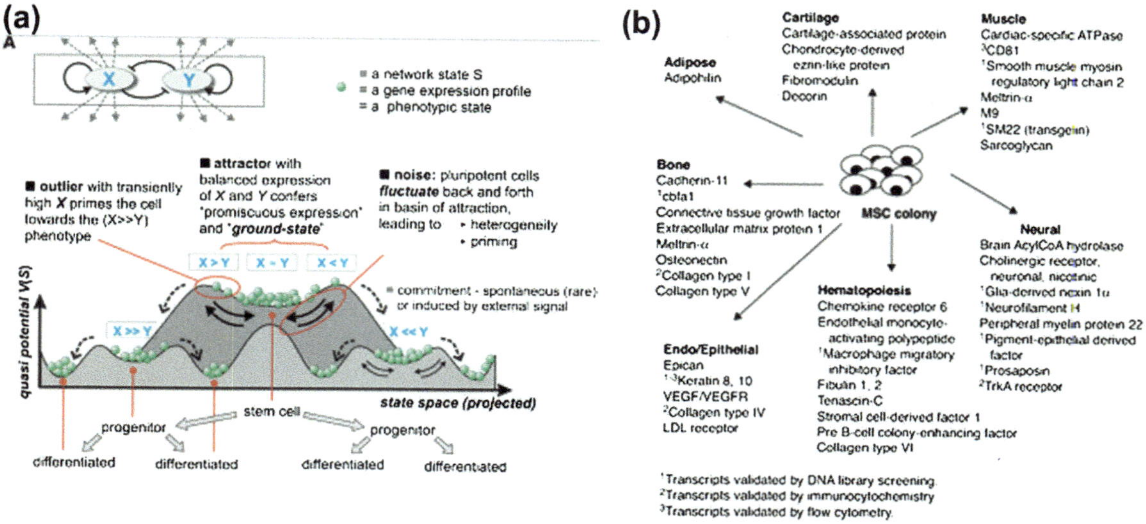

FIGURE 2.10

Stem cells can express a range of genes without losing identity or differentiating. (a) The figure shows a hypothetical epigenetic landscape with a field of multipotential cells. A population of stem cells, even if clonally derived, can express a range of genes, and the levels of gene and protein expression can oscillate, although some states are more common or "preferred". X and Y may represent master genes leading to different pathways; in the case of MSCs, X = PPARγ (adipocyte pathway) and Y = BMP2 (osteocyte pathway). (a) Drawing reprinted with permission from Figure 3A in Huang (2009). (b) Using micro-Serial Analysis of Gene Expression (microSAGE), the genes expressed from a single colony derived from a single MSC can be analyzed and are found to contain transcripts from many different lineages, ready to respond to a variety of environments. *Reprinted from Tremain et al. (2001).*

a fluid-filled cavity called the blastocoel, and an ICM. Embryonic stem cells were first derived from 3.5-day-old mouse embryos (Evans and Kaufman, 1981). The embryos are transferred to a culture dish, and after dissection, the ICM is isolated from the rest of the embryo by aspiration into a pipette. The ICM is then transferred to a new dish, and examined for undifferentiated morphology (Figure 2.11).

Human embryonic stem cells are obtained from surplus embryos generated from in vitro fertilization by fertility clinics. These embryos are usually normal but will be unused by the patients who, with informed consent, donate these cells for generating stem cells for research that may one day help others. Most human ES cells are derived from the ICM of 5-day-old blastocyst consisting of 30–34 cells, and separated from the trophectoderm by microsurgery or immunosurgery (using antibodies that break down the trophectoderm). Others have also shown that human ES cells can be derived at much earlier stages of embryogenesis, such as the morula stage (Chung et al., 2008). Parthenogenesis is the process in which an egg is able to undergo development into an embryo in a process that mimics early fertilization without a sperm. While lower vertebrates can asexually reproduce this way, the parthenotes of higher vertebrates

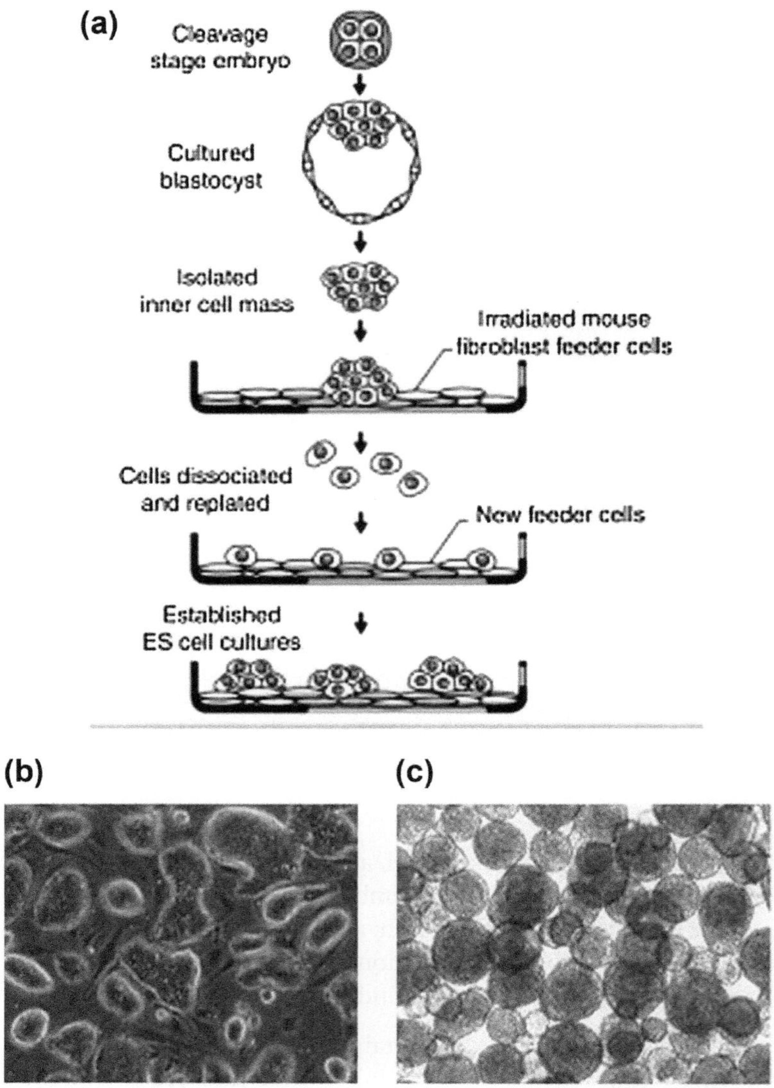

FIGURE 2.11

(a) Derivation of human ES cell lines. Human blastocysts were grown from cleavage-stage embryos produced by in vitro fertilization. ICM cells were separated from trophectoderm by immunosurgery, plated onto a fibroblast feeder substratum in a medium containing fetal calf serum. *Reproduced with permission from Odorico et al. (2001). (b) The onset of ES cell differentiation. Left: Mouse embryonic stem cells in colonies attached to culture plastic. (c) Embryoid bodies floating in the culture medium. Image from first edition of this book.*

do not survive (Lengerke et al., 2007). The parthenotes are alternatives to the destruction of a viable embryo. Guidelines for the detailed characterization of human ES cells have been reviewed by researchers in the field and published (Adewumi et al., 2007) (Figure 2.12).

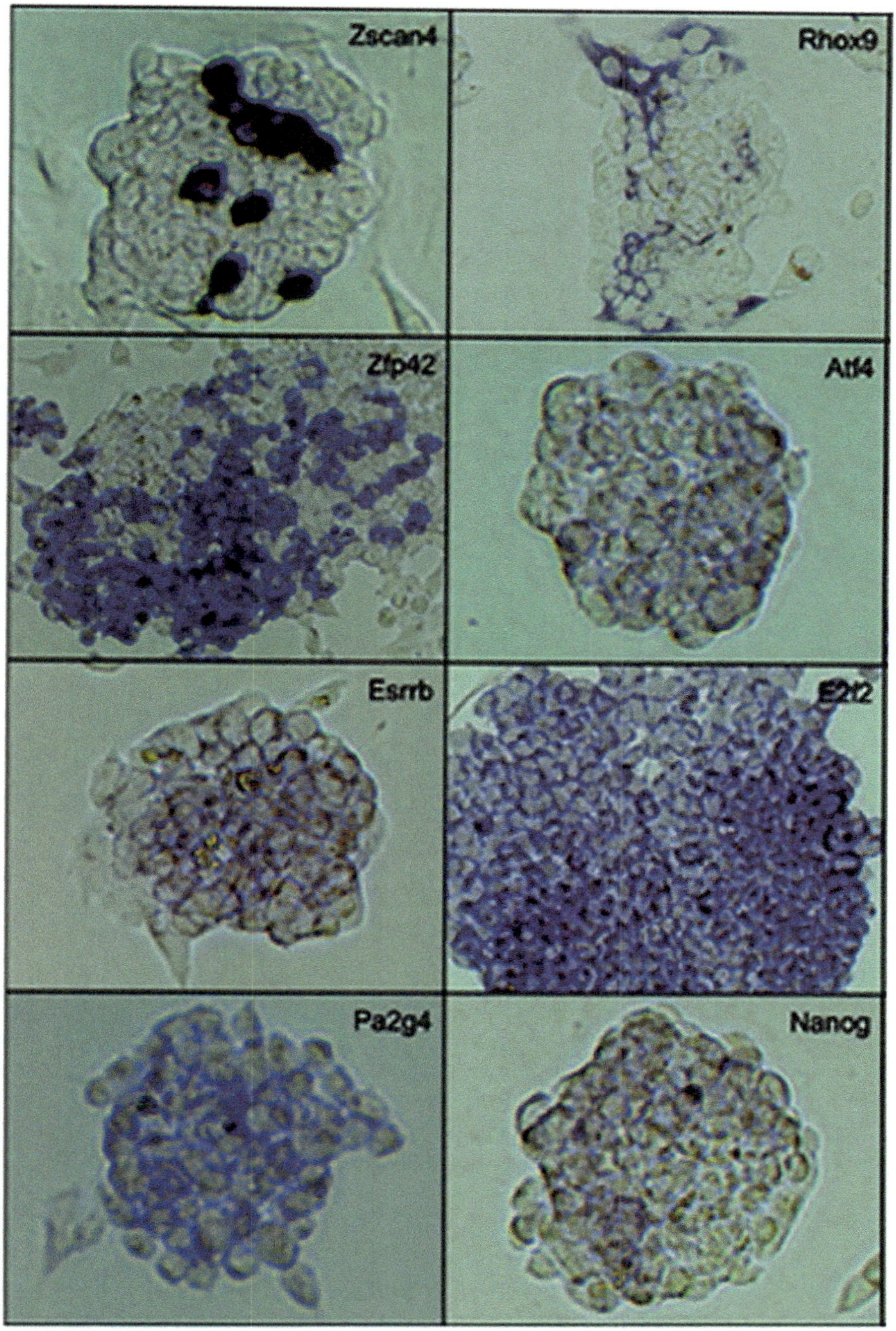

FIGURE 2.12

Microheterogeneity is a characteristic of stem cells. Mouse embryonic stem cells from cell line 129.3 were cultured without feeder cells on multiwell plastic plates for 3 days to allow small colonies to form. The cells were fixed in 4% paraformaldehyde in saline. Each well of the culture plate was hybridized with a single complimentary gene probe for genes of interest, here Zscan4, Rhox9, Zfp42, and so forth, and the probes detected by the antibody to the digoxigenin-labeled probes and colorigenic NBT/BCIP buffer. Results show that even in the colonies derived from a single cell, some daughter cells express different genes more strongly than other cells. *Reprinted with permission from Figure 3b (left 2 columns) from paper by Carter et al. (2008).*

2.6.2 Germ Cell-derived Stem Cells

Pluripotent stem cells can also be derived from germ cells and their progenitors. Embryonal carcinoma cells (ECCs) were the first pluripotent stem cells to be identified in the 1960s, originally from the mouse (Stevens, 1967) and later from human tissue (Fogh, 1975). ECCs are pluripotent cells derived from adult testicular teratomas (benign tumors) or teratocarcinomas (malignant tumors) originating from undifferentiated primordial germ cells (PGCs). PGCs are the progenitors of the germ cell lineage and are unipotent. However, under the appropriate environmental cues, PGCs can revert or dedifferentiate into pluripotent stem cells. This ability is also demonstrated in culture from PGCs isolated from late embryonic or early fetal period which can be used to generate pluripotent stem cells called embryonic germ (EG) cells (Resnick et al., 1992; Shamblott et al., 1998). Pluripotent stem cells have also been generated from adult mouse and human spermatogonial stem cells and referred to as germ-line stem (GS) cells (Guan et al., 2006; Conrad et al., 2008). EC, EG, and GS cells share similar features with iPS and ES cells including growth factor requirements and marker expression with some exceptions. For example, unlike other pluripotent stem cells, EC cells do not grow in colonies and like other cancer cells do not require additional growth factors beyond serum in the culture media. EC cells are cancer stem cells and their karyotype is unstable rendering their use in therapeutic development studies confounded with genetic abnormalities and not appropriate for any potential use in stem cell-based therapies or tissue engineering.

2.6.3 Induced Pluripotent Stem Cells

The discovery by two independent laboratories that human ES and EG cells can be grown in culture revived the notion that human cells could be expanded in large numbers and caused to differentiate, providing an unlimited cell source for treatment of a variety of diseases and tissue injuries. However, technical (immune rejection of transplanted cells) and ethical concerns (use of human embryos) regarding the use of these cells provided the impetus to find alternative somatic cell sources for pluripotent stem cells.

It was already known that somatic cells can be reprogrammed into an ES cell-like state by a technique called somatic cell nuclear transfer. Using this technique the first mammal, Dolly the sheep, was cloned by replacing the nucleus from an egg with the nucleus of a somatic cell and implanting the new cell in its mother for normal development (Wilmut et al., 1997). Using a similar approach, an ES cell was fused with a somatic cell and produced viable offspring (Cowan et al., 2005; Tada et al., 2001). These experiments provided the proof of principle that ES cell cytoplasm can cause reversion of a somatic cell nucleus into an embryonic state from which it can develop normally.

Using a novel approach, Yamanaka et al. (Takahasi et al., 2006) demonstrated that by expressing genes known to promote the ES phenotype in somatic cells, they could convert them into ES-like cells. These cells are called induced pluripotent stem cells and they were originally generated by infecting a somatic cell with retroviruses carrying the four coding sequence of transcription factors Oct4 and Sox2, which promote pluripotency in early embryos and ES cells, and the oncogenes c-Myc and Klf-4 which were also known to contribute to the ES pluripotency and self-renewal. By retroviral infection with these four transcription factors in mouse fibroblasts, the differentiated fibroblasts were reprogrammed into a pluripotent state.

Since their conception, iPS cells have been generated by multiple laboratories from a variety of species and also from cells of patients with specific diseases to allow for detailed study of the diseased state. To date, most iPS cells have been generated by adenoviruses and lentiviruses that express genes regulating pluripotency—Oct4, Nanog, and Sox2—in addition to oncogenic factors such as c-Myc and Klf4 (Daley et al., 2009). These studies show that pluripotent genes such as Oct4 and Sox2 and/or Nanog are critical for reprogramming while c-Myc and Klf4 expression, though not required, significantly increases reprogramming efficiency. Klf4 or c-Myc can also be replaced with small molecules that modify the epigenome such as histone deacetylase inhibitor (Huangfu et al., 2008) or by the SV40 large T antigen (Mali et al., 2008). Another issue is the low efficiency of converting somatic cells into iPS cells and most laboratories report 0.001–10% efficiency rates. This inefficiency of iPS cell generation is also dependent on the cell source where less differentiated cell types (such as MSCs or fetal cells) generate iPS cells at higher efficiency than adult or more mature cell types (Yu et al., 2007).

An important goal is to generate iPS cells within which the pluripotency genes can be turned off to allow their differentiation into functionally mature cell types, and to prevent the iPS cells from causing tumor formation after transplantation. To alleviate these concerns, great effort is being spent to develop technology that will allow the expression of pluripotent factors without the integration of foreign DNA into the host genome. These strategies include: (1) transfections of the proteins themselves (protein transduction) or mRNAs for the reprogramming factors; (2) small molecules (drugs) that are able to facilitate expression of the reprogramming factors; and by (3) the use of nonintegrating vectors (DNA "footprint free" technology) such as episomal plasmids that transiently express the reprogramming factors.

In the laboratory, after transfection of the reprogramming factors into cells, stable iPS cell lines can be detected in 2–4 weeks. At 2 weeks, individual colonies with tight compact morphology are carefully selected, dissociated, and used to generate stable cell lines. Selection of bonafide iPS colonies is critical as many colonies will never become pluripotent stem cells completely. For this purpose, iPS cell lines may be FACS sorted for cell surface marker expression such as TRA-1-60, TRA-1-81, and SSEA4 used to isolate human iPS cells. Although iPS cells display similar morphology and marker expression as ES cells, their genetic and epigenetic expression patterns are not identical and iPS cells retain some characteristic markers of their cell of origin, and must be further characterized.

2.6.4 Differentiation of Pluripotent Stem Cells

Human pluripotent stem cells might be an ideal cell source for tissue engineering and regenerative medicine, because of their indefinite proliferation capacity and pluripotency. Pluripotent stem cells are often mentioned as a promise for the cure of Parkinson's disease, diabetes, and cardiovascular diseases. This inevitably means that these cells have to be differentiated into, respectively, neurons, insulin-producing cells, or cardiomyocytes. In many published articles, the in vitro and in vivo differentiation of human pluripotent stem cells has been described. Useful cell types such as neurons (Carpenter et al., 2001; Reubinoff et al., 2001; Zhang et al., 2001), cardiomyocytes (He et al., 2003; Mummery et al., 2003), hepatocytes (Lavon et al., 2004; Rambhatla et al., 2003), pancreatic beta cells (Assady et al., 2001), endothelial cells (Levenberg et al., 2002), blood cells (Chadwick et al., 2003; Kaufman et al., 2001), and chondrocytes (Vats et al., 2006) have all been successfully derived in the laboratory. Levenberg et al. (2003) demonstrated the differentiation of human ES cells on polymeric scaffolds into three-dimensional structures with characteristics of developing neural tissues, cartilage, liver, or blood vessels.

2.6.5 Applications of Pluripotent Stem Cells

2.6.5.1 Application of Mouse Pluripotent Stem Cells

It is obvious that mouse pluripotent stem cells will never find clinical applications. However, the knowledge gained from mouse pluripotent stem cells was used for the isolation, growth, and differentiation and thus possible future applications of human pluripotent stem cells. Furthermore, mouse ES and iPS cells are used as a model system to study early embryonic development and differentiation.

Another valuable feature of mouse ES cells is the ability to create genetically modified mice. When genes are introduced into an ES cell, and these ES cells contribute to the germ cells (eggs or sperm) of the chimeric progeny, it is possible to breed a line of genetically modified mice. In a knock-out mouse, the function of a specific gene is interrupted, and the biological function of the gene can be studied. When a mutation is introduced that is known to be the cause of a human genetic disease, the mice may serve as a model for this human disease. Much of the knowledge of stem cell self-renewal is also based on the use of these knock-out, knock-in, and gene over-expressing mouse models.

2.6.5.2 Application of Human Pluripotent Stem Cells

Human ES and iPS cells can be used as a model system to study human embryonic development and differentiation. While mouse models have provided critical information regarding our current understanding of human biology, the ability to study human cells and their development is needed to fill gaps in our understanding of human biology. In addition, human iPS cells have been generated from patients with specific diseases including Parkinson's, multiple sclerosis, sickle-cell anemia, and diabetes. Thus iPS cell lines make it possible to study the development and characteristics of these diseases in vitro.

However, to use human ES or iPS cells in therapeutic applications, further optimization of their expansion and differentiation into function cell types is required. The current methods for deriving these cell lines and maintaining them in culture is time-consuming, labor-intensive, and their chemically defined media is expensive. This is despite current efforts to generate affordable feeder-free growth conditions.

The ability to derive and grow these cells without animal products is also necessary for their clinical use. After the discovery of prion diseases, such as Creutzfeldt–Jakob disease (Bovine Spongiform Encephalopathy or mad cow disease), the risk of infection by nonhuman pathogens is well recognized. It is not just the animal sera used in culture that contaminate the human ES cell cultures. In 2005, it was discovered that human ES cells take up nonhuman sialic Neu5Gc (Martin et al., 2005) from the growth media and feeder cells. This uptake and expression of Neu5Gc on the surface of any tissue developed from human ES cells could induce an immune response upon transplantation. The use of serum-replacement media instead of serum in the media does not solve this problem, since there are still animal proteins in the serum replacement. The extracellular matrix proteins that may replace the feeder cells are also usually from animal sources. Thus, the only solution that could alleviate these concerns would be to culture on human feeders, using human serum and supplements.

Immune rejection can be expected from any human ES cell line that is not derived from the patient's own cells. Like patients with organ transplants, patients injected with human stem cells would have to be given immunosuppressive drugs for the rest of their lives. Possibly, ES cells could be genetically modified to escape host immune responses. Alternatively, the use of the patient-derived iPS cells is a possible solution. However, it has also been shown that iPS cells transplanted back into the same strain of inbred mice elucidated a similar immune response as ES cells derived from the same strain (Zhao et al., 2011). Thus, it is currently unclear whether patient-specific iPS transplants will solve the problem of immune rejection.

Another challenge will be the purification of differentiated cell types from the heterogeneous cell populations that normally result from most current methods of differentiation. The complex signaling pathways and growth factor requirements to differentiate ES and iPS cells into mature functional cell types are not fully understood. As a result, even though specific growth factors are added to the culture medium, the outcome is usually a low proportion of differentiated cells in a mixed population of cells. Purification of a cell population can be done by selecting for tissue-specific surface markers, or by transducing the cells using a tissue-specific promoter to drive a selectable marker such as antibiotic resistance. In the latter way, only cells of the desired tissue type will survive treatment with antibiotics.

Besides selecting for the differentiated cells, it will also be important to negatively select for the remaining undifferentiated population of pluripotent stem cells. These cells, once implanted into the patient, may have the capacity to form teratomas. Although these tumors are generally benign, it is a highly undesired side effect.

Another major milestone will be the demonstration of treatment efficacy in large animal models. ES and iPS cells have been isolated from nonhuman primates, which are widely used as experimental animal models for human diseases. These pluripotent stem cell lines could be a useful resource for preclinical stem cell research and developing ES and iPS cell-based transplantation therapies.

In recent years, there has been tremendous progress in the isolation procedures and differentiation of human ES cells, but therapeutic application is still premature. The first clinical trial using human pluripotent stem cells involved human embryonal carcinoma cells differentiated into neural cells to treat stroke (Wechsler, 2009). While this study showed significant signs in patient improvement early after transplantation and no adverse effects, it did not show long-term benefits. Currently, there is a clinical study underway in the United States testing retinal cells derived from human embryonic stem cells and a second trial using iPS cells to treat macular degeneration (Schwartz et al., 2012).

2.6.6 Political and Ethical Concerns

Embryonic stem cell research is controversial because of the use of human embryos. One of the main questions in ES cell isolation is whether destroying the embryo means destroying life. To answer that question, one has to define when life begins. For some, life starts with the fertilization of the egg, for others after 40 days, or after 120 days, or sometimes even at birth.

In some countries, research on human ES cell is completely forbidden; in other countries, research is allowed but it is illegal to create new cell lines that would require the destruction of an embryo. Not all countries have policies covering embryonic stem cell research and many countries have a moratorium on ES cell research. Laws and guidelines generally describe:

- The use of embryos for research.
- The production of embryos for research.
- The use of excess embryos from IVF treatment.
- The creation of new ES lines.
- The use of existing established human ES lines.
- Somatic Cell Nuclear Transfer in ES cells and therapeutic cloning.
- Reproductive cloning.

In Europe, ES cell research is allowed, for example, in the United Kingdom, the Netherlands, Sweden and Belgium, whereas it is forbidden in Lithuania and Austria. In Germany, only embryonic stem cells that date from before 2007 can be imported. Australia and Israel allow ES cell isolation from excess IVF embryos. In the United States, embryonic stem cell research is not forbidden, but only federally approved human ES cell lines are allowed when using federal funds. These lines are registered at the National Institutes of Health. Private funding of research through foundations does allow further isolation of human ES cells. In contrast, there are no similar restrictions for research on human iPS or germ cell-derived stem cell lines. Instead, iPS cells have been promoted as an alternative cell source to replace ES cells for research and therapeutic purposes to alleviate the ethical concerns involving the use of embryos.

2.7 MULTIPOTENT STEM CELLS

Multipotent stem cells come from a variety of adult and fetal tissues. They are restricted in the lineages that they may differentiate into, but some of these are very useful lineages. Here we will outline a few sources that have been considered for tissue engineering and therapeutic use.

2.7.1 Hematopoietic Stem Cells

From before birth to the time of death the blood cells are continually replaced. Hematopoietic stem cells are the precursor cells for all of the blood lineage cells including the lymphoid progenitor cells that differentiate into the immune system T cells, B cells, and natural killer (NK) cells, as well as the myeloid progenitor cells that differentiate to monocytes/macrophages, basophils, eosinophils, red cells, and platelet-releasing megakaryocytes. The importance of HSCs to the study of stem cells cannot be overstated, as many of the stem cell concepts and definitions derive from the scientific and medical study of these stem cells. HSCs, however, are difficult to grow in culture and much effort has been expended to improve their in vitro expansion. The current challenges are many but in vitro expansion that can maintain the stem cell phenotype, along with reliable engraftment in patients needing a bone marrow transplant are primary areas of research effort. The in vitro study of HSCs really began in the 1960s with the experiments of James Till and Ernest McCulloch who studied the effects of radiation on bone marrow in mice at the Ontario Cancer Center in Toronto (Becker et al., 1993). They were able to develop methods to save animals from the effects of radiation by transplanting donor bone marrow that repopulated all of the blood cell lineages. They noticed that many of the transplanted donor cells were then found at an early stage in the spleen where they developed as colonies derived from a single unit, perhaps a single donor stem cell. These colonies could be counted and the stem cells were termed colony forming unit-spleen (CFU-S). Methods were then developed to optimize the engraftment of donor marrow cells and increase the number of CFU-S (Figure 2.13). While there are HSCs engrafted elsewhere in the recipient bone marrow, it is simpler to count the colonies easily visible in the excised spleen as an assay (Figure 2.14).

Through innumerable studies, the blood cell lineages have been traced from the adult cell types back to the blood stem cell—the HSC (Figure 2.3). After many years of study, an understanding of the relationship of HSCs and their progeny at cellular and molecular levels, and the genes and signals involved in stepping from the stem cell to the fully differentiated adult cell is possible. Still, there are many details to be studied further.

Hematopoietic stem cells can be isolated from the bone marrow, but since it is an invasive procedure, this becomes increasingly infrequent. Today, clinical mobilization of HSCs from the bone marrow is accomplished by treating the donor with the HSC growth factor granulocyte-colony stimulating factor (G-CSF) for several days which results in more circulating HSCs in the peripheral blood, and collecting the HSCs by the process of plasmaphoresis, wherein the cells are collected and the plasma fluid fraction is returned to the donor. While the G-CSF induces the HSCs to proliferate in the bone marrow, they are partly held in the stem cell niche by HSC–stromal cell interactions of their surface receptors, notably stromal cell-derived factor-1 (SDF-1) and its receptor on the HSC, chemokine CXCR4. To help displace the proliferating HSCs from their bone marrow niche a compound such as AMD3100 may be given; this compound displaces the SDF-1/CXCR4 interaction and allows greater release of HSCs from bone marrow into the circulation (Figure 2.15).

From the tissue engineering perspective, there are a number of areas that need further study including: (1) improvements in clinical HSC isolation, (2) in vitro propagation of HSCs, and (3) gene-modified HSCs for many inborn error of metabolism diseases. Blood from the placenta and umbilical cord is an alternative rich source for hematopoietic stem cells. It is an appealing cell source, since these tissues are usually discarded after birth.

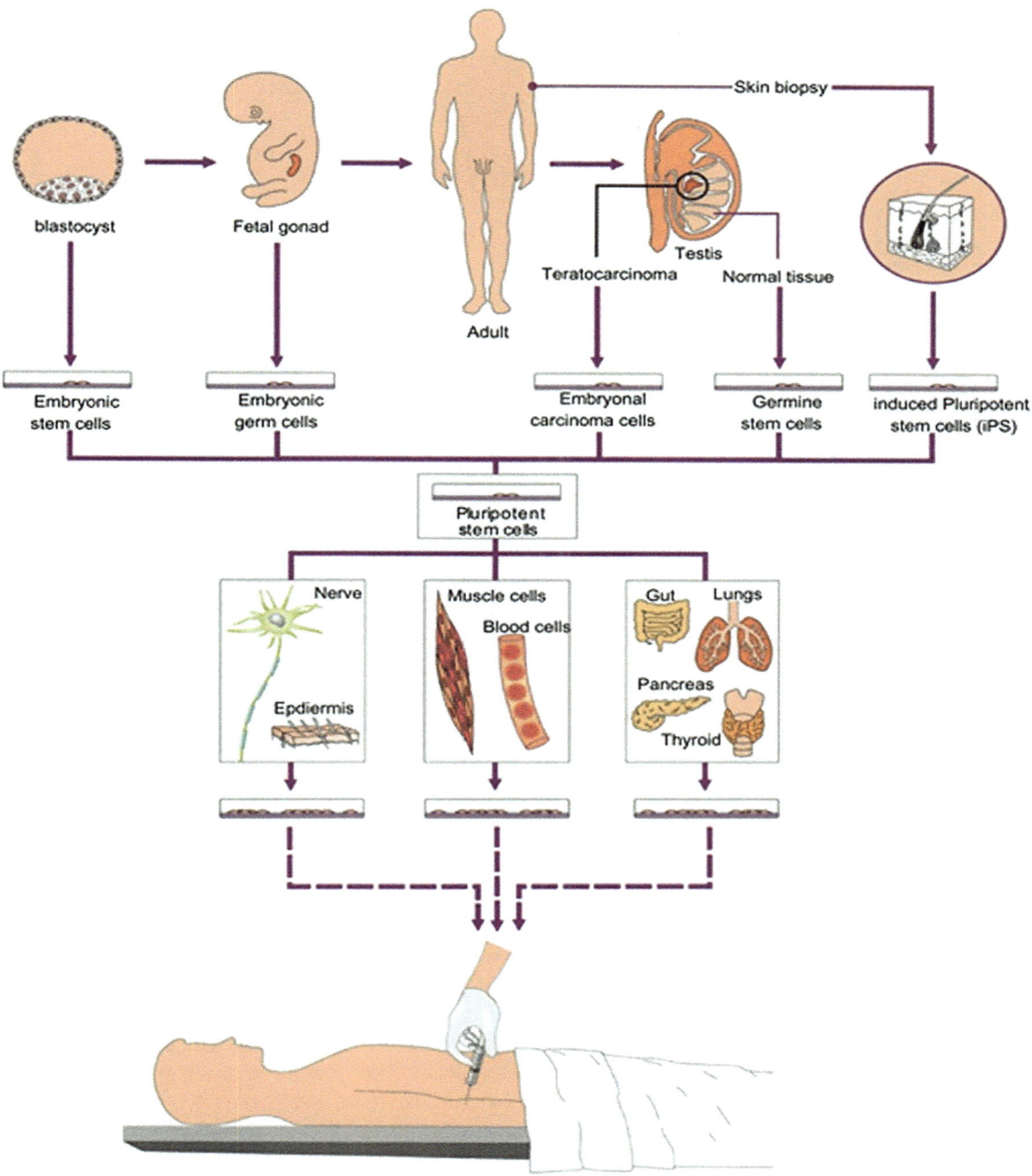

FIGURE 2.13

Human pluripotent stem cells include embryonic stem cells cultured from cells of the inner cell mass of normal or parthenogenetic blastocysts, embryonic germ cells generated from primordial germ cells in late embryonic development, embryonal carcinoma cells isolated from adult teratocarcinomas, germline stem cells derived from spermatogonia, and induced pluripotent stem cells generated by reprogramming differentiated adult cells. Pluripotent stem cells exhibit the potential to produce all cell types of the body. Thus, directed differentiation of these cells holds promise for treating a wide variety of diseases and injuries. *Image from Swelstad and Kerr (2009).*

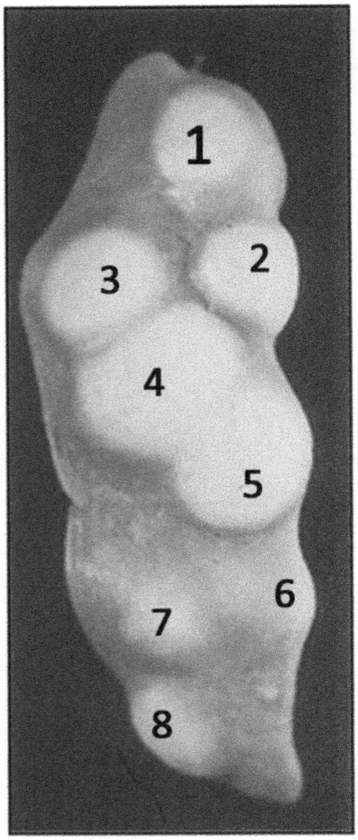

FIGURE 2.14

Mouse spleen displaying ~8 clonal outgrowths from transplanted hematopoietic stem cells. Isolated marrow cells (1–2 x 10^6 cells) containing presumptive stem cells were delivered intravenously to a recipient mouse and the spleen harvested and fixed at 12 days. Waves of hematopoietic repopulation are seen in the spleens of irradiated recipients after transplantation of normal bone marrow cells. At low transplant cell numbers, each wave can be identified as macroscopic colonies. Six to seven days after transplantation, small colonies can be seen. These transient colonies are entirely erythropoietic, and are characteristic of fetal or rapidly regenerating hematopoiesis. These early colonies disappear, and are replaced by days 9–11 after transplantation by larger colonies that contain the three myelopoietic lineages. These also disappear and by days 12–14 when trilineage colonies are seen, with hematopoietic stem-cell progenitors. Assays of colonies of the first two waves are seldom used, since their progenitors may be studied more effectively by tissue culture methods. However, the capacity to form spleen colonies from day 12 to day 14 remains an important criterion for the identification of myelopoietic stem cells. *Reprinted with permission from McCulloch (2001).*

2.7.2 Mesenchymal Stem Cells

Mesenchymal stem cells are a class of multipotent adult stem cells that can be isolated from many tissues but are frequently isolated from bone marrow. Bone marrow is thought to be a renewable source of these cells, as it is for HSCs, and the harvesting of most other tissues would mean the sacrificing tissue that is "in

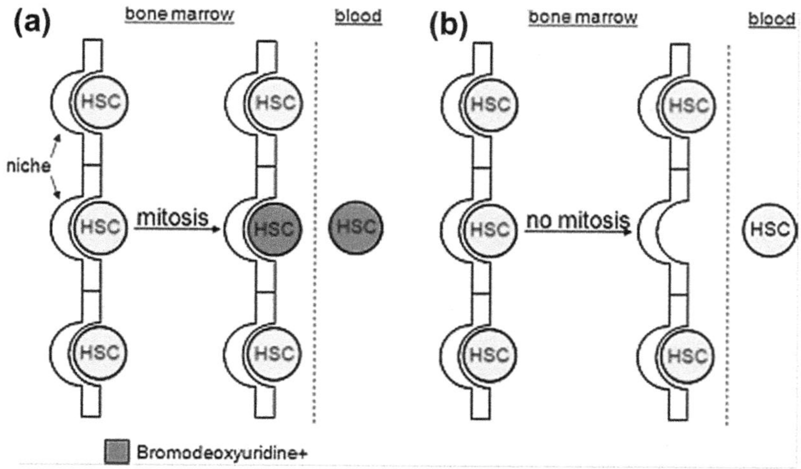

FIGURE 2.15
Hematopoietic stem cells continually replicate and enter the blood stream. Egress from the stem cell niche can be division dependent (a) or division independent (b). The DNA of dividing cells can be labeled by the nucleotide analog bromodeoxyuridine (BrdU) and analyzed. In this experiment, it was found that only a minority of the circulating HSCs were labeled at any time point and thus, replication was not needed for HSCs to exit their bone marrow stem cell niche. *Reprinted with permission from Figure 1 in Bhattacharya et al. (2009).*

use." It is common to designate MSCs by their tissue of origin such as BM-MSCs for bone marrow-derived MSCs. For isolation of human MSCs, a bone marrow aspirate is taken from the iliac crest with a local anesthetic, such as lidocaine. The aspirate sample is subjected to density centrifugation and the density layer containing the nucleated cells is collected with a pipette, washed in saline to remove the density medium, and placed into culture medium. After 10–14 days in the incubator, colonies of proliferating cells are visible under the microscope. MSCs have been shown to differentiate to several important lineages in vitro and in vivo such as bone, cartilage, and fat with strong fidelity, that is, without the appearance of other unwanted lineages, a problem that is common with pluripotent stem cells (Pittenger et al., 1999) (Figure 2.16).

In the 1960s, Alexander Friedenstein isolated cells from guinea pig bone marrow that could be cultured in vitro to form colonies of fibroblast-like cells that he termed colony forming units-fibroblastic or CFU-F. The cells were placed in sealed chambers with a dialysis membrane that allowed the inflow of nutrients and the chambers were imbedded under the skin of host animals for several weeks. When excised and examined histologically, some of the chambers contained bone and cartilage, indicating differentiation of the implanted cells (Friedenstein et al., 1970). Similar work was conducted with rabbit bone marrow cells in the 1980s, a time when the hierarchy of blood lineages was now understood, and Maureen

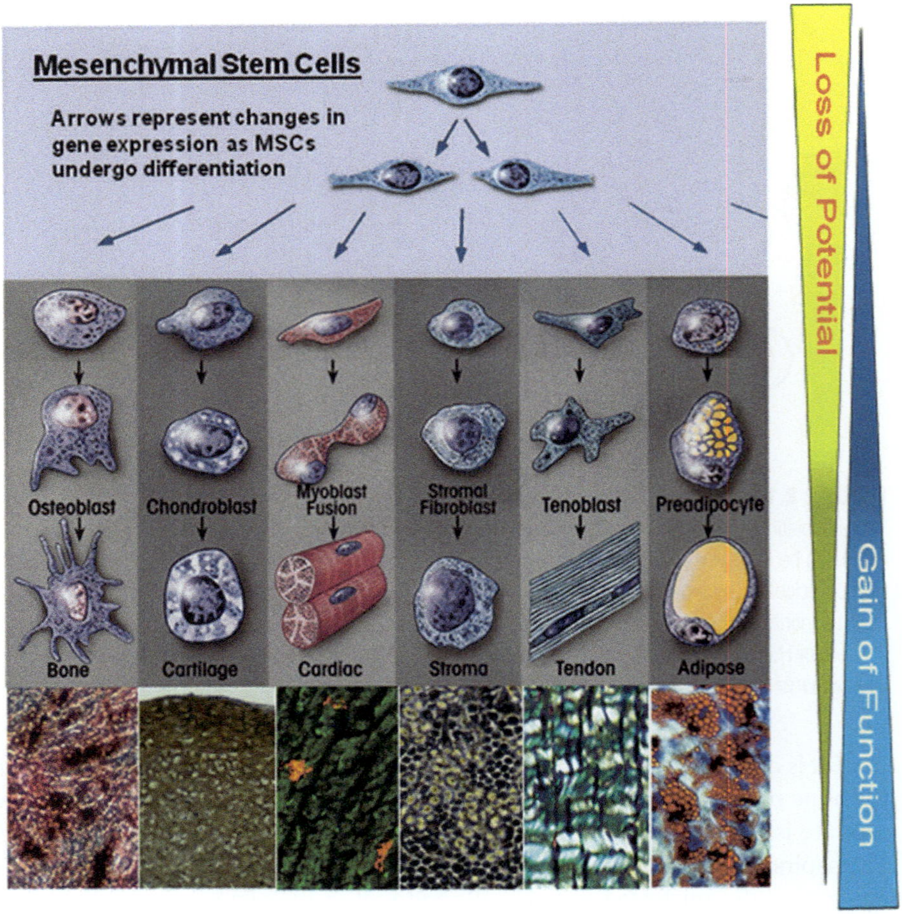

FIGURE 2.16

MSCs can proliferate in culture and retain their ability to differentiate but as they differentiate they stop proliferating. MSCs from bone marrow were cultivated and then subjected to in vitro or in vivo differentiation conditions: (1) osteogenic differentiation in vitro is revealed after 7 days by alkaline phosphatase expression (red) and silver-stained calcium nodules (black clusters); (2) chondrogenic differentiation in vitro is shown by the strong extracellular staining with antibody to collagen type II (brown DAB reaction product) after 21 days and characteristic lacunae histologically; (3) when MSCs are injected into healthy hearts, the ones that are found here shown labeled with red membrane dye Di I are positive for desmin (green) similar to surrounding cardiomyocytes (red + green = yellow); (4) MSCs produce growth factors and support "cobblestones" of hematopoietic stem cells growing on top of a monolayer of MSCs; (5) in tissue-engineered collagen I constructs containing MSCs in a rabbit tendon repair model, the MSCs provide structural strength and a characteristic crimp pattern under polarized microscopy; (6) MSCs induced to differentiate to adipocytes in vitro display large lipid droplets here stained with the lipophilic dye oil red O and counterstained with eosin to show nuclei.

Owen proposed a similar lineage diagram for the CFU-F cells from bone marrow (Luria et al., 1987; Owen and Friedenstein, 1988). Thus, there were two strong lines of evidence of at least two multipotential cells from bone marrow. Arnold Caplan and Steve Haynesworth isolated these multipotential cells from human bone marrow and Caplan coined the term mesenchymal stem cells to describe these cells (Caplan, 1991; Haynesworth et al., 1992). Although publications on MSCs started slowly, over 20,000 papers have been published in the last 20 years, most using bone marrow-derived MSCs.

CLASSICAL EXPERIMENT

Multilineage Potential of Adult Human Mesenchymal Stem Cells

The paper by Pittenger and coworkers in the *Science* issue of April 1999 (Pittenger et al., 1999) marked the broad introduction of the term "mesenchymal stem cells" (MSCs) for a population of cells originally identified in guinea pigs by Friedenstein in the 1960s as adherent colony forming units-fibroblast (CFU-F) isolated from bone marrow (Friedenstein et al., 1970) and the human equivalent were isolated by Caplan (Haynesworth et al., 1992). Pittenger and coworkers obtained 50 bone marrow aspirates from 19- to 57-year-old donors and selected the mononuclear fraction from the aspirate using a density gradient. Most of the cells from this fraction belong to the hematopoietic lineage and will not adhere when brought into culture. A small percentage of the cells (0.001–0.01%) did adhere and developed into symmetric colonies of cells. Flow cytometry analysis revealed that all cells were consistently positive for markers like CD29, CD44, CD71, CD73 (SH3), CD90, CD105 (SH2), and CD106, but negative for the hematopoietic marker CD34. Despite this apparent homogeneity, MSCs, like other stem cells, display some heterogeneity with respect to growth rate, phenotypic plasticity, and colony morphology. Efficient expansion was achieved, with 50–375 million hMSCs within two cell passages (6–10 population doublings). The researchers further demonstrated efficient differentiation of hMSCs into the adipogenic, osteogenic, and chondrogenic lineage by varying culture conditions.

This landmark paper has been cited over 18,000 times since publication, and advocates for human MSCs as a model system for questions of cell biological nature, such as cell fate decision, plasticity, and senescence as well as a readily available source of cells for tissue engineering purposes. However, the name "mesenchymal stem cell" raised an ongoing debate on the nomenclature for mesenchymal cells with multipotentiality isolated from various parts of the body including bone marrow and fat. The senescent phenotype of hMSCs in culture argues against their stemness. Therefore, many researchers rather use terms like marrow stromal cells or mesenchymal progenitor cells to describe a population of cells, which cannot be discriminated from hMSCs on basis of marker gene expression or differentiation potential.

Isolated marrow-derived stem cells differentiate to mesenchymal lineages. Cultured cells from donors were tested for the ability to differentiate in vitro to multiple lineages. Donors (a through c) 158, (d through f) 177, and (g through i) 260 were each shown to differentiate appropriately to the adipogenic (Adipo), chondrogenic (Chondro), and osteogenic (Osteo) lineages. Adipogenesis was indicated by the accumulation of neutral lipid vacuoles that stain with oil red o (a, b, and g), and such changes were not evident (j) with Hs27 newborn skin fibroblasts or (M) with 1087Sk adult mammary tissue fibroblasts. Chondrogenesis was shown by staining with the C4F6 monoclonal antibody to type II collagen and by morphological changes (b, e, and h), which were not seen by similarly culturing (k) Hs27 or (n) 1087Sk cells. Osteogenesis was indicated by the increase in alkaline phosphatase (c, f, and i) and calcium deposition, which was not seen in the (l) Hs27 or (o) 1087Sk cells.

Reproduced with permission from Pittenger et al. (1999).

CLASSICAL EXPERIMENT *Continued*

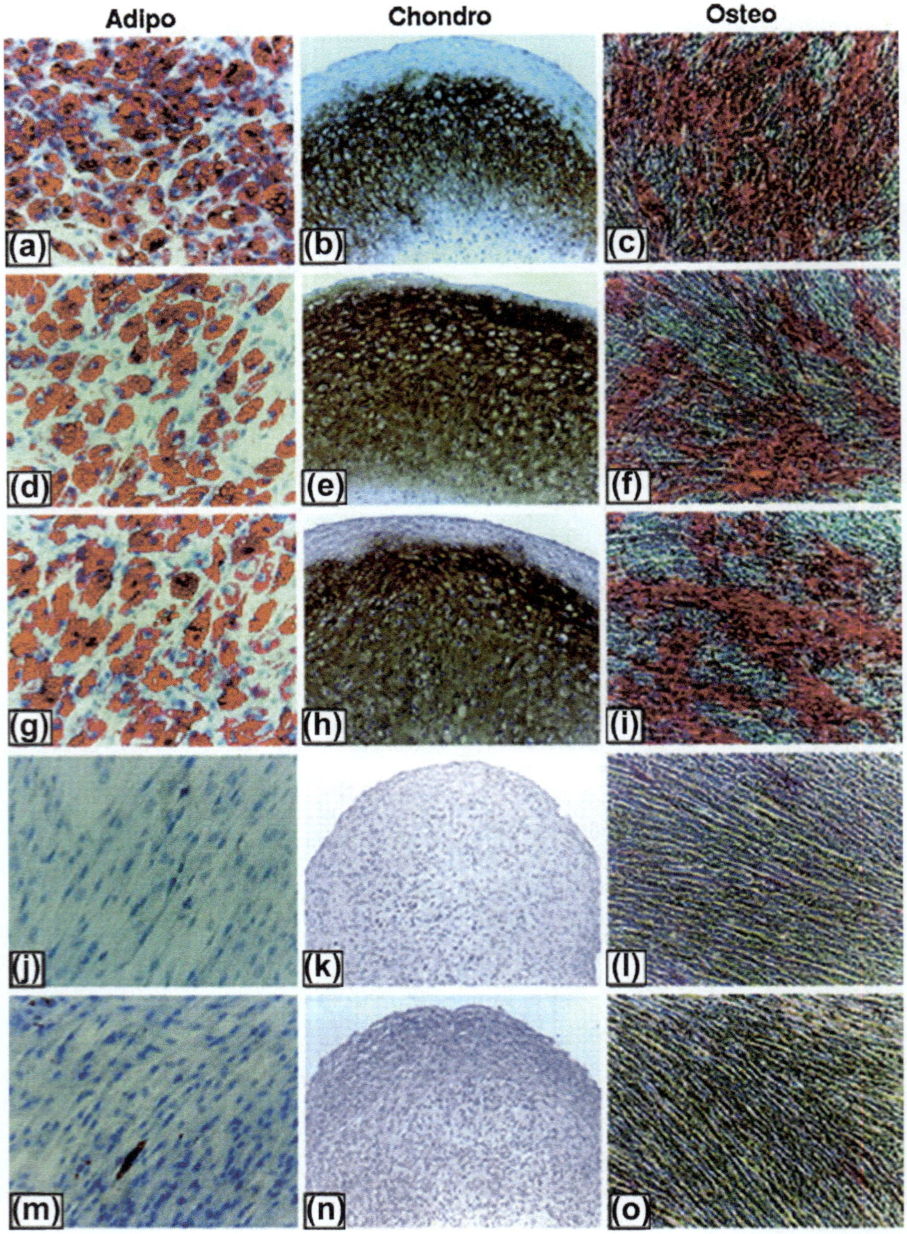

Another common source of MSC-like cells is adipose tissue that is harvested by liposuction. The adipose tissue is highly vascularized and MSCs are part of the vascular network as stromal adventitia along the many vessels (Zuk et al., 2001; Halvorsen et al., 2001). These cells are also isolated by a centrifugation procedure and may be used immediately as an uncharacterized but enriched fraction of progenitor cells termed the stromal vascular fraction (SVF). As an alternative, the cell samples can be cultured to obtain an expanded population of cells for characterization and experimental use, and the cultured cells are termed adipose-MSCs (Ad-MSCs) or adipose tissue-MSCs (AT-MSCs).

We mentioned above that MSC-like cells can be found along blood vessels in the population of microvascular pericytes that are contractile and aid in moving blood along small vessels. The pericytes have cell surface markers similar to BM-MSCs and this indicates that all tissues have "MSCs" as part of their capillary network (Covas et al., 2008). This also suggests that these peripheral MSCs would be ideal locations to quickly respond to tissue injury! This is an attractive idea but experience would indicate that these MSCs are inadequate to repair most injuries of adult tissues, where scarring and fibrosis are usually the result. However, young children have a remarkable regenerative capacity for scarless healing.

MSCs can be isolated from a bone marrow aspirate and expanded 20–30 population doublings and provide enough cells for most regenerative medicine purposes. However, MSCs cannot be propagated in an unlimited fashion in vitro. Whether this is due to an intrinsic limitation in the lifespan of an MSC or due to a limitation in the in vitro culture conditions is not clear. A recent study of MSCs that grew onto the culture dish from umbilical cord tissue pieces cultured for 15 days or 190 days gave rise to similar numbers of MSCs with near identical properties (Otte et al., 2013). This finding suggests that the aging clock only started after the tissue pieces were removed from the outgrowing cells on the culture dish, and that tissue matrix or tissue matrix containing viable cells has properties that promote continued proliferation, furthering the stem cell potential of these outgrowth cells.

2.7.2.1 Are MSCs True Stem Cells?

There is a long standing debate about the best nomenclature for MSCs and other names have been suggested—multipotential stromal cells (MSCs), mesenchymal stromal cells (MSCs), mesenchymal progenitor cells (MPCs), skeletal stem cells (SkSCs), and others. However, as the cells cannot be propagated indefinitely, some researchers prefer not to use "stem" in the naming of these cells, reserving that for the indisputable cell that gives rise to all mesenchymal lineages—when it is found. A quick search on PubMed continues to show that mesenchymal stem cell remains the name most commonly used (three times more frequent than mesenchymal stromal cell) from year to year and the abbreviation MSC also covers most of the other common names. It is likely the name will be used unless a better, bona fide mesenchymal stem cell is identified.

2.7.2.2 MSCs: The Inflammation and Injury Drug Store

One of the surprising features of MSCs and perhaps one of the most useful is their ability to modulate immune responses and inflammation. When lymphocytes, which contain a large number of T cells, from different individuals are mixed, there is a burst of proliferation as each population responds to the foreign cells. This mixed lymphocyte reaction (MLR) is a common test of compatibility and immune suppression of this response is necessary to avoid rejection anytime donor and recipient are not identical. (Autologous (self) transplantation and transplantation from an identical twin does not require immune suppression.) However, MSCs do not stimulate allogeneic lymphocytes in vitro and are not rapidly rejected when transplanted. It has also been reported that differentiated MSCs likewise do not stimulate the MLR assay (Le Blanc et al., 2003a,b). This property of MSCs has been extensively studied with different immune cell types (Aggarwal and Pittenger, 2005). It is known that MSCs produce many factors that suppress lymphocyte activation. Each factor can play an important role in avoiding rejection, from T cells, dendritic cells, NK cells, or B cells, and an increase in regulatory T cells is common as well (Figure 2.17). All of the reduction in the inflammatory response might suggest that greater levels of infection may accompany MSC delivery; however, this is not the case as MSCs also produce factors that inhibit sepsis including PDE-2 (Németh et al., 2009) and the antibacterial peptide LL-37 (Krasnodembskaya et al., 2010).

The MSCs migrate to sites of injury, and it has become common to deliver MSCs intravenously for a variety of diverse disorders including heart ischemia and graft versus host disease. This homing to injury effect has been widely studied and the release of factors at the site of injury such as SDF-1 has been studied as the inducer of MSC homing, although leaking from the capillaries at the site of injury cannot be ruled out (Marquez-Curtis and Janowska-Wieczorek, 2013). Either way, MSCs accumulate at the injury site, reducing inflammation, releasing angiogenic factors, and growth factors that accelerate injury repair and preserve neighboring healthy tissue (Caplan and Dennis, 2006). Many studies have demonstrated that MSCs can increase angiogenesis, both in the infarcted heart (Kraitchman et al., 2003; Williams et al., 2013) and in models of peripheral artery disease (Kedziorek et al., 2012; Yang et al., 2013).

2.8 STEM CELLS IN SKIN EPITHELIA

The skin is the largest organ in the body by weight and constantly exposed to insults and injuries from the environment. The healing of dermal wounds is familiar to all of us, but it is an efficient, complex process that keeps the external bacteria-rich environment outside at bay. Hair and sweat glands are also important components of the skin. There is more than one type of stem cell found in skin and its structures but the epithelial stem cells found in the basal layer of

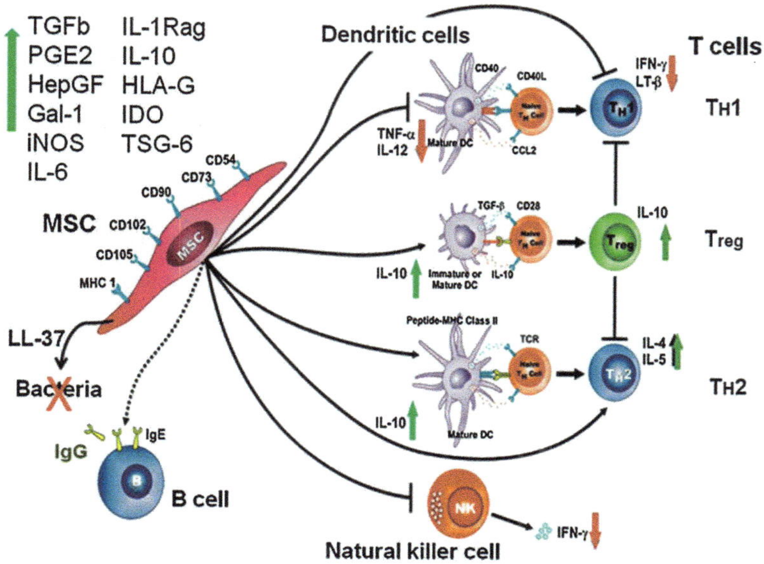

FIGURE 2.17

MSCs produce a variety of factors that can suppress rejection from immune cells. Lymphocytes were fractionated into individual populations of immune cells including dendritic cells, T cells, natural killer cells, and B cells. From these isolated immune cells, the mature proinflammatory DC1 and T_H1 were produced as well as the anti-inflammatory DC2, T_H2, and T_{Reg} and each population was tested separately in the presence of MSCs. The factors identified to date produced by MSCs that reduce the inflammatory response include TGFβ (transforming factor β), PGE2 (prostaglandin E2), HGF (hepatocyte growth factor), Gal-1 (galectin-1), iNOS (inducible nitric oxide synthase), IL-6 (interleukin 6), IL-10, interleukin 1 receptor antagonist (IL-1Rag), HLA-G (human leukocyte antigen-G), IDO (indoleamine 2,3-dioxygenase), and TSG-6 (TNF-α-stimulated gene 6 protein). While the reduction in immune cell responses would seem to open the individual to other infections such as bacteria, the MSCs also produce antibacterial peptides such as LL-37 that inhibit their growth. *Modified from Aggarwal and Pittenger (2005).*

the epidermis are responsible for the daily maintenance of skin. These epithelial stem cells are unipotent. The epidermal basal cells also provide the long-lasting epithelial replacement cells, although other stem cells from the bulge region of the hair follicle may aid rapid reepithelialization after injury. Because of the importance of skin integrity, burn victims constitute a high mortality risk, but now commercial skin graft products are available that greatly speed up the process of closure of the integumentary system (Figure 2.18).

2.9 STEM CELLS IN THE INTESTINE

The small intestine undergoes continuous renewal. The microvilli where adsorption of nutrients occurs are continuously sloughed off and replaced from the intestinal crypts at the base of the microvilli. The crypts contain

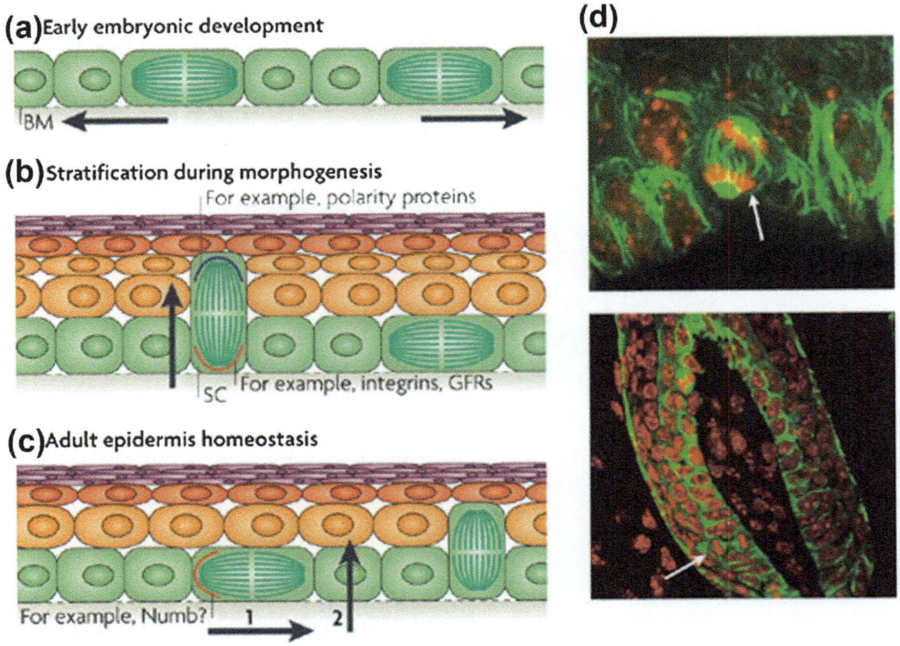

FIGURE 2.18

(a) During the early stages of embryonic skin development, most cell divisions are symmetric and parallel to the basement membrane (BM), which ensures the growth of the surface of the developing embryo and maintains the epithelium as a single layer. (b) During epidermal stratification, 70% of the cell divisions become asymmetric, such that the mitotic spindle is perpendicular to the basement membrane. This allows the development of suprabasal cells that terminally differentiate and establish the skin barrier (indicated by the arrow). In this mode of asymmetric cell division, different cell fate determinants are segregated unequally between the two daughter cells. The basal cell segregates the integrins and growth factor receptors (GFRs), which provide survival and proliferative cues to the stem cell (SC). Polarity proteins are concentrated to the apical surface and presumably become preferentially distributed to the suprabasal cell. (c) During epidermal homeostasis in the adult tail skin, asymmetric cell divisions occur with the plane of division parallel to the BM, such that only one daughter cell inherits a cell fate determinant, such as Numb, and remains an SC (step 1), whereas the other becomes committed to terminal differentiation, and probably undergoes delamination to reach the suprabasal layers (step 2). (d) Examples of asymmetric spindle orientations (white arrow) in the oral epithelium, which are similar to those seen in the embryonic epidermis (upper panel) and in the adult hair bulb (lower panel). Nuclei are shown in red and microtubules in green. *Images in part (d) are courtesy of T. Lechler and E. Fuchs, Rockefeller University, New York, USA. From Blanpain and Fuchs (2009).*

Paneth cells, part of the immune system which release defensin molecules that fight bacteria, and intestinal stem cells. The stem cell-to-microvilli cell renewal in the small intestine is the fastest of all tissues in the body! The location of the intestinal cells as they migrate from the base to the tip represents the life cycle of intestinal stem cells. These stem cells are small cycling cells identified

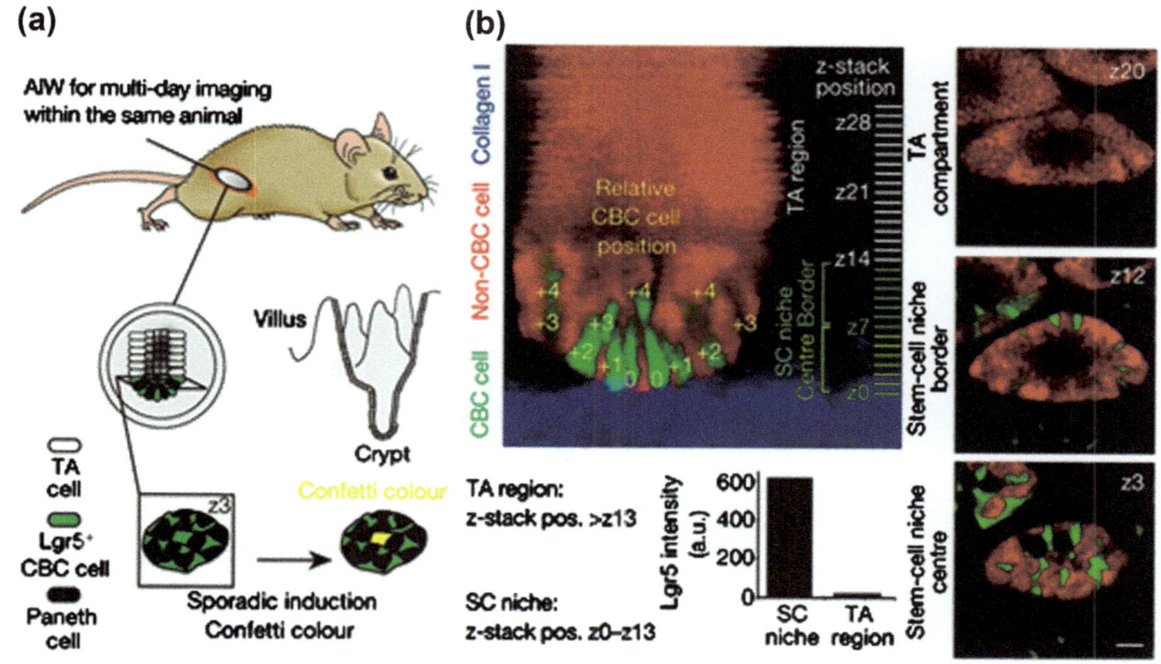

FIGURE 2.19

(a) Cartoon showing an Lgr5$^{eGFP\text{-}Ires\text{-}CreERT2}$/R26R-Confetti mouse with an AIW to visualize intestinal Lgr5^{+} crypt base columnar (CBC) stem cells and their Confetti progeny over multiple imaging sessions. (b) Lateral projection of an z-stack and representative xy images of a crypt at indicated z-stack positions (pos.). The stem cell (SC) niche (z0–z13) is defined by Lgr5-GFP fluorescence. The relative position of CBC cells to the most basal cell (row 0) determines location in the central (row 0 to +2, which translates to z0–z6) or border region (row +3 to +4, which translates to z7–z13) of the stem cell niche. a.u. = arbitrary units. Scale bar, 20 μm. *From Ritsma et al. (2014).*

by the expression of the gene Lgr5 and located between the supporting Paneth cells at the base of the microvilli. Thus, the Lgr5+ cells and the Paneth cells represent the niche for the stem cells. As the cells begin to migrate away from the crypt, they become transit amplifying (TA) cells and as they mature they remain associated with the villus of origin. Once division of the intestinal stem cell begins the replicative nature of the cells of the microvillus is one of symmetrical division, with identical daughter cells as both move out of the stem cell niche (Figure 2.19).

2.10 STEM CELLS IN THE CENTRAL NERVOUS SYSTEM

The central nervous system contains stem cells. Several sites in the brain have been found to produce cells that can proliferate and differentiate to neuronal precursors or glial precursors (to form astrocytes and oligodendrocytes). The anatomical location that is the source of these stem cells is a small region in each hemisphere near the hippocampus known as the subventricular zone (SVZ) and a second site known as the dentate gyrus (Ericksson et al., 1998; Alvarez-Buylla et al., 2002). From studies that label replicating cells, and then visualized their migration to sites of brain injury, it has been shown that some cell replacement, repair, and recovery of function are possible. It remains difficult to access these internal sites to optimize central nervous system

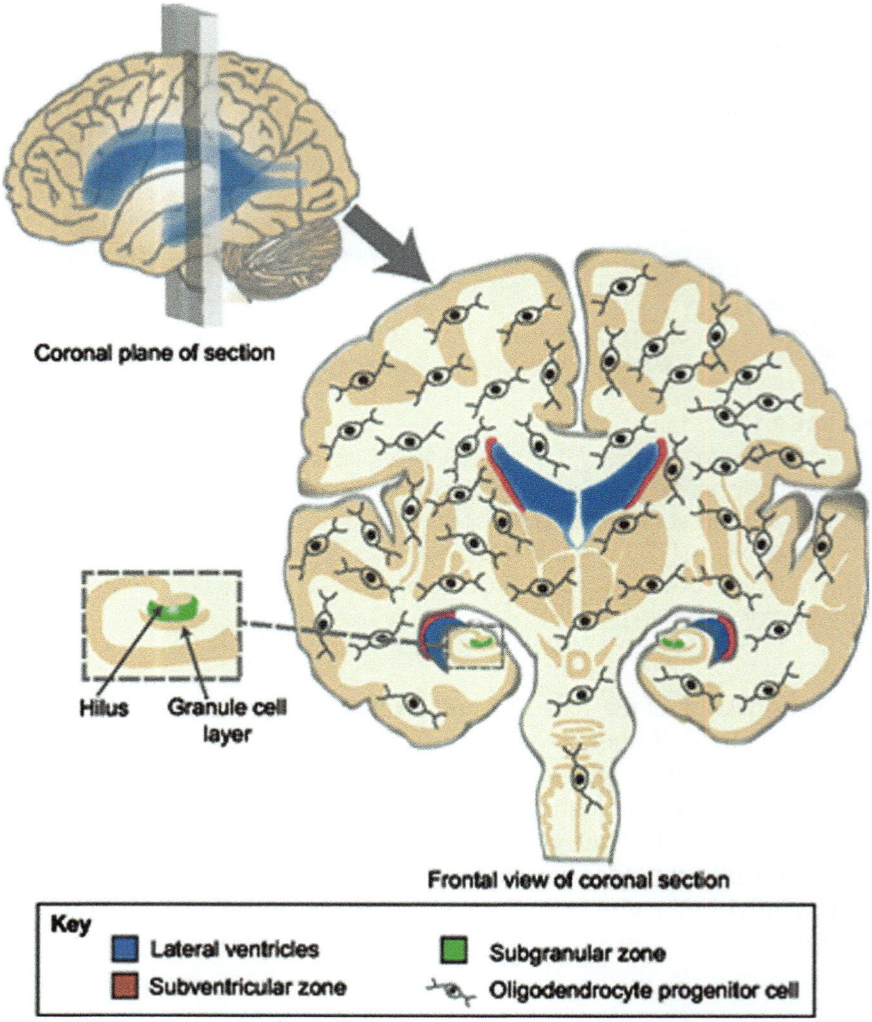

FIGURE 2.20

Location of CNS stem and progenitor cell niches. The subventricular zones line the walls of the lateral ventricles, whereas the subgranular zones reside between the hilus and granule cell layer of each hippocampus. Oligodendrocyte progenitor cells are widely distributed in the white and gray matter of the brain and spinal cord. *From van Wijngaarden and Franklin (2013).*

regeneration. Over the last 20 years, significant research has been performed on the spinal cord, the other half of the central nervous system. Following spinal cord injury, there is not a recognized source of endogenous repairing stem cells as in the brain. Some regeneration following spinal cord injury has been demonstrated by placing exogenous progenitor/stem cells into the injury site, usually with the aid of a bioengineered matrix material. Like most injuries, lesions in the central nervous system seem to respond best to early treatment, before significant fibrosis has occurred (Figure 2.20).

2.11 FUTURE PERSPECTIVES

An incredible amount of new research is now being performed on stem cells, including adult stem cells, induced stem cells, and embryonic stem cells. Many new studies in tissue engineering involve new materials in combination with stem cells but much remains to be discovered. The next decade or two will see greater understanding and implementation of stem cells for tissue engineering. Clinical trials that test effects of stem cell treatment are underway in most medical areas. The US government site that tracks most trials throughout the world (www.clinicaltrials.gov) lists over 4000 clinical trials utilizing stem cells of one type or another, but not all cell types are equally represented—searching hematopoietic stem cells find 2200 trials while the much newer MSCs find 390 trials. Some are ongoing trials while others are completed or not recruiting. Nevertheless, the enthusiasm to utilize stem cells as the latest tool to address unmet medical needs is clear and seems to be growing.

2.12 SUMMARY

- Stem cells are defined by the properties of self-renewal, proliferation, and differentiation.
- Continually renewing tissues such as blood, skin, and intestinal epithelia have abundant stem cells, and these tissues usually repair quickly. Complex large structural tissues such as heart and brain that are slow to repair or regenerate have few stem cells, or, if present, they are difficult to activate for repair purposes.
- Telomeres are the DNA sequence caps on the ends of chromosomes that protect the coding DNA from nucleases. Telomerase is the specialized enzyme that adds and maintains the capping nucleotides. Loss of telomerase activity is associated with DNA damage.
- Apoptosis is the process of programmed cell death from within, while necrosis is cell damage caused by external factors such as trauma or cell membrane rupture.
- Stem cells are found in stem cell niches and interactions with neighboring cells may maintain them in a primitive state.

- Stem cells may exist in metastable states, capable of responding to environmental factors such as growth factors, cytokines, or morphogenetic factors.
- Pluripotent stem cells include embryonic stem (ES) cells derived from inner cell mass of the early embryo, embryonal carcinoma (EC) cells derived from undifferentiated primordial germ cells, and the induced pluripotent stem (iPS) cells for which the 2012 Nobel Prize was awarded to Yamanaka and Gurdon.
- Bone marrow contains several useful adult stem cells: hematopoietic stem cells (HSCs) and mesenchymal stem cells (MSCs).
- HSCs may be the best studied adult stem cells and are the cells used for bone marrow transplantation. However, they are difficult to grow in culture and maintain their ability to differentiate. MSCs are 10-fold less abundant than HSCs in bone marrow. However, they are relatively easy to grow in culture and provide useful precursor cells for many connective tissues including bone and cartilage.
- Adult stem cells and iPS cells are being widely used for tissue engineering.

STATE OF THE ART

The Dark Side: Cancer Stem Cells

We have discussed the excitement and promise of stem cells for tissue healing but as with any powerful technology there are potential problems to understand. Stem cells share many properties with cancer cells. Cancer cells proliferate and some, such as leukemic cells, become a predominate cell in the tissue in which they are growing. Many anticancer treatments are designed to interfere with proliferation, such as nucleoside analogs that mimic the normal A, T, G, C nucleotides and once incorporated into the replicating DNA cause lethal chain termination. Thus, certain cancers respond very well to certain drugs. However, much like the stem cell in its niche that divides only infrequently, a cancer stem cell may reside in an unknown niche location and produce daughter cells infrequently. It is these daughter cells that may amplify their numbers rapidly and be diagnosed and treated but it is thought to be the cancer stem cell that is the source of the problem. Cancer cells that form solid tumors lack contact inhibition and continue to grow, although angiogenesis may not keep pace with tumor growth and the tumor may become self-limiting unless new blood vessels form.

In this recent paper, van Rheenen and colleagues (Zomer et al., 2013) used a genetic Cre/LoxP recombinase cassette system that is capable of labeling individual cells with one of four colors (confetti labeling or Brainbow). This is possible because there are four distinct LoxP sequences and when the Cre recombinase causes excision, the intervening DNA sequence is removed, leaving only the one color adjacent to the promoter to be expressed. The authors labeled mouse mammary cancer cells by this technique and visualized the same region of cells in the living mouse tissue by intravital microscopy over two weeks. The collected data shows that some clonal cancer stem cells expand in an uncontrolled manner, while others proliferate and then regress the clonal colony size, or have a delayed onset, or only show a regressive phenotype. The data also indicates that a clone may remain small but then be triggered to continue growing at a later time, as is common in cancer relapse.

Video of authors comments http://bcove.me/vbkksy19.

STATE OF THE ART *Continued*

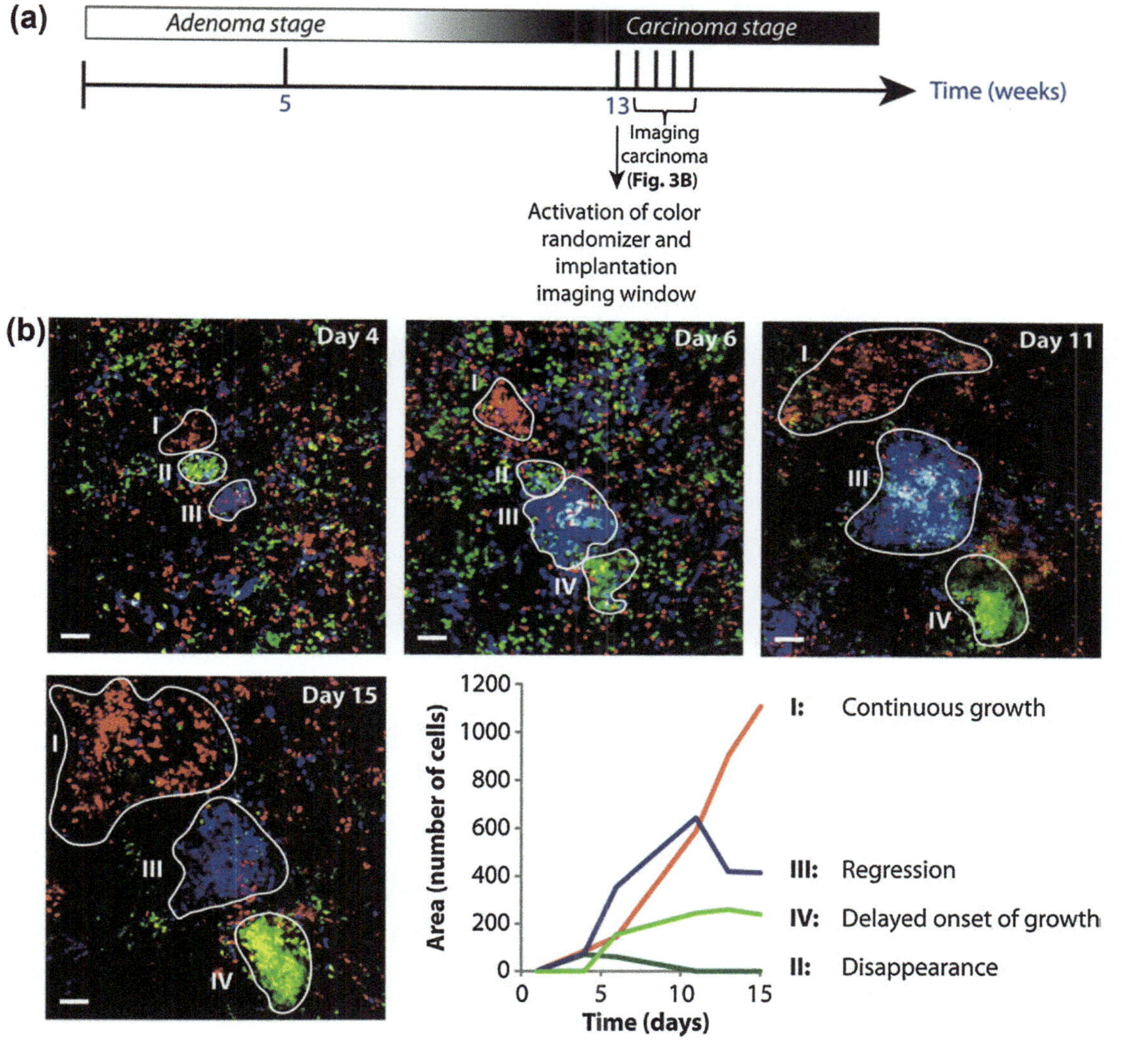

RECOMMENDED READING

Becker, A.J., McCulloch, C.E., Till, J.E., 1993. Cytological demonstration of the clonal nature of spleen colonies derived from transplanted mouse marrow cells. Nature 197, 452–454.

Bergmann, O., Bhardwaj, R.D., Bernard, S., Zdunek, S., Barnabé-Heider, F., Walsh, S., Zupicich, J., Alkass, K., Buchholz, B.A., Druid, H., Jovinge, S., Frisén, J., 2009. Evidence for cardiomyocyte renewal in humans. Science 324, 98–102.

Caplan, A.I., Correa, D., 2011. The MSC: an injury drugstore. Cell Stem Cell 9 (1), 11–15.

Georgantas, R.W., Hildreth, R., Morisot, S., Alder, J., Liu, C.G., Heimfeld, S., Calin, G.A., Croce, C.M., Civin, C.I., 2007. CD34+ hematopoietic stem-progenitor cell microRNA expression and function: a circuit diagram of differentiation control. Proc. Natl. Acad. Sci. U.S.A. 104 (8), 2750–2755.

Huang, S., 2009. Reprogramming cell fates: reconciling rarity with robustness. Bioessays 31 (5), 546–560.

Pittenger, M.F., Mackay, A.M., Beck, S.C., Jaiswal, R.K., Douglas, R., Mosca, J.D., Moorman, M.A., Simonetti, D.W., Craig, S., Marshak, D.R., 1999. Multilineage potential of adult human mesenchymal stem cells. Science 284 (5411), 143–147.

Takahashi, K., Yamanaka, S., 2006. Induction of pluripotent stem cells from mouse embryonic and adult fibroblast cultures by defined factors. Cell 126 (4), 663–676.

REFERENCES

Adcock, I.A., Ford, P., Ito, K., Barnes, P.J., 2006. Epigenetics and airways disease. Respir. Res. 7, 21.

Adewumi, O., Aflatoonian, B., Ahrlund-Richter, L., Amit, M., Andrews, P.W., et al., 2007. Characterization of human embryonic stem cell lines by the International Stem Cell Initiative. Nat. Biotechnol. 25 (7), 803–816.

Aggarwal, S., Pittenger, M.F., 2005. Human mesenchymal stem cells modulate allogeneic immune cell responses. Blood 105 (4), 1815–1822.

Alvarez-Buylla, A., Seri, B., et al., 2002. Identification of neural stem cells in the adult vertebrate brain. Brain Res. Bull. 57 (6), 751–758.

Assady, S., Maor, G., Amit, M., Itskovitz-Eldor, J., Skorecki, K.L., Tzukerman, M., 2001. Insulin production by human embryonic stem cells. Diabetes 50 (8), 1691–1697.

Bhattacharya, D., Czechowicz, A., Ooi, A.G., Rossi, D.J., Bryder, D., Weissman, I.L., 2009. Niche recycling through division-independent egress of hematopoietic stem cells. J. Exp. Med. 206 (12), 2837–2850.

Bianconi, E., Piovesan, A., Facchin, F., Beraudi, A., Casadei, R., Frabetti, F., Vitale, L., Pelleri, M.C., Tassani, S., Piva, F., Perez-Amodio, S., Strippoli, P., Canaider, S., 2013. An estimation of the number of cells in the human body. Ann. Hum. Biol. 40 (6), 463–471.

Blanpain, C., Fuchs, E., 2009. Epidermal homeostasis: a balancing act of stem cells in the skin. Nat. Rev. Mol. Cell Biol. 10, 207–217.

Caplan, A.I., 1991. Mesenchymal stem cells. J. Orthop. Res. 9 (5), 641–650.

Caplan, A.I., Dennis, J.E., 2006. Mesenchymal stem cells as trophic mediators. J. Cell. Biochem. 98 (5), 1076–1084.

Carpenter, M.K., Inokuma, M.S., Denham, J., Mujtaba, T., Chiu, C.P., Rao, M.S., 2001. Enrichment of neurons and neural precursors from human embryonic stem cells. Exp. Neurol. 172 (2), 383–397.

Carter, M.G., Stagg, C.A., Falco, G., Yoshikawa, T., Bassey, U.C., Aiba, K., Sharova, L.V., Shaik, N., Ko, M.S., 2008. An in situ hybridization-based screen for heterogeneously expressed genes in mouse ES cells. Gene Expr. Patterns 8 (3), 181–198.

Chadwick, K., Wang, L., Li, L., Menendez, P., Murdoch, B., et al., 2003. Cytokines and BMP-4 promote hematopoietic differentiation of human embryonic stem cells. Blood 102 (3), 906–915.

Chung, Y., Klimanskaya, I., et al., 2008. Human embryonic stem cell lines generated without embryo destruction. Cell Stem Cell 2 (2), 113–117.

Conrad, S., Renninger, M., et al., 2008. Generation of pluripotent stem cells from adult human testis. Nature 456 (7220), 344–349.

Covas, D.T., Panepucci, R.A., Fontes, A.M., Silva Jr., W.A., Orellana, M.D., Freitas, M.C., Neder, L., Santos, A.R., Peres, L.C., Jamur, M.C., Zago, M.A., 2008. Multipotent mesenchymal stromal cells obtained from diverse human tissues share functional properties and gene-expression profile with CD146+ perivascular cells and fibroblasts. Exp. Hematol. 36 (5), 642–654.

Cowan, C.A., Atienza, J., Melton, D.A., Eggan, K., 2005. Nuclear reprogramming of somatic cells after fusion with human embryonic stem cells. Science (New York, NY) 309 (5739), 1369–1373.

Daley, G.Q., Lensch, M.W., et al., 2009. Broader implications of defining standards for the pluripotency of iPSCs. Cell Stem Cell 4 (3), 200–201 Author reply 202.

Eriksson, P.S., Perfilieva, E., et al., 1998. Neurogenesis in the adult human hippocampus. Nat. Med. 4 (11), 1313–1317.

Evans, M.J., Kaufman, M.H., 1981. Establishment in culture of pluripotential cells from mouse embryos. Nature 292 (5819), 154–156.

Fogh, J., Trempe, G., 1975. New Human Tumor Cell Lines. Plenum Press, New York.

Friedenstein, A.J., Chailakhjan, R.K., Lalykina, K.S., 1970. The development of fibroblast colonies in mono-layer cultures of guinea-pig bone marrow and spleen cells. Cell Tissue Kinet. 3 (4), 393–403.

Greider, C.W., Blackburn, E.H., 1985. Identification of a specific telomere terminal transferase activity in Tetrahymena extracts. Cell 43 (2 Pt 1), 405–413.

Guan, K., Nayernia, K., et al., 2006. Pluripotency of spermatogonial stem cells from adult mouse testis. Nature 440 (7088), 1199–1203.

Halvorsen, Y.D., Franklin, D., Bond, A.L., Hitt, D.C., Auchter, C., Boskey, A.L., Paschalis, E.P., Wilkison, W.O., Gimble, J.M., December 2001. Extracellular matrix mineralization and osteoblast gene expression by human adipose tissue-derived stromal cells. Tissue Eng. 7 (6), 729–741.

Hayflick, L., 1965. The limited in vitro lifetime of human diploid cell strains. Exp. Cell Res. 37, 614–636.

Hayflick, L., 1998. A brief history of the mortality and immortality of cultured cells. Keio J. Med. 47, 174–182.

Haynesworth, S.E., Goshima, J., Goldberg, V.M., Caplan, A.I., 1992. Characterization of cells with osteogenic potential from human marrow. Bone 13 (1), 81–88.

He, J.Q., Ma, Y., Lee, Y., Thomson, J.A., Kamp, T.J., 2003. Human embryonic stem cells develop into multiple types of cardiac myocytes: action potential characterization. Circ. Res. 93 (1), 32–39.

Huangfu, D., Osafune, K., et al., 2008. Induction of pluripotent stem cells from primary human fibroblasts with only Oct4 and Sox2. Nat. Biotechnol. 26 (11), 1269–1275.

Kaufman, D.S., Hanson, E.T., Lewis, R.L., Auerbach, R., Thomson, J.A., 2001. Hematopoietic colony-forming cells derived from human embryonic stem cells. Proc. Natl. Acad. Sci. U.S.A. 98 (19), 10716–10721.

Kedziorek, D.A., Hofmann, L.V., Fu, Y., Gilson, W.D., Cosby, K.M., Kohl, B., Barnett, B.P., Simons, B.W., Walczak, P., Bulte, J.W., Gabrielson, K., Kraitchman, D.L., 2012. X-ray-visible microcapsules containing mesenchymal stem cells improve hind limb perfusion in a rabbit model of peripheral arterial disease. Stem Cells 30 (6), 1286–1296.

Kraitchman, D.L., Heldman, A.W., Atalar, E., Amado, L.C., Martin, B.J., Pittenger, M.F., Hare, J.M., Bulte, J.W., May 13, 2003. In vivo magnetic resonance imaging of mesenchymal stem cells in myocardial infarction. Circulation 107 (18), 2290–2293.

Krasnodembskaya, A., Song, Y., Fang, X., Gupta, N., Serikov, V., Lee, J.W., Matthay, M.A., 2010. Antibacterial effect of human mesenchymal stem cells is mediated in part from secretion of the antimicrobial peptide LL-37. Stem Cells 28 (12), 2229–2238.

Lavon, N., Yanuka, O., Benvenisty, N., 2004. Differentiation and isolation of hepatic-like cells from human embryonic stem cells. Differ. Res. Biol. Divers. 72 (5), 230–238.

Le Blanc, K., Tammik, C., Rosendahl, K., Zetterberg, E., Ringdén, O., October 2003a. HLA expression and immunologic properties of differentiated and undifferentiated mesenchymal stem cells. Exp. Hematol. 31 (10), 890–896.

Le Blanc, K., Tammik, L., Sundberg, B., Haynesworth, S.E., Ringdén, O., 2003b. Mesenchymal stem cells inhibit and stimulate mixed lymphocyte cultures and mitogenic responses independently of the major histocompatibility complex. Scand. J. Immunol. 57 (1), 11–20.

Lengerke, C., Kim, K., et al., 2007. Differentiation potential of histocompatible parthenogenetic embryonic stem cells. Ann. N.Y. Acad. Sci. 1106, 209–218.

Levenberg, S., Golub, J.S., Amit, M., Itskovitz-Eldor, J., Langer, R., 2002. Endothelial cells derived from human embryonic stem cells. Proc. Natl. Acad. Sci. U.S.A. 99 (7), 4391–4396.

Levenberg, S., Huang, N.F., Lavik, E., Rogers, A.B., Itskovitz-Eldor, J., Langer, R., 2003. Differentiation of human embryonic stem cells on three-dimensional polymer scaffolds. Proc. Natl. Acad. Sci. U.S.A. 100 (22), 12741–12746.

Mali, P., Ye, Z., et al., 2008. Improved efficiency and pace of generating induced pluripotent stem cells from human adult and fetal fibroblasts. Stem Cells 26 (8), 1998–2005.

Luria, E.A., Owen, M.E., Friedenstein, A.J., Morris, J.F., Kuznetsow, S.A., 1987. Bone formation in organ cultures of bone marrow. Cell Tissue Res. 248 (2), 449–454.

Marquez-Curtis, L.A., Janowska-Wieczorek, A., 2013. Enhancing the migration ability of mesenchymal stromal cells by targeting the SDF-1/CXCR4 axis. Biomed. Res. Int. 2013, 561098.

Martin, M.J., Muotri, A., Gage, F., Varki, A., 2005. Human embryonic stem cells express an immunogenic nonhuman sialic acid. Nat. Med. 11 (2), 228–232.

McCulloch, E.A., December 16, 2001. CFU-S: an assay for pluripotent myelopoietic stem cells. Hematopoietic Stem Cell Protocols Series: Methods in Molecular Medicine, vol. 63, pp. 153–160.

Morrison, S.J., Kimble, J., 2006. Asymmetric and symmetric stem-cell divisions in development and cancer. Nature 441 (7097), 1068–1074.

Mummery, C., Ward-van Oostwaard, D., Doevendans, P., Spijker, R., van den Brink, S., et al., 2003. Differentiation of human embryonic stem cells to cardiomyocytes: role of coculture with visceral endoderm-like cells. Circulation 107 (21), 2733–2740.

Németh, K., Leelahavanichkul, A., Yuen, P.S., Mayer, B., Parmelee, A., Doi, K., Robey, P.G., Leelahavanichkul, K., Koller, B.H., Brown, J.M., Hu, X., Jelinek, I., Star, R.A., Mezey, E., 2009. Bone marrow stromal cells attenuate sepsis via prostaglandin E(2)-dependent reprogramming of host macrophages to increase their interleukin-10 production. Nat. Med. 15 (1), 42–49.

Odorico, J.S., Kaufman, D.S., Thomson, J.A., 2011. Multilineage differentiation from human embryonic stem cell lines. Stem Cells 19, 193–204.

Otte, A., Bucan, V., Reimers, K., Hass, R., 2013. Mesenchymal stem cells maintain long-term in vitro stemness during explant culture. Tissue Eng. Part C Methods 19 (12), 937–948.

Owen, M., Friedenstein, A.J., 1988. Stromal stem cells: marrow-derived osteogenic precursors. Ciba Found. Symp. 1988 (136), 42–60.

Rambhatla, L., Chiu, C.P., Kundu, P., Peng, Y., Carpenter, M.K., 2003. Generation of hepatocyte-like cells from human embryonic stem cells. Cell Transplant. 12 (1), 1–11.

Resnick, J.L., Bixler, L.S., et al., 1992. Long-term proliferation of mouse primordial germ cells in culture. Nature 359 (6395), 550–551.

Reubinoff, B.E., Itsykson, P., Turetsky, T., Pera, M.F., Reinhartz, E., et al., 2001. Neural progenitors from human embryonic stem cells. Nat. Biotechnol. 19 (12), 1134–1140.

Ritsma, L., Ellenbroek, S., Zomer, A., Snippert, H., de Sauvage, F., Simons, B., Clevers, H., van Rheenen, J., 2014. Intestinal crypt homeostasis revealed at single-stem-cell level by *in vivo* live imaging. Nature 507, 362–365.

Schwartz, S.D., Hubschman, J.P., et al., 2012. Embryonic stem cell trials for macular degeneration: a preliminary report. Lancet 379 (9817), 713–720.

Shamblott, M.J., Axelman, J., et al., 1998. Derivation of pluripotent stem cells from cultured human primordial germ cells. Proc. Natl. Acad. Sci. U.S.A. 95 (23), 13726–13731.

Stevens, L.C., 1967. Origin of testicular teratomas from primordial germ cells in mice. J. Natl. Cancer Inst. 38 (4), 549–552.

Swelstad, B.B., Kerr, C.L., 2009. Current protocols in the generation of pluripotent stem cells: theoretical, methodological and clinical considerations. Stem Cells Cloning 22 (3), 13–27.

Tada, M., Takahama, Y., Abe, K., Nakatsuji, N., Tada, T., 2001. Nuclear reprogramming of somatic cells by in vitro hybridization with ES cells. Curr. Biol. 11 (19), 1553–1558.

Tremain, N., Korkko, J., Ibberson, D., Kopen, G.C., DiGirolamo, C., Phinney, D.G., 2001. MicroSAGE analysis of 2,353 expressed genes in a single cell-derived colony of undifferentiated human mesenchymal stem cells reveals mRNAs of multiple cell lineages. Stem Cells 19 (5), 408–418.

Ueda, H., Fujita, R., Yoshida, A., Matsunaga, H., Ueda, M., 2007. Identification of prothymosin-alpha1, the necrosis-apoptosis switch molecule in cortical neuronal cultures. J. Cell Biol. 176 (6), 853–862.

Van Wijngaarden, P., Franklin, R., 2013. Ageing stem and progenitor cells: implications for rejuvenation of the central nervous system. Development 140, 2562–2575.

Vats, A., Bielby, R.C., Tolley, N., Dickinson, S.C., Boccaccini, A.R., et al., 2006. Chondrogenic differentiation of human embryonic stem cells: the effect of the micro-environment. Tissue Eng. 12 (6), 1687–1697.

Wagner, W., Wein, F., Seckinger, A., Frankhauser, M., Wirkner, U., Krause, U., Blake, J., Schwager, C., Eckstein, V., Ansorge, W., Ho, A.D., 2005. Comparative characteristics of mesenchymal stem cells from human bone marrow, adipose tissue, and umbilical cord blood. Exp. Hematol. 33 (11), 1402–1416.

Wechsler, L.R., 2009. Clinical trials of stroke therapy: which cells, which patients? Stroke 40 (3 Suppl.), S149–S151.

Williams, A.R., Suncion, V.Y., McCall, F., Guerra, D., Mather, J., Zambrano, J.P., Heldman, A.W., Hare, J.M., May 17, 2013. Durable scar size reduction due to allogeneic mesenchymal stem cell therapy regulates whole-chamber remodeling. J. Am. Heart Assoc. 2 (3), 1–8.

Wilmut, I., Schnieke, A.E., McWhir, J., Kind, A.J., Campbell, K.H., 1997. Viable offspring derived from fetal and adult mammalian cells. Nature 385 (6619), 810–813.

Yang, S.S., Kim, N.R., Park, K.B., Do, Y.S., Roh, K., Kang, K.S., Kim, D.I., May 2013. A phase I study of human cord blood-derived mesenchymal stem cell therapy in patients with peripheral arterial occlusive disease. Int. J. Stem Cells 6 (1), 37–44.

Yu, J., Vodyanik, M.A., et al., 2007. Induced pluripotent stem cell lines derived from human somatic cells. Science 318 (5858), 1917–1920.

Zhang, S.C., Wernig, M., Duncan, I.D., Brustle, O., Thomson, J.A., 2001. *In vitro* differentiation of transplantable neural precursors from human embryonic stem cells. Nat. Biotechnol. 19 (12), 1129–1133.

Zhao, T., Zhang, Z.N., et al., 2011. Immunogenicity of induced pluripotent stem cells. Nature 474 (7350), 212–215.

Zomer, A., Ellenbroek, S.I., Ritsma, L., Beerling, E., Vrisekoop, N., Van Rheenen, J., March 2013. Intravital imaging of cancer stem cell plasticity in mammary tumors. Stem Cells 31 (3), 602–606.

Zuk, P.A., Zhu, M., Mizuno, H., Huang, J., Futrell, J.W., Katz, A.J., Benhaim, P., Lorenz, H.P., Hedrick, M.H., April 2001. Multilineage cells from human adipose tissue: implications for cell-based therapies. Tissue Eng. 7 (2), 211–228.

CHAPTER 3

Tissue Formation during Embryogenesis

Marcel Karperien[1], Bernard A.J. Roelen[2], Rob E. Poelmann[3], Adriana C. Gittenberger-de Groot[3], Beerend P. Hierck[3], Marco C. DeRuiter[4], Dies Meijer[5], Susan Gibbs[6]

[1]Department of Developmental BioEngineering, University of Twente, Enschede, The Netherlands; [2]Department of Farm Animal Health, Faculty of Veterinary Medicine, Utrecht University, Utrecht, The Netherlands; [3]Department of Anatomy and Embryology, Leiden University Medical Centre, Leiden, The Netherlands; [4]Department of Clinical and Experimental Anatomy, Leiden University Medical Centre, Leiden, The Netherlands; [5]Centre for Neuroregeneration, University of Edinburgh, Edinburgh, UK; [6]Department of Dermatology, VU Medical Centre, Amsterdam, The Netherlands

LEARNING OBJECTIVES

- To appreciate that tissue engineering strategies rely on the merger of developmental biology with the fields of engineering in a strategy nowadays known as developmental (re)engineering
- To understand the origin of pluripotent stem cell populations in the embryo and to realize that each organ is build out of stem cells from endodermal, mesodermal, ectodermal, and neural crest origin
- Be able to recapitulate some of the basic principles by which cellular differentiation and specification is induced during organogenesis
- To appreciate that the subsequent, timely and proper dosing of signaling events, for example, initiated by cell migration, cell–cell interactions, para- and intracrine signaling of growth factors, epithelial-to-mesenchyme transitions, and environmental factors define cell specification and shape organs
- To understand the importance of paracrine signaling between mesenchymal and epithelial cells, for example, between smooth muscle cell and endothelial cells in blood vessel development or dermal fibroblasts and keratinocytes in skin development and that this paracrine signaling is instrumental to obtain functional tissue
- To understand the role of adaptive signaling in tissue formation in response to environmental factors like pulsatile fluid flow and wet-to-air transitions in cell specification and tissue architecture

Tissue Engineering.
Copyright © 2015 Elsevier Inc. All rights reserved.

- Be able to recapitulate the sequential steps that can be recognized in skin formation and how these steps are reiterated in skin tissue engineering
- To understand the difference between intramembranous and endochondral bone formation and the role of the osteoblasts, chondrocytes, and osteoclasts in both processes
- To appreciate that members of the Bone Morphogenetic Protein (BMP)/ Transforming Growth Factorβ (TGFβ) superfamily, the Fibroblast Growth Factor (FGF) family, and the Wingless-Int (Wnt) family are involved in organogenesis of almost all organs
- To realize that key concepts from developmental biology such as modularity and robustness can be recapitulated in engineering strategies to generate functional organs.

3.1 INTRODUCTION

Tissues in the human body are the result of millions-of-years of evolution. This process has resulted in the natural selection of a tissue structure that is optimally adapted to its function in the body. Tissue engineering aims at repairing or replacing damaged tissue that cannot or insufficiently be repaired by the intrinsic repair mechanisms present in almost all organs of an individual. In most instances, these intrinsic repair mechanisms recapitulate many of the processes involved in the formation of an organ during embryogenesis. Incorporation of these mechanisms in tissue engineering procedures will significantly contribute to the success of the construct in the body. Thus, important lessons for tissue engineering can be learned from the formation of organs during embryogenesis. For example, on (1) the origin of cells that contribute to the formation of a particular organ, (2) the growth factors and their interrelationship in the formation of an organ, (3) the mechanisms by which undifferentiated stem or precursor cells are induced to specialize into an organ of specific cell type, (4) the subsequent steps in organ formation, and (5) the interaction between cells and their environment consisting of both the extracellular matrix and neighboring cells. Ideally, inclusion of all these aspects in the tissue engineering procedure will provide the best opportunities for success. Indeed, one of the basic concepts of developmental biology, i.e., the modularity of the tissue architecture according to which intermediates in tissue development constitute semiautonomous entities, is nowadays incorporated in many tissue engineering strategies to optimize the product.

The past decade has shown a rapid increase in our understanding of the cellular and molecular basis of tissue formation. In addition, our knowledge about stem cell renewal and maintenance of pluri- and multipotency has substantially grown. In fact, it is now possible to generate pluripotent stem cells from a variety of tissue types in a process called reprogramming. It still remains a challenge to combine this knowledge into successful strategies for the engineering of functional tissues that can replace lost or worn out organs. The tissue engineering strategy in which the principles of developmental biology are merged with typical engineering disciplines like chemical, biomechanical, and biophysical engineering is nowadays known as developmental engineering or developmental (re)engineering (Lenas et al., 2009; Tonnarelli et al., 2014).

This chapter aims at providing a background on the formation of various organs during embryogenesis that are subject of tissue engineering techniques, like the heart, blood vessels, peripheral nerves, skin, bone, and cartilage. It starts with providing an overview of the formation of pluri- and multipotent cell populations and their positioning in the overall body plan. Subsequently, some of the basic mechanisms by which these cell populations are recruited in organogenesis and are induced to specify are discussed in greater detail.

3.1.1 The Formation of Pluripotent Cells

All tissues in the adult mammal originate from the fertilized oocyte, called zygote. The first divisions of the fertilized egg are not accompanied by cellular growth and are therefore referred to as cleavage divisions and arising cells are called blastomeres. At the early cleavage stages, each blastomere is still totipotent and thus is capable of forming an entire organism.

The first differentiation occurs at the morula stage after the third series of cleavage divisions (8 cell stage). The cells that are on the outside of the morula differentiate into the trophectoderm, a cell layer mediating implantation and formation of extraembryonic structures such as the placenta. The cells on the inside of the morula will give rise to the inner cell mass. In the next stage, the embryo is called a blastocyst and contains besides the trophectoderm and a fluid-filled cavity or

CLASSICAL EXPERIMENT 1: GENERATION OF THE FIRST EMBRYONIC STEM CELL LINES

Already in the "50s of the twentieth century, it was recognized that the 129Sv strain of mice exhibited a high incidence in the formation of spontaneous testicular tumors, the so-called teratocarcinomas. Teratocarcinomas can also be induced in mice by the grafting of early embryos into extrauterine sites (Stevens, 1967). Individual cells from these tumors could be transplanted to other mice where they formed new tumors, demonstrating the existence of cells with self-renewal capacity or stem cells in the tumors (Kleinsmith and Pierce, 1964). Teratocarcinomas contain a wide variety of differentiated cells that were derivatives of the stem cells. The exact potency of these stem cells remained unclear until chimeric mice were created by injecting few of the tumor cells into blastocyst stage embryos. The tumor cells started to participate in the development of many cells and tissues, including germ cells. This demonstrated that the teratocarcinoma cells were pluripotent (Mintz and Illmensee, 1975; Papaioannou et al., 1975). The cell lines that were derived from teratocarcinomas were called embryonal carcinoma cell lines. The pluripotency of the embryonal carcinoma cells raised many questions. These cells were after all derived from embryos. It was hypothesized that the early embryo contained cells that divided and remained pluripotent unless they received signals for differentiation, as normally occurred during embryogenesis. It seemed logical to assume that these cells could be directly isolated from the embryo and kept in culture.

Continued...

CLASSICAL EXPERIMENT 1: GENERATION OF THE FIRST EMBRYONIC STEM CELL LINES *Continued*

Martin Evans and Matthew Kaufman, then from the University of Cambridge, were the first to directly generate progressively growing cultures from preimplantation mouse embryos. By delaying the implantation of blastocyst stage embryos, they obtained embryos with a bigger-than-normal inner cell mass. Using their knowledge of culture media obtained from working with embryonal carcinoma cells, they were able to culture cells from these large blastocysts. The cells strongly resembled embryonal carcinoma cells; they too formed teratomas when injected into mice and differentiated in vitro. Evans and Kaufman published their seminal paper on the derivation of pluripotent embryonic cells in 1981 (Evans and Kaufman, 1981). In the same year, Gail Martin from the University of California independently established pluripotent cell lines from mouse blastocysts (Martin, 1981). Instead of using delayed implantation embryos, Martin was able to successfully culture the embryonic cells by using a feeder layer of fibroblasts and conditioned medium from embryonal carcinoma cells. Martin coined the term embryonic stem (ES) cells for these embryo-derived cell lines. The pluripotency of the ES cells was convincingly demonstrated by the formation of healthy germ-line chimaras (Bradley et al., 1985).

The aim of the underlying studies was now at reach, namely to use mutation-carrying cells for the generation of mice bearing these mutations to assess their functional consequence. Gene-targeting by homologous recombination in ES cells was developed (Smithies et al., 1985), which proved very fruitful. Indeed, the production of many knockout mice and the knowledge of gene function by studying these mice would not have been possible without ES cells.

blastocoel also an inner cell mass. The inner cell mass consists of pluripotent cells that will ultimately form the entire embryo. Cells can be isolated from the inner cell mass and under the right conditions kept into culture. They are then called embryonic stem (ES) cells. ES cells have the capacity to self-renew and are truly pluripotent as they can differentiate into every cell type of the embryonic and adult organism. ES cells cannot differentiate into extra-embryonic tissue. Their capacity to self-renew and their pluripotency makes ES cells an excellent cell source for tissue engineering purposes. In mouse genetics, ES cells are an important tool for the generation of genetically modified animals.

The first stages of embryonic development occur when the zygote travels through the oviduct prior to implantation in the uterus. Under the appropriate culture conditions, this process can be mimicked in vitro. Thus, isolated oocytes (human or from other mammals) can be fertilized in vitro and cultured until the early blastocyst stage. From these blastocysts, it is possible to generate ES cells that could be applied in regenerative medicine and tissue engineering. Thus, left over embryos from in vitro fertilization procedures could be used for the establishment of new human ES cell lines. Further development of the embryo requires, however, implantation in the uterus and the formation of the germ layers by gastrulation.

3.1.2 Preimplantation Development and Gastrulation

When the embryo has reached the uterus, at around 3–4 days after fertilization in mice and 5–7 days in humans, it is ready to implant in the uterine tissue.

Subsequent steps in embryonic development aim at establishing the proper structure in which specification of the three germ layers, the ectoderm, endoderm, and mesoderm, can occur by a process called gastrulation.

During gastrulation, the three definitive germ layers that construct the adult organism are established: the outer ectoderm, the inner endoderm, and the interstitial mesoderm. In the mouse, gastrulation starts at about 6.5 days after fertilization and in the human at about 13–15 days. The ectoderm forms the outer part of the skin, brain cells, nerve cells, parts of the eye like the lens, epithelial structures of the mouth and anus, the pituitary gland, parts of the adrenal glands, and pigment cells. The endoderm forms the lining of the gastrointestinal and respiratory tracts, plus the liver, pancreas, thyroid gland, thymus, and the lining of the bladder. The mesoderm gives rise to skeletal muscle, heart and blood vessels, connective tissue, kidney, urethra, gonads, bone marrow, blood, bone, cartilage, and fat. During gastrulation, extensive cell movements take place through which cells acquire new positions and new neighbors with which to interact. These cell movements are accompanied by an increase in cell number and a decrease in cell doubling time so that in the mouse, the cell number increases from about 660 at the start of gastrulation to 15,000 at the end of 24 h.

3.1.3 Establishment of the Body Plan by Morphogen Signaling

During gastrulation, not only are the germ layers specified but cells will also receive instructions on their future position and role in the developing fetus by morphogen signaling. An important factor in this process is Nodal, which is a member of the bone morphogenetic protein (BMP)/transforming growth factorβ (TGFβ) superfamily of growth factors. Nodal is first expressed in the proximal part of the epiblast, the part of the embryo that will give rise to all embryonic structures, and is expressed in the node, hence its name. The node organizes gastrulation. Nodal signaling is essential for the formation of mesoderm and definitive endoderm. The highest levels of Nodal are found in the node and with increasing distance to the node, the concentration of Nodal will gradually decrease establishing a so-called morphogen gradient. Cells along this gradient are exposed to different levels of Nodal, providing the cell with important instructions on its future position and role in the fetus. Morphogen gradients are important and recurrent mechanisms by which undifferentiated cells receive instructions on their future position and role in the body.

3.1.4 Neural Crest Cells

Besides the three germ layers that are established during gastrulation, a fourth pluripotent cell population is required for organogenesis. This population consists of neural crest cells (NCCs) (Le Douarin and Klacheim, 1999). Although NCCs are derived from the ectoderm, they are sometimes called the fourth germ layer because of their importance. NCCs arise relatively late in embryonic

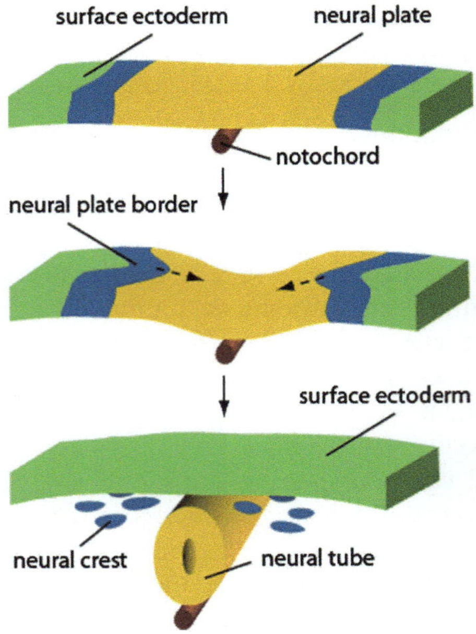

FIGURE 3.1

Formation of the neural crest. Midline ectoderm differentiates into neural cells to form a neural plate (yellow) by signals from the underlying notochord (red). Induction of neural crest (blue) cells takes place at the border between surface ectoderm (green) and the neural plate by signals from the mesoderm and the surface ectoderm. The sides of the neural plate fold together and close to form a neural tube. The neural crest cells delaminate from the ectoderm and migrate to their destinations.

development after establishment of the general body plan and their formation is tightly linked with the development of the central nervous system. The differentiation of the central nervous system starts with formation of the neural tube from the neural plate. The neural plate is a layer of ectodermal cells in the dorsal midline of the embryo. Signals from the underlying notochord (the future nucleus pulposa) instruct cells in the neural plate to proliferate and adapt a tube-like shape that will eventually generate the entire central nervous system. NCCs are generated at the interface of the neuroectoderm with the surface ectoderm, the so-called neural plate border (Figure 3.1).

From their source of origin, NCCs migrate extensively to specific places in the embryo where they generate a diverse group of differentiated cells that can be divided into four distinct subgroups: cranial, cardiac, vagal, and trunk cells that together give rise to sympathetic and parasympathetic neurons, glia cells, fat, cardiac mesenchyme, melanocytes, skin, connective tissue of salivary, thymus, adrenal, thyroid, and pituitary glands, smooth muscle cells (SMCs) of arteries, tooth, and bone and cartilage particularly of the face and cranium. The NCCs therefore have extreme plasticity and are truly pluripotent giving rise to cells from the endoderm, mesoderm, and ectoderm lineage. Due to this extreme plasticity, pluripotent NCCs would provide an excellent cell source for tissue engineering and regenerative medicine particularly for restoration of cranial and facial birth defects. However, NCCs are extremely difficult to isolate due to their high mobility and to the fact that they are derived from a transient structure. The neural crest disappears soon after the neural tube is closed and the neural plate border has disappeared. Consequently, limited studies have been done with NCC as cell source for tissue engineering purposes.

It is thought that the surface ectoderm together with the underlying mesoderm induces the neural plate to form NCCs by signals that include BMPs, in particular BMP4 and BMP7, fibroblast growth factors (FGFs), and members of the Wingless-Int (Wnt) family of morphogens. As they arise, NCCs undergo an epithelial-to-mesenchymal transition (EMT) characterized by expression of members of the *Snail* family of zinc finger transcription factors. Migration is probably initiated by a decrease in the amount of N-cadherin, which is a cell adhesion molecule typically expressed in epithelial cells. EMT is an important mechanism by which cells can dramatically change their fate. It plays an essential role not only in organogenesis but also during tumorigenesis, e.g., the transformation of a carcinoma in situ to an invasive, metastatic tumor.

3.2 CARDIAC DEVELOPMENT

Rebuilding the injured heart after infarction is a major challenge for tissue engineering. Presently, various cell-based approaches are used. Initially, this work started with committed myoblasts, the precursor cells of cardiac muscle, but now a variety of undifferentiated cells are used, such as endothelial progenitors, mesenchymal stem cells, resident cardiac stem cells, and ES cells. Various cell types from different origins are involved in heart formation. Important lessons can be learned from the complex interplay between these cell types that eventually give rise to the four-chambered cyclically beating heart during embryonic development. The heart, being the first functionally active organ, must continuously adapt to the growing needs of the embryo, whereas in postnatal life it has to support the ever-changing activity levels of the body. The development of the heart is modular in which intermediates in tissue development constitute semiautonomous entities. Figure 3.2 provides a time table of the sequential formation of different parts of the heart during embryogenesis. The proper and subsequent integration of all these processes is of crucial importance. Disturbance will result in embryonic heart failure or congenital heart deformities.

Here a framework is provided for understanding the multiple origins of the cardiac cell populations, including the cardiomyocytes, fibroblasts, and smooth muscle and nerve cells. Furthermore, the increasing complexity of the heart's architecture will be explained, in particular chamber formation, cardiac septa

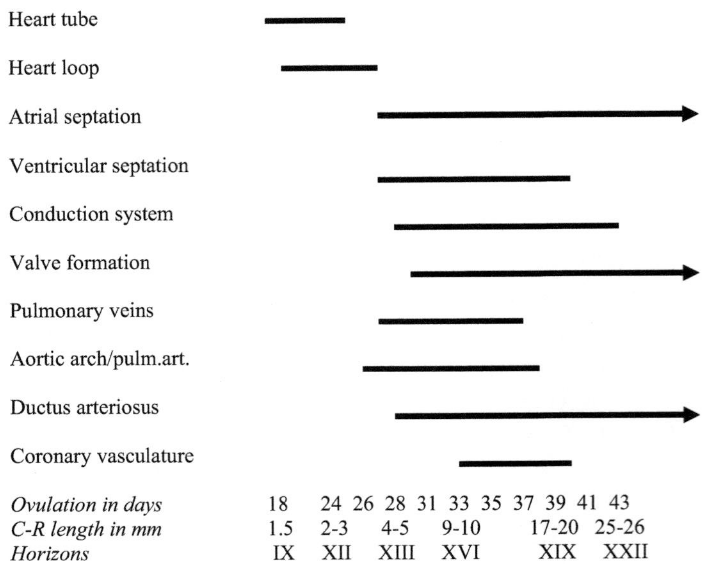

FIGURE 3.2
Time table of human heart development during embryogenesis. Note the different developmental windows for the various parts of the heart. Days since ovulation as a measure of developmental stage is compared with crown-rump length (C-R) and with the human stages of development indicated by Streeter's Horizons.

and valves, the coronary circulation, and the conduction system. Finally, from the myriad of growth factors and transcription factors important during the building of the heart a brief selection will be provided.

3.2.1 Cell Interactions

Cardiac formation starts in the so-called cardiogenic plates. The cardiogenic plates develop at the bilateral anterior tips of the unsegmented splanchnic mesoderm. The splanchnic mesoderm refers to cells of the inner lining of the body cavity. The transcriptional machinery for defining the cardiomyocyte is still largely unclear, and seems more complicated than, for instance, for skeletal muscle, in which the transcription factor MyoD is elemental and sufficient for muscle cell differentiation. Evidence exists that the transcription factor myocardin and signaling via BMPs govern cardiomyocyte differentiation. Myocardin is not specific for cardiac muscle, since it is also present at the onset of SMC differentiation. The cardiogenic plates fuse in the midline (Figure 3.3(a)) and the precursor cells differentiate into cardiomyocytes and endocardial cells (see Gittenberger-de Groot et al., 2005, for a schematic overview). The cardiomyocytes become the contracting part of the heart. The space between myocytes and endocardial cells, which have adapted an epithelial phenotype, becomes filled by cardiac jelly produced by the myocytes. The endocardium gives rise to mesenchymal cushion cells by a

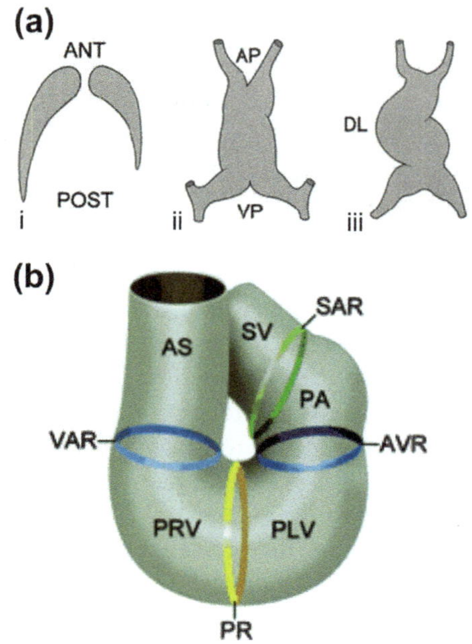

FIGURE 3.3

The first stages in heart development. (a) Schematic representation of (i) the bilateral formation of the mesoderm-derived cardiogenic plates, which are not quite symmetric. (ii) After fusion, these form a straight heart tube with an arterial pole (AP) and a venous (VP) pole. (iii) Thereafter, the tube starts its rightward (dextral) looping (DL). (b) Schematic drawing of the looped heart tube with the cardiac chambers and the transitional zones. Following the blood flow from venous to arterial, we can distinguish the sinus venosus (SV), the sinoatrial ring (SAR), the primitive atrium (PA), the atrioventricular ring (AVR), encircling the atrioventricular canal, the primitive left ventricle (PLV), the primary fold or ring (PR), the primitive right ventricle (PRV), the outflow tract ending at the ventriculoarterial ring (VAR), and the aortic sac (AS). *Adapted from Gittenberger-de Groot et al. (2005).*

process known as EMT (Hay, 2005) and, as a consequence, the cardiac jelly will transform into the thick endocardial cushions containing extracellular matrix and mesenchymal cells. Part of these cells is also recruited from the neural crest (see below). This is a complex process orchestrated by various growth factors and many transcription factors. The atrioventricular cushions reside in the transition zone of the primitive atrium and ventricle, the so-called atrioventricular (AV) canal, whereas the outflow tract (OFT) cushions are found in the conotruncal transition between ventricle and aortic sac (Figure 3.3(b)). Both cushion systems are in close proximity in the inner curvature of the heart. The OFT cushions are important in separation of the OFT into the aorta and the pulmonary trunk, while the AV cushions will play an important role in AV septation by heterologous fusion with the interventricular or primary fold and with the intra-atrial spina vestibuli (see Figure 3.3(b) and Gittenberger-de Groot et al., 2005).

3.2.2 Extracardiac Cell Populations

Two extracardiac cell populations are very important in heart formation: the cardiac neural crest cells (CNCCs) and the proepicardial organ (PEO). Both primordia have in common that their original epithelium (being neuroectoderm and splanchnic epithelium, respectively) transforms by (EMT into migratory mesenchyme. Subpopulations of NCCs migrate into the pharyngeal arches (to form, e.g., the arterial vessel wall: adventitial fibroblasts and SMCs), and into the secondary heart field, a part of the dorsal body wall adjacent to the heart tube (*vide infra*), subsequently migrating into and contributing to the OFT and the AV cushions. The PEO starts as a small grape-like structure extending from the inner lining of the body cavity, the splanchnic mesoderm. It expands and migrates as an epithelial sheet over the outer surface of the myocardium to form the epicardium. After complete coverage of the heart tube, EMT will give rise to subepicardial mesenchyme. A large number of these so-called epicardial-derived cells (EPDCs) migrate into the heart and differentiate into cardiac fibroblasts, SMCs, and adventitial fibroblasts of the coronary vessels (Gittenberger-de Groot et al., 1998). In bulk number the fibroblasts comprise the largest cell population, providing for the extracellular matrix evolving into the fibrous heart skeleton. This system mechanically supports the cardiac valves and the tendinous apparatus during the various phases of life. After cardiac infarction, the EPDC-derived fibroblasts provide for the scar tissue giving integrity to the cardiac wall, although it is, of course, not contractile. However, in early embryonic development, it is responsible for the induction of the architecture of the cardiac wall. Experimental ablation of the PEO results in an aberrant coronary vasculature feeding a paper thin myocardium almost devoid of trabeculations. This resembles a failing heart, eventually leading to embryonic death (Gittenberger-de Groot et al., 2000).

3.2.3 Looping and Chamber Formation

Initially, the heart is formed as a single almost strait tube. By a series of geometrical changes this tube finally transforms into a double pump, separately serving lungs and body. Various mechanisms involve looping of the primary tube, wedging of the OFT between left and right atrium, expansion of the chambers, atrial and ventricular septum formation, valve differentiation, and formation of the pharyngeal arterial system. Recent descriptions of the morphological changes important for the understanding of congenital anomalies have been provided elsewhere (Gittenberger-de Groot et al., 2005), while part of the genomic coding underlying chamber formation has been reviewed (Moorman and Christoffels, 2003) (Figure 3.4). It is evident that one simple transcriptional code for all cardiomyocytes does not exist, as left and right ventricular, atrial, and OFT myocytes present with a variety of different gene expression patterns in successive time windows.

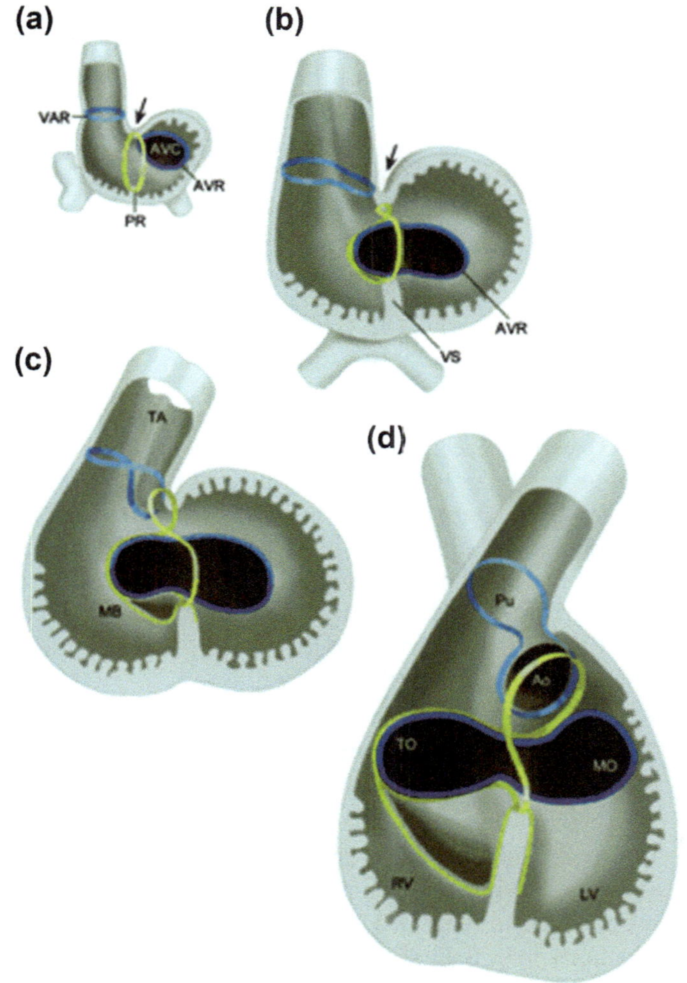

FIGURE 3.4

Transformation of a single heart tube into a four-chambered pumping heart. Schematic representation of the remodeling of the cardiac chambers and the transitional zones at the ventricular level. (a) Internal view (compare with Figure 3.6(b)) of the looped heart tube. The transitional zones are, going from the venous to the arterial pole, the atrioventricular ring (*AVR, dark blue*), the primary ring (*PR, yellow*), and, at the distal end of the myocardial outflow tract, the ventriculoarterial ring (*VAR, bright blue*). (b) During looping, with tightening of the inner curvature (*arrow*), the right part of the AVR moves to the right of the ventricular septum (*VS*). (c) Start of formation of the inflow tract of the right ventricle by excavation of the PR. The lower border is formed by the moderator band (*MB*). (d) Completion of the process with formation of a tricuspid orifice (*TO*) above the right ventricle (*RV*) and the aortic orifice (*Ao*) and the mitral orifice (*MO*) above the left ventricle (*LV*). It is easily appreciated that there is aortic-mitral continuity, whereas the distance between the TO and the pulmonary orifice (*Pu*) is marked. *Adapted from Gittenberger-de Groot et al. (2005).*

3.2.4 Septation and Valve Formation

Septation of the OFT requires the coordinated interactions of the OFT cushions, containing also NCCs, with the surrounding cardiomyocytes. AV canal septation requires fusion of the AV cushions, harboring also NCCs and EPDCs. As a final step in septum formation, cardiomyocytes penetrate parts of the fused cushions. Differentiation of the OFT semilunar valves, and AV mitral and tricuspid valves and their tendinous apparatus, involves coordinated interactions between cells of various origins such as migrated NCCs in conjunction with the surrounding cardiomyocytes (Gittenberger-de Groot et al., 2000). Replacement of malfunctioning valves, which is a highly relevant clinical problem, by tissue-engineered substitutes is an active field of research. The complexity in the cellular origin of the valves next to their unique biomechanical properties makes tissue engineering of these structures challenging.

3.2.5 Conduction

Contractions of the heart tube start as a peristaltic movement (de Jong et al., 1992) with a pacemaker functioning at the venous pole, but this will soon change in a cyclic base-to-apex contraction that in time will be followed by an apex-to-base constriction in a time-specific manner. In yet to establish time windows, these alterations in contraction mode are paralleled by differentiation of the myocardial entity acquiring fast and slow conduction properties. The formation of the conduction system probably requires the involvement of NCC, for the induction or insulation of the central conduction system, and EPDC for the peripheral Purkinje fibers.

3.2.6 Concluding Remarks

The formation of the heart is a complex, modular, and time-consuming process, which is not finished before birth. The multiple transcriptional interactions and molecular and cellular contributions framed in exact developmental windows provide an overwhelming myriad of possibilities, of which only a few combinations are adequate to result in a normally structured and functioning heart. Many combinations prove to be embryonic lethal or result in dramatic congenital malformations. Reconstructing and engineering the heart for major diseases must be based on thorough knowledge of the multiple interactions at all levels of development, function, and maintenance of the heart. The different origin of various cell populations involved in heart formation still provides a challenge for tissue engineering.

3.3 BLOOD VESSEL DEVELOPMENT

A clear need for a suitable arterial replacement has prompted researchers to look beyond autologous and synthetic tissue replacements toward the engineering of vessels using mesenchymal, hematopoietic, and embryonic stem cells. In embryonic development, the formation of the vasculature is tightly linked with heart development. Thus, when the heart starts to beat (about 4 weeks after conception) a network of endothelial-lined vessels through which the blood will be transported to and from the yolk sac, is in place. This almost two-dimensional extraembryonic plexus will rapidly grow out and gives rise to organ-specific vasculatures within the fast growing embryo. Except for the capillaries each vessel will be enveloped by one or more layers of specialized mural cells, such as SMCs, largely depending on their position and function within the vasculature.

Here the embryonic origin and development of the multiple cell populations involved in vessel formation, such as endothelial cells (ECs), SMCs, fibroblasts, and pericytes, are related to their contribution to the various types of vessels (muscular and elastic arteries, capillaries, and veins). Recent studies lift a tip of the veil clouding the molecular pathways involved in embryonic endothelial and SMC differentiation and the role of environmental factors like hemodynamic forces in the morphogenesis of the vessel wall. Although the combination of these factors will determine the ultimate phenotype of the vascular cells, one has to keep in mind that a single endothelial or a single SMC phenotype does not exist. In fact, there is a large heterogeneity in endothelial and smooth muscle phenotypes reflecting their multiple origins and their site and tissue-specific role in controlling homeostasis of the body.

Endothelial specification is an excellent example of a tissue structure that develops and specifies under blood pressure. For example, the pressure in the great arteries sets specific demands for the architecture of this vessel, which are not needed in small arteries penetrating in areas of a particular organ. To accomplish this, environmental cues activate biochemical pathways in cells which subsequently orchestrate tissue architecture, i.e., the alignment of cells and extracellular matrix.

3.3.1 Origin of the Endothelial Cells

Like the endocardium, the ECs of the blood vessels differentiate in situ within the so-called splanchnic mesoderm that faces the endoderm (Figure 3.5 and Section 3.2.1). This process is called vasculogenesis. The endothelial precursors line up, connect, lumenize, and form a plexus that gives rise to both the embryonic and extraembryonic vessels in the yolk sac and later in the placenta. Vascular endothelial growth factor (VEGF), TGFβ, and FGF signaling pathways are key players in the initial establishment of this endothelial network. This endothelial plexus in the splanchnic mesoderm expands into the somatic mesoderm by the formation of new sprouts. This process is also termed angiogenesis. Recent studies demonstrate that even before the heart starts to beat, an imprinting of arterial and venous identity exists within this capillary plexus. Arterial identity is characterized by the expression of genes like *ephrin-B2 (ephB2)*, *neuropilin-1*, *Notch3*, *DLL4*, and *gridlock*. *EphB4* and *neuropilin-2* are indicative for a venous identity. Blood flow is necessary for

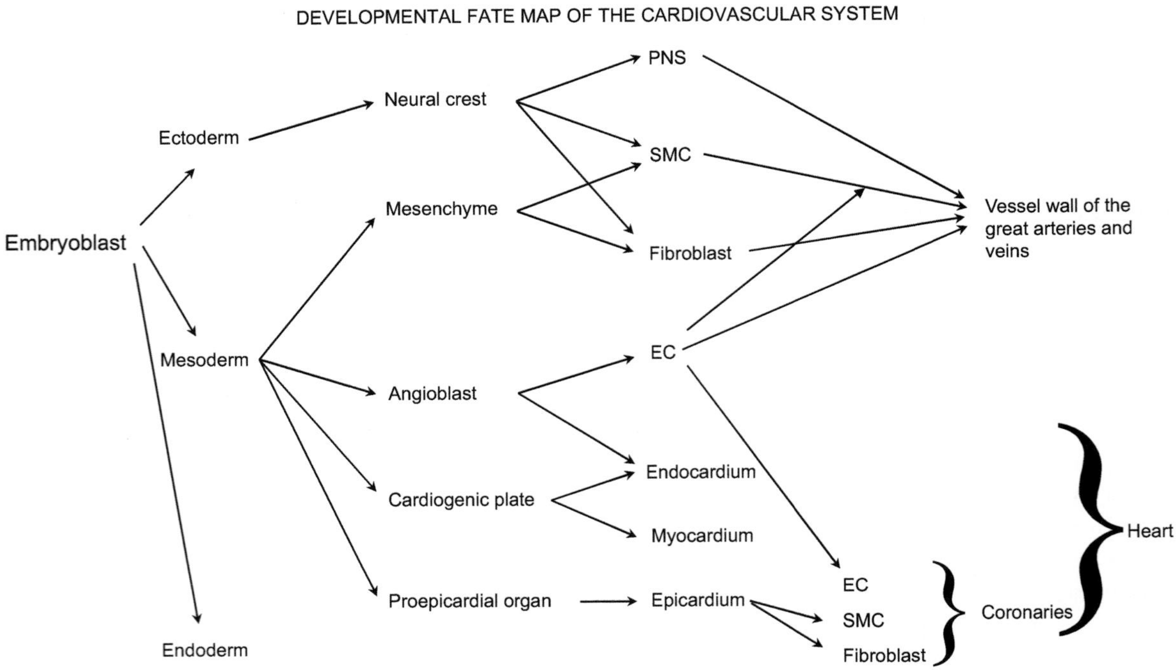

FIGURE 3.5
Developmental origin of cells contributing to the cardiovascular system. The mesodermal compartment of the three-layered embryo gives rise to the major part of the mural cells within the embryo. Angioblasts are the first vascular cells that differentiate by vasculogenesis within the mesoderm. Then the crescent-shaped cardiogenic plate arises giving rise to the endocardium and myocardium of the heart. The remaining mesenchyme will later on differentiate into SMC and fibroblasts of most of the organs. The vascular SMC and fibroblasts in the head and trunk, however, are derived from the mesectodermal neural crest population. Although the endothelial lining of the coronaries are derived by angiogenesis within the septum transversum, the SMC and fibroblasts differentiate from a specialized region within the splanchnic mesothelial lining of the coelomic cavity. Abbreviations: Endothelial cell (EC); Peripheral nerve system (PNS); Smooth muscle cell (SMC).

ongoing differentiation and shaping of the vasculature. Blood flow induces shear stress in as well as cyclic expansion and relaxation of ECs. Both processes are known to activate specific biomechanical signaling pathways that can specify the differentiation process and orchestrate tissue build up. Experimentally changing the blood flow can induce a phenotypic shift of the initial identity demonstrating the plasticity of the ECs in the early stages of vasculogenesis (Le Noble et al., 2005).

3.3.2 Origin of the Smooth Muscle Cells

The next step in vessel wall differentiation is the recruitment of SMCs (arteries and veins) and pericytes (precapillaries) to the preformed endothelial scaffolding. The location of the vessel within the embryo determines the origin of the SMCs (Figure 3.5). In the greater part of the body the SMCs have a mesodermal origin. In the head and the thoracic region the arteries and veins recruit their SMCs from the NCCs. The SMCs around the coronary vessels are derived from the PEO. Moreover, even differentiated ECs of the dorsal aorta can contribute to SMCs by rapidly changing their cellular phenotype in a process known as transdifferentiation. Sharp borderlines unmark the segments of the vasculature that are composed of SMCs from different origins. Cell tracing studies have demonstrated that within elastic arteries these borderlines coincide with a change in morphology and functionality. Origin-related SMC differences are, e.g., reflected in the differences in glycosylation of the TGFβ type II receptor resulting in a different response to TGFβ1 in the aortic arch and descending aorta. Vascular abnormalities like Marfan (disrupted fibrillin-1), Williams' syndrome (disrupted elastin), Loeys–Dietz syndrome (TGFβ receptor 1 and receptor 2 mutations), and Char syndrome (mutations in the transcription factor AP-2) have preferential sites of occurrence that probably reflect the embryonic origin of the SMCs.

3.3.3 Stabilization of the Vessel Wall

The vessels stabilize with the recruitment of SMCs from the surrounding mesoderm or NCCs. This process is in part governed by the growth factor platelet-derived growth factor (PDGF)-B, which is expressed by the angiogenic ECs. PDGF-B initiates the recruitment of PDGF receptor-expressing mural cells (pericytes and SMCs) toward the newly formed vessel (Hellström et al., 1999). In turn these cells produce angiopoietin 1, which after binding the Tie2 receptor on the ECs downregulates the angiogenic activities of the ECs. This is an example of the intensive communication between the ECs and underlying pericytes and SMCs and vice versa. Cross talk between ECs and underlying mesenchymal cells is a recurrent theme in cell type specification. Pericytes have recently attracted a lot of attention in the field of tissue engineering as they possess stem cell-like characteristics and are considered to be one of the origins of adult mesenchymal stem or stromal cells in the adult organism (Lv et al., 2014).

3.3.4 Hemodynamics and Vessel Formation

From the onset of circulation, the blood exerts mechanical forces on the ECs and the surrounding undifferentiated mesenchyme which appear to be essential epigenetic and environmental factors in vascular differentiation. It is known that the expression of many genes is regulated by shear stress inherently linked to fluid flow inside a blood vessel. Flow depends, among others, on the diameter of blood vessels, a larger vessel diameter allowing for more flow. Among many others, the expression of TGFβ, endothelial nitric oxide synthase (eNOS), and VEGF is regulated by shear stress. Furthermore, pulsatile and relatively high levels of shear stress protect against apoptosis in which nitric oxide (NO) produced by the enzyme eNOS is a potent intermediate. In contrast, low shear stress areas and turbulent flow promote apoptosis. For these reasons, preconditioning of blood vessels grown in the lab by simulating pulsatile blood flow through the construct is an essential element in tissue engineering of functional blood vessels. The preconditioning helps in defining the identity of the ECs. In addition, it helps in the adaptation and orientation of the underlying SMCs and extracellular matrix in order to withstand the mechanical forces exerted by the pulsatile flow. Such tissue maturation process is needed to avoid rupture and hence failure of tissue-engineered constructs shortly after transplantation.

3.4 DEVELOPMENT OF PERIPHERAL NERVE TISSUE

Driven by an enormous clinical need, peripheral nerve regeneration has become a prime focus within the field of tissue engineering. The various tissue components that make up the peripheral nerve originate from different germ layers. The neurons and glial or Schwann cells are derived from either the neuroectoderm or NCC, while the nerve sheath, which protects the nerves, is derived from the mesoderm (Le Douarin and Kalcheim, 1999). The development and organization of the peripheral nerves with their myelinated and nonmyelinated axonal fibers and their protective epithelial layer involves complex cellular interactions and molecular mechanisms.

3.4.1 Development of the Schwann Cell Lineage

The peripheral nerve contains two types of Schwann cells: myelin-forming Schwann cells that ensheath large caliber axons (diameter larger than 1 μm) and nonmyelinating Schwann cells that accommodate multiple lower caliber axons in cytoplasmic cuffs (Figure 3.6). Both cell types are derived from an immature Schwann cell, which itself is the product of a common precursor cell: the Schwann cell precursor. From studies in the mouse and rat, it was found that Schwann cell precursors are formed early during embryonic nerve development, around embryonic days 12–13 (E12-E13 mouse). At this time, NCCs have populated the outgrowing axon bundles. The cellular interactions and cues that divert the fate of the NCC to the Schwann cell lineage at the eventual exclusion of other fate options are still largely unknown. However, it is known that the survival and ensheathment of axons by Schwann cell precursors critically depend on the axonally derived neuregulin1 protein. Migrating NCCs express the neuregulin1 receptor ErbB3/ErbB2 which depends on the Sry box transcription factor Sox-10. Both Sox-10 and ErbB3/ErbB2 are essential

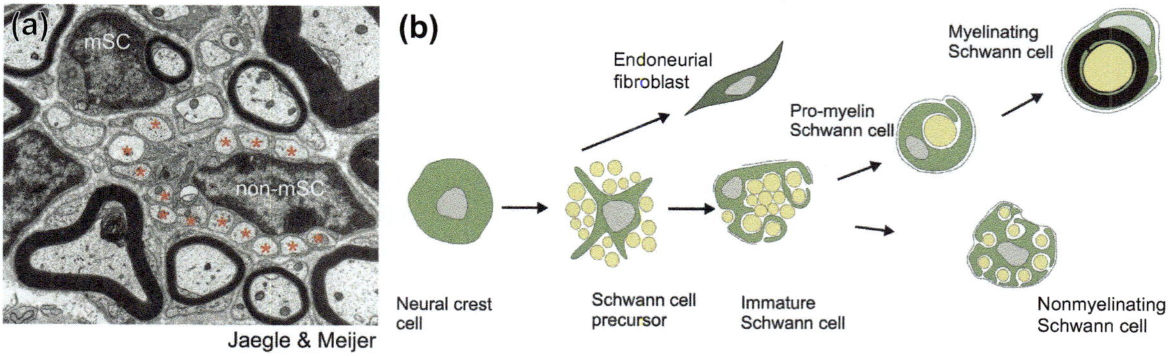

FIGURE 3.6

Development of the Schwann cell. The mature Schwann cell phenotypes that can be distinguished in the nerve differentiate through a number of intermediate stages from the neural crest. (a) This electron micrograph of a transverse section of a typical mixed nerve (containing both sensory and motor fibers) shows myelinating Schwann cells associated with a single larger axon and a nonmyelinating Schwann cell that ensheaths multiple small axons (asterisks) accommodating them in cytoplasmic cuffs. (b) The schematic drawing illustrates the different intermediate stages that can be distinguished in the development of a Schwann cell from the neural crest. Lineage tracing studies in mice have indicated that the endoneurial fibroblasts also are derived from the neural crest cells that populate the embryonic nerves. The basal lamina that surrounds the axon Schwann cell units is produced by the Schwann cell and is here illustrated by a stippled line. Immature Schwann cells have not produced a continuous basal lamina yet. For further details see text.

for survival and proliferation of Schwann cell precursors. Lack of embryonic Schwann cells results in massive sensory and motor neuron cell death in these embryos, demonstrating that neurons not only rely on target-derived trophic support but also critically depend on Schwann cell support.

Schwann cell precursors migrate, proliferate, and ensheath axons and differentiate into immature Schwann cells. Immature Schwann cells populate the fetal nerves and differ from Schwann cell precursors in several ways. While the survival of Schwann cell precursors depends strictly upon axonal contact, immature Schwann cells have acquired the ability to survive in the absence of axonal contact by establishing an autocrine survival pathway (Meier et al., 1999). Insulin-like growth factor (IGF1), Neurotrophin-3 (NT-3), and PDGFβ are important components of this pathway. The survival of immature and mature Schwann cells in the absence of axonal contact is of physiologic importance as successful regeneration of injured peripheral nerves greatly depends on axonal contact with living Schwann cells in the denervated nerve stump. Indeed, these so-called reactive Schwann cells secrete a range of neurotrophic factors and cytokines, such as nerve growth factor (NGF), ciliary neurotrophic factor (CNTF), leukemia inhibitory factor (LIF), and interleukin-6 (Il-6), which stimulate and direct axonal growth (Scherer and Salzer, 2001). Consequently, successful tissue engineering or even regeneration of peripheral nerves requires the presence of a healthy Schwann cell population which can support and direct nerve out growth.

3.4.2 Radial Sorting and Myelination of Peripheral Nerves

In mouse embryos from mid-gestation onwards, the migration, proliferation, and ensheathment of axonal fibers continue until around birth, when the number of Schwann cells and axons are eventually matched. The process of radial sorting and ensheathment of axons relies on interactions between the Schwann cell and its basal lamina (Colognato et al., 2005). In particular laminin, which is a major component of the basal lamina, is of importance for proper axonal sorting by Schwann cells. During the process of radial sorting, individual Schwann cells select larger caliber axons in a 1:1 relationship. These Schwann cells will exit the cell cycle and initiate myelin formation. In contrast, groups of lower caliber axons (mainly the fibers of nociceptive and most autonomic neurons) remain associated with a single Schwann cell. These so-called non-myelin-forming Schwann cells accommodate and ensheath the individual axons in cytoplasmic intentions or cuffs. These populations of axons-glia are called Remak fibers (see Figure 3.6(a)).

The fate decision to become either a myelin-forming or a non-myelin-forming cell is governed by axonal cues. The exact nature of these cues has long remained elusive and is still not understood in full detail. The extent of myelination, that is, the number of layers of compact myelin and the length of the internode, the distance between two successive nodes of Ranvier, which play an essential role in action potential conduction, correlates with the diameter of the axon. Thus, thicker axons have thicker myelin and longer internodes.

3.4.3 Structure of the Peripheral Nerve Sheath

While nerve tracts in the central nervous system are protected by the rigid bony structure of the skull and vertebral column, peripheral nerves lack these protective structures. Instead, they have a compact connective tissue matrix which protects the peripheral nerves. Three layers can be distinguished in this compacted tissue, and these have been termed epineurium, perineurium, and endoneurium. The epineurial and perineurial cells of the nerve sheath are derived from the mesoderm (Bunge et al., 1989). The majority of endoneurial cells are derived, like the Schwann cells, from the NCC (Joseph et al., 2004).

While the epineurium is a condensation of connective tissue surrounding the peripheral nerve, the perineurium is a multilayered cellular sheath surrounding individual fascicles of the nerve. The layers consist of concentric sleeves of flattened epithelial cells that interdigitate and are connected by tight junctions at their margins.

The endoneurium mainly consists of a thick layer of collagen fibrils running parallel with the axon—Schwann cell units. Axons, Schwann cells, and fibroblasts are the major cellular components of the intrafascicular space.

3.4.4 Development of the Peripheral Nerve Sheath

The peripheral nerve sheath develops from mesenchymal cells surrounding the embryonic nerves. From E15 onwards, mesenchymal cells assemble concentrically around the developing nerve, and around birth a recognizable perineurium is present. At this stage the perineurium is still leaky, but becomes impermeable a few weeks after birth (Sharghi-Namini et al., 2006).

Genetic experiments have demonstrated that the formation of a structurally and functionally normal perineurium largely depends on the Schwann cell-derived signaling molecule desert hedgehog (Dhh). Dhh is one of three members of the hedgehog family of intercellular signaling molecules that play key roles in embryonic pattern formation and organogenesis (Hammerschmidt et al., 1997). The fact that still a thin layer of perineurial cells is formed in the absence of Dhh signaling suggests that Dhh is not required for the initial recruitment of mesenchymal cells. Since perineurial-like sheaths are formed in genetic mutants that lack Schwann cells, this suggests that the initial signal might be derived from the axon.

3.4.5 Summary and Perspective

The development of the peripheral nerve tissue is orchestrated through a number of cellular and molecular interactions that are now understood in some detail. In addition to providing essential support for neurons, Schwann cells orchestrate the development of the protective perineurial sheath. Ideally, artificial nerve grafts that aim to guide and support the regenerating nerve into a distal, denervated nerve stump should include autologous Schwann cells. However, to obtain autologous Schwann cells healthy nerves need to be sacrificed. Much effort is put in place to obtain Schwann cells from various stem cell populations such as embryonic stem cells, induced pluripotent stem cells, stem cells isolated from the hair follicle bulge, and mesenchymal stromal cells. The successful application of these cells in a clinical setting will require a detailed understanding of the biology of these stem cells and the development of procedures to derive and expand Schwann cells and their subsequent incorporation in a tissue engineering strategy.

3.5 EMBRYONIC SKIN DEVELOPMENT

Tissue-engineered living skin substitutes have found their way into clinical practice for treatment of severe skin defects. While the application of these in vitro generated skin substitutes have greatly improved treatment outcome of severe burn defects, many issues remain to be resolved to end up with an aesthetic and functional equivalent that matches the natural skin. Here the sequential steps and factors involved in embryonic skin development are summarized and correlated with the formation of skin substitutes in vitro. Skin

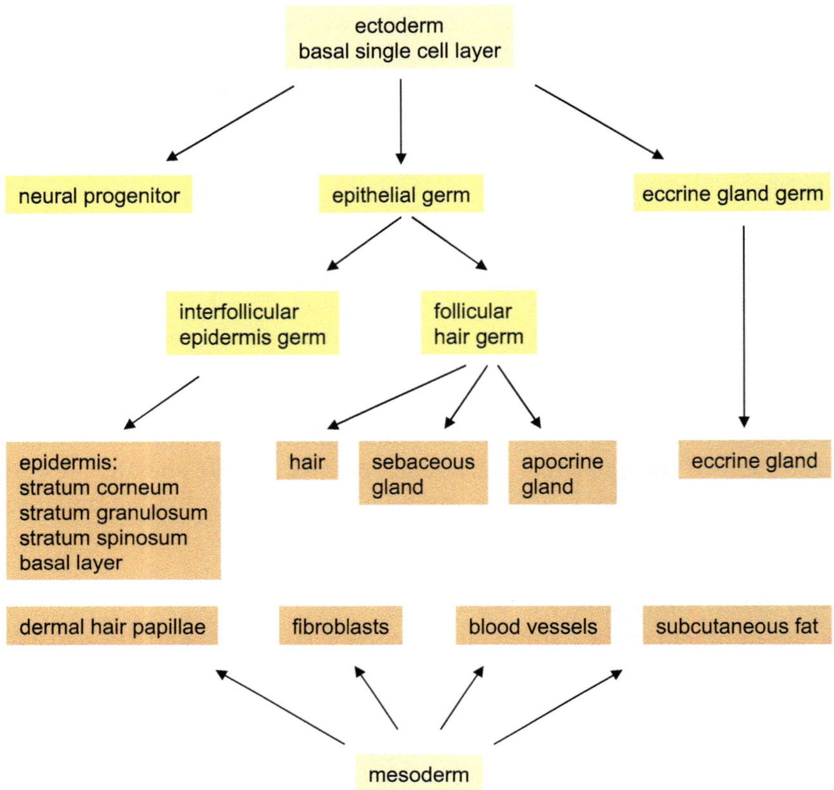

FIGURE 3.7
Schematic diagram of the cellular origin of the skin.

tissue engineering is one of few examples in which developmental engineering has resulted in clinically applicable tissue replacements.

Embryonic skin develops from the ectoderm and mesoderm via a complex process of cell proliferation, differentiation, migration, and apoptosis (Figure 3.7; Table 3.1). The ectoderm is the progenitor of the follicular and interfollicular epidermis and is also the neural progenitor (Moreau and Leclerc, 2004). The mesoderm is the progenitor of the dermis and subcutaneous fat.

3.5.1 Interfollicular Epidermis

In 5-week gestational skin, the epidermis consists of a single layer of ectodermal cells with intermittent areas containing a second suprabasal layer—the periderm (Figures 3.8(a) and 3.9(e)). After 7 weeks, Langerhans cells (immune surveillance cells of the skin), derived from bone marrow cells, begin to appear in the developing epidermis. By the 10th week, the stratum intermedium forms between the basal cell layer and the periderm through upward movement of

Table 3.1 Timelines for Embryonic Skin Development

Gestation (weeks)	Fetal Length (cm)	Developing Structure
5	0.14	Ectoderm: basal single cell layer and intermittent periderm
7		Langerhans cell precursors from hematopoietic origin
8		Mesoderm: loosely arranged mesenchymal cells embedded in ground substance; Merkel cell differentiation from ectoderm
10	5.0	Epithelial germ: basal cell layer, stratum intermedium, periderm
12		Epithelial germ: melanocyte precursors in intermediate layer; early dermis: argyrophilic reticulum fibers (collagen III), abundant fibroblasts
13		Eccrine glands start to develop
17		Hair and sebaceous gland development
19	17	Epidermis: basal layer, 2–3 intermediate layers, flattened periderm
20		Apocrine glands; fat cells mainly in brown fat (very little white fat)
22		Dermis: elastic fibers
23	19	Epidermis: keratinization in stratum intermedium (keratohyalin granules), premature stratum corneum
24		Formation of vernix caseosa
32		Papillary and reticular dermis
36	40–50	Fully developed immune competent skin

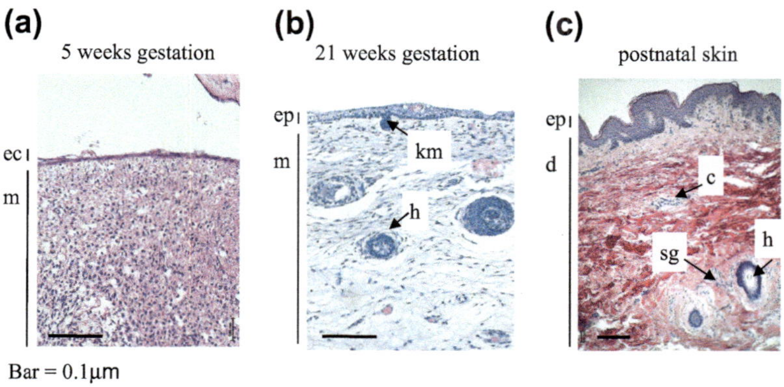

FIGURE 3.8

Developing embryonic skin. (a) Embryonic skin at 5 weeks gestation. Note the single-layered ectoderm, ec, containing sporadic periderm cells, the cell dense mesenchyme, m, and the absence of capillaries. (b) Embryonic skin at 21 weeks gestation. Note the multilayered developing epidermis, ep, lacking a stratum corneum. The epidermal keratinocyte mass, km, that develops into the hair follicle and sebocytes can be seen budding into the dermis. Mesenchymal cells are becoming less dense in the dermal matrix and precursor hair follicles, h, are visible. (c) Fully developed new born skin. Note the stratified epidermis, ep, endothelial capillaries, c, and low density of fibroblasts (single cells) in the dermis, d. Fully developed hair follicles, h, and sebaceous glands, sg, are visible deep in the dermis. Bar = 100 μm.

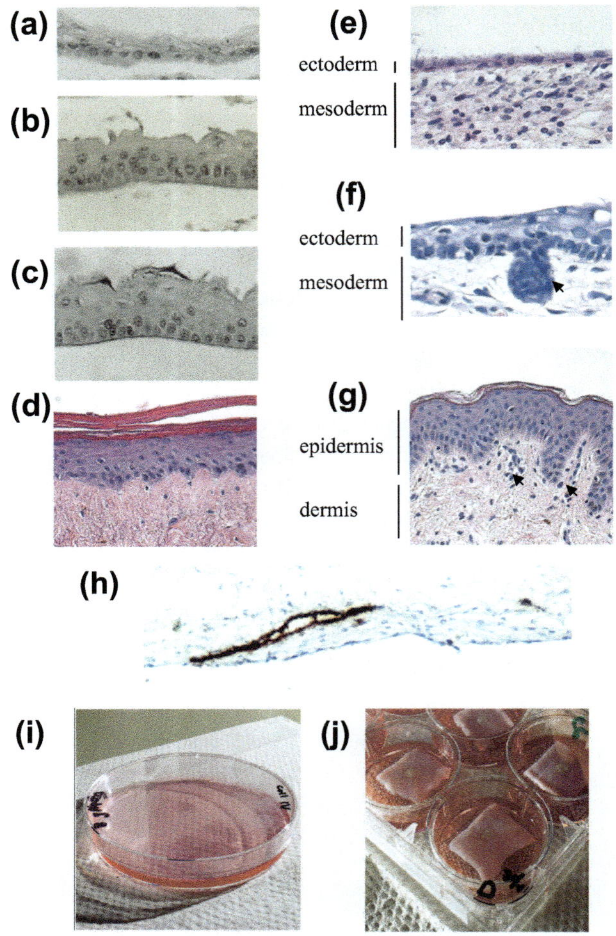

FIGURE 3.9

In vitro versus in vivo developing skin. (a) Cross section of a keratinocyte sheet cultured submerged under the culture medium on a feeder layer of irradiated 3T3 mouse fibroblasts. Fibroblasts are washed away before harvesting and therefore are not visible. Note the similarity with 5-week gestational ectoderm (E). (b) Keratinocyte sheet cultured as described in (a) but with addition of retinoic acid to the culture medium which stimulates keratinocyte proliferation. Note the resemblance to 21-week gestation skin (f), only the follicular bud is absent (arrow). (c) Keratinocyte sheet as described in (a) but with the addition of vitamin D as well as retinoic acid to the culture medium. Retinoic acid stimulates proliferation and stratification whilst vitamin D stimulates differentiation. Note the keratohyalin-like granules in the upper layers making this culture resemble developing epidermis at 23 weeks gestation. (d) Reconstructed epidermis cultured on fibroblast populated human dermis. As the cultures are grown at the air–liquid interface, with nutrients from the medium diffusing in through the dermis, complete epidermal differentiation occurs. Note the fibroblasts populate the dermis at a similar frequency to that observed in (g) but significantly less than in (e). Also, no endothelial capillaries are present. (h) Fibroblast–endothelial cell coculture system stimulates endothelial capillary formation (blue shows hematoxylin staining of fibroblast nuclei; red shows EN4 immunostaining of endothelial cells) and deposition of extracellular matrix from fibroblasts. (i) Petri dish showing experimental setup for culturing under submerged conditions as described in (a), (b), (c), and (h). (j) Experimental setup for culturing at the air–liquid interface as in (d).

cells from the basal cell layer. The cells of the periderm become large and protrude into the amniotic cavity. Merkel cells (nerve/touch cells) differentiate from the ectoderm between weeks 8 and 12 of gestation. Around 12 weeks, melanoblasts migrate from the NCC into the ectoderm (Hirobe, 2005). These cells later develop into melanocytes—the pigment forming cells in the epidermis. The appearance of melanoblasts takes place in a craniocaudal direction following the "Lines of Blaschko" (Figure 3.10) (Bolognia et al., 1994). After 21 weeks of gestation, fetal skin becomes thickened due to increasing numbers of stratum intermedium cells and the periderm becomes flattened (Figures 3.8(b) and 3.9(f)). At 23 weeks, keratinization takes place in the upper cell layers and small keratohyalin granules form. The cells of the periderm are shed into the amniotic fluid, leaving only fragments of degenerated periderm cells above the keratinized cells of the newly formed stratum corneum. The epidermis continues to form by upward movement of differentiating keratinocytes from the proliferating basal layer until a fully differentiated, stratified epidermis is formed (Figures 3.8(c) and 3.9(g)) (Mack et al., 2005). During the last trimester of gestation (>24 weeks), the newly formed stratum corneum of the fetus is protected from the amniotic fluid by a white, greasy biofilm called vernix caseosa. Vernix caseosa consists of water-containing corneocytes embedded in a lipid matrix and the basic structure shows certain similarities with the stratum corneum (Rissmann et al., 2006). After birth, the vernix caseosa is absorbed and the epidermis forms a competent barrier to the environment preventing infection and dehydration of the infant.

3.5.2 Follicular Epidermis

The ectoderm basal cell layer proliferates and differentiates into the keratinizing interfollicular epidermis as described above. Additionally, the ectoderm forms the eccrine glands (sweat) and hair germs which then further differentiate into hair, sebaceous glands, and apocrine glands (Figure 3.7) (Fu et al., 2005). In early gestation (13 weeks), when fetal skin is composed of two or three epidermal layers, eccrine glands begin to form. Eccrine gland islands composed of epidermal cells gradually migrate down into the dermis to form the juvenile sweat glands at weeks 18–20. Hair follicles begin to form at 17 weeks. The formation of these follicles represents a proteotypic interaction between the neuroectoderm and mesoderm that is provided by three different stem cell sources: epidermal, neural crest, and mesenchyme. The epidermal keratinocyte mass that differentiates into hair follicles and sebocytes buds into the deeper layers of the dermis (Figures 3.8(b) and 3.9(f)). Beneath each bud lies a group of mesenchymal cells (fibroblasts) from which the dermal hair papillae and connective tissue sheath are later formed. NCCs give rise to the melanocytes of the hair follicle pigmentary unit. After 21 weeks of gestation, the structures of skin appendages (eccrine glands, hair follicles, and sebaceous

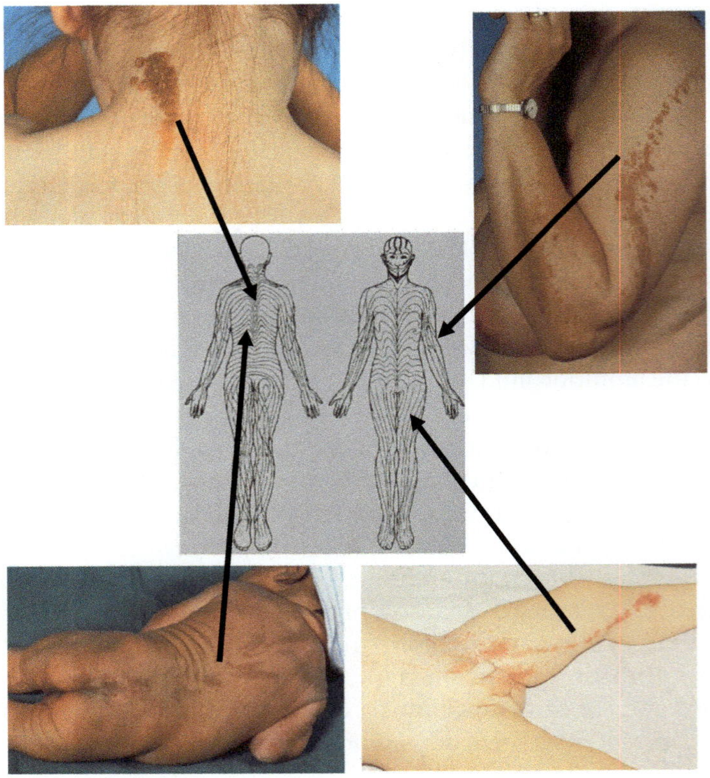

FIGURE 3.10

Lines of Blaschko. More than a century ago, the lines of Blaschko were described by a dermatologist named Alfred Blaschko (Happle and Assim, 2001; Rott, 1999; Traupe, 1999; Bolognia et al., 1994). The lines were based on the observation of patterned skin lesions that were linear on the extremities, S-shaped on the anterior trunk and V-shaped on the back. The lines (or mosaicism) result when a postzygotic mutation occurs leading to a subject with two or more genetically different populations of cells that originate from the genetically homozygous zygote. In most cases, the linear patterns of abnormal skin are surrounded by normal skin. Blaschko's lines are thought to trace pathways of ectodermal development. Several studies have demonstrated that the genotypes of keratinocytes and underlying fibroblasts do not correlate in mosaic skin conditions, confirming the different developmental patterns of ectoderm and mesoderm. The epidermis and its appendageal structures, the melanocytes, the vascular system, and the subcutaneous fat, separately or in combination may be involved in the morphological manifestations which follow Blaschko's lines. The central figure illustrates the Lines of Blaschko originating from the neural crest. Upper left/right and lower right show subjects with epidermal nevi; lower left shows a baby with "linear and whorled nevoid hypermelanosis." *Photographs are supplied by Dr. M. Wintzen, dermatologist, VU university medical center, NL.*

glands) can be detected (Figure 3.8(b)). After 28 weeks of gestation, as fetal skin becomes even thicker, the number of eccrine glands and hair follicles is increased and the structure of the appendages becomes mature. At birth, the sebaceous glands are considerably larger than they are in infancy and secrete the vernix caseosa.

3.5.3 Dermis

In 5-week gestational skin, the dermis consists of loosely arranged mesenchymal cells that are embedded in a ground substance (Figure 3.8(a)). At 12 weeks, argyrophilic reticulum fibers appear. As these fibers increase in number and thickness, they arrange themselves in bundles. Simultaneously, mesenchymal cells develop into fibroblasts and ECs (Figure 3.7). The fetal dermis shows many more fibroblasts than adult dermis (compare Figure 3.8(a) and (b) with Figure 3.9(c)). Also, the dermis of the fetus contains a large amount of collagen type III, in contrast to adult skin which contains a large amount of collagen type I. Elastic fibers appear in the dermis at 22 weeks. As gestation continues, elastic fibers increase in number until at 32 weeks, a well-developed network is formed in the reticular and papillary dermis which is indistinguishable from that found in new born infants.

Fat cells begin to develop in the subcutaneous tissue after 20 weeks. Three types of cell can be found: (1) spindle shaped, lipid-free mesenchymal precursor cells; (2) young type fat cells containing two or more small lipid droplets; and (3) mature fat cells possessing one large lipid droplet.

3.5.4 Tissue Engineering of Embryonic and New Born Skin

The various stages of skin development can be mimicked in vitro by adjusting and finely tuning the culture conditions under which the cells are grown. One should not just consider the tissue itself (in this case skin) but also the environment around the tissue. For example, embryonic ectoderm is in a wet environment engulfed by amniotic fluid on one side and the developing mesoderm rich in fibroblasts on the other side. This provides an extremely wet and nutrient-rich environment. Upon birth, the environment changes significantly. The skin is now exposed to the air. Nutrients reach the epidermis by diffusion from blood vessels in the underlying dermis. Therefore, conditions switch from a wet to a dry environment and also from a nutrient-rich to a nutrient-poor environment. This transition is thought to trigger the final stage in the formation of fully developed skin, thus forming a competent barrier to the environment. Therefore when trying to generate different stages of skin development, for example, the ectoderm transition to epidermis, one should try to copy these environmental changes. Cultures that resemble early ectoderm can be generated from newborn or adult keratinocytes by culturing the keratinocytes on a dense feeder layer of lethally irradiated fibroblasts completely submerged in culture medium (Figure 3.9(a) and (i)) (Gibbs et al., 1996). The fibroblasts provide a contact point for the initially seeded keratinocytes to nestle up to, but importantly the fibroblasts are metabolically active resulting in the secretion of growth factors and cytokines essential for keratinocyte viability and proliferation. Again this is an example of the importance of cross talk between, in this case, epidermal cells and the underlying mesenchyme which directs tissue formation and maturation. The result is a keratinocyte sheet which consists of approximately two layers of unkeratinized epidermis hardly showing any similarity to new born skin. The keratinocyte sheet cultured in this way visibly resembles early ectoderm at 5 weeks gestation (compare Figure 3.9(a) with (e)).

In order to progress to the next stage of embryonic epidermal development in vitro, additional nutrients are required. Addition of, for example, retinoic acid (a factor known to be involved in healthy skin development) to the primitive ectoderm-like culture stimulates cell division and upward migration of keratinocytes. This increases the number of cell layers and results in cultures which visibly resemble embryonic skin at 21 weeks gestation (compare Figure 3.12(b) and (f)). Addition of vitamin D in combination with retinoic acid further increases the number of layers, epidermal differentiation, keratinization, and the appearance of small keratohyalin-like granules in the upper layers (Figure 3.12(c)) (Gibbs et al., 1996). These cultures visibly resemble embryonic skin at approximately 23 weeks gestation.

In order to mimic the final stage in epidermal development, the culture environment has to be changed significantly—copying the environmental change at birth. The most profound change is exposure of the skin to the air in place of the warm and wet environment of the amniotic fluid. In vitro this can be mimicked by raising the keratinocyte cultures to the air–liquid interface (Figure 3.12(j)) (Gibbs et al., 1997). The transition from submerged culture conditions to air exposed culture conditions stimulates the final step in skin development—the formation of the stratum corneum.

The developing dermis can be copied in vitro by coculturing fibroblasts and ECs (Ponec et al., 2004). Cell–cell interactions and optimal culture conditions stimulate individual ECs to migrate toward each other to form capillary-like tubes (Figure 3.12(h)). Fibroblasts synthesize extracellular matrix components thus forming a human collagen–elastin rich dermal matrix similar to in vivo.

Whereas the development of the interfollicular epidermis and dermis can be mimicked in vitro, development of the follicular epidermis is still challenging although in the past years much progress has been made.

3.5.5 Cell—Cell Interactions and Growth Factors

The skin development and thereafter the maintenance of skin integrity (homeostasis) is dependent on a complex interplay between cell types within the ectoderm (epidermis) and mesoderm (dermis). Cell homing to the correct location, proliferation, and differentiation is regulated by growth factors, cytokines, and chemokines that act in autocrine and paracrine loops. The result is the formation of a fully developed skin consisting of keratinocytes and fibroblasts with constant ratios of keratinocytes:melanocytes (36:1) and keratinocytes:Langerhans cells (53:1) (Hoath and Leahy, 2003). The formation of cell units can be mimicked in vitro and give an insight into embryonic skin development. Melanocytes cocultured with keratinocytes under submerged conditions (ectoderm-like culture) maintain a constant ratio for up to three passages (3 weeks) and when coseeded onto acellular dermis and cultured at the air–liquid interface form a differentiated epidermis interdispersed with melanocytes in the basal layer (Gibbs et al., 2000a). Melanosomes formed in the melanocytes are transferred to keratinocytes and cap the nuclei just as in vivo in order to protect the dividing keratinocyte population from harmful ultraviolet irradiation after birth. It has been reported that fetal keratinocytes possess a greater stimulatory effect on proliferation of melanocytes than neonatal keratinocytes, as fetal keratinocytes produce and release more mitogens than newborn keratinocytes. Similar to melanocytes, Langerhans cell precursors ($CD34^+$ cord blood-derived) can be introduced into submerged keratinocyte cultures which when air-exposed, develop into fully immunocompetent Langerhans cells in the epidermis in ratios similar to those found in fully developed skin (Facy et al., 2005). Cocultures of keratinocytes and fibroblasts (irradiated feeder layer) maintain the keratinocyte stem cell population in vitro and enable large amounts of keratinocytes, including stem cells, to be amplified for transplantation as sheets onto large burns wounds (Green et al., 1979). Upon transplantation, the ectoderm-like culture develops into a differentiated adult epidermis as the environmental conditions change (submerged, nutrient-rich environment changes to air-exposed, nutrient-poor environment). Soluble factors secreted by keratinocyte–fibroblast interactions are involved in formation of the basement membrane (El Ghalbzouri and Ponec, 2004). The basement membrane is essential for the attachment of the epidermis to the dermis and heritable defects can result in blistering (bullous disease). Soluble factors secreted by keratinocytes result in homing of fibroblasts into the dermis (Gibbs et al., 2006), whereas soluble factors secreted by fibroblasts stimulate formation of the epidermal layers (basal layer, spinous layer, granular layer, and stratum corneum) (El Ghalbzouri et al., 2002). One of these factors is keratinocyte growth factor (KGF or FGF7) (Figure 3.11). Supplementation of FGF7 to the culture medium induces keratinocyte proliferation and

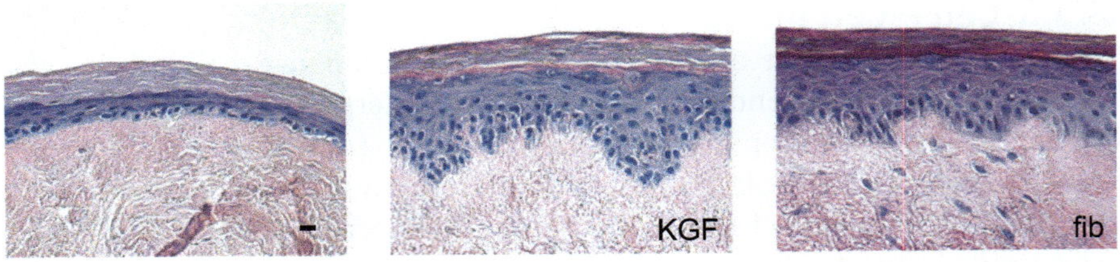

FIGURE 3.11
Keratinocyte growth factor (KGF) can be used to replace fibroblasts (fib) in stimulating epidermal morphogenesis. In the absence of KGF or fibroblasts (fib), the stratified layers of the epidermis are not properly formed. KGF is a growth factor made and secreted by fibroblasts which stimulates keratinocyte proliferation and differentiation and is used as a substitute for fibroblasts (left panel) in coculture experiments with human keratinocytes.

Table 3.2 Growth Factors Secreted by Keratinocytes and Fibroblasts Which Are Involved in Skin Development

Growth Factor	Secreted by	Target Cell	Property
Bone morphogenic protein	?	Ectoderm/KC	Differentiation of ectoderm and epidermis
Noggin	Mesoderm	Ectoderm	Neural differentiation, hair follicle formation, suppression of ectoderm
Epiregulin	KC	KC	Proliferation
Epidermal growth factor	Hair follicle?	KC/Fib	Proliferation, migration
Keratinocyte growth factor	Fib	KC	Proliferation, differentiation
Tumor growth factor-α	KC/Fib	KC/Fib	Differentiation, migration, extracellular matrix synthesis
Vascular endothelial growth factor	KC	Endo	Angiogenesis
Tumor necrosis factor-α	KC/Fib	KC/Fib/LCp	KC/Fib proliferation, LCp differentiation to LC
Tumor growth factor-β	KC	Fib/LCp	Fibroblast differentiation to myofibroblast, LCp differentiation to LC
Granulocyte macrophage-colony stimulating factor	KC	LCp/MC	LCp differentiation to LC, MC proliferation, melanogenesis, dendritogenesis
Basic fibroblast growth factor	KC	MC/Endo/Fib	Proliferation, angiogenesis
Hepatocyte growth factor	KC/Fib	MC	Proliferation
α-MSH	KC	MC	Melanogenesis, dendritogenesis
Nerve growth factor	KC	MC	Melanogenesis, dendritogenesis
Endothelin 1	KC	MC	Proliferation, melanogenesis, dendritogenesis
Stem cell factor	KC/Fib	MC	Proliferation, melanogenesis, dendritogenesis

KC: keratinocyte; Fib: fibroblast; LCp: Langerhans cell precursor; LC: Langerhans cell; MC: melanocyte; Endo: endothelial cell.

differentiation and can be used to replace living fibroblasts in the dermal matrix (Figure 3.14) (Gibbs et al., 2000b). Table 3.2 summarizes the properties of some keratinocyte- and fibroblast-derived growth factors involved in skin development.

3.5.6 Summary

The development of embryonic skin involves environmental, matrix, and cell–cell interactions. Inspired by developmental biology these processes can be mimicked in vitro to engineer skin substitutes. Very important in this engineering process is the fine tuning of differential and sequential secretion of soluble factors which regulate cell growth and differentiation. In vitro, a number of phases of embryonic skin development can be mimicked, such as development of the interfollicular epidermis including keratinocytes, melanocytes, and Langerhans cells. Also a capillary network in a fibroblast-populated dermal matrix can be generated from a mesenchymal cell mix. However, the formation of appendages such as hair follicles and eccrine glands, which involves extensive dermal and epidermal interactions, is still challenging.

3.6 SKELETAL FORMATION

The skeleton is a dynamic living tissue largely consisting of an extracellular matrix combined with matrix forming and degrading cells. The extracellular matrix can be divided into two types, bone and cartilage, which differ in their matrix composition and physical properties in line with their respective functions. Three cell types are involved in skeletal formation: the cartilage producing chondrocyte, the bone forming osteoblast, and the cartilage and bone resorbing osteoclast. Major challenges for tissue engineering of the skeleton are (i) the repair of damaged articular cartilage covering the distal ends of the bone in the joints; (ii) the repair of nonunion fractures, and (iii) the improvement of the union of synthetic prostheses with native bone. The purpose of this section is to provide a general overview of skeletogenesis during fetal development including the origin of the bone forming and degrading cells and the basic principles underlying the maintenance of skeletal integrity during adulthood.

The adult human skeleton contains 213 bones. Depending on their location in the body, these bones have one or more specific functions. They provide structural support and integrity to the body and protect vital internal organs such as the brain, heart, and lungs. Bones play an essential role in body movement by providing attachment sites for muscles. They are also involved in the regulation of mineral homeostasis. Furthermore, they provide the appropriate environment for the development of the hematopoietic system inside the bone marrow cavity. Bones are formed either by endochondral ossification or by intramembranous ossification.

Three cell lineages are responsible for the formation of the skeleton during embryonic development. The mesodermal paraxial and lateral plate mesoderm and the neuroectoderm-derived cranial NCC. The lateral plate mesoderm will form the long bones in the limbs. The paraxial mesoderm will form the axial skeleton (vertebral column and rib cage) and together with the NCC will shape the craniofacial bones. The formation of bones starts during the fourth week of human development and begins with the formation of the bones at the base of the skull. New bones are formed in an anterior–posterior direction. With exception of the bones in the jaws, the formation of the craniofacial bones begins later. The shape of each bone is specifically adapted to its function, which depends on its place in the skeleton. Even before the onset of skeletogenesis, cells at the site of the future bone already contain information on the shape and structure of the bone they will form.

3.6.1 Skeletal Precursor Cells

Osteoblasts and chondrocytes are both derived from the multipotent mesenchymal stromal cells or also often referred to as mesenchymal stem cells (MSCs). These cells also give rise to a variety of related mesodermal cell lineages such as fibroblasts, muscle cells, tendons, and adipocytes. The MSC is derived from either the mesoderm or the cranial NCC. During fetal development, MSCs can be isolated from virtually any tissues in relatively high quantities. In adulthood, most tissues possess only a limited number of MSCs, except for the bone marrow in which relatively high amounts of MSCs reside throughout life. These cells play an important role in the continuous cycle of bone formation and degradation by providing a source for new osteoblasts. Signaling by the Wnt family of morphogens plays an important role in the maintenance of the undifferentiated stem cell population (Hartmann, 2006). Wnts stimulate stem cell proliferation and renewal and simultaneously block the initiation of differentiation by increasing the nuclear localization and the transcription potential of β-catenin, an essential and critical intracellular mediator of the Wnt signal transduction pathway.

At sites of future bones, the MSC will give rise to a skeletal precursor cell which can differentiate into either an osteoblast or a chondrocyte. The direction of differentiation depends on the activity of two transcription factors: Sox-9 and RunX2. Knockout studies in mice have shown that Sox-9 is indispensable for chondrocyte differentiation, while RunX2 is indispensable for osteoblast differentiation. The skeletal precursor cell expresses both transcription factors. Intracellular levels of β-catenin controlled by Wnt signaling determine the direction of differentiation. High levels of β-catenin inhibit Sox-9 and potentiate RunX2 activity resulting in osteoblast differentiation. In contrast, low levels of β-catenin result in unopposed Sox-9 activity and thus in the formation of chondrocytes (Figure 3.12) (Hill et al., 2005; Day et al., 2005). In embryogenesis, chondrocyte differentiation appears to be the default differentiation route of the MSC.

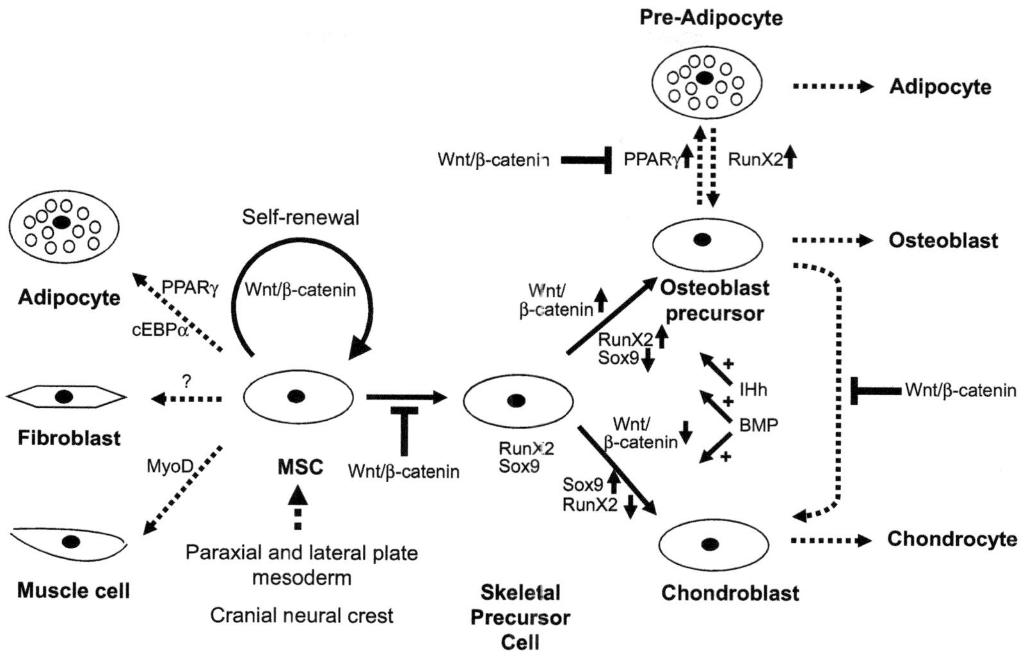

FIGURE 3.12

The formation of skeletal precursor cells. Skeletal precursor cells are derived from the mesenchymal stromal cells (MSCs) that also give rise to adipocytes, tendonocytes, and fibroblasts. Wnt/β-catenin is responsible for stem cell proliferation and renewal. The MSC differentiates into a skeletal precursor cell that expresses the transcription factors for osteoblast differentiation (RunX2) and chondrocyte differentiation (Sox-9). Wnt/β-catenin signaling directs the precursor cell into the osteoblast lineage and prevents transdifferentiation into chondrocytes by augmenting RunX2 activity and repressing Sox-9. Also Indian Hedgehog (IHh) and BMP signaling are involved in this process. The osteoblast precursor is still able to differentiate into adipocytes. The balance between the transcription factors PPARγ (adipocyte specific) and RunX2 (osteoblast specific) determines the fate of the cell. See text for further details.

3.6.2 Endochondral Ossification

The axial and appendicular skeleton is formed by endochondral ossification, i.e., the skeletal elements are preshaped in a cartilaginous mold which is subsequently replaced by bone (Kronenberg, 2003). This process starts with the condensation of loosely connected mesenchymal cells at sites of the future bones (Figure 3.13). This is accompanied with a rearrangement of the vasculature resulting in an avascular condensed mesenchyme surrounded by blood vessels. The transcription factor Sox-9 is responsible for the condensation of the skeletal precursor cells. The combination of increased cell–cell contacts, resulting in signaling of, amongst others, cadherins, low oxygen conditions due to the absence of vessels and low levels of Wnt signaling, is believed to trigger the initiation of chondrocyte differentiation. It is characterized by the production of an abundant extracellular matrix consisting predominantly of collagen type 2a1 and glycosaminoglycans, which are typical cartilage markers. In the middle of the cartilage anlage, chondrocytes start

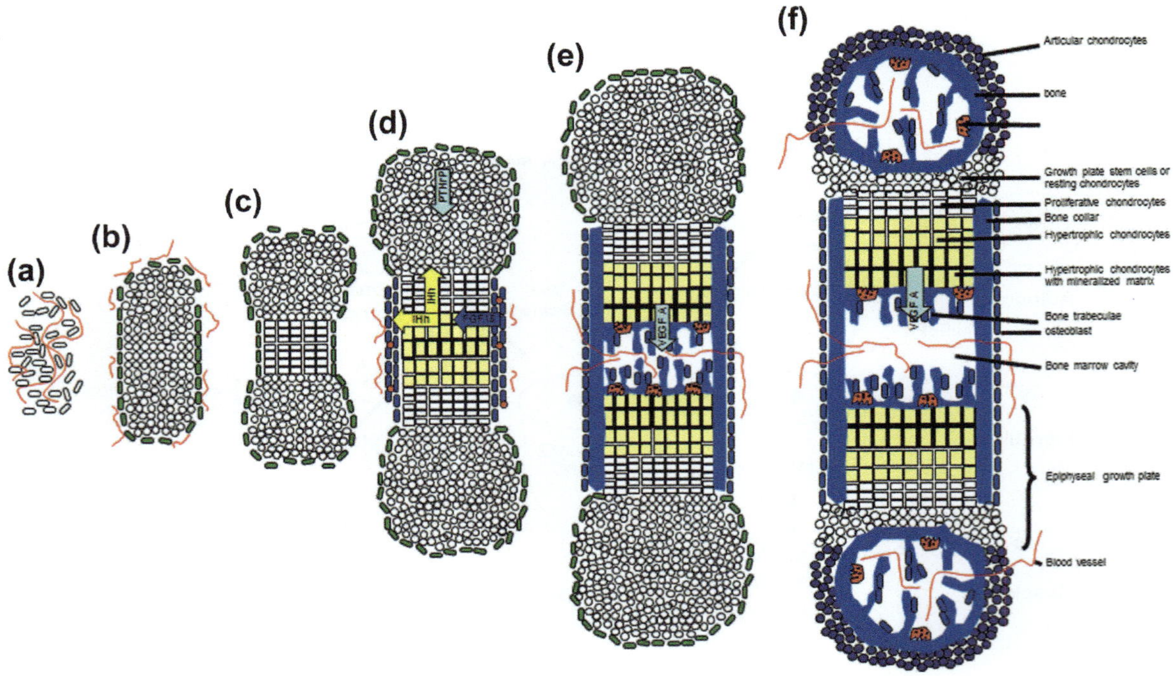

FIGURE 3.13

Endochondral bone formation. Schematic presentation of the subsequent stages of endochondral ossification. It starts with loosely connected mesenchyme invaded with blood vessels (red) (a). The mesenchymal cells start to condense and the vasculature is rearranged. The condensed cells differentiate into chondrocytes that are surrounded by a perichondrium (green) (b). Cells in the middle of the cartilage anlage arrange into columns and start to proliferate (c) and subsequently start to hypertrophy (yellow) (d). Chondrocytes in the transition zone secrete Indian Hedgehog (IHh) that stimulates the formation of a bone collar (blue), chondrocyte proliferation, and the expression of PTHLH in the periarticular region of the long bone. Osteoblasts secrete FGF18 that inhibits chondrocyte proliferation. Terminally hypertrophic chondrocytes start to mineralize their matrix (black lines). At this time point mononuclear osteoclast precursors (red dots) reside in the perichondrium. These cells differentiate into multinucleated osteoclasts that together with blood vessels and osteoblasts (blue) invade the hypertrophic cartilage and start to resorb the mineralized cartilage matrix. VegfA expressed due to the hypoxic conditions in the cartilage plays an important role in the attraction of blood vessels. This results in the formation of a bone marrow cavity. The osteoblasts replace the chondrocyte matrix with a bone matrix (e). Somewhat later this process is repeated in the secondary ossification centers in the cartilaginous heads of the long bones. The formation of the secondary ossification centers divides the cartilage into articular chondrocytes (dark blue) covering the bone surface in the joints and in epiphyseal growth plate cartilage that is responsible for bone elongation (f).

to arrange in typical columns and display high proliferative activity. Somewhat later these chondrocytes stop proliferating and differentiate further into hypertrophic chondrocytes. This is characterized by a dramatic increase of the cell's volume and a reshuffling of the extracellular matrix, which now contains collagen type 10 instead of collagen type 2. The hypertrophic chondrocytes start to mineralize their matrix forming the primary ossification center and subsequently die by apoptosis. At the same time, the chondrocytes signal to the surrounding perichondrial cells in which osteoblast precursors reside by producing Indian hedgehog (IHh). IHh initiates osteoblast differentiation and the formation of the bone collar. It is the most important coupling factor between chondrocyte and osteoblast differentiation during endochondral ossification (Kronenberg, 2003). The hypoxic conditions in the center of the bone anlage induce the expression of the transcription factor hypoxia inducible factor1α (Hif1α), an essential transcription factor for the expression of VEGF. VEGF recruits blood vessels from the lateral perichondrium (Schipani et al., 2006). With these factors in growing vessels, both osteoclasts and osteoblasts enter the bone resulting in the resorption of cartilage, replacement by newly formed bone and the formation of the bone marrow cavity. In this process a cartilaginous matrix containing collagen type 2, collagen type 10, and glycosaminoglycans is replaced by a mineralized bone matrix predominantly consisting of collagen type 1. Somewhat later, chondrocytes start to hypertrophy in the center of the cartilaginous heads at the distal ends of the bones and mineralize their extracellular matrix resulting in the formation of the secondary ossification center. This is followed by ingrowth of vessels, osteoclasts, and osteoblasts. The secondary ossification center demarcates the separation of two types of chondrocytes: articular chondrocytes that cover the distal ends of the bones and have a relatively low metabolic activity, and epiphyseal growth plate chondrocytes that have a high metabolic activity and are solely responsible for bone elongation. The signaling mechanisms involved in this process are largely unknown. Experimental evidence suggests that specification of the articular versus the growth plate chondrocyte occurs already at the first stages of endochondral ossification.

3.6.3 Intramembranous Ossification and Osteoblast Differentiation

The flat bones of the skull and the flat part of the clavicle are formed by intramembranous ossification. In this process cells directly deposit a mineralized bone matrix without a cartilage intermediate. This process also starts with condensation of mesenchymal cells. In contrast to endochondral bone, these condensations occur in vascularized regions of mesenchyme. The condensed cells directly differentiate into osteoblasts which start to produce a mineralized bone matrix. The sequential activation of two transcription factors is essential for the formation of osteoblasts from skeletal precursor cells. This differentiation route is initiated by RunX2 and requires relatively high levels of Wnt signaling. The subsequent activation of a second transcription factor, Osterix (Osx), is needed for further maturation of the cells and the production of a mineralized bone matrix. Indeed, knockout mice lacking either RunX2 or Osx lack bone formation due to the absence of mature osteoblasts. In Osx knockout mice, osteoblasts are arrested in a somewhat later differentiation stage in comparison to the RunX2 knockout mice (Day et al., 2005; Hill et al., 2005).

Other transcription factors like Msx2, ATF4, deltaFosB, and Fra1 and 2 further facilitate and cooperate with RunX2 and Osx in osteoblast differentiation and production of a mineralized bone matrix (Karsenty and Wagner, 2002; Nakashima and de Crombrugghe, 2003).

The life cycle of the osteoblast can be divided into various phases. In the first phase, the number of osteoprogenitors is increased by rapid proliferation. Subsequently, the cells begin to secrete large quantities of extracellular matrix predominantly consisting of collagen 1 followed by a maturation phase in which the matrix is prepared for matrix mineralization. Each of these phases is characterized by the expression of typical markers (Figure 3.14). At the end of the life span, the osteoblast has three choices: it can die by apoptosis, it can become fully embedded in the extracellular bone matrix and differentiate further into an osteocyte, or it can become a quiescent bone lining cell that covers the bone surface. Osteocytes are single cells fully surrounded by a mineralized bone matrix. They are in close contact with each other and the bone surface with a network of cell extensions. These cell protrusions are localized in small channels, the caniculae. The lining cells are quiescent cells which are no longer involved in bone formation. Upon the appropriate signals these cells can, however, resume their activity and start participating again in the formation of a bone matrix.

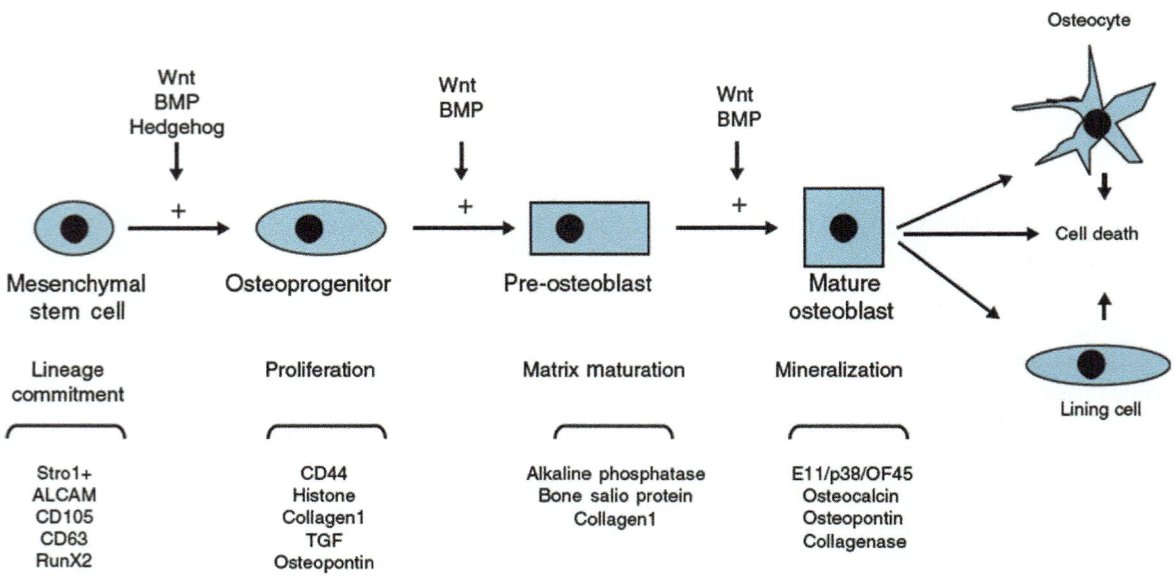

FIGURE 3.14

Osteoblast differentiation. Schematic illustration of osteoblast lineage cells and frequently used markers of the stages of maturation. Wnt and BMPs have positive effects on osteoblast differentiation in all stages of development. The actions of hedgehog are limited to the earliest phases of differentiation from mesenchymal stromal cells.

A large number of growth factors are involved in the regulation of osteoblast differentiation. Members of the TGFβ/BMP superfamily, particularly the BMPs, IHh, and Wnt family members can mediate initiation of osteoblast differentiation from uncommitted precursors. These factors act, amongst others, by inducing the expression of RunX2. Other factors like insulin-like growth factor I (IGF-I), TGFβ, FGFs to name a few play a role in osteoblast proliferation, matrix production, and mineralization. Of particular interest are members of the Wnt family, which are involved in successive stages of osteoblast differentiation such as in the initiation of osteoblast differentiation, matrix production, and osteoblastic cell death. Furthermore, they are involved in the regulation of the coupling between bone formation and bone resorption. Besides control by locally produced growth factors, bone formation by osteoblasts is controlled by systemic factors, such as sex steroids, growth hormone and parathyroid hormone, and by the hypothalamus via the sympathic nervous system (Karsenty and Wagner, 2002; Manolagas, 2000; Ducy et al., 2000).

3.6.4 Osteoclast Differentiation

The osteoclast is a multinucleated highly specialized cell specifically equipped for the resorption of a mineralized matrix (Figure 3.15). Osteoclasts are derived from the hematopoietic stem cell. This cell gives rise to erythrocytes, granulocytes, mast cells, megakaryocytes, lymphocytes, and macrophages. The transcription factor PU-1 and the growth factor macrophage-colony stimulating factor (M-CSF) are required for the establishment of the macrophage lineage. The derivation of osteoclasts from this lineage requires a direct interaction between the osteoclast precursor and the osteoblast involving the expression of receptor activator of nuclear factor kappa β (RANK) by the osteoclast precursor and the membrane associated ligand (RANKL) by the osteoblast. The interaction between RANK and RANKL is required for all subsequent stages of osteoclastogenesis including the fusion of mononucleated precursor cells into a multinucleated functional osteoclast. The subsequent stages of osteoclast development are characterized by specific marker genes (Figure 3.15) (Roodman, 2006).

The interaction between RANK and RANKL can be antagonized by the decoy receptor osteoprotegerin (OPG). The major source of RANKL and OPG is cells from the osteoblast lineage although also T-cells appear to be involved. A large variety of stimuli like cytokines (e.g., IL-1, IL-6, TNFα) and hormones (vitamin D, parathyroid hormone, estrogens) are involved in the regulation of the osteoblastic expression of RANKL and OPG. The balance between these two proteins determines whether osteoblasts can support osteoclastogenesis or not. The formation of osteoclasts is tightly linked with osteoblast differentiation, since the expression of RANKL is controlled by RunX2. Furthermore, the expression of OPG is repressed by Wnt signaling which simultaneously increases osteoblast activity (Glass and Karsenty, 2006).

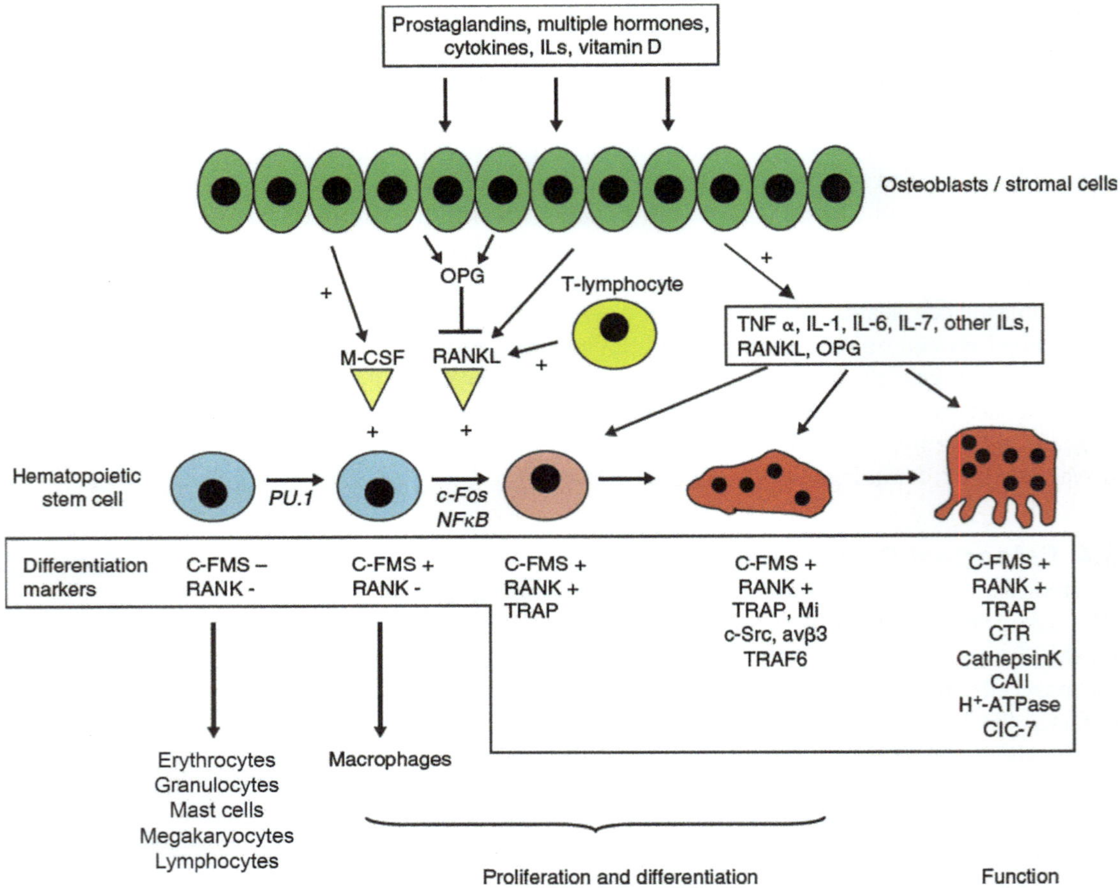

FIGURE 3.15

Osteoclast differentiation. Osteoclasts are derived from the hematopoietic stem cell. These cells become committed to the myeloid lineage under the influence of the transcription factor PU-1 and start to express c-FMS and RANK, the receptors for *m*-CSF and RANKL, respectively. In subsequent differentiation steps, the transcription factors c-Fos and NFκB are of critical importance. Osteoclastogenesis can be divided into various stages, that are characterized by the expression of typical marker genes. The formation of osteoclasts is tightly controlled by osteoblasts and stromal cells that express *m*-CSF and RANKL. Furthermore they express a negative regulator of RANKL, osteoprotegerin (OPG). The RANKL/OPG ratio is a critical component in the regulation of osteoclast formation and activity. Besides RANKL and OPG, osteoblasts and stromal cells express a variety of other growth factors and cytokines that are either pro- or antiosteoclastogenic. The expression of all these factors is tightly controlled by prostaglandins, a variety of hormones like vitamin D3 and estrogens, interleukins (ILs), and other cytokines. Besides osteoblasts and stromal cells, T-lymphocytes also express RANKL and can thus influence osteoclastogenesis and bone resorption.

3.6.5 Bone Remodeling

While the majority of organs in the human body are static, i.e., do not undergo major changes once the formation of the adult organ is completed, bone is continuously renewed by a process called remodeling. It starts as soon as the first bone is formed. In this process, old fatigued bone is replaced by new mechanically sound bone. It is an essential mechanism in the preservation of bone strength and in skeletal adaptation to changing environmental conditions such as in load-bearing of the skeleton. Remodeling is largely orchestrated by the osteocytes and osteoblasts and is controlled by environmental, hormonal, and hypothalamic functions. For example, weight bearing of the skeleton induces a fluid shear stress in the caniculae which is sensed by the osteocytes. The osteocytes signal to the osteoblasts at the bone surface. Amongst others, the secreted Wnt antagonist SOST is an osteocyte expressed gene that appears to play an important role in this process (van Bezooijen et al., 2004). Depending on the signal, osteoblastic bone formation is either activated or repressed. In addition, these signals may influence the balance between OPG and RANKL expression thereby modulating osteoclast formation and bone resorption. Disturbances in the balance between bone formation and resorption are at the basis of all skeletal disorders.

3.6.6 Tissue Engineering of Bone

MSCs derived from adult bone marrow are the most frequently used cell source to engineer bone. In most strategies, their capacity to differentiate into bone forming osteoblasts in a process that resembles intramembranous ossification is exploited. In the treatment of critical size bone defects, limited clinical successes have been achieved. Based on lessons from bone formation in the embryo, this could have been predicted. Bone fractures normally heal, with a few exceptions, in a process resembling endochondral ossification. Recapitulation of this process in a tissue engineering strategy to treat large bone defects implies that one should not aim at directly making bone out of MSCs but one should first make a cartilage anlage (Box State-of-the-Art Experiment). After implantation, this cartilage will be replaced by bone through endochondral ossification which is organized by the tissue itself through interaction with host tissue. Implantation of tissue-engineered cartilage instead of bone has various advantages. For example, cartilage is an avascular tissue and chondrocytes are resilient to low oxygen concentrations. In contrast, bone is highly vascularized and metabolically very active. Lack of oxygen as a consequence of the absence of a vasculature in many tissue-engineered constructs is the main reason for transplant failure. In endochondral ossification, the cartilage self-organizes the ingrowth of blood vessels, which is the first step leading to cartilage degradation, and its replacement by bone. This is an example how tissue engineers have started to exploit the insights from developmental biology into an engineering strategy using the modularity and self-organizing principles of developing tissue to obtain a robust strategy for treatment of bone defects.

3.7 FUTURE DIRECTIONS

Over the past years the tissue engineering field is gradually changing. While at its infancy, tissue engineering was mainly based on trial and error, this new discipline is rapidly developing into a technology-based discipline comparable to other branches of engineering. Reiteration of processes in developmental biology is now exploited to build in vitro tissues. Concepts such as modularity, robustness, and self-organization, which are at the basis of the formation of organs, will be more and more incorporated in tissue engineering strategies. This trend will continue in the coming years resulting in maturation of a new discipline in the formation of tissues called developmental engineering.

3.8 SUMMARY

- Recapitulation of developmental processes in combination with typical engineering disciplines is known as developmental (re)engineering and forms the basis of most, if not all, tissue engineering procedures.
- Tissue formation depends on the timely and subsequent interaction between multiple stem or progenitor cells derived from various origins. Cell migration, cell–cell contact, auto- and paracrine signaling and also environmental cues like pulsatile flow or exposure to air are some of the processes that are used for the specification of cells into a functionalized tissue.
- The inner cell mass in the blastocyst contains pluripotent cells that give rise to all tissues in the adult body. When brought into culture these cells are called embryonic stem cells that retain pluripotency and self-renewal in vitro.
- In gastrulation, the three germ layers ectoderm, mesoderm, and endoderm are formed by a complex process involving cell–cell interactions, cell movements, and gradients of morphogens by which the body plan is established.
- Pluripotent neural crest cells originate from the neural plate border. They migrate via clearly defined patterns to all tissues in the body where they actively participate in tissue formation. They give rise to cell types of ectodermal, mesodermal, and endodermal origin.
- The heart is formed by mesoderm- and ectoderm-derived cell types. The mesodermal cells are present in the cardiogenic plates and will give rise to cardiomyocytes and endocardial cells. The cardiac neural crest, which

will contribute to formation of, e.g., the arterial vessel wall, and the proepicardial organ, which will form the epicardium, originate from the ectoderm by epithelial-to-mesenchyme transition.

- The formation of the heart is a complex and time-consuming process. Initially the heart is formed as a single beating tube that by a series of geometrical changes transforms into a four-chambered septated beating heart.
- In blood vessel formation, endothelial cells are derived from the splanchnic mesoderm. The smooth muscle cells and pericytes are derived from the mesoderm or the neural crest.
- Blood flow and communication between endothelial cells and surrounding smooth muscle cells or pericytes will shape and define the identity of the vessel. In the absence of blood flow, vessels will not develop properly or even deteriorate.
- In peripheral nerve tissue, the neurons and glial cells (also called Schwann cells) are derived from either the neuroectoderm or neural crest, while the nerve sheath is derived from the mesoderm.
- Neurons can only survive in close contact with Schwann cells. Vice versa, mature Schwann cells can only survive in close contact with neurons. Only immature Schwann cells can survive without axonal contact.
- In adult skin, the follicular and interfollicular epidermis is derived from the ectoderm. The mesoderm is the progenitor of the dermis and subcutaneous fat.
- The development of embryonic skin involves environmental, matrix, and cell–cell interactions. These processes can be mimicked to some extent in vitro by changing the culture conditions from a wet to air-exposed environment and by coculture of dermal fibroblasts and keratinocytes.
- The major bone forming cells, the osteoblast and chondrocyte, are derived from the paraxial and lateral plate mesoderm and the neuroectoderm-derived cranial neural crest. The bone-degrading osteoclasts are derived from the mesenchymal hematopoietic cell lineage.
- Bone is an active tissue that is continuously remodeled. It is formed either by intramembranous ossification or by endochondral ossification. In intramembranous ossification, mesenchymal stem cells directly differentiate into a bone matrix depositing osteoblast. In endochondral bone formation, skeletal elements are preformed in a cartilaginous mold that is replaced by bone.
- Osteoblasts and chondrocytes have a common precursor, the multipotent MSC. The differentiation into either cell lineage is governed by a complex array of signaling molecules including BMPs and Wnts of which the intracellular Wnt target β-catenin appears to play a central role.

STATE-OF-THE-ART EXPERIMENT

Realizing that most bones in our body are formed and healed by endochondral ossification, Martin and coworkers started to explore whether this developmental biological phenomenon could be used to generate bone. They started with human mesenchymal stromal cells. While others focused on directly making bone via intramembranous ossification, Martin started to make cartilage first. The cartilage templates were subsequently implanted ectopically in a mouse. Indeed the cartilage templates were efficiently replaced by functional bone containing a mineralized matrix. The efficacy depended on the maturation state of the cartilage. Efficient replacement by bone was shown only in cartilage constructs containing hypertrophic cartilage. The ectopic bone was vascularized and contained a bone marrow compartment in which hematopoiesis took place. Hypertrophic cartilage was capable to recruit the ingrowth of blood vessels from the host tissue. In addition, the cartilage was able to recruit osteoblast precursors from the host which started to deposit bone. Cells from the bone marrow compartment, such as the osteoclasts, were also derived from the host. Some human MSCs also differentiated into osteoblasts particularly at the pericortical region. The bone in this region was thus derived from both implanted human cells and host cells. Importantly, the underlying morphogenetic process was structurally and molecularly similar to the temporal and spatial progression of long bone development in the limbs. The study by Martin and coworkers provides a model for fundamental and translational studies on bone morphogenesis and regeneration by invoking a developmental engineering paradigm.

REFERENCES

Bolognia, J.L., Orlow, S.J., Glick, S.A., 1994. Lines of Blaschko. J. Am. Acad. Dermatol. 31, 157–190.

Bradley, A., Evans, M., Kaufman, M.H., Robertson, E., 1985. Formation of germ-line chimaeras from embryo-derived teratocarcinoma cell lines. Nature 309, 255–256.

Bunge, M.B., Wood, P.M., Tynan, L.B., Bates, M.L., Sanes, J.R., 1989. Perineurium originates from fibroblasts: demonstration in vitro with a retroviral marker. Science 243, 229–231.

Colognato, H., ffrench-Constant, C., Feltri, M.L., 2005. Human diseases reveal novel roles for neural laminins. Trends Neurosci. 28, 480–486.

Day, T.F., Guo, X., Garrett-Beal, L., Yang, Y., 2005. Wnt/beta-catenin signaling in mesenchymal progenitors controls osteoblast and chondrocyte differentiation during vertebrate skeletogenesis. Dev. Cell. 8, 739–750.

de Jong, F., Opthof, T., Wilde, A.A., Janse, M.J., Charles, R., Lamers, W.H., Moorman, A.F., 1992. Persisting zones of slow impulse conduction in developing chicken hearts. Circ. Res. 71, 240–250.

Ducy, P., Schinke, T., Karsenty, G., 2000. The osteoblast, a sophisticated fibroblast under central surveillance. Science 289, 1501–1504.

El Ghalbzouri, A., Ponec, M., 2004. Diffusible factors released by fibroblasts support epidermal morphogenesis and deposition of basement membrane components. Wound. Repair Regen. 12, 359–367.

El Ghalbzouri, A., Gibbs, S., Lamme, E., Van Blitterswijk, C.A., Ponec, M., 2002. Effect of fibroblasts on epidermal regeneration. Br. J. Dermatol. 147, 230–243.

Evans, M.J., Kaufman, M.H., 1981. Establishment on culture of pluripotential cells from mouse embryos. Nature 292, 154–156.

Facy, V., Flouret, V., Regnier, M., Schmidt, R., 2005. Reactivity of Langerhans cells in human reconstructed epidermis to known allergens and UV radiation. Toxicol. In Vitro 19, 787–795.

Fu, X., Li, J., Sun, X., Sun, T., Sheng, Z., 2005. Epidermal stem cells are the source of sweat glands in human fetal skin: evidence of synergetic development of stem cells, sweat glands, growth factors, and matrix metalloproteinases. Wound Repair Regen. 13, 102–108.

Gibbs, S., Backendorf, C., Ponec, M., 1996. Regulation of keratinocyte proliferation and differentiation by all-trans-retinoic acid, 9-cis-retinoic acid and 1,25-dihydroxy vitamin D3. Arch. Dermatol. Res. 288, 729–738.

Gibbs, S., Vicanova, J., Bouwstra, J., Valstar, D., Kempenaar, J., Ponec, M., 1997. Culture of reconstructed epidermis in a defined medium at 33 degrees C shows a delayed epidermal maturation, prolonged lifespan and improved stratum corneum. Arch. Dermatol. Res. 289, 585–595.

Gibbs, S., Murli, S., De Boer, G., Mulder, A., Mommaas, A.M., Ponec, M., 2000a. Melanosome capping of keratinocytes in pigmented reconstructed epidermis–effect of ultraviolet radiation and 3-isobutyl-1-methyl-xanthine on melanogenesis. Pigment Cell Res. 13, 458–466.

Gibbs, S., Silva Pinto, A.N., Murli, S., Huber, M., Hohl, D., Ponec, M., 2000b. Epidermal growth factor and keratinocyte growth factor differentially regulate epidermal migration, growth, and differentiation. Wound Repair Regen. 8, 192–203.

Gibbs, S., Hoogenband van den, H.M., Kirschig, G., Richters, C.D., Spiekstra, S.W., Breetveld, M., Scheper, R.J., Boer de, E.M., 2006. Autologous full-thickness skin substitute for healing chronic wounds. Br. J. Dermatol. 155 (2), 267–274.

Gittenberger-de Groot, A.C., Bartelings, M.M., DeRuiter, M.C., Poelmann, R.E., 2005. Basics of cardiac development for the understanding of congenital heart malformations. Pediatr. Res. 57, 169–176.

Gittenberger-de Groot, A.C., Vrancken Peeters, M.-P.F.M., Bergwerff, M., Mentink, M.M.T., Poelmann, R.E., 2000. Epicardial outgrowth inhibition leads to compensatory mesothelial outflow tract cuff and abnormal septation and coronary formation. Circ. Res. 260, 373–377.

Gittenberger-de Groot, A.C., Vrancken Peeters, M.-P.F.M., Mentink, M.M.T., Gourdie, R.G., Poelmann, R.E., 1998. Epicardial derived cells, EPDCs, contribute a novel population to the myocardial wall and the atrioventricular cushions. Circ. Res. 82, 1043–1052.

Glass, D.A., Karsenty, G., 2006. Canonical Wnt signaling in osteoblasts is required for osteoclast differentiation. Ann. N. Y. Acad. Sci. 1068, 117–130.

Green, H., Kehinde, O., Thomas, J., 1979. Growth of cultured human epidermal cells into multiple epithelia suitable for grafting. Proc. Natl. Acad. Sci. U.S.A 76, 5665–5668.

Hammerschmidt, M., Brook, A., McMahon, A.P., 1997. The world according to hedgehog. Trends Genet. 13, 14–21.

Happle, R., Assim, A., 2001. The lines of Blaschko on the head and neck. J. Am. Acad. Dermatol. 44, 612–615.

Hartmann, C., 2006. A Wnt canon orchestrating osteoblastogenesis. Trends Cell Biol. 16, 151–158.

Hay, E.D., 2005. The mesenchymal cell, its role in the embryo, and the remarkable signaling mechanisms that create it. Dev. Dyn. 233, 706–720.

Hellström, M., Kalén, M., Lindahl, P., Abramsson, A., Betsholtz, C., 1999. Role of PDGF-B and PDGFR-beta in recruitment of vascular smooth muscle cells and pericytes during embryonic blood vessel formation in the mouse. Development 126, 3047–3055.

Hill, T.P., Spater, D., Taketo, M.M., Birchmeier, W., Hartmann, C., 2005. Canonical Wnt/beta-catenin signaling prevents osteoblasts from differentiating into chondrocytes. Dev. Cell 8, 727–738.

Hirobe, T., 2005. Role of keratinocyte-derived factors involved in regulating the proliferation and differentiation of mammalian epidermal melanocytes. Pigment Cell Res. 18, 2–12.

Hoath, S.B., Leahy, D.G., 2003. The organization of human epidermis: functional epidermal units and phi proportionality. J. Invest. Dermatol. 121, 1440–1446.

Joseph, N.M., Mukouyama, Y.S., Mosher, J.T., Jaegle, M., Crone, S.A., Dormand, E.L., Lee, K.F., Meijer, D., Anderson, D.J., Morrison, S.J., 2004. Neural crest stem cells undergo multilineage differentiation in developing peripheral nerves to generate endoneurial fibroblasts in addition to Schwann cells. Development 131, 5599–5612.

Karsenty, G., Wagner, E.F., 2002. Reaching a genetic and molecular understanding of skeletal development. Dev. Cell 2, 389–406.

Kleinsmith, L.J., Pierce, G.B., 1964. Multipotentiality of single embryonal carcinoma cells. Cancer Res. 24, 1544–1551.

Kronenberg, H.M., 2003. Developmental regulation of the growth plate. Nature 423, 332–336.

Le Douarin, N., Kalcheim, C., 1999. The Neural Crest. Cambridge University Press, Cambridge.

Le Noble, F., Fleury, V., Pries, A., Corvol, P., Eichmann, A., Reneman, R.S., 2005. Control of arterial branching morphogenesis in embryogenesis: go with the flow. Cardiovasc. Res. 65, 619–628.

Lenas, P., Moos, M., Luyten, F.P., 2009. Developmental engineering: a new paradigm for the design and manufacturing of cell-based products. Part I: from three-dimensional cell growth to biomimetics of in vivo development. Tissue Eng. Part B Rev. 15 (4), 381–394.

Lv, F.J., Tuan, R.S., Cheung, K.M., Leung, V.Y., 2014. The surface markers and identity of human mesenchymal stem cells. Stem Cells 32, 1408–1419.

Mack, J.A., Anand, S., Maytin, E.V., 2005. Proliferation and cornification during development of the mammalian epidermis. Birth Defects Res. C Embryo Today 75, 314–329.

Manolagas, S.C., 2000. Birth and death of bone cells, basic regulatory mechanisms and implications for the pathogenesis and treatment of osteoporosis. Endocr. Rev. 21, 115–137.

Martin, G.R., 1981. Isolation in culture of a pluripotent cell line from early mouse ebmryos cultured in medium conditioned by teratocarcinoma stem cells. Proc. Natl. Acad. Sci. U.S.A. 78, 7634–7638.

Meier, C., Parmantier, E., Brennan, A., Mirsky, R., Jessen, K.R., 1999. Developing Schwann cells acquire the ability to survive without axons by establishing an autocrine circuit involving insulin-like growth factor, neurotrophin-3, and platelet-derived growth factor-BB. J. Neurosci. 19, 3847–3859.

Mintz, B., Illmensee, K., 1975. Normal genetically mosaic mice produced from malignant teratocarcinoma cells. Proc. Natl. Acad. Sci. U.S.A. 72, 3585–3589.

Moorman, A.F.M., Christoffels, V.M., 2003. Cardiac chamber formation: development, genes and evolution. Physiol. Rev. 83, 1223–1267.

Moreau, M., Leclerc, C., 2004. The choice between epidermal and neural fate: a matter of calcium. Int. J. Dev. Biol. 48, 75–84.

Nakashima, K., de Crombrugghe, B., 2003. Transcriptional mechanisms in osteoblast differentiation and bone formation. Trends Genet. 19, 458–466.

Papaioannou, V.E., McBurney, M., Gardner, R.L., Evans, M.J., 1975. The fate of teratocarcinoma cells injected into early mouse embryos. Nature 258, 70–73.

Ponec, M., El Ghalbzouri, A., Dijkman, R., Kempenaar, J., van der, P.G., Koolwijk, P., 2004. Endothelial network formed with human dermal microvascular endothelial cells in autologous multicellular skin substitutes. Angiogenesis 7, 295–305.

Rissmann, R., Groenink, H.W., Weerheim, A.M., Hoath, S.B., Ponec, M., Bouwstra, J.A., 2006. New insights into ultrastructure, lipid composition and organization of vernix caseosa. J. Invest. Dermatol. 126, 1823–1833.

Roodman, G.D., 2006. Regulation of osteoclast differentiation. Ann. N. Y. Acad. Sci. 1068, 100–109.

Rott, H.D., 1999. Extracutaneous analogies of Blaschko lines. Am. J. Med. Genet. 85, 338–341.

Scherer, S.S., Salzer, J.L., 2001. Axon-Schwann cell interactions during peripheral nerve degeneration and regeneration. In: Jessen, K.R., Richardson, W.D. (Eds.), Glial Cell Development. Bios Scientific Publishers Ltd, Oxford, pp. 299–330.

Schipani, E., 2006. Hypoxia and HIF-1alpha in chondrogenesis. Ann. N. Y. Acad. Sci. 1068, 66–73.

Sharghi-Namini, S., Turmaine, M., Meier, C., Sahni, V., Umehara, F., Jessen, K.R., Mirsky, R., 2006. The structural and functional integrity of peripheral nerves depends on the glial-derived signal desert hedgehog. J. Neurosci. 26, 6364–6376.

Smithies, O., Gregg, R.G., Boggs, S.S., Korawelski, M.A., Kucherlapati, R.S., 1985. Insertion of DNA sequences into the human chromosomal b-globin locus by homologous recombination. Nature 317, 230–234.

Stevens, L.C., 1967. The biology of teratomas. Adv. Morphog. 6, 1–31.

Traupe, H., 1999. Functional X-chromosomal mosaicism of the skin, Rudolf Happle and the lines of Alfred Blaschko. Am. J. Med. Genet. 85, 324–329.

Tonnarelli, B., Centola, M., Barbero, A., Zeller, R., Martin, I., 2014. Re-engineering development to instruct tissue regeneration. Curr. Top. Dev. Biol. 108, 319–338.

Van Bezooijen, R.L., Roelen, B.A., Visser, A., et al., 2004. Sclerostin is an osteocyte-expressed negative regulator of bone formation, but nog a classical BMP antagonist. J. Exp. Med. 199, 805–814.

CHAPTER 4

Cellular Signaling

Vanessa L.S. LaPointe[1], Jan de Boer[1], Adam J. Engler[2]

[1]*Department for Cell Biology-inspired Tissue Engineering, Merlin Institute, Maastricht University, Maastricht, The Netherlands;* [2]*Department of Bioengineering, University of California, San Diego, USA*

LEARNING OBJECTIVES

- To describe the paradigm of cell signaling, which incorporates signal initiation, transduction, and gene activation
- To appreciate the complexity of signaling and give examples of deviations from the paradigm
- To use the nomenclature to annotate the classes of molecules involved in cell signaling
- To describe how integrins and G-protein-coupled receptors transduce extracellular signals
- To know the major steps in BMP and Rho GTPase signaling
- To be able to describe some major signaling pathways in skin, vascular, and bone regeneration
- To be able to describe the structures involved in transmitting forces in skeletal muscle
- To value the amount of effort and time taken in the discovery of signaling pathways and their implementation in tissue engineering
- To understand how tissue engineers can use knowledge of cellular signaling to improve their designs
- To be able to give examples of when tissue engineers used knowledge of cell signaling to influence cell behavior

4.1 GENERAL INTRODUCTION

4.1.1 Paradigm of Cellular Signaling

Tissue engineering, in analogy with development and wound healing, is a dynamic process requiring the right concentration of a signal to be received by the right cells with spatial and temporal precision in order to establish a normally functioning tissue. Communication between cells, their local environment, and the

Tissue Engineering.
Copyright © 2015 Elsevier Inc. All rights reserved.

rest of the body plays a central role in the coordination of cell number, position, and function. The ways in which the cells function in a tissue environment will be discussed in other chapters of this book. In this chapter, we will discuss the molecular mechanisms by which cells communicate with each other and respond to their environments, referred to as cellular signaling, and we will show how tissue engineers can control signaling to influence cell behavior. In the first section, some general principles of cell signaling will be outlined and further illustrated using specific examples from major classes of signaling pathways. Examples will be given of tissue engineering strategies that target these pathways in order to evoke specific responses from cells. We will then discuss the signaling requirements of several tissue types that are of particular interest in tissue engineering—namely, skin, blood vessels, bone, and skeletal muscle. It is important to note that due to the sheer number of molecules involved in signaling and the complexity of their interactions, there will be many exceptions to the general rules outlined here.

Generally speaking, cell signaling is initiated by the generation of a ligand, which is secreted by a sending cell to bring about a change in the physiology of a responding cell (Figure 4.1(a)). The cellular hardware, or the repertoire of proteins expressed by the cell, determines whether and how it responds to a certain signal. This typically involves the binding of the ligand to a receptor on the receiving cell to initiate a cascade of biochemical changes that result in immediate changes to proteins or end in the transcription of genes. Possible responses include altered proliferation, differentiation, migration, cytokine production, metabolism, cytoskeletal rearrangement, and extracellular matrix production. Considering the diversity of both the signals and the cell responses, it should be clear that the mechanisms of cell signaling can be

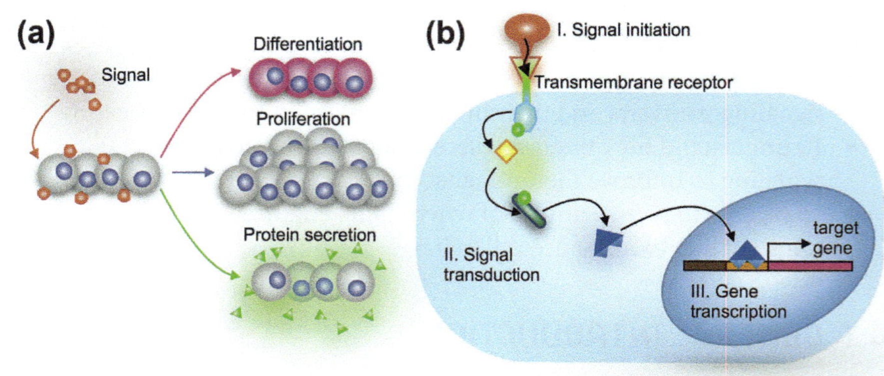

FIGURE 4.1
(a) Cell signaling leads to changes in the responding cell, such as differentiation, proliferation, or altered protein secretion or activity. (b) Although many variations on this theme exist, signaling events often adhere to the paradigm of signal initiation by an extracellular ligand followed by signal transduction into the nucleus, where the change in gene transcription underlies physiological adaptations of the cell.

extremely complex. Researchers are still a long way from generating a complete description of the response of a cell to a given ligand, never mind the more dynamic and complex events in development, cancer, and tissue regeneration. Nonetheless, our present understanding is detailed enough to be applicable by tissue engineers aiming to direct cell behavior, and many studies have shown that mimicking the ligand–receptor interactions that exist in native tissue can improve the behavior of cells in a tissue-engineered construct.

A fundamental lesson to take from this chapter is that cell signaling is astonishingly complex. For instance, since the gene and protein expression profile are different for every cell type, cells will respond differently to certain signals. Examples abound, just one of which is the cellular response to the glucocorticoid hormones, which trigger cell death in lymphocytes, but stimulate osteogenic differentiation in mesenchymal stem cells. Furthermore, cells respond differently to ligands at different concentrations and different rates of change. They also respond differently depending on other aspects of their environment and their individual history. To add to this complexity, cells in the body are not exposed to a single signal at a single concentration, but to a cocktail of hormones, cytokines, and growth factors. And finally, not all signaling begins with a ligand–receptor interaction. Of particular relevance to tissue engineering, cell signaling can also be initiated by physical cues, such as the elasticity of the substrate upon which a cell resides.

That being said, a central paradigm can be recognized in most events of cellular signaling, consisting of three distinct steps (Figure 4.1(b)):

1. Signal initiation: An extracellular ligand binds to a receptor on the surface of the cell. Ligand binding changes the activity or conformation of the receptor, thus transducing the signal to the inside of the cell.
2. Signal transduction: The activated receptor triggers a signal transduction cascade in which intracellular proteins are activated, ultimately leading to activation of a transcription factor in the nucleus.
3. Gene activation: The transcription factor binds to regulatory sequences in target genes, resulting in gene activation, protein synthesis, and changed cellular physiology.

In the following section, the main types of signals, receptors, and the machinery for gene activation are described.

4.1.2 Signal Initiation

The first step in this paradigm of cell signaling is the interaction of a ligand with its receptor. A wide variety of molecules in the body serve as ligands and these can be incorporated in regenerative medicine strategies to influence cell behavior. Irrespective of their identity, most ligands share the characteristic that they physically interact with their ligand, which is usually a transmembrane protein that binds with the signal molecule in the extracellular space and transduces the signal into the cell. This is certainly the case for hydrophilic ligands or molecules that cannot pass through the plasma membrane, but some small or hydrophobic molecules can diffuse through the membrane and bind targets in the cytosol or the nucleus.

Generally, ligands can be categorized into:

1. Extracellular matrix proteins
2. Membrane-bound proteins
3. Diffusible molecules

When an extracellular matrix protein acts as a ligand, it enables a cell to sense and respond to the composition of its environment, an event that primarily occurs via a class of receptors known as the integrins (Section 4.1.2.1). Many extracellular matrix proteins are ligands for specific receptors. For example, cells can detect whether they are surrounded by collagen and produce the correct enzymes to remodel their surroundings. More information on the role of the extracellular matrix is discussed in Chapter 5.

By detecting membrane-bound proteins, cells can also determine whether they are in contact with each other, and gain some information about the identity of an adjacent cell. An impressive example of the power of membrane-bound proteins to influence cell behavior is the way by which cadherins enable similar cells to stick together and segregate themselves from other cell types. It even appears that cells can organize based on both a quantitative (i.e., the level of a cadherin) and qualitative (i.e., the member of the cadherin family) basis. For example, if cells from two tissues such as the heart (expressing N-cadherin) and the skin epithelium (expressing E-cadherin) are dissociated and mixed together, they will form two distinct populations. Epithelial cells expressing high and low E-cadherin would be expected to do the same. Cell–cell contact can inform diverse changes, such as telling cells within a tissue when to stop growing (the Hippo pathway) or initiating an immune response.

Diffusible molecules are produced when a sending cell produces a protein (e.g., vascular endothelial growth factor) or organic substance (e.g., a steroid hormone) and secretes it into its environment. Other diffusible molecules include amino acids, nucleotides, and dissolved gases. Some diffusible molecules can also bind to receptors within the cytoplasm. Nitric oxide, for example, can be released by endothelial cells lining the inside of a vessel in response to regulation by the nervous system or local mechanical forces. The nitric oxide passes through the plasma membrane of the adjacent smooth muscle cells lining the wall and causes them to relax, thereby dilating the vessel. Hydrophobic compounds, such as steroid hormones, can also pass through the membrane and initiate signaling from inside the cell. The synthesis of vitamin D, discussed in Section 4.4.3, is an example of this mechanism.

The action range of diffusible ligands varies. Long range, or endocrine, action occurs for the parathyroid hormone, which is produced in the parathyroid glands but affects bone remodeling throughout the body. Short range, or paracrine activity, occurs as interleukins produced in the dermis of the skin control epidermal biology at a distance of a few layers of cells. Finally, in some cases, a cell will produce and retain a signaling ligand intracellularly or at its own surface membrane, where it activates signaling within the producing cell, a phenomenon known as autocrine stimulation.

In addition to developing strategies that provide ligands to cell receptors, tissue engineers can also improve upon the activity of a ligand. For example, insulin-like growth factor-1 (IGF-1) was fused with a protein fragment that improved its incorporation into a fibrin matrix. When implanted in a bladder lesion in rats, the engineered growth factor acted to increase the proliferation of the smooth muscle cells to improve regeneration (Lorentz et al., 2012).

4.1.2.1 Integrins

As an example of a major class of receptors, integrins are heterodimeric transmembrane proteins consisting of α- and β-subunits that serve both as adhesive molecules and signaling receptors. The α- and β-subunits combine to form one of 24 known integrins in humans. Each is thought to have a specific and unique function, despite a significant amount of sequence homology and cross reactivity between integrins and their ligands. When an integrin binds its ligand, it establishes physical contact between the cell and its extracellular matrix or an adjacent cell (Table 4.1). Integrins transmit information in both directions across the plasma membrane (bidirectional signaling) to integrate the intracellular and extracellular environments. They do this by having a ligand binding site in their extracellular domain and sites for linkage to the cytoskeleton and major signaling pathways on their cytoplasmic domain. Outside-in signaling occurs when an extracellular ligand binds to the extracellular domain of an integrin and initiates a cascade of intracellular signaling events. Inside-out signaling occurs when cytoplasmic events result in a conformational change of the integrin, thereby changing its affinity for extracellular ligands. Integrin signaling often occurs in coordination with other receptors and pathways. For example, it was recently found that Gα13, a G-protein subunit (Section 4.1.2.2), binds integrin β3 in a mutually exclusive, but temporally regulated wave-like pattern to promote outside-in signaling (Shen et al., 2013).

An important property of integrins is that they can readily modulate their propensity to bind a ligand. They do this through a conformational switch that affects both their extracellular and intracellular domains. When a ligand binds, the extracellular domain adapts an active, extended conformation, which also improves its affinity for the ligand (Takagi et al., 2002). In what is known as outside-in activation, the cytosolic domains of the α- and β-subunits, which

Table 4.1 Integrin α-β Subunit Combinations and Ligands

α1β1	Collagen, laminin
α2β1	Collagen, laminin, thrombospondin, E-cadherin, tenascin
α3β1	Laminin, thrombospondin
α4β1	Fibronectin, thrombospondin, osteopontin, MAdCAM-1, VCAM-1, ADAM family, ICAM-4
α4β7	Fibronectin, osteopontin, MAdCAM-1, VCAM-1
α5β1	Fibronectin, osteopontin, ADAM family, COMP, fibrillin, CD171, thrombospondin
α6β1	Laminin, ADAM family, CCN1, thrombospondin
α6β4	Laminin
α7β1	Laminin
α8β1	Fibronectin, osteopontin, vitronectin, tenascin, LAP TGF-β, nephronectin
α9β1	Osteopontin, VCAM-1, tenascin, ADAM family, VEGF-C, VEGF-D
α10β1	Collagen, laminin
α11β1	Collagen
αvβ1	Fibronectin, osteopontin, LAP TGF-β, CD171
αvβ3	Fibronectin, thrombospondin, osteopontin, BSP, MFG-E8, Del-1, vitronectin, vWF, tenascin, PECAM-1, LAP TGF-β, fibrillin, fibrinogen, ADAM family, COMP, CCN1, CCN2, ICAM-4, MMP-2, CD171
αvβ5	Osteopontin, BSP, MFG-E8, Del-1, vitronectin, LAP TGF-β
αvβ6	Fibronectin, osteopontin, LAP TGF-β, ADAM family
αvβ8	LAP TGF-β
αIIbβ3	Fibronectin, thrombospondin, vitronectin, vWF, fibrinogen, CCN2, Cyr61, ICAM-4, CD171
αDβ2	VCAM-1, ICAM, CCN1, fibrinogen, plasminogen, vitronectin
αEβ7	E-cadherin
αLβ2	ICAM-1, ESM-1, ICAM-4
αMβ2	ICAM-1, fibrinogen, Factor X, iC3b, ICAM-4, heparin, negative charges in denatured proteins
αXβ2	ICAM, fibrinogen, iC3b, collagen, ICAM-4, heparin, negative charges in denatured proteins

Adapted from Humphries et al. (2006).

were previously bound to each other, move apart. Then the binding site for talin, one of the proteins responsible for linking integrins to the actin cytoskeleton, is exposed, and an intracellular tensile force to act against the integrin–ligand binding is generated.

Integrins also participate in inside-out activation. When the binding affinity of talin is increased by molecules such as phosphoinositide, PIP_2, talin can compete with the intracellular domain of the α-subunit for a binding domain on the β-subunit. Upon binding, the conformational change causes an increase in affinity for the extracellular ligand. Inside-out signaling also protects the host from excessive integrin-mediated cell adhesion, which could have pathological consequences (Qin et al., 2004). The involvement of integrin αIIbβ3 in platelet aggregation has been used as a model of inside-out signaling, which was found to cause a conformation change in the integrin that allows it to bind von Willebrand factor and fibrinogen, two components essential for platelet

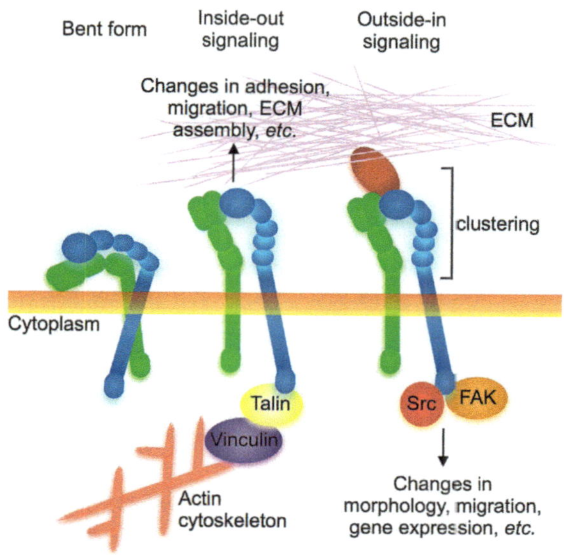

FIGURE 4.2
Schematic overview of integrin-mediated activation leading to inside-out and outside-in signaling. In their resting state, integrins reside in a bent conformation. In inside-out activation, the β-subunit (blue) binds an intracellular activator to induce a conformation change thereby increasing integrin affinity to extracellular ligands. In outside-in signaling, a ligand binds to the extracellular domain to initiate signaling cascades to change cell behavior. The clustering that follows stabilizes the interaction.

aggregation (Zou et al., 2007). Sequential activation of cytosolic proteins Pkg, p38, and ERK regulate the activation of integrin αIIbβ3 in platelets (Li et al., 2006). Then, as platelets begin to aggregate, integrins cluster and outside-in signaling recruits cytoplasmic proteins that stabilize the interactions and promote platelet spreading, a phenotypic change.

The adhesion of a cell to a surface is an example of integrins acting as both adhesive proteins and ligand receptors. Firstly, when an integrin binds an extracellular matrix ligand, it recruits additional integrins and supporting proteins to stabilize the interaction, ultimately forming a focal adhesion, which securely attaches the cell to a particular substrate. However, the role of the integrins does not end with adhesion (Wolfenson et al., 2013). The subsequent signaling components that become activated include enzymes (e.g., focal adhesion kinase (FAK), Ras and Rho GTPases) and adapters (e.g., p130Cas/Crk, Paxillin, Talin, α-actinin) that assemble within focal adhesions (Figure 4.2). These signals cooperate with others to regulate cell behavior in a complex tissue microenvironment. For example, transforming growth factor-β (TGF-β) signaling (Section 4.1.3.1) can depend on integrin activation. TGF-β is produced in a latent complex with a latency-associated peptide tethering it to the extracellular matrix. Integrins are central to the mechanisms of releasing these key growth factors from the extracellular matrix, which they do by binding to the latency-associated peptide and coordinating matrix metalloproteases and other molecules at the cell surface to cleave the peptide and release the growth factor (Munger and Sheppard, 2011).

Understanding the role of integrins is central to tissue engineering, where both cell adhesion and signaling are of the highest importance for generating a functioning tissue construct. Of interest for tissue engineers, many integrins are able to bind a peptide derived from an extracellular matrix protein. The most validated of these interactions is the amino acid sequence arginine-glycine-aspartic acid (RGD), which was first derived from fibronectin, but is also present in other proteins known to adhere cells. The cells can bind this sequence of three amino acids, something tissue engineers can easily take advantage of by designing their scaffolds to present a short peptide with the RGD sequence. Integrins can also be targeted with other peptides. For example, when a synthetic peptide (LLP2A-Ale) with a high affinity for integrin α4β1 was immobilized on hydroxyapatite, it promoted the homing of mesenchymal stem cells and improved bone healing in a mouse model with bone loss related to age and hormone deficiency (Yao et al., 2013).

4.1.2.2 G-Protein-Coupled Receptors

G-protein-coupled receptors (GPCRs) are the largest family of cell surface receptors. This family contains many different receptors that play an important role in physiology, including the adrenergic receptor, light and odor receptors, and receptors for small peptide hormones, such as vasopressin and glucagon, and neuropeptides such as serotonin. All GPCRs have a common structure: they consist of a single amino acid chain containing an N-terminal extracellular domain, a transmembrane domain in which the amino acid chain traverses seven times through the cell membrane, and a cytoplasmic C-terminus. The N-terminus and the extracellular amino acids of the transmembrane domain are involved in binding of the ligand, while the cytoplasmic amino acids of the transmembrane domain and the C-terminus are involved in signal transduction (Rosenbaum et al., 2009).

All G-protein-coupled receptors transduce their signal via heterotrimeric G-proteins, consisting of an α-, β-, and γ-subunit. In the absence of a ligand, the α-subunit, a GTPase, binds GDP (guanosine diphosphate) and the G-protein is inactive. Binding of a ligand induces a conformational change in the receptor, resulting in the release of GDP from the α-subunit and the binding of GTP in its place. What happens next is still not completely understood, but it appears that activation causes a dissociation of the α-subunit from the β- and γ-subunits. Once dissociated, the α-subunit and the βγ complex are active and available for signaling. The second mechanism of action is thought to be that the conformational change results in the exposure of previously hidden binding sites residing between the α-subunit and the βγ complex, which are targets for enzymes or ion channels.

The parathyroid hormone-like hormone, PTHLH, is a small peptide hormone that binds with high affinity to the type 1 parathyroid hormone receptor (PTH1R; Figure 4.3), a member of the GPCR superfamily. GPCRs generally transduce their signal through one class of α-subunits: Gαs, Gαq, Gαi, or Gα12/13, but some receptors, like PTH1R, can activate two classes of α-subunits (Gαs and Gαq). Gαs-subunits are involved in the activation of cyclases, which are anchored in the cell membrane. Cyclases convert nucleotide triphosphates (NTPs) into cyclic nucleoside monophosphates (cNMPs), which act as second messengers (intracellular signaling molecules). Activation of the PTH1R results in the release of the Gαs-subunit and the activation of adenylate cyclase, which converts ATP into cAMP. Subsequently, cAMP binds to the regulatory subunit of cAMP-dependent protein kinase A (PKA), thereby removing its inhibitory

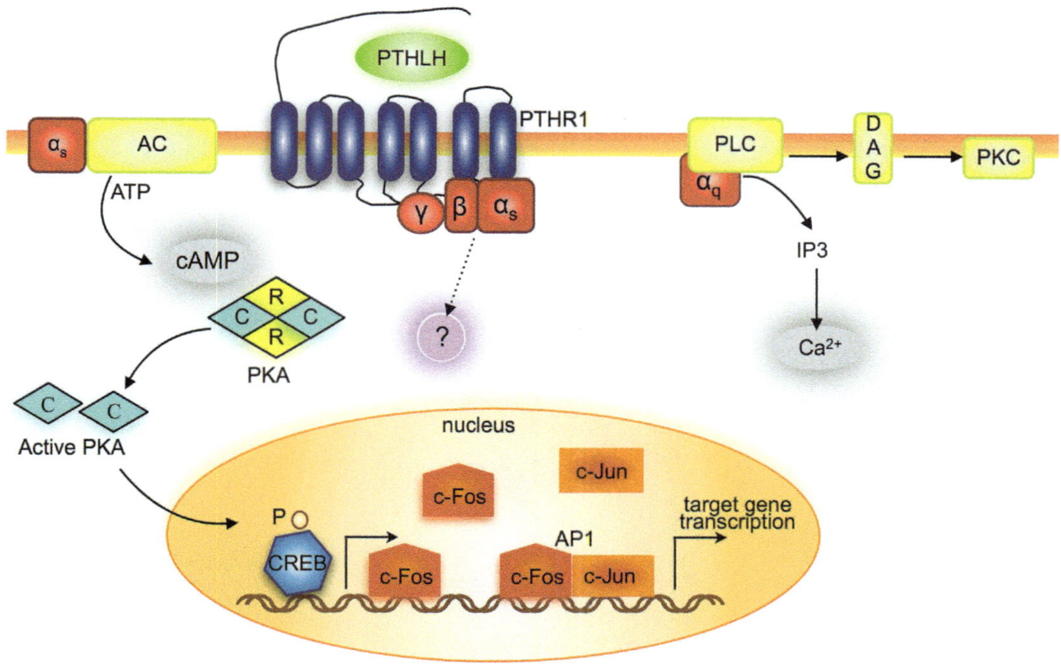

FIGURE 4.3

G-protein-coupled receptor signaling. Binding of the PTH1R by PTHLH results in the activation of two classes of G-protein α-subunits. Activation of Gαs converts ATP into cAMP by the membrane-bound enzyme adenylate cyclase (AC). cAMP binds to the regulatory (R) subunits of protein kinase A (PKA). This releases the catalytic subunits (C), which phosphorylate cAMP response element binding (CREB) protein resulting in the activation of transcription of target genes, such as c-Fos, which together with c-Jun is part of the AP1 family of transcription factors.

actions on the catalytic subunit of PKA. The newly available catalytic subunit of PKA then phosphorylates serine residues on proteins. One of the main nuclear targets of PKA is the transcription factor cAMP response element binding (CREB) protein. This transcription factor is bound in an inactive form to specific DNA sequences present in promoters of a wide variety of genes until PKA activates it by phosphorylation of serine-133, resulting in the start of gene transcription. Of interest for regenerative medicine, the cAMP/PKA pathway can be activated by various chemical substances such as forskolin, which is an activator of adenylate cyclase, and isobutyl methyl xanthine (IBMX), which is an inhibitor of cAMP degradation.

4.1.3 Signal Transduction

Cell signaling ultimately results in biochemical changes to proteins (Figure 4.4). When these changes occur to proteins that are already present in the cell, such as many signals resulting in altered migration, they occur quickly. However, signals that require changes in gene expression and translation of new proteins occur more slowly and often involve a more complex series of

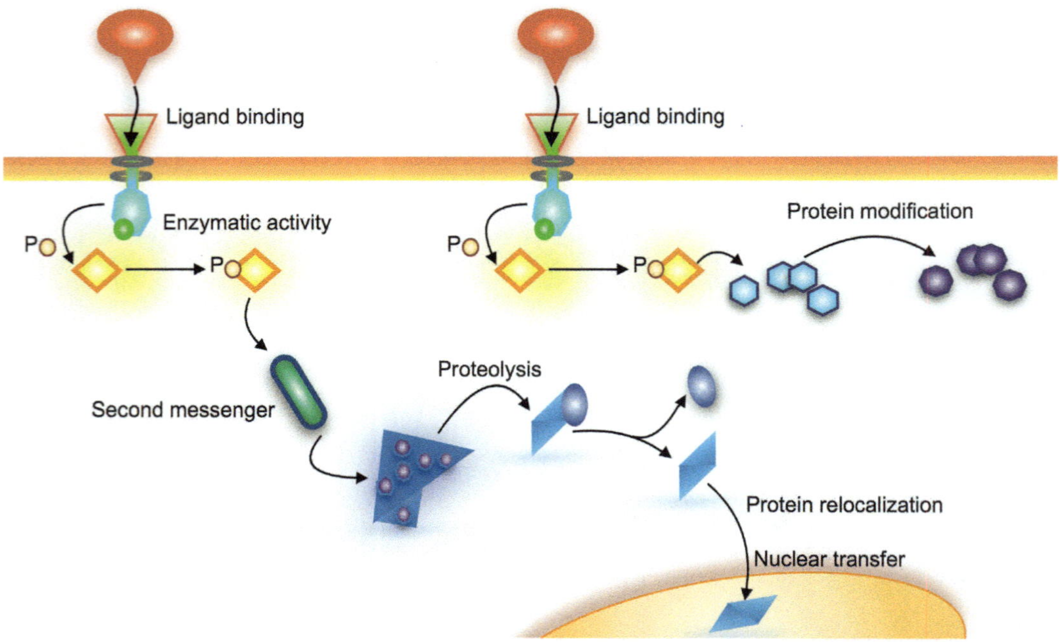

FIGURE 4.4
A hypothetical signaling pathway with some of the most relevant molecular mechanisms involved in signal transduction. The event of a ligand binding to a receptor can lead to diverse downstream events. For example, when the enzymatic activity of a receptor is activated, it can initiate a signaling cascade leading to the translocation of a protein from the cytoplasm to the nucleus or it can act directly on proteins in the cytoplasm to produce phenotypic changes.

events. The complexity and specificity of signaling pathways are both largely due to the cascades they initiate. These cascades allow different ligand–receptor interactions to converge on the same pathway, or for the same ligand–receptor interaction to initiate different pathways depending on factors like the cell type or the extracellular environment.

When the endpoint of signaling is transcription, the signal initiated by the ligand–receptor interaction has to be transduced from the membrane to the nucleus. Different ligand–receptor combinations can trigger different cascades of molecular interactions in the cytoplasm, resulting in the relay of the signal. In many cases, the signal is conveyed by direct protein–protein interactions and subsequent modification of target proteins, leading to their activation. For example, binding of BMP2 (bone morphogenetic protein 2) to its receptor leads to activation of the receptor's serine/threonine kinase activity and phosphorylation of target proteins. Phosphorylation is a common biochemical event that acts as a cellular signal. Some of the most prevalent protein modifications are summarized in Table 4.2. Alternatively, activated proteins drive the production or relocalization of the so-called second messengers, which are small molecules such as cyclic AMP, phospho-inositol-2-phosphate (PIP2), calcium, and nitric oxide, which act as ligands for target proteins, hence the name second messenger.

A single signaling cascade will typically contain many types of proteins. It might begin with a ligand–receptor interaction that changes the conformation of the receptor and causes it to recruit enzymes. The next steps may involve posttranslational modifications, amplification steps, the convergence or divergence of the cascade, and ultimately result in a transcription factor entering the nucleus and binding DNA in consort with cofactors or other transcriptional modulators. Specific examples of pathways follow in Sections 4.2–4.5.

Table 4.2 Enzymatic Activities Involved in Signal Transduction

Enzyme	Activity	Putative Effect
Kinase	Protein phosphorylation	Changed enzymatic activity or protein stability
Phosphatase	Protein dephosphorylation	Changed enzymatic activity or protein stability
Ubiquitin ligase	Protein ubiquitination	Protein degradation
Proteases	Protein site-specific cleavage	Degradation, activation
Fatty acid esterase	Palmitoyl/myristoyl addition to protein	Relocalization to the plasma membrane
SUMO ligase	Protein SUMO modification	Protein degradation
Lipase	Cleavage of phospholipids	Generation of second messengers, such as PIP2 and IP3
Cyclase	Forms a cyclic compound	Generation of second messenger cAMP and cGMP

Generally, protein activity can be controlled in several ways. The enzymatic activity of the protein can be switched on, as with BMP receptor activation. This is one mechanism of signal amplification, for a single activated protein can catalyze the modification of many others. Furthermore, proteins can be sequestered away from their site of action as seen, for instance, with the transcription factor NFκB, which is held in the cytoplasm by the inhibitory protein IκB until upstream signaling events allow its translocation to the nucleus (Section 4.2.2). Another way to control protein activity is its abundance. Cell signaling can, in some cases, inhibit the constitutive proteolytic degradation of a certain protein, which then accumulates and exerts its function (Section 4.2.2).

Importantly, signaling cascades have feedback loops, which can attenuate or amplify the signal. In a pathway with positive feedback, the outcome of one step increases its own production, thereby increasing the response to the signal. Alternatively, in a pathway with negative feedback, the output inhibits its own production, thereby preventing a response to a ligand. In reality, these feedback loops are more complicated, and are responsible for phenomena such as the oscillation of an enzyme in response to sudden changes in the concentration of a ligand.

4.1.3.1 TGF-β Superfamily

TGF-β signaling is a central pathway in both tissue development and homeostasis. The TGF-β superfamily consists of various subfamilies, including TGF-βs, BMPs, activins, inhibins, and the anti-Müllerian hormone. TGF-β signaling displays some remarkably complex regulation. For example, when a TGF-β receptor is bound to its ligand, it is endocytosed and either further activated or inactivated depending on the means of endocytosis. Activation depends on clathrin-coated vesicles and the formation of endosomes, inside of which, the Smad-dependent pathway continues. Inactivation depends on caveolae, and is followed by ubiquitination of the receptor leading to its degradation (Di Guglielmo et al., 2003).

Here we focus on the major group of the TGF-β superfamily, the BMPs, which are secreted growth factors that were originally identified by their ability to induce ectopic bone and cartilage formation (Urist, 1965). The BMP subfamily consists of more than 20 members. BMPs exert their effects through distinct combinations of two different types of serine/threonine kinase receptors: type I receptors and type II receptors. Receptor activation involves ligand-induced hetero-oligomerization of two sequentially acting kinases, with the type I receptor acting as a substrate for the type II receptor kinase (Figure 4.5). The activated, phosphorylated type I receptor then propagates the signal through phosphorylation of receptor Smads (R-Smads-1, -5, and -8 for BMP and -2 and -3 for TGF) that assemble into heteromeric complexes with Co-Smad4 and translocate to the nucleus, where they regulate transcription of target genes. In addition, BMPs stimulate other pathways that are distinct from the Smad pathway, such as activation of Jnk and p38 MAP kinase pathways (Guicheux et al., 2003).

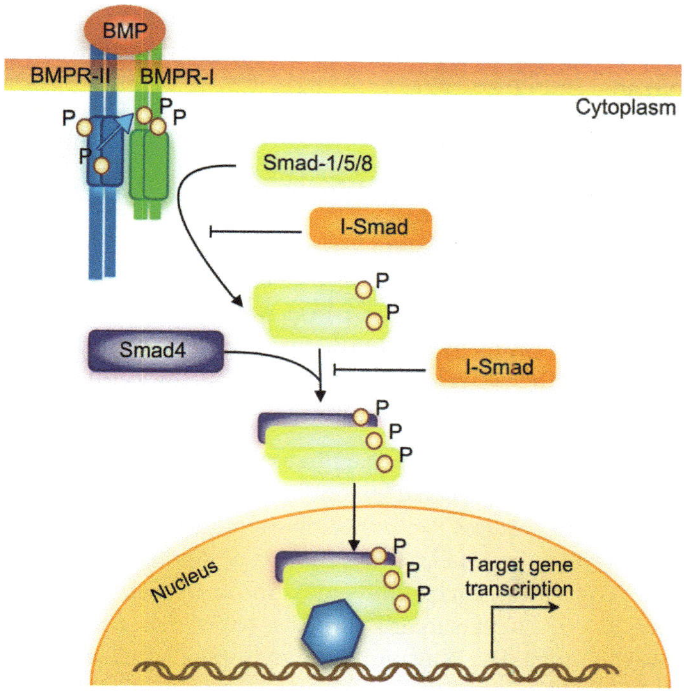

FIGURE 4.5

A model for activation of bone morphogenetic protein (BMP) receptors. BMP-mediated heteromeric complex formation of two type I and two type II BMP receptors (BMPRs) induces activation of the BMPR-I by the BMPR-II kinase. The activated BMPR-I propagates the signal downstream through phosphorylation (P) of R-Smads1/5/8. These phosphorylated Smads assemble into heteromeric complexes with Co-Smad4 and translocate into the nucleus where they regulate gene transcription.

Negative regulation of BMP activity occurs at nearly every step in the BMP/Smad pathway. Intracellularly, inhibitory Smads (I-Smad6 and 7) antagonize R-Smads by competing with them for interaction with activated receptors or heteromeric complex formation with Co-Smad4 (Figure 4.5). I-Smads also prevent activation of R-Smads by recruitment of ubiquitin ligases, like Smad ubiquitination regulatory factor 1 (Smurf-1) and Smurf-2, which add a ubiquitin moiety to Smad proteins, earmarking them for destruction by the proteasome. Extracellularly, soluble pseudoreceptors and antagonists inhibit BMP activity. Extracellular antagonists include Noggin, Chordin, DAN, Cerberus, and Gremlin, which are thought to operate by blocking binding of BMPs to their receptor (Canalis et al., 2003).

In tissue engineering, strategies can be designed to control BMP signaling activity at each of the regulatory steps, both outside and inside the cell. Human embryonic stem cells could be differentiated into chondrocytes by induction with BMP-7 and TGF-β1 (Nakagawa et al., 2009), and the controlled release of

TGF-β1 from a hydrogel scaffold activated Smad and ERK1/2 pathways in mesenchymal stem cells and improved their chondrogenic differentiation both in vitro and after implantation (Re'em et al., 2012).

4.1.3.2 Rho Kinase Signaling

Cell migration requires coordinated signaling between the protruding front of the cell and the retracting rear. It is the Rho family of GTP-binding proteins (GTPases) that play key roles in mediating cytoskeletal arrangement and migration, namely the three family members Rho, Rac, and Cdc42. Rho regulates the banding of actin filaments into stress fibers and helps in the formation of focal adhesions by clustering integrins. Rac promotes lamellipodia extension through the polymerization of actin at the cell periphery. Finally, Cdc42 mediates filopodia formation.

Like other GTPases, the Rho family proteins are molecular switches, residing in an inactive state when bound to GDP and an active state when bound to GTP. Forward cell movement requires Rac activation to induce protrusions at the leading edge of the cell (Nobes and Hall, 1999). These lamellipodia are actin-rich adhesive structures that concentrate high affinity integrins into focal adhesions to facilitate their binding to extracellular matrix proteins, while also actively advancing the leading edge of the cell. Rac activation is an example of a positive feedback loop, since it produces PI 3-kinase which leads to further activation of Rac. Rho affects stress fiber formation by activating formins and Rho kinases (ROCKs). ROCKs act by both inhibiting a phosphatase and by phosphorylating various downstream targets including myosin II light chain (MLC), leading to cell contraction.

For tissue engineering, the Rho-GTPase pathway has been implicated in the phenomenon known as mechanotransduction, which refers to the fact that some cell types can sense the elasticity of their substrates. The seminal paper in this field demonstrated that mesenchymal stem cells could be induced to differentiate to a specific tissue lineage when grown on a substrate that matched the elasticity of tissue from that lineage (Engler et al., 2006). Cells grown on a material with elasticity similar to brain, muscle, and bone osteoid differentiated into neurogenic, myogenic, and osteogenic cells, respectively. This was found to depend on nonmuscle myosin II and differentiation did not occur when blebbistatin, a nonmuscle myosin II inhibitor, or ML7, a myosin light chain kinase inhibitor, were used.

4.1.4 Gene Activation

The last step in the signal transduction cascade is gene activation (Figure 4.6). Gene activity is mainly controlled by the rate of transcription of the gene, the process in which a messenger RNA is produced by RNA polymerase. Signal transduction initiates the transcription process by activating the so-called transcription factors, which are sequence-specific DNA-binding proteins. The transcription factor is either retained in the cytoplasm and is translocated upon

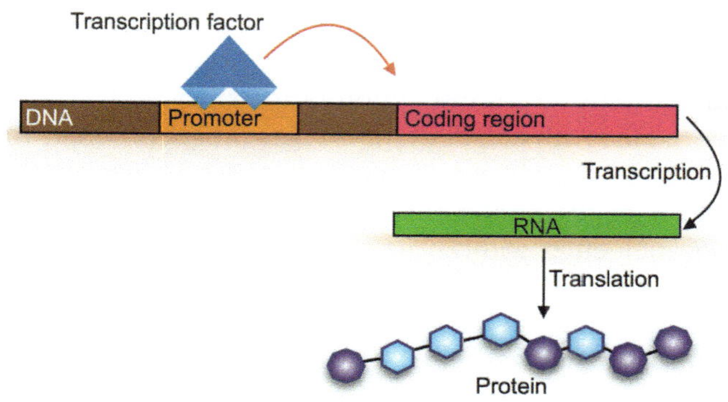

FIGURE 4.6
The activation of gene transcription results in protein translation and a physiological response.

activation or the transcription factor is already present in an inactive form in the nucleus where it becomes activated. Once actively present in the nucleus, they bind to unique DNA sequences in the promoter region of target genes. The promoter is the DNA sequence typically adjacent to the protein-encoding DNA sequence, and contains the regulatory elements that control gene transcription. Every gene has a unique promoter sequence and therefore recruits one or more different transcription factors. For instance, the Runx2 gene, which is a master regulator of osteogenic differentiation, contains binding sites for the Smad transcription factors, which are activated by the BMP signal transduction cascade. Transcription, like the entire signaling cascade, does not operate as a simple on/off switch. For example, it was recently found that while the short Smad binding element (CAGAC) is very common in the genome and in the promoter region of genes, the presence of the sequence does not necessarily mean Smad signaling will occur. Indeed, only a subset of these genes are activated by BMP signaling in any given cell type. Recent work has shed light on the molecular mechanisms underlying the diverse, cell type-specific responses to a single, generic signal (e.g., Smad activation). It was found that the recruitment of Smads to their binding sites on gene promoters was determined by tissue-specific transcription factors already present in the cell (Mullen et al., 2011; Trompouki et al., 2011). Thus, although a majority of genes may contain potential Smad binding sites, transcriptional activation is restricted to genes appropriate to the cell type—a finding that may apply to other signaling cascades. Once bound to the promoter, the transcription factor activates the RNA polymerase complex, resulting in transcription of the gene into a messenger RNA, which will be translated into a protein in the cytoplasm.

4.1.5 Variations on a Theme

The paradigm of cellular signaling described earlier is a conceptually good starting point, but it should be noted that many signal transduction pathways use variations on this theme or use only a part of the route to convey a signal from outside the cells to evoke a biological response. First of all, some ligands directly bind and activate a transcription factor, such as vitamin D3 (Section 4.4.3). Moreover, the path length of the cascade differs strikingly from a single step as for instance in BMP signaling (Section 4.1.3.1) to an elaborate cascade seen with integrin signaling (Section 4.1.2.1). Some signal transduction pathways exert their effect without activation of gene transcription. One of the well-established examples of this is the family of Rho kinases, which is downstream of some of the main signal transduction pathways and controls cytoskeletal organization (Section 4.1.3.2). However, more and more examples of this are beginning to emerge, again highlighting the complexity of protein interactions within cells. For example, it has recently been demonstrated that the long cytoplasmic tail of the type II BMP receptor—which is unique within this family of receptors and whose function remains largely undetermined—interacts directly with LIM kinase (LIMK), thus directly affecting cytoskeletal organization (Foletta et al., 2003). Likewise, the vascular endothelial growth factor (VEGF) receptor directly phosphorylates profilin, another cytoskeletal regulator (Fan et al., 2012). It seems inevitable that numerous examples are yet to be discovered.

Another important phenomenon in cellular signaling is cross talk, in which the signals of two different ligands converge on the same effector molecule. Cross talk may occur at several levels in the signaling cascade. In some cases, coactivation of a signal transduction molecule by two signaling pathways is required. For instance, CREB activation requires phosphorylation at serine-133 and serine-129, which are downstream events of G-protein-coupled receptor signaling and insulin signaling, respectively. In other cases, the transcription factors downstream of two or even more signal transduction pathways have to bind simultaneously to a promoter in order to activate gene transcription. Finally, many signaling molecules are not uniquely activated by a single signal transduction pathway. For instance, mitogen-activated protein kinase activation can be downstream of receptor tyrosine kinases but also of integrins.

In the following section, we discuss a number of signal transduction pathways involved in specific tissues to illustrate their complexity and also to point out how detailed knowledge on signaling can help in the rational design of tissue engineering approaches.

4.2 CELLULAR SIGNALING IN SKIN BIOLOGY

4.2.1 Skin Signaling Overview

Wound healing in skin is characterized by an orderly sequence of events beginning immediately after injury. The healing process can be broadly classified into four phases: hemostasis, inflammation, proliferation, and remodeling. These phases proceed in a systematic fashion, in which specific signaling routes are responsible for proper, organized, and controlled healing (Figure 4.7). Tissue engineers can develop strategies that more closely mimic the process of wound healing and apply them toward the generation of new skin matrices to cover epidermal and dermal wounds.

In the hemostasis and inflammation phases, which typically last for 24–48 h following injury, a fibrin clot forms and the wound is a hypoxic environment. Macrophages, lymphocytes, neutrophils, and platelets have to kill bacteria and clear the wound bed. TGF-β facilitates polymorphonuclear neutrophil migration from the surrounding blood vessels, from which they extrude themselves. These cells cleanse the wound and clear it of debris. After this process, the neutrophils will release factors known

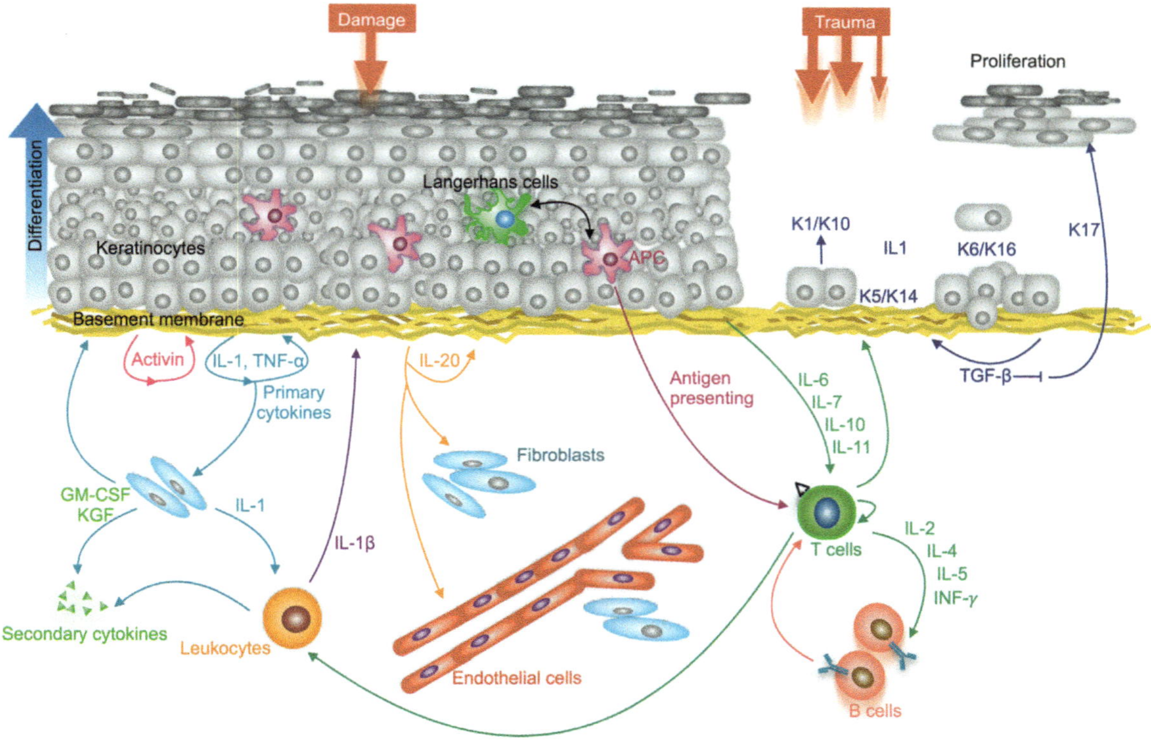

FIGURE 4.7

When keratinocytes are activated by an injury, they will express a number of specific keratins and interleukins. The homeostasis of the skin is regulated by an exquisitely finely tuned exchange of signaling factors among the dermal and epidermal cells. APC, antigen presenting cells; IL, interleukins; K, keratins; TNF-α, tumor necrosis factor-α; GM-CSF, granulocyte monocyte-colony stimulating factor; KGF, keratinocyte growth factor; TGF-β, transforming growth factor-β; INF, interferon.

Table 4.3 Signals Involved in the Recruitment of Cells to the Wound Bed

Cell Type	Signals
Keratinocytes	EGF, PDGF, IL-1, IL-6, IGF-1
Fibroblasts	PDGF, platelet factor 4 (PF4), TGF-β, IL-4, hypoxia, fragments of fibronectin and types I and III collagen
Endothelial cells	bFGF, TNF-α, IL-8, hypoxia, heparin, fibronectin
Monocytes/Macrophages	Monocyte chemoattractant protein-1 (MCP-1), CCL2, platelet-derived endothelial cell growth factor (PD-ECGF), thrombin, TGF-β, fibronectin, and elastin
Neutrophils	Fibrinopeptides, fibrin lysis products, C5a, formyl methionyl peptides (from bacterial proteins), platelet activating factor (PAF), TNF-α, PDGF, PF4

as chemotactic agents to attract various cell types to the site (Table 4.3). In the proliferative and remodeling phases, cells synthesize and release growth factors to regulate wound repair. Dermal fibroblasts serve multiple important purposes during this process, one of which is to produce the collagen that provides strength and integrity to the healed wound and a path to guide the keratinocytes. While the extracellular matrix proteins are produced by fibroblasts, other cells such as keratinocytes are active in producing cytokines, including tumor necrosis factor (TNF)-α, various interleukins, growth factors (TGF-α, TGF-β, VEGF, IGF) and signaling polypeptides to restore the epithelium (Barrientos et al., 2008). Vascular endothelial growth factor-A (VEGF-A) and basic fibroblast growth factor (FGF2 or bFGF) are the key growth factors in angiogenesis. The major cytokine involved in keratinocyte and fibroblast proliferation is IL-1α/β. IL-1 triggers release of keratinocyte growth factor (KGF) from fibroblasts, and in this way keratinocytes indirectly promote their own proliferation. IL-1α is constitutively expressed in human keratinocytes while in other cells it is released under stimulation. The final remodeling phase is characterized by synthesis and degradation of collagen and depends on a balance between factors that endorse synthesis of extracellular matrix components and enzymes that degrade these components, such as matrix metalloproteases (MMPs). For instance, MMP-11 is involved in remodeling of the extracellular matrix and may function by releasing growth factors from the extracellular matrix.

Recent studies, in particular those using genetically manipulated animal models, have highlighted the impact of TGF-β on various aspects of wound healing, and surprisingly, not all of its effects are conducive to optimal healing. Intriguingly, TGF-β1 mutations, or mutations in the cell signaling intermediate Smad3, lead to normal or even accelerated cutaneous wound healing responses (Ashcroft et al., 1999). Controlling the

TGF-β signaling is a strategy being used to prevent excessive wound healing, which leads to hypertrophic scars. Since increased levels of TGF-β were found in these wounds, antagonist molecules aimed at inhibiting TGF-β signaling are promising treatments to reduce scarring and improve healing (Gurtner et al., 2008).

4.2.2 Interleukin and NFκB Signaling

NFκB proteins are known as latent regulatory proteins, referring to the fact that they are already present in most cells, but are activated in response to stress. They are central to the inflammatory and innate immune responses, which, in the case of wound healing, are initiated by the binding of interleukin-1 to its receptor.

Interleukins (ILs) are a class of proteins (cytokines and lymphokines) that are released by cells of the immune system and act as intercellular mediators in developmental regulation, tissue repair, hemopoiesis, inflammation, and specific and nonspecific immune responses. The relevance of interleukin signaling is highlighted in hyperactivated skin (e.g., caused by chemical insults) or in psoriasis, where the IL balance is shifted toward overproduction of IL-1α or IL-1β. The misbalance in IL-1 expression has recently been used as an indicator for skin irritation, and the cosmetic industry is testing thousands of chemicals and substances for IL-1 expression. For this, tissue-engineered skin is used as an animal substitute and IL-1 is used as a marker for discriminating nonirritants from irritants.

Interleukin-1 promotes keratinocyte and fibroblast proliferation and induces the expression of intercellular adhesion molecules in endothelial cells and fibroblasts. The signal transduction events begin with the binding of IL-1 to its receptor, IL-1RI. This induces the recruitment and interaction of IL-1RacP to IL-1RI, resulting in a high affinity complex for IL-1 (Figure 4.8). IL-1RI and IL-1RacP recruit MyD88, a member of the IRAK (IL-1 receptor-associated protein kinases) family, and TRAF-6 (tumor necrosis factor receptor-associated factor-6). Together, they activate the IκK (inhibitory κB) complex, consisting of the inhibitory protein IκK and the transcription factor NF-κB. The TRAF-6 signal phosphorylates the IκB subunit, thereby releasing the NF-κB dimer from the complex. NF-κB then translocates to the nucleus where it acts as a transcription factor by binding to κB binding sites in target genes such as IL-6 and IL-8. IL-1 signaling can also lead to the activation of c-jun N-terminal kinase (Jnk) and other mitogen-activated protein kinases (MAPK) that result in the phosphorylation and activation of AP-1, another transcription factor that causes expression of genes related to wound healing (Lawrence, 2009).

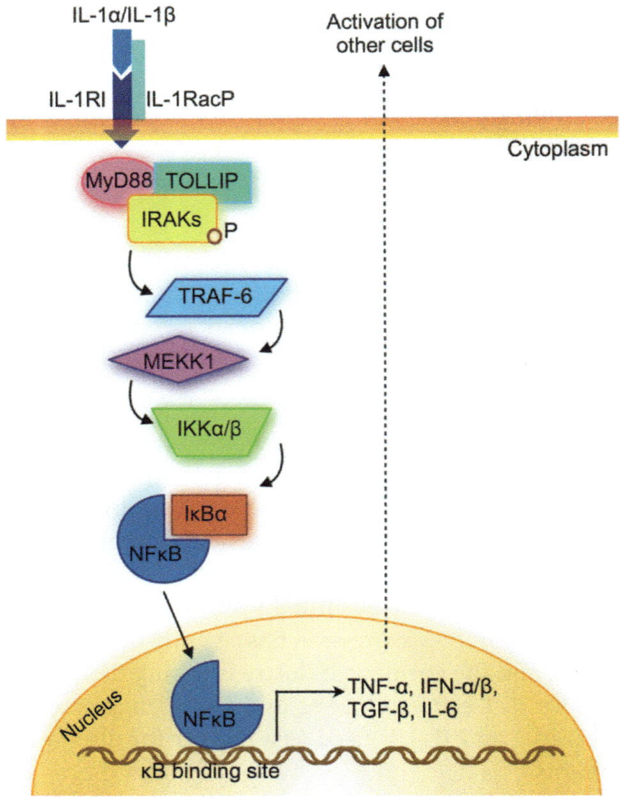

FIGURE 4.8

The interleukin (IL-1) signal transduction pathway in keratinocytes. After binding, recruitment of MyD88, TOLLIP, and IRAKs will follow. Thereafter TRAF-6 will interact, which causes activation of several protein kinases such as MEKK1. This finally results in the activation of several transcription factors, such as NFκB. IRAK, IL-1 receptor-associated protein kinases; TRAF6, tumor necrosis factor receptor-associated factor-6; IκB, inhibitory κB; IKK, IκB kinase; MEKK1, MEK kinase; TOLLIP, Toll-interacting protein.

4.3 CELLULAR SIGNALING IN VASCULAR BIOLOGY

4.3.1 Vascular Signaling Overview

The development of blood vessels is an important area for tissue engineering and is one of the major barriers to overcome to establish a functioning tissue construct. Apart from engineering blood vessels for bypass surgeries, blood vessel formation becomes an important issue for other engineered tissues after implantation. In the initial phase after implantation, engineered tissues have to rely on diffusion for the delivery of oxygen and nutrients and the

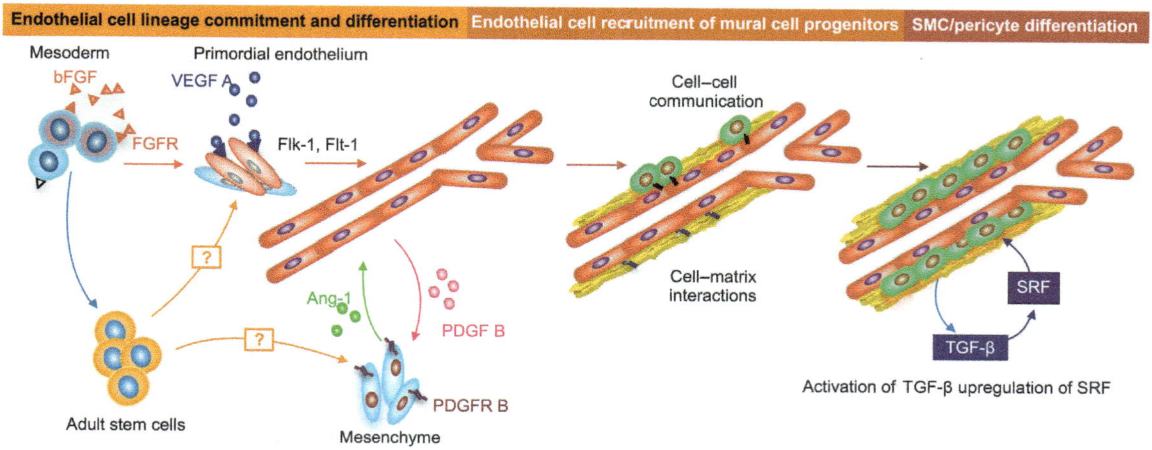

FIGURE 4.9

Blood vessel assembly can be divided into different stages. First, endothelial progenitor cells are attracted to the site and stimulated to differentiate into endothelial cells. VEGF is the key factor governing these processes. When endothelial cells have organized into a vascular structure, they secrete PDGF-B to recruit mural cell progenitors. Communication between the endothelial cells and the mural cell progenitors is important for the differentiation of the mural progenitors into pericytes or smooth muscle cells (depending on the vessel type) and the proper remodeling of the vessel. TGF-β is amongst the factors that are important for this process. EC, endothelial cell; SMC, smooth muscle cell; bFGF, basic fibroblast growth factor; FGFR, fibroblast growth factor receptor; VEGF-A, vascular endothelial growth factor-A; Flk-1, vascular endothelial growth factor receptor 2; Flt-1, vascular endothelial growth factor receptor 1; Ang-1, angiopoietin 1; Tie-2, angiopoietin 1 receptor; PDGF-B, platelet derived growth factor B; PDGFRβ, platelet derived growth factor receptor-β; TGF-β, transforming growth factor-β; SRF, serum response factor.

removal of waste products. Over time, blood vessels from the host will grow into the engineered tissue, allowing for a better transport of these components. If the process of vascularization is too slow, however, it can lead to nonoptimal integration or even partial death of the engineered tissue due to nutrient limitations. There has been significant progress in understanding the molecules and mechanisms involved in the assembly, remodeling and maturation of blood vessels. The factors that mediate these various phases are delivered to the endothelial cells in a paracrine or juxtacrine manner, making vessel formation a locally controlled event (Figure 4.9). The formation of vessels can occur via vasculogenesis, a process in which endothelial precursors coalesce in situ to form vessels, or via angiogenesis, in which new vessels arise by budding from preexisting vessels. Signaling pathways stimulated by growth factors such as VEGF and FGF2 (bFGF) are known to play critical roles in the initial stages of vascular development. The major steps in blood vessel assembly and the signaling events involved are depicted in Figure 4.9.

4.3.2 Receptor Tyrosine Kinase Signaling

Receptor tyrosine kinases are a major class of transmembrane receptors that are responsible for binding many extracellular and cell surface proteins and initiating the subsequent signaling pathways. The extracellular ligands include many relevant proteins in morphogenesis and regeneration, such as VEGF, FGFs, platelet-derived growth factor (PDGF), nerve growth factor (NGF), and insulin. The cell surface proteins are primarily from the ephrin family, which are especially important for morphogenesis in development. Receptor tyrosine kinases generally work by autophosphorylation upon ligand binding, which causes dimerization of the receptor such that the kinase domain on one receptor can phosphorylate its neighbor.

Several tyrosine kinase receptors act during vascular morphogenesis, including the VEGF receptors, VEGFR-1 (Flt-1) and VEGFR-2 (Flk-1), PDGF receptor-β (PDGFRβ), and the angiopoietin receptor, Tie-2. The VEGF receptors are primarily involved in the early stages of blood vessel assembly. Their activation upon ligand binding leads to the migration, proliferation, and eventually the differentiation of endothelial cells to form a new, lumen-containing vessel.

The binding of VEGF to VEGFR-2, a receptor tyrosine kinase, can lead to the activation of several signaling pathways involved in angiogenesis (Figure 4.10), one of which is the MAPK pathway. The binding activates the tyrosine kinase activity of the cytoplasmic domain of the receptor, providing a binding site for the adapter protein Grb2 that in turn localizes Sos to the plasma membrane. Sos activates Ras by replacing GDP with GTP. Ras-GTP binds directly to a serine-threonine kinase, Raf, forming a transient membrane-anchoring signal. Active Raf kinase phosphorylates a dual specificity kinase, mitogen-activated kinase (MEK), and activates it. The activated MEK phosphorylates ERK1/ERK2, which subsequently translocates to the nucleus, where it can phosphorylate the transcription factor Ets-1. Ets-1 then regulates the expression of multiple angiogenic genes like MMP-1, MMP-9, integrin β3, and VE-cadherin (Watanabe et al., 2004).

Endothelial cells within the primitive capillary network govern the subsequent acquisition of mural cells (pericytes or smooth muscle cells) that will make up the surrounding vessel wall. Proliferating endothelial cells recruit mural cell precursors via secretion of PDGF-B that acts as a chemoattractant and mitogen for multipotent progenitors derived from the mesenchyme surrounding the endothelial tubes. Conversely, the mesenchymal progenitors secrete soluble factors, Angiopoietin-1 (Ang-1) and -2 (Ang-2), which both bind the receptor Tie-2 that is expressed by endothelial cells. Ang-1 and Ang-2 have opposing roles in vessel assembly. Ang-1 consistently actives Tie-2 to stimulate vessel remodeling and angiogenesis, unlike Ang-2, which appears to antagonize this response. However, both have an important role in vessel formation. Ang-1 most likely provides a basal signal, but the vessel destabilization promoted by

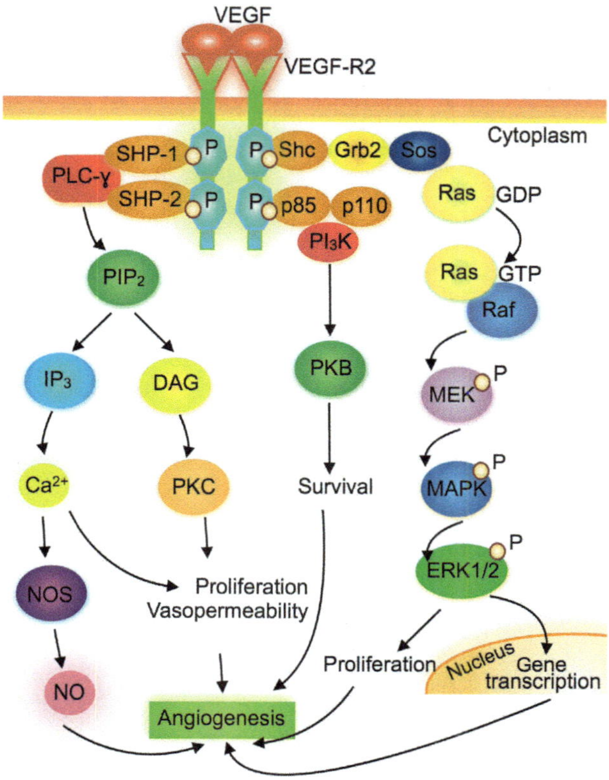

FIGURE 4.10

Binding of VEGF to its receptor leads to the activation of several signaling pathways that are involved in angiogenesis. Activation of these pathways leads to endothelial cell proliferation, differentiation, and migration. One of the pathways involved is the MAPK pathway. The MAPK pathway is activated through a Ras/Raf complex, but it should be noted that an alternative route via PKC also plays an important role in endothelial cells. VEGF, vascular endothelial growth factor; VEGF-R2, vascular endothelial growth factor receptor 2; SHP, SH phosphatase; Grb2, growth factor receptor-bound protein 2; PI3K, phosphoinositide-3 kinase; PLC, phospholipase C; PIP2, phosphatidylinositol bisphosphate; IP3, phosphoinotide-3,4,5-triphosphate; DAG, diacylglycerol; PKC, protein kinase C; NOS, nitric oxide synthase; NO, nitric oxide; PKB, protein kinase B; MEK, MAPK/ERK kinase; MAPK, mitogen-activated protein kinase; ERK, extracellular signal-regulated kinase.

Ang-2 in the presence of VEGF allows the proliferation and migration of endothelial cells (Brindle et al., 2006).

The addition of growth factors like VEGF to engineered tissues could initiate faster vessel formation after implantation. However, in order to get stable, functional vascular structures, other factors are necessary as well. For example, a sequential release approach was taken to supply two growth factors to improve vascularization. PDGF-BB, which signals smooth muscle cells to stabilize vessel walls, was combined with VEGF-A inside a hydrogel. The two growth factors had two different release profiles, a strategy that improved cardiac function after a myocardial infarction in rats (Fischbach and Mooney, 2007; Hao et al., 2007). Therefore, a thorough understanding of receptor tyrosine kinase signaling pathways involved both in initial vessel formation and vessel maturation is important to be able to enhance the vascularization of engineered tissues after implantation.

4.4 CELLULAR SIGNALING IN BONE BIOLOGY

4.4.1 Bone Signaling Overview

Cellular signaling in bone occurs mainly between four cell types that are specific for bone: the lining cells, osteoclasts, osteoblasts, and osteocytes, as well as their hematopoietic and mesenchymal progenitors (Figure 4.11). The cascades of events in bone remodeling are strictly orchestrated by communication between the different bone cells within a basic multicellular

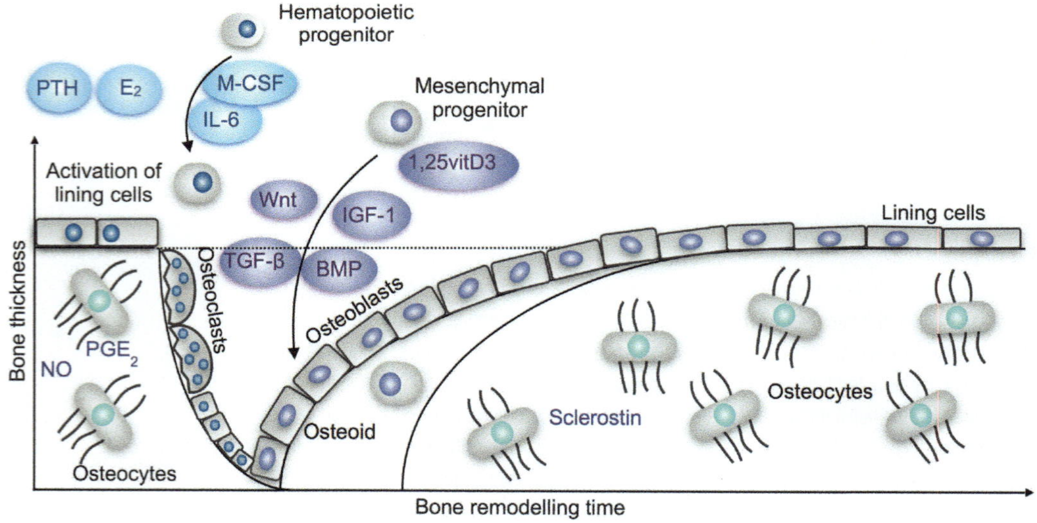

FIGURE 4.11

Bone cells within a basic multicellular unit interact in a time-dependent manner during bone modeling. The process is initiated when the activation of the bone lining cells residing on the bone surface receive signals from osteocytes within the bone and cells in the bone marrow, including nitric oxide (NO) and prostaglandin E2 (PGE_2). The osteoclasts are affected by systemic factors such as parathyroid hormone (PTH) and estrogen (E_2) via the activated lining cells. Following osteoclast resorption, growth factors such as transforming growth factor-β (TGF-β) are released, stimulating bone formation by osteoblasts and inhibiting bone resorption. Bone formation is also regulated by many local factors, including wingless-related integration site family of proteins (Wnts), insulin-like growth factor-1 (IGF-1), and systemic factors such as 1,25 dihydroxyvitamin D3 (1,25-$(OH)_2$D3). As time progresses, the osteoblasts are incorporated into bone (osteocytes), become resting lining cells, or die by apoptosis. Upon mineralization of osteoid, the incorporated osteocytes secrete sclerostin to provide negative feedback for bone formation to prevent overfilling of the resorption pit.

unit. Osteocytes activate lining cells on the bone surface that subsequently produce cytokines to recruit osteoclast progenitors to the bone surface. Differentiation into and stimulation of mature osteoclasts results in the formation of a resorption pit, during which factors are released that contribute to the recruitment, proliferation, and differentiation of osteogenic progenitors into mature osteoblasts. Active osteoblasts produce an unmineralized matrix called osteoid, in which some osteoblasts are embedded and become osteocytes after mineralization of the matrix. The remaining osteoblasts either become lining cells or die by apoptosis. Some of the key locally secreted growth factors are BMPs (Section 4.1.3.1), Wingless-related integration site family of proteins (Wnts) (Section 4.4.2), Hedgehogs, and IGFs. Systemic factors, such as parathyroid hormone (PTH), play an important role in bone metabolism. The gender steroids estrogen and androgen also affect bone metabolism, but they primarily act via a different mechanism, instead binding to intracellular receptors that then interact with binding elements in promoter regions of bone-related genes. Deficiency or decreased levels of these gender steroids are one of the causes of osteoporosis, mainly by affecting the expression of membrane-bound and secreted signaling molecules that regulate bone resorption. Another systemic factor acting on nuclear receptors and with pronounced effects on bone is vitamin D (Section 4.4.3).

In addition to diffusible molecules produced for the purpose of signaling, some small molecules and ions such as calcium, glucose, and oxygen can also trigger a physiological response by acting as a ligand. Extra- and intracellular levels of calcium regulate a multitude of biological processes by interacting with an enormous variety of proteins and pathways. Just one of its myriad roles is to act as a ligand for receptors on the surface of pituitary cells. These cells sense serum levels of calcium, and in response to low calcium levels, they secrete parathyroid hormone, which affects the bone remodeling process throughout the body.

4.4.2 Wnt Signaling

Like BMPs, Wnts (pronounced as "wint") are secreted signaling molecules with pivotal rules in a variety of cellular activities including cell fate determination, proliferation, differentiation, migration and polarity, and have been implicated in cancer. Wnt signaling is an example of a pathway that depends on regulated proteolysis, a mechanism that enables cells to respond rapidly to an extracellular signal. The canonical Wnt signaling pathway is important for bone homeostasis and the pathway represents a therapeutic target for patients at risk for fractures related to aging and disease (Baron and Kneissel, 2013).

There are 19 Wnts in humans that all bind to a receptor of the Frizzled family, sometimes in conjunction with other membrane-associated proteins, such as low-density lipoprotein receptor-related proteins (LRPs). Frizzled then recruits Dishevelled, a scaffold protein that initiates the signaling cascade. There are three major signaling pathways that Wnts can activate: (1) β-catenin (also known as canonical Wnt signaling), (2) planar cell polarity pathways, and (3) Ca^{2+} pathway (Figure 4.12). In canonical Wnt signaling,

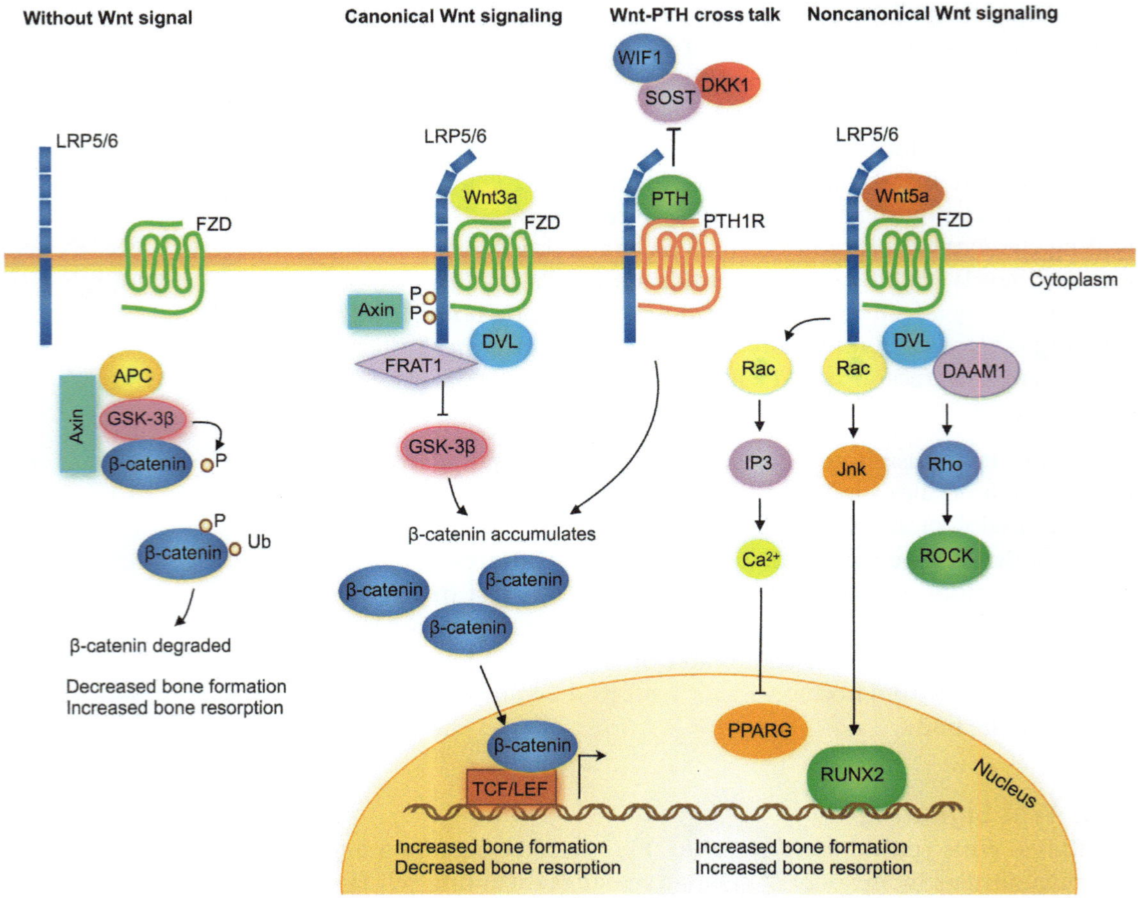

FIGURE 4.12

A simplified view of Wnt signaling in bone homeostasis. In the absence of Wnt, the levels of β-catenin in the nucleus are low because it gets targeted by the destruction complex, leading to phosphorylation, ubiquitination, and degradation. In the canonical Wnt pathway, the binding of Wnt to its receptor complex initiates Wnt/β-catenin signaling. When β-catenin accumulates in the cytoplasm, it translocates to the nucleus, where it associates with members of the TCF/LEF transcription factor family to control gene transcription. The PTH1 receptor can activate β-catenin signaling in the absence of Wnt. Bone formation is increased and bone resorption is decreased with β-catenin signaling. The noncanonical Wnt pathways (planar cell polarity and Ca^{2+}) do not act through β-catenin. They also affect homeostasis by increasing both bone formation and resorption. *Adapted from Baron and Kneissel (2013).*

LRP5 and LRP6 act as coreceptors and upon Wnt binding, intracellular β-catenin is stabilized and thereafter enters the nucleus where it interacts with transcription factors in the TCF/LEF family. The key to this latent regulation mechanism is that without Wnt signaling, a group of proteins form a degradation complex that binds and phosphorylates β-catenin, marking it for ubiquitination and degradation in proteasomes. In the planar cell polarity pathway, activation of Frizzled regulates the polarity of cells by affecting their cytoskeletal organization. In the Ca^{2+} pathway, activation of a heterotrimeric G-protein results in an increase in intracellular Ca^{2+} and activation of calcium/calmodulin-regulated kinase II (CamKII) and protein kinase C (PKC). Extracellularly, the Wnt inducible factor 1 (Wif1) and the secreted Frizzled-related proteins (sFRPs) antagonize Wnt signaling by interfering with the functional interaction between Wnts and their Frizzled receptors (Kawano and Kypta, 2003). Using another mechanism, Dickkopf (Dkk) proteins compete for the LRP5 and LRP6 co-receptors to prevent activation of the canonical signaling pathway.

4.4.3 Vitamin D

Vitamin D signaling exemplifies the deviation of the paradigm of cell signaling because the vitamin D receptor resides in the cytoplasm and acts directly as a transcription factor once bound to vitamin D. Vitamin D is essential for the development and maintenance of a mineralized matrix and enhancement of calcium absorption. Strictly speaking, vitamin D is not a vitamin but should rather be considered as a hormone produced by the skin that is metabolized into more active compounds in peripheral tissues (Figure 4.13). The compound 1,25-$(OH)_2$D3 is the active form of vitamin D with the highest binding affinity for the vitamin D receptor (VDR), a member of the steroid receptor superfamily. These nuclear receptors act as ligand-activated transcription factors that interact with coregulators and the transcriptional preinitiation complex to regulate gene transcription. Binding of 1,25-$(OH)_2$D3 induces heterodimerization of the VDR with the retinoid X receptor (RXR). This heterodimer binds with high affinity to vitamin D response elements (VDREs) in the promoters of target genes such as osteocalcin.

The activity of 1α-hydroxylase is highly regulated to ensure appropriate plasma levels of active 1,25-$(OH)_2$D3 to maintain calcium homeostasis. Dietary calcium intake can regulate 1α-hydroxylase levels directly through serum calcium levels and indirectly through PTH. Like calcium, dietary phosphate also increases renal 1α-hydroxylase activity, but seems to mediate this via systemic hormones. 1α-hydroxylase activity is also expressed by other tissues than bone, and contributes to local 1,25-$(OH)_2$D3 needs in a paracrine way.

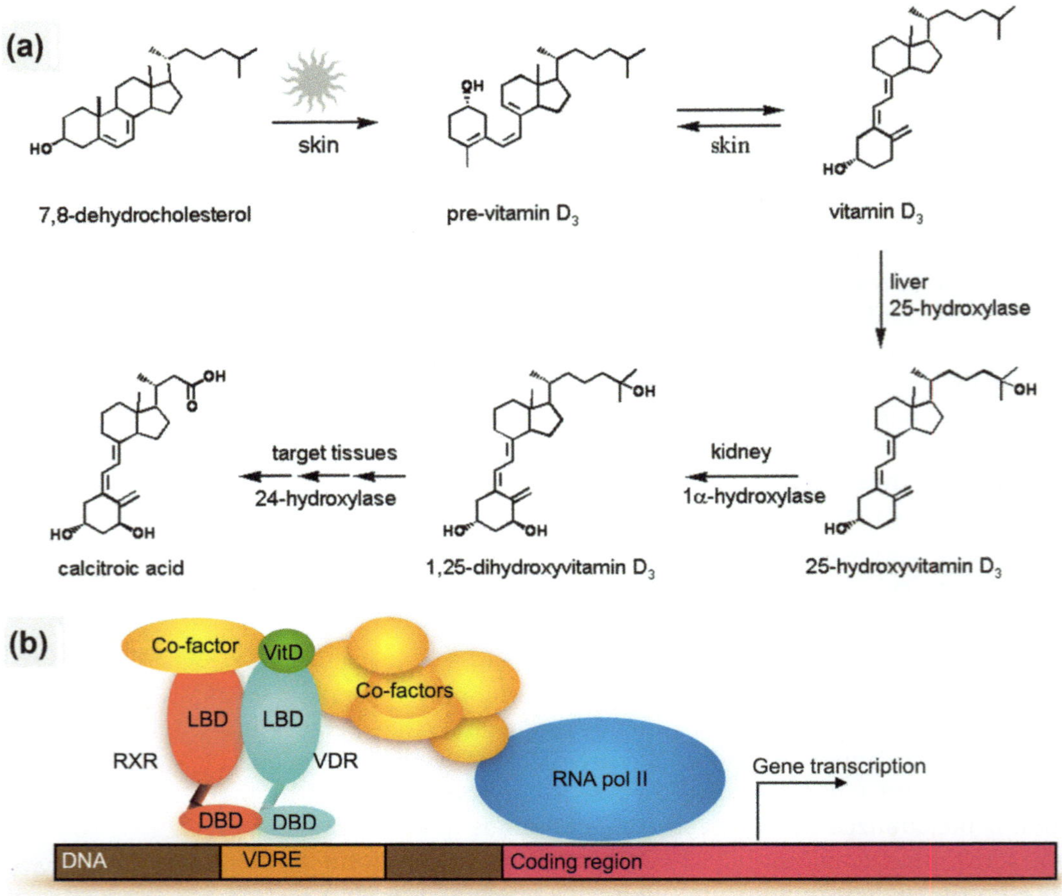

FIGURE 4.13

(a) Vitamin D3 synthesis, activation, and catabolism. Vitamin D3 is produced in the skin by the photolytic cleavage of 7-dehydrocholesterol followed by thermal isomerization. Vitamin D is transported to the liver via the serum vitamin binding protein, where it is converted to 25-hydroxyvitamin D3, the major circulating metabolite of vitamin D3. The final step, 1α-hydroxylation, occurs primarily, but not exclusively, in the kidney, forming 1,25-dihydroxyvitamin D3, the hormonal form of the vitamin. Catabolism inactivation is carried out by 24-hydroxylase, which catalyzes a series of oxidation steps resulting in side chain cleavage. (b) 1,25-dihydroxyvitamin D3 regulates gene transcription. The VDR-RXR heterodimer recruits co-factors and binds to VDRE using the DNA binding domains. Complex formation with basal transcription machinery and histone modifiers enables activation of gene transcription. DBD, DNA binding domain containing the two zinc fingers; LBD, ligand binding domain; VDR, vitamin D receptor; VDRE, vitamin D responsive element; VitD, 1,25-dihydroxyvitamin D3. *From Dusso et al. (2005).*

4.5 CELLULAR SIGNALING IN SKELETAL MUSCLE

4.5.1 Muscle Signaling Overview

As with bone, muscle is a highly vascularized and dynamic tissue. Mammals have skeletal muscle, cardiac muscle, smooth muscle, and myoepithelial cells, but only skeletal muscle will be discussed in this section. In order to understand

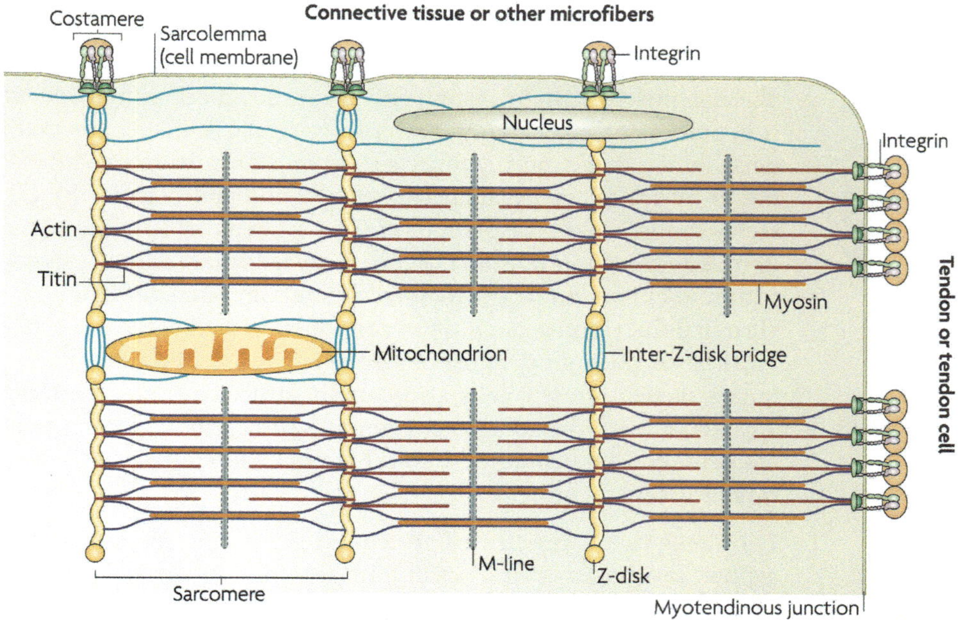

FIGURE 4.14

A schematic of skeletal muscle, showing two myofibrils composed of actin, myosin, and associated proteins aligned at the Z-disc. The myofibrils are connected to costameres by intermediate filament proteins (blue) and cytoskeletal proteins (yellow circles). Costameres connect the Z-discs to both the surrounding tissue and other myofibers. *From Sparrow and Schöck (2009).*

how signaling occurs in skeletal muscle, it is important to understand its structure. Skeletal muscles are composed of long fibers, which are each a large multinuclear cell that formed after the fusion of many cells. The cytoplasm of the myofibers contains many myofibrils, which are chains with a 1–2 μm diameter that often span the entire length of the fiber, which is composed of many subunits called sarcomeres. The sarcomeres are the contractile portion of the cell and are made up of overlapping thin actin filaments and think myosin filaments, the latter of which slides over the former to generate the contractile force for which muscles are known. A single contraction of a myofiber is a highly coordinated event, with tropomyosin and troponins bound to thin filaments rotating out of position to expose a myosin head-binding site to facilitate binding, ATP hydrolysis to ADP, the power stroke, and uncoupling (Figure 4.14).

Muscle contraction occurs when a motor nerve signals an action potential, which leads to an influx of Ca^{2+} into the cytosol. Tropomyosin and troponin, a complex of three polypeptides (troponin T, I, and C), are responsible for the contractile response to Ca^{2+} influx. In resting cells, troponin I and T pull the tropomyosin bound to the actin filaments into a position that interferes with the binding of myosin. At high levels of Ca^{2+}, troponin C inhibits the action of troponin I, thereby allowing tropomyosin to move into a position on actin that permits the action of myosin.

Skeletal muscle is an interesting system to study mechanotransduction because it is both a force-generating and a force-sensitive tissue. For example, it has been shown that culturing muscle stem cells on substrates with elasticity similar to muscle promotes their self-renewal and ability to contribute to muscle regeneration upon transplantation (Gilbert et al., 2010). The signaling events in muscle mechanotransduction primarily take place at the Z-disc, which is the region at either end of a sarcomere where the thin actin filaments are bound via α-actinin. Upon generation of a mechanical signal, events such as the translocation or phosphorylation of sarcomeric-associated proteins occur. Importantly, skeletal muscle cells also contain structures called costameres, which are related to focal adhesions containing integrins in nonmuscle cells, and are believed to transmit mechanical forces by interacting with the Z-discs and other sarcomeric proteins.

Titin, zeugmatin, muscle LIM protein (MLP), α-actinin-associated LIM protein, cypher, myopodin, and α-actinin are recruited to the Z-discs where they are involved in signaling. The role of the Z-discs in both signaling and maintaining tissue structure is revealed by a number of knockdown studies. For example, when MLP, a protein containing a series of kinase binding sites that could initiate signaling cascades is knocked down, the muscle has lower passive elastic properties (Knöll et al., 2002). It is important to note that muscle has many redundant signaling pathways in the Z-discs. For example, mutations in myotilin, a protein that stabilizes Z-discs and signals to costameres, do not significantly alter sarcomere assembly or function (Moza et al., 2007).

4.5.2 Costameres

Costameric proteins found at the cell–extracellular matrix interface transmit both extracellular and muscle-generated forces to mediate signaling and can be considered analogous to focal adhesions in nonmuscle cells (Ervasti, 2003). As a hallmark of costameres, vinculin was first identified in the striated patterns of muscle adhesions and has been directly implicated as a molecular strain sensor regulating the differentiation of stem cells (Anastasi et al., 2009; Holle et al., 2013). Beyond ubiquitous adhesive proteins such as vinculin, costameres contain at least 23 distinct proteins that could act as mechanical sensors in the spectrin-ankyrin and dystroglycan complexes. In the spectrin complexes, ankyrin connects spectrin filaments to the membrane, and both proteins are very extensible, capable of transducing significant force (Krieger et al., 2011; Lee et al., 2006). At their other end, ankyrins bind to $Na^{+}K^{+}$ ATPases, and cannot transmit force as effectively. The dystrophin-associated protein complex (DAPC) directly transduces sarcomeric forces via dystrophin, which connects Z-disc α-actinin-bound γ-actin to the DAPC and α-syntrophin (Ehmsen et al., 2002). Dystrophin is mutated and truncated in muscular dystrophies, where increasing severity of the mutation correlates to decreased mechanical function and force transduction. Signaling at costameres is clearly important for normally functioning muscle. Some recent strategies to cure muscular dystrophies have included the use of a fragment of a Wnt protein with higher solubility without sacrificing activation of the signaling pathway (von Maltzahn et al., 2013) and by mimicking events from development to differentiation of embryonic stem cells into muscle progenitors (Darabi et al., 2008).

4.6 FUTURE DEVELOPMENTS

The molecular processes underlying cellular signaling show a high degree of complexity, as demonstrated by the content of this chapter. To fully understand the molecular ins and outs of a biological process, countless numbers of proteins have to be considered as candidate players. High throughput analytical methods are increasingly used to simultaneously analyze thousands of molecules, such as micro-array and next generation sequencing technology for gene expression profiling and mass spectrometry for protein analysis. Critical to this development is improved software to assist the interpretation the data and to model gene regulatory networks. Another challenge for the future lies in the extrapolation of the molecular knowledge to tissue engineering. Many signaling pathways have been delineated using model cell lines, but cells are exposed to a dynamic three-dimensional environment in tissue engineering. We anticipate that principles of wound repair and developmental biology will be increasingly incorporated. Because the signaling events that play a role in tissue engineering should ideally be studied in situ, molecular imaging techniques such as bioluminescent imaging, magnetic resonance imaging, and positron emission tomography will be increasingly used to noninvasively image gene expression and enzymatic activity of proteins. These technological developments will increase our knowledge of the molecular basis of tissue engineering. Once critical cellular signaling events have been identified in a certain tissue engineering problem, the next challenge lies in implementation of the knowledge for which a whole toolbox is currently being developed. For instance, ligands involved in tissue regeneration can be incorporated into biomaterials. Extracellular matrix proteins or peptides mimicking their action can be used to coat biomaterials, whereas growth factors, cytokines, hormones, and small molecules can be released from scaffolds. The same compounds can also be used to control the culture medium of expanding and differentiating stem cells. Furthermore, materials can be designed in such a way that they elicit the desired response. Both the chemical composition and micro-architecture have an influence on the biological behavior of the cells growing on them. Finally, tissue formation can be regulated by systemic administration of pharmacological agents. As biologists continue to unravel the complex signaling mechanisms inherent in tissue development and regeneration, more targets for tissue engineers will become apparent.

4.7 SNAPSHOT SUMMARY

- Most cell signaling events follow the three-step paradigm of signal initiation, signal transduction, and gene transcription.
- The presence of a ligand will not necessary have the same or any effect on different cell types. The cellular response depends on its gene and protein expression, the extracellular environment, and many other factors.
- Cell signaling is a complex process that is not completely understood, but there are many targets for tissue engineers to manipulate cell behavior.
- Manipulation of signal transduction for tissue engineering purposes should be considered in its in vivo context with phenomena such as cell type and cross talk with other signaling mechanisms.
- Intricate signaling cross talk between different cell types and pathways exists in skin, vessels, bone, and muscle.
- Signaling pathways are regulated at many different levels: extracellularly, in the cytoplasm, and in the nucleus. This provides engineers multiple means to manipulate the cell.
- The intricate details of many signaling pathways have yet to be discovered and may provide new tools for tissue engineering in the future.
- Activating pathways, such as Notch, that are affected in aging cells can restore the regeneration potential of tissues.
- Manipulating signaling through the delivery of microRNAs (miRNAs) can cause cells to proliferate and contribute to tissue regeneration.

STATE-OF-THE-ART EXPERIMENT

It is now known that cardiomyocytes have some limited capacity for renewal, but their regenerative potential is no match for the damage caused by cardiac ischemia. Nonetheless, following a heart attack, there is an increase in cardiomyocyte proliferation at the bordering region of the infarct, suggesting that the injury promotes a small regenerative response. Given that a number of small molecules can also promote cardiomyocyte proliferation, Eulalio et al. hypothesized that a mechanism must exist that normally prevents cell division and sought to discover whether miRNAs might be responsible.

MicroRNAs (miRNAs) are small RNA molecules that function by binding to mRNAs. This often results in degradation of the mRNA, thereby silencing protein translation, or it affects translation directly. Research into the role of these noncoding RNAs has revealed another level of complexity to the regulation of gene and protein expression. For regenerative medicine, miRNAs represent a new therapeutic target since they are implicated in many disease states.

Eulalio et al. used an unbiased high-throughput fluorescence microscopy-based screening method for miRNAs that could more than double the rate of cardiomyocyte proliferation (Eulalio et al., 2012). Briefly, they transfected rat cardiomyocytes with a library of 875 miRNA mimics. After 72 h, cells were stained with sarcomeric α-actinin to identify the cardiomyocytes and either Ki-67 (a proliferation antigen) or 5-ethynyl-2′-deoxyuridine (EdU; a uridine analogue incorporated into newly synthesized DNA) was added to the cultures. High-content fluorescent imaging was used to identify which miRNAs caused more than a twofold increase in cardiomyocyte proliferation.

Two miRNAs, hsa-miR-590 and hsa-miR-199a, were then tested in vivo in a rodent model for cardiac infarction. When combined with a viral vector used in gene therapy, the miRNAs increased cardiomyocyte proliferation following the injury and resulted in improved heart repair and function.

Researchers then went on to identify the molecular targets of these miRNAs to better understand their role. This is not simply a matter of identifying complementary mRNA sequences, as miRNA–mRNA binding depends on surrounding sequences. To overcome this, researchers performed a second high-throughput screening and used deep sequencing to find the gene transcripts affected by either hsa-miR-590 or hsa-miR-199a. They found approximately 700 transcripts downregulated by the miRNAs and 1000 transcripts upregulated, underlining the complex mechanisms by which the regulation occurs. The researchers knocked down 601 of those genes using short interfering RNAs (siRNAs), which only partly reproduced the effect of the miRNAs, revealing the cooperative and complicated nature by which the miRNAs maintain quiescence.

There are still major questions to answer before these miRNAs could be used in therapeutic settings. For example, more information is needed to determine whether overriding an endogenous mechanism of quiescence might have a harmful effect on cardiac function. Nonetheless, this study aids in the development of new therapies for cardiac regeneration by identifying new molecular targets for drugs or gene therapy (Figure 4.15).

Continued...

STATE-OF-THE-ART EXPERIMENT *Continued*

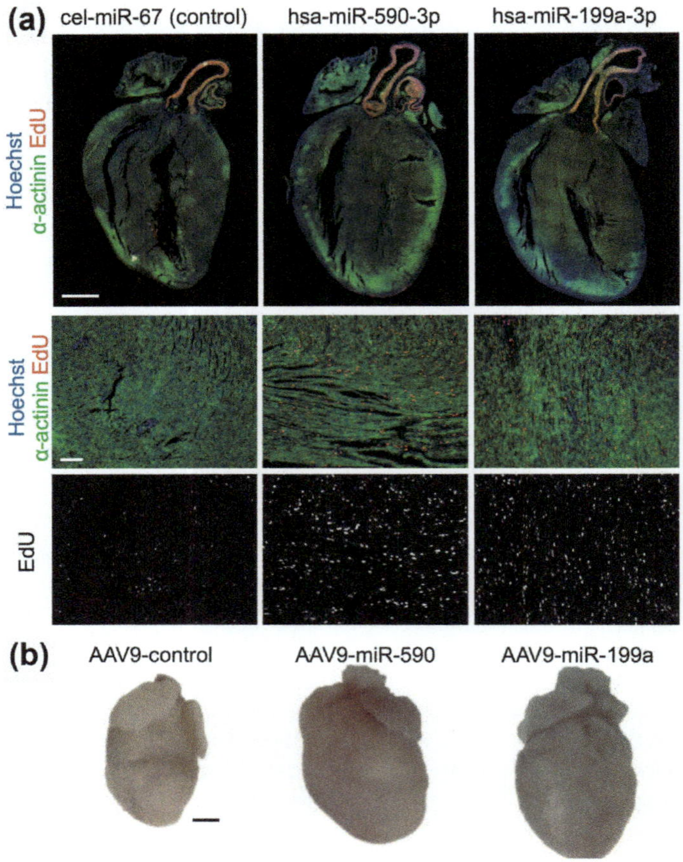

FIGURE 4.15

(a) EdU and α-actinin staining of mouse hearts injected with hsa-miR-590 and hsa-miR-199a show a marked increase of EdU+ (proliferating) cardiomyocytes compared to the control. Scale: 1 mm (top), 100 μm (bottom). (b) 12 days after injection of a viral vector expressing the miRNAs, mouse hearts were morphologically normal, but significantly enlarged. Scale: 1 mm.

CLASSICAL EXPERIMENT

A landmark study by Conboy et al. showed that mimicking Notch signaling can be used to improve muscle regeneration (Conboy et al., 2003).

Satellite cells are the resident stem cell population that is responsible for the regeneration of muscle following injury. Like the regeneration of many tissues, the efficiency of this process diminishes with age. Indeed, when the hindlimb muscles were injured in young (2–3 months) mice, the satellite cells produced new myoblasts and myofibers. However, in aged (23–24 months) mice, there was only 20–25% contribution from the satellite cells and the regeneration was impaired. This could not be explained by a changing number of satellite cells, so researchers hypothesized that the dramatic decline of myoblast generation in aging mice might be due to the ability of the satellite cells to be activated in response to injury.

Signaling through the Notch receptor was known to be critical for satellite cell activation. When Notch binds its ligand, Delta, which is present on adjacent cells, a protease in the plasma membrane cleaves the cytosolic tail of Notch, which then translocates to the nucleus to activate Notch target genes. Conboy et al. used fluorescence-activated cell sorting (FACS) to detect the expression of Delta-1 and a Notch inhibitor, Numb, on the surface of satellite cells in their resting state and after injury. In resting cells from mice of all ages, low levels of Delta-1 and high levels of Notch were detected. In injured cells, Delta-1 was upregulated in young and adult mice, but not in old mice, and this increase was also met by a decrease in Numb. These results were confirmed in vivo, where Delta expression was upregulated adjacent to the injury site in young mice, but was almost unchanged in old mice. When a Notch inhibitor, Jagged-Fc, was introduced at the site of injury, the regenerative potential of young mice was inhibited.

To test whether these findings could be used to improve muscle regeneration, an antibody that binds to the extracellular domain of Notch and thereby mimics its activation was introduced at the injury site. Remarkably, by restoring Notch activation, the aged mice were able to regenerate injured muscle with the same efficiency as young mice (Figure 4.16).

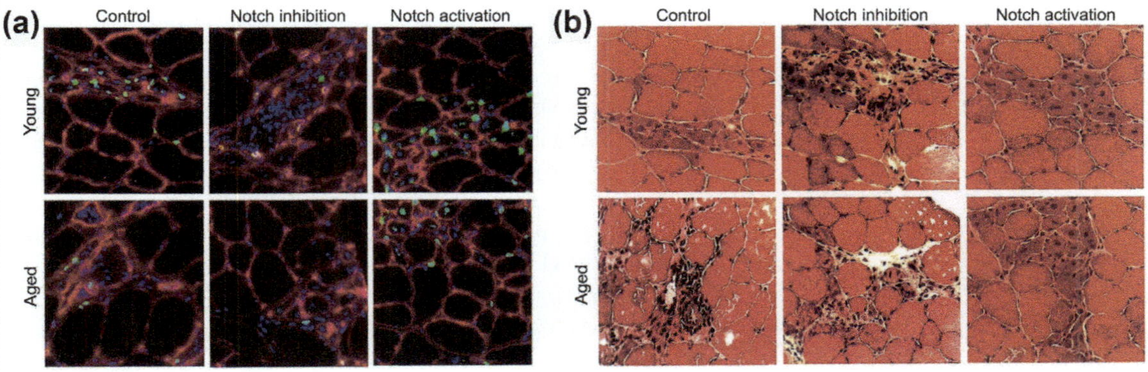

FIGURE 4.16

(a) Muscles of young and aged mice were injected with either a Notch inhibitor or activator. Two days later, BrdU was injected and muscles were excised 5 days after injury. Immunofluorescence using antibodies against BrdU (green), showing proliferating cells, and laminin (red), marking the boundaries of myofibers, showed that Notch inhibition reduced myofiber formation, while Notch activation promoted regeneration in old muscle. (b) Hematoxylin and eosin staining confirmed the immunofluorescence results, showing the negative effect of Notch inhibition on muscle regeneration and the positive effect of Notch activation on regeneration in aged mice.

RECOMMENDED READING

Hynes, R.O., 2002. Integrins: bidirectional, allosteric signaling machines. Cell 110, 673–687.

Nelson, W.J., Nusse, R., 2004. Convergence of Wnt, β-catenin, and cadherin pathways. Science 303, 1483–1487.

REFERENCES

Anastasi, G., Cutroneo, G., Gaeta, R., Di Mauro, D., Arco, A., Consolo, A., Santoro, G., Trimarchi, F., Favaloro, A., 2009. Dystrophin-glycoprotein complex and vinculin-talin-integrin system in human adult cardiac muscle. Int. J. Mol. Med. 23, 149–159.

Ashcroft, G.S., Yang, X., Glick, A.B., Weinstein, M., Letterio, J.L., Mizel, D.E., Anzano, M., Greenwell-Wild, T., Wahl, S.M., Deng, C., Roberts, A.B., 1999. Mice lacking Smad3 show accelerated wound healing and an impaired local inflammatory response. Nat. Cell Biol. 1, 260–266.

Baron, R., Kneissel, M., 2013. Wnt signaling in bone homeostasis and disease: from human mutations to treatments. Nat. Med. 19, 179–192.

Barrientos, S., Stojadinovic, O., Golinko, M.S., Brem, H., Tomic-Canic, M., 2008. Growth factors and cytokines in wound healing. Wound Repair Regen. 16, 585–601.

Brindle, N.P.J., Saharinen, P., Alitalo, K., 2006. Signaling and functions of angiopoietin-1 in vascular protection. Circ. Res. 98, 1014–1023.

Canalis, E., Economides, A.N., Gazzerro, E., 2003. Bone morphogenetic proteins, their antagonists, and the skeleton. Endocr. Rev. 24, 218–235.

Conboy, I.M., Conboy, M.J., Smythe, G.M., Rando, T.A., 2003. Notch-mediated restoration of regenerative potential to aged muscle. Science 302, 1575–1577.

Darabi, R., Gehlbach, K., Bachoo, R.M., Kamath, S., Osawa, M., Kamm, K.E., Kyba, M., Perlingeiro, R.C.R., 2008. Functional skeletal muscle regeneration from differentiating embryonic stem cells. Nat. Med. 14, 134–143.

Di Guglielmo, G.M., Le Roy, C., Goodfellow, A.F., Wrana, J.L., 2003. Distinct endocytic pathways regulate TGF-β receptor signalling and turnover. Nat. Cell Biol. 5, 410–421.

Dusso, A.S., Brown, A.J., Slatopolsky, E., 2005. Vitamin D. Am. J. Physiol. Renal Physiol. 289, F8–F28.

Ehmsen, J., Poon, E., Davies, K., 2002. The dystrophin-associated protein complex. J. Cell Sci. 115, 2801–2803.

Engler, A.J., Sen, S., Sweeney, H.L., Discher, D.E., 2006. Matrix elasticity directs stem cell lineage specification. Cell 126, 677–689.

Eulalio, A., Mano, M., Dal Ferro, M., Zentilin, L., Sinagra, G., Zacchigna, S., Giacca, M., 2012. Functional screening identifies miRNAs inducing cardiac regeneration. Nature 492, 376–381.

Ervasti, J.M., 2003. Costameres: the Achilles' heel of Herculean muscle. J. Biol. Chem. 278, 13591–13594.

Fan, Y., Arif, A., Gong, Y., Jia, J., Eswarappa, S.M., Willard, B., Horowitz, A., Graham, L.M., Penn, M.S., Fox, P.L., 2012. Stimulus-dependent phosphorylation of profilin-1 in angiogenesis. Nat. Cell Biol. 14, 1046–1056.

Fischbach, C., Mooney, D.J., 2007. Polymers for pro- and anti-angiogenic therapy. Biomaterials 28, 2069–2076.

Foletta, V.C., Lim, M.A., Soosairajah, J., Kelly, A.P., Stanley, E.G., Shannon, M., He, W., Das, S., Massagué, J., Bernard, O., Soosairaiah, J., 2003. Direct signaling by the BMP type II receptor via the cytoskeletal regulator LIMK1. J. Cell Biol. 162, 1089–1098.

Gilbert, P.M., Havenstrite, K.L., Magnusson, K.E.G., Sacco, A., Leonardi, N.A., Kraft, P., Nguyen, N.K., Thrun, S., Lutolf, M.P., Blau, H.M., 2010. Substrate elasticity regulates skeletal muscle stem cell self-renewal in culture. Science 329, 1078–1081.

Guicheux, J., Lemonnier, J., Ghayor, C., Suzuki, A., Palmer, G., Caverzasio, J., 2003. Activation of p38 mitogen-activated protein kinase and c-Jun-NH2-terminal kinase by BMP-2 and their implication in the stimulation of osteoblastic cell differentiation. J. Bone Miner. Res. 18, 2060–2068.

Gurtner, G.C., Werner, S., Barrandon, Y., Longaker, M.T., 2008. Wound repair and regeneration. Nature 453, 314–321.

Hao, X., Silva, E.A., Månsson-Broberg, A., Grinnemo, K.-H., Siddiqui, A.J., Dellgren, G., Wärdell, E., Brodin, L.A., Mooney, D.J., Sylvén, C., 2007. Angiogenic effects of sequential release of VEGF-A165 and PDGF-BB with alginate hydrogels after myocardial infarction. Cardiovasc. Res. 75, 178–185.

Holle, A.W., Tang, X., Vijayraghavan, D., Vincent, L.G., Fuhrmann, A., Choi, Y.S., del Álamo, J.C., Engler, A.J., 2013. In situ mechanotransduction via vinculin regulates stem cell differentiation. Stem Cells 31, 2467–2477.

Humphries, J.D., Byron, A., Humphries, M.J., 2006. Integrin ligands at a glance. J. Cell Sci. 119, 3901–3903.

Kawano, Y., Kypta, R., 2003. Secreted antagonists of the Wnt signalling pathway. J. Cell Sci. 116, 2627–2634.

Knöll, R., Hoshijima, M., Hoffman, H.M., Person, V., Lorenzen-Schmidt, I., Bang, M.-L., Hayashi, T., Shiga, N., Yasukawa, H., Schaper, W., McKenna, W., Yokoyama, M., Schork, N.J., Omens, J.H., McCulloch, A.D., Kimura, A., Gregorio, C.C., Poller, W., Schaper, J., Schultheiss, H.P., Chien, K.R., 2002. The cardiac mechanical stretch sensor machinery involves a Z disc complex that is defective in a subset of human dilated cardiomyopathy. Cell 111, 943–955.

Krieger, C.C., An, X., Tang, H.-Y., Mohandas, N., Speicher, D.W., Discher, D.E., 2011. Cysteine shotgun-mass spectrometry (CS-MS) reveals dynamic sequence of protein structure changes within mutant and stressed cells. Proc. Natl. Acad. Sci. 108, 8269–8274.

Lawrence, T., 2009. The nuclear factor NF-κB pathway in inflammation. Cold Spring Harbor Perspect. Biol. 1, a001651.

Lee, G., Abdi, K., Jiang, Y., Michaely, P., Bennett, V., Marszalek, P.E., 2006. Nanospring behaviour of ankyrin repeats. Nature 440, 246–249.

Li, Z., Zhang, G., Feil, R., Han, J., Du, X., 2006. Sequential activation of p38 and ERK pathways by cGMP-dependent protein kinase leading to activation of the platelet integrin $\alpha_{IIb}\beta_3$. Blood 107, 965–972.

Lorentz, K.M., Yang, L., Frey, P., Hubbell, J.A., 2012. Engineered insulin-like growth factor-1 for improved smooth muscle regeneration. Biomaterials 33, 494–503.

von Maltzahn, J., Zinoviev, R., Chang, N.C., Bentzinger, C.F., Rudnicki, M.A., 2013. A truncated Wnt7a retains full biological activity in skeletal muscle. Nat. Commun. 4, 2869.

Moza, M., Mologni, L., Trokovic, R., Faulkner, G., Partanen, J., Carpén, O., 2007. Targeted deletion of the muscular dystrophy gene myotilin does not perturb muscle structure or function in mice. Mol. Cell. Biol. 27, 244–252.

Mullen, A.C., Orlando, D.A., Newman, J.J., Lovén, J., Kumar, R.M., Bilodeau, S., Reddy, J., Guenther, M.G., DeKoter, R.P., Young, R.A., 2011. Master transcription factors determine cell-type-specific responses to TGF-β signaling. Cell 147, 565–576.

Munger, J.S., Sheppard, D., 2011. Cross talk among TGF-β signaling pathways, integrins, and the extracellular matrix. Cold Spring Harbor Perspect. Biol. 3, a005017.

Nakagawa, T., Lee, S.Y., Reddi, A.H., 2009. Induction of chondrogenesis from human embryonic stem cells without embryoid body formation by bone morphogenetic protein 7 and transforming growth factor beta1. Arthritis Rheumatol. 60, 3686–3692.

Nobes, C.D., Hall, A., 1999. Rho GTPases control polarity, protrusion, and adhesion during cell movement. J. Cell Biol. 144, 1235–1244.

Qin, J., Vinogradova, O., Plow, E.F., 2004. Integrin bidirectional signaling: a molecular view. PLoS Biol. 2, e169.

Re'em, T., Kaminer-Israeli, Y., Ruvinov, E., Cohen, S., 2012. Chondrogenesis of hMSC in affinity-bound TGF-β scaffolds. Biomaterials 33, 751–761.

Rosenbaum, D.M., Rasmussen, S.G.F., Kobilka, B.K., 2009. The structure and function of G-protein-coupled receptors. Nature 459, 356–363.

Shen, B., Zhao, X., O'Brien, K.A., Stojanovic-Terpo, A., Delaney, M.K., Kim, K., Cho, J., Lam, S.C.T., Du, X., 2013. A directional switch of integrin signalling and a new anti-thrombotic strategy. Nature 503, 131–135.

Sparrow, J.C., Schöck, F., 2009. The initial steps of myofibril assembly: integrins pave the way. Nat. Rev. Mol. Cell Biol. 10, 293–298.

Takagi, J., Petre, B.M., Walz, T., Springer, T.A., 2002. Global conformational rearrangements in integrin extracellular domains in outside-in and inside-out signaling. Cell 110, 599–611.

Trompouki, E., Bowman, T.V., Lawton, L.N., Fan, Z.P., Wu, D.-C., DiBiase, A., Martin, C.S., Cech, J.N., Sessa, A.K., Leblanc, J.L., Li, P., Durand, E.M., Mosimann, C., Heffner, G.C., Daley, G.Q., Paulson, R.F., Young, R.A., Zon, L.I., 2011. Lineage regulators direct BMP and Wnt pathways to cell-specific programs during differentiation and regeneration. Cell 147, 577–589.

Urist, M.R., 1965. Bone: formation by autoinduction. Science 150, 893–899.

Watanabe, D., Takagi, H., Suzuma, K., Suzuma, I., Oh, H., Ohashi, H., Kemmochi, S., Uemura, A., Ojima, T., Suganami, E., Miyamoto, N., Sato, Y., Honda, Y., 2004. Transcription factor Ets-1 mediates ischemia- and vascular endothelial growth factor-dependent retinal neovascularization. Am. J. Pathol. 164, 1827–1835.

Wolfenson, H., Lavelin, I., Geiger, B., 2013. Dynamic regulation of the structure and functions of integrin adhesions. Dev. Cell 24, 447–458.

Yao, W., Guan, M., Jia, J., Dai, W., Lay, Y.-A.E., Amugongo, S., Liu, R., Olivos, D., Saunders, M., Lam, K.S., Nolta, J., Olvera, D., Ritchie, R.O., Lane, N.E., 2013. Reversing bone loss by directing mesenchymal stem cells to bone. Stem Cells 31, 2003–2014.

Zou, Z., Chen, H., Schmaier, A.A., Hynes, R.O., Kahn, M.L., 2007. Structure-function analysis reveals discrete beta3 integrin inside-out and outside-in signaling pathways in platelets. Blood 109, 3284–3290.

CHAPTER 5

Extracellular Matrix as a Bioscaffold for Tissue Engineering

Brian M. Sicari[1,2], Ricardo Londono[1,4], Stephen F. Badylak[1,2,3]

[1]McGowan Institute for Regenerative Medicine, University of Pittsburgh, Pittsburgh, PA, USA; [2]Department of Surgery, University of Pittsburgh, Pittsburgh, PA, USA; [3]Department of Bioengineering, University of Pittsburgh, Pittsburgh, PA, USA; [4]School of Medicine, University of Pittsburgh, Pittsburgh, PA, USA

LEARNING OBJECTIVES

- To understand the composition and organization of native extracellular matrix (ECM)
- To recognize the utility of scaffolds derived from ECM in the field of tissue engineering
- To understand the process by which ECM scaffolds are manufactured
- To understand the meaning of "constructive tissue remodeling"
- To identify mechanisms behind constructive tissue remodeling by ECM bioscaffolds
- To recognize recent clinical translations of ECM bioscaffolds

5.1 INTRODUCTION

Tissue engineering and regenerative medicine strategies for the reconstruction of injured or missing tissues have included the use of implantable scaffold materials. Such scaffolds are composed of synthetic or naturally occurring materials and are typically designed to be degradable in vivo. There are advantages and disadvantages of both synthetic and naturally occurring biomaterials. The present chapter will focus solely upon biologic materials, specifically, biologic scaffold materials composed of extracellular matrix (ECM).

ECM biologic scaffolds are manufactured by the decellularization of source tissues such as dermis (Reing et al., 2010), urinary bladder (Freytes et al., 2004), small intestinal submucosa (SIS) (Badylak et al., 1989; Turner et al., 2010), and pericardium (Yang et al., 2009), among others, and by the decellularization/demineralization of bone (Einhorn, 1984; Harakas, 1984). Chemical, enzymatic, and/or mechanical methods are used to achieve decellularization of source tissues and such methods have marked effects upon

Tissue Engineering.
Copyright © 2015 Elsevier Inc. All rights reserved.

the mechanical and biologic properties of the resultant bioscaffold and the induced in vivo remodeling outcome. Such scaffold materials have been commonly used in both preclinical and clinical applications including the reinforcement and reconstruction of musculotendinous structures (Lai et al., 2003; Derwin et al., 2010), lower urinary tract (Fiala et al., 2007), esophageal structures (Badylak et al., 2011), skin (Lindberg and Badylak, 2001), upper airway (Haykal et al., 2012), bone (An, 1995), and myocardium (Kelly et al., 2009), among others. The remodeling outcomes have varied from excellent (Turner et al., 2010) to unsatisfactory (Doede et al., 2009). The reasons for these disparate outcomes are associated with differences in decellularization and processing methods, the use of chemical cross-linking agents, and/or appropriate clinical application.

There are currently more than 40 commercially available products with a broad range of indications for use. In the past decade, more than 3 million patients have been treated by surgical implantation of ECM scaffolds, most of which are xenogeneic (typically porcine) in origin. Documented reports of infectious disease transmission or immune-mediated rejection are nonexistent and are testament to the safety of these materials. ECM bioscaffolds are typically regulated as devices by the Food and Drug Administration (FDA) and therefore are subject to the same safety testing requirements as other medical devices.

The mechanisms by which ECM scaffolds facilitate a constructive remodeling response in tissues are partially understood and include relatively rapid and complete degradation, the release of bioactive molecules that are resident within normal mammalian ECM, and the release of bioactive cryptic peptide molecules with diverse activities promoting the recruitment of endogenous stem and progenitor cells, and modulation of the host innate immune response. Such events alter the default response of mammalian tissue to injury. Stated differently, the typical injury response consisting of inflammation leading to deposition of nonfunctional scar tissue is altered by the surgical placement of ECM to a response consisting of constructive, site-appropriate, and functional tissue formation (Figures 5.1–5.10).

5.2 NATIVE EXTRACELLULAR MATRIX

The ECM represents the secreted products of cells in all tissues and organs. ECM composition is generally conserved across tissue types; however, there exists a unique, tissue-specific, and three-dimensional ultrastructure. The structural and functional molecules that comprise the ECM provide the mechanical properties necessary for proper functioning of each tissue, and facilitate signal transduction between adjacent cells and between cells and the ECM itself. The ECM fluctuates in response to changes in microenvironmental cues including cellular activity and mechanical loading, and therefore is considered to be in a

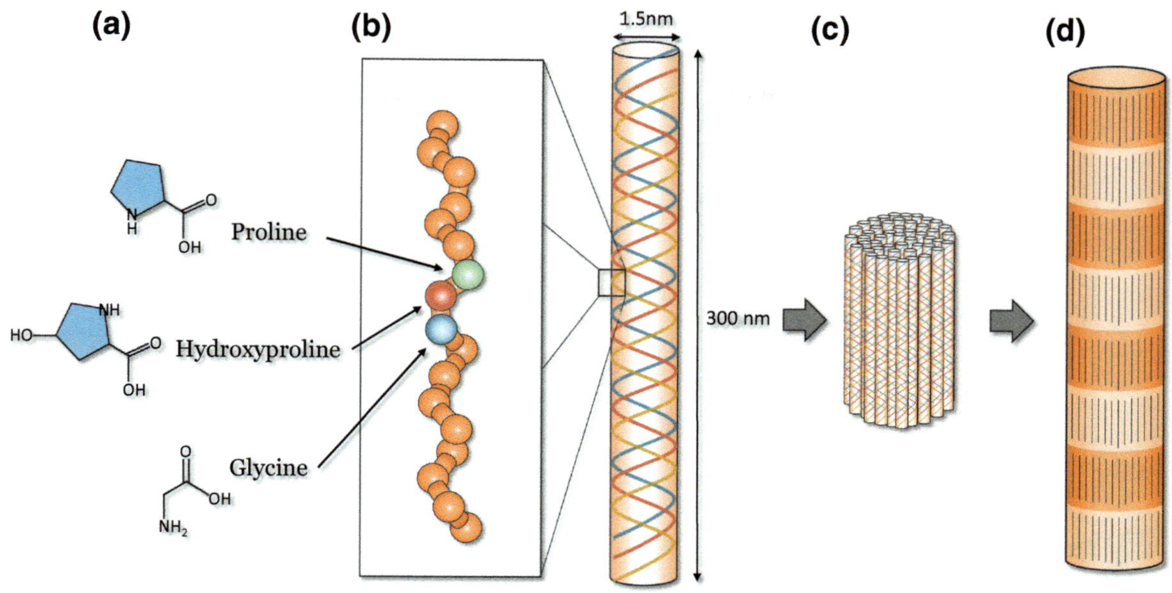

FIGURE 5.1

Schematic representation of the assembly of collagen I. (a) Individual collagen polypeptide chains. The polypeptides consist of the repeating sequence Gly-X-Y, where X is often proline and Y is often hydroxyproline; (b) collagen is made up of three polypeptide strands which are all left-handed helixes and twist together to form a right-handed coiled coil. The polypeptide strands are synthesized as precursor chains with propeptides (globular extensions) on the C and N ends. The propeptides are cleaved into short nonhelical telopeptides; (c) collagen molecules self-assemble into collagen fibrils; (d) collagen fibers are formed by end-to-end and lateral assembly of collagen fibrils, resulting in a regular banding pattern that is characteristic of collagen.

state of dynamic reciprocity with the cells and factors that influence the ECM (Bissell and Aggeler, 1987).

5.2.1 ECM Composition

The ECM is composed of a complex combination of proteins including structural molecules such as collagen, fibronectin, and laminin, as well as functional molecules such as glycosaminoglycans (GAGs) and growth factors all arranged in a tissue-specific three-dimensional ultrastructure (Badylak, 2005). In addition to molecules present within the native ECM, low molecular weight "matricryptic" peptides are liberated upon matrix proteolytic degradation that are biologically active yet distinct from native parent molecules.

5.2.1.1 Collagen

Collagen is the most abundant protein within mammalian ECM and comprises more than 90% of its dry weight (Van Der Rest and Garrone, 1991). More than 25 distinct types of collagen have been identified, each with

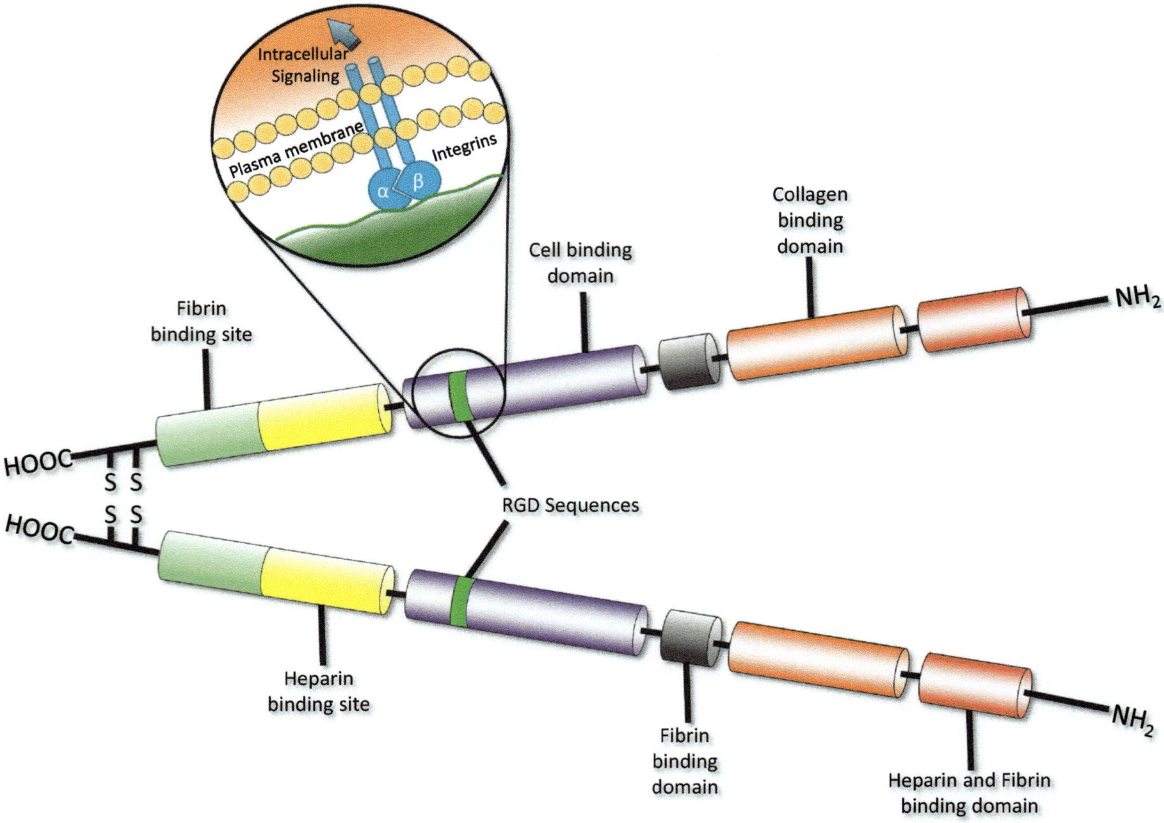

FIGURE 5.2
Fibronectin is a dimeric molecule joined by two disulfide bonds at the carboxyl end. Each domain of the fibronectin molecule has binding sites for cell receptors and ECM molecules.

a unique biologic function. Type I collagen is the most abundant structural protein present within tissues and is highly conserved across both animal and plant kingdoms.

Large quantities of Type I collagen are present within the ECM of tendons and ligaments, providing the necessary uniaxial and multiaxial mechanical strength required to meet the demands of these connective tissues.

Other collagen types are present in the ECM of most tissues but typically in much lower quantities. These collagen types provide tissue-specific physical and mechanical characteristics and allow for tissue-specific cell–ECM interactions. For example, less rigid, type III collagen is present in the submucosal layers of tissues that require increased flexibility and compliance such as the urinary bladder (Barnard and Gathercole, 1991). Type IV collagen has ligand

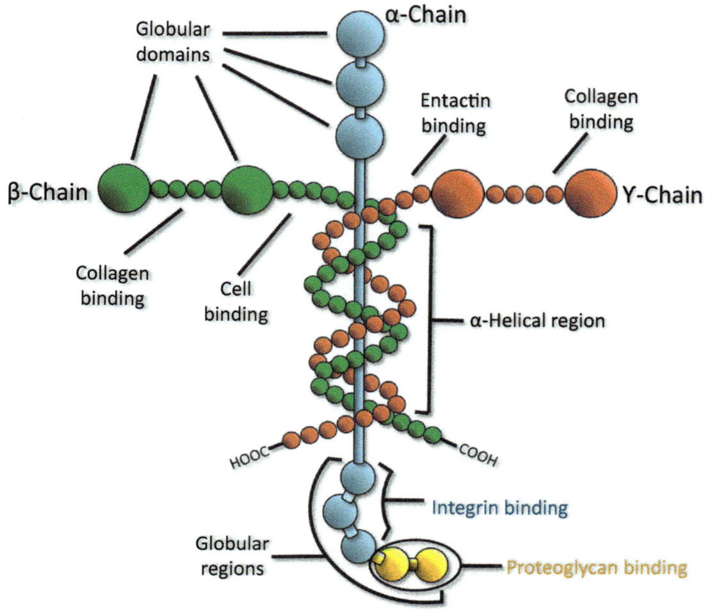

FIGURE 5.3
Laminin is composed of three polypeptide chains (α, β1, and β2) organized into the shape of a cross. Laminin has binding domains for heparin, collagen IV, heparin sulfate, and cells.

affinity for endothelial cells and is present within the basement membrane of most vascular structures (Yurchenco and O'rear, 1994). Type VII collagen is present within the basement membrane of the epidermis and facilitates fibril anchoring to protect overlying keratinocytes from shear stress (Yurchenco and O'rear, 1994). Each of these types of collagen is present within most of the ECM bioscaffold materials used for constructive tissue remodeling.

5.2.1.2 Fibronectin

Fibronectin is the second most abundant ECM molecule present in both soluble and tissue isoforms in submucosal, basement membranes, and interstitial tissues (Hodde et al., 2002). The fibronectin molecule is rich in ligands that promote adhesion of diverse cell types including the Arg-Gly-Asp (RGD) subunit which facilitates cell–ECM adhesion via the alpha5 beta1 integrin. Because of these characteristics, fibronectin has been used as a coating for synthetic scaffold materials to promote host biocompatibility.

5.2.1.3 Laminin

Laminin is another adhesion molecule found in the ECM, especially within basement membranes. This trimeric cross-linked polypeptide exists in numerous forms depending upon the combination of peptide chains. Laminin has

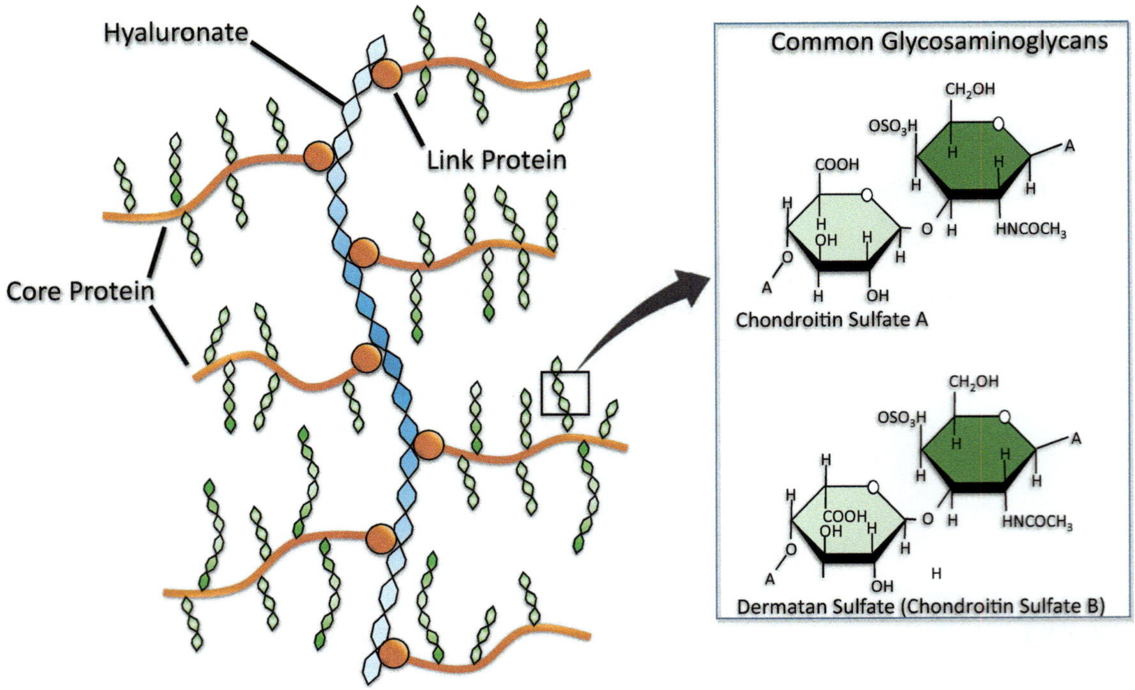

FIGURE 5.4
Glycosaminoglycans are long unbranched polysaccharides that are composed of a repeating disaccharide unit. The disaccharide unit is composed of one of two modified sugars, either *N*-acetylgalactosamine or *N*-acetylglucosamine.

been shown to play a prominent role in the maintenance of vascular structures and mesenchymal stem cell–ECM interactions (Ponce et al., 1999).

5.2.1.4 Glycosaminoglycans

GAGs are long unbranched polysaccharides that are composed of a repeating disaccharide unit. GAGs possess a variety of biologic activities including the ability to bind growth factors and cytokines and promote water retention. The GAGs present in ECM include heparin, heparan sulfate, chondroitin sulfate A and B, and hyaluronic acid (HA). HA retains significant biologic activity and directly promotes cell proliferation, migration, and differentiation (Hodde et al., 1996). The concentration of HA within ECM is highest in fetal and newborn tissues and is therefore associated with enhanced healing properties.

5.2.1.5 Growth Factors

Intact ECM is composed of a combination of sequestered biologically active molecules present in a unique and natural spatial distribution. Although some cytokines and growth factors are present within ECM in low concentrations,

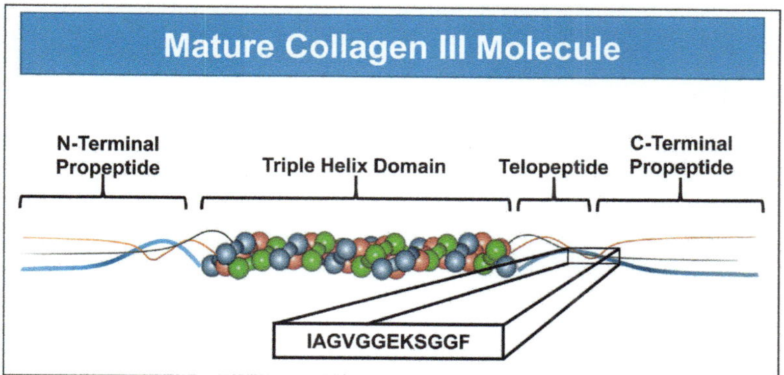

Species	Sequence	GenBank Accession	GeneID
Peptide	IAGVGGEKSGGF	-	-
Rat	IAGVGGEKSGGF	AAA40942	84032
Rabbit	IAGVGGEKSGGF	XP_002712379	100009177
Mouse	IAGVGGEKSGGF	AAH28248	12825
Human	IAGIGGEKAGGF	AAH28178	1281
Monkey	IAGVGGEKAGGF	XP_002712379	100009177
Chimpanzee	IAGIGGEKAGGF	XP_001163292	459815
Orangutan	IAGIGGEKAGGY	XP_002812713	100457897
Bovine	IAGVGAEKAGGF	DAA32863	510833

FIGURE 5.5

Peptides of urinary bladder matrix (UBM) were fractionated via ammonium sulfate precipitation and subsequent size exclusion, ion exchange, and reverse phase chromatography. Analysis showed that a chemotactic fraction consisted predominantly of a single peptide with the amino acid sequence IAGVGGEKSGGF, a sequence identical to a string of amino acids form the C-terminal telopeptide of collagen IIIα. A BLAST search for the isolated peptide sequence showed over 75% homology with the Collagen IIIα molecule over eight separate species.

they act as potent modulators of cell behavior. The list of growth factors found within ECM is extensive and includes vascular endothelial growth factor (VEGF), the fibroblast growth factor (FGF) family, stromal-derived growth factor (SDF-1), epithelial cell growth factor (EGF), transforming growth factor B (TGF-B), keratinocyte growth factor (KGF), hepatocyte growth factor (HGF), platelet-derived growth factor (PDGF), and bone morphogenetic protein (BMP), among others (Voytik-Harbin et al., 1997; Hodde et al., 2001; Badylak et al., 2009).

Therapies delivering enriched and purified forms of growth factors, such as VEGF to promote angiogenesis or BMP to promote bone formation, have been associated with limitations including dosage and route of administration, among others. An advantage of utilizing the ECM in its intact state as

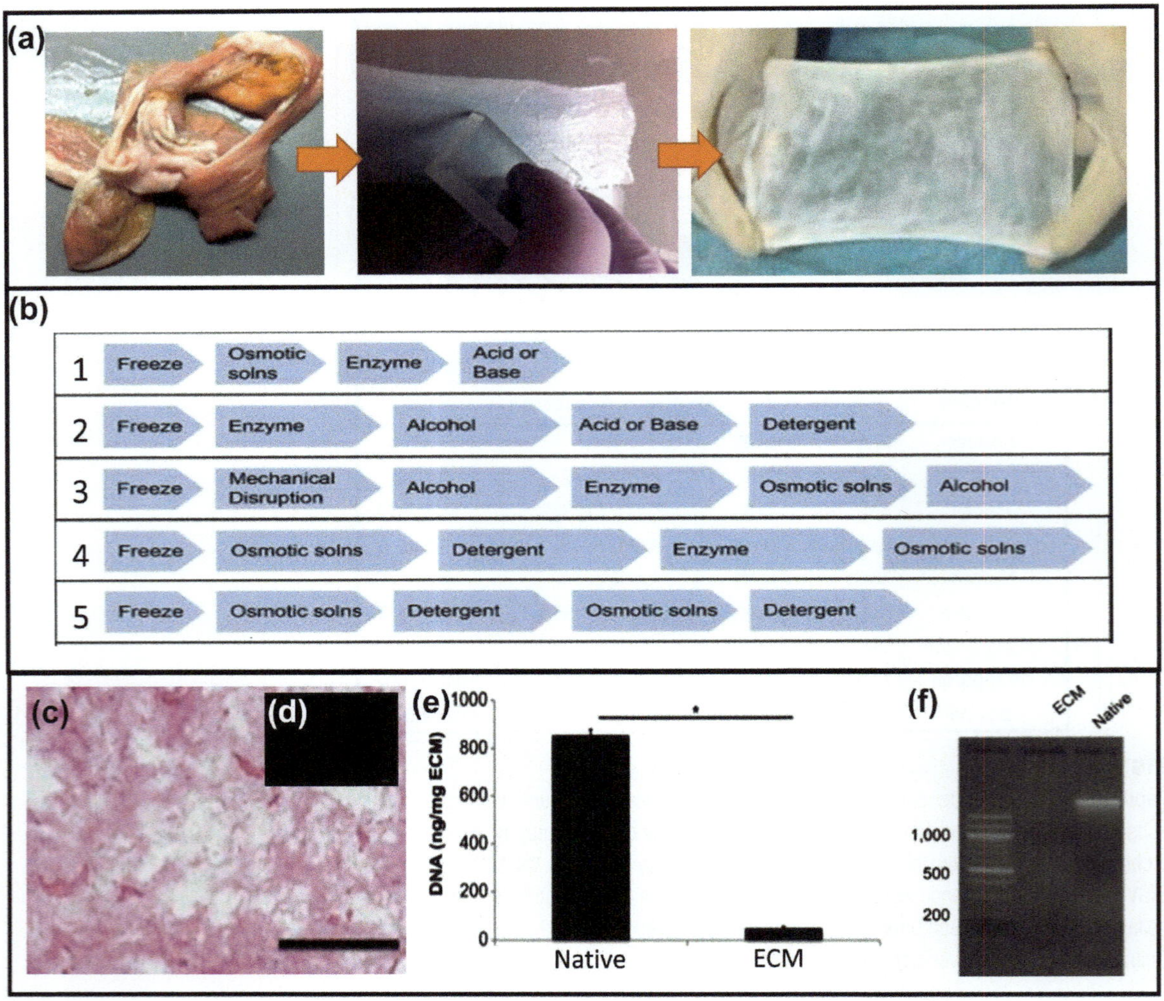

FIGURE 5.6

(a) Urinary bladder matrix (UBM) is mechanically decellularized by scarping and chemically decellularized by peracetic acid. The resultant acellular material is white in appearance. (b) Examples of decellularization protocols for (1) thin laminates such as pericardium, (2) thicker laminates such as dermis, (3) fatty, amorphous tissues such as adipose, (4) composite tissues or whole simple organs such as trachea, and (5) whole vital organs such as liver. Arrow lengths represent relative exposure times for each processing step. The criteria for efficient scaffold material decellularization suggests the following. (c) The decellularized material should lack the presence on nuclei upon H&E histologic and (d) 4,6-diamidino-2-phenylindole (DAPI) staining. (e) The material should contain less than 50 ng/mg dsDNA dry weight. (f) Remaining DNA fragments should be less than 200 bp in length. *Figure adapted from Crapo et al. (2011).*

a scaffold for tissue engineering is the presence of all the attendant growth factors in the same relative amounts and three-dimensional ultrastructure that exist in nature. The ECM protects these growth factors from degradation and efficiently presents them to resident or migrating cells. Because the ECM is

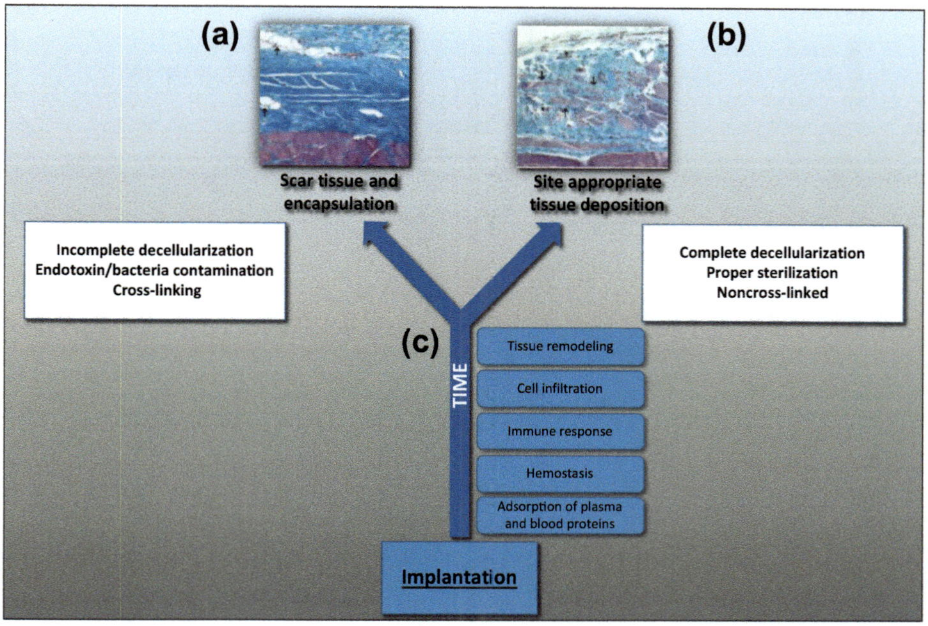

FIGURE 5.7

The remodeling outcome associated with biologic scaffolds depends upon a variety of factors including the effectiveness of the scaffold decellularization process (Keane et al., 2012), and whether or not the scaffold material has been chemically cross-linked (Valentin et al., 2006), among others. Efficient decellularization removes proinflammatory cellular material and is therefore necessary for a constructive tissue remodeling response. Conversely, cross-linking prevents scaffold degradation and as a result decreases the release of beneficial matricryptic peptides and growth factors within the ECM. (a) Poor decellularization, bacterial contamination, and/or cross-linking may induce a host response consistent with a foreign body reaction and results in scar tissue formation (blue staining: Collagen) and some adipose deposition (arrows). (b) Thoroughly decellularized and sterilized, noncross-linked scaffolds promote site-appropriate tissue remodeling (arrows: skeletal muscle fibers). (c) The scaffold remodeling process occurs through a series of events following scaffold implantation.

in a state of dynamic reciprocity with resident cells, these sequestered growth factors are liberated from the ECM in a controlled released manner dependent upon cellular activity.

5.2.1.6 Matricryptic Peptides

Biologic scaffolds composed of ECM are rapidly infiltrated by host polymorphonuclear and mononuclear cells in vivo, which initiate the scaffold degradation process. Degradation of ECM generates cryptic oligopeptides with potent in vitro and in vivo bioactivity not present in the parent ECM molecules. These "matricryptic" peptides have shown chemotactic activity for a number of stem/progenitor cells in vitro and have been shown to induce the endogenous recruitment of host progenitor cells in vivo (Agrawal et al., 2010; Tottey et al., 2011b). A large number of these multipotential

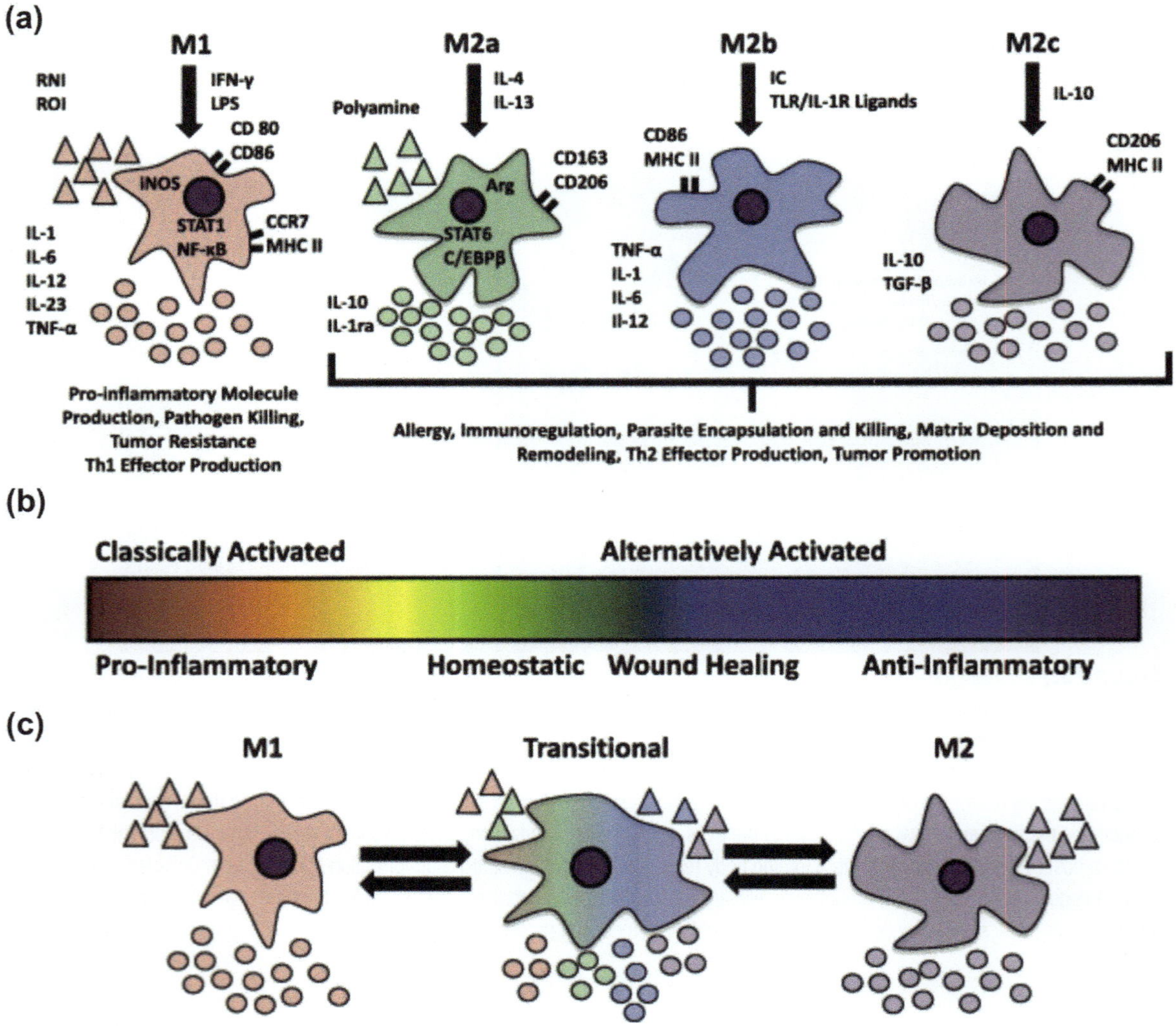

FIGURE 5.8

Macrophages have typically been recognized as phagocytic and antigen presenting cells responsible for propagating the proinflammatory response of the mammalian innate immune system. It is now accepted that macrophages are capable of diverse phenotypes and also function as regulators of tissue homeostasis during the host response to disease and tissue injury. (a) Common activators, associated surface molecules, transcription factors, effector molecules, and characteristics of recognized macrophage phenotypic subsets. (b) Macrophage polarization occurs along a spectrum between M1 and M2 extremes. (c) Macrophage phenotype is plastic and can change in response to paracrine and autocrine signals. Triangles represent secreted reactive species and circles represent secreted cytokines. *Figure from Brown and Badylak (2013).*

cell types influenced by the degradation products of ECM scaffold, both in vitro and in vivo, have the ability to direct cell phenotype. For example, a low molecular weight peptide derived through proteolytic degradation of the collagen IIIα molecule has been shown to possess osteogenic potential (Agrawal et al., 2011).

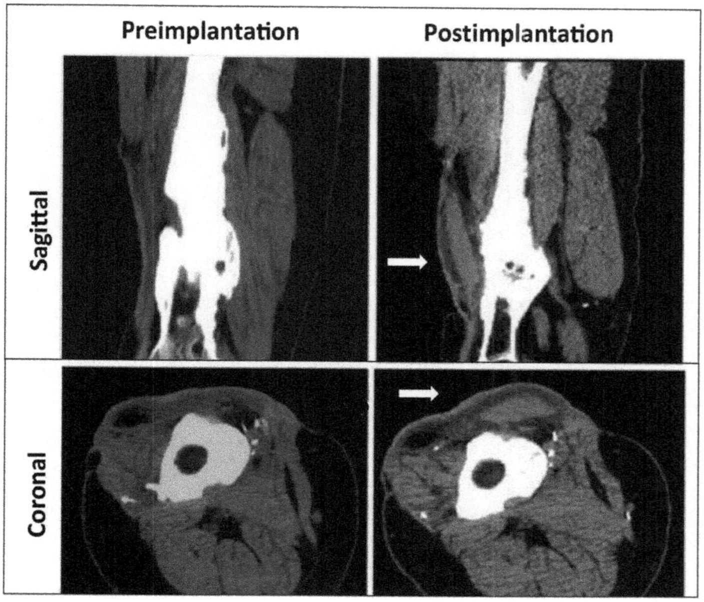

FIGURE 5.9

A marine suffering from VML of the quadriceps femoris muscle resulting from injuries sustained in battle was treated with surgical placement of a biologic scaffold composed of porcine small intestinal submucosa (SIS) ECM. CT scan showed the presence of radiopaque tissue formation (arrows), consistent with skeletal muscle, at the site of ECM bioscaffold implantation, at 4 months postsurgery. *Figure adapted from Mase et al. (2010).*

5.3 ECM SCAFFOLD PREPARATION

The preparation of an ECM scaffold derived from native mammalian tissue requires a combination of chemical and mechanical processing steps that result in efficient decellularization of the source tissue. As stated earlier, ECM scaffolds have been prepared from a variety of tissues including dermis, small intestine, urinary bladder, pericardium, liver, skeletal muscle, and adipose tissue, among others.

Most source tissue material used for derivation of ECM bioscaffolds are xenogeneic or allogeneic in nature, and therefore require the efficient removal of cellular antigens to prevent foreign body recognition by the host following in vivo implantation. Depending upon the tissue of interest, mechanical methods such as manual or automated delamination of muscle and mucosal layers is typically followed by a combination of physical, chemical, and enzymatic methods to achieve complete decellularization. The goal of any decellularization process is the removal of all cellular and nuclear material while maintaining the composition, microstructure, mechanical properties, and biologic activity of the remaining native ECM. A biologic scaffold material is considered

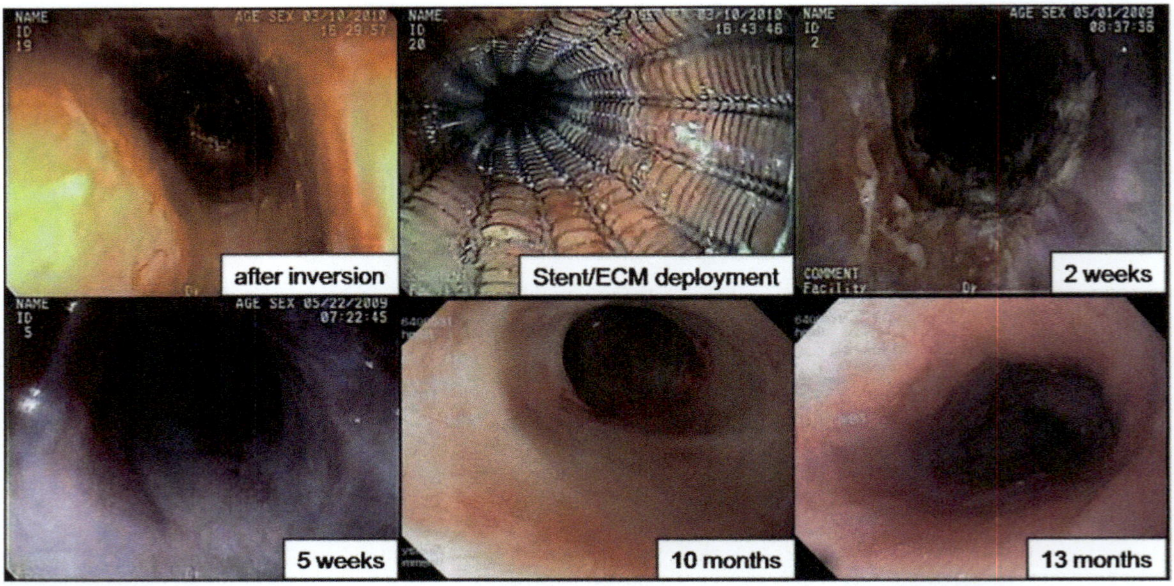

FIGURE 5.10

Representative endoscopic images of ECM bioscaffold placement in a patient suffering from superficial esophageal adenocarcinoma. Upper Row: Left, Complete circumferential, 8 cm long, muscosectomy and inversion exposed the underlying esophageal muscularis externa. Middle, Stent deployment facilitated placement and gentle compression of a tubularized ECM bioscaffold against the muscularis externa. Right, Endoscopy at 2 weeks, immediately after stent removal, showed the ECM scaffold material firmly placed against the area of resection. Bottom Row: Left, The resected area showed the presence of squamous epithelium and no ECM scaffold material was detectable after 5 weeks. Middle, The esophageal mucosa appeared normal with short segment circumferential strictures after 10 months. Right, The resected area was covered by normal esophageal epithelium without stricture formation at 13 months.
Figure from Badylak et al. (2011).

to be efficiently decellularized when it meets the following criteria: (1) lack of nuclei present within the scaffold material following histologic or chemical nuclear staining; (2) total scaffold material dsDNA content must be less than 50 ng/mg; (3) any remaining DNA remnants present in the scaffold material must be less than 200 base pairs in nucleotide length (Crapo et al., 2011). Physical treatments for decellularization include sonication, scraping and/or shaking, or subjection to freeze–thaw cycles. Such methods facilitate the disruption of cell membranes and removal of cell contents from the native ECM. Enzymatic treatments with detergents or ionic solutions further denature cell membranes and disrupt intercellular bonds. Because the ECM of different tissues varies with regard to composition, structure, and density, the combination of methods used to achieve efficient decellularization varies across tissues.

Following decellularization, ECM intended for use as a biologic scaffold must be sterilized prior to surgical placement. Sterilization of biologic materials, including ECM scaffolds, involves unique considerations such as the shrink temperature of collagen, potential cross-linkage of bioactive components, and potential changes in surface characteristics, among others. Despite these considerations, numerous methods for the terminal sterilization of ECM scaffold materials have been identified. These techniques include ionizing radiation such as gamma and electron beam irradiation, and exposure to ethelyne oxide (ETO) gas, among others.

5.4 CONSTRUCTIVE TISSUE REMODELING

5.4.1 Default Mammalian Wound Healing versus Constructive Remodeling by ECM

The default mammalian response to injury is highly conserved across tissue types and occurs in three overlapping yet distinct phases: the inflammatory phase, the proliferative phase, and the remodeling phase (Singer and Clark, 1999). The inflammatory phase occurs immediately after tissue damage and is marked by activation of the coagulation cascade to promote hemostasis and by the subsequent influx of innate inflammatory cells that facilitate the removal of cellular debris and microbicidial activity for the prevention of infection. Specifically, hemostasis is achieved through the deposition of a provisional fibrin matrix which acts as a scaffold for infiltrating neutrophils and macrophages (Clark et al., 1982). The proliferative phase is characterized by angiogenesis and the migration, proliferation, and differentiation of different cell types (Nissen et al., 1998). During the remodeling phase, unnecessary macrophages and endothelial cells undergo apoptosis while fibroblasts and myofibroblasts produce and deposit a collagenous extracellular matrix. This host-derived fibrous tissue matrix represents the precursor of scar tissue, which replaces site-appropriate functional tissue within the injury site.

In contrast, constructive tissue remodeling is a process by which the default healing response to injury is modified from one that results in scar tissue formation to one that results in the formation of site-appropriate functional tissue. Mechanisms behind an ECM scaffold-mediated constructive remodeling response to injury are discussed in the following section and include the participation of an activated host progenitor cell population and a predominant M2/Th2 immune response among others (Badylak et al., 2008; Beattie et al., 2009).

5.4.2 Mechanisms behind ECM-Mediated Constructive Tissue Remodeling

While the exact mechanisms behind the ability of ECM-based bioscaffolds to promote constructive tissue remodeling remain only partially understood, there are a number of processes which are believed to be key contributors. Among these processes are the ability of the scaffold material to degrade rapidly with concomitant release of bioactive matricryptic peptides; the ability of the scaffold material to recruit and activate stem/progenitor cells; and the ability of the scaffold material to elicit a immunomodulatory host response which does not result in a foreign body response or immunologic rejection of the implant. Each of these processes is discussed in the following section, and explored in further detail in the studies which comprise the present chapter.

5.4.2.1 Scaffold Degradation

Recent studies suggest an important role for ECM-derived biologically active molecules, including growth factors and matricryptic peptides as discussed earlier, in the tissue response to injury. The release of ECM-derived bioactive molecules is dependent upon ECM degradation. When used as a surgically placed biologic scaffold material, ECM degradation is a critical and necessary event for constructive tissue remodeling. Macrophages, in particular, are key facilitators of ECM bioscaffold degradation, and their suppression prevents the degradation of ECM scaffolds (Valentin et al., 2009). Quantitative studies of 14C-labeled ECM scaffolds show that approximately 50% of the ECM scaffold is degraded after 28 days and virtually all of the ECM is replaced with host tissue by 60–90 days postimplantation (Gilbert et al., 2007), a process partially dependent upon the anatomic site of placement. Chemically cross-linked ECM scaffolds that are unable to be degraded fail to induce constructive remodeling effects and, in contrast, promote a foreign body response. Bioactive molecules released upon ECM scaffold degradation have been shown to direct stem/progenitor cell chemotaxis, mitogenesis, and influence cell fate. These molecules have also been shown to modulate the host immune response and promote antimicrobial activity.

5.4.2.2 Endogenous Cell Therapy by ECM Bioscaffolds

Degradation of the ECM by physical and chemical methods creates low molecular weight peptides with chemoattractant activity. These peptides, derived from ECM, range in size from 5 to 16 kDa and have demonstrated chemotactic activity for primary endothelial cells (Li et al., 2004), perivascular stem cells (Crisan et al., 2008), and neural progenitor cells (Crapo et al., 2014), among others.

Primitive $Sox2^{+}$ cells were shown to be present at the site of ECM implantation in an adult mammalian model of digit injury (Agrawal et al., 2010, 2012), and multipotent bone marrow-derived cells were shown to participate in the remodeling of ECM scaffold-mediated Achilles tendon repair (Zantop et al., 2006). Specifically, in the preclinical Achilles tendon injury model, ECM scaffold explants promoted chemotaxis of progenitor cells after 3, 7, and 14 days of in vivo remodeling (Beattie et al., 2009). The results of the study showed greater migration of progenitor cells toward tendons repaired with ECM scaffolds compared to tendons repaired with autologous tissue and uninjured normal contralateral tendon. These results support the concept that ECM scaffolds are capable of recruiting progenitor cells to the site of implantation to participate in constructive tissue remodeling over the long term.

5.4.2.3 Modulation of the Host Immune Response by ECM Bioscaffolds

The host innate immune system and specifically macrophages play a pivotal role in the host response to ECM bioscaffold implantation. Macrophages have been shown to respond to many different implanted biomaterials, including those composed of metals (Takebe et al., 2003), ceramics (Lu et al., 2002), polymers (Labow et al., 2005), and biologic proteins such as collagens and xenogeneic ECM (Valentin et al., 2009). Macrophage-mediated foreign body giant cell formation and proinflammatory cytokine production are frequently associated with the implantation of nondegradable or synthetic biomaterials. However, several recent studies suggest that immunomodulatory macrophages can facilitate constructive and site-appropriate tissue remodeling in response to ECM bioscaffold implantation (Valentin et al., 2009; Sicari et al., 2012).

5.4.2.3.1 Macrophage Heterogeneity

Macrophages have been characterized as mononuclear phagocytes responsible for microbicidal activity and antigen presentation, the removal of cellular debris following acute tissue injury, and propagation of a type-I proinflammatory immune response (Kedzierska et al., 2003; Savill et al., 1989). Macrophages are now recognized as a heterogeneous cell population capable of immunomodulation and tissue repair (Brown and Badylak, 2013; Mantovani et al., 2013; Martinez et al., 2008; Mosser and Edwards, 2008). Specifically, following the Th1/Th2 nomenclature, macrophages have been categorized according to their functional properties as either M1 or M2 (Mantovani et al., 2004). M1 macrophages are proinflammatory or "classically activated" cells stimulated by proinflammatory factors (e.g., IFNγ, LPS). M1-activated macrophages secrete toxic reactive oxygen species, nitric oxygen intermediates, and inflammatory cytokines such as IL-1β, IL-6, and TNF-α. These cells are responsible for microbicidal activity, debris clearance, and the propagation of a proinflammatory and Th1 response. Conversely, "alternatively activated" and constructive M2 macrophages are stimulated by immunomodulatory and antiinflammatory factors (e.g., IL-4, IL-10). M2-activated macrophages express IL-10, have high levels of scavenger and mannose receptors, and produce arginase in the place of inducible nitric oxide synthase (iNOS), subsequently producing ornithine and polyamines. These cells are responsible for propagation of an immunomodulatory or Th2 response, matrix deposition, and tissue repair and reconstruction (Mantovani et al., 2005). Most relevant to the present chapter is the phenomenon of surgically placed ECM bioscaffolds to be associated with the constructive M2 macrophage phenotype. Moreover, the phenotype of responding macrophages has been found to be an important determinant factor in the success of an implanted biomaterial scaffold and its ultimate remodeling outcome (Badylak et al., 2008; Brown et al., 2009).

5.4.2.3.2 ECM Bioscaffolds Promote a Constructive Macrophage Phenotype

ECM scaffold materials which promoted a downstream constructive tissue remodeling outcomes were associated with a predominant M2 macrophage phenotype during the initial host innate immune response. Specifically, studies have shown that ECM bioscaffolds promote a transition from a predominantly M1 macrophage cell population immediately following surgical placement to a predominant M2 cell population by 7–14 days postimplantation (Badylak et al., 2008; Brown et al., 2009). As stated earlier, the phenotype of host macrophages which respond to biologic scaffold materials upon implantation has been shown to serve as a statistical predictor of the resultant tissue remodeling outcome. Specifically, synthetic scaffold materials (i.e., polypropylene mesh) as well as chemically cross-linked ECM scaffold materials, which do not degrade and are therefore unable to liberate bioactive matricryptic peptides, are associated with a predominant M1 macrophage phenotype at early and later time points and ultimately result in a chronic inflammatory response. Conversely, noncross-linked ECM bioscaffolds, capable of degradation and liberation of bioactive cryptic peptides, are associated with an early M1 phenotype, which rapidly transitions to a predominant M2 macrophage phenotype and results in constructive remodeling including the deposition of site-appropriate and functional tissue. These results suggest that interactions of host cells with degradation products from ECM bioscaffolds may affect macrophage phenotype. While the exact bioactive component(s) of the matrix responsible for this phenomenon remains unknown, a better understanding of macrophage–scaffold interaction has potential to affect future biomaterial design and application.

5.4.2.4 Antimicrobial Properties of ECM Bioscaffolds

Naturally occurring antimicrobial peptides are commonplace within both the plant and animal kingdoms. It has been shown that components of ECM scaffolds contain antibacterial activity against both Gram negative and positive bacteria (Sarikaya et al., 2002). Furthermore, this antibacterial activity is conserved across several tissue types. Specifically, degradation products from ECM scaffold materials derived from small intestine, bladder, and liver showed bactericidal activity and that degradation of the scaffold material is necessary for the inhibition of bacterial growth (Brennan et al., 2006).

Scaffolds composed of ECM are resistant to bacterial infections in both preclinical and clinical applications. This resistance is evident in clinical applications at a high risk for bacterial infection as well as in deliberate infection preclinical models (Badylak et al., 2003). In one study, dogs were subjected to segmented aortic replacement with either a synthetic graft or with an ECM scaffold, and were deliberately challenged with *Staphyloccus aureus* at the time of implantation. Dogs implanted with the synthetic graft had positive bacterial cultures and showed the presence of persistent infection after 30 days. However, dogs implanted with ECM had negative bacterial cultures and were free of an adverse inflammatory response after 30 days (Badylak et al., 1994).

The acute antibacterial effects of ECM degradation may prevent immediate postimplantation infection before the host innate immune system becomes activated. Additionally, as the ECM continues to degrade over time, the continued release of bioactive molecules may provide sustained antibacterial effects.

5.5 CLINICAL TRANSLATION OF ECM BIOSCAFFOLDS

As stated earlier, ECM bioscaffolds have been surgically implanted in more than 3 million patients during the past decade. ECM scaffolds are decellularized, disinfected, and terminally sterilized according to strict QC and GMP standards and are most commonly implemented to promote tissue ingrowth for hernia repair, pelvic floor reinforcement, and tendon reconstruction.

5.5.1 Skeletal Muscle Reconstruction

Adult mammalian skeletal muscle tissue retains remarkable regenerative capacity following minor trauma such as exercise-induced injury. Specialized skeletal muscle progenitor cells called satellite cells are responsible for this regenerative plasticity. Following injury, signals from the injured skeletal muscle microenvironment stimulate resident satellite cells to proliferate and give rise to myoblasts, which go on to fuse and form contractile skeletal muscle myofibers. In the case of minor injury, these regenerated myofibers are able to recapitulate the injured skeletal muscle tissue.

When penetrating soft tissue injuries result in a massive loss of skeletal muscle tissue, the resulting defect is unable to be compensated for by the regenerative response of skeletal muscle. Such devastating trauma, referred to as volumetric muscle loss (VML), is commonplace on modern battlefields and results in significant cosmetic and functional deficits for wounded soldiers. Limited treatment options for VML, including autologous tissue transfer, are associated with complications such as donor site morbidity and failure of graft integration. The constructive tissue remodeling associated with surgically placed ECM bioscaffolds suggests a potential therapeutic application for VML treatment.

In vitro studies showed degradation products from ECM scaffold materials were chemotactic for myogenic skeletal muscle myoblasts and perivascular stem cells (Wolf et al., 2012; Crisan et al., 2008). In a mouse model, surgical placement of an ECM bioscaffold within sites of VML promoted the formation of islands of skeletal muscle cells after 56 days. Furthermore, in a preclinical canine model, an ECM bioscaffold remodeled into contractile skeletal muscle tissue after 6 months when placed within a VML defect affecting gastrocnemius skeletal muscle and associated Achilles tendon musculotendinous tissue (see State-of-the-Art Experiment textbox).

A recent clinical case study highlighted the ability of an acellular ECM scaffold to promote skeletal muscle constructive remodeling within a wounded soldier suffering from VML of the quadriceps femoris muscle (Mase et al., 2010). At 16 weeks postimplantation, the patient showed a 30% increase in muscle function that was concomitant with the presence of soft tissue consistent with skeletal muscle on computed tomography (CT) scan.

Furthermore, a clinical study is currently in progress, which evaluates the effect of ECM implantation at the site of VML in 80 patients. The outcome measures of this trial include mechanical strength and function, the cell populations involved in the process, quality of life in these patients, as well as to examine the cellular composition of the remodeled tissue. Preliminary results show the presence of multipotent perivascular cells, myosin heavy chain + skeletal muscle cells, neovascularization, and regenerating skeletal muscle tissue widely distributed throughout the remodeling ECM. Furthermore, at 24 weeks postoperatively these patients have exhibited at least a 25% increase in strength compared to presurgical values (Unpublished data). Taken together, these results demonstrate the potential clinical efficacy of ECM scaffolds in promoting the recruitment of perivascular progenitor cells and facilitating restoration of structure and function in volumetric skeletal muscle defects.

5.5.2 Esophageal Mucosa Reconstruction

The default response to injury in the esophagus is characterized by inflammation resulting in rapid, robust scar tissue deposition with clinical stricture. Recent advancements in therapies including stepwise radiofrequency ablation and radical endoscopic resection require numerous interventions and can lead to high incidence of metachronous lesions (Shaheen et al., 2009). As a result, the standard of care for many esophageal diseases, especially overt cancer and its precursor Barrett's disease with high grade dysplasia (HGD), involves watchful waiting and eventually performing radical esophagectomy, a procedure associated with complication rates that approach 50% (Orringer et al., 2001).

Regenerative medicine strategies that utilize scaffold-based approaches to tissue repair provide a viable alternative for esophageal tissue replacement. Early results of a bioactive ECM scaffold-based approach have been promising. In a preclinical dog model, full thickness esophageal defects that included approximately 40–50% of the circumference and 5 cm of length can be repaired with ECM facilitating a constructive, nonstenotic healing response with formation of all layers of the esophageal wall (Badylak et al., 2000). However, when the same scaffolds are used to attempt reconstruction of complete circumferential full thickness defects, occurrence of severe clinical stricture was uniformly observed (Badylak et al., 2005). When the complete circumferential defects do not include the full thickness of the esophagus and the lesions are limited to the mucosa, then placement of an ECM scaffold upon the subjacent muscularis externa promotes the constructive remodeling of a functional mucosa without clinical stricture (Nieponice et al., 2009).

These studies, during the past 12 years, have shown that the surgical placement of xenogeneic ECM bioscaffolds can modulate the default esophageal injury response toward a constructive remodeling outcome (Nieponice et al., 2009). Results from these preclinical studies laid the foundation for the successful clinical translation of this regenerative medicine approach in five patients with adenocarcinoma. These patients were nonsurgical candidates as a result of comorbidities. Complete circumferential mucosectomy of the diseased tissues in these patients followed by placement of an SIS-ECM bioscaffold induced the formation of a neomucosa free of disease (Badylak et al., 2011). This regenerative medicine approach has a potential to change the practice of medicine for patients affected with esophageal adenocarcinoma, the cancer type showing the greatest rate of increase in recent years (Enzinger and Mayer, 2003).

5.6 COMMERCIALLY AVAILABLE SCAFFOLDS COMPOSED OF ECM

Scaffold materials composed of ECM are currently being used in several tissue engineering applications. In addition to the clinical studies discussed earlier, ECM scaffolds are being used to treat chronic dermal wounds such as nonhealing diabetic ulcers, and for the reconstruction and reinforcement of abdominal wall, pelvic floor, and orthopedic musculotendinous tissue (Table 5.1).

Table 5.1 Commercially Available Products Composed of Intact Extracellular Matrix

Product	Manufacturer	Source Tissue
AlloDerm	Lifecell	Human skin
AlloPatch®	Musculoskeletal Transplant Foundation	Human fascia lata
Axis™ dermis	Mentor	Human dermis
Bard® Dermal Allograft	Bard	Cadaveric human dermis
CopiOs®	Zimmer Inc.	Bovine pericardium
CorMatrix ECM™	CorMatrix Cardiovascular	Porcine small intestine
CuffPatch™	Arthrotek	Porcine small intestinal submucosa (SIS)
DurADAPT™	Pegasus Biologicals	Horse pericardium
Dura-Guard®	Synovis Surgical	Bovine pericardium
Durasis®	Cook SIS	Porcine small intestinal submucosa (SIS)
Durepair®	TEI Biosciences	Fetal bovine skin
Epic™	St. Jude Medical Inc.	Porcine heart valve

Continued...

Table 5.1 Commercially Available Products Composed of Intact Extracellular Matrix *Continued*

Product	Manufacturer	Source Tissue
FasLata®	Bard	Cadaveric fascia lata
FortaFlex®	Organogenesis Inc.	Porcine small intestine
Freestyle®	Medtronic Inc.	Porcine heart valve
Graft Jacket®	Wright Medical Tech.	Human skin
Hancock® II	Medtronic Inc.	Porcine heart valve
IOPatch™	IOP Inc.	Human pericardium
Matristem®	Acell	Porcine urinary bladder
Meso Biomatrix™	Kensey Nash Corp.	Porcine mesothelium
Mosaic®	Medtronic Inc.	Porcine heart valve
NeoForm™	Mentor Worldwide LLC	Human dermis
Oasis®	Healthpoint	Porcine small intestinal submucosa (SIS)
OrthADAPT™	Pegasus Biologicals	Horse pericardium
Pelvicol®	Bard	Porcine dermis
Peri-Guard®	Synovis Surgical	Bovine pericardium
Permacol™	Tissue Science Laboratories	Porcine skin
PriMatrix™	TEI Biosciences	Fetal bovine skin
Prima™ Plus	Edwards Lifesciences LLC	Porcine heart valve
Restore™	DePuy	Porcine small intestinal submucosa (SIS)
SJM Biocor®	St. Jude Medical Inc.	Porcine heart valve
Stratasis	Cook SIS	Porcine small intestinal submucosa (SIS)
Strattice™	Lifecell Corp.	Porcine dermis
SurgiMend™	TEI Biosciences	Fetal bovine skin
Surgisis®	Cook SIS	Porcine small intestinal submucosa (SIS)
TissueMend®	TEI Biosciences	Fetal bovine skin
Veritas®	Synovis Surgical	Bovine pericardium
Xelma™	Molnlycke	ECM protein, PGA, water
Xenform™	TEI Biosciences	Fetal bovine skin
Zimmer Collagen Patch	Tissue Science Laboratories	Porcine dermis

5.7 FUTURE CONSIDERATIONS

Tissue engineering and regenerative medicine strategies include the use of cell-based therapies, scaffold-based approaches, and/or the use of bioactive molecules. Optimal approaches will undoubtedly differ for each clinical application. However, it is clear that ECM-based biologic scaffolds have a prominent place in the regenerative medicine armamentarium of therapeutic applications. The recent use of three-dimensional scaffolds composed of whole organs, which have been decellularized, represents a potential major advance for patients with end stage organ failure (Badylak et al., 2012). In summary, by taking advantage of Mother Nature's ideal scaffold, ECM, tissue engineering, and regenerative medicine approaches can have an immediate impact upon the practice of medicine.

5.8 SUMMARY

- Scaffolds composed of ECM are useful in many tissue engineering applications.
- The ECM is composed of structural and functional molecules arranged in a tissue-specific three-dimensional structure.
- The main components of the ECM are collagen, fibronectin, laminin, and GAGs.
- Following isolation from various tissues, the ECM maintains structural and functional proteins that support constructive tissue remodeling.
- Constructive tissue remodeling is distinct from the default scar forming injury response and results in the formation of site-appropriate and functional tissue.
- ECM bioscaffold-mediated constructive tissue remodeling is dependent upon scaffold degradation.
- ECM bioscaffolds are associated with endogenous cell therapy including the recruitment of progenitor cells and modulation of the innate immune response. Specifically, ECM scaffolds are associated with a constructive M2 macrophage phenotype.
- ECM bioscaffolds promote constructive remodeling of soft tissues and have been translated from preclinical models to clinical implementation in patients suffering from VML and esophageal adenocarcinoma.

CLASSICAL EXPERIMENT

An assay for the identification of chemotaxis induced by degradation products from extracellular matrix bioscaffolds.

Tissue engineering strategies have utilized the transplantation of exogenous stem/progenitor cells, referred to as cell therapy, for the treatment of injured or missing tissues. These strategies have been associated with limitations including immunologic rejection of the cells and failure of the cells to incorporate within host tissue. ECM scaffold materials have been shown to release cryptic bioactive molecules upon their degradation in vivo. These matricryptic peptides recruit endogenous host stem/progenitor cells to sites of ECM bioscaffold surgical placement. The implementation of acellular bioscaffolds composed of ECM may obviate the limitations associated with exogenously delivered cell therapy by recruiting endogenous host-derived stem/progenitor cells to sites of tissue injury. The present study uses a chemotaxis chamber assay to examine the ability of degradation products from ECM bioscaffolds to recruit perivascular stem cells (PVSC).

Biologic scaffolds composed of porcine urinary bladder matrix (UBM) were digested as previously reported (Freytes et al., 2008), resulting in the creation of bioactive degradation products including matricryptic peptides. The potential chemotaxis of PVSCs toward degradation products of UBM was examined using a chemotaxis chamber assay as previously described (Agrawal et al., 2010). Briefly, UBM digests were placed in the bottom wells of a chemotaxis chamber while PVSCs were placed in the top. A membrane containing 8 μm sized pores, which trapped migrating cells, separated UBM from cells. Cells were allowed to migrate for 3 h at 37 °C. At the end of the experiment, nonmigrated cells are washed from the membrane while migrated cells are counted and chemotaxis is quantified. The assay showed that degradation products from biologic scaffolds composed of UBM are chemotactic for multipotent PVSCs and that low oxygen conditions potentiated this effect.

Continued...

CLASSICAL EXPERIMENT *Continued*

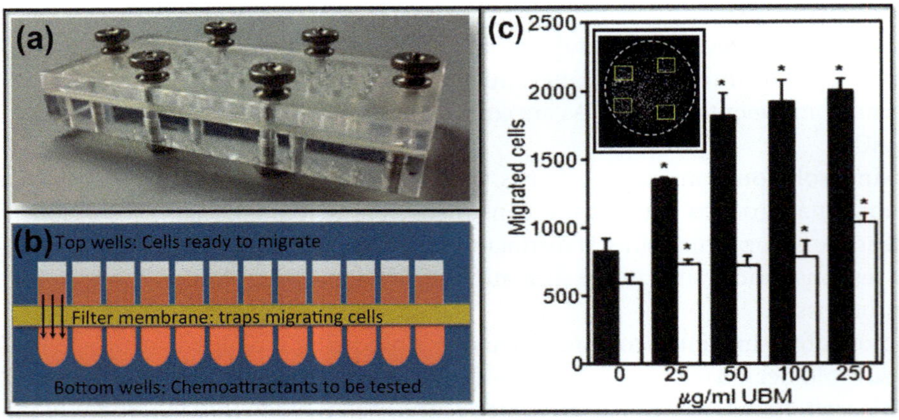

(a) The present study utilized a chemotaxis chamber. (b) PVSCs were placed in the top wells while degradation products from UBM were placed in the bottom wells. (c) Migrated cells were trapped in an 8 μm pore membrane and stained with a nuclear stain for counting and subsequent quantification. (d) The present study showed that degradation products from UBM promoted the chemotaxis of PVSCs in a dose-dependent manner (white bars), and that low oxygen conditions (black bars) potentiated this effect (Tottey et al., 2011a). *Figure adapted from Tottey et al. (2011a).*

STATE-OF-THE-ART EXPERIMENT

The massive loss of skeletal muscle tissue, with inability to regenerate, resulting from traumatic penetrating soft tissue injury is referred to as volumetric muscle loss (VML). Biologic scaffolds composed of extracellular matrix (ECM) have been shown to recruit stem/progenitor cells with myogenic potential and promote site-appropriate formation of muscle tissue in several tissue engineering applications. The present study examines the ability of ECM bioscaffolds to promote musculotendinous junction (MTJ) reconstruction in a canine model of MTJ VML injury (Turner et al., 2010).

Porcine small intestinal submucosa (SIS) was prepared as previously described. The scaffold material was then milled to produce particulate SIS, which was vacuum pressed and encapsulated between SIS sheets for formation of a pillow construct.

Eleven female dogs were subjected to a model of VML injury consisting of unilateral resection of the distal third of the gastrocnemius muscle and proximal half of the associated Achilles tendon bundle. The resected tissue was replaced with the SIS-ECM pillow construct and secured by nondegradable marker sutures placed at the proximal and distal anastomoses. The time points for evaluation were 2, 4, and 6 months. The study showed that vascularized, innervated skeletal muscle, which was similar to normal muscle tissue, had formed within the ECM bioscaffold implantation site at 6 months postsurgery. Furthermore, this de novo skeletal muscle tissue generated 48% of the contractile force of native contralateral musculotendinous junction.

STATE-OF-THE-ART EXPERIMENT *Continued*

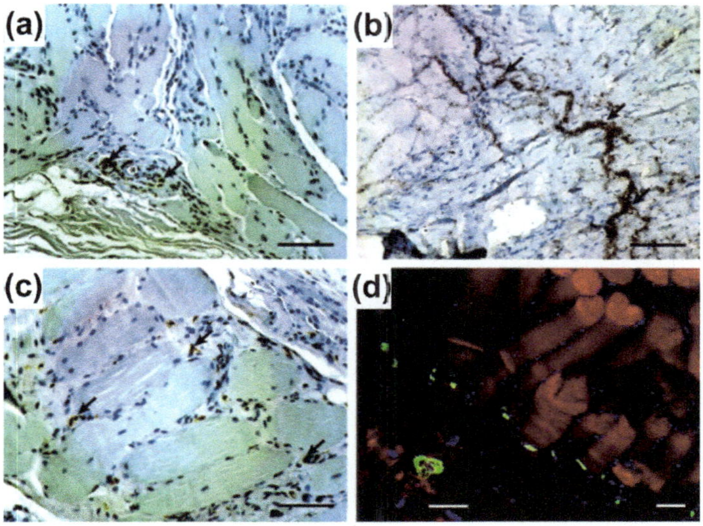

Remodeled ECM bioscaffolds are innervated. (a) Tissue sections from the scaffold implantation site showed β-III tubulin positive neurons (brown staining, arrows) in close proximity to blood vessels at 2 months postsurgery. (b) Neurons could be identified extending toward de novo skeletal muscle tissue after 4 months. (c) Nerves could be identified in close proximity to new skeletal muscle fibers after 6 months. (d) α-Bungarotoxin positive motor end plates (green staining) were shown within the remodeled skeletal muscle tissue (red staining) at 6 months postsurgery. *Figure from Turner et al. (2010).*

REFERENCES

Agrawal, V., Johnson, S.A., Reing, J., Zhang, L., Tottey, S., Wang, G., Hirschi, K.K., Braunhut, S., Gudas, L.J., Badylak, S.F., 2010. Epimorphic regeneration approach to tissue replacement in adult mammals. Proc. Natl. Acad. Sci. USA 107, 3351–3355.

Agrawal, V., Siu, B.F., Chao, H., Hirschi, K.K., Raborn, E., Johnson, S.A., Tottey, S., Hurley, K.B., Medberry, C.J., Badylak, S.F., 2012. Partial characterization of the Sox2+ cell population in an adult murine model of digit amputation. Tissue Eng. Part A 18, 1454–1463.

Agrawal, V., Tottey, S., Johnson, S.A., Freund, J.M., Siu, B.F., Badylak, S.F., 2011. Recruitment of progenitor cells by an extracellular matrix cryptic peptide in a mouse model of digit amputation. Tissue Eng. Part A 17, 2435–2443.

An, H.S., Simpson, J.M., Glover, J.M., Stephany, J., 1995 Oct 15. Comparison between allograft plus demineralized bone matrix versus autograft in anterior cervical fusion: a prospective multicenter study. Spine 20 (20), 2211–2216. PubMed PMID: 8545714.

Badylak, S.F., 2005. Regenerative medicine and developmental biology: the role of the extracellular matrix. Anat. Rec. B New Anat. 287, 36–41.

Badylak, S.F., Coffey, A.C., Lantz, G.C., Tacker, W.A., Geddes, L.A., 1994. Comparison of the resistance to infection of intestinal submucosa arterial autografts versus polytetrafluoroethylene arterial prostheses in a dog model. J. Vasc. Surg. 19, 465–472.

Badylak, S.F., Freytes, D.O., Gilbert, T.W., 2009. Extracellular matrix as a biological scaffold material: structure and function. Acta Biomater. 5, 1–13.

Badylak, S.F., Hoppo, T., Nieponice, A., Gilbert, T.W., Davison, J.M., Jobe, B.A., 2011. Esophageal preservation in five male patients after endoscopic inner-layer circumferential resection in the setting of superficial cancer: a regenerative medicine approach with a biologic scaffold. Tissue Eng. Part A 17, 1643–1650.

Badylak, S.F., Lantz, G.C., Coffey, A., Geddes, L.A., 1989. Small intestinal submucosa as a large diameter vascular graft in the dog. J. Surg. Res. 47, 74–80.

Badylak, S.F., Meurling, S., Chen, M., Spievack, A., Simmons-byrd, A., 2000. Resorbable bioscaffold for esophageal repair in a dog model. J. Pediatr. Surg. 35, 1097–1103.

Badylak, S.F., Valentin, J.E., Ravindra, A.K., Mccabe, G.P., Stewart-akers, A.M., 2008. Macrophage phenotype as a determinant of biologic scaffold remodeling. Tissue Eng. Part A 14, 1835–1842.

Badylak, S.F., Vorp, D.A., Spievack, A.R., Simmons-byrd, A., Hanke, J., Freytes, D.O., Thapa, A., Gilbert, T.W., Nieponice, A., 2005. Esophageal reconstruction with ECM and muscle tissue in a dog model. J. Surg. Res. 128, 87–97.

Badylak, S.F., Weiss, D.J., Caplan, A., Macchiarini, P., 2012. Engineered whole organs and complex tissues. Lancet 379, 943–952.

Badylak, S.F., Wu, C.C., Bible, M., Mcpherson, E., 2003. Host protection against deliberate bacterial contamination of an extracellular matrix bioscaffold versus dacron mesh in a dog model of orthopedic soft tissue repair. J. Biomed. Mater. Res. B Appl. Biomater. 67, 648–654.

Barnard, K., Gathercole, L.J., 1991. Short and long range order in basement membrane type IV collagen revealed by enzymic and chemical extraction. Int. J. Biol. Macromol. 13, 359–365.

Beattie, A.J., Gilbert, T.W., Guyot, J.P., Yates, A.J., Badylak, S.F., 2009. Chemoattraction of progenitor cells by remodeling extracellular matrix scaffolds. Tissue Eng. Part A 15, 1119–1125.

Bissell, M.J., Aggeler, J., 1987. Dynamic reciprocity: how do extracellular matrix and hormones direct gene expression? Prog. Clin. Biol. Res. 249, 251–262.

Brennan, E.P., Reing, J., Chew, D., Myers-irvin, J.M., Young, E.J., Badylak, S.F., 2006. Antibacterial activity within degradation products of biological scaffolds composed of extracellular matrix. Tissue Eng. 12, 2949–2955.

Brown, B.N., Badylak, S.F., 2013. Expanded applications, shifting paradigms and an improved understanding of host-biomaterial interactions. Acta Biomater. 9, 4948–4955.

Brown, B.N., Valentin, J.E., Stewart-akers, A.M., Mccabe, G.P., Badylak, S.F., 2009. Macrophage phenotype and remodeling outcomes in response to biologic scaffolds with and without a cellular component. Biomaterials 30, 1482–1491.

Clark, R.A., Lanigan, J.M., Dellapelle, P., Manseau, E., Dvorak, H.F., Colvin, R.B., 1982. Fibronectin and fibrin provide a provisional matrix for epidermal cell migration during wound reepithelialization. J. Invest. Dermatol. 79, 264–269.

Crapo, P.M., Gilbert, T.W., Badylak, S.F., 2011. An overview of tissue and whole organ decellularization processes. Biomaterials 32, 3233–3243.

Crapo, P.M., Tottey, S., Slivka, P.F., Badylak, S.F., 2014. Effects of biologic scaffolds on human stem cells and implications for CNS tissue engineering. Tissue Eng. Part A 20, 313–323.

Crisan, M., Yap, S., Casteilla, L., Chen, C.W., Corselli, M., Park, T.S., Andriolo, G., Sun, B., Zheng, B., Zhang, L., Norotte, C., Teng, P.N., Traas, J., Schugar, R., Deasy, B.M., Badylak, S., Buhring, H.J., Giacobino, J.P., Lazzari, L., Huard, J., Peault, B., 2008. A perivascular origin for mesenchymal stem cells in multiple human organs. Cell Stem Cell 3, 301–313.

Derwin, K.A., Badylak, S.F., Steinmann, S.P., Iannotti, J.P., 2010. Extracellular matrix scaffold devices for rotator cuff repair. J. Shoulder Elbow Surg. 19, 467–476.

Doede, T., Bondartschuk, M., Joerck, C., Schulze, E., Goernig, M., 2009. Unsuccessful alloplastic esophageal replacement with porcine small intestinal submucosa. Artif. Organs 33, 328–333.

Einhorn, T.A., Lane, J.M., Burstein, A.H., Kopman, C.R., Vigorita, V.J., 1984 Feb. The healing of segmental bone defects induced by demineralized bone matrix: a radiographic and biomechanical study. J. Bone Joint Surg. Am. 66 (2), 274–279. PubMed PMID: 6693455.

Enzinger, P.C., Mayer, R.J., 2003. Esophageal cancer. N. Engl. J. Med. 349, 2241–2252.

Fiala, R., Vidlar, A., Vrtal, R., Belej, K., Student, V., 2007. Porcine small intestinal submucosa graft for repair of anterior urethral strictures. Eur. Urol. 51, 1702–1708. Discussion 1708.

Freytes, D.O., Badylak, S.F., Webster, T.J., Geddes, L.A., Rundell, A.E., 2004. Biaxial strength of multilaminated extracellular matrix scaffolds. Biomaterials 25, 2353–2361.

Freytes, D.O., Martin, J., Velankar, S.S., Lee, A.S., Badylak, S.F., 2008. Preparation and rheological characterization of a gel form of the porcine urinary bladder matrix. Biomaterials 29, 1630–1637.

Gilbert, T.W., Stewart-akers, A.M., Simmons-byrd, A., Badylak, S.F., 2007. Degradation and remodeling of small intestinal submucosa in canine achilles tendon repair. J. Bone Joint Surg. Am. 89, 621–630.

Harakas, N.K., 1984 Sep. Demineralized bone-matrix-induced osteogenesis. Clin. Orthop. Relat. Res. (188), 239–251. PubMed PMID: 6380863.

Haykal, S., Soleas, J.P., Salna, M., Hofer, S.O., Waddell, T.K., 2012. Evaluation of the structural integrity and extracellular matrix components of tracheal allografts following cyclical decellularization techniques: comparison of three protocols. Tissue Eng. Part C Methods 18, 614–623.

Hodde, J., Record, R., Tullius, R., Badylak, S., 2002. Fibronectin peptides mediate HMEC adhesion to porcine-derived extracellular matrix. Biomaterials 23, 1841–1848.

Hodde, J.P., Badylak, S.F., Brightman, A.O., Voytik-harbin, S.L., 1996. Glycosaminoglycan content of small intestinal submucosa: a bioscaffold for tissue replacement. Tissue Eng. 2, 209–217.

Hodde, J.P., Record, R.D., Liang, H.A., Badylak, S.F., 2001. Vascular endothelial growth factor in porcine-derived extracellular matrix. Endothelium 8, 11–24.

Keane, T.J., Londono, R., Turner, N.J., Badylak, S.F., 2012. Consequences of ineffective decellularization of biologic scaffolds on the host response. Biomaterials 33, 1771–1781.

Kedzierska, K., Azzam, R., Ellery, P., Mak, J., Jaworowski, A., Crowe, S.M., 2003. Defective phagocytosis by human monocyte/macrophages following HIV-1 infection: underlying mechanisms and modulation by adjunctive cytokine therapy. J. Clin. Virol. 26, 247–263.

Kelly, D.J., Rosen, A.B., Schuldt, A.J., Kochupura, P.V., Doronin, S.V., Potapova, I.A., Azeloglu, E.U., Badylak, S.F., Brink, P.R., Cohen, I.S., Gaudette, G.R., 2009. Increased myocyte content and mechanical function within a tissue-engineered myocardial patch following implantation. Tissue Eng. Part A 15, 2189–2201.

Labow, R.S., Sa, D., Matheson, L.A., Santerre, J.P., 2005. Polycarbonate-urethane hard segment type influences esterase substrate specificity for human-macrophage-mediated biodegradation. J. Biomater. Sci. Polym. Ed. 16, 1167–1177.

Lai, J.Y., Chang, P.Y., Lin, J.N., 2003. Body wall repair using small intestinal submucosa seeded with cells. J. Pediatr. Surg. 38, 1752–1755.

Li, F., Li, W., Johnson, S., Ingram, D., Yoder, M., Badylak, S., 2004. Low-molecular-weight peptides derived from extracellular matrix as chemoattractants for primary endothelial cells. Endothelium 11, 199–206.

Lindberg, K., Badylak, S.F., 2001. Porcine small intestinal submucosa (SIS): a bioscaffold supporting in vitro primary human epidermal cell differentiation and synthesis of basement membrane proteins. Burns 27, 254–266.

Lu, J., Descamps, M., Dejou, J., Koubi, G., Hardouin, P., Lemaitre, J., Proust, J.P., 2002. The biodegradation mechanism of calcium phosphate biomaterials in bone. J. Biomed. Mater. Res. 63, 408–412.

Mantovani, A., Biswas, S.K., Galdiero, M.R., Sica, A., Locati, M., 2013. Macrophage plasticity and polarization in tissue repair and remodelling. J. Pathol. 229, 176–185.

Mantovani, A., Sica, A., Locati, M., 2005. Macrophage polarization comes of age. Immunity 23, 344–346.

Mantovani, A., Sica, A., Sozzani, S., Allavena, P., Vecchi, A., Locati, M., 2004. The chemokine system in diverse forms of macrophage activation and polarization. Trends Immunol. 25, 677–686.

Martinez, F.O., Sica, A., Mantovani, A., Locati, M., 2008. Macrophage activation and polarization. Front. Biosci. 13, 453–461.

Mase Jr., V.J., Hsu, J.R., Wolf, S.E., Wenke, J.C., Baer, D.G., Owens, J., Badylak, S.F., Walters, T.J., 2010. Clinical application of an acellular biologic scaffold for surgical repair of a large, traumatic quadriceps femoris muscle defect. Orthopedics 33, 511.

Mosser, D.M., Edwards, J.P., 2008. Exploring the full spectrum of macrophage activation. Nat. Rev. Immunol. 8, 958–969.

Nieponice, A., Mcgrath, K., Qureshi, I., Beckman, E.J., Luketich, J.D., Gilbert, T.W., Badylak, S.F., 2009. An extracellular matrix scaffold for esophageal stricture prevention after circumferential EMR. Gastrointest. Endosc. 69, 289–296.

Nissen, N.N., Polverini, P.J., Koch, A.E., Volin, M.V., Gamelli, R.L., Dipietro, L.A., 1998. Vascular endothelial growth factor mediates angiogenic activity during the proliferative phase of wound healing. Am. J. Pathol. 152, 1445–1452.

Orringer, M.B., Marshall, B., Iannettoni, M.D., 2001. Transhiatal esophagectomy for treatment of benign and malignant esophageal disease. World J. Surg. 25, 196–203.

Ponce, M.L., Nomizu, M., Delgado, M.C., Kuratomi, Y., Hoffman, M.P., Powell, S., Yamada, Y., Kleinman, H.K., Malinda, K.M., 1999. Identification of endothelial cell binding sites on the laminin gamma 1 chain. Circ. Res. 84, 688–694.

Reing, J.E., Brown, B.N., Daly, K.A., Freund, J.M., Gilbert, T.W., Hsiong, S.X., Huber, A., Kullas, K.E., Tottey, S., Wolf, M.T., Badylak, S.F., 2010. The effects of processing methods upon mechanical and biologic properties of porcine dermal extracellular matrix scaffolds. Biomaterials 31, 8626–8633.

Sarikaya, A., Record, R., Wu, C.C., Tullius, B., Badylak, S., Ladisch, M., 2002. Antimicrobial activity associated with extracellular matrices. Tissue Eng. 8, 63–71.

Savill, J.S., Wyllie, A.H., Henson, J.E., Walport, M.J., Henson, P.M., Haslett, C., 1989. Macrophage phagocytosis of aging neutrophils in inflammation: programmed cell death in the neutrophil leads to its recognition by macrophages. J. Clin. Invest. 83, 865–875.

Shaheen, N.J., Sharma, P., Overholt, B.F., Wolfsen, H.C., Sampliner, R.E., Wang, K.K., Galanko, J.A., Bronner, M.P., Goldblum, J.R., Bennett, A.E., Jobe, B.A., Eisen, G.M., Fennerty, M.B., Hunter, J.G., Fleischer, D.E., Sharma, V.K., Hawes, R.H., Hoffman, B.J., Rothstein, R.I., Gordon, S.R., Mashimo, H., Chang, K.J., Muthusamy, V.R., Edmundowicz, S.A., Spechler, S.J., Siddiqui, A.A., Souza, R.F., Infantolino, A., Falk, G.W., Kimmey, M.B., Madanick, R.D., Chak, A., Lightdale, C.J., 2009. Radiofrequency ablation in Barrett's esophagus with dysplasia. N. Engl. J. Med. 360, 2277–2288.

Sicari, B.M., Johnson, S.A., Siu, B.F., Crapo, P.M., Daly, K.A., Jiang, H., Medberry, C.J., Tottey, S., Turner, N.J., Badylak, S.F., 2012. The effect of source animal age upon the in vivo remodeling characteristics of an extracellular matrix scaffold. Biomaterials 33, 5524–5533.

Singer, A.J., Clark, R.A., 1999. Cutaneous wound healing. N. Engl. J. Med. 341, 738–746.

Takebe, J., Champagne, C.M., Offenbacher, S., Ishibashi, K., Cooper, L.F., 2003. Titanium surface topography alters cell shape and modulates bone morphogenetic protein 2 expression in the J774A.1 macrophage cell line. J. Biomed. Mater. Res. A 64, 207–216.

Tottey, S., Corselli, M., Jeffries, E.M., Londono, R., Peault, B., Badylak, S.F., 2011a. Extracellular matrix degradation products and low-oxygen conditions enhance the regenerative potential of perivascular stem cells. Tissue Eng. Part A 17, 37–44.

Tottey, S., Johnson, S.A., Crapo, P.M., Reing, J.E., Zhang, L., Jiang, H., Medberry, C.J., Reines, B., Badylak, S.F., 2011b. The effect of source animal age upon extracellular matrix scaffold properties. Biomaterials 32, 128–136.

Turner, N.J., Yates Jr., A.J., Weber, D.J., Qureshi, I.R., Stolz, D.B., Gilbert, T.W., Badylak, S.F., 2010. Xenogeneic extracellular matrix as an inductive scaffold for regeneration of a functioning musculotendinous junction. Tissue Eng. Part A 16, 3309–3317.

Valentin, J.E., Badylak, J.S., Mccabe, G.P., Badylak, S.F., 2006. Extracellular matrix bioscaffolds for orthopaedic applications: a comparative histologic study. J. Bone Joint Surg. Am. 88, 2673–2686.

Valentin, J.E., Stewart-akers, A.M., Gilbert, T.W., Badylak, S.F., 2009. Macrophage participation in the degradation and remodeling of extracellular matrix scaffolds. Tissue Eng. Part A 15, 1687–1694.

Van Der Rest, M., Garrone, R., 1991. Collagen family of proteins. FASEB J. 5, 2814–2823.

Voytik-harbin, S.L., Brightman, A.O., Kraine, M.R., Waisner, B., Badylak, S.F., 1997. Identification of extractable growth factors from small intestinal submucosa. J. Cell. Biochem. 67, 478–491.

Wolf, M.T., Daly, K.A., Reing, J.E., Badylak, S.F., 2012. Biologic scaffold composed of skeletal muscle extracellular matrix. Biomaterials 33, 2916–2925.

Yang, M., Chen, C.Z., Wang, X.N., Zhu, Y.B., Gu, Y.J., 2009. Favorable effects of the detergent and enzyme extraction method for preparing decellularized bovine pericardium scaffold for tissue engineered heart valves. J. Biomed. Mater. Res. B Appl. Biomater. 91, 354–361.

Yurchenco, P.D., O'rear, J.J., 1994. Basement membrane assembly. Methods Enzymol. 245, 489–518.

Zantop, T., Gilbert, T.W., Yoder, M.C., Badylak, S.F., 2006. Extracellular matrix scaffolds are repopulated by bone marrow-derived cells in a mouse model of achilles tendon reconstruction. J. Orthop. Res. 24, 1299–1309.

CHAPTER 6

Degradation of Biomaterials

Noel L. Davison[1,2], Florence Barrère-de Groot[2], Dirk W. Grijpma[1,3]
[1]*MIRA Institute for Biomedical Technology and Technical Medicine, and Department of Biomaterials Science and Technology, University of Twente, Enschede, The Netherlands;* [2]*Xpand Biotechnology BV, 3723 MB Bilthoven, The Netherlands;* [3]*Department of Biomedical Engineering, University of Groningen, University Medical Center Groningen, Groningen, The Netherlands*

LEARNING OBJECTIVES

- To understand what defines degradable biomaterials and why they are needed
- To identify the most important degradation mechanisms of biomaterials
- To understand the effect of tissue response on the (in vivo) degradation rate
- To understand the requirements for a degradable biomaterial to be used as a cell scaffold for tissue engineering
- To learn why degradable bioceramics are used in bone regeneration
- To learn how degradable polymers can be synthesized
- To recognize examples of degradable polymers broadly used in tissue engineering

6.1 DEGRADABLE BIOCERAMICS

Inorganic nonmetallic solids can be classified as ceramic, glass, or glass-ceramics.

The word ***ceramic*** is derived from the Greek word keramikos, "having to do with pottery." The American Society for Testing and Materials defines a ceramic article as "an article having a ... body of crystalline or partly crystalline structure, or of glass, which body is produced from essentially inorganic, nonmetallic substances and either is formed from a molten mass which solidifies on cooling, or is formed and simultaneously or subsequently matured by the action of the heat."

For several decades, bioceramics (i.e., calcium phosphates (CaPs) and bioactive glasses (BGs)) have been used clinically as bone-regenerative materials,

Tissue Engineering.
Copyright © 2015 Elsevier Inc. All rights reserved.

particularly in nonload-bearing dental and orthopedic applications since they are not mechanically strong (Temenoff and Mikos, 2008). Current biomedical applications of bioceramics include repair for periodontal disease, maxillofacial reconstruction, augmentation and stabilization of the jawbone, spinal fusion, and bone fillers after tumor surgery. On the horizon, bioceramics show promise as effective carriers for drug delivery, growth factors, bioactive peptides, and various types of cells for tissue engineering purposes (Williams, 2014). Because of their excellent biocompatibility and ability to regenerate bone, integration of these bioceramics with stronger biomaterials (e.g., polymers) is another research focus to improve the mechanical properties of bone substitutes. For example, a titanium hip implant can be coated with bioceramics to aid integration with native bone or bioceramic particles can be blended with polymers to create composites with enhanced mechanical properties (Dorozhkin, 2011).

Since 1920, clinicians have used CaPs as bone graft substitutes with success. Over the past two decades, CaPs consisting of hydroxyapatite (HA) and β-tricalcium phosphate (β-TCP) phases have been commonly used as bone void fillers because they are both osteoconductive, i.e., able to support bone formation directly at their surface (Dorozhkin, 2011). However, because of their different crystalline structure, HA and β-TCP degrade at significantly different rates in physiologic solution given by their thermodynamic solubility constant (K_S): 6.62×10^{-126} and 2.07×10^{-33} for HA and β-TCP, respectively (Elliot, 1994). Hence, HA bioceramics are considered to be nondegradable and β-TCP bioceramics are considered to be degradable. Their distinct degradation profiles are apparent in preclinical studies, where HA and β-TCP were implanted in bony locations and analyzed over time (Figure 6.1). Similarly in clinical studies, β-TCP ceramics can totally degrade, while HA ceramics remain almost unaltered several years after implantation (Horch et al., 2006; Mastrogiacomo et al., 2005).

One way to modulate the degradation profile of CaP bioceramics is by mixing less soluble HA and highly soluble β-TCP, or incorporating other biocompatible CaP phases. In addition, other CaPs with various solubilities and degradation kinetics are also used clinically to regenerate bone, such as dicalcium phosphate dihydrate (DCPD) and amorphous CaP (ACP). Table 6.1 summarizes CaPs of current biological and clinical interest.

BGs represent another class of degradable bioceramics. Glass is an amorphous inorganic solid. Glass results from the very rapid cooling of a viscous molten material to a solid state without crystallization. Silica (SiO_2) is the common base for glass. A slower, more controlled cooling protocol can induce the formation of glass-ceramics, which are partly crystalline and partly glassy. An important change in the glass microstructure, which usually precedes crystallization, is the glass-in-glass phase separation. After phase separation, the glass

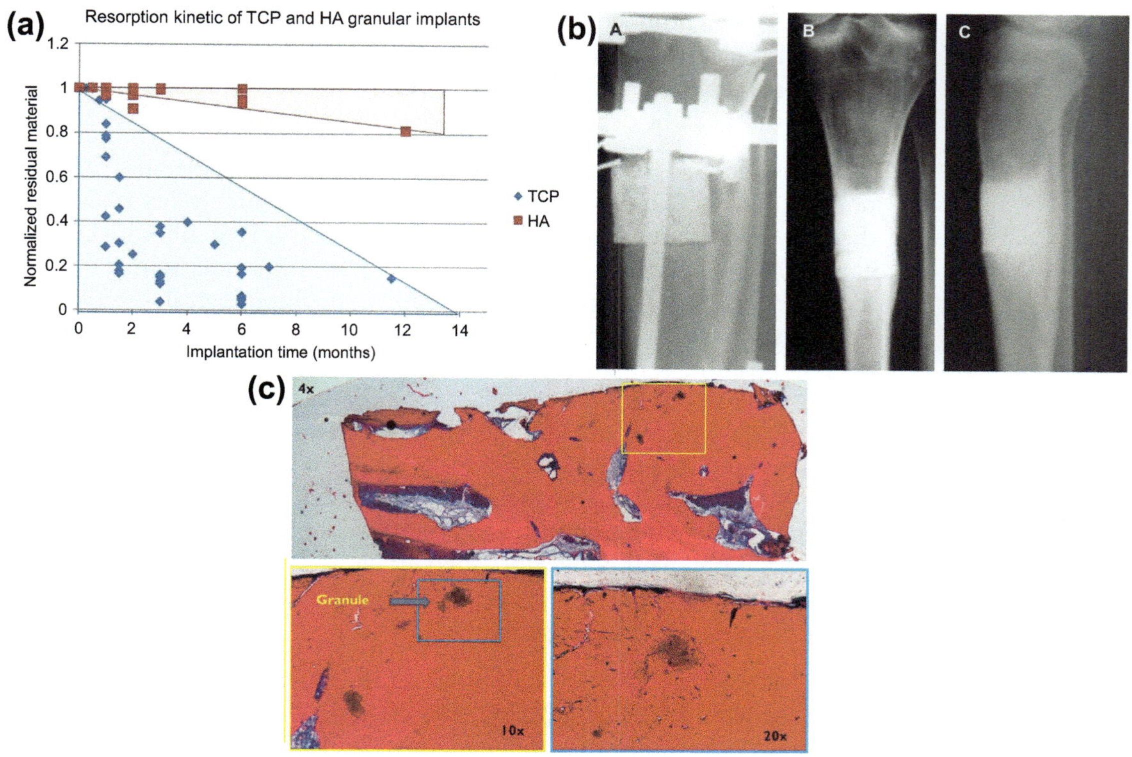

FIGURE 6.1

(a) Degradation kinetics of HA and β-TCP ceramics implanted in preclinical studies. Residual material was determined by histological analysis on undecalcified sections. Material was normalized across publications: the starting amount of material at the time of implantation corresponds to 1; no residual material corresponds to 0 (Nyangoga et al., 2010; Walsh et al., 2008; Hing, 2005; Von Doernberg et al., 2006; Jensen et al., 2009; Wiltfang et al., 2002; Bodde et al., 2007; Gonda et al., 2009; Peltola et al., 2001; Peltola et al., 2003; Liu et al., 2000); (b) Tibia repair by a tissue engineering approach in a human subject. Radiographs obtained immediately post (A) 18 months and (B) 5.5 years after surgery (Mastrogiacomo et al., 2005); (c) Histological evidence of β-TCP resorption after 1 year implantation (de Ruijter et al., 2013).

Table 6.1 Calcium Phosphates of Biological Interest

Name	Formula	Abbreviation	Ca/P
Dicalcium phosphate anhydrate	$CaHPO_4$	DCPA	1.00
Dicalcium phosphate dihydrate	$CaHPO_4.2H_2O$	DCPD	1.00
Octacalcium phosphate	$Ca_8(PO_4)_4(HPO_4)_2{\cdot}5H_2O$	OCP	1.33
Tricalcium phosphate	$Ca_3(PO_4)_2$	TCP	1.50
Hydroxyapatite	$Ca_{10}(PO_4)_6(OH)_2$	OHAp	1.67

no longer has a homogenous composition, but rather consists of two or more immiscible glassy phases of different chemical constituents. Unlike crystallization, phase separation in glass might not be visible by optical microscopy and in most cases can be detected only by electron microscopy.

BGs consist of SiO_2, CaO, Na_2O, and P_2O_5 in specific proportions. These components form a two-dimensional glass network in which silica (SiO_2) forms the network and the alkali metal (e.g., Na, K, Ca, Mg) modifies it (Ducheyne et al., 1994). For instance, the ratio of silica network former to metal network modifier in a BG determines its solubility in physiological solutions, and hence its bioactivity and resorbability. For example, BG45S5, the first tested BG composition contains (in weight percentage, wt%) 45% of SiO_2, 24.5% CaO, 24.5% Na_2O, and 6% P_2O_5. This glass was first studied in the late 1960s and early 1970s (Hench, 1991). Later, glass-ceramics were developed in order to increase the mechanical properties of these BGs. To this end, BGs are subjected to thermal treatments that alter the material's microstructure. Although mechanisms were not understood, it was suggested that this glass type formed a chemical bond with bone. Since then, it has been established that glasses do not bond to bone if the Ca/P ratio is substantially lower than 5:1, as seen in Figure 6.2, relating to its degradability.

The degradation kinetics of bioceramics is an important factor in bone-regenerative medicine. First, the degradation profile orchestrates the available space to allow new bone to grow. To illustrate this, the implantation of two macroporous β-TCP bioceramics (TCP-1 and TCP-2) of identical composition but made by different synthesis methods demonstrated significantly different in vivo degradation profiles and consequently significantly different

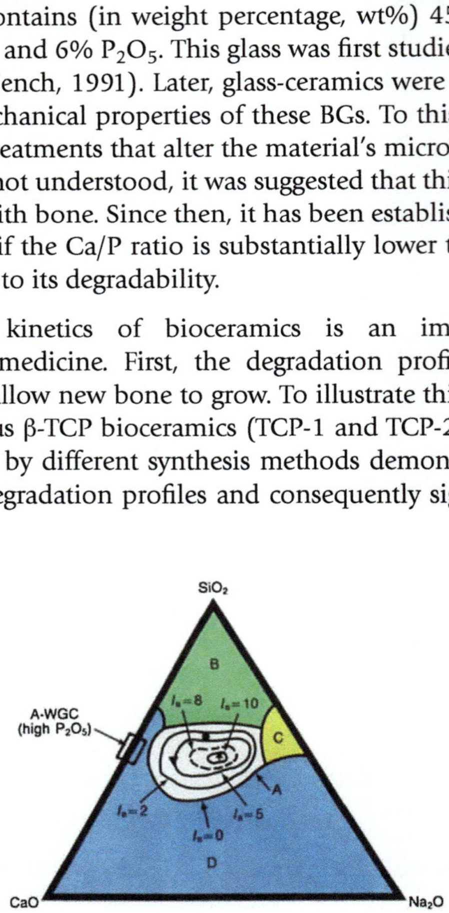

FIGURE 6.2
Compositional dependence (in wt%) of bone and soft tissue bonding of BGs and glass-ceramics. All compositions in region A are bioactive and bond to bone. Compositions in region B are bioinert and lead to formation of a nonadherent fibrous capsule. Compositions in region C are resorbable. Region D is restricted by technical factors. Region E (soft tissue bonding) is inside the dashed line (Hench, 1991).

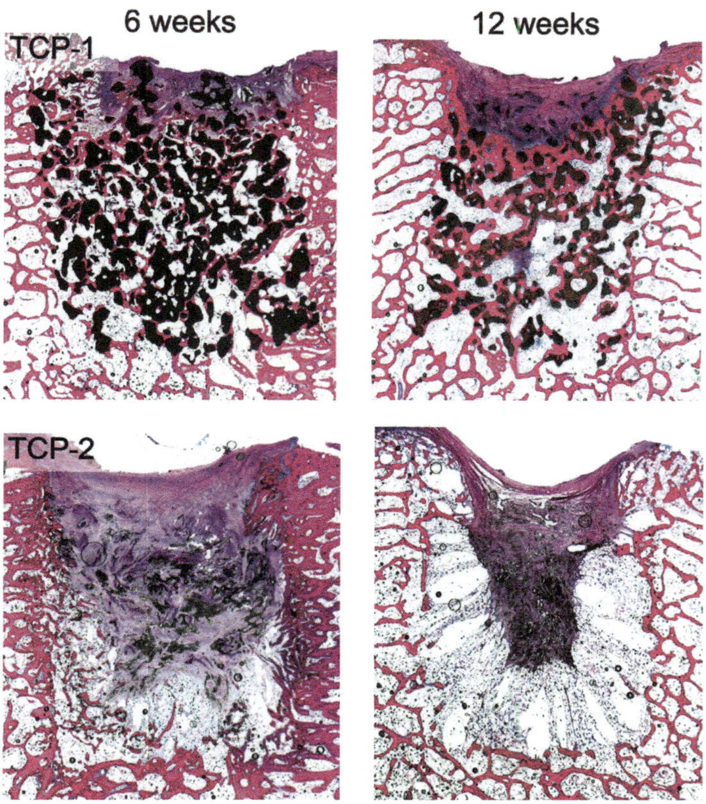

FIGURE 6.3

Light microscopy pictures of macroporous CaPs implanted in goat's femoral condyle (magnification: 2×). This figure represents the tunable nature between degradation and bone formation of two β-TCP ceramics with similar composition but made by different synthesis methods. TCP-1 degrades slower than TCP-2 and supports continuous bone ingrowth from the host bone to the center of the implant, reflecting full bone regeneration in the defect. TCP-2 degrades too fast and does not support bone ingrowth; instead, the defect is filled with fibrous tissue. *Courtesy of Dr. Huipin Yuan.*

bone-regenerative potential in the femoral condyle of goats (Figure 6.3). TCP-2, which was resorbed too rapidly, produced only fibrous tissue growth in the bone defect, while TCP-1, which was resorbed much slower, produced significant bone growth in the defect. In addition to the kinetics of degradation, it is theorized that the biodegradation mechanisms and products have a direct (positive or negative) impact on cells and tissues, thus directing tissue formation.

6.1.1 Degradation Mechanisms of Bioceramics

Ceramic materials are held together by either ionic or covalent bonds. They can be crystalline or amorphous. The crystalline and amorphous states are typical solid states which represent the degree of order between ions, atoms, or molecules. The ***crystalline state*** is characterized by a well-ordered, defined internal molecular structure. This structure is periodically repeated in a specific tri-dimensional pattern (a). The cohesion in crystals is insured by the binding energy between the atoms in the case of covalent solids, or by electrostatic forces existing between anions and cations in the case of ionic solids. On the contrary, ***amorphous state*** is characterized by no order between ions, atoms, or molecules (b); therefore the values of binding energy between entities greatly vary within the solid. The dissolution of ceramics is strongly dependent on these binding energies.

In physiologic and artificial aqueous environments, bioceramics can degrade via multiple mechanisms: (1) physico-chemical dissolution accompanied by possible phase transformation, (2) cellular degradation mediated by multinucleated cells (MNC), and (3) mechanical fragmentation due to a loss of structural integrity resulting from mechanisms (1) and (2). In biological systems, degradation of bioceramics reflects nonequilibrium processes that occur simultaneously or in competition with each other. Because bioceramics alone are not used in load-bearing conditions, the mechanical degradation mechanism will not be addressed in this chapter.

6.1.1.1 Physico-Chemical Degradation of Bioceramics

The physico-chemical degradation of bioceramics can be described as a dissolution–reprecipitation phase transformation process, which is the result of ion exchange at the solid–liquid interface. The kinetics of this degradation process and the surface transformations resulting from ionic exchange depend on the intrinsic properties of the bioceramics and the nature of the aqueous environment.

From a thermodynamic point of view, CaPs are sparingly soluble in water but readily dissolve in acids. To study and compare the dissolution kinetics of CaPs in vitro, physiologic saline solutions at different pH are generally used; in these simplified solutions, solubility kinetics correlate with the thermodynamic solubility constant of the CaP. However, more complex supersaturated solutions can be used to more closely mimic the in vivo environment. Supersaturated solutions, such as blood, contain higher concentrations of ions than could be dissolved in water under equilibrium conditions; thus, the supersaturation state is inherently unstable. Figure 6.4(a) shows the solubility of various CaP phases as a function of Ca^{2+} concentration of the solution and pH under equilibrium conditions at 37 °C. In supersaturation conditions, crystal growth has not yet occurred, but this can be initiated by introducing seeds, i.e., CaP substrates (Figure 6.4(b)). Supersaturated solutions frequently used to mimic physiologic solution include cell culture media supplemented with serum or simulated body fluid (SBF) (Bohner and Lemaitre, 2009).

Although not always the case, it can be generally stated that the physico-chemical dissolution behavior of CaP ceramics in vitro and in vivo follows

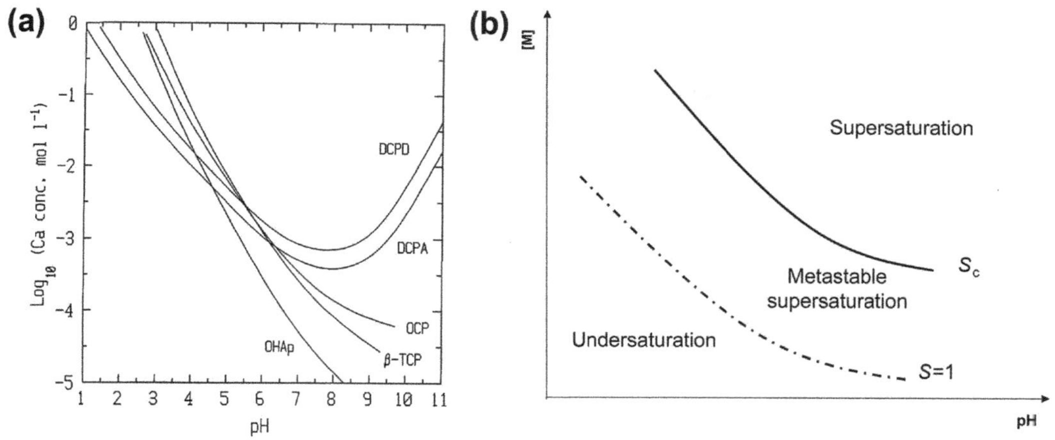

FIGURE 6.4
(a) Solubility isotherms of calcium phosphates at 37°C (Elliot, 1994). (b) Solubility isotherms of salts in solution as a function of the pH. This diagram represents the three different saturation states of an ionic solution susceptible to precipitation.

their *thermodynamic solubility*, i.e., HA<TCP<OCP<DCPD, from least soluble to the most soluble. Still, additional factors can also affect the physico-chemical dissolution kinetics:

- *The crystalline features*. The higher the crystallinity, the lower the degradation kinetics. Eventually, amorphous biomaterials dissolve faster than crystalline ones.
- *The presence of additives*. The presence of some additives of mineral origins within the CaP structure can affect the crystal lattice, and therefore can accelerate the dissolution, e.g., carbonate, silicate, or strontium added to HA.
- *The physical structure and geometry (e.g., porosity, granules vs blocks)*. The larger the exposed surface to the solution environment, the faster the biomaterial dissolves, simply because more exchanges can take place.

Crystalline solids as bioceramics are different from the theoretical crystals. First, these crystals generally contain imperfections: (1) additional entities are inserted in the crystal lattice (c), or (2) some entities can be absent in the crystal, i.e., vacancies (d). Second, crystalline solids are composed of several crystals packed together in a roughly organized bulk structure. These imperfections result from the materials' processing (e.g., impurities, sintering) and can affect their solubility.

From a surface reactivity point of view, physico-chemical dissolution can be seen as ionic transfer from the solid phase to the aqueous liquid via surface hydration of Ca^{2+} and phosphate species (PO_4^{3-}, HPO_4^{2-}, $H_2PO_4^{-}$) as bioceramic material interacts with other ions, proteins, and sugars present in the biological fluid. On the one hand, this dissolution process is highly dependent on the nature of the

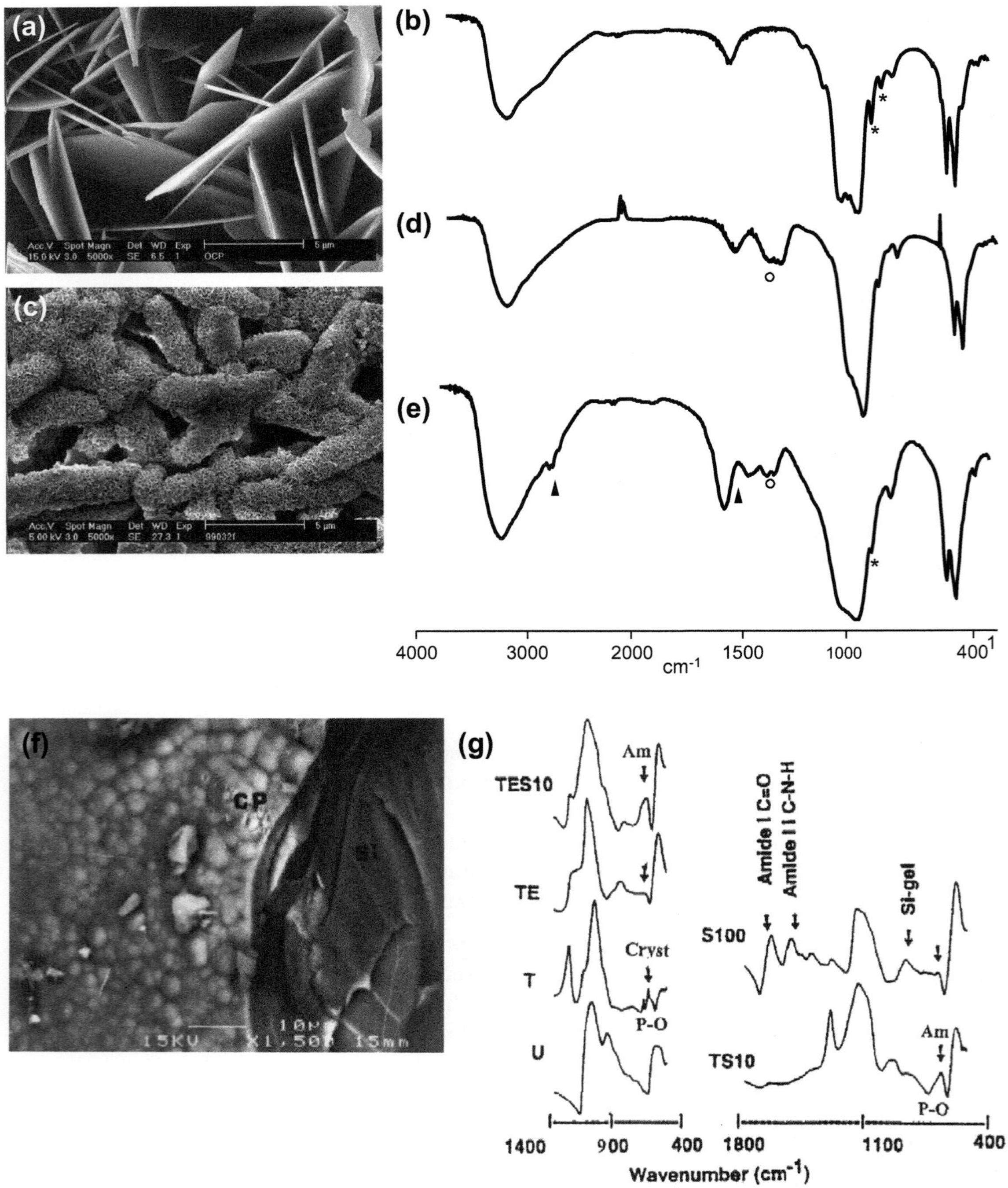
(a)
5 μm
OCP
(b)
(c)
5 μm
(d)
(e)
4000
3000
2000
cm-1
1500
1000
400
(f)
CP
Si
10μm
15KV X1,500 15mm
(g)
TES10
Am
TE
T
Cryst
P-O
U
Amide I C=O
Amide I I C-N-H
Si-gel
S100
Am
TS10
P-O
1400
900
400
1800
1100
400
Wavenumber (cm-1)

CaP bioceramics and on the other hand, it is dependent on the nature of the environment, i.e., chemical composition, supersaturation, and pH. Whether the environment is in vivo, in culture medium, or in SBF, the nature of the precipitate is a poorly crystallized carbonated apatite. When organic compounds are present in the surrounding milieu, they generally slow down the formation of this new apatitic layer, and they can be incorporated therein. However, not all CaP ceramics without known bone-regenerative properties in vivo allow the growth of such a newly formed apatite layer in vitro (Bohner and Lemaitre, 2009).

The physico-chemical degradation of BGs can be generalized as three processes including leaching, dissolution, and precipitation. Briefly, sodium ions leach from the surface and are replaced by hydrogen ions through an ion-exchange reaction. This depletion of sodium leads to the formation of a silica-rich layer. An ACP layer forms on the silica-rich layer. Calcium is present in the solid state, but is equally drawn from solution. In simple electrolyte solutions (which are not good simulations of in vivo behavior), the amorphous layer crystallizes to form carbonated HA (c-Ap) with properties akin to the mineral phase of bone. This simple reaction sequence that brings about the deposition of c-Ap was first proposed by Hench (1971).

Hench described five reaction stages:

Stage 1: Leaching and formation of silanols
Stage 2: Loss of soluble silica and formation of silanols
Stage 3: Polycondensation of silanols to form a hydrated silica gel
Stage 4: Formation of an amorphous calcium phosphate layer
Stage 5: Crystallization of c-Ap layer (Figure 6.5(a)).

FIGURE 6.5

Morphology and structure of an octacalcium phosphate (OCP) coating (a and b) and the subsequent changes due to the dissolution/reprecipitation process in vitro after immersion in culture medium for 2 weeks (c and d), and in vivo after 4 weeks of subcutaneous implantation in rat (e). The morphology of the substrate is shown in scanning electronic microscopy (SEM) (a and c). The structure is characterized by Fourier Transform InfraRed (FTIR) spectroscopy (b, d, and e). The initial OCP substrate is composed of sharp elongated crystals (a) and its crystal structure is characterized by sharp phosphate (P–O) bands between 1100 and 1000 cm^{-1} and at ~600 and 560 cm^{-1} (b). Two typical OCP bands can be noted at 906 and 852 cm^{-1} (*). In vitro, the substrate exhibits smaller crystals that seem to grow along the initial OCP crystals (c), the FTIR structure of these crystals have lost their OCP features and exhibit broader and less defined bands (d). In addition, new bands typical of carbonate groups at 1460 and 1416 cm^{-1} (O) have appeared. This FTIR spectrum indicates a carbonate apatitic structure (d). In vivo (subcutaneous implantation in rats for 4 weeks), the FTIR structure of the coating has evolved in a similar way as in vivo: the coating transformed into a carbonated apatite (e). In addition, organic compounds (▲) were found in the transformed substrate. Phase transformations of bioglass (BG) in vitro (f). SEM of a fractured granule showing the Ca–P (CP) surface layer and underlying silica (Si) (Radin et al., 2000) (g). FTIR spectra of unreacted BG 45S5 prior to (U) and after immersion for 3 h into 0.05 M Tris buffer (T), T+plasma electrolyte (TE), TE+10% serum (TES-10), T+10% serum (TS-10), and 100% serum (S-100). Notice the appearance of a split P–O bend indicating formation of a crystalline phase (Cryst) after immersion in T and undivided P–O bend indicating amorphous (Am) phosphate phase after immersion in other solutions (Radin et al., 2000).

The dissolution of BGs has been studied for over 40 years. Many different factors influence the mechanism and rate of the dissolution of BGs, such as glass composition, glass particle size, and glass powder type.

6.1.1.2 Cellular Degradation of Bioceramics

The cellular degradation of bioceramics is mediated by MNC, which resorb the material surface and phagocytose degraded material particulate. Osteoclasts are specialized MNC uniquely responsible for the resorption of bone through the acid dissolution of bone mineral and enzymatic cleavage of bone protein matrix. Osteoclasts also play a critical role in regulating new bone formation through bone coupling mechanisms. Bone coupling describes the tendency of osteoblasts to lay down new bone precisely where it was first resorbed by osteoclasts in tight synchronicity.

In vitro and in vivo, it has been observed that osteoclasts can degrade CaPs in a similar way as bone mineral: osteoclasts adhere to the CaP surface and create resorption pits, presumably through acid dissolution of the sealed microenvironment formed between the osteoclast and the material substrate (the "sealing zone"). In vitro, osteoclasts readily adhere and resorb several CaP ceramics including HA and TCP. In vivo, multinucleated osteoclast-like cells form microconcavities on the ceramic surface adjacent to newly formed bone and degraded material particulate (Figure 6.6.). In vitro experiments identified specific physico-chemical parameters beyond the primary CaP composition that influence cellular degradation.

- *The physico-chemical dissolution kinetics of the CaP ceramics.* Depending on the synthesis methods, dissolution kinetics can vary widely among CaPs with the same chemical composition. The release of Ca^{2+} influences osteoclastic activity; for instance, above a certain Ca^{2+} concentration, osteoclastic resorption is inhibited. As previously discussed, CaP structure and crystallinity also play a role in dissolution kinetics and therefore may also dictate osteoclast activity.
- *The presence of additives.* The incorporation of carbonate or other ions in CaP may stimulate the carbonic anhydrase activity known to promote the osteoclastic acidic secretion in vitro. On the other hand, the presence of zinc and fluoride in CaP can inhibit osteoclastic resorption in vitro and in vivo.
- *The surface features.* Surface energy was found to modulate the osteoclastic adhesion in vitro. Surface roughness and microporosity appear to enhance osteoclastic attachment and activity (Barrère et al., 2006).

Osteoclast resorption of BGs has not been extensively studied. Both in vivo and in vitro studies demonstrate that active osteoclasts are only detected on the

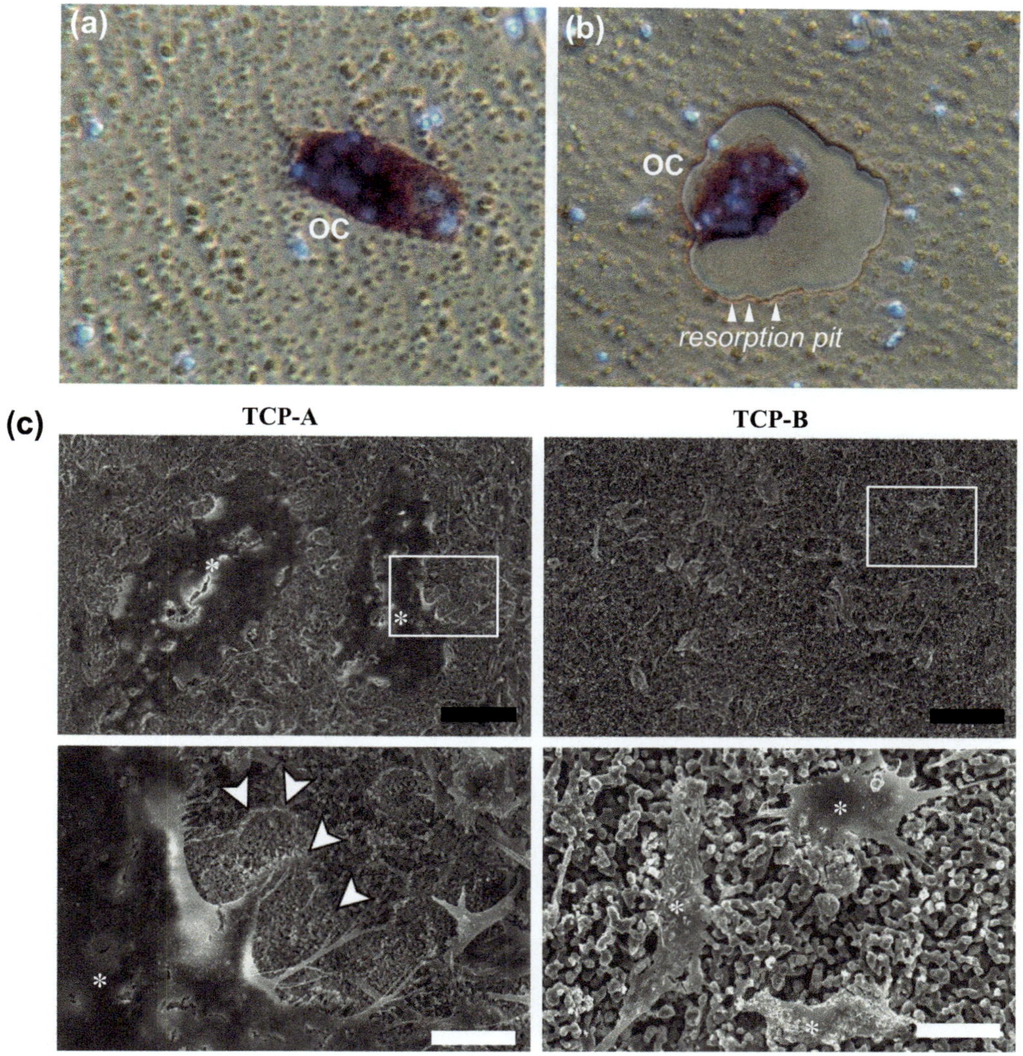

FIGURE 6.6

Osteoclasts (OCs) on CaPs in vitro (a–e) and in vivo (f–m). OCs stain positive for tartrate-resistant acid phosphatase (TRAP) (red) in both (a) nonresorbing and (b) resorbing states. Nuclei staining (blue) shows that OCs are multinucleated. *(Courtesy Dr. S.G. Perez.)*
(c) OC (white arrows) formation occurs differently on two TCP ceramics with different surface structure (black scale bars = 100 μm; white scale bars = 20 μm). After removing OCs, extensive resorption pits were evident on TCP-A (lighter areas) but not on TCP-B (d) (white scale bars = 500 μm; yellow scale bars = 50 μm). (e) Fluorescent microscopy of an OC resorbing TCP. Actin rings (white arrows)—where osteoclast resorption takes places—are evident near the cell membrane border (Green = F-actin, Red = cell membrane, Blue = cell nuclei). (f and g) TRAP-positive OCs (→, stained red) resorbing a CaP implant (*) and bone (B), next to bone marrow (BM). At high magnification, many nuclei (N) in one resorbing OC can be seen ((f), scale bar = 50 mm; (g), 10 mm) (Wenisch et al., 2003). (h and i) Consecutive histological sections from a biopsy after 6 months implantation of TCP granules for sinus lift augmentation: (h) CaP granules in contact with bone (B) are surrounded by OCs (→) stained positively for TRAP (red). *(Courtesy of Dr. J.W. Hoekstra.)* (j) Transmission electronic micrograph of an OC that is closely associated with a ceramic surface. Several cell features are noted: the osteoclast nuclei (N), ruffled border (r), sealing zone (s), dorsal microvilli (mv), golgi complexes (G), vacuoles filled with longer slender CaP crystals (→), and denser particles (▼) (scale bar = 2 μm) (Wenisch et al., 2003) (k). Ultrastructural details of the osteoclastic ruffled border at the implant surface showing internalized CaP particles in the cellular vacuoles (scale bar = 1 μm).

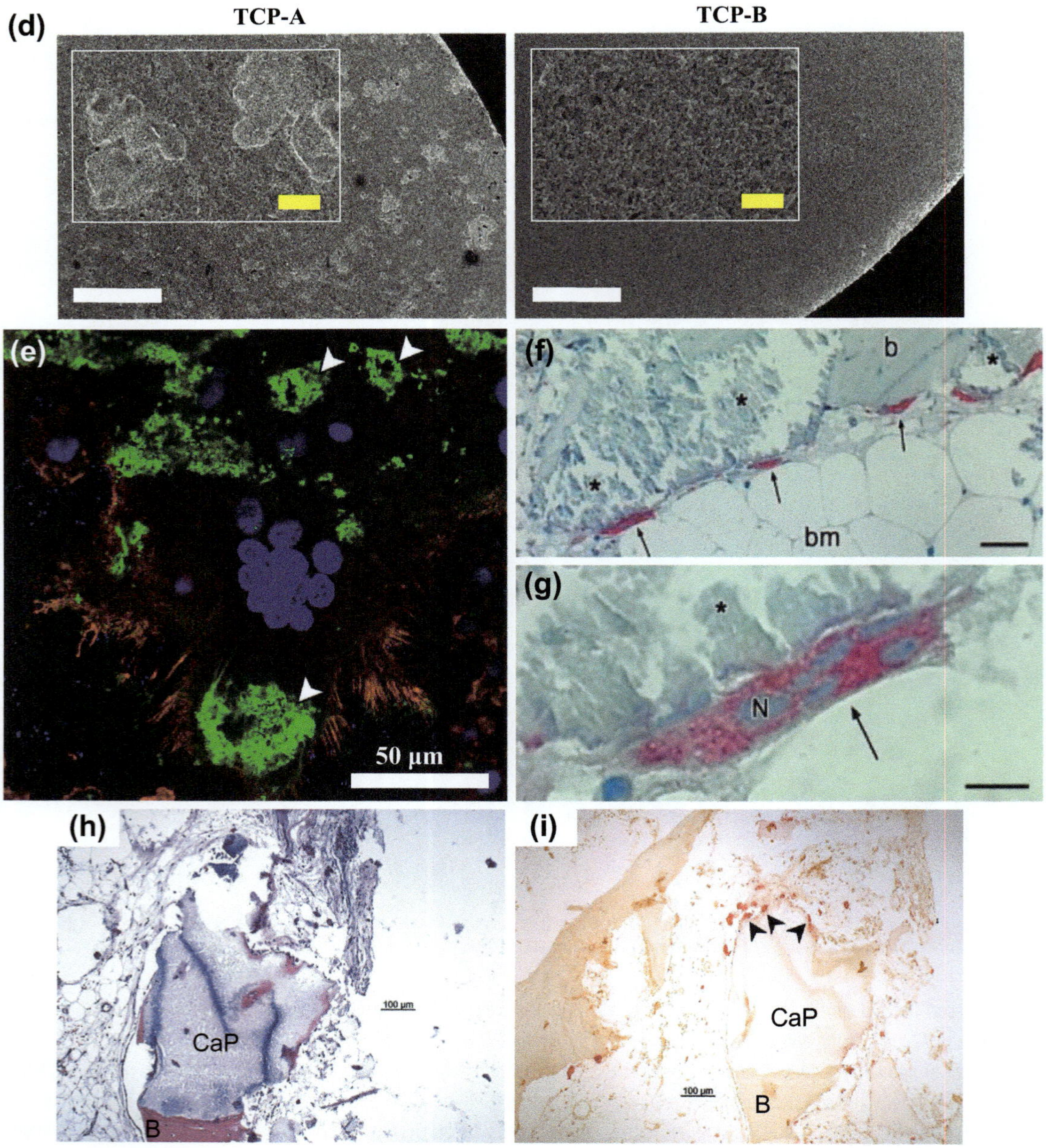

FIGURE 6.6 ***Continued***

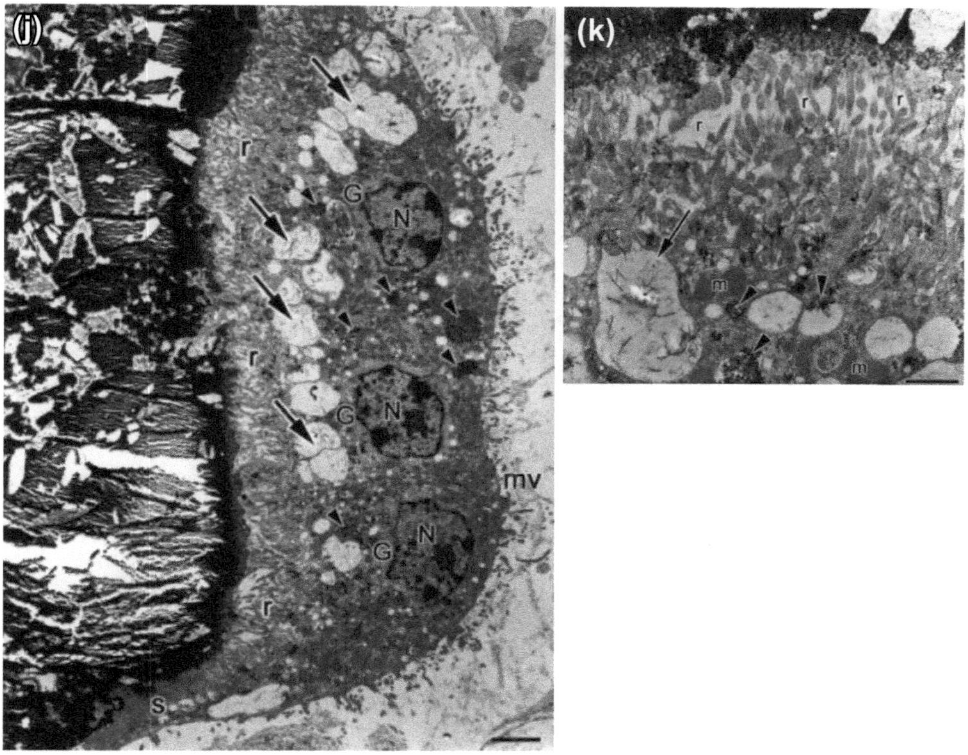

FIGURE 6.6 ***Continued***

CaP layer that can precipitate on the BG surface. Osteoclasts may be responsible for resorbing this formed CaP layer; however, there is no direct evidence of cellular degradation of silicon. Regardless of the mode of degradation, the silicon content of BG can be safely excreted in urine when implanted in vivo (Lai et al., 2005).

6.1.2 Translation of Degradable Bioceramics to Bone Tissue Engineering Systems

6.1.2.1 Bone Bonding and Stimulation of Bone Formation

The dissolution/precipitation phenomena are associated with bioactivity related to bone bonding, i.e., the formation of an interfacial mineralized layer between bioceramics and bone tissue that ensures their cohesion. Structurally, this layer is comparable to the films grown in vitro by dissolution–precipitation mechanisms, i.e., nanocrystals of carbonated apatite in SBF that mimic the mineral composition of blood plasma (Figure 6.7). When formed in the presence of osteogenic cells, this mineralized layer growing on CaP ceramics is comparable to the cement lines present in bone. In vitro, modification of Ca^{2+}

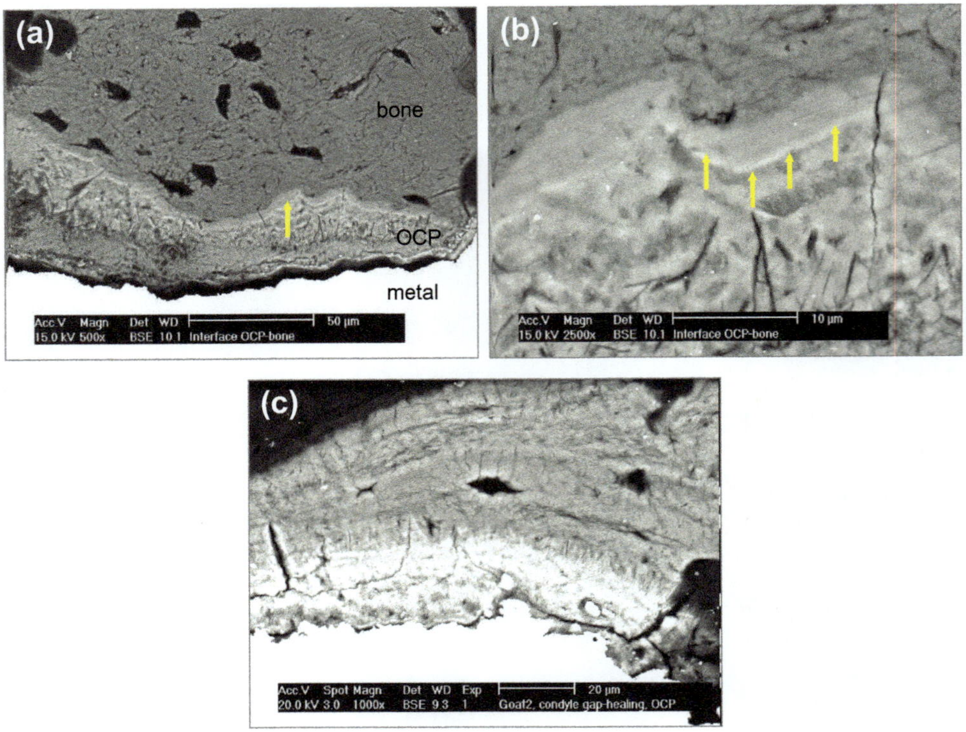

FIGURE 6.7

Back-scattering electronic microscopy picture of an octacalcium phosphate (OCP) coating on a metallic porous scaffold implanted for 12 weeks in the femoral condyle (goat) at different magnification. (a and b) Between the OCP coating and the newly formed bone, an interfacial phase (arrow) that can be attributed to superficial phase transformation is clearly visible. However, on the same implant, this interfacial layer is not always visible (c).

and inorganic phosphate (P_i) concentrations in cell culture medium due to incubation with CaPs are reported to affect cell proliferation, differentiation, and may even induce cell death. In the absence of CaPs, increasing or decreasing Ca^{2+} and P_i concentrations in culture medium directly affect osteoblast activity (Barrère et al., 2008). Human mesenchymal stem cells (hMSC) cultured on various biphasic calcium phosphate (BCP) ceramics, ranging from 100% HA to 100% β-TCP, showed higher osteogenic differentiation on the ceramics composed of more β-TCP (i.e., 20% HA/80% β-TCP, "20/80 HA/β-TCP") potentially due to increased Ca^{2+}/P_i release from β-TCP dissolution. In a mouse model of ectopic bone formation, hMSC cultured on the same range of BCP also induced more bone formation and at a faster rate on the 20% HA/80% β-TCP (20/80 HA/β-TCP) versus the other, more stable formulations (e.g., 100% HA, 76/24, 63/37, and 56/44 HA/β-TCP), or 100% TCP, which may have degraded too quickly (Arinzeh et al., 2005). Degradable β-TCP had

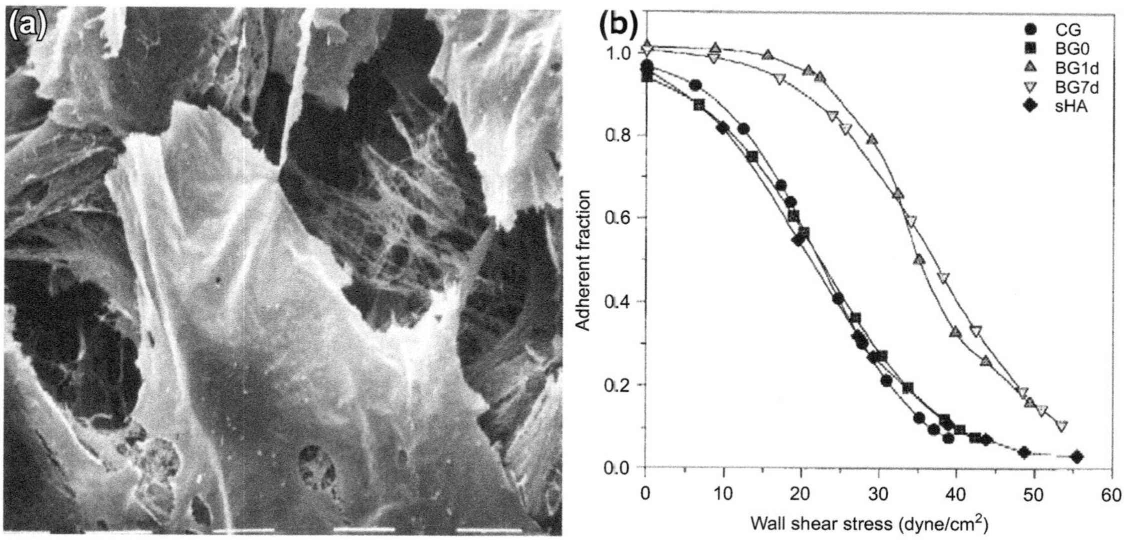

FIGURE 6.8

(a) Scanning electron micrograph of a BG surface cultured with rat calvaria osteoblasts for 1 week. The glass surface was covered by collagen fibrils and bone-like tissue (magnification = 1000×) (Ducheyne et al., 1994); (b) cell detachment profiles for substrates coated with 0.1 mg/mL fibronectin (10 dyne cm^{-2} = 1 N m^{-2}). Cell adhesion was significantly higher on reacted bioactive (BG1d, BG7d) than on unreacted bioactive (BG0) and control (CG) glasses and stoichiometric hydroxyapatite (sHA) (García et al., 1998).

a positive influence on osteogenic differentiation of hMSC compared to the nondegradable HA (Barradas et al., 2013).

From in vitro cell culture studies, BGs have shown stimulatory effects on the differentiation of bone-forming osteoblasts from precursor cells. Ducheyne et al. (1994) reported the characteristics of the BG substrate that augment extracellular matrix secretion by osteoblast lineage cells. After 1 week in culture with calvarial osteoblasts, SEM (Figure 6.8(a)) shows that the BG surface was covered by collagen fibrils and bone-like tissue. This and many other studies showed and confirmed the ability of BGs to stimulate the commitment and bone formation of osteoblasts and their precursors. It is thought that the dissolution products of BGs play an important role in stimulating osteogenesis (Xynos et al., 2000); thus the degradation of BG is likely to be crucially important to its performance.

6.1.2.2 Osteoinduction and Degradable Bioceramics

Osteoinductive biomaterials have intrinsic ability to induce ectopic bone formation in nonosseous environments, i.e., where bone cells are initially absent, such as muscle or subcutis. Osteoinductive bioceramics perform better in bone defects than those that are only osteoconductive in preclinical models (Yuan et al, 2010). They may be regarded as the most simple but effective bone tissue engineering approache because an osteoinductive material can induce stem cell differentiation and tissue formation in situ without the need to seed and culture cells on the material or add growth factors prior to implantation. Although the performance of osteoinductive biomaterials has been reported for over 20 years, the biological mechanism of how an osteoinductive biomaterial stimulates de novo bone formation is largely unknown. Three-dimensional architecture, concavities, interconnected pores, and surface microstructure have all been shown to promote bone formation. Increased ion and protein reactivity resulting from larger specific surface area are thought to play a role in this enhanced bone-forming capacity. For instance, increased specific surface area may lead to increased precipitation/release of Ca^{2+}/P_i or increased absorption of bone morphogenetic proteins (BMPs) (Barradas et al., 2011). Therefore the degradation rate of these ceramics may be crucial for their performance.

Osteoclasts that resorb CaPs may also contribute to osteoinduction by CaPs, again underlining the importance of bioceramic degradation and functionality. For instance, osteoclasts have been found to precede de novo bone formation by osteoinductive β-TCP in the dorsal muscle of dogs by 4 weeks (Kondo et al., 2006), and osteoclast inhibitors have been shown to block osteoinduction by CaP (Tanaka et al., 2010). Though osteoclasts are normally only found in osseous sites or in pathologies related to heterotopic ossification, the natural inflammatory response following the implantation of a bioceramic is thought to provide all of the necessary components for osteoclastogenesis—both osteoclast precursors and osteoclast differentiating cytokines. This potential link between cellular degradation of CaPs and osteoinduction is illustrated in Figure 6.9.

6.2 BIODEGRADABLE POLYMERS

Scaffolds for tissue engineering have been prepared from biologically derived polymers, such as collagen, hyaluronic acid, and chitosan. However, the number of natural polymers that can be used, as well as the possible modifications to improve their mechanical properties and degradability, are limited. Natural polymers are often difficult to process, and pathogenic risks may be associated with materials of animal or human origin. Alternatively, the use of synthetic polymers of different chemistries enables the design of scaffolds with specific mechanical and biological properties, and degradation rates. Synthetic polymers can be produced cheaply and reproducibly, and can be easily processed into devices of virtually any shape and form.

Much of the research on degradable polymers in tissue engineering has been focused on hybrid cell/scaffold constructs using degradable polymers such as poly(lactic acid) (PLA), poly(glycolic acid) (PGA), and their copolymers (PLGA), which are well known in the medical field. These polymers degrade via hydrolysis of the main chain ester bonds and are resorbable in vivo as their

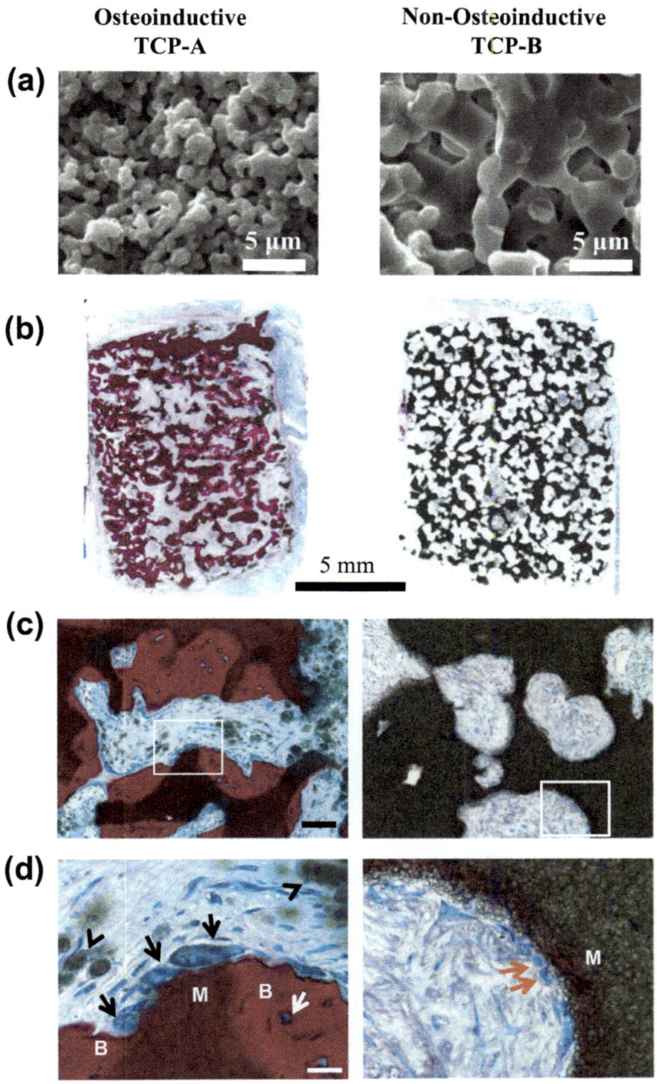

FIGURE 6.9

Osteoinductive and resorbable β-TCP (TCP). TCP with different surface microstructures were synthesized and implanted in the dorsal muscle of dogs for 12 weeks. (a) Scanning electron micrographs show the different surface structures of TCP-A and TCP-B. (b) Histological sections were stained with basic fuchsin and methylene blue for bone. Overview images show ectopic bone formation (bright red/pink) by TCP-A (reddish brown) and not by TCP-B (brown). (c) In TCP-A, extensive bone formation is bonded to the material and degraded material particulate is present in the loose connective tissue space (light brown spots), but no bone or material particulate is apparent in TCP-B (scale bar = 100 μm). (d) At high magnification, osteoclast-like cells (black arrows) resorb the TCP-A material (M) surface between stretches of mature bone (B) containing osteocytes (white arrow) in characteristic lacunae. In the loose connective tissue, material particulate has been internalized by other cells (arrow heads). In contrast, mononuclear cells (red arrows) are mainly found on TCP-B and no material particulate is evident in the loose connective tissue space (bottom row: scale bar = 25 μm). *Adapted from Davison et al. (2014).*

degradation products (lactic- and glycolic acid) are part of the Krebs metabolic cycle. Besides PLA, PGA, and their copolymers, which are considered the golden standard in tissue engineering and regenerative medicine applications, novel polymers are continuously being developed, characterized, and used to fabricate scaffolds and medical devices. Some of these polymers will be outlined in the following paragraphs of this chapter. From a degradation mechanism point of view, it is also important to differentiate between the resorption behavior of polymers from that of ceramics: in the former case soluble compounds are generated upon degradation and scission of the polymer chain, while in the latter case the scaffolds resorb by slow dissolution of the constituent ions into intracellular or extracellular fluids. In this part, we will discuss how polymers are synthesized and how their properties can be characterized and controlled. Particular emphasis will be put on degradation mechanisms of these biomaterials.

6.2.1 Synthesis and Properties of Polymers

6.2.1.1 Polymer Synthesis

Polymers are long-chain, high molecular weight macromolecules formed by reaction of monomers. A polymer chain consists of many covalently bound repeating units, formed upon sequential reactions of the monomers (Odian, 2004). These monomers are multifunctional molecules that can react with each other to form homopolymers or with monomers of a different type to form copolymers. Important reaction mechanisms through which polymers are prepared are *addition polymerizations*, *step-growth polymerizations*, and *ring-opening polymerizations*.

In *addition polymerizations*, double bond containing monomers such as ethylene, vinyl chloride, styrene, and methyl methacrylate are polymerized by action of radicals or ionic species. In these polymerizations, distinct steps can be discerned: initiation and the formation of reactive radicals or ionic compounds, chain propagation, and chain termination. Once a reactive species is formed, successive additions of large numbers of monomers will lead to the formation of a polymer chain. Each monomer addition regenerates a reactive center. This chain reaction will continue until termination occurs. Overall, typical addition polymerizations can be presented as:

$$\underset{\text{Ethylene}}{n\ H_2C{=}CH_2} \rightarrow \underset{\text{Polyethylene (or PE)}}{-[H_2C-CH_2]_n-}$$

and

$$\underset{\text{Methyl methacrylate}}{n\ H_2C{=}C(CH_3)COOCH_3} \rightarrow \underset{\text{Poly(methyl methacrylate) (or PMMA)}}{-[H_2CC(CH_3)COOCH_3]_n-}$$

As a rule, addition polymers are not considered to be degradable polymers, as no labile bonds are present in the main chain. However, prolonged implantation of PE and PMMA and other such polymers will result in some chain scission due to cellular action in which radical species are formed.

Step-growth polymerizations are characterized by the stepwise reaction of the functional groups of the reactants. These reactions are analogous to the simple reactions of functional groups. The size of the polymer molecule increases relatively slowly as dimers, trimers, tetramers, pentamers, etc., are formed until eventually large-sized molecules are formed. These reactions occur between any of the different-sized species present in the reaction system.

In *step-growth polymerization* reactions, often a condensation reaction takes place in which a small molecule is liberated. For example, lactic acid is a molecule that contains two functional groups: a hydroxyl group (–OH) and a carboxylic group (–COOH). Upon removal of water by heating and application of vacuum, lactic acid can be polymerized to form a polyester:

$$n\ HOCH(CH_3)COOH \rightarrow -[OCH(CH_3)CO]_n- + (n-1)\ H_2O$$

Lactic acid — Poly (lactic acid) (or PLA)

But in the *step-growth* preparation of polyurethanes, in which a diisocyanate reacts with a (polymeric) diol, a small molecule is not liberated and condensation does not take place:

$$OCN-(CH_2)_6-NCO + HO-[CH_2CH_2O]_m-H \rightarrow -[O(CH_2CH_2O)_mCONH(CH_2)_6NHCO]-$$

Hexamethylene diisocyanate — Polyethylene glycol — A polyurethane

A third mechanism by which polymers can be prepared is *ring-opening polymerization* (ROP) of cyclic monomers. As an example, the ring opening of ethylene oxide is shown:

$\rightarrow$ $-[CH_2CH_2O]_n-$

Ethylene oxide — Poly(ethylene oxide) (or PEO)

Besides the monomer used and the way in which it has been polymerized, the molecular weight and the molecular weight distribution of the polymer obtained are important factors that determine its properties.

6.2.1.2 Polymer and Copolymer Properties

The nature of the monomer(s) used to prepare the polymer will determine the repeating units within the polymer chain, and, therefore, will determine to a large extent the physical, chemical, and biological properties of the polymeric material as an implant or tissue engineering scaffold.

Depending on the intended application, polymeric materials with widely differing physical properties (e.g., thermal properties, mechanical properties, and hydrophilicity) and degradation characteristics are required (Ratner et al., 2013). Parameters that often need to be optimized are the melting and glass transition temperatures, the tensile strength, the elastic modulus or stiffness, and the (surface) hydrophilicity. Therefore, copolymerization, i.e., the preparation of polymers from two or more types of monomers, is often employed to tune the material properties (Odian, 2004).

Copolymers can be classified as *random-*, *alternating-*, *block-* or *graft copolymers*. *Random copolymers* have a statistical arrangement of different monomer units in their backbone. In *alternating copolymers*, the different monomer units are arranged in such a way that they alternate along the polymer chain. The physical and degradation properties depend strongly on the nature and composition of monomers. *Block copolymers* are composed of connected polymeric or oligomeric segments built up of different monomer units. Examples are diblock copolymers, triblock copolymers, and multiblock copolymers. In *graft copolymers*, a comb-like structure is obtained by attaching (co)polymer segments to a polymer backbone at various places. Several types of block copolymer architectures can be synthesized for use as biomaterials.

An important class of block copolymers is that of phase-separated multiblock copolymers containing alternating soft- and hard segments. Frequently, a hard phase (with high melting temperature or glass transition temperature) dispersed in a continuous soft phase (with a glass transition temperature below room temperature) can be distinguished. These hard segments form physical cross-linkages between the polymeric chains adding strength and toughness to a flexible and elastic thermoplastic elastomeric material.

Well-known examples of multiblock copolymers are segmented poly(urethane)s. Building blocks of segmented poly(urethane)s include diisocyanates, polyols, and chain extenders. Usually the hard segment originates from the diisocyanate and chain extender molecules, and the soft segment contains the polyol (a hydroxyl group terminated polymer or oligomer) moiety. A general structure of such a poly(urethane) is shown in Figure 6.10.

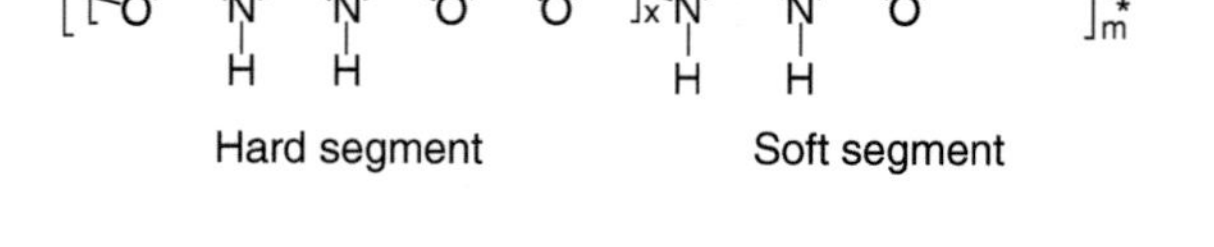

FIGURE 6.10

General structure of a segmented poly(urethane) prepared from a diisocyanate, OCN-R-NCO; chain extender, HO–R′–OH; and polyol building blocks.

6.2.1.3 Chemically Cross-Linked Polymer Networks

Besides the *physical* cross-linkages between the polymeric chains as described earlier, polymer networks can be obtained by *chemical* cross-linking. Bifunctional monomers such as methyl methacrylate and lactic acid yield linear polymers (which are often soluble in solvents and can be processed at elevated temperatures above their glass transition temperature or melting temperature). Monomers with a functionality greater than 2, e.g., glycerol which contains three reactive hydroxyl groups, can be used to prepare branched polymers and cross-linked network structures when polymerized with a diacid monomer.

In chemically cross-linked polymers a network structure is created by the formation of covalent linkages between the polymer chains (Odian, 2004). In contrast to linear and branched polymers, chemically cross-linked polymers do not dissolve in solvents, but only swell in them. The extent of the polymer network swelling by a solvent is dependent on the density of cross-linkages: the higher the cross-linking density, the lower the degree of swelling. As a chemically cross-linked polymer is in fact a single giant molecule, it cannot be processed at temperatures above its melting temperature or glass transition temperature either.

Important examples of cross-linked polymers are hydrogels. These polymers are based on hydrophilic, water soluble oligomeric or polymeric chains that have been reacted with each other to form a covalent network. In one preparation route, polyethylene glycol (PEG) is functionalized at both termini with acrylic groups that in a subsequent stage are linked by radical photopolymerization reactions (Sawhney et al., 2003). To prepare a degradable and resorbable polymeric scaffold for tissue engineering, it will be necessary to have cleavable bonds in the polymer main chain or in the cross-linkages between the polymer chains.

6.2.2 Biodegradable Polymers: Applications and Synthesis

Significant efforts have been devoted to prepare biodegradable polymers and use them in medical applications as implants, drug delivery devices, and tissue engineering scaffolds (Ratner et al., 2013; Uhrich et al., 1999). Polymer degradation and decrease of its molecular weight occurs through scission of the main chain. Although degradation reactions can occur during processing at elevated temperatures or upon sterilization with high-energy radiation, for a polymeric material to resorb in the body it is necessary that the main chain of the macromolecule contains labile bonds. Such bonds can then be scissioned by hydrolysis or oxidation reactions (also through enzymes and cellular activity) to yield (soluble) compounds of lower molecular weight (Göpferich, 1996; Tamada and Langer, 1993). Figure 6.11 gives an overview of important cleavable bonds that can be present in the main chain of biodegradable polymers.

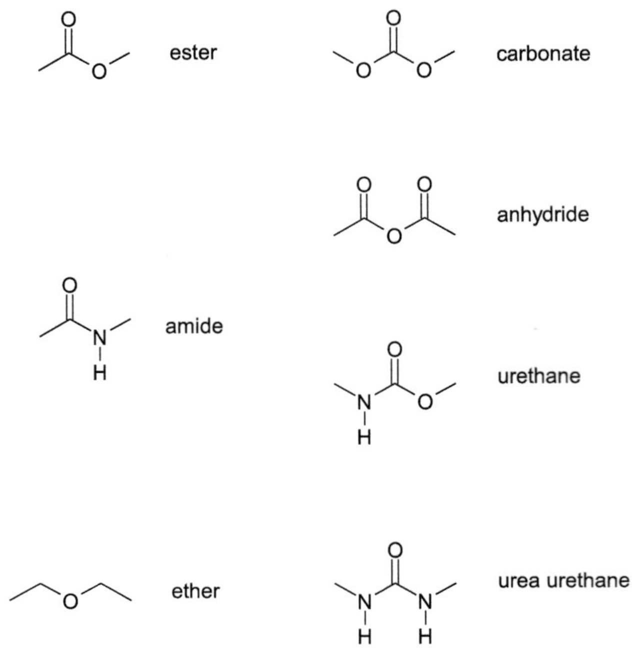

FIGURE 6.11
Overview of important cleavable labile bonds in the main chain of biodegradable polymers.

6.2.2.1 Biodegradable Polymers in Tissue Engineering

Of the currently employed biomaterials, synthetic polymers possess most characteristics required to fabricate scaffolds for tissue engineering applications. The most often used synthetic polymers and their areas of applications are summarized in Table 6.2.

Of the synthetic polymers, linear aliphatic polyesters like PLA, PGA, and PLGA are very well known in the medical field and have therefore been broadly used in tissue engineering. In general, they elicit a minimal or mild foreign body reaction, and as such are considered to be biocompatible. By varying their molecular weight and in case of PLGA also their copolymer ratio, the biodegradation rate and the mechanical properties can be tailored. They are suited for tissue engineering applications as their degradation products (lactic and glycolic acids) obtained resulting from hydrolysis are normally present in the metabolic pathways of the human body. However, their bulk degradation behavior (see Section 6.2.3) may lead to the formation and accumulation of relatively large amounts of acidic degradation products that cannot be easily disposed of. This can result in local inflammation (Bostman et al., 1989; Fu et al., 2000). Furthermore, in semicrystalline polymers, the crystalline residues may take a long time to resorb completely. Another linear aliphatic polyester commonly used in

Table 6.2 Overview of Synthetic Polymeric Biomaterials Used in Tissue Engineering Applications

Degradable Polymer	Abbreviation	Tissue Engineering Applications
Poly(lactic acid)	PLA	Skin, cartilage, bone, ligament, tendon, vessels, nerve, bladder, liver
Poly(glycolic acid)	PGA	Skin, cartilage, bone, ligament, tendon, vessels, nerve, bladder, liver
Poly(lactic-co-glycolic acid)	PLGA	Skin, cartilage, bone, ligament, tendon, vessels, nerve, bladder, liver
Poly(ε–caprolactone)	PCL	Skin, cartilage, bone, ligament, tendon, vessels, nerve
Multiblock copolymers comprising poly(ethylene oxide) and butylene terephthalate units	PEOT/PBT	Skin, cartilage, bone, muscle
Polyphosphoesters	PPEs	Cartilage, bone, nerve, liver
Polyphosphazenes	PPAs	Bone, nerve
Polyanhydrides	PAs	Bone
Polyorthoesters	POEs	Bone
Poly(propylene fumarate)-diacrylate networks	PPF-DA	Bone, vessels
Poly(ethylene glycol)-diacrylate hydrogel networks	PEG-DA	Skin, vessels, nerve

tissue engineering is poly(ε-caprolactone) (PCL). This polymer has found many applications as a result of its good biocompatibility and favorable mechanical properties. It degrades at a much lower rate than PLA, PGA, and PLGA, making it attractive when long-term implants are desired (Hutmacher et al., 2001).

Another family of thermoplastic polymers that has been developed recently for tissue engineering and drug delivery are multiblock copolymers comprising poly(ethylene oxide) (PEO) and poly(butylene terephthalate) (PBT) (PEOT/PBT)). They are prepared by polycondensation of (dimethyl) terephtalate (T), butanediol (B), and polyethylene glycol (PEG). These poly(ether ester) copolymers exhibit good physical properties like elasticity, toughness, and strength in combination with easy processability (Bezemer et al., 1999). Their properties result mainly from a phase separated morphology in which soft, hydrophilic PEO segments are physically cross-linked by hard, semicrystalline PBT segments. In contrast to chemically cross-linked materials, these cross-links are reversible and will be disrupted at temperatures above their glass transition- or melting point, giving the polymer material its good processability.

The interest arisen in tissue engineering applications is due to the fact that by varying the molecular weight of the starting PEG segments and the weight ratio of PEOT and PBT blocks, it is possible to tailor properties, such as wettability, swelling, biodegradation rate, protein adsorption, and mechanical properties (Bezemer et al., 1999; Deschamps et al., 2002).

An alternative to the preparation of tissue engineering scaffolds by processing polymers from melts or solutions is making use of injectable polymers. This class of polymeric materials is very attractive as they can be used in minimally invasive surgery such as arthroscopy, which is highly beneficial to the patient. Furthermore, they can fill irregularly shaped defects (Burdick et al., 2001), and cells and bioactive agents can easily be incorporated (Elbert et al., 2001; Mann et al., 2001). In particular, photopolymerizable systems based on poly(propylene fumarate)-diacrylates (PPF-DA) and poly(ethylene glycol)-diacrylates (PEG-DA), have been much investigated. They even can be hardened transdermally by applying light (He et al., 2000). Despite the advantages in using these polymers, some issues may still rise from their low mechanical properties and possible cytotoxicity remains due to the acrylic groups.

Biodegradable polyesters synthesized by ROP in the past years, biodegradable aliphatic polyesters, such as poly(lactide) [poly(lactic acid), PLA], poly(glycolide) [poly(glycolic acid), PGA], poly PCL, and their copolymers, have attracted much attention (for reviews see Albertsson and Varma, 2002; Uhrich et al., 1999).

Although polyesters can be synthesized by polycondensation of hydroxyl acids such as lactic acid, it is difficult to achieve high molecular weights and control the molecular weight, molecular weight distribution, and architecture of the polymer. In most cases, biodegradable polyesters are synthesized by ROP of cyclic esters (or lactones). Structures of some of these lactones often used are shown in Figure 6.12.

Ring-opening polymerization of cyclic esters proceeds through cationic polymerization, anionic polymerization, or coordination-insertion reactions. The coordination-insertion reactions (Kowalski et al. (2000); Kricheldorf et al., 2000; Nijenhuis et al., 1992) are often initiated by metal alkoxides, which contain a covalent metal–oxygen bond and have a weak Lewis acid character. The propagation proceeds by coordination of the monomer to the metal alkoxide, followed by the insertion of the monomer into the metal–oxygen bond. A scheme of the propagation step of this mechanism is shown in Figure 6.13.

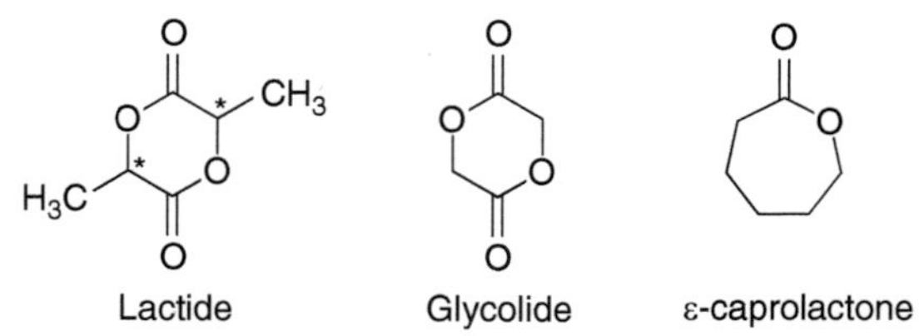

FIGURE 6.12
Structures of cyclic esters used in ring-opening polymerizations. Lactide has two optically active carbon atoms in the ring.

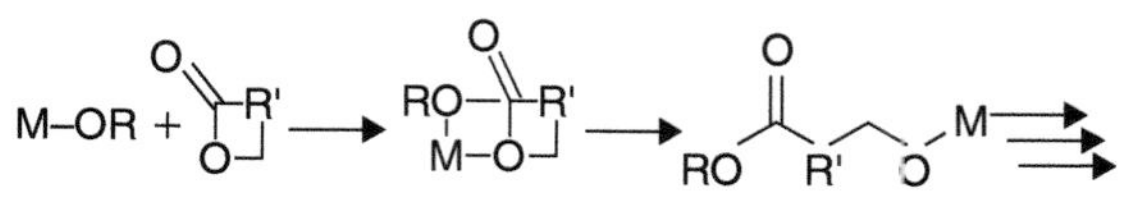

FIGURE 6.13
Scheme illustrating the propagation step of lactone ring-opening polymerization via a coordination-insertion mechanism.

It should be noted that the growing chains contain the metal–oxygen bond during the propagation, i.e., they remain active.

Tin(II) 2-ethylhexanoate (stannous octoate or $SnOct_2$) is the most often used catalyst for the ROP of cyclic esters. $SnOct_2$ is a highly active catalyst that is soluble in the monomer melts, allowing the ROP to be performed in bulk. Although there is some concern about the cytotoxicity of $SnOct_2$, the American Food and Drug Administration has approved many degradable implants in which the polymer has been prepared using small amounts of this catalyst.

The mechanism of the ROP process using $SnOct_2$ as a catalyst has been studied in depth by Kowalski et al. (2000) and Kricheldorf et al. (2000). For the ROP of L-lactide and ε-CL, Kowalski and coworkers proposed the following mechanism shown in Figure 6.14: first, a tin–alkoxide bond is formed via a reversible reaction between $SnOct_2$ and a compound containing a hydroxyl group (a). The monomer is then coordinated to and inserted into the formed tin–alkoxide bond. Subsequent coordination and insertion of a next monomer then propagates the polymerization (b). Chain transfer reactions between the growing polymer chains (in which the tin–alkoxide bonds are still present) and compounds containing hydroxyl groups can also occur, leading to an increase in the molecular weight distribution (c).

The compounds containing hydroxyl groups (shown in Figure 6.14) can have very small amounts of impurities present in the monomer, catalyst, or reaction environment. If stringent purification and reaction conditions are applied, very

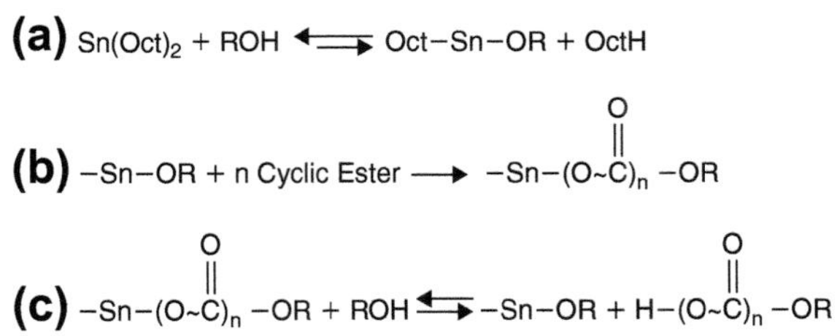

FIGURE 6.14
Ring-opening polymerization of cyclic esters using $SnOct_2$ as a catalyst: (a) formation of a tin–alkoxide bond by reaction of an alcohol with $SnOct_2$; (b) propagation by monomer coordination and insertion; (c) chain transfer reaction.

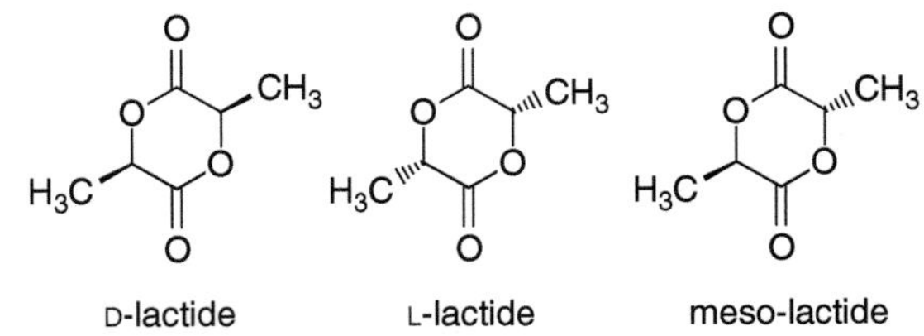

FIGURE 6.15
Structures of lactide stereoisomers. D,L-lactide is an equimolar racemic mixture of LLA and DLA.

high molecular weight polyesters (such as PLA with M_n of 310 × 10^3g/mol) can be synthesized. These compounds can also be added on purpose to control molecular weight, composition, and architecture of the polymer that is to be synthesized. When, e.g., monomethoxy-PEG (mPEG) or a PEG-diol is used together with $SnOct_2$, the corresponding diblock or triblock copolymers of PEG and polyesters can be synthesized.

6.2.2.2 Lactides and Polylactides

Of the degradable polyesters used in medical applications, PLA has received special interest due to the presence of asymmetrical carbon atoms in the lactide monomers. Lactide compounds exist in three different stereoisomeric forms (Figure 6.15): D-(R-) lactide (DLA), L-(S-) lactide (LLA), and meso-lactide. It should be noted that with D,L-lactide (DLLA), an equimolar racemic mixture of LLA and DLA isomers is meant. The structures of these different compounds are shown in the following figure.

In the ROP with $SnOct_2$, the asymmetric carbon atoms in the lactide monomers can retain their conformation. Therefore, PLA polymers with different stereoregularities (isotactic poly(LLA) (PLLA), poly(DLA) (PDLA), and atactic poly(DLLA) (PDLLA)) can be prepared.

Poly(LLA) and poly(DLA) are semicrystalline polymers with a glass transition temperature (T_g) of approximately 60 °C and a peak melting temperature (T_m) of approximately 180 °C. PDLLA is amorphous with a T_g of approximately 55 °C. Both semicrystalline PLLA and amorphous PDLLA polymers are rigid materials. Their modulus of elasticity and stress at break (σ_{break}) values are close to, respectively, 3.5GPa and 65MPa. However, these polymers are relatively brittle with an elongation at break (ε_{break}) less than 6%.

Interestingly, mixtures of PLLA and PDLA can form the so-called stereocomplexes with even higher melting temperatures of up to 230 °C.

Both polymers degrade by hydrolysis in which naturally occurring lactic acid is formed. Degradation of the polymers starts with water uptake, followed by random cleavage of the ester bonds in the polymer chain. The degradation takes place throughout the bulk of the material. Upon degradation, the number of carboxylic end groups increases, which leads to a decrease in pH and autocatalytic hydrolytic degradation (Li et al., 1990). During the degradation of semicrystalline PLLA, crystallinity of the residual material increases as hydrolysis preferentially takes place in the amorphous domains. In general, the rate of degradation and erosion of amorphous PDLLA is faster than those of PLLA.

6.2.3 Mechanisms of Polymer Degradation and Erosion

6.2.3.1 Definitions

In polymer degradation, chain scission occurs, and oligomers, monomers, and other low molecular weight species are formed (Tamada and Langer, 1993). The degradation of polymers can be induced by thermal activation, oxidation, photolysis, radiolysis, or hydrolysis processes (Gopferich, 1996). When degradation is affected by the biological environment, the term biodegradation can be used. As a consequence of degradation, *erosion* of the material can occur. Erosion specifically refers to the loss of material by monomers and oligomers leaving the polymer mass. Obviously, if polymer erosion and resorption is to occur, first soluble components need to be formed.

6.2.3.1.1 Chain Scission and Polymer Degradation

For polymeric biomaterials, the most important degradation reaction is hydrolysis. There are several factors that influence the rate of this reaction: *the nature of the chemical bond, the pH, the copolymer composition, and the extend of water uptake* are the most relevant. Of these factors, it is mainly the type of bond in the polymer backbone that determines the rate of hydrolysis. Anhydrides and *ortho*-ester bonds are the most reactive ones. In water, anhydrides are more reactive than esters and amides. Such ranking must be viewed with circumspection. However, reactivity of the bonds is much dependent on catalysis and steric or electronic effects.

In organic chemistry, ester hydrolysis of organic compounds occurs by reaction of water with the ester bond. However, in the presence of acidic or basic environments, the hydrolysis reactions are catalyzed and the rates are much higher. The hydrolytic degradation of poly(CL), poly(DLLA), and related aliphatic polyesters involves the generation of carboxylic end groups that are also able to catalyze the hydrolysis reaction. Based on a first-order kinetic model, Pitt et al. (1981) related the rate of chain scission of an aliphatic polyester autocatalyzed by the generated carboxylic acid end groups to the decrease of the average number molecular weight in time:

$$\ln\left(\overline{M}_n\right) = \ln\left(\overline{M}_n^0\right) - kt$$

where $\overline{M}_n$ is the average number molecular weight at the start of the hydrolysis, k is the rate constant, and t is the degradation time.

If the hydrolysis is not autocatalytic, chain cleavage will follow the rate law (Pitt and Gu, 1987):

As discussed ealier, the composition can have a large influence on the chemical and physical properties of a (co)polymer. The rate of degradation of a polymer is dependent on the ease of hydrolysability as well as on the accessibility of the cleavable main chain bonds to enzymes and water. Besides the nature of these labile bonds, the hydrophilicity of the material, the morphology and crystallinity of the polymer, and its molecular weight are important parameters in determining the degradability of a polymer.

In the hydrolysis of a polyester, which initially is hydrophobic and not soluble in water, the average molecular weight of the polymer will decrease in time. It will require a certain degree of main chain hydrolysis before appreciable amounts of water soluble oligomers and low molecular compounds are formed.

6.2.3.2 Bulk- and Surface-Erosion

Polymer degradation processes can result in erosion, i.e., loss of mass of the materials, when degradation products diffuse and dissolve into the degradation environment. Polymer erosion can be a more complex process as it depends on many other processes besides degradation, such as morphological changes and characteristics of the oligomers formed (Göpferich, 1996).

In polyesters there are four main factors that determine the erosion diffusion and dissolution phenomena (Vert, 2005):

1. The hydrolysis rate constant of the ester bond.
2. The diffusion coefficient of water within the polymer matrix.
3. The diffusion coefficient of chain fragments within the polymer matrix.
4. The solubility of the degradation products (oligomers) in the surrounding medium.

Erosion of polymers can be classified as a bulk erosion process or surface erosion process (Göpferich, 1993, 1996; Von Burkersroda et al., 2002). In bulk erosion, the polymer chain scission occurs throughout the specimen. The molecular weight and the mechanical strength of the specimens decrease in time. The decrease in molecular weight occurs essentially from the beginning of the degradation process, whereas loss of mass is much delayed. The external dimensions of the polymer specimens remain essentially unchanged, until the specimens disintegrate at a critical time point. It should be noted that the loss in mechanical properties of the material precedes the loss in mass. The process of bulk erosion is schematically described in Figure 6.16(a).

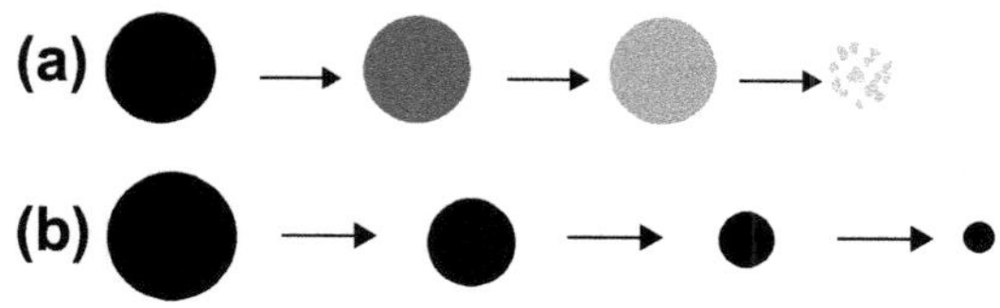

FIGURE 6.16
Scheme illustrating degradation of materials by bulk erosion (a) and surface erosion (b) processes.

In surface erosion, loss of material is confined to the surface of the polymer device only. Size and mass of the device decrease in time, whereas molecular weight and mechanical properties of the polymer device remain unchanged. In surface erosion, the rate of mass loss is proportional to the surface area. Surface erosion is schematically depicted in Figure 6.16(b).

Most biodegradable polyesters that are currently available degrade by a bulk erosion process that predominantly involves simple hydrolysis of main chain ester bonds (Middleton and Tipton, 2000). First, small amounts of water diffuse into the bulk of the material preferentially attacking the ester bonds present in the amorphous domains of the polymer. Initially, this decrease in molecular weight will not affect the mechanical properties of the device as physical cross-linkages due to entanglements and regions of crystallinity maintain the structure of the material. At a certain point, however, the reduction in molecular weight will lead to a decrease in the mechanical properties of the material. As more carboxylic acid end groups are formed, the rate of the hydrolysis reaction will increase and loss of mass will occur when the degradation products become soluble in water. Figure 6.17(a) shows a graphic representation of these effects.

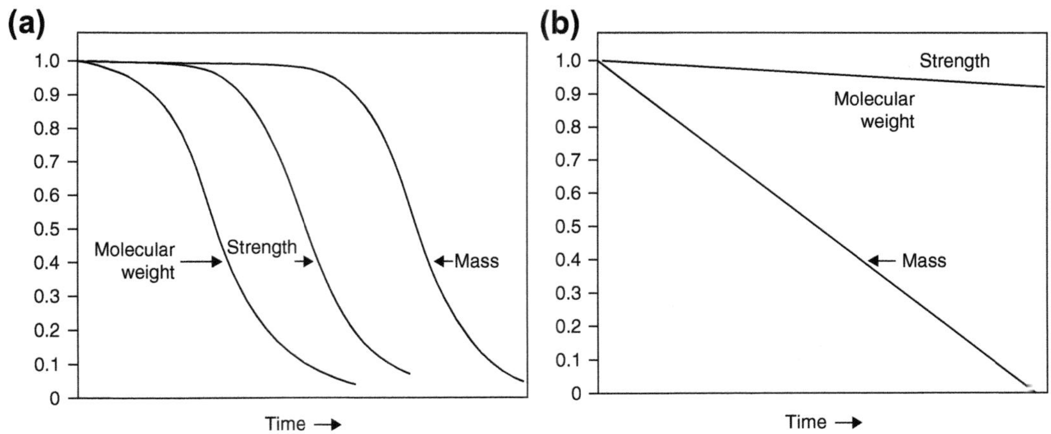

FIGURE 6.17
(a) Representation of the bulk erosion process of degradable polymers; (b) representation of the surface erosion process of degr‑dable polymers.

In a surface erosion process (Figure 6.17(b)), the rate of hydrolysis of labile bonds is relatively fast in comparison to the diffusion rate of water into the bulk, and the degradation reactions are limited to the surface of the polymer material. The molecular weight of the polymer, and therefore its mechanical properties, remain more or less unchanged with time, while the mass of the device decreases at an appreciable rate. A surface-eroding polymer could provide constant and well controlled release rates in drug delivery applications, but only few polymers show surface erosion characteristics: examples are polyanhydrides (copolymers of sebacic acid, 1,3-bis(*p*-carboxyphenoxy) propane and 1,3-bis (*p*-carboxyphenoxy)hexane poly(adipic anhydride)) and poly(orthoesters). These polymers have very hydrolytically labile bonds in their main chains, which react with water rapidly. Therefore, the degradation process occurs at the polymer surface resulting in a surface eroding material.

6.2.3.3 In vivo Degradation

In the development of a novel degradable polymer, most degradation experiments are most often conducted in vitro by incubating the polymeric specimen in phosphate buffered saline (PBS) at body temperature (37°C). However, in vitro degradation behavior can considerably differ from in vivo degradation. In most cases, in vivo degradation is faster than in vitro degradation (Mainil-Varlet et al., 1997). The faster in vivo degradation can be attributed to the tissue response (Stoll et al., 2001). Acute inflammation, due to the injury of implantation, and foreign body reaction, due tont site (Anderson, 1994; Anderson and Shive, 1997). Reactive species, like super oxide and hydrogen peroxide species the presence of the implant, induce the migration of polymorphonuclear leucocytes and macrophages to the impla, produced by these inflammatory cells can oxidize the polymer chains (Deschamps et al., 2002). Therefore, the in vivo decrease in molecular weight values is not only due to hydrolytic degradation caused by the extracellular fluid, but probably also by the influence of the oxygen free radicals and of the other species generated by the inflammatory cells (Ali et al., 1994). In some cases, however, such as in hydrogel-like polymer systems, a slower in vivo degradation has been reported compared to the in vitro degradation in phosphate buffer (Changez et al., 2005). The presence of less fluid near the implant site retards swelling and subsequently inhibits degradation. For copolymers that degrade only by hydrolysis in the bulk, the in vitro degradation will mimic the in vivo degradation (Deschamps et al., 2003).

Besides the increased chain scission rate, the erosion rate in vivo differs from the in vitro situation. Mass loss in vivo can be based on either *passive transport* of monomers and oligomers by dissolution and diffusion (similar to in vitro) or by *active transport* by phagocytes or a combination of these processes. The solubility of the oligomers may be increased in vivo due to the presence of lipids, resulting in a faster mass loss. The dissolved degradation products will be removed from the implantation site via the lymphatic system and subsequently secreted from the body by the kidneys. The in vivo mass loss can be further increased by mechanical stresses and cellular activity. Various studies demonstrated the fragmentation of a biomaterial in time. Implants based on PEOT/PBT copolymers have been evaluated extensively in various implantation sites and extensive fragmentation of the implants was visible within 4weeks, thereby significantly enlarging the implant surface (Lu et al., 2000). Higher fragmentation rates and smaller average particle sizes were observed at intramuscular implantation sites as compared to subcutaneous sites. This could be attributed to the higher stresses and higher extents of tissue vascularization in muscular tissue. Ultimately, the implant breaks down into particles smaller than 10 μm, which undergo phagocytosis by macrophages (Anderson and Shive, 1997). Transmission electron microscopy (TEM), Raman spectroscopy, and imaging showed phagocytes containing intracellular polymeric particles.

Besides the site of implantation, other parameters that affect the in vivo degradation are the *size* and the *shape* of the implant. The tissue response to a biomaterial depends to a large extent on the shape of the implant (Anderson, 1994; Matlaga et al., 1976). Being a surface related process, a higher surface area (due to fragmentation or porosity) can affect the tissue response. In addition, sharp angular shapes can induce a higher response than more rounded shapes (Matlaga et al., 1976). The formation of a fibrous capsule as part of the foreign body response may also affect the rate of degradation. Furthermore, one would expect that larger implants require longer degradation times. However, this may depend upon auto or normal catalysis of the degradation process. In the case of poly(lactide-co-glycolide) (PLGA), for example, the in vivo degradation is autocatalyzed due to formation of acidic degradation products (see Classical Experiment). Various studies showed that the autocatalytic effect can be affected by the size of the implant (Grizzi et al., 1995). Comparison of PLGA implants of various sizes indicated that a critical thickness can be determined earlier, which hydrolytic degradation is accelerated due to accumulation of acidic degradation products in the implant.

In conclusion, the in vivo degradation of a biodegradable polymer is affected by many factors, which are influenced by the implant characteristics. This complex situation stresses the importance of performing relevant in vivo experiments when selecting a biodegradable polymer for a specific tissue engineering application.

6.3 FUTURE PERSPECTIVES FOR DEGRADABLE BIOMATERIALS IN TISSUE ENGINEERING

Current research on degradable biomaterials in tissue engineering focuses on balancing the ability to support tissue growth with the degradation rate. A degradable biomaterial should disappear without leaving inert residues behind which may lead to mechanical instability. The optimal biomaterial degradation characteristics and resorption rates are specific for each specific application. Future research on degradable biomaterials will not only focus on the supporting task of these materials but harness degradation characteristics to instruct tissue regrowth. For example, CaPs can be used not only as scaffolds to support cells but can also be designed to incorporate and release growth factors such as growth hormone, BMPs, and insulin-like growth factor to stimulate bone formation or drugs such as bisphosphonates to limit bone resorption (Chapter 11). In order to achieve a sustained release of these molecules, more research is needed to carefully tune the degradation rate while maintaining structural rigidity to support new tissue formation. Trace ions can also be integrated into the crystal structure to stimulate the differentiation of osteoblasts and inhibit bone resorption of osteoclasts. Calcium phosphate nanoparticles are even being researched to deliver cell-death-inducing genes to tumor cells for cancer therapy (Bose and Tarafder, 2012).

There is much interest in the development of responsive degradable polymeric biomaterials in tissue engineering. Such polymers can be made to degrade on demand by action of the cells that grow into or surround the implanted scaffold or supporting hydrogel structure. As cells infiltrate and degrade the temporary cell-supporting scaffold, e.g., by secretion of proteolytic enzymes, newly formed tissue will infiltrate the structure and tissue remodeling can occur (Dispinar et al., 2012; Gustafson et al., 2013; Jo et al., 2010). In addition, responsive degradable scaffolds can be designed to have thermosensitive properties as well. Polymer solutions that are liquid at room temperature can be designed to gel upon implantation at body temperature. In this manner, suspensions of cells can be injected in a minimally invasive manner and form a semisolid cell-loaded construct in situ (Galperin et al., 2010; Jeong et al., 2009).

6.4 SUMMARY

- Degradable biomaterials are suitable as scaffold material for tissue engineering as interference with the development and growth of new tissue and unwanted long-term reactions are prevented.

- Degradable biomaterials for use in tissue engineering should be biocompatible, and the resulting degradation products should be nontoxic. The time required to complete degradation resorption depends on the intended application and preferably matches the formation of functional tissue.
- The in vivo degradation of biomaterials combines physico-chemical degradation (chain scission and dissolution in wet environment), enzymatic activity, and cellular degradation (inflammation, foreign body response).
- The soluble biomaterial degradation products are transported from the implantation site via the lymphatic system to the kidneys, which excrete them from the body.
- Polymers are long-chain macromolecules formed by covalent coupling of monomers. Copolymerization, i.e., the preparation of polymers from two or more types of monomers, is often employed to tune their properties.
- CaPs and BGs have unique bone-bonding properties related to their superficial phase transformation mechanisms involving dissolution and precipitation.
- The bioceramics are (partially) crystallized minerals resembling bone mineral. Their degradation profile is partly dictated by the thermodynamic solubility of the crystalline phase.

The nonresorbable HA and the resorbable β-TCP are the most commonly used ceramics in clinics, and they can be combined to tailor their degradation rate.

- The most important degradable polymers are polyesters. In the presence of water, chain scission occurs by hydrolysis of ester bonds, which lowers the molecular weight. The resulting oligomers dissolve in the surrounding environment or undergo further hydrolysis.
- More advanced degradable biomaterials will have an optimal balance between mechanical properties and degradation rate. In addition, incorporated functional groups will attract cells and/or growth factors required for tissue formation.

CLASSICAL EXPERIMENT: THE DISCOVERY OF BIOACTIVE GLASSES

While attending a US Army Materials Research Conference in 1967, Larry Hench was horrified by the stories of Vietnam War battlefield casualties that were causing thousands of amputations in young soldiers. "Instead of making materials to destroy people," Hench took the challenge to "make materials to repair people." Indeed, in the 1960s, bone implants made out of metals and plastics were not suitable for bone repair, as they were rejected too often by the body. His idea was to develop a material that would be accepted by the host bone. Because bone is composed of a mineral phase made up of calcium and phosphate, he hypothesized that an implant containing calcium and phosphate in the right proportions would not be rejected by the body. He designed three compositions of glass that contained calcium and phosphate, silicon dioxide to hold the structure together, and soda to stimulate the melting of glass in the furnace. These samples were then implanted into the thighbones of rats. Larry Hench was called by the surgeon who was taking care of the animal experiments and reports his conversation:

Six weeks later Ted called me and yelled "Larry, what is that stuff you gave me?" He was so excited, I thought for sure the glasses had killed the rats. So I quickly replied, "Calm down, Ted. They're only the first tries. There are lots of other compositions I can make." I'll never forget his answer. He said "Larry, you don't need to make any other glasses. The first ones work. Those

CLASSICAL EXPERIMENT: THE DISCOVERY OF BIOACTIVE GLASSES *Continued*

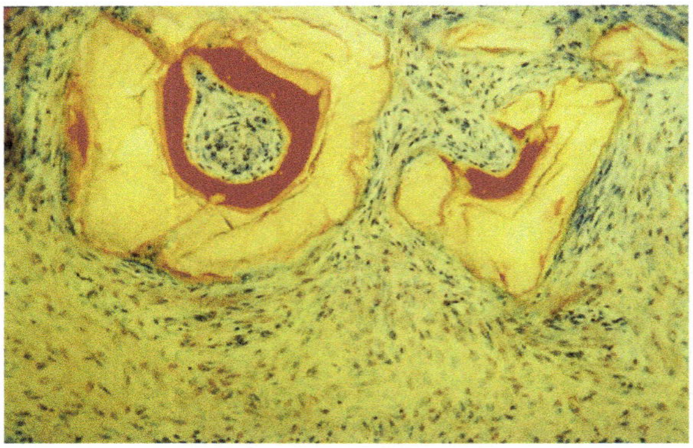

FIGURE CE1

Bone tissue formation in excavated bioactive glass particles of narrow size range. Bone tissue is stained red. Also note the channels connecting the interior of the particle with the surrounding milieu (Ducheyne and Qiu, 1999).

implants won't come out. They are bonded to the bone. I've never seen anything like it before." He was right. The glass implants did not come out ... A bond formed that was as strong as bone. In contrast, control implants of other materials slipped easily out of the bone because of the scar tissue formed at their interface (Figure CE1).

Several BGs were derived from the first Bioglass 45S5 composition by varying the percentage of SiO_2, CaO, Na_2O, and P_2O_5. Their ability to form bone bonding with time was evaluated in vivo. This thorough study described the relationship between glass composition, their dissolution/reprecipitation properties, and their bone-bonding ability. Along with the development of new biological assays, it has also been found that the dissolution products of BGs stimulated osteogenic differentiation. Although these bioceramics are relatively "old" biomaterials, their ability to trigger bone formation are incomparable with other biomaterials. Nowadays, they benefit from new technologies allowing, on one hand, to investigate the reason behind their bone-bonding ability, and on the other hand to process them into high-performance scaffolds for bone tissue engineering.

CLASSICAL EXPERIMENT: DEGRADATION BEHAVIOR OF POLY-DL-LACTIC ACID

Amorphous, noncrystallizable copolymers based on L-lactide, D-lactide, and glycolide show a surface to center differentiation upon hydrolysis, resulting in a hollowing out of the specimens during degradation (Grizzi et al., 1995; Li et al., 1990). This phenomenon is related to the autocatalytic hydrolysis of the ester bonds in the main chain.

Until the publications of Vert and coworkers, the extent of degradation of these polymers both in vitro and in vivo was investigated by visual observation, viscometry, and changes in specimen mass. Polymer degradation was considered to be a homogeneous process, where water absorption throughout the polymer is followed by hydrolysis reactions. However, careful analysis of the molecular weight of the degrading polymer in 2 mm thick specimens showed a bimodal molecular weight distribution. It was found that in such poly(lactide) specimens the

CLASSICAL EXPERIMENT: DEGRADATION BEHAVIOR OF POLY-DL-LACTIC ACID *Continued*

inner part of the specimen degrades at a higher rate than the outside (Grizzi et al., 1995; Vert, 2005).

When the specimen is placed in the aqueous medium, water penetrates into the material. Hydrolytic cleavage of the polymer chains occurs and as carboxylic acid groups are generated, autocatalysis occurs as well. Within the specimen, degradation of the still insoluble polymer chains proceeds homogenously via an autocatalytic mechanism. As soon as the molecular weight of the oligomers formed becomes low enough to allow solubility in the surrounding aqueous medium, these oligomers diffuse to the surface of the specimen and into the surrounding medium as they continue to degrade. This process, which combines hydrolytic degradation, diffusion, and solubilization, results in a differentiation between the rates of degradation at the surface and at the interior of the polymer specimen. As a result, hollow specimens can be formed during degradation, as shown in Figure CE2.

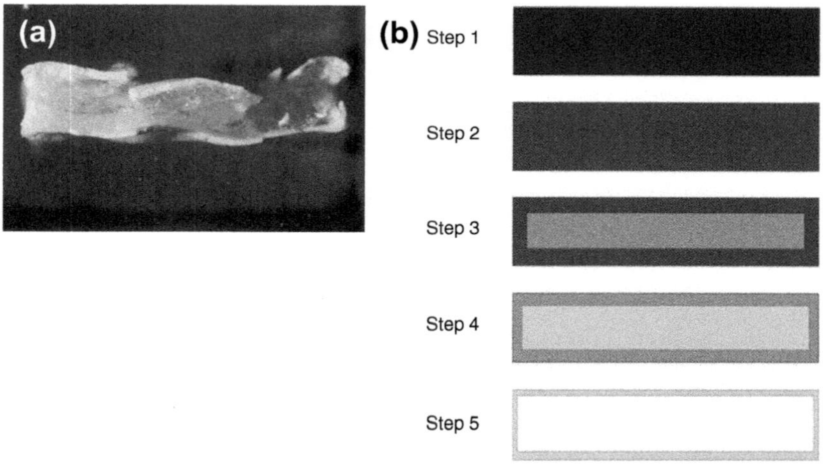

FIGURE CE2

(a) Cross-section of a PDLLA specimen degraded for 5 weeks in saline buffer. (b) Schematic representation of the different steps of the degradation of PDLLA specimens in aqueous medium. *(Reproduced with permission from Li et al. (1990).)* Step 1: initial specimen; Step 2: water absorption, start of ester bond cleavage, and decrease in molecular weight; Step 3: differentiation between surface and center, with dramatic decrease in molecular weight in inner part of the specimen; Step 4: diffusion of oligomers through thinning surface when molecular weight is low enough to allow solubilization in the medium; Step 5: hollow shell remaining after release of oligomers and slow degradation of the shell. *(Reproduced with permission from Li et al. (1990).)*

REFERENCES

Albertsson, A.-C., Eklund, M., 1995. Influence of molecular structure on the degradation mechanism of degradable polymers: in vitro degradation of poly-(trimethylenecarbonate), poly(trimethylenecarbonate-co-caprolactone), and poly(adipic anhydride). J. Appl. Polym. Sci. 57, 87–108.

Albertsson, A.-C., Varma, I.K., 2002. Polyesters: synthesis, properties and applications. Adv. Polym. Sci. 157, 1–40.

Ali, S.A.M., Doherty, P.J., Williams, D.F., 1994. Molecular biointeractions of biomedical polymers with extracellular exudates ad inflammatory cells and their effects on the biocompatibility, in vivo. Biomaterials 15 (10), 779–785.

Anderson, J.M., 1994. In vivo biocompatibility of implantable delivery systems and biomaterials. Eur. J. Pharm. Biopharm. 40, 1–8.

Anderson, J.A., Shive, M.S., 1997. Biodegradation and biocompatibility of PLA and PLGA microspheres. Adv. Drug. Deliv. Rev. 28, 5–24.

Arinzeh, T.L., et al., 2005. A comparative study of biphasic calcium phosphate ceramics for human mesenchymal stem-cell-induced bone formation. Biomaterials 26 (17), 3631–3638.

Barradas, A.M.C., et al., 2013. Molecular mechanisms of biomaterial-driven osteogenic differentiation in human mesenchymal stromal cells. Integr. Biol. 5 (7), 920–931.

Barradas, A.M.C., et al., 2011. Osteoinductive biomaterials: current knowledge of properties, experimental models and biological mechanisms. Eur. Cells Mater. 21, 407–429.

Barrère, F., et al., 2008. Advanced biomaterials for skeletal tissue regeneration: instructive and smart functions. Mater. Sci. Eng. Rep. 59 (1–6), 38–71.

Barrère, F., van Blitterswijk, C.A., de Groot, K., 2006. Bone regeneration: molecular and cellular interactions with calcium phosphate ceramics. Int. J. Nanomed. 1 (3), 317–332.

Bezemer, J.M., Grijpma, D.W., Dijkstra, P.J., et al., 1999. A controlled release system for proteins based on poly(ether ester) block-copolymers: polymer network characterization. J. Control. Release. 62 (3), 393–405.

Bodde, E.W.H., et al., 2007. Bone regeneration of porous b-tricalcium phosphate (Conduit TM TCP) and of biphasic calcium phosphate ceramic (Biosel 1) in trabecular defects in sheep. J. Biomed. Mater. Res. A 82, 711–722.

Bose, S., Tarafder, S., 2012. Calcium phosphate ceramic systems in growth factor and drug delivery for bone tissue engineering: a review. Acta Biomater. 8, 1401–1421.

Bostman, O., Hirvensalo, E., Vainionpaa, S., et al., 1989. Ankle fractures treated using biodegradable internal fixation. Clin. Orthop. Relat. Res. 238, 195–203.

Burdick, J.A., Peterson, A.J., Anseth, K.S., 2001. Conversion and temperature profiles during the photoinitiated polymerization of thick orthopaedic biomaterials. Biomaterials 22 (13), 1779–1786.

Von Burkersroda, F., Sched, L., Göpferich, A., 2002. Why degradable polymers undergo surface erosion or bulk erosion. Biomaterials 23, 4221.

Bohner, M., Lemaitre, J., 2009. Can bioactivity be tested in vitro with SBF solution? Biomaterials 30 (12), 2175–2179.

Changez, M., Koul, V., Krishna, B., et al., 2005. Studies on biodegradation and release of gentamcin sulphate from interpenetrating network hydrogels based on poly(acrylic acid) and gelatin: in vitro and in vivo. Biomaterials 25 (1), 139–146.

Davison, N.L., Luo, X., Schoenmaker, T., Everts, V., Yuan, H., Barrère-de Groot, F., de Bruijn, J.D. (2014). Submicron-scale surface architecture of tricalcium phosphate directs osteogenesis in vitro and in vivo. Eur. Cell. Mater. 27, 281–297; discussion 296–297.

Von Doernberg, M.-C., et al., 2006. In vivo behavior of calcium phosphate scaffolds with four different pore sizes. Biomaterials 27 (30), 5186–5198.

Deschamps, A.A., Claase, M.B., Sleijster, W.J., et al., 2002. Design of segmented poly(ether ester) materials and structures for the tissue engineering of bone. J. Control. Release. 78 (1–3), 175–186.

Deschamps, A.A., Van Apeldoorn, A.A., De Bruijn, J.D., et al., 2003. Poly(ether ester amide)s for tissue engineering. Biomaterials 24, 2643–2652.

Dispinar, T., Van Camp, W., De Cock, L.J., De Geest, B.G., Du Prez, F.E., 2012. Redox-responsive degradable PEG cryogels as potential cell scaffolds in tissue engineering. Macromol. Biosci. 12, 383–394.

Dorozhkin, S., 2011. Medical application of calcium orthophosphate bioceramics. BIO 1 (1), 1–51.

Ducheyne, P., El-Ghannam, A., Shapiro, I., 1994. Effect of bioactive glass templates on osteoblast proliferation and in vitro synthesis of bone-like tissue. J. Cell. Biochem. 56 (2), 162–167.

Ducheyne, P., Qiu, Q., 1999. Bioactive ceramics: the effect of surface reactivity. Biomaterials (20), 2287–2303.

Elbert, D.L., Hubbell, J.A., 2001. Conjugate addition reactions combined with free-radical cross-linking for the design of materials for tissue engineering. Biomacromolecules 2 (2), 430–441.

Elliot, J.C., 1994. Structure and Chemistry of the Apatites and Other Calcium Orthophosphates. Elsevier, Amsterdam.

Li, S.M., Garreau, H., Vert, M., 1990. Structure-property relationships in the case of the degradation of massive poly(a-hydroxy acids) in aqueous media. J. Mater. Sci. Mater. Med. 1, 198.

Lu, L., Peter, S.J., Lyman, M.D., et al., 2000. In vitro and in vivo degradation of porous poly(DL-lactic-co-glycolic acid) foams. Biomaterials 21, 1845–1873.

Fu, K., Pack, D.W., Klibanov, A.M., Langer, R., 2000. Visual evidence of acidic environment within degrading poly(lactic-co-glycolic acid) (PLGA) microspheres. Pharm. Res. 17 (1), 100–106.

Galperin, A., Long, T.J., Ratner, B.D., 2010. Degradable, thermo-sensitive poly(N-isopropyl acrylamide)-based scaffolds with controlled porosity for tissue engineering applications. Biomacromolecules 11, 2583–2592.

García, A.J., Ducheyne, P., Boettiger, D., 1998. Effect of surface reaction stage on fibronectin-mediated adhesion of osteoblast-like cells to bioactive glass. J. Biomed. Mater. Res. 40 (1), 48–56.

Gonda, Y., et al., 2009. Stimulatory effect of hydrothermally synthesized biodegradable hydroxyapatite granules on osteogenesis and direct association with osteoclasts. Biomaterials 30 (26), 4390–4400.

Göpferich, A., 1996. Mechanisms of polymer degradation and erosion. Biomaterials 17, 103.

Göpferich, A., Langer, R., 1993. Modeling of polymer erosion. Macromolecules 26, 4105.

Grizzi, I., Garreau, H., Li, S., Vert, M., 1995. Hydrolytic degradation of devices based on poly(DL-lactic acid) size dependence. Biomaterials 16 (4), 305–311.

Gustafson, J.A., Price, R.A., Frandsen, J., Henak, C.R., Cappello, J., Ghandehari, H., 2013. Synthesis and characterization of a matrix-metalloproteinase responsive silk-elastinlike protein polymer. Biomacromolecules 14, 618–625.

He, S., Yaszemski, M.J., Yasko, A.W., et al., 2000. Injectable biodegradable polymer composites based on poly(propylene fumarate) cross-linked with poly (ethylene glycol)-dimethacrylate. Biomaterials 21 (23), 2389–2394.

Hench, L.L., et al., 1971. Mechanisms of interfacial bonding between ceramics and bone. J. Biomed. Mater. Res. Symp. 2, 117–141.

Hench, L.L., 1991. Bioceramics: from concept to clinic. J. Am. Ceram. Soc. 74 (7), 1487–1510.

Hing, K.A., 2005. Bioceramic bone graft Substitutes: influence of porosity and chemistry. Int. J. Appl. Ceram. Technol. 2 (3), 184–199.

Horch, H.-H., et al., 2006. Synthetic, pure-phase beta-tricalcium phosphate ceramic granules (Cerasorb) for bone regeneration in the reconstructive surgery of the jaws. Int. J. Oral. Maxillofac. Surg. 35 (8), 708–713.

Hutmacher, D.W., Schantz, T., Zein, I., et al., 2001. Mechanical properties and cell cultural response of polycaprolactone scaffolds designed and fabricated via fused deposition modeling. J. Biomed. Mater. Res. 55 (2), 203–216.

Jensen, S.S., et al., 2009. Comparative study of biphasic calcium phosphates with different HA/TCP ratios in mandibular bone defects. A long-term histomorphometric study in minipigs. J. Biomed. Mater. Res. B 90 (1), 171–178.

Jeong, Y., Joo, M.K., Bahk, K.H., Choi, Y.Y., Kim, H.-T., Kim, W.-K., Lee, H.J., Sohn, Y.S., Jeong, B., 2009. Enzymatically degradable temperature-sensitive polypeptide as a new in-situ gelling biomaterial. J. Control. Release. 137, 25–30.

Jo, Y.S., Rizzi, S.C., Ehrbar, M., Weber, F.E., Hubbell, J.A., Lutolf, M.P., 2010. Biomimetic PEG hydrogels crosslinked with minimal plasmin-sensitive tri-amino acid peptides. J. Biomed. Mater. Res. A 93, 870–877.

Kondo, N., et al., 2006. Osteoinduction with highly purified beta-tricalcium phosphate in dog dorsal muscles and the proliferation of osteoclasts before heterotopic bone formation. Biomaterials 27 (25), 4419–4427.

Kowalski, A., Duda, A., Penczek, S., 2000. Kinetics and mechanism of cyclic esters polymerization initiated with tin(II) octoate. 3. Polymerization of l,l-dilactide. Macromolecules 33, 7359.

Kricheldorf, H.R., Kreiser-Saunders, I., Stricker, A., 2000. Polylactones 48. SnOct2-initiated polymerizations of lactide: a mechanistic study. Macromolecules 33, 702.

Lai, W., et al., 2005. Excretion of resorption products from bioactive glass implanted in rabbit muscle. J. Biomed. Mater. Res. A 75 (2), 398–407.

Liu, Y.L., et al., 2000. Bone healing in porous implants: a histological and histometrical comparative study on sheep. J. Mater. Sci. Mater. Med. 11 (11), 711–717.

Mainil-Varlet, P., Curtis, R., Gogolewski, S., 1997. Effect of in vivo and in vitro degradation on molecular and mechanical properties of various low-molecular-weight polylactides. J. Biomed. Mater. Res. 36, 360–380.

Mann, B.K., Gobin, A.S., Tsai, A.T., et al., 2001. Smooth muscle cell growth in photopolymerized hydrogels with cell adhesive and proteolytically degradable domains: synthetic ECM analogs for tissue engineering. Biomaterials 22 (22), 3045–3051.

Mastrogiacomo, M., et al., 2005. Tissue engineering of bone: search for a better scaffold. Orthod. Craniofac. Res. 8 (4), 277–284.

Matlaga, B.F., Yasenchak, L.P., Salthouse, T.N., 1976. Issue response to implanted polymers: the significance of sample shape. J. Biomed. Mater. Res. 10, 391–397.

Middleton, J.C., Tipton, A.J., 2000. Synthetic biodegradable polymers as orthopedic devices. Biomaterials 21, 2335–2346.

Nijenhuis, A.J., Grijpma, D.W., Pennings, A.J., 1992. Lewis acid catalyzed polymerisation of L-lactide. Kinetics and mechanism of esters of bulk polymerization. Macromolecules 25, 6419–6424.

Nyangoga, H., et al., 2010. A non-steroidal anti-inflammatory drug (ketoprofen) does not delay beta-TCP bone graft healing. Acta Biomater. 6 (8), 3310–3317.

Odian, G., 2004. Principles of Polymerization, fourth ed. Wiley Interscience, Hoboken.

Peltola, M.J., et al., 2003. Frontal sinus and skull bone defect obliteration with three synthetic bioactive materials. A comparative study. J. Biomed. Mater. Res. B 66 (1), 364–372.

Peltola, M.J., et al., 2001. In vivo model for frontal sinus and calvarial bone defect obliteration with bioactive glass S53P4 and hydroxyapatite. J. Biomed. Mater. Res. 58 (3), 261–269.

Pitt, C.G., Gu, Z., 1987. Modification of the rates of chain cleavage of poly(e-caprolactone) and related polyesters in the solid state. J. Control. Release. 4, 283–292.

Pitt, C.G., Chasalow, F.I., Hibionada, D.M., Klimas, D.M., Schindler, A., 1981. Aliphatic polyesters. I. The degradation of poly(ε-caprolactone) in vivo. J. Appl. Polym. Sci. 26, 3779–3787.

Radin, S., et al., 2000. In vitro transformation of bioactive glass granules into Ca–P shells. J. Biomed. Mater. Res. 49 (2), 264–272.

Ratner, B.D., Hoffman, A.S., Schoen, F.J., Lemons, J.E. (Eds.), 2013. Biomaterials Science, an Introduction to Materials in Medicine, third ed. Academic Press, Elsevier, Amsterdam.

de Ruiter, A., Dik, E., van Es, R., van der Bilt, A., Janssen, N., Meijer, G., Koole, R., Rosenberg, A., 2014. Micro-structured calcium phosphate ceramic for donor site repair after harvesting chin bone for grafting alveolar clefts in children. J. Craniomaxillofac Surg. 42(5), 460–468.

Sawhney, A.S., Pathak, C.P., Hubbell, J.A., 1993. Bioerodible hydrogels based on photopolymerized poly(ethylene glycol)-co-(alpha-hydroxy acid) diacrylate macromers. Macromolecules 26, 581–587.

Stoll, G.H., Nimmerfall, F., Acemoglu, M., et al., 2001. Poly(ethylene carbonate)s, part II: degradation mechanisms and parenteral delivery of bioactive agents. J. Control. Release. 76, 209–225.

Tamada, A., Langer, R., 1993. Erosion kinetics of hydrolytically degradable polymers. Proc. Natl. Acad. Sci. USA 90, 552–556.

Tanaka, T., Saito, M., Chazono, M., et al., 2010. Effects of alendronate on bone formation and osteoclastic resorption after implantation of beta-tricalcium phosphate. J. Biomed. Mater. Res. A 93 (2), 469–474.

Temenoff, J.S., Mikos, A.G., 2008. Biomaterials: The Intersection of Biology and Materials Science. Pearson-Prentice Hall, Upper Saddle River, USA.

Uhrich, K.E., Cannizzaro, S.M., Langer, R.S., Shakesheff, K.M., 1999. Polymeric systems for controlled drug release. Chem. Rev. 99, 3181–3198.

Vert, M., 2005. Aliphatic polyesters: great degradable polymers that cannot do everything. Biomacromolecules 6, 538–546.

Walsh, W.R., et al., 2008. Beta-TCP bone graft substitutes in a bilateral rabbit tibial defect model. Biomaterials 29 (3), 266–271.

Wenisch, S., et al., 2003. In vivo mechanisms of hydroxyapatite ceramic degradation by osteoclasts: fine structural microscopy. J. Biomed. Mater. Res. A 67 (3), 713–718.

Williams, D.F., 2014. The biomaterials conundrum in tissue engineering. Tissue Eng. A 20, 1129–1131.

Wiltfang, J., et al., 2002. Degradation characteristics of alpha- and beta-tri-calcium-phosphate (TCP) in minipigs. J. Biomed. Mater. Res. 63, 115–121.

Xynos, I.D., Hukkanen, M.V., et al., 2000. Bioglass 45S5 stimulates osteoblast turnover and enhances bone formation in vitro: implications and applications for bone tissue engineering. Calcif. Tissue. Int. 67 (4), 321–329.

Yuan, H., et al., 2010. Osteoinductive ceramics as a synthetic alternative to autologous bone grafting. Proc. Natl. Acad. Sci. USA 107 (31), 13614–13619.

CHAPTER 7

Cell–Material Interactions

Penelope M. Tsimbouri, Laura E. McNamara, Enateri V. Alakpa, Matthew J. Dalby, Lesley-Anne Turner

Centre for Cell Engineering, Institute of Molecular, Cell and Systems Biology, College of Medical, Veterinary and Life Sciences, Joseph Black Building, University of Glasgow, Glasgow, Scotland, UK

LEARNING OBJECTIVES

- Understand the importance of the cell–material interface for tissue engineering
- Appreciate that the adhesome is a specialized and highly complex cell structure, currently thought to consist of well over 150 components with approximately 690 possible interactions
- To understand the biochemical nature of mechanotransduction
- Know that cytoskeleton-mediated biochemical signaling may confer advantages in terms of targeted mechanotransduction and faster signal transduction times compared to signals relayed by diffusion
- Know that solid-state signals from the interface are integrated with other signals both locally and globally resulting in a whole-cell response to materials
- Appreciate the dynamic and complex structure of the extracellular matrix and its ability to respond and interact with attached cells in a bidirectional manner, making it an "active" part of the body, rather than the inactive and passive support role that it has traditionally been assigned
- Know that the roles of extracellular cues can be dissected by engineering synthetic surfaces with alterations in topography, chemistry, or mechanics (stiffness)
- Understand that physical cues from topography, chemistry, and substrate mechanics have strong effects on cells (including stem cells) altering growth and phenotype
- Appreciate that cell–material interface information can be used to generate new stem cell–specific tissue engineering scaffold materials

Tissue Engineering.
Copyright © 2015 Elsevier Inc. All rights reserved.

7.1 INTRODUCTION

This chapter will discuss how cells interact and understand their extracellular environments. This is critical to developing new materials as our ability to engineer surface chemistry, topography, and mechanics is improving apace and it is clear that these signals have large effects on, e.g., stem cells. Development of understanding in this area will allow for the development of novel biomaterial and tissue engineering scaffolds.

7.1.1 Cell–Material/Extracellular Matrix Interactions

Although important for tissue engineering, soluble chemical interactions at the cell–material interface (such as those mediated by soluble growth factors) will not be discussed in detail as they are reviewed in Chapter 6, rather we will focus on solid-state interactions. At the cell–material interface, solid-state interactions confer effects through physical distortion, chemical signaling, adhesion points, or a combination.

Physical distortion results from force transmission between cell and material due to spatial proximity. For example, extracellular matrix (ECM)-induced distortion of the cell membrane can cause activation of stress-sensitive ion channels triggering downstream biochemical signaling cascades that result in force-dependent changes in cell behaviors. Interactions mediated through chemical means follow similar mechanisms as soluble chemical signaling; for example, growth hormone and cytokine ligand binding to receptors resulting in activation/repression of signaling cascades. In solid-state interactions, these chemical factors are held in place through interactions with insoluble structures, such as the cell cytoskeleton or ECM fibers. If fact, most growth factors in vivo are bound to ECM components or are part of membrane complexes. Adhesion points at the cell–material interface are physicochemical connections between specialized subcellular sites at the cell membrane and specific components of the material surface. There is often not a clear distinction between interaction "types" (i.e., distortion, chemical signaling, and adhesion points); for example, adhesion points enable cells to physically distort the ECM using traction when cells "pull" on collagen fibers during wound healing, resulting in alignment of ECM fibrils.

7.1.2 Adhesion Sites

Adhesion formation and dynamics occur at the nanoscale; the plasma membrane is approximately 10 nm thick, the glycocalyx up to hundreds of nanometers thick, and membrane proteins tens of nanometers in diameter. Although membrane and glycocalyx composition varies between cell types, meaning that the specifics of interactions vary, common adhesion structures and mechanisms are observed. A number of potential cell adhesion molecules have been identified, which include integrins, CD44, syndecans, and discoidin domain receptors (DDRs), although syndecans and DDR's "true" status as adhesion receptors remains to be clarified (Schmidt and Friedl, 2010). Integrins and CD44 molecules are thought to provide high affinity adhesions, binding predominantly extracellular proteins and carbohydrate polymers such as hyaluronic acid, respectively (Schmidt and Friedl, 2010). Integrins in particular are thought to be important for cell adhesion to synthetic materials. Adhesions can be enhanced by cell binding to additional ECM structures/motifs; this mechanism is called synergistic binding. Syndecans and DDRs are thought to synergistically cooperate with integrins to mediate adhesions (Vogel et al., 2006; Morgan et al., 2007).

7.1.3 Integrins

Integrins are transmembrane heterodimeric proteins (with one alpha and one noncovalently bound beta subunit) that act as mechanotransducers; that is, they are able to translate extracellular mechanical stimulations into intracellular responses. Three domains make up integrins: a large extracellular domain, a membrane spanning region, and, typically, a short intracellular cytoplasmic domain (reviewed in Hynes (1999)). Integrins exist in "affinity states"—that is, their affinity for ECM ligands varies and is determined by both extracellular factors (such as divalent cations) and intracellular processes involving small GTPases (Keely et al., 1998; Parise et al., 2000). In higher affinity states, integrins are able to bind specific extracellular ligands resulting in changes (conformational and relative movement of subunits) that alter their intracellular domains and thus functional state. Common to all integrins is their need for stimulation by an external divalent cation (Ca^{2+}or Mg^{2+} specific to the individual integrin) in order to bind extracellular ligands. Intracellular domains interact with a number of cytoplasmic structures and signaling cascades, dependent on the type of integrin, ligand, and the molecules and structures nearby. Unbound integrins (independent of affinity state) diffuse freely around the plasma membrane and do not tend to initiate signaling pathways.

Due to their range of intracellular interactions, integrin activation can result in a number of downstream effects, one of which is integrin clustering (multimerization), whereby other integrins are recruited to the initial attachment site. Clustering is dependent on ligand density, e.g., clustering only happens in osteoblasts, fibroblasts, and melanocytes, if a maximum ligand spacing of less than 73 nm between integrin attachment sites is presented (Arnold et al., 2004). Integrin clustering may be followed by structural and signaling protein recruitment and connection with the cell cytoskeleton, which leads to the formation of larger adhesion structures. Many adhesion structures have been described, including focal complexes, focal adhesions, podosomes, fibrillar adhesions, and three-dimensional (3D) matrix adhesions. All of these are dense, complicated structures and although heterogeneous, they share common features. One of the best-defined adhesions is the focal adhesion, composed of integrins, structural proteins, adaptor proteins, signaling molecules, and cytoskeletal components.

7.1.4 Integrin Mediated Adhesion Structures

Within seconds, the initial connection between integrins and the cytoskeleton results in the build up of tractional forces on the surface (Choquet et al., 1997), which are required in order to counterbalance internal forces produced as a result of increased cytoskeletal tension. In response to these forces, further integrin clustering takes place and molecules are recruited to the adhesion site, resulting in lateral adhesion growth and ultimately maturation into focal adhesions (mechanostimulation). It should be noted that force is not always required for initial integrin clustering, indeed integrins can cluster in the absence of external forces and then connect to the cytoskeleton (Yu et al., 2011; Bakker et al., 2012). Integrin-mediated adhesions vary in size and, as such, are often categorized as focal complexes (approximately 1 μm long), focal adhesions (2–10 μm long) or fibrillar adhesions (1–10 μm long) (Stricker et al., 2011) and (Abercrombie and Dunn, 1975; Davies et al., 1994) (Figure 7.1). Fibrillar adhesions, sometimes described as super-mature adhesions, are typically seen in cells in vivo and in vitro and have been implicated in osteogenesis (Biggs et al., 2009b). Typically, in vitro these structures are found at the leading edge of lamellipodia (the ruffling membrane found at the leading edge of cells where adhesions are rapidly remodeled during locomotion), and around the periphery and central areas of cells.

Molecules recruited to the adhesion site during maturation include structural and adaptor proteins, signaling molecules, and cytoskeletal elements. There are currently thought to be 156 components of the adhesome and due to the multiple interaction capabilities of many components, approximately 690 interactions are possible (Zaidel-Bar et al., 2007) (Figure 7.2). The most abundant of the adhesome molecules function to form its physical structure and can be categorized as adaptor proteins, adhesion receptors (such as integrins), and actin regulators. Adaptor proteins link focal adhesions to the cytoskeleton, e.g., vinculin and tensin reinforce early links. Of the other recruited molecules, most are signaling proteins that signal to other regions of the cell and regulate assembly and turnover of the adhesome itself. In addition to this structural/signaling classification, proteins of the adhesome can be considered as intrinsic or associated; for some proteins (intrinsic) the vast majority of all their known interactions are in the adhesome complex (e.g., vinculin), whereas others (associated) have many additional interactions outside the adhesome (e.g., protein kinases and phosphatases involved in signaling). During maturation protein cross-linking also takes place, e.g. alpha-actinin cross-links F-actin fibrils to form stress fibers in the cytoskeleton. In focal adhesions integrin and actin are separated by approximately 40 nm and the proteins making it are segregated into multiple

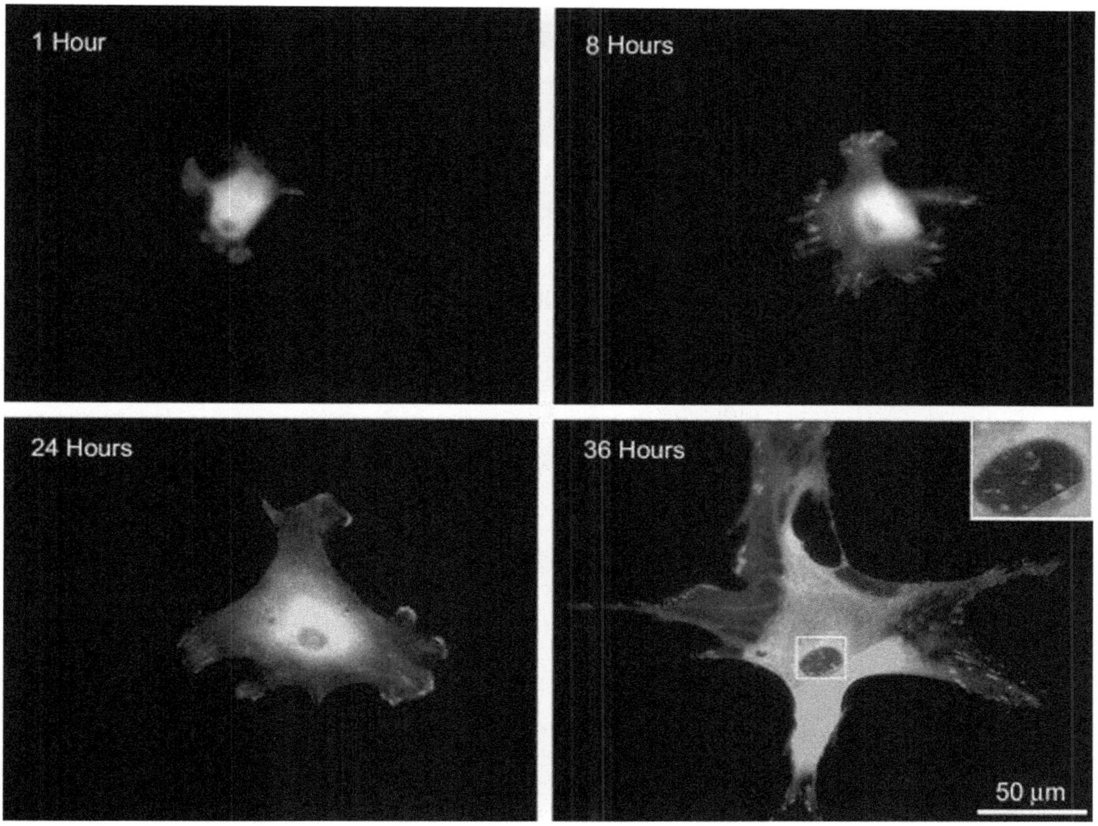

FIGURE 7.1

Fluorescent microscopy images of vinculin staining in human osteoblastic cells cultured on planar poly(methyl methacrylate) substrates at increasing time points. As cells spread, focal complexes present 8 h after seeding were slowly replaced by focal adhesions and then super-mature adhesions by 36 h. Subnuclear adhesion formation on the cellular ventral membrane was evident (boxed area). *Image reprinted from Biggs et al., 2009b. The use of nanoscale topography to modulate the dynamics of adhesion formation in primary osteoblasts and ERK/MAPK signaling in STRO-1+ enriched skeletal stem cells. Biomaterials 30(28), 5094–5103, copyright 2009, with permission from Elsevier.*

protein-specific strata, consisting of an integrin signaling layer (integrin cytoplasmic tails, focal adhesion kinase, and paxillin), an intermediate force transduction layer (talin and vinculin), and an actin-regulatory layer (zyxin, vasodilator-stimulated phosphoprotein, and alpha-actinin) (Kanchanawong et al., 2010). As vinculin is found in all classes of integrin-mediated adhesion and its recruitment correlates with increasing adhesion size, it is often the protein of choice for staining by researchers interested in observing cell adhesion.

Initial adhesions are transient and maturation into focal adhesions is thought to be mediated by changes in cellular tension, either intracellularly or as a result of the extracellular environment. Adhesion size appears to be an important factor

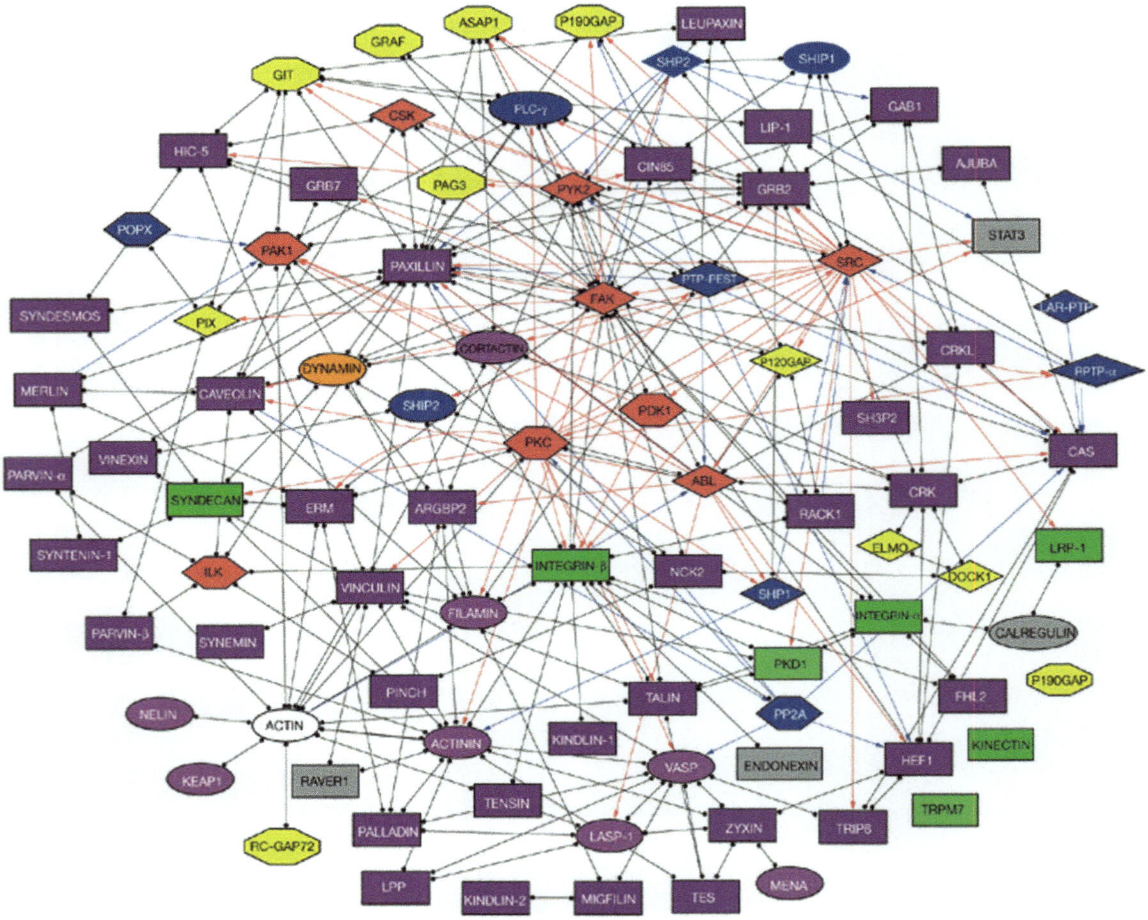

FIGURE 7.2

Diagram illustrating the complexity of the adhesome, showing interactions between intrinsic components. Black lines and full circles: nondirectional binding interactions; blue arrows: directional inhibition (e.g., dephosphorylation); red arrows: directional activation (e.g., phosphorylation). *Reprinted by permission from Macmillan Publishers Ltd: Nature Cell Biology (Zaidel-Bar et al., 2007), copyright (2007).*

for cell mechanics; increasing adhesion size has been found to correlate with increased cellular tension (Engler et al., 2006). Recruitment of molecules and cross-linking expands and strengthens the adhesion as well as decreases the force generated/experienced per molecule (reviewed in (Schoen et al., 2013)), thus feeding back into the force generated signaling controlling adhesion maturation.

Physically, focal adhesions provide a link between extracellular materials and intracellular components of the cytoskeleton and signaling molecules (Figure 7.3). Functionally, focal adhesions provide a route by which extracellular mechanical signals and the physical environment can be translated into

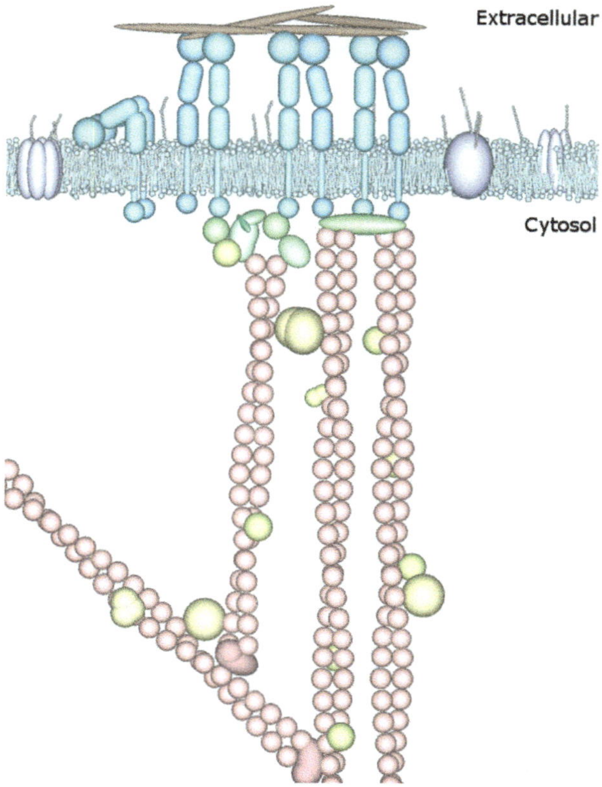

FIGURE 7.3

Schematic illustration of a selection of the molecular components of the adhesome; integrins (blue) connect extracellular fibronectin (brown) with an intracellular complex ccntaining, amongst other structures, adaptor proteins (green). Connectivity between these and signaling molecules (yellow) and/or the cytoskeleton (pink) enables mechanical forces to be converted into intracellular signals.

an intracellular message that cells can bidirectionally communicate with, be that respond to and/or be influenced by. Where cells are already attached to a material surface, and focal adhesions are mature, mechanical stresses can induce new integrin activations and trigger associated changes in focal adhesion dimensions and function.

7.1.5 Mechanotransduction

Upon ligand binding intracellular integrin domains undergo conformational change, which, through signaling protein activation (such as focal adhesion kinase) and/or physical propagation through structural and cytoskeletal proteins (such as talin and alpha-actinin), results in signal propagation through the cell. This process of signal propagation as a result of force is called mechanotransduction (with sensors such as integrins termed mechanosensors) and results in downstream effects as diverse as growth, apoptosis, differentiation, morphology, and proliferation. Mechanotransduction is thought to propagate through cells via either, or a combination of, biochemical and physical pathways.

Both direct signaling protein activation and cytoskeletal protein interactions can result in activation of downstream biochemical signaling cascades. Mechanotransduction via direct signaling protein activation follows mechanisms similar to signaling cascades generated by soluble chemicals such as growth factors—once a mechanotransducer is activated, multiple pathways can cause inhibition or enhancement of any one particular signaling cascade. As described earlier, integrins undergo structural change when activated resulting in activation of adjacent molecules. Adhesion-related signaling cascades are complex and may lead to a variety of cell behaviors depending on the proteins involved. A well-described example of such a cascade is integrin mechanostimulation of focal adhesion kinase (FAK), which activates the ERK/MAPK (extracellular signal-related kinase/mitogen-activated protein kinase) pathway and is known to affect various cellular responses such as proliferation and differentiation (Cattaruzza et al., 2004; Miyamoto et al., 1995; Zhu and Assoian, 1995).

Many molecules and structures are immobilized on the insoluble cytoskeleton anchored to adhesion sites; the cytoskeleton is thought to act as both a scaffold for solid-state biochemical reactions and as a reservoir for sequestered soluble factors (e.g., signaling molecules with cytoskeletal binding domains such as SH3 domains (Bar-Sagi et al., 1993; Koch et al., 1991)). It therefore follows that if integrin signaling is sufficient to influence cytoskeletal molecular conformation, thus modifying cytoskeleton shape, tensions, structure, or kinetics, any of the structures associated with the cytoskeleton may also be influenced (Ingber, 2003). Where these structures are signaling molecules, if sequestered molecules are released or binding sites exposed or hidden, downstream biochemical signaling cascades may be enhanced or inhibited following mechanisms similar to biochemical signaling cascades initiated by soluble chemicals, e.g., cytokines and growth factors (Negrutskii and Deutscher, 1992). In this way, mechanical force can be transduced through the cytoskeleton and using solid-state chemistry can be converted into biochemical reactions.

7.1.6 Tensegrity Theory

The cell cytoskeleton can be described as a tensegrity structure. Tensegrity architecture is the name given to a form of tensile architecture, whereby physical structures are held in place by the balance of tensile and compression forces acting upon them (Ingber, 1993) (Figure 7.4). The opposing forces act in unison to maintain the shape of the structure, whilst providing strength and resilience. Within the cytoskeleton, actin microfilaments in combination with myosins are contractile elements, which pull cell membranes inwards.

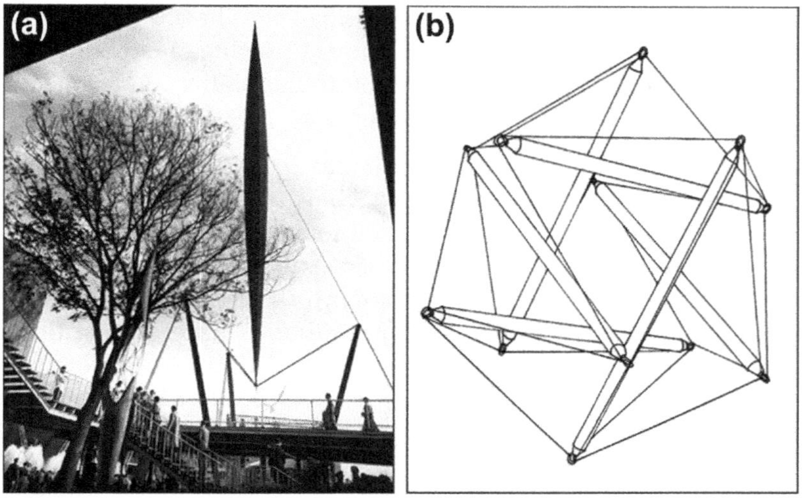

FIGURE 7.4
Images show tensegrity structures: (a) The Skylon Tower erected on the South Bank of the River Thames, London, as part of the 1951 Festival of Britain (designed by Moya, Powell, and Samuely/fabricated by the Painter Brothers of Hereford, England). Tensegrity architecture gives the impression of a tower floating above the ground. (b) A line drawing showing structures in a tensegrity architecture icosahedron. *Images produced with permission (a) "In 16695" by British Official Photograph - Museum of London website (http://www.museumoflondon.org.uk/). Licensed under Public domain via Wikimedia Commons - https://commons.wikimedia.org/wiki/File:In16695.jpg#mediaviewer/File:In16695.jpg. (b)'Tensegrity Icosahedron' 1994, Bob Burkhardt, Design first exhibited by Buckminster Fuler in 1949. Licensed under Public domain via Wikimedia Commons - http://commons.wikimedia.org/wiki/File:Tensegrity_Icosahedron.png.*

They are under tension and as such exert an overall compression force onto cells (Ingber, 2003). These compression forces are resisted intracellularly by stiff microtubules that act like compression struts. Intermediate filaments provide lateral support to microtubules, thus helping them to resist buckling through compression.

When adhered to other cells or substrates, microfilament contractile forces are also resisted extracellularly by transmembrane cell adhesion receptors (e.g., integrins). As such, receptors resist the contractile microfilament forces through opposing tension. Cells are therefore maintained in a prestressed state and are in equilibrium under a balance of forces (Ingber, 2003). That is, the cytoskeletal network is under isometric tension which removes any slack in the system; "lack of slack" means that any mechanical stresses applied to the system (cytoskeleton) are immediately sensed. Even though cytoskeletal structures undergo continuous remodeling at the molecular level, they are stable enough as interconnected units to resist mechanical stress and maintain structural integrity for longer than the turnover time of individual protein monomers. The tensegrity architecture model therefore provides a mechanism whereby mechanical stimulation of the cell cytoskeleton via attachment proteins, at a point on a cell's membrane, causes the cell to react as a whole and results in integrated, global changes in cytoskeletal structure (Maniotis et al., 1997).

Even though cytoskeletal structures undergo continuous remodeling at the molecular level, they are stable enough as interconnected units to resist mechanical stress and maintain structural integrity for longer than the turnover time of individual protein monomers. Stresses tend to dissipate less in stiffer structures, thus the stiff, prestressed cytoskeleton is able to concentrate and focus stresses and facilitates longer distance force propagation compared to less stressed structures. Stiffness differentials between the cytoskeleton and nucleus (the nucleus is the stiffest organelle being approximately 9 times stiffer than the surrounding cytoskeleton (Maniotis et al., 1997)) are thought to facilitate force propagation over the relatively large distances between membranes and nuclear envelope (Wang et al., 2009). Cytoskeleton-mediated mechanotransduction may thus confer a number of advantages above soluble biochemical mechanotransduction conferred via diffusion. The quick response system means that mechanical forces can cause very quick effects, more so than biochemical signaling that takes diffusion time. In addition to quick response times, cytoskeletal-mediated biochemical signaling channels forces along discrete fibers of the cytoskeleton providing a mechanism by which to concentrate stresses on specific molecules at particular locations, whilst protecting other cellular components from these mechanical forces, meaning that only a subset of molecules will experience force levels strong enough to alter their activities (Ingber, 2006).

7.1.7 Extracellular Matrix

The diverse, heterogeneous ECM structure that surrounds and supports cells in tissues is composed of both soluble and insoluble components. Soluble factors include sequestered cytokines and growth factors; insoluble components, of which there are currently thought to be over 100 (Davies, 2001), are largely composed of glycoproteins and proteoglycans able to bind onto specific sites on other ECM molecules, forming a dense mesh like structure (summarized in (Rovensky, 2012)). The importance of the ECM in tissue regeneration is demonstrated by the application of decellularized tissues; for example, the trachea of a cadaver was decellularized leaving only ECM before being repopulated with recipient cells and then successfully transplanted without the need for immunosuppressant drugs (Macchiarini et al., 2008).

Extracellular matrix glycoproteins can be roughly classified as mainly structural (e.g., collagens and elastins) or adhesive (Rovensky, 2012). Adhesive glycoproteins, which include fibronectin, laminin, tenascin, and fibulin, have different domains and each binds to specific molecules, the characteristic domains being those that bind cell surface receptors, e.g., integrins. Many interactions are possible between adhesive glycoproteins and other cell and ECM structures. Although the structures of individual ECM components vary widely, many share common structural motifs, e.g., the arginine-glycine-aspartic acid (RGD) motif found on the hydrophilic loops of a number of ECM molecules including fibronectin, vitronectin, and tenascin. RGD is a particularly widespread motif, and is required for cell binding to fibronectin as well as for binding of many (although not all) integrins. For further discussion on components of the ECM, please refer to Chapter 5.

7.1.8 Bidirectional Signaling

Cell adhesion sites enable a route by which bidirectional signals can be relayed at the cell–material interface. Outside-in signaling (i.e., from ECM to cell) is critical for cell viability; cell adhesion and mechanical stiffness cues from the ECM are mechanotransduced into intracellular signals that influence a host of cell behaviors, as discussed previously. In the same way inside-out signaling (i.e., from cell to ECM) is important in maintenance of the ECM, for example, in fibronectin fibril assembly. Fibronectin is secreted from cells as a dimer that does not bind other fibronectin dimers. However, excreted dimerized fibronectin binds activated (high affinity) integrins, which results in the generation of tension via the actin cytoskeleton. This tension stretches the bound fibronectin, which results in exposure of cryptic self-association binding sites enabling other fibronectin molecules to bind and thus a chain reaction of self-assembly and fibril formation (fibrillogenesis) ensues (Baneyx et al., 2002; Mao and Schwarzbauer, 2005). Cell signaling to the ECM through adhesion points can also result in cells "pulling" on ECM structures, causing cell-mediated remodeling. For example, cells forming and maintaining bone can orient collagen fibers and trabeculae in the best direction to withstand the applied stress.

7.1.9 Synthetic Materials

It is these cell–ECM interactions that the tissue engineer needs to understand and exploit when designing synthetic materials to take the role of the ECM. It is important to note that cells do not tend to interact directly with synthetic materials. Whether in vitro or in vivo, when cells encounter synthetic surfaces a layer of proteins, from culture media or plasma, respectively, adheres to the surface within milliseconds (Atala et al., 2008) (Figure 7.5). Protein adsorption is affected by the properties of the material, such as surface charge, wettability (hydrophobicity), and topography. It is this adsorbed layer of protein that is thought to influence cell interactions with the underlying material (Wilson et al., 2005). The remainder of this chapter will review ways in which the material surface can be engineered to enhance cell interaction and gain control of phenotype.

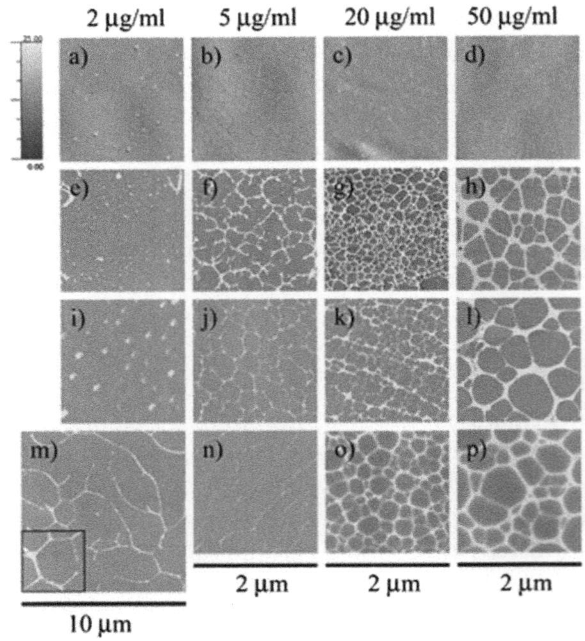

FIGURE 7.5

Phase magnitude atomic force microscopy images of fibronectin adsorption on four different materials, poly(methyl acrylate), poly(butyl acrylate), poly(ethyl acrylate), and poly(hexyl acrylate) (PMA, PEA, PBA, and PHA, respectively), after 10 min of incubation in different solution concentrations. *Images reproduced in part from Guerra et al. (2010) with permission of The Royal Society of Chemistry.*

7.2 SURFACE CHEMISTRY

First, we will consider the role and exploitation of chemistry at the cell–material interface. Given that integrins recognize motifs such as RGD and isoleucine-lysine-valine-alanine-valine (IKVAV) in proteins of the ECM, the surface chemistry of a material substrate will affect cell attachment and spreading. Surface chemistry refers to the chemical properties of the surface of the material in addition to any other modifications that are added to the surface; both of these aspects modulate cellular responses. The chemistry of the materials surface modulates the ability of adhesive proteins to deposit onto it and thus affect cell adhesion, and the presentation or concealment of additional modified chemical or peptide cues can directly affect the engagement of cell adhesion molecules such as integrins. Chemical surface patterning offers an advanced platform to study the contribution of different chemical cues to processes such as integrin clustering, cell adhesion, and stem cell differentiation (please see Chapter 2 for further discussion surrounding stem cells).

7.2.1 Hydrophobicity and Hydrophilicity

Surfaces can be hydrophobic ("water-hating"), which repel water on the surface, or hydrophilic ("water loving"), which attract water and polar molecules. A bead of water on a hydrophobic surface will become rounded, a feature that can be measured using a technique called contact angle goniometry, which assesses the angle between the surface and the droplet of water. Hydrophobic surfaces have high water contact angles (>90°), while a water droplet on a hydrophilic surface will have a low contact angle (<90°) because the droplet is able to spread more on this surface. Extremely hydrophobic surfaces (known as super-hydrophobic surfaces) can have contact angles >150°.

Hydrophobic surfaces are not generally conducive to cell culture. The chemistry of some surfaces can be modified using plasma treatment, where a flow of oxygen plasma is used to introduce polar groups on to the surface in order to increase its hydrophilicity (Tan et al., 2010). This can be a temporary effect as some materials, such as the polymer poly(dimethylsiloxane) (PDMS), can recover their hydrophobicity with time. Typically increasing the hydrophilicity of a surface promotes protein adsorption. For in vitro tissue culture in the laboratory, these proteins are generally provided by fetal bovine serum in the culture medium, which possesses a complex, undefined mix of growth factors and proteins. Once such proteins are adsorbed to the surface, cells can begin to attach and spread on the substrate—a result of, e.g., integrin ligation and initiation of signaling cascades that instruct the cells to attach and proliferate. In the absence of these cues, noncancerous cells undergo anoikis, where they fail to attach and eventually become apoptotic.

7.2.2 Patterning Using Surface Chemistry

In the laboratory, surface chemistry can be harnessed to selectively pattern zones on materials where cells are permitted to attach to the substrate and other areas where cell adhesion is prevented. The polymer poly(ethylene glycol) (PEG) is often used to block adhesion, since PEG is considered to be a "nonfouling" surface which discourages protein adsorption, and thus precludes cell spreading. The development of effective nonfouling surface chemistries is also of interest for coating medical devices, to prevent the accumulation of pathogenic organisms. Surface chemistries can be used to pattern devices, but compared with, e.g., topographies (as will be discussed) suffer from the limitation that they are generally more rapidly denuded from the surface. PEG derivatives, where PEG is attached to other polymers (known as "PEG grafts"), are being developed with the aim of extending the persistence time of the surface chemistry. For mammalian cell research, combinations of PEGylated regions and hydrophilic regions coated with adhesive proteins can be used to generate tightly zoned areas for cell adhesion or nonadhesive sites.

This approach has been employed to study processes such as adhesion-mediated cell survival. The results of one such study indicated that apoptosis was initiated when the cells were confined to small adhesive islands, as the cells were unable to spread sufficiently to gain the appropriate integrin-mediated survival signals (Chen et al., 1997).

Defined regions of surface chemistry can also be used to affect stem-cell fate. Changes in cytoskeletal tension can modulate fate determination, which was illustrated by the confinement of mesenchymal stem cells (MSCs) to small adhesive squares, which predisposed the cells to adipogenesis, while larger adhesive squares tended to stimulate osteogenesis (McBeath et al., 2004). These effects were dependent on cell shape and tension-sensitive G-protein Rho A and its effector kinase ROCK (Rho kinase) (McBeath et al., 2004)—for a more detailed review of this classical manuscript please see Box "State of the Art Manuscript,". This observation was extended in an elegant study where MSCs were cultured on proadhesive surface chemistries patterned into round or angular features and coated with fibronectin. The surrounding regions were treated with PEG to reduce protein adsorption and define shape boundaries. On the angular features greater cytoskeletal tension was generated by the production of large actin stress fibers, and the stem cells tended to undergo osteogenesis, differentiating along the bony lineage (Kilian et al., 2010). In contrast, when the MSCs were cultured on rounded adhesive features, cytoskeletal tension was reduced and the cells tended to undergo adipogenesis, becoming fat (Kilian et al., 2010) (Figure 7.6). Adhesive chemical patterns with defined shapes were also used to examine

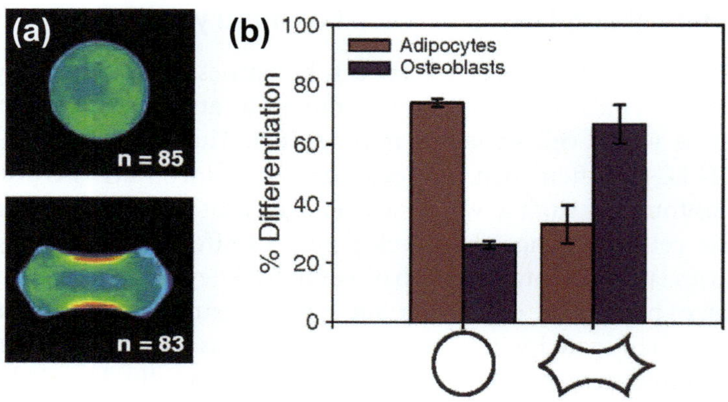

FIGURE 7.6
An example of the use of adhesive patterned shapes to confine stem cell morphology, showing the reduced cytoskeletal tension on the rounded shape, and increased tension (myosin II staining) on the angular surface in (a), and the predisposition for adipogenesis on the circular shape and osteogenesis on the angular "holly" shape in (b). *Reproduced with permission from Kilian et al. (2010).*

changes in the organization of cytoskeletal actin microfilament and vimentin intermediate filaments by tightly confining cell morphologies to different patterns (Shabbir et al., 2014). Such adhesive features offer the opportunity to investigate cellular architecture and physiology in response to morphological changes in a tightly controlled manner.

7.2.3 Effect of Chemistry on Cell Behavior

Studies into the effects of distinctions in the presentation of a number of surface chemistries on planar glass surfaces found that different lineages could be induced by culture of the MSCs on distinct substrate chemistries, with presentation of –OH groups predisposing cells toward chondrogenesis, $-NH_2$ groups promoting osteogenesis, and $-CH_3$ groups assisting with the retention of stem cell markers (Curran et al., 2005). In addition, modifying the chain length and side chains of $-CH_3$ groups could further modulate the MSC phenotype, with particular presentations of $-CH_3$ groups offering potentially useful surfaces for the selection of MSCs from bone marrow extracts and retention of stem cell characteristics (Curran et al., 2005; Curran et al., 2011). A study of hydrogels functionalized with simple chemistries suggested that presentation of phosphate predisposed stem cells to osteogenesis, while presentation of hydrophobic t-butyl groups encouraged the cells to undergo adipogenesis (Benoit et al., 2008). The development of substrates with surface chemistry capable of controlling stem cell phenotype would have important implications for the supply of stem cells for clinical applications.

7.2.4 Ligand Spacing

Due to the nature of integrin interactions, which require engagement of multiple integrins in order to initiate the receptor clustering required for the induction of integrin-mediated signaling, the density and spacing of adhesive ligands presented on a surface affect the induction of downstream signaling. The use of tightly controlled spacing and arrangements of surface chemistries, such as self-assembled monolayers (SAMs, e.g., self-assembling arrays of gold nanoparticles), has enabled this process to be studied in more detail. Using gold nanoparticle SAMs conjugated to adhesive RGD ligands, the maximum inter-feature density capable of inducing integrin clustering in a number of cell types was shown to be 70 nm, as has been mentioned by Arnold et al. (2008). Using dip-pen nanolithography, a technique which enables the patterning of specific arrangements of chemical surface modifications using chemical "inks" to print arrays of various moieties (functional groups such as $-CH_3$ and –OH) shown to affect stem-cell fate, it has been demonstrated that modulating the arrangement and spacing of these chemistries also affected the cell response (Curran et al., 2010).

7.2.5 Dynamic Chemistry

Surface chemistries that are capable of being triggered to convert to different states in response to stimuli such as light (DeForest and Anseth, 2012; Mosiewicz et al., 2013), temperature (da Silva et al., 2007), and changes in electrical potential (Ng et al., 2012) have also been synthesized. Although most surfaces are designed to change states in an irreversible manner, a number of surfaces have been engineered to have a reversible switch. These platforms offer an intriguing research platform to study the mechanosignals required to switch stem-cell fate, and offer the potential to influence processes such as cell migration and stem-cell fate with a level of "on demand" control.

One study employed photocleavable "caging" of an enzyme substrate. The enzyme substrate was not accessible to the enzyme while masked by the caging group, but could be exposed by using UV laser light to remove the caging group (Mosiewicz et al., 2013). In the presence of the cross-linking enzyme transglutaminase factor XVIII, the substrate could be cross-linked to other biomolecules, enabling region-specific patterning of substrates. The authors demonstrated that specific patterning could also be achieved in 3D by using the system to attach –RGD ligands to defined regions of hydrogels, enabling the migration of stem cells to be controlled within the gel (Mosiewicz et al., 2013).

Other dynamic chemistries have been engineered such that the adhesive ligand is removed in response to an externally applied stimulus. An example of this was the use of an external light stimulus to photocleave adhesive –RGD sequences from chemically modified hydrogels, which enhanced chondrogenesis in the gels relative to –RGD modified gels (Salinas and Anseth, 2008). Initially, the –RGD motifs allowed the cells to adhere, then the cleavage acted to reduce the intracellular tension experienced by the stem cells encapsulated within the gels, stimulating them to undergo chondrogenesis (Salinas and Anseth, 2008). Most of the stimulus-responsive changes in dynamic chemistries are irreversible, but some can be reversed. One study examined surface chemistry produced using "click" chemistry (modular reactions that generate larger molecules from smaller units), e.g., which exposed or concealed adhesive –RGD sites depending on the applied electrical potential. Altering the conformation of sulphonate (negatively charged) and ammonium (positively charged) terminated molecules in response to applied positive or negative electrical potentials facilitated the switch between exposure or concealment of adjacent –RGD sites, and thus allowed cells to migrate onto certain regions of the pattern only under the permissive conditions when –RGD was exposed (Ng et al., 2012) (Figure 7.7). Thus, dynamic surfaces can be used to modulate cell adhesion via exposure and concealment of adhesive ligands.

Dynamic surfaces have also been used to promote switching between attachment and detachment of cells. The intrinsic thermoresponsive properties of the polymer

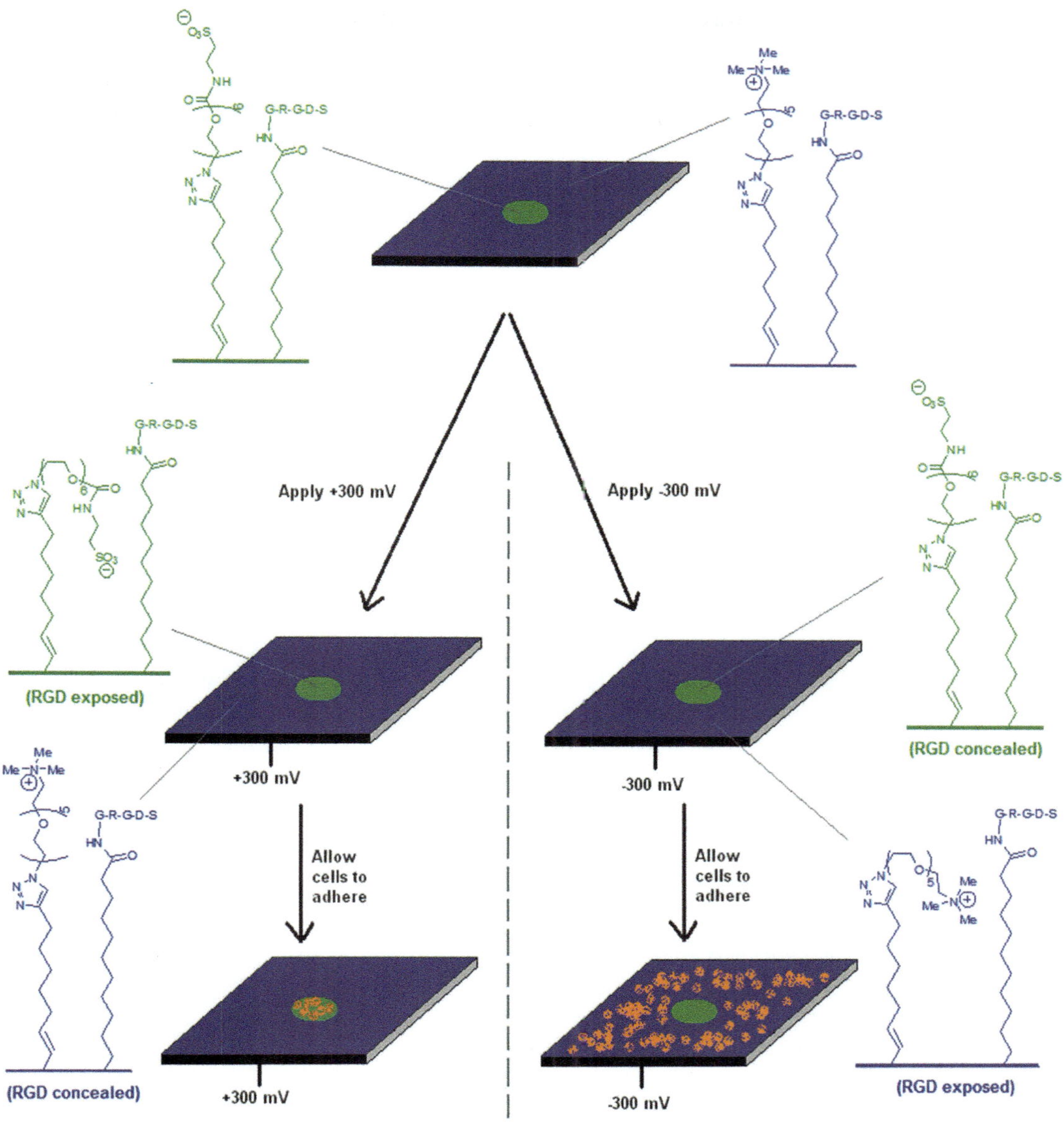

FIGURE 7.7

Illustration of the surface chemistry produced to enable reversible switching of defined areas between a proadhesive and antiadhesive state in response to applied positive and negative electrical potential, in order to restrict cell migration. Surfaces were patterned to possess regions with a nonfouling chemistry terminated with a sulphonate (anionic; green) or ammonium (cationic; blue) moiety with adjacent –RGD groups. Adhesive –RGD sites were exposed under conditions where the sulphonate or ammonium groups were attracted to the surface (positive and negative potentials of ±300 mV, respectively), thus favoring selective cell adhesion in distinct regions in response to electrical potential-induced surface switching. *Reproduced with permission from Ng et al. (2012).*

poly(*N*-isopropylacrylamide) (PNIPAAm) have been harnessed to harvest cell sheets, as the surface chemistry switches from hydrophilic to hydrophobic with increasing temperature, leading to the detachment of the cell sheet (da Silva et al., 2007). The harvest of such cell sheets could be beneficial for therapeutic applications. Click chemistry was used to generate an adhesive surface that could be changed to a nonadhesive state using UV light, which allowed selective detachment of cells from microregions of hydrogels (DeForest and Anseth, 2012). As discussed for hydrogels by DeForest and Anseth, region-specific dynamic approaches offer the potential for the harvest of cells from particular areas of the substrate, which could make comparative analyses of the cellular physiology from different areas of the same surface more feasible (DeForest and Anseth, 2012).

To better mimic the stem cell niche and other complex biological sites, the physicochemical environment of in vitro culture systems has to reflect the subtleties of the situation in the body. In conjunction with other approaches, chemical patterning offers the opportunity to investigate the interactions between different factors in influencing stem-cell fate. The stiffness of hydrogels can be tuned for particular research applications, and it is now possible to have spatial control over the zoning of different chemical modifications within the gels. Using a two-photon microscope to induce photo-cross-linking of the gel with laser light, distinct regions of gels could be patterned with different growth factors (Wylie et al., 2011). Gradients of soluble factors have also been produced in hydrogels using microfluidic approaches, which utilize microchannels to deliver small volumes of fluid in a controlled manner. This approach could potentially be combined with fixed chemistries to better recapitulate the niche by studying the effects of distinct combinations and concentrations of fixed and soluble factors on cell fate determination (Lutolf et al., 2009).

To conclude, engineered surface chemistries offer distinct functionalities for modulating cell responses, including tuneable surfaces to control processes such as cell attachment, signaling, migration, and stem cell differentiation.

7.3 SURFACE TOPOGRAPHY

Next, we shall consider physical patterning of the material surface—topography. Topography is a useful tool that provides a controlled and well-defined stimulus for influencing cell responses. It is possible to create a variety of pattern types with specified sizes for the noninvasive modulation of cellular organization and hence gene expression and cell behavior. These characteristics make topography an important tool for fundamental biological research and translational studies. Fabrication methods are not the focus of this chapter; they are reviewed in detail in Chapter 9. We will instead, after brief introduction, focus on the range of effects that topography can have on cells—both stem cells and differentiated cells.

7.3.1 Differentiated Cell Guidance by Nanostructures

The fact that cells guide to topographical features has been known for over 100 years (Harrison, 1911). The term contact guidance for cells aligning along topographical features was coined in the 1950s (Weiss and Garber, 1952), and in 1964 Curtis and Varde showed that fibroblast cells can position themselves parallel to silica fibers with diameters of 10–30 μm (Curtis and Varde, 1964). Since then, micron size features have continued to be central in topographical research. However, over the last 10 years research focus has shifted to nanotopographical features and their effects on cells. Nanoscale features are present in both cells and the ECM that surrounds them. Research has shown that nanotopography affects several cellular processes including adhesion, morphology, gene expression, and proliferation. The cellular response to different nanotopographies, such as pits, pillars, and grooves, has been studied using a range of cell types such as neurons, fibroblasts, smooth muscle cells, osteoblasts, and MSCs and will be discussed below.

Nanopit arrays with different degrees of disorder have been shown to influence the adhesion and morphology of fibroblasts (Biggs et al., 2009a), osteoblasts (Biggs et al., 2007), and MSCs (Tsimbouri, 2013a,b). Osteoblasts and MSCs grown on nanotopographies showed marked differences in cell adhesion and morphology when compared to cells on flat materials, with differences in the geometry of the nanopits also affecting cell behavior (Biggs et al., 2007; Tsimbouri, 2013a,b). In both studies it was shown that cells on nanopits with a highly ordered square pattern (SQ) developed smaller adhesions compared to those cultured on nanopit substrates with a small degree of disorder (NSQ50), which demonstrated formation of fibrillar adhesions. In a different study, which looked at the effect of variable nanopit depth on integrin-mediated focal adhesion formation, nanopits <29 nm deep enhanced cell adhesion, increased the expression of adhesion proteins such as integrins and phosphorylated FAK (P-FAK), in comparison to flat controls (Lim et al., 2007).

Pillar height has been shown to be important in cell responses to topographies. By changing the height of nanopillars over a range of between 13 and 95 nm changes in adhesion were noted; nanopillars with the smallest height increased cell adhesion whereas the taller nanopillars reduced adhesion. In agreement, MSCs cultured on 15 nm titanium nanopillars showed increased adhesion in comparison to flat controls (Sjostrom et al., 2013). These results provide evidence that topographical effects translate between different materials (the original studies were performed on patterns in biocompatible polymers such as polycarbonate, polystyrene, polymethylmethacrylate, and polycaprolactone).

The effect of different pillar diameters was examined and showed that smaller diameter pillars resulted in an increase in cell adhesion, however, with a marked reduction in proliferation. In contrast, nanopillars with larger diameters resulted in reduced cell adhesion but with increased proliferation when compared to flat controls (Milner and Siedlecki, 2007).

Nanogrooves are probably the best-studied topographies and their effects on cell responses have been relatively well characterized. Different studies, involving nanogrooved substrates and many different cell types, have shown that fibroblasts, osteoblasts, epithelial cells, endothelial cells, smooth muscle and neural cell cultures adopt an elongated morphology when cultured on them (Loesberg et al., 2007; Teixeira et al., 2003; Barbucci et al., 2003; Biggs et al., 2008; Yim et al., 2005; Johansson et al., 2006; Tsimbouri et al., 2013b). It has been shown that an increase in cell alignment along grooves is inversely proportional to groove width (Biggs et al., 2009a; Biela et al., 2009).

7.3.2 Nanotopography and Stem Cell Differentiation

The study of embryonic stem (ES) cell–material surfaces is complicated by the need for a feeder layer of mouse embryonic fibroblasts (MEFs) for their culture or by the need to use, e.g., matrigel (derived from murine basal lamina). The presence of the feeder cell layer is likely to mask any response of ES cells to nanotopography. A few researchers have, however, managed to study both differentiation and retention of the self-renewal abilities of ES cells in response to topographic stimuli. In one of these studies, the effects of topography on ES cell differentiation, in both the presence and absence of a feeder layer, was investigated. The results showed that in the presence of a feeder layer, the differentiation rate of ES cells cultured on grooved substrates was significantly reduced in comparison to flat controls. In contrast, higher levels of spontaneous differentiation were observed in ES cells cultured on nanogrooves without feeder layers compared to those grown on flat controls with a feeder layer (McFarlin et al., 2006). These results are suggestive of a synergistic role between nanotopography and feeder layers in the promotion of ES cell self-renewal. In a recent study using nanopits to assess ES cell differentiation, it was found that ES cells incubated on nanotopographical substrates had enhanced expression of mesenchymal or stromal markers and expression of early osteogenic progenitors at levels above those detected in cells on planar substrates (Kingham et al., 2013).

In comparison to ES cells, MSCs have been extensively studied for topographical interaction effects.

The involvement of nanotopography on MSCs differentiation down the osteogenic lineage has been closely studied for approximately the last seven years. In one study, the authors examined MSC differentiation on a range of nanopit topographies with different degrees of order (Biggs et al., 2007). Osteogenic differentiation of MSCs, indicated by increased levels of bone cell markers, was observed when grown on a disordered topography (as discussed earlier in relation to adhesion; 120 nm diameter pits with 100 nm depth and 300 nm center–center spacing in a square arrangement but with up to ±50 nm offset from pit center position) (Dalby et al., 2007). This result was repeated using the same nanotopographies but in different materials: PCL (McMurray et al., 2011) or polycarbonate (Tsimbouri et al., 2013a; Tsimbouri et al., 2012). It appears that the osteogenic effect is transferable between biocompatible materials and relies on the ability of the MSCs to form fibrillar adhesions that support increased levels of intracellular tension and hence biochemical signaling events. In contrast, if offset was reduced to zero, adhesion reduced and the nanopit topography then supported self-renewal of MSCs (McMurray et al., 2011).

Nanopillars and their involvement in the differentiation of MSCs into osteoblast cells has also been investigated. MSCs cultured on carbon nanotubes with diameters of 100 nm pillar topographies with diameters ranging between 30 and 100 nm showed osteogenic differentiation (Oh et al., 2009) in comparison to the 30 nm nanotubes where no differentiation was observed. Cell adhesion was inversely proportional to the diameter of the nanotubes. Similarly to the nanopits, there is a close relationship between cell adhesion and differentiation and the balance of it controls cell fate. MSCs cultured on 15 nm titania (titanium oxide) nanopillars showed osteogenic marker expression (Sjostrom et al., 2013). These data are interesting as the ability to pattern osteogenic nanopatterns onto metals such as titanium allows translation of fundamental observations, e.g., on plastic, into the real world such as patterning of metallic orthopedic implants where bone-implant integration is of key importance.

MSCs may be able to differentiate into neural cells in the presence of growth factors (Song and Tuan, 2004). In the absence of differentiation factors, MSC-neural differentiation potential using nanogrooved surfaces has also been demonstrated in MSCs cultured on 350 nm deep grooves as indicated by expression of neural markers MAP2 (microtubule-associated protein 2) and β-tubulin III at higher levels than the positive control (1 or 30 μM retinoic acid) (Yim et al., 2007). We note that some caution should be applied to MSCs with regard to neural differentiation as this represents a transdifferentiation; it may be that the MSCs can be coaxed to express neural proteins but a functional role is yet to be demonstrated.

7.3.3 Neural Stem Cells

Neural stem cells can differentiate into neurons and two nonneural, but neural support, cell types (glia): astrocytes and oligodendrocytes.

Nanofibers produced by electrospinning have been used in the study of neural stem cell differentiation into neurons. The effect of varying fiber dimensions from the nanoscale to microscale and their orientation, i.e., aligned or random on neural stem cell differentiation, was studied (Yang et al., 2005). Higher differentiation rates were observed on the aligned and random nanofibers in comparison to aligned and random microfibers. In another study, the effect of fiber diameter (283 ± 45 nm, 749 ± 153 nm, and 1452 ± 312 nm) on the differentiation of neural stem cells was investigated (Christopherson et al., 2009). Similarly to the previous study, the authors found that neural stem cell differentiation was supported on fiber diameters within the nanoscale rather than microscale range. Both studies suggest that it is the fiber diameter that plays a key role in directing the differentiation of neural stem cells rather than fiber orientation.

To conclude this section, topography has broad effects on cell activity and is a powerful surface cue. See Figure 7.8 for further details.

7.4 MATERIAL MECHANICS (STIFFNESS)

Within the mammalian body, tissue and organ types show considerable variation in their make up, as is suited to their structural and functional integrity. This diversity is inclusive of the type of cells native to each tissue, extracellular constituents, physical formation (structure), and ultimately their primary function. These qualities change considerably from development to maturity. As part of this, the relative deformability or stiffness (generally expressed as the Young's modulus in the unit Pascals) is also a characteristic that has considerable variation from one tissue type to the next, such as the relatively soft adipose tissue, which measures an approximate Young's modulus of 1 kPa (Elkin et al., 2007) as compared to that of muscles, tendon and bone tissue, which are much stiffer, measuring in the range of 6 MPa–40 GPa (Discher et al., 2005; Reilly and Burstein, 1974). Specialized cells housed within tissue niches are adapted to exist as such and as a consequence the relative stiffness of a tissue type has considerable importance on how the cells interpret their surrounding matrix, facilitating tissue functionality and efficiency.

7.4.1 Mimicking Tissue Stiffness In Vitro

The manner in which matrix rigidity affects overall tissue integrity and signaling events in cells indicates that the manner in which cells "feel" and respond to mechanical cues presented in their substrate is an important consideration when guiding cell behavioral responses in vitro. The need for high precision and reproducibility when fabricating substrates that vary in stiffness has led to the widespread use of synthetic polymers such as polyacrylamide (PA), PDMS, and polylactic acid (PLLA) whose resultant stiffnesses can be controlled by careful manipulation of the extent to which the polymer molecules are allowed to form cross-links with one another. This, in general practice, is achieved by subtle alterations to polymer concentrations or tuning of the pH

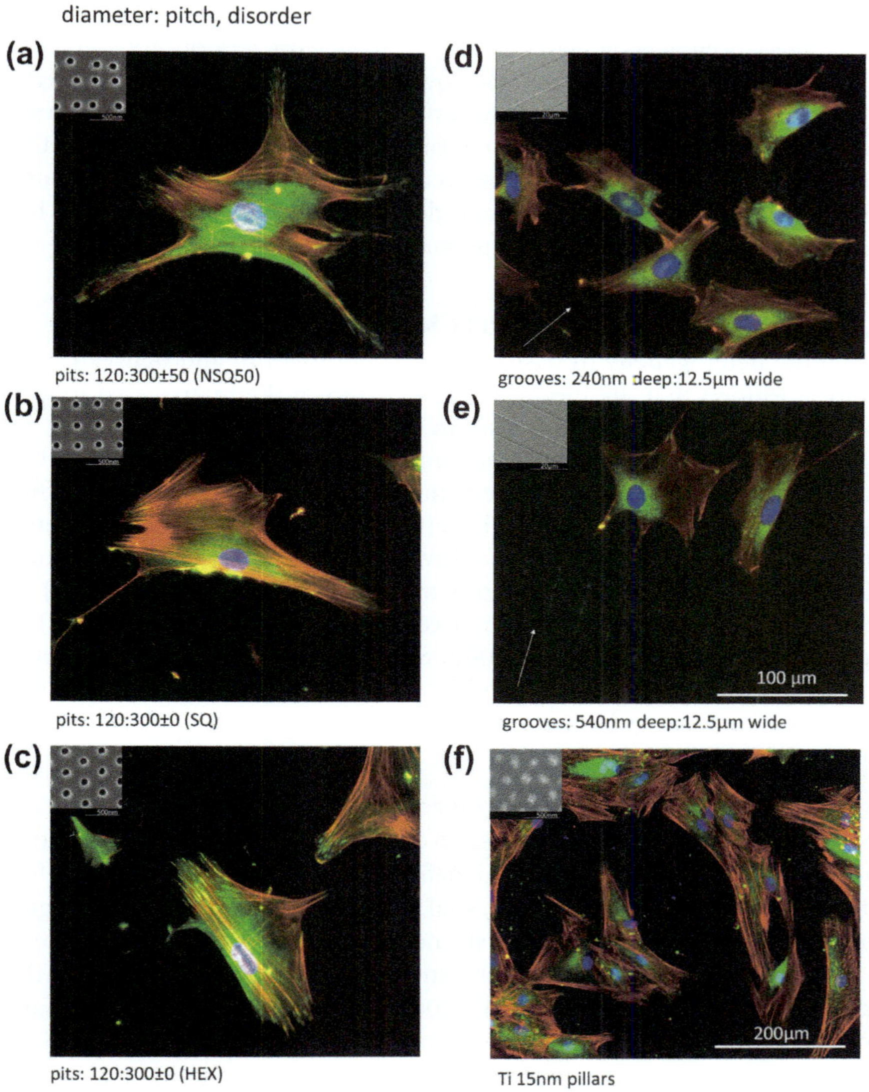

FIGURE 7.8

Scanning electron microscopy (insets) and fluorescent microscopy images of mesenchymal stem cell (MSC) morphology and attachment on different nanotopographies. MSCs grown on nanopits (panels a–c) and nanopillars (f) and mouse embryonic fibroblasts (MEFs) grown on nanogrooves (panels d and e). (a) On NSQ50 disordered pit topographies, MSCs were spread with large lamellae and displayed an elongated shape; cells developed stress fibers with long adhesions. (b) On SQ ordered pits, cells had a less spread morphology and adopted a stellate shape with fewer stress fibers and adhesions. (c) MSCs on ordered hexagonal pits showed a less spread morphology with smaller adhesions. (d) MEFs grown on 240 nm deep grooves were elongated and aligned along the direction of the nanogrooves—alignment was more pronounced on (e) the 540 nm grooves. (f) MSCs on 15 nm nanopillars were spread with a well-organized cytoskeletal architecture and polygonal cell morphology. Actin (red), vinculin (green), and DNA (blue).

and/or temperature during the cross-linking process. A drawback with the use of these polymers, however, is their characteristic lack of biocompatibility; this is usually tempered, however, by the use of proteins, such as collagen, laminin and fibronectin, to coat the surface. Naturally occurring compounds, e.g., collagen or matrigel, have also been used to study the effects of substrate stiffness on cell behavior; it is, however, less preferable as the much sought after qualities of precision and reproducibility during fabrication are hindered as batch sources are known to vary considerably.

7.4.2 Cell Behavior and Matrix Mechanics

Given the heterogeneity of tissue elasticity both within and between tissue types, it follows that the mechanical properties of the matrix used to grow cells in vitro elicit varied responses in different cell types and as such, the use of standard tissue culture plastic may affect behavior responses that are outwith what is naturally expected. In adult cell types, substrate stiffness has been shown to have considerable effect on the survival and proliferation rate of the particular cell types in cultivation. For example, in vitro cultures of neural cells on relatively soft substrates (between 100 and 500 Pa) are known to thrive more when compared to the same cell type cultured on rigid surfaces (Flanagan et al., 2002; Saha et al., 2008). Liver hepatocyte and stellate cells have a higher proliferation rate when cultured on stiff substrates and undergo quiescence or growth arrest when cultured on soft substrates (Wells, 2008).

Cell sensing and response to matrix mechanics, however, is not a wholly precise phenomenon. A measure of overlap exists in cell interpretation as opposed to the substrate having a restrictive or defined effect on cell behavior. Therefore, while cell types such as neutrophils, glial, and myocardial cells are able to grow and survive on a range of substrate stiffnesses, myocyte functionality, however, is most increased when the substrate stiffness matches that of innate cardiac tissue (Engler et al., 2004; Saha et al., 2008; Wells, 2008). This increased functionality is also attributed to an innate mechanical coupling of the cells mechanosensory adhesions to that of its surrounding matrix (Hoerning et al., 2012); not unlike enhanced enzyme activity at physiological body temperature.

The ability of adherent cells to migrate is an important factor with regard to facilitating numerous physiological processes such as immunoresponsive behavior, wound healing, morphogenesis, and development. Cells cultured on substrates with uniform chemistry but alternate rigidity exhibit varied motility, indicating that rigidity sensing by the cell plays an influential role in guiding directional movement—a phenomenon referred to as mechanotaxis. In fibroblast cells, e.g., directional movement is based on the relative deformability of the substrate with migration occurring preferentially from soft to rigid regions (Cortese et al., 2009; Gray et al., 2003; Lo et al., 2000) (Figure 7.9).

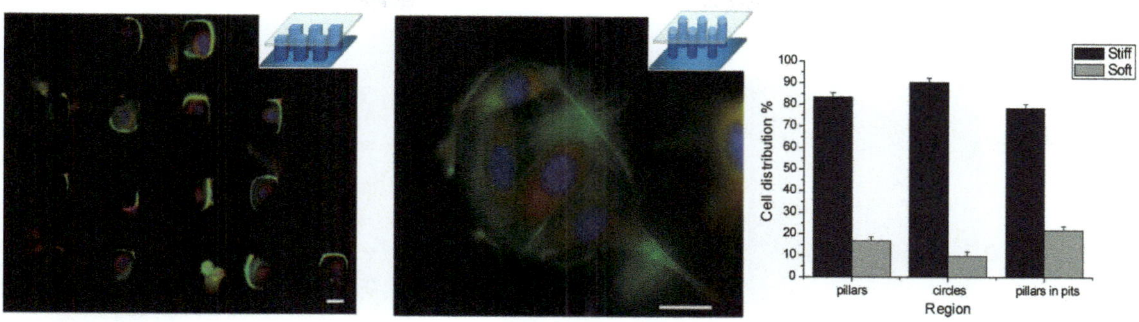

FIGURE 7.9

The effect of matrix stiffness on cell migration and morphology. Fluorescence images of NIH 3T3 fibroblast cells cultured on a uniform PDMS substrate sheet placed above pillars so that the areas under the pillars are stiff while areas with no pillars underneath are soft. This shows that cells preferably migrate toward the stiffer regions supported by the pillars. *Reproduced with permission from Cortese et al. (2009).*

These examples suggest that substrate stiffness acts as far more than a simple by-product of cell activity, determining not only tissue structure and function but also playing a key role in facilitating cell survival, function, migration, and proliferation.

7.4.3 Stem Cell Differentiation and Substrate Stiffness

Interactions between pluripotent cells and matrix stiffness may be a crucial factor in determining early stage development. ECM mechanics play a considerable role in guiding differentiation of ES cells during organogenesis, as well as acting as cues for maintaining their pluripotent state (Evans et al., 2009). The lack of matrix mechanics as a cue results in misguided cell development leading to spontaneous differentiation. In vitro studies using multipotent or adult stem cells also show that matching substrate stiffness to that of naturally occurring tissue types is a powerful tool in influencing stem cell differentiation along a particular lineage.

The use of matrix mechanics to guide stem cell differentiation also has much more influence over coaxing cell development when compared to the use of chemical cues alone (Engler et al., 2006). In order to maintain such organization, cellular sensing of changes in ECM stiffness is mediated through a constant cycle of mechanosensing and responses mediated through focal adhesions that generate intracellular cytoskeletal tensional states (Choquet et al., 1997; Yamada and Geiger, 1997). The extent of resistance or deformation of a substrate experienced by a cell results in increased or decreased intracellular tension within it, facilitated by cytoskeletal contractility. This invariably has an effect on physical characteristics such as cell morphology, spreading, and adhesion strength. Loss of cytoskeletal tension on soft substrates, e.g., leads to cells adopting a small rounded morphology while stiffer substrates, able to withstand higher traction forces by the cell, lead to a higher tensile state of the cytoskeleton and thus the cell

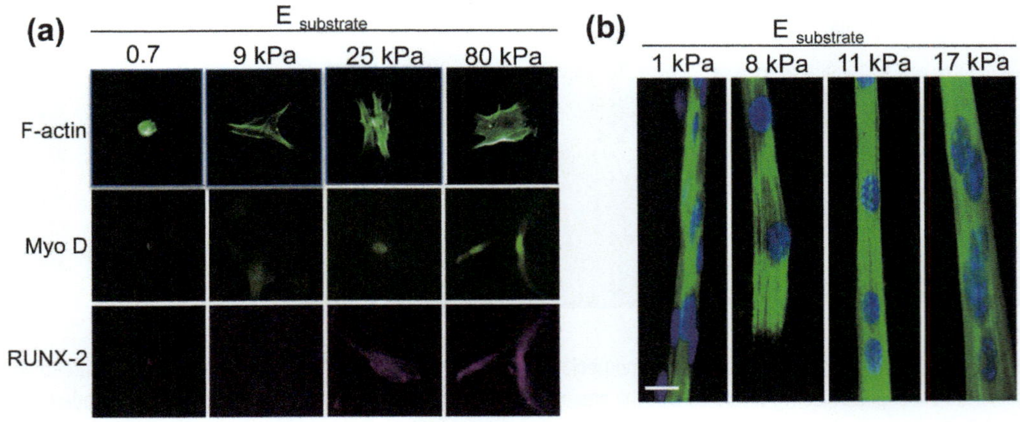

FIGURE 7.10

(a) Mesenchymal stem cells cultured on fibronectin coated PA substrates with varying Young's moduli show different cytoskeletal (F-actin) organization and morphology. Cells cultured on 9–80 kPa substrates show differentiation into myoblast cells (MyoD, myogenic differentiation 1, a transcription factor involved in myogenesis) at all stiffnesses and osteoblast differentiation (RUNX-2) only on the stiffest substrates. (b) Myoblast cells are viable on a range of substrate stiffness but they only undergo striation (showing differentiation indicative of functionality) when cultured on substrates that bear matrix stiffness similar to muscle tissue (8 and 11 kPa). *Image (a) is reproduced with permission from Rowlands et al. (2008) and Engler et al. (2004) respectively; image (b) © 2004 Rockefeller University Press. Originally in Journal of Cell Biology. 166:877–887. doi:10.1083/jcb.200405004.*

is able to adopt a larger and more spread morphology (Figure 7.10). These differences in physical characteristics brought on by cytoskeletal contractility, in stem cells, influence changes in broad signaling events culminating in gene expression effecting differentiation (Kilian et al., 2010; McBeath et al., 2004). The formation of small rounded stem cells on soft substrates culminates in the expression of phenotypically relevant proteins resulting in differentiation into low tensile neuronal and adipose cell types. In a similar vein, MSCs cultured on substrates that have a Young's modulus similar to that of nonmineralized bone (40 kPa) show an upregulation of transcription factor RUNX-2 (runt-related transcription factor 2, a main osteogenic phenotype initiator) and early-stage signaling initiator BMPR-2 (bone morphogenic protein 2) which facilitate the formation of osteoblasts (Engler et al., 2006; Place et al., 2009; Trappmann et al., 2012).

7.5 SUMMARY

This chapter has reviewed how cells interact with and derive mechanotransductive information from chemical, topographical, and mechanical cues. Understanding how cells convert this information into biological processes has particular significance in the field of stem cells. This is important as use of terminally differentiated cells will be central to tissue engineering of simple tissues, but in order to engineer complexity we need to think about control of

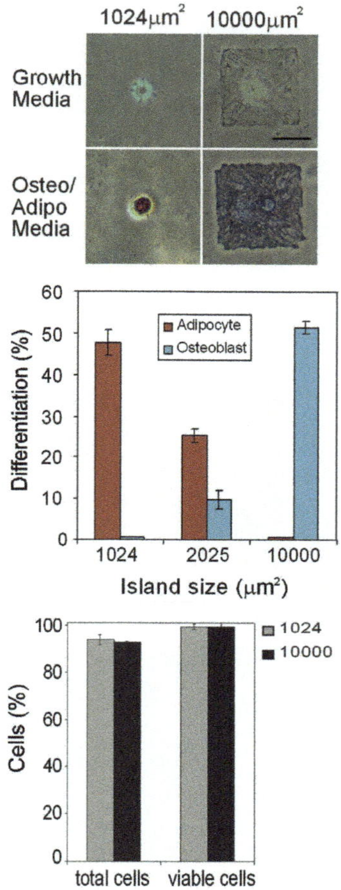

FIGURE 7.11

Cell shape determines MSC differentiation. (a) Brightfield images of MSCs cultured on island of different sizes after 1 week in culture after staining for lipids (red) and alkaline phosphatase (blue). Scale bar = 50 μm. (b) Percentage differentiation of MSCs plated on different-sized islands after 1 week of culture. (c) Percentage of MSCs remaining viable after 1 week culture on different-sized islands. *Reprinted from McBeath et al. (2004). Cell shape, cytoskeletal tension, and RhoA regulate stem cell lineage commitment. Developmental Cell, 6(4), 438–495, copyright (2004), with permission from Elsevier.*

stem cells—because these are the cells that engineered complexity during our embryogenesis and development.

7.5.1 Snapshot Summary

- Cells interact with the extracellular environment through cell adhesion molecules, e.g., integrins.
- This interaction affects the formation of cell adhesions and the cytoskeleton.

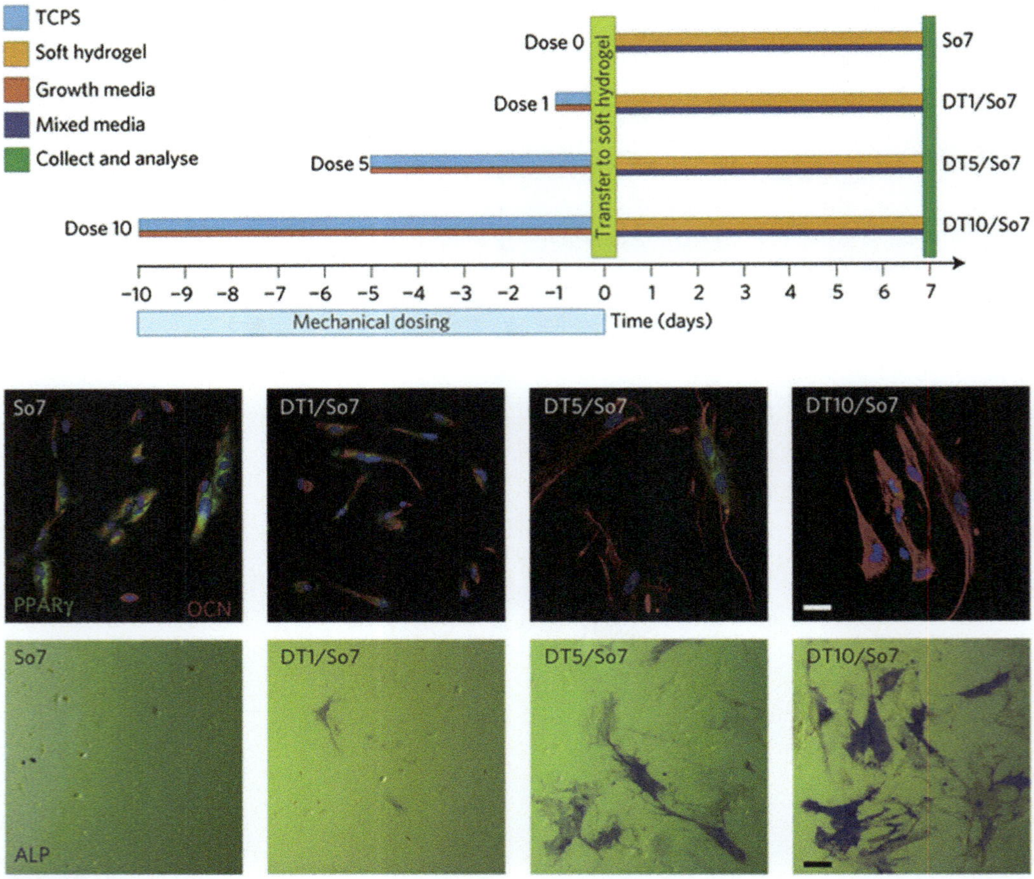

FIGURE 7.12

Immunocytochemistry (b and c) of MSCs cultured under various regimes (a) highlights PPARγ (green—a fat cell marker) and osteocalcin (red—a bone cell marker) in (b) and alkaline phosphatase (c) showing a switch from fat to bone expression with longer rigid substrate pretreatment times. Scale bars are 20 μm. *Reprinted by permission from Macmillan Publishers Ltd: Nature materials (Yang et al., 2014), copyright 2014.*

- Embedded within adhesions and the cytoskeleton are signaling molecules that will affect biochemical pathways depending on how cells adhere.
- We can engineer synthetic substrates that can, through cell adhesion interactions, mimic changes in the chemistry, topography, and elasticity of the native extracellular environment.
- It has been shown that these changes in substrate affect cell adhesion, migration, growth, and gene expression.

- This has implications for stem cell biology where it has been shown that these changes can provide control over differentiation.
- These materials thus provide tools to understand stem cell signalling mechanisms (e.g., mechanotransduction).
- Most data so far has been with mesenchymal stem cells from bone marrow.
- We are starting to understand the interactions of other stem cells (neural, embryonic, etc.) with their physical environment.
- This will open up our ability to use stem cells in tissue engineering (Figure 7.11) (Figure 7.12).

CLASSICAL MANUSCRIPT: SHAPE-MEDIATED SIGNALING

McBeath et al. (2004) and Pittenger et al. (2002) have demonstrated that the seeding density at which mesenchymal stem cells (MSCs) are plated influences their lineage commitment, with high and low densities stimulating adipogenic and osteogenic differentiation, respectively. In this paper the authors identified why this density-regulating lineage commitment happens and the mechanisms that control it, namely mechanical forces.

Using microcontact printing the authors fabricated an array of adhesive fibronectin islands surrounded by a nonadhesive substrate, which forced the cells cultured upon it to grow individually (one on each island). Cell shape and the degree to which these individual cells spread were defined by controlling the size of the islands. These experiments allowed the effects of cell shape to be isolated from other factors of the local tissue microenvironment (such as cell–cell contact inhibition). The authors reported that cell shape alone can regulate MSC lineage commitment, with flattened spread cells committing to an osteogenic lineage and unspread, round cells an adipogenic lineage.

The mechanisms regulating this shape-mediated signaling were found to involve the actin cytoskeleton and RhoA GTPase activity, with both an intact cytoskeleton and RhoA being central to the process—neither of them alone was sufficient to shift lineage commitment. RhoA is a central regulator of cell contractility in many cells and has a number of effectors, including some that influence cytoskeleton structure and function. RhoA activity was found to be greater in spread than unspread cells and thus corresponds with the authors' complementary observation that continually active RhoA stimulated osteogenic lineage commitment while inactive RhoA decreased osteogenic commitment. Remarkably, RhoA activity was found to be a stronger stimulus than traditionally used chemically mediated differentiation signals—demonstrating the power of cell shape to control stem cell differentiation. RhoA is known to play a central role in soluble factor signaling (Sordella et al., 2003) and is thus identified as a potential integrator of structural and soluble cues.

These experiments were fundamental to our understanding of the power by which mechanical forces, embodied by cell shape, cytoskeletal tension, and RhoA signaling, can control stem cell differentiation.

STATE-OF-THE-ART MANUSCRIPT: STEM CELLS HAVE A MECHANICAL MEMORY

Yang et al. (2014) investigated the phenomenon of mesenchymal stem cell (MSC) mechanical memory, or mechanical dosing—the potential of cells to remember past mechanical environments that may influence their long-term fate. Tissue culture plastic, a plastic which is routinely used to culture cells in vitro, and poly(ethylene glycol) (PEG) hydrogels, which have tuneable stiffness, were used to expose cells to a range of mechanical stiffnesses.

The transcription coactivators/factors Yes-associated protein (YAP) and transcriptional co-activator with PDZ-binding motif (TAZ) were found to be key regulators of mechanical memory, thought to act as intracellular mechanical rheostats. In environments where YAP and TAZ would be transiently activated, continual exposure leads to constitutive activation. A threshold exists, after which this constitutive activation continues even after removal of the original mechanical stimulus. Importantly for laboratory-based work, tissue culture plastic was shown to mechanically dose MSCs toward an osteogenic lineage through upregulation of the pre-osteogenic transcription factor RUNX2.

This research highlights the importance of the temporal component of mechanotransduction and has important implications for traditional laboratory-based cell culture.

REFERENCES

Abercrombie, M., Dunn, G.A., 1975. Adhesions of fibroblasts to substratum during contact inhibition observed by interference reflection microscopy. Exp. Cell Res. 92 (1), 57–62.

Arnold, M., et al., 2004. Activation of integrin function by nanopatterned adhesive interfaces. Chemphyschem 5 (3), 383–388.

Arnold, M., et al., 2008. Induction of cell polarization and migration by a gradient of nanoscale. Nano Lett. 8 (7), 2063–2069.

Atala, A., et al., 2008. Principles of Regenerative Medicine. Elsevier.

Bakker, G.J., et al., 2012. Lateral mobility of individual integrin nanoclusters orchestrates the onset for leukocyte adhesion. Proc. Natl. Acad. Sci. 109 (13), 4869–4874.

Baneyx, G., Baugh, L., Vogel, V., 2002. Fibronectin extension and unfolding within cell matrix fibrils controlled by cytoskeletal tension. Proc. Natl. Acad. Sci. U.S.A. 99 (8), 5139–5143.

Barbucci, R., et al., 2003. Micro and nano-structured surfaces. J. Mater. Sci. Mater. Med. 14 (8), 721–725.

Bar-Sagi, D., et al., 1993. SH3 domains direct cellular localization of signaling molecules. Cell 74 (1), 83–91.

Benoit, D.S.W., et al., 2008. Small molecule functional groups for the controlled differentiation of human mesenchymal stem cells encapsulated in poly(ethylene glycol) hydrogels. Nat. Mater. 7, 816–823.

Biela, S., et al., 2009. Different sensitivity of human endothelial cells, smooth muscle cells and fibroblasts to topography in the nano-micro range. Acta Biomater. 5 (7), 2460–2466.

Biggs, M.J., et al., 2008. Adhesion formation of primary human osteoblasts and the functional response of mesenchymal stem cells to 330nm deep microgrooves. J. R. Soc. Interface 5 (27), 1231–1242.

Biggs, M.J., et al., 2009a. Interactions with nanoscale topography: adhesion quantification and signal transduction in cells of osteogenic and multipotent lineage. J. Biomed. Mater. Res. Part A 91 (1), 195–208.

Biggs, M.J., et al., 2007. The effects of nanoscale pits on primary human osteoblast adhesion formation and cellular spreading. J. Mater. Sci. Mater. Med. 18 (2), 399–404.

Biggs, M.J., et al., 2009b. The use of nanoscale topography to modulate the dynamics of adhesion formation in primary osteoblasts and ERK/MAPK signalling in STRO-1+ enriched skeletal stem cells. Biomaterials 30 (28), 5094–5103.

Cattaruzza, M., Lattrich, C., Hecker, M., 2004. Focal adhesion protein zyxin is a mechanosensitive modulator of gene expression in vascular smooth muscle cells. Hypertension 43 (4), 726–730.

Chen, C.S., et al., 1997. Geometric control of cell life and death. Science 276 (5317), 1425–1428.

Choquet, D., Felsenfeld, D.P., Sheetz, M.P., 1997. Extracellular matrix rigidity causes strengthening of integrin-cytoskeleton linkages. Cell 88 (1), 39–48.

Christopherson, G., Song, H., Mao, H., 2009. The influence of fiber diameter of electrospun substrates on neural stem cell differentiation and proliferation. Biomaterials 30 (4), 556–564.

Cortese, B., Gigli, G., Riehle, M., 2009. Mechanical gradient cues for guided cell motility and control of cell behavior on uniform substrates. Adv. Funct. Mater. 19 (18), 2961–2968.

Curran, J.M., et al., 2010. Introducing dip pen nanolithography as a tool for controlling stem cell behaviour: unlocking the potential of the next generation of smart materials in regenerative medicine. Lab. Chip 10, 1662–1670.

Curran, J.M., et al., 2011. The use of dynamic surface characteristics to control msc isolation and function. Biomaterials 32, 4753–4760.

Curran, J.M., Chen, R., Hunt, J.A., 2005. Controlling the phenotype and function of mesenchymal stem cells in vitro by adhesion to silane-modified clean glass surfaces. Biomaterials 26, 7057–7067.

Curtis, A., Varde, M., 1964. Control of cell behaviour: topological factors. J. Natl. Cancer Inst. 33, 15–26.

Dalby, M.J., et al., 2007. The control of human mesenchymal cell differentiation using nanoscale symmetry and disorder. Nat. Mater. 6, 997–1003.

da Silva, R.M.P., Mano, J.F., Reis, R.L., 2007. Smart thermoresponsive coatings and surfaces for tissue engineering: switching cell-material boundaries. Trends Biotechnol. 25, 577–583.

Davies, J.A., 2001. Extracellular Matrix, in Encyclopedia of Life Sciences. Nature Publishing Group.

Davies, P.F., Robotewskyj, A., Griem, M.L., 1994. Quantitative studies of endothelial cell adhesion. Directional remodeling of focal adhesion sites in response to flow forces. J. Clin. Invest. 93 (5), 2031–2038.

DeForest, C.A., Anseth, K.S., 2012. Photoreversible patterning of biomolecules within click-based hydrogels. Angew. Chem. Int. Ed. 51, 1816–1819.

Discher, D.E., Janmey, P., Wang, Y.L., 2005. Tissue cells feel and respond to the stiffness of their substrate. Science 310 (5751).

Elkin, B.S., et al., 2007. Mechanical heterogeneity of the rat hippocampus measured by atomic force microscope indentation. J. Neurotrauma 24 (5).

Engler, A.J., et al., 2006. Matrix elasticity directs stem cell lineage specification. Cell 126 (4), 677–689.

Engler, A.J., et al., 2004. Myotubes differentiate optimally on substrates with tissue-like stiffness: pathological implications for soft or stiff microenvironments. J. Cell Biol. 166 (6), 877–887.

Evans, N.D., et al., 2009. Substrate stiffness affects early differentiation events in embryonic stem cells. Eur. Cells Mater. 18.

Flanagan, L.A., et al., 2002. Neurite branching on deformable substrates. Neuroreport 13 (18).

Gray, D.S., Tien, J., Chen, C.S., 2003. Repositioning of cells by mechanotaxis on surfaces with micropatterned Young's modulus. J. Biomed. Mater. Res. Part A 66A (3), 605–614.

Guerra, N.B., et al., 2010. Subtle variations in polymer chemistry modulate substrate stiffness and fibronectin activity. Soft Matter 6 (19), 4748–4755.

Harrison, R.G., 1911. On the stereotropism of embryonic cells. Science 34 (870), 279–281.

Hoerning, M., et al., 2012. Rigidity matching between cells and the extracellular matrix leads to the stabilization of cardiac conduction. Biophys. J. 102 (3), 379–387.

Hynes, R.O., 1999. Cell adhesion: old and new questions. Trends Cell Biol. 9 (12), M33–M37.

Ingber, D.E., 2006. Cellular mechanotransduction: putting all the pieces together again. Faseb J. 20 (7), 811–827.

Ingber, D.E., Tensegrity, I., 2003. Cell structure and hierarchical systems biology. J. Cell Sci. 116 (7), 1157–1173.

Ingber, D.E., 1993. The riddle of morphogenesis: a question of solution chemistry or molecular cell engineering? Cell 75 (7), 1249–1252.

Johansson, F., et al., 2006. Axonal outgrowth on nano-imprinted patterns. Biomaterials 27 (8), 1251–1258.

Kanchanawong, P., et al., 2010. Nanoscale architecture of integrin-based cell adhesions. Nature 468 (7323), 580–584.

Keely, P., Parise, L., Juliano, R., 1998. Integrins and GTPases in tumour cell growth, motility and invasion. Trends Cell Biol. 8 (3), 101–106.

Kilian, K.A., et al., 2010. Geometric cues for directing the differentiation of mesenchymal stem cells. Proc. Natl. Acad. Sci. U.S.A. 107 (11), 4872–4877.

Kingham, E., et al., 2013. Nanotopographical cues augment mesenchymal differentiation of human embryonic stem cells. Small 9 (12), 2140–2151.

Koch, C.A., et al., 1991. SH2 and SH3 domains: elements that control interactions of cytoplasmic signaling proteins. Science 252 (2006), 668–674.

Lim, J., et al., 2007. The regulation of integrin-mediated osteoblast focal adhesion and focal adhesion kinase expression by nanoscale topography. Biomaterials 28 (10), 1787–1797.

Lo, C.M., et al., 2000. Cell movement is guided by the rigidity of the substrate. Biophys. J. 79 (1), 144–152.

Loesberg, W., et al., 2007. The threshold at which substrate nanogroove dimensions may influence fibroblast alignment and adhesion. Biomaterials 28 (27), 3944–3951.

Lutolf, M.P., Gilbert, P.M., Blau, H.M., 2009. Designing materials to direct stem-cell fate. Nature 462 (7272), 433–441.

Macchiarini, P., et al., 2008. Clinical transplantation of a tissue-engineered airway. Lancet 372 (9655), 2023–2030.

Maniotis, A.J., Chen, C.S., Ingber, D.E., 1997. Demonstration of mechanical connections between integrins, cytoskeletal filaments, and nucleoplasm that stabilize nuclear structure. Proc. Natl. Acad. Sci. U.S.A 94 (3), 849–854.

Mao, Y., Schwarzbauer, J.E., 2005. Fibronectin fibrillogenesis, a cell-mediated matrix assembly process. Matrix Biol. 24 (6), 389–399.

McBeath, R., et al., 2004. Cell shape, cytoskeletal tension, and RhoA regulate stem cell lineage commitment. Dev. Cell 6 (4), 483–495.

McFarlin, D.F., et al., 2006. Nanoscale topography modulates human embryonic stem cell self-renewal. In: The American Society for Cell Biology 46th Annual Meeting.

McMurray, R.J., et al., 2011. Nanoscale surfaces for the long-term maintenance of mesenchymal stem cell phenotype and multipotency. Nat. Mater. 10 (8), 637–644.

Milner, K., Siedlecki, C., 2007. Submicron poly(L-lactic acid) pillars affect fibroblast adhesion and proliferation. J. Biomed. Mater. Res. Part A 82 (1), 80–91.

Miyamoto, S., et al., 1995. Integrin function: molecular hierarchies of cytoskeletal and signaling molecules. J. Cell Biol. 131 (3), 791–805.

Morgan, M.R., Humphries, M.J., Bass, M.D., 2007. Synergistic control of cell adhesion by integrins and syndecans. Nat. Rev. Mol. Cell Biol. 8 (12), 957–969.

Mosiewicz, K.A., et al., 2013. In situ cell manipulation through enzymatic hydrogel photopatterning. Nat. Mater. 12, 1072–1078.

Negrutskii, B.S., Deutscher, M.P., 1992. A sequestered pool of aminoacyl-tRNA in mammalian cells. Proc. Natl. Acad. Sci. U.S.A. 89 (8), 3601–3604.

Ng, C.C.A., et al., 2012. Using an electric potential to reversibly switch surfaces between two states for dynamically controlling cell adhesion. Angew. Chem. Int. Ed. 51, 7706–7710.

Oh, S., et al., 2009. Stem cell fate dictated solely by altered nanotube dimension. Proc. Natl. Acad. Sci. U.S.A. 106 (7), 2130–2135.

Parise, L.V., Lee, J., Juliano, R.L., 2000. New aspects of integrin signaling in cancer. Semin. Cancer Biol. 10 (6), 407–414.

Pittenger, et al., 2002. Stem cell culture: mesenchymal stem cells from bone marrow. In: Atala, A., Lanza, R.P. (Eds.), Methods of Tissue Engineering. Academic Press, San Diego, CA, pp. 461–469.

Place, E.S., Evans, N.D., Stevens, M.M., 2009. Complexity in biomaterials for tissue engineering. Nat. Mater. 8 (6).

Reilly, D.T., Burstein, A.H., 1974. Review article - mechanical-properties of cortical bone. J. Bone Joint Surg. Am. A 56 (5).

Rovensky, Y.A., 2012. Adhesive Interactions in Normal and Transformed Cells. Humana Press.

Rowlands, A.S., George, P.A., Cooper-White, J.J., 2008. Directing osteogenic and myogenic differentiation of MSCs: interplay of stiffness and adhesive ligand presentation. Am. J. Physiol. Cell Physiol. 295 (4), C1037–C1044.

Saha, K., et al., 2008. Substrate modulus directs neural stem cell behavior. Biophys. J. 95 (9), 4426–4438.

Salinas, C.N., Anseth, K.S., 2008. The enhancement of chondrogenic differentiation of human mesenchymal stem cells by enzymatically regulated RGD functionalities. Biomaterials 29, 2370–2377.

Schmidt, S., Friedl, P., 2010. Interstitial cell migration: integrin-dependent and alternative adhesion mechanisms. Cell Tissue Res. 339 (1), 83–92.

Schoen, I., Pruitt, B.L., Vogel, V., 2013. the yin-yang of rigidity sensing: how forces and mechanical properties regulate the cellular response to materials. Annu. Rev. Mater. Res. 43, 589–618.

Shabbir, S.H., et al., 2014. Geometric control of vimentin intermediate filaments. Biomaterials 35, 1359–1366.

Sjostrom, T., et al., 2013. Titanium nanofeaturing for enhanced bioactivity of implanted orthopedic and dental devices. Nanomedicine (Lond) 8 (1), 89–104.

Song, L., Tuan, R., 2004. Transdifferentiation of human mesenchymal stem cells derived from bone marrow. FASEB J. 18 (9), 980–982.

Sordella, et al., 2003. Modulation of Rho GTPse signalling regulates a switch between adipogenesis and myogenesis. Cell 113, 147–158.

Stricker, J., et al., 2011. Spatiotemporal constraints on the force-dependent growth of focal adhesions. Biophys. J. 100 (12), 2883–2893.

Tan, S.H., et al., 2010. Oxygen plasma treatment for reducing hydrophilicity of a sealed polydimethylsiloxane microchannel. Biomicrofluidics 4.

Teixeira, A., et al., 2003. Epithelial contact guidance on well-defined micro- and nanostructured substrates. J Cell. Sci. 116 (10), 1881–1892.

Trappmann, B., et al., 2012. Extracellular-matrix tethering regulates stem-cell fate. Nat. Mater. 11 (7).

Tsimbouri, P.G., et al., 2013a. Nanotopographical effects on mesenchymal stem cell morphology and phenotype. J. Cell. Biochem. 115, 380–390.

Tsimbouri, P.M., et al., 2013b. Genomic analysis of the role of transcription factor C/EBPdelta in the regulation of cell behaviour on nanometric grooves. Biomaterials 34 (8), 1967–1979.

Tsimbouri, P.M., et al., 2012. Using nanotopography and metabolomics to identify biochemical effectors of multipotency. ACS Nano 6 (11), 10239–10249.

Vogel, W.F., Abdulhussein, R., Ford, C.E., 2006. Sensing extracellular matrix: an update on discoidin domain receptor function. Cell. Signal. 18 (8), 1108–1116.

Wang, N., Tytell, J.D., Ingber, D.E., 2009. Mechanotransduction at a distance: mechanically coupling the extracellular matrix with the nucleus. Nat. Rev. Mol. Cell Biol. 10 (1), 75–82.

Weiss, P., Garber, B., 1952. Shape and movement of mesenchyme cells as functions of the physical structure of the medium: contributions to a quantitative morphology. Proc. Natl. Acad. Sci. U.S.A. 38 (3), 264–280.

Wells, R.G., 2008. The role of matrix stiffness in regulating cell behavior. Hepatology 47 (4), 1394–1400.

Wilson, C.J., et al., 2005. Mediation of biomaterial-cell interactions by adsorbed proteins: a review. Tissue Eng. 11 (1–2), 1–18.

Wylie, R.G., et al., 2011. Spatially controlled simultaneous patterning of multiple growth factors in three-dimensional hydrogels. Nat. Mater. 10 (10), 799–806.

Yamada, K.M., Geiger, B., 1997. Molecular interactions in cell adhesion complexes. Curr. Opin. Cell. Biol. 9 (1).

Yang, F., et al., 2005. Electrospinning of nano/micro scale poly(L-lactic acid) aligned fibers and their potential in neural tissue engineering. Biomaterials 26 (15), 2603–2610.

Yim, E., et al., 2005. Nano-pattern induced changes in morphology and motility of smooth muscle cells. Biomaterials 26 (26), 5405–5413.

Yim, E.K., Pang, S.W., Leong, K.W., 2007. Synthetic nanostructures inducing differentiation of human mesenchymal stem cells into neuronal lineage. Exp. Cell Res. 313 (9), 1820–1829.

Yu, C.H., et al., 2011. Early integrin binding to Arg-Gly-Asp peptide activates actin polymerization and contractile movement that stimulates outward translocation. Proc. Natl. Acad. Sci. U.S.A. 108 (51), 20585–20590.

Zaidel-Bar, R., et al., 2007. Functional atlas of the integrin adhesome. Nat. Cell Biol. 9 (8), 858–867.

Zhu, X., Assoian, R.K., 1995. Integrin-dependent activation of MAP kinase: a link to shape-dependent cell proliferation. Mol. Biol. Cell 6 (3), 273–282.

Yang, et al., 2014. Mechanical memory and dosing influence stem cell fate. Nat. Mater 13, 645–652.

FURTHER READING

Gilbert, P.M., et al., 2010. Substrate elasticity regulates skeletal muscle stem cell self-renewal in culture. Science 329 (5995), 1078–1081.

Kloxin, A.M., et al., 2009. Photodegradable hydrogels for dynamic tuning of physical and chemical properties. Science 324 (5923), 59–63.

Stevens, M.M., George, J.H., 2005. Exploring and engineering the cell surface interface. Science 310 (5751), 1135–1138.

Turner, L.A., Dalby, M.J., 2014. Nanotopography: Potential relevance in the stem cell niche. Biomater. Sci. 2, 1574–1594.

CHAPTER 8

Materiomics: A Toolkit for Developing New Biomaterials

Andrew L. Hook[1], Morgan R. Alexander[1], Dave A. Winkler[2]
[1]Laboratory of Biophysics and Surface Analysis, University of Nottingham, Nottingham, UK;
[2]CSIRO Material Science & Engineering, Parkville, Australia

LEARNING OBJECTIVES

- To understand what materiomics is and why it is required
- To become familiar with the various approaches used to design materiomic experiments
- To learn what a polymer microarray is, what it is used for, and how it is produced
- To appreciate the complexity of material–biological interactions
- To become familiar with computational modeling methods as applied to biomaterials
- To gain an insight into how materiomics has and will continue to benefit tissue engineering

8.1 INTRODUCTION: WHAT IS MATERIOMICS?

Materials science is a well-established discipline that is familiar to many. It can be characterized as the study of how a material's structure determines its properties (structure–function relationship) and how this understanding can be used to achieve desirable performance through appropriate processing/synthesis. An example is the control of the mechanical properties of a metal through the manipulation of the microstructure using appropriate heating, cooling, and mechanical deformation. This is a process practiced from the formation of the first wrought iron items such as nails by blacksmiths through to advanced aerospace alloys. Composites then emerged slowly over centuries to give us our existing toolkit of high-performance structural materials. The accumulated knowledge in these material systems now allows the aerospace engineer to "dial up" the materials from their database to design whole plane or individual component on a computer. In tissue engineering of scaffolds, in addition to mechanical property information we are interested in cellular adhesion, immune compatibility, cell adhesion density, phenotype, and so forth. This information is not available for the researcher

Tissue Engineering.
Copyright © 2015 Elsevier Inc. All rights reserved.

who would like to "dial up" the appropriate materials to develop a device that is capable of, for example, manipulating cells to develop into a vascularized functioning organ.

The emerging field of *materiomics* offers a holistic approach to material systems that aims to merge materials science and biological methodologies in order to open new opportunities to discover novel and improved materials with reduced development times (Table 8.1). Materiomics studies the relationships between molecular, physicochemical, and/or processing properties of materials and their characteristics and functions. For example, the process of bone tissue growth on a calcium phosphate scaffold under controlled conditions is a materials science and biological problem, whereas understanding how bone can be grown on any material is a materiomics problem (Groen et al., 2013).

Table 8.1 A List of Key Terminology Associated with Materiomics and the Development of New Biomaterials

The Terminology	Definition
Biomaterials	A synthetic material for application in biology
Combinatorial space	A range of materials created by combining a smaller number of base components at various ratios
Computational modeling in materiomics	Using statistical methods or computer-based methods to simulate and predict the behavior of materials and biological systems such as proteins
High-throughput screening	A method of experimentation whereby experiments are parallelized, usually through the automation of experimental procedures, such that several tests can be achieved in the time it would conventionally take to perform a single test
Materials discovery	Screening a materials space (often a combinatorial space) to discover a material optimal for a particular application
Materials or chemical space	All or a subset of materials of chemistries that exist or could theoretically be produced
Mathematical descriptor	A mathematically derived number that describes a molecular, compositional, or physical property of a material
Ontologies	Controlled vocabularies that precisely describe in terms computers can understand, multiple aspects of materials, such as composition, synthesis, processing, and properties
Polymer library	A collection of different polymers
Polymer microarray	An experimental platform that allows hundreds to thousands of unique materials to be expressed on one support, e.g., a microscope slide, at addressable locations
Synergistic effects	Whereby a combination of components produces a property that could not be predicted from assessing the properties of the individual components

8.2 WHY DO WE NEED NEW BIOMATERIALS

Current biomaterials have revolutionized the aspects of medicine, yet they still routinely fail in the clinic due to immune rejection, infection, and mechanical failure and often require debilitating drug regimens such as blood thinners to prevent clotting. This is because the materials used in these applications were often chosen from "off-the-shelf" existing materials due usually to suitable mechanical properties but with little to no consideration to how biology interacts with these materials. For example, silicone rubbers were initially developed as electrical insulators (Kim et al., 1992) and not urinary catheters, yet millions of such devices are in use in healthcare today. While silicone rubbers have many desirable properties such as flexibility, cost, and chemical resistance, their surface properties are far from optimal as can be seen from the high rates of catheter-associated urinary tract infections (Costerton et al., 1999).

Consequently, these materials are simply not good enough in their biomedical applications and leave little scope for development of new treatments. Tissue engineering constantly requires new biomaterials to realize therapies based on cellular delivery or organ generation (Lutolf and Hubbell, 2005). Within regenerative medicine there is a strong drive toward translation of prospective therapies into the clinic (see also Chapter 22). Often this means that materials choice is dominated by materials, such as poly(lactic acid) and poly(lactide-*co*-glycolide), already licensed for use in man. However, these existing materials are not always suitable for a particular cell type or desired cellular response; so it is essential to have the ability to develop new materials with properties tailored for a particular need (see also Chapter 7).

An increasing number of research groups are investigating the modification of existing materials to improve their properties, and there is an increasing interest in identifying new materials that are tailored toward a particular biological response, such as resistance to bacterial attachment and biofilm formation, or the encouragement of stem cell differentiation to a particular lineage. The emerging paradigm of materiomics promises to shorten the development path of existing and new materials (Cranford et al., 2013; Kohn, 2004).

8.3 THE SIZE OF CHEMICAL SPACE

Captain James T Kirk of the USS Starship Enterprise (Paraphrased) Chemical Space: the final frontier, our mission is to explore strange new chemistries, to seek out new properties, to boldly go where no man has gone before... If we are to accept the challenge of exploring chemical space thoroughly, we are going to need more than a lot of experiments.

It has been estimated using the laws of chemical valence that 10^{100} different molecules could be synthesized with molecular weights below 1000 Da. Such a large number is difficult to comprehend; it is greater than the estimated number of particles of matter in the observable universe, approximately 10^{80} atoms!

Experimentally, high-throughput microarray-based methods allow hundreds to thousands of materials to be screened in parallel (Hook et al., 2010), which is at least 100 times faster than conventional single material measurements but still vastly too slow to explore even a fraction of the possible materials that could be synthesized. It is clear that we need a faster method to navigate the enormous regions of molecular space. We envisage our ever-increasing computational power to describe and navigate these huge spaces *in silico,* making it feasible to explore and identify valuable regions of materials space many orders of magnitude faster than real experiments. The following sections of ontology and modeling deal with the separate challenges of physical and computational experimentation. Computational modeling acts as a bridge between the materials description and the performance, for example, the surface chemistry and the cell attachment. The generation of the data is described in sections on high-throughput experimentation.

8.4 DESIGN OF EXPERIMENTS/GENETIC EVOLUTION/ PARALLELS TO DRUG DISCOVERY

The most advanced robotic systems for materials synthesis and property measurement cannot create more than a minute fraction of the materials that could theoretically be produced, both now and in the future. Consequently, mathematical, statistical, and computational methods are being adopted to allow researchers to create "virtual materials" and experiments that can achieve several useful outcomes.

Materials scientists and tissue engineers are fortunate in that some of the required high-throughput experimental methods and many of the mathematical modeling tools have been developed previously by the pharmaceutical industry. In the early 1990s, methods for rapid synthesis of potential drug candidates using *combinatorial chemistry* were developed (Lowe, 1995). This allowed hundreds or thousands of new molecules to be made at a time instead of one at a time as had been done previously. At the same time, very fast and efficient methods for measuring the biological effects of these large libraries of drugs were also developed (Lowe, 1995).

This paradigm shift in the way drugs were discovered also increased by orders of magnitude the amount of data that resulted, requiring new methods of data processing, storage, and most importantly modeling to be developed. The skills developed in the pharmaceutical industry are transferable to materials and tissue engineering meaning that efforts in high-throughput discovery in this new research domain are accelerating rapidly. The mathematical and statistical tools that were developed to make sense of the vast amount of data generated by the pharmaceutical industry are also very much applicable to data generated by materials scientists and tissue engineers, with one important difference. The complex materials employed in materials science and tissue engineering are not single, precisely defined molecules like drugs. They are often complex polymers with distributions of chain length, often cross-linked in random ways, or containing mixtures (blocks) of multiple polymer segments derived from two or more monomers. In tissue engineering specifically, they could be complex biological polymers with a mixture of components with distribution properties at the molecular scale. Complex materials like these often have properties that are sensitive to the way the materials are made and processed (their *provenance* or history), in particular in terms of their surface properties which are often the dominant factors in cellular response. Developing analytical methodologies to measure the actual surface chemistry and adapting computational modeling methods to deal with this heterogeneity of structure and provenance dependence is one of the most important research issues in materiomics.

Once the complex material can be described mathematically and the processes that were used to create it are encoded, it is possible to build very effective models relating these mathematical "descriptors" to the materials property of interest using statistical and machine learning methods. These methods are known generically as quantitative structure–property relationships (QSPR) modeling methods (Katritzky et al., 1995). They may be simple linear regression models relating the intrinsic molecular, structural, physical, and process properties of a material to some useful property that the material exhibits. In many cases, the relationship between the molecular or process properties and the useful overt properties is very complex and nonlinear. In these instances, machine learning methods like neural networks (discussed in Section 8.6.3) can be very effective at finding this complex relationship and making accurate predictions of useful properties of new materials.

Additionally, scientists are now starting to explore genetic methods for materials design. In this approach, a family of potential materials exhibiting the desired property are synthesized and tested. The structural and process properties of these materials are encoded into a fingerprint of material genome. The members of this initial pool of materials that have properties closest to those required are then mutated to create a new pool of materials. These are tested again and the best of these are mutated to create a third pool of materials. As this process is repeated, like with evolution of species, the materials gradually become "fitter," that is, have properties closer to those required. The advantage of these genetic methods is that they can explore very large areas of materials space to identify regions containing useful properties. Their disadvantage is that they cannot search for the very best possible material in the entire space of possibilities, but find instead materials that have locally optimal properties.

The bottle neck of material discovery studies is being able to experimentally test individual materials. A technique called Design of Experiments (DoE) allows experimental scientists to carry out a smaller number of materials synthesis or characterization experiments that are representative of a much larger number (Gullberg et al., 2004). For example, full factorial designs encode independent parameters such as structural components, synthesis and process parameters into a small number of discrete values (e.g., high/low, polar/nonpolar, negative/positive). For example, if seven parameters (k) each take three levels (l) (e.g., negative/neutral/positive), there are $l^k = 3^7$ or 2187 experiments if all the possible combinations of these parameters are explored experimentally. Usually this number of experiments is too costly or time consuming. A partial factorial design carries out a smaller number of experiments l^{k-p} depending on the parameter p, the size of the fraction of the full factorial design used. The principle is illustrated by a simple example in Figure 8.1. If the experiment varied three factors, and each factor could take either of two values, there would be $2^3 = 8$ possible experiments in the full factorial design. A partial factorial design with $p = 1$ would require $2^{3-2} = 4$ experiments. If the experiments were correctly chosen, each factor would occur at the high and low levels the same number of times (balanced design).

Mathematical or statistical models can then make predictions of the properties of the materials that were not synthesized or tested in the partial factorial design, that is, the models can "fill in the gaps." These or existing experiments on materials, particularly those done using high-throughput robotic methods

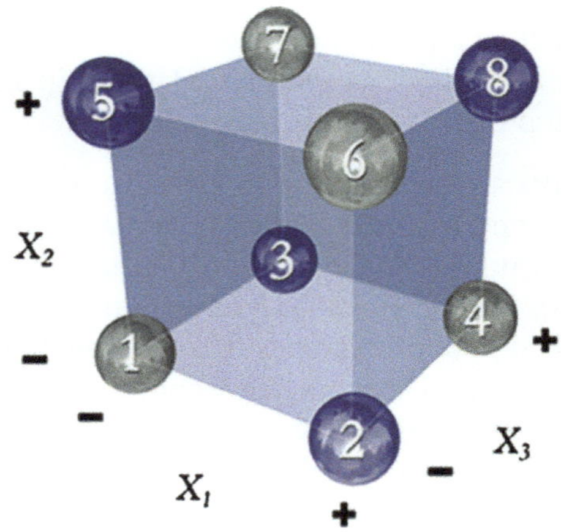

FIGURE 8.1
A full factorial design is represented by the eight spheres at the vertices of the cube. A balanced partial factorial design would carry out four experiments involving the combinations of the three factors represented by the four purple spheres.

that generate large amounts of data, can be used to generate sophisticated mathematical models that relate material composition, structure, processing, and so on, to their performance to modulate cell activity as biomaterials, tissue engineering scaffolds, or solution-borne molecules. This enables the most to be learned from a small number of experiments that can represent that largest possible materials space.

8.5 HIGH-THROUGHPUT EXPERIMENTAL METHODS

Any search for a new material has to begin with experimental observations of the performance of a set of candidate materials (Kell and Oliver, 2003). In some cases, there is sufficient knowledge of biology's response to a material such that candidate materials can be judiciously chosen to test a specific hypothesis. Where the current understanding of those interactions is not sufficiently developed to allow *ab initio* design, the set of initial candidate materials must be selected to represent the largest material diversity in order to maximize the understanding gained by the initial experiment. Taking a "blind" approach to materials discovery can also be useful in cases where the biological–material interaction is well understood because it enables the possibility for the discovery of an entirely new class of materials that perform well in a given role and that would not have been predicted from the present knowledge of a given biological response to a material (Hook et al., 2012) The success of any further "chase" for the ideal material depends on the quality of the measurements made, the number of different measurements made, and how representative the initial candidate materials were for the chemical space being screened. It is, therefore, advantageous in an initial screen to have as many measurements as possible, and the polymer microarray format is ideally suited to achieve this (Hook et al., 2010). The aim of a materials microarray is to present a library of materials in a format that enables the library to be rapidly screened, normally in a parallel cell exposure format. Screening exposure in parallel is of benefit not just from a throughput perspective but also in terms of reducing interexperimental biological variance. This sample format is the key enabling tool for experimental high-throughput materials discovery projects. Typical libraries consist of hundreds to thousands of unique members formed, e.g., from combinations of tens to hundreds of individual monomers combined to form copolymers, e.g., 24 monomers combined pairwise at 70:30 yield 576 unique copolymer compositions (Anderson et al., 2004, 2005; Tourniaire et al., 2006). More recently, Celiz et al. have expanded this approach to include 141 different acrylate and acrylamide monomers thereby significantly increasing the chemical space that is experimentally accessible (Celiz et al., 2013). The advantage of this approach is the ability to study the interaction of materials with biological systems.

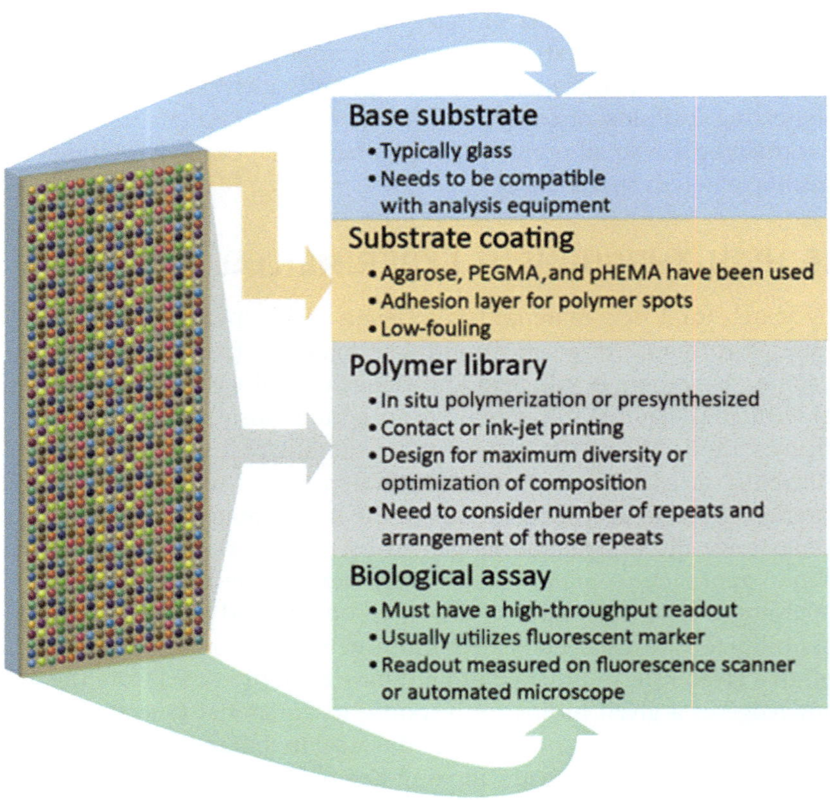

FIGURE 8.2

The key parameters that must be considered for the creation of a polymer microarray. PEGMA, poly(ethylene glycol) methacrylate; pHEMA, poly(hydroxyethyl methacrylate). *From Hook (2014) with permission.*

The design of a polymer microarray is directed by the biological assay it will be used for and consideration must be given to:

1. the supporting substrate and its coating,
2. the members of the polymer library and their synthesis (on-slide or before),
3. the biological assay which can couple with the microarray (Figure 8.2).

It is possible not only to explore chemistry but also to systematically investigate other contributors to the cell response to materials, broadly collected into four groups: specific biochemical ligand interactions (e.g., surface peptides), surface chemistry (see also Chapter 6), surface topography, and material stiffness (Engler et al., 2006; Flaim et al., 2005; Hook et al., 2012; Unadkat et al., 2011). These key factors are summarized in Figure 8.3.

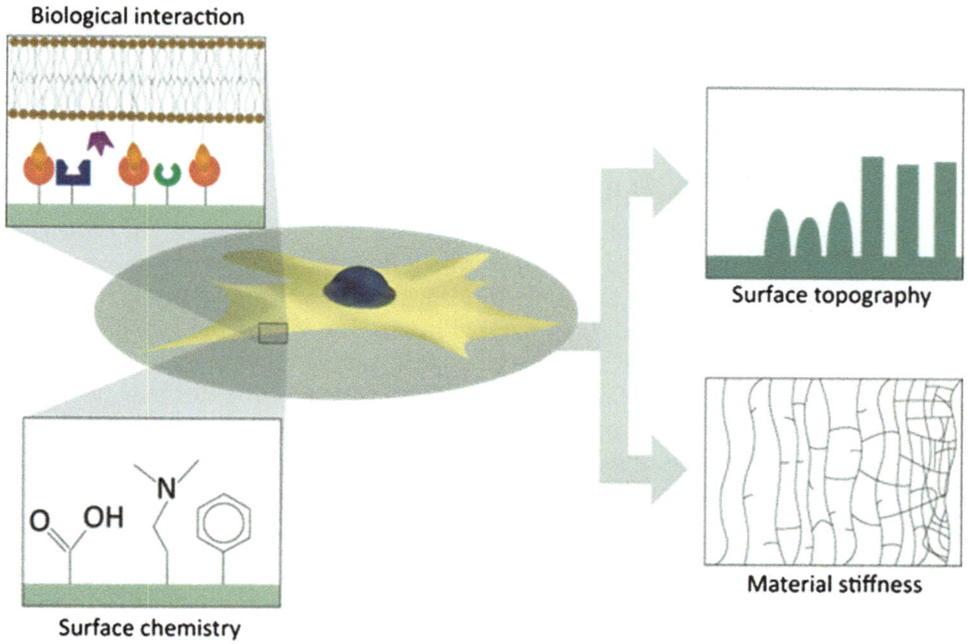

FIGURE 8.3
The key aspects of biological–material interactions. *Based upon figure in Hook (2014), with permission.*

8.5.1 The Supporting Substrate

The key requirement of the array support is that it can be applied easily to the biological assay and associated high-throughput readout. The substrate of choice is a glass microscope slide due to the existing range of stage holders, scanners, and microscopes that have been adapted to this format. Furthermore, the slide is transparent, which allows cells growing on the slide to be easily assessed using light microscopy. The glass slide is also widely compatible with most cell culture methods, enabling a diverse range of biological assays to be applied. However, polymer microarrays can equally be produced on other materials, for example, directly onto polystyrene-based tissue culture ware, which may enable the polymer microarray format to be accessible to a greater number of biological assays.

The surface chemistry of the underlying substrate plays an important role in the formation of a microarray as well as the success of subsequent bioassays. The substrate coating must be both adherent to the materials printed onto it and resistant to the attachment of biomolecules and living cells (nonfouling) in order to optimize the signal-to-noise ratio of any biological assay and prevent cross talk from one spot to another. A number of different surface coatings have been developed with these needs in mind, with emphasis generally placed on using cheap and robust coating methodologies. Examples include poly(hydroxyethyl methacrylate) (Anderson et al., 2004, 2005), agarose (Tourniaire et al., 2006), and poly(ethylene glycol) based coatings (Kurkuri et al., 2009).

8.5.2 The Library of Materials

The creation of material arrays requires a combinatorial library of materials to be generated and positioned at addressable locations on a substrate surface. A key factor that will influence the design of a polymer microarray will be which biological–material interaction will be probed. Biology will respond to the material's surface chemistry, topography, hardness, and the presence of any bioactive factors. Some or all of these properties may be included within the materials space represented on an array depending on the anticipated biological response to the materials. The ultimate screen of a biological–material interaction would ensure all elements of materials are included.

Polymer microarrays are typically formed by either contact or ink-jet printing, although other microfabrication methodologies are possible (see Chapter 9). Contact printing involves the use of a robot moving a metallic pin, which is dipped into a solution and then spotted onto the substrate surface by making contact. For ink-jet printing, a nozzle is used to draw up then eject polymer solution onto the substrate at a defined droplet volume.

The library with a diverse **chemistry** can be generated prior to the synthesis of the array, for example, by printing presynthesized polymers (Tourniaire et al., 2006) or by printing monomer solutions that are polymerized in situ (Anderson et al., 2004). The in situ polymerization (Anderson et al., 2004) of a polymer library can be achieved using contact printing in five steps: mixing monomers at various ratios in a source plate, printing acrylate monomers with an initiator, activation of initiator upon UV irradiation, polymerization of the monomer, and finally removal of the solvent (Figure 8.4). This approach decreases the time required for polymer library synthesis and microarray formation by combining these processes and also allows materials that cannot be printed, such as cross-linked polymers, to be included in the polymer library. This allows materials with variable hardness to be included. However, the polymer synthesis conditions are likely to differ when materials are produced on a larger scale.

An alternative approach is to print a library of presynthesized materials (Anderson et al., 2005; Tourniaire et al., 2006), which allows standard polymer characterization techniques such as gel permeation chromatography, nuclear magnetic resonance, and differential scanning calorimetry to be utilized without having to extract minute amounts of polymer from the slide. However, this method is restricted to polymers that can be solvated, and is considerably more laborious for the initial creation of the polymer library.

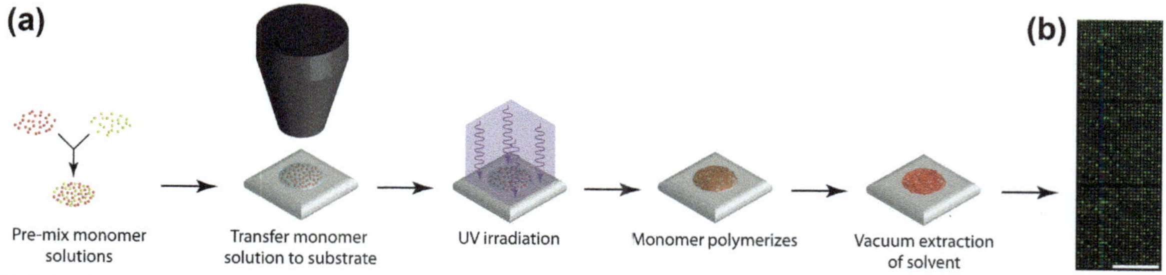

FIGURE 8.4

(a) The five steps required for the in situ polymerization of a polymer library. This leads to the production of a polymer microarray. (b) Fluorescence image of a polymer microarray after incubation with fluorescent bacteria, scale bar is 10 mm. *From Hook et al. (2010) with permission.*

Protein readily adsorbs to the surfaces of materials from media, thus, in most cell culture conditions the attachment of cells is regulated by the intermediate adsorbed protein layer (Wilson et al., 2005). This means that very few assays truly probe material–cell interactions, but rather the surface chemistry of the materials controls the protein adsorption that is then the determining factor on the behavior of cells at the surface (Mei et al., 2010). Exposure of a protein containing media or body fluid will rapidly cause the surface to be coated with a large assortment of various proteins that can be difficult to understand due to the many and complex interactions. To study this in a more controlled manner, microarrays have been preadsorbed with proteins such as laminin, fibronectin, albumin, and vitronectin.

High-throughput screens of material–cell interactions are not limited to synthetic polymers but can be applied to **biological** molecules such as proteins directly (Flaim et al., 2005; Gobaa et al., 2011; Kurkuri et al., 2009). The proteins of greatest interest are those whose biological function is associated with cell attachment. Within the body, cells secrete and bind to the extracellular matrix (ECM), which is composed of an assortment of proteins including collagen, laminin, vitronectin, and fibronectin (see also Chapter 5). These proteins and combinations of these proteins can be printed directly onto glass substrates to form a high-throughput platform that can be applied to assess the response of cells (Flaim et al., 2005). When applied to mouse embryonic stem (mES) cells, combinations of collagen I and Fn were best able to instigate the differentiation of ES cells to an early hepatic fate. A key outcome of this (Flaim et al., 2005) and other work (Hook et al., 2012) is the observation of synergistic effects, where the combination of two components induce a biological response that would not be predicted by assessing the response to the components alone. The occurrence of such events is a key validation of the combinatorial approach to material library design and also highlights the need for high-throughput assays to uncover these unique and unpredictable interactions.

The **topography** of a surface plays a key role in the attachment and behavior of cells; contact guidance of cells along grooves has been observed down to nanoscale features (Johansson et al., 2006; Tessierlavigne and Goodman, 1996), altering the roughness of prosthetic implants has been a key parameter to improve bone-implant contact (Cochran et al., 2002; Martin et al., 1995; Price et al., 2004), and it has also been demonstrated that surface roughness can influence the differentiation of preosteoblast cells (Schneider et al., 2004). Thus, surface topography is a key parameter to be included in biomaterial development programs (Adler et al., 2011; Unadkat et al., 2011).

Similar to chemistry, when considering the different angles, curvature, spacing, and feature densities possible that would alter topography the possibilities are endless. Thus, the challenge remains to produce a subset of materials that is representative of the materials' topography space. In one strategy, topographical units were derived from very simple geometric shapes—circles, isosceles triangles, and rectangles arranged within an imaginary square. The three base shapes are good representative base units as they allow the inclusion of large smooth areas (circles), angles (triangles), and stretched elements (rectangles) within the resultant topographical features. By this arrangement there is still an infinite number of combinations that are available; however, in one study 2176 possible units were selected at random, produced by photolithography, and used to emboss poly(lactic acid) films to produce a TopoChip (Figure 8.5). These films were used to study the influence of topography on the bioactivity of human mesenchymal stem cells (hMSCs). Varying levels of alkaline phosphatase, which is a marker for early osteogenic differentiation (Unadkat et al., 2011), were observed for cells grown on the varied topographical patterns, demonstrating that various topographies can induce cellular differentiation.

8.5.3 The Biological Assay

The final consideration for applying polymer microarrays to the discovery of biomaterials is the design of a suitable bioassay that is compatible with the high-throughput format. To achieve this, the key requirement is the inclusion of a suitable readout that can be assessed in an automated fashion. Typically, a fluorescence marker is used that can be measured using an automated fluorescence microscope or a fluorescence scanner. However, fundamentally each biological application requires the development of a suitable biological assay that is unique and specialized for each case. Polymer microarrays have been successfully used to probe a diverse range of biological systems, including supporting stem cell outgrowth (Anderson et al., 2004; Anglin et al., 2010a; Fernandes et al., 2010; Flaim et al., 2005; Gobaa et al., 2011; King et al., 2010; Mei et al., 2009, 2010; Unadkat et al., 2011; Yang et al., 2010; Zhang et al., 2013),

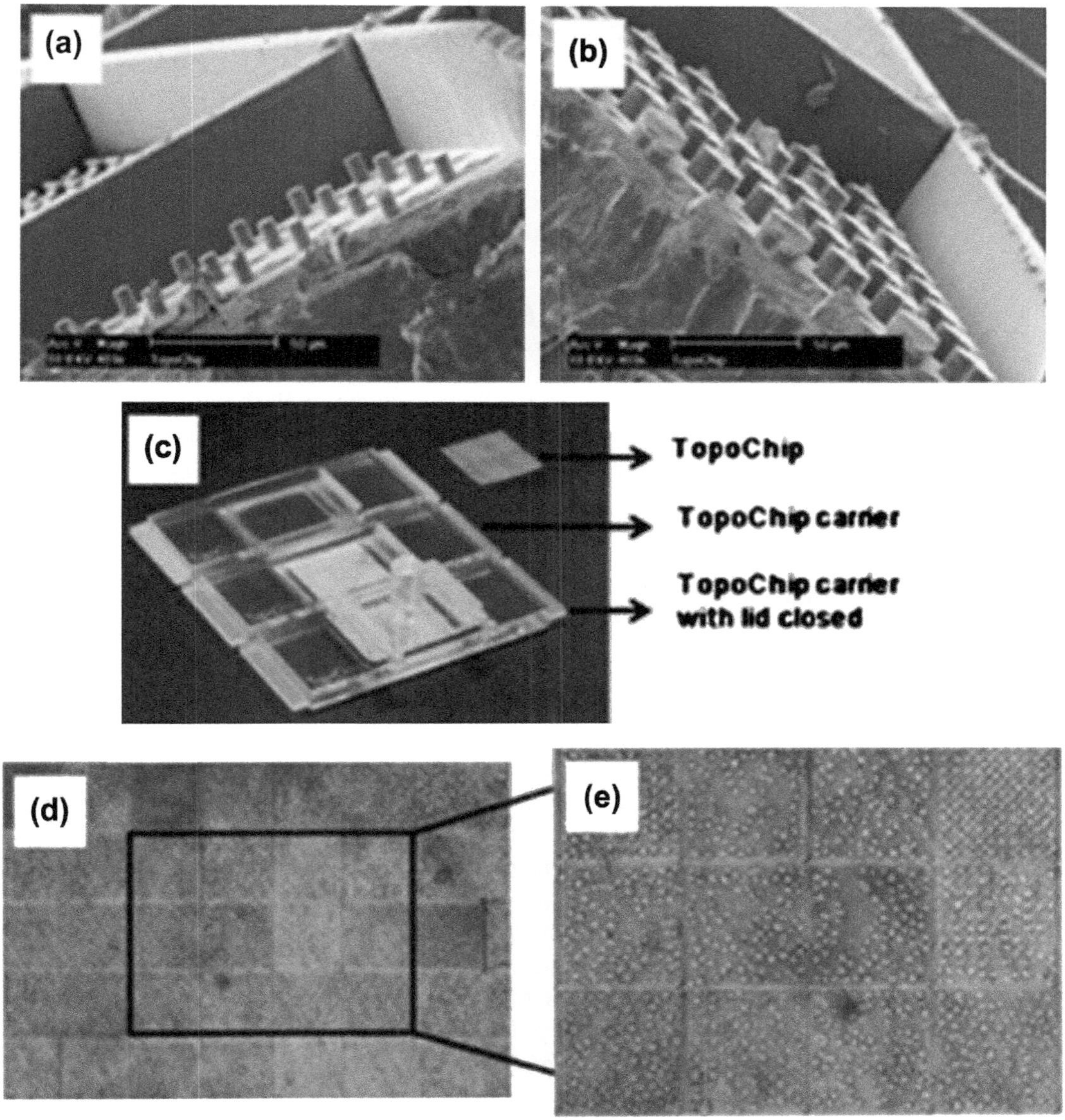

FIGURE 8.5

A platform for high-throughput screening of topographical patterns. (a) and (b) Scanning electron microscopy images of sections of TopoChips displaying accurate feature replication on the TopoChip. (Scale bar: 50 μm.) (c) The TopoChip carrier, lid, and chip assembly. This chip carrier can even be used as a microbioreactor for perfusion culture of cells, or with a second set of attachment (not shown) for static cell culture. (d) and (e) Light microscopic images of cells seeded using the chip carrier displaying homogeneity in cell distribution within and between TopoUnits. *From Unadkat et al. (2011) with permission.*

dendritic cell maturation and phagocytosis (Mant et al., 2006), materials resistant to bacteria (Hook et al., 2012, 2013a; Pernagallo et al., 2011), switchable materials (Hook et al., 2013b, 2014; Zhang et al., 2009, 2013), platelet activation (Hansen et al., 2011), cell sorting (Anglin et al., 2010a,b; Tare et al., 2009), hepatocytes and toxicity models (Flaim et al., 2005; Hay et al., 2011), human skeletal cell attachment (Khan et al., 2010; Tare et al., 2009), endothelization (Pernagallo et al., 2012), *Giardia lamblia*–material interactions (Pickering et al., 2012), cell transfection (Adler et al., 2011; Hook et al., 2009a,b), and *Cryptosporidium parvum*–material interactions (Wu et al., 2012).

8.6 COMPUTATIONAL MODELING

Despite advances in high-throughput materials screening and the rapid creation of material libraries, the number of materials that can feasibly be screened using experimental methods is limited in the order of thousands of materials. As we have illustrated earlier, this is still vastly smaller than the estimated size of materials space, materials that could in principle be synthesized. Computational modeling provides the bridge between the experimental limitations and the vastness of chemical and materials spaces. The large number of biological–material interactions that can be assessed using polymer microarrays provides an ideal basis for the creation of models.

8.6.1 Correlating Measured Material Properties with Cellular Attachment

Computational modeling attempts to predict the biological performance of a material from its known composition, molecular, and process properties. Living cells will interact with the surfaces of materials; thus, it is the surface properties of the materials that provide the defining parameter which can be used to describe their biological response (see also Chapter 7). It is therefore useful to be able to measure various surface properties of materials concurrently with biological measurements. Where large libraries of materials have been used, it is necessary to have a high-throughput surface analysis method that is capable of taking individual measurements from the hundreds to thousands of unique materials. High-throughput measurements of material surfaces have been achieved by X-ray photoelectron spectroscopy (XPS), time-of-flight secondary ion mass spectrometry (ToF-SIMS) (Urquhart et al., 2007, 2008), water contact angle (WCA) (Taylor et al., 2007), atomic force microscopy (AFM) (Hook et al., 2011), surface plasmon resonance (SPR) (Hook et al., 2009c), and force measurements (Tweedie et al., 2005). Once material properties have been measured, the challenge remains to connect them in a meaningful and quantitative way to a biological response. This is difficult due to the complex and multifaceted interactions of cells with materials, whereby cells respond to a material's surface chemistry, hardness, topography, and so forth. (Figure 8.3). Simple correlations with properties that are commonly implicated in cell–material interactions, such as WCA or roughness, typically do not work. An important step in deriving a correlation between biological response and a measured surface property has been the use of partial least square (PLS) regression to correlate a univariate property, such as cell numbers, with a multivariate dataset, such as the hundreds of secondary ions produced in a ToF-SIMS spectra (Taylor et al., 2009). Modeling a multivariate dataset enables the inclusion of multiple aspects of the materials. This method was initially validated by linking ToF-SIMS spectra with WCA (Taylor et al., 2009), and has been successfully applied to predict the frequency of colony formation of stem cells from the chemical information represented in ToF-SIMS spectra (Figure 8.6) (Mei et al., 2010; Yang et al., 2010). Similarly, bacterial attachment was predicted for *Pseudomonas aeruginosa* and *Staphylococcus aureus* (Hook et al., 2012). The successful construction of these models likely results from the diverse material properties that are represented within ToF-SIMS spectra. Secondary ions emitted from a surface can be assigned to chemical functionalities that are associated with surface charge, hydrophobicity, aromatic and cyclic groups, and material stiffness (Hook et al., 2012, 2013b; Mei et al., 2010; Taylor et al., 2009; Urquhart et al., 2008). A powerful application of this approach would be the design of improved materials based upon the models produced. Limiting this approach is the extensive surface analysis required to gather the relevant chemical/physical properties of the materials being studied. Moreover, the number of materials that can be included is limited to the number of different samples that can feasibly be screened. Although this is significantly expanded by the use of polymer microarrays to the order of hundreds to thousands of materials, this is still very small compared with the large number of varied polymer chemistries that are possible.

To overcome this constraint, modeling based upon molecular descriptors rather than measured properties has been explored (Hook et al., 2013c). Using this method, materials do not need to be produced and analyzed, but rather a virtual library of materials can be created where the size of the library is limited by computational power and the materials space in which the model is applicable, rather than experimental requirements. This approach has been used to predict defects within polymers from their molecular descriptors using a PLS regression model (Hook et al., 2013c). This work was expanded to enable the prediction of the adhesion of human embryoid body cells (hEBs) to materials using only calculated molecular descriptors (Epa et al., 2012). Combining such modeling methods with polymer microarray screens, and specifically the design of polymer libraries, enables exciting opportunities to assess a larger material property space than could be accessed by experimental methods alone.

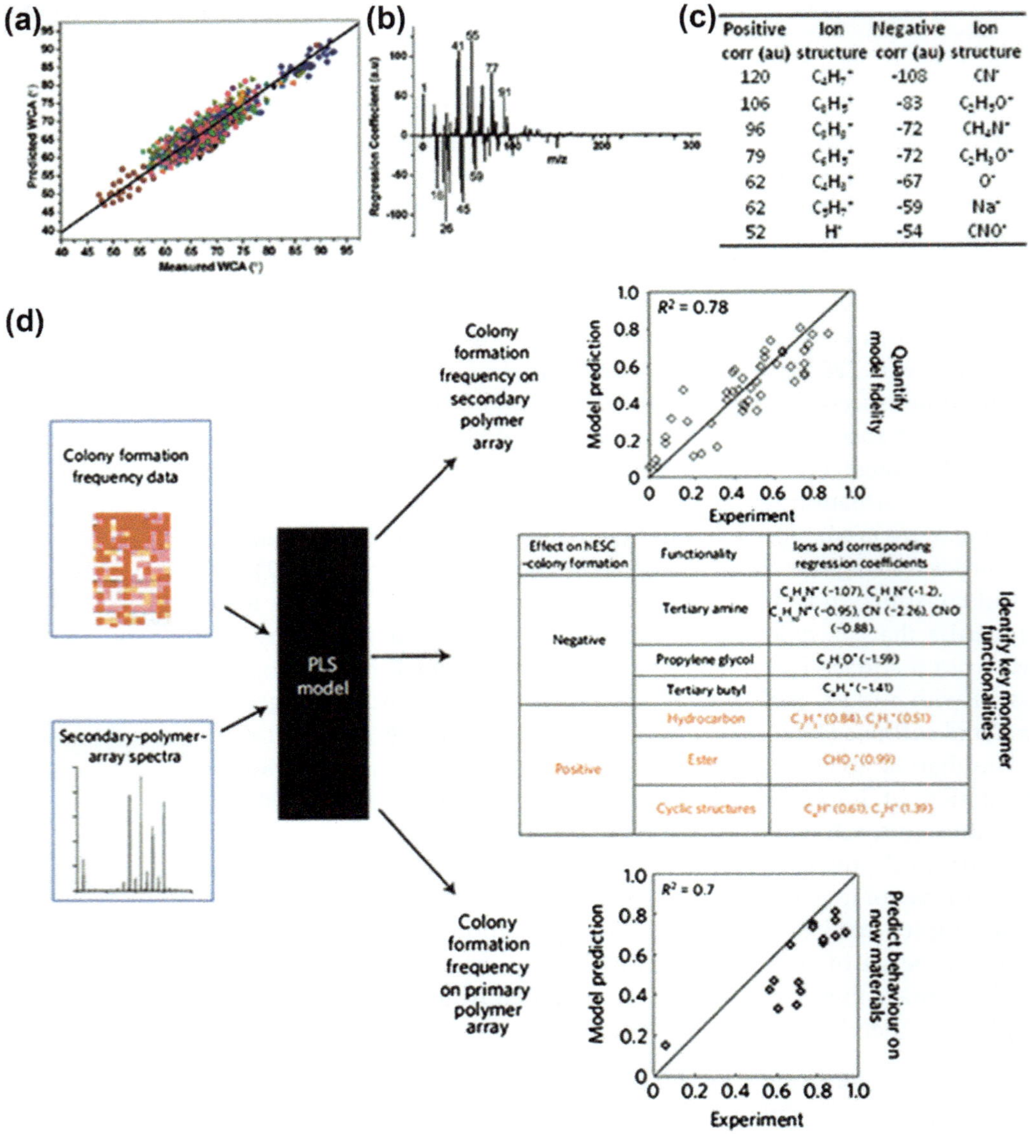

FIGURE 8.6

Utilizing PLS modeling for materials development. (a)–(c) To validate the technique PLS modeling was first used to model WCA and surface chemistry—a system that is well understood. (a) The predicted WCA against the measured WCA for a library of polymers (linear fit is shown, $R^2 = 0.94$). (b) The regression coefficients produced by the PLS model that informs how each variable within the multivariate dataset (in this case ToF-SIMS ions) correlates with the univariate dataset (WCA). The regression coefficients are plotted versus the peak *m/z*. (c) Table showing the ion structures for the ions with the largest positive or negative regression coefficients for the PLS model, confirming that aliphatic carbon groups correlate positively with WCA and oxygen and nitrogen containing groups correlate negatively with WCA. (d) Having been validated, PLS modeling could be applied to a system that is not well understood—predicting colony formation frequency of stem cells from material chemistry represented by ToF-SIMS ions. Top: Each point in the figure represents one of 48 different polymers in the secondary array, and the inserted line represents the ideal situation when the prediction matches the experiment completely. Middle table: Functionalities and their associated characteristic ions that either correlate positively or negatively with human embryonic stem (hES) cells colony formation. Bottom: The model produced from the secondary array was validated by applying to a primary array, thus, successfully predicting the colony formation frequency for materials not used to initially produce the PLS model. *From Urquhart et al. (2008) and Mei et al. (2010) with permission.*

8.6.2 Ontologies

The full mathematical description of materials must be achieved in order to computationally model the biological response they induce. This must include the multiple aspects of materials that define their performance, including both composition and synthesis. Ontologies are controlled vocabularies that are becoming increasingly important in all "omic" fields. They allow genes, chemicals, materials, and so forth, to be precisely defined in terms of composition, relationships to other entities, and history (provenance). Much work in ontology is carried out within the framework of the semantic web, a movement that aims to convert the current World Wide Web (consisting of largely unstructured documents) into a *web of data*. This will allow computers, instruments, and so forth, to understand and exchange data easily, without the need for translation between formats and human interpretation of elements of the data, because the information about the data is embedded in the data. This is achieved using ontology languages like OWL (Web Ontology Language) that allow knowledge representation. The data described by an ontology is interpreted as a set of entities (such as a specific type of polymer) and a set of *property assertions* (e.g., "is a homopolymer") that describe the relationships between the entities. An ontology is a set of axioms (premises or starting points for reasoning) that place constraints on sets of entities (or classes) and the types of relationships that are permitted between them. These axioms allow systems to infer additional information from these data without human intervention. Ontologies have been used for a considerable time in biology (e.g., gene ontology), but are beginning to be developed for chemistry and materials (Taylor et al., 2006). An example of chemical ontology is ChEBI. ChEBI consists of ~30,000 terms and acts as a reference for chemical entities, from subatomic particles to polymers, across multiple research domains (Hill et al., 2013). Materials-specific ontologies are now being developed. One of the first examples is the polymer documentation system of the international data corporation (IDC) with inclusion of analytical and synthetic concept relations (POLIDCASYR), a controlled vocabulary of hierarchically ordered terms (Figure 8.7) (Adams, 2010).

A simple organization of concepts related to materials science might contain four groups of core ontologies: Substance, Process, Property, and Environment. These ontologies give structured definitions of terms, names, and vocabularies for the basic concepts of each core domain. Figure 8.8 shows a component of the Property ontology, Property class, which has six attributes. It corresponds to the fundamental data structure of materials property (Ashino, 2010).

8.6.3 Modeling Methodologies

As highlighted in Section 8.4, computational tools like QSPR, initially developed to model large datasets derived from high-throughput

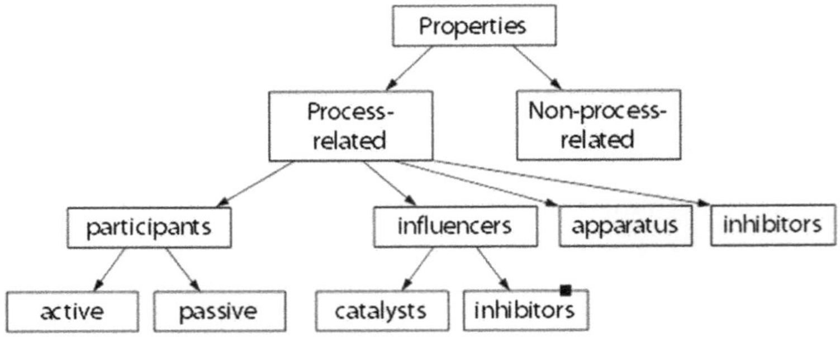

FIGURE 8.7
The POLIDCASYR semantic categories. *From Adams (2010) with permission.*

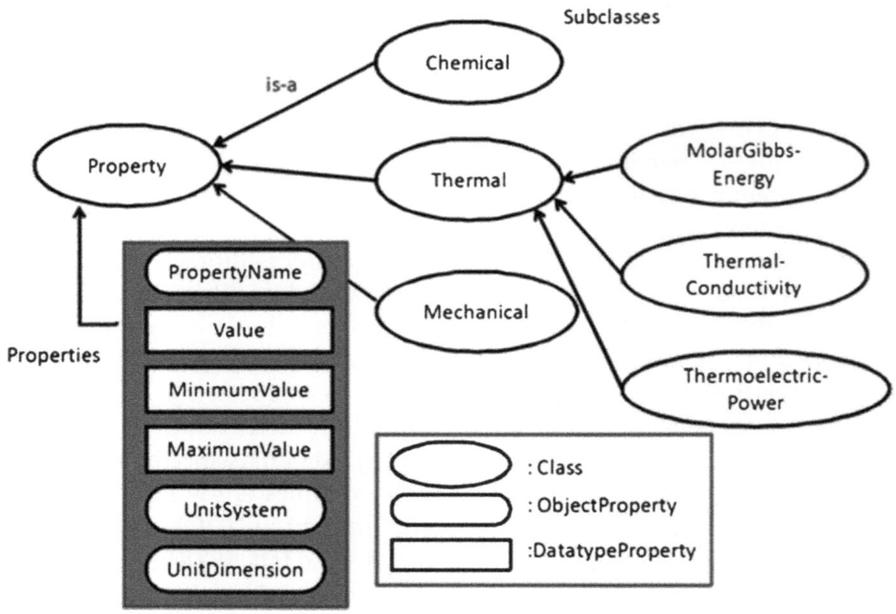

FIGURE 8.8
Structure of the Property ontology. The Property class has the properties: PropertyName, Value, and UnitSystem. The subclasses, chemical, thermal, and mechanical, correspond to the taxonomy of materials property. *From Ashino (2010) with permission.*

pharmaceutical research, have been shown to be very useful for materials design and discovery. Although this field of modeling is relatively at an embryonic stage compared to its use in the pharmaceutical and related fields, a number of key proof of concept studies have demonstrated its applicability.

The key steps in QSPR modeling of complex materials are similar to those required for molecule or drug modeling:

- The relevant molecular, structural, physicochemical, and process parameters of the materials need to be encoded into mathematical *descriptors*, numbers that capture these properties accurately and succinctly.
- As there are many ways of encoding these microscopic properties or parameters, methods need to be employed that choose the best subset of descriptors that are most likely to influence the materials property, such as cell adhesion, that is being modeled.
- The selected subset of mathematical descriptors is then used to train a statistical or machine learning model that is quite often nonlinear. Machine learning methods called artificial neural networks are commonly used. They are computational models inspired by the brain that are capable of learning complex relationships and pattern recognition.
- The quality, robustness, predictive power, and domain of applicability (the region of property and descriptor space where the model was generated) of the model must be assessed.

A comprehensive review of the application of statistical and machine learning methods to materials, including biomaterials discovery, was published recently (Le et al., 2012).

We illustrate QSPR modeling using two examples: adhesion of embryoid body cells and cell expansion performance of hematopoietic (blood) stem cell bioreactors.

8.6.4 Attachment of Embryoid Body Cells

Tissue engineering applications require materials that give a known and controlled cell response, and as such further progress in this field requires the development of new materials that can manipulate cells in novel ways. In particular, concerns about using animal-derived products to culture cells for human transplantation have stimulated research on designing chemically defined, serum-free, feeder-free synthetic substrates and media to support robust self-renewal of pluripotent cells (see Chapter 2). This is being increasingly undertaken using high-throughput synthesis and assessment methods (Figure 8.9) (Anderson et al., 2004; Epa et al., 2012; Hansen et al., 2013; Mei et al., 2010; Yang et al., 2010). A model that relates cell adhesion with material properties would both enhance the understanding of the cell–material interaction as well as direct future materials design. Robust, predictive, quantitative, and purely computational models of the adhesion of hEBs to the surfaces of a 496-member polymer microarray library have been developed. Linear regression modeling and a special kind of neural network was used to construct models relating the surface chemistry of the polymers to their ability to attach and grow hEB cells (Epa et al., 2012). The structures of the polymers were converted into mathematical descriptors that captured relevant polymer properties

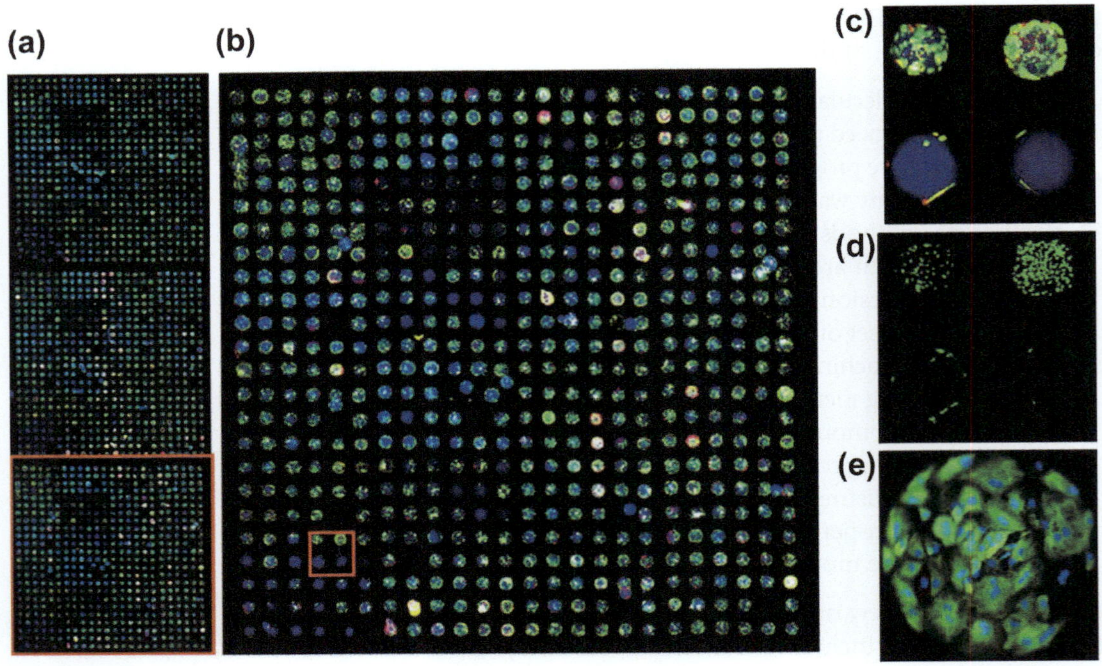

FIGURE 8.9

The application of human embryonic stem (hES) cells onto a polymer microarray for the high-throughput discovery of materials that support cell attachment. (a)–(c) Six million hES cells were added on the polymer microarray for 6 days and then stained with cytokeratin 7 (green—an intermediate filament protein found in most glandular and transitional epithelia) and vimentin (red—an intermediate filament protein common in cells of mesenchymal origin). Polymer spots can be identified by blue fluorescence. (d) Nuclei were also stained (green) (not shown in other images to simplify presentation). (e) Typical cytokeratin 7 positive spot. *From Anderson et al. (2004) with permission.*

like the ability to form hydrogen bonds, the wettability of the surface, the dipole moment, the polymer flexibility, the atomic composition, the number and types of functional groups. The neural network models could predict quantitatively the hEB adhesion for the polymers used to train it, and also was effective at predicting hEB adhesion for other polymers not used to generate the model (Figure 8.10). The ability of the model to predict the performance of materials not used to produce the model is the key validation of the robustness of the model. The next step then is to use the model to make predictions about the likely performance of additional polymers not yet synthesized, which could increase the efficiency of materials discovery by identifying polymers not worth pursuing.

8.6.5 Hematopoietic Stem Cell Bioreactor Performance

Computational modeling is not limited to materiomic applications but can be applied to any system where multiple factors combine to produce a particular result. This is true for many tissue engineering applications, for example, in the design of cell culture protocols.

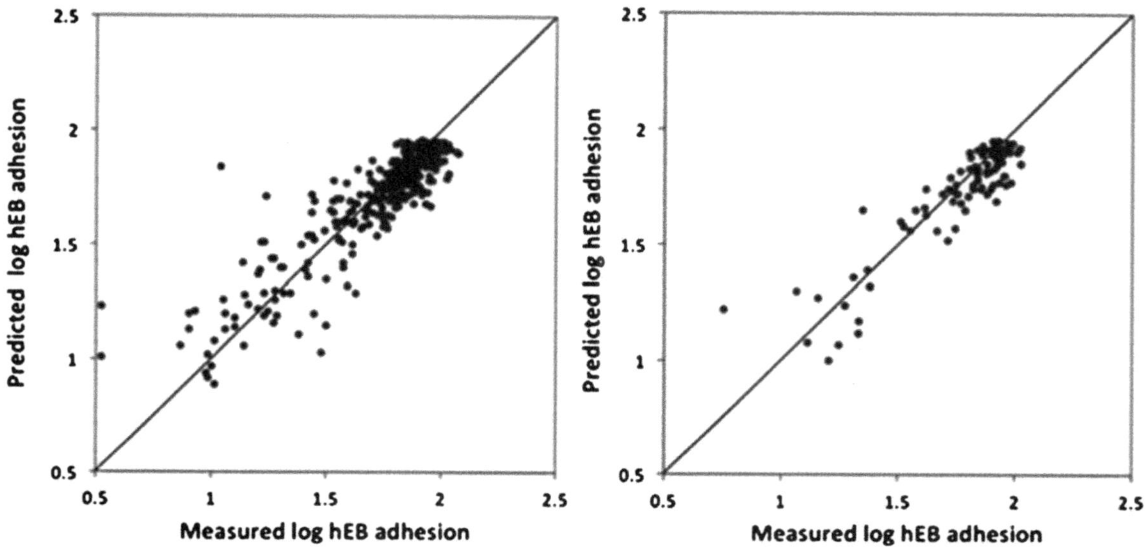

FIGURE 8.10
Predictions of the log human embryoid body cell (hEB) adhesion on the polymers for the training (left) and test (right) sets for the nonlinear Bayesian neural net model. *From Epa et al. (2012) with permission.*

Ex vivo expansion of stem cell populations is a very active and important area of stem cell research (see Chapter 13). While expansion of ES cells can be carried out routinely in many laboratories, significant ex vivo expansion of adult stem cells is proving more difficult to achieve. The shortage of hematopoietic stem cells (HSCs) in particular poses serious clinical problems and limits the clinical application of HSC therapies such as bone marrow transplants. Development of bioreactors, manufactured or engineered devices or systems that create a biologically active environment for cell culture, is consequently a very active research area. A myriad of bioreactor configurations, growth factor regimens, and culture media conditions have been reported in literature publications. Many of these bioreactor parameters interact with each other to influence the number of phenotypes of the cells the bioreactors generate. Clearly, it is important to understand these interactions so that the bioreactors can be controlled to produce the maximum expansion of cells of the desired type.

Recent studies have shown that it is feasible to model quantitatively the expansion of cells in bioreactors and to understand which bioreactor parameters contribute most to efficient expansion of given cell phenotypes (Winkler and Burden, 2012). Both linear regression and nonlinear neural network methods have generated very good prediction for the expansion of six different types of cells as a function of bioreactor operating parameters. These included: the inoculate density; the concentrations of growth factors such as interleukins,

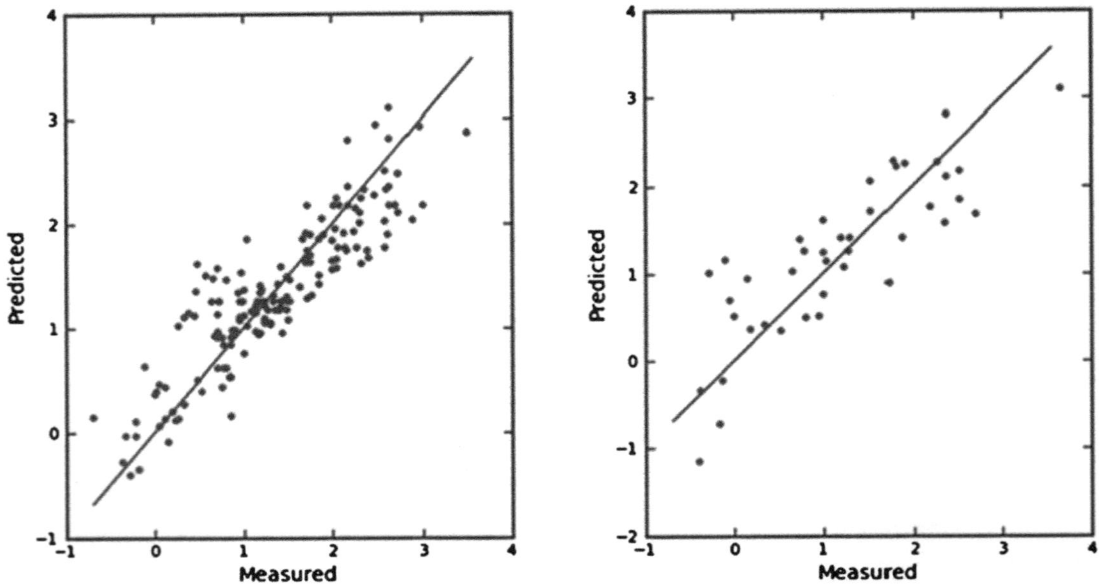

FIGURE 8.11
Performance of BRANNGP models with two hidden layer nodes to predict the expansion of nucleated cells. Prediction of the expansion of experiments used to train the model (left), and of the expansion of asset of bioreactors not used in model building (right). *From Winkler and Burden (2012) with permission.*

colony stimulating factors, stem cell factors, and erythropoietin; presence of serum or stromal layers; and bioreactor size, shape, and geometry. The performance of a neural network model to predict the expansion of cells is illustrated in Figure 8.11.

8.7 FUTURE PERSPECTIVE

All the tools for achieving a materiomics approach to biomaterials development are in place. Experimental platforms have now been developed that enable the biological response to a range of materials properties to be assessed, including material surface chemistry, topography, and hardness. Coupled with this has been the successful computational modeling of material–biological interactions based upon mathematical descriptors that could easily be applied to virtual polymers. The final step that remains is to integrate computational modeling to direct a high-throughput materials discovery program. This could be achieved using the following approach. Initially, a diverse library of polymers can be screened to produce a "learning" set that would enable the construction of an experimental model that relates the biological response to material composition. This model can then be used to screen virtual materials and thus direct the design of the next polymer library generation. By repeating this process an ideal material composition can be identified while minimizing the number of materials that actually need to be synthesized, and in doing so minimizing the experimental bottle neck associated with a high-throughput materials discovery approach.

One challenge remains in how best to assess the complex nature of polymers. Whereas in drug discovery there really is only one key parameter to consider, the chemical structure of the drug, for materials, in addition to the chemical structure of its constituents, the polymer's molecular weight and polydispersity can significantly alter the polymer's properties. Furthermore, in the case of copolymers the type of copolymerization can play a key role. The way in which a polymer is made and subsequently processed can also play a key role in its property and function, and the challenge is to both be able to characterize this in high throughput and include these parameters when necessary within computational models. While this has been done through a property high-throughput characterization approach, for example, modulus being determined by nano indentation (Mei et al., 2010), it would still be beneficial to be able to make measurements such as green fluorescing protein (GFP) on the small volume of polymer represented by a microarray spot. This requires further development in both experimental methods and within computational modeling.

8.8 SUMMARY

- Current materials that are used for tissue engineering applications are limited and often have suboptimal interactions with biology.
- Polymer microarrays are a key enabling tool for high-throughput materials discovery.
- Synergistic effects are often observed for materials, whereby two components together will produce an effect that would not be predicted from observing the two components individually.
- The number of potential materials that could be applied to tissue engineering applications is huge (10^{100} different molecules could be synthesized with molecular weights below 1000 Da).
- The most advanced high-throughput experimental techniques cannot effectively screen the huge materials space.
- Computational modeling of material–biological interactions offers the potential to screen a larger materials space by producing a model from known interactions and applying the model to a virtual library of materials.
- High-throughput material screens offer the perfect starting point for producing robust computational models.
- Bacterial attachment and stem cell attachment to materials have been successfully modeled using only mathematical descriptors derived from monomer/polymer structure.
- Rapid and effective identification of optimal materials for a given tissue engineering application will be achieved when high-throughput materials discovery methodologies are used in tandem with computational modeling.

CLASSIC MANUSCRIPT 1

One of the first studies that assessed the biological properties of a library of polymers in order to establish a structure–property relationship was conducted by Brocchini et al. (1998). To achieve this, initially a library of 112 unique polymers was formed by copolymerizing 14 different diphenols (the A component) with 8 different diacids (the B component) in all possible combinations to form alternating (polyAB) polymers. These were not assessed on an array, but rather as coatings on glass coverslips. The large number of materials included within this study enabled the relationship between the polymers' composition and properties, such as the WCA and glass transition temperature (T_g), to be explored. For assessing the biological response to the materials, a subset of 42 copolymers were selected and produced as coatings that were then exposed to fibroblast cells that attached and proliferated on the materials. The proliferation of cells was compared with the WCA of materials, and although proliferation was highest at the lowest WCA, this study highlighted the complex nature of cell interactions with materials—by changing the backbone of the polymer to include oxygen cell proliferation was increased despite the WCA being the same. This suggests that cells respond to specific chemistries in addition to general surface properties such as surface energy.

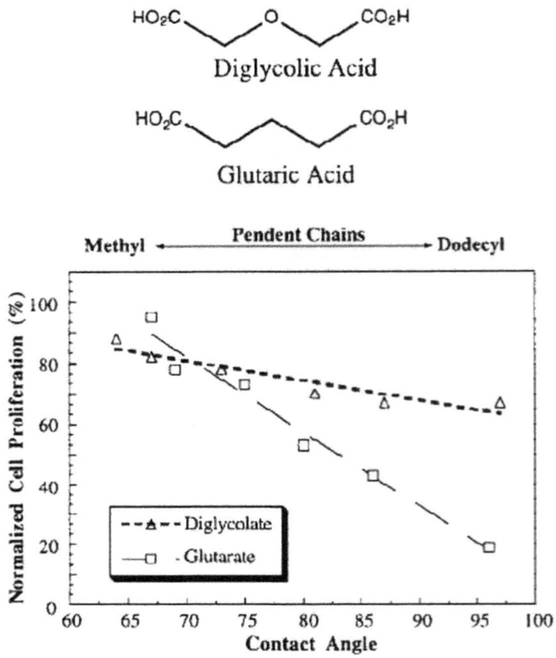

CLASSIC MANUSCRIPT 2

Kohn heralded the beginning of biomateriomics with his classic commentary paper of 2004 and identified the key obstacles for materiomics materials discovery to be realized. Initially, Kohn discussed the growing field of tissue engineering and particularly the need to have new materials that could be used as tissue scaffolds. He identified a lack of diversity in materials used to create biodevices and the need to develop new biomaterials that have greater control over biological systems. Rather than relying on the good judgment of scientists, Kohn suggested that a far more effective approach would be to apply computational methods to screen virtual materials and direct the material compositions that are actually experimentally screened.

Kohn's commentary highlighted the challenges associated with the development of materiomics (although he did not use this term). He stated "the capability of synthesizing large numbers of polymers is of little utility unless it is integrated into an overall experimental work flow that includes effective screening and characterization steps and advanced methods of computational modelling and data mining." Furthermore, the challenge of modeling material systems was discussed, focusing on the many parameters that determine the functionality of a polymeric material, including its chemical composition and the way it was prepared. The challenge of how to characterize polymers within a library in a high-throughput manner remained.

STATE-OF-THE-ART MANUSCRIPT

Infection by pathogenic bacteria on implanted and indwelling medical devices is a major cause of device failure and hospital-acquired infection with significant costs to hospitals and increased patient morbidity and mortality (Costerton et al., 1999). Attempts to minimize this important medical issue have included the development of materials with inherently low bacterial attachment, but have been limited by the poor understanding of how bacteria interact with material surfaces. However, application of a high-throughput materials discovery strategy enabled a new class of materials that are resistant to bacterial attachment to be discovered (Hook et al., 2012). Polymer microarrays exposed to three clinical important pathogens, *Pseudomonas aeruginosa* (PA), *Staphylococcus aureus* (SA), and uropathogenic *Escherichia coli* (UPEC) enabled the interaction of bacteria with thousands of materials to be assessed. In a state-of-the-art manuscript, Epa et al. demonstrated the union of computational modeling with high-throughput experimental methods. The bacterial attachment data for all three strains on the polymer library was used to derive robust and predictive machine learning models of bacterial attachment (Epa et al., 2014). Mathematical descriptors encoding polymer properties were used to generate the models. These included the numbers and types of chemical functional groups in the polymers, descriptors relating to the ability of the polymers to form hydrogen binds (relevant for pathogen biofilm formation), the dipole moments, surface wettability, molecular shape, and complexity. The models could predict pathogen attachment for the polymer library quantitatively with accuracy (see figure). As the materials had been modeled using mathematical descriptors, the models produced could be easily applied to predict pathogen attachment for a second-generation library that was different from the library initially used to generate the model. Furthermore, by retrospectively assessing the model polymer surface, chemistries that enhance or diminish pathogen attachment can be identified.

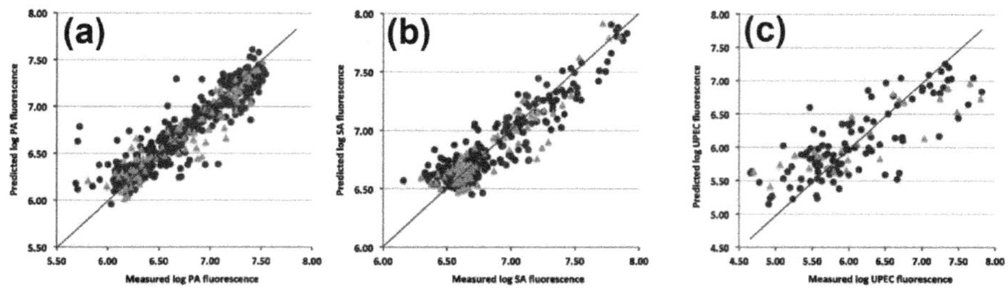

Predicted and measured attachment of (a) PA, (b) SA, and (c) UPEC attachment (fluorescent intensity) to a polymer library for the training set (black circles) and test set (gray triangles). The successful prediction of the bacterial attachment to materials outside those used to construct the model validates the robustness of the model and suggests the model may be useful to screen virtual libraries of polymers. *From Epa et al. (2014) with permission.*

REFERENCES

Adams, N., 2010. Polymer informatics. In: Meier, M.A.R., Webster, D.C. (Eds.), Polymer Libraries, Springer.

Adler, A.F., Speidel, A.T., Christoforou, N., Kolind, K., Foss, M., Leong, K.W., 2011. High-throughput screening of microscale pitted substrate topographies for enhanced nonviral transfection efficiency in primary human fibroblasts. Biomaterials 32, 3611–3619.

Anderson, D.G., Levenberg, S., Langer, R., 2004. Nanoliter-scale synthesis of arrayed biomaterials and application to human embryonic stem cells. Nat. Biotechnol. 22, 863–866.

Anderson, D.G., Putnam, D., Lavik, E.B., Mahmood, T.A., Langer, R., 2005. Biomaterial microarrays: rapid, microscale screening of polymer-cell interaction. Biomaterials 26, 4892–4897.

Anglin, E., Davey, R., Herrid, M., Hope, S., Kurkuri, M., Pasic, P., Hor, M., Fenech, M., Thissen, H., Voelcker, N.H., 2010a. Cell microarrays for the screening of factors that allow the enrichment of bovine testicular cells. Cytometry Part A 77A, 881–889.

Anglin, E.J., Salisbury, C., Bailey, S., Hor, M., Macardle, P., Fenech, M., Thissen, H., Voelcker, N.H., 2010b. Sorted cell microarrays as platforms for high-content informational bioassays. Lab Chip 10, 3413–3421.

Ashino, T., 2010. Materials ontology: an infrastructure for exchanging materials information and knowledge. Data Sci. J. 9, 54–61.

Brocchini, S., James, K., Tangpasuthadol, V., Kohn, J., 1998. Structure-property correlations in a combinatorial library of degradable biomaterials. J. Biomed. Mater. Res. 42, 66–75.

Celiz, A.D., Hook, A.L., Scurr, D.J., Anderson, D.G., Langer, R., Davies, M.C., Alexander, M.R., 2013. ToF-SIMS imaging of a polymer microarray prepared using ink-jet printing of acrylate monomers. Surf. Interface Anal. 45, 202–205.

Cochran, D.L., Buser, D., Ten Bruggenkate, C.M., Weingart, D., Taylor, T.M., Bernard, J.P., Peters, F., Simpson, J.P., 2002. The use of reduced healing times on ITI (R) implants with a sandblasted and acid-etched (SLA) surface: early results from clinical trials on ITI (R) SLA implants. Clin. Oral Implants Res. 13, 144–153.

Costerton, J.W., Stewart, P.S., Greenberg, E.P., 1999. Bacterial biofilms: a common cause of persistent infections. Science 284, 1318–1322.

Cranford, S.W., De Boer, J., Van Blitterswijk, C., Buehler, M.J., 2013. Materiomics: an -omics approach to biomaterials research. Adv. Mater. 25, 802–824.

Engler, A.J., Sen, S., Sweeney, H.L., Discher, D.E., 2006. Matrix elasticity directs stem cell lineage specification. Cell 126, 677–689.

Epa, V.C., Yang, J., Mei, Y., Hook, A.L., Langer, R., Anderson, D.G., Davies, M.C., Alexander, M.R., Winkler, D.A., 2012. Modelling human embryoid body cell adhesion to a combinatorial library of polymer surfaces. J. Mater. Chem. 22, 20902–20906.

Epa, V.C., Hook, A.L., Chang, C., Yang, J., Langer, R., Anderson, D.G., Williams, P., Davies, M.C., Alexander, M.R., Winkler, D.A., 2014. Modelling and prediction of bacterial attachment of polymers. Adv. Funct. Mater. 24, 2085–2093.

Fernandes, T.G., Kwon, S.J., Bale, S.S., Lee, M.Y., Diogo, M.M., Clark, D.S., Cabral, J.M.S., Dordick, J.S., 2010. Three-dimensional cell culture microarray for high-throughput studies of stem cell fate. Biotechnol. Bioeng. 106, 106–118.

Flaim, C.J., Chien, S., Bhatia, S.N., 2005. An extracellular matrix microarray for probing cellular differentiation. Nat. Methods 2, 119–125.

Gobaa, S., Hoehnel, S., Roccio, M., Negro, A., Kobel, S., Lutolf, M.P., 2011. Artificial niche microarrays for probing single stem cell fate in high throughput. Nat. Methods 8, 949–955.

Groen, N., Cranford, S.W., De Boer, J., Buehler, M.J., Van Blitterswijk, C.A., 2013. Introducing materiomics. In: De Boer, J., Van Blitterswijk, C.A. (Eds.), Materiomics: High-Throughput Screening of Biomaterial Properties. Cambridge University Press, Cambridge.

Gullberg, J., Jonsson, P., Nordstrom, A., Sjostrom, M., Moritz, T., 2004. Design of experiments: an efficient strategy to identify factors influencing extraction and derivatization of *Arabidopsis thaliana* samples in metabolomic studies with gas chromatography/mass spectrometry. Anal. Biochem. 331, 283–295.

Hansen, A., Mcmillan, L., Morrison, A., Petrik, J., Bradley, M., 2011. Polymers for the rapid and effective activation and aggregation of platelets. Biomaterials 32, 7034–7041.

Hansen, A., Mjoseng, H.K., Zhang, R., Kalloudis, M., Koutsos, V., De Sousa, P.A., Bradley, M., 2013. High-density polymer microarrays: identifying synthetic polymers that control human embryonic stem cell growth. Adv. Healthcare Mater. 3, 848–853. http://dx.doi.org/10.1002/adhm.201300489.

Hay, D.C., Pernagallo, S., Diaz-Mochon, J.J., Medine, C.N., Greenhough, S., Hannoun, Z., Schrader, J., Black, J.R., Fletcher, J., Dalgetty, D., Thompson, A.I., Newsome, P.N., Forbes, S.J., Ross, J.A., Bradley, M., Iredale, J.P., 2011. Unbiased screening of polymer libraries to define novel substrates for functional hepatocytes with inducible drug metabolism. Stem Cell Res. 6, 92–102.

Hill, D.P., Adams, N., Bada, M., Batchelor, C., Berardini, T.Z., Dietze, H., Drabkin, H.J., Ennis, M., Foulger, R.E., Harris, M.A., Hastings, J., Kale, N.S., De Matos, P., Mungall, C.J., Owen, G., Roncaglia, P., Steinbeck, C., Turner, S., Lomax, J., 2013. Dovetailing biology and chemistry: integrating the gene ontology with the ChEBI chemical ontology. BMC Genomics 14.

Hook, A.L., Creasey, R., Thissen, H., Hayes, J.P., Voelcker, N.H., 2009a. Laser-based patterning for transfected cell microarrays. Biofabrication 1, 045003.

Hook, A.L., Thissen, H., Voelcker, N.H., 2009b. Advanced substrate fabrication for cell microarrays. Biomacromolecules 10, 573–579.

Hook, A.L., Thissen, H., Voelcker, N.H., 2009c. Surface plasmon resonance imaging of polymer microarrays to study protein-polymer interactions in high throughput. Langmuir 25, 9173–9181.

Hook, A.L., Anderson, D.G., Langer, R., Williams, P., Davies, M.C., Alexander, M.R., 2010. High throughput methods applied in biomaterial development and discovery. Biomaterials 31, 187–198.

Hook, A.L., Yang, J., Chen, X., Roberts, C.J., Mei, Y., Anderson, D.G., Langer, R., Alexander, M.R., Davies, M.C., 2011. Polymers with hydro-responsive topography identified using high throughput AFM of an acrylate microarray. Soft Matter 7, 7194–7197.

Hook, A.L., Chang, C., Yang, J., Luckett, J., Cockrayne, A., Atkinson, S., Mei, Y., Bayston, R., Irvine, D.J., Langer, R., Anderson, D.G., Williams, P., Davies, M.C., Alexander, M.R., 2012. Combinatorial discovery of polymers resistant to bacterial attachment. Nat. Biotechnol. 30, 868–875.

Hook, A.L., Chang, C., Yang, J., Atkinson, S., Langer, R., Anderson, D.G., Davies, M.C., Williams, P., Alexander, M.R., 2013a. Discovery of novel materials with broad resistance to bacterial attachment using combinatorial polymer microarrays. Adv. Mater. 25, 2542–2547.

Hook, A.L., Scurr, D.J., Anderson, D.G., Langer, R., Williams, P., Davies, M., Alexander, M.R., 2013b. High throughput discovery of thermo-responsive materials using water contact angle measurements and time-of-flight secondary ion mass spectrometry. Surf. Interface Anal. 45, 181–184.

Hook, A.L., Scurr, D.J., Burley, J.C., Langer, R., Anderson, D.G., Davies, M.C., Alexander, M.R., 2013c. Analysis and prediction of defects in UV photo-initiated polymer microarrays. J. Mater. Chem. B 1, 1035–1043.

Hook, A.L., 2014. Polymer microarrays for high throughput biomaterials discovery. In: Palmer, E. (Ed.), Cell Based Microarrays, Springer.

Hook, A.L., Chang, C., Scurr, D.J., Langer, R., Anderson, D.G., Williams, P., Davies, M.C., Alexander, M.R., 2014. Thermally switchable polymers achieve controlled *Escherichia coli* detachment. Adv. Healthcare Mater. http://dx.doi.org/10.1002/adhm.201300518.

Johansson, F., Carlberg, P., Danielsen, N., Montelius, L., Kanje, M., 2006. Axonal outgrowth on nano-imprinted patterns. Biomaterials 27, 1251–1258.

Katritzky, A.R., Lobanov, V.S., Karelson, M., 1995. QSPR – the correlation and quantitative prediction of chemical and physical-properties from structure. Chem. Soc. Rev. 24, 279–287.

Kell, D.B., Oliver, S.G., 2003. Here is the evidence, now what is the hypothesis? The complementary roles of inductive and hypothesis-driven science in the post-genomic era. BioEssays 26, 99–105.

Khan, F., Tare, R.S., Kanczler, J.M., Oreffo, R.O.C., Bradley, M., 2010. Strategies for cell manipulation and skeletal tissue engineering using high-throughput polymer blend formulation and microarray techniques. Biomaterials 31, 2216–2228.

Kim, S.H., Cherney, E.A., Hackam, R., 1992. Hydrophobic behavior of insulators coated with RTV silicone-rubber. IEEE Trans. Electr. Insul. 27, 610–622.

King, W.J., Jongpaiboonkit, L., Murphy, W.L., 2010. Influence of FGF2 and PEG hydrogel matrix properties on hMSC viability and spreading. J. Biomed. Mater. Res. Part A 93A, 1110–1123.

Kohn, J., 2004. New approaches to biomaterials design. Nat. Mater. 3, 745–747.

Kurkuri, M.D., Driever, C., Johnson, G., Mcfarland, G., Thissen, H., Voelcker, N.H., 2009. Multifunctional polymer coatings for cell microarray applications. Biomacromolecules 10, 1163–1172.

Le, T., Epa, V.C., Burden, F.R., Winkler, D.A., 2012. Quantitative structure-property relationship modeling of diverse materials properties. Chem. Rev. 112, 2889–2919.

Lowe, G., 1995. Combinatorial chemistry. Chem. Soc. Rev. 24, 309–317.

Lutolf, M.P., Hubbell, J.A., 2005. Synthetic biomaterials as instructive extracellular microenvironments for morphogenesis in tissue engineering. Nat. Biotechnol. 23, 47–55.

Mant, A., Tourniaire, G., Diaz-Mochon, J.J., Elliott, T.J., Williams, A.P., Bradley, M., 2006. Polymer microarrays: identification of substrates for phagocytosis assays. Biomaterials 27, 5299–5306.

Martin, J.Y., Schwartz, Z., Hummert, T.W., Schraub, D.M., Simpson, J., Lankford, J., Dean, D.D., Cochran, D.L., Boyan, B.D., 1995. Effect of titanium surface-roughness on proliferation, differentiation, and protein-synthesis of human osteoblast-like cells (MG63). J. Biomed. Mater. Res. 29, 389–401.

Mei, Y., Gerecht, S., Taylor, M., Urquhart, A., Bogatyrev, S.R., Cho, S.W., Davies, M.C., Alexander, M.R., Langer, R.S., Anderson, D.G., 2009. Mapping the interactions among biomaterials, adsorbed proteins, and human embryonic stem cells. Adv. Mater. 21, 2781–2786.

Mei, Y., Saha, K., Bogatyrev, S.R., Yang, J., Hook, A.L., Kalcioglu, Z.I., Cho, S.W., Mitalipova, M., Pyzocha, N., Rojas, F., Van Vliet, K.J., Davies, M.C., Alexander, M.R., Langer, R., Jaenisch, R., Anderson, D.G., 2010. Combinatorial development of biomaterials for clonal growth of human pluripotent stem cells. Nat. Mater. 9, 768–778.

Pernagallo, S., Wu, M., Gallagher, M.P., Bradley, M., 2011. Colonising new frontiers-microarrays reveal biofilm modulating polymers. J. Mater. Chem. 21, 96–101.

Pernagallo, S., Tura, O., Wu, M., Samuel, K., Diaz-Mochon, J.J., Hansen, A., Zhang, R., Jackson, M., Padfield, G.J., Hadoke, P.W.F., Mills, N.L., Turner, M.L., Iredale, J.P., Hay, D.C., Bradley, M., 2012. Novel biopolymers to enhance endothelialisation of intra-vascular devices. Adv. Healthcare Mater. 1, 646–656.

Pickering, H., Wu, M., Bradley, M., Bridle, H., 2012. Analysis of *Giardia lamblia* interactions with polymer surfaces using a microarray approach. Environ. Sci. Technol. 46, 2179–2186.

Price, R.L., Ellison, K., Haberstroh, K.M., Webster, T.J., 2004. Nanometer surface roughness increases select osteoblast adhesion on carbon nanofiber compacts. J. Biomed. Mater. Res. Part A 70A, 129–138.

Schneider, G.B., Zaharias, R., Seabold, D., Keller, J., Stanford, C., 2004. Differentiation of preosteoblasts is affected by implant surface microtopographies. J. Biomed. Mater. Res. Part A 69A, 462–468.

Tare, R.S., Khan, F., Tourniaire, G., Morgan, S.M., Bradley, M., Oreffo, R.O.C., 2009. A microarray approach to the identification of polyurethanes for the isolation of human skeletal progenitor cells and augmentation of skeletal cell growth. Biomaterials 30, 1045–1055.

Taylor, K.R., Essex, J.W., Frey, J.G., Mills, H.R., Hughes, G., Zaluska, E.J., 2006. The semantic grid and chemistry: experiences with CombeChem. J. Web Semantics 4, 84–101.

Taylor, M., Urquhart, A.J., Zelzer, M., Davies, M.C., Alexander, M.R., 2007. Picoliter water contact angle measurement on polymers. Langmuir 23, 6875–6878.

Taylor, M., Urquhart, A.J., Anderson, D.G., Langer, R., Davies, M.C., Alexander, M.R., 2009. Partial least squares regression as a powerful tool for investigating large combinatorial polymer libraries. Surf. Interface Anal. 41, 127–135.

Tessierlavigne, M., Goodman, C.S., 1996. The molecular biology of axon guidance. Science 274, 1123–1133.

Tourniaire, G., Collins, J., Campbell, S., Mizomoto, H., Ogawa, S., Thaburet, J.F., Bradley, M., 2006. Polymer microarrays for cellular adhesion. Chem. Commun., 2118–2120.

Tweedie, C.A., Anderson, D.G., Langer, R., Van Vliet, K.J., 2005. Combinatorial material mechanics: high-throughput polymer synthesis and nanomechanical screening. Adv. Mater. 17, 2599–2604.

Unadkat, H.V., Hulsman, M., Cornelissen, K., Papenburg, B.J., Truckenmuller, R.K., Carpenter, A.E., Wessling, M., Post, G.F., Uetz, M., Reinders, M.J.T., Stamatialis, D., Van Blitterswijk, C.A., De Boer, J., 2011. An algorithm-based topographical biomaterials library to instruct cell fate. Proc. Natl. Acad. Sci. U.S.A. 108, 16565–16570.

Urquhart, A.J., Anderson, D.G., Taylor, M., Alexander, M.R., Langer, R., Davies, M.C., 2007. High throughput surface characterisation of a combinatorial material library. Adv. Mater. 19, 2486–2491.

Urquhart, A.J., Taylor, M., Anderson, D.G., Langer, R., Davies, M.C., Alexander, M.R., 2008. TOF-SIMS analysis of a 576 micropatterned copolymer array to reveal surface moieties that control wettability. Anal. Chem. 80, 135–142.

Wilson, C.J., Clegg, R.E., Leavesley, D.I., Pearcy, M.J., 2005. Mediation of biomaterial-cell interactions by adsorbed proteins: a review. Tissue Eng. 11, 1–18.

Winkler, D.A., Burden, F.R., 2012. Robust, quantitative tools for modelling ex-vivo expansion of haematopoietic stem cells and progenitors. Mol. Biosyst. 8, 913–920.

Wu, M., Bridle, H., Bradley, M., 2012. Targeting *Cryptosporidium parvum* capture. Water Res. 46, 1715–1722.

Yang, J., Mei, Y., Hook, A.L., Taylor, M., Urquhart, A.J., Bogatyrev, S.R., Langer, R., Anderson, D.G., Davies, M.C., Alexander, M.R., 2010. Polymer surface functionalities that control human embryoid body cell adhesion revealed by high throughput surface characterization of combinatorial material microarrays. Biomaterials 31, 8827–8838.

Zhang, R., Liberski, A., Sanchez-Martin, R., Bradley, M., 2009. Microarrays of over 2000 hydrogels – identification of substrates for cellular trapping and thermally triggered release. Biomaterials 30, 6193–6201.

Zhang, R., Mjoseng, H.K., Hoeve, M.A., Bauer, N.G., Pells, S., Besseling, R., Velugotla, S., Tourniaire, G., Kishen, R.E.B., Tsenkina, Y., Armit, C., Duffy, C.R.E., Helfen, M., Edenhofer, F., De Sousa, P.A., Bradley, M., 2013. A thermoresponsive and chemically defined hydrogel for long-term culture of human embryonic stem cells. Nat. Commun. 4. http://dx.doi.org/10.1038/ncomms2341.

CHAPTER 9

Microfabrication Technology in Tissue Engineering

Chaenyung Cha[1], Francesco Piraino[2], Ali Khademhosseini[3]

[1]School of Materials Science and Engineering, Ulsan National Institute of Science and Technology, Ulsan, Korea; [2]Institute of Bioengineering, École Polytechnique Fédérale de Lausanne, Lausanne, Switzerland; [3]Harvard-MIT Division of Health Sciences and Technology, Brigham and Women's Hospital, Harvard Medical School, Cambridge, MA, USA

LEARNING OBJECTIVES

- To learn the history of microfabrication and its impact on science, technology, and society
- To learn the advantages of microfabrication technology over macroscale approaches
- To identify major microfabrication techniques used in tissue engineering (TE)
- To understand the concept and working principle of each microfabrication process
- To understand how each microfabrication technique is utilized in various TE applications
- To learn the strategies used to assemble microstructures into larger, hierarchical constructs

9.1 INTRODUCTION

9.1.1 Background

In the last half century, the advent and development of microfabrication technologies have contributed tremendously to advancing many facets of science and engineering. Microfabrication, by definition, is a process by which a structure with micrometer dimensions and resolution is fabricated (Madou, 2002). Often, an overall microfabrication process consists of several pertinent techniques in order to create the final desired structure. Developing structures in micrometer scale clearly offers several advantages over macroscale structures. First, miniaturization can substantially decrease the cost per experiment by reducing the amount of materials needed for development. When fully realized, the microfabrication process can be achieved using automated systems,

Tissue Engineering.
Copyright © 2015 Elsevier Inc. All rights reserved.

which results in significant increase in productivity. Miniaturization also makes it possible to integrate numerous structures or devices in a confined space, therefore enabling efficient high-throughput experiments.

In a historical perspective, microfabrication was conceptualized and came into fruition in the realm of electronics; the fabrication of integrated circuits involves a series of micrometer-scale techniques such as thin film deposition, lithography, etching, and doping on silicon substrates in semiconductor manufacturing (Madou, 2002). Needless to say, electronics has had and will continue to have a vital and revolutionary impact on our society, completely transforming our everyday lives. More recently, the same techniques are being used to fabricate devices for Micro-Electro-Mechanical Systems (MEMS) for such applications in sensors and actuators that have already been commercialized and widely used (Madou, 2002; Grayson et al., 2004; Paul, 2006). For example, pressure transducers are used as tire pressure monitors in automobiles. Inkjet printers also utilize a piezoelectric device which undergoes vibrating motion upon applied electricity, resulting in a continuous formation of ink droplets.

Microfluidics has emerged as the latest innovative microfabrication technology that allows high-throughput and precision analytics (Whitesides, 2006). Microfluidics is a technology that controls the flow of fluids in amounts ranging from tens to hundreds of microliters. Therefore, manipulating various fluid flows within a microfluidic device allows for highly sensitive and accurate separation and detection of analytes, using only small amounts of sample reagents. Microfluidic concept is also combined with MEMS-based analytical devices, termed Micro Total Analysis System (μTAS), with the goal of fully automating the analytical process (Reyes et al., 2002).

Microfabricated materials and devices have been traditionally made using hard substrates such as metal, silicon, and glass. However, polydimethylsiloxane (PDMS)-based elastomers have become a popular material of choice in recent years for several reasons (McDonald and Whitesides, 2002). First, PDMS elastomers are mechanically stable and yet not brittle, as they possess elasticity. Second, they are easily molded and fabricated by simply mixing a liquid polymer base with a curing agent. Third, they are optically transparent, which makes it possible to visualize any experimental process with a microscope. Fourth, they are permeable to gas, chemically inert, and biocompatible for cell culture, therefore highly applicable for biological experiments.

9.1.2 Microfabrication Meets Tissue Engineering

Microfabrication consists of the design, characterization, production, and application of structures, devices, and systems at the micrometer scale. The properties of materials at such scales are significantly different from those at a macroscopic scale, and for this purpose, microfabrication has been receiving much attention in recent years. Developing materials used in biology and medicine can benefit greatly from the evolution of microfabrication, as it can optimize the interactions occurring at the interface between materials and biological systems with micrometer-scale precision. In addition, by providing the possibility to tailor detailed topography as well as overall structure, these novel technologies allow for the development of new materials possessing unique properties.

In the field of tissue engineering (TE), many strategies often involve the use of various biomaterials as artificial extracellular matrices (ECMs) to engineer tissues or organs, in order to provide suitable microenvironments for cells or tissues to function properly (Lutolf et al., 2005). There is a substantial need to engineer these artificial tissue constructs in a micrometer scale, in order to manipulate the architecture to control cellular behavior and pattern the inner structure (Bhatia et al., 1999). In addition, micrometer-sized biomaterials could be useful as injectable delivery vehicles of therapeutic molecules in clinical applications. Naturally, several microfabrication techniques that are currently available in electronics, MEMS, and microfluidics are increasingly adopted to engineer materials used in TE to control their various physical properties: size, overall shape, spacing, architectural details, and porosity (Khademhosseini et al., 2006; Zorlutana et al., 2012).

In this chapter, we will examine the concept, working principle, and advantages of microfabrication technologies that are actively employed in various TE applications, including photolithography, rapid prototyping (RP), and soft lithography. In addition, strategies to assemble micrometer structures into hierarchical constructs (i.e., self-assembly and templated assembly) will also be discussed.

9.2 MICROFABRICATION TECHNIQUES IN TISSUE ENGINEERING

9.2.1 Photolithography

Photolithography is one of the most extensively utilized technologies due to its usage in fabricating microchips in the semiconductor industry. In essence, this technique involves the fabrication of a thin film on top of a substrate (usually a silicon wafer), in which UV light is passed through a photomask that has a desired micropattern. The UV light induces cross-linking reaction of a liquid-based, photosensitive monomer, resulting in a micropatterned, solidified material.

Photolithography has also become a common tool to develop micrometer-scale tissue-engineered constructs not only for its popularity as an emblematic technique of microfabrication, but also for its compatibility with photo-cross-linkable polymers that are extensively used in TE (Figure 9.1(a)). A wide array of both natural and synthetic polymers can be rendered photo-cross-linkable by employing chemical modification schemes to conjugate vinyl groups, which undergo radical polymerization upon photoreactive initiators (Ifkovits et al., 2007).

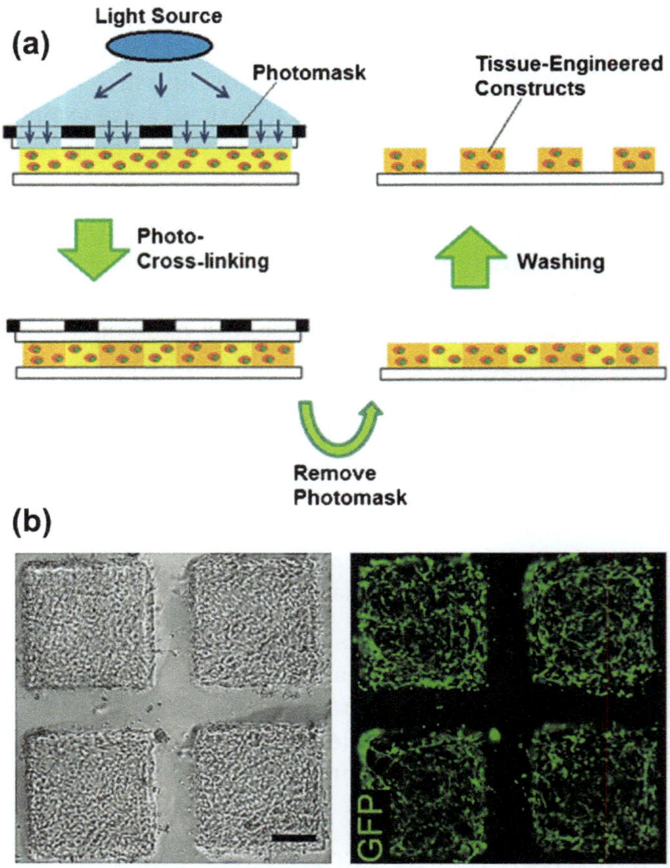

FIGURE 9.1

(a) Schematic illustration of photolithographic technology to fabricate tissue constructs. (b) Gelatin methacrylate microgels laden with endothelial cells were fabricated via photolithography (scale bar: 200 μm). *Reprinted with permission from Nichol et al. (2010). Copyright 2010 Elsevier.*

For example, poly(ethylene glycol) (PEG), a widely used polymer in various biomedical applications, is often functionalized with (meth)acrylate groups (Ifkovits et al., 2007). The resulting photo-cross-linkable PEG hydrogel is one of the most widely used biomaterials in TE applications. Pishko and coworkers demonstrated that PEG hydrogels with varying micrometer scales and spacing could be successfully fabricated on a silicon substrate (Revzin et al., 2001). Natural polymers modified with photo-cross-linkable groups are also widely used for this purpose. For example, Khademhosseini and coworkers used a photolithographic technique to fabricate cell-laden gelatin methacrylate hydrogels with varying length scales (Figure 9.1(b)) (Nichol et al., 2010). Another benefit of using photolithography is the ability to create three-dimensional (3D) materials efficiently using layer-by-layer method; multilayered 3D materials can be easily developed by simply repeating the fabrication process to create a layer on top of the previous layer (Liu Tsang and Bhatia, 2004).

To engineer micropatterned hydrogels by photolithography a number of requirements must be met. First, the light source must be able to adequately initiate the chemical reaction of the precursor solution so that the gelation occurs efficiently, while not damaging the encapsulating species (e.g., cells and proteins). The most commonly used light source is UV, as UV-activated initiators are often used to start the polymerization process of vinyl-based molecules. More recently, there are efforts to use light source in the visible range that is well known to activate certain initiators (e.g., camphorquinone, 1-phenyl-1,2-propadione, Irgacure®819, VA-086) instead of UV, due to the potential harmful effect of UV on encapsulated cells (Arikawa et al., 2009). Second, the fidelity of the micropattern printed on a photomask used to transmit light to the gel precursor solution must be robust enough such that the intended resolution of the pattern should translate into that of the fabricated material. Modern high-resolution printers can now be used to generate darkened patterns on the photomask with sub-ten-micrometer resolution. Therefore, this allows the fabrication via photolithography to achieve intended architecture and resolution with high fidelity.

More recently, photolithography based on Digital Micromirror Device™ (DMD) called DMD Projection Printing (DMD-PP) has become the latest RP technology that is adopted in TE (Figure 9.2(a)) (Lu et al., 2006; Grogan et al., 2013; Gauvin et al., 2012). In DMD-PP, the DMD chip replaces the role of a photomask of the traditional photolithography. In each DMD chip, there are hundreds of thousands of micromirrors, each of which can be tilted to either −10° or +10° with respect to the surface using a CAD-type modeling. This will act as "on" or "off" switch, in which only the light reflected off one angle goes to the sample stage and induces photo-cross-linking reaction (on), and the light reflected off the other angle is collected by a light absorber (off). In other words, the DMD chip acts as a "digital photomask" to control the micropatternable area. In addition, the DMD chip can also be moved laterally to expand the fabrication area. Furthermore, the same layer-by-layer approach can be applied to create larger 3D constructs. Recent works by Chen and coworkers highlight the efficacy of utilizing DMD-PP technology in fabricating tissue constructs, in which cell-encapsulated photo-cross-linkable hydrogels, such as PEG and gelatin methacrylate, in micrometer-scale resolutions are successfully fabricated (Lu et al., 2006; Grogan et al., 2013; Gauvin et al., 2012).

9.2.2 Rapid Prototyping

Aided by the remarkable development in computer technology and robotics, many manufacturing processes have adopted automated systems. Obvious

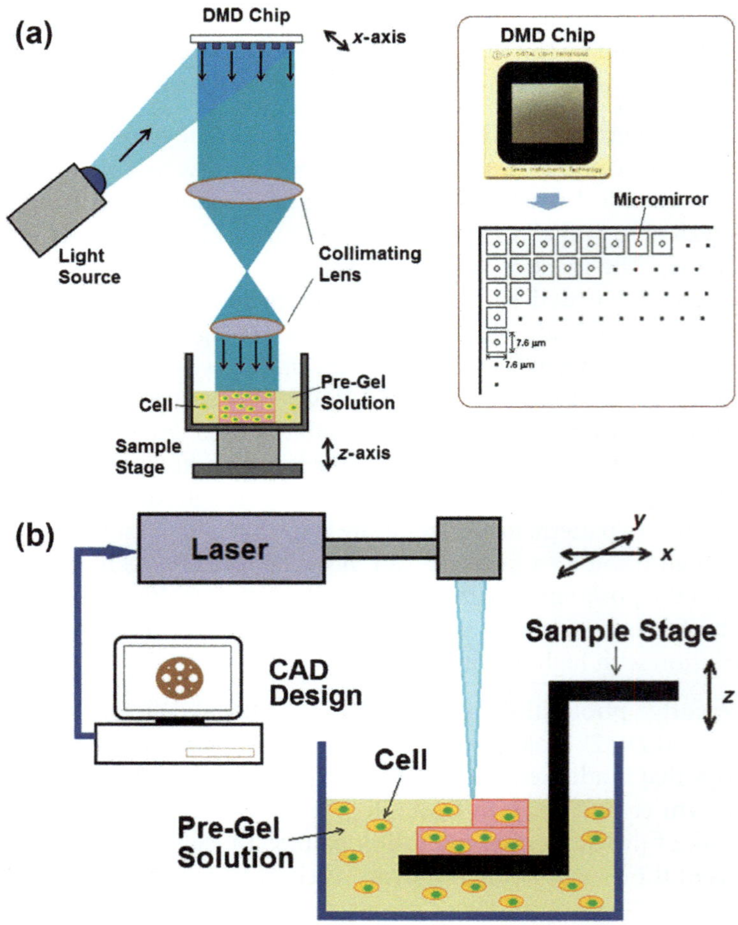

FIGURE 9.2

Schematic illustrations of (a) DMD projection printing and (b) stereolithographic technologies to fabricate tissue constructs.

advantages of automation include increased productivity and quality control, and reduced cost. Much of our everyday lives are run by various machines that are, at least in part, automated; they work simply with "the touch of a button."

Nowadays, many microfabrication processes also largely rely on automated systems. For example, most of the microfabrication processes in semiconductor manufacturing, including lithography, etching, cleaning, and doping, are now fully automated. RP has especially benefited various materials research efforts, because the automated material fabrication process can bypass exhaustive and time-consuming experiments by using computer-based design, and efficient and accurate fabrication using automated systems (Chua et al., 2010; Hutmacher et al., 2004).

RP technology to engineer a 3D material typically involves: (1) computer-aided design (CAD) for modeling, (2) application of base material to a fabrication apparatus, (3) automated material fabrication step based on the CAD model, and (4) layer-by-layer fabrication or component assembly to build the final 3D structure. With the increasing popularity of RP technology within the last decade, several RP-based microfabrication techniques are utilized in the field of TE to develop 3D tissue-engineered constructs to control their spatial resolution and overall architecture in micrometer scale (Hutmacher et al., 2004; Billiet et al., 2013).

9.2.2.1 Stereolithography

RP in the realm of material fabrication began with the invention of stereolithography by Charles W. Hull in 1986 (Jacobs, 1992). In essence, stereolithography is an automated version of the photolithography technique using laser as a photocurable light source (Figure 9.2(b)). Unlike the traditional photolithography which uses a photomask to create a micropattern, stereolithography utilizes laser to selectively fabricate the desired area. The pattern design is done with the CAD-based modeling, and the movement of the laser is controlled automatically based on the CAD model. For developing 3D structures, a layer-by-layer approach is used in which the sample stage is lowered and the next layer is fabricated on top of the previously developed layer. Like conventional photolithography, tissue-engineered constructs based on photo-cross-linkable materials can be efficiently fabricated in micrometer resolutions. However, stereolithography often does not involve photomasks, as the fabrication is achieved by the automatically controlled laser. This aspect becomes especially advantageous when developing 3D structures with multiple layers; if each layer requires a distinctive design, then a different photomask is required for each layer in traditional photolithography. In stereolithography, the entire 3D structure can be easily designed and constructed using the CAD modeling. For these advantages, the stereolithography is increasingly used to create tissue-engineered constructs with more complex micrometer dimensions and resolutions. For example, Bashir and coworkers successfully demonstrated the use of stereolithography in engineering tissue constructs by encapsulating various types of cells into photo-cross-linkable hydrogels (Chan et al., 2010; Zorlutana et al., 2011).

9.2.2.2 Extrusion Deposition Printing

In the last few years, RP technology has gained tremendous amount of interest in the general public, with the introduction of commercially available "3D printers." The working concept of 3D printers is based on the deposition of a material via extrusion; a molten, thermoplastic material which solidifies upon cooling is extruded out of a nozzle and deposited in a desired pattern (Yan et al., 1996; Khalil et al., 2005; Dimitrov et al., 2006). Like any other RP technology, the extrusion and the position of material deposition are automatically controlled by CAD-based modeling, and the layer-by-layer approach can be used to develop 3D constructs.

However, the innate drawback of extrusion deposition printing technology is the limitation on the choice of materials. The materials must be in liquid state during the extrusion, and they must be solidified immediately upon deposition in order to maintain the intended resolution and overall architecture. Therefore, thermoplastic materials, which melt at high temperature and solidify upon cooling, are generally used.

In TE, the application of the extrusion deposition printing has been limited due to the aforementioned reason. For example, using gel precursor solutions commonly used to fabricate hydrogels on extrusion deposition printing is not feasible; since an additional cross-linking step (e.g., photo-cross-linking) is required to fabricate the material, it is not possible to solidify the solution immediately. As a result, the extruded solution would disperse before fabrication and the intended resolution would be lost. In addition, it is impossible to encapsulate cells in situ due to the high temperature required to melt the material for extrusion. Therefore, the extrusion deposition printing so far has been limited to fabricating tissue constructs based on thermosetting materials (e.g., collagen, agarose) or hard scaffolds for bone TE applications (Leukers et al., 2005; Hollister, 2005). However, a recent work by Lewis and coworkers has expanded the use of extrusion-based printing in TE by utilizing the technology to create a microchannel network within a biomaterial for vascularization (Figure 9.3) (Wu et al., 2011). In this study, they printed a biocompatible, sacrificial ink in a desirable pattern within a gel precursor solution. Eventual cross-linking reaction followed by the removal of ink resulted in a hydrogel embedded with microchannels.

9.2.3 Soft Lithography

Soft lithography refers to a collection of techniques that enable the fabrication of structures or the transfer of materials using elastomeric "soft" masters (e.g., stamps, molds, and photomasks). Generally, PDMS elastomers are used as masters due to their elasticity and their ease of fabrication. The standard procedure for fabricating a PDMS master is as follows:

1. A standard photolithography technique is used to fabricate a desired pattern on a silicon substrate: Photoresist (e.g., SU-8 2075 by MicroChem Co.) is coated on top of a silicon wafer with a desired thickness, and cured by UV passing through a photomask. The patterned surface acts as a template for PDMS fabrication.

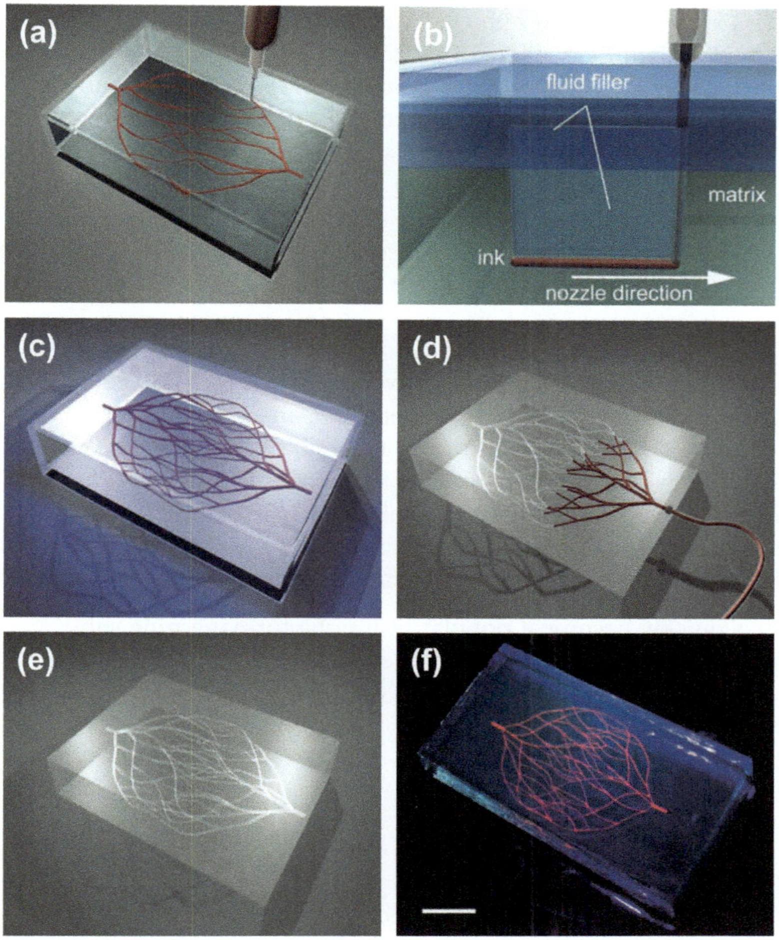

FIGURE 9.3

Extrusion-based printing of 3D microchannel network within a hydrogel. (a) Deposition of a fugitive ink into a pregel solution. (b) Voids left by the moving nozzle are replenished with the pregel solution. (c) Subsequent photopolymerization of the pregel solution results in hydrogel formation. (d, e) The ink is removed via vacuum to expose the microvascular channels. (f) Fluorescent image of the constructed hydrogel with microchannels (scale bar: 10 mm). *Reprinted with permission from Wu et al. (2011). Copyright 2011 Wiley-VCH Verlag GmbH & Co.*

2. PDMS elastomer (e.g., SYLGARD® 184 by Dow Corning Inc.) is cured on top of the silicon template: After detaching from the template, the surface of the PDMS elastomer will have an intended patterned surface which can be used as masters for soft lithography applications.

There are several advantages of utilizing soft lithography as compared to standard lithographic techniques (Brittain et al., 1998). First, it is easier and less expensive to develop PDMS elastomers which can be repeatedly fabricated as long as there is a template (the microfabricated silicon wafer) in place. Also, it is easier to develop PDMS elastomers with well-defined 3D architecture. In addition, due to their elasticity and hydrophobicity, PDMS elastomers are able to make tighter contact with other substrates, and therefore better suited as molds and stamps for accurate patterning. In TE, such soft lithographic techniques as micromolding, microcontact printing, and microfluidics are commonly used since they provide a facile route for controlling the shape and the pattern of micrometer-scale tissue constructs.

9.2.3.1 Microcontact Printing

It is often desirable to pattern bioactive molecules, such as proteins, peptides, and DNA, on biomaterials in micrometer scale in order to control the organization of cells within a tissue construct in a desirable pattern. Microcontact printing (μCP) is widely used for this purpose (Bernard et al., 2000). μCP is essentially a stamp printing technique, in which a microfabricated stamp is coated with a bioactive molecule (Figure 9.4). The coated stamp is then pressed against a surface, resulting in a transfer of the coated material from the stamp to the biomaterial.

μCP has quickly become a popular technique for patterning substrates for its obvious simplicity and cost-effectiveness. However, despite its widespread use,

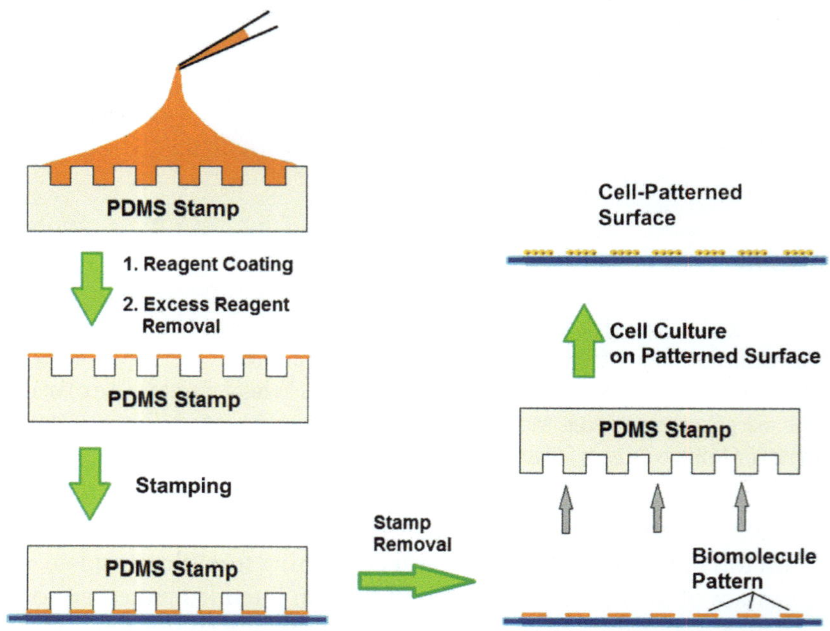

FIGURE 9.4

Schematic illustration of microcontact printing (μCP) technology to create a pattern of biological molecules on a substrate in order to pattern cells.

this μCP method suffers from various limitations, which greatly affect its accuracy and reproducibility. First, stamping conditions can easily vary the patterned area, as the stamping force can deform the stamp itself. In addition, due to the hydrophobicity of PDMS elastomer, the coating of a material depends greatly on its hydrophobicity/hydrophilicity. Furthermore, it is difficult to control the amount of material being transferred by printing.

This technique has been used in TE applications such as the control of cell adhesion by printing various cell-adhesive molecules on biomaterial substrates. For example, Chen and coworkers demonstrated that μCP of ECM proteins and subsequent cellular organization on a variety of substrates (e.g., glass, silicone, polystyrene) could be effectively controlled by the surface wettability (Tan et al., 2004). Similarly, Bizios and coworkers were also able to control the osteoblast organization by patterning cell-adhesive peptides (e.g., Arg-Gly-Asp (RGD) and Lys-Arg-Ser-Arg (KRSR)) using μCP (Hasenbein et al., 2002).

9.2.3.2 Micromolding

Using a mold having desired size and shape to fabricate materials is a highly efficient and reproducible method. The size and shape of a material can be easily adjusted by simply using different types of molds. In addition, this technique is not bound by the type of cross-linking strategy, as the mold only acts as a vessel in which the material is fabricated. Therefore, this technique can be applied to fabricate any material of choice. Creating the right mold is the most crucial aspect of this technology. Molds can be fabricated by various microfabrication techniques such as lithography and laser cutting that enable the fabrication of structures with micrometer resolution (Qin et al., 2010). As long as the mold is not reactive toward biomaterials and does not have any deleterious effect on the cells, micromolded tissue constructs can be developed using any fabrication schemes.

Traditionally, the mold is made by cutting or milling glass or plastic materials. More recently, photolithography is widely used to generate more accurate and higher resolution micromolds. For example, the micropatterned photoresist on a silicon wafer is used as a micromold by itself, or a template to fabricate PDMS elastomer-based micromolds (Figure 9.5) (Qin et al., 2010). Using the template has the advantage of easily reproducing identical molds. For example, micrometer-scale hyaluronic acid and PEG hydrogels encapsulated with cells have been fabricated using PDMS micromolds (Yeh et al., 2006). The size and shape of the hydrogels could be conveniently controlled by using different types of the PDMS molds.

9.2.3.3 Microfluidics

Microfluidics is defined as science and technology that control the flow of fluids using channels with micrometer dimensions (Whitesides, 2006). Microfluidic devices allow for the manipulation of multiple fluids in highly complex channel

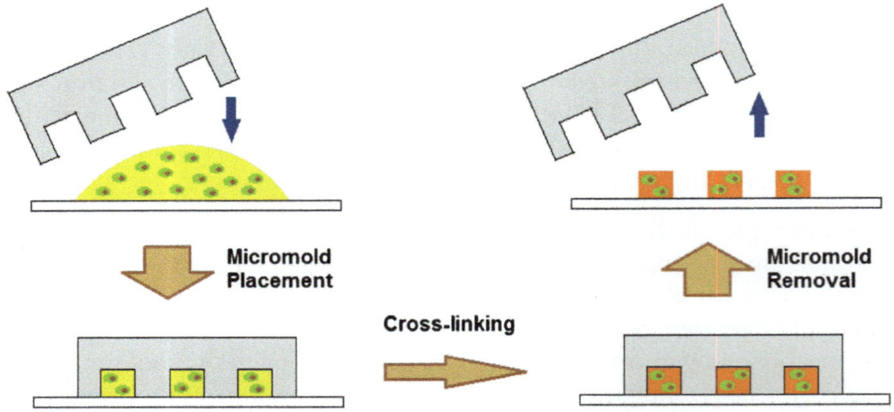

FIGURE 9.5
Schematic illustration of micromolding technology to fabricate tissue constructs.

configurations, as well as having obvious benefits of any microscale devices (e.g., miniaturization, high-throughput, cost-effectiveness). Therefore, microfluidics has not only contributed significantly to advancing the physical science of fluid dynamics, but is also actively being pursued in several applications, including chemical and biological analytics and MEMS. In addition, extensive research efforts are being made to combine microfluidics with other types of microsystems such as microscale sensors and actuators to increase the functionality of these systems.

Like other soft lithographic techniques, PDMS elastomers are generally used to make microfluidic devices. The standard fabrication procedure for a microfluidic device is similar to that of PDMS molds and stamps used in micromolding and μCP. However, there are additional steps required after the fabrication of PDMS elastomers with desired channel geometry (Figure 9.6) (McDonald and Whitesides, 2002; Qin et al., 2010).

1. Fluid inlets and outlets for the microfluidic channels are created in the PDMS elastomer: holes can be easily created using a handheld punch, which is a clear advantage over other hard substrates such as plastic and glass.
2. A glass plate is irreversibly attached to the bottom (the side with printed grooves) of the PDMS elastomer to finally develop a microfluidic device: treating the PDMS elastomer and a glass plate with oxygen plasma generates reactive oxygen species (ROS) on both surfaces. When the two come in contact with each other, ROS on both sides react with each other resulting in the irreversible bonding.

In TE, the ability to control the direction and quantity of fluid flows using micrometer-scale channels is utilized in several applications, including cellular patterning, perfusion bioreactors, and droplet-based biomaterials.

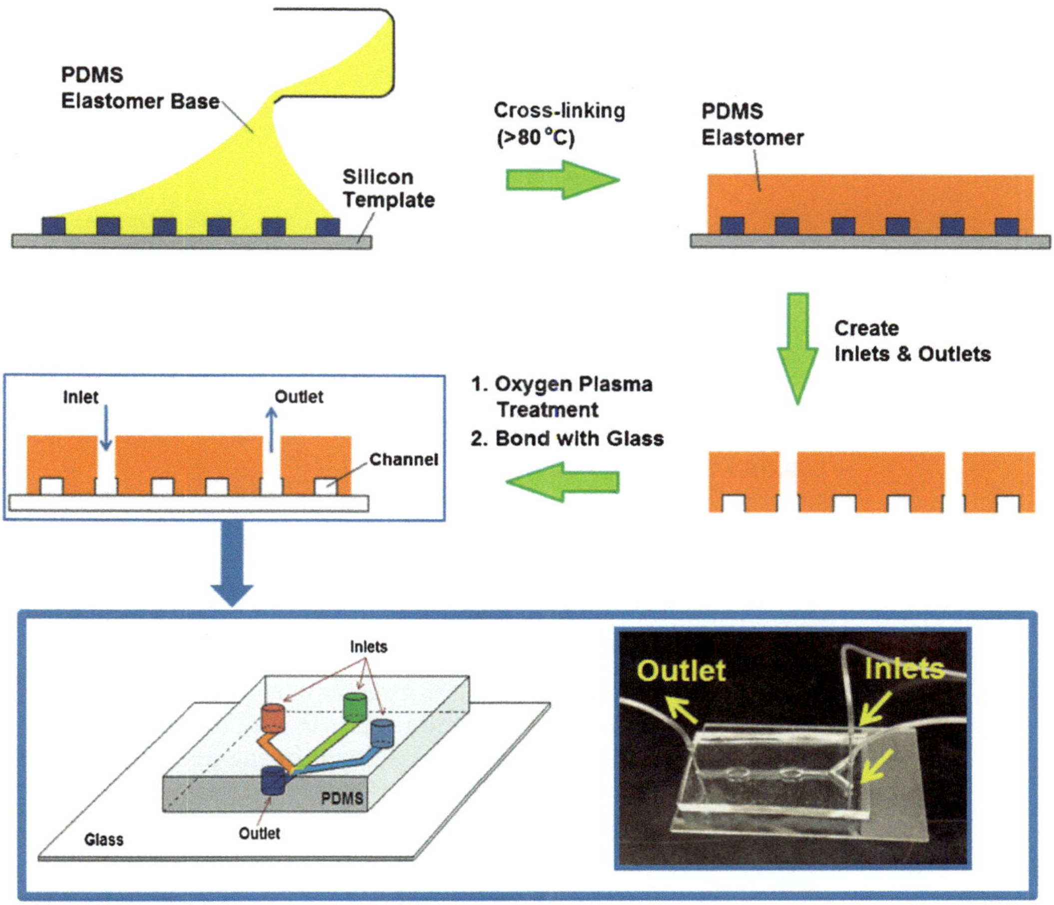

FIGURE 9.6

Schematic illustration of a soft lithographic process to fabricate a microfluidic device. The bottom image is an example of a microfluidic device.

Patterning cells: To be able to pattern cells within a material in a desirable and controllable complexity is a requisite technique for designing tissue constructs with micrometer-scale architecture and resolution. Microfluidics has shown to be a highly effective tool for this purpose, since the cells can be delivered to a desired area within a construct via microchannels (Chiu et al., 2000). The patterning of cells using a microfluidic device is commonly accomplished by the following procedure (Figure 9.7):

1. Reversible assembly of a microfluidic device on a substrate.
2. Delivery of cell-responsive molecules (e.g., proteins and peptides) via microchannels: The molecules become conjugated to the substrate.

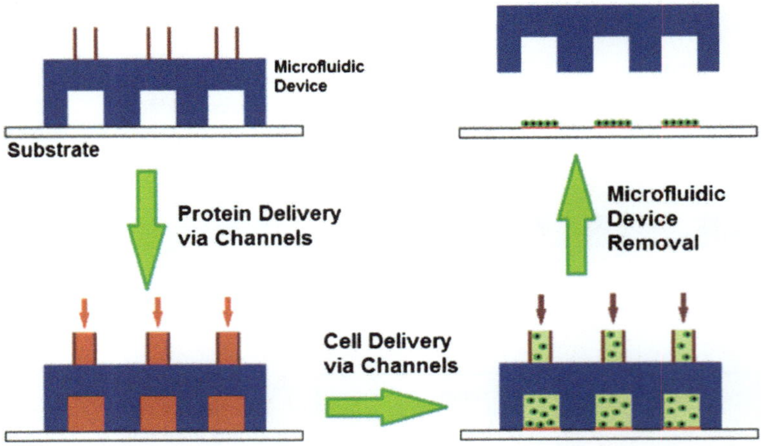

FIGURE 9.7
Schematic illustration of patterning cells on a substrate using a microfluidic device.

3. Delivery of cells via microchannels: The cells adhere to the surface by the interaction with the cell-responsive molecules.
4. Removal of the microfluidic device from the substrate.

Bioreactors: Acquiring an adequate number of cells is a prerequisite for their use in any TE application. In addition, it is vital to maintain the viability and physiological functions of the cells within engineered tissue constructs, which is a significant challenge for developing larger tissue constructs beyond their diffusion limit that severely compromises the long-term cellular viability and functions. Therefore, continuous supply of oxygen and nutrients to the cells is a crucial factor in designing biomaterials for TE applications.

Microfluidic devices are increasingly used as perfusion bioreactors to provide cell culture media to tissue constructs (Leclerc et al., 2003; Gómez-Sjöberg et al., 2007). The ability to manipulate the flow of fluids through microchannels is utilized to supply the culture media to cells or tissues grown within a microfluidic device. Typically, cells are directed into and cultured in chambers within a microfluidic device, in which the culture medium is precisely supplied via channels (Figure 9.8). Using a microfluidic device as a bioreactor for cell culture has several advantages. First, it is not only possible to precisely control the supply of cell culture media in spatiotemporal fashion, but also in varying combinations of multiple media types. In addition, high-throughput experiments can be made possible by using a microfluidic device having multiple cell culture chambers as the experimental platform. It is also possible to create 3D cell culture conditions by stacking multiple device layers connected via channels or membranes.

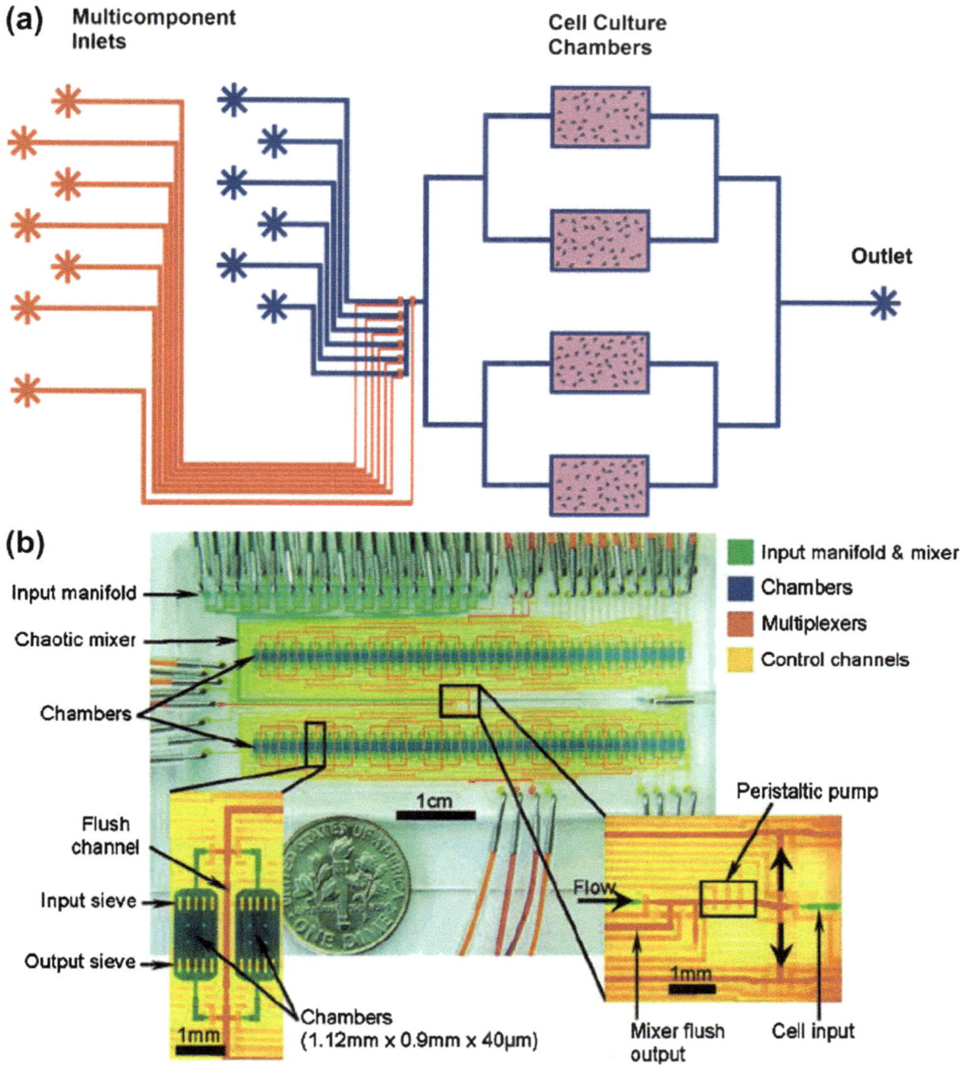

FIGURE 9.8

(a) Schematic illustration of a microfluidic cell culture system. (b) A photograph of a complex microfluidic cell culture system. *Reprinted with permission from Chiu et al. (2000). Copyright 2007 American Chemical Society.*

Rather than using microfluidic devices, other microfabrication techniques such as stereolithography, extrusion-based printing, and micromolding are used to create microchannels within tissue-engineered constructs, with the goal of increasing media diffusion or inducing vascularization within the constructs for their long-term survival (Figure 9.9) (Golden et al., 2007; Kaully et al., 2009; Borenstein et al., 2002; Laschke et al., 2006; Du et al., 2011).

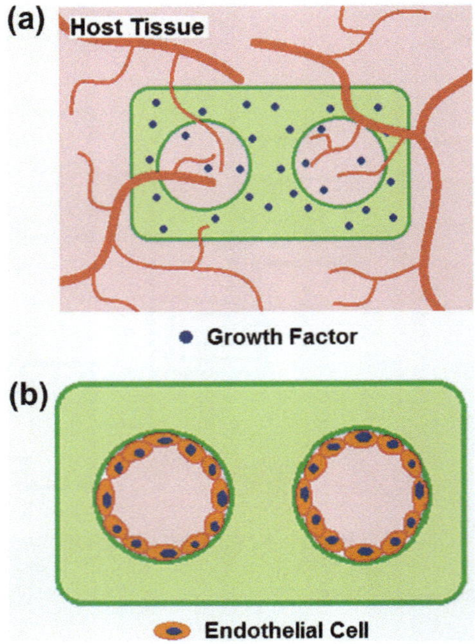

FIGURE 9.9

Schematic illustrations of the strategies of inducing vascularization within microchannels of a tissue construct. (a) Inclusion of growth factors to promote angiogenesis in vivo. (b) In vitro vascular cell culture on the channel walls to form vasculature.

In order to induce vascularization inside the microchannels of a tissue-engineered construct, the construct is often encapsulated with growth factors or cytokines which become released and induce angiogenesis within the channels upon implantation into a host tissue (Sun et al., 2005). Conversely, vascular cells such as endothelial cells and smooth muscle cells are often delivered through the channels, so the channel wall becomes covered with the cells to induce vascularization in vitro (Fidkowski et al., 2005).

Droplet-based biomaterials: Polymeric microparticles are widely used in biomedical applications, predominantly as injectable carriers of biological entities like cells, proteins, DNA, and so on (Freiberg and Zhu, 2004). They are commonly fabricated by emulsification, in which high energy (i.e., sonication) is applied to the mixture of two immiscible phases to generate emulsions. However, this type of bulk emulsification despite its simplicity suffers from polydispersity of droplet size and damage to encapsulating species, which severely hampers its use in biological applications. Microfluidics provides a highly controllable, biocompatible, and versatile route to creating emulsions, in which discrete "droplets" of a fluid are generated within an

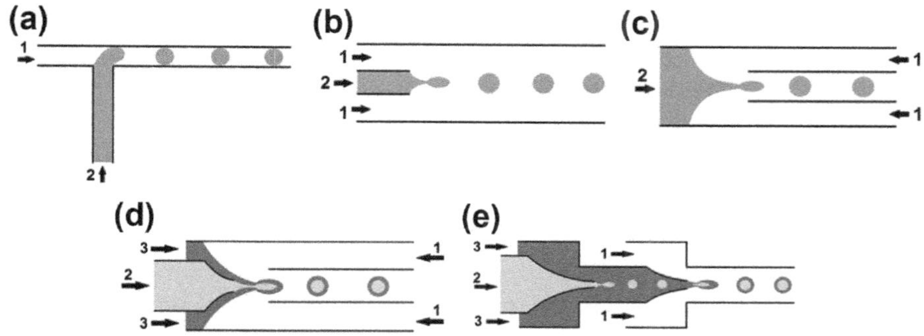

FIGURE 9.10

Microfluidic geometries used to generate droplets. (a) T-junction, (b) co-flow, (c) flow-focusing, (d) second-order co-flow, and (e) second-order flow-focusing. (1: continuous phase, 2: discontinuous phase, and 3: second discontinuous phase).

immiscible continuous phase by tuning microchannel geometry. The microfluidics-assisted emulsification offers several distinct advantages, as compared to the conventional emulsification (Shah et al., 2008; Christopher and Anna, 2007). First, monodisperse droplets can be created by using a steady flow rate that applies a consistent stress for droplet generation. Second, the size of the droplets is easily tuned by controlling the flow rates. Third, higher order droplets can be generated and manipulated by changing the channel geometry: size, number of orders, and number of encapsulated droplets. Fourth, it is possible to generate anisotropic droplets (Janus particles) using microfluidics (Shepherd et al., 2006).

There are three different types of channel geometries used to generate droplets: T-junction, co-flow, and flow-focusing (Figure 9.10). T-junction is the simplest and earliest microfluidic channel to generate droplets, in which two immiscible fluids meet at an intersection (two inlets, one outlet) (Garstecki et al., 2006). On the other hand, the flow of the continuous phase intersects that of the discontinuous phase from two opposing channels (three inlets, one outlet) in co-flow and flow-focusing geometries, which allows for a more robust droplet generation. They also allow for the fabrication of multiple-order droplets in a sequential fashion, in which the first set of droplets is in turn delivered to the next round, and so on. The major difference between co-flow and flow-focusing geometries is the directions of the continuous and discontinuous phases; two fluids are in the same direction in co-flow geometry, whereas they are in the opposite directions in flow-focusing geometry. Since more stress can be applied to the discontinuous phase using the flow-focusing geometry, it can be used to generate droplets of a more viscous fluid.

Regardless of the channel geometry, they all possess the same overarching mechanism for droplet generation; the flow of a continuous phase in one channel applies a consistent force on the flow of a discontinuous phase in the other channel at their intersection, resulting in droplet formation. In addition, the size of the droplets can be controlled by changing the flow rates; the higher the flow rate of a continuous phase, the greater is the force applied to a discontinuous phase resulting in smaller droplets.

More recently, the microfluidic droplets are used as precursors to fabricate microgels (i.e., hydrogels in micrometer scale) as TE scaffolds (Tumarkin and Kumacheva, 2009). The aqueous droplets contain monomers that undergo a cross-linking reaction to form a hydrogel network. Therefore, a cross-linking step is applied to the droplets generated from a microfluidic device in situ to fabricate microgels. For example, droplets containing vinyl-containing monomers and initiators can be cross-linked by exposing them to UV (Shepherd et al., 2006; Cha et al., 2013). Alternatively, droplets with certain polysaccharides that undergo cross-linking reactions with divalent ions (e.g., alginate cross-linked with calcium ions) are also used to fabricate microgels (Choi et al., 2007).

9.2.4 Hierarchical Assembly of Tissue Microstructures

Microfabrication technologies generally develop desired structures via top-down approach; larger structures are conceptualized and developed before their sublevel structures. On the other hand, there have been recent efforts geared toward utilizing bottom-up strategies, in which structures in micrometer scale can be used as components that are assembled into larger constructs. Due to their small sizes, it is extremely difficult to assemble them individually into a desired overall structure, or to direct them into intended target sites. Therefore, several strategies have been developed to intelligently derive a hierarchical and complex assembly of these microstructures.

9.2.4.1 Self-assembly

In order to induce their self-assembly, microstructures are often designed to present certain functional groups or molecules that can induce physical interactions (e.g., hydrophobic interaction, surface tension, capillary force), which drive their movements into intended rearrangement. For example, microgels suspended in hydrophobic solution could be assembled into larger constructs due to high surface tension of aqueous-based microgel suspension (Figure 9.11(a)) (Du et al., 2008).

For its high specificity and strong attractive interaction, DNA base pairing has been demonstrated as a highly useful tool for inducing self-assembly. Khademhosseini and coworkers conjugated complementary DNA strands on

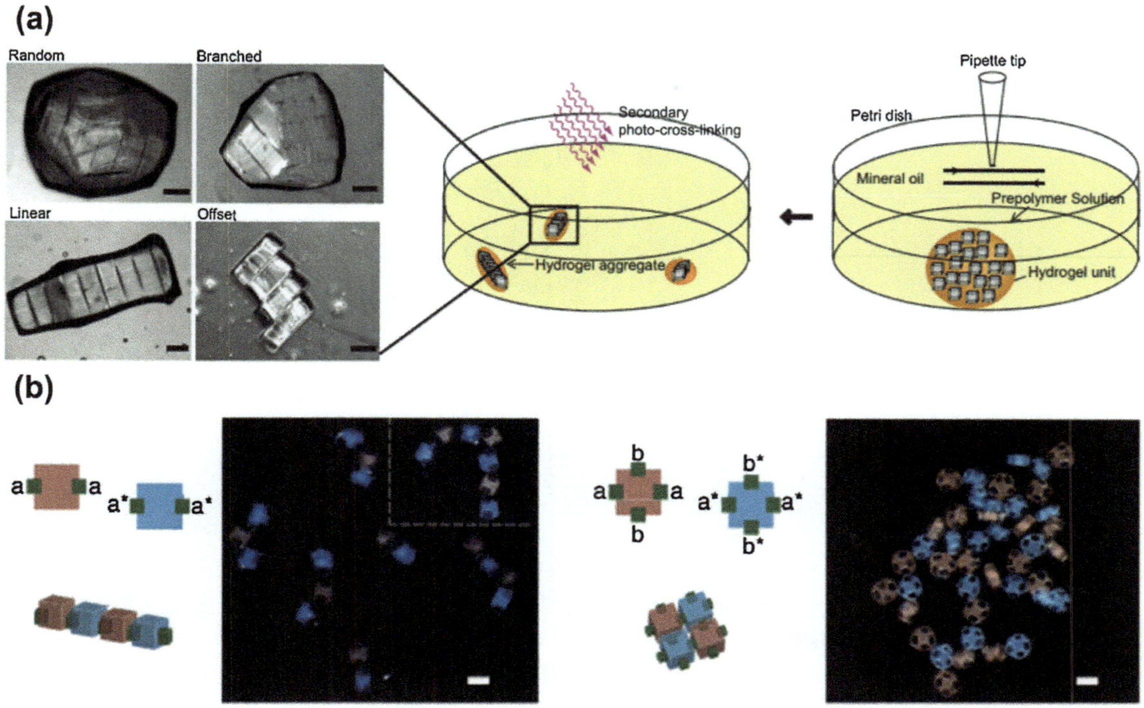

FIGURE 9.11

Self-assembly of tissue microstructures to form larger tissue constructs. (a) Microgels suspended in hydrophobic solution coalesce to form various macrostructures due to high surface tension of aqueous-based suspension. *Reprinted with permission from Du et al. (2008). Copyright 2008 National Academy of Sciences, USA.* (b) Two sets of microgels were conjugated with complementary DNA strands (a and a′, b and b′), which act as "glue" to guide their assembly via strong DNA base pairing (scale bar: 200 μm). *Reprinted with permission from Qi et al. (2013). Copyright 2013 Nature Publishing Group.*

two different sets of cell-laden microgels. These DNA strands were used as a "glue" to induce strong attraction between those two microgels to engineer larger aggregate tissue structures (Figure 9.11(b)) (Qi et al., 2013).

9.2.4.2 Templated Assembly

Microstructures can be assembled with high specificity simply by matching their size and shape. It is analogous to fitting a peg to a hole (or a key to a lock). Current microfabrication technology allows for the fabrication of structures with micrometer-scale precision and complexity. Therefore, highly specific microstructures with matching size and shape can be used as building blocks to assemble larger, hierarchical structures. For example, Vunjak-Novakovic and coworkers induced the assembly of microgels with varying shapes, developed via micromolding technique, into a larger mold containing matching holes

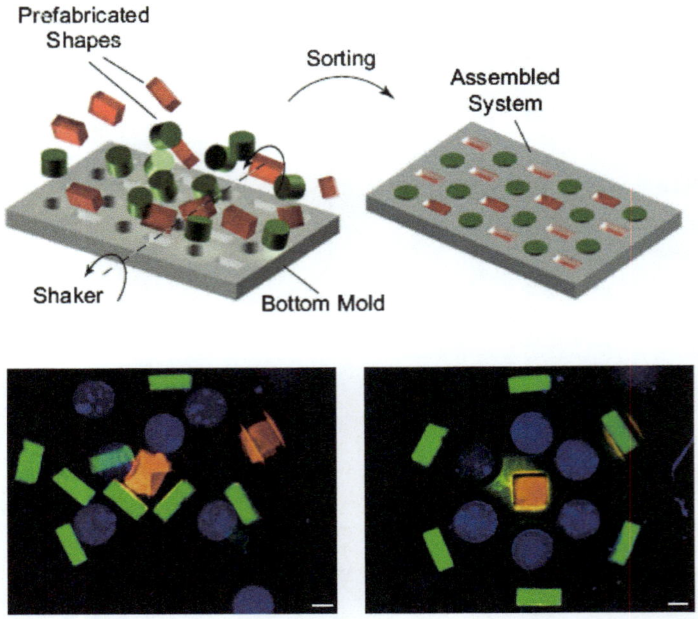

FIGURE 9.12
"Docking" microgels into a mold by matching size and shape. *Reprinted with permission from Eng et al. (2013). Copyright 2013 National Academy of Sciences, USA.*

(Figure 9.12) (Eng et al., 2013). The cell-seeded microgels were "docked" into the mold in a desired geometric pattern, which was eventually used to study various cellular behaviors.

Microfluidic devices are also being used to assemble microstructures. For example, Whitesides and coworkers used microfluidic chambers where tissue constructs were delivered and assembled into larger structures (Bruzewicz et al., 2008). Kwon and coworkers used microfluidic channels as "rails" that guide microstructures into designated areas to assemble complex structures (Figure 9.13) (Chung et al., 2008). They demonstrated that by changing the channel pattern, the types of microstructures, and the order of delivery, a variety of complex structures could be engineered.

More recently, RP techniques are also being utilized to assemble larger tissues (Karoly et al., 2010). Commonly referred to as "bioprinters," these automated printers are designed so that cell-seeded microstructures can be deposited on a predefined template. Due to the cells' innate nature to form aggregates via cell–cell adhesion (e.g., cadherins, a class of cell-surface proteins, are involved in cell junction formation), these deposited structures would eventually become fused over time leading to larger tissue constructs. For example, Gabor and

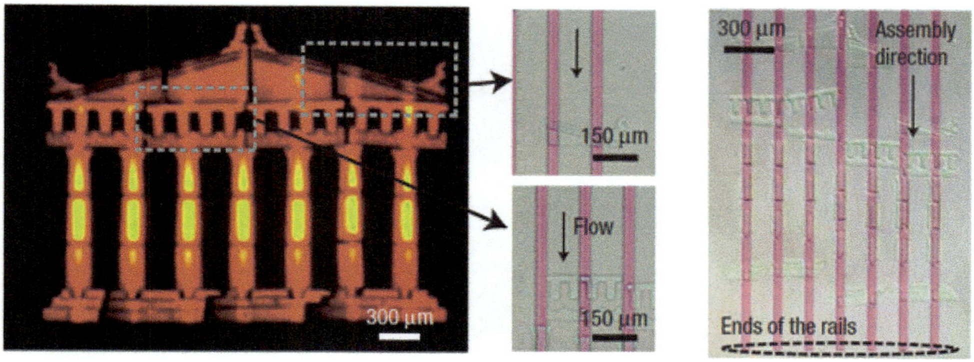

FIGURE 9.13

Microfluidic assembly of microstructures. Microfluidic channels were used as "rails" to guide microstructures to desired areas to develop complex structures. *Reprinted with permission from Chung et al. (2008). Copyright 2008 Macmillan Publishers Limited.*

coworkers utilized this bioprinting approach to deposit tissue microstructures on a desired 3D pattern, and successfully created vascular-like tubular tissue structures (Figure 9.14) (Karoly et al., 2010).

9.3 CONCLUSION AND FUTURE PERSPECTIVE

The ability to manipulate the fabrication and the control of a system in a micrometer scale is a highly innovative technology that is greatly coveted in all areas of research and industry. It offers several advantages as compared with bulk, macroscale systems: efficiency, reproducibility, high resolution control, cost-effectiveness, miniaturization, and high-throughput. The microfabrication technology was first driven by the remarkable technological advancements in microelectronics and its applications in semiconductor industry, fundamentally revolutionizing our way of life.

Many microfabrication processes involved in microelectronics such as μCP and photolithography have been the basis for burgeoning affluence of current microfabrication technology, because these same processes are utilized to create tools needed to fabricate other devices. For example, the majority of MEMS devices are made by processing silicon wafers. In addition, PDMS elastomer-based microfluidic devices are made by fabricating PDMS on the silicon master by photolithography. RP technology is also largely an extension of automated lithography processes. More recently, extensive research efforts are invested to combine several microfabrication technologies to develop more complex systems, such as μTAS by combining MEMS with microfluidics.

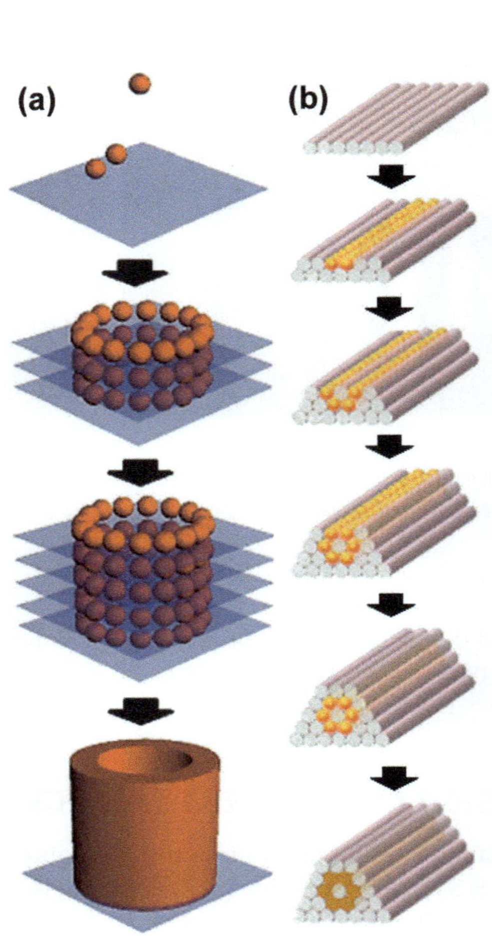

FIGURE 9.14

Rapid prototyping technology to pattern and assemble tissue microstructures. (a) Layer-by-layer deposition. (b) Horizontal layer-by-layer deposition to create tubular structures. *Reprinted with permission from Karoly et al. (2010). Copyright 2010 IOP Publishing, Ltd.*

The ultimate goal of TE is to artificially engineer tissues or entire organs by encompassing the principles of biology, chemistry, and materials engineering. Like any other research area involving material fabrication, TE has benefited tremendously from microfabrication technology in recent years. It is now well established that cells and tissues are regulated in micrometer scales. Therefore, controlling the spacing, orientation, and overall architecture of a tissue-engineered construct in a micrometer scale would allow for the enhanced regulation of various cellular phenotypes. For this reason, microfabrication technologies such as photolithography, microfluidics, and RP, which have been originally developed for microelectronics and analytics, are now extensively investigated for a wide range of TE applications.

As the field of TE itself is considered by many to be still in its infancy, the utilization of microfabrication technology so far has been limited, and therefore not yet reached its true potential. As the TE continues to expand and likely moves toward tissue, organ, and patient-specific strategies, this trend of utilizing microfabrication technology is expected to continue and eventually become a fixture in the future. This will be further fueled by the technological advancement that will render many microfabrication processes less expensive and more accessible, and therefore become more popular; do not look any further than the recent meteoric rise and popularity of "3D printers."

9.4 SUMMARY

- Microfabrication was conceptualized and came into fruition in the realm of electronics, fabrication of integrated circuits in semiconductor industry.
- Microfabrication technology is increasingly utilized in various TE applications, as it offers several advantages over conventional bulk methods: efficiency, reproducibility, high resolution control, cost-effectiveness, miniaturization, and high-throughput.
- Photolithography is used to fabricate photo-cross-linkable tissue-engineered constructs in micrometer-scale resolution and dimensions by controlling the patternable area with a photomask.
- RP is an automated manufacturing technology fueled by remarkable advances in computer-aided design and control system; stereolithography and extrusion-based printing are widely used in TE.
- The size and shape of tissue-engineered constructs can be easily tuned by using a patterned substrate as a micromold.
- A desired pattern of biological molecules (e.g., proteins, peptides, and DNA) can be transferred onto a substrate using microcontact printing (μCP).
- Microfluidic technology allows the control of small amounts of fluids through micrometer-scale channels, which is increasingly utilized in various TE applications such as perfusion bioreactors, patterning cells, and fabrication of droplet-based biomaterials.
- Microfabricated structures can be used as building blocks that are assembled into larger, hierarchical structures ("bottom-up" approach).

CLASSICAL EXPERIMENT

In order to engineer micrometer-scale tissue-engineered constructs, Khademhosseini and coworkers devised a simple yet highly versatile and efficient micromolding approach for generating cell-laden 3D hydrogels of controlled sizes and shapes in micrometer scale in the form of harvestable free units (Yeh et al., 2006). In this work, they encapsulated fibroblasts and murine embryonic stem cells within photo-cross-linkable hyaluronic acid or PEG hydrogels using a PDMS micromold. This study was carried out following three key phases of microfabrication: fabrication of PDMS micromold, hydrogel fabrication using the mold, and retrieval of micromolded hydrogels.

Templates of various shapes were fabricated by curing PDMS on silicon wafers previously patterned with negative photoresist. The fabricated PDMS to be used as micromolds consisted of various shapes (squares, circles, and rectangles) and sizes (from 50 to 400 μm). The PDMS micromolds were rendered hydrophilic by plasma treatment for easy removal of hydrogels, while nontreated ones were used as covers to reversibly seal the patterns into individual volumes during the procedure.

Cells were suspended in a gel precursor solution, which was then pipetted onto the PDMS mold. A PDMS cover was then placed on top and sealed. The assembly was then exposed to UV light to induce the photo-cross-linking reaction to fabricate hydrogels. After removing the cover, the hydrogels could be easily collected by hydrating them with a buffer solution.

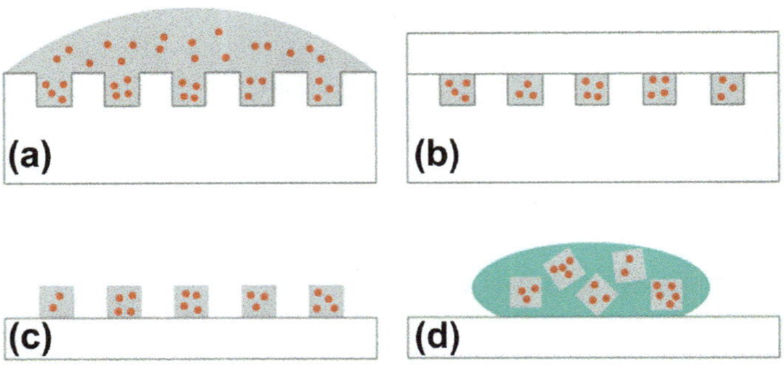

Cell-laden hydrogel formation via micromolding: (a) The cells within a gel precursor solution are deposited onto a plasma-treated PDMS template. (b) A PDMS cover is placed on top and irradiated with UV to induce photo-cross-linking. (c) The PDMS mold is removed. (d) Hydrogels are hydrated and harvested.

STATE-OF-THE-ART EXPERIMENT

Stereolithography is an RP microfabrication technology that allows for the automated control of fabricating tissue-engineered constructs. This technology relies on a computer-aided design (CAD) for modeling and a laser as a light source to selectively fabricate the desired area. 3D structures can be easily developed by repeating the fabrication process ("layer-by-layer" approach).

Chen and coworkers devised a new stereolithography technique that is different from the traditional laser-based one to fabricate tissue-engineered constructs (Gauvin et al., 2012). They used a DMD as a digital photomask, in which each DMD chip contains micromirrors that can be controlled using the same CAD modeling to define the patternable area. A single light exposure reflected off a DMD chip results in the fabrication of an entire layer, which is a clear advantage over the traditional stereolithography that takes much longer time to fabricate a single layer because the laser must move along the entire pattern. This aspect is especially beneficial for fabricating tissue constructs, as minimizing fabrication time significantly enhances the viability of encapsulating cells.

They successfully demonstrated the efficacy of the DMD projection system by encapsulating endothelial cells into photo-cross-linkable gelatin hydrogels having various micrometer-scale architectures. In addition, like a conventional stereolithography, this technique could be repeated to create a layer on top of the previously fabricated layer ("layer-by-layer" approach) to create larger 3D constructs, in which cellular activities were well maintained.

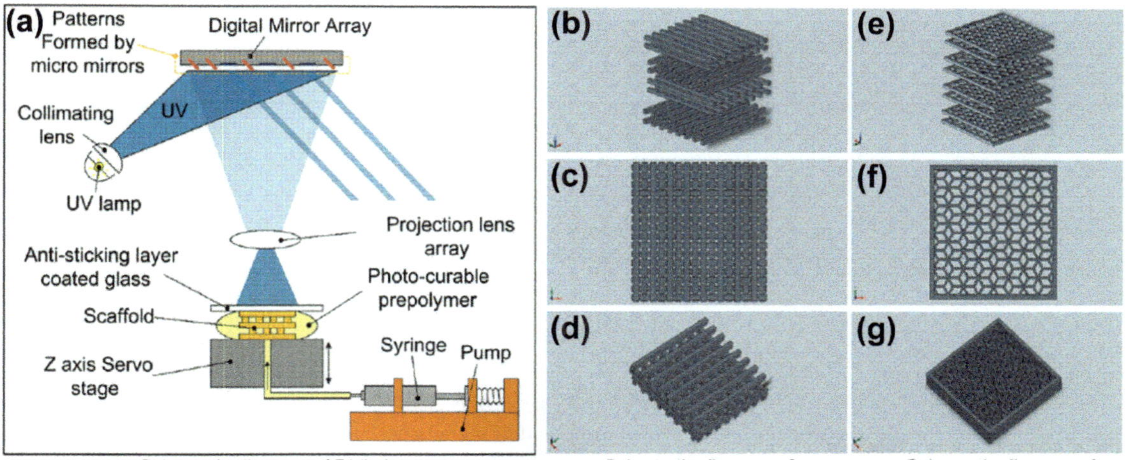

Schematic diagram of Digital Mirror Device (DMD) system

Schematic diagram of woodpile scaffold (b)–(d)

Schematic diagram of haxagonal scaffold (e)–(g)

Digital Micromirror Device (DMD) projection printing system to fabricate tissue-engineered constructs. (a) Schematic illustration of a DMD system. (b)–(g) Examples of microfabricated constructs using the DMD system.

RECOMMENDED READING

Chu, L.-Y., Utada, A.S., Shah, R.K., Kim, J.-W., Weitz, D.A., 2007. Controllable monodisperse multiple emulsions. Angew. Chem. Int. Ed. 46 (47), 8970–8974.

Jakab, K., Norotte, C., Marga, F., Murphy, K., Vunjak-Novakovic, G., Forgacs, G., 2010. Tissue engineering by self-assembly and bio-printing of living cells. Biofabrication 2 (2), 022001.

REFERENCES

Arikawa, H., Takahashi, H., Kanie, T., Ban, S., 2009. Effect of various visible light photoinitiators on the polymerization and color of light-activated resins. Dent. Mater. J. 28 (4), 454–460.

Bernard, A., Renault, J.P., Michel, B., Bosshard, H.R., Delamarche, E., 2000. Microcontact printing of proteins. Adv. Mater. 12 (14), 1067–1070.

Bhatia, S.N., Chen, C.S., 1999. Tissue engineering at the micro-scale. Biomed. Microdevices 2 (2), 131–144.

Billiet, T., Vandenhaute, M., Schelfhout, J., Vlierberghe, S., Dubruel, P., 2013. Exploring the future of hydrogels in rapid prototyping: a review on current trends and limitations. In: Antoniac, I. (Ed.), Biologically Responsive Biomaterials for Tissue Engineering. Springer, New York, pp. 201–249.

Borenstein, J., Terai, H., King, K., Weinberg, E.J., Kaazempur-Mofrad, M.R., Vacanti, J.P., 2002. Microfabrication technology for vascularized tissue engineering. Biomed. Microdevices 4 (3), 167–175.

Brittain, S., Paul, K., Zhao, X.-M., Whitesides, G., 1998. Soft lithography and microfabrication. Phys. World 11 (5), 31–36.

Bruzewicz, D.A., McGuigan, A.P., Whitesides, G.M., 2008. Fabrication of a modular tissue construct in a microfluidic chip. Lab Chip 8, 663–671.

Cha, C., Oh, J., Kim, K., et al., 2013. Microfluidics-assisted fabrication of gelatin-silica core–shell microgels for injectable tissue constructs. Biomacromolecules 15 (1), 283–290.

Chan, V., Zorlutuna, P., Jeong, J.H., Kong, H., Bashir, R., 2010. Three-dimensional photopatterning of hydrogels using stereolithography for long-term cell encapsulation. Lab Chip 10 (16), 2062–2070.

Chiu, D.T., Jeon, N.L., Huang, S., et al., 2000. Patterned deposition of cells and proteins onto surfaces by using three-dimensional microfluidic systems. Proc. Natl. Acad. Sci. U.S.A. 97 (6), 2408–2413.

Choi, C.-H., Jung, J.-H., Rhee, Y., Kim, D.-P., Shim, S.-E., Lee, C.-S., 2007. Generation of monodisperse alginate microbeads and in situ encapsulation of cell in microfluidic device. Biomed. Microdevices 9 (6), 855–862.

Christopher, G.F., Anna, S.L., 2007. Microfluidic methods for generating continuous droplet streams. J. Phys. D Appl. Phys. 40 (19), R319–R336.

Chua, C.K., Leong, K.F., Lim, C.S., 2010. Rapid Prototyping: Principles and Applications. World Scientific Publishing, London.

Chung, S.E., Park, W., Shin, S., Lee, S.A., Kwon, S., 2008. Guided and fluidic self-assembly of microstructures using railed microfluidic channels. Nat. Mater. 7 (7), 1476–122.

Dimitrov, D., Schreve, K., de Beer, N., 2006. Advances in three dimensional printing – state of the art and future perspectives. Rapid Prototyping J. 12 (3), 136–147.

Du, Y., Lo, E., Ali, S., Khademhosseini, A., 2008. Directed assembly of cell-laden microgels for fabrication of 3D tissue constructs. Proc. Natl. Acad. Sci. U.S.A. 105 (28), 9522–9527.

Du, Y., Ghodousi, M., Qi, H., Haas, N., Xiao, W., Khademhosseini, A., 2011. Sequential assembly of cell-laden hydrogel constructs to engineer vascular-like microchannels. Biotechnol. Bioeng. 108 (7), 1693–1703.

Eng, G., Lee, B.W., Parsa, H., et al., 2013. Assembly of complex cell microenvironments using geometrically docked hydrogel shapes. Proc. Nat. Acad. Sci. U.S.A. 110 (12), 4551–4556.

Fidkowski, C., Kaazempur-Mofrad, M.R., Borenstein, J., Vacanti, J.P., Langer, R., Wang, Y., 2005. Endothelialized microvasculature based on a biodegradable elastomer. Tissue Eng. 11 (1–2), 302–309.

Freiberg, S., Zhu, X.X., 2004. Polymer microspheres for controlled drug release. Int. J. Pharm. 282 (1–2), 1–18.

Grayson, A.C.R., Shawgo, R.S., Johnson, A.M., et al., 2004. A BioMEMS review: MEMS technology for physiologically integrated devices. Proc. IEEE 92 (1), 6–21.

Garstecki, P., Fuerstman, M.J., Stone, H.A., Whitesides, G.M., 2006. Formation of droplets and bubbles in a microfluidic T-junction-scaling and mechanism of break-up. Lab Chip 6 (3), 437–446.

Gauvin, R., Chen, Y.-C., Lee, J.W., et al., 2012. Microfabrication of complex porous tissue engineering scaffolds using 3D projection stereolithography. Biomaterials 33 (15), 3824–3834.

Golden, A.P., Tien, J., 2007. Fabrication of microfluidic hydrogels using molded gelatin as a sacrificial element. Lab Chip 7 (6), 720–725.

Gómez-Sjöberg, R., Leyrat, A.A., Pirone, D.M., Chen, C.S., Quake, S.R., 2007. Versatile, fully automated, microfluidic cell culture system. Anal. Chem. 79 (22), 8557–8563.

Grogan, S.P., Chung, P.H., Soman, P., et al., 2013. Digital micromirror device projection printing system for meniscus tissue engineering. Acta Biomater. 9 (7), 7218–7226.

Hasenbein, M.E., Andersen, T.T., Bizios, R., 2002. Micropatterned surfaces modified with select peptides promote exclusive interactions with osteoblasts. Biomaterials 23 (19), 3937–3942.

Hollister, S.J., 2005. Porous scaffold design for tissue engineering. Nat. Mater. 4 (7), 518–524.

Hutmacher, D.W., Sittinger, M., Risbud, M.V., 2004. Scaffold-based tissue engineering: rationale for computer-aided design and solid free-form fabrication systems. Trends Biotechnol. 22 (7), 354–362.

Ifkovits, J.L., Burdick, J.A., 2007. Review: photopolymerizable and degradable biomaterials for tissue engineering applications. Tissue Eng. 13 (10), 2369–2385.

Jacobs, P.F., 1992. Rapid Prototyping & Manufacturing: Fundamentals of Stereolithography, first ed. Society of Manufacturing Engineers, Dearborn.

Kaully, T., Kaufman-Francis, K., Lesman, A., Levenberg, S., 2009. Vascularization – the conduit to viable engineered tissues. Tissue Eng. 15 (2), 159–169.

Karoly, J., Cyrille, N., Francoise, M., Keith, M., Gordana, V.-N., Gabor, F., 2010. Tissue engineering by self-assembly and bio-printing of living cells. Biofabrication 2 (2), 022001.

Khademhosseini, A., Langer, R., Borenstein, J., Vacanti, J.P., 2006. Microscale technologies for tissue engineering and biology. Proc. Nat. Acad. Sci. U.S.A. 103 (8), 2480–2487.

Khalil, S., Nam, J., Sun, W., 2005. Multi-nozzle deposition for construction of 3D biopolymer tissue scaffolds. Rapid Prototyping J. 11 (1), 9–17.

Laschke, M.W., Harder, Y., Amon, M., et al., 2006. Angiogenesis in tissue engineering: breathing life into constructed tissue substitutes. Tissue Eng. 12 (8), 2093–2104.

Leclerc, E., Sakai, Y., Fujii, T., 2003. Cell culture in 3-dimensional microfluidic structure of PDMS (polydimethylsiloxane). Biomed. Microdevices 5 (2), 109–114.

Leukers, B., Gülkan, H., Irsen, S., et al., 2005. Hydroxyapatite scaffolds for bone tissue engineering made by 3D printing. J. Mater. Sci. Mater. Med. 16 (12), 1121–1124.

Lu, Y., Mapili, G., Suhali, G., Chen, S., Roy, K., 2006. A digital micro-mirror device-based system for the microfabrication of complex, spatially patterned tissue engineering scaffolds. J. Biomed. Mater. Res. Part A 77A (2), 396–405.

Lutolf, M.P., Hubbell, J.A., 2005. Synthetic biomaterials as instructive extracellular microenvironments for morphogenesis in tissue engineering. Nat. Biotechnol. 23 (1), 47–55.

Liu Tsang, V., Bhatia, S.N., 2004. Three-dimensional tissue fabrication. Adv. Drug Delivery Rev. 56 (11), 1635–1647.

Madou, M.J., 2002. Fundamentals of Microfabrication: The Science of Miniaturization. CRC Press, Boca Raton.

McDonald, J.C., Whitesides, G.M., 2002. Poly(dimethylsiloxane) as a material for fabricating microfluidic devices. Acc. Chem. Res. 35 (7), 491–499.

Nichol, J.W., Koshy, S.T., Bae, H., Hwang, C.M., Yamanlar, S., Khademhosseini, A., 2010. Cell-laden microengineered gelatin methacrylate hydrogels. Biomaterials 31 (21), 5536–5544.

Paul, O., 2006. Microtransducer operation. In: Korvink, J.G., Paul, O. (Eds.), MEMS: A Practical Gruide to Design, Analysis and Applications. William Andrew, Inc., New York, pp. 1–47.

Qi, H., Ghodousi, M., Du, Y., et al., 2013. DNA-directed self-assembly of shape-controlled hydrogels. Nat. Commun. 4.

Qin, D., Xia, Y., Whitesides, G.M., 2010. Soft lithography for micro- and nanoscale patterning. Nat. Protoc. 5 (3), 491–502.

Revzin, A., Russell, R.J., Yadavalli, V.K., et al., 2001. Fabrication of poly(ethylene glycol) hydrogel microstructures using photolithography. Langmuir 17 (18), 5440–5447.

Reyes, D.R., Iossifidis, D., Auroux, P.-A., Manz, A., 2002. Micro total analysis systems: 1. Introduction, theory, and technology. Anal. Chem. 74 (12), 2623–2636.

Shah, R.K., Shum, H.C., Rowat, A.C., et al., 2008. Designer emulsions using microfluidics. Mater. Today 11 (4), 18–27.

Shepherd, R.F., Conrad, J.C., Rhodes, S.K., et al., 2006. Microfluidic assembly of homogeneous and janus colloid-filled hydrogel granules. Langmuir 22 (21), 8618–8622.

Sun, Q., Chen, R., Shen, Y., Mooney, D., Rajagopalan, S., Grossman, P.M., 2005. Sustained vascular endothelial growth factor delivery enhances angiogenesis and perfusion in ischemic hind limb. Pharm. Res. 22 (7), 1110–1116.

Tumarkin, E., Kumacheva, E., 2009. Microfluidic generation of microgels from synthetic and natural polymers. Chem. Soc. Rev. 38 (8), 2161–2168.

Tan, J.L., Liu, W., Nelson, C.M., Raghavan, S., Chen, C.S., 2004. Simple approach to micropattern cells on common culture substrates by tuning substrate wettability. Tissue Eng. 10 (5–6), 865–872.

Whitesides, G.M., 2006. The origins and the future of microfluidics. Nature 442 (7101), 368–373.

Wu, W., DeConinck, A., Lewis, J.A., 2011. Omnidirectional printing of 3D microvascular networks. Adv. Mater. 23 (24), H178–H183.

Yan, X., Gu, P., 1996. A review of rapid prototyping technologies and systems. Comput. Aided Des. 28 (4), 307–318.

Yeh, J., Ling, Y., Karp, J.M., et al., 2006. Micromolding of shape-controlled, harvestable cell-laden hydrogels. Biomaterials 27 (31), 5391–5398.

Zorlutuna, P., Jeong, J.H., Kong, H., Bashir, R., 2011. Stereolithography-based hydrogel microenvironments to examine cellular interactions. Adv. Funct. Mater. 21 (19), 3642–3651.

Zorlutuna, P., Annabi, N., Camci-Unal, G., et al., 2012. Microfabricated biomaterials for engineering 3D tissues. Adv. Mater. 24 (14), 1782–1804.

CHAPTER 10

Scaffold Design and Fabrication

Dietmar W. Hutmacher[1], Tim B.F. Woodfield[2], Paul D. Dalton[3]

[1]Institute of Health and Biomedical Innovation, Queensland University of Technology, Brisbane, Australia; [2]Department of Orthopaedic Surgery and Centre for Bioengineering & Nanomedicine, University of Otago, Christchurch, Christchurch, New Zealand; [3]Department of Functional Materials in Medicine and Dentistry, University of Würzburg, Germany

LEARNING OBJECTIVES

- To outline the sequence of events in the formation of a tissue-engineered construct and the changes in important factors with time
- To understand the key rationale, characteristics, and process parameters of the currently used scaffold fabrication techniques
- To understand which scaffold processing techniques would allow for the direct control over the design and fabrication of 3D pore architecture
- To provide a comprehensive list of various scaffold techniques, with the most relevant references
- To demonstrate the differences between pore size and pore interconnectivity and its physical and biological implications
- To learn how various types of scaffolds can be prepared and how different configurations form various scaffolding materials
- To assist research teams with their choice of a specific additive manufacturing processing technology to produce scaffolds

10.1 INTRODUCTION

Scaffold-based tissue engineering (TE) combines viable cells, biomolecules, and a structural scaffold into a "tissue engineering construct" (TEC) to promote the repair and/or regeneration of tissues. The construct is intended to support cell migration, growth, and differentiation, and guide tissue development and organization into a mature and healthy state. The science in the field is still developing and various approaches and strategies are under experimental investigation. It is by no means clear what defines ideal scaffold/cell or scaffold/neotissue constructs, even for a specific tissue type. The considerations are complex and include architecture, structural mechanics, surface properties, degradation products, composition of biological components, and

Tissue Engineering.
Copyright © 2015 Elsevier Inc. All rights reserved.

the changes of these factors with time in vitro and/or in vivo (Hutmacher et al., 2004; Woodfield et al., 2004; Hollister, 2005).

However, TE scaffolds do have certain minimum requirements for biochemical as well as chemical and physical properties. Scaffolds must provide sufficient initial mechanical strength and stiffness to substitute for the mechanical function of the diseased or damaged tissue, which it aims at repairing or regenerating. Scaffolds may not necessarily be required to provide complete mechanical equivalence to healthy tissue, but stiffness and strength should be sufficient to at least support and transmit forces to the host tissue site in the context. For example, in skin TE, the construct should be able to withstand the wound contraction forces. In case of bone engineering, external and internal fixation systems might be applied to take support of the majority load-bearing forces until the bone has matured (Henkel and Hutmacher, 2013).

Cell and tissue remodeling is the key for achieving stable biomechanical conditions and vascularization at the host site. Hence, the three-dimensional (3D) scaffold/tissue construct should maintain sufficient structural integrity during the in vitro and/or in vivo growth and remodeling process. The degree of remodeling depends on the tissue itself (e.g., skin 4–6 weeks, bone 4–6 months) and its host anatomy and physiology. Scaffold architecture has to allow for initial cell attachment and subsequent migration into and through the matrix, mass transfer of nutrients and metabolites, and provision of sufficient space for development and later remodeling of organized tissue. The degradation and resorption kinetics of the scaffold need to be designed based on the relationships of mechanical properties, molecular weight, mass loss, and tissue development (including vascularization) described in Figure 10.1 for bone TE.

With the essentials of mechanics and geometry achieved, a suitable construct will possess surface properties which are optimized for the attachment and migration of cell types of interest (depending on the targeted tissue). The external size and shape of the construct must also be considered, especially if the construct is customized for an individual patient (Henkel and Hutmacher, 2013). Using discrete bioactive regions to control cell fate in 3D systems, multiphasic scaffolds, and dynamically responsive materials are the three recent trends in scaffold design.

Practical considerations of manufacture arise, even after a holistic TE strategy for scaffold performance. From a clinical point of view, a scaffold must be manufactured under GMP (Good Manufacturing Practice) conditions in a reproducible and quality-controlled fashion at an economic cost and speed (Hollister, 2009a, 2009b; Spector, 2014). To move the current TE practices to the next frontier, some manufacturing processes will have to accommodate the incorporation of cells and/or growth factors during the scaffold fabrication process. To address this, novel manufacturing processes such as robotic assembly and machine- and computer-controlled 3D cell encapsulation are under

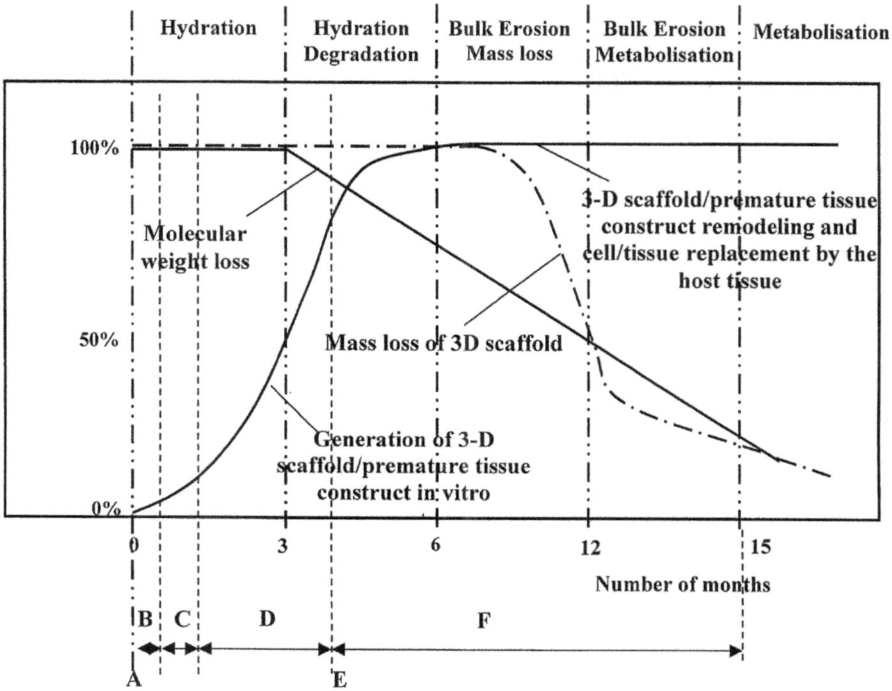

FIGURE 10.1

A timeline of the molecular weight loss and mass loss for a polymeric scaffold for bone TE including (A) Fabrication of the scaffold; (B) cell seeding of the osteoblast/cartilage; (C) growth of cells in a dynamic environment; (D) maturation of tissue in a physiologic environment (bioreactor); (E) in vivo implantation and (F) in vivo modeling and assimilation. Vascularization occurs over several weeks and new bone forms around the scaffold struts over many months. The graphical illustration shows that the scaffold material must be selected and/or designed with a degradation and resorption rate such that the strength of the scaffold is retained until the tissue-engineered transplant is fully remodeled by the host tissue and can assume a progressively increasing structural role. Mechanical stability of the scaffold carrying cells, proteins, genes, or growth factors once deposited within the defect is a key requirement. If the mechanical integrity is compromised, cell viability may be jeopardized under constant biomechanical loading and micromotion as well as fibrous encapsulation may result in failure of the construct to truly regenerate the defect side. Hence, a key requirement of the scaffold design must be that the physical support by the scaffold itself is maintained until the engineered tissue within the interconnected pore architecture has been remodeled and built-up sufficient mechanical integrity to support the major loading profile in the former defect area. *Image reproduced with permission from Hutmacher (2000).*

development, so that the tissue-engineered construct not only has a controlled spatial distribution of cells and growth factors, but also a versatility of scaffold materials and microstructure within one construct. There are well-defined steps to producing a TE product for clinical use, and these are reviewed well elsewhere (Hollister, 2009a, 2009b; Hollister and Murphy, 2011; Spector, 2014).

A number of fabrication technologies have been applied to process biodegradable and bioresorbable polymers (Nair and Laurencin, 2007) into 3D scaffolds of high porosity and surface area. From a scaffold design and function point of view, each processing methodology has its pros and cons. It is beyond the scope of this chapter to cover all scaffold fabrication techniques. Hence, the authors aim at providing the reader with an overview of the methods which are currently most relevant for scaffold-based TE and to present this data in a comprehensive form. While we have attempted to sequentially introduce concepts of various scaffold manufacturing processes in this chapter, in reality different approaches are being increasingly combined. Therefore, some scaffold fabrication approaches may be mentioned prior to their introduction later in the chapter.

10.2 SCAFFOLD DESIGN

It is important to emphasize, at the outset, that the field of scaffold-based TE is still developing, with many different approaches under experimental investigation. Thus, it is by no means clear as to what defines an ideal scaffold/cell or scaffold/neotissue construct, even for a specific tissue type. Indeed, since some tissues perform multiple functional roles, it is unlikely that a single scaffold would serve as a universal foundation for the regeneration of even a single tissue. Hence, the considerations for scaffold design are complex and they include material composition, porous architecture, structural mechanics, surface properties, degradation properties and products, together with the composition of any added biological components, and, of course, the changes in all of these factors over time. Hollister (2009a) described scaffold design as a 4F concept; the scaffold should have a suitable *form*, temporarily support tissue *function*, enhance tissue *formation*, and all delivered in a manner so that a surgeon can perform rapid *fixation*. Recent research in scaffold design has also focused on immune modulating materials to improve tissue integration (Lewis et al., 2014).

As TE scaffolds will eventually be implanted in the human body, it should be nonantigenic, noncarcinogenic, nontoxic, nonteratogenic, and possess high cell/tissue biocompatibility so that they will not trigger pathological reactions after implantation. Scaffolds may be required to provide sufficient initial mechanical strength and stiffness to substitute for the mechanical function of the diseased or damaged tissue, which the TEC aims at repairing or regenerating. Scaffolds may not necessarily be required to provide complete mechanical equivalence to healthy tissue, indeed the variability in architecture for a single tissue type is so extensive that it is inconceivable that a single TEC would serve universal applications for even a single tissue. Nevertheless, stiffness and strength should be sufficient to at least either permit requisite cell seeding of the scaffold in vitro without compromising scaffold architecture, or support and transmit forces in an in vivo healing site. Thus, in the context of skin TE, the scaffold material should be sufficiently robust to not only resist change in shape as a result of the introduction of cells into the scaffold, each of which would be capable of exerting tractional forces, but also the wound contraction forces which will be invoked during tissue healing in vivo. The same general rules would apply to bone engineering, although the in vivo phase external and internal fixation systems, or other supports or restrictions on patient activity in early stage recovery, may reduce such mechanical roles of the intended scaffold.

Tissue remodeling is important for achieving stable biomechanical conditions and vascularization at the host site. Hence, the scaffold/tissue construct should maintain sufficient structural integrity during the in vitro and/or in vivo growth and remodeling process. Scaffold architecture has to allow for initial cell attachment and subsequently migration into and through the matrix as well as mass transfer of nutrients and metabolites and the provision of sufficient pore space for development and later remodeling of the organized tissue. The porosity and internal space within a degradable scaffold will increase with time, allowing for tissue to develop and/or remodel.

10.2.1 Morphology/Architecture

Scaffolds can have a broad range of morphological properties, and examples are shown in Figure 10.2. Porous materials can be classified into three groups: interconnecting (open pores), nonconnecting (closed pores), or a combination of both. In all but the most biomechanically challenging applications, it is likely that the test for the scaffold engineer will be to achieve sufficient stiffness and strength in a highly porous structure to provide adequate mechanical integrity. One of the most demanding applications will be the repair and generation of musculoskeletal tissues, particularly bone, where scaffolds need to have a high elastic modulus in order to provide temporary mechanical support without showing symptoms of fatigue or failure, to be retained in the space they were designated for, and to provide the tissue with adequate space for growth (Henkel and Hutmacher, 2013). One of the fundamental challenges of scaffold design and materials selection is that to achieve sufficient strength and stiffness, the scaffold material must have both a sufficiently high interatomic and intermolecular bonding and/or a physical and chemical structure which allows for hydrolytic attack and breakdown. The production of scaffold structures based upon mathematic models and unit structures (Figure 10.3(a)) is one fascinating recent advance (Melchels et al., 2010), as is the fabrication of degradable magnesium scaffolds consisting of complex architectures based on optimized unit structures (Figure 10.3(b)) (Nguyen et al., 2011).

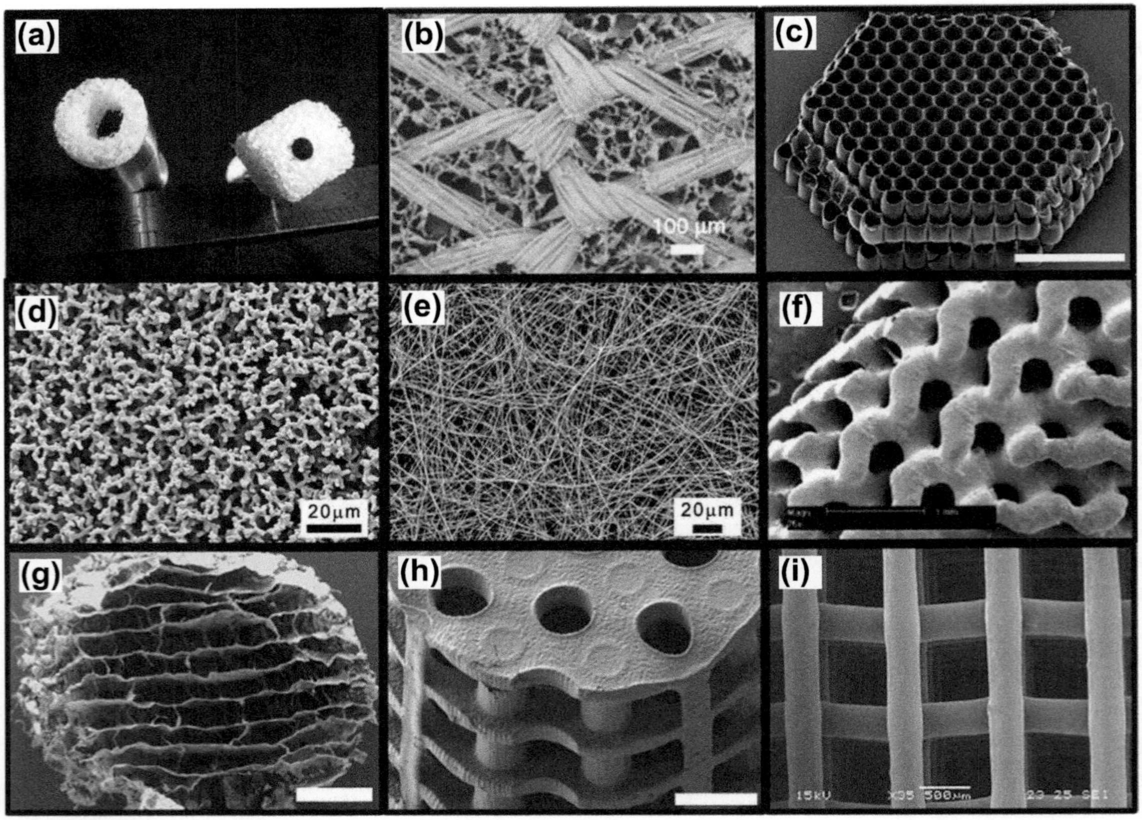

FIGURE 10.2
Selection of random examples of scaffold designs for different tissue engineering applications. (a) Porogen leaching. (b) Braided yarns macropore architecture combined with a micropore filling collagen matrix (Chen et al., 2005). (c) Mixed lithography/micromolding approaches (Koroleva et al., 2012). (d) Randomly organized pore morphology of a scaffold made via polymerization-induced phase separation (Li et al., 2013). (e) A nonwoven fabricated via solution electrospinning (Li et al., 2013). (f) Gyroid scaffolds built using stereolithography (Schuller-Ravoo et al., 2013). (g) Channeled pores made via uniaxial freezing (Bozkurt et al., 2012). (h) Structurally supported scaffold built via stereolithography (Kim et al., 2011). (i) Fused deposition modeling (FDM). *All Figures reproduced with permission.*

10.2.2 Porosity

A pore can be defined as a void space within a scaffold, whereas porosity can be considered as a collection of pores (Figure 10.4). Pore size and porosity are two of the most important scaffold parameters and are reviewed in depth elsewhere (Loh and Choong, 2013). Macro-pores (i.e., >50 μm) are of a scale to influence tissue function. For example, pores greater than 300 μm in size are typically suggested for bone ingrowth in relation to vascularization of the construct. Micropores (i.e., <50 μm) are of a scale to influence cell function (e.g., cell attachment) given that mammalian cells typically are 10–20 μm in size. Nanoporosity refers

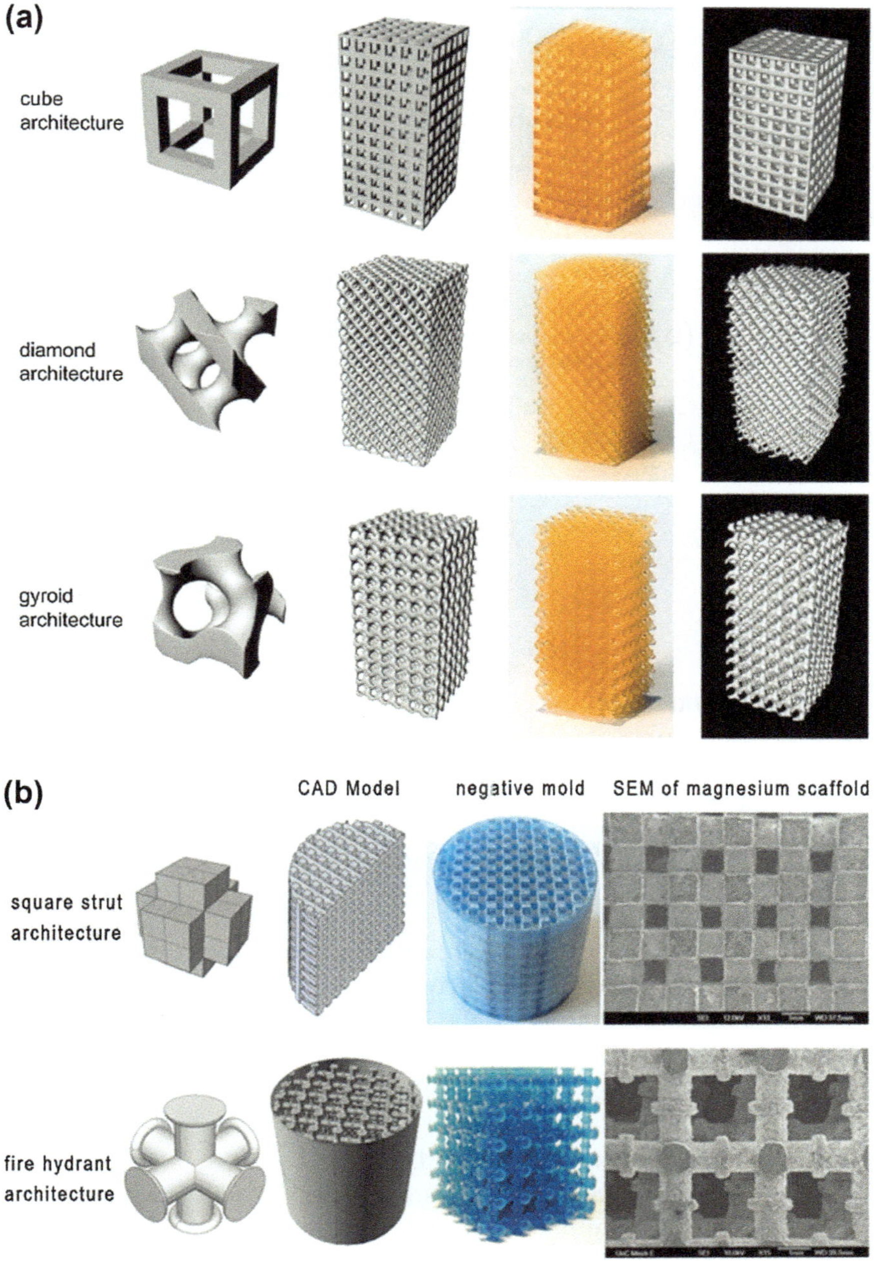

FIGURE 10.3

(a) Various scaffold architectures fabricated with stereolithography, using mathematically derived structural units. Each column (from left to right) illustrates the structural unit; a computer-aided design model of the scaffold; the scaffold fabricated via stereolithography; and a microcomputed tomography (μCT) reconstruction of the scaffold. (b) Similar approaches using unit structures have been adopted to fabricate degradable magnesium scaffolds consisting complex architectures using indirect additive manufacturing techniques. *Images reproduced from Melchels et al. (2010) and Nguyen et al. (2011).*

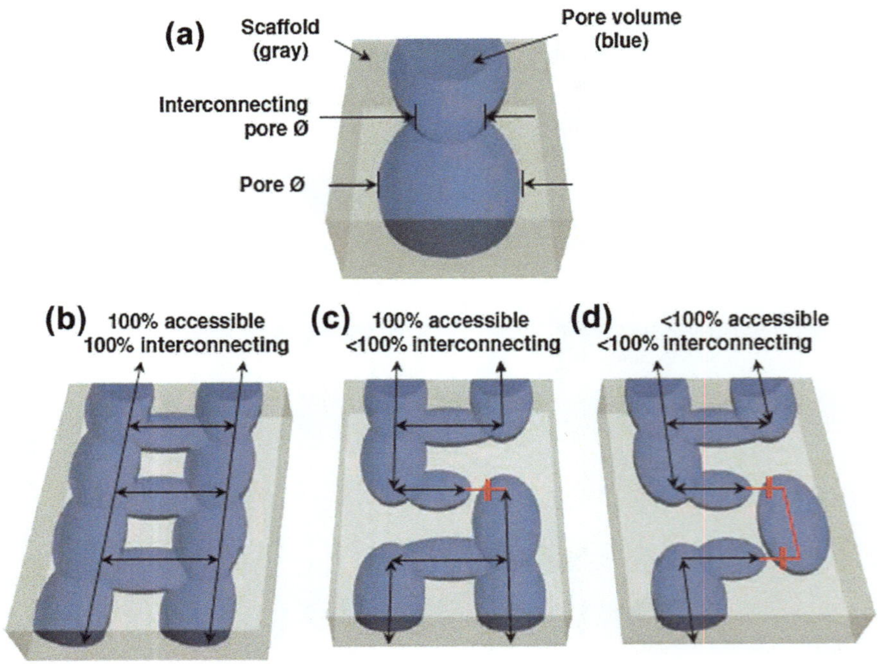

FIGURE 10.4

Ingenuous graphical illustration which clarifies the relationship between pore morphology and interconnectivity. Imaging of a section through a scaffold shown in (a) (gray) and corresponding pore geometries in relation to overall accessible pore volume (blue). (b) Pore volume 100% accessible and 100% interconnecting. (c) Pore volume 100% accessible, yet not 100% interconnecting. (d) Pore volume neither 100% accessible nor 100% interconnecting. *Original image from 1st edition.*

to pore architectures or surface textures on a nanoscale (i.e., 1–1000 nm). There is often a compromise between porosity and scaffold mechanical properties. High porosity (e.g., 90%) may provide a greater pore volume for cell infiltration and extracellular matrix (ECM) formation, but conversely decreases mechanical properties in accordance to a power–law relationship.

10.2.3 Interconnectivity

Pore interconnectivity (Figure 10.4) is a critical factor, and surprisingly overlooked in scaffold design and characterization. A scaffold may be porous, but unless the pores are interconnecting (i.e., voids linking one pore to another), they serve no purpose and become superfluous in a scaffold for TE. The size of interconnections between pores should be suitably large to support cell migration and proliferation in the initial stages and consequent ECM infiltration of desired tissue and, therefore, is more critical than pore size. It is preferable that scaffolds for TE have 100% interconnecting pore volume, thereby also maximizing the diffusion and exchange of nutrients (e.g., oxygen) throughout the entire scaffold pore volume.

10.2.4 Pore Characterization

As a measure of pore interconnectivity, the accessible pore volume, or permeability, of a scaffold can be measured. Accessible pore volume can be defined as the total volume of pores which can be infiltrated from all peripheral borders to the interior of the scaffold (Miot et al., 2005). Scaffold permeability can be measured by determining the flow rate of fluid flow through interconnecting pores. However, this technique is not suitable in scaffolds with large, 100% interconnected pore volumes as the scaffold provides no resistance to fluid flow. Alternatively, accessible pore volume, as well as vol% porosity, pore size distribution, and scaffold surface area to volume ratio (i.e., volume fraction) can be characterized using techniques such as mercury intrusion porosimetry, microcomputed tomography (μCT), or image analysis (Ruegsegger et al., 1996; Muller and Ruegsegger, 1997; Odgaard, 1997; Laib et al., 2001). Mercury porosimetry is a popular technique which is based on the principle that the pressure required to force a nonwetting liquid such as mercury into pores, against the resistance of liquid surface tension, is indicative of the pore size, assuming the pores are cylindrical in shape. However, the resolution of the technique is severely limited in scaffolds with large pore sizes (>500 μm) where low mercury intrusion pressures are necessary, and it has limitations when applied to materials that have irregular pore geometries. Alternative techniques such as μCT, which utilize 3D CT imaging to generate computer models of porous materials, have been developed for analyzing bone architecture and more recently scaffold architecture (Figure 10.5). Using 3D μCT techniques, considerably greater information can be obtained to characterize pore architectures containing features ranging from 6 μm to >1500 μm, without the physical limitations associated with mercury porosimetry (Ho and Hutmacher, 2006).

10.2.5 Vascularization

It has become increasingly clear that scaffolds used for repair of large volume defects in vascularized tissue require a distinct vascularization strategy. There are a few exceptions to this statement, for instance, with cartilage which does not have a vasculature. It is not sufficient to assume that rapid vascularization will occur and cells within the TEC will have sufficient nutrients. Vascularization is such a key issue in scaffold-based engineering that there is a separate chapter dedicated to this topic in this textbook. Figure 10.6 shows how microvessels can be produced in vitro, and how large pores designed for vascular networks have been introduced into a matrix. Microfluidics has become a key aspect of studying vascularization, reviewed elsewhere (Wong et al., 2012).

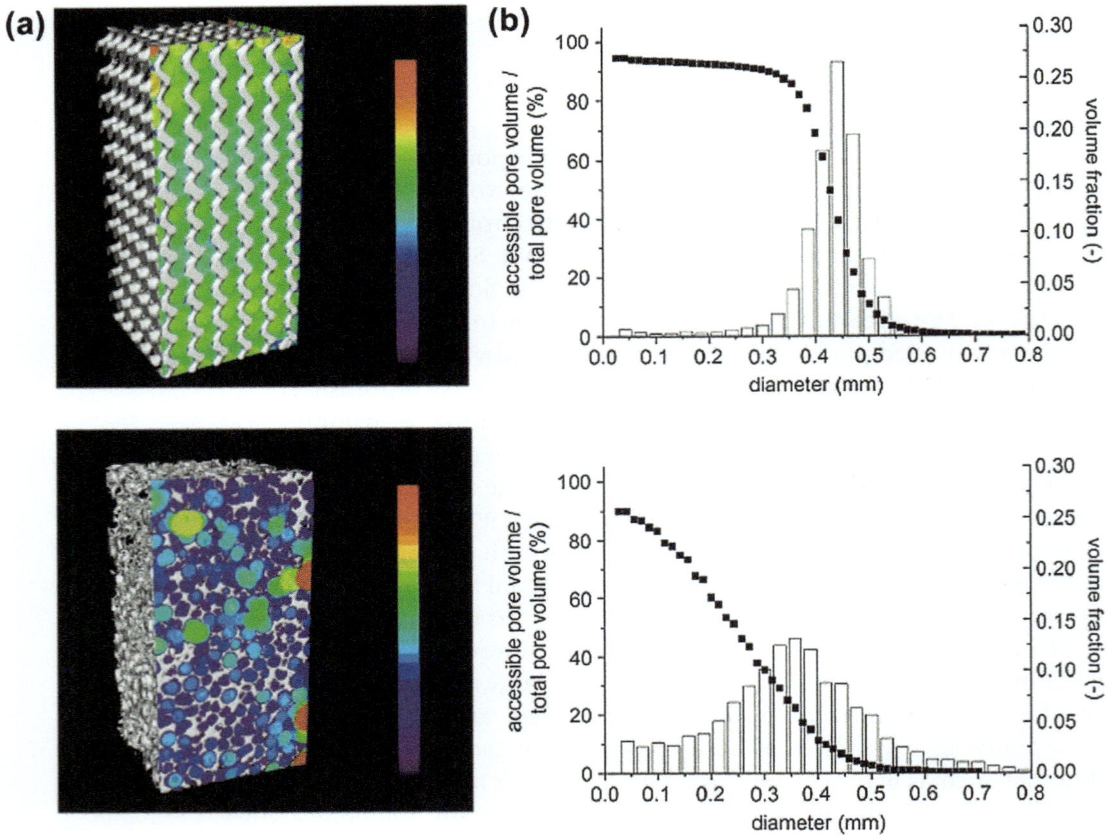

FIGURE 10.5

Microcomputed tomography (μCT) analysis combined with sophisticated image analysis allows detailed characterization from both a qualitative and quantitative point of view of scaffold morphology. Comparison of well-organized (gyroid architecture shown in upper image) and randomly organized (salt leached morphology shown in lower image) scaffolds. Software coding allows for better visualization based on color coding of varying pore sizes, while the right image shows in chart form the accessible pore volume and volume fraction taken up with spheres with increasing diameter. *Images reproduced from Melchels et al. (2010).*

10.2.6 Multiphasic and Multimodal Scaffolds

Since tissue is rarely isotropic, the scaffold architecture also does not need to be uniform. Furthermore, there are often instances where a gradient of mechanical properties is desired (Seidi et al., 2011) for "hard to soft" TE, including for materials in dentistry. The increased utility of additive manufacturing (described later in the chapter) permits complex scaffold manufacture, including gradients (Melchels et al., 2010), and these can be used to replicate the natural anatomic structure of tissue.

When a scaffold manufacturing approach such as solution electrospinning (described later) does not produce desired porosities, then it can be

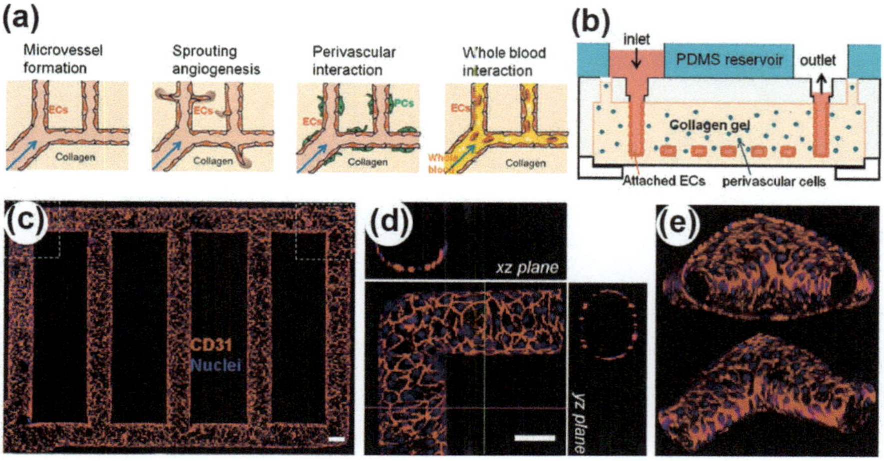

FIGURE 10.6

(a) Fast and sufficient vascularization plays an important role in scaffold-based tissue engineering concepts. This illustration shows several in vitro microvessel formation strategies. (b) Using lithography, a fluid channel bioreactor for endothelial cells is established within a collagen hydrogel seeded with pericytes. (c)–(e) Within the lumen of the channels, an endothelial cell layer is formed. *Reproduced from Zheng et al. (2012).*

combined with another scaffold manufacturing approach. These scaffolds are termed multimodal and commonly include submicron elements. As shown in Figure 10.2(d), solution electrospun scaffolds have low porosity; however, the submicron scale fibers are known to provide some important signals to cells, such as for adhesion and differentiation (Liu et al., 2012; Mammadov et al., 2013). Multimodal scaffolds combine these submicron elements with micron-scale scaffold manufacturing approaches such as direct writing (Figure 10.7(a)–(e)) (Park et al., 2008) and melt electrospinning (Figure 10.7(f)–(g)) (Kim et al., 2010). This results in scaffolds with high surface area due to the submicron fibers (promoting cell attachment and assisting in higher seeding of the scaffold), as well as a suitable porosity and pore size due to the microscale elements. There are many more possibilities for generating multimodal scaffolds, and undoubtedly their fabrication will be the focus of future research.

10.3 CLASSICAL SCAFFOLD FABRICATION TECHNIQUES

There are many different methods described in the literature to design and fabricate porous scaffolds. However, a significant number of these techniques did not yet make significant impact and/or have no relevance for bench to bedside translation. Many scaffold fabrication approaches have been used for decades and, while still having utility, are subjected to less research as more advanced

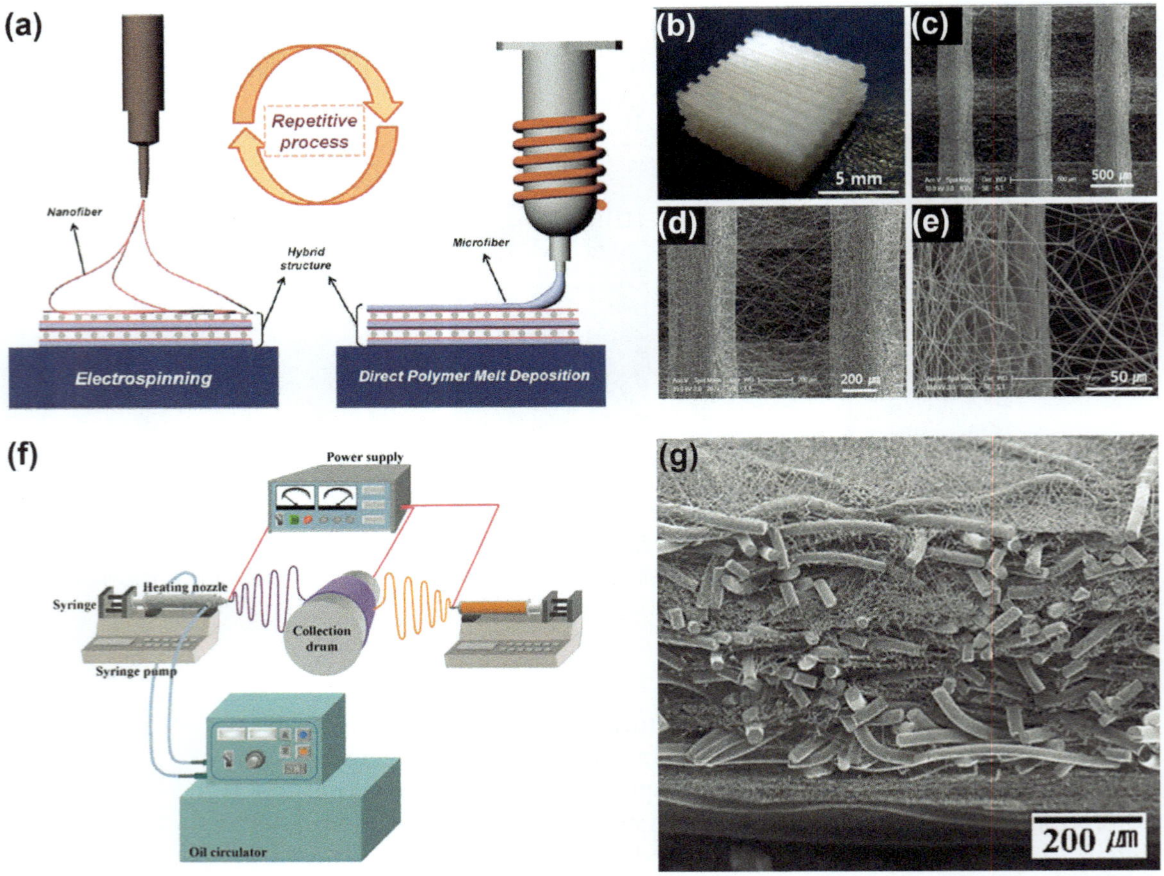

FIGURE 10.7

(a) The most current literature shows an increasing trend toward the engineering of tissue interfaces and hence multimodal scaffold designs are en vogue. (b)–(e) This schematic illustrates a process for combining 3D extrusion and electrospinning to design and fabricate a multimodal scaffold architecture which has different scale elements. (f)–(g) In an innovative fabrication strategy, melt and solution electrospinning of the same polymer have been combined to build a multimodal scaffold. *(a) Reproduced from Kim et al. (2008); (b) Reproduced from Kim et al. (2010).*

or flexible processing technologies emerge such as additive manufacturing or biofabrication (i.e., the printing of scaffolds + cells). This could be a temporary issue—a breakthrough in processing within an area can rapidly change the TE community perspective on a fabrication process.

10.3.1 Porogen Leaching

Porogen leaching was used in the early days of TE as it is one of the oldest polymer processing technologies to make porous products. It is based on dispersing a template (particles, etc.) within a polymeric or monomeric solution, gelling or

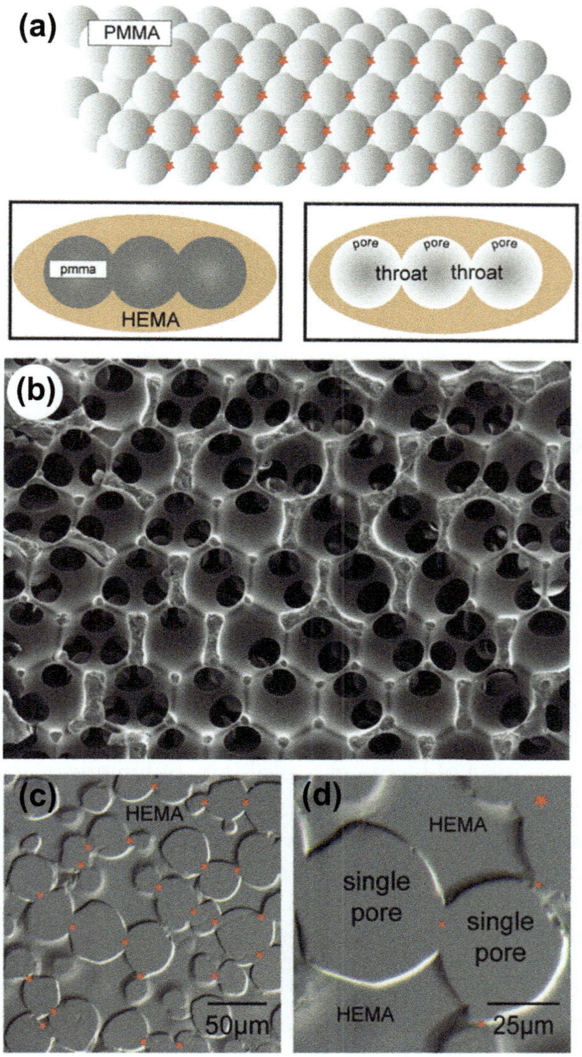

FIGURE 10.8

With the goal to enhance vascularization and tissue integration, scaffolds have also been designed by using hydrogels as building materials. (a) By applying a porogen leaching technique, similar diameter poly(methyl methacrylate) (PMMA) beads are shaken into a compact arrangement, then cross-linked pHEMA is formed around the spheres. (b–d) After PMMA dissolution, the scaffold has regular, defined pores with good interconnectivity. When subcutaneously implanted, macrophages (not shown) penetrate and adopt a more antiinflammatory phenotype. Good vascularization occurs when the interconnecting pores are approximately 35 μm with reduction in fibrous capsule formation. *Reproduced from Bryers et al. (2012) and Fukano et al. (2006).*

fixing the structure, and removal of the template to result in a porous architecture (Figure 10.2(a)). Recent improvements with combining shaking of regular sized spherical porogens, produced extremely well-defined scaffolds (Figure 10.8) (Bryers et al., 2012). Furthermore, the defined morphology was able to alter the inflammatory reaction to the scaffold. This "stealth-like" ability for a scaffold to avoid negative inflammation is a property that allows it to be used in a broad range of tissue/biomaterial applications. This manufacturing process is applicable across various polymer platforms, for example, zwitterionic scaffolds based on polymers and cross-linking agents of carboxybetaine have been investigated (Zhang et al., 2013). Subcutaneous implantation of this low-fouling material did not result in a fibrous capsule for at least 3 months after subcutaneous implantation. It also promoted angiogenesis in the tissue surrounding the implant, possibly due to a switch to prohealing phenotype by the invading macrophages. These porogen leached materials are part of an important class of biomaterials that harness and modulate the immune system (Lewis et al., 2014).

10.3.2 Phase Separated Scaffolds

Phase separation has been employed for decades for producing hollow fiber membranes and porous structures. Thermally induced phase separation (TIPS) in particular has produced a range of scaffolds for TE. The morphological nature of the scaffold is variable and, due to the self-assembling nature of the construct, may be combined with porogen leaching to generate a macroarchitecture within the scaffold. The TIPS technique for producing scaffolds is based upon the reduction of polymer solubility when the temperature is lowered, or when the polymer is frozen out of solution. These two types of TIPS are termed liquid–liquid and solid–liquid phase separation, respectively. The preparation of hydrogel scaffolds by polymerization-induced phase separation (Figure 10.2(c)) is another type of approach and is described in the Classical Experiment (boxed text). An extensive review on liquid–liquid phase separation was published by Elbert (2011).

CLASSICAL EXPERIMENT: PHEMA SCAFFOLDS

The modern era of TE dates back to the work described by Rheinwald and Green for skin TE and then by Langer and Vacanti in the 1980s, although investigations on scaffolds and matrices were performed in earlier decades. As one example, the synthesis and polymerization of 2-hydroxyethyl methacrylate (HEMA) by Otto Wichterle (1913–1998) and Drahoslav Lím (1925–2003) in Prague in the early 1960s initiated a series of successful experiments and applications for this newly discovered synthetic hydrogel. The contact lens industry was the major beneficiary of the manufacture of transparent pHEMA. Otto Wichterle rapidly demonstrated the benefits of such hydrogels as a contact lens; pHEMA was so successful in this role that it is still a viable commercial product over 40 years later. In no small part, pHEMA helped propel contact lenses into a billion-dollar industry; Bausch and Lomb sold the first commercial pHEMA contact lens in 1971. The rapid recognition of the potential for pHEMA reflects the exceptional polymer research performed in Prague during this period.

Also in the early 1960s, polymer chemists in the former Czechoslovakia demonstrated that porous opaque materials are formed after polymerization of HEMA in excess water. This system phase separates due to an increasing insolubility of the growing polymeric

CLASSICAL EXPERIMENT: PHEMA SCAFFOLDS *Continued*

radical. When the two liquid phases are separated into a bicontinuous system, the propagating radicals in the polymer rich phase result in gelation. Such hydrogels were proposed as soft tissue replacements and cellular penetration into the pores was observed, when implanted subcutaneously. These cell-invasive pHEMA scaffolds were termed "sponges," with their first clinical uses being breast augmentation and nasal cartilage replacement. The pHEMA scaffolds withstood autoclaving, were easy to form into different shapes, and felt like soft tissue to the surgeon with their "spongy" physical properties. pHEMA scaffolds were hailed by Czech scientists and surgeons as a soft tissue replacement with a multitude of potential biomedical applications. Clinical studies, initiated in the early 1970s, included numerous women acquiring cell-invasive pHEMA breast implants. However, long-term in vivo calcification of the scaffolds prevented the development of pHEMA as a successful implantable material. In the 1990s, after a surge in interest regarding cell-invasive materials, pHEMA scaffolds were developed in the Chirila laboratory as a porous rim to an artificial cornea that became a clinical product (Jiraskova et al., 2011). Cell-invasive pHEMA scaffolds may have one of the simplest and inexpensive experimental protocols in TE. The solvent is water, and particulate leaching is unnecessary; the components are simply mixed and polymerized and the porosity results from phase separation of the propagating radical. Figure 10.2(c) shows the morphology of a phase separated pHEMA scaffold.

10.3.3 Uniaxial Freezing

The freezing of hydrogels has been increasingly used to induce channels and pores. Various hydrogels have been transformed into scaffolds by uniaxial freezing including agarose, cross-linked polyethylene glycol (PEG) (Schulte et al., 2013), chitosan (Jana et al., 2013), and collagen (Bozkurt et al., 2012). Channels in the hydrogels reflect the ice crystal structures formed and can guide cell and/or tissue growth in the scaffolds. The size and scale of the pores is controlled by the temperature gradient, with larger ice crystals producing larger pores. They have been investigated primarily for tissue such as the peripheral nerve, spinal cord, and muscle repair (Jana et al., 2013). An example of a scaffold made by uniaxial freezing is shown in Figure 10.2(e).

10.3.4 Micromolding

The molding of porous objects is increasingly relevant for TE, particularly as the resolution levels now enter the nanoscale. Furthermore, the complexity of structures is constantly improving and this process will have a more significant impact on TE in the future. While additive manufacturing is acknowledged as being highly flexible in producing complex scaffold architectures of varying size and shape (Figure 10.2(h)), there remain limits in large-scale production. Therefore, once a scaffolding structure has been decided upon using additive manufacturing, molding offers an attractive approach to produce tens of thousands of such samples. Figure 10.2(h) shows a scaffold made via micromolding, which is used for glia culture (Koroleva et al., 2012). An excellent in vitro study on microvessels used micromolding to produce high fidelity structures for understanding blood vessel formation (Zheng et al., 2012). Another approach is to use additive manufacturing techniques to fabricate a complex structure that is then part of a micromolded matrix (Koroleva et al., 2012; Miller et al., 2012).

10.3.5 Gas Foaming

Gas foaming of biodegradable polymers found its original application in the biomedical sciences in drug delivery applications during the 1980s. It is a scaffold fabrication technique that permits solvent-free formation of porous materials through generation of gas bubbles within a polymer. Molded polymers may be pressurized with a gas, typically CO_2, until the polymer is saturated. The release of pressure results in nucleation and growth of the air bubbles up to 100 μm; however, interconnectivity is still limited and is often combined with particulate leaching to obtain improved interconnectivity between pores. This process was recently reviewed by Dehghani and Annabi (2011).

10.3.6 Classical Nonwoven Textiles

Cell-invasive fiber-based scaffolds can be produced using methodologies developed for the textile industry. TE textiles have a relatively high surface area and their "value-added" application, from an industry with established techniques, is an advantage. Additionally, textiles are typically formed into sheets and the permeability is high, allowing the necessary nutrients to reach the seeded cells. Many of the recent advances in fiber-based technologies have involved hydrogel fibers (Tamayol et al., 2013) and "living fiber" systems. The latter involves the fibers being used as a carrier for cells (Akbari et al., 2014).

10.3.7 Knitted and Braided Textiles

Individual fibers, or multifilament yarns, may be woven, knitted, or braided into patterns with variable pore sizes. Extremely complex fabric structures can result, and the relatively large pores between the fibers can be filled with a secondary scaffold, such as collagen gel or electrospun fibers (Figure 10.2(b)) (Chen et al., 2005). The fabric structure is, in this instance, a mechanical support for a secondary, interstitial scaffold that may be otherwise easily damaged. These approaches are often ideal for forming scaffold sheets or flexible rope-like scaffolds with excellent tensile properties for skin and ligament repair but not necessarily applicable to forming thick scaffolds used in application for compressive loading.

10.4 ELECTROSPINNING

Electrospinning is a relatively inexpensive manufacturing technique for submicron and micron diameter fibers from polymer solutions or melts. Although it is a process known since the 1930s, the past decade has seen dramatic expansion of its use in TE since reintroduction by the Reneker laboratory in the 1990s.

Electrospinning has become an important process for scaffold fabrication; however, the next step for this technology is to take TE products to the clinic.

10.4.1 Electrospinning Principles

Electrospinning is commonly performed with polymeric solutions; however, there has been recent interest in melt electrospinning, particularly as a direct writing approach (see boxed text). Typically, a high (positive or negative) voltage is applied to a polymer solution or melt that is pumped to a spinneret (small orifice, or flat-tipped needle) facing an earthed target, or collector (Figure 10.9). Upon reaching a critical voltage, the surface tension of the polymer at spinneret tip is counterbalanced by localized charges generated by the electrostatic force,

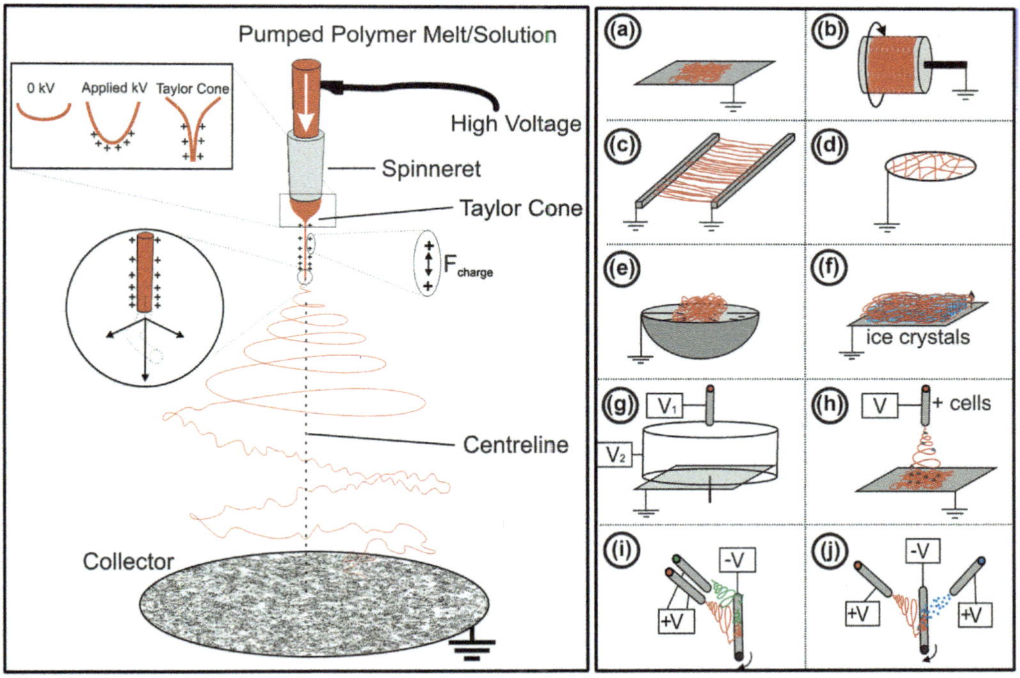

FIGURE 10.9

Left: Detailed graphical illustration of the electrospinning process by using different collectors and spinning modes. An electric field results in accumulated surface charges on the droplet until a polymer jet is ejected directly toward the collector when the applied voltage is positive and collector is grounded (inset). The surface charge density increases on the polymer jet until distortions and displacements occur from the centerline. This initial instability results in a spiraling jet that is eventually balanced by the surface tension of the polymer solution/melt. Further bending instabilities result in a chaotic path prior to landing on the collector. Right: Selected collection schemes for electrospun fibers (red). (a) Single ground, (b) rotating single ground, (c) dual bar, (d) single horizontal ring, (e) needle containing hemisphere, (f) electrospinning in vitro onto ice crystals, (g) electrospinning with circular auxiliary electrode and guiding pin, (h) electrospinning cells with polymer, (i) electrospinning with two different spinnerets, and (j) simultaneous electrospinning/electrospraying. *Original image modified from 1st edition.*

and the droplet elongates and stretches into a (Taylor) cone where a continuous jet is ejected. Initially, the polymer jet travels directly toward the target.

Once the Taylor cone is formed, the droplet contains fewer surface charges than the polymer jet. The surface charge density increases with distance from the spinneret (toward the collector), until the surface charges overcome surface tension, and the polymer jet deviates from the most direct path to the collector, or the centerline. The instabilities of the jet result in spiraling and further stretching. Solvent evaporation or cooling of the polymer prior to landing at the collector results in electrospun fibers with sizes typically between 200 nm and 5 μm, depending on the conditions. Fridrikh et al. (2003) and Reneker and Yarin (2008) provide excellent perspectives on electrospinning physics and fiber formation.

10.4.2 Collection Systems

Depending on the collection configuration, the morphology of the scaffold can be vastly different. The majority of electrospinning systems use a single collection plate to collect the fibers (Figure 10.9(a)), while grounding a single rotating cylindrical collector can also collect fibrous mats (Figure 10.9(b)). Oriented fibers can be collected between grounded elements (Figure 10.9(c)–(d)); in this configuration, the fibers are suspended in a space between the two grounds. An inverted hemisphere containing wire filaments has also been used as a collector to increase the porosity of the electrospun material (Figure 10.9(e)) (Blakeney et al., 2011), while collecting onto ice crystals while they are growing is another method to increase porosity (Figure 10.9(f)) (Simonet et al., 2007). An interesting approach to control deposition was performed with auxiliary electrodes and a point collector below a dielectric plate (Figure 10.9(g)) (Lee et al., 2012). Cells can also be added to the electrospinning media (Figure 10.9(h)) (Eddaoudi et al., 2010). A significant proportion of advances within electrospinning techniques have been achieved with new collector concepts.

10.4.3 Spinneret Configurations

The spinneret can be a metal or an insulator, and can potentially have a variety of shapes—including cylinders that rotate through a solution and electrospin off the cylinder surface. Slit and line spinnerets have also been used with success to produce multiple Taylor cones. Dual spinnerets have been used to form tubular constructs with different fiber types, including bimodal scaffolds (Kim et al., 2010). Spinnerets can also be designed to deliver multiple fluids to the Taylor cone. This approach is called coaxial electrospinning and it results in hollow electrospun fibers, and can contain multiple channels within the fiber.

10.4.4 Electric Field Variations

The electric field is what drives the polymer jet to the collector(s), and variations affect the deposition pattern of the electrospun fibers. The gap method of alignment has electric field force vectors separating close to the collector. As the fibers deposit from one collector to another, suspended oriented fibers result across the gap. The electric field is related to the collector shape; however, external manipulations are possible without changing the collector. For example, single collectors have uniform electric field vectors, whilst auxiliary electrodes have been placed between the spinneret and collector to alter the electric field and guide and orient fibers collected upon a collector (Arras et al., 2012).

10.4.5 Cell/Electrospun Scaffold Interactions

Cells often adopt significantly different morphologies on electrospun fabrics, when compared to flat surfaces. Many cells extend and orient along the electrospun fibers and form tissue constructs, which are typically low volume materials. Fibroblasts are particularly motile on electrospun fibers and stretch their cell body into "spindle-like" morphologies. However, the surface properties of electrospun fibers are important for cell adhesion and spreading. Fluid flow and bioreactor perfusion will also have a significant effect—for example, Wallin et al. (2012) performed lithography on photosensitive electrospun fibers and then placed them in microfluidic channels. Surface modification of electrospun fibers to promote adhesion can be achieved either post-processing using chemical or physical adsorption, or during processing. For example, the chemical deposition of calcium phosphate (CaP) onto the surface of a poly(ε-caprolactone) (PCL) fiber can be performed. Surface modification can also occur with thermodynamically driven surface segregation approaches (Grafahrend et al., 2008; Grafahrend et al., 2011) that are also resistant to protein adhesion. More complex is the issue of differentiation due to electrospun fibers, as some studies are performed in media that also promote differentiation. The body of literature describing cell/electrospun fiber interactions is vast, and there are many reviews available to gain an understanding on differentiation of cells upon electrospun substrates (Brown et al., 2011; Liu et al., 2012; Mammadov et al., 2013).

One short-coming of solution electrospinning is that the very small fibers produced are often not conducive for cell penetration (Blakeney et al., 2011). The lack of pore size and interconnectivity resulting from the chaotic deposition of electrospun nanofibers creates a barrier to cell invasion. Important fundamental research by Pham et al. (2006) investigated the pore size generated when random fibers of different diameters are deposited upon each other. For random fibers, a critical fiber diameter of 4 μm was necessary for pores of 20 μm and higher (Pham et al., 2006). Therefore, approaches to expand the pores of electrospun fibers have been developed (Simonet et al., 2007; Blakeney et al., 2011).

10.4.6 Melt Electrospinning

Many tissue engineers aim to incorporate cells with electrospun material for 3D tissue constructs. The recent interest with electrospun fibers generated a wealth of literature, and yet includes only a handful of melt electrospinning approaches.

Electrospinning polymers without solvents (via the melt), however, may be attractive for applications and systems where solvent accumulation or toxicity is a concern. With a lack of solvents for preparation, melt electrospinning can be performed without a ventilation system, and once produced the scaffolds can be used immediately. Therefore, melt electrospinning may allow for new TE approaches, overcoming technical restrictions governed by solvent accumulation and toxicity. For more information on melt electrospinning writing, refer to the boxed text.

10.5 ADDITIVE MANUFACTURING

Automated fabrication of scaffolds with well-defined architectures has been increasing, and will increase further in the future. Described as rapid prototyping or solid freeform fabrication (SFF) technologies, the latest American Society for Testing and Materials (ASTM) standards defined them as additive manufacturing techniques. Additive manufacturing has also been incorrectly termed as 3D printing; however, the latter term is becoming so widespread that they are becoming interchangeable. In this chapter, we will abide by current ASTM standards and use the term additive manufacturing. The goal of additive

STATE OF THE ART—MELT ELECTROSPINNING WRITING

Next to laser-based systems (mentioned earlier) considerable effort to develop new forms of "ink" based on colloids and polyelectrolyte solutions have indeed produced high-resolution and complex scaffold architectures; however, from a translational biomedical research point of view their pathway through regulatory approval for clinical applications is challenging.

One barrier for melt-based additive manufacturing techniques is to extrude smaller diameter struts and fibers below 50 µm. A straightforward approach is to use electrostatics to draw molten filaments, rather than pushing them through small diameter nozzles. There are instances when the electrified jet in electrospinning is predictable, without electrical instabilities (Fridrikh et al., 2003). The high viscosities and low conductivities of many polymer melts in particular make them a prime candidate for such electrostatic direct writing approaches. Integrating an *x–y* stage as a collector with melt electrospinning, Brown et al. (2011) demonstrated the linear deposition of fibers over a range of diameters. For each flow rate, there was a critical translation speed, where the fiber deposition changed from coiled to linear. Importantly, the coiled morphology of the fibers collected was not due to "electrical bending instabilities," but compressive forces associated with a liquid thread impinging onto a flat surface (Figure 10.10(a)). With an approach similar and as simple as "writing with honey," the PCL filament was placed onto the flat surface with control. This approach has subsequently been described as melt electrospinning writing (MEW).

The accurate writing of melt electrospun fibers is one important aspect and another is to stack the electrostatically deposited fibers upon each other. This is a key property for this approach performed in an additive manufacturing mode (Dalton et al., 2013). Indeed, this stacking property of MEW enables the production of cell invasive and highly porous scaffolds (Figure 10.10(b), (c), (e)). Adopting MEW and using much closer distances, Wei and Dong (2013) produced closely placed fibers (Figure 10.10(d)). MEW can be performed onto cylindrical collectors to fabricate tubular scaffolds (Brown et al., 2012) that were seeded with osteoblasts and placed subcutaneously in nude mice to investigate metastases of human breast cancer cells (Thibaudeau et al., 2014) (Figure 10.10(f)).

STATE OF THE ART—MELT ELECTROSPINNING WRITING *Continued*

FIGURE 10.10

State of the art: (a) With an elongated initial straight jet for melt electrospinning, the coiling of the fiber is due to buckling associated with a viscous thread; the faster speeds will result in a straight filament. (b) A poly(ε-caprolactone) melt electrospinning writing (MEW) scaffold with a porosity of 87%. (c) A MEW scaffold produced using 120° turns. (d) Densely compacted fibers made with lower voltages and shorter collection distance. (e) Fibroblasts seeded onto a scaffold as shown in (b). (f) Tubular scaffolds made by melt electrospinning writing onto a cylindrical collector and used for in vivo metastasis studies. *(a), (b), (d)–(f) reproduced from Brown et al. (2011), Farrugia et al. (2013), Wei and Dong (2013), Farrugia et al. (2013), and Thibaudeau et al. (2014) respectively. (c) is a previously unpublished image provided by Toby Brown.*

manufacturing in TE is to not only produce scaffolds with precise morphologies, but also combine medical imaging techniques to fabricate anatomically shaped (custom) implants. There are increasing numbers of additive manufacturing techniques with an extensive list shown in Figure 10.11. Some excellent reviews on additive manufacturing for TE are available in the literature (Billiet et al., 2012; Melchels et al., 2012; Seliktar et al., 2013).

Additive manufacturing for TE can be classified into three different approaches: (1) laser-based, (2) nozzle-based, and (3) printer-based (Billiet et al., 2012). They all selectively add material, layer-by-layer, as controlled by a computer program. Each layer represents the shape of the cross section of the computer-aided design (CAD) model at a specific level (Figure 10.12). Today, additive manufacturing is

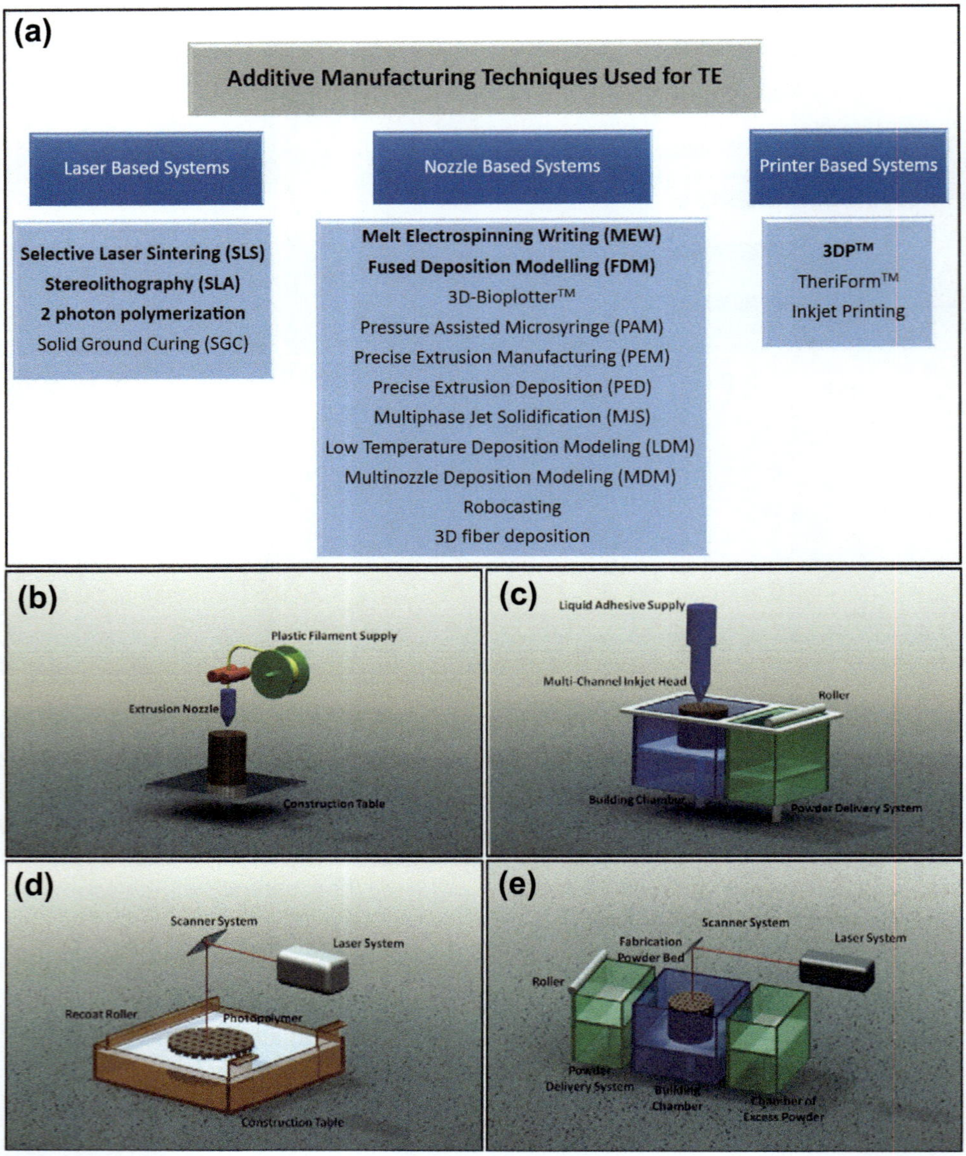

FIGURE 10.11

(a) Summary of selected additive manufacturing technologies used for fabricating scaffolds. Technologies that are leading the field and henceforth highlighted in bold are covered in this chapter. Schematics of four additive manufacturing approaches: (b) Melt extrusion, (c) 3D Printing (3DP™), (d) Stereolithography (SLA), and (e) Selective laser sintering (SLS). *(a) Modified from Billiet et al. (2012); (b)–(e) Reproduced from Melchels et al. (2012).*

viewed as a highly potent technology for the generation of TE scaffolds. One of the potential benefits offered by additive manufacturing technology is the ability to create parts with compositional variation across the entire structure due to the computer controlled nature of fabrication. Another strength of additive manufacturing is that control of scaffold structure and morphology can be discretely and systematically controlled for rational scaffold design.

10.5.1 Direct Writing and Extrusion

Melt extrusion of strands/filaments is one of the most affordable additive manufacturing approaches. A typical machine consists of a heated liquefier head attached to a carriage moving in the horizontal x–y plane. Polymer is forced through the liquefier/heating zone to a nozzle to fabricate the scaffold following a programmed path which is based on CAD model and the slice

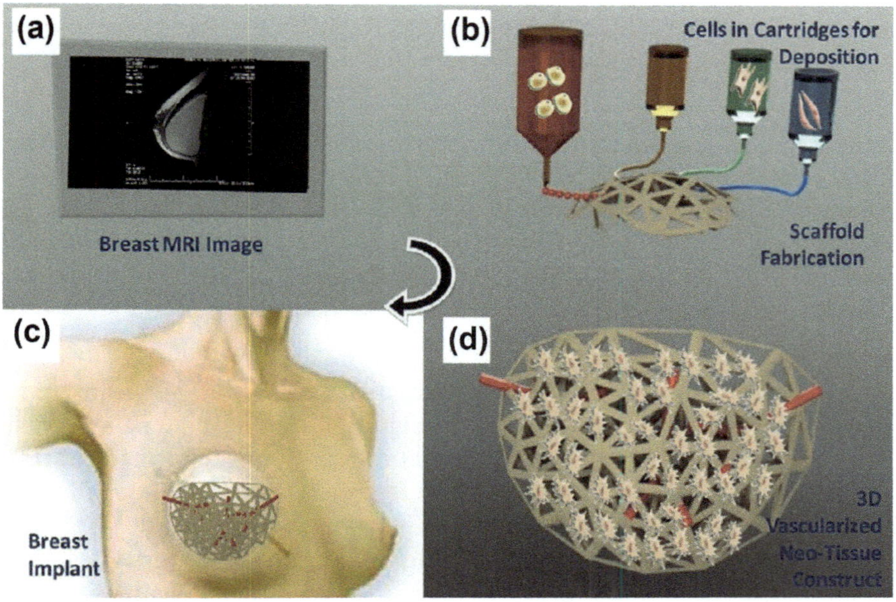

FIGURE 10.12
The combination of high-resolution clinical imaging, computer-aided design (CAD), and additive manufacturing allows one to develop custom-made and/or patient-specific scaffolds and/or tissue engineering constructs (TECs). Overall these types of scaffold-based tissue engineering strategies are inherited to personalized medicine. This figure shows the concept involved in breast tissue engineering (Melchels et al., 2012). (a) High-resolution magnetic resonance imaging of the organ (breast) allows to obtain the exact shape and volume. Powerful software packages such as Mimics and Amira allow generation of breast tissue in silico which can be manipulated into different scaffold shapes and internal architectures using CAD. (b) Using multiple printing systems that include scaffolding materials as well as different cell types allows TECs to be fabricated, which can form breast neotissue and a microvascular network in vitro in a bioreactor. The TEC is then implanted (c) and supported by good vascularization (d) in vivo. *Reproduced from Melchels et al. (2012).*

parameters. Once a layer is built, the platform moves down one step in the z-direction to deposit the next layer. Parts are fabricated layer-by-layer with the layer thickness varying in proportion to the nozzle diameter chosen. This high temperature process is restricted to the use of thermoplastic materials with good melt viscosity properties; cells or other thermosensitive biological agents cannot be encapsulated into the scaffold matrix during the fabrication process.

A design limitation by using melt extrusion systems is that the pore size for the scaffolds may not be consistent in all three dimensions—as the melt requires a surface to be deposited upon, and there are challenges with overhanging structures. The pores in both the x- and y-directions are formed in between the intersection of material strands/filaments and are determined by user-defined settings for fiber spacing. However, the pore size in the z-direction is formed from voids created by the stacking of material layers and hence their sizes are restricted to the thickness (diameter) of the deposited polymer fibers. By far the most widely used polymer for this approach is PCL (Woodruff and Hutmacher, 2010).

One technical limitation of mechanically extruding polymer melts is the large diameters of fibers produced, typically above 100 μm. The addition of a solvent to the polymer will permit smaller nozzle use, and therefore smaller filaments. However, the fidelity of the filament requires rapid gelling after passing through the nozzle. Ultra-violet (UV) cross-linking is an often used approach, and this has been adopted to cross-link pHEMA solutions in a direct writing capacity (Figure 10.13(a)–(b)) (Shepherd et al., 2011).

The requirement to rapidly form a solid after passing through a nozzle has resulted in significant research into printing shear thinning materials, polyelectrolyte complexes, and colloids. Direct writing can also be performed under the surface of a liquid (termed a coagulation reservoir), with the liquid aiding the ink gelation to attain a high structural fidelity (Gratson et al., 2004). This has been used to great effect to produce high-resolution scaffolds for TE and other applications. These colloids/polyelectrolyte complexes should have defined viscoelastic properties so that they can pass through the nozzle and form a solid rapidly. This allows the "suspension" of a discrete filament without sagging or the need to place it on support surface. Polyacrylic acid rich complexes were found to be some of the most suitable polyelectrolyte inks for direct-write assembly (Figure 10.13(c)–(d)) (Gratson et al., 2004).

There are various approaches to hydrogel printing, including robotic dispensing through nozzles, inkjet printing, and laser-induced forward transfer. The inclusion of cells within the fibers of a scaffold is an exciting concept; however, it is restricted by the limitations of the hydrogel (or "bio-ink") in which cells are both suspended and printed. This is highlighted by the "biofabrication

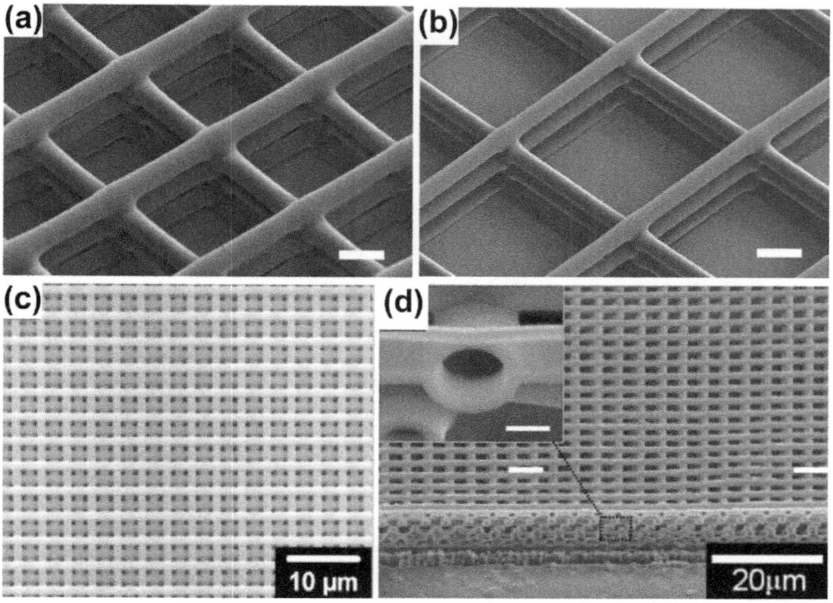

FIGURE 10.13

PHEMA scaffolds made using direct writing and photopolymerization with fiber spacing of (a) 60 µm and (b) 80 µm. Scale bar for A and B is 20 µm. A top view of a polyelectrolyte scaffold (c) made with direct writing. This morphology is commonly referred to as a "woodpile" structure. Such direct written filaments may be hollow, with the lumen exposed in (d) using a focused ion beam. *(a)–(b) Reproduced from Shepherd et al. (2011); (c)–(d) Reproduced from Gratson et al. (2006).*

window" outlined in Figure 10.14(a), whereby the scaffold fidelity is often compromised by cell requirements (and vice versa). The process of printing can damage cells due to the shear stress, while gel toxicity (including exposure to UV light) and nutrient restrictions can cause cell death. Furthermore, it is not sufficient to simply print a viable cell. It must also perform the intended function after printing, including establishing cell–cell contact, migration, and expression of certain factors which may not be possible with a nonoptimized hydrogel. Malda et al. (2013) recently reviewed such cell/hydrogel printing in depth and described the challenges of the biofabrication window as well as printing hydrogels (Figure 10.14(b)).

10.5.2 Inkjet/Powder Systems (3D Printing)

While 3D printing is today a generic term to describe additive manufacturing, this term was used to describe a specific technology developed in the 1990s at the Massachusetts Institute of Technology. It involved creating a solid object by selective inkjet printing into a binder and repeating this in sequential layers. Each layer is created by spreading a thin layer of powder over the previous

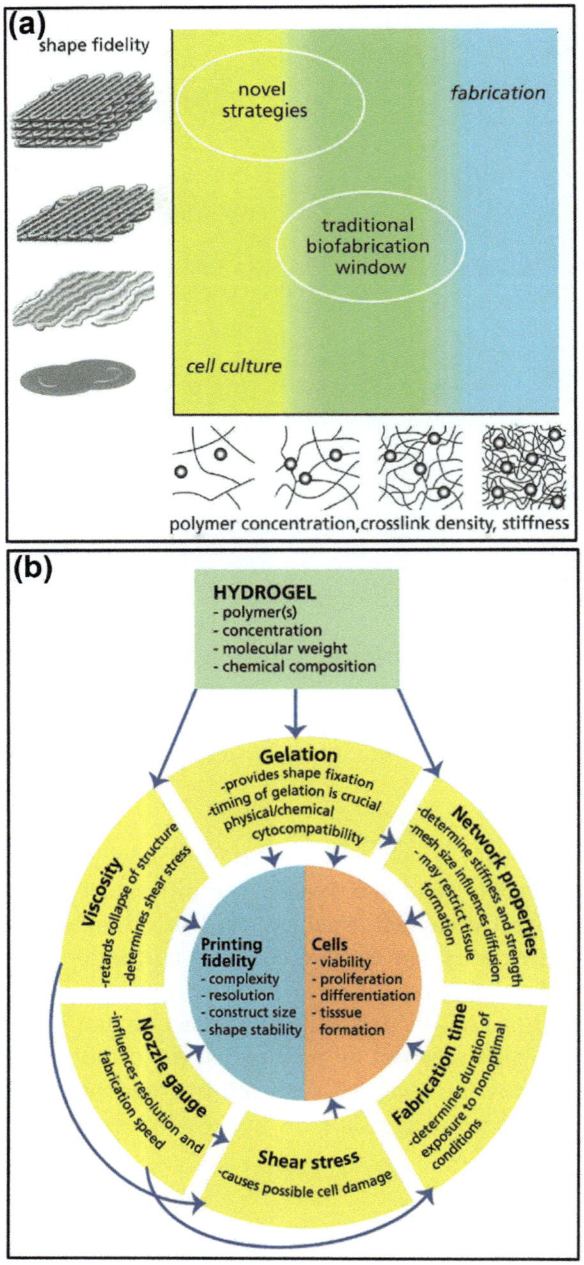

FIGURE 10.14

The "Biofabrication" window is shown in (a), showing the challenges faced with producing cell-laden hydrogels. To produce a scaffold with a certain fidelity, hydrogel properties are often altered that deleteriously affect the cell. Key to the future of hydrogel printing are new types of "ink" that allow cell function as well as structural placement (identified by the "novel strategies" window). These challenges are further described in (b), showing how the printing of cells is greatly affected by many of the parameters in the hydrogel printing system. *Images reproduced from Malda et al. (2013).*

printed layer. The powder bed is supported by a piston which descends upon the powder spreading and printing each layer (or, conversely, the inkjets and spreader are raised after printing of each layer and the bed remains stationary). The unbound powder supports temporarily unconnected portions of the component as the scaffold is built but is removed after completion of printing. After the binder has dried in the powder bed, the finished component can be retrieved and unbound powder removed for post-processing, if necessary.

The solvent drying rate is an important variable in the production of scaffolds by this method. Very rapid drying of the solvent tends to cause warping of the printed component; however, much of this can be eliminated by choosing a solvent with a low vapor pressure. It is often an advantage to combine solvents to achieve minimal warping and adequate bonding between the biomaterials particles. This process is capable of overcoming the limitations in manufacturing overhanging structures. As the layers are spread, there is always a supporting platform of powder for printing and binding to take place. However, one drawback is that the surface roughness and the aggregation of the powdered materials affect the efficiency of powder removal. The resolution of the printer is also limited by the specification of the nozzle size and the position controller that defines the print head movement. Another factor is the particle size of the powder used, which simultaneously determines the layer thickness. The versatility of using a powdered material is both an advantage and a constraint of the inkjet/powder process. Many biomaterials do not come in a powder form and need special processing conditions to get a powder which fulfills the printing requirements.

10.5.3 Stereolithography

Stereolithography (SLA) is often considered to be the pioneer of the additive manufacturing industry with the first commercial system introduced in 1988 by 3D Systems Inc. SLA is based on the use of a focused UV laser which is vector scanned over the top of a liquid bath of a photopolymerizable material. The UV laser causes the bath to polymerize where the laser beam strikes the surface of the bath, resulting in the creation of a first solid plastic layer at and just below the surface. The solid layer is then lowered into the bath and the laser-generated polymerization process is repeated for the generation of the next layer, and so on, until a plurality of superimposed layers forming the desired scaffold architecture is obtained. The most recently created layer in each case is always lowered to a position for the creation of the next layer slightly below the surface of the liquid bath. Once the scaffold is complete, the platform rises out of the vat and the excess resin is drained. The scaffold is then removed from the platform, washed of excess resin, and then placed in a UV oven for a final curing. Skoog et al. (2014) provide an excellent review on SLA, including two-photon lithography.

The photopolymer resins used for industrial applications are mixtures of simple low-molecular-weight monomers capable of chain-reacting to form solid long-chain polymers when activated by radiant energy within specific range of wavelength. The commercial materials used by SLA equipment are epoxy-based or acrylate-based resins that offer strong, durable, and accurate parts/models. However, this material cannot be used as scaffold material due to lack of biocompatibility and biodegradability. Hence, limited selection of photopolymerizable biomaterials is a major constraint for the use of the SLA technique in the design and fabrication of scaffolds for TE applications. Curable bioadhesives or injectable materials may provide suitable materials while gelatin methacrylamide has shown utility in SLA (Gauvin et al., 2012). Additionally, monomers based upon 1, 3-trimethylene carbonate have been investigated (Figure 10.2(f)) (Schuller-Ravoo et al., 2013). One advantage with this form of scaffold is the low variation (less than 5%) in pore size and interconnecting pores.

10.5.4 Two-Photon Polymerization

This method is a subset of SLA; however, the use of a multiphoton femtosecond laser produces results that are quite unique (Greiner et al., 2012; Ovsianikov et al., 2012). The resolutions are exceptional (Figure 10.15(a)–(c)), with nanoscale tolerances and the ability to produce very small pores and highly detailed structures (Hribar et al., 2014). Similar to that of SLA, a photocurable resin is required and currently only small objects can be produced and the resins are often toxic (Figure 10.15(d)–(f)). However, these limitations are well known and there are groups working intensively to overcome these challenges. Ovsianikov et al. (2012) demonstrated 3D photografting of a PEG-based matrix to produce functionalized regions within the hydrogel. Laza et al. (2012) also performed two-photon polymerization within lamellar fluid flow (Figure 10.15(g)–(i)). Without turbulence, the precise flow allows a small structure such as a spring to be produced in a continuous process. While not built as an entire scaffolding structure, fabricating elements that later can be combined into a scaffold is an exciting prospect.

10.5.5 Selective Laser Sintering

Selective Laser Sintering (SLS) also uses a focused laser beam, but to sinter areas of a loosely compacted powder. Reviewed by Duan and Wang (2011), this method begins with a thin layer of powder spread evenly onto a flat surface with a roller mechanism. The powder is then raster-scanned with a high-power laser beam. The powder material that is struck by the laser beam is fused, while the other areas of powder remain dissociated. Successive layers of powder are deposited and raster-scanned, one on top of another, until an entire part is complete. Each layer is sintered deeply enough to bond it to the preceding layer. However, SLS poses significant material constraints for scaffold fabrication and is currently mainly used to make CaP-based scaffolds for bone engineering. SLS has the disadvantage that incorporation of sensitive

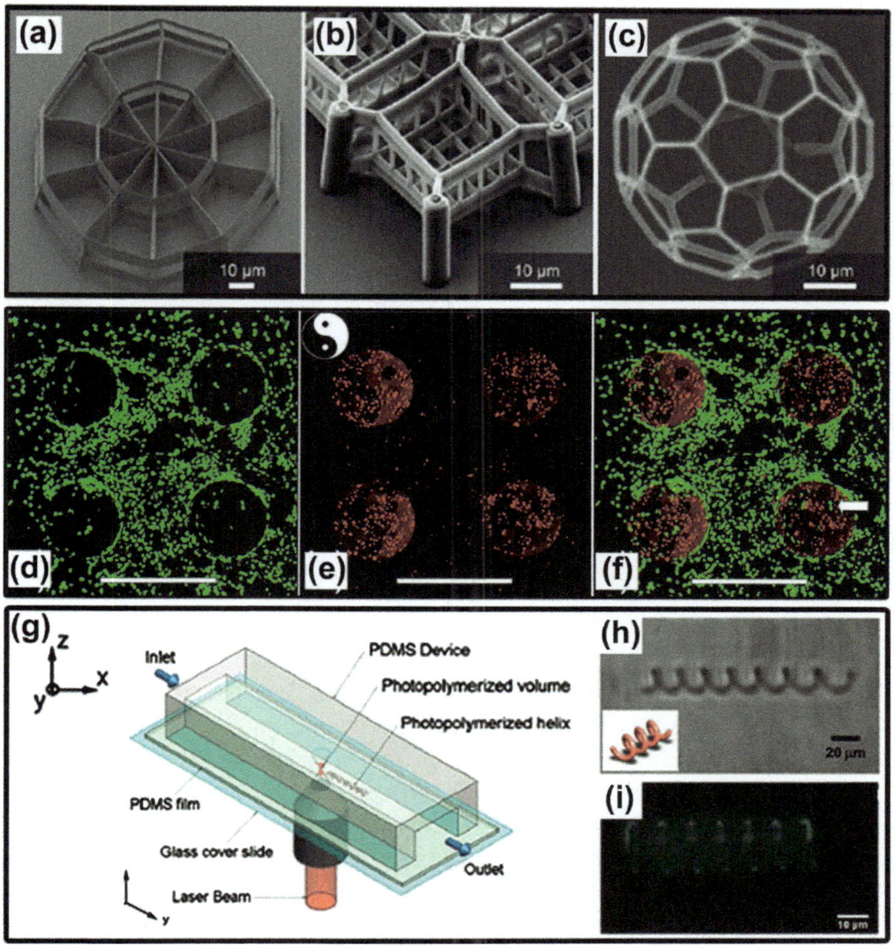

FIGURE 10.15

(a)–(c) SEM images of various structures made using two-photon polymerization show superior resolution of scaffolds fabricated. (d)-(f) Encapsulation of cells within a gelatin-based hydrogel is challenging due to toxicity. Live/dead staining of cells shows significant cell death, however some cells do survive (arrowed in (f)). Scale bar represents 500 μm. (g)–(i) is an example of continuous flow lithography, where two-photon polymerization is performed in a microfluidics channel to produce a discrete object (e.g., a spring). *(a)–(c) Reproduced from Greiner et al. (2012); (d)–(f) Reproduced from Ovsianikov et al. (2014); (g)–(i) Reproduced from Laza et al. (2012).*

biomolecules is difficult because of the need to locally heat the powder layer so as to sinter it.

10.5.6 Indirect SFF

Indirect SFF relies on a molding step after fabricating the master pattern by additive manufacturing. Indirect SFF adds further versatility and detail in scaffold design and fabrication (Figure 10.16). The previous restriction on casting was the inability of molds to produce complex geometry and internal

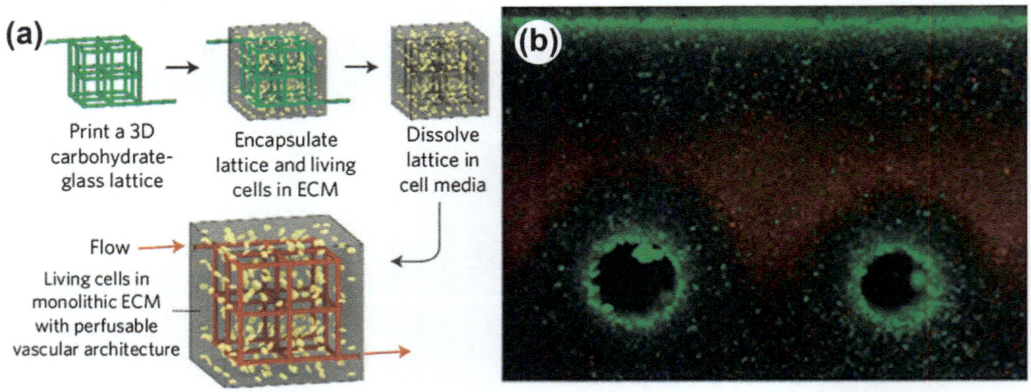

FIGURE 10.16

(a) One example for indirect SFF scaffold manufacture are these collagen matrices seeded with cells, which are receiving nutrients through channels previously produced by 3D printing a sugar lattice. (b) After leaching out of the sugar channels matrices are seeded with human umbilical vein endothelial cells shown by the intracellular enhanced green fluorescence protein reporter. *Reproduced from Miller et al. (2012).*

architecture. Now with indirect SFF, traditional casting processes with these SFF molds can meet the specific TE requirements, including mechanical integrity and customized shapes (Nguyen et al., 2011). Some highlighted advantages of indirect SFF include cost savings as the materials required for mold casting is substantially less and need not be processed into a dedicated form for any particular SFF process, such as processing into a powder for SLS and 3D printing. In addition, indirect SFF allows the usage of a wide range of materials or a combination of materials (composites or copolymers). However, some significant drawbacks still revolve around this method including the resolution of the SFF method, as the cast model would inherit the errors and defects from the mold, such as cracks and dimensional changes. Also, a mold removal method must be developed to remove the mold while preserving the as-cast scaffold intact and desired properties undisturbed.

10.6 CONCLUSION AND FUTURE DIRECTIONS

Scaffolds are of great importance for TE since they enable the fabrication of functional living implants using patient's cells following in vitro cell culture or delivery of cells harvested during surgery. In addition to considerations of scaffold performance based on a holistic TE strategy, workable considerations of manufacture arise. For wider and routine clinical applications, it must be possible to manufacture scaffolds in a reproducible, controlled fashion at an economic cost and speed, and that meet necessary regulatory standards. The fabrication process in many instances needs to accommodate the presence of biological components to allow cell seeding or incorporation of biomolecules, which may be soluble or heat labile. TECs by reason of their very complexity, particularly over small length scales, have stimulated the exploration of innovative and novel manufacturing techniques.

In general, the scaffolds require individual form and well-defined interior pore architecture with interconnected porosity to host most cell types. From a biological point of view, the designed matrix should serve at least the following small set of functions: (1) an immobilization site for transplanted cells and/or host cells, (2) formation of a protective space to prevent unwanted tissue growth into the wound bed and allow healing with differentiated tissue, (3) directing migration or growth of cells via surface properties of the scaffold, and (4) directing migration or growth of cells and ECM via release of soluble molecules such as growth factors, hormones, and/or cytokines. Future work has to provide further evidence that some of these techniques offer the right balance of capability versus practicality to be suitable for fabrication of materials in sufficient quantity and quality to allow holistic TE technology platforms to enter clinical application.

Solution electrospinning has been acknowledged for a decade as a process with potential for TE, yet there remains a barrier with solvent toxicity being a major factor hindering development toward translation from bench to bedside. One can envision that melt electrospinning will be a more straightforward process for the design and fabrication of implants and medical devices. Melt electrospinning is also aided by its ability to be used in a direct writing mode which allows to manufacture custom-made and patient-specific scaffolds.

While in the past, there was a reliance for self-organization of cells within scaffolds, there is a clear trend toward precision placement and control of the cell environment to direct and instruct the tissue formation better. A key challenge of vascularization remains throughout the entire scaffold volume, while ensuring cells are distributed in the required locations and are sufficiently nourished. Furthermore, there exist two possibilities of incorporating cells into the scaffolds: (1) seeding of cells onto the scaffold surface subsequent to scaffold fabrication and (2) the incorporation of cells into the scaffold fabrication process. This second approach is of interest especially when incorporating cells into the scaffold material; however, scaffold fabrication processes require cellular compatibility at the stage of inclusion. Another popular option, not covered in this chapter, is microtissue or the aggregate assembly of cells.

In summary, additive manufacturing is primed to have an extensive impact on scaffold-based TE design over the next decade. Certain tissues such as bone are currently well supported in a broad range of functional scaffolds; however, the next decade should see advances in soft TE as a result of higher resolution additive manufacturing. The diversity in applications for scaffold use is set to increase, as in vitro models for injury and disease are refined and increasingly used. Key to the success of additive manufacturing in TE is the effective translation from concept to medical product, or positive in vitro model validation.

RECOMMENDED READING

Melchels, et al., 2012. Review on Additive Manufacturing.

Billiet, et al., 2012. Review on Additive Manufacturing.

Zheng, et al., 2012. Engineering Microvessels.

Melchels, et al., 2010. Scaffolds Built with Mathematically Derived Sub-units.

Malda, et al., 2013. Hydrogel Printing Review.

Dalton, et al., 2013. Review on Electrospinning and Additive Manufacturing Overlap.

Brown, et al., 2011. Melt Electrospinning Writing.

Skoog, et al., 2014. Review on SLA & 2-photon Polymerization.

Hollister, 2009a. Taking TE to the Clinic.

REFERENCES

Akbari, M., Tamayol, A., Veronique, L., Annabi, N., Najafabadi, A.H., Khademhosseini, A., Juncker, D., 2014. Composite living fibers for creating tissue constructs using textile techniques. Adv. Funct. Mater.1–8.

Arras, M.M.L., Grasl, C., Bergmeister, H., Schima, H., 2012. Electrospinning of aligned fibers with adjustable orientation using auxiliary electrodes. Sci. Technol. Adv. Mater. 13 (3).

Billiet, T., Vandenhaute, M., Schelfhout, J., Van Vlierberghe, S., Dubruel, P., 2012. A review of trends and limitations in hydrogel-rapid prototyping for tissue engineering. Biomaterials 33 (26), 6020–6041.

Blakeney, B.A., Tambralli, A., Anderson, J.M., Andukuri, A., Lim, D.J., Dean, D.R., Jun, H.W., 2011. Cell infiltration and growth in a low density, uncompressed three-dimensional electrospun nanofibrous scaffold. Biomaterials 32 (6), 1583–1590.

Bozkurt, A., Lassner, F., O'Dey, D., Deumens, R., Bocker, A., Schwendt, T., Janzen, C., Suschek, C.V., Tolba, R., Kobayashi, E., Sellhaus, B., Tholl, S., Eummelen, L., Schugner, F., Damink, L.O., Weis, J., Brook, G.A., Pallua, N., 2012. The role of microstructured and interconnected pore channels in a collagen-based nerve guide on axonal regeneration in peripheral nerves. Biomaterials 33 (5), 1363–1375.

Brown, T.D., Dalton, P.D., Hutmacher, D.W., 2011. Direct writing by way of melt electrospinning. Adv. Mater. 23 (47), 5651.

Brown, T.D., Slotosch, A., Thibaudeau, L., Taubenberger, A., Loessner, D., Vaquette, C., Dalton, P.D., Hutmacher, D.W., 2012. Design and fabrication of tubular scaffolds via direct writing in a melt electrospinning mode. Biointerphases 7 (1–4), 13.

Bryers, J.D., Giachelli, C.M., Ratner, B.D., 2012. Engineering biomaterials to integrate and heal: the biocompatibility paradigm shifts. Biotechnol. Bioeng. 109 (8), 1898–1911.

Chen, G.P., Sato, T., Ohgushi, H., Ushida, T., Tateishi, T., Tanaka, J., 2005. Culturing of skin fibroblasts in a thin PLGA-collagen hybrid mesh. Biomaterials 26 (15), 2559–2566.

Dalton, P.D., Vaquette, C., Farrugia, B.L., Dargaville, T.R., Brown, T.D., Hutmacher, D.W., 2013. Electrospinning and additive manufacturing: converging technologies. Biomater. Sci. 1 (2), 171–185.

Dehghani, F., Annabi, N., 2011. Engineering porous scaffolds using gas-based techniques. Curr. Opin. Biotechnol. 22 (5), 661–666.

Duan, B., Wang, M., 2011. Selective laser sintering and its application in biomedical engineering. MRS Bull. 36 (12), 998–1005.

Eddaoudi, A., Townsend-Nicholson, A., Timms, J.F., Schorge, S., Jayasinghe, S.N., 2010. Molecular characterisation of post-bio-electrosprayed human brain astrocytoma cells. Analyst 135 (10), 2600–2612.

Elbert, D.L., 2011. Liquid-liquid two-phase systems for the production of porous hydrogels and hydrogel microspheres for biomedical applications: a tutorial review. Acta Biomater. 7 (1), 31–56.

Farrugia, B.L., Brown, T.D., Upton, Z., Hutmacher, D.W., Dalton, P.D., Dargaville, T.R., 2013. Dermal fibroblast infiltration of poly(epsilon-caprolactone) scaffolds fabricated by melt electrospinning in a direct writing mode. Biofabrication 5 (2), 025001.

Fridrikh, S.V., Yu, J.H., Brenner, M.P., Rutledge, G.C., 2003. Controlling the fiber diameter during electrospinning. Phys. Rev. Lett. 90 (14).

Fukano, Y., Tavakkol, Z., Isenhath, S.N., Usui, M.L., Underwood, R.A., Irvin, C.A., Marshall, A.J., Hauch, K.D., Ratner, B.D., Fleckman, P., Olerud, J.E., 2006. Epidermal and dermal integration into implanted percutaneous biomaterials in mice. J. Invest. Dermatol. 126, 31.

Gauvin, R., Chen, Y.C., Lee, J.W., Soman, P., Zorlutuna, P., Nichol, J.W., Bae, H., Chen, S.C., Khademhosseini, A., 2012. Microfabrication of complex porous tissue engineering scaffolds using 3D projection stereolithography. Biomaterials 33 (15), 3824–3834.

Grafahrend, D., Calvet, J.L., Klinkhammer, K., Salber, J., Dalton, P.D., Moller, M., Klee, D., 2008. Control of protein adsorption on functionalized electrospun fibers. Biotechnol. Bioeng. 101 (3), 609–621.

Grafahrend, D., Heffels, K.H., Beer, M.V., Gasteier, P., Moller, M., Boehm, G., Dalton, P.D., Groll, J., 2011. Degradable polyester scaffolds with controlled surface chemistry combining minimal protein adsorption with specific bioactivation. Nat. Mater. 10 (1), 67–73.

Gratson, G.M., Garcia-Santamaria, F., Lousse, V., Xu, M.J., Fan, S.H., Lewis, J.A., Braun, P.V., 2006. Direct-write assembly of three-dimensional photonic crystals: conversion of polymer scaffolds to silicon hollow-woodpile structures. Adv. Mater. 18 (4), 461.

Gratson, G.M., Xu, M.J., Lewis, J.A., 2004. Microperiodic structures – direct writing of three-dimensional webs. Nature 428 (6981), 386.

Greiner, A.M., Richter, B., Bastmeyer, M., 2012. Micro-engineered 3D scaffolds for cell culture studies. Macromol. Biosci. 12 (10), 1301–1314.

Henkel, J., Hutmacher, D.W., 2013. Design and fabrication of scaffold-based tissue engineering. BioNanoMaterials 14 (3–4), 171–193.

Ho, S.T., Hutmacher, D.W., 2006. A comparison of micro CT with other techniques used in the characterization of scaffolds. Biomaterials 27 (8), 1362–1376.

Hollister, S.J., 2005. Porous scaffold design for tissue engineering. Nat. Mater. 4 (7), 518–524.

Hollister, S.J., 2009a. Scaffold design and manufacturing: from concept to clinic. Adv. Mater. 21 (32–33), 3330–3342.

Hollister, S.J., 2009b. Scaffold engineering: a bridge to where? Biofabrication 1 (1).

Hollister, S.J., Murphy, W.L., 2011. Scaffold translation: barriers between concept and clinic. Tissue Eng. Part B Rev. 17 (6), 459–474.

Hribar, K.C., Soman, P., Warner, J., Chung, P., Chen, S.C., 2014. Light-assisted direct-write of 3D functional biomaterials. Lab Chip 14 (2), 268–275.

Hutmacher, 2000. Biomaterials 21, 2529–2543.

Hutmacher, D.W., Sittinger, M., Risbud, M.V., 2004. Scaffold-based tissue engineering: rationale for computer-aided design and solid free-form fabrication systems. Trends Biotechnol. 22 (7), 354–362.

Jana, S., Cooper, A., Zhang, M., 2013. Chitosan scaffolds with unidirectional microtubular pores for large skeletal myotube generation. Adv. Healthcare Mater. 2, 557–561.

Jiraskova, N., Rozsival, P., Burova, M., Kalfertova, M., 2011. AlphaCor artificial cornea: clinical outcome. Eye 25 (9), 1138–1146.

Kim, G., Son, J., Park, S., Kim, W., 2008. Hybrid process for fabricating 3D hierarchical scaffolds combining rapid prototyping and electrospinning. Macromol. Rapid Commun. 29 (19), 1577–1581.

Kim, K., Dean, D., Wallace, J., Breithaupt, R., Mikos, A.G., Fisher, J.P., 2011. The influence of stereolithographic scaffold architecture and composition on osteogenic signal expression with rat bone marrow stromal cells. Biomaterials 32 (15), 3750–3763.

Kim, S.J., Jang, D.H., Park, W.H., Min, B.M., 2010. Fabrication and characterization of 3-dimensional PLGA nanofiber/microfiber composite scaffolds. Polymer 51 (6), 1320–1327.

Koroleva, A., Gill, A.A., Ortega, I., Haycock, J.W., Schlie, S., Gittard, S.D., Chichkov, B.N., Claeyssens, F., 2012. Two-photon polymerization-generated and micromolding-replicated 3D scaffolds for peripheral neural tissue engineering applications. Biofabrication 4 (2).

Laib, A., Kumer, J., Majumdar, S., Lane, N., 2001. The temporal changes of trabecular architecture in ovariectomized rats assessed by MicroCT. Osteoporosis Int. 12 (11), 936–941.

Laza, S.C., Polo, M., Neves, A.A.R., Cingolani, R., Camposeo, A., Pisignano, D., 2012. Two-photon continuous flow lithography. Adv. Mater. 24 (10), 1304–1308.

Lee, J., Lee, S.Y., Jang, J., Jeong, Y.H., Cho, D.W., 2012. Fabrication of patterned nanofibrous; mats using direct-write electrospinning. Langmuir 28 (18), 7267–7275.

Lewis, J.S., Roy, K., Keselowsky, B.G., 2014. Materials that harness and modulate the immune system. MRS Bull. 39 (1), 25–34.

Li, H.Y., Fuhrmann, T., Zhou, Y., Dalton, P.D., 2013. Host reaction to poly(2-hydroxyethyl methacrylate) scaffolds in a small spinal cord injury model. J. Mater. Sci. Mater. Med. 24 (8), 2001–2011.

Liu, T., Houle, J.D., Xu, J.Y., Chan, B.P., Chew, S.Y., 2012. Nanofibrous collagen nerve conduits for spinal cord repair. Tissue Eng. Part A 18 (9–10), 1057–1066.

Loh, Q.L., Choong, C., 2013. Three-dimensional scaffolds for tissue engineering applications: role of porosity and pore size. Tissue Eng. Part B Rev. 19 (6), 485–502.

Malda, J., Visser, J., Melchels, F.P., Jungst, T., Hennink, W.E., Dhert, W.J.A., Groll, J., Hutmacher, D.W., 2013. 25th Anniversary Article: engineering hydrogels for biofabrication. Adv. Mater. 25 (36), 5011–5028.

Mammadov, B., Sever, M., Guler, M.O., Tekinay, A.B., 2013. Neural differentiation on synthetic scaffold materials. Biomater. Sci. 1 (11), 1119–1137.

Melchels, F.P.W., Bertoldi, K., Gabbrielli, R., Velders, A.H., Feijen, J., Grijpma, D.W., 2010. Mathematically defined tissue engineering scaffold architectures prepared by stereolithography. Biomaterials 31 (27), 6909–6916.

Melchels, F.P.W., Domingos, M.A.N., Klein, T.J., Malda, J., Bartolo, P.J., Hutmacher, D.W., 2012. Additive manufacturing of tissues and organs. Prog. Polym. Sci. 37 (8), 1079–1104.

Miller, J.S., Stevens, K.R., Yang, M.T., Baker, B.M., Nguyen, D.H.T., Cohen, D.M., Toro, E., Chen, A.A., Galie, P.A., Yu, X., Chaturvedi, R., Bhatia, S.N., Chen, C.S., 2012. Rapid casting of patterned vascular networks for perfusable engineered three-dimensional tissues. Nat. Mater. 11 (9), 768–774.

Miot, S., Woodfield, T.B.F., Daniels, A.U., Suetterlin, R., Peterschmitt, I., Heberer, M., van Blitterswijk, C.A., Riesle, J., Martin, I., 2005. Effects of scaffold composition and architecture on human nasal chondrocyte redifferentiation and cartilaginous matrix deposition. Biomaterials 26 (15), 2479–2489.

Muller, R., Ruegsegger, P., 1997. Micro-tomographic imaging for the nondestructive evaluation of trabecular bone architecture. Stud. Health Technol. Inf. 40, 61–79.

Muschler, G.F., Nakamoto, C., Griffith, L.G., 2004. Engineering principles of clinical cell-based tissue engineering. J. Bone Joint Surg Am. 86-A (7), 1541–1558.

Nair, L.S., Laurencin, C.T., 2007. Biodegradable polymers as biomaterials. Prog. Polym. Sci. 32 (8–9), 762–798.

Nguyen, T.L., Staiger, M.P., Dias, G.J., Woodfield, T.B.F., 2011. A novel manufacturing route for fabrication of topologically-ordered porous magnesium scaffolds. Adv. Eng. Mater. 13 (9), 872–881.

Odgaard, A., 1997. Three-dimensional methods for quantification of cancellous bone architecture. Bone 20 (4), 315–328.

Ovsianikov, A., Mironov, V., Stampfl, J., Liska, R., 2012. Engineering 3D cell-culture matrices: multiphoton processing technologies for biological and tissue engineering applications. Expert Rev. Med. Devices 9 (6), 613–633.

Ovsianikov, A., Mühleder, S., Torgersen, J., Li, Z., Qin, X-H., Van Vlierberghe, S., Dubruel, P., Holnthoner, W., Redl, H., Liska, R., Stampfl, J., 2014. Laser photofabrication of cell-containing hydrogel constructs. Langmuir 30 (13), 3787–3794.

Park, S.H., Kim, T.G., Kim, H.C., Yang, D.Y., Park, T.G., 2008. Development of dual scale scaffolds via direct polymer melt deposition and electrospinning for applications in tissue regeneration. Acta Biomater. 4 (5), 1198–1207.

Pham, Q.P., Sharma, U., Mikos, A.G., 2006. Electrospun poly(epsilon-caprolactone) microfiber and multilayer nanofiber/microfiber scaffolds: characterization of scaffolds and measurement of cellular infiltration. Biomacromolecules 7 (10), 2796–2805.

Reneker, D.H., Yarin, A.L., 2008. Electrospinning jets and polymer nanofibers. Polymer 49 (10), 2387–2425.

Ruegsegger, P., Koller, B., Muller, R., 1996. A microtomographic system for the nondestructive evaluation of bone architecture. Calcif. Tissue Int. 58 (1), 24–29.

Schuller-Ravoo, S., Teixeira, S.M., Feijen, J., Grijpma, D.W., Poot, A.A., 2013. Flexible and elastic scaffolds for cartilage tissue engineering prepared by stereolithography using poly(trimethylene carbonate)-based resins. Macromol. Biosci. 13 (12), 1711–1719.

Schulte, V.A., Alves, D.F., Dalton, P.P., Moeller, M., Lensen, M.C., Mela, P., 2013. Microengineered PEG hydrogels: 3D scaffolds for guided cell growth. Macromol. Biosci. 13 (5), 562–572.

Seidi, A., Ramalingam, M., Elloumi-Hannachi, I., Ostrovidov, S., Khademhosseini, A., 2011. Gradient biomaterials for soft-to-hard interface tissue engineering. Acta Biomater. 7 (4), 1441–1451.

Seliktar, D., Dikovsky, D., Napadensky, E., 2013. Bioprinting and tissue engineering: recent advances and future perspectives. Israel J. Chem. 53 (9–10), 795–804.

Shepherd, J.N.H., Parker, S.T., Shepherd, R.F., Gillette, M.U., Lewis, J.A., Nuzzo, R.G., 2011. 3D microperiodic hydrogel scaffolds for robust neuronal cultures. Adv. Funct. Mater. 21 (1), 47–54.

Simonet, M., Schneider, O.D., Neuenschwander, P., Stark, W.J., 2007. Ultraporous 3D polymer meshes by low-temperature electrospinning: use of ice crystals as a removable void template. Polym. Eng. Sci. 47 (12), 2020–2026.

Skoog, S.A., Goering, P.L., Narayan, R.J., 2014. Stereolithography in tissue engineering. J. Mater. Sci. Mater. Med. 25 (3), 845–856.

Spector, M., 2014. The tissue engineer's toolbox manifesto. Biomed. Mater. 9 (1), 010201.

Tamayol, A., Akbari, M., Annabi, N., Paul, A., Khademhosseini, A., Juncker, D., 2013. Fiber-based tissue engineering: progress, challenges, and opportunities. Biotechnol. Adv. 31 (5), 669–687.

Thibaudeau, L., Taubenberger, A., Holzapfel, B.M., Quent, V.M., Fuehrmann, T., Hesami, P., Brown, T.D., Dalton, P.D., Power, C.A., Hollier, B., Hutmacher, D.W., 2014. A tissue engineered humanized xenograft model of human breast cancer metastasis to bone. Dis. Models Mech. 7 (2), 299–309.

Wallin, P., Zandén, C., Carlberg, B., Hellström Erkenstam, N., Liu, J., Gold, J., 2012. A method to integrate patterned electrospun fibers with microfluidic systems to generate complex microenvironments for cell culture applications. Biomicrofluidics 6, 024131.

Wei, C., Dong, J.Y., 2013. Direct fabrication of high-resolution three-dimensional polymeric scaffolds using electrohydrodynamic hot jet plotting. J. Micromech. Microeng. 23 (2).

Wong, K.H.K., Chan, J.M., Kamm, R.D., Tien, J., 2012. Microfluidic models of vascular functions. Annu. Rev. Biomed. Eng. 14, 205–230.

Woodfield, T.B.F., Malda, J., de Wijn, J., Peters, F., Riesle, J., van Blitterswijk, C.A., 2004. Design of porous scaffolds for cartilage tissue engineering using a three-dimensional fiber-deposition technique. Biomaterials 25 (18), 4149–4161.

Woodruff, M.A., Hutmacher, D.W., 2010. The return of a forgotten polymer-polycaprolactone in the 21st century. Prog. Polym. Sci. 35 (10), 1217–1256.

Zhang, L., Cao, Z.Q., Bai, T., Carr, L., Ella-Menye, J.R., Irvin, C., Ratner, B.D., Jiang, S.Y., 2013. Zwitterionic hydrogels implanted in mice resist the foreign-body reaction. Nat. Biotechnol. 31 (6), 553.

Zheng, Y., Chen, J.M., Craven, M., Choi, N.W., Totorica, S., Diaz-Santana, A., Kermani, P., Hempstead, B., Fischbach-Teschl, C., Lopez, J.A., Stroock, A.D., 2012. In vitro microvessels for the study of angiogenesis and thrombosis. Proc. Natl. Acad. Sci. USA 109 (24), 9342–9347.

CHAPTER 11

Controlled Release Strategies in Tissue Engineering

Jeffrey J. Rice[1], Mikaël M. Martino[2], Evan A. Scott[3], Jeffrey A. Hubbell[4]

[1]Department of Chemical Engineering, Tennessee Technological University, Cookeville, Tennessee, USA; [2]World Premier International Immunology Frontier Research Center, Osaka University, Suita, Osaka 565-0871, Japan; [3]Department of Biomedical Engineering, Northwestern University, Evanston, Illinois, USA; [4]Institute of Bioengineering, Ecole Polytechnique Fédérale de Lausanne, CH-1015 Lausanne, Switzerland

LEARNING OBJECTIVES

- To understand the methods employed in controlled morphogen release
- To gain an understanding of the clinical need for controlled release strategies of morphogens in tissue engineering
- To understand the influence of the physicochemical properties of the incorporated bioactive agent on its controlled release
- To be able to explain the classes of morphogens used in tissue engineering, and the implications of their mode of action on appropriate controlled release strategies
- To be able to relate physicochemical and molecular characteristics of materials with their resorption behavior and controlled release characteristics
- To appreciate examples of current success and future challenges of controlled morphogen release in tissue engineering therapies
- To be able to list out important morphogens used in tissue engineering and relate their biological function to therapeutic targets

11.1 INTRODUCTION

Tissue engineering strategies often depend heavily on provision of bioactive signals, be they bound or diffusible growth factor signals, or substrate-bound cell adhesion signals. As such, controlled release of the signaling molecules play an integral role in the overall task of tissue engineering. In tissue engineering, one often uses biomaterial scaffolds or matrices; here the word *scaffold* refers to a macroporous or microporous material used to provide a structural support

Tissue Engineering.
Copyright © 2015 Elsevier Inc. All rights reserved.

(e.g., a polymer sponge) and the word *matrix* refers to a nanoporous continuous material for the same purpose (e.g., a hydrogel). Cells can be transplanted within such scaffolds and matrices, for example, cells have been cultured and placed there in vitro. In many cases, one must coax the cells to behave in desired ways by providing them with bioactive signals in a controlled manner. It is sometimes even hoped that, by providing bioactive signals within a matrix, we may succeed in coaxing cells, including progenitor and stem cells, to migrate into the matrix from the surrounding tissues, to obviate the need for cell transplantation, and the logistical complexities that are associated with these. Thus, controlled release strategies are needed to provide signals to cells in the vicinity of the scaffolds and matrices and within the constructs to behave in desired ways.

In vitro, when cells are cultured within a scaffold or matrix, controlled release may not be necessary: the bioactive signals may simply be added to the medium in the bioreactor. Controlled release may be beneficial, even if not necessary, to reduce the amounts of very expensive signals, such as growth factors, by providing them only locally within the scaffold or matrix. In vivo, however, it makes little sense to administer such factors to the whole system. Systemic delivery could mean that very little and possibly even no growth factor would actually find its way to the implant, being cleared by other pathways in the body before it reaches that site. Thus, in vitro, controlled release may be beneficial, and in vivo it is usually essential.

The term *controlled release* is often taken to refer to release of a bioactive molecule from the bound state to a freely diffusible state, whereupon it can exert its biological activity. In tissue engineering, however, many of the bioactive molecules of interest are active in their immobilized state, and some, such as growth factors, can be active either immobilized or freely diffusible. As such, in this chapter, we shall use the term controlled release to refer to the provision of a bioactive molecule over time in a manner such that its biological activity can be productively harnessed. Thus, controlled release more precisely refers to the controlled provision of a biologically active molecule in its most appropriate state. For a molecule promoting cell adhesion, controlled release would refer to immobilization of that ligand and presentation on a surface for a prolonged duration. For a growth factor that signals intracellularly, it would refer to release of the factor into the diffusible state, so that it could be bound by the corresponding receptor and internalized.

The overall goal of the controlled release of signaling molecules is to present a biologically active factor to the tissues at a concentration that is optimal—not so low as to be ineffective, and not so high as to display toxic side effects; this concentration range is known as the *therapeutic window*, as illustrated in Figure 11.1. If simply injected into the tissue site as a bolus, without incorporation into a biomaterial scaffold or matrix, release will usually occur as a sudden burst, perhaps well above the upper limit of the therapeutic window, and then

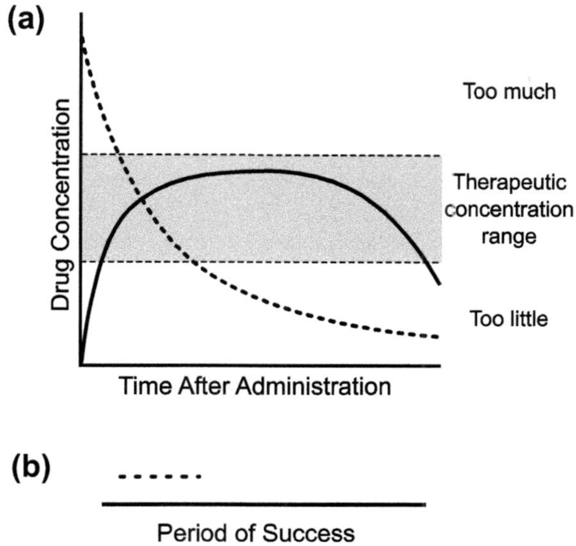

FIGURE 11.1

The goal of controlled release is to provide effective concentrations of a drug to a tissue space for an effective period of time. Most drugs are active in a so-called therapeutic window (gray box), above which concentration the drug demonstrates toxic side effects and below which concentration the drug is ineffective. In the absence of controlled release (dashed line in (a)), the drug is injected at too high of a concentration, rapidly disperses in the tissue, passes through the therapeutic window, only within the effective concentration range for a brief period of time (dashed line in (b)), and then rapidly declines to too low a concentration to be effective. With controlled release (bold line in (a)), the drug rises in tissue concentration after implantation up to the therapeutic window, and is maintained within an effective range for a desired duration, and eventually is depleted from the implant This places the drug within the therapeutic window for a much longer period of time (bold line in (b)).

rapidly decline to concentrations that are too low to be effective. If released slowly from a scaffold, a typical case in tissue engineering, the concentration of the drug in the vicinity of the implant would rise to within the therapeutic window, remain there for a prolonged duration while the desired signaling occurs, and then gradually decline to zero as the factor is depleted from the scaffold.

A number of strategies have been explored and can be envisioned for controlled release of biological signals in tissue engineering (Table 11.1). These include (1) entrapping factors within gel networks. (2) entrapping factors within hydrophobic degradable polymers that either make up the matrix, or that are used to form microparticles that are themselves entrapped within the matrix, (3) binding factors within a matrix that has an affinity for the factors, (4) binding factors to affinity sites naturally present in the matrix or to affinity sites conjugated to the matrix, and (5) binding the factors directly to the matrix material.

Table 11.1 Drug Delivery Strategies Commonly Encountered in Tissue Engineering

Case	Binding or Affinity	Example
Physical mixtures	None or physicochemical	Growth factors in collagen
Gel networks	Physical entrapment	Growth factors in synthetic polymer hydrogels
Polymer fibers and particles	Physical entrapment	Growth factors or small molecules in microparticles
Specific affinity mixtures	Biochemical	Growth factors in heparin-containing matrices
Chemical immobilization	Covalent	Adhesion factors on scaffolds

11.1.1 Bioactive Molecules of Interest in Tissue Engineering

Before considering the approaches to controlled release that are likely to be successful in tissue engineering, it is important to consider what types of molecules might be important to release. For example, release of low-molecular-weight hydrophobic drugs, as might be used in contraception, inflammation reduction, or cancer, requires vastly different material types than might be used in the release of proteins drugs, as might be used in diabetes or growth hormone deficiency. In tissue engineering, it is these latter biomolecules that are of more interest, such as bioactive proteins. These proteins include growth factor morphogens, such as vascular endothelial growth factors (VEGFs), bone morphogenetic proteins (BMPs), transforming growth factor (TGF)-β, and interleukins; adhesion factor morphogens, such as whole and domains of fibronectin, vitronectin, and laminin; and peptide mimics of these proteins. These molecules are large (molecular weight 5000–2,000,000 Da), are water soluble, and are very sensitive to their environment. Even small changes in their environment, such as exposure to hydrophobic surfaces, can result in protein denaturation and loss of biological activity. Perhaps even more importantly, the fraction of the protein that becomes denatured may expose sites that are immunogenic and lead to the formation of antibodies against the protein that is being therapeutically delivered. The unintended generation of antibodies may not only neutralize the favorable biological interactions between the protein and its target but could also generate a highly undesirable immunological response. As such, one must pay a great deal of attention to the details of the released molecule, to ensure that it is not only released but that it is also released in a native form and with the desired kinetics.

Growth factors are by far the drugs of the highest current interest in tissue engineering. These molecules are ligands for cell-surface receptors that turn on complex cascades of signal transduction within the cell. Many, but not all, of the growth factors are dimers and bind to receptor tyrosine kinases. Typically, each monomer of a dimerized growth factor binds to the extracellular domain of a receptor, inducing dimerization of the receptor, which leads to phosphorylation of the intracellular domains of the receptor. The resulting phosphorylation of the cytoplasmic region of the receptors leads to altered binding of other intracellular proteins, which then triggers the signal transduction cascade.

The use of tissue-engineered constructs for immunological purposes is a burgeoning area of research in the biomedical field. The proteins that play a crucial role for the signaling of immune cells are cytokines and chemokines. The signaling occurs in much the same way as growth factors, but cytokines, in general, stimulate immune cell activation, differentiation, and proliferation, while chemokines direct cell migration. The methods used to control the delivery of growth factors can also be applied for cytokines and chemokines.

Examples of growth factors that are important in tissue engineering are BMPs, TGF-βs, the platelet-derived growth factors (PDGFs), the fibroblast growth factors (FGFs), the neurotrophins, VEGFs, and the insulin-like growth factors (IGFs)—a nonexhaustive list. The plural is used in the list for the names, because the names represent families of proteins, for example, the angiogenic growth factors of the VEGFs consist of VEGF-A, VEGF-B, VEGF-C, and VEGF-D, and moreover, several isoforms of these growth factors exists, such as VEGF-A_{110}, VEGF-A_{121}, VEGF-A_{165}, and VEGF-A_{189}. The different family members are related by partial sequence homology and sometimes by overlap in receptor specificity, but may have vastly different function. For example, VEGF-A primarily induces hemangiogenesis (i.e., the formation of new blood vessels from existing ones), whereas VEGF-C primarily induces lymphangiogenesis (i.e., the formation of new lymphatic vessels). The different isoforms are typically different splice variants of the same parent sequence and may also have a different signaling capacity. For example, VEGF-A_{165} binds to both VEGF receptors 1 and 2 (among other receptors) as well as to several extracellular matrix (ECM) proteins and cell-surface heparan sulfate, whereas VEGF-A_{121} binds to VEGF receptors 1 and 2, but not to ECM proteins and cell-surface heparan sulfate, which leads to very different biological activities between the two isoforms.

Some of the growth factors mentioned earlier have already been turned into commercial tissue engineering products, for example, PDGF-BB into a product for healing chronic wounds in the skin (Regranex®, from Johnson & Johnson) and bone morphogenetic protein-2 (BMP-2) for healing surgical and traumatic defects in bone (e.g., INFUSE®, from Medtronic Sofamor Danek). In both cases, prolonged delivery of the growth factor is key to efficacy. In the case of Regranex®, frequent readministration is needed for long-term delivery; in the case of INFUSE®, this is partially achieved by affinity between the growth factor and a biomaterial matrix, as will be discussed further in the following text. However, the clinical translation of these growth factors has been disappointing (Lee et al., 2011; Rice et al., 2013), and it additionally faces issues related to safety and cost effectiveness (Woo, 2012; Lo et al., 2012; Solchaga et al., 2012). For example, PDGF-BB has not been widely adopted as a main therapy, and the United States Food and Drug Administration (USFDA) recently issued a warning regarding its use related to cancer risk (USFDA, 2008). These problems derive most likely from the fact that the growth factors are used at vastly supraphysiological levels, which may relate to their safety issues. Therefore, design of controlled release strategies for dose reduction and localization is an ongoing challenge.

Adhesion factors are also important candidates for drug delivery in tissue engineering. These molecules are normally components of the ECM and serve a role of mechanical signaling between the cell's cytoskeleton and the ECM. For more detailed information about ECM proteins, see Chapter 5 within the text. The extracellular domains of transmembrane receptors, typically integrins, bind to the adhesion factors, and the intracellular domains interact, via several other intermediary proteins, with the cytoskeleton. Since mechanical signaling is the ultimate function of receptor ligation, clearly the adhesion factor must be immobilized in the extracellular milieu. As such, these factors function as agonists (i.e., inducing a function in its desired form) in the immobilized state, and as competitive antagonists in their diffusible state. Thus, in the case of these molecules, controlled release means controlled immobilized presentation.

Proteins that contain an array of adhesion factors include fibronectin, vitronectin, fibrinogen, osteopontin, the laminins, and thrombospondin—also a very nonexhaustive list. In addition to the whole proteins, specific integrin-binding domains have been identified within these proteins, and these smaller domains can be recombinantly produced and employed within a controlled release matrix or scaffold. This has been extensively done with fibronectin, which contains multiple independently folding protein domains, many of which bind to integrins (Petrie et al., 2006; Martino et al., 2009, 2011). Moreover, still smaller receptor-binding peptides sequences within those protein domains have been identified and successfully used as adhesion factors. Among these, the most widely studied in tissue engineering is the tripeptide sequence Arg-Gly-Asp, or RGD in the one-letter code, from fibronectin and a number of other proteins, which bind to the integrin family of adhesion receptors (Hersel et al., 2003). Although the very small RGD sequence binds to a number of integrins without a great deal of specificity between integrins, extending the sequence with flanking amino acids or conformationally constraining it by cyclizing the sequence can both increase its affinity for the receptor and recreate some of the specificity between different integrin receptors that exist in the proteins themselves.

From the perspective of current tissue engineering, fibrinogen and fibronectin are by far the most important adhesion factors. Fibrinogen is the basis for fibrin, a material widely commercially used as a surgical sealant and adhesive, as well as experimentally in a number of cell transplantation applications. Here, fibrinogen forms the entire basis for the biomaterial scaffold (the fibrin scaffold is formed by polymerization of fibrinogen after its cleavage by the coagulation protease thrombin) and thus presents key adhesion factors throughout the scaffold. This fibrinogen is purified from blood plasma by a cryoprecipitation procedure, which also copurifies substantial amounts of fibronectin. This is why it is also listed as a commercially important protein at this date, for promoting cell adhesion and migration within the fibrin scaffold. A number of other adhesion factors are under commercial development in tissue engineering at this point in time, including members of the laminin family, vitronectin, and peptides based on the RGD sequence.

11.1.2 Modes of Controlled Release

Drugs may be incorporated within biomaterials in a homogeneous or a heterogeneous manner, as shown in Figure 11.2(a). When incorporated in a homogeneous manner, an implicit requirement is that the drug be somehow soluble in the matrix material, or at least soluble in its precursors. In this way, the drug can be homogeneously dissolved in the biomaterial, and, although a solid, the matrix acts as a solvent for the drug; the entire complex of drug and matrix is referred to as a *solid solution* of the drug within the matrix biomaterial. Given that the drugs of interest in tissue engineering are typically water soluble and macromolecular, this situation usually occurs only when the biomaterial is very hydrophilic. Such biomaterials then absorb large amounts of water and are referred to as hydrogels. Hydrogels may either be microporous, for example, composed of hydrophilic nanoscopic fibrils or cellular structures, such as a fibrin hydrogel, or they may be nanoporous, composed of a homogeneous network of water-soluble synthetic polymer chains crosslinked into a solid by covalent or physicochemical bonds.

When hydrophobic polymers are used, that is, those not forming hydrogels, the drugs that are of most interest in tissue engineering are not soluble within them and they form a heterogeneous structure. Domains of the drug are dispersed throughout the biomaterial scaffold heterogeneously as nanodomains or microdomains. This situation would exist with polymers such as the degradable polyesters poly(lactic acid), poly(glycolic acid), and poly(ε-caprolactone) and their copolymers. If the protein drug were incorporated within the polymer scaffold as a dry powder, such as produced by spray drying or by lyophilization, then pools of dry or highly concentrated protein would exist dispersed throughout the scaffold. It is also possible to incorporate the protein drug as an aqueous solution dispersed throughout the polymer scaffold, using an emulsion process; this results in a polymer scaffold with pools of dissolved protein drug dispersed heterogeneously throughout as shown in Figure 11.2(a) (see Section 4, for more detail on this method and others).

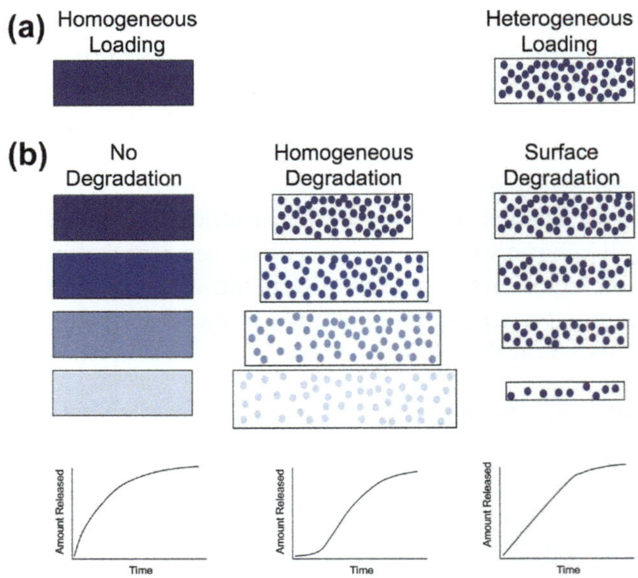

FIGURE 11.2

(a) Protein drugs can be loaded within hydrophilic polymer matrices in a homogeneous manner (left), but within a hydrophobic polymer matrix they are loaded in a heterogeneous manner of protein particles or highly concentrated protein in pools of buffer in the polymer (right). (b) Protein may be released from polymer matrices by simple diffusion in the absence of polymer degradation, which can occur with hydrophilic gel matrices. In this manner, the rate of diffusion of protein controls the rate of release (left). Alternatively, the polymer may degrade homogeneously, swelling as it degrades, releasing drug either homogeneously (not shown) or heterogeneously (shown) dispersed throughout the matrix. In this manner, the rate of degradation controls the rate of release. Homogeneous degradation often occurs with hydrophilic gel matrices (center). With hydrophobic degradable polymers, surface degradation can occur, which releases the heterogeneous particles or pools of protein as they are exposed to the surface of the polymer implant (right). (c) With diffusion-controlled release, release rates are not linear (left), however with homogeneous degradation (center) and surface degradation (right) near linear release profiles can be obtained.

Just like there is more than one mode of drug incorporation (homogeneous, heterogeneous), there is also more than one mode of drug release, depending upon the mode of drug interaction with the biomaterial and the mode of material degradation, as shown in Figure 11.2(b). A first mode of release may be found in the case of hydrogel matrices that have mesh sizes that are sufficiently large as to not present a substantial resistance to drug diffusion, for example, mesh sizes larger than approximately 100 nm. In the case of highly porous hydrogels, the drug is not entrapped within the matrix, and drug release thus proceeds independently of biomaterial degradation; the biomaterial may degrade quickly, or slowly, or not at all, but drug release proceeds more or less independently of these considerations. Such a situation is referred to as *diffusion-controlled release*. In this case, the diffusion coefficient of the drug in the hydrogel matrix is constant over time, since the hydrogel structure does not change during the process of release. Since the concentration of drug remaining in the hydrogel matrix decreases over time, due to the release process, the rate of release decreases over time, and a constant rate of release is unachievable. When relatively short-term release characteristics are desired, this approach may be appropriate. The biomaterial plays its main role as a vehicle for local, if transient, administration of the drug. The biomaterial may also play a second role, as a matrix for cell transplantation or cell invasion, independent of its role in drug release.

A second mode of release may be found in the case of hydrogel matrices with smaller mesh sizes. The polymer network that composes the gel may efficiently entrap the protein drug and hinder its diffusion from the material. In this case, the drug cannot escape from the matrix until the matrix begins to degrade. As hydrolysis occurs in the material, the architecture of the polymer network is cleaved, and results in an increase in the mesh size and an increase in the overall swelling of the material. As this occurs, the drug becomes less entrapped and may begin to diffuse from the matrix. Given that hydrolysis occurs throughout the material and that drug release then ensues from the entire material, this situation is referred to as *homogeneous degradation*. Such situations would exist in a uniform hydrolytically degradable hydrogel, such as a hydrolytically degradable polyethylene glycol (PEG) (Zustiak and Leach, 2010). Wherever water is in the hydrogel structure, hydrolysis occurs. Since the hydrogel is highly swollen with water, water is equally present throughout the hydrogel matrix, and as such degradation proceeds equally fast throughout the matrix. This leads to porosity of the hydrogel network that increases over time, and then diffusive release of drug that correspondingly increases over time. In this case, the diffusion coefficient of the drug in the hydrogel matrix starts out at a very low value, too low to demonstrate substantial release. As the gel degrades, the diffusion coefficient increases over time, and this leads to increasing release rates. However, as release occurs, the concentration of the drug remaining in the hydrogel matrix decreases over time, somewhat offsetting this increasing release rate. When these two effects are balanced, release characteristics can be obtained that are nearly *zero order*, that is, they are nearly constant over time. The extent to which this is true depends upon the extent to which the two opposing effects (diffusion coefficient increasing over time and concentration in the matrix decreasing over time) are appropriately counterbalanced. When prolonged release is needed, for example, for slow morphogenetic processes, this mechanism of controlled release can be very appropriate.

A third mode of release may be found with the use of more hydrophobic materials. In the case of such hydrophobic polymers used as biomaterial scaffolds, water will naturally have much better access to the surface of the scaffold than the internal domains of the scaffold; this leads to polymer hydrolysis in a heterogeneous manner, first on the surface, and then deeper in as the surface erodes and exposes new polymer to the surface, much like a bar of hand soap dissolving. Such degradation is referred to as *surface degradation*. With surface degradation, the drug is likewise released from the biomaterial scaffold layer by layer, and the rate of drug release depends strongly on the rate of polymer degradation and also on the morphology of the implant. For example, the surface area of a polymer film does not change very much as the film degrades, and thus, the rate of release in such morphologies is close to constant over time, which is referred to as *zero-order release*. In cases of release by surface degradation, it is easier to achieve zero-order release than with other release mechanisms.

Finally, a fourth mode of controlled drug release may be found in the situation in which the bioactive molecule is not desired to be released at all, for example, of *biomolecule immobilization*. The principle example of when this is desired is with adhesion proteins and peptides, which are active in their bound state, rather than free. Here, both surface and bulk immobilization must be considered. If an adhesion molecule is being immobilized on the surface of a surface-degrading scaffold, to support cell attachment or to trigger cell differentiation, it must clearly remain on the surface for the desired duration. If the surface layer is removed from the scaffold too rapidly, then the biomolecule will not be present sufficiently long to carry out the desired signaling function. In this case, it would have to be immobilized throughout the bulk of the material, so that when one polymer surface layer is removed, new adhesion molecules are exposed. If the scaffold degrades only slowly, compared with the time scale of cell differentiation and tissue formation within the scaffold, then these considerations are less important. Likewise, if the scaffold degrades by bulk degradation, then the rate of removal of the adhesion molecule from the polymer surface will likewise be slow, even if it is not immobilized throughout the bulk. Thus, the necessity of immobilization throughout the bulk, or the satisfactory nature of surface immobilization, depends on the details of the application and the material.

Having now introduced the topic to consider what types of drugs are interesting to release (mostly proteins and peptides), and the different modes of controlled release that may be encountered (diffusive release, release with bulk or surface degradation, no release at all), we can now turn our attention to specific examples, as indicated in Table 11.1.

11.2 BIOACTIVE FACTORS ADMIXED WITH MATRICES

In some cases, the proteins that form biomaterial matrices and scaffolds may serve directly as a carrier for controlled delivery of a bioactive molecule. A number of possibilities exist by which the scaffolds may prolong release of diffusible protein drugs such as growth factors. The scaffold or matrix may merely serve to localize the placement of the drug as a diffusion source, as described earlier in regard to diffusion-controlled release. Alternatively, even a relatively small affinity between the biomaterial matrix and the factor may result in a useful prolongation of release, as has been observed with gelatin and collagen matrices (Shi et al., 2011). Finally, for drugs that display a relatively poor solubility, the matrix or scaffold can contain a continuous source of slowly dissolving factor, and also provide sustained release for an effective duration.

Type-I collagen makes up a large fraction of connective tissues, such as skin and tendon, and bovine skin from closed and controlled herds makes up a useful supply of type-I collagen for use as biomaterial matrices in tissue engineering (Koide, 2005; Lee et al., 2001; Clark et al., 1989). The material may be used in granular form, as is used in the commercial product OP-1® Implant in the United States and in Europe (Stryker Biotechnology), which consists of osteogenic protein-1 (OP-1, also known as bone morphogenetic protein-7 or BMP-7) dispersed in granules of type-I collagen. This product has been approved for clinical use in tibial nonunions after failed autograft. Bovine type-I collagen may also be used in the form of a macroporous sponge as a drug delivery and cell ingrowth matrix. Such sponges may be placed in bony defects, as well as within medical devices such as spinal fusion cages, which are hollow structural devices used to maintain the spacing between the vertebral bodies when the intervertebral disc is removed and bony fusion between the two vertebrae is sought. BMP-2 has been developed as a product for such bone growth applications, in the product INFUSE® in the United States and InductOs® in Europe (Medtronic Sofomor Danek). BMP-2 in such collagen sponges has been shown to remain at the implantation site in vivo for at least 2 weeks, using radiolabeled protein (Seeherman and Wozney, 2005).

BMPs in collagen matrices have gone forward to clinical testing in humans, after substantial testing in preclinical animal models (De Biase and Capanna, 2005; Seeherman et al., 2002; Seeherman and Wozney, 2005). For example, BMP-2 in bovine type-I collagen sponges was evaluated in two clinical studies (Boden et al., 2000; Burkus et al., 2003) applied within cylindrical interbody spinal fusion cages. This compared the cage plus the BMP-2 in a collagen sponge with the standard of care at the time, which was the spinal fusion cage with bone autograft taken from the iliac crest. Bone formation in and around the cage is illustrated in Figure 11.3. The BMP-2-treated group performed as well in terms of fusion between the two vertebrae as the group treated with standard of care. The controlled release of BMP-2 from collagen sponges demonstrated a substantial advantage by negating the need to harvest bone from the iliac crest, a procedure that is associated with significant morbidity. BMP-2 in collagen sponges was also evaluated in treatment of open tibial fractures with

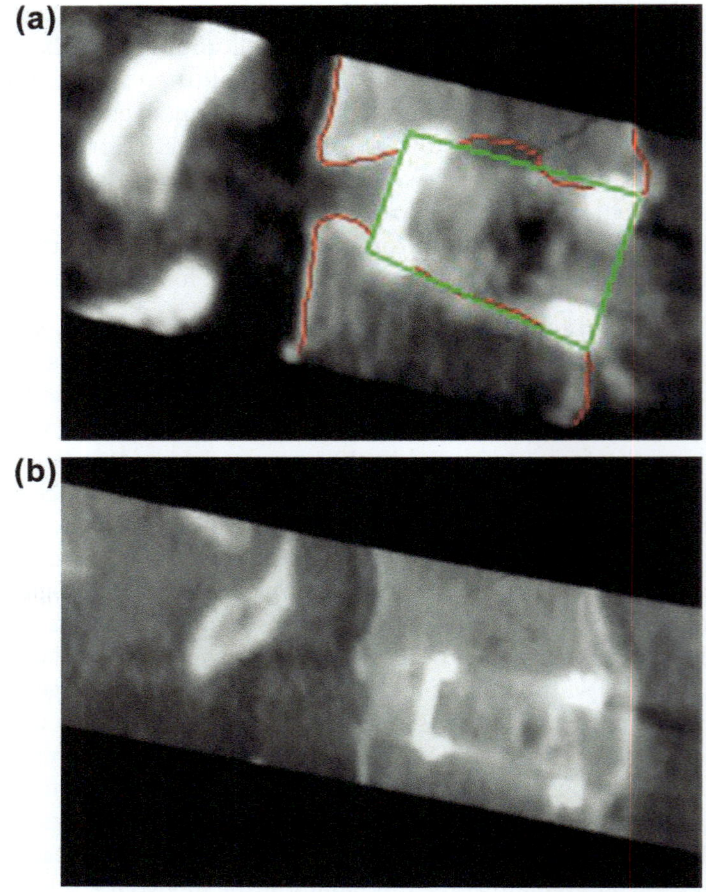

FIGURE 11.3

Radiographic image of vertebrae and the intervertebral space in a human patient after treatment with a spinal fusion cage filled with BMP-2 contained within an absorbable collagen sponge. Thin-cut computed tomography (1 mm) sagittal reconstructions are shown. Mineralized tissue is white, and the metal cage is dense white. (a) A section 2 days after implantation of the cage. For clarity, the edges of the two vertebrae are outlined in red, and the edges of the fusion cage are outlined in green. No bone is present within the cage, or on either end of it. (b) A section 2 years after implantation of the cage. Bone is present both within the cage and at both ends. *From Burkus et al. (2003), used with permission.*

an intermedullary nail, compared to treatment with the nail alone and soft tissue management in both cases (Govender et al., 2002). The group treated with BMP-2 in a collagen sponge was demonstrated to heal faster and with fewer complications than the group treated with the standard of care. BMP-7 in granular collagen matrices has been compared to bone autograft in both spinal fusion indications with interbody fusion cages (Johnsson et al., 2002) and in treatment of tibial nonunions (i.e., fractures that failed to heal and resulted in a chronic lack of union between the proximal and distal tibial portions) (Friedlaender et al., 2001), and in both cases, the BMP-7 in collagen granules demonstrated efficacy that was statistically equivalent to bone autograft. These clinical studies have led to approval for use of BMP-2 in treatment of open tibial fractures and in spinal fusion and of BMP-7 in treatment of tibial nonunions after failed autograft.

In the earlier examples, we see that bovine type-I collagen can be used successfully as a carrier of growth factors in tissue engineering, such as BMPs in bone tissue repair. However, due to the poor affinity between BMPs and type-I collagen, the factors are not retained in the matrix for very long, a fact that requires the use of supraphysiological doses of the growth factors in the successful therapies. The use of very high doses has caused safety issues, for example, the safety of BMP-2 in INFUSE® for bone repair has been called into question (Woo, 2012), and the use of BMP-2 for posteriolateral spinal fusion showed clinical incidences of cancer that were of substantial concern to the USFDA (2010). In addition to the lack of highly prolonged delivery, there may be immunological consequences of the use of bovine proteins. For example, in the clinical evaluation of BMP-7 in bovine type-I collagen granules, 5% of the patients developed antibodies for the bovine collagen, and 10% of the patients developed antibodies against the recombinant human BMP-7 (Friedlaender et al., 2001). One could hypothesize that the bovine collagen acted as an adjuvant to enhance immune reactions against the recombinant human protein, and these worries create a driving force for the development of either human-derived or synthetic matrices that would function as well.

Human-derived matrix proteins are being developed as an alternative to bovine-derived proteins. Although human collagen is available for clinical use (e.g., purified from placentas), its expense is such that it is not widely used for implants that require high masses of protein. Rather, human fibrin is widely available and broadly utilized, purified from human blood plasma donations (Albala and Lawson, 2006). Human fibrin has the advantage that it can be injected as a liquid: double-barrel syringes are usually used, in which a fibrinogen stream from one syringe is mixed via a static mixer with a thrombin stream from the second syringe, and coagulation into a gel occurs over a period of seconds to minutes, depending on the concentration of thrombin in the final mixture (Albala, 2003).

As an example of another sustained-release strategy for these admixtures in which the gel does not provide a substantial permeability barrier (the network structure of fibrin gels, as well as collagen, is very open compared to the size of the proteins being released), proteins of low solubility can be used. We present another example with BMP-2, this time contained within a fibrin matrix, to illustrate this point. Recombinant BMP-2 when produced in mammalian cells is glycosylated, whereas BMP-2 produced in engineered bacterial cells is not glycosylated. The glycosylation of BMP-2 (and other proteins) has an effect to increase its solubility, and nonglycosylated bacterially produced BMP-2 is much less soluble than mammalian cell-produced BMP-2. Delayed release due to this limited solubility is illustrated in Figure 11.4. This has been utilized to produce BMP-2 that is retained in fibrin matrices for prolonged periods relative to mammalian cell-produced BMP-2, which demonstrated in clinical studies in dogs (veterinary patients) healing in bone fusions that was statistically equivalent to that obtained with bone autografts (Schmoekel et al., 2004). A typical example of healing in these animals is illustrated in Figure 11.5. This principle of controlled release would presumably also work in other highly porous biological matrices, such as type-I collagen matrices.

As an alternative to native protein matrices being used as admixtures with protein drugs such as growth factors, denatured protein matrices can also be used. The main example of this is bovine-derived gelatin. Gelatin is produced by denaturation of type-I collagen, and an unfolded protein ensemble that self-assembles into a gel by hydrophobic interactions is obtained. The gel that so forms, identical to that in edible gelatin, is not very stable at physiological temperature, and one can achieve much stabler gels by chemical crosslinking

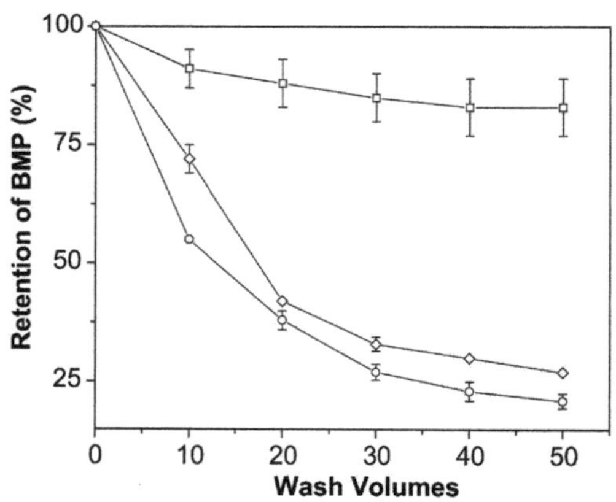

FIGURE 11.4

Nonglycosylated BMP-2 (produced in bacteria) demonstrates much less solubility than glycosylated BMP-2 (native, or produced in mammalian cells, for example). This limited solubility has been used to prolong release of protein from fibrin surgical matrices. Release in vitro is shown. Release of the nonglycosylated BMP-2 (squares) is very slow compared to release of the glycosylated protein (diamonds). When the nonglycosylated protein was complexed with heparin (circles), its release was accelerated and became as fast as the glycosylated protein. The limited solubility of certain drugs can be used as a tool to prolong release from matrices. *From Schmoekel et al. (2004), used with permission.*

with bifunctional agents, such as glutaraldehyde, which reacts with amine functionalities in the denatured protein chains coming mostly from lysine residues. Crosslinked gelatin matrices have been widely used in animal studies with proteins such as BMP-2, both with and without inorganic particles of tricalcium phosphate to assist in stability during bone formation (Takahashi et al., 2005; Yamamoto et al., 2006). In this case, it is likely that the controlled release of the growth factor within the gelatin matrix is due to many low-affinity interactions due to electrostatic attraction with the gelatin matrix. This has been more clearly demonstrated in a study with two hepatocyte growth factor (HGF) isoforms interacting with gelatin. In this study, native HGF was demonstrated to be released very slowly from crosslinked gelatin matrices, almost as slowly as material degradation. By contrast, when a highly charged region of the protein was deleted to make a less charged variant protein, that variant was released by simple diffusion from the matrix, much faster than

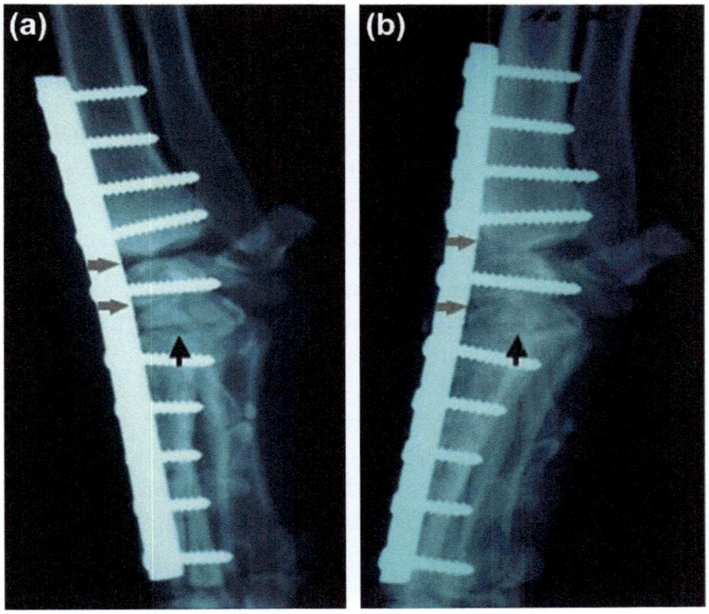

FIGURE 11.5

Nonglycosylated BMP-2 was used in fibrin surgical matrices to fuse the carpal joint complex in canine patients in pancarpal arthrodesis procedures. Normally, the joint is stabilized with a surgical plate (shown on the radiographs), and autologous bone is implanted within the joint space to induce joint fusion. Here, fibrin with nonglycosylated BMP-2 was injected in the joint space, rather than bone autograft. (a) Immediately after the procedure, no bone is visible within the three joints of the complex (arrowheads). (b) After healing, the joints were completely fused, as demonstrated by radio-opacity within the joint space (arrowheads). In a study comparing nonglycosylated BMP-2 in fibrin, arthrodesis was as effective as in a control cohort treated with both autograft (Schmoekel et al., 2004).

the native isoform (Ozeki and Tabata, 2006). This demonstrates an additional principle for controlled release in highly porous biopolymer matrices and the use of numerous, but low-affinity electrostatic interactions between the protein drug and the matrix to obtain sustained release.

In addition to low-affinity interactions, some biopolymeric matrices have the potential to exhibit high-affinity biophysical interactions for some protein drugs, especially some growth factors. This principle can be illustrated with fibrin. Fibrin forms in the body at sites of injury, and accordingly some growth factors have evolved to be immobilized in such fibrin clots (Mosesson, 2005). Cells such as blood platelets are naturally present within fibrin clots, and these cells release large quantities of stored growth factors to induce healing. A number of growth factors such as placenta growth factor-2 (PlGF-2) and fibroblast growth factor-2 (FGF-2), which contain very high affinity binding sites for fibrinogen and fibrin, with a dissociation constant <10 nM (Peng et al., 2004; Sahni et al., 2003; Martino et al., 2013). For example, this affinity immobilizes the FGF-2 within fibrin matrices, protects it from proteolytic degradation (Sahni et al., 2000), and may potentiate its activity in biological processes as complicated as induction of angiogenesis (Sahni et al., 1999). Fibrin is not the only biological matrix molecule with activity to bind certain growth factors: several growth factors have been reported to bind fibronectin (Martino and Hubbell, 2010), vitronectin (Upton et al., 1999), tenascin-C (De Laporte et al., 2013), osteopontin (Martino et al., 2014), type-I collagen (Kanematsu et al., 2004a), and type-II and type-IV collagen (Sieron et al., 2002; Scheufler et al., 1999).

In this section, we have seen that relatively simple approaches to controlled release of protein growth factors as mixtures in biopolymer matrices can be very useful. Examples of such biopolymers are type-I collagen granules and sponges, crosslinked gelatin matrices, and fibrin. The scaffold or matrix may merely localize the drug, or it may serve to delay release of the drug by subtle electrostatic affinity. In the case of particular matrix-growth factor pairs, such as fibrin and PlGF-2, naturally evolved high-affinity interactions may result in prolonged retention of the drug within the matrix. Admixtures of growth factors in biopolymer matrices have already resulted in commercial products in tissue engineering, such as the BMPs used in bone repair.

11.3 BIOACTIVE FACTORS ENTRAPPED WITHIN GEL MATRICES

It is possible to make hydrogel matrices that have characteristics and porosities such that protein permeation is directly inhibited by entrapment. This may be done in two ways: either the network structure of the hydrogel matrix may be sufficiently tight as to resist protein diffusion, or the polymer chains in the network may drive the protein out of solution. Both concepts are described in this section.

Homogeneous hydrogels have a structure not unlike a three-dimensional (3D) fishing net, with polymer chains connecting crosslinks within the network, which may be covalent or physical (Martens et al., 2004; Anseth et al., 2002; Kissel et al., 2002). The molecular weight of the chains that connect the crosslinks and the basic structure of the polymer chain determine the distance between the crosslinks and thus the average pore size within the material (Lustig and Peppas, 1988). These pores should not be thought of as static pores with rigid walls, rather they are formed dynamically and represent a time average, like the porosity created by the strings of a fishing net under water, sometimes looser and sometimes tighter as the polymer chains self-diffuse. When the average distance between crosslinks within the hydrogel network becomes of the same order of magnitude as the diameter of the protein drug, diffusion of the drug becomes significantly restricted, and the drug becomes entrapped within the hydrogel. This physical restriction can be useful in controlled release in tissue engineering.

Protein can also be entrapped within small hydrogel particles, and these particles can be distributed throughout a matrix or scaffold to provide a sustained source of a diffusible bioactive factor such as a growth factor. One such hydrogel structure is obtained when water-soluble polymer chains are attached to a polymerizable species at their termini, and then, this macromonomer is further polymerized. An example of this is with PEG, a polymer that is highly soluble in water. This chain can be reacted with ester monomers, so as to make a still water-soluble ABA block copolymer, where the A block is an oligomer of, for example, lactic acid, and the B block is the central PEG chain, as illustrated in Figure 11.6. This polymer is still water soluble, and as such, it is not useful in sustained release (unless the A blocks are sufficiently large to drive the entire polymer out of solution). If the chain termini are attached to polymerizable acrylate groups (Ac), to form Ac-oligo(lactic acid)-poly(ethylene glycol)-oligo(lactic acid)-Ac, then, the resulting macromonomer can be polymerized to form a crosslinked hydrogel (the crosslinks, or knots in the fishing net, are chains of Ac groups attached to each other) connected by PEG chains (the strings in the fishing net), with a hydrolytically sensitive link between the crosslinks (the oligo(lactic acid) blocks). Polymerization requires a crosslinker that contains at least three reactive groups to achieve network formation. With only two reactive groups, polymerization would result in only a long linear chain instead of a crosslinked network. The Ac groups can each react with two other Ac groups during photopolymerization, resulting in four reactive groups per ABA block and a highly crosslinkable polymer. These polymers have been created (Sawhney et al., 1993), and the polymerization can be carried out in water under sufficiently mild conditions to not degrade proteins incorporated in the hydrogels. By tuning the molecular weight of the PEG blocks, and thus the length between the crosslinks within the matrix, the release rate of the entrapped proteins can be modulated (Mason et al., 2001).

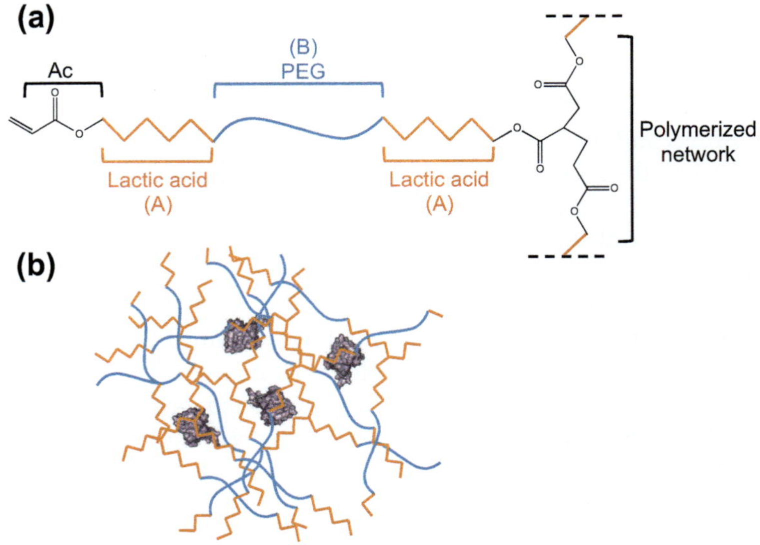

FIGURE 11.6

A number of hydrogel systems have been devised to create nanoporous networks for entrapment of protein drugs. One material is formed by polymerization of a water-soluble macromonomer (a), consisting of a central large chain of poly(ethylene glycol), flanked by short degradable regions of oligo(lactic acid), which are then end-caped with polymerizable acrylate groups. The central chain provides solubility and creates the strings in the meshwork, the flanking regions make the strings degradable, and the end-capping groups create knots in the three-dimensional network to tie the strings together. Such a network is shown in (b). When the distance between crosslinks is of the same order as the diameter of the protein, the protein drug may be entrapped. As the strings in the network degrade, this distance between crosslinks increases, and the protein may escape.

The incorporation of proteins within a 3D network can be challenging. With a 2D net, such as a fishing net, the net can be wrapped around the objects within—the net can be preformed, and then deformed around the objects. However, with a 3D net, somehow the objects must be placed within the net; however, if they are too large (relative to the distance between crosslinks) to get out, then they are also too large to get in. In the earlier example, this dilemma was solved by constructing the net around the objects—the protein was dissolved in the macromonomer solution, and the macromonomer was polymerized to form the net in situ. Other chemistries have also been developed to accomplish these ends. For example, PEG has been derivatized to form a chemically reactive polymer; PEG can be branched, like a four-arm or an eight-arm starburst, and when these branched reactive polymers are crosslinked by the addition of a corresponding chemically reactive group, a crosslinked hydrogel will form (van de Wetering et al., 2005; Metters and Hubbell, 2005; Schoenmakers et al., 2004; Elbert et al., 2001). Here, the knots in the network

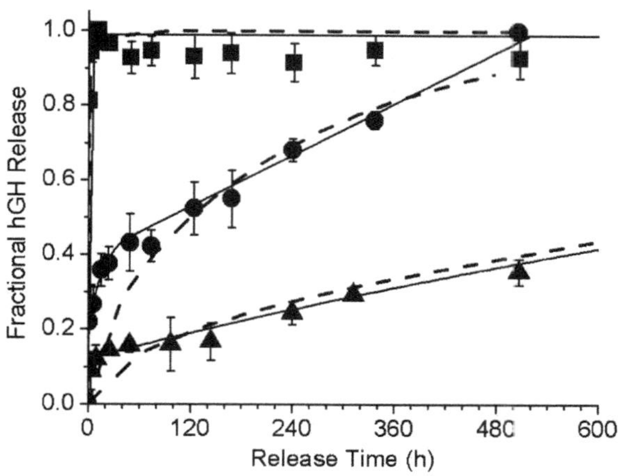

FIGURE 11.7

The network architecture of crosslinked PEG hydrogels can be adjusted by using different molecular weight PEGs. Here, release of human growth hormone (hGH) from PEG gels formed from an eight-arm, 2 kDa PEG octa-acrylate (triangles—a very tight network), an eight-arm 10 kDa PEG octa-acrylate (squares—a very open network), and 50%:50% blend of the two (circles—an intermediate network). The PEG multiacrylates were crosslinked by reaction with equivalent dithiothreitol. The solid lines represent fitting of the experimental data, and the dashed lines represent results of a predictive model based on measured network characteristics. It is seen that the protein release rate can be controlled by careful selection of the network architecture. *From van de Wetering et al. (2005), used with permission.*

are the central portions of the starburst polymers, and the strings between the knots consist of one PEG chain, the crosslinking molecule, and the second PEG chain. In the example cited, the polyethylene chains are reacted with acrylate groups, and these are crosslinked to small molecular weight dithiols such as dithiothreitol or a low-molecular-weight PEG diacrylate. When the networks are very loose, the incorporated protein is released by simple diffusion; however, when the meshwork is tighter, because the molecular weight of the arms on the PEG chains is shorter, degradation-controlled release can be obtained. Control over release rate via meshwork architecture is illustrated in Figure 11.7.

In the earlier examples, sustained release was obtained when the network size was of the same order of magnitude as the protein diameter. If this is the case, of course, the network porosity is very, very small relative to the size of a cell, and cells are unable to penetrate the hydrogel network. This is why these sorts of delivery vehicles would commonly be presented as microparticles dispersed through another scaffold or matrix. An alternative would be to make a microporous scaffold out of these nanoporous materials, such that cells could migrate through the micropores (say, of a diameter of 10–100 μm) while the proteins diffused through the nanopores. A second alternative would be to somehow endow the nanoporous network with an ability to be degraded by cells as they migrate. When

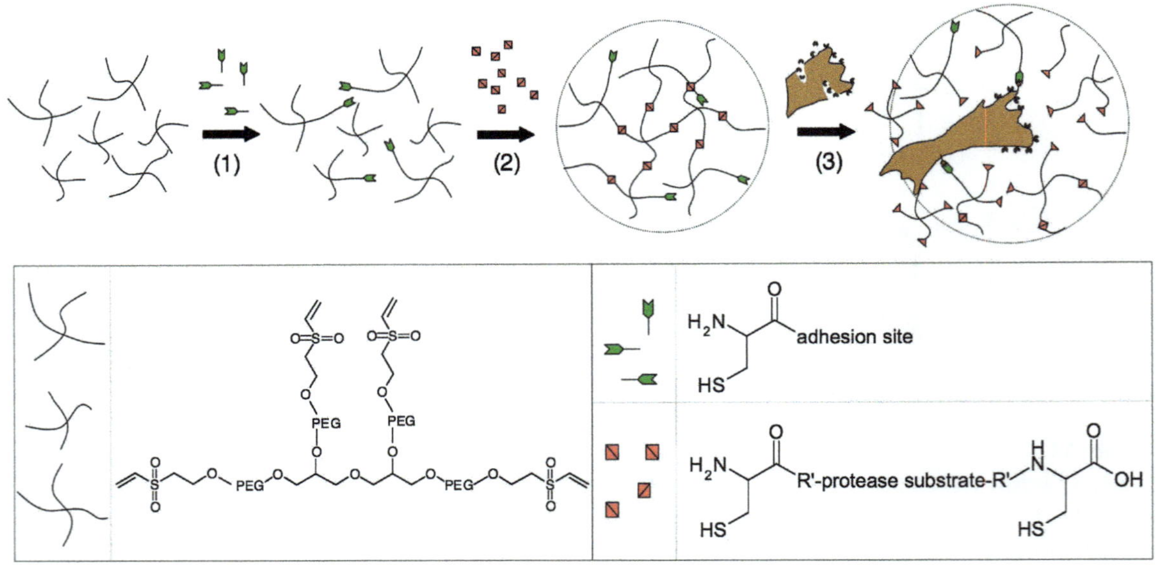

FIGURE 11.8

Biomimetic gel matrices can be constructed by crosslinking of reactive branched PEGs, e.g., terminated with vinylsulfone moieties, and di-thiol peptides (red squares), with a cysteine residue on each end of the peptide. Because the PEG chain contains more than two reactive groups, and the peptide contains two, a crosslinked network forms in step 2 of the reaction. In step 1, an adhesion peptide, such as the RGD peptide, can be incorporated into the network as a peptide that contains a single cysteine residue. If the crosslinking peptide is designed as a substrate for cell-expressed enzymes, such as matrix metalloproteinases, then cells can locally degrade the matrix and migrate through it, as in step 3 of the reaction. Given that the enzymes are localized near the surfaces of the cells, matrix degradation occurs only at the surface of the cells as well, degrading the matrix as the cells migrate through it. *From Lutolf et al. (2003b), used with permission.*

cells migrate, they activate a number of proteolytic cascades at their surface (Lutolf and Hubbell, 2005). If one could somehow construct the hydrogel network to be sensitive to these enzymes, cells could create their own micropores in the matrix as they migrate, allowing the protein to be released from the nanoporous material or to be released from it as the matrix is degraded by the cell-associated enzymes.

This concept of a cell-degradable, synthetic, nanoporous matrix has been achieved (Lutolf et al., 2003a,c; Gobin and West, 2002; West and Hubbell, 1999). For example, peptide substrates can be used as crosslinkers for correspondingly reactive PEG chains. These peptides being designed to be enzyme substrates for cell-activated enzymes such as plasmin (Pratt et al., 2004) or matrix metalloproteinases (Lutolf et al., 2003a), as illustrated in Figure 11.8 (Lutolf et al., 2003b). These hydrogel matrices offer a great deal of design flexibility: if peptides are designed to be fast-degrading substrates, then cell migration through materials can be very fast, and if slower-degrading substrates are used, then cell migration and material degradation, and thus protein release, are slower. In one example, when BMP-2 was incorporated into such matrices and implanted in bone defects

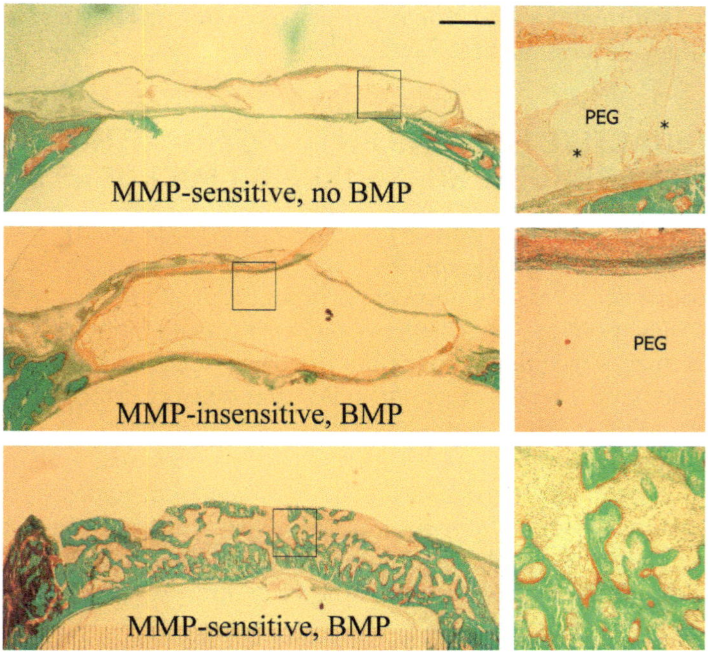

FIGURE 11.9

Matrix metalloproteinase-sensitive biomimetic gels, as illustrated in Figure 11.8, were used to release BMP-2 in calvarial defects in rats. (Upper panels) A negative control in which an MMP-sensitive gel was implanted, however without BMP-2, demonstrated no bone formation. (Middle panels) A negative control in which BMP-2 was implanted, however in a gel that was insensitive to MMPs, demonstrated no bone formation. The gel was not degraded or infiltrated by cells. (Lower panels) The positive, in which a gel was used that was sensitive to MMPs and that contained BMP-2, demonstrated matrix infiltration and good bone formation (green in the histology). All gels contained the RGD sequences as an adhesion moiety. All images are at 3 weeks postoperatively. *From Lutolf et al. (2003c), used with permission.*

in rats, bone repair was induced (Lutolf et al., 2003c), as shown in Figure 11.9. In the earlier example of BMP-2 release from an enzymatically sensitive gel, the protein was incorporated throughout the gel in an entrapped state. As cells migrated into the gel, degrading it enzymatically as they proceeded, they also liberated the growth factor into the diffusible state, micrometer by micrometer (Lutolf et al., 2003c). Such a release concept can be very advantageous for morphogenesis into a matrix, in that the rates of release of the bioactive agent are automatically linked to the rate of cell infiltration into the matrix.

In this section, we have seen that proteins, which are essentially hydrophilic in nature, can be entrapped within hydrophilic networks. The network structure, if sufficiently small, can resist the diffusion of the protein. The controlled release of proteins from these networks can be achieved through severely restricted diffusion or through degradation of the network. The network degradation can be hydrolytic or it can be proteolytic.

11.4 BIOACTIVE FACTORS ENTRAPPED WITHIN HYDROPHOBIC SCAFFOLDS OR MICROPARTICLES

Degradable hydrophobic polymers such as poly(lactic acid), poly(glycolic acid), poly(ε-caprolactone), and their copolymers have been widely explored for the controlled release of proteins, for applications that include tissue engineering. These polymers can display distinct advantages for many applications; these advantages include the long duration over which controlled release may be sustained and the ability to integrate drug delivery directly into a structural scaffold built from the same polymer. Since the polymer is hydrophobic and the drug is usually hydrophilic, the drug is distributed heterogeneously throughout the polymer and it escapes as the polymer degrades. In the following paragraphs, we describe several means by which protein drugs have been incorporated into hydrophobic polymers, either as microparticles distributed throughout a scaffold or within a scaffold itself directly.

One of the easiest ways to incorporate a protein drug within a polymeric scaffold is by simple adsorption to the scaffold surface. Although this does not allow great control over the release characteristics of the drug, useful sustained release and efficacy can sometimes be obtained when the therapeutic window of the drug is relatively wide. A good example has been provided in scaffolds intended for bone regeneration (Borden et al., 2004; Kofron et al., 2004; Borden et al., 2002). Polylactide-*co*-glycolide microparticles, 600–700 μm in diameter, were compressed within a metal mold, which was then heated to a temperature in excess of the polymer's glass transition temperature. At such temperatures, molecular mobility within the polymer is enabled, to allow chains from one microparticle to diffuse into the surface of neighboring microparticles. When the system is cooled, this gives a sintered solid, with porosity remaining from the interstitial space between the microparticles. This creates a very high surface area within the implant per unit volume. This surface can be used to adsorb protein, at surface concentrations up to 100–500 ng/cm^2. Given that many growth factors are active at amounts from 10 to 100 ng, this can represent a large quantity of the growth factor. One favorable feature of such porous solids is that they can be very strong, and thus can be used as structural implants in the repair of load-bearing tissues, such as bone. When BMP-7 was incorporated into such porous sintered solids and used for bone repair in rabbits, very effective bone healing was observed. When cells from the bone marrow were also seeded within the scaffolds, even more effective bone healing was observed. This example demonstrates that growth factor can be incorporated by simple means such as adsorption (because very little of the growth factor is in principle required), that the drug delivery vehicle can be the same as the scaffold, and that such approaches can be useful for cells invading a matrix as well as in differentiation of cells that are transplanted into a matrix.

More sophisticated approaches for drug incorporation can also be undertaken, with correspondingly better control of the release characteristics of the construct. For example, the protein drug can be formulated as a dry powder in a finely distributed state, such as by lyophilization into a fine cake and then micronizing that cake into small particles in a ball mill. Alternatively, the protein can be spray dried directly into fine particles. Processes have been described for the incorporation of such powders directly into polymer scaffolds (Lu et al., 2003)—see also the text box describing the seminal work of Robert Langer on the topic. BMP-7 was suspended in degradable polyester polymer in an organic solvent, films were cast, and the solvent was evaporated to result in fine particles of the protein drug distributed throughout the polymer films. When these films with BMP-7 were seeded with cells isolated from muscle, the growth factor enhanced their osteogenic differentiation, and these cell-seeded polymer films were useful in repair of bone in rabbit models. This example demonstrates that protein drug can be loaded as a particulate discontinuous phase within the degradable hydrophobic polymers, and that these constructs can be useful in cell differentiation and tissue repair.

Methods such as those described earlier can be used with a variety of polymeric scaffolds in tissue repair, even in rather complicated conditions. For example, angiogenic growth factors such as VEGF-A can be incorporated into polymeric scaffolds formed by a gas foaming method (Sun et al., 2005). When the VEGF-A was incorporated into the surface of the polymer, it was released fairly quickly from the implant. However, if VEGF-A was preencapsulated within degradable hydrophobic polyester microparticles, which were then incorporated into the scaffold, release was substantially delayed (Ennett et al., 2006). Such composite polymer systems showing distinct release kinetics for growth factors are useful, since tissue repair and regeneration involve the sequential signaling of multiple growth factors (Figure 11.10). This difference in release rates has been strategically used to promote angiogenesis (Richardson et al., 2001). The growth factor VEGF-A induces angiogenesis; however, the newly formed blood vessels are often immature, as they are hyperpermeable and unusually branched in architecture. Other growth factors, such as PDGF, are known to induce maturation of blood vessels, by associating the vessels with other cells that stabilize and reduce the permeability of the newly formed vessels. This two-stage release method was used to release VEGF-A quickly, which is incorporated into the surface of the scaffold, and PDGF more slowly, which is preincorporated into microparticles that were then incorporated into the scaffold. When angiogenesis was studied in animal models, substantially more mature angiogenesis was obtained with the dual growth factor release system than with the single

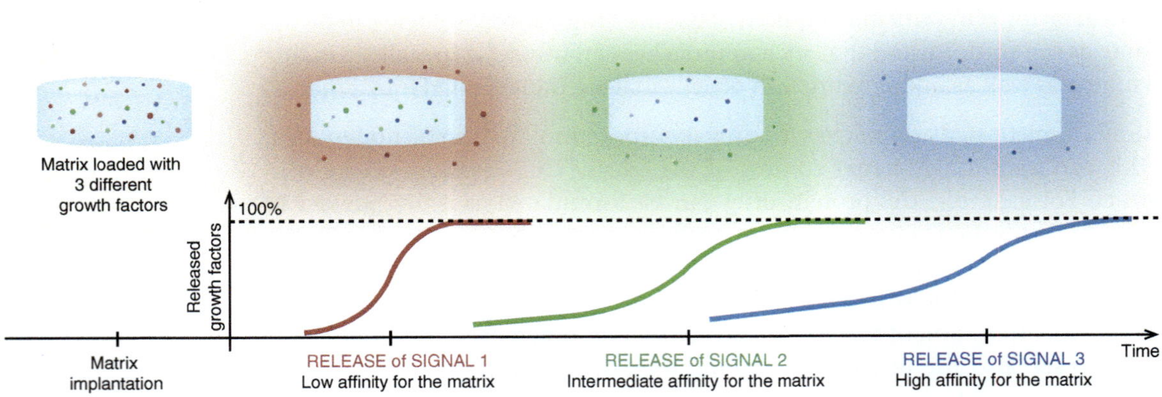

FIGURE 11.10

Properly engineering matrices for control release of biomolecules enable spatiotemporal delivery of multiple growth factors. This can be achieved by creating biomaterial systems showing distinct release kinetics for growth factors. The matrix system can be designed to display different affinities or physical barriers for distinct growth factors or the growth factors are engineered to display different affinities for the matrix producing different release profiles. The system controls the temporal release of distinct growth factors resulting in sequential waves of signaling, which can control various stages of the healing process, beginning with the initial inflammatory response and cell recruitment, leading to cell differentiation and tissue morphogenesis, and finally maturation of the tissue. *From Rice et al. (2013), used with permission.*

growth factor release system releasing VEGF-A (Richardson et al., 2001). The difference in the maturity of newly formed vessels with dual release is illustrated in Figure 11.11. Similarly, in models of myocardial infarction, sequential delivery of VEGF and PDGF-BB (Hao et al., 2007), or IGF-1 and HGF (Ruvinov et al., 2011), from alginate hydrogel systems induced the formation of mature vessels and improved cardiac function more efficiently than each factor did separately. These examples demonstrate that separately controlled release rates can be achieved for growth factors in hydrophobic degradable polymer scaffolds, and that this can be used to obtain high-level biological function in tissue responses.

It is instructive to think about the details of how protein drugs can be incorporated into polymer microparticles. The most common method is referred to as the water-in-oil-in-water method (Cohen et al., 1991). The final structure consists of a polymer microparticle with dispersed domains of protein-containing water droplets within it. It is also possible in some circumstances to further dry the watery domains so that they consist of merely dry protein dispersed throughout the polymer microparticle.

The construct is made as follows: in one preparation, the protein is dissolved in a buffer, which is usually optimized for long-term stability of the protein to be encapsulated. In another preparation, the polymer is dissolved in a water-immiscible organic solvent (the "oil"); the solvent is selected to be highly volatile, to facilitate its removal, and CH_2Cl_2 is a common choice. The protein

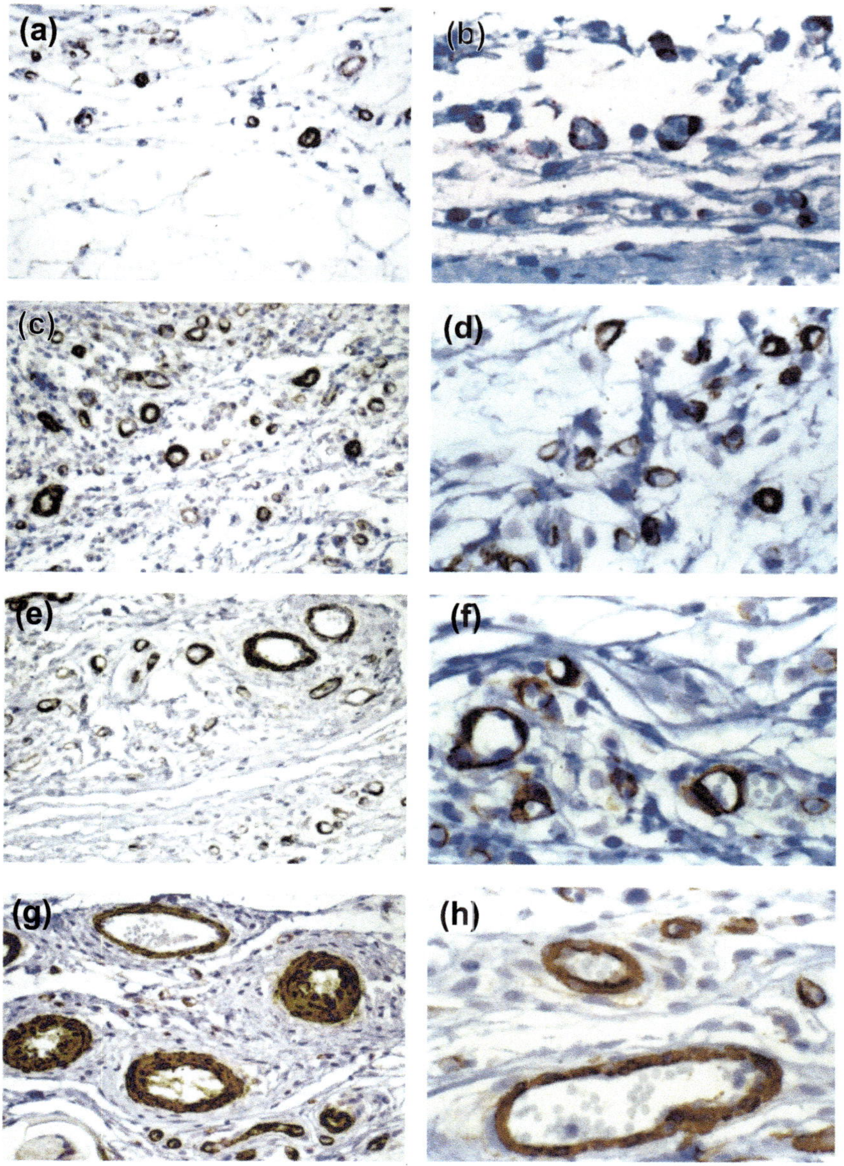

FIGURE 11.11

Physiological angiogenesis involves more than one growth factor. Dual growth factor release systems have been explored as a means by which to induce mature angiogenesis, using VEGF as an angiogenesis-inducing factor and PDGF as a vascular maturation factor, to induce cladding of the new vessels with stabilizing smooth muscle cells. Here, VEGF was incorporated within a scaffold and was released quickly, whereas PDGF was incorporated within microparticles contained in the scaffold and was released slowly. α-Smooth muscle actin was stained as an indication of vessel stabilization by smooth muscle cells. (a, b) Angiogenesis within scaffolds containing no growth factor was minimal. (c, d) Angiogenesis within scaffolds containing only VEGF. (e, f) Angiogenesis within scaffolds containing only microencapsulated PDGF. (g, h) Angiogenesis within scaffolds with dual-release of VEGF and PDGF demonstrated a vascular response that was clearly more mature than with either growth factor alone, with statistically more vessels that were positive for both endothelial cell and smooth muscle cell markers. Magnification in (a, c, e, and g) is 400× and in (b, d, f, and h) is 1000×. *From Richardson et al. (2001), used with permission.*

solution and the polymer solution are mixed; since water is not miscible in the organic solvent, an emulsion forms, and the size of the watery droplets in the polymer continuous phase may be controlled via the energy used in preparing the mixture. Now what exists is a polymer solution in organic solvent, with microdroplets of protein solution in aqueous buffer dispersed throughout. How does one make microparticles from this? This emulsion is further mixed with water that contains an emulsion stabilizer, such as poly(vinyl alcohol), and again energy is added, although at a lower level so as to obtain larger droplets. Following the mixing is a watery continuous phase, with microdroplets of polymer solution in the organic solvent, which in turn have smaller droplets within them made up of aqueous buffer containing the bioactive protein—a so-called water-in-oil-in-water double emulsion, that is, an emulsion within an emulsion. This double emulsion can be subjected to low pressures to evaporate off the organic solvent, during which the polymer solution merely becomes a solid polymer, containing the watery droplets of protein throughout—microparticles of polymer with smaller microdroplets and nanodroplets of protein in buffer. This method has been used to develop a number of commercial products in protein drug delivery.

One difficulty with the use of hydrophobic polyesters is that they degrade to yield acidic products, and the pH of the interior of the microparticle may become very low (Ding et al., 2006; Li and Schwendeman, 2005). This can lead to chemical changes in the protein, such as deamidation of glutamine and asparagine residues to glutamic acid and aspartic acid residues, respectively. Since this unfavorable side reaction is due to the acidic microenvironment within the particle, approaches have been developed to coencapsulate bases, such as $Mg(OH)_2$, a common antacid (Zhu et al., 2000). This has been demonstrated to dramatically improve the stability of the encapsulated protein, and has been shown to enhance the delivered activity of protein growth factors such as FGF-2 and BMP-2.

The examples in this section demonstrate that hydrophobic degradable polymers, in the form of microparticles or complete scaffolds, can be used to incorporate protein drugs and control their release. These scaffolds can be used to incorporate multiple growth factors with various release rates.

11.5 BIOACTIVE FACTORS BOUND TO AFFINITY SITES WITHIN MATRICES

In synthetic and biomolecular scaffolds that are highly permeable, it is difficult to retain protein drugs for prolonged periods. It was mentioned earlier, in the context of gelatin matrices, for example, that low-affinity electrostatic interactions only partially remedy this. It was further mentioned that some biomolecular complexes between growth factors and heparan sulfate proteoglycans or between growth factors and ECM proteins have an evolved high-affinity interaction. This evolved affinity can be further engineered to prolong sustained release, and this forms the topic of the present section.

Most growth factors of interest in tissue engineering are referred to as heparin-binding growth factors. They contain positively charged domains on their surfaces (Maxwell et al., 2005) that bind with a relatively high affinity to heparan sulfate-containing proteoglycans in the ECM. This affinity for heparin has been used in engineering systems for the controlled release of growth factors. Early work utilized heparin bound into hydrogel matrices, where the heparin then served to bind the growth factors. Microparticles of such biofunctional gels have been used in controlled release of FGF-2 (Laham et al., 1999; Sellke et al., 1998). Chemically modified heparin has been bound within poly(ethylene glycol) hydrogels to bind FGF-2 for use in inducing differentiation of mesenchymal stem cells (Benoit and Anseth, 2005b).

Quite sophisticated delivery matrices have been developed based on this principle of heparin affinity. For example, fibrin derivatives have been constructed, which contain heparin-binding peptides (Maxwell et al., 2005; Taylor et al., 2004; Sakiyama-Elbert and Hubbell, 2000a). These peptides then bind heparin, which can then bind heparin-binding growth factors. The higher the ratio of bound heparin to incorporated growth factor, the more delayed is the release of the heparin-binding growth factor (Sakiyama-Elbert and Hubbell, 2000b). Although the overall interactions are a bit indirect (to fibrin is coupled a peptide, which then binds heparin, which then binds the growth factor), the approach has been demonstrated to lead to a dramatically prolonged release of many growth factors, including nerve growth factor (NGF), FGF-2, TGF-β1, VEGF, and BMP-2 (Jeon et al., 2011; Sakiyama-Elbert and Hubbell, 2000a). The control release of VEGF from a heparin functionalize alginate matrix is shown in Figure 11.12.

Nonproteoglycan ECM proteins that do not contain highly negatively charged sugar chains, such as fibrinogen and fibronectin, can also be used to bind a number of growth factors (Macri et al., 2007; Martino and Hubbell, 2010; Kanematsu et al., 2004a,b). Smaller domains within the ECM proteins are responsible for the growth factor-binding activity and can be used instead of the full-length proteins, since the latter are often more difficult to translate into the clinic due to cost and safety reasons. For example, the main growth factor-binding domain of fibronectin has been recombinantly produced and incorporated into fibrin, to sequester VEGF-A165, PDGF-BB, and BMP-2 (Martino and Hubbell, 2010; Martino et al., 2011). Similarly, the growth factor-binding domain of fibrin was incorporated within a PEG hydrogel in combination with a cell adhesion peptide. The synthetic matrix, intended to mimic fibrin, was loaded with FGF-2 and/or PlGF-2 and implanted in full thickness

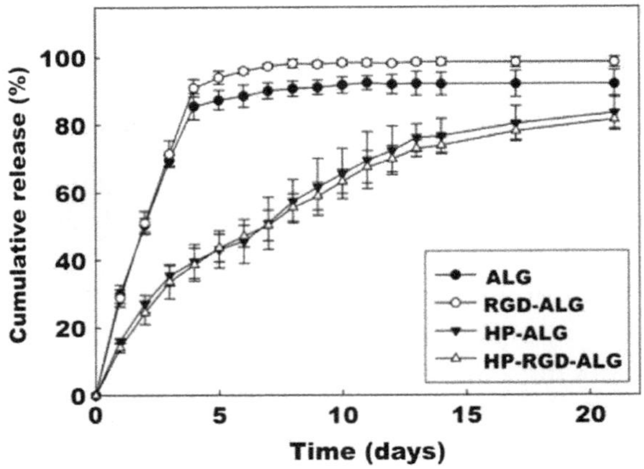

FIGURE 11.12

Many growth factors used in tissue engineering bind heparin. Here, a heparin (HP)-functionalized alginate (ALG) hydrogel prolongs the release of VEGF as compared to the hydrogel without heparin. A peptide with the RGD integrin-binding motif is also incorporated within a set of hydrogels, the addition of the peptide does not affect the release rate of VEGF. *From Jeon et al. (2011), used with permission.*

skin defects in a mouse presenting impaired wound healing. This scaffold incorporating the growth factor-binding domain and growth factors induced wound healing comparable to the one observed when the growth factors were delivered within a natural fibrin matrix.

Besides the fact that the ECM binds growth factors, the cooperative interactions between ECM proteins, growth factors, adhesion receptors, and growth factor receptors can be exploited (Kim et al., 2011; Comoglio et al., 2003; Hynes, 2009). For instance, ECM/growth factor complexes can act as a cooperative signaling molecule and induce the formation of clusters between growth factor receptors and integrins. Because several molecules involved in the signaling machinery of growth factors and integrins are common, the formation of such clusters can enhance and prolong the signaling of growth factor receptors. Exploiting the synergistic signaling between the growth factor receptors and adhesion receptors is particularly interesting, because it allows strong cellular signaling yet with a lower dose of growth factor. For example, a multifunctional recombinant fragment of fibronectin has been engineered to comprise both the majors integrin-binding and growth factor-binding domains. This fragment was able to mediate synergistic signaling between the integrin $\alpha_5\beta_1$ and the growth factor receptors for VEGF-A165, PDGF-BB, and BMP-2. In a model of chronic wounds in diabetic mice and in a model of bone repair in rat, codelivery of low doses of growth factors with the multifunctional fibronectin fragment was able to induce skin repair and bone regeneration at doses at which the growth factors delivered without the fragment had no significant effects (Martino et al., 2011).

The strategies discussed earlier always involve the use of a biomaterial for delivering growth factors. Therefore, when the growth factor is released from the biomaterial or when the biomaterial is degraded, the growth factor quickly diffuses and signals within the surrounding tissue or may enter the circulatory system. This is especially true for growth factors that do not display a strong binding to the ECM. In contrast, some growth factors have evolved to strongly bind ECM components. For example, PlGF-2 binds exceptionally strongly and promiscuously to ECM proteins, such as fibrin, fibronectin, and osteopontin, the binding is mediated by a 22-amino acid domain within PlGF-2 (Martino et al., 2014). By fusing this peptide sequence of PlGF-2 to VEGF-A, PDGF-BB, and BMP-2, improved ECM-binding variants have been engineered displaying 2- to 100-fold greater affinities to ECM proteins and heparan sulfate. Therefore, these ECM superaffinity growth factors could stay at the site of delivery for an extended period, and they could induce repair in rodent models of chronic wounds and bone defects that were greatly enhanced as compared to treatment with wild-type growth factors. In the case of VEGF-A, it was possible to obtain superior activities from the growth factor, as highlighted in the text box.

The examples in this section demonstrate that biologically evolved affinities between protein drugs and biomolecules can be exploited in engineering systems for sustained release, and that these affinities can be endowed upon both synthetic and biological matrices for tissue engineering.

11.6 BIOACTIVE FACTORS COVALENTLY BOUND TO MATRICES

As mentioned in the Introduction, controlled release in tissue engineering does not always imply that the bioactive agent is actually released from the support; a desired controlled release outcome may mean that the bioactive molecule is retained on the surface of a polymer matrix or scaffold. This is true especially for adhesion factors as drugs, but may be true as well for growth factors. We begin with the obvious case of the adhesion factor, and then proceed to the less obvious case of the growth factor, to make some cases in point.

In order for an adhesion factor to carry out its biological function of transmitting stress between a substrate outside the cell and the cytoskeleton within the cell, the adhesion factor must be immobilized to the substrate. In the case of biopolymeric scaffolds such as collagen and fibrin, there exist many adhesion ligands that are built into the scaffold material by evolution. In the case of synthetic matrices, it is necessary to incorporate exogenous ligands within the scaffold to enable biospecific cell adhesion.

A number of chemical approaches have been developed by which to endow materials with exogenous cell adhesion sites (Hubbell, 1995; Lutolf and

Hubbell, 2005), and a detailed coverage of these methods is outside the scope of this chapter. For example, with hydrogels, adhesion peptides may be covalently coupled to an immobilization site through a linking polymer, such as a poly(ethylene glycol) chain (Hern and Hubbell, 1998). Such approaches have been used to direct cell migration on hydrogel surfaces (DeLong et al., 2005), for example, by incorporating gradients of the adhesion peptide on the biomaterial. By carefully selecting the peptide that is incorporated within the overall nonadhesive hydrogel platform, one may select between different adhesion receptors on the surface of the cell to induce different functions, for example, between different β_1 and β_3 integrins dependent upon the identity of the peptide and the protein that it mimics (Gonzalez et al., 2004). Such approaches can be used to create biomaterials that demonstrate selectivity to specific cell types, such as endothelial cells (Hubbell et al., 1991; Jun and West, 2005). One can also trigger specific cell functions with immobilized adhesion ligands, such as differentiation of precursor cells (Patel et al., 2005) and osteoblasts (Benoit and Anseth, 2005a).

Even biopolymeric matrices that already have cell adhesion characteristics can benefit from the incorporation of additional cell adhesion domains. This is because the cell adhesion peptides and proteins do more than just provide adhesion; they also provide specific cell differentiation cues much like growth factors do. Interesting examples can be taken in the context of engineering of fibrin scaffolds. Peptides and proteins can be incorporated into fibrin during coagulation using the enzymology of coagulation. During coagulation, an enzyme, the transglutaminase factor XIIIa, links glutamine residues to lysine residues to crosslink the fibrin gel into a strong and stable matrix. It is possible to synthesize exogenous peptides or to express recombinant proteins so as to incorporate a substrate site for factor XIIIa (Schense and Hubbell, 1999). The net effect of this is that the exogenous peptide or protein will be covalently incorporated into the fibrin network during coagulation. This has been explored with adhesion peptides from the protein laminin, a protein that is involved in a number of morphogenetic processes, including nerve growth. When four laminin peptides were covalently coupled into fibrin matrices and the derivatized fibrin was used to repair resected nerves in the rat, the number of myelinated axons bridging the resection site was nearly doubled compared to unmodified fibrin (Schense et al., 2000), as illustrated in Figure 11.13. Fibrin was already adhesive for nerve cells, but the laminin-derived peptides bound to other adhesion receptors and signaled greater regeneration than with the native fibrin alone. Similar effects have been seen with other adhesion proteins linked into fibrin and inducing complicated morphogenetic processes such as angiogenesis (Hall and Hubbell, 2004).

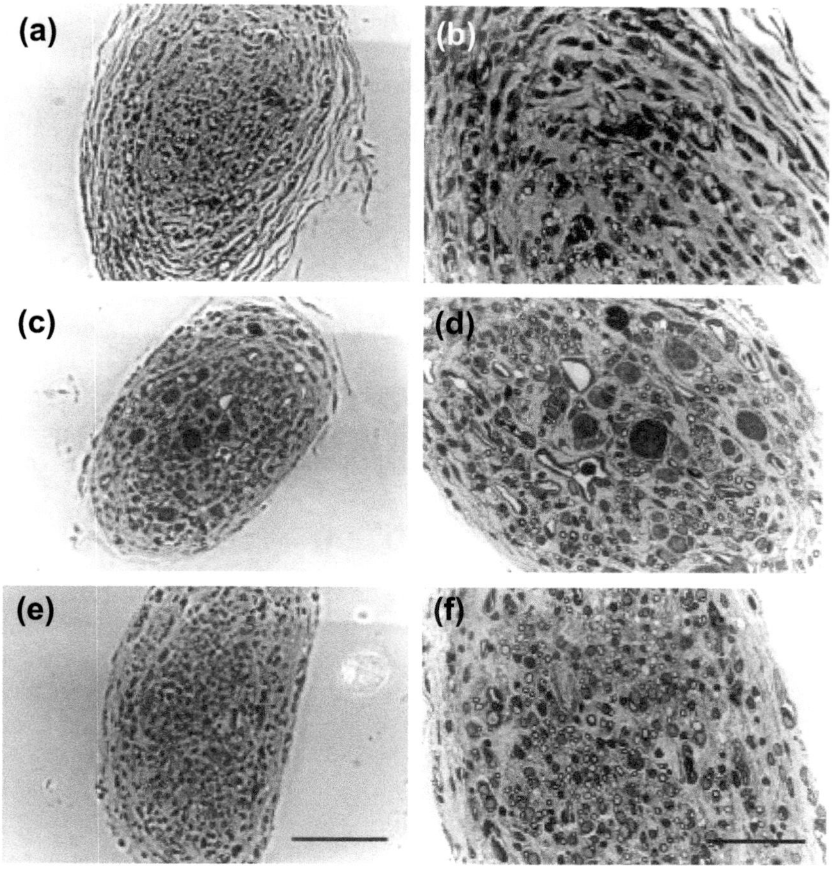

FIGURE 11.13

Exogenous adhesion ligands can be covalently incorporated within fibrin matrices during coagulation by appending the adhesion ligands to the peptide sequence recognized by factor XIIIa, which covalently crosslinks proteins during coagulation. This was explored in regeneration of peripheral nerves in rat models, using resections of the dosal root. (a, b) Transmission electron microscopy (TEM) showing nerve structure in the middle of gaps left empty within a polymer tube. Only a small number of myelinated axons were visible. (c, d) TEM in the middle of a gap treated with unmodified fibrin. Here also, only a small number of myelinated axons were visible. (e, f) TEM in the middle of a gap treated with fibrin containing four peptides derived from laminin, namely peptides containing RGD, YIGSR, IKVAV, and RNIAEIIKDI sequences. The number of myelinated axons in these samples was increased by 85% compared to nerves treated with unmodified fibrin. (a, c, and e), length bar is 50 μm. (b, d, and f), length bar is 25 μm. *From Schense et al. (2000), used with permission.*

Growth factors are also interesting drugs to consider for immobilization within biomaterial matrices and on the surfaces of biomaterial scaffolds. One of the earliest demonstrations that substrate-bound growth factors could be biologically active was carried out with epidermal growth factor, as highlighted in the text box, with work on cell differentiation in vitro (Kuhl and Griffith-Cima, 1996). One of the earliest schemes to construct biomaterials for implantation as matrices was by derivatization of type-I collagen with the growth factor TGF-β2. The growth factor was linked to collagen via a reactive bifunctional linker and was useful as an injectable biomaterial scaffold for cell infiltration (Bentz et al., 1998).

The scheme described earlier in this section for functionalization of fibrin using the transglutaminase factor XIIIa has also been used for chemical linking of growth factors within fibrin. Growth factors are recombinantly expressed as a fusion protein to include a sequence that is coupled by factor XIIIa to fibrin, with an intervening amino acid sequence that is cleaved by cell-activated enzymes. This scheme has been carried out with VEGF-A, BMP-2, NGF, TGF-β, and IGF-I to demonstrate high activity in vitro and in vivo (Schmoekel et al., 2005; Ehrbar et al., 2004; Sakiyama-Elbert et al., 2001; Liang and Andreadis, 2011; Lorentz et al., 2012). Additionally, these substrate-bound growth factors can interact with the physical aspects of their environment, such as flow near the surface of cells (Helm et al., 2005).

11.7 MATRICES USED FOR IMMUNOMODULATION

Similar to wound healing, the development of an immune response is a tightly orchestrated process involving multiple cell populations, soluble factors, and cell adhesion molecules. Likewise, tissue-engineered scaffolds for immunomodulation hold immense therapeutic potential, and contribute to strategies for vaccination, cancer, autoimmunity, inflammation, and a variety of diseases stemming from dysregulated immunity. The design of such scaffolds incorporate the same methods of controlled release previously discussed for wound healing and bone regeneration, and one could argue that an even higher level of controlled delivery is necessitated. In this section, we shall discuss the up-and-coming field of immunoengineering, which applies tissue engineering principles to the design of materials intended to modulate the immune system.

Depending on whether an immune response is mediated by molecular or cellular mechanisms, immunity can be, respectively, divided into two categories: innate and adaptive. Adaptive immune responses require an intricate series of cellular interactions that are well suited for manipulation and guidance by synthetic microenvironments. The first step in cell-mediated immunity is the activation of professional antigen-presenting cells (APCs), which include dendritic cells, macrophages, and B cells. APC activation is dependent on the presence of immunogenic molecules, or antigens, as well as immunostimulatory molecules, referred to as adjuvants, which can be incorporated into an engineered scaffold or matrix. The mechanisms by which APCs are activated determine whether an inflammatory or tolerogenic response is elicited.

The fibrin scaffold that forms within a wound following tissue injury is a natural example of how both antigen and adjuvant can be properly displayed for an appropriate immunological response. Fibrin has also been extensively used in tissue engineering applications as described early in this chapter. It naturally contains adhesion motifs that specifically bind to cell-surface receptors on leukocytes to enhance their infiltration into the scaffold matrix. In particular, the leukocyte $\alpha_M\beta_2$/MAC-1 integrin receptor interacts with the peptide sequence NRLSIGE within the fibrinogen gamma chain and is critical to the induction of in vivo inflammatory responses (Flick et al., 2004). In additional to integrin-binding motifs, key soluble proteins, that is, cytokines and chemokines, need to be present to help control the immune response. They are produced by recruited inflammatory cells or could be delivered within the hydrogel scaffold. Cytokines are factors responsible for immune cell activation, differentiation, and proliferation, while chemokines direct cell migration. As mentioned earlier, adjuvants are soluble factors that play a fundamental role in the activation of APCs. APCs possess specialized membrane-bound and intracellular toll-like receptors (TLRs) that bind and detect evolutionarily conserved adjuvants from microbes and pathogens. TLR agonists are pathogen-associated molecular patterns (PAMPS) that include lipopolysaccharides, viral and bacterial DNA and RNA, as well as synthetic small molecules such as imiquimod derivatives and monophosphoryl lipid A. Further, APCs process soluble antigenic components, such as foreign proteins and carbohydrates present within the matrix, and when properly activated by PAMPS, they are able to activate effector cells for the subsequent generation of antigen-specific cytotoxic and antibody responses.

Following exposure to antigen, adjuvant, cytokines, and chemokines, APCs are induced to migrate out of the scaffold into secondary lymphoid organs, which are specialized microenvironments optimized for leukocyte activation and function. Thus, immunomodulatory scaffolds must not often only recruit specific cell populations but they must also direct their exit from the scaffold for the induction of subsequent effector functions. Secondary lymphoid organs contain the effector cells of the immune system, lymphocytes, whose function is directed by the activated APCs. Lymphocytes comprise T cells and B cells, which form a unique cell–cell junction with activated APCs referred to as the immunological synapse. The types and concentrations of cytokines and chemokines present during formation of the immunological synapse can drastically change the induced cellular responses. Thus, controlled delivery of soluble factors is again a critical design criterion. Further, the formation of the immunological synapse is enhanced by cell adhesion motifs and the 3D architecture of lymphoid organs. The supportive stromal cells that comprise the connective tissues within lymphoid organs are specialized to express chemokines and cell adhesion ligands that, respectively, attract specific leukocytes and promote their adhesion within separate domains. The lymph node, for example, is divided into B and T cell zones that greatly enhance lymphocyte interactions with APCs and promote the generation of antibodies and cytotoxic T cells that are precisely tailored to address a particular infection as necessary. A detailed description of the immunological processes that induce these responses is beyond the scope of this text, but we intend to call attention to the lessons that can be learned from natural microenvironments when designing immunomodulatory tissue engineering constructs.

As described earlier, in vivo microenvironments have evolved to present immunostimulatory molecules to immune cells as well as to encourage essential leukocyte cell–cell interactions, and thus, tissue-engineered immunomodulatory scaffolds are frequently designed to mimic the natural microenvironments of fibrin matrices and secondary lymphoid organs. Key design parameters to consider include the selection and rate of delivery of multiple cytokines and chemokines, the type and surface concentration of cell adhesion molecules, and the overall 3D structure of the microenvironment. The majority of therapeutic scaffolds have therefore been composed of either poly(lactide-*co*-glycolide) (PLGA), which is advantageous for controlled delivery of soluble factors, or natural fibrinogen or collagen matrices that contain important cell adhesion motifs. The Mooney group fabricated PLGA scaffolds that contained synthetic TLR agonists, cancer antigens and the chemokine GM-CSF, which attracts dendritic cells (Ali et al., 2009). The scaffold could thus recruit dendritic cells via a chemokine gradient, and stimulate them with cancer antigens in the presence of a potent adjuvant. The dendritic cells were thus induced to leave the scaffold and elicit tumor-specific lymphocyte responses within secondary lymphoid organs. The Irvine group developed an injectable alginate hydrogel that contained preactivated dendritic cells (Hori et al., 2008). After injection, T cells were recruited to this artificial "node" and were directed to perform effector functions after interacting with the activated APCs. Several attempts have been made to better mimic the 3D structures of secondary lymphoid organs to better control or model elicited immune responses. Kuzin et al. developed a bioreactor containing scaffolds composed of macroporous cellulose microspheres (Kuzin et al., 2011). By culturing primary human tonsil cells within this in vitro system, T cells and B cells were found to be enriched in separate zones and antigen-specific antibody responses could be generated. In vivo generation of lymphoid tissue has also been achieved. The Watanabe group implanted collagenous "sponge-like" scaffolds that were seeded with lymphoid stromal cells (Suematsu and Watanabe, 2004). The scaffolds generated lymphoid-tissue-like organoids that could elicit strong antibody responses. The incorporation of both lymphoid stromal cells as well as activated dendritic cells was found to be essential for the clustering of T cells and B cells into separate zones within the secondary lymphoid organ-mimicking structures (Figure 11.14).

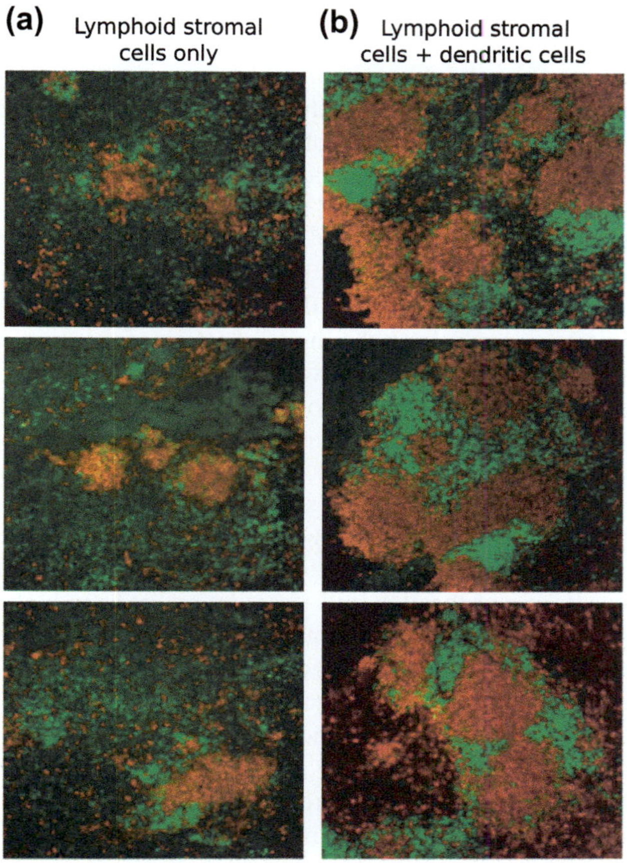

FIGURE 11.14

Inclusion of activated dendritic cells (DCs) in collagenous scaffold in which TEL-2-LTa stromal cells are embedded accelerates lymphocyte accumulation in implants. (a, b) Two types of transplants were prepared. TEL-2-LTa cells embedded in collagenous scaffolds were transplanted under the capsule of left kidney in mice (a), and TEL-2-LTa plus activated DCs embedded in collagenous scaffolds were transplanted to the right kidney (b). Three weeks later, transplants were collected and 5-mm-thick frozen sections were immunostained. T cells are stained in red with anti-Thy1.2 monoclonal antibody (mAb), and B cells are stained in green with anti-B220 mAb. *From Suematsu and Watanabe (2004), used with permission.*

In this section, we have seen that tissue-engineered constructs can be used for immunomodulation by applying methods of controlled release that were initially developed for tissue healing and regeneration. Tailoring matrices to present antigens, adjuvants, cytokines, and chemokines in a fashion that mimics natural microenvironments can allow precise control over cellular migration and the immune response.

11.8 SUMMARY

- Most drugs of interest for controlled release in tissue engineering are proteins and peptides. These include adhesion factors and growth factors. The hydrophilic nature of these drugs imposes many constraints on the controlled release strategies that can be employed.
- In tissue engineering, controlled release sometimes means that the drug is released from the surface or bulk of an implant slowly over long periods of time, and sometimes, it means that the drug is immobilized on the surface of or throughout the material and is not released at all. The difference depends on the mode of action of the drug. For example, adhesion factors are only active when immobilized.
- Hydrophilic drugs may be incorporated homogeneously through hydrophilic matrices, but only heterogeneously through hydrophobic matrices. In hydrophobic matrices, they may be incorporated as solid drug particles through solvent processes, or as aqueous pools rich in protein, for example, through water-in-oil-in-water double emulsion processes.
- Protein drugs may be released from hydrophilic matrices through diffusive release or through bulk degradation, or from hydrophobic matrices through surface degradation. With diffusive release, it is not possible to obtain constant releases rates over time. With bulk degradation, this can be obtained if certain features of the network structure are appropriately balanced. It is easier to achieve this with surface degradation.
- Many growth factors bind to other biological molecules, such as heparin and heparan sulfate proteoglycan. This affinity can be exploited in heparin-containing systems for controlled release. Some growth factors have evolved affinities for ECM proteins, such as fibrin, fibronectin, and collagen. This can also be exploited in controlled release.
- Hydrogels resemble 3D fishing nets, and protein drugs can be efficiently entrapped within hydrogels if the distance between crosslinks within the network is sufficiently close to the diameter of the protein being released.
- Degradable polyesters, such as poly(lactic acid), poly(glycolic acid), poly(ε-caprolactone), and their copolymers are useful in controlled release from tissue engineering scaffolds.
- The internal pH in degradable polyesters is often very acidic, which may cause problems in the long-term stability of protein drugs being released. This can be corrected by coencapsulating pharmaceutically acceptable bases, such as $Mg(OH)_2$.
- The creation of tissue-engineered constructs will play an increasing role in immunomodulation by allowing for cell migration into environments tailored with antigens, adjuvants, cytokines, and chemokines.

CLASSICAL EXPERIMENT—CONTROLLED PROTEIN RELEASE FROM HYDROPHOBIC POLYMERS

In the 1970s, many investigators were working in the delivery of hydrophobic drugs from hydrophobic materials; however, very few investigators were working on the controlled release of protein drugs. In a seminal paper in 1976 in Nature (Langer and Folkman, 1976), this limitation was decisively broken. Robert Langer, now a professor and world-renowned researcher at the Massachusetts Institute of Technology and then a postdoctoral fellow working with Prof. Judah Folkman at Harvard Medical School, shared his results on controlled release of proteins from hydrophobic polymers, both in vitro and in vivo. The method was simple and elegant, and has been used by countless researchers around the world and in numerous clinical products to date. The method takes advantage of the fact that many polymers are soluble in organic solvents, but proteins are generally not. If a protein is dissolved in a buffer, and then a water-miscible organic solvent (such as ethanol) is mixed into the water, denaturation of the protein typically occurs and perhaps permanently loses some or all of its biological activity. However, if the protein is mixed as a solid particle directly into an organic solvent, it cannot be denatured, because it has no molecular mobility in the solid state. In this method, protein particles are mixed into an organic solution of a hydrophobic polymer, such as ethylene–vinylacetate copolymer in CH_2Cl_2. The solvent is then evaporated to form a polymeric film or some other form. If the protein particles within the film are sparse and are completely surrounded by the polymer, then little protein release occurs. However, if the protein particles are denser and make contact with each other, then a slowly dissolving tortuous path of protein may exist within the polymer, allowing access of water outside the sample to the protein-rich interstices within the polymer. This in turn allows slow release. Release of up to 100 days was observed, and the rate of release depended on the molecular weight, and thus the diffusion coefficient, of the test protein, as illustrated in the figure. To test this concept in vivo, Langer encapsulated an angiogenic protein within ethylene–vinylacetate pellets and implanted these in the corneas of rabbits, a tissue in which vessels can be easily visualized. The protein released from these pellets-induced angiogenesis from the sclera into the cornea in a manner that was dependent on the protein being released. The method is easy to carry out and to scale into an industrial process, which has led to its success.

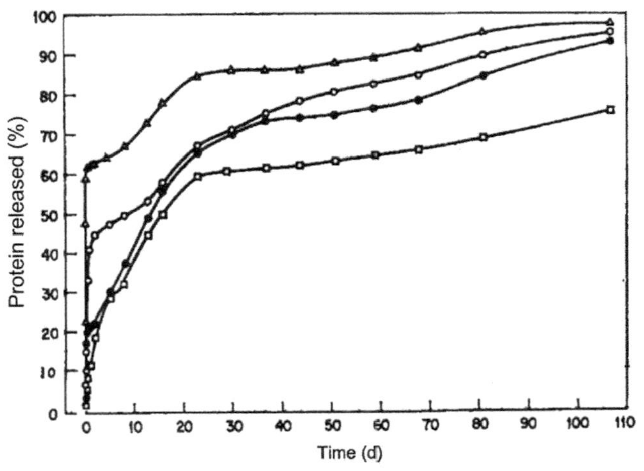

Proteins released from ethylene vinylacetate copolymer demonstrated that long-term release could be obtained and that the rate of release depended upon the molecular weight, and thus the diffusion coefficient, of the probe protein. Lysozyme (14 kDa, triangle), soybean trypsin inhibitor (21 kDa, open circle), alkaline phosphatase (88 kDa, closed circle), and catalase (250 kDa, square) were used. *From Langer and Folkman (1976), used with permission.*

CLASSICAL EXPERIMENT—SURFACE IMMOBILIZATION OF GROWTH FACTORS

Growth factors are typically thought to be active in their freely diffusible state. It is certainly the case that many growth factors bind to components of the ECM; however, the ECM is enzymatically remodeled, and this results in mobilization and release of the growth factors. A very early study on permanently bound growth factors was performed by Philip Kuhl, who was then a graduate student at the Massachusetts Institute of Technology with Prof. Linda Griffith. The study was published in Nature Medicine in 1996 (Kuhl and Griffith-Cima, 1996). The experiment was testing whether epidermal growth factor retained and could exert its biological activity when permanently immobilized upon the surface of a biomaterial. Kuhn and Griffith worked with mouse epidermal growth factor, which contains no lysine residues; this means that the only amine on the protein is at the N-terminus, and if an amine-reactive chemistry is used to immobilize the protein, the protein can only be immobilized in one conformation and orientation, by the N-terminus. Glass surfaces were derivatized to graft a multiarmed star poly(ethylene glycol), to which the mouse epidermal growth factor was immobilized, as illustrated in the figure. The surface-bound growth factor was found to be competent in inducing cell proliferation and morphological changes as effectively as controls with soluble growth factor. This demonstrates that the factor merely needed to ligate its receptors on the cell surface, that is, that internalization of the growth factor and recycling were not needed to induce the positive signaling of the growth factor. This concept set the stage for a number of other investigators to consider surface immobilization of growth factors as an approach to controlled release, that is, an approach to providing the sustained activity of a growth factor in the vicinity of a biomaterial matrix or scaffold.

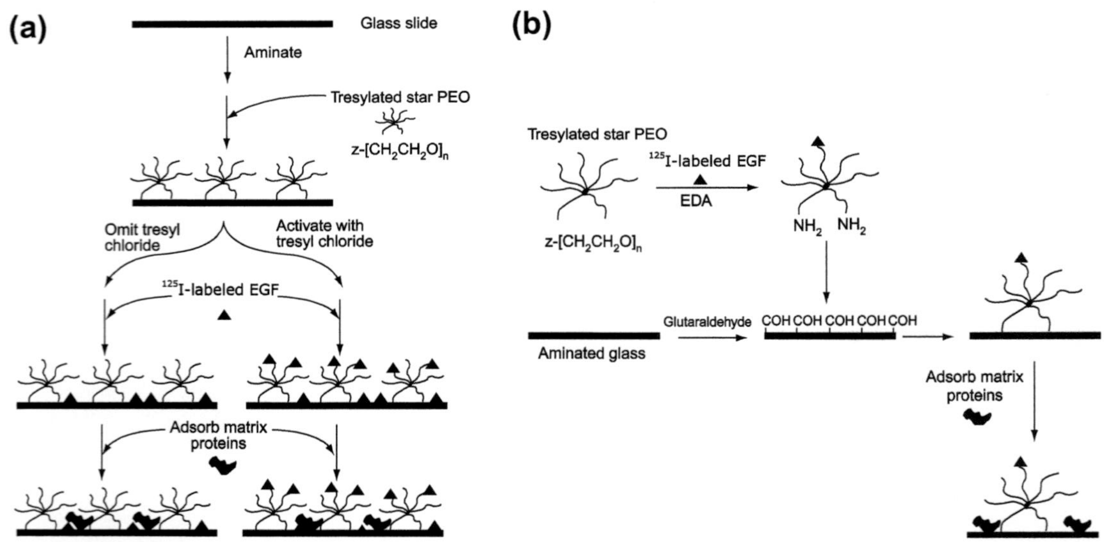

Surface-immobilized epidermal growth factor (EGF) was demonstrated to be biologically active in hepatocyte cultures. EGF was bound to a branched star-PEG, either after grafting the PEG to the surface (a) or prior to grafting to the surface (b). The PEG served to tether the growth factor to the surface in a manner that retained its biological activity and ability to ligate the EGF receptor.
From Kuhl and Griffith-Cima (1996), used with permission.

STATE-OF-THE-ART EXPERIMENT—MATRIX IMMOBILIZATION OF VEGF-A IN THERAPEUTIC ANGIOGENESIS

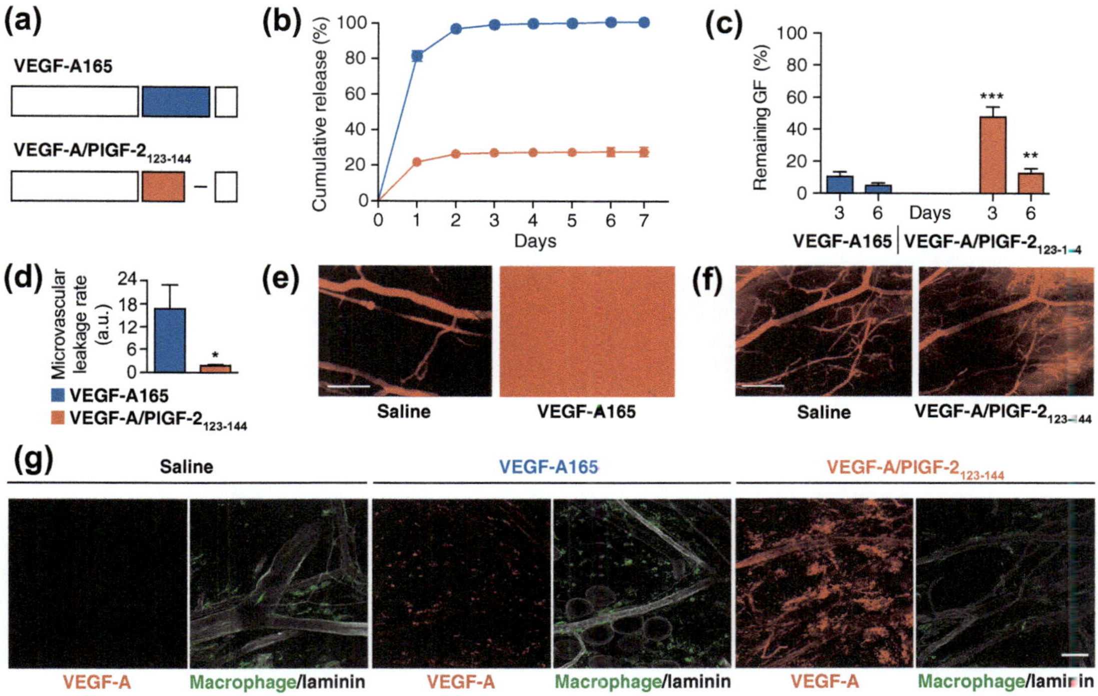

Under most conditions of release, VEGF-A induces a pathological morphology of angiogenesis, with vessels being tortuous and hyperpermeable. When VEGF-A was engineered to bind to the ECM more tightly, a very low and continuous supply of VEGF-A was generated. This resulted in an improvement of its angiogenic effect in a mouse model of skin repair, while reducing a major problem that has arisen in translating VEGF-A to clinical use, i.e., induction of vascular permeability. (a) In order to generate, VEGF-A/PIGF-$2_{123\text{-}144}$, the heparin-binding domain of VEGF-A165 (blue box) is substituted with PIGF-$2_{123\text{-}144}$ (red box). (b) Retention of wild-type versus PIGF-$2_{123\text{-}144}$-fused VEGF-A within a fibrin matrix. VEGF-A165 is quickly released, while VEGF-A/PIGF-$2_{123\text{-}144}$ remains sequestered in the matrix. (c) Retention of wild-type versus VEGF-A/PIGF-$2_{123\text{-}144}$ within a tissue. VEGF-A165 and VEGF-A/PIGF-$2_{123\text{-}144}$ were injected intradermally in the back of mice, then the growth factors still present in the tissue surrounding the injection site were detected at different time points, using enzyme-linked immunosorbent assay. The results demonstrate that wild-type VEGF-A disappears rapidly from the injection site, while the ECM-binding VEGF-A stays for an extended period. (d–f) Because VEGF-A/PIGF-$2_{123\text{-}144}$ has a strong affinity for the ECM, the engineered growth factor induces less vascular permeability than the same dose of wild-type VEGF-A165. (d) The graphs show measurement of vascular permeability in the mouse ear skin. The rate of leakage due to application of VEGF-A/PIGF-$2_{123\text{-}144}$ is only 10% of that induced by application of wild-type VEGF-A. (e, f) Representative images of the mouse ear skin vasculature 25 min after saline or growth factor application. Permeability is visualized by red-labeled dextran leaking from the blood vessels. (g) Localization of growth factor delivered in the ear skin flap. Skin flaps were incubated with saline or growth factor for 16 h, then, the growth factor was detected by immunohistochemistry. VEGF-A165 is slightly detected as small clusters localizing to some extent with macrophages (in green). In contrast, providing evidences of robust binding to tissue, VEGF-A/PIGF-$2_{123\text{-}144}$ is strongly detected bound to interstitial fibers, basement membrane, and macrophages and their surrounding matrix. Specifically, VEGF-A/PIGF-$2_{123\text{-}144}$ colocalizes with collagen-like fibers and with the basement membrane of blood vessels (in white). Representative confocal 20 μm optical sections are shown. Scale bar = 50 μm.

From Martino et al. (2014), used with permission.

STATE-OF-THE-ART EXPERIMENT—MATRIX IMMOBILIZATION OF VEGF-A IN THERAPEUTIC ANGIOGENESIS *Continued*

It is often hoped that a new growth factor lies just around the corner, waiting to be discovered to solve important tissue engineering problems. But not all the magic lies in the growth factor itself, rather the details with how the growth factor is delivered—the controlled release strategy—may matter as much as the growth factor itself in determining efficacy. An example is found in controlled release strategies with VEGF-A. VEGF-A induces angiogenesis, yet it also induces blood vessels to form in a relatively immature manner. Here, we show that this problem can be solved without using any additional factors. In research led by Mikaël M. Martino, Priscilla S. Briquez, and Jeffrey A. Hubbell at EPFL, a recombinant VEGF-A variant was made, namely, VEGF-A/PlGF-$2_{123\text{-}144}$ (Martino et al., 2014). The ECM-binding domain of VEGF-A165 (called heparin-binding domain) has been substituted with the 22-amino acid sequence of PlGF-2, which binds exceptionally strongly and promiscuously to ECM proteins. Thus, insertion of PlGF-$2_{123\text{-}144}$ sequence conferred superaffinity toward ECM proteins and heparan sulfate. Binding affinities of VEGF-A/PlGF-$2_{123\text{-}144}$ for ECM proteins were 5- to 50-fold higher compared to those of wild-type VEGF-A165. Consequently, VEGF-A/PlGF-$2_{123\text{-}144}$ could be strongly retained in a fibrin matrix and by native ECM molecules, in vivo. In addition, the growth factor could be released from the matrix by the protease plasmin, since PlGF-$2_{123\text{-}144}$ contains cleavage sites for plasmin. Thus, cells migrating into the ECM encounter a field of bound VEGF-A, which can be liberated by local proteolytic remodeling. This controlled release strategy of VEGF-A/PlGF-$2_{123\text{-}144}$ was more potent in the induction of angiogenesis and skin wound healing in animal models. Importantly, the approach could also resolve a major problem that has arisen in translating VEGF-A to clinical use. VEGF-A has been shown to rapidly induce vascular permeability, which leads to systemic hypotension and edema. This phenomenon has been the dose-limiting toxic response in peripheral and cardiovascular applications, due to the fact that VEGF-A, as a free soluble molecule, rapidly activates its receptor. In contrast, because VEGF-A/PlGF-$2_{123\text{-}144}$ tightly binds to the ECM, the number of receptors activated within the same period of time will be lower. Therefore, because vascular permeability is directly linked to the number of activated VEGF receptors, VEGF-A/PlGF-$2_{123\text{-}144}$ induces less vascular permeability than wild-type VEGF-A does. For example, in a model of dextran extravasation from vessels in the skin of the mouse, the rate of leakage due to the application of VEGF-A/PlGF-$2_{123\text{-}144}$ was only 10% of that induced by application of wild-type VEGF-A.

REFERENCES

Albala, D.M., 2003. Fibrin sealants in clinical practice. Cardiovasc. Surg. 11 (Suppl. 1), 5–11.

Albala, D.M., Lawson, J.H., 2006. Recent clinical and investigational applications of fibrin sealant in selected surgical specialties. J. Am. Coll. Surg. 202, 685–697.

Ali, O.A., Huebsch, N., Cao, L., Dranoff, G., Mooney, D.J., 2009. Infection-mimicking materials to program dendritic cells in situ. Nat. Mater. 8, 151–158.

Anseth, K.S., Metters, A.T., Bryant, S.J., Martens, P.J., Elisseeff, J.H., Bowman, C.N., 2002. In situ forming degradable networks and their application in tissue engineering and drug delivery. J. Control. Release 78, 199–209.

Benoit, D.S.W., Anseth, K.S., 2005a. The effect on osteoblast function of colocalized RGD and PHSRN epitopes on PEG surfaces. Biomaterials 26, 5209–5220.

Benoit, D.S.W., Anseth, K.S., 2005b. Heparin functionalized PEG gels that modulate protein adsorption for hMSC adhesion and differentiation. Acta Biomater. 1, 461–470.

Bentz, H., Schroeder, J.A., Estridge, T.D., 1998. Improved local delivery of TGF-β 2 by binding to injectable fibrillar collagen via difunctional polyethylene glycol. J. Biomed. Mater. Res. 39, 539–548.

Boden, S.D., Zdeblick, T.A., Sandhu, H.S., Heim, S.E., 2000. The use of rhBMP-2 in interbody fusion cages – definitive evidence of osteoinduction in humans: a preliminary report. Spine 25, 376–381.

Borden, M., Attawia, M., Khan, Y., El-Amin, S.E., Laurencin, C.T., 2004. Tissue-engineered bone formation in vivo using a novel sintered polymeric microsphere matrix. J. Bone Joint Surg. Br. 86B, 1200–1208.

Borden, M., Attawia, M., Laurencin, C.T., 2002. The sintered microsphere matrix for bone tissue engineering: in vitro osteoconductivity studies. J. Biomed. Mater. Res. 61, 421–429.

Burkus, J.K., Dorchak, J.D., Sanders, D.L., 2003. Radiographic assessment of interbody fusion using recombinant human bone morphogenetic protein type 2. Spine 28, 372–377.

Clark, D.P., Hanke, C.W., Swanson, N.A., 1989. Dermal implants – safety of products injected for soft-tissue augmentation. J. Am. Acad. Dermatol. 21, 992–998.

Cohen, S., Yoshioka, T., Lucarelli, M., Hwang, L.H., Langer, R., 1991. Controlled delivery systems for proteins based on poly(lactic glycolic acid) microspheres. Pharm. Res. 8, 713–720.

Comoglio, P.M., Boccaccio, C., Trusolino, L., 2003. Interactions between growth factor receptors and adhesion molecules: breaking the rules. Curr. Opin. Cell Biol. 15, 565–571.

De Biase, P., Capanna, R., 2005. Clinical applications of BMPs. Injury 36, 43–46.

De Laporte, L., Rice, J.J., Tortelli, F., Hubbell, J.A., 2013. Tenascin C promiscuously binds growth factors via its fifth fibronectin type III-like domain. PLoS One 8, e62076.

DeLong, S.A., Gobin, A.S., West, J.L., 2005. Covalent immobilization of RGDS on hydrogel surfaces to direct cell alignment and migration. J. Control. Release 109, 139–148.

Ding, A.G., Shenderova, A., Schwendeman, S.P., 2006. Prediction of microclimate pH in poly(lactic-*co*-glycolic acid) films. J. Am. Chem. Soc. 128, 5384–5390.

Ehrbar, M., Djonov, V.G., Schnell, C., Tschanz, S.A., Martiny-Baron, G., Schenk, U., Wood, J., Burri, P.H., Hubbell, J.A., Zisch, A.H., 2004. Cell-demanded liberation of VEGF(121) from fibrin implants induces local and controlled blood vessel growth. Circ. Res. 94, 1124–1132.

Elbert, D.L., Pratt, A.B., Lutolf, M.P., Halstenberg, S., Hubbell, J.A., 2001. Protein delivery from materials formed by self-selective conjugate addition reactions. J. Control. Release 76, 11–25.

Ennett, A.B., Kaigler, D., Mooney, D.J., 2006. Temporally regulated delivery of VEGF in vitro and in vivo. J. Biomed. Mater. Res. 79, 176–184.

Flick, M.J., Du, X., Witte, D.P., Jirouskova, M., Soloviev, D.A., Busuttil, S.J., Plow, E.F., Degen, J.L., 2004. Leukocyte engagement of fibrin(ogen) via the integrin receptor alphaMbeta2/Mac-1 is critical for host inflammatory response in vivo. J. Clin. Invest. 113, 1596–1606.

Friedlaender, G.E., Perry, C.R., Cole, J.D., Cook, S.D., Cierny, G., Muschler, G.F., Zych, G.A., Calhoun, J.H., Laforte, A.J., Yin, S., 2001. Osteogenic protein-1 (bone morphogenetic protein-7) in the treatment of tibial nonunions – a prospective, randomized clinical trial comparing rhOP-1 with fresh bone autograft. J. Bone Joint Surg. Am. 83A, S151–S158.

Gobin, A.S., West, J.L., 2002. Cell migration through defined, synthetic extracellular matrix analogues. FASEB J. 16.

Gonzalez, A.L., Gobin, A.S., West, J.L., Mcintire, L.V., Smith, C.W., 2004. Integrin interactions with immobilized peptides in polyethylene glycol diacrylate hydrogels. Tissue Eng. 10, 1775–1786.

Govender, S., Csimma, C., Genant, H.K., Valentin-Opran, A., 2002. Recombinant human bone morphogenetic protein-2 for treatment of open tibial fractures – a prospective, controlled, randomized study of four hundred and fifty patients. J. Bone Joint Surg. Am. 84A, 2123–2134.

Hall, H., Hubbell, J.A., 2004. Matrix-bound sixth Ig-like domain of cell adhesion molecule L1 acts as an angiogenic factor by ligating alpha v beta 3-integrin and activating VEGF-R2. Microvasc. Res. 68, 169–178.

Hao, X., Silva, E.A., Mansson-Broberg, A., Grinnemo, K.H., Siddiqui, A.J., Dellgren, G., Wardell, E., Brodin, L.A., Mooney, D.J., Sylven, C., 2007. Angiogenic effects of sequential release of VEGF-A165 and PDGF-BB with alginate hydrogels after myocardial infarction. Cardiovasc. Res. 75, 178–185.

Helm, C.L.E., Fleury, M.E., Zisch, A.H., Boschetti, F., Swartz, M.A., 2005. Synergy between interstitial flow and VEGF directs capillary morphogenesis in vitro through a gradient amplification mechanism. Proc. Natl. Acad. Sci. USA 102, 15779–15784.

Hern, D.L., Hubbell, J.A., 1998. Incorporation of adhesion peptides into nonadhesive hydrogels useful for tissue resurfacing. J. Biomed. Mater. Res. 39, 266–276.

Hersel, U., Dahmen, C., Kessler, H., 2003. RGD modified polymers: biomaterials for stimulated cell adhesion and beyond. Biomaterials 24, 4385–4415.

Hori, Y., Winans, A.M., Huang, C.C., Horrigan, E.M., Irvine, D.J., 2008. Injectable dendritic cell-carrying alginate gels for immunization and immunotherapy. Biomaterials 29, 3671–3682.

Hubbell, J.A., 1995. Biomaterials in tissue engineering. Biotechnology 13, 565–576.

Hubbell, J.A., Massia, S.P., Desai, N.P., Drumheller, P.D., 1991. Endothelial cell-selective materials for tissue engineering in the vascular graft via a new receptor. Biotechnology 9, 568–572.

Hynes, R.O., 2009. The extracellular matrix: not just pretty fibrils. Science 326, 1216–1219.

Jeon, O., Powell, C., Solorio, L.D., Krebs, M.D., Alsberg, E., 2011. Affinity-based growth factor delivery using biodegradable, photocrosslinked heparin–alginate hydrogels. J. Control. Release 154, 258–266.

Johnsson, R., Stromqvist, B., Aspenberg, P., 2002. Randomized radio stereometric study comparing osteogenic protein-1 (BMP-7) and autograft bone in human noninstrumented posterolateral lumbar fusion – 2002 Volvo Award in clinical studies. Spine 27, 2654–2661.

Jun, H.W., West, J.L., 2005. Endothelialization of microporous YIGSR/PEG-modified polyurethaneurea. Tissue Eng. 11, 1133–1140.

Kanematsu, A., Marui, A., Yamamoto, S., Ozeki, M., Hirano, Y., Yamamoto, M., Ogawa, O., Komeda, M., Tabata, Y., 2004a. Type I collagen can function as a reservoir of basic fibroblast growth factor. J. Control. Release 99, 281–292.

Kanematsu, A., Yamamoto, S., Ozeki, M., Noguchi, T., Kanatani, I., Ogawa, O., Tabata, Y., 2004b. Collagenous matrices as release carriers of exogenous growth factors. Biomaterials 25, 4513–4520.

Kim, S.H., Turnbull, J., Guimond, S., 2011. Extracellular matrix and cell signalling: the dynamic cooperation of integrin, proteoglycan and growth factor receptor. J. Endocrinol. 209, 139–151.

Kissel, T., Li, Y.X., Unger, F., 2002. ABA-triblock copolymers from biodegradable polyester A-blocks and hydrophilic poly(ethylene oxide) B-blocks as a candidate for in situ forming hydrogel delivery systems for proteins. Adv. Drug Deliv. Rev. 54, 99–134.

Kofron, M.D., Li, X.D., Laurencin, C.T., 2004. Protein- and gene-based tissue engineering in bone repair. Curr. Opin. Biotechnol. 15, 399–405.

Koide, T., 2005. Triple helical collagen-like peptides: engineering and applications in matrix biology. Connect. Tissue Res. 46, 131–141.

Kuhl, P.R., Griffith-Cima, L.G., 1996. Tethered epidermal growth factor as a paradigm for growth factor-induced stimulation from the solid phase. Nat. Med. 2, 1022–1027.

Kuzin, I., Sun, H., Moshkani, S., Feng, C., Mantalaris, A., Wu, J.H., Bottaro, A., 2011. Long-term immunologically competent human peripheral lymphoid tissue cultures in a 3D bioreactor. Biotechnol. Bioeng. 108, 1430–1440.

Laham, R.J., Sellke, F.W., Edelman, E.R., Pearlman, J.D., Ware, J.A., Brown, D.L., Gold, J.P., Simons, M., 1999. Local perivascular delivery of basic fibroblast growth factor in patients undergoing coronary bypass surgery – results of a phase I randomized, double-blind, placebo-controlled trial. Circulation 100, 1865–1871.

Langer, R., Folkman, J., 1976. Polymers for sustained-release of proteins and other macromolecules. Nature 263, 797–800.

Lee, C.H., Singla, A., Lee, Y., 2001. Biomedical applications of collagen. Int. J. Pharm. 221, 1–22.

Lee, K., Silva, E.A., Mooney, D.J., 2011. Growth factor delivery-based tissue engineering: general approaches and a review of recent developments. J. R. Soc. Interface 8, 153–170.

Li, L., Schwendeman, S.P., 2005. Mapping neutral microclimate pH in PLGA microspheres. J. Control. Release 101, 163–173.

Liang, M.S., Andreadis, S.T., 2011. Engineering fibrin-binding TGF-beta1 for sustained signaling and contractile function of MSC based vascular constructs. Biomaterials 32, 8684–8693.

Lo, K.W., Ulery, B.D., Ashe, K.M., Laurencin, C.T., 2012. Studies of bone morphogenetic protein-based surgical repair. Adv. Drug Deliv. Rev. 64, 1277–1291.

Lorentz, K.M., Yang, L., Frey, P., Hubbell, J.A., 2012. Engineered insulin-like growth factor-1 for improved smooth muscle regeneration. Biomaterials 33, 494–503.

Lu, H.H., Kofron, M.D., El-Amin, S.F., Attawia, M.A., Laurencin, C.T., 2003. In vitro bone formation using muscle-derived cells: a new paradigm for bone tissue engineering using polymer-bone morphogenetic protein matrices. Biochem. Biophys. Res. Commun. 305, 882–889.

Lustig, S.R., Peppas, N.A., 1988. Solute diffusion in swollen membranes. 9. Scaling laws for solute diffusion in gels. J. Appl. Polym. Sci. 36, 735–747.

Lutolf, M.P., Hubbell, J.A., 2005. Synthetic biomaterials as instructive extracellular microenvironments for morphogenesis in tissue engineering. Nat. Biotechnol. 23, 47–55.

Lutolf, M.P., Lauer-Fields, J.L., Schmoekel, H.G., Metters, A.T., Weber, F.E., Fields, G.B., Hubbell, J.A., 2003a. Synthetic matrix metalloproteinase-sensitive hydrogels for the conduction of tissue regeneration: engineering cell-invasion characteristics. Proc. Natl. Acad. Sci. USA 100, 5413–5418.

Lutolf, M.P., Raeber, G.P., Zisch, A.H., Tirelli, N., Hubbell, J.A., 2003b. Cell-responsive synthetic hydrogels. Adv. Mater. 15, 888–892.

Lutolf, M.R., Weber, F.E., Schmoekel, H.G., Schense, J.C., Kohler, T., Muller, R., Hubbell, J.A., 2003c. Repair of bone defects using synthetic mimetics of collagenous extracellular matrices. Nat. Biotechnol. 21, 513–518.

Macri, L., Silverstein, D., Clark, R.A., 2007. Growth factor binding to the pericellular matrix and its importance in tissue engineering. Adv. Drug Deliv. Rev. 59, 1366–1381.

Martens, P.J., Bowman, C.N., Anseth, K.S., 2004. Degradable networks formed from multifunctional poly(vinyl alcohol) macromers: comparison of results from a generalized bulk-degradation model for polymer networks and experimental data. Polymer 45, 3377–3387.

Martino, M.M., Briquez, P.S., Guc, E., Tortelli, F., Kilarski, W.W., Metzger, S., Rice, J.J., Kuhn, G.A., Muller, R., Swartz, M.A., Hubbell, J.A., 2014. Growth factors engineered for super-affinity to the extracellular matrix enhance tissue healing. Science 343, 885–888.

Martino, M.M., Briquez, P.S., Ranga, A., Lutolf, M.P., Hubbell, J.A., 2013. Heparin-binding domain of fibrin(ogen) binds growth factors and promotes tissue repair when incorporated within a synthetic matrix. Proc. Natl. Acad. Sci. USA 110, 4563–4568.

Martino, M.M., Hubbell, J.A., 2010. The 12th-14th type III repeats of fibronectin function as a highly promiscuous growth factor-binding domain. FASEB J. Off. Publ. Fed. Am. Soc. Exp. Biol. 24, 4711–4721.

Martino, M.M., Mochizuki, M., Rothenfluh, D.A., Rempel, S.A., Hubbell, J.A., Barker, T.H., 2009. Controlling integrin specificity and stem cell differentiation in 2D and 3D environments through regulation of fibronectin domain stability. Biomaterials 30, 1089–1097.

Martino, M.M., Tortelli, F., Mochizuki, M., Traub, S., Ben-David, D., Kuhn, G.A., Muller, R., Livne, E., Eming, S.A., Hubbell, J.A., 2011. Engineering the growth factor microenvironment with fibronectin domains to promote wound and bone tissue healing. Sci. Transl. Med. 3, 100ra89.

Mason, M.N., Metters, A.T., Bowman, C.N., Anseth, K.S., 2001. Predicting controlled-release behavior of degradable PLA-b-PEG-b-PLA hydrogels. Macromolecules 34, 4630–4635.

Maxwell, D.J., Hicks, B.C., Parsons, S., Sakiyama-Elbert, S.E., 2005. Development of rationally designed affinity-based drug delivery systems. Acta Biomater. 1, 101–113.

Metters, A., Hubbell, J., 2005. Network formation and degradation behavior of hydrogels formed by Michael-type addition reactions. Biomacromolecules 6, 290–301.

Mosesson, M.W., 2005. Fibrinogen and fibrin structure and functions. J. Thromb. Hemost. 3, 1894–1904.

Ozeki, M., Tabata, Y., 2006. Affinity evaluation of gelatin for hepatocyte growth factor of different types to design the release carrier. J. Biomater. Sci. Polym. Ed. 17, 139–150.

Patel, P.N., Gobin, A.S., West, J.L., Patrick, C.W., 2005. Poly(ethylene glycol) hydrogel system supports preadipocyte viability, adhesion, and proliferation. Tissue Eng. 11, 1498–1505.

Peng, H., Sahni, A., Fay, P., Bellum, S., Prudovsky, I., Maciag, T., Francis, C.W., 2004. Identification of a binding site on human FGF-2 for fibrinogen. Blood 103, 2114–2120.

Petrie, T.A., Capadona, J.R., Reyes, C.D., Garcia, A.J., 2006. Integrin specificity and enhanced cellular activities associated with surfaces presenting a recombinant fibronectin fragment compared to RGD supports. Biomaterials 27, 5459–5470.

Pratt, A.B., Weber, F.E., Schmoekel, H.G., Muller, R., Hubbell, J.A., 2004. Synthetic extracellular matrices for in situ tissue engineering. Biotechnol. Bioeng. 86, 27–36.

Rice, J.J., Martino, M.M., De Laporte, L., Tortelli, F., Briquez, P.S., Hubbell, J.A., 2013. Engineering the regenerative microenvironment with biomaterials. Adv. Healthcare Mater. 2, 57–71.

Richardson, T.P., Peters, M.C., Ennett, A.B., Mooney, D.J., 2001. Polymeric system for dual growth factor delivery. Nat. Biotechnol. 19, 1029–1034.

Ruvinov, E., Leor, J., Cohen, S., 2011. The promotion of myocardial repair by the sequential delivery of IGF-1 and HGF from an injectable alginate biomaterial in a model of acute myocardial infarction. Biomaterials 32, 565–578.

Sahni, A., Altland, O.D., Francis, C.W., 2003. FGF-2 but not FGF-1 binds fibrin and supports prolonged endothelial cell growth. J. Thromb. Hemost. 1, 1304–1310.

Sahni, A., Baker, C.A., Sporn, L.A., Francis, C.W., 2000. Fibrinogen and fibrin protect fibroblast growth factor-2 from proteolytic degradation. Thromb. Hemost. 83, 736–741.

Sahni, A., Sporn, L.A., Francis, C.W., 1999. Potentiation of endothelial cell proliferation by fibrin(ogen)-bound fibroblast growth factor-2. J. Biol. Chem. 274, 14936–14941.

Sakiyama-Elbert, S.E., Hubbell, J.A., 2000a. Controlled release of nerve growth factor from a heparin-containing fibrin-based cell ingrowth matrix. J. Control. Release 69, 149–158.

Sakiyama-Elbert, S.E., Hubbell, J.A., 2000b. Development of fibrin derivatives for controlled release of heparin-binding growth factors. J. Control. Release 65, 389–402.

Sakiyama-Elbert, S.E., Panitch, A., Hubbell, J.A., 2001. Development of growth factor fusion proteins for cell-triggered drug delivery. FASEB J. 15, 1300–1302.

Sawhney, A.S., Pathak, C.P., Hubbell, J.A., 1993. Bioerodible hydrogels based on photopolymerized poly(ethylene glycol)-*co*-poly(alpha-hydroxy acid) diacrylate macromers. Macromolecules 26, 581–587.

Schense, J.C., Bloch, J., Aebischer, P., Hubbell, J.A., 2000. Enzymatic incorporation of bioactive peptides into fibrin matrices enhances neurite extension. Nat. Biotechnol. 18, 415–419.

Schense, J.C., Hubbell, J.A., 1999. Cross-linking exogenous bifunctional peptides into fibrin gels with factor XIIIa. Bioconjug. Chem. 10, 75–81.

Scheufler, C., Sebald, W., Hulsmeyer, M., 1999. Crystal structure of human bone morphogenetic protein-2 at 2.7 angstrom resolution. J. Mol. Biol. 287, 103–115.

Schmoekel, H., Schense, J.C., Weber, F.E., Gratz, K.W., Gnagi, D., Muller, R., Hubbell, J.A., 2004. Bone healing in the rat and dog with nonglycosylated BMP-2 demonstrating low solubility in fibrin matrices. J. Orthop. Res. 22, 376–381.

Schmoekel, H.G., Weber, F.E., Schense, J.C., Gratz, K.W., Schawalder, P., Hubbell, J.A., 2005. Bone repair with a form of BMP-2 engineered for incorporation into fibrin cell ingrowth matrices. Biotechnol. Bioeng. 89, 253–262.

Schoenmakers, R.G., Van De Wetering, P., Elbert, D.L., Hubbell, J.A., 2004. The effect of the linker on the hydrolysis rate of drug-linked ester bonds. J. Control. Release 95, 291–300.

Seeherman, H., Wozney, J., Li, R., 2002. Bone morphogenetic protein delivery systems. Spine 27, S16–S23.

Seeherman, H., Wozney, J.M., 2005. Delivery of bone morphogenetic proteins for orthopedic tissue regeneration. Cytokine Growth Factor Rev. 16, 329–345.

Sellke, F.W., Laham, R.J., Edelman, E.R., Pearlman, J.D., Simons, M., 1998. Therapeutic angiogenesis with basic fibroblast growth factor: technique and early results. Ann. Thorac. Surg. 65, 1540–1544.

Shi, C., Chen, W., Zhao, Y., Chen, B., Xiao, Z., Wei, Z., Hou, X., Tang, J., Wang, Z., Dai, J., 2011. Regeneration of full-thickness abdominal wall defects in rats using collagen scaffolds loaded with collagen-binding basic fibroblast growth factor. Biomaterials 32, 753–759.

Sieron, A.L., Louneva, N., Fertala, A., 2002. Site-specific interaction of bone morphogenetic protein 2 with procollagen II. Cytokine 18, 214–221.

Solchaga, L.A., Hee, C.K., Roach, S., Snel, L.B., 2012. Safety of recombinant human platelet-derived growth factor-BB in Augment((R)) Bone Graft. J. Tissue Eng. 3. 2041731412442668.

Suematsu, S., Watanabe, T., 2004. Generation of a synthetic lymphoid tissue-like organoid in mice. Nat. Biotechnol. 22, 1539–1545.

Sun, Q.H., Chen, R.R., Shen, Y.C., Mooney, D.J., Rajagopalan, S., Grossman, P.M., 2005. Sustained vascular endothelial growth factor delivery enhances angiogenesis and perfusion in ischemic hind limb. Pharm. Res. 22, 1110–1116.

Takahashi, Y., Yamamoto, M., Tabata, Y., 2005. Enhanced osteoinduction by controlled release of bone morphogenetic protein-2 from biodegradable sponge composed of gelatin and beta-tricalcium phosphate. Biomaterials 26, 4856–4865.

Taylor, S.J., Mcdonald, J.W., Sakiyama-Elbert, S.E., 2004. Controlled release of neurotrophin-3 from fibrin gels for spinal cord injury. J. Control. Release 98, 281–294.

Upton, Z., Webb, H., Hale, K., Yandell, C.A., Mcmurtry, J.P., Francis, G.L., Ballard, F.J., 1999. Identification of vitronectin as a novel insulin-like growth factor-II binding protein. Endocrinology 140, 2928–2931.

USFDA, 2008. Safety Warning on Becaplermin in Regranex®. Safety Warning. Food and Drug Administration, Silver Spring.

USFDA, 2010. Panel Executive Summary for P050036 Medtronic's AMPLIFY™ rhBMP-2 matrix. Orthopaedic and Rehabilitation Devices Advisory Panel. Food and Drug Administration, Silver Spring.

van de Wetering, P., Metters, A.T., Schoenmakers, R.G., Hubbell, J.A., 2005. Poly(ethylene glycol) hydrogels formed by conjugate addition with controllable swelling, degradation, and release of pharmaceutically active proteins. J. Control. Release 102, 619–627.

West, J.L., Hubbell, J.A., 1999. Polymeric biomaterials with degradation sites for proteases involved in cell migration. Macromolecules 32, 241–244.

Woo, E.J., 2012. Recombinant human bone morphogenetic protein-2: adverse events reported to the Manufacturer and User Facility Device Experience database. Spine J. Off. J. North Am. Spine Soc. 12, 894–899.

Yamamoto, M., Takahashi, Y., Tabata, Y., 2006. Enhanced bone regeneration at a segmental bone defect by controlled release of bone morphogenetic protein-2 from a biodegradable hydrogel. Tissue Eng. 12, 1305–1311.

Zhu, G.Z., Mallery, S.R., Schwendeman, S.P., 2000. Stabilization of proteins encapsulated in injectable poly(lactide-*co*-glycolide). Nat. Biotechnol. 18, 52–57.

Zustiak, S.P., Leach, J.B., 2010. Hydrolytically degradable poly(ethylene glycol) hydrogel scaffolds with tunable degradation and mechanical properties. Biomacromolecules 11, 1348–1357.

CHAPTER 12

Bioreactors: Enabling Technologies for Research and Manufacturing

M. Adelaide Asnaghi[1], Timothy Smith[2], Ivan Martin[1], David Wendt[1]

[1]Departments of Surgery and of Biomedicine, University Hospital Basel, Basel, Switzerland;
[2]Octane Medical Group, Kingston, Ontario, Canada

LEARNING OBJECTIVES

- To define the concept of a bioreactor
- To discuss the main functions and benefits of bioreactors in tissue engineering, from basic research through tissue manufacturing
- To illustrate the role of bioreactors as three-dimensional (3D) in vitro model systems for both fundamental research and therapeutic strategies
- To discuss the relevance of bioreactors to facilitate regulatory compliance of manufacturing processes in tissue engineering
- To discuss the impact of automated bioreactor-based manufacturing systems on the economic viability of engineered tissue products

12.1 INTRODUCTION

The term "bioreactor" may initially conjure up images of a fermentation tank with a mixed suspension of bacteria, possibly for the production of an antibiotic. In this classical application, the main functions of the bioreactor are to control the environmental conditions (e.g., pH, temperature, and pressure) and the nutrient/product concentrations during the bioprocess. The level of control, reproducibility, and automation that an optimized bioreactor system enables is essential to manufacture products that must meet specific regulations and criteria regarding efficacy, safety, and quality, in addition to being cost effective.

In the context of tissue engineering, the term bioreactor was initially used to describe little more than simple mixing of a Petri dish. Over time, bioreactors used in tissue engineering have advanced in sophistication to overcome many of the limitations associated with conventional cell and tissue culture methods and/or to direct cell function (Figure 12.1). Various bioreactors have been developed to induce convective mixing of the culture medium around the scaffolds to improve transport of nutrients and wastes to/from the constructs. Perfusion bioreactors have been developed to perfuse the

Tissue Engineering.
Copyright © 2015 Elsevier Inc. All rights reserved.

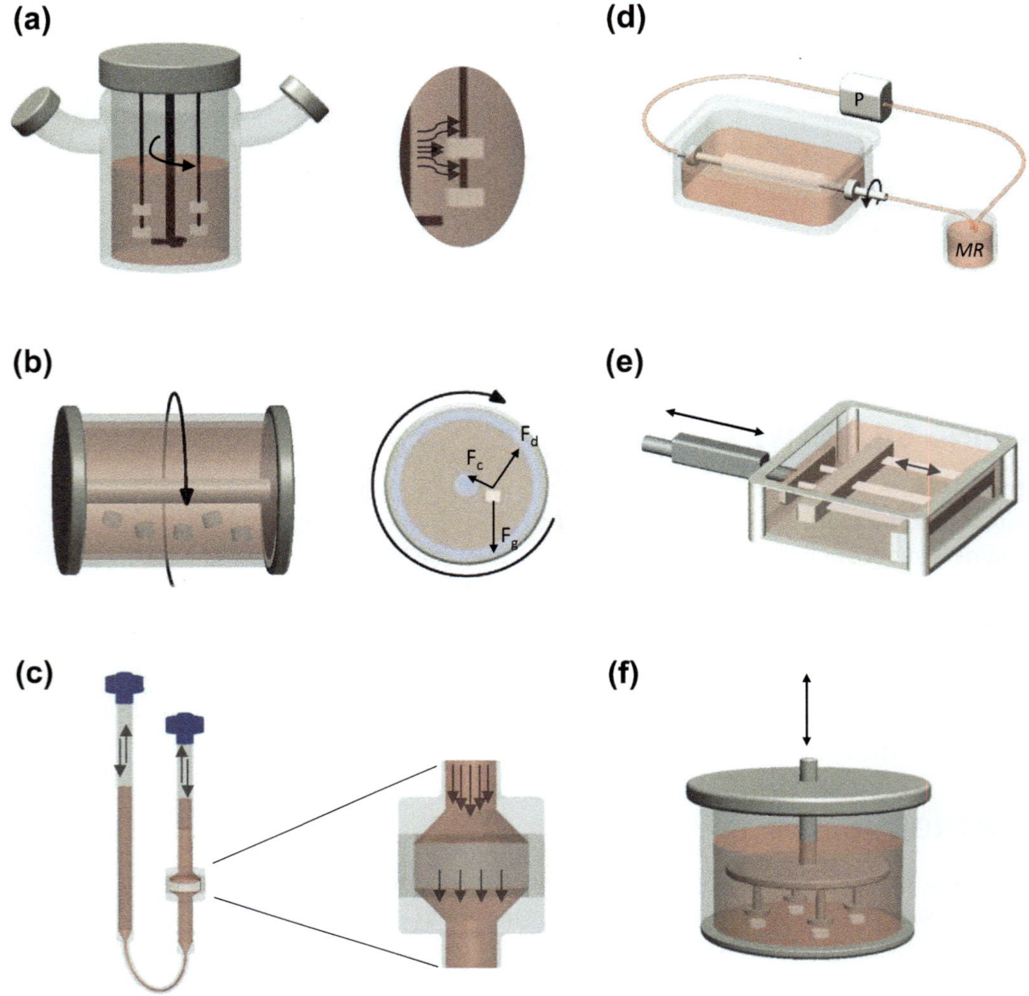

FIGURE 12.1

Representative bioreactors for tissue-engineering applications. (a) In spinner flask bioreactors, the cell suspension/culture medium is mixed around the scaffolds, which are statically suspended from the lid. (b) Appropriate rotational speed of the external wall in rotating-wall vessel bioreactors enables the construct to remain in a state of free fall through the culture medium, thanks to the balance of drag (Fd), centrifugal (Fc), and net gravitational (Fg) forces. (c) Perfusion bioreactors exploit the direct flow of the medium through the pores of the 3D scaffold. (d–f) Specialized bioreactors have been developed to apply controlled mechanical conditioning to engineered constructs in different applications, such as (d) radial strain induced by pulsatile luminal flow (P = pump, MR = medium reservoir), (e) cyclic stretching, and (f) dynamic compression. *Schematics generated by C. Turrisi.*

medium directly through the pores of a scaffold; this therefore better supports cell survival and uniform matrix accumulation. A variety of bioreactors have been developed to apply specific physiological regimes of physical stimulation (e.g., dynamic compression and cyclic stretching) to modulate extracellular matrix (ECM) deposition, cell differentiation, and cell/tissue structural organization.

In Section 2, we will discuss how bioreactors, which provide a high level of control over the physicochemical culture parameters, provide the means to drive the development of structurally defined and functionally effective 3D engineered constructs. Moreover, as discussed in Section 3, by enabling to precisely regulate the cellular microenvironment and to provide spatial and temporal control of multiple regulatory signals (e.g., growth factors, hydrodynamic, mechanical, or electrical stimuli), bioreactors provide well-defined model systems that support controlled investigations on cell function and tissue development in 3D environments. However, despite the impressive scientific progress achieved by researchers in the field of tissue engineering, the need for safe and clinically effective autologous tissue substitutes still remains unsatisfied. In Section 4, we discuss how innovative and low-cost bioreactor-based platforms that automate, standardize, and scale up the production processes will be *central* to establish regulatory compliant and cost-effective manufacturing strategies, and will thus play a *key* role in the successful exploitation of a tissue-engineered product for widespread clinical use.

12.2 ENABLING TOOLS FOR TISSUE ENGINEERS

In this section, we describe how bioreactors provide crucial tools to initiate, maintain, and direct cell cultures and tissue development in a 3D, physicochemically defined, tightly controlled, aseptic environment. Specifically, state-of-the-art devices may offer the possibility to (1) dynamically seed cells within 3D scaffolds, (2) overcome mass transport limitations, in 3D cell-seeded constructs and in tissue/organ decellularization approaches, and (3) physically stimulate the developing constructs (Figure 12.2).

12.2.1 Cell Seeding on 3D Scaffolds

The first step to the in vitro generation of 3D engineered tissue constructs is the seeding of porous scaffolds with cells appropriate to the specific application. Scaffold seeding determines the initial number of cells in the construct, as well as their spatial distribution throughout the matrix. Several studies report a direct significant influence of seeding parameters on subsequent cellular processes, including proliferation, migration, and differentiation, which determine the ultimate structure, composition, and functionality of the engineered tissues. Seeding cells into scaffolds with a uniform spatial distribution has been related to enhanced cell proliferation (Li et al., 2001) and more homogeneous ECM deposition (Holy et al., 2000). In addition, high initial cell densities have been associated with improved biochemical composition and structural stability of engineered cartilage and cardiac tissues (Freed et al., 1994; Carrier et al., 1999), increased mineralization of osteogenic constructs (Holy et al., 2000), increased rate and extent of development of engineered tendon (Awad et al., 2000), and

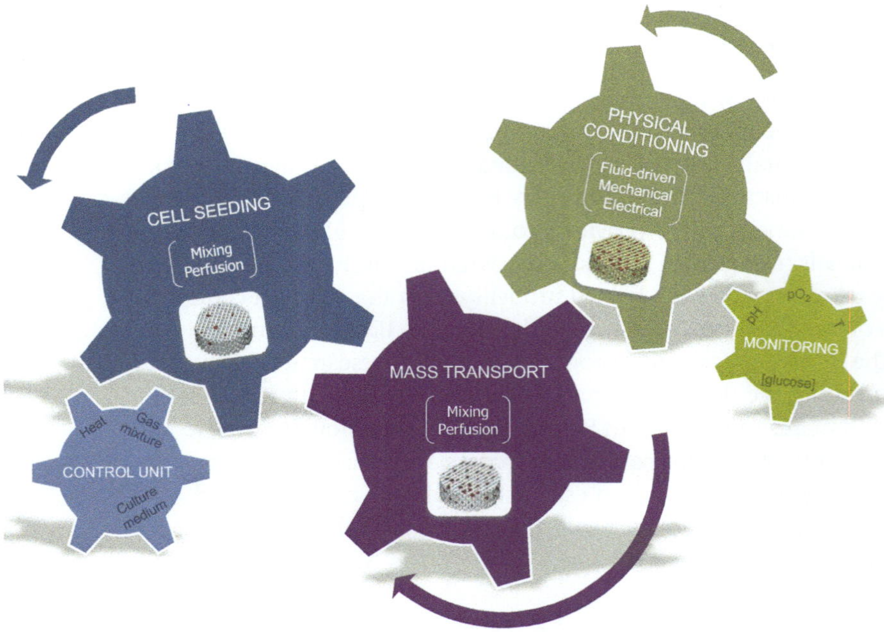

FIGURE 12.2

Schematic representation of the key functions of bioreactors in tissue-engineering applications. (1) Cell seeding of 3D matrices: bioreactors can maximize the cell utilization, control the cell distribution, and improve the reproducibility of the cell seeding process. (2) Increased mass transport: bioreactors can enhance transfer of nutrients and removal of waste products to/from the construct, and support cell survival and uniform ECM accumulation. (3) Physical conditioning of cell/scaffold constructs: bioreactors that apply physiological regimes of physical stimulation can improve the structural and functional properties of engineered tissues. (4) Maintenance of a controlled culture environment: bioreactors that monitor and control culture parameters can enhance the reproducibility and overall quality of engineered tissues and can provide well-defined model systems to investigate the fundamental aspects of cell function and tissue development in a 3D environment (T = temperature, pO_2 = oxygen partial pressure, [ ] = concentrations).

greater biosynthesis and mechanical integrity in engineered fibrocartilage (Wang et al., 2009). Therefore, since human cells are often available in short supply, it is crucial to optimize the efficiency of seeding to maximize the utilization of cells that can be obtained from rather limited tissue biopsies and to limit the extent of in vitro cell expansion.

Manually pipetting a concentrated cell suspension onto the surface of a 3D scaffold (i.e., "static cell seeding") has been the most prevalent seeding method used to establish tissue cultures. By relying on gravity as the leading principle for cell settlement and subsequent adhesion to the scaffold, such a seeding method typically leads to a poor efficiency and a nonuniform cell distribution within the scaffold (Holy et al., 2000; Li et al., 2001; Wendt et al., 2003; Alvarez-Barreto et al., 2007), besides being scarcely reproducible due to marked intraoperator and interoperator variabilities.

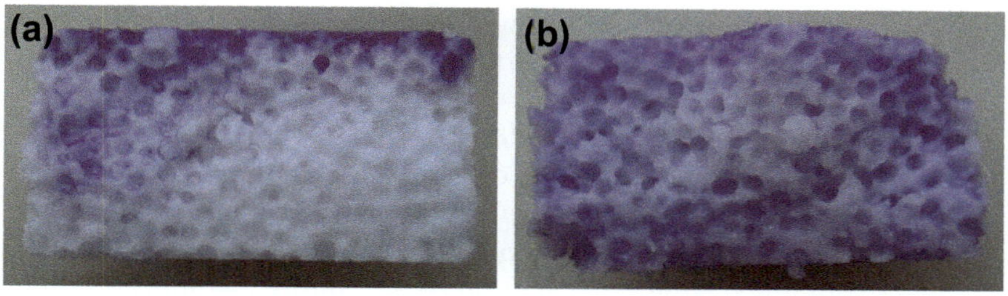

FIGURE 12.3

Cell seeding distribution into 3D scaffolds. MTT staining of living cells seeded onto ChronOS™ porous ceramic. Ceramic seeded statically (a) displayed predominant staining at the surface, with the central region essentially unstained. Ceramic seeded by perfusion (b) was relatively homogeneously stained throughout its depth and diameter (Wendt et al., 2003).

As compared to the conventional static seeding method, dynamic cell seeding in bioreactors has proven to reproducibly provide a higher efficiency and more uniform distributions of cells throughout 3D scaffolds. Spinner flasks (Vunjak-Novakovic et al., 1998), rotating wall vessels (RWVs) (Freed et al., 1997), and wavy-walled bioreactors (Bueno et al., 2007) are just a few of the devices found in the literature; these highlight the contribution of convection for improved cell seeding.

Bioreactors that perfuse a cell suspension directly through the pores of a 3D scaffold, either with a unidirectional or alternating flow, can lead to an even greater efficiency (>90%), and more uniform cell distributions than do static and spinner flasks for a wide variety of scaffolds and cell types (Alvarez-Barreto et al., 2007; Janssen et al., 2006; Shvartsman et al., 2009; Timmins et al., 2007; Wendt et al., 2003; Zhao and Ma, 2005) (Figure 12.3).

However, optimization of a seeding protocol (i.e., cell concentration of the seeding suspension, medium flow rate, flow direction, and timing of the perfusion pattern) is inherently dependent on the specific setup and the result of complex interactions between many different parameters, including cell type, biomaterial physicochemical properties, 3D scaffold architecture, and related fluid dynamics. Although most of the studies reported in the literature rely upon experimental trial-and-error investigations to optimize a seeding protocol, which requires a large number of experimental runs, theoretical models could offer a more systematic alternative for optimization. Li et al. used a mathematical model to evaluate the maximum seeding density achievable within matrices of different porosities when using cell filtration at controlled flow rates (Li et al., 2001). With the support of computational fluid dynamics (CFD) models, the velocity profiles of different perfusion bioreactor configurations could be simulated for effective fine-tuning of a design capable to support uniform cell seeding over large-diameter scaffolds (Santoro et al., 2010). CFD modeling of flow through the pores of scaffolds showed a relationship between the distribution of local wall shear stresses in the scaffolds and the distribution of seeded cells obtained experimentally (Melchels et al., 2011).

12.2.2 Increase of Mass Transport in Cell-Seeded Scaffolds

Unlike 2D monolayer cell cultures, supplying sufficient nutrients and oxygen to cells within a 3D tissue construct is a central challenge that must be addressed. When culturing 3D constructs under conventional static conditions (i.e., without any mixing or circulation of the culture medium), nutrient transport and waste removal only occur by diffusion. The implications of diffusional mass transfer can often be observed within statically cultured constructs, which are frequently reported to be inhomogeneous in structure and composition and to contain a necrotic central region. With increasing construct thickness, more pronounced gradients in tissue quality can be observed; this limits the maximum dimensions of constructs that can be cultured under static conditions (typically <1 mm). Although the limiting nutrient or species is not decisively known, steep oxygen gradients in statically cultured constructs were shown to directly correlate with the inhomogeneous development of engineered cardiac (Radisic et al., 2006), cartilage (Malda et al., 2004), and bone tissues (Volkmer et al., 2008) with viable cells and tissue at the construct periphery and mostly apoptotic cells in the bulk of the constructs.

Several different bioreactor systems have been developed to reduce diffusional limitations. Both spinner flasks and RWVs use convection to ensure that culture media *around* the scaffolds is well mixed; this thus leads to enhanced nutrient transport. The beneficial effects of such systems on cell survival and tissue development compared to static culture conditions have been demonstrated in cartilage (Vunjak-Novakovic et al., 1999), cardiac (Carrier et al., 1999), and bone tissue-engineering applications (Sikavitsas et al., 2002; Stiehler et al., 2009). However, as construct thickness increases and permeability decreases (through ECM deposition), cell growth and tissue formation are restricted to the outer regions of the constructs; this highlights the fact that *internal* mass transfer limitations are not overcome simply by convective mixing around the construct (Sikavitsas et al., 2002).

Bioreactors that perfuse culture medium directly through the pores of a scaffold reduce mass transport limitations both at the construct periphery and within its internal structure. Compared to conventional static culture, perfusion bioreactors have improved (1) proliferation, albumin secretion, and specific gene expression by hepatocytes (Shvartsman et al., 2009); (2) viability, proliferation, and differentiation of cardiomyocytes (Radisic et al., 2004); (3) uniform ECM synthesis and accumulation by chondrocytes (Wendt et al., 2006); (4) proliferation and/or distribution and amount of mineralized matrix by mesenchymal stem cells (MSCs) (Grayson et al., 2010) (Figure 12.4). Again, although the particular nutrient(s) or chemical species leading to the above

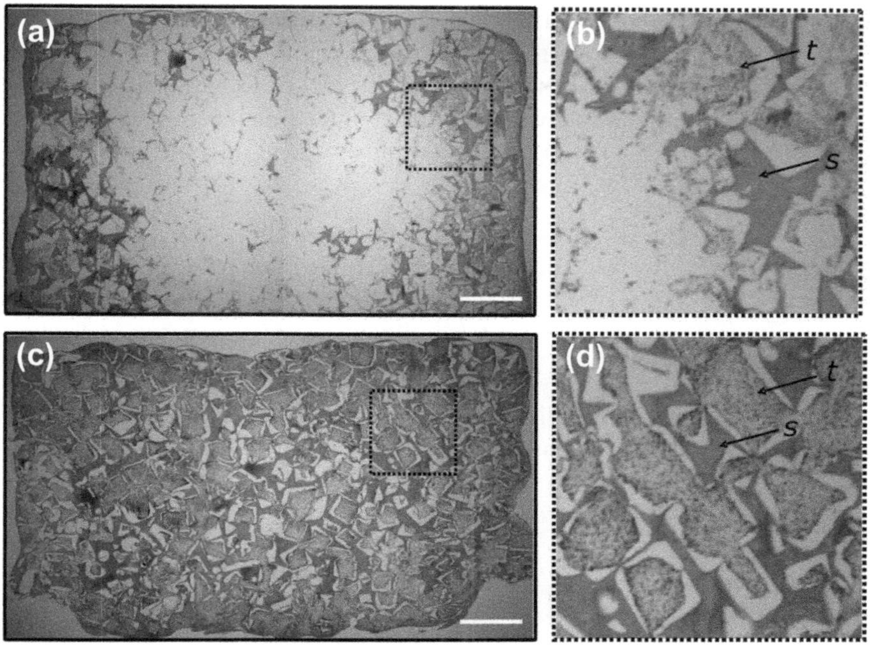

FIGURE 12.4

Implications of mass transfer on cell and matrix distributions. Human chondrocyte–foam constructs following perfusion cell seeding and two weeks of culture. (a, b) Statically cultured constructs; (c, d) perfusion cultured constructs; (a, c) low magnification images show the tissue distribution throughout the entire cross-section (scale bar = 1 mm); (b, d) higher magnification images identify the tissue "*t*" and scaffold "*s*" within the cross-sections. Statically cultured constructs contained cells and matrix only at the construct surface, and reached a depth of approximately 1 mm into the scaffold (note: the central region of the foam scaffold did not adhere to histological slide due to the absence of cells). In contrast, perfusion cultured constructs were highly homogeneous, and contained a uniform distribution of cells and matrix throughout the cross-section (Wendt et al., 2006).

results is not definitively known, increased oxygen transfer within perfused 3D constructs has been confirmed experimentally. When a hypoxia-activated luciferase promoter was used to monitor cellular hypoxia in 3D constructs, a drastically lower luciferase activity was reported in the interior of the grafts cultured in a perfusion bioreactor than under conventional static culture (Liu et al., 2010). Sensor-based measurements also confirmed higher oxygen levels within 3D constructs when cultured under perfusion compared to static conditions, which was related to improved cell viability (Volkmer et al., 2008).

A greater understanding of the basic mechanisms underlying perfusion-associated cell proliferation/differentiation and matrix production is challenging to achieve since the relative effects of perfusion-induced mechanical stresses acting on cells and enhanced mass transfer of chemical species, or a combination of the factors, cannot be easily discerned experimentally. As a result, similar to perfusion seeding parameters, optimization of perfusion culture conditions is commonly achieved by means of an experimental trial-and-error approach. Both the design of new perfusion bioreactors and the optimization of their operating conditions could be more efficient by using computational models aimed at predicting biochemical species concentrations within the pores of 3D scaffolds, as well as fluid velocity and shear profiles (Cioffi et al., 2008).

A more comprehensive strategy that could help to elucidate and decouple the effects of mechanical stimuli and specific species should (1) combine theoretical and experimental approaches (Figure 12.5) and (2) make use of sensing technologies to monitor and control key parameters indicative of the culture progression.

By the use of online sensors in perfusion bioreactors, it is possible to quantitatively measure the metabolite consumption of cultured cells (e.g., oxygen and glucose), and to monitor trends in cell growth online (Janssen et al., 2006); this thus supports/substitutes subjective, qualitative conclusions traditionally

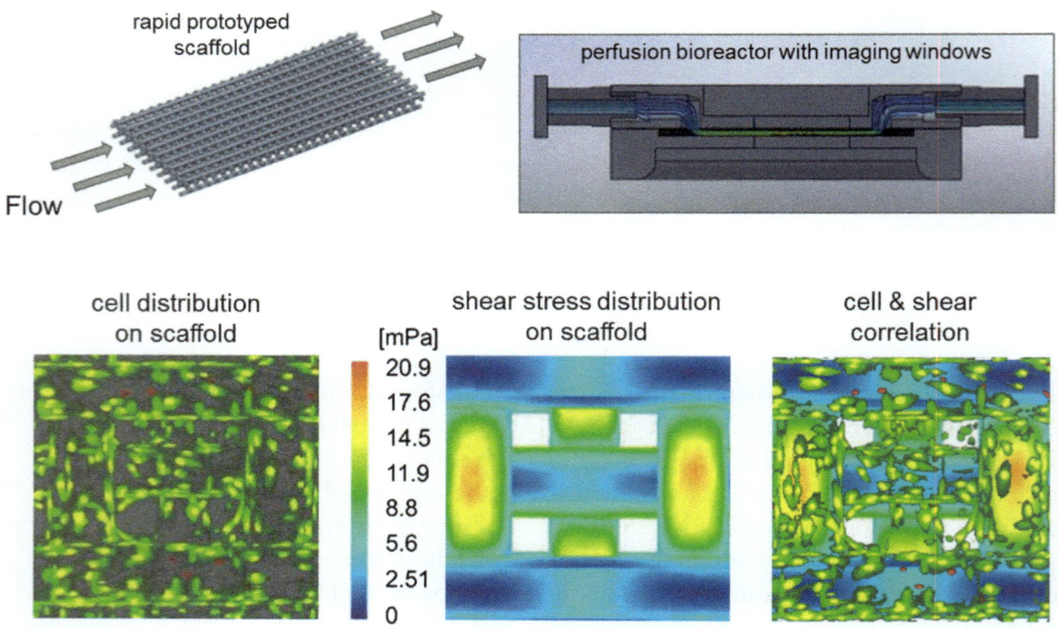

FIGURE 12.5

An integrated system to study the effects of hydrodynamic conditions on cell behavior over time, combining computational flow analysis and a microbioreactor designed for 3D scaffolds. Scaffolds were fabricated by rapid prototyping to obtain a controlled and well-defined architecture to dramatically simplify the computational modeling and facilitate comparison to the experimental data. A perfusion bioreactor was developed with optical windows for noninvasive imaging the cultured scaffold via confocal microscopy. Integrating the computational and experimental data, *local fluid conditions* can be correlated one to one with *local cell behavior* throughout the course of culture (Raimondi et al., 2010).

derived by simply observing the color of the medium. Online measurements of oxygen could be used as a nondestructive method to estimate the number of chondrocytes within an engineered cartilage graft in a perfusion bioreactor. The drop in oxygen across the perfused construct was linearly correlated with the number of cells per construct. The method was valid for a wide range of cell numbers, including cell densities currently used in the manufacture of cartilage grafts for clinical applications (Santoro et al., 2012) (Figure 12.6). Since cells within specific regions of a porous scaffold could be subjected to lower oxygen levels than those measured in the bulk medium, computational models can be used in conjunction with sensor data to predict *local* oxygen levels within the pores of a scaffold (Cioffi et al., 2008) and thus to fine-tune perfusion conditions to assure adequate supply of oxygen to cells throughout the scaffold volume. In addition to sensor technology, metabolic assays (e.g., alamarBlue®) could potentially be used as a nondestructive online method to monitor cell proliferation in engineered constructs (Zhou et al., 2013).

12.2.3 Increase of Mass Transport for Decellularization and Repopulation of Native Tissues/Organs

In recent years, a promising approach for functional organ replacement has emerged, based on the decellularization of allogeneic or xenogeneic donor tissues and organs to provide acellular naturally occurring 3D biologic scaffolds, which can subsequently be seeded with either differentiated or progenitor cells. The most common decellularization method has been to submerge the tissue statically in one or more treatment solutions or expose the tissue to some form of agitation to devitalize the tissue and remove the cell debris while retaining the native composition and structure of the associated matrix (see Chapter 5). These strategies, which rely on passive diffusion, often lead to

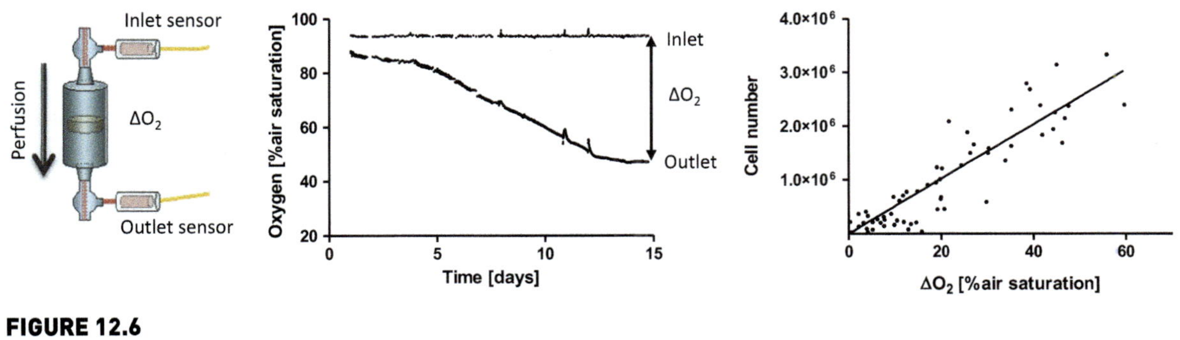

FIGURE 12.6

Online measurements of oxygen concentrations to estimate the number of cells in a perfusion bioreactor. Dissolved oxygen in the culture medium was measured across the perfused construct via disposable chemooptic microsensors. Measurements at the inlet port remained constant throughout the 2-week culture period; this indicated oxygen saturation of the culture medium within the perfusion loop, while oxygen levels measured at the outlet decreased over time due to oxygen consumption by proliferating cells. The oxygen drop (ΔO_2) across the culture chamber could be correlated with the number of cells within the construct (Santoro et al., 2012).

suboptimal efficiency and inconsistent results especially when decellularizing whole organs, due to the low permeability of the complex ECM. Bioreactors have been used to assist decellularization both by improving the transport of devitalizing solutions and rinsing the cellular material away. To improve upon a previous ineffective static process for decellularizing porcine heart valves, Karim et al. developed a rotating and perfusable bioreactor to help release and wash out the cells from the ECM (Karim et al., 2006). For decellularizing human umbilical veins, a modified vascular perfusion bioreactor also proved to be significantly more effective than traditional techniques (Montoya and McFetridge, 2009). Acellular heart scaffolds with patent and perfusable vessel conduits could be produced through perfusion-based decellularization (Ott et al., 2008). Perfusion also supported successful decellularization of rat lung (Ott et al., 2010), and assisted in the scale-up for the generation of clinically relevant acellular lung scaffolds from porcine and human donors with preserved ECM composition and architectures (Gilpin et al., 2014).

Similar to the challenges encountered when seeding conventional 3D scaffolds, reseeding of decellularized tissues/organs is highly challenging due to their complex and tortuous architecture. Rotating bioreactors were efficient in reseeding heart valves (Karim et al., 2006) and trachea (Asnaghi et al., 2009), whereas perfusion seeding has been reported to enable the appropriate cell repopulation of the heart (Ott et al., 2008), lung (Ott et al., 2010), liver (Uygun et al., 2010), and pancreas (Goh et al., 2013).

To circumvent the problem of the limited availability of native donor tissues, engineered tissues could be generated in vitro to possess particular structural, biochemical, and/or biomechanical cues of a native tissue, and then subsequently decellularized. This approach would offer the opportunity to deliver *off-the-shelf* bioactive materials with ECM-presented cues, which can be potently instructive toward tissue regeneration, but without immunomatching requirements. As an example, engineered blood vessels grown from allogeneic cells in a bioreactor could then be efficiently decellularized to preserve the structure and composition of the engineered tissues. When implanted in a large animal model after seeding with cells from the intended recipient, the vascular grafts resisted clotting and intimal hyperplasia (Quint et al., 2011).

12.2.4 Physical Conditioning

Physical forces have been shown to be potent regulators of cell physiology in vivo, and modulate ECM synthesis and organization at various hierarchical levels—from molecules to whole tissues and organs—at all stages of life (development, growth, and remodeling), under both normal and pathological conditions. It has long been hypothesized that the specific mechanical cues experienced by cells in vivo could also modulate the biological response of these cells when applied in vitro. Therefore, a wide spectrum of specialized bioreactors has been developed to provide controlled physiological regimes of biophysical stimuli to 3D constructs in an effort to direct cell differentiation, stimulate ECM synthesis, improve cell/tissue structural organization, and/or enhance or accelerate functional tissue formation (Figure 12.7).

Mechanically generated signals appear to play a role in the *differentiation* of progenitor cells into different lineages, including myofibroblast (pulsatile flow: Hoerstrup et al., 2002), chondrogenic (reviewed in: O'Conor et al., 2013), epidermal (stretching: Powell et al., 2010), ligament (translational and rotational strain: Altman et al., 2002), cardiomyocytic (cyclic stretch: Tiburcy et al., 2011), and fibroblastic phenotype (vibrations: Tong et al., 2013).

Direct mechanical stimulation can stimulate *ECM deposition* in engineered cartilage (dynamic compression: Démarteau et al., 2003), bone (cyclic bending: Mauney et al., 2004), and heart valves (pressure and pulsatile flow: Flanagan et al., 2007).

Physical conditioning has been shown to be an effective means to improve the *cell/tissue structural organization* (see Chapter 3): cyclic strain modulated the morphology of engineered skeletal muscle (Powell et al., 2002),

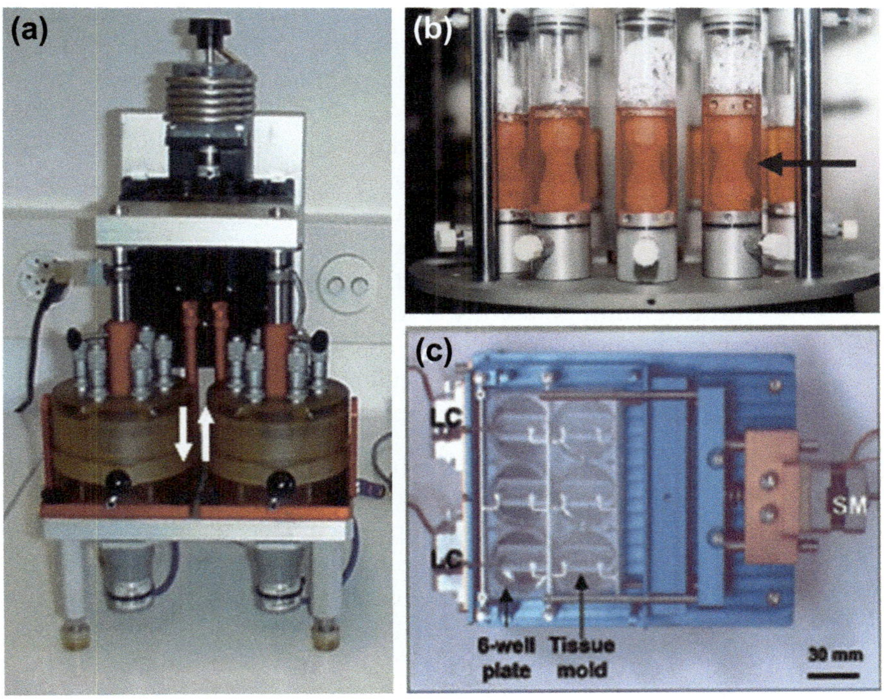

FIGURE 12.7

Bioreactors for mechanical conditioning of engineered constructs. Examples of bioreactor systems designed to apply (a) dynamic compression (Démarteau et al., 2003), (b) torsion and tension (Altman et al., 2002), or (c) cyclic stretching (Powell et al., 2002).

cardiac (Zimmermann et al., 2002), and vascular tissues (Weidenhamer and Tranquillo, 2013); electrical stimulation promoted elongation and ultrastructural organization in cardiomyocytes (Tandon et al., 2009).

Numerous mechanical loading protocols have been tested to enhance a *specific tissue function*: surface motion applied to engineered cartilage increased the expression and release of lubricating proteins (Grad et al., 2005); cyclic stretch guided collagen fiber formation and improved tensile strength of engineered tendons (Issa et al., 2011); pulsatile flow increased rupture and suture retention strength in engineered vascular constructs (Niklason et al., 1999).

Although much has been learned regarding the role of biochemical and biophysical cues on cell function and tissue growth, our understanding of the biological processes is limited by traditional analyses and mechanical testing methods, which require termination of the culture and destruction of the sample. To overcome these limitations, much effort is being devoted in recent years to the development of quantitative noninvasive and nondestructive techniques for online monitoring of the construct developmental state. Online force measurements allowed to monitor over time the progression of the mechanical properties of bioreactor-cultured grafts in the case of cartilage (stiffness under compression: Preiss-Bloom et al., 2009) and vascular grafts (burst pressure in a pulsatile flow bioreactor: Syedain et al., 2011), whereas advanced imaging techniques have enabled the online monitoring of the morphological properties of engineered constructs, in terms of amount, composition, and distribution of the ECM, which is being deposited throughout a scaffold during bioreactor culture. Optical coherence tomography has been successfully employed in the context of tendon (Bagnaninchi et al., 2007) and blood vessels (Gurjarpadhye et al., 2013), whereas the use of microcomputed tomography has been reported in the context of bone engineering (Porter et al., 2007). The integration of quantitative noninvasive and nondestructive technologies will aid in better understanding the effects of physiological stimuli on cells/tissues not only for the goal of engineering more functional tissues but it could also help to understand the response of an engineered tissue when implanted in vivo into a load-bearing site.

In this context, the effective need to generate mature grafts is still controversial: indeed, beyond a few cases for which graft functionality might need to be highly developed prior to implantation, such as for engineered blood vessels (Niklason et al., 1999) and heart valves (Hoerstrup et al., 2002), it is becoming increasingly clear that tissue maturation could be more efficiently induced by the physiological, biochemical, and mechanical cues of the body as an "in vivo bioreactor," provided that a graft of "sufficient" quality is implanted. Bioreactors therefore have the role of actively modulating the growth of engineered grafts to the level required to autonomously progress to engraftment, remodeling, and tissue regeneration once in vivo.

12.3 BIOREACTOR-BASED IN VITRO MODEL SYSTEMS

It has become increasingly clear that conventional 2D culture systems (Petri dishes, culture flasks) cannot adequately recapitulate the microenvironment experienced by cells in vivo. Cell differentiation and tissue development in vivo are strongly dependent on cell spatial arrangements and directional cues, which simply cannot be simulated in 2D (see Chapter 13). Therefore, 3D model systems, which have been developed to mimic specific aspects of the actual cellular microenvironment that exists in vivo, will be vital to gain a greater understanding of basic cell and tissue functions within the native milieu (Figure 12.8). In the previous sections, we have highlighted how controlled bioreactor systems can be used to elucidate the roles of specific biochemical and biophysical stimuli on cell/tissue response, generally in the context of engineering implantable 3D tissues grafts. In this section, we will illustrate that bioreactors can also play a central role as in vitro model systems

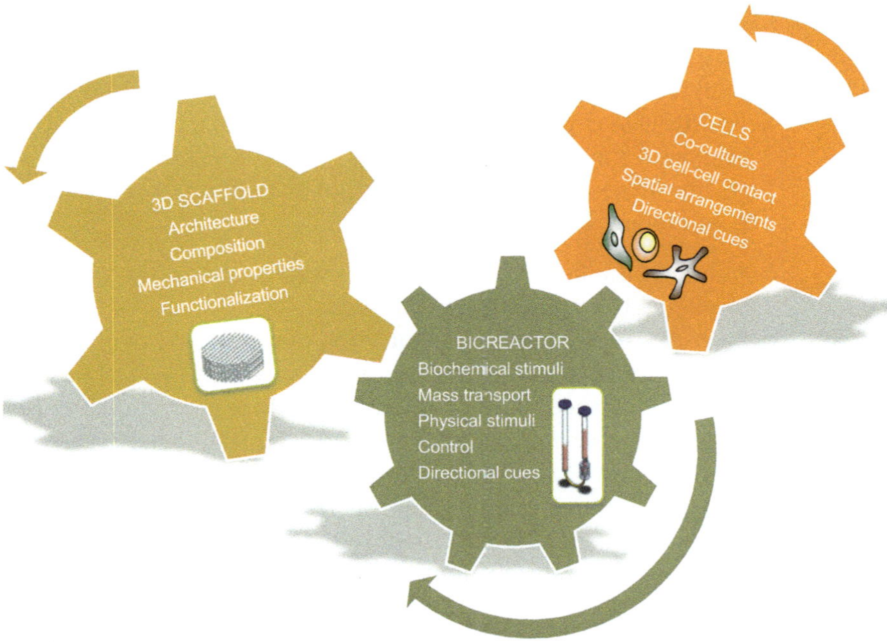

FIGURE 12.8

Schematic illustration of the key parameters that can be modulated in a 3D bioreactor-based in vitro model system. Bioreactors can support cell cultures on 3D scaffolds, which can be adapted to resemble the in vivo 3D microenvironment of a specific cell type and relative cell–matrix interactions (see Chapters 6 and 10). The high degree of flexibility and control over various culture parameters is crucial to replicate the physiological complexity of a native tissue and/or to establish the formation of a functional niche.
3D, three-dimensional.

to investigate the response of engineered tissues upon implantation and for fundamental research in the areas of stem cell biology, disease pathogenesis and drug screening, and oncology.

At the end of Section 2, the point was raised that the need to generate fully mature grafts is still controversial, begging the question, "how mature/functional should a graft be before it can be implanted into the repair site?" Bioreactors can help address this question by providing controlled technological platforms to test and predict how engineered tissue grafts would respond once implanted in vivo and exposed to the associated physical forces. Using a dynamic compression bioreactor, engineered cartilage constructs at different stages of development were subjected to a loading regime resembling a mild postoperative rehabilitation. The response of engineered tissues was correlated with the amount of glycosaminoglycans (GAGs) in the constructs prior to loading, with potential implications in defining functional quality controls for engineered tissues or identifying the most appropriate regimes of physical rehabilitation for a specific graft (Démarteau et al., 2003). Bioreactors have also been used to determine whether cells, which are not normally subjected to physiological loading, could respond to physical stimuli. Engineered constructs generated from nasal chondrocytes were subjected to physical forces resembling joint loading using different bioreactors and were found to respond positively as engineered constructs from articular chondrocytes. This suggested that nasal chondrocytes, which are not subjected to loading in vivo, could represent an appropriate cell source for the repair of articular cartilage defects (Candrian et al., 2008).

Although the field of tissue engineering has primarily focused on engineering implantable grafts for potential clinical applications, the technologies that have been developed offer powerful tools that can potentially revolutionize other areas of research in the biomedical sciences. A number of factors including soluble signals, specific cell–cell and cell–matrix interactions, as well as structural cues and physical forces, are known to be crucial for stem cell maintenance and function (see Chapter 2). Although in vivo studies have been instrumental to elucidate some of the mechanisms by which bone marrow niches exert their function on hematopoietic stem cells (HSCs), the establishment of an in vitro model that recapitulates the fundamental interactions of the niche components in a controlled setting would be of great benefit. An in vitro model system was developed in which freshly harvested bone marrow-derived cells, including both stromal and hematopoietic fractions, were cultured within porous scaffolds in a perfusion bioreactor system (Braccini et al., 2005). The system allowed to bypass the standard 2D expansion phase on plastic, typically associated with the selection of a restricted subset of mesenchymal populations, and resulted in the formation of an organized 3D stromal-like niche, which was able to support hematopoiesis. Based on this model, a perspective was subsequently described as to how the bioreactor-based model system could be further exploited to improve stem cell-based therapies and extended toward the engineering of other specialized stromal niches (Figure 12.9) (Di Maggio et al., 2011).

By replicating key aspects of native microenvironments, bioreactors provide more physiological in vitro models also for studies of disease pathogenesis and drug screening. A hollow fiber bioreactor-based model system was developed for studying cell stroma-dependent hematopoietic/leukemic cell functions and ex vivo expansion. Localization of leukemic cells to stromal cells was significantly enhanced in the bioreactor, along with sustaining prolonged cell growth and inhibited spontaneous erythroid differentiation, mimicking physiological conditions in the bone marrow (Usuludin et al., 2012). Dynamic 3D perfusion culture of adipose-derived stem cells in a multicompartment hollow fiber bioreactor allowed the generation of 3D adipose tissue, and better

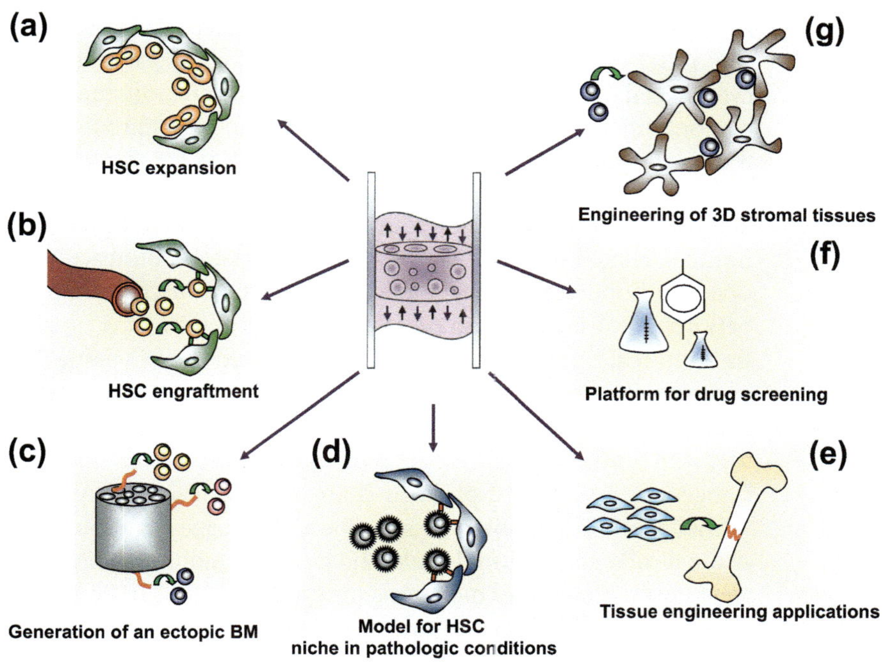

FIGURE 12.9

Perspectives of 3D bioreactor-based culture model systems in the context of stem cell biology and regenerative medicine therapies. The proposed in vitro model could allow one (a) to efficiently expand HSCs for bone marrow transplantation; (b) to study the role of accessory cells in HSC homing and engraftment; (c) to generate an ectopic bone marrow that could provide a healthy environment in cases in which the niche is damaged; (d) to dissect the mechanisms related to stem cell maintenance in pathologic conditions; (e) to develop cell-based engineered grafts for tissue repair; (f) to identify and test new drugs for cancer and bone-related disorders; (g) to reproduce the 3D spatial organization of specialized stromal cells, fundamental to support stem cell function in several tissues (e.g., the thymus) (Di Maggio et al., 2011). HSC, hematopoietic stem cell.

resembled the architecture of native tissue compared to cells grown statically in 2D. Reproducible metabolic activity of the engineered tissue in the bioreactor was demonstrated (increased and inhibited glucose uptake in response to insulin and tumor necrosis factor-alpha), which makes it a potential powerful tool to study adipose tissue function and drug therapies (Gerlach et al., 2012). A bioreactor-based model of normal and diseased kidney tissue could sustain functional kidney tubules and cysts with relevant in vivo-like complexities for prolonged periods of time, an important step for the in vitro investigation of pathophysiology and drug screening (Subramanian et al., 2010). Human arteries mimicking the structural and functional characteristics of native tissues were engineered in a bioreactor and tested as an in vitro model to investigate the fundamental processes involved in atherosclerosis. The accumulation of lipoproteins in the intima, as well as the binding and transmigration of monocytes under dynamic pulsatile flow conditions, demonstrated the suitability of the model (Robert et al., 2013).

A small yet growing number of cancer researchers are also moving toward the use of 3D cell culture systems where physiological patterns of cell adherence, cytoskeletal organization, migration, signal transduction, morphogenesis, proliferation, differentiation, and response to therapeutic stimuli could better mimic the in vivo situation. In this context, bioreactors are essential to sustain long-term culture of uniform 3D tissues and can provide additional technical means to perform controlled studies aimed at understanding specific biological phenomena related to tumor biology and metastases. A bioreactor originally developed for bone engineering applications has been used to study the interaction between metastatic human breast cancer cells and in vitro engineered bone-like tissue in real time. Breast cancer cells colonized the 3D tissue in the bioreactor and recapitulated events observed in clinical samples of bone metastasis from breast cancer (Krishnan et al., 2011). Human bone stromal cells, after 3D coculture with human prostate cancer cells in a RWV bioreactor, underwent permanent cytogenetic and gene expression changes, with reactive oxygen species serving as mediators. These observations collectively suggest coevolution of cancer and stromal cells that occurred under 3D growth conditions; this ultimately accelerates cancer growth and metastasis (Sung et al., 2008). Using a decellularization chamber and a bioreactor, human lung cancer cell lines could be grown on a decellularized rat lung matrix in a manner that mimics human lung cancer, and provides a valuable model to improve understanding of the tumor microenvironment and angiogenesis in lung tumor development (Mishra et al., 2012). Using in vitro bioreactor models, it was shown that physiologically relevant interstitial fluid flow is an active regulator of glioma invasion. This is an important finding, as current therapies often lead to edema and increased flow, and particular experimental treatments have been aimed at increasing the delivery of therapeutics by enhanced convective flow (Munson et al., 2013).

The reported examples highlight how 3D bioreactor-based culture systems, supported by continuous progress in material- and tissue-engineering fields, together with cell and molecular biology, could play a crucial role in unraveling the mechanisms regulating cell physiology in normal and pathological conditions, and on the development/validation of new therapeutic strategies for a variety of clinical needs.

12.4 BIOREACTORS AS TISSUE MANUFACTURING DEVICES

A major challenge to bring a tissue-engineered product into routine clinical practice (see Chapter 22: Clinical translation) is the translation of research-scale production models into clinically applicable manufacturing designs that are compliant with good manufacturing practice (GMP) requirements and are economically sustainable. In this section, we discuss how innovative and low-cost bioreactor systems that automate, standardize, and scale up the production process will be essential to reach this ambitious goal (Figure 12.10).

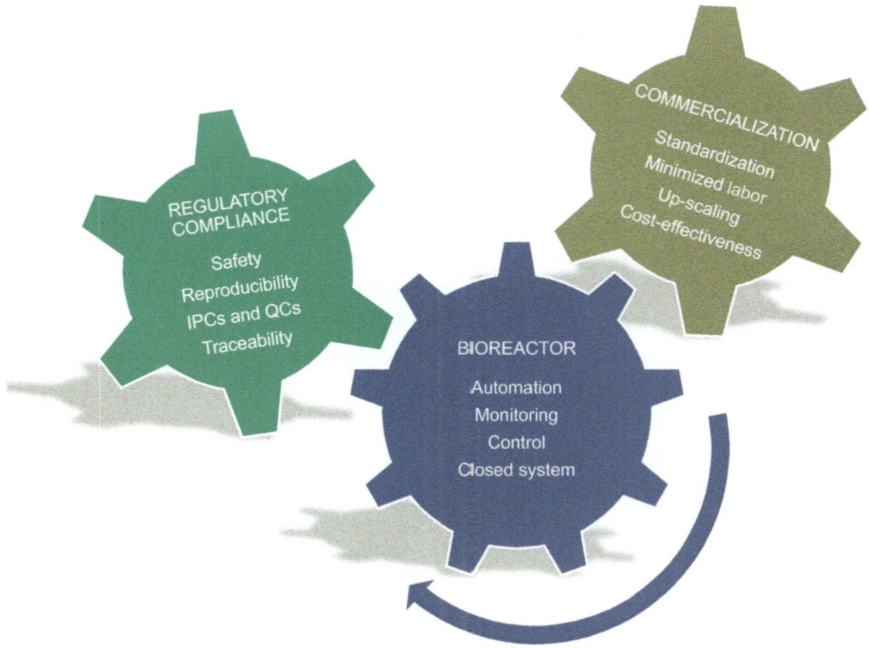

FIGURE 12.10

Schematic representation of the role of bioreactors in the clinical translation of engineered tissue products. Bioreactor-based manufacturing systems offer great opportunities for both regulatory compliance and commercialization (IPC = in-process control, QC = quality control).

12.4.1 Regulatory Aspects

In this section, we illustrate the regulatory challenges (see Chapter 21) that face conventional manufacturing models and discuss how bioreactor-based platforms can address these challenges through (1) automation, (2) monitoring, (3) feed-back control, and (4) process streamlining.

The manufacturing of engineered tissues has traditionally been based on conventional bench top cell and tissue culture techniques, which require a large number of manual and labor-intensive manipulations, and thus impose significant challenges in terms of regulatory compliance. Bench-level production of cell-based therapies in preclinical and clinical stages requires skilled and highly trained personnel working for long hours in stringent clean room conditions. Maintaining an adequate level of consistency and quality to satisfy regulatory requirements is highly challenging due to both interoperator and intraoperator variabilities and operator errors. Moreover, the number of manual manipulations poses a significant threat to the safety of the product due to risks of contamination. Automated bioreactor-based production systems can minimize the requirements for operator handling, and can thus greatly improve process standardization, ensure more reproducible product quality, and guarantee safety of the product by mitigating contamination risks.

In traditional production systems, the process of monitoring the progression of the culture often relies on purely qualitative observations (e.g., cell confluence and morphology in 2D cultures, and color of the culture medium), which implies strict operator dependence. Periodic off-line measurements of biochemical factors in the culture medium are possible, but are affected by intrinsic risks associated with manual sampling (e.g., inappropriate sample storage/processing, and contamination), and reduced accuracy and significance due to lag-time and possible artifacts. Bioreactors, which implement sensing and online monitoring, offer great advantages to provide noninvasive and quantitative data relevant to the process and the engineered graft. Typical environmental factors could be routinely monitored, as well as biochemical parameters correlated with cell number, phenotype, and metabolism. For example, strategies to monitor cell proliferation via online measurements of dissolved oxygen in the culture medium have been reported (Janssen et al., 2006; Santoro et al., 2012); these provide meaningful IPCs, data on graft quality (QC), and traceability data. As technologies mature, the integration of innovative techniques for online assessment of the structural and functional properties of the developing tissue (Section 2.4) will enable novel noninvasive approaches for IPCs and product release criteria; this provides a higher level of traceability of key manufacturing data, as well as improved process robustness and product safety. Together with software-based data management platforms, online IPC/QC strategies will help reduce the workload and risks of error associated with manual traceability and facilitate compliance with GMP regulations.

Additional benefits of bioreactor-based manufacturing processes rely on the use of sensing technologies not only to monitor the culture progression but also to optimize the control of culture parameters at predefined levels and to adapt them to the evolving needs of the culture/engineered tissue in a feedback-controlled loop. Such an approach would considerably aid to improve process robustness and standardization for reduced process and product variability. Feedback-controlled medium exchange aimed at maintaining homeostatic parameters to predefined set points (e.g., oxygen and pH) has been recently implemented in different bioreactors (Hülsmann et al., 2013; Li et al., 2014). The flow rate in a perfusion bioreactor could also be controlled and automatically adjusted according to oxygen concentrations and pH measurements in the center of cultured constructs (Volkmer et al., 2012).

Advanced control systems and automation, along with simplified and streamlined processes, would facilitate the development of a fully closed single-unit bioreactor system; this further minimizes operator-dependent procedures, with increased compliance to safety regulations. As an example, development and preclinical testing of such an approach are being undertaken in the context of cartilage tissue engineering by a European community-funded consortium, BIO-COMET (www.biocomet.eu), gathering up different core competences from academic, clinic, and industrial partners. A fully automated and closed bioreactor-based manufacturing system could potentially run under less stringent clean room conditions as compared to the manual manufacturing processes of today. Operation in Class-C clean rooms would considerably simplify the overall management and flow of the manufacturing process and could dramatically reduce costs. However, in this regard, ambiguous and incomplete regulatory guidelines currently hinder the definition of clear and precise bioreactor design requirements to ensure compliance to GMP requirements.

12.4.2 Manufacturing Costs

We have discussed how bioreactors can improve the quality and reproducibility of engineered tissues, as well as facilitating compliance to regulatory guidelines. Here, we highlight how automated bioreactor-based manufacturing systems can ultimately improve the economic viability of engineered tissue products.

As previously underlined, the manufacturing of cell-based grafts has traditionally been based on conventional manual bench top cell and tissue culture techniques. It is generally recognized that the use of these labor-intensive processes results in complex production protocols, extensive operator training, continuous QC oversight, and high infrastructure costs related to facilities and consumables. While these economic factors are generally absent or minimal in many clinical fields, they are quite significant in the field of tissue engineering. It is imperative that every effort is made to manage production costs; otherwise, the clinical delivery costs of the engineered product will overshadow and undermine the clinical gains relative to the current standard of care.

As an alternative to costly conventional methods, bioreactors that automate the individual steps of a graft manufacturing process can minimize the amount of manual processing labor required; this has a profound impact on production costs. Moreover, in-process monitoring techniques and data management systems, which automatically collect and archive the process data, could significantly reduce labor costs associated with QC and quality assurance.

Although most bioreactors used in tissue engineering to date have been designed and used for cell seeding and/or tissue culture, ideally a bioreactor-based manufacturing system would control and automate all bioprocesses, from the initial cell isolation through the production of the implantable graft. The development of such a closed system would not only have great benefit in terms of regulatory compliance, but it would also have the potential to improve the cost effectiveness of a manufacturing process. The use of a closed system increases the safety of the process, and thereby minimizes product loss due to contamination. As crosscontamination risks would be minimized, closed-system bioreactors also allow greater throughput within a single controlled manufacturing area, and increase the efficiency of costly facility infrastructure. Moreover, although currently all cell-based TE products must be manufactured within costly Class-B and Class-A clean room facilities (see Chapter 21), closed-system bioreactors could potentially be operated under less stringent clean room standards in Class-C. This would have considerable impact on production costs by reducing the building, maintenance, and operating costs of the GMP facility (~10-fold higher for Class-B vs Class-C), labor costs (e.g., time for sterile dressing), and capital equipment costs (clothes, equipment, and environmental controls).

In the early stages of product commercialization, when production volumes remain low, these labor and facility requirements may not appear that formidable. However, as product demand increases, since the production of autologous-based grafts must be performed separately and as independent batches for each patient, a manufacturing process based on manual methods faces pragmatic limits in terms of labor and infrastructure. Alternatively, production systems based on automated and closed bioreactor systems are ideally suited to progressive upscaling by adding more units to the production as product demand increases (i.e., scale-out). Failing to integrate a scalable process design from an early phase of development can lead to late stage failure since revisions to an established process require comparability testing, multiple regulatory agencies reviews, and the potential need for additional clinical trial confirmations.

Engineered tissues are currently manufactured within *centralized* GMP production facilities (i.e., within a company's own GMP facility). As an alternative, fully automated closed bioreactor systems would make a *decentralized* manufacturing strategy plausible; this allows multiple off-site graft production locations, for example, within the confines of competent hospitals. While manufacturing a product within a central location has the clear advantage of enabling close supervision and control over the entire production process, the logistics (i.e., packaging, shipping, and tracking) of transporting living biopsies and engineered grafts (under defined and monitored transport conditions, within defined timeframes) can become significantly complex and costly.

Whether manufacturing is centralized or decentralized, automating and controlling each of the required bioprocesses throughout the engineering procedure will necessitate state-of-the-art technology. Therefore, to minimize capital and operating costs for the system, it is crucial to invest efforts in characterizing the processes and products so as to understand precisely which parameters are critical for the product performance. The essential parameters to monitor and control, together with operating limits, should be identified to standardize production and provide meaningful quality/traceability data, and at the same time minimize risks, costs, and user complexity. In this context, complex and costly systems, which essentially mimic established manual procedures, will unlikely be able to demonstrate real cost benefit by replacing manual cell culture techniques in a manufacturing process. Alternatively, manufacturing processes could be dramatically simplified and costs could be significantly reduced if the conventional tissue-engineering paradigms were replaced with novel concepts and techniques that *streamline* the individual bioprocessing steps. Simplified tissue-engineering processes could be key to future manufacturing strategies by requiring only a minimal number of processes; this facilitates simplified and compact bioreactor designs with limited automation requirements, with the likely result of lower product development and operating costs.

The advantages of a bioreactor-based manufacturing system appear convincing, yet bioreactor technology has not been widely adopted. This will require a change in the mindset to break away from the well-established manual cell culture methods that are used in virtually all biological laboratories. The commercial availability of user-friendly bioreactors, which have slowly begun to emerge onto the market, could help automated systems gain greater acceptance, given that complex and technically challenging systems may discourage their initial introduction during the research phase. Moreover, a sound analysis that critically assesses the potential cost effectiveness of a bioreactor-based production system could help to attract the investments necessary to develop and/or implement sophisticated manufacturing technologies.

In the following, we assess the economic factors of tissue-engineering manufacturing processes, and examine the potential for automation to drastically reduce clinical delivery costs. While the economic gains attributable to automation will be specific to the details of any given process, it is possible to define a generic model to assess the overall magnitude of economic impact. In the generic model shown in Table 12.1, key cost centers for the production process are assessed for manual versus automated production platforms. Each step in the conventional culturing process can be analyzed to determine automation-related cost reductions.

Table 12.1 Cost Analysis of Manual versus Automated Manufacturing

Cost Categories for Clinical GMP Production	Reference	Duration of Culture							
		Manual Cost per Treatment				Automated Cost per Treatment			
		1 Week	2 Weeks	3 Weeks	4 Weeks	1 Week	2 Weeks	3 Weeks	4 Weeks
Labor	Scaling (%)	100%	120%	200%	400%	30%	40%	50%	60%
Process technician	$160 per hour	1600	1920	3200	6400	480	640	800	960
Quality control	$160 per hour	1600	1920	3200	6400	480	640	800	960
Quality assurance	$240 per hour	1920	2400	3360	4800	960	1200	1680	2400
Consumables	Scaling (%)	100%	200%	500%	800%	50%	100%	250%	400%
Reagents		1000	2000	5000	8000	500	1000	2500	4000
Plasticware		300	600	1500	2400	150	300	750	1200
Bioreactor disposables		0	0	0	0	1500	1500	1500	1500
Capital equipment	Scaling (%)	100%	200%	500%	800%	20%	60%	180%	200%
Laboratory equipment		280	560	1400	2240	56	168	504	560
Bioreactor automation unit		0	0	0	0	400	800	1200	1600
Facility	Scaling (%)	100%	200%	500%	800%	20%	60%	180%	200%
Clean room		580	1160	2900	4640	116	348	1044	1160
Certifications		100	200	300	400	20	60	180	200
Total		$7,380	$10,760	$20,860	$35,280	$4662	$6,656	$10,958	$14,540

GMP, good manufacturing practice.

The analysis compares projected costs per implant for manual versus automated production of cell therapeutics and engineered tissue grafts. The cost categories are those typically considered in business models and are subdivided to highlight differences related to manual and automated processing. The model is based on generic assumptions applicable to GMP production within the cell therapy industry; as such, strict application of the conclusions must be adjusted in keeping with therapy-specific conditions. The model utilizes scaling factors that recognize process complexity for increasing production periods. The key assumptions underlying the scaling factors are as follows:

1. Clinical deployment of cell therapeutics and engineered grafts require cell populations in the range of 10 million to 1 billion cells per dose, depending on the clinical application.
2. Production of 10 million cells without secondary differentiation and tissue formation would be representative of relatively short culture periods, whereas higher dose treatments require significant population doublings and associated extended durations.
3. Implementation of tissue-engineering practices involving cell differentiation and ECM deposition also require extended culture periods.
4. The reference for all scaling is based on one week of manual culture (this is defined as 100%).
5. An increase in scaling percentage reflects the additional resources required based on accommodating increased cell populations and all related workflow activities.
6. A decrease in scaling percentage, as shown for automation, reflects the economy attainable in each cost category based on the efficiency of automated operation as highlighted below.

The generic assumptions in each cost category are as follows:

1. The labor rate includes salaries, benefits, the extensive initial training to operate in a GMP facility, and ongoing testing of the operators' techniques.
2. Estimated labor time includes all steps performed by the Process Technician (e.g., biopsy preparation, cell isolation, media changes, and final implant preparation for transport), and gowning time to move between clean room zones.
3. The role of the QC personnel is to provide continuous operator oversight consistent with regulatory policy.
4. The role of quality assurance personnel is to manage the implementation of quality systems related to batch production and release.
5. Reagents and consumables have been estimated at \$1000 and \$300, respectively, for a one week manual process; however, practical implementation for different clinical indications can result in very significant deviations from this estimate based on the use of specialized culture additives such as growth factors.

6. Capital costs of laboratory equipment (e.g., incubators, biological safety cabinets, microscopes, and fridges) and bioreactor automation units have been amortized over the equipment lifetime and divided by the number of cultures that would utilize this equipment.
7. Facility costs have been calculated based on Class-B clean room rental rates, including support staff to maintain the facility and equipment (note that this is similar to the cost of GMP facility construction and validation, amortized >10 years).

When considering the implementation of automated bioreactors for the production of cells and tissues, a number of economic factors influence cost of goods. The factors described below become increasingly influential during scale-up of cell and tissue production from, for example, 1000–10,000 units per year:

1. The role of automation is to fundamentally reduce operator engagement, and thereby reduce overall labor time down to 30% for week one with minimal on-going manual involvement required as the culture process progresses.
2. The requirement for QC personnel is similarly reduced in keeping with the level of activity of the process technician.
3. Automation allows a 50% reduction in the quality assurance labor that is required for in-process and product release testing, due to the ability to incorporate in-process monitoring, automate sample collection, and provide automatic process documentation.
4. The specifications and delivery of process reagents can be optimized through the use of automation due to the ability to provide precise quantities according to ideal temporal conditions (e.g., delivery of media based on measured cell number).
5. Adoption of bioreactors within a cell production process reduces dependency on secondary capital equipment as substantially all processing occurs within closed, dedicated, and cell processing modules.
6. An advantage of closed-system processing within bioreactors for cell production is the ability to contain and monitor the process within an efficient physical envelope; this thereby enables a high process throughput within given facility constraints;

The results of the model indicate that the adoption of automated bioreactors provides economic benefits in all business cost centers. Labor requirements are dramatically lower, as would be expected, but also there are less obvious advantages related to the ability to improve the efficiency of biological processes through systematic optimization only available in automated environments. For example, the use of reagents can be optimized to minimize waste, maximize shelf life, and only deliver reagents when required with the specified volume. With the decreased need for ongoing monitoring and manual manipulation, the use of laboratory equipment per process also decreases. Further, on the basis that the automated bioreactors provide closed-system operation, crosscontamination risks are addressed; this allows greater throughput within a single controlled area and thereby increases the efficiency of facility infrastructure.

Automation strategies involving the controlled environment available within bioreactors offer unique opportunities to substantially improve the economic viability of cell therapy and tissue engineering. The use of these techniques not only provides critical cost reductions in an emerging field of clinical practice but it also underpins gains in product quality and product sophistication unavailable under current manual processing conditions.

12.5 CONCLUSIONS AND FUTURE PERSPECTIVES

The ex vivo generation of living tissue grafts has presented new biological and engineering challenges for establishing and maintaining cells in 3D cultures, therefore necessitating the development of new biological models as compared to those long established for traditional cell culture. In this context, bioreactors represent a key tool in all processes involved in the engineering of 3D tissue products and offer much potential for improving the quality of engineered tissues. The possibility of reproducible and controlled 3D cultures also support fundamental studies aimed at investigating mechanisms of cell function/differentiation and tissue development in biologically relevant models. Understanding the role of specific biochemical or physical factors in tissue development will in turn not only support a more efficient in vitro engineering of grafts but also set the stage for the application of those same factors directly at the site where tissue has to be regenerated and opens up new perspectives for the development/validation of novel therapeutic strategies in different medical fields.

Moreover, fundamental knowledge gained at the research level will be essential to define, optimize, and streamline the key processes required for efficient manufacturing models, essential for the successful exploitation of engineered tissue products for widespread clinical use. As opposed to conventional manual production techniques, automation and control in bioreactor systems will enable safe, standardized, and quality-assured manufacturing processes, and facilitate compliance to relevant regulatory frameworks. Innovative, user-friendly, and low-cost closed bioreactor systems will maximize the potential for smooth upscaling and cost-effective production of engineered tissue products. As a result of improved quality, regulatory compliance and economic viability of tissue-engineered products, the implementation of bioreactor-based manufacturing processes will facilitate broader commercialization and clinical implementation of novel and powerful therapeutical approaches.

Close collaboration among academic institutions, industrial partners, and regulatory agencies will be crucial to expedite the route to successful clinical translation of engineered tissue products. Joined forces from academia and companies would provide solutions to open technical challenges and support industrialization of the processes to be user-friendly, cost-effective, and compliant with regulatory criteria. Considering that specific and clear regulatory requirements remain to be well defined for Advanced Therapy Medicinal Products, close interactions with regulatory bodies will be essential to clarify and shape the current framework.

12.6 SNAPSHOT SUMMARY

- Bioreactors can improve the quality (i.e., efficiency and spatial distribution) and reproducibility of the process of cell seeding into 3D porous scaffolds.
- Convection in bioreactor systems can enhance mass transport of nutrients and waste products to and from the cells within engineered constructs. Perfusion bioreactors have great potential to minimize mass transfer limitations and support cell viability and uniform matrix accumulation within large 3D constructs.
- Improved mass transport in bioreactors can efficiently assist in the decellularization of native tissues/organs, and in subsequent cell repopulation.
- Physical conditioning in controlled bioreactor systems has the potential to modulate cell differentiation, and improve the structural and functional properties of engineered tissues.
- By recapitulating key aspects of the in vivo cellular microenvironment, bioreactors can provide in vitro model systems to investigate cell function and tissue development in 3D environments.
- Technological platforms to noninvasively and nondestructively monitor the culture progression in real-time support the optimization and efficient use of bioreactor-based 3D model systems, in both scientific research and clinically compliant tissue manufacturing.
- Automated, standardized, and closed bioreactor-based manufacturing processes possess great benefits in terms of safety, traceability, and regulatory compliance.
- By automating and streamlining the individual steps of a graft manufacturing process, bioreactors provide economic benefits in all business cost centers.
- Regulatory compliant and cost-effective manufacturing strategies will be central for the successful exploitation of engineered tissue products for widespread clinical use.

CLASSICAL EXPERIMENT

Articular cartilage derives its mechanical function from its matrix, which consists of tissue fluid and a framework of structural macromolecules (collagens, proteoglycans, and noncollagenous proteins and glycoproteins). During the development, maintenance, and remodeling of cartilage, chondrocytes synthesize appropriate types and amounts of macromolecules and assemble them into a highly organized matrix. As adult articular cartilage has a limited capacity for self-repair, significant efforts have been made in an attempt to restore tissue composition, structure, and function, including the development of engineered cartilage. Cartilaginous constructs have been grown in vitro by using isolated cells, biodegradable polymer scaffolds, and bioreactors. In this classical study, the effect of hydrodynamic conditions on the composition and mechanical properties of engineered cartilage was investigated by culturing bovine calf auricular chondrocytes on fibrous polyglycolic acid scaffolds in three different culture environments: static conditions, spinner flasks, and RWV bioreactor. After 6 weeks of cultivation, the composition, morphology, and mechanical properties of the constructs all depended on the conditions of in vitro cultivation. Static culture yielded small and fragile constructs, whereas turbulent flow in spinner flasks yielded constructs with fibrous outer capsules: both environments resulted in constructs with poor mechanical properties. The constructs that were cultured freely suspended in a dynamic laminar flow field in RWV bioreactors were the largest, contained cartilage-like ECM with the highest fractions of GAG and collagen, and had the highest mechanical properties. This study presents a classical experimental approach and early findings suggesting that the hydrodynamic conditions in bioreactors can modulate the composition, morphology, and mechanical properties of in vitro engineered cartilage (Vunjak-Novakovic et al., 1999).

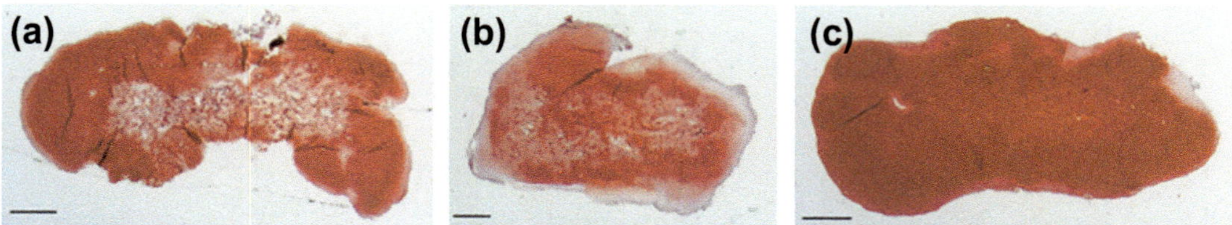

Safranin-O-stained cross-sections of engineered cartilage tissues after six weeks of culture under different hydrodynamic conditions (glycosaminoglycan (GAG) is stained red). (a) Statically cultured constructs contain central regions with negligible GAG deposition. (b) Spinner flask cultured constructs stain more intensely for GAG in their central regions but are encapsulated by fibrous tissue at their periphery. (c) Rotating-wall vessel (RWV) cultured constructs stain intensely and homogeneously for GAG throughout their cross-sectional area (scale bars: 1 mm).

STATE-OF-THE-ART EXPERIMENT

Diseases that compromise large airways seriously threaten the duration and quality of life of many patients due to altered breathing, speech, and swallowing. Most tracheal lesions can now be resected and primary reconstruction safely performed. Unfortunately, the resectable length is restricted to approximately 40–50% of the total length of the organ in adults and one-third in children, and any treatment beyond that requires the development of a safe and functional tracheal replacement. Attempts to provide autologous, allogeneic,

STATE-OF-THE-ART EXPERIMENT *Continued*

or synthetic grafts have been disappointing so far. Tissue engineering has emerged as the most promising technique to generate functional tracheal substitutes, with the first clinical application successfully performed in a patient with end-stage airway disease (Macchiarini et al., 2008). For the scaffold, a detergent-enzymatic decellularization method was applied to a human donor trachea to produce a nonimmunogenic long tubular scaffold with suitable structural integrity. Autologous bone marrow-derived MSCs (then differentiated into chondrocytes) and respiratory epithelial cells were isolated, characterized, and expanded for cell repopulation of the external hyaline cartilage framework and internal epithelial covering, respectively. To provide different cell culture conditions on either side of the organ wall and adequate mass transport throughout the critically sized construct, a rotating double-chamber bioreactor was designed, which allowed confined seeding and culturing on either surfaces of the tubular matrix and included rotatory movement of the scaffold around its longitudinal axis (Asnaghi et al., 2009). An analytical model was also developed to predict oxygen profiles in the bioreactor-cultured construct and in the culture media, and to compare representative culture configurations and operating conditions. The two cell types could be efficiently seeded and cultured on either side of the long tubular scaffold in the bioreactor, with minimal construct manipulation and contamination risks. The engineered construct was successfully implanted in the patient as a replacement for the left main bronchus and immediately provided a functional airway and improved quality of life.

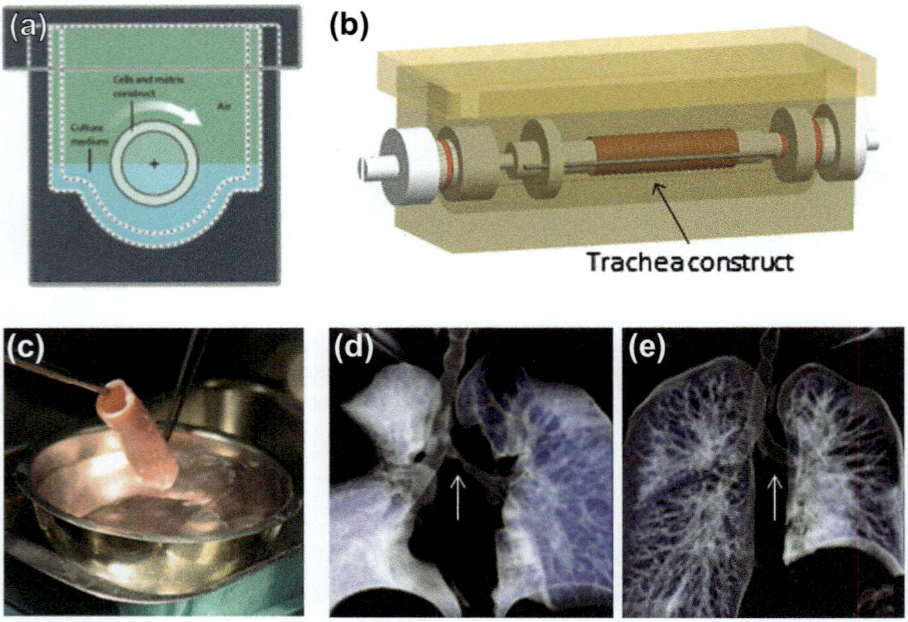

(a) Schematic lateral view and (b) 3D rendering of the bioreactor. By immersing half of the construct in media at any one moment, rotating air-medium interfaces are provided in both luminal and outer compartments. (c) The final graft immediately before surgical implantation. (d, e) Volume-rendering computed tomography before and 1 month after engraftment: terminal long segment narrowing is completely reversed after surgery (arrows).

REFERENCES

Altman, G.H., Horan, R.L., Martin, I., Farhadi, J., et al., 2002. Cell differentiation by mechanical stress. FASEB J. 16 (2), 270–272.

Alvarez-Barreto, J.F., Linehan, S.M., Shambaugh, R.L., Sikavitsas, V.I., 2007. Flow perfusion improves seeding of tissue engineering scaffolds with different architectures. Ann. Biomed. Eng. 35 (3), 429–442.

Asnaghi, M.A., Jungebluth, P., Raimondi, M.T., Dickinson, S.C., et al., 2009. A double-chamber rotating bioreactor for the development of tissue-engineered hollow organs: from concept to clinical trial. Biomaterials 30 (29), 5260–5269.

Awad, H.A., Butler, D.L., Harris, M.T., Ibrahim, R.E., et al., 2000. In vitro characterization of mesenchymal stem cell-seeded collagen scaffolds for tendon repair: effects of initial seeding density on contraction kinetics. J. Biomed Mater. Res. 51 (2), 233–240.

Bagnaninchi, P.O., Yang, Y., Zghoul, N., Maffulli, N., et al., 2007. Chitosan microchannel scaffolds for tendon tissue engineering characterized using optical coherence tomography. Tissue Eng. 13 (2), 323–331.

Braccini, A., Wendt, D., Jaquiery, C., Jakob, M., et al., 2005. Three-dimensional perfusion culture of human bone marrow cells and generation of osteoinductive grafts. Stem Cells 23 (8), 1066–1072 (Dayton, Ohio).

Bueno, E.M., Laevsky, G., Barabino, G.A., 2007. Enhancing cell seeding of scaffolds in tissue engineering through manipulation of hydrodynamic parameters. J. Biotechnol 129 (3), 516–531.

Candrian, C., Vonwil, D., Barbero, A., Bonacina, E., et al., 2008. Engineered cartilage generated by nasal chondrocytes is responsive to physical forces resembling joint loading. Arthritis Rheum. 58 (1), 197–208.

Carrier, R.L., Papadaki, M., Rupnick, M., Schoen, F.J., et al., 1999. Cardiac tissue engineering: cell seeding, cultivation parameters, and tissue construct characterization. Biotechnol. Bioeng. 64 (5), 580–589.

Cioffi, M., Küffer, J., Ströbel, S., Dubini, G., et al., 2008. Computational evaluation of oxygen and shear stress distributions in 3D perfusion culture systems: macro-scale and micro-structured models. J. Biomech. 41 (14), 2918–2925.

Démarteau, O., Wendt, D., Braccini, A., Jakob, M., et al., 2003. Dynamic compression of cartilage constructs engineered from expanded human articular chondrocytes. Biochem. Biophys. Res. Commun. 310 (2), 580–588.

Di Maggio, N., Piccinini, E., Jaworski, M., Trumpp, A., et al., 2011. Toward modeling the bone marrow niche using scaffold-based 3D culture systems. Biomaterials 32 (2), 321–329.

Flanagan, T.C., Cornelissen, C., Koch, S., Tschoeke, B., et al., 2007. The in vitro development of autologous fibrin-based tissue-engineered heart valves through optimised dynamic conditioning. Biomaterials 28 (23), 3388–3397.

Freed, L.E., Marquis, J.C., Langer, R., Vunjak-Novakovic, G., et al., 1994. Composition of cell-polymer cartilage implants. Biotechnol. Bioeng. 43 (7), 605–614.

Freed, L.E., Langer, R., Martin, I., Pellis, N.R., et al., 1997. Tissue engineering of cartilage in space. Proc. Natl. Acad. Sci. U.S.A 94 (25), 13885–13890.

Gerlach, J.C., Lin, Y.-C., Brayfield, C.A., Minteer, D.M., et al., 2012. Adipogenesis of human adipose-derived stem cells within three-dimensional hollow fiber-based bioreactors. Tissue Eng. Part C, Methods 18 (1), 54–61.

Gilpin, S.E., Guyette, J.P., Gonzalez, G., Ren, X., et al., 2014. Perfusion decellularization of human and porcine lungs: bringing the matrix to clinical scale. J. Heart Lung Transplant. 33 (3), 298–308.

Goh, S.-K., Bertera, S., Olsen, P., Candiello, J.E., et al., 2013. Perfusion-decellularized pancreas as a natural 3D scaffold for pancreatic Ktissue and whole organ engineering. Biomaterials 34 (28), 6760–6772.

Grad, S., Lee, C.R., Gorna, K., Gogolewski, S., et al., 2005. Surface motion upregulates superficial zone protein and hyaluronan production in chondrocyte-seeded three-dimensional scaffolds. Tissue Eng. 11 (1–2), 249–256.

Grayson, W.L., Fröhlich, M., Yeager, K., Bhumiratana, S., et al., 2010. Engineering anatomically shaped human bone grafts. Proc. Natl. Acad. Sci. U.S.A 107 (8), 3299–3304.

Gurjarpadhye, A.A., Whited, B.M., Sampson, A., Niu, G., et al., 2013. Imaging and characterization of bioengineered blood vessels within a bioreactor using free-space and catheter-based OCT. Lasers Surg. Med. 45 (6), 391–400.

Hoerstrup, S.P., Kadner, A., Melnitchouk, S., Trojan, A., et al., 2002. Tissue engineering of functional trileaflet heart valves from human marrow stromal cells. Circulation 106 (12 Suppl 1), I143–I150.

Holy, C.E., Shoichet, M.S., Davies, J.E., 2000. Engineering three-dimensional bone tissue in vitro using biodegradable scaffolds: investigating initial cell-seeding density and culture period. J. Biomed. Mater. Res. 51 (3), 376–382.

Hülsmann, J., Aubin, H., Kranz, A., Godehardt, E., et al., 2013. A novel customizable modular bioreactor system for whole-heart cultivation under controlled 3D biomechanical stimulation. J. Artif. Organs 16 (3), 294–304.

Issa, R.I., Engebretson, B., Rustom, L., McFetridge, P.S., et al., 2011. The effect of cell seeding density on the cellular and mechanical properties of a mechanostimulated tissue-engineered tendon. Tissue Eng. Part A 17 (11–12), 1479–1487.

Janssen, F.W., Hofland, I., van Oorschot, A., Oostra, J., et al., 2006. Online measurement of oxygen consumption by goat bone marrow stromal cells in a combined cell-seeding and proliferation perfusion bioreactor. J. Biomed. Mater. Res. Part A 79A (2), 338–348.

Karim, N., Golz, K., Bader, A., 2006. The cardiovascular tissue-reactor: a novel device for the engineering of heart valves. Artif. Organs 30 (10), 809–814.

Krishnan, V., Shuman, L. A., Sosnoski, D.M., Dhurjati, R., et al., 2011. Dynamic interaction between breast cancer cells and osteoblastic tissue: comparison of two- and three-dimensional cultures. J. Cell. Physiol. 226 (8), 2150–2158.

Li, S., Liu, Y., Zhou, Q., Lue, R., et al., 2014. A novel axial-stress bioreactor system combined with a substance exchanger for tissue engineering of 3D constructs. Tissue Eng. Part C, Methods 20 (3), 205–214.

Li, Y., Ma, T., Kniss, D. a, Lasky, L.C., et al., 2001. Effects of filtration seeding on cell density, spatial distribution, and proliferation in nonwoven fibrous matrices. Biotechnol. Prog. 17 (5), 935–944.

Liu, J., Barradas, A., Fernandes, H., Janssen, F., et al., 2010. In vitro and in vivo bioluminescent imaging of hypoxia in tissue-engineered grafts. Tissue Eng. Part C. Methods 16 (3), 479–485.

Malda, J., Rouwkema, J., Martens, D.E., Le Comte, E.P., et al., 2004. Oxygen gradients in tissue-engineered PEGT/PBT cartilaginous constructs: measurement and modeling. Biotechnol. Bioeng. 86 (1), 9–18.

Macchiarini, P., Jungebluth, P., Go, T., Asnaghi, M.A., et al., 2008. Clinical transplantation of a tissue-engineered airKway. Lancet 372 (9655), 2023–2030.

Mauney, J.R., Sjostorm, S., Blumberg, J., Horan, R., et al., 2004. Mechanical stimulation promotes osteogenic differentiation of human bone marrow stromal cells on 3-D partially demineralized bone scaffolds in vitro. Calcif. Tissue Int. 74 (5), 458–468.

Melchels, F.P.W., Tonnarelli, B., Olivares, A.L., Martin, I., et al., 2011. The influence of the scaffold design on the distribution of adhering cells after perfusion cell seeding. Biomaterials 32 (11), 2878–2884.

Mishra, D.K., Thrall, M.J., Baird, B.N., Ott, H.C., et al., 2012. Human lung cancer cells grown on acellular rat lung matrix create perfusable tumor nodules. Ann. Thorac. Surg. 93 (4), 1075–1081.

Montoya, C.V., McFetridge, P.S., 2009. Preparation of ex vivo-based biomaterials using convective flow decellularization. Tissue Eng. Part C, Methods 15 (2), 191–200.

Munson, J.M., Bellamkonda, R.V., Swartz, M.A., 2013. Interstitial flow in a 3D microenvironment increases glioma invasion by a CXCR4-dependent mechanism. Cancer Res. 73 (5), 1536–1546.

Niklason, L.E., Gao, J., Abbott, W.M., Hirschi, K.K., et al., 1999. Functional arteries grown in vitro. Science 284 (5413), 489–493.

O'Conor, C.J., Case, N., Guilak, F., 2013. Mechanical regulation of chondrogenesis. Stem Cell Res. Ther. 4 (4), 61.

Ott, H.C., Clippinger, B., Conrad, C., Schuetz, C., et al., 2010. Regeneration and orthotopic transplantation of a bioartificial lung. Nat. Med. 16 (8), 927–933.

Ott, H.C., Matthiesen, T.S., Goh, S.-K., Black, L.D., et al., 2008. Perfusion-decellularized matrix: using nature's platform to engineer a bioartificial heart. Nat. Med. 14 (2), 213–221.

Porter, B.D., Lin, A.S.P., Peister, A., Hutmacher, D., et al., 2007. Noninvasive image analysis of 3D construct mineralization in a perfusion bioreactor. Biomaterials 28 (15), 2525–2533.

Powell, C.A., Smiley, B.L., Mills, J., Vandenburgh, H.H., 2002. Mechanical stimulation improves tissue-engineered human skeletal muscle. Am. J. Physiol. Cell Physiol. 283 (5), C1557–C1565.

Powell, H.M., McFarland, K.L., Butler, D.L., Supp, D.M., et al., 2010. Uniaxial strain regulates morphogenesis, gene expression, and tissue strength in engineered skin. Tissue Eng. Part A 16 (3), 1083–1092.

Preiss-Bloom, O., Mizrahi, J., Elisseeff, J., Seliktar, D., 2009. Real-time monitoring of force response measured in mechanically stimulated tissue-engineered cartilage. Artif. Organs 33 (4), 318–327.

Quint, C., Kondo, Y., Manson, R.J., Lawson, J.H., et al., 2011. Decellularized tissue-engineered blood vessels as an arterial conduit. Proc. Natl. Acad. Sci. U.S.A. 108 (22), 9214–9219.

Radisic, M., Malda, J., Epping, E., Geng, W., et al., 2006. Oxygen gradients correlate with cell density and cell viability in engineered cardiac tissue. Biotechnol. Bioeng. 93 (2), 332–343.

Radisic, M., Yang, L., Boublik, J., 2004. Medium perfusion enables engineering of compact and contractile cardiac tissue. Am. J. Physiol. Heart Circ. Physiol. 286 (2), H507–H516.

Raimondi, M.T., Bridgen, D.T., Laganà, M., Tonnarelli, B., et al., 2010. Integration of experimental and computational microfluidics in 3D tissue engineering. In: Berthiaume, F., Morgan, J.R. (Eds.), Methods in Bioengineering: 3D Tissue Engineering. Artech House.

Robert, J., Weber, B., Frese, L., Emmert, M.Y., et al., 2013. A three-dimensional engineered artery model for in vitro atherosclerosis research. PloS ONE 8 (11), e79821.

Santoro, R., Krause, C., Martin, I., Wendt, D., 2012. On-line monitoring of oxygen as a non-destructive method to quantify cells in engineered 3D tissue constructs. J. Tissue Eng. Regener. Med. 6 (9), 696–701.

Santoro, R., Olivares, A.L., Brans, G., Wirz, D., et al., 2010. Bioreactor based engineering of large-scale human cartilage grafts for joint resurfacing. Biomaterials 31 (34), 8946–8952.

Shvartsman, I., Dvir, T., Harel-Adar, T., Cohen, S., 2009. Perfusion cell seeding and cultivation induce the assembly of thick and functional hepatocellular tissue-like construct. Tissue Eng. Part A 15 (4), 751–760.

Sikavitsas, V.I., Bancroft, G.N., Mikos, A.G., 2002. Formation of three-dimensional cell/polymer constructs for bone tissue engineering in a spinner flask and a rotating wall vessel bioreactor. J. Biomed. Mater. Res. 62 (1), 136–148.

Stiehler, M., Bünger, C., Baatrup, A., Lind, M., et al., 2009. Effect of dynamic 3-D culture on proliferation, distribution, and osteogenic differentiation of human mesenchymal stem cells. J. Biomed. Mater. Res. Part A 89 (1), 96–107.

Subramanian, B., Rudym, D., Cannizzaro, C., Perrone, R., et al., 2010. Tissue-engineered three-dimensional in vitro models for normal and diseased kidney. Tissue Eng. Part A 16 (9), 2821–2831.

Sung, S., Hsieh, C., Law, A., Zhau, H.E., et al., 2008. Coevolution of prostate cancer and bone stroma in three-dimensional coculture: implications for cancer growth and metastasis. Cancer Res. 68 (23), 9996–10003.

Syedain, Z.H., Meier, L.A., Bjork, J.W., Lee, A., et al., 2011. Implantable arterial grafts from human fibroblasts and fibrin using a multi-graft pulsed flow-stretch bioreactor with noninvasive strength monitoring. Biomaterials 32 (3), 714–722.

Tandon, N., Cannizzaro, C., Chao, P.H., Maidhof, R., et al., 2009. Electrical stimulation systems for cardiac tissue engineering. Nat. Protoc. 4 (2), 155–173.

Tiburcy, M., Didié, M., Boy, O., Christalla, P., et al., 2011. Terminal differentiation, advanced organotypic maturation, and modeling of hypertrophic growth in engineered heart tissue. Circ. Res. 109 (10), 1105–1114.

Timmins, N.E., Scherberich, A., Früh, J.-A., Heberer, M., et al., 2007. Three-dimensional cell culture and tissue engineering in a T-CUP (tissue culture under perfusion). Tissue Eng. 13 (8), 2021–2028.

Tong, Z., Duncan, R.L., Jia, X., 2013. Modulating the behaviors of mesenchymal stem cells via the combination of high-frequency vibratory stimulations and fibrous scaffolds. Tissue Eng. Part A 19 (15–16), 1862–1878.

Usuludin, S.B.M., Cao, X., Lim, M., 2012. Co-culture of stromal and erythroleukemia cells in a perfused hollow fiber bioreactor system as an in vitro bone marrow model for myeloid leukemia. Biotechnol. Bioeng. 109 (5), 1248–1258.

Uygun, B.E., Soto-Gutierrez, A., Yagi, H., Izamis, M., et al., 2010. Organ reengineering through development of a transplantable recellularized liver graft using decellularized liver matrix. Nat. Med. 16 (7), 814–820.

Volkmer, E., Drosse, I., Otto, S., Stangelmayer, A., et al., 2008. Hypoxia in static and dynamic 3D culture systems for tissue engineering of bone. Tissue Eng. Part A 14 (8), 1331–1340.

Volkmer, E., Otto, S., Polzer, H., Saller, M., et al., 2012. Overcoming hypoxia in 3D culture systems for tissue engineering of bone in vitro using an automated, oxygen-triggered feedback loop. J. Mater. Sci. Mater. Med. 23 (11), 2793–2801.

Vunjak-Novakovic, G., Martin, I., Obradovic, B., Treppo, S., et al., 1999. Bioreactor cultivation conditions modulate the composition and mechanical properties of tissue-engineered cartilage. J. Orthop. Res. 17 (1), 130–138.

Vunjak-Novakovic, G., Obradovic, B., Martin, I., Bursac, P.M., et al., 1998. Dynamic cell seeding of polymer scaffolds for cartilage tissue engineering. Biotechnol. Prog. 14 (2), 193–202.

Wang, L., Seshareddy, K., Weiss, M.L., Detamore, M.S., 2009. Effect of initial seeding density on human umbilical cord mesenchymal stromal cells for fibrocartilage tissue engineering. Tissue Eng. Part A 15 (5), 1009–1017.

Weidenhamer, N.K., Tranquillo, R.T., 2013. Influence of cyclic mechanical stretch and tissue constraints on cellular aKnd collagen alignment in fibroblast-derived cell sheets. Tissue Eng. Part C, Methods 19 (5), 386–395.

Wendt, D., Marsano, A., Jakob, M., Heberer, M., et al., 2003. Oscillating perfusion of cell suspensions through three-dimensionaKl scaffolds enhances cell seeding efficiency and uniformity. Biotechnol. Bioeng. 84 (2), 205–214.

Wendt, D., Stroebel, S., Jakob, M., John, G.T., et al., 2006. Uniform tissues engineered by seeding and culturing cells in 3D scaffolds under perfusion at defined cxygen tensions. Biorheology 43 (3–4), 481–488.

Zhao, F., Ma, T., 2005. Perfusion bioreactor system for human mesenchymal stem cell tissue engineering: dynaKmic cell seeding and construct development. Biotechnol. Bioeng. 91 (4), 482–493.

Zhou, X., Holsbeeks, I., Impens, S., Sonnaert, M., et al., 2013. Noninvasive real-time monitoring by alamarblue(®) during in vitro culture of three-dimensional tissue-engineered bone constructs. Tissue Eng. Part C, Methods 19 (9), 720–729.

Zimmermann, W.H., Schneiderbanger, K., Schubert, P., Didié, M., et al., 2002. Tissue engineering of a differentiated cardiac muscle construct. Circ. Res. 90 (2), 223–230.

CHAPTER 13

Clinical Grade Production of Mesenchymal Stromal Cells

Jessica J. Alm[1], Hong Qian[2], Katarina Le Blanc[1]

[1]Center for Hematology and Regenerative Medicine and Division of Clinical Immunology and Transfusion Medicine, Department of Laboratory Medicine, Karolinska Institutet, Stockholm, Sweden; [2]Center for Hematology and Regenerative Medicine, Department of Medicine, Karolinska Institute, Stockholm, Sweden

LEARNING OBJECTIVES

- To recognize current definition and identification of multipotent mesenchymal stromal cells (MSCs)
- To be familiar with isolation procedures for human bone marrow MSCs
- To be familiar with culture expansion of MSCs for research and clinical therapy
- To understand the process and effects of cryopreservation of MSCs
- To recognize basic protocols for trilineage differentiation of MSCs
- To be aware of changes induced in MSCs upon culture expansion
- To get an overview of current MSC therapies and future therapeutic strategies

13.1 INTRODUCTION

13.1.1 Definition and Properties of Mesenchymal Stromal Cells

Multipotent mesenchymal stromal cells (MSCs) constitute a heterogeneous subset of stromal precursor cells present at low frequencies in connective tissue compartments. MSCs have been extensively studied for their multiple differentiation, repair, and immunomodulatory properties that make them attractive as treatment of tissue regeneration, autoimmunity, and inflammation. The mechanisms of action and molecular basis for clinical benefits of MSCs are still not fully understood but are progressively revealed by preclinical and clinical studies. As will be described in this chapter, MSCs can be isolated from bone marrow (BM) or other tissue sources by their ability to adhere to culture plastic, and their capacity to proliferate in culture allow for in vitro expansion for obtaining a sufficient cell number for therapeutic, tissue engineering, or research applications (Figure 13.1).

Tissue Engineering.
Copyright © 2015 Elsevier Inc. All rights reserved.

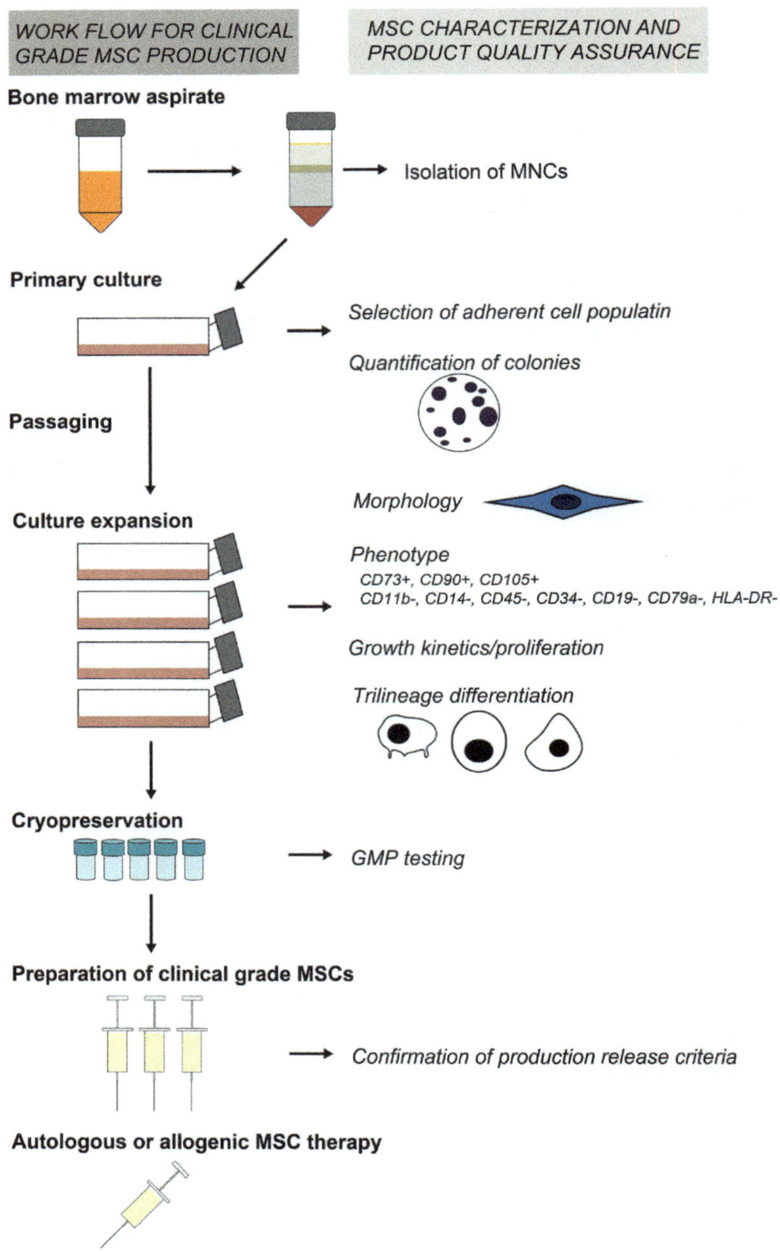

FIGURE 13.1

Schematic overview of basic principles for isolation, culture expansion, and characterization of MSCs for clinical applications. The same principles are applied for research applications of MSCs. MSCs, multipotent mesenchymal stromal cells; MNCs, mononuclear cells.

MSCs are characterized by their ability to differentiate into several cell types of mesodermal origin, including osteoblasts, chondrocytes, and adipocytes. In addition, other lineage differentiations, including skeletal and cardiac muscle, hepatocytes, neural cells, and endothelial and epithelial cells, have been reported, but remain controversial. Culture-isolated MSC populations are heterogeneous with coexisting subsets of cells with differing phenotypes and differentiation potentials (Pittenger et al., 1999; Friedenstein et al., 1974). In addition to regenerative and immunomodulatory properties, MSCs exert a stromal supportive and regulatory role in the hematopoietic stem cell (HSC) niche (Mendez-Ferrer et al., 2010). The true in vivo location and nature of MSCs still remain unclear.

Extensive attempts have been made to (1) enrich cultures for multipotent MSCs and (2) identify a specific surface marker or marker profile. MSCs express a broad range of nonspecific surface markers (Table 13.1), but no unique marker has yet been found despite vigorous investigations. The lack of universal criteria to define MSCs has been a challenge to the field. In an attempt to overcome the heterogeneity in the literature, International Society for Cellular Therapy (ISCT) proposed minimal phenotypic and functional criteria for identifying MSCs (Dominici et al., 2006) (indicated in Figure 13.1).

In spite of extensive MSC research, the original isolation protocol is still largely in use. Depending on the application, different properties of the MSCs might be important and desired. These properties are affected by methods of choice at each step along the process of isolation, expansion, banking, and handling. In addition, donor characteristics influence MSC properties (Box 13.1).

13.1.2 Sources of MSCs

BM, adipose tissue, and umbilical cord are the most common sources for MSCs, mainly due to high frequency, ease of harvest, and availability of tissue. Additional sources are synovium, periosteum, and skeletal muscle. MSC-like cells can also be derived from gingiva, periodontal ligament, and exfoliated deciduous teeth. These sources seem rich in MSCs possessing a high in vitro expansion capacity and low senescence. However, since MSCs from dental tissues are of neural crest origin, their phenotype and differentiation potential differ from classical MSCs of mesodermal origin.

MSCs can also be isolated from fetal tissues, including umbilical cord, cord blood, placenta, and amniotic fluid. Although cord tissue and placenta seem to be rich in MSCs of a high proliferative capacity, low senescence, and high multilineage differentiation potential, the frequency of MSCs in cord blood and amniotic fluid is lower (Thirumala et al., 2013).

BM-MSCs can be isolated from iliac crest aspirates, vertebrae, and marrow and trabecular bone fragments from proximal femur during hip replacement surgery, whereas BM from discarded femoral heads present more variable potential. In this chapter, we focus on isolation of adult MSCs from BM aspirates.

Table 13.1 Overview of Useful Markers for Phenotyping of Culture-Expanded MSCs

Basic Panel According to ISCT Minimal Criteria			
Positive Markers	**Expression and Note**	**Negative Markers**	**Expression and Note**
CD73	97–99% positive	CD11b	Monocyte and macrophage markers
CD90	97–99% positive	CD14	
CD105	97–99% positive	CD34	Hematopoietic progenitor and endothelial cell marker Positive on adipose-derived MSCs during early passages
		CD45	Leukocyte marker
		CD19	B cell markers
		CD79a	
		HLA-DR	

Additional Markers of Interest for MSC Phenotyping	
Positive Markers	**Expression and Note**
CD44	Negative in fresh MSCs, but acquired upon culturing
CD146	~99% positive in fresh BM-MSCs and in nontoxic culture
CD271	~99% positive in fresh BM-MSCs Gradually disappear during culture
CD63	>90%
CD13	>99%
CD29/integrin β	100% Adhesion molecule
CD49 d/VLA-4	~50% of BM-MSCs adhesion molecule
CD54/ICAM-1	>90% Adhesion molecule
CD106/VCAM-1	~40–50% Expression level changes with passaging
CD166/ALCAM	~99%, adhesion molecule
CXCR4	10–40% by FACS Adhesion molecule
CXCR5	~82% adhesion molecule,
CCR1	70%, reduced with passaging Adhesion molecule
CXCR6	~25–40%, adhesion molecule
TLR3	>50%
TLR4	<50% expression not clear in resting MSCs

MSCs, multipotent mesenchymal stromal cells; FM, bone marrow; FACS, fluorescence-activated cell sorting; HLD-DR, human leukocyte antigen D-related.

BOX 13.1 THE NAME OF THE GAME

MSCs were initially termed "fibroblast colony forming cells" (FCFCs) and CFU-Fs (Friedenstein et al., 1974). Since then, the name of the cells has been subject for debate, as the requirement for "stem cell" properties are not always fulfilled. Friedenstein eventually called them *osteogenic stem cells* (Friedenstein), while *marrow stromal stem cells* were proposed to emphasize the supportive stromal capacity (Owen et al., 1987). In 1991, Caplan introduced the term *mesenchymal stem cells* (Caplan, 1991). Due to the unclear true stem cell nature of mesenchymal stem cells, the name *mesenchymal progenitor cells* was introduced (Dennis et al., 1999). To clarify the nomenclature for MSCs and to avoid the inaccurate use of the term stem cell, The ISCT proposed the term *multipotent mesenchymal stromal cells (MSCs)* (Horwitz et al., 2005).

13.1.3 MSC Therapy

Differentiation capacity of MSCs gained early interest for the development of tissue engineering applications, especially for bone and cartilage reconstruction. Much effort has been put into finding cell-based methods for the treatment of degenerative diseases, but the original idea to replace damaged tissue with transplanted MSCs has proven challenging. Besides directly contributing to new tissue formation, MSCs secrete bioactive molecules, such as cytokines, trophic factors, antioxidants, and proangiogenic factors, that mediate repair by limiting stress responses and apoptosis; this controls early inflammation after injury, and recruits immune and reparative cells to the damaged tissue (Doorn et al., 2012; Maumus et al., 2013) (Figure 13.2).

The well-documented immunoregulatory properties of MSCs have contributed to the development of therapeutic applications in immune-based disorders (Le Blanc and Mougiakakos, 2012; Bernardo and Fibbe, 2013). Currently, MSC therapy is under investigation as treatment of a broad spectrum of chronic and degenerative diseases, including autoimmune diseases (e.g., multiple sclerosis, type 1 diabetes mellitus, and Crohn's disease), myocardial repair, liver disease, kidney failure, skeletal tissue disorders including osteogenesis imperfecta and osteoarthritis, and regeneration (e.g., bone, cartilage, tendons, and muscle). So far, the most frequent application has been in HSC transplantation (HSCT), especially for the prevention and treatment of graft-versus-host disease after HSCT, or enhancement of HSC engraftment (Le Blanc et al., 2004; Le Blanc et al., 2008; Lazarus et al., 1995; Koc et al., 2000).

Although MSC therapy is promising in several applications, the field still has challenges and desired improvements to address (Table 13.2). Collectively, clinical trials have provided encouraging outcomes and an excellent safety record, but there are still challenges and desired improvements to address including variability in protocols, and the unknown fate and mechanism of action of MSCs (Table 13.2). MSC migration and engraftment into injured tissue are poorly understood.

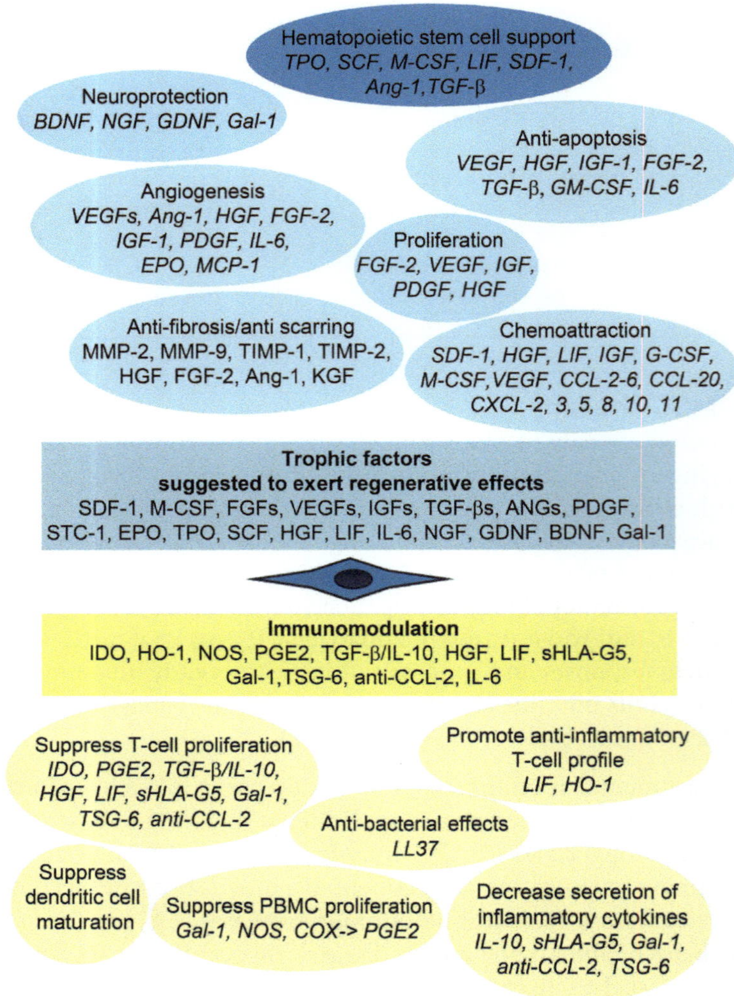

FIGURE 13.2

Various factors released by MSCs, suggested to exert their regenerative (trophic) or immunomodulatory actions. Factors secreted by MSCs may have different effects and functions on other cells depending on the particular microenvironment. MSCs, multipotent mesenchymal stromal cells. *Modified from Doorn et al. (2012) and Maumus et al. (2013).*

The osteogenic potential of MSCs in vitro has brought extensive hope and research interest for MSC-based repair of bone and cartilage. Initial animal studies on culture-expanded MSCs in diffusion chambers or HA scaffolds proved the in vivo osteogenic differentiation of MSCs. Bioscaffolds with autologous MSCs have been used for repairing segmental bone defects in animal models and in proof-of-concept clinical studies. However, the clinical use of MSCs in orthopedic settings remains challenging, and the development of MSC-based solutions

Table 13.2 Current Promises and Challenges of MSC Therapeutics

Strengths and Promises of MSC Therapy	Limitations and Challenges
Proven safe	Heterogeneity in protocols and cell preparations between laboratories
MSCs remain genetically stable through ex vivo expansion (in contrast to other types of stem cells)	Limited understanding of MSC biology in vitro and in vivo
Can be used autologously or allogenically	Large donor variability
MSCs relatively easy to isolate and expand	Clinical grade large-scale MSC production is resource requiring
Encouraging clinical outcomes	Improved clinical effect desirable
Strong trophic effects of MSCs	Limited engraftment, unknown fate of infused and transplanted MSCs
Do not need to be a pure cell population to be clinically useful	Unknown identity and heterogeneous cell population
	Identity of clinically efficient MSCs to be defined
	Need for assays to screen for high-quality therapeutic MSCs

MSCs, multipotent mesenchymal stromal cells.

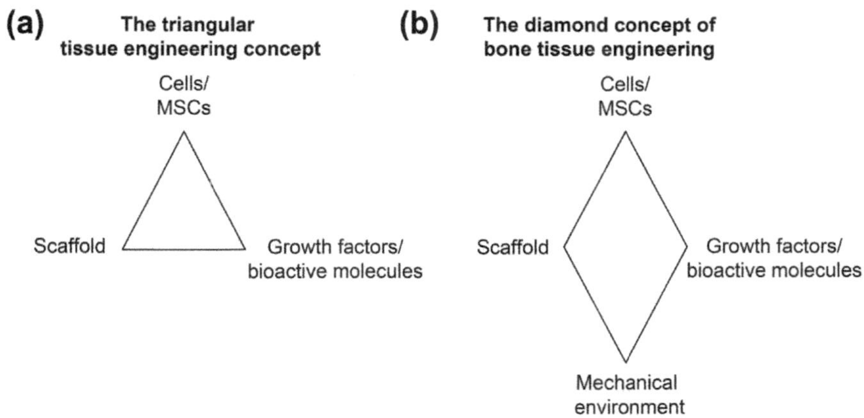

FIGURE 13.3
(a) The traditional triangular tissue engineering concept and (b) the diamond concept of bone tissue engineering. *Modified from Giannoudis et al. (2007).*

has been slow. Clinical studies on stem cell therapies in orthopedics are so far mostly observational, with small patient numbers and limited design regarding controls and outcomes (Gomez-Barrena et al., 2011). For bone healing strategies, a further development of the traditional triangular tissue engineering concept (cells, scaffold, and growth factors) has been proposed (Figure 13.3). Taking the biomechanical environment into account as a fourth factor, the "diamond concept" was presented (Giannoudis et al., 2007), with a promising first clinical outcome for treatment of fracture nonunions (Giannoudis et al., 2013).

13.2 ISOLATION OF BM-MSCs

Most isolation protocols rely on plastic adherence of the MSC population for further expansion, and provide enrichment rather than isolation of specific cells. Differences in isolation protocols select different subpopulations of cells with varying phenotypic and functional properties. These variances in cell products are one reason for discrepancies in outcome of clinical trials (Bara et al., 2014). Therefore, it is important to recognize the methods used in order to properly interpret and compare results.

13.2.1 Isolation Techniques

A BM aspirate contains hematopoietic cells, adipocytes, endothelial progenitors, fibroblastic cells, and osteogenic cells of different maturation stages. Hence, the composition of the cell population obtained is highly dependent on the isolation methods. The three major approaches for MSC isolation are (1) "direct plating of whole marrow," (2) "density gradient separation of the MNC population," and (3) "selection of cellular subfractions based on phenotypic profile" (prospective isolation).

13.2.1.1 Isolation of MSCs from Unprocessed Whole BM

In the original work by Friedenstein and subsequent early studies, MSCs were isolated from whole unprocessed BM. Although density centrifugation for MNC selection is the most common approach for isolating human MSCs today, the original method is also applied for both research and clinical applications.

There are two main approaches for isolating MSCs from whole BM. The BM sample can either be diluted with three equal volumes of growth media before being distributed into culture flasks (~0.1 mL/cm^2) or alternatively concentrated by centrifugation before resuspension. The concentration of nucleated cells is determined by counting in a diluent that lyses the red blood cells (RBCs, e.g., Türk solution), and the marrow sample is seeded based on nucleated cell densities (ranging from 10^4 to 5×10^6 MNCs/cm^2).

Contamination with RBCs significantly reduces the efficiency of MSC isolation by interfering with colony formation, and impairs MSC function. The RBC fraction can be removed by hemolysis treatment, density gradient centrifugation, or by negative immunoselection with magnetic bead separation. Another concern with the whole marrow approach is the high presence of undesired cell types in the adherent cell fraction. If the nonadherent hematopoietic cells are left in a culture plated at a high density, they will eventually attach to the adherent layer of stromal cells and a microenvironment resembling the hematopoietic niche is created (Wagner et al., 2007). This can interfere with the desired MSC phenotype and properties by priming MSCs toward hematopoietic supportive cells. By using a low plating density and removing the nonadherent fraction within 72 h, contamination with other cell types is decreased. Contaminating cell populations are subsequently removed through passaging, since only MSCs will proliferate, detach, and survive passaging. Advantages of using the whole marrow approach include an easier and faster processing procedure.

13.2.1.2 Isolation of MSCs through Density Gradient Centrifugation

For isolation of a more homogenous initial cell population, the MNC fraction is separated by density gradient centrifugation. The method separates cells based on density, irrespective of their size. A density gradient medium is placed in a centrifugal tube, and the BM is loaded onto the gradient (Figure 13.4(a) and (b)). Through centrifugation at 400–600 × g for 30 min, cells sediment according to their density. RBCs and granulocytes have a high density and sediment through the gradient to the bottom of the tube, while the light density monocytes including MSCs sediment on top of the gradient (Figure 13.4(c)). The MNC cell fraction is collected with a pipette and plated in the growth medium. Commonly used gradient media include Ficoll (paque, hypaque, histopaque, and paque premium), Lymphoprep, and Percoll. Ficoll and Lymphoprep are polysaccharide gradients, usually applied at a density of 1.077 g/mL, and are traditionally used for isolation of MNCs and lymphocytes from peripheral blood and BM. Percoll is a suspension of colloidal silica particles used at different densities for experimental isolation of subcellular particles. For MSC isolation, a density of 1.073 g/mL is usually applied.

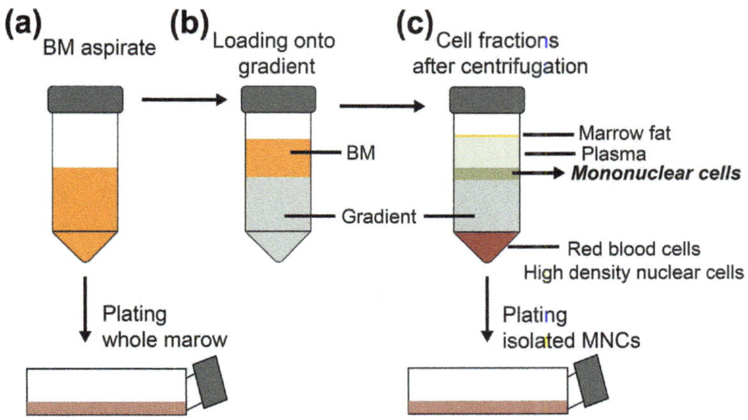

FIGURE 13.4

Two approaches for isolation of MSCs from bone marrow (BM). (a) Initial isolation culturing can be established by direct plating of whole unprocessed BM. More commonly, the mononuclear cell (MNCs) fraction is first isolated by density gradient centrifugation. (b) The BM sample is loaded onto a gradient medium (Percoll, Ficoll, or Lympoprep), (c) after centrifugation at 400–600 g for approximately 30 min cells sediment according to their density, and the MNC fraction can be collected for seeding.

Noticeable, density centrifugation reduces the number of MNCs yielded to approximately 15–30%. Ficoll-based density separation has been associated with the loss of cells, with high regenerative potential, but is reported to be superior to the Percoll-based procedure for isolating cells with osteogenic potential. Compared to density centrifugation, unprocessed whole marrow seems to yield a higher colony forming unit (CFU) frequency and an MSC population with longer telomeres, indicating a loss of more naive MSCs with the extra isolation step. The proportion of committed osteogenic cells is larger in whole marrow preparations.

13.2.1.3 Cell-Sorting Methods for Prospective Selection of Subfractions

With cell-sorting methods, more purified and specific cell populations are isolated based on cell surface marker expression. This can be conducted using either magnetic-activated cell sorting or fluorescence-activated cell sorting (FACS). In the former, cells are labeled with antibodies against desired cell surface markers conjugated with magnetic beads, whereby the marker-expressing positive cells are separated from the negative cells by exposure to a strong magnetic field. For FACS, cells are labeled with fluorescence-conjugated antibodies and separated in an electronic flow cytometer as the fluorescent cells pass through laser beams.

STATE-OF-THE-ART EXPERIMENT

Prospective isolation of MSC subfractions from fresh BM using flow cytometry There are several studies on the positive selection of MSCs based on a few antibodies. In early studies, MSCs were identified based on the expression of CD105, CD73 (Haynesworth et al., 1992), or Stro-1 (Gronthos et al., 1994). Current strategies include the selection of cells based on a surface marker profile combining several positive and negative markers. Briefly, the mononuclear fraction isolated by density centrifugation can be subsequently applied to magnetic sorting to enrich for nonhematopoietic ($CD45^-$, $CD235a^-$) cells by negative selection using CD45 (a panmarker for all white blood cells) and CD235a (a marker for erythrocytes) microbeads. Cells can then be directly plated, or be further purified by staining with a set of flurochrome-conjugated antibodies to sort a highly purified MSC population. There are a range of surface molecules used in different combinations for sorting MSCs. We have previously identified CD44 as a negative marker for human MSCs from BM (Qian et al., 2012). For exclusion of residual hematopoietic cells and endothelial cells, CD31, CD45, and Glycophorin A/CD235a should be included. Thereafter, CD44 can be used as a negative marker to select MSCs. In addition, CD271, CD146, CD106, and CD73 can be added to further purify MSCs and check the specificity of the CD44 MSCs, since all of them are established positive makers for human BM-MSCs (Tormin et al., 2011). Dead cells can be excluded by propidium iodide (PI) staining. Using flow cytometry, the $CD44^+$ and $CD44^-$ stromal cells can be sorted as separate populations. One representative FACS sorting profile and the sorting strategy are illustrated in the figure. Collectively, recent studies indicate that a $CD271^+$/$CD146^+$/$CD44^-$ phenotype on primary nonhematopoietic BM cells might represent a more immature and potent subset of MSCs (Tormin et al., 2011; Qian et al., 2012).

Continued...

STATE-OF-THE-ART EXPERIMENT *Continued*

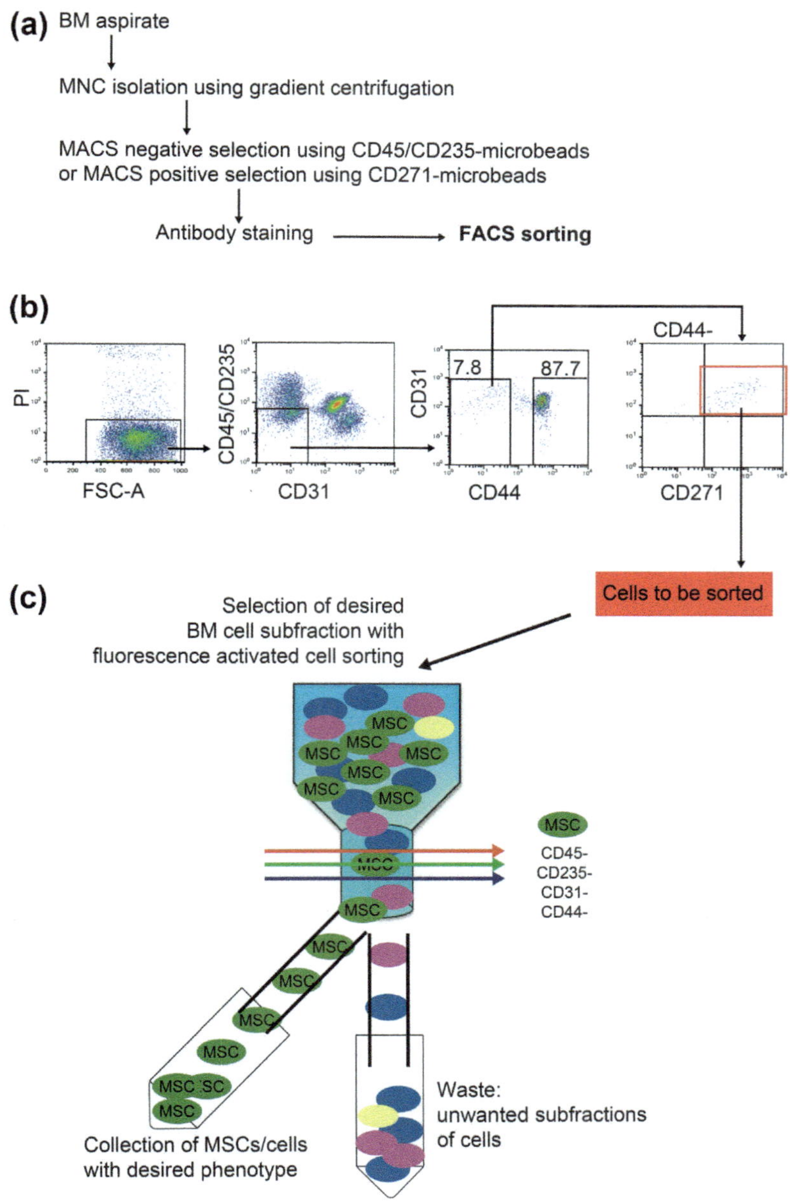

(a) Workflow for prospective MSC isolation from fresh BM. (b) Gating strategy for selection of the desired cell population. (c) This strategy is for selecting the CD44⁻ population for sorting and collection (Qian et al., 2012).

Sorting methods are hardly used for clinical MSC isolation due to the lack of a specific MSC marker, and the fact that it is not yet known what subpopulation is optimal for clinical applications. Further, cell-sorting methods are associated with a risk of cell damage, prolonged isolation procedure, challenges in ensuring absolute aseptic conditions, and higher costs. Current solutions have a limited ability of large-scale applications for clinical MSC manufacturing.

13.2.2 The Impact of Seeding Parameters on MSC Isolation

The initial seeding density of whole marrow or isolated MNCs is an important parameter for efficient MSC yield since this affects adherence of both MSCs and other cell types, and MSC proliferation. A variety of initial plating densities can be found in the literature. For clinical expansion, MNC seeding varies between 10^3 and 10^6 cells/cm^2, with $1.5–1.6 \times 10^5$ being the most commonly used value (Ikebe and Suzuki, 2014).

Several factors contribute to a higher MSC yield with a lower seeding density ($1–200 \times 10^3$ MNCs/cm^2) (Sotiropoulou et al., 2006). At diluted concentrations, soluble factors produced by non-MSC cell fractions are limited, and this thereby reduces their influence on MSCs.

Time for cell adherence influences the cell population(s) selected during primary culture. With increased incubation times, additional cell populations with weaker adhesion capacities attach through proteins derived from culture serum or produced by the cells, or attach to the adherent cell layer. The optimal time point for first media change is difficult to define, since early removal has a risk of eliminating late adherent MSC subpopulations, and late removal increases the impact of undesired cells. In the early work by Castro-Malaspina et al., time-dependent adhesion from 10 to 90 min indicated the complete adherence of desired cells within 60 min (Castro-Malaspina et al., 1980). Today, commonly used adherence times are 24, 48, and 72 h. Further, the initial nonadherent cell population seems to contain an MSC fraction capable of forming CFUs upon serial culturing in the presence of fibroblast growth factors (FGFs) (Di Maggio et al., 2012).

13.2.3 CFU Assay and Expected Frequency of MSCs

BM aspiration technique, volume, and location impact affect MSC yield (Muschler et al., 1997). The number of cells with a CFU capacity is presumed to represent the frequency of MSCs (Friedenstein et al., 1974; Castro-Malaspina et al., 1980). The CFU assay is conducted either by plating a small volume of whole marrow, or isolated MNCs at a low density. For whole marrow CFU assay, 1.7–3.2 μL BM/cm^2 (80–100 μL in 100-mm dishes or T25 flasks) are seeded. For isolated MNC populations, protocols vary greatly. Seeding densities ranging from very low (50–1000 MNCs/cm^2), low

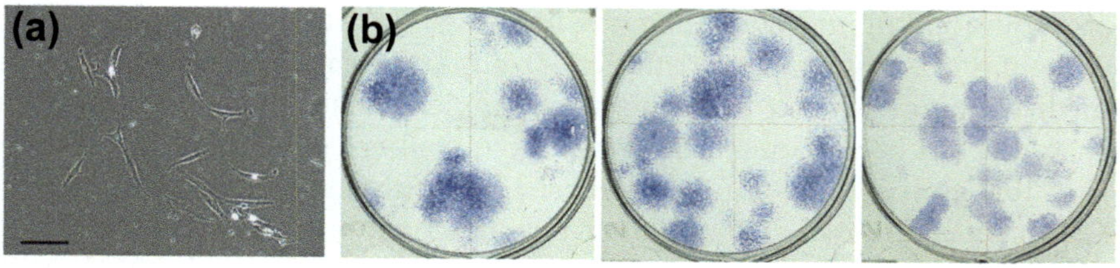

FIGURE 13.5
Primary colony formation assay. When seeded at low densities, MSCs form colonies referred to as colony forming units (CFUs). (a) Within days, small colonies of a few cells can be detected. (b) After culturing for 10–21 days (depending on the setup), colonies can be fixed and stained with crystal violet (or corresponding staining) for visualization and quantification of the number and size of CFUs. BM samples from individual donors usually demonstrate differences in both CFU number and size (b). Scale bar = 20 μm. MSCs, multipotent mesenchymal stromal cells; BM, bone marrow.

(8000–20,000 MNCs/cm^2), to medium low (40,000–120,000 MNCs/cm^2) are commonly used. Colonies can be quantified after 10–21 days by fixing and staining the cells. Colonies having a diameter ≥ 2 mm or consisting of > 40–50 cells are counted (Figure 13.5).

Reported estimates of CFUs range from 10,000 to 100,000 nucleated cells, with frequencies of 1 in 26,000–40,000 common in adult BM, depending on age and possible disease state of the donor (Castro-Malaspina et al., 1980, Stolzing et al., 2008). Caplan and colleagues reported a dramatic change with every age decade, with a 10-fold drop in MSC frequency from birth to teenage, and another 10-fold drop up to old age (Caplan, 2007).

13.3 CULTURE EXPANSION

Both research and therapy application of MSCs require culture expansion to obtain sufficient cell numbers. The choice of medium, serum, and potential supplements affect MSC proliferation, gene expression, and differentiation, and hence the cellular product.

13.3.1 Culture Media

Culture expansion of MSCs is carried out in a defined basal medium. For human MSCs, commonly used culture media are Dulbecco's Modified Eagle's Medium (DMEM or DMEM F12) and alpha-Minimum Essentials Medium (α-MEM). Basal media contain amino acids, glucose, and ions (calcium, magnesium, potassium, sodium, and phosphate) essential for cell survival and growth. L-glutamine is an amino acid that is essential for protein and nucleic acid synthesis and energy production in cell culture. Most media either contain L-glutamine or are substituted with it at the time of use. However, L-glutamine is unstable at physiological pH in aqueous solutions and is easily degraded into ammonium and pyroglutamate that possibly interfere with cell growth. Glutamax is a more stable substitute for L-glutamate that can be added to basal media. Sodium pyruvate is another basal supplement added to culture media as an additional energy source, but is not a requirement for MSC culture. DMEM with low glucose (1.0 g/L) is the most commonly used medium for human MSCs, in accordance with original recommendations (Haynesworth et al., 1992), and also for clinical MSC expansion (Ikebe and Suzuki, 2014). However, in a comparison of eight different basal media, α-MEM supplemented with Glutamax or L-glutamin, and 10% fetal bovine serum (FBS) produced the highest MSC expansion (Sotiropoulou et al., 2006). Alpha-MEM has been demonstrated to better preserve and support osteogenic properties of MSCs.

13.3.2 Serum Supplementation

Serum supplementation of basal medium is a key aspect in obtaining favorable conditions. Serum contains nutrients and growth factors that are indispensable for survival and proliferation. FBS is the gold standard serum for cell cultures, since it is rich in required factors. Human blood-derived products as alternatives to FBS have been extensively studied. These include autologous (AS) and allogenic human serum (HS), platelet lysate (PL), and platelet-rich plasma (PRP). Comparisons between serum alternatives are presented in Table 13.3. Alternatively, serum-free (SF) conditions with defined selected supplements are continuously investigated, but have so far proven to be challenging in fully supporting MSC expansion.

13.3.2.1 FBS—The Traditional Supplement

FBS, commonly added at a final concentration of 10%, has cell growth and trilineage differentiation supportive properties. However, the large batch-to-batch variation of FBS requires tedious and expensive pretesting for selection of optimal batches supporting proliferation. Alternatively, commercially available MSC-qualified serum can be purchased at a considerable cost. The lack of detailed definition of the content of FBS causes quality differences between MSC expansion lots and experiments.

Compared to MSCs grown in human-derived serum, FBS-cultured cells display significant changes in gene expression (Tonti and Mannello, 2008). Genes associated with trilineage differentiation and cell-cycle inhibition are upregulated; this indicates that FBS-cultured cells are more committed than cells cultured in human-derived counterparts. On the other hand, MSCs cultured in FBS show upregulation of several genes associated with prostaglandin synthase, which is of interest for clinical applications that focus on immunosuppressive properties of MSCs (Shahdadfar et al., 2005).

The use of FBS might evoke unfavorable immune responses toward xenogeneic proteins in MSC recipients (Sundin et al., 2007). It carries a potential risk of

Table 13.3 Serum Supplements Used for Human MSC Expansion

Serum Supplement	Commonly Used Concentrations	Clinical Grade Availability	Advantages	Disadvantages	Selected Reading
Fetal bovine serum (FBS)	10%, 15%, 20%	Yes	Abundantly available Commercially available Proven clinically safe in numerous trials Well-established cell growth support Rich in growth factors Contain neuroprotective factors Upregulate prostaglandin synthase genes Maintained differentiation potential	High batch-to-batch variability Content not specified Risk for xenogenic immune response Risk for transmission of bacteria, fungi, virus, prions Prolonged cell cycle Premature replicative senescence Upregulation of trilineage differentiation genes → more committed cells (can be an advantage in certain applications)	Shahdadfar et al. (2005), Tonti and Mannello (2008) Sundin et al. (2007) Hemeda et al. (2014)
Autologous serum (AS)	10%	Yes	No immune reactions No risk for contamination with animal or human pathogens No xenogeneic immune reactions High proliferation Long-term culture stability Maintained differentiation potential Slower differentiation Maintain more primitive cell stage Preservation of high cell motility Upregulation of antiapoptotic genes Support stable transcriptome compared to FBS	Limited availability Limited quality in some donors	Shahdadfar et al. (2005) Tonti and Mannello (2008)

Continued...

Table 13.3 Serum Supplements Used for Human MSC Expansion *Continued*

Serum Supplement	Commonly Used Concentrations	Clinical Grade Availability	Advantages	Disadvantages	Selected Reading
Allogenic human serum (HS)	10%, 5%	Yes	No risk for contamination with animal pathogens No xenogeneic immune reactions AB-serum support proliferation and preserve immunosuppressive properties, phenotype, and differentiation potential	Variability in content Limited batch volumes Variable results, depends on preparation and starting material Reduced MSC yield Low proliferation Early cell death	Tonti and Mannello (2008), Hemeda et al. (2013), Le Blanc et al. (2007) Spees et al. (2004)
Platelet lysate (PL) Platelet rich plasma (PRP)	10%, 5% 5%	Yes Yes	No risk for contamination with animal pathogens No xenogeneic immune reactions Rich in growth factors Increased MSC yield (CFU number and size) Strong proliferative stimulation Preserved trilineage differentiation capacity Smaller cell size Accelerated expansion without loss in basic properties Genetic stability	Batch-to-batch variability, can be reduced by pooling No standardized protocol for production Limited batch volumes Variable growth factor content Limited distribution Possible immune reactions in allogenic settings Risk for transmission of human viruses *Mycoplasma* contaminations needs to be excluded Change in gene expression related to adhesion and migration Concerns for chromosomal aberration and cell exhaustion	Bieback (2013), Griffiths et al. (2013), Hemeda et al. (2014) Tonti and Mannello (2008) Flemming et al. (2011), Abdelrazik et al. (2011), Copland et al. (2013)

MSCs, multipotent mesenchymal stromal cells; CFU, colony forming unit.

bacterial, viral, fungal, or prion transmission and contains a relatively high level of endotoxins. Expansion in FBS was reported in 77% of recently reviewed trials, with few side effects reported (Ikebe and Suzuki, 2014). Manufacturing of animal-derived products for therapeutic applications is highly controlled, and follows regulations for veterinary control set by the European Union and the United States Department of Agriculture. Preferably, FBS should be derived from countries without incidences of spongiform encephalopathy. As for now, the use of FBS is still tolerated by most regulatory authorities, but increased restrictions are expected.

13.3.2.2 Human Serum—Autologous (AS) and HS

Both AS and HS (usually a pool of human AB serum) can substitute FBS (Table 13.3). Although favorable alternatives, the restricted availability of AS and HS due to the need for large blood volumes limits their large-scale use for clinical MSC expansion, and depending on the donor profile serum quality may not always be optimal. Expansion in HS has provided mixed results (Le Blanc et al., 2007; Spees et al., 2004) probably due to large variations between serum preparations. HS is associated with low proliferation, early cell death, reduced attachment, and limited colony formation (Hemeda et al., 2014). Superiority of bovine serum over human allogenic serum is due to the fetal origin of FBS. Compared to adult serum, the fetal-derived serum is enriched in growth factors, although it is poor in antibodies. With human AB serum, proliferation is higher, and phenotypic and immunosuppressive properties are similar to FBS cultures (Le Blanc et al., 2007).

13.3.2.3 Human Platelet-Derived Supplements—PL and PRP

Platelet-derived products have been extensively explored as alternative human-derived supplements. Platelets are rich in growth factors known to stimulate MSC proliferation. MSCs grown in PRP or PL are smaller than FBS- or HS-cultured cells (Tonti and Mannello, 2008). There is no evidence that the increase in the growth rate of MSCs in platelet-derived conditions carries a risk for chromosomal aberrations and cellular exhaustion, and changes in both gene expression and DNA methylation are similar to those for FBS-cultured cells. Proliferative cessation and senescence state of MSCs expanded in FBS can be "rescued" by switching to PL culture conditions (Griffiths et al., 2013). MSCs cultured in the presence of platelet products retain their trilineage differentiation potential, while reports on the immunosuppressive activity are more conflicting and seem to be dependent on the PL preparation (Copland et al., 2013).

PL is prepared from plasma enriched for platelets (PRP) by freeze-thawing procedures and centrifugation. PRP is often used at a final concentration of 5%, whereas PL is used at 10%. Both products require the addition of heparin to avoid gel formation. The heparin concentration is critical, since higher concentrations impair MSC expansion and differentiation. Differences between batches of platelet products and variations in growth factor content can be partly overcome by pooling several units (Hemeda et al., 2014). Limitations related to PL and PRP include the low definition of the content, and the restricted commercial distribution. As with any blood-derived products, concerns are related to the risk of contamination with pathogens.

13.3.2.4 Growth Factor Supplementation and SF Alternatives

A range of growth factors, in particular FGF2 (basic FGF, FGFb), platelet-derived growth factor (PDGF), epidermal growth factor (EGF), transforming growth factor beta (TGF-β), and insulin-like growth factor 1 (IGF-1) have gained interest as MSC-promoting agents, either in combination with serum or in SF conditions. FGF2 is widely used in combination with FBS to accelerate expansion. MSCs grown in the presence of FGF2 maintain differentiation potential, but upregulate human leukocyte antigen D-related (HLA-DR) and downregulate CD146 (Sotiropoulou et al., 2006).

To minimize risks of contamination, transmission of pathogens, and possible immunological reactions, development of SF culture conditions is a future goal. Results are so far diverse and an optimal solution has not yet been defined. Requirement of a broad range of growth factor and cytokine supplies to reach expansion levels comparable to serum conditions is challenging, in finding both the optimal composition and concentrations, and in cost effectiveness. SF conditions also require the addition of adhesion proteins for proper cell attachment. Exogenous proteins are expensive and may carry the risk for pathogen transmission and immune reactions. Although commercially available ready-to-use SF culture media for MSC expansion are available, formulations of these are often not disclosed, and evaluation of each separate product is needed to find the optimal product for each specific purpose. It should also be noted that MSCs cultured in SF conditions differ from those in FBS-cultured cells regarding phenotype, clonogenicity, and differentiation potential, and may therefore not be comparable clinical products.

13.3.3 Plating Density for MSC Expansion

When cell cultures have reached near confluence, MSCs are typically detached by Trypsin and replated at lower densities. Lower plating density favors rapid expansion and reduces contact inhibition. The exponential growth phase lasts longer and thereby allows for more population doublings (PD). Low-density conditions do not seem to affect MSC phenotype or in vitro trilineage differentiation. It is recommended to harvest MSCs when cultures are ≤80% confluent. Despite higher expansion rates with extremely low seeding densities, these protocols are not realistic for large-scale clinical cell production, since the number of culture vials easily exceeds what is manageable in terms of practical handling and cost effectiveness. Hence, MSCs are typically seeded at 1000 and 4000 cells/cm^2 (Ikebe and Suzuki, 2014; Bara et al., 2014).

13.3.4 Harvest of Adherent Cells

To harvest adherent MSCs, detachment is usually facilitated by enzymatic cleavage of adhesion proteins. Trypsin is the most commonly used detachment agent, at varying concentrations (0.25%, 0.05%, and 0.025%). During the procedure, most surface proteins are destroyed and excessive trypsinization can lead to irreversible cell damage. In contrast, insufficient trypsinization leads to incomplete cell harvest. Detachment conditions can be adjusted in terms of the concentration of enzyme, duration, and temperature (usually 37 °C). Several alternative products have been developed to offer animal-free and/or softer detachment. TrypLE is a recombinant bacterial-derived animal-free trypsin product. Compared to trypsin, cell detachment is gentler and does not need to be actively inactivated with serum, just diluted with phosphate buffered saline (PBS) (Carvalho et al., 2011). TrypZean is a comparable corn-derived product. Most products contain ethylenediamine tetra-acetic acid, a chelating agent that promotes cell detachment by sequestering calcium ions necessary for cell attachment through calcium-dependent adhesion molecules.

Accutase is a crab-derived product that contains a mixture of enzymes with both proteolytic and collagenolytic activity; hence, both are mammalian-free and bacterial-free. Compared to other detachment products, Accutase provides a gentle detachment, which leaves most surface proteins intact and is therefore good when harvesting cells for surface marker analysis; it also preserves high cell viability. Up to 50 min of treatment can be done without affecting cell viability. The enzymes are inactive after 1 h at 37 °C, and no active inactivation is required after detachment.

13.3.5 Scale-up Manufacturing and Alternatives to Traditional 2D Culture Systems

Many clinical applications require a large number of MSCs, varying between one million to several billions per treatment depending on the application. To meet these demands, production must be cost effective while adhering to regulatory guidelines. Traditional 2D culture systems are challenging for production of MSCs at a large scale. The large number of flasks required easily exceeds what is practically manageable. The use of multilayered flasks and cell stackers provides limited improvements not only in terms of cell yield but also in terms of labor and cost effectiveness. With an increasing number of culture flasks, the risk for heterogeneity in cell product quality, as well as risk for contaminations, increases. Different bioreactor systems provide ways to overcome these limitations. Suspension reactor systems has proven challenging for adherence-dependent MSCs, for which suspension adaptation can cause undesired cellular changes. The biggest challenges with bioreactor systems are optimization of critical parameters such as pH, oxygen tension, nutrient concentrations, and share stress. In addition, safe and reproducible manufacturing under good manufacturing practice (GMP) conditions requires aseptic operation, automated processing steps, and easy monitoring.

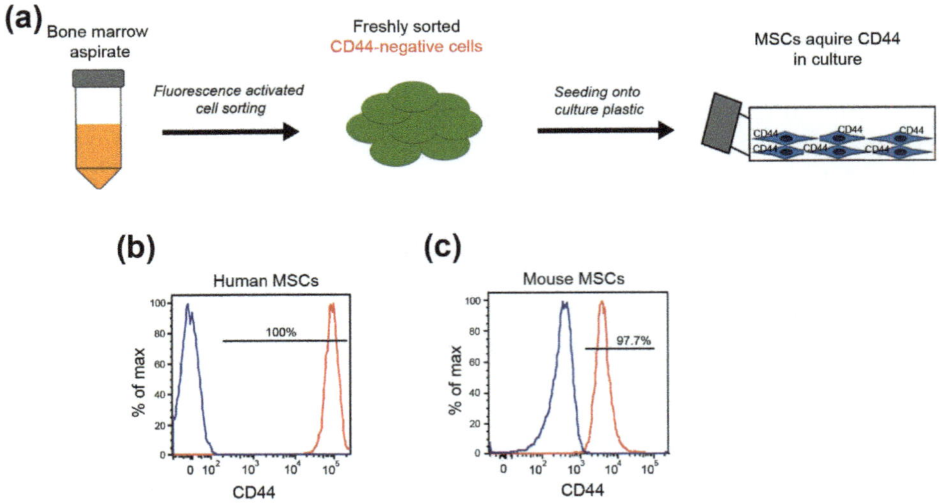

FIGURE 13.6

In vitro cultivation induces CD44 expression in human MSCs. Culturing induces extensive changes in human MSCs. The acquisition of CD44 is one striking example. (a) The MSCs are CD44 negative when they are sorted from fresh human BM aspirates by FACS. However, when seeded in culture, CD44 expression is dramatically upregulated. (b) and (c) Flow cytometry analysis of the cultured cells show that almost 100% of the cells are positive for CD44 already after primary culture in both human (b) and mouse (c) MSCs (Qian et al., 2012). MSCs, multipotent mesenchymal stromal cells; BM, bone marrow; FACS, fluorescence-activated cell sorting.

13.3.6 Culture-Induced Changes in MSCs

Much of the current understanding about MSCs is based on culture-selected cells, which may not predict the true in vivo properties/identity of MSCs. Recent developments of FACS-based protocols open up the possibility to better characterize MSC subpopulations in BM. So far, there is no clear strategy available to distinguish multipotent MSCs from other precursor cells, and phenotyping is still mainly used for characterization (Qian et al., 2012; Tormin et al., 2011). Long-term cultures cause significant changes in MSC proliferative capacity, morphology, differentiation, and global gene expression profile. Changes in MSC surface markers and gene expression patterns occur during the primary culture and early passages (p1–3) (Qian et al., 2012) (Figure 13.6). Microarray analyses have revealed a >2-fold differential expression of >2700 genes in cultured MSCs compared to that in freshly harvested MSCs, mostly in genes associated with cell adhesion and proliferation. These findings emphasize the importance of prospective isolation of MSCs to uncover the nature and therapeutic potentials of the cells.

13.4 CHARACTERIZATION OF CULTURE-EXPANDED MSCs

Since culture expansion is associated with a range of changes in MSCs, characterization of the expanded cells and monitoring of critical parameters are of

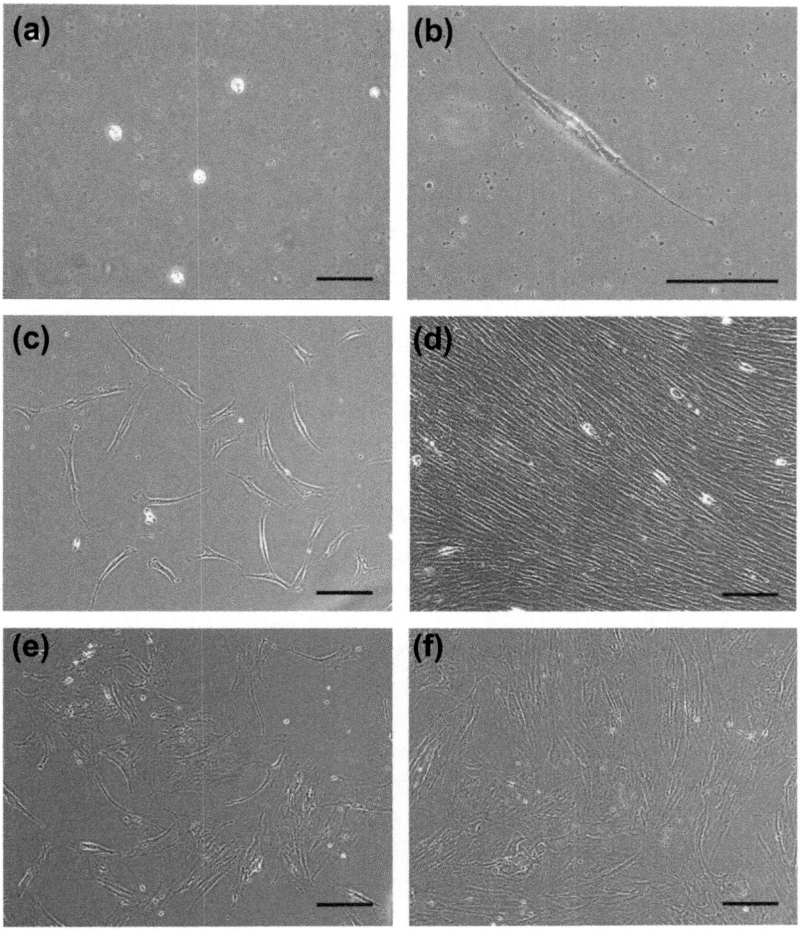

FIGURE 13.7

Morphology of BM-MSCs isolated by plastic adherence. (a) Adherent cells are visible 12–24 h after the initial plating of BM-MNCs. (b) Within 72 h, adherent cells started to present MSC morphology, and (c) increasing numbers of adherent MSC-like cells are visible after additional days. (d) Passaged cells reach confluence within 7–21 days, depending on individual cells and culturing conditions. At higher passages, MSCs acquire a flattened morphology as they approach senescence (e) and (f). Scale bar = 20 μm. MNCs, mononuclear cells; BM, bone marrow; MSCs, multipotent mesenchymal stromal cells.

importance to ensure a product of high quality and safety. These parameters might vary depending on the application. For clinical grade MSCs, release criteria usually include the absence of infectious contamination and high viability (>90%) in addition to the minimal criteria. Monitoring of cell morphology is one of the basic parameters for evaluating cells both during isolation and expansion (Figure 13.7).

13.4.1 Clonogenicity and Growth Kinetics of Expanded MSCs

Maintenance of clonogenic capacity of MSCs through repeated passaging can be monitored by performing a CFU assay at each passage. This type of CFU assay is performed by plating approximately 10 MSCs/cm^2 and culturing for 21 days, following the evaluation as described in Section 13.2.3. The CFU capacity declines with increased passage and can be used as a quality measure to estimate the remaining PD capacity.

Counting the number of cells at harvest compared to cells seeded is a common way of monitoring cell expansion at separate passages. However, from the cell biological perspective, the number of PDs is of greater importance. PDs are calculated using the formula log N/log 2, where N is the number of cells yielded at trypsinization divided by the number of cells plated. Growth kinetics of MSCs through several passages can then be expressed by calculating cumulative PDs.

Both time in culture and number of PDs are critical factors that affect the quality of MSCs. With increased passaging, MSCs lose their spindle-shaped morphology and acquire a flattened morphology as they approach senescence (Figure 13.7(e) and (f)). Long-term growth kinetics vary widely and reported maximum PDs for adult human MSCs range between 4 and 25, 30 and 40, or 50 and 60 PDs, depending on donor profile and culture conditions. Human MSCs retain their phenotype up to approximately 20 passages, but extensive passaging increases the risk of chromosomal instability. The genetic stability is retained in human MSCs at least up to passage 4–5, where after random chromosome alterations can be detected, which has been linked to senescence in the absence of malignant transformation (Tarte et al., 2010).

CLASSICAL EXPERIMENT: CLASSICAL ISOLATION, EXPANSION, AND CHARACTERIZATION OF MSCs

MSCs were discovered in the 1970s by Friedenstein and colleagues, as FCFCs, found in monolayer cultures of explanted cells from guinea pig BM and spleen. They isolate the cells by plating whole BM on tissue culture plastic and washing away the nonadherent cells after a few hours. After 4–5 days, the adherent cells started to proliferate and form colonies (Friedenstein et al., 1970). Based on early animal studies on freshly isolated BM cells, and in vitro cultured FCFCs (Friedenstein et al., 1974), a population of CFU-F was detected, which was capable of forming bone tissue when transplanted in vivo by injection under the renal capsule or in diffusion chambers (a). These cells were further explored by many investigators (Castro-Malaspina et al., 1980; Haynesworth et al., 1992; Kuznetsov et al., 1997; Pittenger et al., 1999), and it was established that the plastic adherent, clonal cells from the BM were multipotent and could differentiate into osteoblasts, adipocytes, and chondrocytes, as well as additional cells of the mesenchymal lineage (b). Taken together, these early studies on BM-MSCs constitute the backbone of the traditional protocols for isolation, expansion, and characterization of MSCs, summarized into the minimal criteria presented by ISCT in 2006 (Dominici et al., 2006) (c), and still widely used.

Continued...

CLASSICAL EXPERIMENT: CLASSICAL ISOLATION, EXPANSION, AND CHARACTERIZATION OF MSCs *Continued*

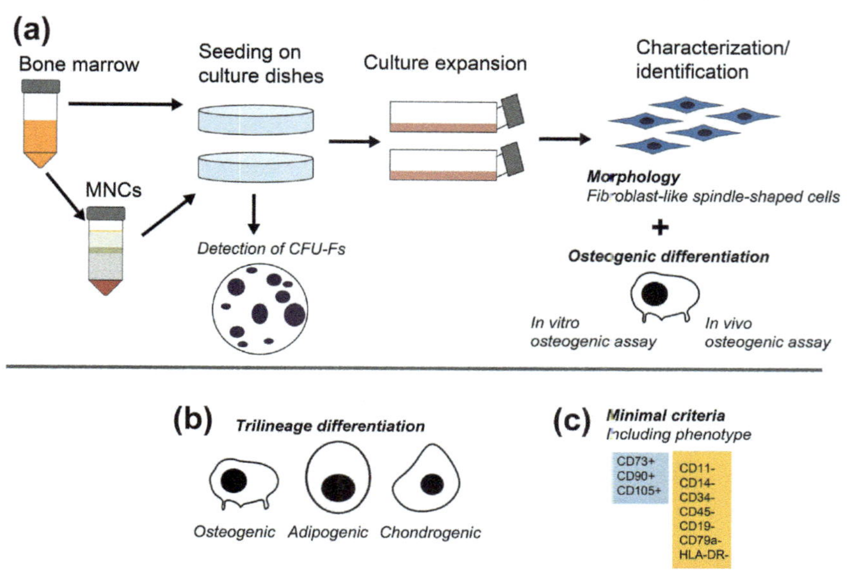

13.4.2 Phenotyping of Culture-Expanded MSCs

Basic phenotyping of culture-expanded MSCs can be performed according to the ISCT minimal criteria. The phenotypic definition requires expression of CD73, CD90, and CD105, together with a lack of expression of monocyte and macrophage markers (CD11b or CD14), a hematopoietic progenitor and endothelial cell marker (CD34), a leukocyte marker (CD45), B cell markers (CD19 or CD79a), and HLA-DR. To characterize the culture-selected MSCs or control the quality of the therapeutic MSCs, FACS is required to phenotypically analyze the culture-expanded MSCs. Since CD19+ cells are also CD45+, CD19 staining can be omitted. With current techniques, multicolor flow cytometry can be performed to obtain a complete phenotypic profile based on concomitant expression of the markers in the basic panel. An example of multicolor staining for simultaneous detection of several positive and negative markers is given in Table 13.4. Besides the basic panel of surface markers, additional markers can be of interest for MSC phenotyping, depending on the application (Table 13.1).

13.4.3 Trilineage Differentiation of MSCs

Differentiation of MSCs into bone, adipose, and cartilage cells is usually required to distinguish MSCs from other stromal cells, and are part of the minimal criteria. Differentiation involves key lineage-specific transcription factors (TFs) and a number of regulatory molecules (Figure 13.8).

Table 13.4 Example of Multicolor Staining Strategy for MSC Phenotyping According to the Basic Panel of Minimal Criteria

Fluorophore/Wave Length Channel for Flow Cytometry	Marker for MSC Phenotyping
1	$CD73^+$
2	$CD90^+$
3	$CD105^+$
4	$CD45^-/CD14^-/CD34^-$
5	$HLA\text{-}DR^-$
6	$HLA\text{-}I^+$
7	PI

MSCs, multipotent mesenchymal stromal cells.

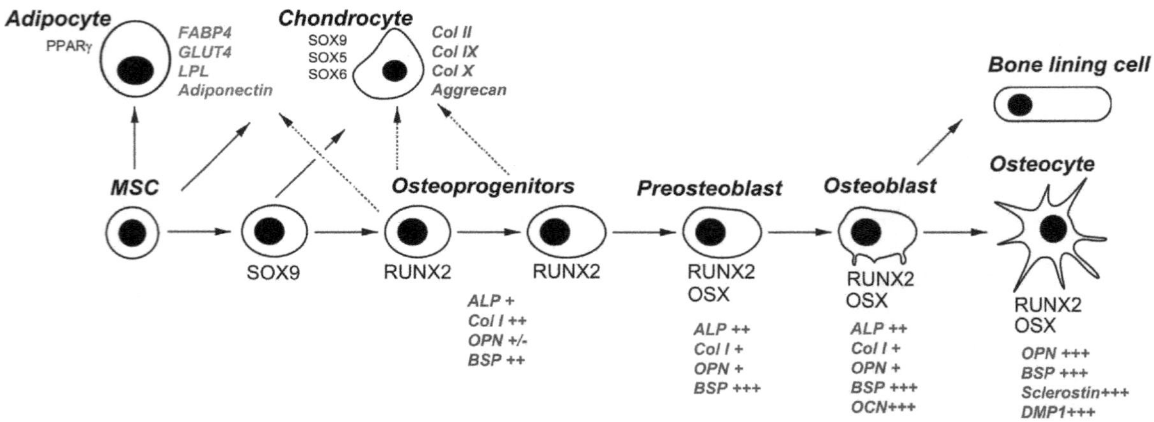

FIGURE 13.8

A simplified overview of trilineage differentiations of MSCs with the corresponding expression of key mediators. The key transcription factors (TFs) responsible for each lineage commitment are indicated in black, whereas crucial lineage specific proteins are indicated in gray. During osteogenic differentiation, expression of TFs, alkaline phosphatase (ALP), and different extracellular matrix proteins follow certain patterns that can be utilized for the detection and estimation of degree of osteogenic maturation. Col, collagen; OPN, osteopontin; BSP, bone sialoprotein; OCN, osteocalcin; MSCs, multipotent mesenchymal stromal cells.

There are several ready-to-use induction media commercially available. The composition and concentrations of crucial inductive agents are not always described. If the purpose of the in vitro differentiation assay is solely to demonstrate trilineage capacity, commercial differentiation media are convenient. If however the differentiation process itself is the subject of investigation, or the aim is to produce fully mature and functional matrix-producing osteoblasts, the composition of inductive agents has to be defined, since these might be needed for fine tuning concentrations, timing, and duration of the supplements depending on the cell source and purpose (Alm et al., 2012) (Boxes 13.2–13.4).

13.5 CRYOPRESENTATION

13.5.1 Cryopreservation and Storage of MSCs

Cryopreservation allows for long-term banking of cells at very low temperatures (down to −196 °C) for varying periods of time while preserving structurally

BOX 13.2 IN VITRO OSTEOGENIC DIFFERENTIATION ASSAY OF MSCs

The classical and well-established protocol for osteoblastic differentiation of human MSCs comprises culturing subconfluent MSCs in the presence of ascorbic acid (AA), β-glycerophosphate (βGP), and dexamethasone (Dex). Additional osteogenic factors, that is, vitamin D and BMPs, have also been used successfully. Depending on experimental settings and donor characteristics, differentiation starts at 7–10 days and can last till 10–21 days. Mineralization can be detected after 3–5 weeks (Aubin et al., 2006).

- *Ascorbic acid* is essential for collagen matrix maturation, as a cofactor for the hydroxylation required for proper collagen fibril assembly, and for stimulation of noncollagenous bone matrix protein synthesis. In human MSCs, ascorbic acid also upregulates cell cycle and mitosis-related genes and stimulates proliferation. Ascorbic acid alone does not stimulate ALP expression, but in its absence, ALP expression is reduced, and calcium accumulation is inhibited. The more stable form ascorbic acid 2-phosphate is used as supplement, usually at concentrations ranging from 50 to 500 μM.
- *β-Glycerophosphate* provides an organic phosphate source for in vitro bone formation, usually applied at concentrations ranging between 5 and 10 mM. It is enzymatically hydrolyzed into inorganic phosphate by ALP. As a consequence, increasing βGP concentrations does not increase mineralization, since the ratio between organic and inorganic phosphate is the critical factor, which is fully dependent on ALP.
- *Dexamethasone* is a synthetic glucocorticoid that is essential for osteogenic differentiation and mineralization of MSCs in vitro. As for other glucocorticoids, effects of Dex on MSCs are complex, and the exact mechanisms are not completely understood. Dex seems to regulate osteogenic differentiation at transcriptional levels through both bone morphogenetic protein (BMP)- and wnt-signaling pathways. For human MSCs, both 10 and 100 nM Dex are commonly used concentrations. By using transient supplementation with 100 nM Dex for the first 7 days of osteogenic induction culturing, differentiation can be enhanced while reducing interindividual and intraindividual donor variabilities.

Upon osteogenic differentiation, MSCs change morphology from spindle-shaped to cuboidal (a) and start to produce a collagen matrix. As the process proceeds, three-dimensional structures (nodules or strings) appear (b) and d), which are later mineralized (e) and (f). Each step is associated with characteristic changes in gene and protein expression (see figure given below). The mature osteoblast is characterized by a cuboidal morphology and strong ALP expression (a)–(c). Differentiation is detected by measurement of ALP activity and visualized by ALP staining. Mature, functional osteoblasts are verified by demonstrating matrix mineralization by staining for phosphate (von Kossa staining) (e) and (f) and calcium (Alizarin Red) deposition. Osteogenic lineage commitment can be detected by analyzing osteoblast-specific gene expression including ALP, Runx2, osterix, collagen type 1, osteocalcin, and osteopontin. At the protein level, collagen I and its synthesis by-products, as well as osteocalcin, in addition to ALP, can be analyzed. Scale bars = 20 μm.

Continued...

BOX 13.2 IN VITRO OSTEOGENIC DIFFERENTIATION ASSAY OF MSCs *Continued*

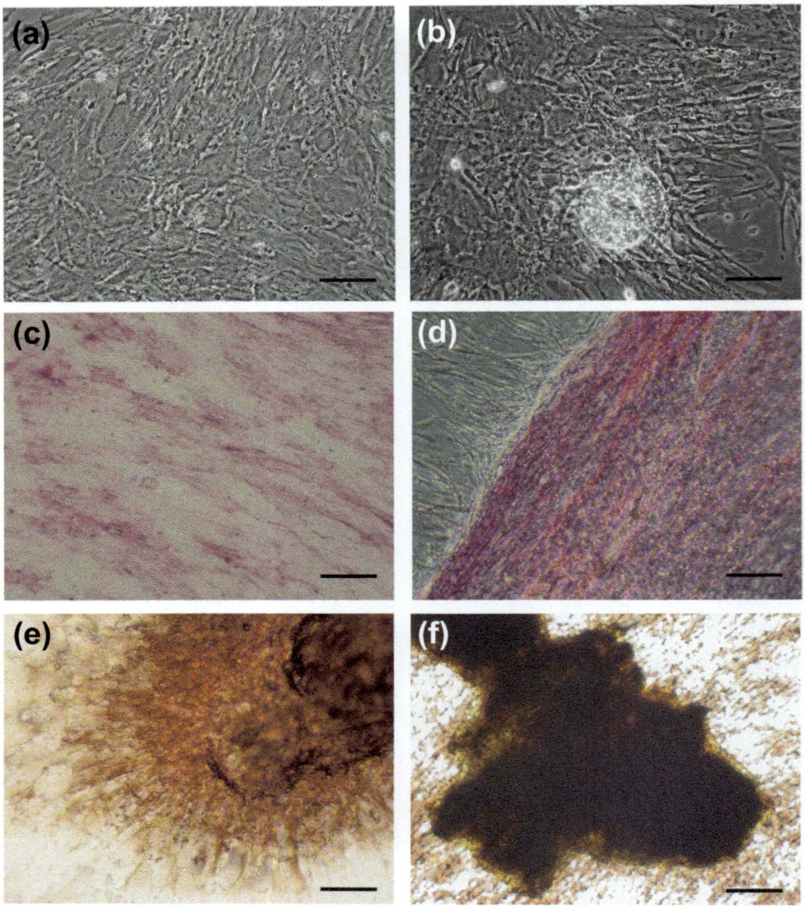

and functionally intact living cells. Cryopreservation and long-term storage of MSCs is a necessity for not only providing "off-the-shelf" availability for clinical application but also for banking for research purposes.

Cryopreservation protocols are developed to prevent, or at least control, intracellular and intercellular ice formation during freezing. Three key factors that determine cell survival following cryopreservation are temperature, rate of temperature change, and freezing media (Leibo and Pool, 2011). The rate at which cells are cooled down from physiological to freezing temperatures is decisive for cell survival. Even more crucial is the thawing rate. The

BOX 13.3 IN VITRO ADIPOGENIC DIFFERENTIATION ASSAY OF MSCs

For adipogenic differentiation, cells are usually cultured for 2 weeks in an induction media supplemented with methyl-isobutylxanthine (0.5 mM), indomethacin (0.1 mM), dexamethasone (1 μM), and insulin (10 μg/mL). This is followed by 1–2 weeks of culturing in adipogenic maintenance media supplemented with only insulin (10 μg/mL). Accumulation of lipid droplets can be observed after 4–5 days of induction culturing, but complete differentiation usually requires 2–3 weeks. Accumulation of lipid droplets is visible under a light microscope (a). Staining with Oil red O can be performed to enhance visualization (b) and (c). However, since Oil red O stains triglycerides, differentiation has to be complete into mature adipocytes for proper staining to occur. Adipogenic lineage commitment is usually verified by analyzing gene expression of TF peroxisome proliferator-activated receptor γ (PPARγ). Mature adipocytes can be identified by the expression of fatty acid-binding protein 4, glucose transporter type 4, leptin, and adiponectin (see figure given below). There is an inverse relationship between differentiation along osteogenic and adipogenic lineages. In culture, many growth factors and hormones promoting adipogenic differentiation have negative effects on osteogenic differentiation, and vice versa. Also wnt signaling serves as an important switcher between the two lineages, with upregulation during osteogenesis, and downregulation during adipogenesis. Scale bars = 20 μm.

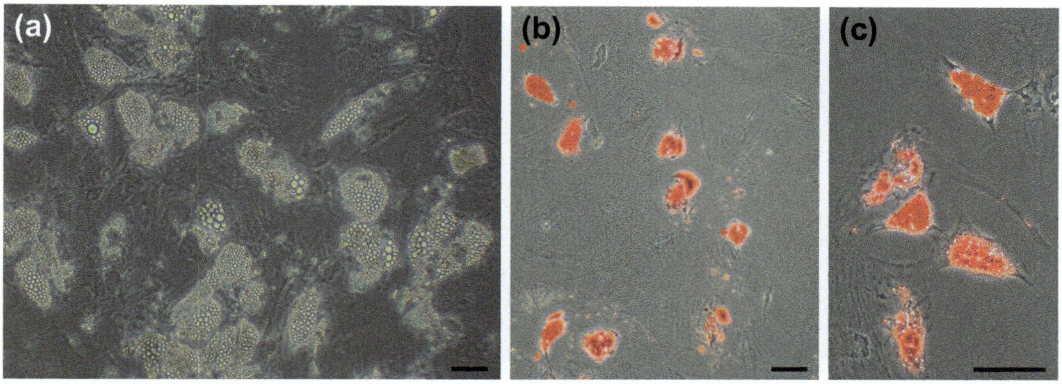

composition of the freezing media and type of cryoprotectant have an additional significant influence on the outcome. Two biophysical events contributing to cell damage during freezing are intracellular ice formation and water transport across the cell membrane. Intracellular ice formation associated with quick freezing causes cell membrane rupture. This is avoided by slow cooling at a controlled rate. However, slow freezing is associated with dehydration injury. As extracellular fluids freeze, there is a build-up of electrolyte concentrations, reaching lethal levels. The increased extracellular solute concentration causes cellular dehydration and shrinking, and potential cell membrane damage (Fowler and Toner, 2005). To prevent this, cryoprotective agents are added to the freezing media. Through their high solubility at low temperatures, cryoprotectants serve two important functions: to lower the

BOX 13.4 IN VITRO CHONDROGENIC DIFFERENTIATION ASSAY OF MSCs

Chondrogenic differentiation assay is usually performed as pellet cultures. For establishing pellet cultures, 200,000 cells are normally washed in 15-mL conical tubes in 1 mL of incomplete chondrogenic medium consisting of high glucose DMEM containing 4.5 g/L of glucose (DMEM-HG) supplemented with 110 mg/L of sodium pyruvate, 50 μg/mL of L-ascorbic acid-2-phosphate, 100 nM dexamethasone, 10 μL/mL of ITS + Premix. The cells are then resuspended in complete chondrogenic media, comprising incomplete chondrogenic media supplemented with10 ng/mL of TGF-β3 and pelleted by centrifugation at 500 g for 8 min. The pellets are maintained in culture with 1 pellet/tube in 1 mL of complete medium. TGF-β3 has to be added fresh whenever the medium is changed. Chondrogenic pellets are usually harvested at 21 days for evaluation of differentiation.

For detection of chondrogenic extracellular matrix formation, harvested pellets are fixed and prepared for histochemistry according to standard protocols, including embedding in paraffin. Paraffin sections can then be stained for specific matrix components. Sulfated proteoglycans are visualized by staining with Toluidine blue (a) and (b), Alcian Blue, or Safranin. Type II collagen can be detected by immunohistochemistry. Chondrogenic lineage commitment can be evaluated by gene expression analysis for sox9, which is the key TF and the earliest chondrogenic marker. Additional genes of interest are sox5, sox6, aggrecan, and collagen types II, IX, and X (see figure given below). Scale bar = 50 μm.

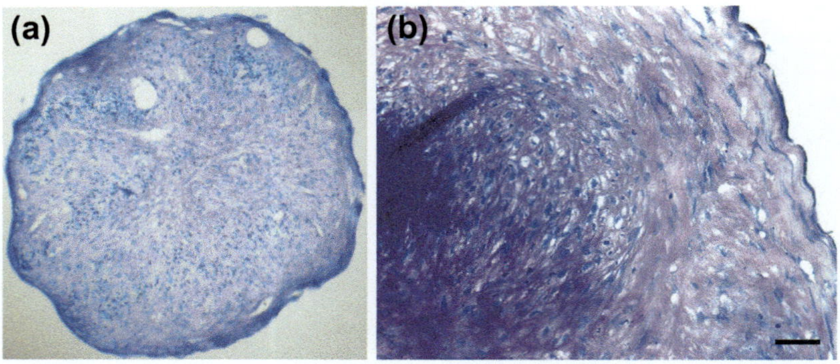

freezing temperature and to enter the cells to prevent shrinking. There are numerous cryoprotectants available, of which dimethylsulfoxide (DMSO) is the most common. DMSO is toxic at temperatures above +4 °C, and therefore, exposure to cells at these temperatures should be minimized. Successful attempts have been made to reduce the commonly used 10% DMSO concentration to 7.5–2.5% (Ginis et al., 2012). Usually serum (FBS) is added at a concentration of ≥10–20%. SF freezing solutions have been tested, with normal or low DMSO levels, or alternative nontoxic cryoprotectants (Fowler and Toner, 2005).

When DMSO is mixed with freezing media, a thermal reaction takes place. Therefore, DMSO should be freshly mixed with freezing media and cooled to 4 °C before use. For optimal cryopreservation, MSCs are frozen at concentrations of 1–5 × 10^6 cells/mL. Cells are kept at 4 °C for 30 min before adding freezing media with DMSO (4 °C). MSCs are cryopreserved through slow-freezing methods at a cooling rate of 1 °C/min. The most widely used methods are conduction-based approaches, using either isopropanol containers for controlled cooling rate or alcohol-free systems where cooling is controlled by utilizing a thermoconductive alloy inside a device made of insulative material. More advanced freezing devices include controlled rate freezers (CRFs), which offer a uniform cooling by precisely controlling and adjusting the temperature of cryogen surrounding the samples in a container using dispersion fans. The system provides complex cooling programs, and the temperature profile of each sample is recorded. However, heterogeneity still exists in a samples exposure to a cryogen during freezing, especially at large-scale preservation; this causes vial-to-vial variation in freezing kinetics and postthaw viability (Thirumala et al., 2013). Although current CRFs are used for clinical cell banking, advanced technologies are desired to achieve a more uniform large-scale freezing.

Cryopreserved cells are usually stored in liquid nitrogen (LN_2). Storage in LN_2 has weaknesses, especially for the preservation of therapeutic cells. Possible shrinking of the cell vials can compromise the closure and allow LN_2 to enter the vial, with the subsequent risk of explosion during thawing as the LN_2 vaporizes at increasing temperatures. Second, there is a risk of contamination between sample vials, as viruses, bacteria, and fungi can be transmitted through N_2 in both liquid and vapor phases. Ultralow temperature freezers (−196 °C) offer an alternative to LN_2 for cryostorage of therapeutic cells.

13.5.2 Thawing of Cryopreserved MSCs

Thawing is more hazardous to the cells than freezing. As temperature increases, the cryoprotectant causes extracellular water to enter the cells, and this causes cell swelling. Therefore, samples need to be diluted with solutions to provide osmotic balance. Cells should be thawed at 37 °C for 1–2 min followed by dilution of the cryoprotectant with prewarmed growth medium containing FCS (or other serum alternative). Cryovials can be thawed in 37 °C heating blocks, water baths, or incubators. To avoid cell damage caused by DMSO upon thawing, relatively large volumes of culture media with serum (10–60%) should be added to reestablish physiological homeostasis (Haack-Sorensen and Kastrup, 2011; Leibo and Pool, 2011). To prevent osmotic shock of the cells during thawing, slow dilution of the cryoprotectant is recommended. Freezing media are discarded by centrifugation, and the cells diluted in fresh growth media are ready to use. Freshly thawed cells should be handled carefully since freeze thawing with cryoprotectants makes the cell membrane fragile. With optimal thawing and exposure of cells to favorable conditions, this can be kept transient.

13.5.3 Effects of Cryopreservation on the Quality of MSCs

The majority of MSC therapeutics uses cryopreserved cells. Banking of expanded and quality controlled MSCs for thawing immediately before use is the most convenient setup for ensuring the availability of sufficient cell numbers, fulfilling regulatory requirements. It is established that cryopreservation of MSCs is safe in terms of retaining their basic characteristics, that is, proliferative, phenotypic, and differentiation capacities, essential for in vitro MSC research. However, preservation of functional properties crucial to their therapeutic efficiency, including immunosuppressive activity, homing, migration, and engraftment, is less investigated. Viability of MSCs is usually good immediately after thawing, and ranges between 70% and 97% (Samuelsson et al., 2009; Ginis et al., 2012). However, compared to freshly detached MSCs, viability of thawed cells decreases significantly if handling and holding times are >2 h (Pal et al., 2008).

For subculturing cryopreserved MSCs, a transient sensitivity and decreased viability are not a problem. The surviving cell fraction recovers in culture and retains proliferative and trilineage differentiation capacity, without changes in the basic phenotype. For clinical applications, however, dying and/or damaged cells potentially reduce the therapeutic efficiency through a range of mechanisms. First, damaged cell surface and subsequent binding of complement activation products promote clearance by various types of phagocytes. In addition to the obvious limitation in cell survival, animal studies have indicated a reduced tolerance. Second, postthaw cytoskeletal and surface molecule deficiency may compromise MSC migration and engraftment. Third, recent data indicate that cryopreservation can affect the immunomodulatory properties of MSCs immediately after thawing, but this seems to be regained after recovery in culture (Francois et al., 2012; Samuelsson et al., 2009). We have demonstrated that MSCs have reduced immunomodulatory and blood regulatory properties directly after thawing, associated with complement-mediated elimination when exposed to whole blood in vitro. These changes seemed to be paired with differences in therapeutic efficiency (Moll et al., 2014).

13.6 PRODUCTION OF CLINICAL GRADE MSCs

Large-scale expansion of clinical grade MSCs has to meet the terms of current good manufacturing practice (cGMP). These include strictly defined conditions and processes (Chapter 21). In addition to phenotypic analysis, quality control of clinical MSCs includes analyses of microbiological safety and genetic stability (Table 13.5). So far, there is no established standardized protocol. The diversity in protocols, study designs, and mechanisms of application among clinical trials limits comparisons between trials, and the translation into clinical practice.

Table 13.5 MSC Product Release Criteria

Cellular Product Testing	Release Criteria
Donor serology for infectious disease markers	Negative
Morphology	Spindle shaped
Phenotypic purity by flow cytometry	Positive for CD73, CD90, CD105 Negative for CD14, CD31, CD34, CD45, HLA-DR
Microbial contamination	Negative
Viability	>70% (Eosin, Trypan-blue, or Türk)
Mycoplasma test of final product	Negative
Appearance of final product	Single cell suspension

MSCs, multipotent mesenchymal stromal cells.

13.6.1 General Considerations

For production of therapeutic doses of autologous MSCs, traditional laboratory scale production and cryopreservation are usually sufficient, but for increasing applications of allogenic MSCs, large-scale expansion and banking are required to meet the higher cell-dose demands and provide stocks of readily available cells.

Documentation of culture variables is important because of the possible impact on the resulting product. Variables of interest are cell density at the time of passage, number of PDs, number of passages, culture medium, and possible growth factors used (Krampera et al., 2013). To ensure a high-quality product, time in culture, number of PDs, and passage number need to be kept limited. This emphasizes the importance of a large volume of high quality starting material to reach the required therapeutic cell dose within a restricted time period and PDs. Although human MSCs have been demonstrated to be genetically stable through repeated culturing, many centers have normal karyotyping as part of release criteria.

13.6.2 Current Good Manufacturing Practice

Regulatory requirements vary between countries, but for clinical procedures, most countries require production protocols that at least adhere to good laboratory practice, while many centers follow the stricter and higher requirements of cGMP (Tolar et al., 2010). For cGMP level manufacturing, all stages of the production has to adhere to preset specifications and protocols, and can only be performed in classified cGMP facilities by trained personnel. The comprehensive cGMP requirements include well-defined and documented methods processes in the form of Standard Operating Procedures, documentation for traceability of all raw materials, reagents, and culture plastic products in use (Chapter 21).

13.7 DONOR VARIABILITY AND DONOR-RELATED PARAMETERS AFFECTING IN VITRO PROPERTIES AND EXPANSION ABILITY OF MSCs

Large individual variability in human MSCs is a well-documented issue. Many baseline characteristics and in vitro properties of human MSCs depend on demographic features of the donor, including age, gender, vitamin D status, smoking, alcohol use, estrogen status, corticosteroid use, immobilization, levels of cytokines and growth factors, and different disease states (Zhou et al., 2012; Hung et al., 2006). Intrinsic alterations in human MSCs with aging contribute to tissue aging (Kassem and Marie, 2011). These aspects are of relevance in the concept of MSC therapy and tissue engineering, since a large proportion of MSC donors (both allogenic and autologous) constitute middle-aged and aged individuals. Age-related alterations of MSCs include a decrease in frequency, proliferative and osteogenic differentiation capacity, function and life span, and increased cell senescence and apoptosis (Stolzing et al., 2008; Alm et al., 2010). The declining number of MSCs is suggested to correspond to different needs for progenitors during growth and adulthood, with corresponding age periods of differing MSC titers (Caplan, 2007).

13.8 RELATIONSHIP BETWEEN IN VITRO ASSAYED MSC PROPERTIES AND THEIR POSSIBLE IN VIVO FUNCTION

Identity and nature of native MSCs and their role in normal tissue turnover and repair remain poorly understood. Current isolation methods and in vitro assays are limited in providing an insight into the "true nature" of MSCs due to heterogeneity of cell populations isolated, culture induced changes, and restricted methods for identifying MSCs. However, correlations between in vitro assayed properties and relevant in vivo functions and parameters can serve other purposes. There are two aspects of interest regarding correlation between in vitro and in vivo function of MSCs: (1) the therapeutic aspect of in vitro evaluation of the cellular product to predict clinical effectiveness in the recipient and (2) the biological aspect of how well in vitro properties of MSCs reflect tissue turnover, quality, and healing capacity, and immunomodulatory capacities in the donors.

13.8.1 Correlations between MSC In Vitro Properties and Therapeutic Effectiveness

For MSC therapy, potency assays that predict clinical efficiency would revolutionize the field. Within tissue engineering, correlation between in vitro differentiation capacity of culture-expanded MSCs and tissue formation after transplantation is essential and often investigated with varying results. The CADM1 gene has been suggested as an in vitro diagnostic marker for predicting bone-forming capacity of MSCs (Mentink et al., 2013). Cell adhesion, migration, and survival are additional parameters considered to be of relevance for therapeutic success, but so far, there are no reliable functional assays for in vivo prediction.

For MSC treatment of immune-based disorders, there are a range of immunological in vitro assays of limited predictive value, including mixed lymphocyte cultures, cocultures with purified T-cells (or other immune cells), or licensing of MSCs by exposure to proinflammatory cytokines. ISCT recently presented a working proposal toward shared guidelines for the characterization of immunological properties of MSCs (Krampera et al., 2013). Inflammatory priming with interferon gamma (IFN-γ) ± tumor necrosis factor-α (TNF-α), followed by response readout by analyzing the expression of indoleamine 2,3-dioxygenase (IDO), interleukin-6, and other possible proinflammatory cytokines upregulated in MSCs following licensing is suggested as a central assay (Bernardo and Fibbe, 2013). However, in vitro immunosuppressive properties of MSCs have so far not showed any correlation with their therapeutic efficacy (von Bahr et al., 2012b).

13.8.2 MSC In Vitro Properties Reflecting Demographic Parameters

Studies of an observational nature have linked in vitro assayed MSC capacities to demographic parameters. For example, MSCs from patients with osteoporosis, osteoarthritis, and osteonecrosis show altered in vitro capacities and changed gene expression (Jiang et al., 2011). The in vitro osteogenic differentiation capacity of MSCs correlates with vitamin D and parathyroid hormone status (PTH), and bone mineral density of the donors (Zhou et al., 2012). In vitro mechanoresponse of MSCs increases with increased serum leptin and estradiol levels, while decreasing with increased body mass index of the donor (Friedl et al., 2009).

Correlation of in vitro capacity of MSCs with donor parameters within the immunoregulatory concept is not well studied. As the number of clinical applications using autologous MSCs for autoimmune disorders (e.g., MS, diabetes, and Crohn's), these types of analyses are important to evaluate whether autologous cells show signs of diminished potential. This is also important when using autologous MSCs for treatment of organ failure (kidney, liver, and heart), since there might be underlying causes of tissue damage that affects the MSC quality. However, due to the lack of "quality assays" predicting the clinical efficiency, evaluation so far relies on comparisons of basic characteristics with MSCs from healthy subjects. Such studies are generally associated with contradictory results regarding disease effects on proliferative, phenotypic, and differentiation capacities of MSCs. The relevance of these in vitro properties for possible roles of MSCs in the specific disease, or suitability of autologous cell therapy, is unclear.

13.9 FUTURE PERSPECTIVES

The safety of MSC treatment is well established for numerous indications. Of outermost importance in current developments of MSC-based therapeutics is further demonstration of beneficial effects, and a better understanding of underlying mechanisms by which MSCs mediate both immune-suppressive and tissue-regenerative actions. Clinical efficiency will gain from increased knowledge, while further clinical trials are needed to elucidate these mechanisms. Diversity in MSC manufacturing and study designs needs to be overcome to obtain more reproducible patient responses. The effect of cell handling and manipulation during manufacturing and storage, as well as impact of donor-related parameters, need further understanding. Future clinical trials need to be designed and conducted in the best possible way to optimize the information that can be obtained.

13.9.1 Practical Issues

Cost-effective production of a sufficient number of clinical grade cells, that is, short in vitro manipulation and presumably high quality, for continuous MSC treatments in an increasing number of patients is one challenge. Another challenge is the development, evaluation, transition, and establishment of animal-free protocols adhering to cGMP standards, a task that requires time and resources before completion. Additional fine-tuning of culturing and handling protocols might be needed for improvement of cell preservation following detachment and cryopreservation.

13.9.2 The Cell Therapy Product

The optimal therapeutic cell might differ in profile depending on the application. Tissue engineering approaches need therapeutic cells for tissue formation and gain of function, while immune-based MSC therapies urge for cells mediating a successful clinical response to improve the (often systemic) health conditions and life expectancy. Several approaches are investigated to improve the cell therapy product, ranging from cell source and donor profile, to isolation and culturing protocols, and even further to refinement of the product by specific manipulations of the cells to reach desired effects (Figure 13.9).

Criteria for recruitment of allogenic MSC donors mainly focus on generally healthy individuals. The improved understanding of important factors for donor selection may increase cell quality. Alternatively, relatively simple measures could be applied to possibly "boost" the MSC compartment in both allogenic and autologous donors, such as monitoring of vitamin D status with subsequent substitution prior to BM harvest if needed, or stricter criteria related to the use of nonprescriptive medication, smoking, or alcohol habits.

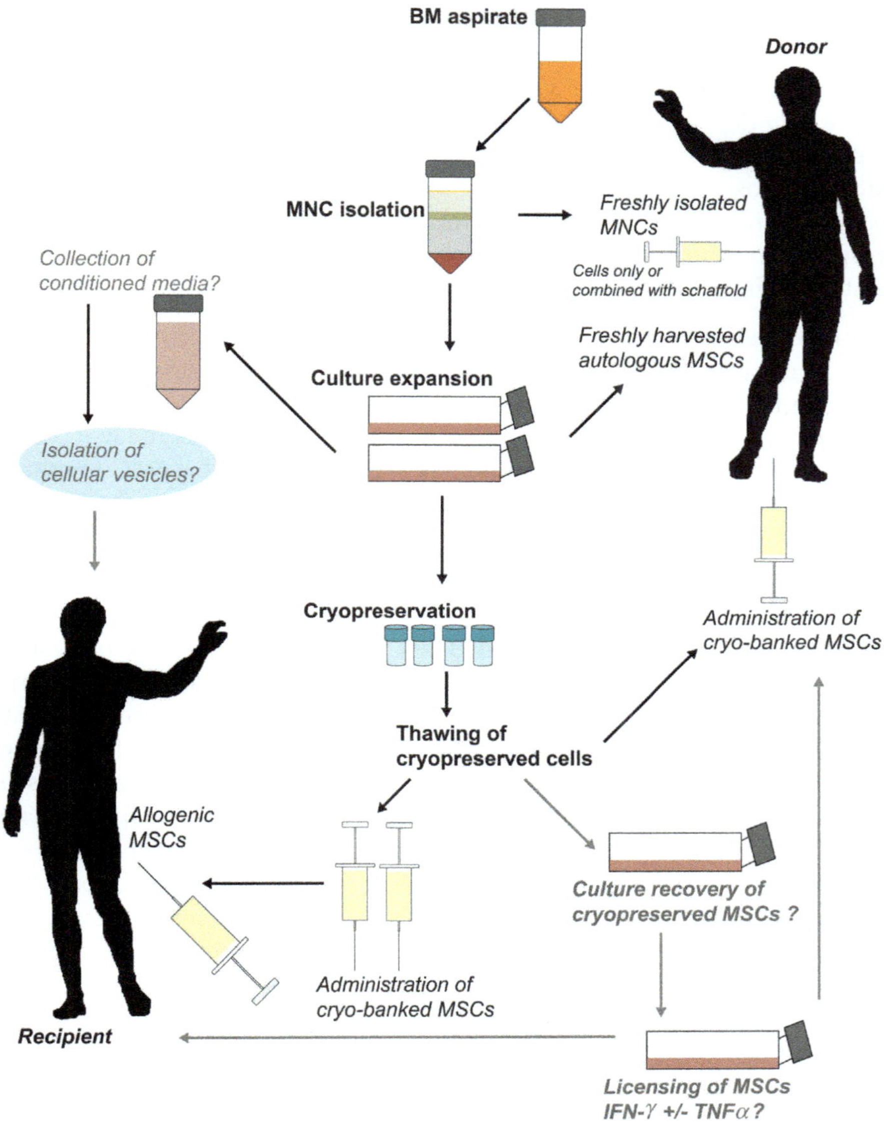

FIGURE 13.9

Current and future perspectives of BM mesenchymal stromal cell-based therapeutics. Current MSC-derived clinical applications encompass both autologous and allogenic solutions, ranging from freshly isolated MNCs, to culture-expanded and cryobanked MSCs. In some cases, freshly detached cultured MSCs are applied. Potential and alternative strategies (indicated with gray) include the application of MSC-produced paracrine factors, culture recovery of cryobanked cells, or licensing of MSC ex vivo prior to transplantation. BM, bone marrow; MSCs, multipotent mesenchymal stromal cells; MNC, mononuclear cells.

13.9.3 Refinement of the MSC Product?

Current research indicates that the immunosuppressive functions of MSCs rely on acquirement of an antiinflammatory phenotype of the MSCs. This is possibly dependent on inflammatory cues in the recipient, which triggers polarization of transplanted MSCs (Bernardo and Fibbe, 2013; Le Blanc and Mougiakakos, 2012), but appropriate responses of the infused MSCs can also be determined by donor-derived factors and postthaw handling. One way to ensure an antiinflammatory phenotype of MSCs is by licensing through stimulation with proinflammatory cytokines (Figure 13.9).

Applications of MSCs for tissue regeneration were originally based on the principle that transplanted cells differentiate in vivo to replace damaged tissue, for example, by cardiomyocyte differentiation after myocardial infarction, or osteogenic differentiation in fracture nonunions or large bone defects. Today, it is clear that long-term persistence of transplanted cells is low and that MSC responses are mediated through paracrine effects. Preclinical studies indicate that engraftment of transplanted MSCs at the site of injury is limited to days (Tolar et al., 2010). We previously investigated the biodistribution of systemically infused MSCs in patients, and found very low engraftment (von Bahr et al., 2012a).

Systemic infusion of culture-expanded MSCs is the main approach for treating immune-based disorders. For cardiac repair, both systemic and local injections of culture-expanded MSCs have been applied with promising results. MSC therapy for musculoskeletal regeneration has proven to be more challenging. Usually, culture-expanded MSCs are injected locally into the damaged tissue, either alone or with a scaffold. Strategies under investigation for improvements of the cellular performance include molecular targeting through crucial signaling pathways, or epigenetic modulations and micro-RNA-based approaches (Baron and Kneissel, 2013; Lian et al., 2012). Another attempt to overcome the limited bone forming potential of MSCs was targeted by high-throughput screening for molecular compounds that impact and enhance the MSC's fate (Alves et al., 2011). With advanced flow cytometry applications, subfractions of BM stromal cells are investigated (see Start-of-the-art experiment). With further characterization of different cellular subgroups, it can be possible to select the most potential cells for specific applications.

13.9.4 BM Mononuclear Cells as a Straightforward Approach for Tissue Regeneration

An alternative approach for stromal cell therapy for tissue regeneration is a local injection of freshly aspirated autologous BM-MNC. This strategy is especially investigated in musculoskeletal regeneration where the tradition of using local BM and bone grafts has provided a natural transition to intraoperative isolation of BM-MNCs (Figure 13.8). Advantages of this approach are that it requires minimal manipulation of cells, it does not require culturing or preservation steps, and there is no need for expensive cGMP facilities. Local application of autologous BM-MNCs has been evaluated in clinical applications such as segmental long-bone defects, fracture nonunions, revision total hip arthroplasty, and as intramuscular injection for enhanced muscle reinnervation and regeneration in patients with traumatic brachial plexus injury. Also, intramyocardial injection of MNC for treatment of chronic myocardial ischemia has proven to be safe, with modest short-term clinical improvements (van Ramshorst et al., 2009).

13.9.5 The MSC Secretome

The plethora of soluble factors produced by MSCs is revealed through analyses of secretome, transcriptome, and proteome, which not only raises challenges but also new possibilities (Figure 13.9). First, donor variability and responses to culture conditions may affect the secretome profile, and reliable read-out assays are required to screen the cells. Second, the ability of MSCs to produce desired and relevant secreted factors in response to the microenvironment may be altered immediately postthaw. Third, it remains unclear whether minimal criteria exist for secreted factors defining clinically efficient cells. Identification of the "optimal secretory profile" of MSCs would allow manufacturing of efficient cell products, possibly designed according to applications. The clinical potency of the MSC secretome has already opened up for new therapeutic solutions using MSC conditioned media, or purified solutions instead of cells themselves, including MSC-derived microvesicles and exosomes.

Increasing applications of MSC-based therapies, in combination with advanced research technologies, will contribute to reveal the mechanism of action of MSCs, and improved therapeutic solutions. Currently, MSC-based therapies develop into two different directions, with one represented by advanced cell selection and/or in vitro licensing before clinical use, whereas the other is represented by freshly aspirated autologous BM-MNCs.

13.10 SNAPSHOT SUMMARY

1. MSCs constitute a heterogeneous subset of stromal precursor cells present at low frequencies in connective tissue compartments, commonly isolated from adult BM and possessing regenerative, stromal supportive, and immunoregulatory capacities.
2. MSCs are characterized by their ability to form colonies (colony-forming unit-fibroblast (CFU-Fs)); proliferate through several passages in culture; and to differentiate into osteoblasts, adipocytes, and chondrocytes.
3. No unique surface markers have been defined for positive selection and identification, but the cells can be isolated directly from BM by the use of combinations of multiple markers and advanced flow

cytometry techniques. However, phenotypic profiles of cultured MSCs are known.

4. MSC therapy is under investigation in a broad spectrum of clinical applications for treatment of chronic and degenerative diseases, with encouraging outcomes, and excellent safety record, although variability in protocols and unknown fate and mechanism of action for clinical benefits of MSCs remain challenges.
5. The differentiation potential of MSCs has brought high expectations for tissue engineering reconstruction of damaged tissues, especially for bone and cartilage repair, but the clinical use of MSCs in these settings remains limited, with survival and engraftment of MSCs being the biggest challenge, followed by a search for optimal cell fraction, dose, and delivery.
6. Methods of choice at each step of the isolation and culturing process significantly affect the MSC population produced, as the isolation method accounts for cell population selected, whereas culture parameters and cryopreservation methods affect the profile and properties of the MSCs.
7. MSCs retain their basic characteristics, that is, proliferative, phenotypic, and differentiation capacities after cryopreservation, but preservation of functional properties essential for therapeutic efficiency (immunosuppressive activity, homing, migration, and engraftment) is less investigated.
8. Both autologous and allogenic MSCs are used for clinical applications for tissue regeneration and immunomodulation. A range of basic demographic parameters (age, gender, and bone quality) have been shown to correlate with in vitro assayed basic MSC capacities, while correlation within the immunoregulatory concept is less studied.
9. Current MSC-based therapies develop in two different directions: (1) advanced cell selection and/or in vitro licensing before clinical use and (2) freshly aspirated autologous BM-MNCs without ex vivo expansion.
10. Further development of isolation and expansion protocols is needed to meet the increasing demands of therapeutic doses of ex vivo expanded MSCs with increasing therapeutic applications.

FURTHER READING

Bianco, P., Robey, P.G., Simmons, P.J., 2008. Mesenchymal stem cells: revisiting history, concepts, and assays. Cell. Stem Cell. 2 (4), 313–319.

Ekdahl, K.N., Hong, J., Hamad, O.A., Larsson, R., Nilsson, B., 2013. Evaluation of the blood compatibility of materials, cells, and tissues: basic concepts, test models, and practical guidelines. Adv. Exp. Med. Biol. 735, 257–270.

Muschler, G.F., Nakamoto, C., Griffith, L.G., 2004. Engineering principles of clinical cell-based tissue engineering. J. Bone Joint Surg. Am. 86-a (7), 1541–1558.

REFERENCES

Abdelrazik, H., Spaggiari, G.M., Chiossone, L., Moretta, L., 2011. Mesenchymal stem cells expanded in human platelet lysate display a decreased inhibitory capacity on T- and NK-cell proliferation and function. Eur. J. Immunol. 41 (11), 3281–3290.

Alm, J.J., Heino, T.J., Hentunen, T.A., Vaananen, H.K., Aro, H.T., 2012. Transient 100 nM dexamethasone treatment reduces inter- and intraindividual variations in osteoblastic differentiation of bone marrow-derived human mesenchymal stem cells. Tissue Eng. Part C Methods 18 (9), 658–666.

Alm, J.J., Koivu, H.M., Heino, T.J., Hentunen, T.A., Laitinen, S., Aro, H.T., 2010. Circulating plastic adherent mesenchymal stem cells in aged hip fracture patients. J. Orthop. Res. 28 (12), 1634–1642.

Alves, H., Dechering, K., Van Blitterswijk, C., De Boer, J., 2011. High-throughput assay for the identification of compounds regulating osteogenic differentiation of human mesenchymal stromal cells. PLoS One 6 (10), e26678.

Aubin, J.E., Lian, J.B., Stein, G.S., 2006. Bone formation: maturation and functional activities of osteoblast lineage cells. In: Favus, M.J. (Ed.), Primer on the Metabolic Bone Diseases and Disorders of Mineral Metabolism, 6th ed. American Society for Bone and Mineral Research, Washington DC, pp. 20–29.

Bara, J.J., Richards, R.G., Alini, M., Stoddart, M.J., 2014. Bone marrow-derived mesenchymal stem cells change phenotype following in vitro culture: Implications for basic research and the clinic. Stem Cells 32 (7), 1713–1723.

Baron, R., Kneissel, M., 2013. WNT signaling in bone homeostasis and disease: from human mutations to treatments. Nat. Med. 19 (2), 179–192.

Bernardo, M.E., Fibbe, W.E., 2013. Mesenchymal stromal cells: sensors and switchers of inflammation. Cell. Stem Cell. 13 (4), 392–402.

Bieback, K., 2013. Platelet lysate as replacement for fetal bovine serum in mesenchymal stromal cell cultures. Transfus. Med. Hemother. 40 (5), 326–335.

Caplan, A.I., 1991. Mesenchymal stem cells. J. Orthop. Res. 9 (5), 641–650.

Caplan, A.I., 2007. Adult mesenchymal stem cells for tissue engineering versus regenerative medicine. J. Cell. Physiol. 213 (2), 341–347.

Carvalho, P.P., Wu, X., Yu, G., Dietrich, M., Dias, I.R., Gomes, M.E., Reis, R.L., Gimble, J.M., 2011. Use of animal protein-free products for passaging adherent human adipose-derived stromal/stem cells. Cytotherapy 13 (5), 594–597.

Castro-Malaspina, H., Gay, R.E., Resnick, G., Kapoor, N., Meyers, P., Chiarieri, D., McKenzie, S., Broxmeyer, H.E., Moore, M.A., 1980. Characterization of human bone marrow fibroblast colony-forming cells (CFU-F) and their progeny. Blood 56 (2), 289–301.

Copland, I.B., Garcia, M.A., Waller, E.K., Roback, J.D., Galipeau, J., 2013. The effect of platelet lysate fibrinogen on the functionality of MSCs in immunotherapy. Biomaterials 34 (32), 7840–7850.

Dennis, J.E., Merriam, A., Awadallah, A., Yoo, J.U., Johnstone, B., Caplan, A.I., 1999. A quadripotential mesenchymal progenitor cell isolated from the marrow of an adult mouse. J. Bone Miner. Res. 14 (5), 700–709.

Di Maggio, N., Mehrkens, A., Papadimitropoulos, A., Schaeren, S., Heberer, M., Banfi, A., Martin, I., 2012. Fibroblast growth factor-2 maintains a niche-dependent population of self-renewing highly potent non-adherent mesenchymal progenitors through FGFR2c. Stem Cells 30 (7), 1455–1464.

Dominici, M., Le Blanc, K., Mueller, I., Slaper-Cortenbach, I., Marini, F., Krause, D., Deans, R., Keating, A., Prockop, D., Horwitz, E., 2006. Minimal criteria for defining multipotent mesenchymal stromal cells. The International Society for Cellular Therapy position statement. Cytotherapy 8 (4), 315–317.

Doorn, J., Moll, G., Le Blanc, K., van Blitterswijk, C., de Boer, J., 2012. Therapeutic applications of mesenchymal stromal cells: paracrine effects and potential improvements. Tissue Eng. Part B Rev. 18 (2), 101–115.

Flemming, A., Schallmoser, K., Strunk, D., Stolk, M., Volk, H.D., Seifert, M., 2011. Immunomodulative efficacy of bone marrow-derived mesenchymal stem cells cultured in human platelet lysate. J. Clin. Immunol. 31 (6), 1143–1156.

Fowler, A., Toner, M., 2005. Cryo-injury and biopreservation. Ann. N. Y. Acad. Sci. 1066, 119–135.

Francois, M., Copland, I.B., Yuan, S., Romieu-Mourez, R., Waller, E.K., Galipeau, J., 2012. Cryopreserved mesenchymal stromal cells display impaired immunosuppressive properties as a result of heat-shock response and impaired interferon-gamma licensing. Cytotherapy 14 (2), 147–152.

Friedenstein, A.J., Deriglasova, U.F., Kulagina, N.N., Panasuk, A.F., Rudakowa, S.F., Luria, E.A., Ruadkow, I.A., 1974. Precursors for fibroblasts in different populations of hematopoietic cells as detected by the in vitro colony assay method. Exp. Hematol. 2 (2), 83–92.

Friedenstein, A.J., Chailakhjan, R.K., Lalykina, K.S., 1970. The development of fibroblast colonies in monolayer cultures of guinea-pig bone marrow and spleen cells. Cell Tissue Kinet. 3 (4), 393–403.

Friedl, G., Windhager, R., Schmidt, H., Aigner, R., 2009. The osteogenic response of undifferentiated human mesenchymal stem cells (hMSCs) to mechanical strain is inversely related to body mass index of the donor. Acta Orthop. 80 (4), 491–498.

Giannoudis, P.V., Ahmad, M.A., Mineo, G.V., Tosounidis, T.I., Calori, G.M., Kanakaris, N.K., 2013. Subtrochanteric fracture non-unions with implant failure managed with the "Diamond" concept. Injury 44 (Suppl. 1), S76–S81.

Giannoudis, P.V., Einhorn, T.A., Marsh, D., 2007. Fracture healing: the diamond concept. Injury 38 (Suppl. 4), S3–S6.

Ginis, I., Grinblat, B., Shirvan, M.H., 2012. Evaluation of bone marrow-derived mesenchymal stem cells after cryopreservation and hypothermic storage in clinically safe medium. Tissue Eng. Part C Methods 18 (6), 453–463.

Gomez-Barrena, E., Rosset, P., Muller, I., Giordano, R., Bunu, C., Layrolle, P., Konttinen, Y.T., Luyten, F.P., 2011. Bone regeneration: stem cell therapies and clinical studies in orthopaedics and traumatology. J. Cell. Mol. Med. 15 (6), 1266–1286.

Griffiths, S., Baraniak, P.R., Copland, I.B., Nerem, R.M., McDevitt, T.C., 2013. Human platelet lysate stimulates high-passage and senescent human multipotent mesenchymal stromal cell growth and rejuvenation in vitro. Cytotherapy 15 (12), 1469–1483.

Gronthos, S., Graves, S.E., Ohta, S., Simmons, P.J., 1994. The STRO-1+ fraction of adult human bone marrow contains the osteogenic precursors. Blood 84 (12), 4164–4173.

Haack-Sorensen, M., Kastrup, J., 2011. Cryopreservation and revival of mesenchymal stromal cells. Methods Mol. Biol. 698, 161–174.

Haynesworth, S.E., Baber, M.A., Caplan, A.I., 1992. Cell surface antigens on human marrow-derived mesenchymal cells are detected by monoclonal antibodies. Bone 13 (1), 69–80.

Hemeda, H., Giebel, B., Wagner, W., 2014. Evaluation of human platelet lysate versus fetal bovine serum for culture of mesenchymal stromal cells. Cytotherapy 16 (2), 170–180.

Hemeda, H., Kalz, J., Walenda, G., Lohmann, M., Wagner, W., 2013. Heparin concentration is critical for cell culture with human platelet lysate. Cytotherapy 15 (9), 1174–1181.

Horwitz, E.M., Le Blanc, K., Dominici, M., Mueller, I., Slaper-Cortenbach, I., Marini, F.C., Deans, R.J., Krause, D.S., Keating, A., 2005. Clarification of the nomenclature for MSC: The International Society for Cellular Therapy position statement. Cytotherapy 7 (5), 393–395.

Hung, C., Nakamoto, C., Muschler, G.F., 2006. Factors affecting connective tissue progenitors and orthopaedic implications. Scand. J. Surg. 95 (2), 81–89.

Ikebe, C., Suzuki, K., 2014. Mesenchymal stem cells for regenerative therapy: optimization of cell preparation protocols. Biomed. Res. Int. 2014, 951512.

Jiang, S.S., Chen, C.H., Tseng, K.Y., Tsai, F.Y., Wang, M.J., Chang, I.S., Lin, J.L., Lin, S., 2011. Gene expression profiling suggests a pathological role of human bone marrow-derived mesenchymal stem cells in aging-related skeletal diseases. Aging (Albany NY) 3 (7), 672–684.

Kassem, M., Marie, P.J., 2011. Senescence-associated intrinsic mechanisms of osteoblast dysfunctions. Aging Cell 10 (2), 191–197.

Koc, O.N., Gerson, S.L., Cooper, B.W., Dyhouse, S.M., Haynesworth, S.E., Caplan, A.I., Lazarus, H.M., 2000. Rapid hematopoietic recovery after coinfusion of autologous-blood stem cells and culture-expanded marrow mesenchymal stem cells in advanced breast cancer patients receiving high-dose chemotherapy. J. Clin. Oncol. 18 (2), 307–316.

Krampera, M., Galipeau, J., Shi, Y., Tarte, K., Sensebe, L., 2013. Immunological characterization of multipotent mesenchymal stromal cells–the International Society for Cellular Therapy (ISCT) working proposal. Cytotherapy 15 (9), 1054–1061.

Kuznetsov, S.A., Krebsbach, P.H., Satomura, K., Kerr, J., Riminucci, M., Benayahu, D., Robey, P.G., 1997. Single-colony derived strains of human marrow stromal fibroblasts form bone after transplantation in vivo. J. Bone Miner. Res. 12 (9), 1335–1347.

Lazarus, H.M., Haynesworth, S.E., Gerson, S.L., Rosenthal, N.S., Caplan, A.I., 1995. Ex vivo expansion and subsequent infusion of human bone marrow-derived stromal progenitor cells (mesenchymal progenitor cells): implications for therapeutic use. Bone Marrow Transpl. 16 (4), 557–564.

Le Blanc, K., Frassoni, F., Ball, L., Locatelli, F., Roelofs, H., Lewis, I., Lanino, E., Sundberg, B., Bernardo, M.E., Remberger, M., Dini, G., Egeler, R.M., Bacigalupo, A., Fibbe, W., Ringden, O., 2008. Mesenchymal stem cells for treatment of steroid-resistant, severe, acute graft-versus-host disease: a phase II study. Lancet 371 (9624), 1579–1586.

Le Blanc, K., Mougiakakos, D., 2012. Multipotent mesenchymal stromal cells and the innate immune system. Nat. Rev. Immunol. 12 (5), 383–396.

Le Blanc, K., Rasmusson, I., Sundberg, B., Gotherstrom, C., Hassan, M., Uzunel, M., Ringden, O., 2004. Treatment of severe acute graft-versus-host disease with third party haploidentical mesenchymal stem cells. Lancet 363 (9419), 1439–1441.

Le Blanc, K., Samuelsson, H., Lonnies, L., Sundin, M., Ringden, O., 2007. Generation of immunosuppressive mesenchymal stem cells in allogeneic human serum. Transplantation 84 (8), 1055–1059.

Leibo, S.P., Pool, T.B., 2011. The principal variables of cryopreservation: solutions, temperatures, and rate changes. Fertil. Steril. 96 (2), 269–276.

Lian, J.B., Stein, G.S., van Wijnen, A.J., Stein, J.L., Hassan, M.Q., Gaur, T., Zhang, Y., 2012. MicroRNA control of bone formation and homeostasis. Nat. Rev. Endocrinol. 8 (4), 212–227.

Maumus, M., Jorgensen, C., Noel, D., 2013. Mesenchymal stem cells in regenerative medicine applied to rheumatic diseases: role of secretome and exosomes. Biochimie 95 (12), 2229–2234.

Mendez-Ferrer, S., Michurina, T.V., Ferraro, F., Mazloom, A.R., Macarthur, B.D., Lira, S.A., Scadden, D.T., Ma'ayan, A., Enikolopov, G.N., Frenette, P.S., 2010. Mesenchymal and haematopoietic stem cells form a unique bone marrow niche. Nature 466 (7308), 829–834.

Mentink, A., Hulsman, M., Groen, N., Licht, R., Dechering, K.J., van der Stok, J., Alves, H.A., Dhert, W.J., van Someren, E.P., Reinders, M.J., van Blitterswijk, C.A., de Boer, J., 2013. Predicting the therapeutic efficacy of MSC in bone tissue engineering using the molecular marker CADM1. Biomaterials 34 (19), 4592–4601.

Moll, G., Alm, J.J., Davies, L.C., von Bahr, L., Heldring, N., Stenbeck-Funke, L., Hamad, O.A., Hinsch, R., Ignatowicz, L., Locke, M., Lönnies, H., Lambris, J.D., Teramura, Y., Nilsson-Ekdahl, K., Nilsson, B., Le Blan, K., 2014. Do cryopreserved mesenchymal stromal cells display impaired immunomodulatory and therapeutic properties? Stem Cell 32 (9), 2430–2442.

Muschler, G.F., Boehm, C., Easley, K., 1997. Aspiration to obtain osteoblast progenitor cells from human bone marrow: the influence of aspiration volume. J. Bone Joint Surg. Am. 79 (11), 1699–1709.

Owen, M.E., Cave, J., Joyner, C.J., 1987. Clonal analysis in vitro of osteogenic differentiation of marrow CFU-F. J. Cell Sci. 87 (Pt. 5), 731–738.

Pal, R., Hanwate, M., Totey, S.M., 2008. Effect of holding time, temperature and different parenteral solutions on viability and functionality of adult bone marrow-derived mesenchymal stem cells before transplantation. J. Tissue Eng. Regen. Med. 2 (7), 436–444.

Pittenger, M.F., Mackay, A.M., Beck, S.C., Jaiswal, R.K., Douglas, R., Mosca, J.D., Moorman, M.A., Simonetti, D.W., Craig, S., Marshak, D.R., 1999. Multilineage potential of adult human mesenchymal stem cells. Science 284 (5411), 143–147.

Qian, H., Le Blanc, K., Sigvardsson, M., 2012. Primary mesenchymal stem and progenitor cells from bone marrow lack expression of CD44 protein. J. Biol. Chem. 287 (31), 25795–25807.

Samuelsson, H., Ringden, O., Lonnies, H., Le Blanc, K., 2009. Optimizing in vitro conditions for immunomodulation and expansion of mesenchymal stromal cells. Cytotherapy 11 (2), 129–136.

Shahdadfar, A., Fronsdal, K., Haug, T., Reinholt, F.P., Brinchmann, J.E., 2005. In vitro expansion of human mesenchymal stem cells: choice of serum is a determinant of cell proliferation, differentiation, gene expression, and transcriptome stability. Stem Cells 23 (9), 1357–1366.

Sotiropoulou, P.A., Perez, S.A., Salagianni, M., Baxevanis, C.N., Papamichail, M., 2006. Characterization of the optimal culture conditions for clinical scale production of human mesenchymal stem cells. Stem Cells 24 (2), 462–471.

Spees, J.L., Gregory, C.A., Singh, H., Tucker, H.A., Peister, A., Lynch, P.J., Hsu, S.C., Smith, J., Prockop, D.J., 2004. Internalized antigens must be removed to prepare hypoimmunogenic mesenchymal stem cells for cell and gene therapy. Mol. Ther. 9 (5), 747–756.

Stolzing, A., Jones, E., McGonagle, D., Scutt, A., 2008. Age-related changes in human bone marrow-derived mesenchymal stem cells: consequences for cell therapies. Mech. Ageing Dev. 129 (3), 163–173.

Sundin, M., Ringden, O., Sundberg, B., Nava, S., Gotherstrom, C., Le Blanc, K., 2007. No alloantibodies against mesenchymal stromal cells, but presence of anti-fetal calf serum antibodies, after transplantation in allogeneic hematopoietic stem cell recipients. Haematologica 92 (9), 1208–1215.

Tarte, K., Gaillard, J., Lataillade, J.J., Fouillard, L., Becker, M., Mossafa, H., Tchirkov, A., Rouard, H., Henry, C., Splingard, M., Dulong, J., Monnier, D., Gourmelon, P., Gorin, N.C., Sensebé, L., Société Française de Greffe de Moelle et Thérapie Cellulaire, 2010. Clinical-grade production of human mesenchymal stromal cells: occurrence of aneuploidy without transformation. Blood 115 (8), 1549–1553.

Thirumala, S., Goebel, W.S., Woods, E.J., 2013. Manufacturing and banking of mesenchymal stem cells. Expert Opin. Biol. Ther. 13 (5), 673–691.

Tolar, J., Le Blanc, K., Keating, A., Blazar, B.R., 2010. Concise review: hitting the right spot with mesenchymal stromal cells. Stem Cells 28 (8), 1446–1455.

Tonti, G.A., Mannello, F., 2008. From bone marrow to therapeutic applications: different behaviour and genetic/epigenetic stability during mesenchymal stem cell expansion in autologous and foetal bovine sera? Int. J. Dev. Biol. 52 (8), 1023–1032.

Tormin, A., Li, O., Brune, J.C., Walsh, S., Schutz, B., Ehinger, M., Ditzel, N., Kassem, M., Scheding, S., 2011. CD146 expression on primary non-hematopoietic bone marrow stem cells correlates to in situ localization. Blood 117 (19), 5067–5077.

van Ramshorst, J., Bax, J.J., Beeres, S.L., Dibbets-Schneider, P., Roes, S.D., Stokkel, M.P., de Roos, A., Fibbe, W.E., Zwaginga, J.J., Boersma, E., Schalij, M.J., Atsma, D.E., 2009. Intramyocardial bone marrow cell injection for chronic myocardial ischemia: a randomized controlled trial. JAMA 301 (19), 1997–2004.

von Bahr, L., Batsis, I., Moll, G., Hagg, M., Szakos, A., Sundberg, B., Uzunel, M., Ringden, O., Le Blanc, K., 2012a. Analysis of tissues following mesenchymal stromal cell therapy in humans indicates limited long-term engraftment and no ectopic tissue formation. Stem Cells 30 (7), 1575–1578.

von Bahr, L., Sundberg, B., Lonnies, L., Sander, B., Karbach, H., Hagglund, H., Ljungman, P., Gustafsson, B., Karlsson, H., Le Blanc, K., Ringden, O., 2012b. Long-term complications, immunologic effects, and role of passage for outcome in mesenchymal stromal cell therapy. Biol. Blood Marrow Transpl. 18 (4), 557–564.

Wagner, W., Wein, F., Roderburg, C., Saffrich, R., Faber, A., Krause, U., Schubert, M., Benes, V., Eckstein, V., Maul, H., Ho, A.D., 2007. Adhesion of hematopoietic progenitor cells to human mesenchymal stem cells as a model for cell-cell interaction. Exp. Hematol. 35 (2), 314–325.

Zhou, S., Glowacki, J., Kim, S.W., Hahne, J., Geng, S., Mueller, S.M., Shen, L., Bleiberg, I., LeBoff, M.S., 2012. Clinical characteristics influence in vitro action of 1,25-dihydroxyvitamin D(3) in human marrow stromal cells. J. Bone Min. Res. 27 (9), 1992–2000.

CHAPTER 14

Vascularization, Survival, and Functionality of Tissue-Engineered Constructs

Marietta Herrmann[1], Matthias W. Laschke[2], Mauro Alini[1], Arnaud Scherberich[3], Sophie Verrier[1]

[1]Musculoskeletal Regeneration, AO Research Institute Davos, Davos, Switzerland; [2]Institute for Clinical & Experimental Surgery, University of Saarland, Homburg, Saarland, Germany; [3]Laboratory of Tissue Engineering, Department of Biomedicine, University Hospital of Basel, Basel, Switzerland

LEARNING OBJECTIVES

- To appreciate the scientific and potential clinical problems associated with the lack of or poor vascularization of tissue-engineered constructs
- To learn about the mechanisms of vasculogenesis and angiogenesis
- To have an overview of the methods currently used in clinics, their advantages, and limitations
- To have an overview of the different strategies used to promote fast neovascularization of tissue-engineered constructs
- To learn about emerging techniques and the latest developments in the area of tissue-engineered construct neovascularization

14.1 INTRODUCTION

Tissue engineering is a fast growing field of research driven by the need for tissue substitutes and transplantable organs. This interdisciplinary field of sciences involves engineering and life sciences toward the development of biological substitutes able to restore, maintain, or improve tissue function or a whole organ (Langer and Vacanti, 1993). Classical tissue engineering approaches are based on the association of cells, tissue-specific cells or stem cells, with a biomaterial scaffold adapted to the tissue to be replaced. Although, even if tremendous progress has been made in material sciences, cell biology, and their interactions, so far, the success of tissue-engineered constructs is limited to avascular tissue (cartilage, cornea) or thin tissues (bladder, skin); this exhibits a slow metabolism and mainly relies on passive oxygen diffusion into the tissue.

Clinically, the demand for engineered tissues with controlled dimensions is enormous. However, in engineering a more complex or larger volume of

Tissue Engineering.
Copyright © 2015 Elsevier Inc. All rights reserved.

tissues, a key challenge is still the inability to provide sufficient blood supply in the initial phase after implantation, which is a prerequisite for the cell viability and long-term construct survival and functionality (Laschke et al., 2006). Upon implantation, and as long as a proper vasculature has not been established, the implant has to rely on diffusion for the supply of oxygen, nutrients, and the removal of metabolites. The diffusion is limited to a distance of about 150–200 μm from the next supplier vessel (Folkman and Hochberg, 1973). This can lead to nutrient limitations, and can result in an improper integration, and eventually to the death of the implant. Thus, the long-term survival of a cell carrying construct strongly depends on its rapid neovascularization (Kannan et al., 2005; Rouwkema et al., 2008).

Neovascularization can occur by two different processes: The formation of new vessels from preexisting blood vessels, referred to as angiogenesis, or by the de novo formation of blood vessels from endothelial progenitor cells (EPCs), referred to as vasculogenesis. Until recently, vasculogenesis was considered to be restricted to embryonic development. With the detection of EPCs in adult circulation, it became evident that these cells can also be home to sites of neovascularization and contribute to vascular regeneration (Alev et al., 2011).

In postnatal life, the main neovascularization process is defined as the formation of new capillaries from preexisting blood vessels. This sprouting angiogenesis was the first identified form of angiogenesis (Risau, 1997). It occurs in several well-orchestrated and characterized stages (Figure 14.1).

First, biological signals known as angiogenic growth factors (Table 14.1) activate receptors expressed on endothelial cells residing in preexisting vessels. Second, the activated endothelial cells begin to release enzymes called proteases that degrade the basement membrane in order to allow endothelial cells to migrate out of the vessel of origin toward the surrounding matrix in which they proliferate and organize in new luminal structures. These sprouts progressively grow and will connect to each other to form new blood-perfused microvascular networks. Third, these newly formed structures are stabilized by the formation of extracellular matrix and the adjunction of stabilizing cells, pericytes, and/or smooth muscle cells. New vessel growth through sprouting is a slow process with an estimated velocity of about 5 μm/h. This enables new vessels to grow across small gaps in the vasculature but is not sufficient to promote vascularization in large three-dimensional (3D) tissue-engineered constructs. In constructs, >400 μm in any dimension, the lack of rapid vascularization will lead to cell death and therefore failure of the tissue-engineered constructs to integrate to the host tissue, which will end with the implants' fibrous encapsulation and increased scar formation (Auger et al., 2013; Scheufler et al., 2008) (Figure 14.2).

In this chapter, we aim to give an overview of the current strategies to improve the host tissue integration of tissue-engineered constructs, with a

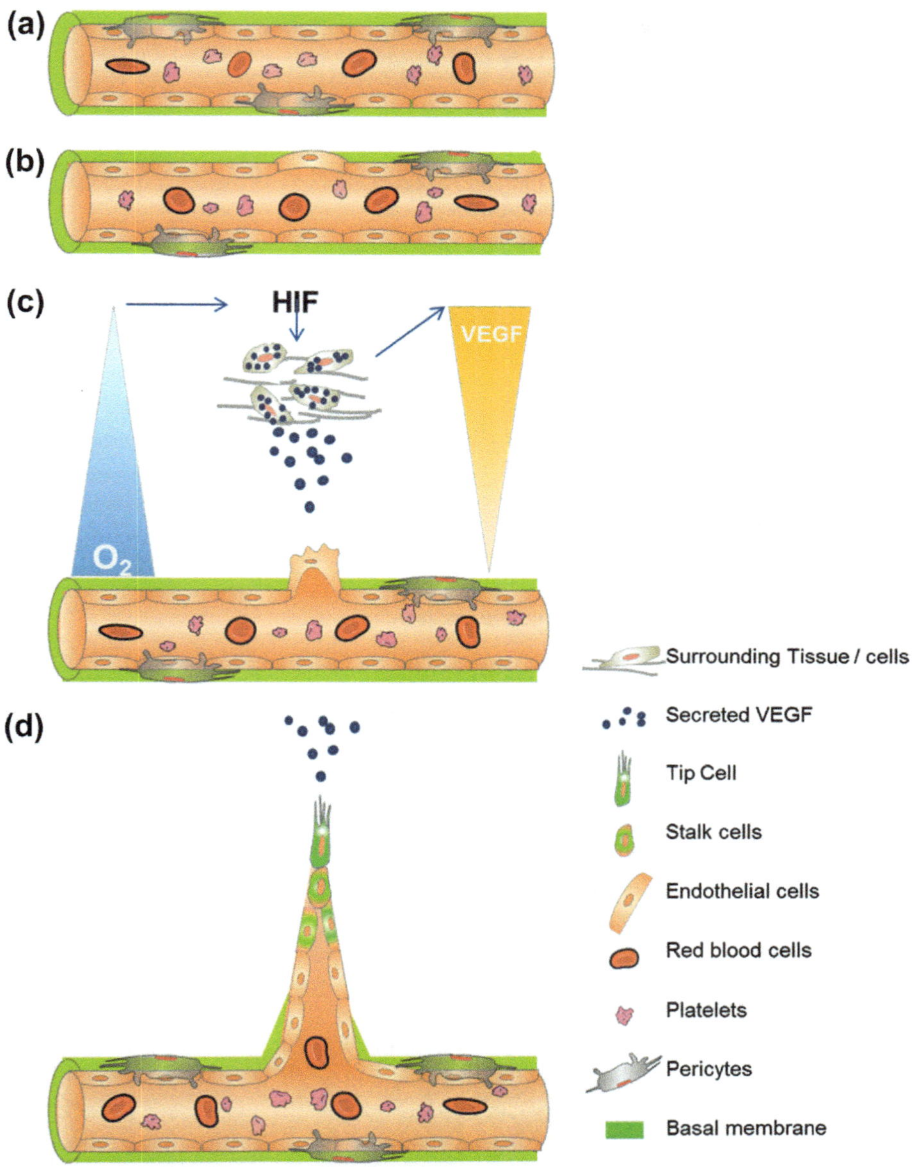

FIGURE 14.1 **Schematic illustration of the different sprouting steps during angiogenesis.**
(a) Capillary. (b) Initiation of the sprouting. Activation of an endothelial cell at the initiation site, and differentiation into a tip cell. (c) Degradation of the basement membrane by the tip cell and migration toward a gradient of angiogenic signals. (d) Generation of the new perfused vessel's trunk by stalk cells. VEGF, vascular endothelial growth factor; HIF, hypoxia inducible factor.

Table 14.1 Important Factors in Angiogenesis

Key Factors in Angiogenesis	Function
VEGF (isoform 1) (Vascular endothelial growth factor)	Endothelial cells proliferation, migration, and differentiation
HIF (Hypoxia inducible factor (α and β))	Stimulation of transcription of VEGF, PDGF receptor (PDGFR), TGF-α EGF, and erythropoietin
PDGF (Platelet-derived growth factor)	VEGF upregulation Recruitment and proliferation of pericytes and smooth muscle cells
Ang-1 Ang-2 (Angiopoietin)	Regulates endothelial cell survival, sprouting, and branching of vessels, pericyte recruitment
TGF-α (Transforming growth factor α)	Induces angiogenesis and cell proliferation
TGF-β (Transforming growth factor β)	Upregulation of angiogenic factors and proteinases, promotes basement membrane formation
EGF (Epithelial growth factors)	Upregulates VEGF production
MMP-1, -2, and -9 (Matrix Metalloproteinase)	Basal lamina degradation, extracellular matrix remodeling
IL-1, -6, -8, and -13 (Interleukins)	Endothelial cell proliferation

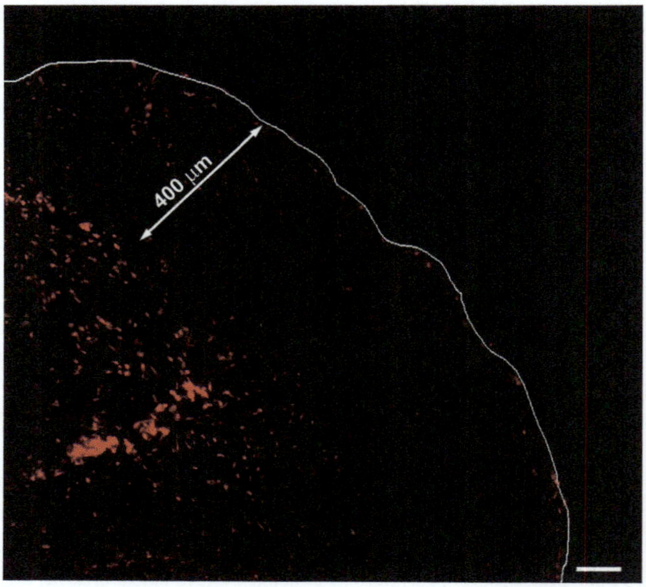

FIGURE 14.2

Cell survival is impaired in the core of 3D tissue-engineered constructs. A disc-shaped hydrogel (diameter 9 mm, height 3 mm) seeded with mesenchymal stem cells (MSCs) was cultured in vitro for 7 days. Cell viability was assessed on cryosections using ethidium heterodimer staining (red fluorescence), which is only taken up by dead cells with a perforated cell membrane. Scale bar = 100 μm.

special emphasis on the promotion of an early and rapid neovascularization (Figure 14.3). In the first part, we will approach several in vitro strategies to promote vascular ingrowth from surrounding tissues; in the second part, we will focus more on methods to produce prevascularized constructs. After having discussed about strategies to increase cell viability in the constructs, we will describe a few methods to study neovascularization in vivo. Finally, the limitation of the current approaches and future directions will be discussed.

14.2 STRATEGIES TO IMPROVE VASCULAR INGROWTH INTO TISSUE-ENGINEERED CONSTRUCTS

14.2.1 Modification of Scaffolds

Scaffolds represent the structural units for the generation of tissue-engineered constructs (see also Chapter 10 "Scaffold design and fabrication"). They serve as a 3D matrix for cells to attach, to proliferate, and to develop into a functional tissue. Accordingly, the interaction of scaffolds with their biological environment crucially determines the extent of construct vascularization. Therefore, the choice of an appropriate scaffold, which adequately promotes vascular ingrowth, is the first important step toward a successful tissue-engineering application.

Several impacting scaffold-related parameters can be identified, which may be modified to support and increase vascularization. These include chemical composition, degradation properties, architecture, porosity, and surface topography. Scaffolds solely differing in their chemical composition or degradation rates have been shown to induce foreign body reactions to a varying extent, which closely correlates with vascular ingrowth (Rucker et al., 2006). Vascular ingrowth is induced by a rapid coating of the implanted biomaterials with adsorbed host proteins, which attract inflammatory cells to the implantation site, where they release various cytokines and angiogenic growth factors (Anderson et al., 2008). In this context, it should be noted that mild proinflammatory properties of scaffolds may be beneficial to promote vascularization (Rucker et al., 2006). However, these properties should of course not negatively influence the biocompatibility of an implanted tissue construct.

Relating to scaffold architecture and porosity, both in silico and in vivo analyses demonstrate that scaffolds with large pore sizes ranging between 250 and 400 μm and high interconnectivity optimally support vascular ingrowth, which can be further improved by increasing the pore size distribution (Bai et al., 2010; Mehdizadeh et al., 2013). Moreover, scaffolds with large pore sizes favor the formation of vascular networks with larger blood vessels and deeper penetration depth (Choi et al., 2013), which is particularly important to guarantee the survival of cells in the center of large 3D tissue constructs.

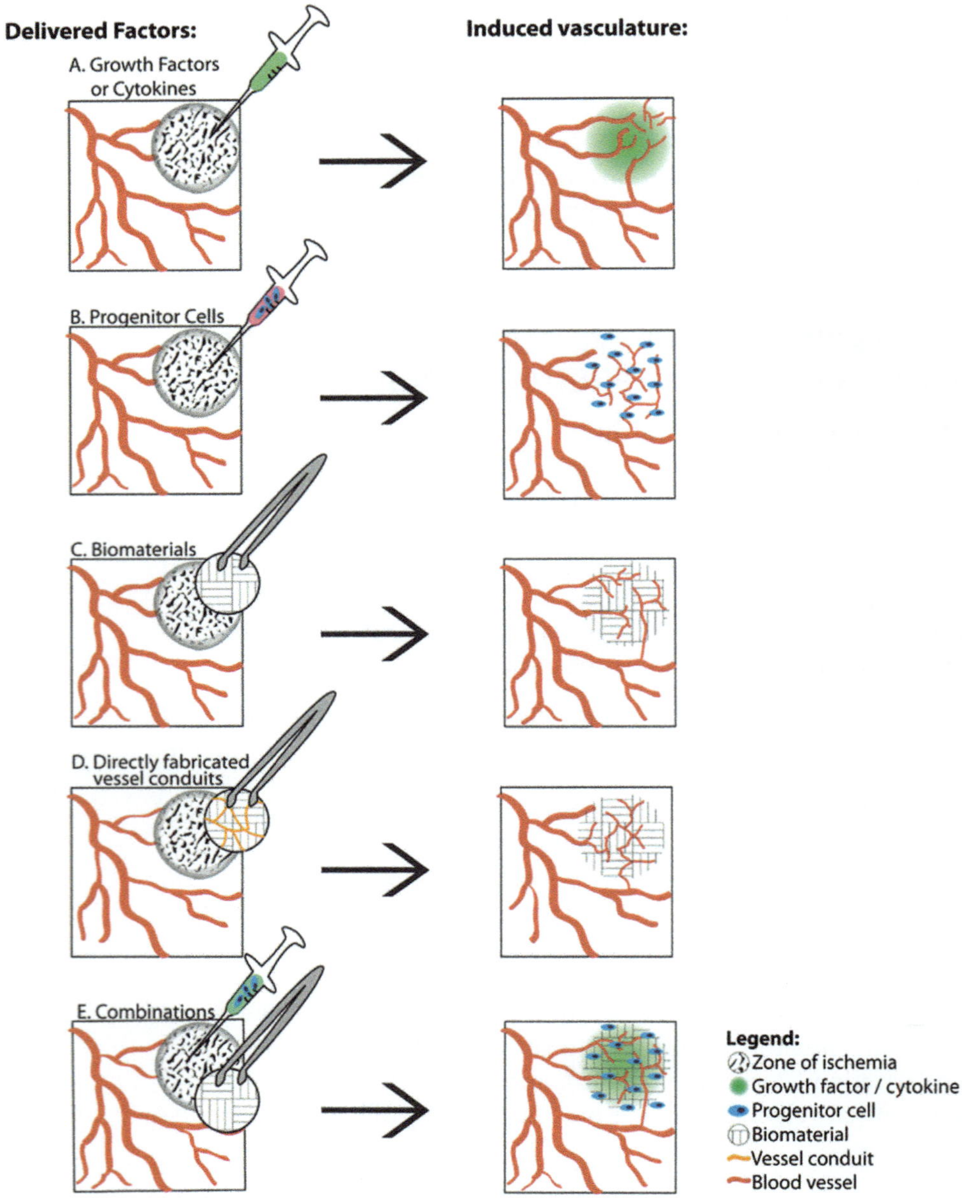

FIGURE 14.3

Current options to induce vascularization in tissue-engineered constructs Phelps and Garcia (2010). *With permission from Elsevier.* (a) Addition of proangiogenic growth factors and cytokines. (b) Seeding of stem cells or progenitor cells such as EPCs or MSCs. (c) Selection of bioactive biomaterials. (d) Addition of preformed microvascular vessels (e.g., adipose-derived microvessels). (e) Combination of several strategies (e.g., cells and growth factors).

Further, sophisticated approaches for the modification of scaffold surfaces offer the possibility to create a proangiogenic environment. For this purpose, heparin or heparin mimetic molecules are attached to the surface of scaffolds as anchoring points for angiogenic growth factors (Leiton et al., 2013). Another option is the treatment of the scaffold surface with glow discharge gas plasma to break covalent bonds and, thus, to change chemical and physical adhesion parameters (Chim et al., 2003).

Finally, rapid progress in the field of scaffold design and production will make it possible in the near future to generate scaffolds with microchannels mimicking the physiological vascular architecture of tissues or scaffolds with clearly defined proangiogenic and antiangiogenic surface areas. Such scaffolds will be able to induce guided vascular ingrowth from the surrounding host tissue with the formation of a highly organized, functional microvasculature inside implanted tissue-engineered constructs.

14.2.2 Incorporation of Growth Factors

Angiogenesis is a highly dynamic multistep process, which involves the activation, migration, proliferation, and tube formation of endothelial cells and the recruitment of perivascular cells, which finally stabilize the wall of the newly formed microvessels. Each of these steps is induced and tightly regulated by cellular release of angiogenic growth factors (Carmeliet, 2003). Many of these factors have been identified and characterized in terms of their protein biochemistry, structure, and biological function. Recombinant forms can be synthesized and are commercially available. Accordingly, the incorporation of angiogenic growth factors into scaffolds is a frequently used strategy to promote vascular ingrowth into tissue engineering constructs. For this purpose, the most commonly used factors (Table 14.2) are vascular endothelial growth factor (VEGF), basic fibroblast growth factor (bFGF), angiopoietin-2 (Ang-2), and platelet-derived growth factor (PDGF).

Table 14.2 Growth Factors Incorporated in Tissue-Engineered Constructs to Promote Vascularization

Factors to Promote Angiogenesis	Factors to Promote Vessel Maturation
VEGF	PDGF
PDGF	TGF-β
bFGF	Ang-1
Ang-2	

VEGF, vascular endothelial growth factor; PDGF, platelet-derived growth factor; bfGF, basic fibroblast growth factor; TGF, transforming growth factor. Ang, angiopoietin.
Soures: Amini et al. (2012) and Brudno et al. (2013).

Listed factors have been applied to promote (1) angiogenesis by inducing the invasion of endothelial cells and vascular sprouting and (2) the stabilization and maturation of newly formed vessels.

There are several possibilities to incorporate angiogenic growth factors into scaffolds. They can be covalently immobilized or loaded on the scaffold surface by collagen coating (Lindhorst et al., 2010). In addition, the pores of scaffolds may be temporarily filled with growth factor-containing gels, which are replaced over time by newly formed tissue. For this approach, platelet-rich plasma (PRP) represents a good, cost-effective candidate, as it can easily be isolated under clinical conditions from blood samples and is a rich, autologous source of numerous growth factors (Eppley et al., 2006). However, its composition is not standardized and may vary considerably from donor to donor. Alternatively, growth factors can be supplied via artificial drug delivery systems such as hydrogels, microspheres, or nanoparticles (Tabata, 2003). These systems bear the advantage that they exhibit clearly defined growth factor concentrations and release kinetics (see also Chapter 11 "Controlled release strategies in tissue engineering"). Accordingly, it is possible to temporally control growth factor signaling and, thus, to optimally support individual stages of the angiogenic process, including vessel formation, maturation, and remodeling (Brudno et al., 2013).

14.2.3 Incorporation of Vascular or Proangiogenic Cells

Incorporation of endothelial cells in a tissue-engineered construct is commonly used to promote its neovascularization. This is critical because the invasion of scaffolds with host endothelial cells is often not fast enough to avoid failure of an implant. Various tissues have been used as a source for endothelial cells. Mature endothelial cells, such as human umbilical vein endothelial cells (HUVECs) or human dermal microvascular endothelial cells, have been isolated from umbilical cord and skin, respectively. In 1997, Asahara and colleagues were the first to isolate a population of EPCs from peripheral blood (Asahara et al., 1997). The use of EPCs may be of advantage for clinical applications as these cells can be obtained autologously and display better proliferation capabilities. Two cell populations referred to as early EPCs and late outgrowth endothelial cells (OECs) have been isolated from peripheral and cord blood. Characterization of these cell populations revealed that early EPCs contribute to neovascularization by paracrine mechanisms, whereas OECs are actively integrated in newly formed vessels (Fuchs et al., 2010). Besides that, different isolation strategies have been developed to select EPCs from blood, bone marrow, and stromal vascular fractions of adipose tissue. These protocols are based on the expression of various combinations of surface markers, such as CD34, CD133, and VEGFR2 (CD309) (Masuda and Asahara, 2003). However, EPCs represent a heterogeneous cell population (Timmermans et al., 2009), which indicates the need for functional characterization of these cells

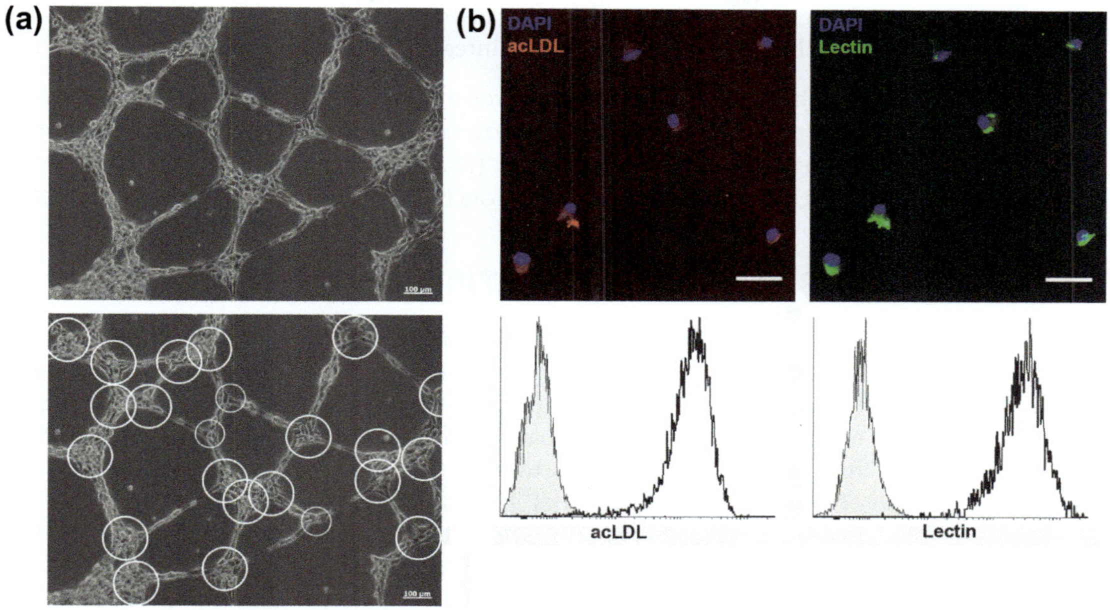

FIGURE 14.4 Functional characterization of endothelial cells.
(a) Endothelial cells are characterized by their capability to form tube-like structures on Matrigel™, a gelatineous protein matrix, constituting components of the basement membrane. In standard protocols for tube formation assays, endothelial cells are seeded on a thin layer of Matrigel™, and tube formation is observed over 24 h. Quantitative analyses include measurements of the total tube length, the mean diameter of tubes, and amount of branching points (indicated by circles in the bottom picture). (b) Uptake of acetylated low-density lipoprotein particles and binding of Ulex europaeus agglutinin (UEA) lectin are typical characteristics of endothelial cells. These can be assessed with fluorescently labeled molecules, which can be easily monitored by fluorescence microscopy (top panel) or flow cytometry (bottom panel, unstained controls are indicated by filled gray curves).

(Figure 14.4). In addition, it should be considered that culture expansion of endothelial cells and EPCs requires specific, recombinant growth factor containing cell culture medium, which is associated with additional costs.

In bone tissue engineering, for instance, MSCs or primary osteoblasts are commonly used as the osteogenic cell source (see also Chapter 16 "Musculoskeletal tissue engineering"). Several publications suggested that cocultures of the two cell types further facilitate neovascularization (Aguirre et al., 2010). This effect is mediated by paracrine mechanisms, including the secretion of proangiogenic factors and by direct cell-to-cell contacts, which drive MSCs toward a phenotype of stabilizing pericytes (Duttenhoefer et al., 2013; Loibl et al., 2014). Perivascular cells have recently attracted great interest in the field of regenerative medicine and are thought to represent an endogenous reservoir of multipotent progenitor cells in adults. Pericytes are localized at microvessels and capillaries, embedded in the basement membrane and can therefore be isolated from virtually all vascularized tissues (Crisan et al., 2008). The first tissue regeneration approaches using transplanted pericytes have confirmed their beneficial effects on tissue integration (James et al., 2012).

Another cell-based strategy involves the implementation of tissue-forming cells genetically manipulated to overexpress proangiogenic factors; this ensures a long-term release of the factor (Akiyama et al., 2010). Although, a too high concentration of angiogenic factors has also been shown to be detrimental.

14.3 PREVASCULARIZATION STRATEGIES

14.3.1 In vitro Prevascularization

In vitro prevascularization of tissue-engineered constructs uses a strategy based on the assembly and prematuration of vascular structures in vitro prior to implantation. This approach is of great interest as anastomosis of vascular fragments with the host vascular system occurs much faster than the formation of new vessels (Laschke et al., 2009). Prevascularization is generally achieved by seeding endothelial cells (see Section 14.2.3) often together with the tissue-specific cells in the construct (Figure 14.5) (Levenberg et al., 2005). Various constructs have been shown to support tube formation and vascularization, including scaffolds, spheroids, and cell sheets.

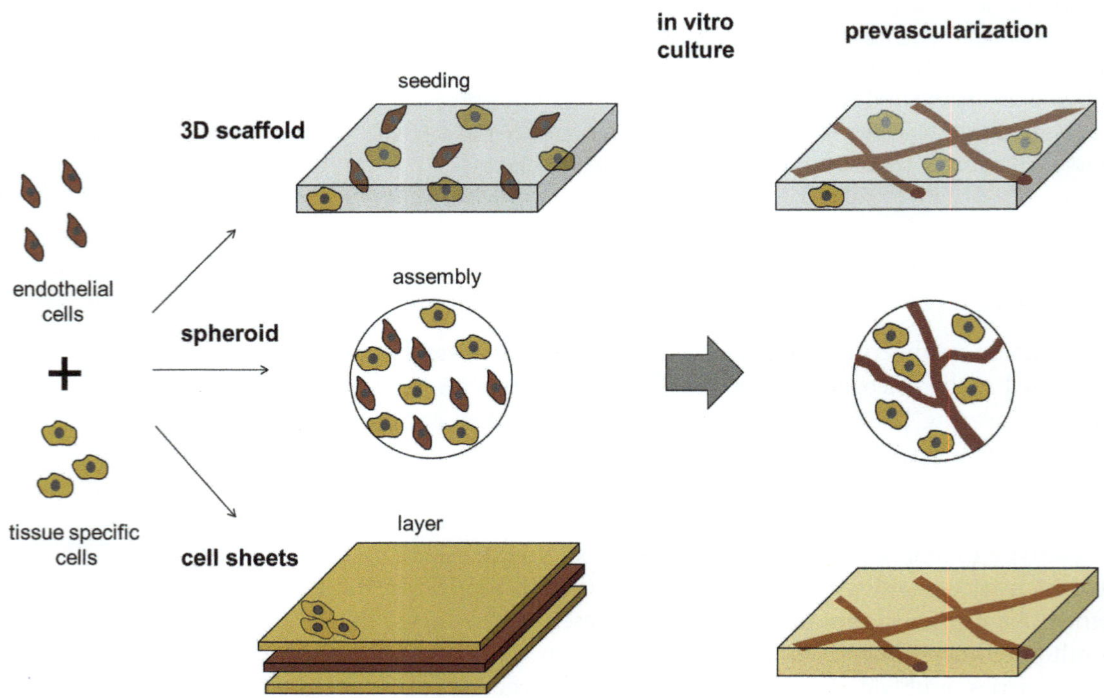

FIGURE 14.5 Schematic representation of common prevascularization strategies.

Prevascularization of scaffold-based constructs has been extensively studied in the field of bone and skin tissue engineering. For bone tissue engineering, cocultures of endothelial cells and osteogenic cells (often MSCs) are coseeded in porous scaffolds (see Section 14.2.1). Cells are often applied in a hydrogel, such as fibrin or PRP, to retain the cells within the construct (Duttenhoefer et al., 2013; Wu et al., 2012).

Multicellular spheroids can be obtained by different methods, such as pellet cultures, spinner cultures, the liquid overlay technique, and microfluidic approaches. These techniques are based on the fact that in the absence of an attachment surface, cells will aggregate and undergo self-assembly. Spheroids are free of any exogenous material and can be also used as endogenous building blocks for more complex structures (Achilli et al., 2012).

The fabrication of cell sheets takes advantage of the extracellular matrix produced by cells themselves. Cell sheets provide another scaffold-free option for engineering 3D constructs. Vascularized cell sheets are generated by sequentially layer cell sheets from endothelial cells and tissue-specific cells (Asakawa et al., 2010). For dermal substitutes, developed for patients with large skin defects, the coculture of OECs and fibroblasts into 3D sheets has shown promising results (Hendrickx et al., 2010).

The time frame of vessel prematuration depends on the construct and cell types. In a 3D scaffold-based approach, tubular structures could be detected after 7 days of in vitro preculture as illustrated in Figure 14.6 (Duttenhoefer

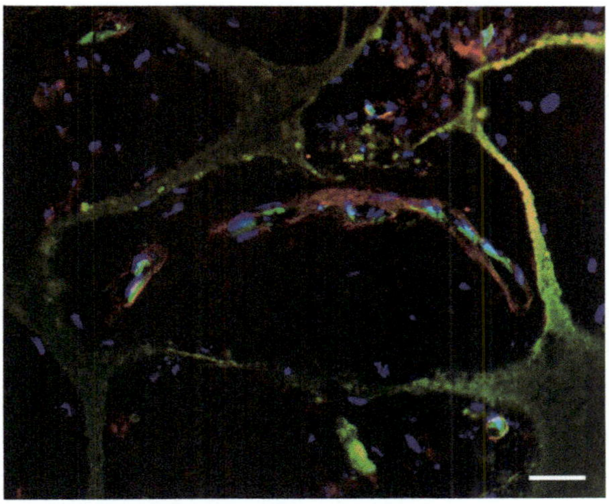

FIGURE 14.6 **In vitro prevascularization of a 3D scaffold.**
HUVECs (green) and MSCs were seeded in a PRP gel and applied to a porous polyurethane scaffold (green autofluorescence). Tube-like structures (red) can be detected after 7 days of coculture. Scale bar = 50 μm.

et al., 2013). The efficient integration of a prevascularized scaffold has been shown in constructs containing OECs and human osteoblasts; this shows perfusion 48 h after subcutaneous implantation (Ghanaati et al., 2011).

Besides the strategies described earlier, decellularized tissues have been used in a similar way as that for scaffolds for various tissue repair approaches. A construct of extracellular matrix is obtained by chemical or enzymatic decellularization of allogeneic tissue and may guide the formation of angiogenic vessels upon engraftment (see also Chapter 5 "Extracellular matrix as bioscaffold in tissue engineering"). In the context of bone tissue engineering, for many years, decellularized native bone matrix has been used as the scaffold of choice for clinical applications. When implanted, such osteoinductive matrices promote direct osteoblastic differentiation of seeded cells, following intramembranous ossification principles (direct bone matrix formation), but fail in large constructs, certainly due to the lack of vascularization. The use of decellularized hypertrophic cartilage matrix has since long been explored for its potential translation orthopedic applications (Holtrop, 1967). This strategy aims to mimic the endochondral ossification, involved in long bone development (see Box 14.1 for a recent example). During endochondral ossification, the chondrogenic cells within the tissue differentiate toward hypertrophy, then vascular ingrowth from surrounding tissue occurs, and eventually a bone matrix gets deposited. This neovascularization strategy is thus based on the natural capacity of chondrocytes to survive in a poor oxygen and nutrient environment.

14.3.2 In situ Prevascularization

A major drawback of the aforementioned in vitro prevascularization strategies is the fact that they involve complex and time-consuming cell isolation, seeding, and cultivation procedures (Koike et al., 2004). Accordingly, the implementation of such strategies into clinical practice is extremely challenging. Moreover, the seeding of scaffolds with cells or cell aggregates does not necessarily result in the development of functional microvascular networks. To overcome these problems, in situ prevascularization strategies have been introduced in the field of tissue engineering. The basic concept of these strategies is to use the human body as a "bioreactor" for the generation of a microvasculature within tissue constructs. This can be achieved in different ways.

In the 1970s, the flap technique was already developed. This consists of wrapping an implant inside a segment of tissue with an independent blood supply (McDowell, 1978) (Figure 14.7).

In this case, the vascular pedicle is still linked to the donor site. On the opposite, in a free flap, the isolated tissue and its vasculature are detached from their site of origin and transferred to a new location with an intact vascular pedicle that will be connected surgically to the vasculature of the new site.

BOX 14.1 TYPICAL EXPERIMENTAL PROTOCOL FOR THE IN VIVO ANALYSIS OF TISSUE CONSTRUCTS IN THE DORSAL SKINFOLD CHAMBER MODEL

1. Use mice with a body weight of 22–25 g (immunocompromised mice are needed for testing constructs of human origin)
2. Anesthesia by intraperitoneal injection of ketamine (75 mg/kg of body weight) and xylazine (15 mg/kg of body weight)
3. Shave and chemical depilation of the back
4. Preparation of the dorsal skinfold chamber (detailed description in Laschke et al. (2011))
5. A 48 h recovery from anesthesia and surgical trauma
6. Implantation of tissue construct into the dorsal skinfold chamber (maximum size: ~3 × 3 × 1 mm)
7. Repetitive intravital fluorescence microscopy of the implant and surrounding chamber tissue on the day of implantation (d0) as well as on d3, d6, d10, and d14
   a. In vivo staining of microvessels: 0.05 mL of fluorescein isothiocyanate (FITC)-labeled dextran (molecular weight: 150,000 Da) i.v. via the retrobulbary space
   b. In vivo staining of leukocytes: 0.05 mL of rhodamine 6G intravenous via the retrobulbary space
8. Euthanasia of the animal after last microscopy by overdose of anesthetics and excision of dorsal skinfold chamber preparation for additional histological and immunohistochemical analyses

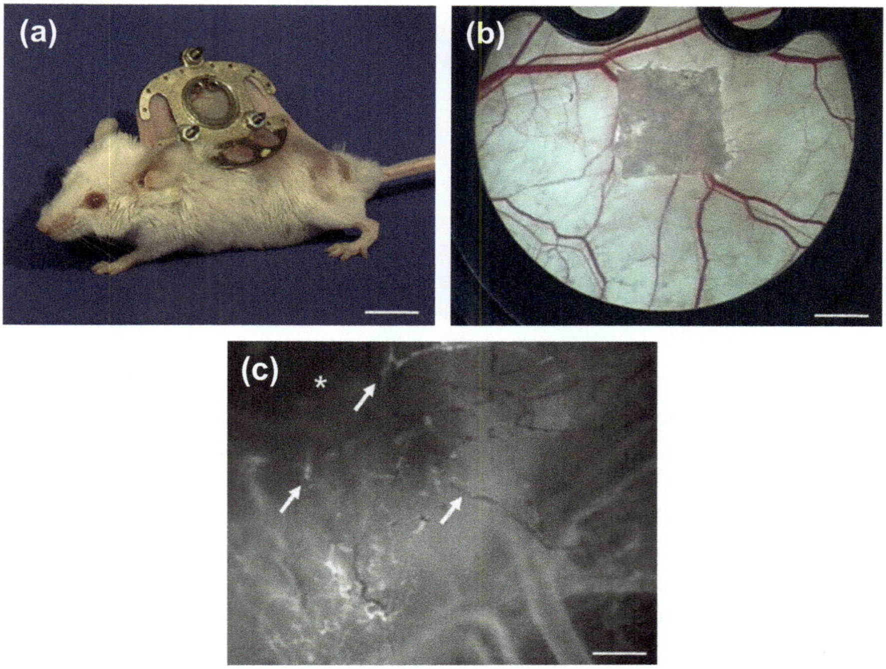

(a) Balb/c mouse equipped with a dorsal skinfold chamber (chamber weight ~2 g). (b) Overview of the observation window of the dorsal skinfold chamber directly after implantation of a porous polyurethane scaffold for bone tissue engineering. Note that the larger microvessels of the chamber (arterioles and venules of the striated muscle and subcutaneous tissue) can be detected macroscopically without the use of high-resolution intravital fluorescence microscopy. (c) Intravital fluorescence microscopy in blue-light epi-illumination with contrast enhancement by 5% FITC-labeled dextran 150,000 i.v. of newly formed microvessels (arrows) growing from the surrounding host tissue into the scaffold (asterisk) at day 14 after implantation. Scale bars: A = 12.5 mm; B = 1.4 mm; and C = 150 μm.

FIGURE 14.7 Schematic representation of a vascular pedicle skin flap.

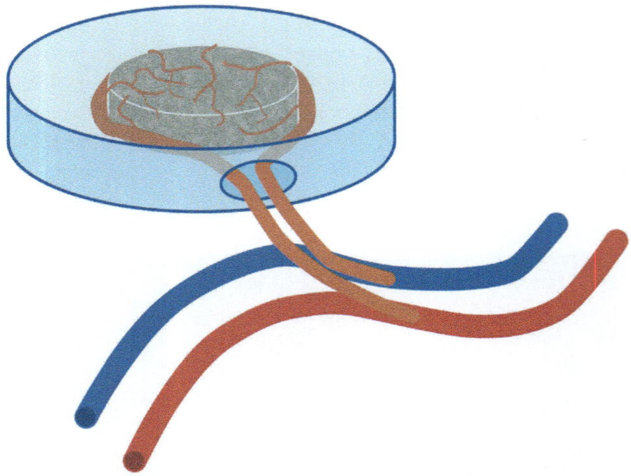

FIGURE 14.8 Schematic representation of an arteriovenous loop chamber.
An artificial vessel is grafted between a vein and an artery. This perfused graft is rolled around the construct, placed in an isolation chamber and implanted ectopically until vascularization is achieved, or directly extended to the ischemic tissue.

However, this involves tissue transfer to another location and therefore the creation of a second lesion site in the body with all the associated potential risks. A rather simple approach is the implantation of scaffolds into an easily accessible and well-vascularized host tissue site of the body, such as subcutis or muscle (Kokemueller et al., 2010). This induces an angiogenic host tissue response and functional, blood-perfused microvessels grow randomly into the implant over time. However, it should be noted that this is also associated with the invasion of the scaffolds with unspecific granulation tissue. Subsequently, these prevascularized scaffolds are excised and transferred to the final defect site. The preformed microvessels will eventually develop interconnections to the surrounding microvasculature via inosculation and get perfused again (Koike et al., 2004). Until then, the cells inside the implant suffer from hypoxia, and there is a high risk that parts of the construct do not survive this critical phase. Therefore, prevascularized tissue constructs may be alternatively generated with a vascular pedicle by flap fabrication and arteriovenous loop technique (Figure 14.8)

(Lokmic et al., 2007; Mian et al., 2000). These constructs can be microsurgically anastomosed to vessels at the implantation site, and can result in an immediate blood perfusion. Nonetheless, irrespective of the approach, all of these in situ prevascularization strategies require three surgical interventions, that is, the implantation of scaffolds for prevascularization, their removal, and their final insertion into the host defect site. Hence, although these strategies are highly promising and already tested under clinical conditions, there are ongoing attempts to further simplify in situ prevascularization. One idea, for instance, is to seed scaffolds intraoperatively with isolated functional microvascular fragments from adipose tissue samples, which are then directly implanted to the defect site in a one-step procedure. Under experimental conditions, these microvascular fragments have already been shown to develop into fully functional microvascular networks inside scaffolds within a short period of time (Laschke et al., 2012).

14.4 STRATEGIES TO IMPROVE CELL SURVIVAL

Beside the above-described methods to promote and accelerate the neovascularization of tissue-engineered constructs, several strategies have been reported to improve the survival of transplanted cells by in vitro preconditioning. On the one hand, these approaches use chemical and biological inhibitors of major cell death pathways. For example, heat-shock treatment of cells induces the expression of cytoprotective proteins. In embryonic stem cell-derived cardiomyocytes, an increased survival of implanted cells was shown after exposure to a heat shock in vitro (Don and Murry, 2013). A novel approach demonstrated key proangiogenic and prosurvival functions of glucose in MSCs upon in vivo transplantation and proposes to include slow-releasable glucose inside the scaffolds to ensure a better cell survival in constructs (Deschepper et al., 2013).

Strategies however aim to enhance the secretion of proangiogenic growth factors, which then subsequently promote the neovascularization of the tissue-engineered construct in vivo. Hypoxic pretreatment has been shown to increase the secretion of proangiogenic factors (e.g., VEGF) by inducing the expression of the transcription factor hypoxia inducible factor 1, alpha (Hif1α) (Skiles et al., 2013).

14.5 IN VIVO MODELS

A functional validation of the different approaches described in this chapter requires in vivo models that are able to assess some specific parameters related to the engraftment of tissue-engineered tissues, in order to compare their relative performance and suitability. In particular, parameters such as (1) the level of vessel ingrowth into the tissue-engineered constructs, (2) the level of survival of the implanted cells inside the constructs, and (3) their functional capacities in vivo can be investigated in such models.

This part does not aim to provide an exhaustive list and description of available animal models used to test tissue-engineered constructs, but rather describes examples of animal models commonly used for such purposes. The present section will also concentrate on small animal models, because of their easy access, lower cost, and higher throughput. Some of those models, involving mouse or rat, further allow investigating the performance of human cell-based tissue-engineered constructs, by selecting immune-incompetent animals such as athymic nude or severe combined immunodeficiency (SCID) rats/mice. Athymic nude mice/rats are T-cell deficient, whereas SCID animals have deficiencies in both T- and B-cells, as well as impaired natural killer cells. In such animals, the xenogeneic in vivo transplantation and testing of human cells is therefore possible with a limited risk of rejection (Sachs et al., 1995).

In general, it is also very important to consider the test tissue to target and its environment when selecting the appropriate animal model. For the specific testing of the vascular ingrowth, only a well-vascularized bed with reproducible angiogenesis is needed. As regards the engraftment and functionality of the generated graft/tissue, ectopic and orthotopic in vivo implantations are available. Ectopic implantation assesses the engraftment and generation of tissue in a general, typically in subcutaneous environment, whereas orthotopic implantation targets an environment similar to the tissue to be formed by the implant (e.g., in a skin defect for tissue-engineered skin substitutes). Engraftment and formation of the desired tissue by the implanted cells can then be assessed either during the in vivo phase or after explantation of the graft.

14.5.1 Assessments of the Vascularity in the Graft

Two different study models to assess the vascularization of a graft are commonly used.

1. The first one is the chicken chorioallantoic membrane, often referred to as chorioallantoic membrane procedure (CAM). The CAM is a vascularized membrane of the chicken embryo reachable through perforation of the egg shell (Ausprunk et al., 1975), in which materials, cell-based tissue-engineered constructs, and also tumor models can be physically implemented and tested. The vascular network around the implant is monitored and the vascular growth can thereby be analyzed. CAM has several advantages. It is technically simple and inexpensive, which allows for large-scale screening. Until day 10 of fertilization, it is not considered to be an animal model in many countries, which dramatically reduces ethics concerns and reduces regulatory requirements. Mammalian xenografts can be implemented thanks to the absence of rejection by the host membrane. It also allows for noninvasive, nondestructive serial observations, but the visualization of new vessel formation can sometimes be challenging. Several features however limit its significance:

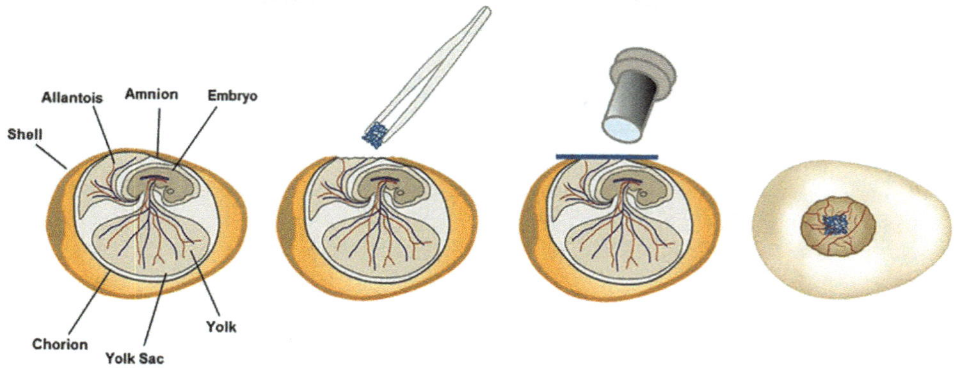

FIGURE 14.9 Schematic representation of the chorioallantoic membrane procedure (CAM).

it is a nonmammalian, embryonic model, in which angiogenesis occurs both by sprouting and intussusceptive (where one vessel splits into two) mechanisms. The model moreover exhibits nonspecific inflammatory reactions poorly predictive of the mechanisms happening in a more clinical relevant scenario (Figure 14.9).

2. The second model is the dorsal skinfold chamber, which allows the assessment of vascularization in implanted scaffolds and tissue constructs. In this model, two symmetrical titanium frames are implanted onto the extended dorsal skinfold, by pinching a double layer of skin (Box 14.2). Then, a 15-mm-diameter circular area of one of the skin folds is excised. The remaining layer consisting of striated muscle, subcutaneous tissue, and skin is therefore apparent and is covered by a coverslip incorporated into the observation window of the titanium frame (Menger et al., 2002). Thereafter, intravital fluorescent staining coupled to multifluorescence microscopy and image analysis techniques allows for the accurate quantification of various parameters such as angiogenic sprouting and network formation, microvessel density, and microhemodynamics (Laschke et al., 2011). Specific immunostaining also permits the study of cellular functions such as apoptosis (Harder et al., 2004). This model is available for studies in rats (Papenfuss et al., 1979), immune-competent mice (Cardon et al., 1970), nude and SCID mice (Lehr et al., 1993), and also in hamsters (Endrich et al., 1980). It is a perfect model to study the early, host-driven inflammatory and angiogenic responses to implanted materials, and its design allows varying specific parameters in the chambers such as temperature or composition of the surrounding milieu. The major limitations related to this model are that it only allows the testing of rather small implants, not >3 × 3 × 1 mm, and that it is not really suitable for long-term studies, that is, a maximum of 2–3 weeks (Laschke et al., 2011).

BOX 14.2 USE OF HYPOXIA-RESISTANT TISSUES TO OVERCOME THE LIMITATIONS OF DELAYED VASCULARIZATION: THE EXAMPLE OF ENDOCHONDRAL OSSIFICATION

Mesenchymal stem/stromal cells (MSCs) are typically used to generate bone tissue by a process resembling intramembranous ossification, that is, by direct osteoblastic differentiation. This process is known to strongly rely on a rapid vascularization of the implanted MSC-containing graft in vivo. However, most bones develop by endochondral ossification, via remodeling of hypertrophic cartilaginous templates. A major advantage of this ossification route is that the implanted construct is made of cartilage, a tissue known to require very low oxygen tensions in a physiological situation, the cellular part of which can therefore survive for prolonged periods in hypoxic conditions before vascular ingrowth at the site of implantation can take place by angiogenic processes from the surrounding tissues.

In a study by Scotti and colleagues, the capacity of expanded human MSCs to generate bone tissue through such an endochondral ossification process, previously demonstrated with embryonic stem cells (Jukes et al., 2008), was tested. To mimic condensation and chondrogenic differentiation of mesenchymal cells typical of endochondral ossification, human bone marrow MSCs were expanded in monolayer cultures and thereafter 5×10^5 cells/insert were cultured in a transwell, scaffold-free system for 1 or 2 weeks in a chondrogenic medium containing transforming growth factor (TGF)-β1 (prechondrogenic, and early hypertrophic stages, respectively), or for 3 weeks in chondrogenic medium followed by 2 weeks in a hypertrophic medium (without TGF-β1 and with beta-glycerophosphate and thyroxine; late hypertrophic stage). The resulting tissues in the form of discs show increasing degrees of chondrogenic differentiation and hypertrophy.

When implanted subcutaneously in nude mice and retrieved after 4 or 8 weeks, the implanted templates exhibited an organized series of events. A cellular condensation and hypertrophic chondrogenesis induced the formation of a bony collar by perichondral ossification. Matrix remodeling by matrix metalloproteases and osteoclastic activity took place and paralleled the vascular ingrowth from the host tissue into the cartilaginous template by angiogenesis. Indeed, capillary vessels positive for CD31 (platelet endothelial cell adhesion molecule-1) were shown to invade the graft starting from the outer matrix, likely to transport host osteoclasts, nutrients, and proapoptotic signals to the internal hypertrophic cartilaginous template (see attached figure). In the vascularized regions of the graft, bone matrix deposition onto the resorbed cartilaginous template could take place and bone tissue was formed, including functional hematopoietic foci, suggesting the formation of a functional bone organ.

The documented endochondral morphogenesis, with essential features of physiological endochondral ossification, is an example of how tissue engineering could harness developmental processes to overcome limitations related to delayed vascularization.

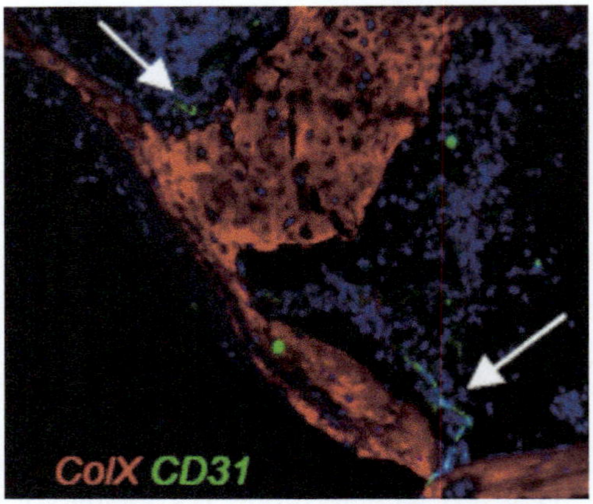

$CD31^+$ endothelial cells (green on the picture) of ingrowing vessels, penetrating the hypertrophic matrix that contains collagen type X (ColX) toward the center of the graft. Those blood vessels are located right next to the cartilaginous areas, which undergo remodeling (arrows). *Adapted from Scotti et al. (2010).*

14.5.2 Assessment of Graft Vascularization and Functionality of Engineered Tissues

How the different vascularization approaches described earlier can improve the engraftment and functionality of tissue-engineered constructs is a key step in their evaluation, comparison, and validation. This can be evaluated in two different situations: either ectopic or orthotopic.

14.5.2.1 Subcutaneous, Ectopic Implantation in Mice or Rats

In this model, the mouse or rat is positioned in a prone position on sterile drapes. An incision in the skin is made by using a surgical knife or scissors under sterile conditions. The length of the incision typically does not exceed 15 mm, and a subcutaneous pocket on the back of the animal is produced by blunt dissection. The tissue-engineered graft is then introduced into the pocket, and the incision is closed by using staples. The duration of the in vivo implantation is thereafter highly variable, and depends on the aim of the experiment. Short time points (e.g., few days to 2 weeks) are selected to study initial vascularization and anastomosis of prevascularized grafts, whereas the study of the formation and maturation of the target tissue may require short, for skin tissue-engineered constructs (Klar et al., 2014), to long implantation for tissues like bone (Guven et al., 2011). After explantation, the constructs can be analyzed by various techniques, essentially by histological means, for specific features with some described in more detail (Section 14.5.2.3). Overall, subcutaneous implantation is a technically simple, informative, and inexpensive in vivo model. Implanted tissue-engineered constructs are typically well tolerated. It is however a rather time-consuming model, in which the grafts are encapsulated by granulation tissue from the host. The subcutaneous tissue is also not a highly relevant site for engraftment and development of several specific tissues, but it allows for the assessment of the intrinsic performance of a graft in a standardized environment and is independent of the influence of trophic factors present in the specific target tissue.

14.5.2.2 Orthotopic Implantation

The basic principle of such in vivo implantations consists in the implementation of a tissue-engineered construct directly into its target tissue. It offers a more physiological and also a much more clinically relevant model for translational approaches. It allows for the actual testing of the functionality of a tissue-engineered graft in the physiological or pathological context of its target environment. There are as many models as there are species (from mouse to goat or horse), tissues, and pathologies, which implies thousands that cannot be detailed here. After explantation, it offers the same opportunities to analyze the implant by various techniques as in ectopic models. In addition, how the graft improved or restored the function of an organ and/or tissue can easily be assessed. For instance, in bone tissue engineering, bone defects/disease that alter the biomechanical properties of the bone can be created by surgical intervention or the use of transgene animal models and the contribution of the tissue-engineered graft can then be assessed also by biomechanical tests such as torsion, compression, or bending tests (Wong et al., 2011).

14.5.2.3 Examples of Techniques for the Detection of Implanted Vascular Cells

A critical issue in the study of the role of vascular cells in the contribution of a tissue-engineered graft to organ/tissue repair is to define the mechanisms by which the implanted vascular cells and tissues contributed to tissue formation and repair in vivo. In vivo CT using application of a vital contrast agent by intravenous injection allows repetitive observation of the vascularization at the injury, implant site. In addition, there are several methods to achieve the identification of implanted cells and to discriminate them from the host-derived cells. Implanted vascular cells can be labeled prior to their implantation with cell staining dyes such as DiI, a lipophilic fluorescent dye that gets fixed by the membrane and diffuses laterally to stain the entire cell, or Carboxyfluorescein succinimidyl ester (Hemmrich et al., 2006). When human vascular cells are implanted into immune-incompetent mice or rats, it is possible to discriminate the human (grafted) vascular cells from those of the mouse or rat (from the host). Several techniques can be used for that purpose. In situ hybridization for species-specific genomic motifs, especially human ALU sequences (Guven et al., 2011) and mouse short interspersed elements (SINE) (Brocher et al., 2013), allow to identify which cell species contributed to the tissue and vascularity present in the explanted graft. Alternatively, immune-histological staining using antibodies directed against human-specific or murine-specific epitopes of vascular-specific proteins, such as CD31 or CD34, allows the specific contribution of implanted and host cells to the generation of blood vessels in the grafts to be determined (Guven et al., 2011; Klar et al., 2014).

14.6 CONCLUSION/OUTLOOK

Clinical success of tissue-engineered constructs is so far limited to thin tissues such as skin, cornea, large diameter blood vessels, bladder, and avascular tissues like cartilage. For more complex 3D tissues, clinical applications of tissue-engineered substitutes are still unsuccessful, due to the limited oxygen diffusion into the graft. Therefore, implanted tissues or tissues substitutes need newly formed microvessels and capillaries to overcome this limitation. Here, we described different strategies that are currently being investigated to promote the cell survival and implant's neovascularization directly after implantation, and as a consequence the long-term survival and full integration of the tissue-engineered constructs. The present clinical approaches of in vivo vascularization methods show a high efficiency in promoting the vascularization of tissue substitutes, and they are not limited in size. Nevertheless, they present non-negligible drawbacks. The main one being the necessity of additional surgeries associated with an increased risk of possible complications and the necessity of microsurgical manipulations. The second one is the time required to achieve a full prevascularization of the scaffold, time during which the patient cannot (or only partially) reintegrate his normal life routine.

To accelerate the host vessels ingrowth (extrinsic vascularization) into the 3D structures, several strategies are considered using tools such as drug release, chemoattractive factors, or functionalized matrices. The second strategy to grow capillaries in an implant aims to promote the formation of new vascular structure inside the scaffold, prior to implantation by in vitro prevascularization, for example. Upon implantation, those newly formed structures will further grow and establish connections with the vasculature of surrounding tissues (intrinsic vascularization). This prevascularization procedure involves the preseeding of endothelial cells or progenitor cells in the presence or absence of tissue-specific cells before implantation, and is the result of a finely tuned cascade of in vitro events. Thus, after implantation, those structures need to undergo a rapid anastomosis with the surrounding microvessels to ensure their viability.

The use of human cells is clearly an advantage with respect to future clinical translation. However, for in vivo evaluation, this requires the use of immune-compromised animals, either mice or rats. This may have an impact on the outcome of the study, as an initial mild inflammation has beneficial effects on the integration of the construct. It has also been shown that the invading myeloid cells are required for the neovascularization of tissue-engineered constructs in vivo (Melero-Martin et al., 2010). Another important consideration during the in vivo evaluation phase is the implantation sites of the construct. Many in vivo studies on neovascularization assess the angiogenic capabilities of tissue-engineered constructs in ectopic sites (e.g., subcutaneous implantation, dorsal skin fold chamber). The tissue environment may however affect the outcome of these studies, and a translation of results to the orthotopic situation is questionable. Moreover, the size of the constructs to be implanted is relatively limited for such ectopic sites.

The need for tissue-engineered constructs, or tissue substitutes, is usually related to the size of the defect to be healed and therefore requires large 3D volumes. After the proof of concept stage in vitro and in vivo in small animals, the scaling up of tissue replacement constructs is mandatory and will still need validation in large animals to justify their potential use in clinics.

The development of proper systems (e.g., bioreactors) that will be organ specific (in terms of mechanical and biological environments) and enable the development of large-scale and/or tailored tissue-engineered constructs will need to be investigated. These bioreactors could be in vivo systems in which the prevascularization will be enhanced at the site of implantation, by adding cells, biological factors, or a combination of different strategies. The "right" cell types to be seeded would need to be better defined. Moreover, an ideal material that favors both the target tissue regeneration and an adequate vascularization has still not been developed (tissues have different functions and therefore different requirements). But one could ask oneself: "Do we need to have proangiogenic cells and tissue-specific cells at the same time?" Upon direct implantation in the damaged tissue, the presence of a functional neovascular network might be sufficient to promote the integration of the implant, and tissue-specific cells will invade the constructs gradually.

From a clinical perspective, a single-step procedure and intraoperative method would be clearly beneficial. Having a better insight into the required cell population(s) and their requirements in the target tissue in vivo would allow a fast isolation and possible direct reimplantation in the presence of an appropriate stimulating factor. Finally, having a better knowledge of the required cells, their tissue of origin (bone marrow, peripheral blood, etc.), and their mode of mobilization toward the injury site, one could enable the exploitation of these cues to prestimulate their recruitment (homing) and the cascade of events leading to a rapid and functional vascularization process of the structure.

14.7 SUMMARY

- Neovascularization of tissue-engineered constructs is required to ensure a sufficient supply with oxygen and nutrients and the removal of metabolites.
- In adult life, neovascularization is mainly driven by angiogenesis, a highly regulated, multistep process driven by proangiogenic growth factors and cell–cell interactions.
- The microstructure of a tissue-engineered construct, particularly the porosity, can guide and promote the neovascularization of the construct.
- Vascular sprouting and invasion of endothelial cells into a tissue-engineered construct can be induced by the incorporation of proangiogenic factors into the construct.
- Incorporation of mature endothelial cells or EPCs may be used to promote intrinsic neovascularization.
- In vitro prevascularized constructs can be produced by seeding vascular and tissue-specific cells into scaffolds, aggregated into multicellular spheres or layered into cell sheets.
- Allogenic decellularized tissues retain the natural architecture of the extracellular matrix and can be used to guide neovascularization.

- In situ prevascularization techniques aim to use the human body as a bioreactor to promote neovascularization in a tissue-engineered construct.
- Various preclinical models have been developed to assess the vascularity of grafts, including the chorioallantoic membrane model, the dorsal skinfold chamber, and the ectopic subcutaneous or orthotopic implantation of the construct in mice or rats.
- The potential translation of strategies into a clinical application has to be considered, particularly with respect to minimal manipulation, scale-up, and cost effectiveness.

REFERENCES

Achilli, T.M., Meyer, J., Morgan, J.R., 2012. Advances in the formation, use and understanding of multi-cellular spheroids. Expert. Opin. Biol. Ther. 12 (10), 1347–1360.

Aguirre, A., Planell, J.A., Engel, E., 2010. Dynamics of bone marrow-derived endothelial progenitor cell/mesenchymal stem cell interaction in co-culture and its implications in angiogenesis. Biochem. Biophys. Res. Commun. 400 (2), 284–291.

Akiyama, H., Ito, A., Kawabe, Y., Kamihira, M., 2010. Genetically engineered angiogenic cell sheets using magnetic force-based gene delivery and tissue fabrication techniques. Biomaterials 31 (6), 1251–1259.

Alev, C., Ii, M., Asahara, T., 2011. Endothelial progenitor cells: a novel tool for the therapy of ischemic diseases. Antioxid. Redox. Signal 15 (4), 949–965.

Amini, A.R., Laurencin, C.T., Nukavarapu, S.P., 2012. Bone tissue engineering: recent advances and challenges. Crit. Rev. Biomed. Eng. 40 (5), 363–408.

Anderson, J.M., Rodriguez, A., Chang, D.T., 2008. Foreign body reaction to biomaterials. Semin. Immunol. 20 (2), 86–100.

Asahara, T., Murohara, T., Sullivan, A., Silver, M., van der Zee, R., Li, T., Witzenbichler, B., Schatteman, G., Isner, J.M., 1997. Isolation of putative progenitor endothelial cells for angiogenesis. Science 275 (5302), 964–967.

Asakawa, N., Shimizu, T., Tsuda, Y., Sekiya, S., Sasagawa, T., Yamato, M., Fukai, F., Okano, T., 2010. Pre-vascularization of in vitro three-dimensional tissues created by cell sheet engineering. Biomaterials 31 (14), 3903–3909.

Auger, F.A., Gibot, L., Lacroix, D., 2013. The pivotal role of vascularization in tissue engineering. Annu. Rev. Biomed. Eng. 15, 177–200.

Ausprunk, D.H., Knighton, D.R., Folkman, J., 1975. Vascularization of normal and neoplastic tissues grafted to the chick chorioallantois. Role of host and preexisting graft blood vessels. Am. J. Pathol. 79 (3), 597–618.

Bai, F., Wang, Z., Lu, J., Liu, J., Chen, G., Lv, R., Wang, J., Lin, K., Zhang, J., Huang, X., 2010. The correlation between the internal structure and vascularization of controllable porous bioceramic materials in vivo: a quantitative study. Tissue Eng. Part A 16 (12), 3791–3803.

Brocher, J., Janicki, P., Voltz, P., Seebach, E., Neumann, E., Mueller-Ladner, U., Richter, W., 2013. Inferior ectopic bone formation of mesenchymal stromal cells from adipose tissue compared to bone marrow: rescue by chondrogenic pre-induction. Stem Cell. Res. 11 (3), 1393–1406.

Brudno, Y., Ennett-Shepard, A.B., Chen, R.R., Aizenberg, M., Mooney, D.J., 2013. Enhancing microvascular formation and vessel maturation through temporal control over multiple pro-angiogenic and pro-maturation factors. Biomaterials 34 (36), 9201–9209.

Cardon, S.Z., Oestermeyer, C.F., Bloch, E.H., 1970. Effect of oxygen on cyclic red blood cell flow in unanesthetized mammalian striated muscle as determined by microscopy. Microvasc. Res. 2 (1), 67–76.

Carmeliet, P., 2003. Angiogenesis in health and disease. Nat. Med. 9 (6), 653–660.

Chim, H., Ong, J.L., Schantz, J.T., Hutmacher, D.W., Agrawal, C.M., 2003. Efficacy of glow discharge gas plasma treatment as a surface modification process for three-dimensional poly (D,L-lactide) scaffolds. J. Biomed. Mater. Res. A 65 (3), 327–335.

Choi, S.W., Zhang, Y., Macewan, M.R., Xia, Y., 2013. Neovascularization in biodegradable inverse opal scaffolds with uniform and precisely controlled pore sizes. Adv. Healthc. Mater 2 (1), 145–154.

Crisan, M., Yap, S., Casteilla, L., Chen, C.W., Corselli, M., Park, T.S., Andriolo, G., Sun, B., Zheng, B., Zhang, L., Norotte, C., Teng, P.N., Traas, J., Schugar, R., Deasy, B.M., Badylak, S., Buhring, H.J., Giacobino, J.P., Lazzari, L., Huard, J., Peault, B., 2008. A perivascular origin for mesenchymal stem cells in multiple human organs. Cell. Stem Cell 3 (3), 301–313.

Deschepper, M., Manassero, M., Oudina, K., Paquet, J., Monfoulet, L.E., Bensidhoum, M., Logeart-Avramoglou, D., Petite, H., 2013. Proangiogenic and prosurvival functions of glucose in human mesenchymal stem cells upon transplantation. Stem Cells 31 (3), 526–535.

Don, C.W., Murry, C.E., 2013. Improving survival and efficacy of pluripotent stem cell-derived cardiac grafts. J. Cell Mol. Med. 17 (11), 1355–1362.

Duttenhoefer, F., Lara de, F.R., Meury, T., Loibl, M., Benneker, L.M., Richards, R.G., Alini, M., Verrier, S., 2013. 3D scaffolds co-seeded with human endothelial progenitor and mesenchymal stem cells: evidence of prevascularisation within 7 days. Eur. Cell. Mater. 26, 49–64.

Endrich, B., Asaishi, K., Gotz, A., Messmer, K., 1980. Technical report—a new chamber technique for microvascular studies in unanesthetized hamsters. Res. Exp. Med. (Berl) 177 (2), 125–134.

Eppley, B.L., Pietrzak, W.S., Blanton, M., 2006. Platelet-rich plasma: a review of biology and applications in plastic surgery. Plast. Reconstr. Surg 118 (6), 147e–159e.

Folkman, J., Hochberg, M., 1973. Self-regulation of growth in three dimensions. J. Exp. Med. 138 (4), 745–753.

Fuchs, S., Dohle, E., Kolbe, M., Kirkpatrick, C.J., 2010. Outgrowth endothelial cells: sources, characteristics and potential applications in tissue engineering and regenerative medicine. Adv. Biochem. Eng. Biotechnol. 123, 201–217.

Ghanaati, S., Fuchs, S., Webber, M.J., Orth, C., Barbeck, M., Gomes, M.E., Reis, R.L., Kirkpatrick, C.J., 2011. Rapid vascularization of starch–poly(caprolactone) in vivo by outgrowth endothelial cells in co-culture with primary osteoblasts. J. Tissue Eng. Regener. Med. 5 (6), e136–e143.

Guven, S., Mehrkens, A., Saxer, F., Schaefer, D.J., Martinetti, R., Martin, I., Scherberich, A., 2011. Engineering of large osteogenic grafts with rapid engraftment capacity using mesenchymal and endothelial progenitors from human adipose tissue. Biomaterials 32 (25), 5801–5809.

Harder, Y., Contaldo, C., Klenk, J., Banic, A., Jakob, S.M., Erni, D., 2004. Improved skin flap survival after local heat preconditioning in pigs. J. Surg. Res. 119 (1), 100–105.

Hemmrich, K., Meersch, M., von, H.D., Pallua, N., 2006. Applicability of the dyes CFSE, CM-DiI and PKH26 for tracking of human preadipocytes to evaluate adipose tissue engineering. Cells Tissues Organs 184 (3–4), 117–127.

Hendrickx, B., Verdonck, K., Van den Berge, S., Dickens, S., Eriksson, E., Vranckx, J.J., Luttun, A., 2010. Integration of blood outgrowth endothelial cells in dermal fibroblast sheets promotes full thickness wound healing. Stem Cells 28 (7), 1165–1177.

Holtrop, M.E., 1967. The potencies of the epiphyseal cartilage in endochondral ossification. Proc. K. Ned. Akad.Wet. C 70 (1), 21–28.

James, A.W., Zara, J.N., Zhang, X., Askarinam, A., Goyal, R., Chiang, M., Yuan, W., Chang, L., Corselli, M., Shen, J., Pang, S., Stoker, D., Wu, B., Ting, K., Peault, B., Soo, C., 2012. Perivascular stem cells: a prospectively purified mesenchymal stem cell population for bone tissue engineering. Stem Cells Transl. Med. 1 (6), 510–519.

Jukes, J.M., Both, S.K., Leusink, A., Sterk, L.M., van Blitterswijk, C.A., de Boer, J., 2008. Endochondral bone tissue engineering using embryonic stem cells. Proc. Natl. Acad. Sci. USA 105 (19), 6840–6845.

Kannan, R.Y., Salacinski, H.J., Sales, K., Butler, P., Seifalian, A.M., 2005. The roles of tissue engineering and vascularisation in the development of micro-vascular networks: a review. Biomaterials 26 (14), 1857–1875.

Klar, A.S., Güven, S., Biedermann, T., Luginbühl, J., Böttchner-Haberzeth, S., Meuli-Simmen, C., Meuli, M., Martin, I., Scherberich, A., Reichmann, E., June 2014. Tissue-engineered dermo-epidermal skin grafts prevascularized with adipose-derived cells. Biomaterials 35 (19), 5065–5078.

Koike, N., Fukumura, D., Gralla, O., Au, P., Schechner, J.S., Jain, R.K., 2004. Tissue engineering: creation of long-lasting blood vessels. Nature 428 (6979), 138–139.

Kokemueller, H., Spalthoff, S., Nolff, M., Tavassol, F., Essig, H., Stuehmer, C., Bormann, K.H., Rucker, M., Gellrich, N.C., 2010. Prefabrication of vascularized bioartificial bone grafts in vivo for segmental mandibular reconstruction: experimental pilot study in sheep and first clinical application. Int. J. Oral Maxillofac. Surg. 39 (4), 379–387.

Langer, R., Vacanti, J.P., 1993. Tissue engineering. Science 260 (5110), 920–926.

Laschke, M.W., Harder, Y., Amon, M., Martin, I., Farhadi, J., Ring, A., Torio-Padron, N., Schramm, R., Rucker, M., Junker, D., Haufel, J.M., Carvalho, C., Heberer, M., Germann, G., Vollmar, B., Menger, M.D., 2006. Angiogenesis in tissue engineering: breathing life into constructed tissue substitutes. Tissue Eng. 12 (8), 2093–2104.

Laschke, M.W., Kleer, S., Scheuer, C., Schuler, S., Garcia, P., Eglin, D., Alini, M., Menger, M.D., 2012. Vascularisation of porous scaffolds is improved by incorporation of adipose tissue-derived microvascular fragments. Eur. Cell Mater. 24, 266–277.

Laschke, M.W., Vollmar, B., Menger, M.D., 2009. Inosculation: connecting the life-sustaining pipelines. Tissue Eng. Part B Rev. 15 (4), 455–465.

Laschke, M.W., Vollmar, B., Menger, M.D., 2011. The dorsal skinfold chamber: window into the dynamic interaction of biomaterials with their surrounding host tissue. Eur. Cell Mater. 22, 147–164.

Lehr, H.A., Leunig, M., Menger, M.D., Nolte, D., Messmer, K., 1993. Dorsal skinfold chamber technique for intravital microscopy in nude mice. Am. J. Pathol. 143 (4), 1055–1062.

Leijon, J., Carlsson, F., Brannstrom, J., Sanchez, J., Larsson, R., Nilsson, B., Magnusson, P.U., Rosenquist, M., 2013. Attachment of flexible heparin chains to gelatin scaffolds improves endothelial cell infiltration. Tissue Eng. Part A 19 (11–12), 1336–1348.

Levenberg, S., Rouwkema, J., Macdonald, M., Garfein, E.S., Kohane, D.S., Darland, D.C., Marini, R., van Blitterswijk, C.A., Mulligan, R.C., D'Amore, P.A., Langer, R., 2005. Engineering vascularized skeletal muscle tissue. Nat. Biotechnol. 23 (7), 879–884.

Lindhorst, D., Tavassol, F., von, S.C., Schumann, P., Laschke, M.W., Harder, Y., Bormann, K.H., Essig, H., Kokemuller, H., Kampmann, A., Voss, A., Mulhaupt, R., Menger, M.D., Gellrich, N.C., Rucker, M., 2010. Effects of VEGF loading on scaffold-confined vascularization. J. Biomed. Mater. Res. A 95 (3), 783–792.

Loibl, M., Binder, A., Herrmann, M., Duttenhoefer, F., Richards, R.G., Nerlich, M., Alini, M., Verrier, S., 2014. Direct cell–cell contact between Mesenchymal Stem Cells and Endothelial Progenitor Cells induces a Pericyte-like Phenotype in vitro. Biomed. Res. Int.

Lokmic, Z., Stillaert, F., Morrison, W.A., Thompson, E.W., Mitchell, G.M., 2007. An arteriovenous loop in a protected space generates a permanent, highly vascular, tissue-engineered construct. FASEB J. 21 (2), 511–522.

Masuda, H., Asahara, T., 2003. Post-natal endothelial progenitor cells for neovascularization in tissue regeneration. Cardiovasc. Res. 58 (2), 390–398.

McDowell, F., 1978. Plastic surgery in the twentieth century. Ann. Plast. Surg. 1 (2), 217–224.

Mehdizadeh, H., Sumo, S., Bayrak, E.S., Brey, E.M., Cinar, A., 2013. Three-dimensional modeling of angiogenesis in porous biomaterial scaffolds. Biomaterials 34 (12), 2875–2887.

Melero-Martin, J.M., De Obaldia, M.E., Allen, P., Dudley, A.C., Klagsbrun, M., Bischoff, J., 2010. Host myeloid cells are necessary for creating bioengineered human vascular networks in vivo. Tissue Eng. Part A 16 (8), 2457–2466.

Menger, M.D., Laschke, M.W., Vollmar, B., 2002. Viewing the microcirculation through the window: some twenty years experience with the hamster dorsal skinfold chamber. Eur. Surg. Res. 34 (1–2), 83–91.

Mian, R., Morrison, W.A., Hurley, J.V., Penington, A.J., Romeo, R., Tanaka, Y., Knight, K.R., 2000. Formation of new tissue from an arteriovenous loop in the absence of added extracellular matrix. Tissue Eng. 6 (6), 595–603.

Papenfuss, H.D., Gross, J.F., Intaglietta, M., Treese, F.A., 1979. A transparent access chamber for the rat dorsal skin fold. Microvasc. Res. 18 (3), 311–318.

Phelps, E.A., Garcia, A.J., 2010. Engineering more than a cell: vascularization strategies in tissue engineering. Curr. Opin. Biotechnol. 21 (5), 704–709.

Risau, W., 1997. Mechanisms of angiogenesis. Nature 386 (6626), 671–674.

Rouwkema, J., Rivron, N.C., van Blitterswijk, C.A., 2008. Vascularization in tissue engineering. Trends Biotechnol 26 (8), 434–441.

Rucker, M., Laschke, M.W., Junker, D., Carvalho, C., Schramm, A., Mulhaupt, R., Gellrich, N.C., Menger, M.D., 2006. Angiogenic and inflammatory response to biodegradable scaffolds in dorsal skinfold chambers of mice. Biomaterials 27 (29), 5027–5038.

Sachs, D.H., Sykes, M., Greenstein, J.L., Cosimi, A.B., 1995. Tolerance and xenograft survival. Nat. Med. 1 (9), 969.

Scheufler, O., Schaefer, D.J., Jaquiery, C., Braccini, A., Wendt, D.J., Gasser, J.A., Galli, R., Pierer, G., Heberer, M., Martin, I., 2008. Spatial and temporal patterns of bone formation in ectopically pre-fabricated, autologous cell-based engineered bone flaps in rabbits. J. Cell Mol. Med. 12 (4), 1238–1249.

Scotti, C., Tonnarelli, B., Papadimitropoulos, A., Scherberich, A., Schaeren, S., Schauerte, A., Lopez-Rios, J., Zeller, R., Barbero, A., Martin, I., 2010. Recapitulation of endochondral bone formation using human adult mesenchymal stem cells as a paradigm for developmental engineering. Proc. Natl. Acad. Sci. USA 107 (16), 7251–7256.

Skiles, M.L., Sahai, S., Rucker, L., Blanchette, J.O., 2013. Use of culture geometry to control hypoxia-induced vascular endothelial growth factor secretion from adipose-derived stem cells: optimizing a cell-based approach to drive vascular growth. Tissue Eng. Part A 19 (21–22), 2330–2338.

Tabata, Y., 2003. Tissue regeneration based on growth factor release. Tissue Eng. 9 (Suppl 1), S5–S15.

Timmermans, F., Plum, J., Yoder, M.C., Ingram, D.A., Vandekerckhove, B., Case, J., 2009. Endothelial progenitor cells: identity defined? J. Cell Mol. Med. 13 (1), 87–102.

Wong, R.C., Tideman, H., Merkx, M.A., Jansen, J., Goh, S.M., Liao, K., 2011. Review of biomechanical models used in studying the biomechanics of reconstructed mandibles. Int. J. Oral Maxillofac. Surg. 40 (4), 393–400

Wu, X., Ren, J., Luan, J., Yao, G., Li, J., 2012. Biochemical, mechanical, and morphological properties of a completely autologous platelet-rich wound sealant. Blood Coagul. Fibrinolysis, 23 (4), 290–295.

CHAPTER 15

Skin Engineering and Keratinocyte Stem Cell Therapy

Simon Myers[1], Harshad Navsaria[1], Nkemcho Ojeh[2]
[1]Centre for Cutaneous Research, Blizard Institute, Barts and The London School of Medicine and Dentistry, Queen Mary, University of London, United Kingdom; [2]Faculty of Medical Sciences, The University of the West Indies, Cave Hill, Bridgetown, Barbados

This chapter is dedicated to the memory of Professor Harshad A. Navsaria, who recently passed away during its revision.

LEARNING OBJECTIVES

- To understand the structure and function of skin
- To know the function of differentiation markers and role of basement membrane
- To understand the concept of keratinocyte cell expansion in vitro
- To know different delivery systems for keratinocyte application
- To understand the significance of dermis for keratinocyte grafting
- To distinguish different types of tissue-engineered dermal templates/products
- To understand the role of keratinocyte stem cells for clinical application
- To explore the potential use of adult stem cells in skin tissue engineering

15.1 INTRODUCTION

Large area skin loss is seen in acute trauma including burn trauma, where skin engineering has particular relevance. Annually, an estimated 6 million patients seek medical treatment for burns. House fires account for only 4% of admissions, but 70% of fatalities are largely due to associated inhalation injury (Hasselgren, 1999). It has been estimated that between 5 and 10% of admitted patients require admission to a specialized burn unit and at least 1% die from their injuries. Although the incidence of life-threatening burn injury has fallen and the mortality improved significantly (Roberts et al., 2012), such burn care remains a significant financial burden. Burn wound depth assessment is the key to wound management. Early wound closure is the key to limiting mortality from sepsis and scar morbidity. The first

Tissue Engineering.
Copyright © 2015 Elsevier Inc. All rights reserved.

attempt to classify burn depth was made by Jean de Vigo (1460–1525) of Italy in 1483, who recognized superficial and deep wounds, but it was not until 1953 that Jackson led current classifications (Jackson, 1953). Jackson's description includes superficial partial skin loss, deep partial skin loss, and full-thickness skin loss, and he went on to note the poor and delayed healing associated with the latter two, recommending skin grafting in those cases (Wallace, 1982).

Chronic leg ulcers, often venous stasis in etiology, represent further skin defects; in these wounds there is altered vascular physiology and features of cell senescence. The cost for treatment of such ulcers is difficult to assess and estimates have varied considerably. Given the chronic nature of the pathology, although relatively small skin defects, the total cost per patient may be enormous (Hankin et al., 2012).

15.2 STRUCTURE OF THE EPIDERMIS

The epidermis is a multilayered cellular structure. The main cell type, making up 95% of the total, is the keratinocyte, a terminally differentiating cell that moves progressively from deep to superficial before being shed from the outer surface as a keratin scale. Morphologically, the epidermis is divided (from deep to superficial) into four layers: the strata basale, spinosum (prickle), granulosum, and corneum (Figure 15.1).

The stratum basale consists of cuboidal cells, usually one cell deep, in a continuous layer and is attached to the basement membrane. The epibasal layer, the stratum spinosum, is distinguished by numerous desmosomes interconnecting the cells to produce a stable network. The desmosomes appear to provide not only a means of intercellular adhesion, but also to form a focus for the internal organization of cytoskeletal components, particularly tonofilaments. The next layer, the stratum granulosum, is characterized by the intracellular deposition of keratohyalin granules, particles of irregular shape and approximately 2 nm in diameter which are arranged in rows or lattices. Another feature of this layer is the Odland body—a lamellated granule of 100–300 nm in size containing lipid for discharge into the extracellular space. It is this lipid that binds cells in the stratum corneum. The cells in this layer (corneocytes) become flattened as they lose their nuclei and organelles, and as the keratin filaments within them polymerize with the formation of disulfide bonds. This occurs under the influence of filaggrin, the protein component of the keratohyalin granule, and results in the formation of a flattened, highly insoluble keratin shell bound to the cell membrane by involucrin. Degradation of the intercellular lipid component and the desmosomes precedes the loss of groups of cells as a flake of skin.

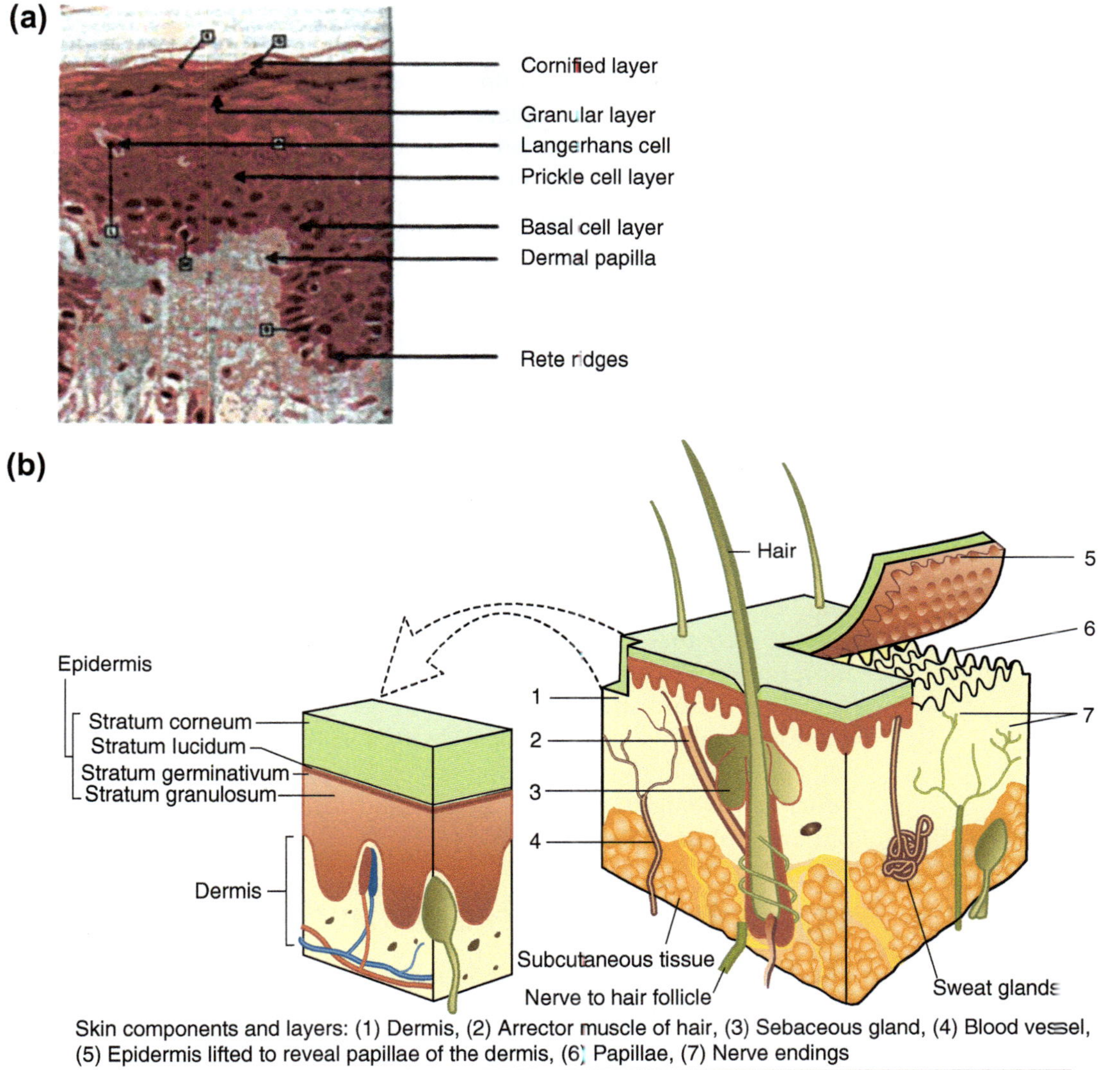

FIGURE 15.1
Structure of normal skin showing the different layers and cell types in the epidermis (a) and dermis (b).

The cell cycle within the epidermis begins with the asymmetric division of a putative stem cell (in the equivalent of the murine bulge region of the outer root sheath of the hair follicle and the deeper rête ridges of glabrous skin). This cell gives rise, by definition, to all keratinocytes in the epidermis including the transient amplifying population which undergoes further limited proliferation. Although the description given earlier presents a linear progression from strata basale to corneum, there are at least two other models recognized from work in animal models.

The kinetics of epidermal proliferation is complicated by the observation that postmitotic cells may be very active metabolically and may increase tissue mass without an increase in cell numbers. The turnover time is the time taken for the whole cell population to replace itself and depends on the length of the cell cycle (the time between successive mitoses) and the growth fraction, i.e., the proportion of cells undergoing mitosis. In normal human skin, the net effect of both parameters is to generate an epidermal turnover time of 52–75 days.

15.3 KERATINS

The cytoskeleton of the keratinocyte is formed from three groups of filaments: actin (7 nm diameter), tubulin (20–25 nm), and intermediate filaments (7–10 nm). The latter group contains several recognized components, including vimentin, desmin, the nuclear laminins, and the keratins, the polypeptide molecules characteristic of all epithelial cells and which account for 65% of the final mass of a corneocyte. The keratin family consists of over 30 individual members and are separated by electrophoresis into basic keratins (numbered 1–8) and the acidic keratins (numbered 9–12), and whose corresponding gene families are termed, respectively, types I and II. The keratins pair into acidic/basic dimers that are specifically expressed according to cell physiology and morphology. For example, simple epithelia are characterized by the keratin pair K8/K18, and stratified epithelia by K5/K14. Further, sole and palm skin express K9, differentiated skin K1/K10, and hyperproliferative or regenerating epidermis K6/K16/K17. This expression provides a convenient way with which to investigate morphology and function in the epidermis with the use of immunohistochemistry (Table 15.1).

15.4 STRUCTURE OF THE DERMOEPIDERMAL JUNCTION

The dermis and epidermis meet at the dermoepidermal junction (DEJ), which forms an intimate and complex attachment that allows not only the physical attachment of the epidermis to deeper structures, but also a physiological link between the two components that is integral to the normal function of both (Figure 15.2). The DEJ extends in an undulating pattern to form interdigitations

Table 15.1 Epithelial Keratin Expression

Keratin	Epidermal Expression
K5/K14	Stratified epithelia (mature)
K1/K10	Differentiated epidermis
K6/K16	Proliferative/regenerating epithelia
K9	Skin of palmoplantar origin

throughout its length and is continuous with epidermal appendages. These interdigitations significantly increase the surface area of contact allowing more intimate communication and preventing shear forces from separating the two layers. The DEJ consists of the cell membranes of those cells superficial to it (i.e., keratinocytes, melanocytes, and Langerhans cells) and those deep to it (fibroblasts, mast cells, etc.) bound to a common membrane—the basement membrane. DEJ and the surrounding complexes are referred to as the basement membrane zone (BMZ).

The basal surface of the keratinocyte attaches to the lamina lucida via hemidesmosomes. These consist primarily of inner plaques on the inner surface of the cell membrane (allowing continuity with tonofilaments within the cell), outer plaques (closely associated with the cell membrane itself), and the sub-basal plate, external to the cell and immediately adjacent (and embedded into) to the basement membrane. The basement membrane itself may be resolved into three laminae by electron microscopy: the outer lucida, intermediate densa, and deep fibroreticularis. Within the lamina lucida are numerous anchoring filaments related to the hemidesmosomes and orientated perpendicular to the cell and basement membranes. The lamina densa appears relatively amorphous, and the fibroreticular layer appears to demonstrate anchoring fibrils,

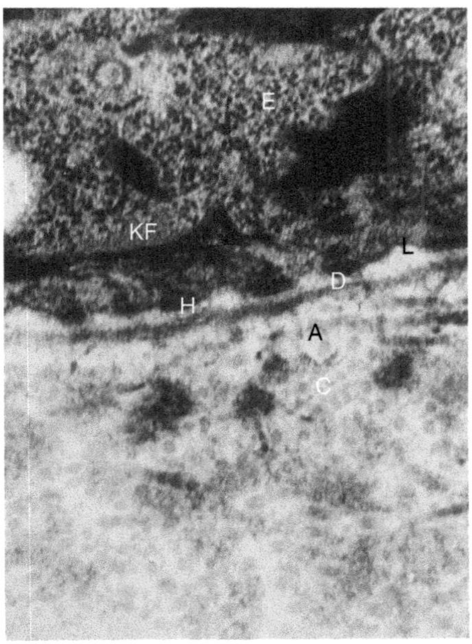

FIGURE 15.2

Electron micrograph of the DEJ after treatment with hyaluronic acid and application of cultured keratinocytes on Laserskin® showing a well-differentiated basement membrane and abundant collagen fibers in the newly formed dermis (×50,000).

Table 15.2 Molecular Components of the Basement Membrane

Component	Location
Bullous pemphigoid antigen-1	Hemidesmosome plaque
Bullous pemphigoid antigen-2	Hemidesmosome plaque and anchoring fibrils
Laminins 5, 6, and 7	Anchoring filaments
Collagen IV	Lamina densa
Chondroitin sulfate	Lamina densa
Collagen VII	Anchoring fibrils

short, curving structures that extend into the dermis either singly or in relation to amorphous bodies and then termed anchoring plaques. In addition to the above electron microscopic description, a number of specific molecular components can be identified within the basement membrane; these allow assessment of maturity by specific immunohistochemical staining (Table 15.2).

The basement membrane is intrinsic to the stability of healing wounds. Rigal et al. (1991) delineated the pattern of basement membrane component deposition with healing over a period of 10 days and showed a near-linear production of most components. Initially, collagen VII is absent, while bullous pemphigoid antigen (BPA) and collagen IV are the predominant components. Collagen VII production only begins after day 3, by which time BPA, laminin, fibronectin, and collagen IV are present in appreciable quantities. By day 7, the levels of all these components are relatively equal and the basement membrane may be described as fully formed.

15.5 IN VITRO KERATINOCYTE CULTURE

Replacement of whole skin relies upon the provision of both the dermis and epidermis. Although split-thickness skin grafting satisfies these criteria, it is not always possible, particularly after a burn injury involving a large total burn surface area. In this instance, two obvious problems present themselves. The first is that there is, by definition, insufficient healthy skin available to resurface the burn area, even with mesh expansion of those grafts. Second, inflicting a further injury to the integument only adds to the physiological challenge facing a patient who may already be critically ill.

In 1975, Rheinwald and Green published their chance discovery that human epidermal cells proliferated under the same conditions as mouse teratoma cells (Rheinwald and Green, 1975). In this instance, it was the provision of a feeder layer of lethally irradiated Swiss 3T3 mouse fibroblasts and of mitogens within the culture media that enabled human epithelial cells to proliferate rapidly, and proved one of the most important breakthroughs in the field (Figure 15.3).

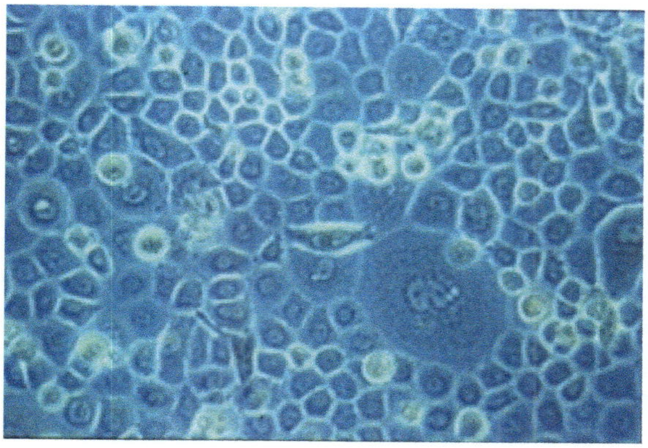

FIGURE 15.3
Cultured keratinocytes on a plastic substrate (×200).

Keratinocyte culture methods have since been modified in a number of ways. Further work (Rheinwald and Green, 1977) optimized the growth conditions resulting in a mean cell cycle time of 22–24 h, much greater than that seen in either normal or hyperproliferative (psoriatic) skin. A number of minor modifications have been made to the culture medium resulting in the conditions used today (Aasen and Izpisúa Belmonte, 2010). Keratinocytes can also be expanded in low calcium, feeder-free, and serum-free conditions (xenobiotic-free culture), which is favorable in removing contaminating cells like fibroblasts, but their proliferative lifespan is reduced (Aasen and Izpisúa Belmonte, 2010). In optimal conditions, a 3-cm^2 piece of skin can be expanded more than 5000-fold within 3–4 weeks, yielding sufficient sheets of keratinocytes to cover the entire surface of an adult human (Atiyeh and Costagliola, 2007).

There are many factors that affect keratinocyte culture, the most important being the age of the donor. Population doublings of cells tend to decrease with age. Moreover, both fibroblasts and keratinocytes have a finite number of population doublings, after which they are unsuitable for culture. Systemic illness in donors may decrease the proliferative capacity of keratinocytes. In addition, the composition of the clonal types, i.e., holoclones, meroclones, or paraclones, generated from the keratinocyte culture is important. Cultures containing more holoclones, which are stem cells, have the greatest proliferative capacity and these stem cells are known to decline with age (Green, 2008).

Within the culture flask, proliferating cells form islands which spread laterally and, to a lesser extent, stratify; growth is predominantly in a horizontal plane, but the daughter cells of those in the center of the island start to stratify. If the entire flask is allowed to stratify, a so-called "Green sheet" is formed, consisting of up to 12 cell layers. The number of layers is determined by the calcium ion concentration in the medium; low calcium solutions promote the formation of a monolayer, and high calcium solutions that of a stratified sheet. The keratinocyte sheet, also known as cultured epithelial autograft (CEA), is commercially available as Epicel and EpiDex that are used in the treatment of partial- and full-thickness burns, and venous and diabetic ulcers. Both of these products are derived from autologous keratinocytes, although EpiDex is made from outer root sheath keratinocytes derived from scalp hair follicles (Greaves et al., 2013). As CEA is very fragile, in order to minimize trauma due to handling, backing materials such as paraffin gauze or polyurethane are used. A variety of delivery systems have been developed to facilitate transplantation, and to limit any delays in in vitro expansion. In this instance, cells are primarily grown on a substrate and transferred without the need for enzymatic dissociation (Myers et al., 2007; Atiyeh and Costagliola, 2007). Cells are plated onto a transfer material (e.g., Laserskin®; Fidia Advanced Biopolymers) in vitro, without the necessity for them to form a confluent sheet—they are transferred while still in individual colonies (Figure 15.4). The lack of confluence in this instance is beneficial with minimal stratification and, therefore, differentiation. When transplanted, the majority of cells presented are of a proliferative phenotype. This method has been shown to be at least as effective as other methods and has the potential benefits of decreasing the time required for culture, or increasing the area that can be covered at a given time point (Shevchenko et al., 2010). Another strategy is to use an aerosol method for keratinocyte delivery where the cells are suspended in fibrin glue and sprayed directly onto the wound bed. Cells have also been harvested from skin using the commercially available ReCell kit from Avita Medical. This allows for immediate delivery of a noncultured mixed cell population consisting of keratinocytes, fibroblasts, Langerhans cells, and melanocytes to the wound (Wood, 2012). Melanocyte numbers decrease rapidly in culture and eventually stabilize at approximately 10% of their initial levels (Staiano-Coico et al., 1990). This means that the end product obtained by current tissue engineering technologies is inferior not only in terms of the lack of skin

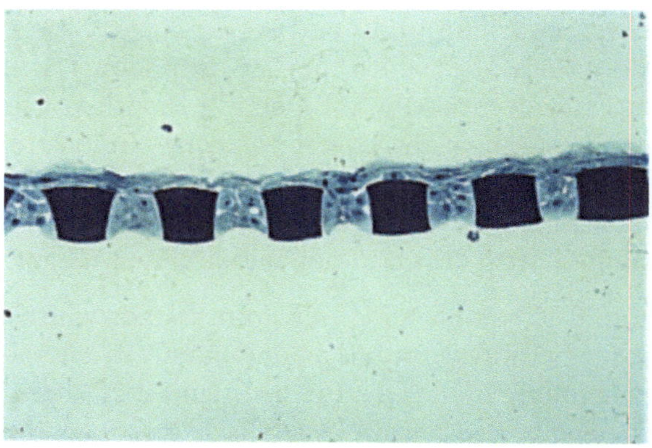

FIGURE 15.4

Cross section of keratinocytes cultured on a Laserskin® membrane (×250).

adnexae, but also in terms of the color match and the potential to pigment on UV exposure. However, with the ReCell kit, the maintenance of the melanocytes allows for appropriate development of pigmentation (Wood, 2012).

Histologically, cultured keratinocyte sheet grafts consist of sheets of cells with much simpler organization than that of native epidermis. The basal layer is relatively well formed, but the suprabasal layers are irregular in thickness, consisting of 2 to 10 layers, with poor organization. Although desmosomes and gap junctions are present, clinically the skin is extremely fragile.

15.6 IMMUNOGENICITY AND CULTURED KERATINOCYTES

It has been established that allografted keratinocytes do not persist in an immunocompetent host (Griffiths et al., 2004). Langerhans cells are epidermal dendritic, antigen-presenting cells. It is possible to deplete Langerhans cell number by in vitro UV irradiation and this may attenuate the rejection response to allograft (Granstein et al., 1987). Langerhans cells proliferate poorly if at all in vitro. Aübock et al. (1988) demonstrated prolonged but not indefinite survival of grafts with Langerhans cell depletion, and postgrafting UV irradiation appeared to prolong allograft survival (Granstein et al., 1987). The latter cannot be repeated indefinitely, however, and would be expected to adversely affect wound healing. Systemic approaches to prevent allograft loss have included Cyclosporin A administration, alone or in conjunction with antibody administration (Ossevoort et al., 1999), as part of a standard immunosuppression protocol used for renal transplantation (Wendt et al., 1994), or as a combination of UV irradiation and 8-methoxypsoralen (Granstein et al., 1987). Ultimately, however, there appears little doubt that transplanted allogeneic keratinocytes are lost from the wound (Griffiths et al., 2004).

15.7 DEVELOPMENT OF IN VIVO SOMATIC KERATINOCYTE STEM CELL GRAFTING

15.7.1 Evolution of Epidermal Replacement

Early keratinocyte culture developed from the 1930s onwards. However, cultured keratinocyte characterization was not performed until 1956. The fact that epidermal cells were able to proliferate in culture and their mitotic activity was clear by 1960. Early keratinocyte differentiation occurred in the absence of fibroblasts (Pruniéras et al., 1965). The classic Rheinwald–Green technique was not published until the mid-1970s (Rheinwald and Green, 1977).

Although keratinocyte expansion in vitro has many potential clinical applications (Shevchenko et al., 2010; Lootens et al., 2013), keratinocyte attachment, proliferation, and differentiation depend on interaction with the dermis. Key to the development of stable epidermal replacement has been the study of the interactions between the epithelial and mesenchymal layers. Initial excitement was dampened by early reports of poor "take," fragile epidermis, and late blistering. Autologous grafting became a clinical reality in 1981 with the grafting of two patients suffering 40% and 80% total body surface area (TBSA) burns (O'Connor et al., 1981). Only a small portion of their total wound areas was grafted, but it appeared that graft "take" was superior on fresh rather than granulating wounds. Reported successful engraftment has been highly variable, between 0 and 95% mean "take." Even experienced teams have been unable to attain the same engraftment rates as with split-thickness autografting. Recently, a case series of 88 burns with a mean graft "take" of 72.7% with a 91% overall patient survival rate has been reported (Sood et al., 2010). In the same study, some of the complications included early blistering and shearing, pruritus and itching, late CEA loss, and wound contractures (Sood et al., 2010).

15.7.2 Evolution of Dermal Replacement

A dermis or equivalent is essential to the "take" and quality of cultured keratinocyte grafts (Burke et al., 1981a). Functional and cosmetic outcomes are superior when more dermis is included in split-thickness or full-thickness skin grafts, limiting scarring and wound contracture (Burke et al., 1981a). In massive burn injury, there is a limited donor area for split-thickness skin graft harvest, and whole allograft is transplanted to provide temporary wound closure (Cleland et al., 2014). It is the epidermal component of the allograft skin that primarily results in rejection. In large burn injuries there is a suppression of adaptive and innate immunity and initial immunological and clinical "take" with vascularization of the graft over some weeks. Removal of the alloepidermis after engraftment and its replacement with cultured autologous keratinocyte sheets was a technique developed to limit rejection. Other forms of dermal replacement have been developed, an early example comprised fibroblasts incorporated into a glycosaminoglycan (GAG) membrane (Lazic and Falanga, 2011).

15.8 POOR KERATINOCYTE "TAKE"

The reasons for cultured keratinocyte sheet graft failure are complex, but possibly the most important factor is the state of the wound bed engrafted. Both the quality of the tissue and the bacterial load are significant factors that adversely affect cultured keratinocyte take. Therefore, early excision of the burn wound reduces bacterial colonization and enhances keratinocyte take. Paradoxically, wound bed preparation with topical antiseptics can damage keratinocytes (Tatnall et al., 1990). Even in those circumstances in which the wound bed was sterile and culture conditions optimal, e.g., the elective excision of congenital nevi, Sood et al. (2009) reported mean take rates of 66%. Take rates are, however, improved by the use of dermal equivalents (Fang et al., 2013).

As early as 1979, Pruniéras et al. (1983) described a method for producing what is now referred to as de-epidermalized dermis (DED) from cadaveric skin (Figure 15.5). The cellular components of whole skin were removed and the remnant was used as a dermal equivalent. This sparked a wave of research, led by Bell and coworkers, into the possibility of creating a bioengineered dermis based upon a simple molecular scaffold. In the same year, Bell et al. (1979) published their first description of a contracted collagen lattice suitable for in vitro research, and which was eventually grafted in vivo by Burke et al. (1981a,b). The initial composition was modified to include chondroitin-6-sulfate for biochemical stability and became marketed commercially as Integra® (Figure 15.6). Alternative strategies have included the use of polyglycolic acid matrices and hyaluronic

FIGURE 15.5
Glycerol-preserved cadaveric skin used for preparing DED.

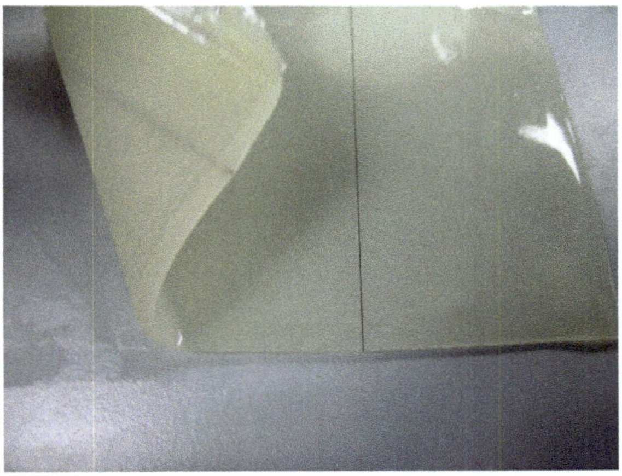

FIGURE 15.6
Integra with the silicone upper layer in situ. *From Anthony et al. (2006).*

acid scaffolds. The latter are of particular note since the degradation products of the macromolecule seem to exert a marked influence on the phenotype of the healing wound (Voigt and Driver, 2012). Most dermal equivalents in current use are presented as "dermal regeneration templates," providing a scaffold for host fibroblasts and endothelial cells to populate and generate a "neodermis."

15.9 ENHANCED DERMAL GRAFTING

A dermal equivalent scaffold can be enhanced by a period of in vitro coculture with dermal fibroblasts, or presentation as a composite graft including an epidermal component. Landeen et al. (1992) described the changes associated with a period of in vitro coculture of a collagen gel with dermal fibroblasts. They demonstrated a rise in extracellular matrix (ECM) components such as hydroxyproline, hyaluronic acid, and chondroitin 4,6-disulfate. These findings suggested that such a period of culture might have profound effects on the dermal equivalent eventually grafted; after 4 weeks in culture the material was clearly much more complex and, in many ways, more closely resembled normal dermis. Since that time, other benefits associated with in vitro coculture have surfaced and are summarized in Table 15.3. Dermal equivalents repopulated with cells may enhance dermal regeneration and improve both keratinocyte adherence and wound healing. Nevertheless, a number of milestones have been achieved with keratinocyte stem cell technology alone, treating a range of diseases at a variety of anatomical sites (Table 15.4).

Composite grafts are generated by combining in vitro a dermal equivalent (with or without fibroblasts) and an epidermis. This may then be grafted as a "whole-skin equivalent," e.g., Apligraf®. The more complex the construct, however, the more significant the development, regulatory, and production costs, and so dermal regeneration templates closed with thin split-thickness autografts remain the focus of the clinical marketplace.

Although the benefits of cellular skin substitutes have been reported, new evidence suggests that with the recognition of two macrophage phenotypes, namely M1 and M2, these cellular substitutes may elicit the proinflammatory

Table 15.3 Reported Benefits Associated with In Vitro Seeding of Dermal Analogues

Proposed Benefits
Improved collagen deposition and remodeling
Improved DEJ formation
Enhanced survival of keratinocytes
Decreased myofibroblast formation
Decreased wound contraction

M1 response that results in scar formation compared to acellular substitutes. Removing the cellular component changes the macrophage response to the immunomodulatory M2 response which facilitates constructive tissue remodeling (Greaves et al., 2013).

15.9.1 Tissue-Engineered Constructs

Constructs may be synthetic or natural (based on allograft dermis), made by various techniques such as electrospinning, phase separation, freeze-drying, self-assembly, and solid free-form fabrication. These may or may not include cells and may vary in cost, engraftment, risk of disease transmission, handling, and storage. See Tables 15.5 and 15.6.

15.9.2 Biobrane and Transcyte

Biobrane is a thin bilaminar dressing with a porous silicone outer layer bonded to an inert nylon fabric that is coated with peptides derived from porcine type I collagen in order to promote adherence. It is a very useful primary dressing for partial-thickness wounds in children that limits the need for painful dressing changes and reduces in-patient stay (Hankins et al., 2006). In a recent in vivo study, where preconfluent autologous keratinocytes were seeded in Biobrane and applied directly to three patients' donor and burn wound sites, healing times were found to be comparable to the use of keratinocyte sheets. Moreover, the preconfluent grafting method was quicker to produce and the keratinocyte delivery system was

Table 15.4 Applications of Cultured Keratinocyte Technology

Site of Use
Autologous
Intraoral
Mastoid cavities
Superficial burn
Full-thickness burn injury
Facial scar revision
Chronic ulcers
Giant congenital nevi
Hypospadias
Separation of Siamese twins
Recessive epidermolysis bullosa
Allogeneic
Chronic venous ulcers
Diabetic leg ulcers
Skin graft donor sites
Burn injuries

Table 15.5 Examples of Replacement Matrices

Dermal Replacements	Epidermal and Dermal Combination Replacements
Cryopreserved allograft	Apligraf™
Glycerol preserved allograft	OrCel™
Alloderm™	
Integra™	
Dermagraft™	
Dermagraft-TC™	
Hyalograft 3D™	
Matriderm®	
Pelnac®	
Ez Derm™	

Table 15.6 Examples of Products for Wound Repair

Product	Epidermis	Dermis	Characteristics
Dermagraft™	N/A	Neonatal allogeneic fibroblasts,absorbable scaffold (polyglycolic acid or polyglactin 910)	Remains metabolically active after cryopreservation and the dermal matrix is reabsorbable; similar to Alloderm™
Integra™	Silicone epidermis	Bovine tendon collagen and shark GAGs	Allows for immediate coverage of large acute wounds reducing bacterial ingress and water egress
Apligraf™	Neonatal foreskin allogeneic human keratinocytes	Human neonatal fibroblasts and bovine type 1 collagen matrix	Requires less donor site so that the split-skin graft size is decreased
Biobrane™	Porous silicone outer layer	Nylon fabric coated with cross-linked peptides derived from porcine type I collagen	The initial adherence is promoted by the collagen-derived peptides on the nylon which bind to the fibrin and collagen on the surface of the wound
Trancyte™	Polymer membrane	Allogeneic human fibroblasts (Figure 15.7)	The tissue matrix and the growth factors are left intact postfreezing
Hyalograft 3D™	N/A	Human fibroblasts and microperforated hyaluronic acid membrane	Provides a hyaluronic acid scaffold
Laserskin™	Autologous human keratinocytes	N/A	Used in combination with Hyalograft 3D™
Cryospray	Allogeneic human keratinocytes	Silicone backing	Accelerate wound closure in acute partial-thickness and refractory chronic wounds

easier to handle (Frew et al., 2013). In a retrospective review, seven pediatric burns cases were reported with the use of Biobrane dressing to cover split-thickness skin grafts and nearby skin graft donor sites or partial-thickness burns. This promoted adherence of the split-thickness skin graft to the wound bed, allowed fluid drainage, and prevented shearing, thus facilitating healing (Farroha et al., 2013). When a similar construct was seeded with allogeneic fibroblasts, an immature ECM of fibronectin, type I collagen, and growth factors was generated. The cellular component was suspended after 17 days to produce Transcyte, formerly Dermagraft-TC (Figure 15.7). It was suggested that Transcyte might limit the extension of stasis burns, but it required storage at −80 °C and its expense was prohibitive.

15.9.3 AlloDerm

AlloDerm is an acellular human allogeneic dermal matrix preserved by freeze-drying. The epidermal basement membrane is preserved, and it can be used to prepare a full-thickness wound bed before application of thin split-skin graft following burn excision. Using an in vivo method such as the Cuono technique (Cuono et al., 1987), the physical removal of the epidermis will tend to remove the basement membrane proteins such as collagen IV and laminin. More recently, AlloDerm has been used in other applications including alloplastic breast reconstruction, abdominal wall reconstruction, rhinoplasty, vaginal repair, and radial forearm free-flap donor site coverage (Jansen et al., 2013). Although this product has found its marketplace as a subcutaneous filler in aesthetic procedures, other similar DED products are increasingly becoming available from tissue banks, e.g., Glyaderm from the Euro Tissue Bank and from UK National Health Service Blood and Transplantation (NHSBT).

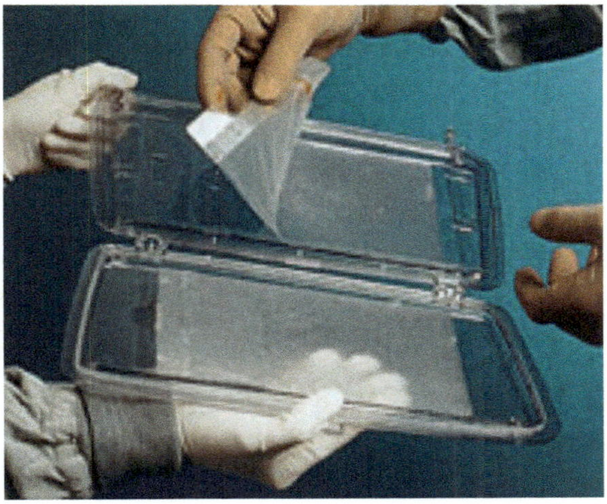

FIGURE 15.7
Commercially available product "Transcyte." *Reproduced with permission from Advanced Biohealing.*

15.9.4 Dermagraft

Dermagraft comprises an absorbable polymer scaffold (polyglycolic acid or polyglactin-910) seeded with neonatal allogeneic fibroblasts (Figure 15.8). As with Transcyte, an ECM is laid down consisting of growth factors and collagens, tenascin, vitronectin, and GAGs, but unlike Transcyte this assembly remains metabolically active after cryopreservation and the dermal matrix is resorbable. Dermagraft has been used more to stimulate wound healing in chronic wounds such as diabetic foot ulcers than resurface full-thickness acute burn wounds (Naughton et al., 1997), but was as successful as split-thickness grafts in the treatment of burns wounds excised to varying depths (Hansbrough et al., 1992). In a recent prospective, multicenter, randomized controlled study, a combination of Dermagraft and standard compression therapy in the treatment of venous leg ulcers did not show a statistically significant improvement over compression therapy alone for overall healing by 12 weeks. However, efficacy seemed to improve as ulcer duration decreased (Harding et al., 2013).

15.9.5 Integra Dermal Regeneration Template

Dermal regeneration templates have been translated to clinical practice. The first to the marketplace is derived from bovine collagen and shark chondroitin sulfate, covered by a silicone surface layer (Figure 15.6). It allows immediate coverage of large, acutely excised wounds, reducing bacterial ingress and water egress (Yannas and Burke, 1980; Burke et al., 1981b). It has also been reported

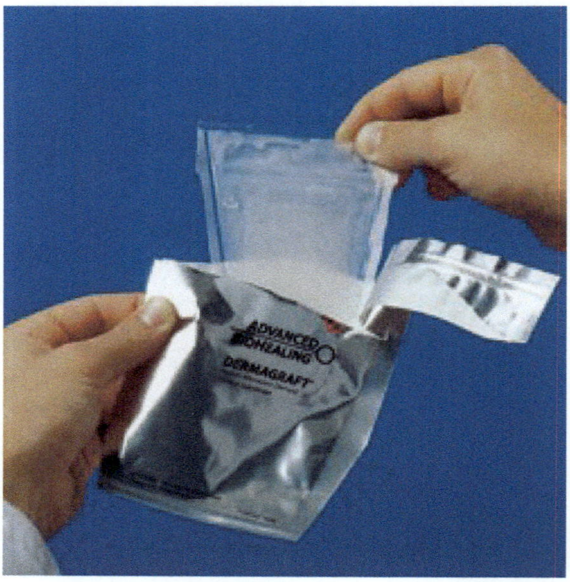

FIGURE 15.8
Commercially available product "Dermagraft." *Reproduced with permission from Advanced Biohealing.*

to be used successfully in the treatment of chronic wounds (Kahn et al., 2011). A pore score of 70–200 μm allows migration of endogenous fibroblasts and endothelial cells into the dermal matrix producing a vascularized neodermis, and Integra is probably the most prevalent "artificial skin." In the original system, there remained a delay of 3 weeks before the definitive epidermal cover was provided in the form of an "ultra-thin" split-skin graft (around 0.15 mm), but the donor site healed more rapidly than a conventional split-thickness graft (around 0.33 mm) and can be reharvested more frequently. Integra provides an excellent bed for sheet split-thickness skin grafts, and at best provides the quality of a full-thickness skin graft from a split-thickness skin graft donor site. A competing similar but cheaper product, Pelnac, is now available (Suzuki et al., 2000).

The 3-week delay while a neodermis formed is similar to the time taken to expand cultured autologous keratinocytes for transplantation in massive burn injury. However, a case report of this approach had a poor result (Pandya et al., 1998). The approach that has evolved combines widely meshed split-thickness autograft with cultured autologous keratinocytes, fibroblasts, or noncultured cell suspension (ReCell) as adjuvants (Wisser and Steffes, 2003; Wood, 2012).

15.9.6 Matriderm

Matriderm is a more recent product with a collagen base that includes elastin, and no silicone surface layer. Although AlloDerm had been used as a single-staged grafting system, Matriderm has been the first to popularize this approach. Matriderm is engrafted to a full-thickness wound bed with an overlying thin split-thickness autograft, in the same procedure (Ryssel et al., 2008). This has significant cost advantages compared to two procedures separated by weeks, and appears to provide acceptable graft take. As a consequence, there is now a single-stage Integra product on the market, and a number of further products in development stages.

15.9.7 Composite Grafts/Cultured Skin Substitutes

A dermal collagen–GAG membrane seeded with autologous keratinocytes and cultured until maturation and differentiation was successfully applied to full-thickness defects. This construct was modified by incorporating autologous fibroblasts in the dermal layer and applied to four patients; of 13 grafts nine took successfully (Hansbrough et al., 1989). Later papers showed similar results and again less successful than the conventional autograft where take rates approach 100% (Boyce et al., 1995). Of note, histological differences included lack of a vascular plexus in the dermis of the cultured skin substitute (CSS) when compared to conventional split-thickness graft, and (or possibly subsequent to) a less keratinized epithelium. Grossly, the CSS was more hyperpigmented but also smoother, lacking any meshed pattern.

Eleven out of 17 wounds grafted with CSS later needed regrafting, but in those that did heal the epidermis was more like normal skin than the blister-prone epithelium from grafting keratinocyte sheets alone. Since then, Boyce et al. (2006) have designed the product Permaderm, an autologous CSS with the purpose of achieving permanent wound closure. Although the product has shown promise in the treatment of full-thickness burns, it is not yet available on the market (Boyce et al., 2006). In a variation of their technique, allogeneic keratinocytes and fibroblasts were used successfully to treat chronic wounds (Boyce et al., 2006). Recently, the same group has explored the generation of hair follicles in CSS in vivo using a combination of human and mice epidermal and dermal cells (Sriwiriyanont et al., 2013).

15.9.8 Cryospray

Cryospray is an example of a screened and expanded allograft keratinocyte cell line produced by the only commercial provider of keratinocyte cultures in the UK, Regenerys Ltd. It is becoming licensed for clinical use in acute partial-thickness and refractory chronic wound healing.

15.9.9 Apligraf/Allo-CSS

Commercially available Apligraf (allo-CSS) is an advanced tissue-engineered composite skin equivalent consisting of bovine type I collagen matrix cultured with allogeneic male neonatal fibroblasts and keratinocytes that resemble normal human skin structure (Figure 15.9). Apligraf is immunologically inert and this is thought to be ascribed to the lack of Langerhans cells. At the same time, this product was shown not to persist long term in a deep dermal wound model but instead acted as a temporary biological dressing, providing growth factors to the wound (Griffiths et al., 2004). In the acute wounds of immunocompetent humans, 15 patients received Apligraf. Although no frank rejection was seen, the Apligraf took on an unusual appearance becoming translucent and jelly-like below the allogeneic epidermis, and biopsy demonstrated marked dermal fibrosis (Eaglstein et al., 1995). Grafted wounds were noted to contract more than full-thickness grafted wounds. No antigens were expressed by the grafted keratinocytes or fibroblasts, nor to any of the xenogenic collagen used could be detected. Apligraf had similar effects to meshed autograft when applied to partial-thickness donor sites, and no classical clinical rejection was observed (Kirsner, 1998). Apligraf combined with autologous meshed split-skin graft produced more favorable results than conventional split-skin grafting alone (Waymack et al., 2000). At present, Apligraf is licensed only for the treatment of diabetic foot and venous leg ulcers. In a diabetic foot ulcer study, more ulcers were healed successfully in the group treated with Apligraf and with no clinical rejection (Brem et al., 2000).

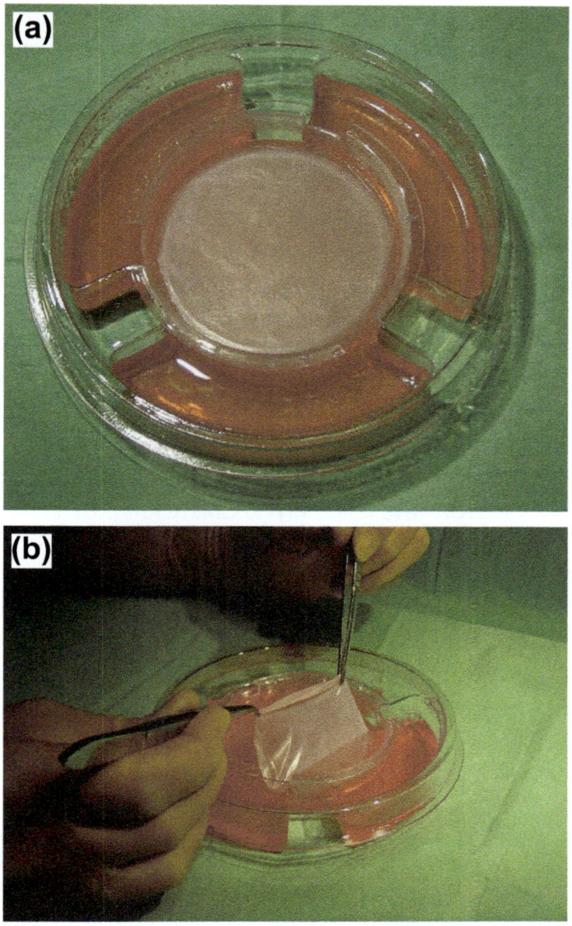

FIGURE 15.9
Apligraf in Transwell (a) and Apligraf being freed from agarose media prior to clinical application (b).

A similar product, OrCel™, is indicated in the treatment of donor sites of recessive dystrophic epidermolysis bullosa, burn, and surgical wounds; however, limited clinical trials data is available on this product (Greaves et al., 2013).

15.10 THE USE OF ADULT STEM CELLS IN TISSUE-ENGINEERED SKIN

The ideal cell source to be used in tissue-engineered skin should be pluripotent, highly proliferative and expandable, readily available in abundance, easily accessible with minimal invasive procedures, and safe for autologous and allogeneic transplantation without provoking an immune response. Stem cells are receiving a lot of attention because of their transgermal and pluripotential nature. Some populations, e.g., hematogenous stem cells, have been well described and clinical therapies developed. The rationale for the potential use of adult stem cells in skin repair and tissue-engineered skin is discussed in the following section.

15.10.1 Keratinocyte Stem Cells

The significant plasticity of somatic stem cells including keratinocytes is becoming more evident, although this raises new questions about the role of epithelial–mesenchymal transition in wound healing and scarring. The keratinocyte stem cell (KSC) (holoclone) cycles slowly, minimizing potential DNA mutations, and probably as a consequence is long lived. Each stem cell is capable of dividing many times (more than 140) into another copy and a separate daughter cell. The resulting transient amplifying cell (meroclone) may divide a number of times providing an explosion in cell numbers, vital in a high turnover environment, before further differentiation into a terminal cell (paraclone) which migrates away from the basal layer. Classically, the epidermis has been thought to be maintained by this epidermal stem cell/transit amplifying cell model. Recently, however, other models have challenged the original model. One model of homeostasis involves only one type of progenitor cell that may undergo an unlimited round of divisions producing postmitotic basal cells in a stochastic process. Another model, the population asymmetry with stem cells model, involves a hierarchical organization consisting of both fast-cycling progenitor cells and slow-cycling stem cells that have different functions during homeostasis and repair of the interfollicular epidermis in adult mice (Li et al., 2013).

The location of the KSC seems to vary. Typically, they are found in patches or clusters related to the DEJ along the basal lamina. There is probably a segmental distribution of progeny related to a stem cell, resulting in their patchy distribution along the DEJ. In glabrous (hairless) skin of the soles and palms, the clusters are found deeper, at the tips of the deep rête ridges. Estimates vary as to how bountiful they are, but probably 5–10% of the basal population is stem cells. The "bulge region" of the hair follicle, a morphologically distinct region in mice but not humans, has been identified as a major reservoir for stem cells in hair-bearing skin. Recently, however, multiple new hair follicle stem cell populations have also been identified outside this anatomical area that have differing abilities to contribute not only to the hair follicle itself but also to the interfollicular epidermis and the sebaceous gland. As there is a distinct pool of KSC in the interfollicular epidermis, hair follicle stem cells are not necessary for interfollicular epidermal homeostasis but contribute during wound healing (Goldstein and Horsley, 2012).

In vivo identification of stem cells relies on their ability to retain labels for long periods because they cycle slowly. Lineage tracing studies, which involve stably integrating a label into the genome of a cell and its progenies, have also helped to identify stem cells and trace their progenies. Stem cells have also been identified and isolated based on the expression of a number of marker molecules. The transcription factor p63, a p53 homologue required for epithelial development, has been shown to be "abundantly expressed by epidermal and limbal (cornea) holoclones," in vitro (Pellegrini et al., 2001). The levels of p63 expressed by meroclones drop dramatically as they leave the stem cell niche.

A population of cells that may remain for the whole life of the animal and has the potential to reconstitute the whole epidermis offers a convenient vehicle for genetic manipulation. Keratinocytes have been cultured in vitro in the last three decades and can be grafted onto wounds via carrier dressings or on skin constructs (Figures 15.10–15.12). A recent study demonstrated a transduction protocol to correct a *LAMA3* gene mutation in primary keratinocytes derived from a patient with the blistering disorder, Herlitz junctional epidermolysis bullosa. The corrected primary keratinocytes were able to express functional laminin-332 heterotrimer proteins and the blistering phenotype was fully reversed both in vitro and in skin grafts (Melo et al., 2014).

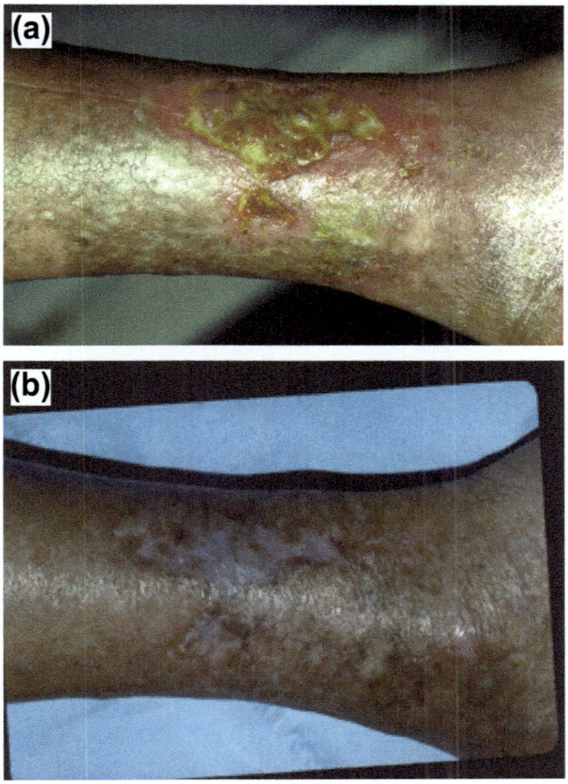

FIGURE 15.10

Cultured keratinocyte grafts used to cover a leg ulcer before (a) and after 3 weeks (b).

FIGURE 15.11
Hyalomatrix 3D consisting of esterified hyaluronic acid with a silicone membrane.

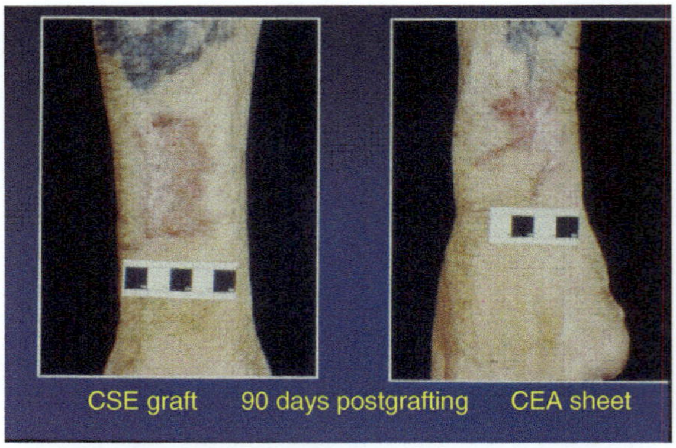

FIGURE 15.12
Cultured skin equivalent (CSE) (keratinocytes grown on DED) and cultured epithelial autograft (CEA) grafted on a patient with tattoo excision 90 days postgrafting. The clinical appearance is superior with CSE.

15.10.2 Induced Pluripotent Stem Cells

Differentiated, adult somatic cells can be reprogrammed to generate induced pluripotent stem (iPS) cells with similar characteristics to embryonic stem cells (ESCs). These cells have been derived from human skin keratinocytes and fibroblasts by exogenous addition of four transcription factors (Oct-3/4, Sox2, c-Myc, and Klf4) and have been shown to generate a wide range of differentiated cell types (Takahashi et al., 2007; Aasen et al., 2008). The use of iPS cells surmounts any moral and ethical issues associated with ESCs. As autologous cells can be used, they can also promote patient-tailored treatment and circumvent any immune rejection problems. iPS cells therefore show promise as a specific source of stem cells that could be used in conjunction with tissue-engineered skin.

15.10.3 Hair Follicle Dermal Stem Cells

In humans, subpopulations of stem cells reside within the hair follicle including epithelial stem cells, melanocyte stem cells, mesenchymal-like stem cells derived from the dermal sheath (DS) and dermal papilla (DP), and nestin-positive stem cells.

The DS and DP comprise the mesenchymal portions of the hair follicle and are important in hair follicle formation, growth, and support. It has been hypothesized that DP and DS cells play a role in dermal repair during wound healing. DS and DP cells have been shown to display hematopoietic activity both in vivo and in vitro as well as adipogenic and osteogenic differentiation in vitro. Multipotent stem cells within the dermis, termed skin-derived precursors, have been shown to originate in the DP and DS (Biernaskie et al., 2009) and can differentiate into neurons, glial cells, smooth muscle cells, and adipocytes (Richardson et al., 2005).

Nestin, a protein marker for neural stem cells, is also expressed in the hair follicle. Nestin-expressing stem cells originate from the bulge region and migrate to the DP and surrounding tissue. These cells have a broad potential to differentiate into neural, hepatic, pancreatic endocrine, and mesenchymal/mesodermal cell lineages. Using green fluorescent protein (GFP) tracing techniques to tag nestin-expressing cells in mice, these cells were shown to form cell colonies, and differentiate into multiple lineages such as neurons, keratinocytes, melanocytes, glial cells, and smooth muscle cells. Implanting GFP-tagged nestin-expressing cells into the gap region of severed sciatic nerve promoted nerve regeneration and restored nerve function. These cells differentiated mostly into Schwann cells, which are known to support neuron regeneration (Amoh et al., 2005).

15.10.4 Mesenchymal Stem Cells

Human mesenchymal stem cells (MSCs) are derived from the stroma of the bone marrow. They have also been isolated from nerve tissue, adipose tissue, umbilical cord blood, and dermis. These progenitor cells are natural key players in tissue regeneration and give rise to many mesenchymal-derived tissues/cells both in vivo and in vitro, some of which include bone, cartilage, osteocytes, and adipocytes.

MSCs play a role in wound healing in both acute and chronic wounds through expression of several growth factors and cytokines that promote angiogenesis, cell proliferation, migration as well as ECM deposition and turnover, and through having anti-inflammatory, antiscarring, antimicrobial, immune-modulative, and immunosuppressive properties (Jackson et al., 2012). In 2007, Falanga and coworkers showed delivery of autologous MSCs to chronic wounds in humans using a fibrin spray system led to the stimulation of elastic fiber production (Falanga et al., 2007). The topical application of autologous GFP-positive MSCs to mouse wounds led to GFP-positive blood vessels, suggesting that the applied cells may persist as well as stimulate the wound repair process (Falanga et al., 2007).

Transdifferentiation of MSCs into other cell lineages has been reported. Some of these include epidermal keratinocytes, vascular endothelial cells, hepatocytes, neural cells, fibroblasts, hair follicle, sweat gland, and sebaceous gland appendage cells. In an in vivo wound healing study, Sasaki et al. (2008) reported that intravenously injecting MSCs derived from GFP transgenic mice resulted in GFP-positive cells that were found at wound sites that expressed markers for keratinocytes, endothelial cells, and pericytes in a murine model and that this occurrence was not as a result of spontaneous fusion as determined by fluorescence in situ hybridization techniques. Even though there is a growing body of evidence for MSC transdifferentiation, more experimental studies are necessary to completely rule out the likelihood of cell fusion. Nevertheless, the properties offered by MSCs make them a potential candidate for allogeneic transplantation and cell-based bioengineered products to repair or replace damaged tissue.

15.10.5 Adipose Stem Cells

Adipose-derived stem cells (ASCs) have gained a lot of attention due to their abundant nature, ease of availability, and rapid expansion in vitro. These cells may be harvested from liposuction aspirates and discarded subcutaneous tissue and when grown in culture are easily expanded and retain their multipotency. ASCs have the ability to differentiate into mesenchymal and nonmesenchymal lineages in vitro. These multipotent cells secrete cytokines that are antioxidative, angiogenic, and immunosuppressive. They also lack human leukocyte antigen (HLA) expression which reduces their immunogenicity and renders them potentially useful for allogeneic applications.

ASCs seeded onto AlloDerm and a fibrin-chitosan scaffold was shown to significantly improve wound healing in mice by releasing angiogenic factors that contributed to the development of a vascular network. These cells also spontaneously differentiated into fibroblastic and epidermal epithelial lineages illustrating their plasticity in cutaneous wound healing (Altman et al., 2008). Autologous ASCs injected intradermally to the subcutaneous level in human nasal tip wounds led to the healing of these difficult-to-treat wounds without the requirement for skin grafting or flap reconstruction (Sung et al., 2012). The successful use of these cells for the generation of a vascularized trilayered skin substitute consisting of epidermis, dermis, and adipocyte-containing hypodermis has also been described (Chan et al., 2012). Other authors have demonstrated that ASCs can contribute to wound healing by promoting hair growth and releasing paracrine factors that stimulate both fibroblast and keratinocyte proliferation and by increasing collagen type I synthesis. The phenotypic plasticity of these cells and the role they play in enhancing wound healing make them a good source of stem cells to be used in skin tissue engineering.

15.10.6 Summary of Use of Adult Stem Cells in Tissue-Engineered Skin

If an adequate population of stem cells has been transfected and carried over through in vitro cultivation then longevity should be assured (Itoh et al., 2011). If the correct signals can be identified and the right environments provided, then it is likely that we could also progress from just stem cells producing regeneration of skin to complex skin constructs containing all skin appendages. The versatility and plasticity of pluripotent stem cells place them at the advancing front of stem cell-based therapies with the potential to generate all components of the skin and replace all functions of the skin when grafted in vivo.

However, there are notes of caution ringing in the euphoria surrounding "stemness." How many steps are there between a pluripotential, slow cycling cell capable of ordered multiple divisions and the unregulated propagation of malignancy? It is felt, for example, that follicular and epidermal stem cells are pivotal in the pathogenesis of skin tumors, being targeted by chemical and viral attack. As with all treatments, potential or otherwise, we must first strive to do no harm.

15.11 THE FUTURE OF TISSUE-ENGINEERED SKIN

Currently, from a small skin biopsy, we can isolate and expand enough keratinocytes in culture to cover the entire body surface area of an adult in 2–3 weeks. Unfortunately, this lengthy period is at odds with the current management of acute burns and is affected by the age of the patient. Moreover, the resultant sheets are difficult to handle, have variable take rates, and are susceptible to shear forces due to the unsophisticated DEJ. Depletion of slow-cycling stem cells in current culture systems can also contribute to late failure of grafts after apparently successful grafting. These shortcomings led to the development of keratinocyte delivery methods in which these cells are delivered on a variety of backing substrates as confluent or preconfluent cells. However, no one ideal delivery method has emerged as the technique of choice as evidenced by the number of different systems described in the literature.

Though epidermal cover will save a life, a substantial dermis is necessary for quality of life, which is lacking if cultured autologous keratinocytes only are used on full-thickness wounds. This is an effective technique for large burns victims and the healed skin is tough and supple. Unfortunately, a delay of 3–4 weeks is required while the neodermis develops. If this process were more prompt, dermal formation would enable quicker grafting and definitive cover.

The classical method of keratinocyte culture, at least in Europe, will need shifting to a more xenobiotic-free culture system to remove the use of mouse 3T3 feeder layer, bovine serum, and other xenobiotic culture products. Even though this culture system is likely to be more costly, it may help reduce public health concerns about prion diseases like bovine spongioform encephalopathy and Creutzfeldt–Jakob. Future research should focus on how to minimize theoretical risk of contamination and disease transmission. Nevertheless, as infection still remains a problem to the successful application of tissue-engineered skin, biomaterials with smart ligands (e.g., defensins) could be constructed and investigated for cell delivery.

It is known that allogeneic keratinocytes survive for a number of weeks in immunocompetent subjects. Genetic modification of HLA antigens could possibly extend the lengthy period of tolerance that these cells experience. Then we may be able to approach the ideal of a living skin substitute consisting of "immunologically inert" keratinocytes and fibroblasts, organized around a dermal template, ideally with adnexal structures acting as reservoirs of stem cells to ensure graft longevity and also a more sophisticated construct.

Despite the significant advances made in the field of tissue-engineered skin, a better understanding of the cellular and molecular mechanisms of other skin cell types and their relation to the keratinocyte (epidermal–mesenchymal interaction) is required to achieve the ideal skin substitute. In addition, adult stem cells now offer exciting opportunities as cell-based bioengineered products. Similarly, the generation of iPS cells offers new experimental tools for developing patient-specific cells for cell replacement therapy. How these stem cells can be used to improve skin constructs will warrant further testing.

15.12 SUMMARY

- The structure of skin is very complex consisting of an epithelial and mesenchymal component. True tissue-engineered skin will need both these components and all the cell types.

- Great advances in skin substitutes have occurred over the last 30 years. Skin is the first organ to be tissue-engineered. This began with the ability to culture and expand keratinocytes in 1975.
- It is now generally accepted that allogeneic keratinocytes do not survive transplantation and autologous cells are required for permanent closure. There is a role for allogeneic cells as a biological dressing.
- A dermal component is required for optimal take and improved cosmesis. An understanding of epithelial–mesenchymal signaling developed over the years will allow for improvements in skin immunology, graft quality, and "take."
- Tissue-engineered skin products are currently available as the epithelial component alone, as a template for dermal regeneration (with or without mesenchymal cells), or dermal component, or both epidermal and dermal components combined together.
- Synthetic dermal replacements usually take the form of part of biological dressings.
- The true future of epidermal and dermal replacements lies in the advances of stem cell technology, identification of embryonic signaling pathways, and taking a holistic approach to tissue-engineered skin based on the structure and function of skin.

CLASSICAL EXPERIMENTS: KERATINOCYTE STEM CELL THERAPY

Rheinwald and Green (1977) performed the classic experiments that founded keratinocyte stem cell therapy as we know it today. The requirement for calcium, trace elements, and hormones such as transferrin, insulin, and hydrocortisone was established. Medium which was specific to keratinocytes and did not require feeders for cell proliferation was developed and produced a single-layered nondifferentiated keratinocyte graft (Tsao et al., 1982). The addition of fetal bovine serum to the cells allowed the cells to stratify rather than to remain as a single-layered sheet. These approaches made keratinocyte grafts more clinically applicable. Another important step was the demonstration that a dermis or a dermal equivalent is required for optimal clinical graft take. The grafting of keratinocyte sheets alone led to poor take rates and graft failure. The significance of dermis was demonstrated by Kangesu et al. (1993) in the porcine model and by Cuono et al. (1987) in humans, although the Cuono technique is challenging clinically. This initiated the development of artificial dermal substitutes and dermal templates. Landmark observations in the field have included: the realization that allogeneic keratinocytes do not survive transplantation, the role of preconfluent delivery systems for autologous keratinocyte grafting to avoid ex vivo delays, the potential for enrichment with stem cells for permanent graft take, the potential for micrografting of hair follicles into dermal regeneration templates when practical, and the potential developments from a new understanding of the signaling pathways involved in epithelial regeneration.

STATE-OF-THE-ART EXPERIMENT: TISSUE-ENGINEERED SKIN

Recessive dystrophic epidermolysis bullosa (RDEB) is a genetic blistering disorder caused by mutations in the COL7A1 gene that encodes collagen type VII, an important component of the anchoring fibrils that constitutes the BMZ. Although no definite treatment exists for RDEB, progress in developing cell- and gene-based therapies has been made through understanding of the molecular basis of RDEB. In a recent study, induced pluripotent stem (iPS) cells, which are generated from adult somatic cells by exogenous expression of defined transcription factors that exhibit similar characteristics to embryonic stem cells (ESCs), were generated from fibroblasts isolated from patients with RDEB (Itoh et al., 2011). The authors reported differentiation of these cells into functional keratinocytes which were used to generate 3D skin equivalents. The same authors were also able to differentiate iPS cells into dermal fibroblasts that could synthesize collagen type VII and reconstruct human skin structure with the use of normal human keratinocytes on the backs of mice (Itoh et al., 2013). Moreover, both iPS-derived fibroblasts and keratinocytes could be used to generate 3D skin equivalents. These studies highlight the potential of iPS cells in the development of gene and stem cell therapy for skin disorders like RDEB, wound healing therapies, and provide useful in vitro models to study skin diseases.

RECOMMENDED READING

Barrandon, Y., Grasset, N., et al., 2012. Capturing epidermal stemness for regenerative medicine. Semin. Cell. Dev. Biol. 23, 937–944.

Dahl, M.V., 2012. Stem cells and the skin. J. Cosmet. Dermatol. 11, 297–306.

Galach, M., Utikal, J., 2011. From skin to the treatment of diseases – the possibilities of iPS cell research in dermatology. Exp. Dermatol. 20, 523–528.

Ghadially, R., 2012. 25 years of epidermal stem cell research. J. Invest. Dermatol. 132, 797–810.

Kamel, R.A., Ong, J.F., et al., 2013. Tissue engineering of skin. J. Am. Coll. Surg. 217, 533–555.

Lewis, C.J., 2013. Stem cell application in acute burn care and reconstruction. J. Wound Care 22, 7–8, 10, 12-6.

Lu, T., Li, Y., et al., 2013. Techniques for fabrication and construction of three-dimensional scaffolds for tissue engineering. Int. J. Nanomed. 8, 337–350.

Van Der Veen, V.C., Boekema, B.K., et al., 2011. New dermal substitutes. Wound Repair Regen. 19 (Suppl. 1), s59–65.

Wong, V.W., Gurtner, G.C., 2012. Tissue engineering for the management of chronic wounds: current concepts and future perspectives. Exp. Dermatol. 21, 729–734.

REFERENCES

Aasen, T., Raya, A., et al., 2008. Efficient and rapid generation of induced pluripotent stem cells from human keratinocytes. Nat. Biotechnol. 26, 1276–1284.

Aasen, T., Izpisua Belmonte, J.C., 2010. Isolation and cultivation of human keratinocytes from skin or plucked hair for the generation of induced pluripotent stem cells. Nat. Protoc. 5, 371–382.

Altman, A.M., Matthias, N., et al., 2008. Dermal matrix as a carrier for in vivo delivery of human adipose-derived stem cells. Biomaterials 29, 1431–1442.

Amoh, Y., Li, L., et al., 2005. Multipotent nestin-positive, keratin-negative hair-follicle bulge stem cells can form neurons. Proc. Natl. Acad. Sci. U.S.A. 102, 5530–5534.

Anthony, E.T., Syed, M., et al., 2006. The development of novel dermal matrices for cutaneous wound repair. Drug Discov. Today 3, 81–86.

Atiyeh, B.S., Costagliola, M., 2007. Cultured epithelial autograft (CEA) in burn treatment: three decades later. Burns 33, 405–413.

Aubock, J., Irschick, E., et al., 1988. Rejection, after a slightly prolonged survival time, of Langerhans cell-free allogeneic cultured epidermis used for wound coverage in humans. Transplantation 45, 730–737.

Bell, E., Ivarsson, B., et al., 1979. Production of a tissue-like structure by contraction of collagen lattices by human fibroblasts of different proliferative potential in vitro. Proc. Natl. Acad. Sci. U.S.A. 76, 1274–1278.

Biernaskie, J., Paris, M., et al., 2009. SKPs derive from hair follicle precursors and exhibit properties of adult dermal stem cells. Cell. Stem Cell. 5, 610–623.

Boyce, S.T., Goretsky, M.J., et al., 1995. Comparative assessment of cultured skin substitutes and native skin autograft for treatment of full-thickness burns. Ann. Surg. 222, 743–752.

Boyce, S.T., Kagan, R.J., et al., 2006. Cultured skin substitutes reduce requirements for harvesting of skin autograft for closure of excised, full-thickness burns. J. Trauma 60, 821–829.

Brem, H., Balledux, J., et al., 2000. Healing of diabetic foot ulcers and pressure ulcers with human skin equivalent: a new paradigm in wound healing. Arch. Surg. 135, 627–634.

Burke, J.F., Quinby Jr., W.C., et al., 1981a. 1981 – approach to burn therapy. Surg. Annu. 13, 1–14.

Burke, J.F., Yannas, I.V., et al., 1981b. Successful use of a physiologically acceptable artificial skin in the treatment of extensive burn injury. Ann. Surg. 194, 413–428.

Chan, R.K., Zamora, D.O., et al., 2012. Development of a vascularized skin construct using adipose-derived stem cells from debrided burned skin. Stem Cells Int. 2012, 841203.

Cleland, H., Wasiak, J., et al., 2014. Clinical application and viability of cryopreserved cadaveric skin allografts in severe burn: a retrospective analysis. Burns 40, 61–66.

Cuono, C.B., Langdon, R., et al., 1987. Composite autologous-allogeneic skin replacement: development and clinical application. Plast. Reconstr. Surg. 80, 626–637.

Eaglstein, W.H., Iriondo, M., et al., 1995. A composite skin substitute (graftskin) for surgical wounds. A clinical experience. Dermatol. Surg. 21, 839–843.

Falanga, V., Iwamoto, S., et al., 2007. Autologous bone marrow-derived cultured mesenchymal stem cells delivered in a fibrin spray accelerate healing in murine and human cutaneous wounds. Tissue Eng. 13, 1299–1312.

Fang, T., Lineaweaver, W.C., Sailes, F.C., Kisner, C., Zhang, F., 2013 July 11. Clinical application of cultured epithelial autografts on acellular dermal matrices in the treatment of extended burn injuries. Ann. Plast. Surg [Epub ahead of print].

Farroha, A., Frew, Q., et al., 2013. The use of Biobrane® to dress split-thickness skin graft in paediatric burns. Ann. Burns Fire Disasters 26, 94–97.

Frew, Q., Philp, B., et al., 2013. The use of Biobrane® as a delivery method for cultured epithelial autograft in burn patients. Burns. doi: 10.1016/j.burns.2012.11.013. [Epub ahead of print].

Greaves, N.S., Iqbal, S.A., et al., 2013. The role of skin substitutes in the management of chronic cutaneous wounds. Wound Repair Regen. 21, 194–210.

Green, H., 2008. The birth of therapy with cultured cells. Bioessays 30, 897–903.

Goldstein, J., Horsley, V., 2012. Home sweet home: skin stem cell niches. Cell. Mol. Life Sci. 69, 2573–2582.

Granstein, R.D., Smith, L., et al., 1987. Prolongation of murine skin allograft survival by the systemic effects of 8-methoxypsoralen and long-wave ultraviolet radiation (PUVA). J. Invest. Dermatol. 88, 424–429.

Griffiths, M., Ojeh, N., et al., 2004. Survival of Apligraf in acute human wounds. Tissue Eng. 10, 1180–1195.

Hankin, C.S., Knispel, J., et al., 2012. Clinical and cost efficacy of advanced wound care matrices for venous ulcers. J. Manag. Care Pharm. 18, 375–384.

Hankins, C.L., Tang, X.Q., et al., 2006. Hot beverage burns: an 11-year experience of the Yorkshire Regional Burns Centre. Burns 32, 87–91.

Hansbrough, J.F., Boyce, S.T., et al., 1989. Burn wound closure with cultured autologous keratinocytes and fibroblasts attached to a collagen–glycosaminoglycan substrate. J. Am. Med. Ass. 262, 2125–2130.

Hansbrough, J.F., Dore, C., et al., 1992. Clinical trials of a living dermal tissue replacement placed beneath meshed, split-thickness skin grafts on excised burn wounds. J. Burn Care Rehabil. 13, 519–529.

Harding, K., Sumner, M., et al., 2013. A prospective, multicentre, randomised controlled study of human fibroblast-derived dermal substitute (dermagraft) in patients with venous leg ulcers. Int. Wound J. 10, 132–137.

Hasselgren, P.O., 1999. Burns and metabolism. J. Am. Coll. Surg. 188, 98–103.

Itoh, M., Kiuru, M., et al., 2011. Generation of keratinocytes from normal and recessive dystrophic epidermolysis bullosa-induced pluripotent stem cells. Proc. Natl. Acad. Sci. U.S.A. 108, 8797–8802.

Itoh, M., Umegaki-Arao, N., et al., 2013. Generation of 3D skin equivalents fully reconstituted from human induced pluripotent stem cells (iPSCs). PLoS ONE 8, e77673.

Jackson, D.M., 1953. The diagnosis of the depth of burning. Br. J. Surg. 40, 588–596.

Jackson, W.M., Nesti, L.J., et al., 2012. Concise review: clinical translation of wound healing therapies based on mesenchymal stem cells. Stem Cells Transl. Med. 1, 44–50.

Jansen, L.A., De Caigny, P., et al., 2013. The evidence base for the acellular dermal matrix AlloDerm: a systematic review. Ann. Plast. Surg. 70, 587–594.

Kahn, S.A., Beers, R.J., et al., 2011. Use of acellular dermal replacement in reconstruction of nonhealing lower extremity wounds. J. Burn Care Res. 32, 124–128.

Kangesu, T., Navsaria, H.A., et al., 1993. Kerato-dermal grafts: the importance of dermis for the *in vivo* growth of cultured keratinocytes. Br. J. Plast. Surg. 46, 401–409.

Kirsner, R.S., 1998. The use of Apligraf in acute wounds. J. Dermatol. 25, 805–811.

Landeen, L.K., Zeigler, F.C., et al., 1992. Characterization of a human dermal replacement. Wounds 4, 167–175.

Lazic, T., Falanga, V., 2011. Bioengineered skin constructs and their use in wound healing. Plast. Reconstr. Surg. 127 (Suppl. 1), 75S–90S.

Li, X., Upadhyay, A.K., et al., 2013. Skin stem cell hypotheses and long term clone survival – explored using agent-based modelling. Sci. Rep. 3, 1904.

Lootens, L., Brusselaers, N., et al., 2013. Keratinocytes in the treatment of severe burn injury: an update. Int. Wound J. 10, 6–12.

Melo, S.P., Lisowski, L., et al., 2014. Somatic correction of junctional epidermolysis bullosa by a highly recombinogenic AAV variant. Mol. Ther. 22, 725–733.

Myers, S.R., Partha, V.N., et al., 2007. Hyalomatrix: a temporary epidermal barrier, hyaluronan delivery, and neodermis induction system for keratinocyte stem cell therapy. Tissue Eng. 13, 2733–2741.

Naughton, G., Mansbridge, J., et al., 1997. A metabolically active human dermal replacement for the treatment of diabetic foot ulcers. Artif. Organs 21, 1203–1210.

O'Connor, N., Mulliken, J., et al., 1981. Grafting of burns with cultured epithelium prepared from autologous epidermal cells. Lancet 317, 75–78.

Ossevoort, M.A., Lorre, K., et al., 1999. Prolonged skin graft survival by administration of anti-CD80 monoclonal antibody with cyclosporin A. J. Immunother. 22, 381–389.

Pandya, A.N., Woodward, B., et al., 1998. The use of cultured autologous keratinocytes with integra in the resurfacing of acute burns. Plast. Reconstr. Surg. 102, 825–858, discussion 829–930.

Pellegrini, G., Dellambra, E., et al., 2001. p63 identifies keratinocyte stem cells. Proc. Natl. Acad. Sci. U.S.A. 98, 3156–3161.

Pruniéras, M., Mathivon, M.F., et al., 1965. Euploid culture of adult epidermal cells in monocellular layers. Ann. Inst. Pasteur (Paris) 108, 149–165.

Pruniéras, M., Régnier, M., et al., 1983. Nouveau procédé de culture des cellules épidermiques humaines sur derme homologue ou hétérologue: préparation de greffons recombinés. Ann. Chir. Plast. (Suppl.), 28s–33s.

Rheinwald, J.G., Green, H., 1975. Formation of a keratinizing epithelium in culture by a cloned cell line derived from a teratoma. Cell 6, 317–330.

Rheinwald, J.G., Green, H., 1977. Epidermal growth factor and the multiplication of cultured human epidermal keratinocytes. Nature 265, 421–424.

Richardson, G.D., Arnott, E.C., et al., 2005. Plasticity of rodent and human hair follicle dermal cells: implications for cell therapy and tissue engineering. J. Investig. Dermatol. Symp. Proc. 10, 180–183.

Rigal, C., Pieraggi, M.-T., et al., 1991. Healing of full-thickness cutaneous wounds in the pig. *I.* Immunohistochemical study of epidermo-dermal junction regeneration. J. Invest. Dermatol 96, 777–785.

Roberts, G., Lloyd, M., et al., 2012. The Baux score is dead. Long live the Baux score: a 27-year retrospective cohort study of mortality at a regional burns service. J. Trauma Acute Care Surg. 72, 251–256.

Ryssel, H., Gazyakan, E., et al., 2008. The use of MatriDerm in early excision and simultaneous autologous skin grafting in burns – a pilot study. Burns 34, 93–97.

Sasaki, M., Abe, R., et al., 2008. Mesenchymal stem cells are recruited into wounded skin and contribute to wound repair by transdifferentiation into multiple skin cell type. J. Immunol. 180, 2581–2587.

Shevchenko, R.V., James, S.L., et al., 2010. A review of tissue-engineered skin bioconstructs available for skin reconstruction. J. R. Soc. Interface 7, 229–258.

Sood, R., Balledux, J., et al., 2009. Coverage of large pediatric wounds with cultured epithelial autografts in congenital nevi and burns: results and technique. J. Burn Care Res. 30, 576–586.

Sood, R., Roggy, D., et al., 2010. Cultured epithelial autografts for coverage of large burn wounds in eighty-eight patients: the Indiana University experience. J. Burn Care Res. 31, 559–568.

Sriwiriyanont, P., Lynch, K.A., et al., 2013. Characterization of hair follicle development in engineered skin substitutes. PLoS ONE 8, e65664.

Staiano-Coico, L., Hefton, J.M., et al., 1990. Growth of melanocytes in human epidermal cell cultures. J. Trauma 30, 1037–1042, discussion 1043.

Sung, H.M., Suh, I.S., et al., 2012. Case reports of adipose-derived stem cell therapy for nasal skin necrosis after filler injection. Arch. Plast. Surg. 39, 51–54.

Suzuki, S., Kawai, K., et al., 2000. Long-term follow-up study of artificial dermis composed of outer silicone layer and inner collagen sponge. Br. J. Plast. Surg. 53, 659–666.

Takahashi, K., Tanabe, K., et al., 2007. Induction of pluripotent stem cells from adult human fibroblasts by defined factors. Cell 131, 861–872.

Tatnall, F.M., Leigh, I.M., et al., 1990. Comparative study of antiseptic toxicity on basal keratinocytes, transformed human keratinocytes and fibroblasts. Skin. Pharmacol. 3, 157–163.

Tsao, M.C., Walthall, B.J., et al., 1982. Clonal growth of normal human epidermal keratinocytes in a defined medium. J. Cell. Physiol. 110, 219–229.

Voigt, J., Driver, V.R., 2012. Hyaluronic acid derivatives and their healing effect on burns, epithelial surgical wounds, and chronic wounds: a systematic review and meta-analysis of randomized controlled trials. Wound Repair Regen. 20, 317–331.

Wallace, A.F., 1982. The Progress of Plastic Surgery. Oxford University Press, Oxford.

Waymack, P., Duff, R.G., et al., 2000. The effect of a tissue engineered bilayered living skin analog, over meshed split-thickness autografts on the healing of excised burn wounds. The Apligraf Burn Study Group. Burns 26, 609–619.

Wendt, J.R., Ulich, T.R., et al., 1994. Indefinite survival of human skin allografts in patients with long-term immunosuppression. Ann. Plast. Surg. 32, 411–417.

Wisser, D., Steffes, J., 2003. Skin replacement with a collagen based dermal substitute, autologous keratinocytes and fibroblasts in burn trauma. Burns 29, 375–380.

Wood, F., 2012. Tissue engineering of skin. Clin. Plast. Surg. 39, 21–32.

Yannas, I.V., Burke, J.F., 1980. Design of an artificial skin. I. Basic design principles. J. Biomed. Mater. Res. 14, 65–81.

CHAPTER 16

Cartilage and Bone Regeneration

Anders Lindahl[1], Mats Brittberg[2], David Gibbs[3], Jonathan I. Dawson[3], Janos Kanczler[3], Cameron Black[3], Rahul Tare[3], Richard OC. Oreffo[3]

[1]Department of Clinical Chemistry and Transfusion Medicine Institute of Biomedicine, The Sahlgrenska Academy University of Gothenburg, Sweden; [2]Department of Orthopaedics, Institute of Clinical Sciences, The Sahlgrenska Academy University of Gothenburg, Sweden; [3]Bone and Joint Research Group, Centre for Human Development, Stem Cells and Regeneration, Institute of Developmental Sciences, University of Southampton, Southampton, United Kingdom

LEARNING OBJECTIVES

- To determine the location and differences between hyaline, fibrous, and elastic cartilage
- To recognize the developmental aspects of bone and cartilage including joint development
- To understand whether cartilage stem cells and stem cell niches exist and to realize the relative importance of the cell layers in hyaline cartilage
- To understand the differences in the repair capacity of articular cartilage between fetal and adult cartilage and to acknowledge that there is an inherent difference in the repair capacity among individuals
- To understand the rational for cell transplantation and the advantages and limitations of other cell sources for cartilage repair
- To recognize the use of scaffolds in cartilage repair and the use of bioreactors in the development of cartilage implants
- To understand the different types of bones and the functions of bone
- To indicate the role and function, limitations, and applications of skeletal stem cells
- To indicate the different biomaterials available and their use and limitations in skeletal regeneration
- To detail the role of growth factors in bone repair

16.1 INTRODUCTION: CARTILAGE

Cartilage tissue is derived from the mesoderm and gives rise to three different types of cartilages where the properties are dependent on the cells (the chondrocytes) and the extracellular matrix (ECM) produced by the

Tissue Engineering.
Copyright © 2015 Elsevier Inc. All rights reserved.

chondrocytes: (1) hyaline cartilage is found in joints, rib cartilage, nose, trachea, and larynx; (2) elastic cartilage is found in the ear, epiglottis, and larynx; and (3) fibrous cartilage is found in intervertebral discs. The long bones initially consist of hyaline cartilage, and ossification gradually takes place in ossification centers where cartilage is gradually replaced by bone (Figure 16.1) as presented later in this chapter.

After bone formation, hyaline cartilage remains at the end of long bones and in the epiphyseal plate cartilage in the ends of long bones. During longitudinal bone growth, the hyaline cartilage of the epiphyseal growth plate constantly adds new cartilage that is gradually replaced with bone and thus length is added until the end of adolescence in humans. In most primates, the epiphyseal growth cartilage is ossified in adulthood.

The cartilage of mesoderm origin is not the only cartilage formed in the body. The formation of the craniofacial skeletal elements and the jaw in particular was an important step in the evolution of higher vertebrates. Most facial bones and cartilage are generated during embryonic development by cranial neural crest cells (NCCs) that are a transient, multipotent population of cells that subsequently delaminate from the dorsal tip of the neural tube. The cells migrate along the long axis of the embryo and give rise to distinct populations. One of

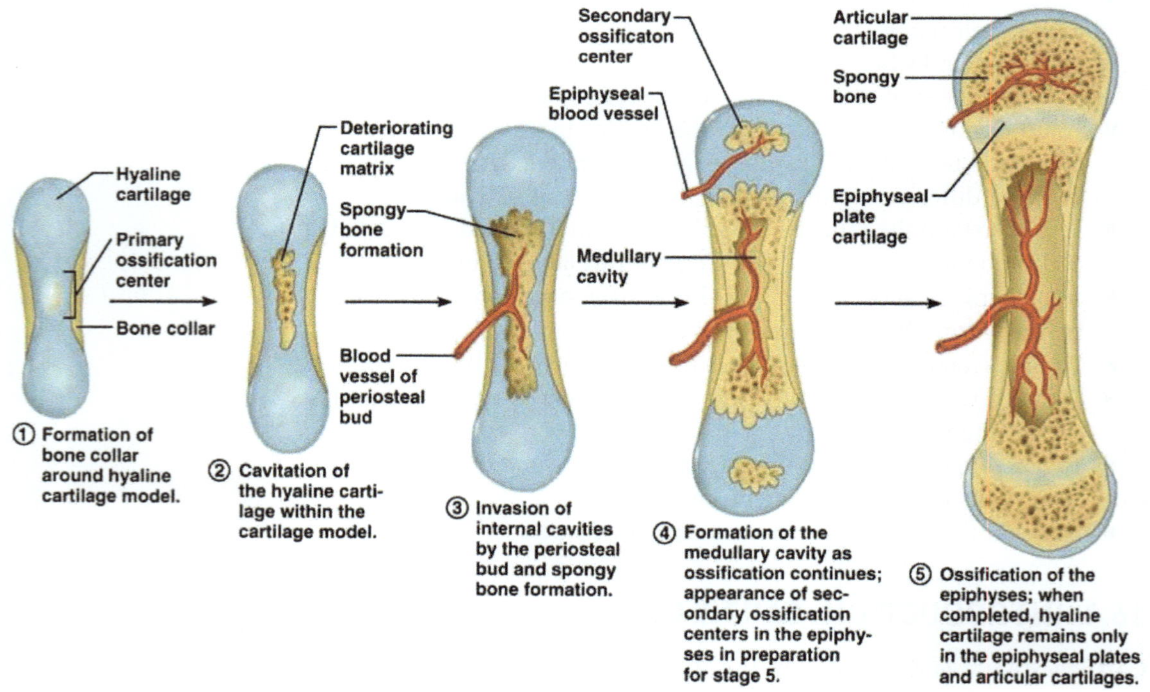

FIGURE 16.1
Cartilage and bone formation.

these cell types are the cranial NCCs that migrate to the level of the first somite and give rise to cartilage and connective tissues of the craniofacial region (Le Douarin and Dupin, 2012).

16.2 CELLULAR STRUCTURES AND MATRIX COMPOSITION OF HYALINE CARTILAGE

Embryonal cartilage gradually matures to meet the needs and demands of the adult organism. The initial high cell-to-matrix ratio (30% cells) in the embryo is reduced to approximately 2% cell content in the adult cartilage. The ECM consists of water (65–80%), and the rest is solid material (20–35%). Of the solid part, about 60% is collagen, 30% proteoglycans, and 10% comprises other proteins. The mechanical function of proteoglycans is to act as shock absorbers, whereas collagens provide resistance to shear stress. The exact composition of the cartilage matrix and the biomechanical properties vary between different locations and ages. The average thickness in loaded areas of the human hyaline cartilage is 2.4 mm.

In young cartilage, only two zones can be identified in articular cartilage: the superficial and growth zones in contrast to mature cartilage where four structurally different zones with different functions and protein composition can be identified: the superficial, transitional, radial, and calcified zones (Figure 16.2)

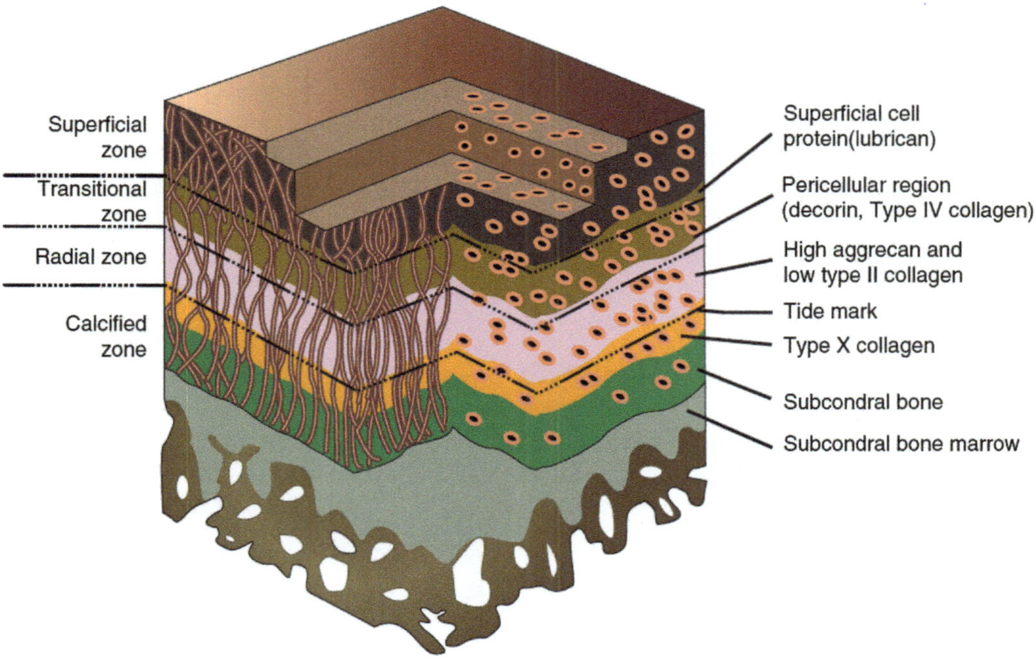

FIGURE 16.2
Structural and functional zones of articular cartilage. *Adapted from Poole et al. (2001).*

(Poole et al., 2001). The superficial zone consists of elongated and flat cells embedded mainly in type I collagen, the zone lines the synovial joint, and its major component lubricin or superficial zonal protein functions as a lubricant and is normally only found in this area of the articular cartilage.

In the transitional zone, the cells are more rounded and pericellularly type IV collagen can be found. The major proteoglycan in this area is decorin. The collagen fibers in these first two regions are horizontally oriented, while in the radial zone, the fibers are vertically oriented. In the radial zone, chondrocytes are more sparsely distributed, and the major matrix proteins are the typical cartilage type II collagen and the proteoglycan, aggrecan. Next to the subchondral bone, the cartilage becomes mineralized, and type X collagen is found in this area.

16.3 COLLAGEN

The ECM in most tissues contains a number of related molecules that interact to form a structured network. The major protein cartilage is the fibril forming type II collagen that gives structure and tensile strength to the cartilage. Collagen is defined as a structural protein of the ECM, and it contains one or more domains having the conformation of a triple helix (van der Rest and Garrone, 1991). Collagen fibrils are composed of aggregates of five tropocollagens (three identical alpha II polypeptide chains assembled in a superhelix) that have a diameter of 20 nm. This unique structure of the fibrils makes collagen insoluble and resistant to attack by degrading enzymes.

In the radial zone, the fibers are vertically oriented in a hexameric structure surrounding the chondrocytes; this thus explains the columnar appearance of chondrocytes typically seen in this zone. The predominant type of collagen is type II collagen, but small amounts of types I, V, VI, IX, X, and XI are also present. Collagens participating in quarter-staggered fibrils are of types II and XI; collagen-forming sheets have type X; collagen-beaded filaments have type VI; and fibrin-associated collagens are of type IX. The principal differences between collagens are outlined in Figure 16.3.

The turnover of collagens in adult normal cartilage is considered to be very slow, for example, the turnover of the whole collagenous component of human femoral head cartilage matrix is estimated to be 400 years. However, in certain disease states, the synthesis can be increased; this indicates an imbalance in tissue homeostasis.

16.4 PROTEOGLYCANS

Two different types of proteoglycans are present in the articular cartilage: nonaggregating and aggregating. The smaller nonaggregating proteoglycans

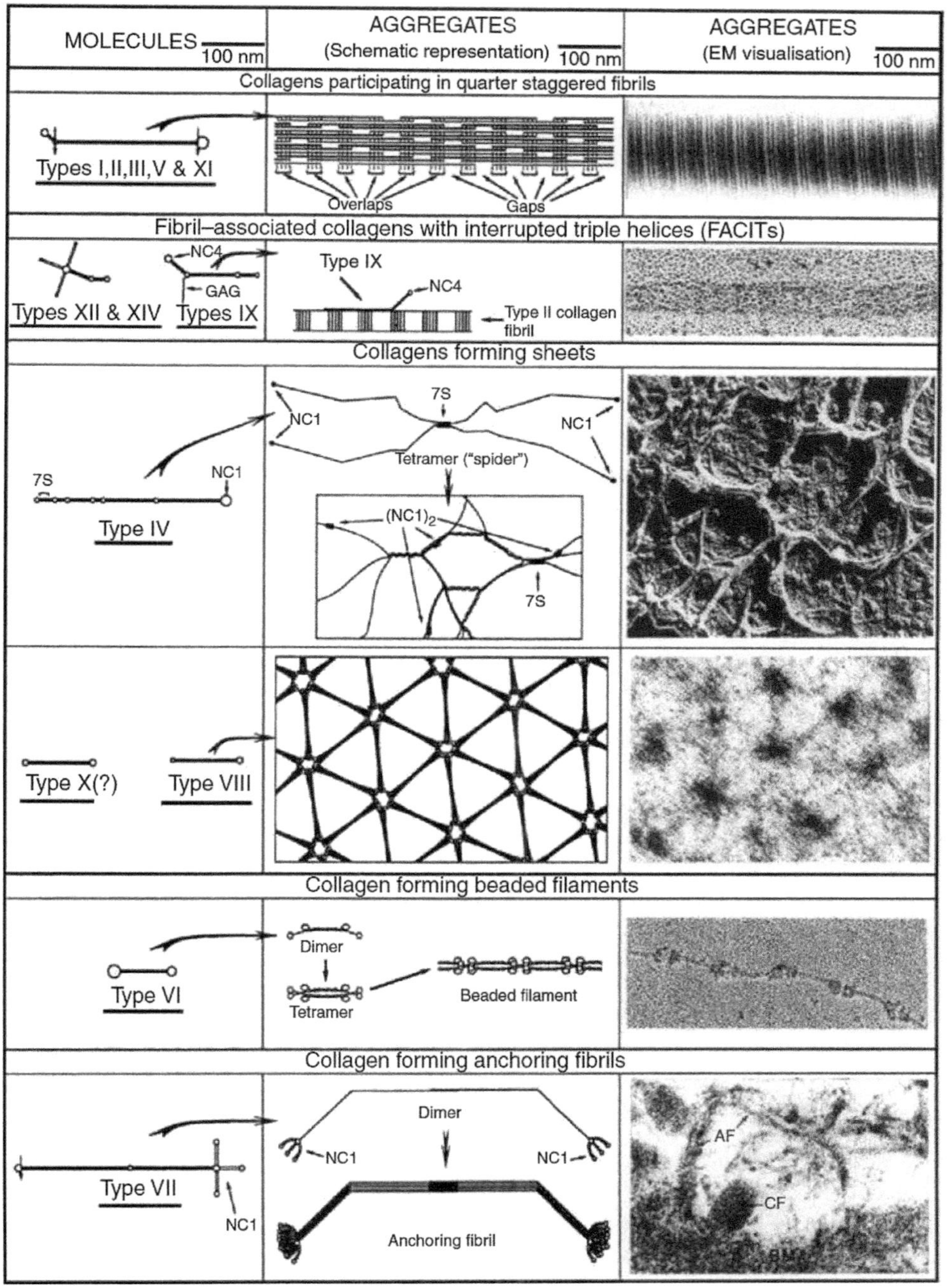

FIGURE 16.3
Molecular structure and supramolecular assemblies of collagens. *From van der Rest and Garrone (1991).*

consist of a core protein to which a large number of sulfated glycosaminoglycans (GAGs) are attached laterally. The GAGs normally found in articular cartilage are keratin sulfate, chondroitin sulfate, and dermatan sulfate.

Larger aggregated proteoglycans are formed by the binding of smaller nonaggregating proteoglycans to hyaluronic acid, which has binding sites for core proteins. The hyaluronic acid is further bound to chondrocytes through the surface receptor CD44 (Figure 16.4). The concentration of hyaluronic acid is higher during the embryonic formation of cartilage, and it has been shown that hyaluronic acids of certain molecular weights are chondroinductive.

Within proteoglycans, negatively charged GAGs are packed around the core protein at a distance of 20–50 nm. Water molecules are bound to GAGs, and when the cartilage is compressed, water is retained within the matrix, which thereby limits the forces of compression. In compression studies with cartilage, it has been shown that dynamic compression selectively stimulates the synthesis of GAGs, which thus indicates an adaptation capacity of the tissue to mechanical force. Decorin and biglycan both interact with collagens through the leucine-rich structures of their core proteins (Figure 16.4).

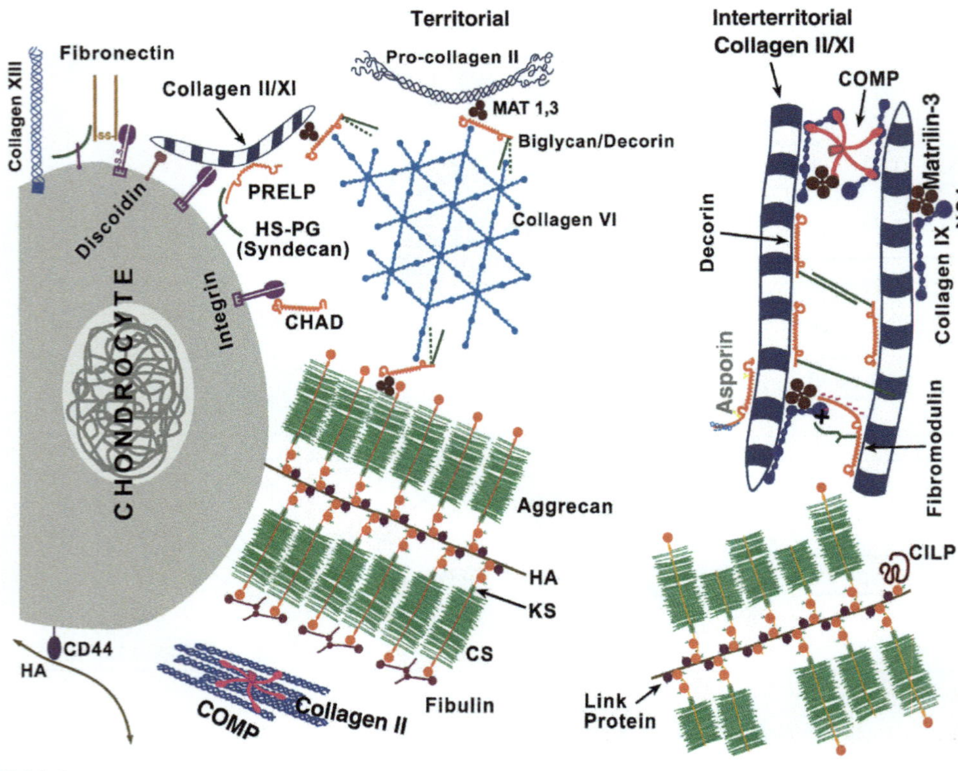

FIGURE 16.4

The molecular organization of normal articular cartilage. *From Heinegard and Saxnae (2011).*

One matrix protein, which was initially identified in association with aggrecan, is a cartilage-derived protein or Matrilin-1. The amount of this protein has been shown to change during the maturation of cartilage and is increased in osteoarthritis (OA). Other proteins involved in macrofibril organization and stabilization of type II collagen include the five-armed member of the thrombospondin family, cartilage oligomeric protein. This protein has been used as a diagnostic marker in serum for determining the progress of matrix degradation in OA.

16.5 THE CHONDROCYTE

The cells of the cartilage are chondrocytes. These cells are characterized by their special ability to produce matrix and their highly developed cytoskeleton of actin filaments. The actin filaments contribute to the biomechanical properties of the chondrocytes as well as connect chondrocytes to the articular matrix. The interactions with the ECM are critical for both the development and maintenance of the cartilage. The cartilage matrix that surrounds chondrocytes in healthy articular cartilage is arranged into three zones depending on the distance from the cell (Figure 16.3). The pericellular matrix lies immediately around cells, and in this microenvironment, molecules are located that interact with cell surface receptors, for example, hyaluronan binds the receptor CD44. Next to the pericellular matrix lies the territorial matrix, and further out is the interterritorial matrix. The types of collagens and the collagen-binding proteins that form the matrices are different in each zone. The chondron is defined as the functional unit of cartilage and consists of several chondrocytes and the immediate microenvironment of the surrounding pericellular matrix. The chondron also functions in insulating and physically separating chondrocytes from direct interaction with the bulk of the load-bearing matrix. The number of chondrocytes per chondron varies from one to eight, with the lowest number being present in the superficial zone and the highest existing in the radial zone (Hunziker et al., 2002). The distribution of chondrocytes varies in different zones from approximately 7000 to 24,000 cells/mm^3. From these figures, the number of chondrocytes beneath a certain area of an articular joint surface can be calculated to be approximately 24,000 cells/mm^2.

16.6 STEM CELLS IN CARTILAGE AND PROLIFERATION OF CHONDROCYTES

Although one of the characteristics of articular cartilage is low cell turnover, there is evidence in the literature to show that proliferation may occur in healthy cartilage, OA cartilage, or in experimentally induced trauma. Generally, proliferating cells have been found either in superficial layers or in deep calcified layers. Due to the fact that proliferative cells are normally associated with some primitive features and stem cell properties, the findings of proliferative cells in these layers also raise the question as to the area from which the cartilage grows, that is, intrinsic or appositional. Based on cell tracing studies in the developing joints of a marsupial (*Monodelphis domestica*), it was demonstrated that the progenitor cells of cartilage are located in the superficial layer, and one explanatory hypothesis is that articular cartilage grows appositionally (Hayes et al., 2001). This hypothesis is further supported by additional studies where cells of the superficial zone were shown to express the cell fate selector gene Notch-1, which is involved in the embryonic development of articular cartilage, especially in the condensation process (Watanabe et al., 2003).

The anatomical structure surrounding the growth plate cartilage consists of a perichondreal ring of connective tissue, and the zone of Ranvier was recently demonstrated to harbor a stem cell niche where cells were demonstrated to migrate into the joint cartilage, a similar structure was also found in the vertebras of rabbits. The results have future implications in understanding the regenerative potential in the joint (Karlsson et al., 2009).

16.7 PATHOPHYSIOLOGY OF CARTILAGE LESION DEVELOPMENT

Cartilage has a limited self-repair capacity, and as far back as the eighteenth century, Hunter stated "If we consult the standard Chirurgical Writers from Hippocrates down to the present Age, we shall find, that an ulcerated Cartilage is universally allowed to be a very troublesome disease ... when destroyed, it is never recovered." This statement is still valid 200 years later even if spontaneous repair of small defects has been reported.

The major explanation for this limited capacity of repair is that cartilage is a unique tissue with a total lack of innervation and vascular supply. These two components normally work together upon an injury to obtain an inflammatory response that attracts phagocytic and multipotent stem cells. Another component limiting the ability of repair is that chondrocytes, even if they could respond to an injury, are trapped within the tight cartilage matrix.

In contrast to the adult cartilage, fetal cartilage has the ability to repair isolated lesions and ulcers (Figure 16.5). This specific self-repairing ability could possibly be explained by the structural and compositional differences between adult and embryonic cartilage (Namba et al., 1998). Fetal cartilage contains higher amounts of hyaluronic acid and has a higher matrix-to-cell ratio. At the matrix level, the embryonic cartilage contains type I collagen and the GAG chondroitin-6 sulfate, while the adult cartilage mainly contains type II collagen and chondroitin-4 sulfate.

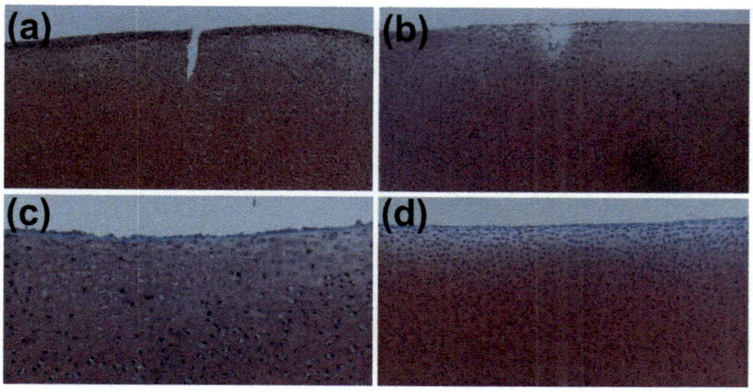

FIGURE 16.5

Spontaneous repair of superficial defects in articular cartilage in a fetal lamb model. (a) Acute injury consisting of a relatively parallel incisional defect that did not violate any blood vessels (×10). (b) Photomicrograph obtained 3 days after the injury, showing dead chondrocytes adjacent to the incisional defect and beneath sites where the footpad of the ophthalmic knife came into contact with the cartilage (×10). (c) Seven days after the injury, the defect is completely filled. The site of the defect and the areas that came into contact with the footpads were hypocellular and contained sparsely populated chondrocytes. Increased mitotic figures were visible adjacent to the zone of injury (×40). (d) Twenty-eight days after the injury, the cellularity, matrix staining, and architectural arrangement of the cells matched those of the sham controls (×10). *Reproduced from Namba et al. (1998).*

Acute or repetitive blunt joint trauma can damage articular cartilage and can result in the formation of isolated defects, and the remaining cells will be subjected to new mechanical forces that may result in cell death or apoptosis. The cellular processes ongoing on the rim of isolated defects have been compared to the process of the development of idiopathic OA. It is therefore thought that even if smaller defects may self-repair, larger isolated lesions caused by trauma that are left untreated may progress into OA. OA is one of the most common forms of musculoskeletal diseases that affects millions of people throughout the world. The clinical symptoms of OA are pain and functional impairment that induce joint stiffness and dysfunction with subsequent impaired performance in daily life. As many as 25% of patients cannot cope with daily activities, and this often results in depression and social isolation. In the European Union and the United States combined, over one million joint replacements are performed each year as a consequence of OA. The main risk factors for OA are age, obesity, and any form of joint trauma where the limited repair capacity of articular cartilage is a confounding factor. OA is often thought of as a disease of the elderly; however, OA is evident in younger patients, with approximately 5% of the population between 35 and 54 years having radiographic signs of OA. Inflammation is a very important factor in the early event of OA and is probably initiated by the traumatic injury to tissue (Heinegard and Saxne, 2011).

16.8 ARTIFICIAL INDUCTION OF CARTILAGE REPAIR

The limited repair capacity of cartilage has forced researchers and surgeons to find methods to induce cartilage repair. Several methods have been used where the general concept has been to use cells or tissues with chondrogenic potential to repair cartilage defects. The most frequently used methods are drilling of the subchondral bone to attract bone marrow cells, and resurfacing with periosteum and perichondrium. However, the repair tissue formed is usually of the fibrous cartilage type unable to withstand the joint load over time.

The use of in vitro expanded chondrocytes for transplantation has been used as an alternative method for cartilage repair, and the first cellular approach in humans for cartilage tissue engineering was initiated in the late 1980s. The treatment was named autologous chondrocyte transplantation (ACT) (Brittberg et al., 1994) and has since become a clinical cell-based therapy for the repair of cartilage lesions worldwide under the alternative name of autologous chondrocyte implantation (ACI). In this procedure, cartilage pieces are harvested arthroscopically from a healthy part of the joint. The cells are isolated by enzymatic digestion, subsequently culture expanded in vitro to 10–30 times the initial amount, and finally implanted into the defect site at a high density (Figure 16.6).

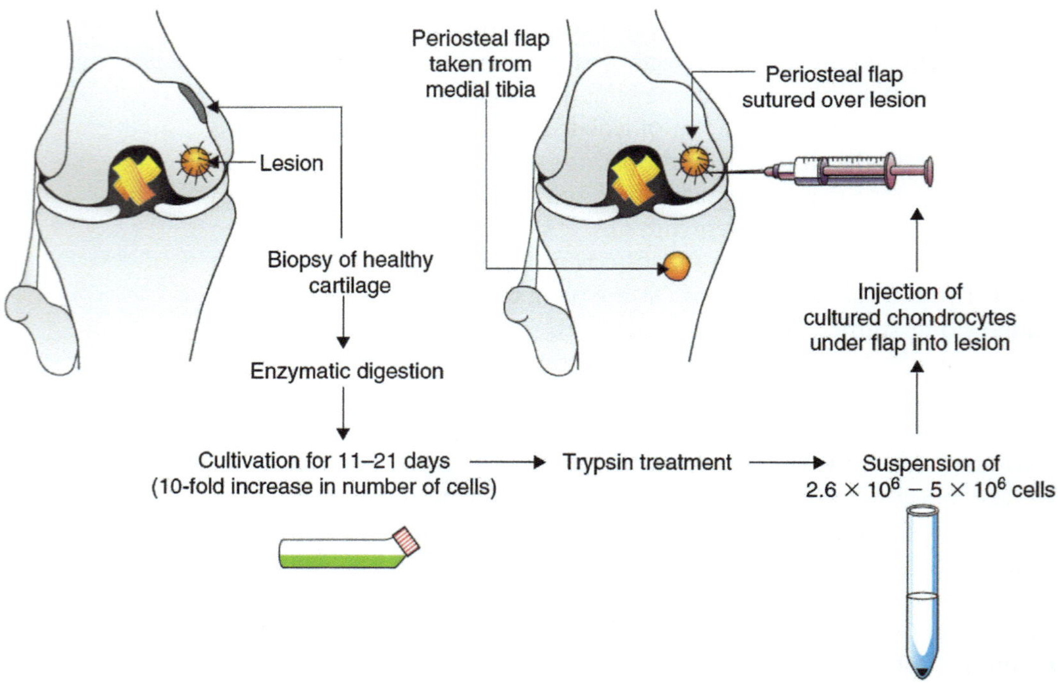

FIGURE 16.6
Autologous chondrocyte transplantation procedure.

16.9 RATIONALE FOR CELL IMPLANTATION

Articular chondrocytes are cells responsible for the unique features of articular cartilage; therefore, it would appear appropriate to use true committed chondrocytes to engineer in vitro or in vivo cartilage to repair a cartilaginous defect. Human articular chondrocytes derived from articular cartilage biopsies have a limited proliferative potential, and the number of cell divisions they undergo in vitro decreases with age. Chondrocytes grown as a monolayer dedifferentiate and lose both their chondrogenic phenotype with type II collagen production and their redifferentiation potential (Figure 16.7). The use of specific growth factors, such as basic fibroblast growth factor (FGF-2) and transforming growth factor (TGF)-β1, stimulates cell proliferation and maintains the cell's ability to redifferentiate upon transfer into a three-dimensional (3D) environment. Expansion of human articular chondrocytes to be used for cell therapy is presently performed in the presence of 10% fetal calf serum or in autologous serum (Brittberg et al., 1994), and their redifferentiation in vitro and in vivo remains a challenge. However, if dedifferentiated chondrocytes are cultured in 3D cultures in gels as well as in porous scaffolds in the presence of chondrogenic inducers (TGF-β1), the cells reexpress phenotypic markers of the articular chondrocyte.

In a mesenchymal tissue defect site, the initial number of multipotential cells that can take part in the repair process is critical (Caplan et al., 1997). The dedifferentiated chondrocytes have similarities to primitive mesenchymal cells, and a high-density cell implantation imitates the prechondrogenic cell condensation seen during limb formation and results in subsequent

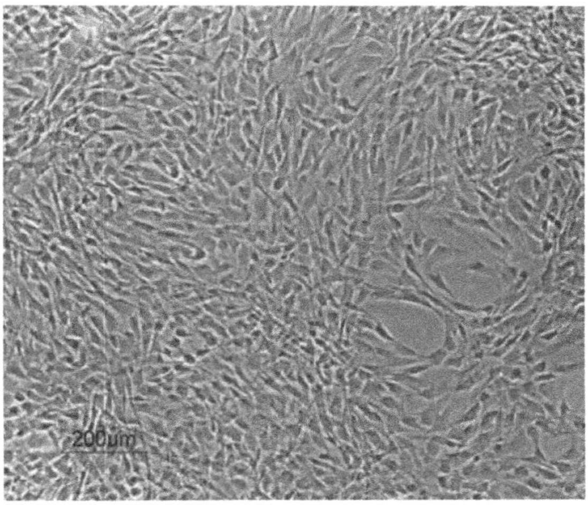

FIGURE 16.7

Human articular chondrocytes in monolayer cultures prior to implantation in the autologous chondrocyte transplantation procedure.

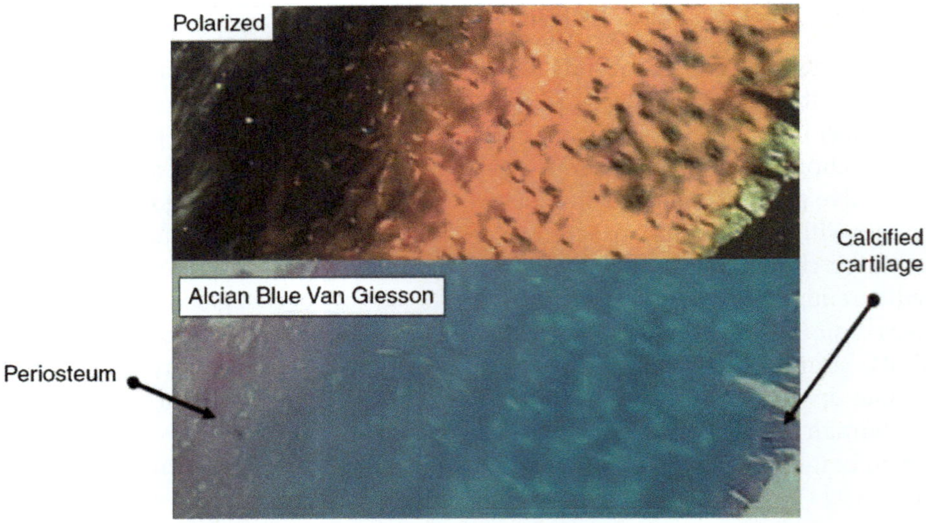

FIGURE 16.8
Regenerated cartilage from the autologous chondrocyte transplantation procedure 2 years after cell implantation.

cartilage formation (Peterson et al., 2002; Figure 16.8). Further, the clinical and functional long-term outcome of ACI remains high even after 10–20 years (Peterson et al., 2010).

16.10 CARTILAGE SPECIMENS FOR IMPLANTATION

Articular cartilage is composed of zones with different mechanical and functional properties; further, different locations in the joint have different cartilage properties, which raises the question of harvesting site. The superficial cells comprise approximately 4% of the total cells obtained from cartilage, and the zone has a higher production capacity of superficial zone proteins (SZPs) (Darling et al., 2004), while the growth zone produces more type II collagen; no difference in aggrecan production is seen between these zones. A surface abrasion technique was used to separate the superficial zone from the underlying growth zone in young articular cartilage. There was no regional variation seen between the femoral groove and medial and lateral condyles, which makes the pooling of zonal cells from different regions an acceptable option for tissue engineering studies. If chondrocytes from the superficial and middle zones of immature bovine cartilage are cultured separately or sequentially, cartilaginous tissue formed from superficial zone chondrocytes exhibits less matrix growth and lower compressive properties than constructs from middle zone chondrocytes, with the stratified superficial/middle constructs exhibiting intermediate properties. Expression of SZPs was the highest at the construct surfaces, with the localization of SZPs in superficial/middle constructs being concentrated at the superficial surface (Klein et al., 2003). Full-thickness, mid-zone and deep-zone, and deep-zone chondrocytes isolated from bovine cartilage cultured on ceramic substrates formed cartilage tissues with different structures. Tissue developed from deep-zone chondrocytes was thicker and had accumulated larger amounts of ECM than did the tissues formed by the full-thickness and mid- and deep-zone chondrocytes. The tissue formed by full-thickness chondrocytes accumulated the greatest amount of collagen, whereas the tissue formed by the mid-zone and deep-zone chondrocytes accumulated significantly more proteoglycans. The mid-zone and deep-zone chondrocytes produced tissue that had compressive mechanical properties up to four times greater than those of the cartilaginous tissues formed by cells from either the full-thickness or deep-zone part of the cartilage (Waldman et al., 2003). A conclusion from these studies is that full-thickness cartilage biopsies are needed for the tissue engineering of articular cartilage.

16.11 CELL SEEDING DENSITY

The mean number of chondrocytes in human cartilage tissue is 9626 cells/mm^3 (Hunziker et al., 2002). With a mean height of the hyaline articular cartilage layer of 2.4 mm, a 1 cm^2 defect would have a volume of 24,000 mm^3 and need approximately 2.4×10^7 cells. In vitro experiments with chondrocytes directly seeded to cartilage explants show a linear relationship between biosynthetic activity and the number of seeded chondrocytes (Chen et al., 1997). However, the number of cells needed for the implantation either as a suspension or in a scaffold has been studied extensively, although no consensus of seeding density has been reached. Seeding of polylactide–polyglycolide scaffolds with a density of <10 million cells/mL resulted in very little cartilaginous material (LeBaron and Athanasiou, 2000), while seeding scaffolds at a density ranging from 20 to 100×10^6 cells/ml resulted in the formation of clinically appropriate cartilage when implanted subcutaneously into nude mice (Puelacher et al., 1994). Chondrocytes seeded in high-density cultures (10^7 cells/ml) in Atelocollagen gel generated a cartilage-like tissue (Iwasa et al., 2003). All the publications that explore the appropriate cell seeding density have used different cell sources, scaffolds cell seeding methods, and growth media. There are no published studies that evaluate the influence of cell-seeding concentration using human articular chondrocytes and clinically validated methods of culture in 3D. However, it is accepted that chondrogenesis is enhanced by increasing the cell seeding density and cell–cell interactions. However, the ideal number of cells needed to start and retain secure chondrogenesis is not known, although the goal of many commercial companies is to deliver 1×10^6 cells/cm^2.

16.12 WHAT TYPE OF CHONDROGENIC CELLS ARE IDEAL FOR CARTILAGE ENGINEERING?

Pure chondrocytes, epiphyseal or mature chondrocytes, and allogeneic or autologous chondrocytes have been used in similar processes; these include ear and nasal chondrocytes and autologous mesenchymal bone marrow stromal cells (BMSCs), as summarized in Table 16.1.

16.13 ALLOGENEIC VERSUS AUTOLOGOUS CELLS

Chondrocytes express transplantation antigens and can theoretically participate in immunological reactions although the matrix produced by chondrocytes protects cells from rejection. Experiments have demonstrated that allogeneic growth plate chondrocyte implantation in cartilage defects of adult rabbits yields neocartilage; however, it is degenerated after a few weeks not only due to the humoral immune response but also because of cell-mediated cytotoxicity (Kawabe and Yoshinao, 1991). When comparing the long-term follow-up of allografted chondrocytes with autograft cells, both treatments gave cartilage regeneration. In humans, however, autologous chondrocytes seem to be preferable since there may be a risk of disease transmission with allografted cells.

16.14 ARTICULAR CHONDROCYTES VERSUS OTHER CELLS

Articular chondrocytes have already been determined by the cartilage lineage, but other cartilage sources could be considered, for example, nose, rib ear, and growth plate chondrocytes. However, the behaviors of these cells are not similar. When porcine auricular, costal, and articular chondrocytes were isolated and mixed with a fibrin polymer, and subsequently implanted into severe combined immunodeficiency (SCID) mice, a new cartilaginous matrix was synthesized by the transplanted chondrocytes in all experimental groups. However, the ratios of dimension and mass for auricular chondrocyte constructs increased by 20–30%, the ratios for costal chondrocyte constructs were equal to the initial values, and the ratios for articular chondrocyte constructs

Table 16.1 Cells with Chondrogenic Potential

Chondrocytes	MSCs
Articular	Bone marrow
Auricular	Adipose cells
Septal nasal	Periosteum
Costal	Synovial membrane

decreased by 40–50%. Further, the biomechanical properties of the engineered cartilage made with auricular or costal chondrocytes were superior to those of cartilage made with articular chondrocytes (Xu et al., 2004). The difference might be explained by the fact that auricular and costal cartilages normally exist in a vascularized area, while articular chondrocytes exist in a nonvascularized area. Human ear, nasal, and rib chondrocytes are equally suited to the generation of autologous cartilage grafts for nonarticular reconstructive surgery, and their regeneration potential could be enhanced using a combination of growth factors (Tay et al., 2004).

Within the mesenchymal tissue, hyaline cartilage chondrocytes are usually considered tissue restricted and without broader differentiation potential (Friedenstein et al., 1970). Culture-expanded mesenchymal stem cells (MSCs) have multipotency, and this gives rise to bone, cartilage, muscle, fat, and stromal cells under defined culture conditions. Hyaline chondrocytes are generally considered to lack multipotency and be restricted to cartilage lineage. However, when human articular chondrocytes are subjected to the same culture techniques and analysis used in the study of phenotypic plasticity of marrow-derived MSCs, chondrocytes have the potential to form cartilage in pellet mass cultures, adipose cells in dense monolayer cultures, and a calcium-rich matrix in an osteogenic assay. In contrast to MSCs, chondrocytes formed only cartilage, and not bone, in an in vivo osteochondrogenic ceramic implantation assay (Tallheden et al., 2003); this indicates a different in vivo default pathway. However, additional work will be necessary to assess whether cells of nonarticular origin can be successfully used for articular cartilage repair.

16.15 EMBRYONIC STEM CELLS AND INDUCED PLURIPOTENT STEM CELLS

In the early 1980s, several laboratories were able to culture cells derived from the inner cell mass of 3.5 day-old mouse embryos. If cells are cultured on feeders or in conditioned media containing inhibitors of differentiation, embryonic stem (ES) cells are able to maintain their undifferentiated stage in cell cultures over numerous cell doublings. Since then ES cells have been derived from several species including the human blastocyst (Thomson et al., 1998). ES cells exhibit all the characteristics of stem cells with properties including self-renewal, pluripotency in differentiation, high levels of telomerase activity, short G1 cycle checkpoint, and ability to initiate DNA replication without external stimulation. When injected into muscles or testis of SCID mice, the cells form teratomas with cells representing all three germ cell layers (Figure 16.9).

Embryonic stem cells cultured in confluent cultures that are devoid of their factors of inhibiting differentiation (leukemia inhibitory factor or continuous expansion of cells) demonstrate aggregate formation. After several days in vitro,

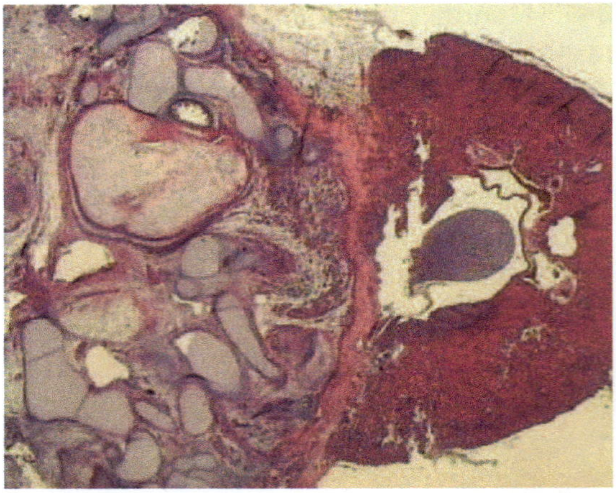

FIGURE 16.9
Teratoma formation of human embryonic stem cells in severe combined immunodeficiency mice.

the cells form large aggregates or embryoid bodies (EBs) due to the presence of many early embryoid derivatives. When subjecting the EBs to differentiation factors, the differentiation fate can be modified. Bone morphogenetic protein (BMP)-2 and -4 have been demonstrated to induce differentiation of chondrogenesis in mouse ES cell (Kramer et al., 2003). In human ES cells, induced differentiation to the hyaline cartilage phenotype has been described if cocultured with irradiated human chondrocytes (Bigdeli et al., 2009) or subjected to a strict culture protocol (Oldershaw et al., 2010).

The Nobel Prize in Physiology or Medicine 2012 was awarded jointly to Sir John B. Gurdon and Shinya Yamanaka "for the discovery that mature cells can be reprogrammed to become pluripotent." The discovery that a limited number of genes (Oct4, Sox2, Klf4, and c-Myc) are able to reprogram adult human cells came as a revolutionary discovery that will have a broad implication on future cell therapies and tissue engineering efforts (Takahashi et al., 2007; Yamanaka and Blau, 2010). Since then, numerous papers have been published in the field, but the use of retroviral vectors cause a change to the induced pluripotent stem (IPS) cell genome and make them unsuitable for cell therapy purpose. However, mRNA transfections with reprogramming factors are able to generate a footprint-free IPS cells that could potentially be used for cell therapy. The technology was recently applied in human chondrocytes where an IPS cell line was derived with a chondrogenic differentiation potential (Borestrom et al., 2014).

16.16 XENOGRAFT CELLS

Another alternative cell source is xenograft cartilage, although the immunologic problem of cross-species transplantation, as well as the problem of cross-species pathogen infectivity, has to be solved if those cell types are to be used instead of true committed chondrocytes (Deschamps et al., 2005).

16.17 DIRECT ISOLATION OF TISSUE

Cell isolation and direct implantation have been described. The method involves a tissue fragmentation step to mince the cartilage tissue mechanically into small tissue fragments before reimplantation. The purpose is to promote outgrowth of embedded chondrocytes through the increased tissue surface area. Using this methodology, researchers established that chondrocytes can effectively grow into adjacently placed scaffold materials and produce neocartilage (Lu et al., 2006). The technique has also been successfully demonstrated in a rabbit model (Frisbie et al., 2009). If one uses such a technique, it might be possible to engineer cartilage by a one-stage procedure, in the future.

16.18 SCAFFOLDS IN CARTILAGE TISSUE ENGINEERING

The primary goal of the tissue engineering approach to full-thickness cartilage defects is to provide engineered cartilage with the emphasis on regeneration rather than on repair. The ideal scaffold should provide an immediate support to cells and have mechanical properties matching those of the tissue being repaired. Gradually, the material then has to degrade, as the cells begin secreting their own ECM, and this thus allows for an optimal integration between newly formed and existing tissue (Freed et al., 1994; Table 16.2).

In cartilage tissue engineering, the scaffold must promote chondrocyte attachment, in vitro proliferation, and cell–cell interactions, as well as must favor the expression and maintenance of a cell-differentiated phenotype. For the development of a suitable polymer scaffold, it is essential to consider the fact that chondrocytes must organize three dimensionally to stimulate the synthesis of ECM molecules and prevent the loss of cell phenotype for a proper cartilage function, once implanted. Ideally, the scaffold should provide a temporary framework until the repaired cells have produced their own matrix. Both

Table 16.2 Criteria for Scaffolds for Tissue Engineering

A	The surface should permit cell adhesion and growth
B	Neither the polymer nor its degradation should provoke inflammation or toxicity when implanted in vivo
C	The material should be reproducible in 3D structures
D	The porosity should be at least 90% to provide a high surface area for cell–polymer interactions and sufficient space for ECM regeneration; minimal diffusional constraints during in vitro cultures
E	The scaffolds should be absorbed once they have served the template function
F	The scaffold degradation should match the rate of tissue regeneration

biologic and synthetic matrices (collagen and fibrin gels, decalcified bone, poly-glycolic acids (PGAs), ceramics, and carbon fibers) have been used for the induced repair of cartilage. The different types of scaffolds can be preseeded by cultured cells before the introduction into the cartilage defect, or the cells may migrate into the scaffold after implantation. Cartilage repair with allogeneic or autologous cells of mesenchymal origin with different types of scaffolds has been tried in different models (Table 16.3). The successful molding of a human ear in PGA fibers subsequently seeded with calf articular chondrocytes and subcutaneously implanted in severe combined immunodeficiency (SCID) mice demonstrated the potential of combination of artificial scaffolds and cells (Cao et al., 1997) (see Classical Experiment).

The use of a matrix for chondrocyte grafting started in the 1980s when embryonic chondrocytes in fibrin gel were used to maintain transferred chondrocytes within prefabricated cartilage defects in roosters (Itay et al., 1987). Biodegradable PGA polymer seeded with chondrocytes was an alternative material used to resurface joint cartilage defects in the rabbit knee (Vacanti et al., 1994). Type I collagen gel seeded with adherent cells from bone marrow or cells from the periosteum grown in culture were both able to regenerate cartilaginous tissue, although with mechanical properties similar to those of hyaline cartilage

Table 16.3 Different Scaffolding Materials Used for Cartilage Repair

Absorbable Fabricated Matrices		
Polygalactin	Vicryl, multifilament yarn	Vacanti et al. (1990)
Poly-L-lactic acid		Wedge et al. (1986)
PGA	Dexon	Freed et al. (1994)
Polyurethane		Klompmaker et al. (1992)
Nonabsorbable Fabricated Matrices		
Polyvinyl alcohol	Ivalon, porous sponge	Cobey (1967)
Nylon		Kuhn (1953)
Polyester	Dacron	Messner and Gillquist (1993)
Polytetrafluoroethylene	Teflon	Messner and Gillquist (1993)
Carbon fiber meshworks		Muckle and Minns (1989)
Polyethylene		Mahmood et al. (2005)
Matrices from Animals or Humans		
Collagen sponges (mainly type I collagen)		Wakitani et al. (1989)
Meniscus		Heatley and Revell (1985)
Decalcified bone		Dahlberg and Kreicberg (1991)
Fibrin polymers		Nixon et al. (1992)
Hyaluronic acid		Robinson et al. (1989)

(Wakitani et al., 1994). Despite numerous experiments (Table 16.2), the combinations of scaffolds and cells have shown variable results, and the clinical use of these scaffolds has been lacking. In the search for the ideal scaffold material, hyaluronic acid has proven to be an ideal molecule for tissue engineering strategies in cartilage repair, given its impressive multifunctional activity in cartilage homeostasis (Chen and Abatangelo, 1999; Figure 16.10).

Nature often gives clues to the researcher in the tissue engineering field, and before starting with a new tissue engineering concept, it is necessary to gather knowledge of the biology of tissue formation during embryogenesis, to possess scientific knowledge to mimic the consistency and architecture of tissues in the adult organism, and to understand tissue OH mechanical properties and their alterations due to the physiological activity. Further, the interactions between cells and the ECM have to be studied since the ECM not only acts as a structural support but it is also involved in informational exchange with the cellular component. Cells produce many kinds of molecules that form a particular environment that provides both a scaffold and guidance during development and throughout adult life. The recapitulation of embryonic events represents a major goal for tissue engineering, but it is important to emphasize that the adult environment is quite different from the one present during the embryonic stage and that tissues respond differently to specific information (Caplan, 2003).

All this is performed through complicated multistep differentiation cascades that implies fine crosstalk between cells and ECM with the synthesis of site-specific specialized molecules. Therefore, the appropriate scaffold should be designed to play this multifunctional role and not be used as a mere vehicle for reparative cells or growth factors. The scaffold has to change its features as the regenerative process takes place—starting with tissue induction in the early events and then supporting tissue formation during differentiation. With

Normal HYAFF Thick HYAFF

FIGURE 16.10

Cartilage graft construct based on chondrocytes cultured in hyaluronic acid.

this notion in mind, the materials used in tissue engineering have to be biocompatible and biodegradable. Physical properties have to match those of the tissue being replaced, and promote cell attachment and ultimately possess the capability of being remodeled by tissue-specific cells.

Hyaluronan is present in all soft tissues of higher organisms and, in particularly high concentrations, in the ECM of articular cartilage and in the mesenchyme of the developing embryo (Toole, 2001). It is a linear polymer consisting of a regular repeating sequence of a nonsulfated disaccharide units, glucuronic acid, and N-acetyl-glucosamine with a molecular weight varying from 4000 to 8×10^6 Da. Hyaluronan is basically unmodified by evolution, which underlies its importance in the physiological environment. Hyaluronan plays different biological roles depending on its physical and chemical properties. It interacts with binding proteins, proteoglycans, and active molecules, such as growth factors that influence matrix structure, water balance lubrication, and cell interaction. Hyaluronan can influence cell movement by its ability to alter the osmotic pressure leading to the formation of hydrated pathways in the tissue.

In the last few years, the use of hyaluronan, in its highly purified form, has become common practice in medicine. Intraarticular administration of hyaluronan is widely being used to relieve pain and improve joint mobility in the nonsurgical treatment of OA. Hyaluronan is a major component of the ECM of embryonic mesenchymal tissues; thus, the use of hyaluronan-based scaffolds will help create an embryonic-like milieu.

The greatest barrier to the successful use of this polymer in tissue engineering is the fact that hyaluronan, in its purified form, has such characteristics that limit its use as a biomaterial. Water solubility, rapid resorption, and short residence time in the tissue along with its poor ductility hamper its possible applications. Crosslinking and coupling reactions are two ways to obtain a modified, stable form of hyaluronan (Campoccia et al., 1998). With the right application, hyaluronan derivatives can be extruded to produce fibers or membranes, lyophilized to obtain sponges, or processed to obtain microspheres. The combination of hyaluronan-based scaffolds and fibers has found its way into the clinic, where several hundreds patients have now been treated (Dickinson et al., 2005; Pavesio et al., 2003).

16.19 BIOREACTORS IN CARTILAGE TISSUE ENGINEERING

Despite the good results of the first-generation technology of ACT, some of the concerns regarding the transplantation of immature chondrocytes in suspension led to the second generation of ACT. This technique involves the transplantation of chondrocytes in a 3D scaffold. This technique has several potential advantages such as the possibility of arthroscopic implantation, predifferentiation of cells in vitro, and implant stability. The nutritional requirements of cells that synthesize ECM increase along the differentiation process. The mass transfer must be increased according to the tissue properties. Bioreactors therefore represent a great tool to accelerate the biochemical and mechanical properties of the engineered tissues, and provide adequate mass transfer and physical stimuli. The tissue engineering challenges in combining scaffolds and cells include the seeding properties where cells usually do not penetrate deep into the scaffold in static cultures. However, if biomaterials and bioreactor culture vessels are successfully combined, a basis for systematic and controlled in vitro tissue growth can be obtained. Such model systems have been created based on chondrocytes, biodegradable PGA scaffolds, and bioreactors where the scaffold induces cell differentiation and degrades at a defined rate, whereas the bioreactor maintains controlled in vitro culture conditions that permit tissue growth and development. The bioreactor has enabled tests for cartilage formation in different gravitations where Earth gravitation was superior to low gravitation in space (Freed et al., 1997) (see State-of-the-Art Experiment).

16.20 GROWTH FACTORS THAT STIMULATE CHONDROGENESIS

The cartilage matrix contains several growth factors that are released at the time of injury from the damaged cartilage matrix as well as from the damaged subchondral bone. Such growth factors include TGF-βs, insulin-like growth factor-I (IGF-I), BMPs, and basic FGF (FGF-2) (Morales and Hascall, 1989). The factors orchestrate different effects. bFGF stimulates chondrocyte cell proliferation, stabilizes phenotypic expression, and inhibits terminal chondrocyte differentiation. TGF-β controls chondrocyte proliferation and differentiation and induces chondrogenic differentiation of mesenchymal cells in vitro and chondrogenesis in vivo (Joyce et al., 1990). It also inhibits chondrocyte terminal differentiation and stimulates chondrocyte proliferation in cooperation with FGF, and it has been shown to promote the repair of damaged cartilage. BMPs initiate cartilage differentiation and maintain the chondrocytic phenotype. IGFs have important anabolic effects on cartilage metabolism with increased matrix production. Little is known about the interactions among these growth factors, and the knowledge of their effect and role in chondrogenesis is limited. However, these findings indicate a potential therapeutic way to use exogenous growth factors to enhance the repair of cartilage defects.

16.21 FUTURE DEVELOPMENTS IN CARTILAGE BIOLOGY

Research in regenerative medicine involves cell and molecular biology, developmental cell biology, immunology, and polymer chemistry. This new direction in medicine will use three strategies: transplantation of cells to form new tissue in the transplant site, implantation of bioartificial tissues constructed in vitro, and induction of regeneration in vivo from healthy tissues next to an injury (Stocum, 2004).

However, regarding the future for cartilage repair in the new century, the idea is to transplant stem/progenitor cells, or their differentiated products, into a cartilage lesion site where they may form new tissue, or the cells could be used to construct a bioartificial tissue in vitro to replace the original tissue or organ. Bioartificial tissues are made by seeding stem cells or differentiated cells into a natural or artificial biomaterial scaffold shaped in the appropriate form and then by implanting the construct in place of the damaged tissue or organ. Theoretically, the use of cartilage stem cells is preferable to the use of differentiated cells harvested directly from a donor because stem cells have the potential for unlimited growth and thus supply. Such so-called uncommitted cells are capable of a broad range of chondrogeneic expression and could provide a regenerative tissue that recreates the embryonic lineage transitions originally involved in joint tissue formation. The technology of IPS cells could also be a potential source of a universal donor tissue cell line. However, the recent research described in this chapter shows that the use of true committed chondrocytes is still reasonable, but more research is needed to know how to make use of them in a more efficient way. We may also come across semibiological approaches where cells are used in combination with artificial implants. Such techniques could be suitable for resurfacing large osteochondral defects and may be one way to develop semibiological articular prostheses.

16.22 INTRODUCTION: BONE—BASIC BONE BIOLOGY: STRUCTURE, FUNCTION, AND CELLS

Bone comprises the largest proportion of the body's connective tissue and serves a number of key functions including (1) structural support, (2) protection of internal organs, (3) locomotion, and (4) maintenance of mineral balance, and as a reservoir for growth factors and for hematopoiesis (Taichman, 2005). Bone is unique and it has the capacity to renew and remodel throughout life. Bone can be separated into two main types: cortical (compact) and cancellous (trabecular or spongy) bone. Eighty to ninety percent of cortical or compact bone is calcified and is located at the diaphysis of long bones, on the surface of flat bones and as a thin shell around the epiphyses and metaphyses of long bones to provide mechanical and protective functions. In contrast, only 15–25% of cancellous or trabecular bone is calcified. Cancellous bone, typically a 3D lattice that comprises bone plates and columns, forms the epiphyseal and metaphyseal regions of long bones and the cortex of the flat bones and provides the metabolic functions in bone (Marieb and Hoehn, 2007).

16.23 BONE COMPOSITION

The ECM of bone consists of mineral, collagen, water, noncollagenous proteins, and lipids. The mineral phase is predominantly hydroxyapatite [$Ca_{10}(PO_4)_6(OH)_2$], with small quantities of carbonate, magnesium, and acid phosphate. This provides bone with its characteristic mechanical rigidity and compressive strength. In contrast, the organic matrix is mostly composed of type I collagen with small amounts of type III, V, and nonfibrillar collagens, with approximately 10–15% of the organic matrix comprising noncollagenous proteins. Noncollagenous proteins include albumin, α2-HS-glycoprotein, and growth factors and noncollagenous proteins synthesized by osteoblasts (proteoglycans, glycosylated proteins, glycosylated proteins, and γ-carboxylated proteins) (Table 16.4). The organic components of bone provide bone with elasticity and flexibility (Clarke, 2008).

Table 16.4 Extracellular Matrix Proteins in Bone

Protein		Function/Location	Disease (Absence or Dysfunction)
Collagen-related proteins	Type I	Binds and orientates other proteins	Osteogenesis imperfecta
	Type X	Present in hypertrophic cartilage	Metaphyseal chondroplasia
	Type III	Trace, may regulate collagen fibril diameter	Ehlers–Danlos syndrome
	Type V	Trace, may regulate collagen fibril diameter	Disrupted fibrils in mouse model
Serum proteins	Albumin	Inhibits HA crystal growth	None
	AHSG	Promotes endocytosis Monocyte chemoattractant Inhibits calcification	Ectopic calcification in mouse knockout
GAG-containing molecules	Aggrecan	Matrix organization and strength Retention of H_2O, Ca, PO_4	Spondyloepiphyseal dysplasia
	Versican	Regulates chondrogenesis Defines space destined to ossify	Wagner syndrome
	Decorin	May regulate collagen fibril size May modulate TGF-β activity	Progeroid form of Ehlers-Danlos syndrome (with double decorin—biglycan knockout)
	Biglycan	Determines peak bone mass May regulate growth factor activity	
	Asporin	Regulates collagen structure	Human polymorphism associated with OA
	Hyaluronan	May define space destined to ossify	None
Glycoproteins	Alkaline phosphatase	Hydrolyzes inhibitors of mineral deposition Increases local PO_4 concentration	Hypophosphatasia—poor growth and mineralization
	Osteonectin	Regulates collagen organization May regulate HA deposition Regulates growth factors	Low mineral content Osteopenia in knockout mouse

Continued...

Table 16.4 Extracellular Matrix Proteins in Bone *Continued*

Protein		Function/Location	Disease (Absence or Dysfunction)
SIBLING proteins	Osteopontin	Binds to cells, controls proliferation May regulate mineralization Inhibits nitric oxide synthase	Poor osteoclast remodeling Increased mineral content in knockout mouse
	Bone sialoprotein	Binds to cells May initiate mineralization	None
	MEPE	Regulation of PO_4 metabolism	Oncogenic osteomalacia
RGD-containing glycoproteins	Thrombospondins	Cell attachment Binds to collagens, platelets, thrombin, fibrinogen, laminin, plasminogen, and PAI	Pseudoachondroplasia
	Fibronectin	Cell attachment Binds to fibrin, heparin, gelatin, collagens	Lethal before skeletal development in knockout mouse
	Vitronectin	Cell attachment Binds: collagens, plasminogen, PAI, heparin	None
	Fibrillin 1 and 2	Regulates elastic fiber formation	Marfan syndrome in Fibrillin 1 mutation
γ-Carboxy glutamic acid-containing proteins	Matrix Gla protein	Inhibitor of mineralization	Keutel syndrome (excessive cartilage mineralization)
	Osteocalcin	Regulates osteoclasts Inhibits mineralization May coordinate bone turnover	Knockout mouse—osteopetrosis
	Protein S	May be synthesized by osteoblasts	Osteopenia

HA = Hydroxyapatite, AHSG = α2HS glycoprotein, H_2O = water, Ca = calcium, PO_4 = phosphate, TGF-β = Transforming Growth Factor Beta, OA = Osteoarthritis, SIBLING = Small Integrin Binding Ligand N-Glycosylated, MEPE = Matrix extracellular phosphoglycoprotein, PAI = Plasminogen Activator Inhibitor, RGD = Arginyl-glycyl-aspartic acid DNA motif.
Adapted from Clarke (2008).

The structure of a long bone (femur) (Figure 16.11) can be separated into distinct regions. The central shaft of a long bone is called the diaphysis, and it has a hollow center, the medullary cavity, containing the bone marrow. In children, long bones are filled with red marrow that is gradually replaced with yellow marrow with age. The diaphysis is predominantly composed of compact bone with only a small component of spongy bone on its inner surface around the medullary cavity. The end zone of the diaphysis is referred to as the metaphysis; and at each end of the long bone is located the epiphysis (proximal/distal).

The metaphysis is the developmental zone between the epiphysis and the diaphysis, known as the "growth plate" or "epiphyseal plate." The growth plate is central in the process of endochondral ossification and enables growth during childhood. The joint surface of each epiphysis is covered by a thin layer of articular cartilage that protects the opposing bone ends during joint movement and

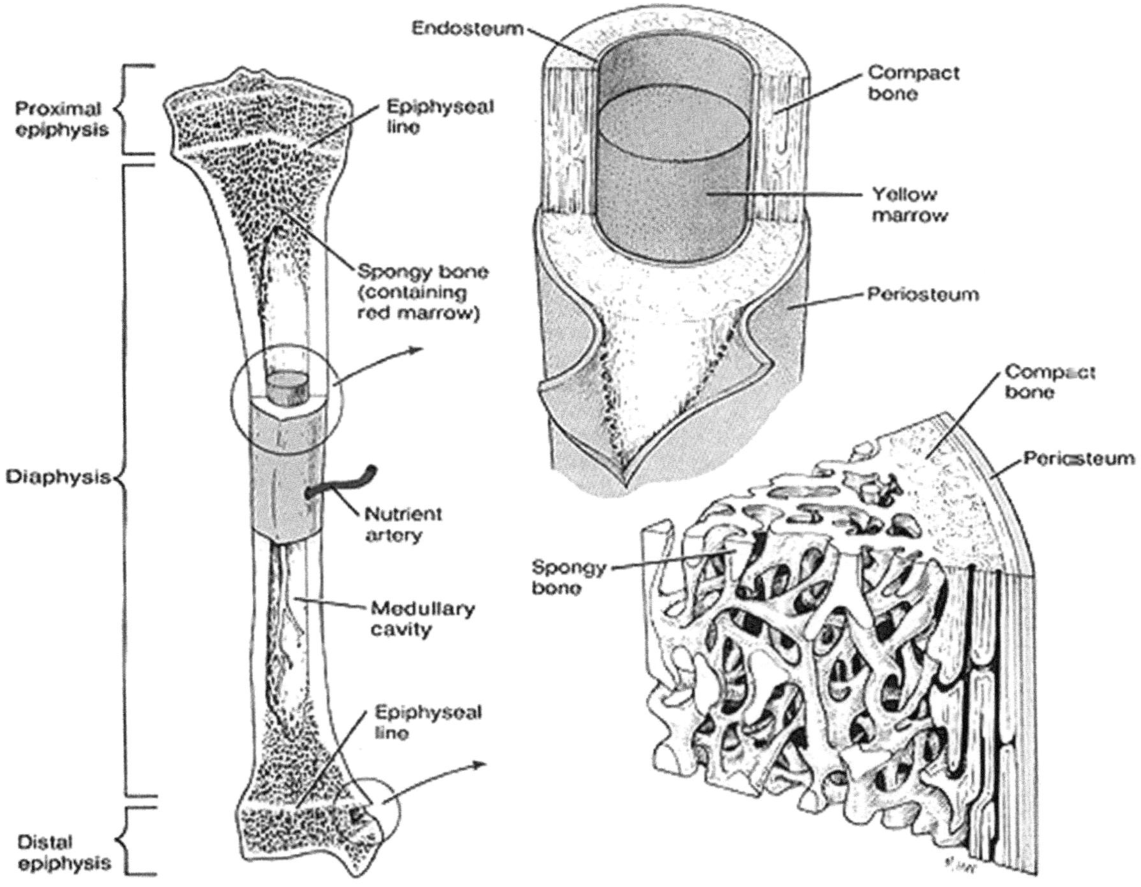

FIGURE 16.11
Diagram of the structure of a long bone. *From Basic Medical Anatomy (Spence, 1991).*

serves to also absorb stress (Rosen et al., 2009). The articular cartilage covers a layer of compact bone, beneath which trabecular bone resides.

Cortical bone is made up of an interconnected group of subunits known as osteons, or the Haversian system (Figure 16.12). A single osteon consists of a hollow, laminated rod of collagen and inorganic calcium phosphates with a hollow core, the Haversian canal that contains a blood and nerve vessel. Surrounding the outer surface of most bones, except at the joints, is the periosteum. The periosteum contains a double-layered membrane—an outer fibrous layer of collagen fibers and an inner layer of skeletal cells including skeletal stem cells (SSCs). The periosteum has a number of blood and lymphatic vessels, as well as nerve fibers, and serves to provide much of the bone nutrients as well as a key role in appositional bone growth and fracture repair.

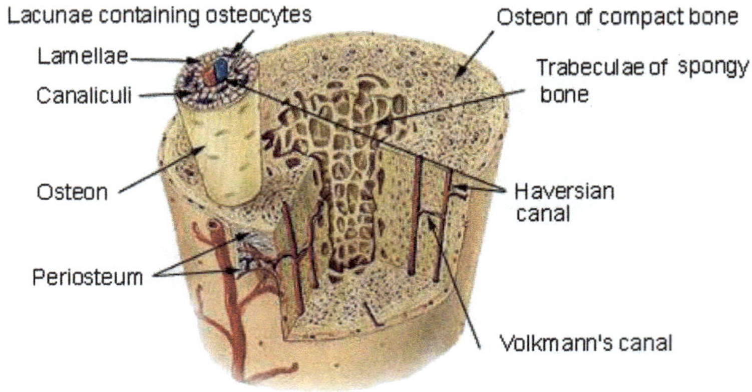

FIGURE 16.12
The microstructure of bone (White and Folkens, 2005).

The periosteum also gives rise to important muscular tendon and ligament insertions making it important for locomotion. The endosteum lines the inner bone surfaces and is composed of a single layer of connective tissue that lines the surface of the medullary cavity.

Flat bones are found where the principal requirement is either protection or provision of broad surfaces for muscular attachment. These bones are typically, broad, flat plates (e.g., cranium, ilium, and scapula) and comprise two thin plates of cortical bone with cancellous bone between the plates. Similarly, short bones, such as the wrist, and irregular bones, such as the spine and hip, also consist of cortical bone surrounding trabecular bone (Marieb and Hoehn, 2007).

16.24 BONE FORMATION

The de novo formation of trabecular and cortical bone involves the differentiation of skeletal cells into osteoblasts that synthesize osteoid (nonmineralized bone matrix) and subsequently mineralize the matrix. It is important to note that skeletal cells need a substrate upon which the bone will be deposited, either a membrane (as in intramembranous ossification), preexisting cartilage spicules (as in endochondral ossification), or preexisting bone.

16.25 INTRAMEMBRANOUS OSSIFICATION

The flat bones of the skull, a number of facial bones, and most of the clavicle form by intramembranous ossification. Indeed, intramembranous ossification is the process responsible for the growth of the majority of the axial skeleton via osteogenic cells of the periosteum. Intramembranous ossification occurs through the direct transformation of mesenchymal cells into osteoblasts and, importantly, is the mechanism through which structural remodeling occurs at the subperiosteal surface of the long bones. Critically, intramembranous ossification is distinct from endochondral ossification in that there is no cartilage template. In intramembranous ossification, bone is laid down by direct apposition within stromal connective tissue. Thus, mesenchymal cells cluster within ossification centers where the stem cells proliferate and then differentiate to form osteoblasts. The osteoblasts secrete matrix, initially as random disorganized collagen fibrils and later as a denser and more homogeneous matrix that is subsequently mineralized by the osteoblast. Growth in the diameter of the shaft of the long bones also occurs intramembranously as the bone increases in length.

16.26 ENDOCHONDRAL OSSIFICATION

The long bones of the skeleton develop predominantly by endochondral ossification (Mackie et al., 2008). In this process, mesenchymal cells cluster to form a cartilage template or "anlage," which acts as a scaffold on which the new bone is formed. The longitudinal growth of the bones requires the development of the growth plate or physis, which has histologically distinct zones. The uppermost of these is the resting zone, which consists of a layer of hyaline cartilage, produced by cartilage cells. Below the resting zone is the proliferative zone where the cartilage cells display a flattened appearance and are typically organized in columns with cartilage cells separated by narrow bands of interstitial tissue. The chondrocytes within the upper hypertrophic zone are metabolically active, and synthesize Type II and Type X collagen and aggrecan and display significantly increased cell size and a less regular columnar arrangement. In the lower hypertrophic zone, cartilage matrix calcification occurs within the longitudinal septae between the columns of chondrocytes, and in essence provides the scaffold upon which bone deposition in the metaphysis can occur. The hypertrophic zone is rich in alkaline phosphatase for mineralization. At the vascular front, where cartilage to bone transition is present, blood vessels invade the connective tissue in this region and penetrate into the vertical columns (Mackie et al., 2008). The longitudinal calcified cartilage septae provide the surface/scaffold for new bone formation. Bone matrix is then laid down on this primary spongiosa to form the secondary spongiosa and thus the first trabeculae. The new metaphyseal woven bone is replaced by modeling via osteoclast activity and osteoblast activity to give mature trabeculae. Bone growth and bone modeling cease when the growth plates close after puberty.

16.27 FRACTURE REPAIR

Normal fracture healing is a complex process that involves a coordinated interplay of cells, growth factors, and ECM. Two basic types of fracture healing exist: Primary bone healing is rare, and it requires absolute contact of the bone fragments, with stability, such that cells within the cortical bone can directly reestablish continuity (Giannoudis et al., 2007). The vast majority of fractures undergo secondary bone healing, characterized by callus formation due to the processes of intramembranous and endochondral ossification. During the initial *inflammatory stage*, soft tissue disruption and bleeding lead to nonspecific wound-healing pathways, stimulated by platelet degranulation and mononuclear inflammatory cell activation. These cells release cytokines and procoagulant factors, which modify the initial hematoma into a fibrinous thrombus and form a positive feedback loop, to activate and recruit further cells. In the following days, the thrombus is reorganized into granulation tissue under the exquisite control of cytokines and growth factors (including TGF-β, platelet-derived growth factor (PDGF), basic FGF-2, vascular endothelial growth factor (VEGF), macrophage colony-stimulating factor, BMPs, interleukin (IL)-1 and IL-6, and tumor necrosis factor-alpha (TNF-α)) produced by the infiltrating cells. These factors are responsible for the recruitment and stimulation of stem cells from bone marrow, periosteum, blood vessels, and surrounding soft tissue.

In the second phase of fracture healing, *endochondral ossification* commences via soft fibrocartilaginous callus formation. Granulation tissue is replaced by fibrocartilage synthesized by chondrocytes and fibroblasts and the eventual formation of a semirigid cartilaginous template that holds the fracture. Hard callus formation is subsequently driven by proangiogenic factors, including VEGF, BMPs, FGF-1, and TGF-β, which stimulate callus vascularization, chondrocyte hypertrophy, and apoptosis. SSCs from the periosteum, bone marrow, vasculature, and surrounding soft tissues aid hard callus formation as bone progenitors differentiate under the actions of factors, such as the BMPs, to form osteoblasts and the generation of mineralized bone matrix resulting in the formation of woven bone that is subsequently remodeled by osteoclast activity tightly coupled to osteoblast activity to restore normal bone structure and function (Giannoudis et al., 2007).

16.28 SKELETAL STEM CELLS

Bone marrow stromal tissue contains a subset of nonhematopoietic progenitor cells that exhibit the two characteristic features of stem cells, namely, capacity for prolonged or unlimited self-renewal under controlled conditions and the potential to differentiate into a variety of mature cell types. The multipotent, self-renewing progenitor cells of bone marrow stroma are widely referred to as MSCs, a term originally coined by Caplan (1991) and extensively used in the literature, notably by Pittenger et al. (1999) (see Chapter 2). The term "skeletal stem cells" was introduced by Bianco and Robey to denote specifically MSCs derived from bone marrow stromal tissue (Dawson et al., 2014). The regenerative capacity inherent to bone can therefore be attributed to its SSC population.

Friedenstein developed the in vitro colony adherence assay to demonstrate the presence of fibroblast-like clonogenic precursor cells (1–10 per 100,000 nucleated cells), referred to as colony-forming unit-fibroblastic (CFU-F), in the tissue culture plastic-adherent nonhematopoietic fraction of fresh bone marrow aspirates (Friedenstein references in Chapter 2). Although the CFU-F population can demonstrate stem cell characteristics, such as self-renewal capacity under clonogenic conditions and the potential to differentiate into multiple stromal lineages, it is not a homogeneous population; nevertheless, cells from these cultures represent the most commonly used population by researchers in the field of bone regeneration and should be referred to as BMSCs rather than as SSCs.

16.28.1 Isolation and Characterization

An array of cell surface markers, namely, the trypsin-resistant antigen recognized by the STRO-1 antibody; CD29 (β1 subunit of integrin family); CD44 (glycoprotein expressed by both HSCs and MSCs); CD49a (Laminin and collagen receptor VLA-α1), CD73 (ecto-5″-nucleotidase); CD90 (Thy-1), CD105 (Endoglin); CD106 (VCAM-1), CD146 (MCAM), CD166 (ALCAM); and CD271 (low-affinity nerve growth factor receptor) are expressed by SSCs, and thereby facilitate their isolation from bone marrow aspirates using a positive immunoselection strategy, or by negative immunoselection for hematopoietic cell surface antigens such as CD34, CD45, CD19, CD14, CD11b, CD79α, and HLA-DR surface molecules (reviewed in Dawson et al. (2014)). The identity of a single unique cell surface marker that specifically recognizes SSCs still remains elusive.

In the field of skeletal science, there are two distinct perspectives regarding the biological nature and function of SSCs. Bianco et al. are of the view that the SSC is a cell with a "precisely defined physical and conceptual entity" and it cannot be described outside of its in situ context. This is based on the results of their seminal work, which provided compelling evidence that the SSCs reside in the perivascular spaces surrounding the vascular sinusoids in the bone marrow in a subendothelial position (Sacchetti et al., 2007). Perisinusoidal SSCs were found to express the STRO-1 antigen, CD146, CD105, alkaline phosphatase, and VCAM-1, and their transcriptome was constituted by genes distinctive of mural cells (i.e., vascular smooth muscle cells and pericytes) and committed osteoprogenitor cells, as well as those related to the HSC niche, but was devoid of endothelial and late osteogenic genes (Sacchetti et al., 2007). Thus, in accordance with the perspective of Bianco et al., upon transplantation in vivo, an SSC is a single cell capable of generating a complete heterotopic bone or bone marrow organ/ossicle (including a compartment of perivascular stromal cells with similar phenotypes and properties as those of the originally explanted cell) in which hematopoiesis from the recipient animal becomes established (Figure 16.13). The capacity to establish, organize, and transfer the hematopoietic niche in vivo is regarded as an important defining feature of SSCs.

In contrast, Caplan and Pittenger are of the view that MSCs are a product of their environment that can exist or be developed in other locations, including the culture dish, that is, in vitro (Caplan, 1991; Caplan and Correa, 2011;

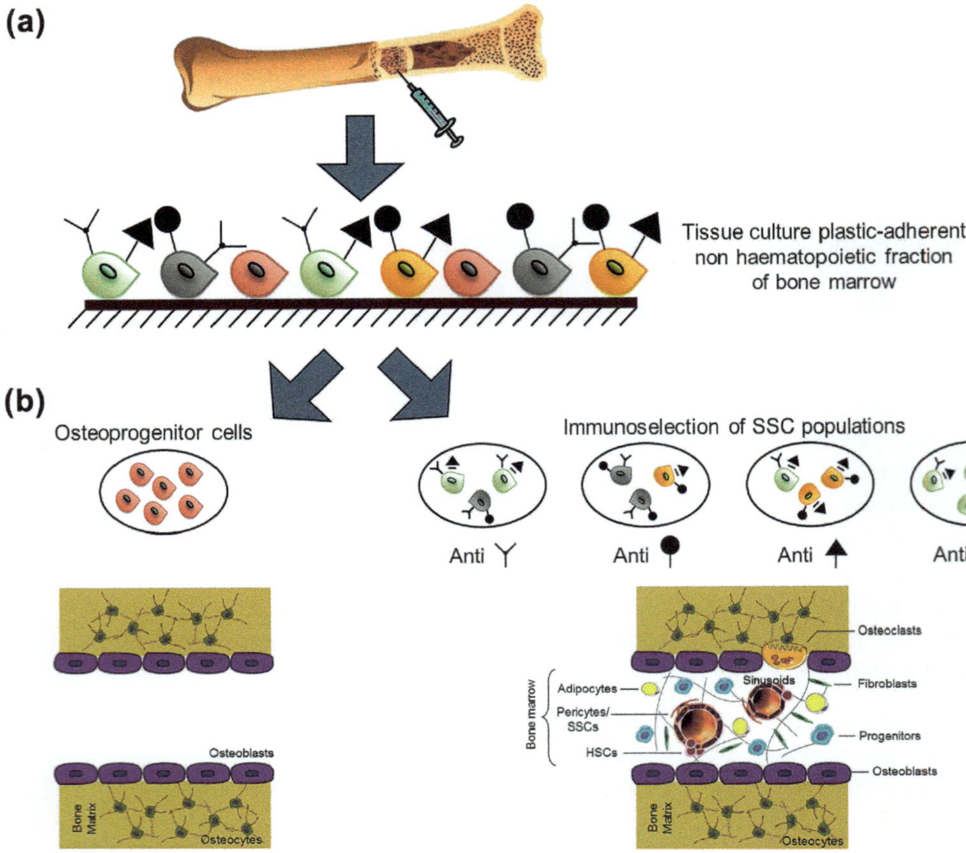

FIGURE 16.13

SSCs establish, organize, and transfer the hematopoietic niche in vivo. *From Dawson et al. (2014).* (a) The tissue culture plastic-adherent nonhematopoietic fraction of bone marrow is heterogeneous as it contains both osteoprogenitors and skeletal stem cells (SSCs). (b) Upon ectopic implantation with a suitable osteoinductive carrier in immunocompromised mice, osteoprogenitor cells are only able to form bone, while relatively homogenous populations of SSCs, immunoselected on the basis of expression of one or more SSC-specific surface markers, are able to regenerate bone and stroma and, critically, establish a hematopoietic microenvironment and, critically, establish a hematopoietic microenvironment.

Pittenger et al., 1999). In accordance with the perspective of Caplan and Correa (2011), MSCs exhibit the following two phenotypes:

- a "constitutive" phenotype in which, as perivascular cells, the population expresses the characteristic cell surface markers both in vivo and ex vivo, and demonstrates functionality through its multipotential ex vivo differentiation capabilities.
- a "regulatory" phenotype in situations of tissue/vessel damage in vivo, where the released pericytes are activated by injury to become MSCs, which secrete a variety of bioactive molecules as part of their local trophic

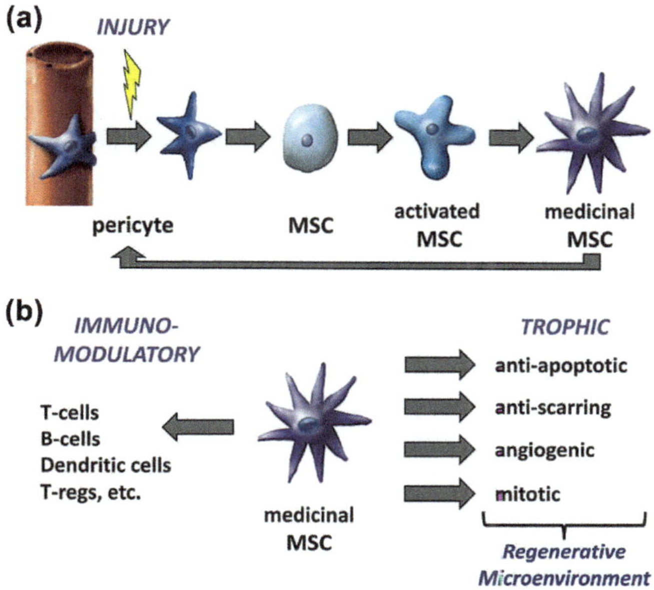

FIGURE 16.14
MSCs are immunomodulatory and trophic. *From Caplan and Correa (2011).* (a) The proposed sequential activation of pericytes as a response to injury. Local vessel damage affects resident pericytes and liberates them from functional contact with blood vessels to become activated mesenchymal stem cells (MSCs). Upon immune activation, these mobilized, "medicinal" MSCs secrete factors that organize a regenerative microenvironment. Subsequent repair is reinforced when activated MSCs reacquire a stabilizing pericyte phenotype in the abluminal space. (b) The bioactive molecules secreted by medicinal MSCs are immunomodulatory and affect a variety of immune cell lineages. Other secreted molecules establish a regenerative microenvironment by establishing a powerful trophic field.

and immunomodulatory activities that help to establish a regenerative microenvironment to support the repair of the injured tissue (Figure 16.14). In this context, MSCs serve as site-regulated, multidrug dispensaries to promote and support the natural regeneration of focal injuries. The natural supply of MSCs needs to be supplemented by local or systemic delivery in situations where the injuries are large or occur in older individuals.

16.28.2 Immunomodulatory Properties

In addition to their stem/progenitor properties, there is considerable interest in the immunomodulatory properties of SSCs and a large body of literature has documented the pleotropic effects of SSCs on cells of the immune system (Caplan and Correa, 2011) (see Chapters 2 and 12). The phenotype of SSCs has been widely described as Major Histocompatibility Complex (MHC) Class I$^+$, MHC Class II$^-$, CD40$^-$, CD80$^-$, and CD86$^-$ (see Chapters 2 and 12). Although MHC Class I molecules have the potential to activate alloreactive T-cells, the absence of

CD80 and CD86, responsible for providing costimulatory signals necessary for T-cell activation and survival, would leave the T-cells anergic. Macrophages are key players of the innate immune system in initiating and controlling inflammation, and SSCs can influence macrophage function depending upon the inflammatory context. Monocytes arriving at an inflammatory environment can develop into activated macrophages, which produce proinflammatory cytokines such as TNF-α and IFN-γ. The resulting proinflammatory microenvironment activates SSCs to adopt an immune-suppressive phenotype and release mediators, which skew the differentiation of monocytes toward antiinflammatory macrophages that secrete high levels of IL-10, TGF-β1 (supports SSC-mediated generation of regulatory T-cells/Tregs that suppress T-cell-mediated immune responses) and low levels of TNF-α, IFN-γ, IL-1, IL-6, and aid tissue regeneration following inflammation (see Chapter 2). Since the precise identity of the regulatory effects of SSCs on immune and inflammatory cells have not been validated conclusively in defined in vivo systems thus far, rigorous in vivo studies on the immunogenicity and immunomodulatory capacity of SSCs in specific pathophysiologic models are needed before the full therapeutic potential of these cells can be realized.

16.29 EXPANSION AND DIFFERENTIATION

SSCs are characterized by limited capacity for proliferation in monolayer cultures and typically exhibit "replicative senescence" as a consequence of repeated subculture. The problem is exacerbated further in SSCs from elderly individuals due to accelerated senescence following ex vivo culture. Development of simple, safe, and efficacious culture techniques that allow ex vivo expansion of SSCs without loss of their functional, immunophenotypic, and cytogenetic characteristics (see Chapter 12) is therefore one of the key requirements for the successful application of SSCs in bone regeneration.

16.30 GROWTH FACTORS FOR BONE REPAIR

Bone formation involves the complex interplay of a number of hormones and growth factors. The growth factors most extensively characterized for their ability to stimulate osteogenesis are the BMPs. Other important local signaling molecules of relevance to bone regeneration include the TGF-β subfamily (TGF-βs), the Wnt proteins (Wnts), IGFs, FGFs, PDGFs, and the VEGFs.

BMPs—Around 20 proteins have been designated as BMPs, though not all are directly osteogenic. With the exception of BMP-1, the BMPs are structurally related to the TGFβ superfamily, which includes TGF-βs, activins, and inhibins. The BMPs, like the TGF-β proteins, are temporarily stored in the bone matrix after synthesis by osteoprogenitor cells, osteoblasts, chondrocytes, and platelets. Clinical interest has been centered particularly on BMP-2, BMP-4, and BMP-7 in the potential treatment of fracture repair, large segmental bone defects, spinal fusion, and in the fixation of prosthetic implants (Even et al., 2012). An ongoing challenge for the clinical application of BMPs is the supraphysiological doses typically required for efficacy and concerns over immunogenicity and the risk of ectopic bone formation associated with high dosage.

TGF-βs—It is a potent multifunctional 25 kD polypeptide that exists in four distinct isoforms (TGF-β1, -β2, -β3, and -β5), sequestered within the bone matrix and has been shown to play a key role in bone formation through the generation of committed osteoblast cells from determined osteogenic precursor cells. It also has a role in the coupling of bone formation and bone resorption.

Wnts—In relation to bone, the class of Wnts associated with β-Catenin-mediated pathways (the canonical Wnts) has been most thoroughly investigated and shown to promote the recruitment and osteogenic differentiation of osteoprogenitor populations (Minear et al., 2010).

IGFs—IGF-I and IGF-II are 7.6 kD polypeptides synthesized by skeletal and nonskeletal tissues and act independently to stimulate osteogenesis and chondrogenesis. Both molecules stimulate osteoblast function and bone matrix deposition, and while IGF-I is more potent, IGF-II is able, unlike IGF-I, to induce the differentiation of MSCs into osteoblasts and stimulate proliferation (Arvidson et al., 2011).

FGFs—FGF-2 is expressed in periosteal cells and osteoblasts and is thought to be of particular relevance to postnatal bone repair and plays a role in both osteoblast differentiation and the stimulation of angiogenesis through endothelial cell recruitment.

PDGFs—It exists as an extracellular factor as one of three biologically active isoforms (AA, AB, BB) composed of a combination of PDGF-A and PDGF-B polypeptide chains. PDGF is expressed during normal bone repair and is deficient in diabetes-induced impaired fracture healing. Preclinical studies have suggested the potential of PDGF-BB and to a lesser extent PDGF-AA to stimulate fracture healing although other studies are conflicting.

VEGFs—The VEGF family includes five members (VEGF-A to VEGF-D, and placental growth factor) with VEGF-A being the most studied for its critical role in angiogenesis (Kanczler and Oreffo, 2008).

16.30.1 Scaffolds for Bone Regeneration

The scaffold provides an extracellular microenvironment for the support and stimulation of stem cell-driven tissue regeneration. It can host transplanted cells, recruit endogenous cells, mediate biological signaling, and provide mechanical support. Along with classic bone graft strategies (autologous and allogeneic bone graft), bone tissue engineering scaffolds can be composed of calcium phosphate-based ceramics, extracted xenogeneic ECM components, and naturally or synthetically derived polymers. These various materials and forms possess intrinsic strengths and weaknesses (Table 16.5), and thus,

Table 16.5 Scaffold Materials and Their Relative Strengths and Weaknesses for Bone Tissue Engineering

	Class of Biomaterial								
				ECM Scaffolds		Non-ECM-Derived Scaffolds			
						Solid-wall Scaffolds			
	Autograft	Allograft	DBM	Fibrous Scaffolds	Hydrogels	Polymer	Ceramic	Fibrous Polymer	Polymer Hydrogel
Manufacture/Clinical Specifications									
Allow economical sourcing and manufacture	--	+	+	+	+	++	++	++	++
Manufacture according to GMP	NA	+	+	+	+	++	++	++	++
Meet FDA approval	NA	+	+	±	±	±	+	±	±
Amenable to sterilization	NA	-	-	-	-	++	++	++	++
Allow facile intraoperative handling/implantation	-	+	-	+	++	+	-	+	++
Nonimmunogenic	++	-	-	++	++	++	++	++	++
Controlled 3D architecture/geometry	-	-	-	+	++	++	±	++	++
Impart provisional weight bearing functionality	++	++	--	--	--	±	++	--	--
Radiologically distinguishable	+	+	-	-	-	-	+	-	-
Scaffolding Specifications									
Biocompatible	++	+	+	+	+	+	+	+	+
Provide mechanical support for surrounding tissue	+	+	--	±	--	+	+	+	±
Facilitate cell delivery	++	-	-	+	++	+	+	+	++
Controlled degradation	++	++	--	+	+	+	+	+	+
Conducive for cell ingrowth/migration	++	+	+	++	±	+	+	++	±

ECM Specifications									
Provision of cell–matrix interactions	++	+	+	+	+	-	+	-	-
Transmission of mechanical signals to cells	++	++	-	+	+	+	-	+	+
Localization of bioactive factors	++	-	+	-	-	-	+	-	+
Extended release of bioactive factors	++	-	+	-	-	-	+	-	±
Preserved bioactivity of bioactive factors	++	-	+	-	-	-	+	-	±

Key: ++ particular strength, + strength, ± variable between specific materials, - weakness, -- particular weakness. Relative strengths and weaknesses are assessed for unmodified scaffold materials. Scaffold modifications and the generation of composite scaffolds overcome many of the weaknesses inherent to the unmodified platform technologies (Table 16.2).

Table 16.6 Modifications to Scaffold Design Associated with Enhanced Bone Formation In Vivo

Class of Modification	Specific Modification	Class of Platform Biomaterial (Table 16.1)
Porosity/pore size	Increased interconnection between pores	Solid-wall polymer, ceramic
	Incorporation of macroscopic channels	Solid-wall polymer
	Increased porosity (within 30–70% range)	Ceramic
	Reduced interconnection between pores (but higher surface area for bone apposition)	Ceramic
Degradation/resorption rate	Controlled degradation via composite ratio	Solid-wall polymer
	Slower resorption via modulation of material chemistry	Solid wall polymer, ceramic, polymer hydrogel
	Slower resorption via polymer coating	DBM
	Enhanced degradation via incorporation of specific enzyme degradable crosslinkers	Polymer hydrogel
	Faster degradation via processing technique	Polymer hydrogel
Calcium-phosphate incorporation	Apatite coating	Solid-wall polymer, fibrous polymer
	Addition of calcium carbonate microspheres	Allograft
ECM protein incorporation	Addition of demineralized bone matrix	ECM fibrous, fibrous polymer
	Collagen and chondroitin sulfate coating	Solid-wall polymer
	Collagen-based hydrogels (over alginate or fibrin)	ECM hydrogel
	Collagen-binding domain surface coating	Ceramic
Surface properties/topography	Nanofiber construction	Fibrous polymer
	Macroscopic trabecular architecture	Solid-wall polymer
	Surface hydrophobicity	Solid-wall polymer, polymer hydrogel
Bioactive factor incorporation	Nanosphere/microsphere encapsulation	Solid-wall polymer
	Heparin binding	DBM, solid-wall polymer, polymer hydrogel
	Hydrogel coating	Ceramic
	Peptide amphiphile gelation	Polymer (peptide) hydrogel

Many additional modifications not listed above have shown improved bone formation in vitro or else offer enhancements not directly assayable by in vivo bone forming assays.

there is an expanding body of research engaged in modifying and optimizing scaffold design to meet the complex of specifications required for clinically relevant bone regeneration (Table 16.6, Figure 16.15; Dawson et al., 2011).

Bioceramics incorporating hydroxyapatite or calcium phosphates are able to integrate well with bone tissue and have a high compressive strength. However, such ceramics do not in themselves stimulate bone formation (osteoinductivity), and while increased porosity facilitates greater penetration of new bone formation, complete resorption of a ceramic scaffold material is rarely observed.

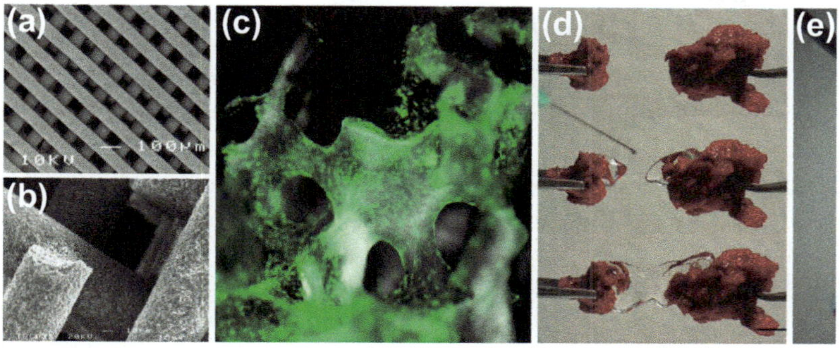

FIGURE 16.15

Scaffold forms: (a) Hydroxyapatite scaffolds generated with a defined 100-μm porosity via the additive manufacture technique extrusion free forming. (b) Along with macropore control via AM microporosity can be controlled via the temperature of sintering. (c) Porous decellularized bone graft seeded with viable HBMSCs. (d), (e) Injectable hydrogels that set in situ allow minimally invasive delivery of cells and growth factors.

Natural polymers, particularly those derived from extracted ECM components, such as collagens, fibrin, or the GAG hyaluronic acid, have the inherent ability to modulate cell behavior via complex cell–matrix interactions. These biologic materials are intrinsically biocompatible, and while their complexity typically precludes mass synthesis and so require a xenogeneic source, biologic materials have been used for many years in humans with minimal adverse immunological outcomes (Badylak, 2007).

Synthetic polymers have also been extensively developed as scaffold materials, due, not least, to their sheer versatility. The physical and chemical properties of these macromolecules can be modulated on the basis of their monomer constituents, the relative ratios of copolymers, the interactions of polymers with one another (i.e., in the mode and degree of crosslinkage) and the incorporation of side chains. In addition, side chains can be functionalized by the covalent immobilization of biologically active molecules, such as adhesion factors and cytokines, that, together with hydrophilic molecules otherwise resistant to cell adhesion and protein absorption, provide a highly defined mode of controlling specific cell responses (Drury and Mooney, 2003).

Hydrogels—They are hydrated polymer chains, while alone they have not been shown to enhance bone formation; they have been used effectively in the delivery of cells and growth factors. Naturally derived hydrogels extensively investigated as cell, growth factor, and drug delivery vehicles in bone regeneration include gelatin, hyaluronic acid, fibroin, collagen, and alginate. Polyethylene glycol and derivatives represent the most commonly used synthetic gels.

16.31 SCAFFOLD BIOCOMPATIBILITY

Perhaps the most basic property required of a bone tissue engineering scaffold is biocompatibility. Naturally occurring polymers and bioceramics tend to display good biocompatibility due to their respective similarity to the organic and inorganic components of natural bone tissue. While biocompatibility is more of a concern with synthetic polymers, biodegradable polymers, such as poly(L-lactic acid), PGA, poly(D,L-lactic-co-glycolic acid), and poly(ε-caprolactone), are already widely used for certain biological applications. Additionally, various other polymers such as polyorthoester, polyanhydrides, and poly(propylene fumarate) are being investigated for tissue engineering, and have shown favorable biocompatibility in animal models (Hutmacher, 2000).

Biodegradability—The degradation of a tissue engineering scaffold is to be carefully controlled so that scaffold resorption and bone formation are tightly coupled. While, on the one hand, persistence of a graft material can result in poor clinical outcomes, especially when bone integration is poor, premature resorption of a scaffold also compromises bone repair by the destabilization of early bone formation and the loss of osteoconductive surfaces for new bone apposition (Hing et al., 2007). Controlled resorption rate is thus a key parameter in bone tissue engineering.

Mechanical properties—A scaffold for bone tissue engineering should approximate the mechanical function of the callus in natural bone repair, and provide a dynamically durable-degradable 3D structure, which, over time, can be replaced by cell-derived tissue function (Dawson and Oreffo, 2008). While weight-bearing functionality may be an ideal in certain contexts, it is not a prerequisite specification of a bone tissue-engineering scaffold (especially if resorbability or mechanical signal transmission is thereby compromised) as support can be achieved via alternative orthopedic techniques (Logeart-Avramoglou et al., 2005).

Porosity—Scaffold porosity (percentage void space in a solid), pore size, and pore interconnectivity impact on the surface area available for cell growth and accessibility for host tissue, including vasculature, to penetrate into the central regions of the scaffold (Logeart-Avramoglou et al., 2005). Increase of porosity and pore size has been found to positively influence bone formation, pore interconnections $<100\ \mu m$ have been found to restrict vascular penetration and supplementation of a porous structure with macroscopic channel-enhanced tissue penetration and bone formation (Silva et al., 2006).

Scaffold as ECM—In addition to providing appropriate mechanical support at the defect site, the tissue-engineering scaffold serves to reestablish a regenerative microenvironment at the site of damage by providing the dynamic extracellular signaling—chemical, biological, and mechanical—that directs progenitor cell recruitment, growth, and differentiation. Bone ECM itself provides a ready example of the dynamic role the ECM plays in biological signaling. Early work in the 1960's identified BMPs as the active factors responsible for the ability of demineralized bone matrix to induce ectopic bone formation upon transplantation (Sampath and Reddi, 1981). However, while recombinant BMPs (rhBMP) have now been clinically approved for the induction fracture repair, BMP-based strategies require milligram doses of the protein for reliable fracture repair in contrast to the micrograms of BMP per kilogram of bone graft.

In addition to controlling the diffusion of chemical and biological signals, the ECM is also associated with directing cell growth and differentiation via direct interaction with cell surface receptors. Thus, for example, Type I collagen, the major organic component of bone ECM is chemotactic to fibroblasts possessing high-affinity cell-binding domains, and type I collagen-specific binding is known to mediate the osteogenic response of human BMSCs. Synthetic polymers may be modified by the attachment of certain peptides derived from ECM proteins. One such peptide, Arg–Gly–Asp (RGD), is derived from the ECM proteins fibronectin and laminin, and is widely studied for promoting biomolecular recognition, cell attachment, and cell function.

16.32 THE FUNCTION OF THE VASCULATURE IN SKELETAL REGENERATION

The vascular system is essential for organ growth and repair and has the formidable task of preserving blood flow to organ systems, yet it retains its ability to continually adapt to various changes within tissues (Baiguera and Ribatti, 2013; Gerber et al., 1999). This adaptive response is acutely apparent in the development and repair of bone. The bone vasculature provides an essential transport conduit for oxygen, nutritional factors, host stem cells, hormones, immunity, and the removal of waste products, and it also provides the essential intrinsic cointegration of vascular cells and skeletal cells to release essential growth factors to modulate skeletal growth and regeneration (Kanczler and Oreffo, 2008).

A number of osteogenic and angiogenic factors are interlinked in their effects on angiogenesis, fetal and adult bone development, and clinical repair of bone fractures (Gerber et al., 1999). The potent angiogenic inductive growth factor VEGF plays an integral role in osteogenesis and bone fracture repair (Gerber et al., 1999). In endochondral skeletal development, VEGF induces angiogenesis, chondrocyte apoptosis, SSC homing, and the ossification of the growth plate (Gerber et al., 1999). VEGF, a prosurvival factor for endothelial cells along with hypoxia can elevate the levels of BMP-2 expression in vascular endothelial cells. Further, bone marrow-derived MSCs that overexpress HIF-1α have the ability to regenerate critically sized bone defects, and bone regeneration rates can be elevated by the upregulation of the HIF-1α/VEGF pathway (Wan et al., 2008).

Conversely, Bone Morphogenic Proteins (BMPs) can enhance the production of VEGF-A in osteoblasts and result in increased angiogenesis. Indeed, BMP-2 combined with FGF-2 and VEGF can enhance endothelial angiogenesis, and the combination of BMP-2/VEGF has the ability to influence stem cell homing and differentiation leading to enhanced bone regenerative outcomes (Zhang et al., 2014). Critically, in the bone repair cascade of fracture models, angiogenesis is induced prior to the onset of programmed osteogenesis; however, if VEGF expression is reduced, vascularization of the fracture is impaired, and this results in nonunion of the bone defect (Hausman et al., 2001).

The microvascular endothelium' ability to communicate with MSCs and osteoprogenitors is fundamental to maintaining the integrity of the skeleton. A number of reports have indicated that this crosstalk and influence on cell primary angiogenesis/osteogenic modification is a result of direct cell contact (Villars et al., 2000). In vivo studies in critical sized bone defects demonstrate that the concomitant delivery of human osteoprogenitors and endothelial cells via an alginate microsphere delivery system can promote significant bone healing (Grellier et al., 2009). Additionally, the introduction of vascular inducing growth factors has led to a substantial increase in the efficacy of regenerating bone in vivo. This has led to the development of smart biocompatible scaffolds that can release active growth factors, such as VEGF seeded with human bone marrow stromal cells (HBMSCs), to successfully enhance the repair of critical sized bone defects (Kanczler et al., 2010). Recent advances in vascular analysis, such as vascular perfusion contrast agents and microcomputed tomography (μCT) and the use of the chorioallantoic membrane (CAM) assay model (Figure 16.16), are now starting to provide us with valuable information in the direct role of the vasculature in bone fracture repair.

Biocomposites with the ability to release dual growth factors (i.e., VEGF and BMP-2) at the same release rate or at staggered release intervals at concentration levels compatible to that expected in vivo bone fracture repair have been demonstrated to enhance fracture repair with increased vascularization (Kanczler et al., 2010). Aside from the usual growth factor suspects promoting angiogenesis and osteogenesis, other pathways have recently been trialed to enhance fracture repair. DJ-1 (a 189-amino acid protein encoded by PARK7); a mediator of the crosstalk between vascular and osteogenic cells has been demonstrated when administered to bone defect sites to enhance bone regeneration in vivo by their modulating effects on angiogenesis and osteogenesis (Kim et al., 2012). Moreover, it must be noted that direct mechanical stimulation can also contribute to the effect on vascular growth and subsequent bone regeneration. Thus, future delivery and success will lie in the capacity to scale up bioengineered cell-scaffold constructs for clinical application is central in the application of bone graft substitutes.

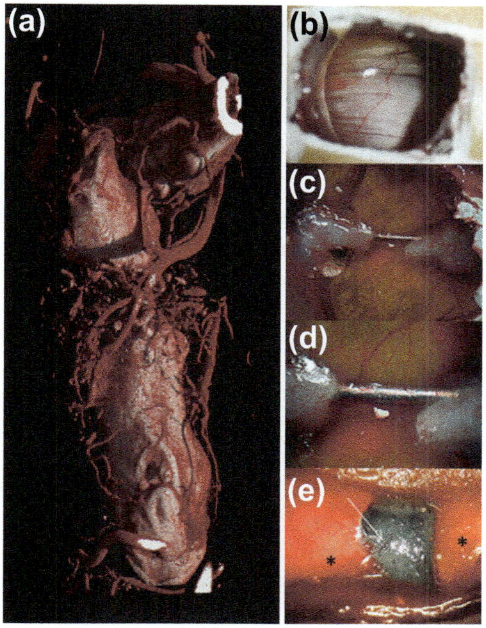

FIGURE 16.16

The three-dimensional image (μCT) of perfused contrast vascular dye (red) depicting the blood vessels in a critical sized mouse femur defect (a). The window of a D18 chick egg depicting the vasculature of the CAM (b). D18 chick femur defect stabilized with a 26 G needle after 8 days of implantation in the CAM: (c) low magnification, (d) high magnification. D18 chick demur defect with an implanted bioscaffold in the defect region (e) (arrow = blood vessel; * = bone) after 8 days of implantation in the CAM.

16.33 ANIMAL MODELS IN BONE TISSUE ENGINEERING

Animal models form an essential link in the development of bone tissue engineering solutions. Animal models can be categorized into three areas of use (Muschler et al., 2010): (1) High throughput "fundamentals" research, which typically uses small mammal laboratory species: mouse, hamster and rat. These species are suitable for large study sizes and are the staple elements for early stage in vivo models; these offer ease of handling, reduced costs, and ready availability through commercial channels. Adversely, short study duration and considerable dissimilarities biomechanically and physiologically limit the clinical relevance of outcomes. (2) Comprehensive, longer-term bioactivity and feasibility studies favor classically medium-sized species such as Rabbit, Dog, and Goat. Studies inform further translation principally through a scale-up in body mass, increase in defect volumes, and marginal scope for incorporating application specific scenarios. However, quadrupedal gait, the lack of analogous secondary bone remodeling, and poor experimental tolerance are issues to consider. Although not without merit, midsized in vivo models cannot inform one to the same degree of accuracy as large mammal studies do and as such are typically considered where facility limitations dictate or as suitable preclinical trial model. (3) Late-translational stage clinical modeling, designed to trial a solution in a scenario as close as is possible to its clinical application. This stage is almost exclusively the domain of ovine models with a limited number of studies using bovine and porcine species. Models in this category are most accurate for assessing clinical efficacy due to analogous species weight, biomechanics, and bone structure with humans. Increased body size and mass allow design protocols matched to human procedural protocol and negative disparities resulting from a quadrupedal gait, although experimentation is typically costly with significant infrastructural requirements often restricting study number (Reichert et al., 2009).

16.34 CURRENT STATUS OF BONE TISSUE ENGINEERING

Clinical experience in Orthopedics—A range of synthetic materials have been used in bone engineering as summarized, with their associated limitations, as in Figure 16.17. A number of randomized controlled clinical trials have evaluated the use of these products alone, and in combination with allograft and bone

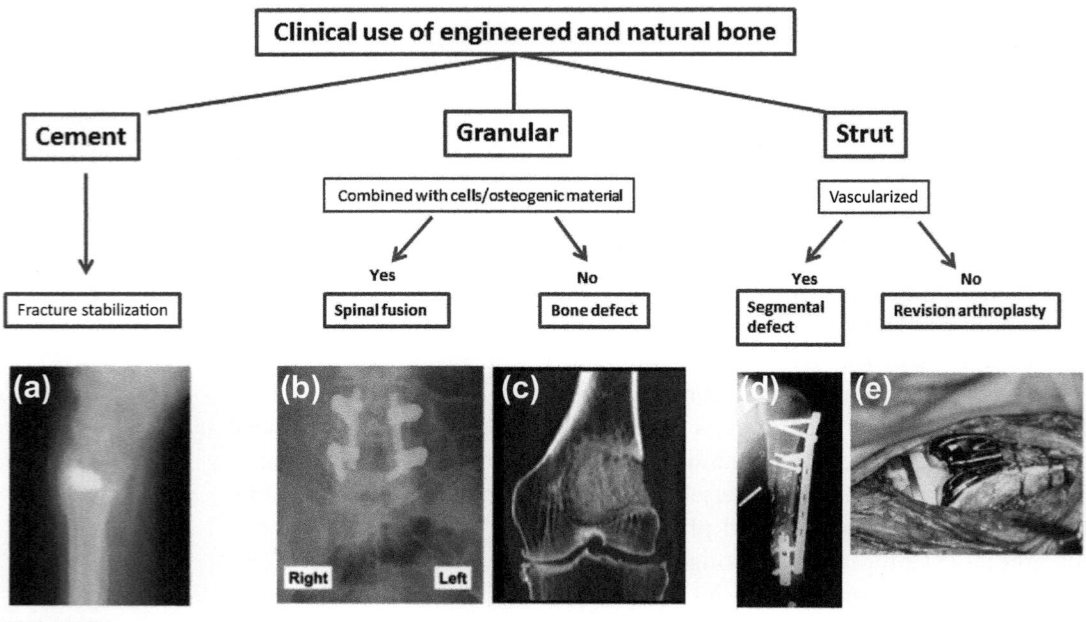

FIGURE 16.17

Clinical use of engineered bone (a, b, and, c), autograft (d), and allograft (e). (a) Cement (Norian SRS) use in stabilization of distal radius fracture (Cassidy et al., 2003). (b, Right): Calcium sulfate (Osteoset) in combination with bone marrow aspirate in posteriolateral lumbar spinal fusion (Niu, 2009). (c) Ultraporous β-TCP in treatment of benign cavity defects (Damron et al., 2013). (d) Reconstruction of long bone defect with free vascularized bone graft (Tu, 2001). (e) The use of structural allograft in revision knee arthroplasty (Clatworthy, 2001).

marrow aspirate as summarized in Table 16.7. Synthetic/or engineered bone has been used successfully in a wide variety of clinical applications including fracture stabilization, arthrodesis, and filling of defects; however, widespread clinical translation is yet to be achieved in situations requiring vascularized and nonvascularized strut grafts, as summarized in Figure 16.18.

Randomized controlled trials have been conducted to investigate the efficacy of injectable calcium phosphate-based cements in the treatment of fractures in the distal radius (Cassidy et al., 2003), calcaneum (Johal et al., 2009), and tibial plateau (Russell and Leighton, 2008). Cassidy et al. found that the use of cement resulted in similar clinical and radiographic outcomes at one year to standard therapy in the management of distal radius fractures. In contrast,

Table 16.7 Clinical Studies Evaluating Synthetic Bone

Randomized controlled trial. $N = 55$ (Damron et al., 2013)	Benign Cavity Lesions, Variety of Bones	β-TCP + BMA (Bone Marrow Aspirate) vs β-TCP alone	Clinical, radiographic mean follow-up 20.2 months	Improvement in both groups. No significant benefit from BMA
Randomized controlled trial. $N = 323$ (Cassidy, 2003)	Distal radius fractures	Closed reduction vs closed reduction + Norian (calcium phosphate cement) *k* wiring in some cases	Clinical, radiographic 1-year follow-up	Radiographic and clinical outcomes similar in both groups
Randomized controlled trial. $N = 52$ (Johal, 2009)	Displaced intraarticular calcaneal fractures	ORIF vs ORIF + α BSM (calcium phosphate cement)	Clinical, radiographic 2-years of follow-up	Less loss of reduction (Bohler's angle) with cement. No clinical difference seen
Randomized controlled trial. $N = 120$ (Russell, 2008)	Tibial plateau fractures	ORIF + autologous graft vs ORIF + α BSM	Clinical, radiographic 1–4 years of follow-up	Subsidence more in autologous group. No clinical difference
Prospective series. $N = 50$ (McNamara, 2010)	Impaction grafting of acetabulum during arthroplasty surgery	1:1 Mixture Allograft and Apapore 60 (Hydroxyapatite 60% porosity)	Clinical (Harris hip score) radiographic mean follow-up 64 months	Mixture allograft and HA comparable to allograft alone in published studies
Retrospective series. $N = 54$ (Oshima, 2012)	Impaction grafting of femur in arthroplasty surgery	Impaction bone grafting with allograft + Hydroxyapatite	Clinical, radiographic scoring. Mean follow-up 92 months	Favorable compared to published data of allograft alone
Retrospective series. $N = 43$ (Whitehouse, 2013)	Impaction grafting of acetabulum in arthroplasty surgery	Bone save (80% tricalcium phosphate and 20% hydroxyapatite)	Clinical, radiographic mean follow-up 4 years	Bone save alone reliable material for impaction grafting in medium term
Randomized controlled trial $N = 256$ (Bae, 2012)	Vertebral compression fractures	Cortoss (resin + ceramic) vs PMMA	Clinical, radiographic 2 years follow-up	Both materials comparable at most time points

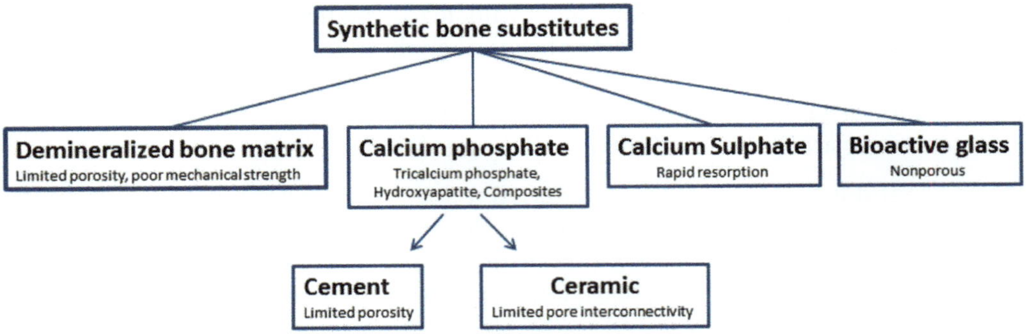

FIGURE 16.18
Variety of materials employed in the production of synthetic bone.

Johal et al. found that the use of bone cement with open reduction internal fixation (ORIF) reduced loss of reduction compared with ORIF alone in the management of displaced intraarticular calcaneal fractures, although the clinical significance of this result was not proven. Russell and Leighton reported less subsidence in tibial plateau fractures managed with ORIF and bone cement compared with cement and autologous bone graft. This is likely to be due to the slower rate of resorption of the cement compared with autologous graft. In general terms, the primary aim of synthetic material in these clinical scenarios was the enhancement of stability to permit early rehabilitation, rather than biological stimulation of healing, as stated by Russell et al. "Primary aim of grafting in a sub articular defect with a tibial plateau fracture is to prevent the collapse of the articular surface" and Johal et al. "primary purpose of the study is to determine whether ORIF plus alpha BSM is superior to ORIF alone in maintaining postoperative calcaneal height and posterior facet reduction." In many cases, clinical outcomes may not correlate with radiographic outcomes, which add further difficulty to quantifying the benefit derived from the engineered bone products. β-Tricalcium phosphate has been shown to be effective in the treatment of benign cavity lesions; however, in a recent randomized controlled trial, the addition of bone marrow aspirate was not found to enhance bone formation (Damron et al., 2013).

Aseptic loosening results from bone resorption and is the most common indication for revision arthroplasty. Impaction bone grafting remains the only reconstructive technique that facilitates reconstitution of bone during revision. Recent case series have demonstrated that hydroxyapatite in combination with allograft and tricalcium phosphate/hydroxyapatite alone (Whitehouse et al., 2013) are capable of producing good medium-term results when used in impaction grafting during revision arthroplasty. Graft material used must be able to withstand the mechanical loading during the impaction, in addition, if polymethyl methacrylate (PMMA) is used, that is, a cemented prosthesis, the graft should be heat stable. In contrast to less vascular areas of skeletal reconstruction, material in impaction grafting is in direct contact with bleeding host bone marrow, and as such is exposed to osteoprogenitor cells, and maybe less likely to benefit from the delivery of additional cells in the scaffold.

Vertebroplasty describes the use of cement to treat fractures secondary to osteoporosis or tumor, and is primarily used to control pain. Cortoss is a nonresorbable bioactive glass that mediates surface deposition of hydroxyapatite, which promotes integration with bone and is radiopaque, a critical property for use in vertebroplasty. Cortoss has a lower exotherm than that of PMMA, and therefore, less necrosis of surrounding biological material would be expected. Bae et al. (2012) demonstrated noninferiority of Cortoss when compared with PMMA in a randomized controlled trial including 256 patients with 2 years of follow-up. The primary endpoint in the study was pain and disability, which reflects the indication for this treatment.

Spinal arthrodesis is used as a treatment for a range of conditions, and is traditionally achieved with the use of autologous bone graft, although the problems associated with harvest of bone graft has prompted the investigation of alternative materials. A recent systematic review explored the use of bone graft alternatives for common spinal arthrodesis procedures (Fischer et al., 2013). The outcome of bone graft alternatives used varied according to the surgical procedure and underlying condition as would be anticipated; however, overall, the use of ceramics was advocated, but not without the use of additional osteoinductive material, such as bone marrow aspirate.

In summary, Clinical demands are diverse, and optimal properties for synthetic bone will vary accordingly. While the use of synthetic engineered bone products in some clinical applications is supported by randomized controlled trials, the development of cell loaded and vascularized construct that could be implanted into a segmental bone defect remains a distant dream. Thus far "low tech" approaches to bone engineering involving the intraoperative application of bone marrow aspirate to synthetic graft material appear to have shown most success in clinical translation.

16.35 FUTURE PERSPECTIVES FOR BONE REGENERATION

In comparison to other cell sources, including pluripotent ES cells and multipotent adult stem cell populations from a range of connective tissues, the relative accessibility of an autologous osteoprogenitor population has led to progress in the application of SSC therapy in the clinic hampered mainly predominantly by our limited understanding of SSC fate, immunophenotype and selection criteria. Future approaches will undoubtedly be centered on the development of simple, safe, and efficacious protocols of cell expansion that meet stringent regulatory body criteria for clinical application together with the application of bioinformatic approaches to generate a phenotypic fingerprint of the skeletal stem can progenitor cell at a single cell resolution, together with the derivation of skeletal cells from pluripotent stem cell sources. It is likely that approaches will include the development and integration of immunoprivileged constructs containing an appropriate scaffold/growth factor(s) composition for autologous and potentially, allogeneic skeletal populations.

For routine and, critically, robust bone formation approaches, the future challenge for regenerative medicine will center on the ability to scale up bioengineered vascularized grafts for clinical applications. The implantation of graft substitutes with temporospatial growth factor release adaptations with seeded stem cells may well be the way forward to promote revascularized bone regeneration, but another strategy where bioartificial bone grafts are prevascularized prior to implantation to the sites of a bone defect could have very positive outcomes (Kokemüller et al., 2013). In summary, the successful repair of large osseous defects or the integration of any other large regenerative tissue constructs will be only as good as the development concomitantly of a functioning blood supply.

16.36 SUMMARY

- Adult articular chondrocytes exhibit a level of phenotypic plasticity that is comparable with that of MSCs.
- The number of chondrocytes needed to start chondrogenesis in a scaffold is $> 20 * 10^6$ cells or at least $1 * 10^6$ cells/cm^2 of cartilage defect area.
- There is a need to study the use of freshly isolated chondrocytes for direct implantation.
- Pooling of zonal chondrocytes from different regions is an acceptable source for chondrocytes to be used for cartilage tissue engineering.
- There is, as yet, no ideal scaffold for cartilage engineering. In small defects in vitro, developed mature osteochondral constructs may be used in the future instead of autologous osteochondral plugs with or without cells. In larger defects, hyaluronan-based scaffolds in combination with chondrocytes is an option that could be further evaluated.
- Small- to medium-sized traumatic defects need to be treated differently compared to large osteoarthritic defects, where the cartilage engineering process has to be directed not only to fill a defect but also to withstand the negative influence of a diseased joint. The tissue engineering approach should address several phenomena including blocking the

production of proinflammatory factors and suppressing the progression of the degenerative process affecting both cartilage and the bone compartment.

- Bone is unique and has a capacity to renew and remodel throughout life. Bone serves a number of key functions that include structural support, protection of internal organs, locomotion, maintenance of mineral balance, and as a reservoir for growth factors and for hematopoiesis.
- Bone marrow contains the SSC that provides the inherent regenerative capacity of bone. These skeletal stem and progenitor cells are typically identified by using a panel of antibodies by applying either positive or negative immunoselection; however, to date, a single unique cell surface maker for the SSC remains elusive. The SSC has been characterized as residing within the perivascular spaces around the vascular sinusoids in the bone marrow and, critically, a single SSC is capable of generating a complete heterotopic bone or bone marrow organ-ossicle including establishment of host hematopoiesis.
- The precise identity of the regulatory effects of SSCs on immune and inflammatory cells remains to be validated conclusively in a defined in vivo system.
- Bone formation involves the exquisite spatiotemporal regulation of recruitment and directed differentiation of bone cell populations under the control of a range of growth factors. A number of osteoinductive growth factors have been identified, including BMP-2 and BMP-7.
- Three-dimensional matrices, or scaffolds, form a vital component of bone tissue engineering. Ideally, the scaffold design and properties should align with the mechanical properties of the bone it will replace and should be resorbable. Cancellous bone autograft provides an osteoconductive, osteoinductive material for orthopedic application; however, limitations in supply and requisite surgery (and complications therein) have led to extensive research to derive alternative materials for clinical applications.
- Critical in any bone tissue regeneration development is a functional vascular system to provide transport of oxygen, nutritional factors, host stem cells, hormones, immunity, and the removal of waste products as well as vascular cells and skeletal cells to aid skeletal growth and regeneration. Development of cell-loaded, vascularized constructs for application in large segmental bone defects or large bone reconstruction remains an unmet challenge.
- The future challenge for regenerative medicine will center on the ability to scale up bioengineered grafts for clinical applications, including an appropriate cell source and vascular supply.

CLASSICAL EXPERIMENT: URIST—BMPS

In the 1960s, Marshall R. Urist (Urist, 1965) investigated decalcified bone that was implanted at various locations, both heterotopic and homotopic, in several types of animals and in humans. These acellular, decalcified bones (i.e., only the organic part of bone tissue) yielded new bone in amounts proportional to the volume of the implant, with over 90% positive experimental results. Evaluation using light microscopy showed that initially (3 weeks postimplantation) these implants were enveloped in highly vascular, inflammatory and fibrous connective tissue. Subsequently, invading cells of the host started the process of resorption of the collagenous matrix, after which new bone deposits became apparent after 4–6 weeks. These early deposits were always found in the interior of well-vascularized excavation chambers. Also, cartilage formation was observed in the closed ends of old, vascular channels, which was gradually replaced by bone tissue between eight and 16 weeks via the process of endochondral ossification. At this point, Urist called this process "osteogenesis by autoinduction." In later experiments, Urist et al. (1976) introduced the term BMP for a proteinaceous compound that was obtained by chromatographic separation of a collagenase digest of bone matrix gelatin. This compound was loaded into diffusion chambers, which were implanted in muscle pouches in adult rats, and demonstrated to induce osteogenesis in the direct vicinity of the diffusion chamber. The purification and characterization of BMPs became established in the late 1980s, as did their amino acid sequence and DNA clones (Wozney, 1992). Currently, many different types of BMPs have been identified, which differ in function (Rengachary, 2002). Although the yield of fresh BMPs out of bone is extremely low (10 kg of bovine bone powder yields 10 μg of BMP), recombinant DNA technologies have allowed now the production of large quantities of the so-called rhBMP for commercial exploitation and clinical use. An example of the in vivo potential of rhBMP-2 on porous CaP implants in rabbits is shown below. Several clinical reports on nonunions in tibia, femur, and the spine have also shown beneficial results for the use of rhBMPs. Spine fusion models, such as an anterior (cage) or posterolateral approach (both with or without instrumentation), have shown additional positive results for the use of both rhBMP-2 and OP-1 (rhBMP-7) as an alternative to the use of autograft bone. Several studies have measured both clinical signs of fusion (such as the OSWERTY pain and disability scoring system) and radiographic evidence for successful fusion. Unfortunately, large quantities of BMPs of up to 20 mg per side (posterolateral fusion model) are needed with subsequent considerable financial consequences.

STATE-OF-THE-ART EXPERIMENT: CLAY NANOPARTICLES FOR REGENERATIVE MICROENVIRONMENTS

The tissue engineering scaffold serves a dual function. On the one hand, the scaffold serves as an implant material with appropriate physical properties to facilitate the delivery of stem cells to the site of damage and provide a template for new tissue formation. At the same time, the scaffold is to provide a developmentally conducive microenvironment that presents, in space and time, the biological cues that direct stem cell behavior. The optimization and marrying of these complex design parameters remain a critical unmet challenge. Recent studies investigating, for the first time, clay nanoparticles for stem cell-based regenerative strategies suggest striking potential to provide new opportunities for regenerative medicine (Dawson and Oreffo, 2013). The well-established utility of clay nanoparticles to interact with polymers as seen in the development of polymer-clay nanocomposites offers exciting possibilities for physical improvement of hydrogels and scaffolds. Significantly improved toughness and compressive properties, along with tuned degradation, have been achieved through composite approaches. Furthermore, clays have been extensively used in the pharmaceutical industry for their ability to interact with drugs and other biological molecules for controlled delivery. The potential of clay–biomolecule interactions for the development of regenerative microenvironments is

STATE-OF-THE-ART EXPERIMENT: CLAY NANOPARTICLES FOR REGENERATIVE MICROENVIRONMENTS *Continued*

an exciting recent development in the field. For example, hydrogels formed through the electrolyte-induced formation of gel capsules from Laponite clay particles have been explored for the potential to localize bioactive molecules with the aim of developing biochemically defined niches for stem cell–mediated regeneration (Dawson et al., 2011). Laponite gels display a high and broad-spectrum affinity for solubilized proteins in cell culture media. Clay gel addition to solutions of model proteins albumin (66.4 kDa, PI, 4.7) and lysozyme (14.4 kDa, PI, 11) induced the active uptake of the two proteins from the media via sorption to the gel capsule surface (Figure 16.19(a)). The utility of this approach to generate niches for tissue regeneration has been demonstrated with the matrix molecules fibronectin and type I collagen and with the angiogenic VEGF. Addition of fibronectin and collagen into clay gels allowed adhesion of cultured cells (Figure 16.19(b)), addition of fibronectin into clay gels was observed to enhance the chondrogenic differentiation of encapsulated MSCs, and active uptake by clay gels of exogenously applied VEGF enhanced angiogenesis when the gels were implanted into a murine femoral defect (Figure 16.19(c)).

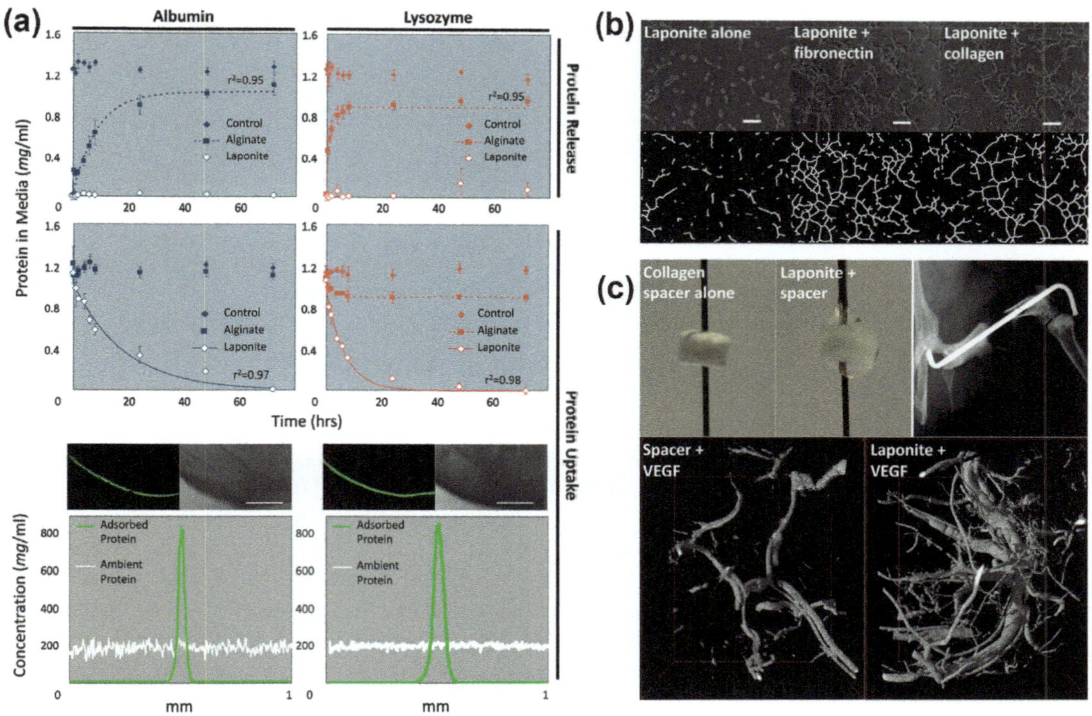

FIGURE 16.19

Biomolecule uptake and delivery in clay gels. (a) Negligible release of premixed model proteins was observed from Laponite gel capsules compared with alginate controls. Addition of clay gels to protein solution resulted in the active uptake of protein from the media and concentrated upon the surface of the gel capsule. Function of adsorbed protein was observed through premixing of the matrix molecule fibronectin and type I collagen into clay gel films to sustain (b) adhesion and network organization of endothelial cells in vitro (scale bars 100 μm) and (c) through preincubation of Laponite gels in VEGF solutions to induce an in vivo angiogenic response in a murine femoral defect. *From Dawson and Oreffo (2013) Copyright 2013, Wiley-VCH.*

REFERENCES

Arvidson, K., Abdallah, B., Applegate, L., Baldini, N., Cenni, E., Gomez–Barrena, E., Granchi, D., Kassem, M., Konttinen, Y., Mustafa, K., 2011. Bone regeneration and stem cells. J. Cell. Mol. Med. 15, 718–746.

Badylak, S.F., 2007. The extracellular matrix as a biologic scaffold material. Biomaterials 28, 3587–3593.

Bae, H., Hatten Jr., H.P., Linovitz, R., Tahernia, A.D., Schaufele, M.K., McCollom, V., Gilula, L., Maurer, P., Benyamin, R., Mathis, J.M., 2012. A prospective randomized FDA-IDE trial comparing Cortoss with PMMA for vertebroplasty: a comparative effectiveness research study with 24 month follow-up. Spine 37, 544–550.

Baiguera, S., Ribatti, D., 2013. Endothelialization approaches for viable engineered tissues. Angiogenesis 16, 1–14.

Bigdeli, N., Karlsson, C., Strehl, R., Concaro, S., Hyllner, J., Lindahl, A., 2009. Coculture of human embryonic stem cells and human articular chondrocytes results in significantly altered phenotype and improved chondrogenic differentiation. Stem Cells 27, 1812–1821.

Borestrom, C., Simonsson, S., Enochson, L., Bigdeli, N., Brantsing, C., Ellerstrom, C., Hyllner, J., Lindahl, A., 2014. Footprint-free human induced pluripotent stem cells from articular cartilage with redifferentiation capacity: a first step toward a clinical-grade cell source. Stem Cells Transl. Med.

Brittberg, M., Lindahl, A., Nilsson, A., Ohlsson, C., Isaksson, O., Peterson, L., 1994. Treatment of deep cartilage defects in the knee with autologous chondrocyte transplantation. N. Engl. J. Med. 331, 889–895.

Campoccia, D., Doherty, P., Radice, M., Brun, P., Abatangelo, G., Williams, D.F., 1998. Semisynthetic resorbable materials from hyaluronan esterification. Biomaterials 19, 2101–2127.

Cao, Y., Vacanti, J.P., Paige, K.T., Upton, J., Vacanti, C.A., 1997. Transplantation of chondrocytes utilizing a polymer-cell construct to produce tissue-engineered cartilage in the shape of a human ear. Plast. Reconstr. Surg. 100, 297–302. Discussion 303–304.

Caplan, A.I., Correa, D., 2011. The MSC: an injury drugstore. Cell. Stem Cell 9, 11–15.

Caplan, A.I., 1991. Mesenchymal stem cells. J. Orthop. Res. 9, 641–650.

Caplan, A.I., 2003. Embryonic development and the principles of tissue engineering. Novartis Found. Symp. 249, 17–25. Discussion 25–33, 170–174, 239–241.

Caplan, A.I., Elyaderani, M., Mochizuki, Y., Wakitani, S., Goldberg, V.M., 1997. Principles of cartilage repair and regeneration. Clin. Orthop. Relat. Res. 254–269.

Cassidy, C., Jupiter, J.B., Cohen, M., Delli-Santi, M., Fennell, C., Leinberry, C., Husband, J., Ladd, A., Seitz, W.R., Constanz, B., 2003. Norian SRS cement compared with conventional fixation in distal radial fractures. A randomized study. J. Bone Jt. Surg. 85, 2127–2137.

Chen, A.C., Nagrampa, J.P., Schinagl, R.M., Lottman, L.M., Sah, R.L., 1997. Chondrocyte transplantation to articular cartilage explants in vitro. J. Orthop. Res. 15, 791–802.

Chen, W.Y., Abatangelo, G., 1999. Functions of hyaluronan in wound repair. Wound Repair Regen. 7, 79–89.

Clarke, B., 2008. Normal bone anatomy and physiology. Clin. J. Am. Soc. Nephrol. 3, S131–S139.

Damron, T.A., Lisle, J., Craig, T., Wade, M., Silbert, W., Cohen, H., 2013. Ultraporous beta-tricalcium phosphate alone or combined with bone marrow aspirate for benign cavitary lesions: comparison in a prospective randomized clinical trial. J. Bone Jt. Surg. Am. 95, 158–166.

Darling, E.M., Hu, J.C., Athanasiou, K.A., 2004. Zonal and topographical differences in articular cartilage gene expression. J. Orthop. Res. 22, 1182–1187.

Dawson, J.I., Oreffo, R.O.C., 2008. Bridging the regeneration gap: stem cells, biomaterials and clinical translation in bone tissue engineering. Archives Biochem. Biophys. 473, 124–131.

Dawson, J.I., Oreffo, R.O.C., 2013. Clay: new opportunities for tissue regeneration and biomaterial design. Adv. Mater.

Dawson, J.I., Kanczler, J.M., Yang, X.B., Attard, G.S., Oreffo, R.O.C., 2011. Clay gels for the delivery of regenerative microenvironments. Adv. Mater. 23, 3304–3308.

Dawson, J.I., Kanczler, J., Tare, R., Kassem, M., Oreffo, R.O., 2014. Concise review: bridging the gap: bone regeneration using skeletal stem cell–based strategies—where are we now? Stem Cells 32, 35–44.

Deschamps, J.Y., Roux, F.A., Sai, P., Gouin, E., 2005. History of xenotransplantation. Xenotransplantation 12, 91–109.

Dickinson, S.C., Sims, T.J., Pittarello, L., Soranzo, C., Pavesio, A., Hollander, A.P., 2005. Quantitative outcome measures of cartilage repair in patients treated by tissue engineering. Tissue Eng. 11, 277–287.

Drury, J.L., Mooney, D.J., 2003. Hydrogels for tissue engineering: scaffold design variables and applications. Biomaterials 24, 4337–4351.

Even, J., Eskander, M., Kang, J., 2012. Bone morphogenetic protein in spine surgery: current and future uses. J. Am. Acad. Orthop. Surg. 20, 547–552.

Fischer, C.R., Cassilly, R., Cantor, W., Edusei, E., Hammouri, Q., Errico, T., 2013. A systematic review of comparative studies on bone graft alternatives for common spine fusion procedures. Eur. Spine J. 1–13.

Freed, L.E., Langer, R., Martin, I., Pellis, N.R., Vunjak-Novakovic, G., 1997. Tissue engineering of cartilage in space. Proc. Natl. Acad. Sci. USA 94, 13885–13890.

Freed, L.E., Vunjak-Novakovic, G., Biron, R.J., Eagles, D.B., Lesnoy, D.C., Barlow, S.K., Langer, R., 1994. Biodegradable polymer scaffolds for tissue engineering. Biotechnology (NY) 12, 689–693.

Friedenstein, A.J., Chailakhjan, R.K., Lalykina, K.S., 1970. The development of fibroblast colonies in monolayer cultures of Guinea-Pig bone marrow and spleen cells. Cell. Tissue Kinet. 3, 393–403.

Frisbie, D.D., Lu, Y., Kawcak, C.E., Dicarlo, E.F., Binette, F., Mcilwraith, C.W., 2009. In vivo evaluation of autologous cartilage fragment-loaded scaffolds implanted into equine articular defects and compared with autologous chondrocyte implantation. Am. J. Sports Med. 37 (Suppl. 1), 71s–80s.

Gerber, H.-P., Vu, T.H., Ryan, A.M., Kowalski, J., Werb, Z., Ferrara, N., 1999. VEGF couples hypertrophic cartilage remodeling, ossification and angiogenesis during endochondral bone formation. Nat. Med. 5, 623–628.

Giannoudis, P.V., Einhorn, T.A., Marsh, D., 2007. Fracture healing: the diamond concept. Injury 38, S3–S6.

Grellier, M., Granja, P.L., Fricain, J.-C., Bidarra, S.J., Renard, M., Bareille, R., Bourget, C., Amédée, J., Barbosa, M.A., 2009. The effect of the co-immobilization of human osteoprogenitors and endothelial cells within alginate microspheres on mineralization in a bone defect. Biomaterials 30, 3271–3278.

Hausman, M., Schaffler, M., Majeska, R., 2001. Prevention of fracture healing in rats by an inhibitor of angiogenesis. Bone 29, 560–564.

Hayes, A.J., Macpherson, S., Morrison, H., Dowthwaite, G., Archer, C.W., 2001. The development of articular cartilage: evidence for an appositional growth mechanism. Anat. Embryol. Berl. 203, 469–479.

Heinegard, D., Saxne, T., 2011. The role of the cartilage matrix in osteoarthritis. Nat. Rev. Rheumatol. 7, 50–56.

Hing, K.A., Wilson, L.F., Buckland, T., 2007. Comparative performance of three ceramic bone graft substitutes. Spine J. 7, 475–490.

Hunziker, E.B., Quinn, T.M., Hauselmann, H.J., 2002. Quantitative structural organization of normal adult human articular cartilage. Osteoarthr. Cartil. 10, 564–572.

Hutmacher, D.W., 2000. Scaffolds in tissue engineering bone and cartilage. Biomaterials 21, 2529–2543.

Itay, S., Abramovici, A., Nevo, Z., 1987. Use of cultured embryonal chick epiphyseal chondrocytes as grafts for defects in chick articular cartilage. Clin. Orthop. Relat. Res. 284–303.

Iwasa, J., Ochi, M., Uchio, Y., Katsube, K., Adachi, N., Kawasaki, K., 2003. Effects of cell density on proliferation and matrix synthesis of chondrocytes embedded in atelocollagen gel. Artif. Organs 27, 249–255.

Johal, H.S., Buckley, R.E., Le, I.L., Leighton, R.K., 2009. A prospective randomized controlled trial of a bioresorbable calcium phosphate paste ([alpha]-BSM) in treatment of displaced intra-articular calcaneal fractures. J. Trauma Acute Care Surg. 67, 875–882.

Joyce, M.E., Roberts, A.B., Sporn, M.B., Bolander, M.E., 1990. Transforming growth factor-beta and the initiation of chondrogenesis and osteogenesis in the rat femur. J. Cell. Biol. 110, 2195–2207.

Kanczler, J.M., Oreffo, R.O., 2008. Osteogenesis and angiogenesis: the potential for engineering bone. Eur. Cell. Mater. 15, 100–114.

Kanczler, J.M., Ginty, P.J., White, L., Clarke, N.M., Howdle, S.M., Shakesheff, K.M., Oreffo, R.O., 2010. The effect of the delivery of vascular endothelial growth factor and bone morphogenic protein-2 to osteoprogenitor cell populations on bone formation. Biomaterials 31, 1242–1250.

Karlsson, C., Thornemo, M., Henriksson, H.B., Lindahl, A., 2009. Identification of a stem cell niche in the zone of Ranvier within the knee joint. J. Anat. 215, 355–363.

Kawabe, N., Yoshinao, M., 1991. The repair of full-thickness articular cartilage defects. Immune responses to reparative tissue formed by allogeneic growth plate chondrocyte implants. Clin. Orthop. Relat. Res. 279–293.

Kim, J.-M., Shin, H.-I., Cha, S.-S., Lee, C.S., Hong, B.S., Lim, S., Jang, H.-J., Kim, J., Yang, Y.R., Kim, Y.-H., Yun, S., Rijal, G., Lee-Kwon, W., Seo, J.K., Gho, Y.S., Ryu, S.H., Hur, E.-M., Suh, P.-G., 2012. DJ-1 promotes angiogenesis and osteogenesis by activating FGF receptor-1 signaling. Nat. Commun. 3, 1296.

Klein, T.J., Schumacher, B.L., Schmidt, T.A., Li, K.W., Voegtline, M.S., Masuda, K., Thonar, E.J., Sah, R.L., 2003. Tissue engineering of stratified articular cartilage from chondrocyte subpopulations. Osteoarthr. Cartil. 11, 595–602.

Kokemüller, H., Jehn, P., Spalthoff, S., Essig, H., Tavassol, F., Schumann, P., Andreae, A., Nolte, I., Jagodzinski, M., Gellrich, N.-C., 2013. En bloc prefabrication of vascularized bioartificial bone grafts in sheep and complete workflow for custom-made transplants. Int. J. Oral Maxillofac. Surg.

Kramer, J., Hegert, C., Rohwedel, J., 2003. In Vitro differentiation of mouse ES cells: bone and cartilage. Methods Enzymol. 365, 251–268.

Le Douarin, N.M., Dupin, E., 2012. The neural crest in vertebrate evolution. Curr. Opin. Genet. Dev. 22, 381–389.

Lebaron, R.G., Athanasiou, K.A., 2000. Ex vivo synthesis of articular cartilage. Biomaterials 21, 2575–2587.

Logeart-Avramoglou, D., Anagnostou, F., Bizios, R., Petite, H., 2005. Engineering bone: challenges and obstacles. J. Cell. Mol. Med. 9, 72–84.

Lu, Y., Dhanaraj, S., Wang, Z., Bradley, D.M., Bowman, S.M., Cole, B.J., Binette, F., 2006. Minced cartilage without cell culture serves as an effective intraoperative cell source for cartilage repair. J. Orthop. Res. 24, 1261–1270.

Mackie, E., Ahmed, Y., Tatarczuch, L., Chen, K.-S., Mirams, M., 2008. Endochondral ossification: how cartilage is converted into bone in the developing skeleton. Int. J. Biochem. Cell Biol. 40, 46–62.

Marieb, E.N., Hoehn, K., 2007. Human Anatomy & Physiology. Pearson Education.

Minear, S., Leucht, P., Jiang, J., Liu, B., Zeng, A., Fuerer, C., Nusse, R., Helms, J.A., 2010. Wnt proteins promote bone regeneration. Sci. Translational Med. 2, 29ra30.

Morales, T.I., Hascall, V.C., 1989. Factors involved in the regulation of proteoglycan metabolism in articular cartilage. Arthritis Rheum. 32, 1197–1201.

Muschler, G.F., Raut, V.P., Patterson, T.E., Wenke, J.C., Hollinger, J.O., 2010. The design and use of animal models for translational research in bone tissue engineering and regenerative medicine. Tissue Eng. Part B Rev. 1, 123–145.

Namba, R.S., Meuli, M., Sullivan, K.M., Le, A.X., Adzick, N.S., 1998. Spontaneous repair of superficial defects in articular cartilage in a fetal lamb model. J. Bone Jt. Surg. Am. 80, 4–10.

Oldershaw, R.A., Baxter, M.A., Lowe, E.T., Bates, N., Grady, L.M., Soncin, F., Brison, D.R., Hardingham, T.E., Kimber, S.J., 2010. Directed differentiation of human embryonic stem cells toward chondrocytes. Nat. Biotechnol. 28, 1187–1194.

Pavesio, A., Abatangelo, G., Borrione, A., Brocchetta, D., Hollander, A.P., Kon, E., Torasso, F., Zanasi, S., Marcacci, M., 2003. Hyaluronan-based scaffolds (Hyalograft C) in the treatment of knee cartilage defects: preliminary clinical findings. Novartis Found. Symp. 249, 203–217 Discussion 229–233, 234–238, 239–241.

Peterson, L., Brittberg, M., Kiviranta, I., Akerlund, E.L., Lindahl, A., 2002. Autologous chondrocyte transplantation. Biomechanics and long-term durability. Am. J. Sports Med. 30, 2–12.

Peterson, L., Vasiliadis, H.S., Brittberg, M., Lindahl, A., 2010. Autologous chondrocyte implantation: a long-term follow-up. Am. J. Sports Med. 38, 1117–1124.

Pittenger, M.F., Mackay, A.M., Beck, S.C., Jaiswal, R.K., Douglas, R., Mosca, J.D., Moorman, M.A., Simonetti, D.W., Craig, S., Marshak, D.R., 1999. Multilineage potential of adult human mesenchymal stem cells. Science 284, 143–147.

Poole, A.R., Kojima, T., Yasuda, T., Mwale, F., Kobayashi, M., Laverty, S., 2001. Composition and structure of articular cartilage: a template for tissue repair. Clin. Orthop. S26–S33.

Puelacher, W.C., Kim, S.W., Vacanti, J.P., Schloo, B., Mooney, D., Vacanti, C.A., 1994. Tissue-engineered growth of cartilage: the effect of varying the concentration of chondrocytes seeded onto synthetic polymer matrices. Int. J. Oral Maxillofac. Surg. 23, 49–53.

Reichert, J.C., Saifzadeh, S., Wullschleger, M.E., Epari, D.R., Schütz, M.A., Duda, G.N., Schell, H., van Griensven, M., Redl, H., Hutmacher, D.W., 2009. The challenge of establishing preclinical models for segmental bone defect research. Biomaterials 30 (12), 2149–2163.

Rengachary, S.S., 2002. Bone morphogenetic proteins: basic concepts. Neurosurg. Focus 13, 1–6.

Rosen, C.J., Compston, J.E., Lian, J.B., 2009. Primer on the Metabolic Bone Diseases and Disorders of Mineral Metabolism. John Wiley & Sons.

Russell, T.A., Leighton, R.K., 2008. Comparison of autogenous bone graft and endothermic calcium phosphate cement for defect augmentation in tibial plateau fractures. A multicenter, prospective, randomized study. J. Bone Jt. Surg. 90, 2057–2061.

Sacchetti, B., Funari, A., Michienzi, S., Di Cesare, S., Piersanti, S., Saggio, I., Tagliafico, E., Ferrari, S., Robey, P.G., Riminucci, M., 2007. Self-renewing osteoprogenitors in bone marrow sinusoids can organize a hematopoietic microenvironment. Cell 131, 324–336.

Sampath, T., Reddi, A., 1981. Dissociative extraction and reconstitution of extracellular matrix components involved in local bone differentiation. Proc. Natl. Acad. Sci. USA 78, 7599–7603.

Silva, M., Cyster, L., Barry, J., Yang, X., Oreffo, R., Grant, D., Scotchford, C., Howdle, S., Shakesheff, K., Rose, F., 2006. The effect of anisotropic architecture on cell and tissue infiltration into tissue engineering scaffolds. Biomaterials 27, 5909–5917.

Spence, A., 1991. Basic Human Anatomy, third ed. Addison-Wesley, Redwood City, CA.

Stocum, D.L., 2004. Tissue restoration through regenerative biology and medicine. Adv. Anat. Embryol. Cell. Biol. 176 (III–VIII), 1–101 (Back Cover).

Taichman, R.S., 2005. Blood and bone: two tissues whose fates are intertwined to create the hematopoietic stem-cell niche. Blood 105, 2631–2639.

Takahashi, K., Tanabe, K., Ohnuki, M., Narita, M., Ichisaka, T., Tomoda, K., Yamanaka, S., 2007. Induction of pluripotent stem cells from adult human fibroblasts by defined factors. Cell 131, 861–872.

Tallheden, T., Dennis, J.E., Lennon, D.P., Sjogren-Jansson, E., Caplan, A.I., Lindahl, A., 2003. Phenotypic plasticity of human articular chondrocytes. J. Bone Jt. Surg. Am. 85-A (Suppl. 2), 93–100.

Tay, A.G., Farhadi, J., Suetterlin, R., Pierer, G., Heberer, M., Martin, I., 2004. Cell yield, proliferation, and postexpansion differentiation capacity of human ear, nasal, and rib chondrocytes. Tissue Eng. 10, 762–770.

Thomson, J.A., Itskovitz-Eldor, J., Shapiro, S.S., Waknitz, M.A., Swiergiel, J.J., Marshall, V.S., Jones, J.M., 1998. Embryonic stem cell lines derived from human blastocysts. Science 282, 1145–1147.

Toole, B.P., 2001. Hyaluronan in morphogenesis. Semin. Cell. Dev. Biol. 12, 79–87.

Urist, M.R., 1965. Bone: formation by autoinduction. Science 150, 893–899.

Urist, M., Nogami, H., Mikulski, A., 1976. A bone morphogenetic polypeptide. In: Calcified Tissues 1975. Springer.

Vacanti, C.A., Kim, W., Schloo, B., Upton, J., Vacanti, J.P., 1994. Joint resurfacing with cartilage grown in situ from cell-polymer structures. Am. J. Sports Med. 22, 485–488.

Van Der Rest, M., Garrone, R., 1991. Collagen family of proteins. FASEB J. 5, 2814–2823.

Villars, F., Bordenave, L., Bareille, R., Amedee, J., 2000. Effect of human endothelial cells on human bone marrow stromal cell phenotype: role of VEGF? J. Cell. Biochem. 79, 672–685.

Wakitani, S., Goto, T., Pineda, S.J., Young, R.G., Mansour, J.M., Caplan, A.I., Goldberg, V.M., 1994. Mesenchymal cell-based repair of large, full-thickness defects of articular cartilage. J. Bone Jt. Surg. Am. 76, 579–592.

Waldman, S.D., Grynpas, M.D., Pilliar, R.M., Kandel, R.A., 2003. The use of specific chondrocyte populations to modulate the properties of tissue-engineered cartilage. J. Orthop. Res. 21, 132–138.

Wan, C., Gilbert, S.R., Wang, Y., Cao, X., Shen, X., Ramaswamy, G., Jacobsen, K.A., Alaql, Z.S., Eberhardt, A.W., Gerstenfeld, L.C., 2008. Activation of the hypoxia-inducible factor-1α pathway accelerates bone regeneration. Proc. Natl. Acad. Sci. USA 105, 686–691.

Watanabe, N., Tezuka, Y., Matsuno, K., Miyatani, S., Morimura, N., Yasuda, M., Fujimaki, R., Kuroda, K., Hiraki, Y., Hozumi, N., Tezuka, K., 2003. Suppression of differentiation and proliferation of early chondrogenic cells by notch. J. Bone Min. Metab. 21, 344–352.

White, T.D., Folkens, P.A., 2005. The Human Bone Manual. Academic Press.

Whitehouse, M.R., Dacombe, P.J., Webb, J.C., Blom, A.W., 2013. Impaction grafting of the acetabulum with ceramic bone graft substitute: high survivorship in 43 patients with a mean follow-up period of 4 years. Acta Orthop. 84, 371–376.

Wozney, J.M., 1992. The bone morphogenetic protein family and osteogenesis. Mol. Reprod. Dev. 32, 160–167.

Xu, J.W., Zaporojan, V., Peretti, G.M., Roses, R.E., Morse, K.B., Roy, A.K., Mesa, J.M., Randolph, M.A., Bonassar, L.J., Yaremchuk, M.J., 2004. Injectable tissue-engineered cartilage with different chondrocyte sources. Plast. Reconstr. Surg. 113, 1361–1371.

Yamanaka, S., Blau, H.M., 2010. Nuclear reprogramming to a pluripotent state by three approaches. Nature 465, 704–712.

Zhang, W., Zhu, C., Wu, Y., Ye, D., Wang, S., Zou, D., Zhang, X., Kaplan, D.L., Jiang, X., 2014. VEGF and BMP-2 promote bone regeneration by facilitating bone marrow stem cell homing and differentiation. Eur. Cells Mater. 27, 1.

CHAPTER 17

Tissue Engineering of the Nervous System

Paul D. Dalton[1], Alan R. Harvey[2], Martin Oudega[3], Giles W. Plant[4]
[1]Department of Functional Materials in Medicine and Dentistry, University of Würzburg, Germany; [2]School of Anatomy, Physiology and Human Biology, The University of Western Australia, Perth, WA, Australia; [3]Departments of Physical Medicine and Rehabilitation, Bioengineering, and Neurobiology, University of Pittsburgh, Pittsburgh, PA, USA; [4]Department of Neurosurgery, Stanford University, Stanford, CA, USA

LEARNING OBJECTIVES

- To understand the different regenerative capacities of the peripheral nervous system (PNS) and central nervous system (CNS), and identify the pros and cons of using different cell types for transplantation into the injured CNS
- To appreciate the critical gap length in peripheral nerves, and how therapies can increase this distance
- To gain a knowledge of different bioengineering strategies being used to promote peripheral nerve repair
- To understand the potential role of genetic modification of cells for use in biohybrid implants in the nervous system, and the limitations of animal models for human neurotrauma and neurodegenerative disease
- To recognize the different types of spinal cord injury (SCI) and how this affects potential treatment strategies
- To understand the importance of using biologically relevant peptide sequences in polymer scaffolds
- To appreciate the unique nature of the optic nerve as a model for CNS regeneration
- To appreciate a key challenge of astrocyte scarring to neuroprosthetic arrays

17.1 INTRODUCTION

Diseases or injuries to the nervous system are devastating to the individual and have substantial societal implications and costs. The retina, cochlea, spinal cord, brain, and peripheral nerve have different molecular and cellular environments and therefore require specific tissue engineering (TE) strategies. For example,

Tissue Engineering.
Copyright © 2015 Elsevier Inc. All rights reserved.

the limited regenerative capacity in the central nervous system (CNS) is primarily due to factors that are both intrinsic and extrinsic to the mature neuron. Depending on the neuropathological condition under investigation, injectable matrices for the CNS can provide a substrate for axonal growth in the inhibitory CNS environment, act as a delivery vehicle, and/or a supportive environment for repair-supporting donor cells of various origins. The primary focus of peripheral nerve TE is the reconnection of severed nerves separated from their targets, or from distal nerve segments. A PNS therapeutic approach may also require the implantation of artificial structures that support and guide regenerating axons.

17.2 PERIPHERAL NERVE

Peripheral nerve injuries (PNIs) can lead to a lifetime loss of function and disfigurement even though the peripheral nerve is capable of a substantial amount of regeneration when crushed or severed. Peripheral nerves contain only the axon part of the neuron and one could consider the peripheral nerve trunk as a protective structure for axons. The cell bodies of sensory neurons are located in structures that are adjacent to the spinal cord (dorsal root ganglion (DRG) cells) or in cranial ganglia, while the cell bodies of motor neurons are within the CNS (spinal cord or brainstem) (Figure 17.1). Regenerating

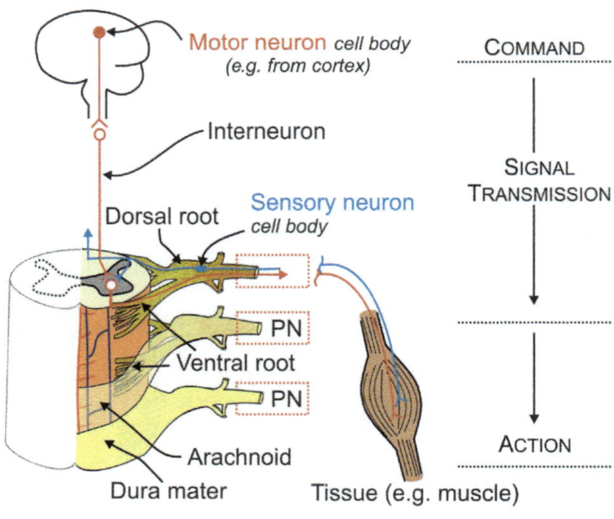

FIGURE 17.1
Origins of PNS axons. The junction between the spinal cord and the PNS, with schematically drawn cell bodies of sensory neurons (blue) and motor neurons (red) and their corresponding neurites. For the length of the peripheral nerve (PN), only axons are present in the tissue. The motor neurons are typically in the brain; however, other intermediate neurons located within the gray matter can play a role in transferring the electrical impulse. The right-hand side text and arrows indicate the direction of the signal for a motor command—a sensory signal is not shown. Dr Ian Galea's assistance with this image is appreciated.

axons have demonstrated guidance over long distances along naturally occurring conduits called Bands of Büngner once the nerve defect is bridged. Small defects may be repaired by suturing nerve stumps together, but for larger nerve gaps, a commonly used approach in peripheral nerve TE involves in vivo implantation of artificial scaffolds and substrates that will bridge the defect and guide naturally regenerating axons to the distal segment. While many experimental in vivo animal models focus on the sciatic nerve in the hindlimb, injuries to cranial nerves can cause permanent facial disfigurement and are a major clinical issue.

17.2.1 Peripheral Nerve Anatomy

Peripheral spinal nerves (Figure 17.2(a)) originate at the dorsal or ventral roots of the spinal cord, while cranial nerves originate from the brainstem. Dorsal roots contain sensory axons which carry signals into the CNS. Ventral roots contains motor axons which carry signals from CNS-originating neurons to muscles and glands (Figure 17.1). Peripheral nerves are discrete fibrocollagenous trunks filled with sensory and motor axons, and support cells such as Schwann cells and fibroblasts. Due to limb movements and the resulting tensile and compressive stresses, the epineurium provides a protective structure to the axons. This fibrocollagenous epineurium binds individual fascicles into one nerve trunk. Within the epineurium are axon-filled fascicles; many but not all the axons are myelinated by supporting glial cells called Schwann cells. Cranial nerves can be purely sensory or motor, or may contain both types of axons. An excellent review of the peripheral nerve extracellular matrix (ECM) is written by Gonzalez-Perez et al.

17.2.2 Peripheral Nerve Injury

Regeneration of severed axons requires that the parent cell bodies survive the initial trauma. Injuries closer to the spinal cord or brainstem are more likely to cause neuronal death and are therefore more complex compared to an injury at the distal periphery. After initial PNI, the proximal nerve stump will swell, but it will undergo minimal damage compared to the distal end. After a nerve injury, the distal portion degenerates due to protease activity termed "Wallerian Degeneration," reviewed by Dubový. In the distal end, the axonal cytoskeleton breaks down, and Schwann cells shed their myelin lipids. Macrophages clear myelin and other debris, a critical event that increases the potential for nerve regeneration. Importantly, Schwann cells in the distal portion of the severed nerve proliferate and begin to secrete specific factors that stimulate axon regrowth. After debris clearance, regeneration begins at the proximal end and continues toward the distal stump. Accurate alignment of the original fascicles is critical to microsurgical repair (Figure 17.2(b)). In the distal nerve, lamina tubules form Bands of Büngner and play a crucial role in successful peripheral nerve regeneration. Axonal regeneration in humans occurs at a rate of about

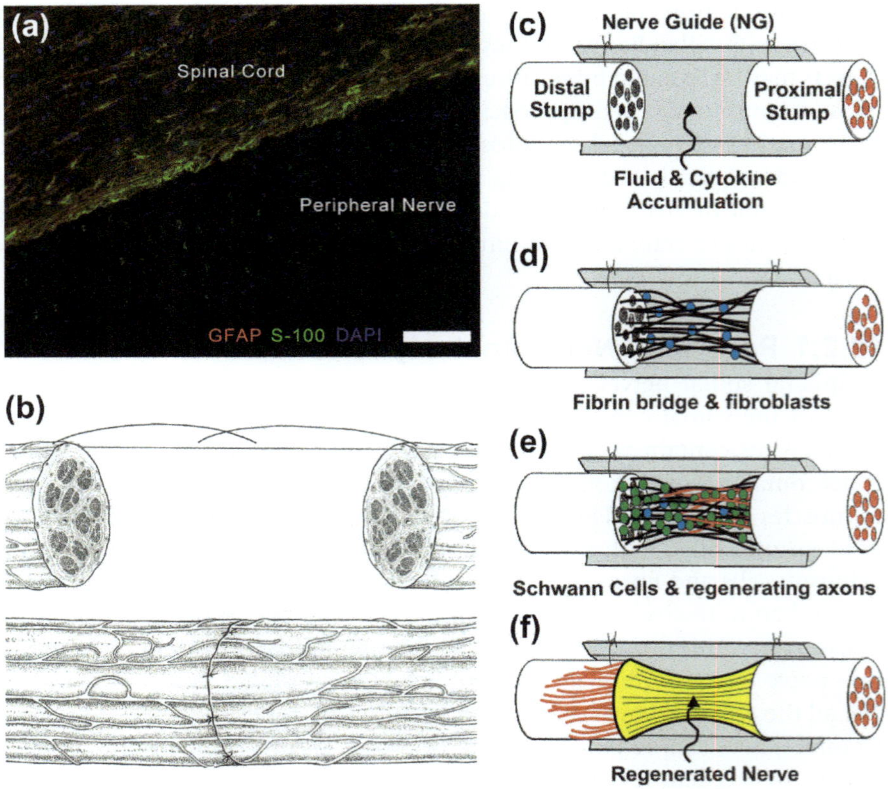

FIGURE 17.2

(a) Immunohistochemistry of the peripheral nerve branching away from the spinal cord. Myelinating Schwann cells are positive to S-100, while astrocytes are positive to both S-100 and GFAP (with the overlap seen as yellow). Scale bar is 100 µm. (b) End-to-end suturing of peripheral nerves. Alignment of the fascicles is critical in successful regeneration, and microsurgical techniques have been developed to optimize this surgery. (c–f) Sequence of events in empty silicone nerve guides. Initially fluid and cytokines fill up the nerve guide (c), followed by the formation of a fibrin matrix inside the lumen which initially supports invading fibroblasts (blue) (d), followed by Schwann cells (green) and axons (red) (e), which eventually penetrate to the distal stump. The regenerated nerve usually is thinner in the middle of the bridge (yellow) as shown in (f). *Figure (b) reprinted from Lundborg (2000).*

2–5 mm/day, and appears to be more effective in younger people. Significant injuries can take many months to heal, and it is uncommon for full restoration of sensory or motor function to result, perhaps because of the gradual loss of neurons or deterioration of the distal nerve stump.

17.2.3 Autologous Nerve Grafts (Autograft)

PNI most commonly results from blunt trauma or from penetrating missiles, such as bullets or other objects, but it is also associated with fractures and fracture dislocations. Crush injuries are therefore more common than nerve transections. When nerve endings are unable to rejoin without tension, a bridging section of nerve is used, and two end-to-end sutures are performed. The crushed section of the nerve is removed and replaced by a nerve taken from another (less essential) site, often the sural nerve; this causes at least some donor site morbidity. With the development and optimization of microsurgical techniques for peripheral nerve repair, the success of end-to-end suturing greatly depends on the alignment of the fascicles. Therefore, the penetration of a sensory axon down a tubule previously filled with an axon from a motor neuron (and vice versa) is an ineffective (but frequent) result. Careful microsurgery, binding the individual fascicles, will maximize the functional recovery; however, axonal misdirection is frequent. Further, donor nerves are often of a small caliber and limited in number. Although the autograft is the gold standard upon which all alternative therapies are judged, these problems drive the search for a TE alternative to this treatment.

17.2.4 Use of Tubes (Nerve Guides) in the PNS

An off-the-shelf alternative for small injury gaps (~3–10 mm) are nerve guides that function by enclosing the therapy area. For nerve gaps up to 10mm, hollow tubes (nerve guides or nerve guidance channels) have successful outcomes comparable to the level of the autograft in various animal models and clinical situations. Many different nerve guides have been implanted and are reviewed elsewhere; however, the silicone nerve guide has been the most frequently implanted. Figure 17.2(c)–(f) schematically shows the general sequence of regeneration within silicone nerve guides. After implantation, fluid and cytokines (including neurotrophic factors) primarily generated by the Schwann cells build up within the nerve guide (Figure 17.2(c)). Within seven days, an oriented fibrin scaffold is formed, and cells—primarily fibroblasts—penetrate into the nerve guide while axon debris is removed by Schwann cells and macrophages in the distal end (Figure 17.2(d)). From 7 to 14 days, regenerating axons and Schwann cells penetrate into the nerve guide by using the newly formed fibrin scaffold as a substrate (Figure 17.2(e)). Schwann cells in the distal component begin to form part of the Bands of Büngner; long cellular tracts from the severed distal nerve ending to the final muscular targets. Between 14 and 56 days, Schwann cells myelinate the regenerating axons that exit the nerve guide, directed by the Bands of Büngner in the distal nerve stump to their targets.

There are three major nerve guides on the medical device market, shown in Figure 17.3: NeuraGen™ (collagen), Neurolac (polylactide (PLA)/caprolactone (PCL)), and Neurotube (knitted polyglycolide). Their performance is reviewed in depth by Deumens et al., and all of the nerve guides are degradable and used in relatively short distance peripheral nerve reconnection. Most research focuses on filling the lumen of nerve guides with matrices, scaffolds, cells, and/or drug delivering therapies.

FIGURE 17.3
SEM images of nerve guides approved for clinical use. (a) Neuragen™ (collagen), (b) Neurolac™ (PLCL), and (c) Neurotube™ (PGA). Scale Bar = 4 mm. *Reproduced from Schlosshauer et al. (2006).*

17.2.5 Critical Gap Length

A critical gap length (L_C) limits successful peripheral nerve regeneration using nerve guides. When the transected peripheral nerve approaches this experimentally determined length, the chances of successful regeneration within nerve guides is diminished. Figure 17.4(a) shows the frequency of successful regeneration (%N) within rat and murine models using silicone nerve guides and L_C (when %N=50) has been determined to be 9.7 ± 1.8 mm for rats and 5.4 ± 1.0 mm for mice. The introduction of matrices, fibers, and cells within nerve guides increases the L_C to varying degrees, which is an important measure of therapeutic success. The shift length (ΔL) due to a therapy is the difference between the L_C of the experimental device and the L_C of the control. The introduction of a therapy will generally increase the L_C; Figure 17.4(b)–(g) are the schematics based upon this approach of assessing additional therapies. Using the L_C normalization of therapies in the PNS, the therapies that significantly increased ΔL, were analyzed and regenerative theories put forward. They theorized that each significant therapeutic approach promoted either microtube synthesis or capsule formation.

17.2.6 Matrices and Scaffolds

Matrices and scaffolds for peripheral nerve TE aim to mimic an ECM environment that promotes increased regeneration after injury, and although most are based on biologically derived polymers, synthetic polymers are increasingly investigated.

17.2.6.1 Matrices

The L_C generally increases for nerve guides when they contain matrices within the lumen (Figure 17.5(b)). Effective matrices have similar mechanical properties; they are all weak, viscoelastic hydrogels with a high water content; however, they cannot be too concentrated as they can obstruct cell and axonal

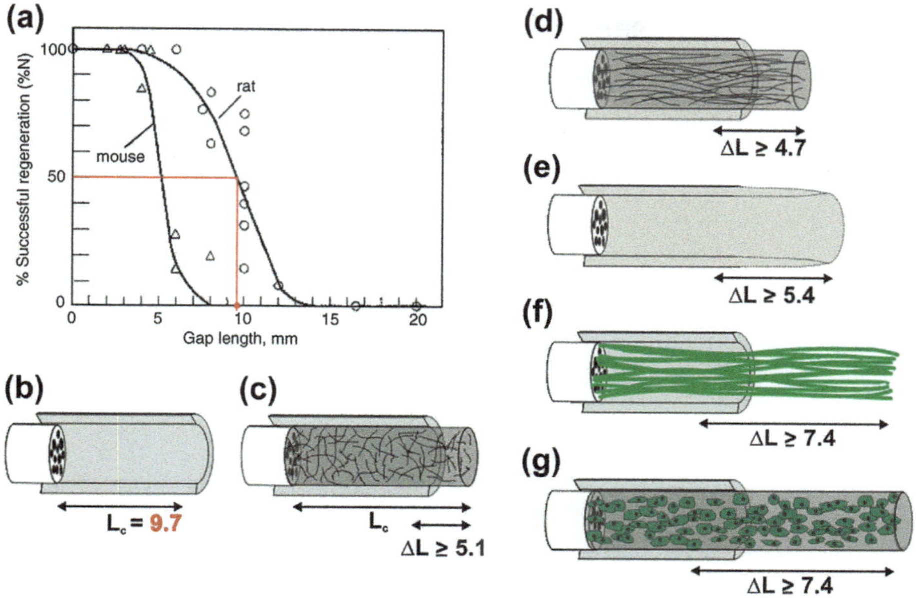

FIGURE 17.4

(a) The chance of successful reinnervation with nerve guides is dramatically reduced, once an injury gap reaches a certain value. This length, termed critical gap length, L_C, is where successful regeneration occurs 50% of the time. Selected examples of the critical gap length, L_C, for the rat sciatic nerve. Empty silicone nerve guides (b) have an L_C of 9.7 mm, and the introduction of a collagen-GAG matrix (c) increases the L_C to at least 14.8 mm with a DL of 5.1 mm. When oriented fibrin matrices are used (d), this results in a DL of at least 4.7 mm *when compared to the same, nonoriented matrices.* Degradable collagen nerve guides (e) have a DL of at least 5.4 mm compared to silicone tubes. When polyamide or collagen fibres are introduced (f), the DL is at least 7.4 mm compared to empty nerve guides, while Schwann cells (g) result in a similar DL value of 7.4 mm, compared to nerve guides filled with PBS. Note: the shift length (DL) is the result of the therapy—*not necessarily compared to empty nerve guides. Figure (a) reprinted from Zhang and Yannas (2005).*

penetration. Matrices also require the capacity to be translated to the clinic. For example, Matrigel™ is a solubilized basement membrane extracted from mouse sarcoma, and while the introduction of Matrigel™ into nerve guides promotes regeneration, it is not a suitable matrix for clinical use. Currently, no ideal matrix promotes the regeneration in a nerve guide beyond the level of the autograft.

17.2.6.2 Oriented Matrices

In vivo experiments with orientated matrices have a higher L_C value than do isotropic matrices of the same material (Figure 17.4(d)). The magnetic alignment of positive or negative diamagnetic anisotropic molecules, such as fibrin and collagen, respectively, elicits guided regeneration in vitro and improved in vivo regeneration. Oriented substratum and techniques to form such structures on a relevant scale are therefore important issues in neural TE. Axons will

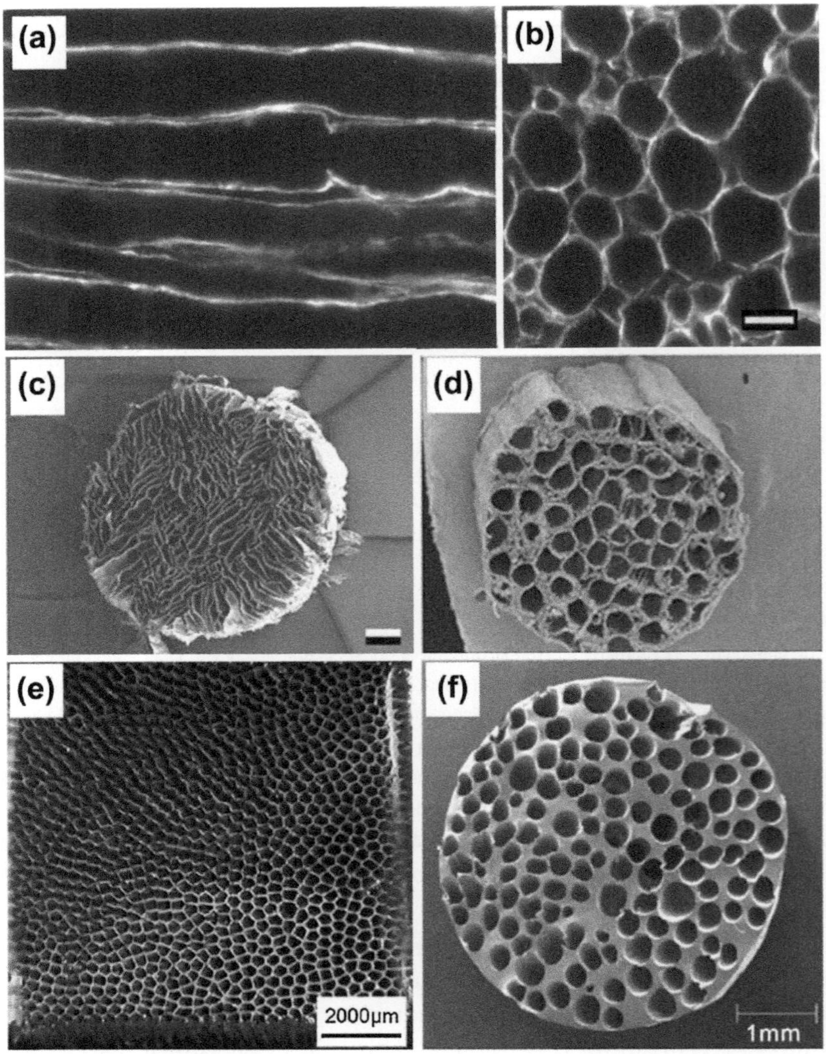

FIGURE 17.5

The Bands of Büngner (a and b) are natural lamina tubules that guide regenerating axons in the distal nerve to their targets (scale bar 10 μm). Uniaxial freezing (c) can induce similar oriented channels within hydrogels, including chitosan (scale bar 200 μm). Using oriented electrospun PCL fibres and a mandrel template, the structure was rolled into a nerve guide with multiple channels. Figure (e) and (f) are of moulded channels formed by bundling then dissolving a fibrous component. *Figures (a) and (b) from Scherer and Salzer (2001) (Oxford Press), (c) Reproduced from Riblett et al. (2012), (d) Reproduced from Jeffries and Wang (2013), (e) Reproduced from Lynam et al. (2011), (f) Reproduced from Yu and Shoichet (2005).*

regenerate significant distances along Schwann cell-containing tubular basal lamina (Figure 17.5(a) and (b)), which are often described as "Schwann cell columns." After PNI, Schwann cells partly clear the myelin while increasing numbers of macrophages in the distal stump complete the task. Throughout such Wallerian degeneration, the microtubules of basal lamina that supports these cells remain and are specific tracts that channel the regenerating axons to the final targets.

17.2.6.3 Scaffolds

Because a successful ECM of the peripheral nerve consists of laminin tubules (Figure 17.5(a) and (b)), it would be expected that similar channel morphologies would result in the directed growth of neurites. One simple technique to induce orientation involves the growth of ice crystals through the hydrogel to impart guided structures (Figure 17.5(c)). Such approaches are applicable to a range of hydrogels, and essentially convert isotropic materials into scaffolds with guidance properties.

Jeffries and Wang produced channeled nerve guides by combining an oriented fiber collection with channel templates; these constructs could be rolled up to produce the channel structures shown in Figure 17.5(d). Simple molding and construct dissolution is another approach to create guidance channels (Figure 17.5(e) and (f)). For such channels, surface modification significantly increases neurite growth in vitro.

Fibers will also guide axons along their length down to the nanoscale, with various oriented electrospun substrates investigated. Figure 17.6(a)–(c) shows the guidance of neurons, glia, or DRGs placed upon to the oriented fiber arrays. Fibers are attractive guidance substrates as they can be bundled together and used as oriented scaffolds (Figure 17.6(d) and (e)). Interestingly, when oriented electrospun fibers are embedded within a hydrogel, a Schwann cell will significantly stretch more than one placed in that hydrogel environment alone (Figure 17.6(f) and (g)). However, the scale of an oriented scaffold is not yet defined, with research demonstrating that much larger channels can regenerate a greater number of axons. In this study, single or folded electrospun membranes were placed within nerve guides over challenging 14-mm nerve gaps (Figure 17.7). The nerve guide that contained a single electrospun membrane had more regenerating axons and better nerve conduction velocity than one containing a folded sheet. While increases in scaffold density could reduce reinnervation, the compartment space within the three-folded sheet nerve guide is much greater than for natural laminin tubules. There remains extensive research to be performed on electrospin fibers for PNI, including correct structures and morphologies.

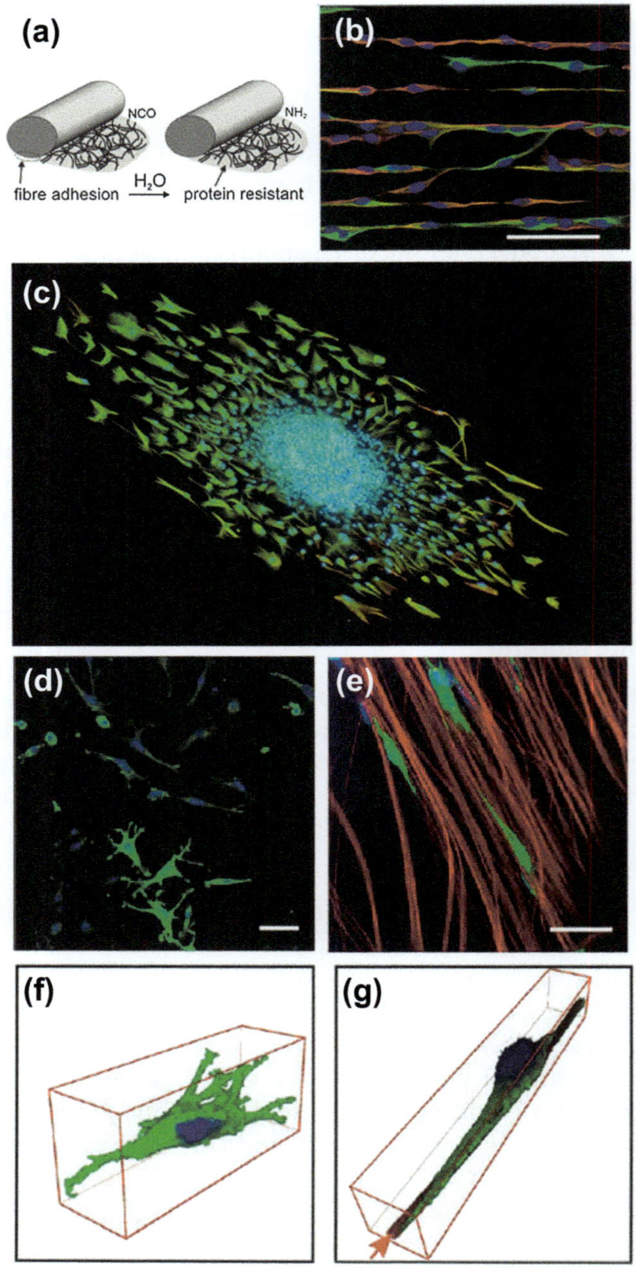

FIGURE 17.6

Electrospun fibers orient neural cells (of either PNS or CNS origin) in vitro in both 2.5 and 3D. (a) Using glass cover slips coated with isocyanate terminated star PEG reactive substrates, oriented electrospun fibers adhere to the surface, which then continues to react with water to become a passive surface. These "dilute" fibers guide U373 glial cells (b) along the length. When a pellet of U373 cells are placed onto the substrate (c), the migration occurs preferentially along the fiber. The embedding of PCL fibers within a 3D collagen gel will also affect cell morphology (d–g) (Kriebel et al., 2013). While morphologies of U373 cells are random when embedded within an agarose hydrogel (d, f), when electrospun fibers are also embedded, the cell will adopt an oriented morphology (e). Colors: (b, c), GFAP (green) and vimentin (red); (d–g), S100 (green) and PCL fibers (red). Scale bar = 100 μm (b) and 50 μm (d–g). *Figures (b) and (c) reproduced from Gerardo-Nava et al. (2009), (d)–(g) reproduced from Kriebel et al. (2014).*

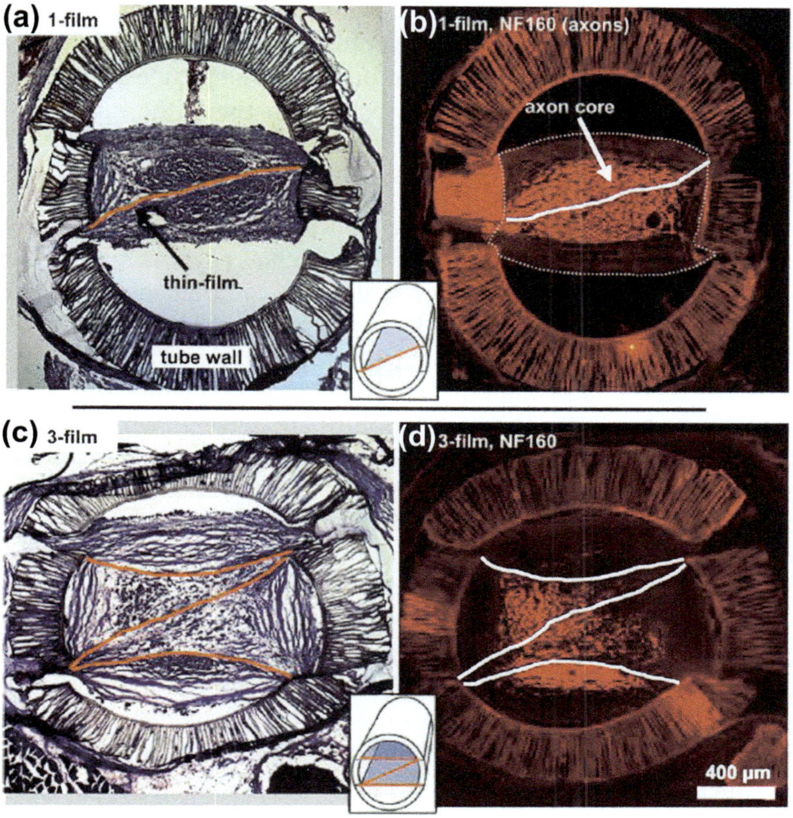

FIGURE 17.7

Cross-sections of nerve guides (a) and (c) 3.5 mm and (b) and (d) 2.5 mm into the 14 mm gap of tibial nerve of a rodent. Inside the nerve guide is an electrospun sheet, with a single sheet resulting in a greater number of axons. *Reproduced from Clements et al. (2009).*

17.2.7 Schwann Cell Grafts

When Schwann cells are included within nerve guides, large shift lengths (ΔL) result, and significant gap lengths are bridged (Figure 17.4(g)). Schwann cells can also be suspended within a matrix, or seeded on decellularized tissue. Schwann cells delivered with Puramatrix™, a commercial self-assembling peptide (SAP) hydrogel, resulted in regeneration similar to that in the autograft and greater than in Schwann cells delivered in an alginate/fibronectin (2%:0.05%) matrix. For cellular nerve graft or cell/biomatrix hybrid structures, the long-term supply of specific neurotrophic factors can be obtained by the targeted genetic modification of Schwann cells using viral vectors in vivo or ex vivo. Depending on the type of neurotrophin that is introduced into grafted Schwann cells, there can be differential effects on graft morphology, the number and type of regenerating axons, myelination, and locomotion. Schwann cell transplantation has great potential as a tool in peripheral nerve TE, although any eventual clinical solution for bridging large injury gaps will have to compete with the relatively low-cost alternative of the autograft.

17.2.8 Neurotrophic Factors

After implantation, the lumen of nerve guides rapidly fills up with cytokines and growth factors, including neurotrophic factors. These soluble molecules assist in neuron survival after axotomy, through complex molecular interactions. Neurotrophic factors include the neurotrophins, neuropoetic cytokines, and fibroblast growth factors. The neurotrophins signal through tyrosine kinase (TrK) receptors in the cell membrane in conjunction with low-affinity receptors such as p75. The neurotrophins are nerve growth factor (NGF), neurotrophin-3 (NT-3), neurotrophin-4/5 (NT-4/5), and brain-derived neurotrophic factor (BDNF). After injury, NGF is elevated in the distal nerve and, acting through TrkA, assists in the survival and outgrowth of a proportion of sensory neurons. Other sensory neurons express different receptors and are supported by NT-3 or by another factor—glial-derived neurotrophic factor (GDNF). BDNF and NT-3, respectively, signal through TrkB and TrkC receptors that are present on motor neurons and increase neuron viability after axotomy. Increased brainstem and spinal cord motoneuron survival has been reported using GDNF. An in vivo study combining GDNF and NGF in delayed release profiles from nerve guides significantly increased the axonal regeneration above either GF alone or controls. Investigations involving growth factors to treat PNI are significant and reviewed elsewhere.

There are various methods of delivering neurotrophic factors in nerve guides. The growth factor can be included within the nerve guide, homogeneously, or specifically at the lumen to inwardly direct the release. The matrix in the lumen of the nerve guide is also a source of growth factors; either simply incorporated, or as part of a noncovalent controlled release system. Heparin-binding growth factors (e.g., NGF, NT-3) can also be locally delivered from matrices as the axons regenerate. The potential permutations and concentrations of neurotrophic factors are many; cocktails of different neurotrophic factors may result in enhanced neuron survival and gradients of such factors may lead to increased axon regrowth. However, more potent could be establishing defined gradients, by using the minimum GF gradient to turn a neurite. Dodla and Bellamkonda recently used incremental increases of LN-1 and NGF within agarose to regenerate across an extremely large 20-mm sciatic nerve gap in rats.

17.2.9 PNS Summary

The clinical outcome after peripheral nerve repair has not significantly improved since the 1980s. Although it is rare for all sensory function to return to a patient, the availability and success of an autograft provide an economic challenge for TE strategies—the ideal therapy needs a relatively low-cost solution.

17.3 CNS: SPINAL CORD

Even the most modest success in treating spinal cord injury (SCI) remains tantalisingly beyond our reach, despite many decades of research. Damage to the spinal cord results in the immediate death of neural cells and disrupts the blood supply to the injured area. A major difference with the PNS is that both axons and neuronal cell bodies are resident within the spinal cord, thus both are affected after injury. The spinal cord also develops a growth-inhibitory glial scar, that impedes potential effective endogenous or treatment-driven regeneration. This scar consists mainly of reactive astrocytes, fibroblasts, and pericytes. These scar cells can express axonal growth-inhibitory molecules, such as chondroitin sulfate proteoglycans and semaphorins, and thus form a physical and chemical barrier for regenerating axons. There is also secondary neuronal and glial degeneration as a delayed consequence of the initial injury that further reduces any residual function. Additionally, with or without injury, CNS white matter, which contains myelin and myelin-associated axonal growth inhibitors, prevents any substantial axonal regenerative response in the adult spinal cord. TE in the spinal cord has focused on cellular substrates that replace lost spinal cells, provide repair-supporting molecules and/or cells, and/or provide a terrain for severed ascending (sensory) or descending (motor) axons to grow across the site of injury. Acellular substrates (scaffolds/matrices) may elicit an axonal regeneration response but are typically less effective than cellular transplants. The mechanisms of reduced cyst formation and astrocytic scarring associated with scaffold implantation are under investigation.

17.3.1 Summary of Anatomy and Injury Response

The middle portion of the spinal cord consists of a butterfly-shaped gray matter containing the cell bodies of neurons and glial cells and numerous blood vessels. The white matter surrounding this structure mainly contains axons and support cells (termed glia) such as oligodendrocytes (providing myelin insulation to the axons), astrocytes, and microglia (immune cells). Traumatic SCI results in the disruption of both sensory and motor axons and the lack of continuity in the spinal cord has devastating consequences for the injured patients. The initial injury response following bruising or transection of the spinal cord involves a primary event in which cells and tissue are crushed or cut by the injury and cells undergo necrosis. After a primary injury, there are secondary events including the influx of Ca^{2+}, release of inflammatory cytokines, oxidative stress, and fluid build-up within the lesion site. Inflammatory and "secondary events" cause neurons in the gray matter, and oligodendrocytes in the white matter, to die from apoptosis. Death of oligodendrocytes and the subsequent loss of the myelin sheath interfere with the speed and reliability of electrical conduction in any surviving axons, and significantly contribute to a sustained and long-lasting loss of function. Successful spinal cord repair requires all these issues to be targeted, which is clearly a complex and difficult task.

17.3.2 SCI Models

There are a number of experimental animal models of SCI and different models are best suited for the measurement of tissue sparing (contusion), axon regeneration (complete transection) or sprouting of intact fibers (hemisection) and mostly involve rodents.

17.3.2.1 Contusion Animal Model

Most human SCI is the result of a compression or contusion injury, which often results in the formation of a cavity or cyst as schematically shown in Figure 17.8(a). Experimentally, such injuries are modeled either by transient compression of the spinal cord using a weight drop method or by displacement of the spinal cord. With a contusion injury, analysis of the axon regeneration response is complex due to the presence of spared axons and their collateral sprouts. To repair the contused spinal cord, cells are injected directly into, or next to, the contusion cavity, in very small volumes. In general, cell grafting into the contusion cavity likely elicits a neuroprotective effect as well as an axon regeneration response that can result in some degree of functional recovery.

17.3.2.2 Hemisection Model

Some instances of human SCI result from a laceration with disruption of the dura matter. This particular type of injury is mimicked in the laboratory using a surgical microknife or microscissors as depicted in Figure 17.8(b). Although with a partial (hemi) section the loss of tissue and function is limited, over time, injury-induced secondary tissue loss may be added to the overall injury zone. With a hemisection SCI model, as with a contusion injury, the presence of spared axons and the formation of axon collaterals make a proper analysis of any transplant-induced regeneration response difficult. To repair the spinal cord following a hemisection, cells/scaffolds can be transplanted as a preformed cable that bridges the injury site or as a suspension/in situ hydrogel that fills the lesion site. An alternative approach is the transplantation of pieces of peripheral nerve. The grafted cells/tissue also elicit axonal growth in a hemisection; however, axons do not exit the transplant without additional interventions such as an increase in the levels of neurotrophic factors beyond the transplant using osmotic minipumps or direct injection of neurotrophic factors (see boxed text) or chondroitinase.

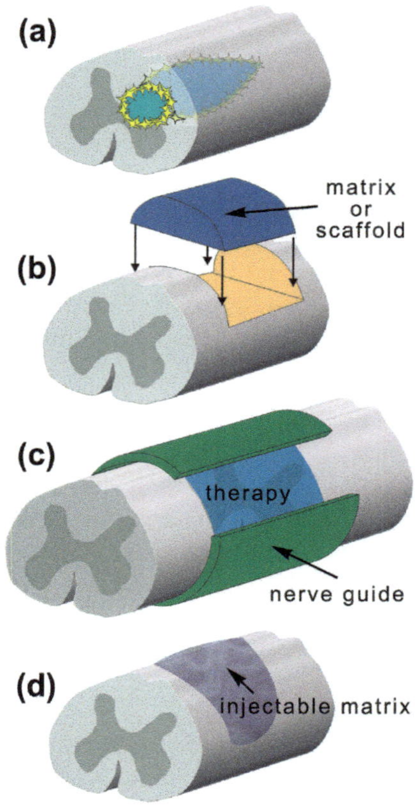

FIGURE 17.8

Contusion (a), hemisection (b), full transection (c) models, and intrathecal delivery (d) for researching SCI models. With contusion injuries, the formation of an astrocytic scar (often termed glial scar) presents a barrier to the axons and lines a fluid-filled cyst. The reactive astrocytes (yellow) lining the cyst are a formidable barrier to regenerating axons. In the surgically created hemisection model, the space can be filled with a matrix/scaffold (dark blue), either injected to fill the lesion or preformed and implanted (shown). With full transection models, the cord is severed and a scaffold is positioned between, or the stumps are inserted into a nerve guide (sectioned to expose lumen) filled with a therapy (light blue). In some species, the spinal cord will retract after full transection, creating a gap to fill with cells/matrix/scaffold. Intrathecal delivery involves an injectable material placed onto the injured spinal cord, inside the sub-arachnoid space.

17.3.2.3 Full Transection Model

Full transection experiments (Figure 17.8(c)) are not common due to surgical complications such as instability of the spinal column and the difficulty in animal care. However, this type of lesion can provide absolute evidence of true axon regeneration. While P(AN-VC) nerve guides have been used for full transection (see boxed text), other investigations using soft, flexible hydrogel nerve guides in the fully transected cord to demonstrate an improved regenerative capacity. However, in addition to the difficulties of maintaining animal care are the problems of bone curvature and instability of the grafts.

17.3.2.4 Intrathecal Delivery

The blood–brain barrier (BBB) prevents many proteins and molecules to pass across the blood vessels within the CNS. The many vessels and capillaries that supply blood to the CNS (the gray matter is highly vascularized in particular) have ordered endothelial cells and tight junctions between them that ensure that the two environments are separately controlled. Intrathecal delivery (Figure 17.8(d)) is one approach to deliver drugs into the CNS without having to cross the BBB.

However, the cerebral spinal fluid flow in the intrathecal space is significant, so if an injectable matrix is to be placed in this environment, it should be engineered to be fast gelling and adhere to the local tissue. A thermoreversible hydrogel matrix blend of hyaluronic acid and methylcellulose (HAMC) was successfully implanted in the intrathecal space to deliver drugs that modulate the severity of SCI. The HAMC injectable and intrathecal delivery approach has flexibility in that cells or drugs can be delivered, to modulate inflammation and limit the extent of injury.

17.3.2.5 Gender and Age

Gender differences can affect the cascade of degenerative processes and functional outcomes after traumatic and ischemic CNS injuries. The presence of estrogens and progestins likely plays an important role in the greater neuroprotection seen in females versus males after injury, but the underlying biological mechanisms are not fully understood. Similarly, age is also an important factor in the pathophysiology of CNS and PNS injuries and in the ability to recover. It is important to realize that animal models do not always reflect these gender and age differences, and it may well be that for a particular type of injury, different animal models should be developed that represent such differences.

17.3.2.6 Species Selection for SCI Models

The selection of the species for SCI studies is an important issue. TE strategies have focused on the mammalian model, whereas invertebrates and nonmammalian vertebrates are aimed at understanding fundamental biochemical and molecular processes associated with cell survival and regeneration. The pathology of rodent SCI appears to be largely similar to that in humans, although important differences have been discerned, including reduced glial scarring, inflammation and demyelination, increased Schwann cell infiltration, and the protracted Wallerian degeneration in humans. Further, there are critical differences in the organization and functional importance of different ascending (sensory) and descending (motor) pathways in human versus rodent brain and spinal cord—differences not always sufficiently taken into account by experimenters. Nevertheless, rodents offer an important and useful model in which to test a variety of SCI repair strategies for their repair potential and ability to influence functional recovery.

Larger species such as pigs, dogs, and cats are used in SCI studies. To assess the efficacy of specific approaches to repair the human spinal cord, it is increasingly accepted that the development and use of primate models are advisable. Besides ethical issues, the limitation with these models is that the maintenance of the injured primate is extremely labor intensive and costly. Due to these high costs, sufficient animal numbers to allow a meaningful statistical analysis are difficult to achieve. Cats are used to model SCI because they offer an important advantage over other models in that they can be trained relatively easily, which is due to sympathetic pathways that are unique to the feline model. One disadvantage of the cat as a model for SCI is that locomotion after injury is strongly influenced by the central pattern generator, which appears to be a crucial difference with humans.

17.3.3 Cell Transplantation

Cell transplants into the spinal cord aim to encourage injured axons to grow through areas of damage toward and into their target areas, and/or as a means to deliver repair-supporting molecules. Various cell types have been tested including stem cells derived from either embryonic or adult tissue.

17.3.3.1 Schwann Cells

Schwann cells have the ability to produce a variety of growth-promoting molecules for providing a growth substrate for injured CNS axons. In the injured spinal cord, Schwann cell transplants elicit axon regeneration (see boxed text). Schwann cells also form myelin sheaths around CNS axons to allow signal conduction in regenerated CNS axons. In addition to axon regeneration, an intraspinal Schwann cell graft has a neuroprotective effect; this reduces secondary injury. Schwann cell transplants in the cystic cavity combined with maintaining an increased level of cyclic adenosine monophosphate results in improved functional recovery.

17.3.3.2 Olfactory Ensheathing Glia

The uninjured adult CNS environment is nonpermissive for axon elongation. However, there is one structure in the adult CNS where axon growth does occur: the olfactory bulb. Throughout adulthood, olfactory neurons generate axons that grow into the olfactory bulb to make their appropriate connections. This particular growth response can occur because of the presence of olfactory ensheathing glia (OEG), which shield off the growing axon from the nonpermissive CNS tissue. When grafted into the injured spinal cord, OEG promote the regrowth of axons beyond areas of injury and into distal spinal tissue (see boxed text). The underlying mechanisms are still unclear, but it is possible that OEG prevent axons from recognizing the nonpermissive adult CNS environment by enfolding them, as they normally do for growing olfactory axons. In addition, OEG may elicit only a moderate astrocytic reaction with low expression of growth-inhibitory molecules.

17.3.3.3 Stem Cells

The application of stem cells for spinal cord repair has to date been limited by technical and ethical concerns and, sometimes, political restrictions. However, different types of stem cells, such as embryonic or adult neural stem cells or bone marrow stromal cells, have been shown to elicit neuroprotective and/or regenerative responses. The mechanisms underlying the repair effects of stem cell transplants are under investigation. The secretion of factors that repair existing blood vessels and/or form new ones are considered important determinants in stem cell-based nervous system repair. Stem-like cells sometimes differentiate into astrocytes, which restrict their effects on axon regeneration and myelination. It may be necessary to differentiate stem cells before/during cell transplantation, as the injury environment does not favor differentiation into neurons or oligodendrocytes. Differentiation of stem cells prior to implantation is accomplished by a variety of factors, although the most efficient combination is yet to be determined experimentally. The potential use of induced pluripotent stem cells to treat SCI is increasingly envisaged and is reviewed well by Kramer et al. In all instances, it is essential to ensure that stem cells do not continue to multiply after transplantation, and potentially form tumor-like structures.

17.3.3.4 Genetically Modified Cells

Gene therapy holds great potential for the treatment of neurotrauma and many types of neurodegenerative CNS disease. Delivery of therapeutic molecules by gene therapy may support the arrest or delay of secondary neurodegeneration and stimulate the regenerative capacity of CNS neurons. This involves the use of different cell types including Schwann cells, fibroblasts, OEG, and neural progenitors.

Gene therapy allows the selective delivery of neurotrophic factors to specific CNS locations. For example, OEG genetically modified to secrete neurotrophins, such as NT-3, BDNF, and GDNF, have stimulated long distance growth in certain spinal cord tracts. In addition, neurotrophin-secreting OEG improves spinal tissue sparing, which is thought to be correlated with improved behavioral outcomes. Genetic engineering of cells can involve the use of different viral systems, and these require hurdles for clinical applications. However, the transfer of genes to transplanted cells enables better monitoring of the cells within the spinal cord or CNS when using reporter genes.

17.3.4 Matrices and Scaffolds

The matrix/substrate must allow axons to penetrate, grow through, and exit so that they can reenter their target regions. After SCI, both descending and ascending axons tracts are disrupted, and each of these tracts has specific locations in spinal cord white matter (Figure 17.9). Since various axon populations have different growth requirements, a future graft may consist of composite bridging materials, each tailored to a particular axon population. Whether matrices or scaffolds are effective due to regenerative properties, or as a result of other factors such as neuroprotection, is a matter still under investigation.

Many biomaterials, either natural (e.g., collagen, alginate, agarose, chitosan, fibrin, fibronectin, and Matrigel™) or synthetic (e.g., methylcellulose, nitrocellulose, PEG, and nanofibers), encourage some degree of tissue preservation and perhaps nerve repair after SCI. The implantation of such scaffolds into the injured spinal cord often results in a significantly different injury response to that commonly observed with a traumatic lesion alone. Cystic cavity formation can be completely absent when scaffolds are implanted, and scarring is often reduced in density. In addition, axons will penetrate

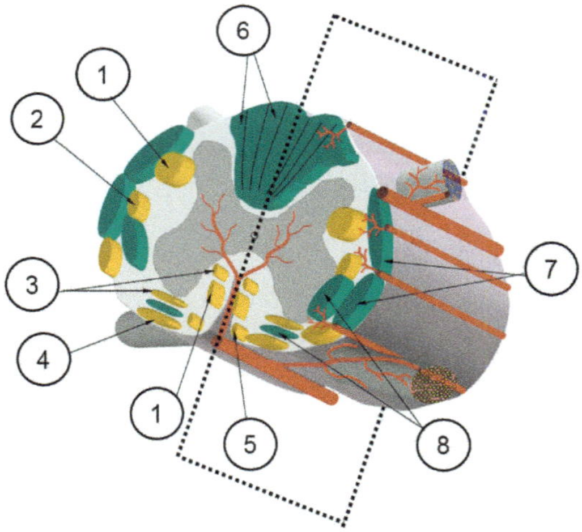

FIGURE 17.9

Selected tracts within the human spinal cord leading to/from the brain. Disruption of tracts will affect the corresponding sensory or motor signals. The tracts are symmetrical along the saggital plane (shown). Motor tracts: (1) corticospinal voluntary, discrete movement, particularly in distal part of limbs; (2) rubrospinal facilitates activity of flexor muscles; (3) reticulospinal facilitate or inhibit voluntary movement or reflex; (4) vestibulospinal facilitates postural activity associated with balance; (5) tectospinal Reflex postural movements in response to vision. Sensory tracts: (6) fasciculus gracilis/cuneatus discriminative touch and conscious perception of body and muscle position; (7) spinocerebellar subconscious perception of body and muscle position; and (8) spinothalamic pathway for crude touch, pain, pressure, and temperature.

into the scaffold in a highly oriented and ordered manner if the scaffold itself has oriented pores. Such observations have been reported from various laboratories, with a range of materials used for the scaffold. For a detailed, recent review on the use of scaffolds and matrices in the spinal cord, refer to Sakiyama-Elbert et al.

17.3.4.1 Relevant Peptide Sequences

For many scaffolds and matrices, there is often a need to direct and target cell/material interactions more specifically than full protein inclusion or adsorption. The conjugation of bioactive peptides with a material is an effective way to achieve controlled interactions, albeit with in vitro environments. Many peptides are various bioactive domains of laminin and include YIGSR, IKVAV, RGDS, and RNIAEIIKDI. The peptide sequence YIGSR is often preferred as a peptide ligand for in vitro applications and is widely proposed as an alternative to RGD for cell adhesion specifically for neural cells.

Another important peptide sequence of laminin is IKVAV; it promotes cell adhesion, neurite outgrowth, experimental metastasis, collagenase IV activity, angiogenesis, plasminogen activator activation, cell growth, tumor growth, and differentiation of neural progenitor cells. Although the aforementioned active peptide sequences are effective in promoting neuron adhesion and axon growth alone, the use of both YIGSR and IKVAV has a synergistic effect on neurite outgrowth. The extended peptides CDPGYIGSR and/or CQAASIKVAV further increase neurite outgrowth.

17.3.4.2 Matrices

The number of in situ gelling matrices to treat SCI is ever increasing. Table 17.1 shows a selected list of different matrices, and the instances in which they are used. There are strong arguments for combining cell therapies with injectable matrices, as they can protect cells from membrane shearing and cell death by "anoikis." In situ gelling hydrogels can be classed as chemically or physically crosslinked, with the latter having a broader range of materials. For extensive descriptions on in situ hydrogels for neural TE, Pakulska et al. and Macaya and Spector are two recommended reviews.

In situ gelling matrices are important for maintaining a noninvasive approach to treating SCI. HAMC is being developed as a flexible injectable technology for both drugs and cells, while SAPs are used frequently in different studies. Fibrin is also commonly used as a matrix, and can be modified with peptide sequences or used to deliver neurotrophic factors with a heparin-binding site. Fibrin also resulted in the greatest number of neurons traced with retrograde labeling in full transection studies, compared to collagen-I, methylcellulose, and Matrigel™. Matrigel™ has also been used in many CNS TE investigations,

Table 17.1 Selected Examples of In situ Gelling Hydrogels for Spinal Cord Injury

Material	Animal Model	Description	References
Agarose	Intrathecal	BDNF delivery; required an external liquid nitrogen cooling system	Jain et al. (2006)
Type I Collagen	Hemisection	DMEM used to buffer with final concentration of 1.9 mg/mL of collagen	Von King et al. (2010)
Collagen/laminin mixtures	Contusion	27% transplanted Schwann cell survival (cf. 14% with media)	Patel et al. (2010)
Fibrin	Full transection	In pHEMA nerve guides; fibrin promoted most regeneration among formulations (methyl cellulose, collagen, Matrigel)	Tsai et al. (2006)
Fibrin	Hemisection	Fibrinogen (10 mg/mL in phosphate buffered saline) mixed with thrombin (final 5.4 U/mL)	Von King et al. (2010)
Fibrin and fibronectin	Hemisection	Blend of 2 mg/mL of fibronectin and 10 mg/mL of fibrinogen—produced best outcome in this comparative study.	Von King et al. (2010)
Fibronectin	Hemisection	PEG-based precipitate of fibronectin to produce viscous gel	Von King et al. (2010)
HAMC	Contusion plus intrathecal		Gupta et al. (2006)
HAMC	Contusion plus intrathecal	Delivered with erythropoietin	Kang et al. (2009)
HAMC	Intrathecal	Delivered with nanoparticles	Baumann et al. (2010)
HAMC	Contusion plus intrathecal	Stem cells plus PGDF direct injection delivery performed 9 days postcontusion	Mothe et al. (2013)
IKVAV-peptide	Contusion	1% peptide solution	Tysseling-Mattiace et al. (2008)
Matrigel	Contusion	36% transplanted Schwann cell survival (cf. 14% with media)	Patel et al. (2010)
Methyl cellulose (5%)	Contusion	2% transplanted Schwann cell survival (cf. 14% with media)	Patel et al. (2010)
QL6 peptide	Contusion	Delivered with neural progenitor cells 2 weeks after SCI	Iwasaki et al. (2014)

HAMC, hyaluronic acid and methylcellulose; pHEMA, poly(2-hydroxyethyl methacrylate); BDNF, brain-derived neurotrophic factor; PEG, polyethylene glycol; PDGF, platelet derived growth factor.

particularly with Schwann cell or OEG transplantation into the spinal cord (see boxed text).

By exploiting the properties of charged peptide sequences, SAPs have proved to be an exciting class of materials for SCI and can encapsulate neurons and other cells. The SAP solution aggregates after mixing the peptide solution with cell suspensions, and such constructs have been used in a number of SCI studies, leading to reduced cavitation and some axon regeneration. Cell therapies delivered with in situ gelling hydrogels can also improve cell survival during/after injection.

17.3.4.3 Scaffolds

Scaffolds made from many different materials have been implanted into the spinal cord and are more extensively reviewed elsewhere. Decellularized tissue has been investigated; however, this is currently limited in scope. Oriented scaffolds, particularly hydrogel scaffolds, demonstrate the ability to stimulate axon growth, and similar to scaffolds used for PNI, controlled uniaxial freezing can generate channels within a polymer solution or hydrogels by using the forces associated with crystal growth. The resulting oriented scaffolds demonstrate neurite and cell penetration into the channels.

Cystic cavity formation can be absent when scaffolds are implanted, and astrocytic scarring is often reduced in density. Short time in vivo studies with poly (2-hydroxyethyl methacrylate) (pHEMA) scaffolds, implanted into the white matter dorsal tracts of rodents, showed a lower expression of glial fibrillary acidic protein (GFAP) compared to that of lesion controls. A notable lack of fluid-filled cysts compared to lesions only has been observed with a range of scaffolds.

17.3.5 Nerve Guides in the CNS

The classical experiment of cell transplantation with nerve guides in the fully transected spinal cord was pioneered in the Bunge group (see boxed text) with P(AN-VC) nerve guides. Collagen nerve guides and pHEMA-based conduits can also be used to promote regeneration in the spinal cord. Hydrogel guides result in the presence of axons within the lumen even when they contain no other therapeutic factor. Regeneration into P(AN/VC) nerve guides occurs with both Matrigel™ and Schwann cells present in the lumen. As is the case with peripheral nerve repair, the inclusion of a matrix into a hydrogel nerve guide increases the number of penetrating axons.

17.3.6 Neuronal and Axon Tracing

Most axon pathways in the brain and spinal cord do not have unique markers that enable easy identification. Also, if present, expression of any such markers may be affected by injury-related processes and/or during the process of regeneration. Thus, neuronal and axon tracers are used to label neurons and axons, respectively, to identify and evaluate the response of fiber tracts to intraspinal transplants. Administration of a tracer near the axon endings allows uptake and retrograde labeling of the neuronal cell bodies of the responding systems. Often-used retrograde fluorescent tracers (back to cell bodies in the brain) are Fluorogold and Fast Blue which require in the rat anywhere between 5 and 14 days for retrograde transport from lower spinal cord levels. These tracers can be injected a few millimeters caudal to, or into, a transplant. When used appropriately and combined with a thorough analysis of the location and number of back-filled neurons in the spinal cord and brain, the response of individual fiber tracts can be assessed. A drawback of this approach is that it is difficult to evaluate how far axons have regenerated. Injection of a tracer at the level of the cell bodies in the brain allows for uptake and anterograde transport toward the axonal endings. Three often-used anterograde tracers are wheat germ-agglutinated horseradish peroxidase, biotinylated dextran amine, and the B subunit of cholera toxin. The first tracer has a relatively fast transport time, but after identification using enzyme histochemistry, the axons are often difficult to quantify. Biotinylated dextran amine has a relatively slow transport time, but its identification using immunocytochemistry produces easily recognizable axons. This is also true for the B subunit of cholera toxin, which in addition has a relative short transport time. In addition to classical tracers, viral vectors have now been developed for the use for neuronal tracing particularly the use of reporter adeno-associated viral vectors (AAV). Only recently have the transduction properties of different AAV serotypes been investigated for the rubrospinal and corticospinal neurons that project to the spinal cord. With future viral vector development, ease or production, transduction efficiency, and better neuronal matching with AAV pseudotypes, we can expect these vectors to become superior to fluorescent tracers. The use of tracers is crucial for the quantitative analysis of the effect of therapeutic strategies on axon regeneration after SCI; this thereby relates anatomical plasticity to any observed behavioral recovery.

In 2013, Chung et al. published a tissue clearing approach—dubbed the CLARITY Protocol—allowing the transformation of an opaque tissue (such as an entire brain) into a clear substance that can be imaged. The CLARITY protocol takes advantage of reactions between monomers and proteins in the ECM and cell membrane. Acrylamide-based monomers are perfused into the animal, after the blood is replaced with saline. The monomers are therefore well distributed throughout the body via the blood vessels and capillaries, and they diffuse and crosslink with the many free amines found in endogenous proteins. Following this tissue crosslinking step, the electrophoretic clearance of lipids is performed using micelle nanoparticles. Combined with solvent mixtures, the tissue is rendered transparent, and immunochemistry can still be performed, while imaging is via fluorescent or confocal microscopy.

This ability to render opaque tissue clear obviates the need to section a sample, which is a particularly time-consuming part of tissue analysis. Further, it provides a quantitative analysis of full tissue, which is inherently more accurate than slicing sections of tissue and extrapolating results. The process allows for single axon tracing, irrespective of the path tortuosity. While the Clarity protocol is not ideal, this development in tissue imaging has a basis in biomedical engineering, and there remains opportunities for improving this technique using advanced polymer and imaging chemistry.

17.3.7 Nanomedicine to Treat SCI

As surgery of the spinal cord brings some risks, minimally invasive approaches are preferred. Nanoparticles are one injectable option and have a high permeability while containing hydrophobic drugs, and can release the drugs in a time-dependent manner. One example of this is the suspension of methylprednisolone-loaded poly(lactic-co-glycolic acid) (PLGA) nanoparticles in agarose and placing this locally after a spinal cord hemisection. For most of the measurements (i.e., improved locomotor ability, reduced scarring, and less invading macrophages), the delivery of loaded nanoparticles in agarose was better than systemic injection, local delivery, or agarose loaded with blank nanoparticles.

Selectively targeting macrophages is another approach to reducing secondary injury, as neuroinflammation is one of the most relevant mechanisms involved in this progression. Recent research from Papa and colleagues has suggested that polymeric nanoparticles can be used to selectively target macrophages because of their specific endocytic/phagocytic activity and their capability to act as a Trojan Horse for targeted drug delivery. Biodegradable polymeric nanoparticles of a 100-nm size were shown to be able to be incorporated into microglia and macrophages and release minocycline into the injured mouse spinal cord to reduce inflammation after trauma.

17.3.8 Approaches to Overcome the Inhibitory Nature of the Injury-Induced Scar

After SCI, a physical and chemical barrier (scar, glial scar) forms at the impact site (Figure 17.10), which prevents the growth of new axon sprouts. This scar consists mainly of reactive astrocytes and meningeal fibroblasts, which express extracellular axonal growth-inhibitory molecules such as chondroitin sulfate proteoglycans and semaphorins. It has become clear that manipulation of this barrier will be an essential component of a successful repair strategy for the injured spinal cord. Such manipulation of the scar may include (1) decreasing the inhibitory nature of the scar, (2) preventing axons from recognizing inhibitory molecules, and (3) enhancing the intrinsic growth ability of axons. The permissiveness of the scar for axons can be increased by reducing receptor-ligand binding, with obstruction of the synthesis of inhibitors, or by degrading biologically active components of the inhibitors. In the laboratory, several approaches have resulted in axon regrowth beyond the scar, but success was usually obtained using injuries that could be considered as "smaller size" injuries, that is, with less formation of scar tissue. The barrier of scarring can be

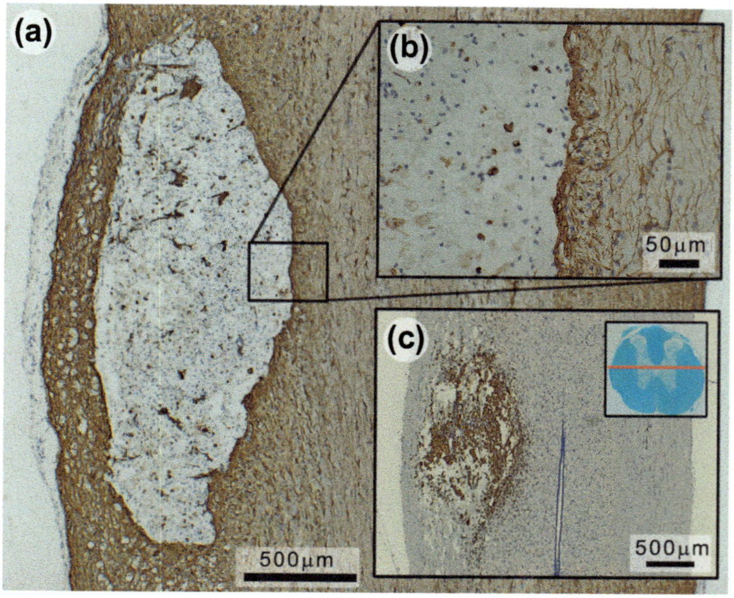

FIGURE 17.10

Sections of adult rat spinal cord six days after contusion injury, showing GFAP positive cells stained with immunohistochemistry (a) and (b). The fluid filled cyst is clearly visible, as is the dense GFAP staining at the border of tissue/cyst (b). This cyst is filled with CD68+ cells, which are activated macrophages/ microglia which should be considered when injecting therapies into this region. Luxol fast blue staining of the spinal cord as an inset in (c), shows the white/gray matter and the plane of sectioning for (a–c). *Adapted and reproduced with permission from Volpato et al. (2013).*

overcome by implantation of OEG into the scar environment, increasing the levels of neurotrophic factors in the nervous tissue using osmotic minipumps or the use of chondroitinase.

17.3.9 Summary

SCI initiates a plethora of destructive events. Some of the cells and molecules present at the site of injury may contribute to protection and repair, but the ultimate outcome is loss of tissue and loss of function in motor and sensory systems. One of the main obstacles for regenerating axons in the injured spinal cord is the presence of a glial scar at the site of injury or, in the case of an intraspinal transplant, at the transplant–host spinal cord interface. Unlike the PNS and the use of peripheral nerve autografts and nerve bridges, there is currently no reproducible, successful clinical therapy for SCI. Cell transplantation, combined with supportive matrices and scaffolds or neurotrophic factors, has been the predominant focus of TE strategies. The emphasis has been on neuroprotection, enhanced tissue sparing, and reestablishment of continuity across an injury site. Future objectives include increasing the number of damaged ascending and descending axons that regenerate, improving the remyelination of these regenerating axons, and the guidance of regenerating axons back to their appropriate target regions in the brain or spinal cord. The development of new techniques involving gene therapy and ex vivo manipulation of cells prior to transplantation will provide additional, and economically sustainable, tools that should improve therapeutic outcomes in injured patients.

17.4 CNS: OPTIC NERVE

Damage to fiber tracts interconnecting different CNS regions causes disruption of neural circuits and often leads to the death of associated neurons. Therapeutic strategies are therefore needed to ensure the viability of affected neurons, promote the regrowth of damaged axons across lesion sites, maintain growth within CNS tissue distal to the injury, and finally ensure that regrowing axons reinnervate appropriate target areas and reconnect with appropriate target neurons.

The visual system is for many a preferred model in which to study regenerative success or failure after nerve fiber damage in the mature CNS:

1. Retinal ganglion cells (RGCs), the neurons that project axons out of the retina and through the optic nerve to central targets in the brain, can be directly targeted via eye injections.
2. There is detailed knowledge of intraretinal and visual brain circuitry.
3. The optic nerve is a discrete, centrally derived tract that is surgically accessible and can be manipulated without affecting other fiber tracts and structures.
4. The extent of RGC survival and axonal regrowth can readily be quantified.

Normally, mature mammalian RGCs do not regrow axons beyond an optic nerve injury and, if the axotomy occurs within the orbit, about 90% of RGCs die within 2 weeks. This lack of regrowth is due to factors both intrinsic and extrinsic to the adult RGC. However, some adult RGCs will survive axotomy and regrow their axons when appropriately stimulated and/or when provided with an appropriate local environment.

17.4.1 Regenerative Therapies

After optic nerve crush, adult RGC survival can be enhanced by intraocular injections of recombinant neurotrophic factors such as BDNF, NT-4/5, ciliary neurotrophic factor, or GDNF or by injections of factors that block intracellular signaling pathways associated with cell death. However, these effects are transient, and viability is in most cases enhanced for only a few weeks. The use of viral vectors, particularly AAV vectors, encoding appropriate growth-promoting genes to transduce and genetically modify RGCs provides more long-term protection.

Some regrowth of RGC axons across an optic nerve crush can be elicited using a number of approaches including the implantation of peripheral nerve fragments into the vitreous, injuring the lens, or recruiting and activating circulating monocytes within the eye. Some regeneration is seen after blockade of growth-inhibitory molecules or after inactivation of Rho, a GTPase that mediates growth cone inhibition. Overall, however, without combinatorial approaches that use several complementary therapies, adult RGC axons fail to regrow more than just a few millimeters past the injury site. Additional approaches include TE strategies to stimulate more sustained regenerative growth of axons from a greater proportion of surviving neurons.

A successful approach that permits substantial long distance regrowth of injured adult RGC axons has been to graft peripheral nerve segments from the same animal onto the cut optic nerve. Under these conditions, regrown axons reinnervate target areas in the brain, and some limited function gets restored. However, there is a functional cost involved in removing the nerves for transplantation. Alternatives are therefore being sought for matrices containing cultured olfactory ensheathing cells and chimeric peripheral nerve grafts containing genetically modified Schwann cells (Figure 17.11). Various types of hydrogel matrices, some containing peptide and aminosugar sequences, have been used in attempts to promote the regeneration of RGC axons after more central optic tract injures in the brain.

The optic nerve injury model provides an optimal system in which to examine combinatorial effects as pharmacological agents, recombinant growth factors, or viral vectors can be injected into the eye to directly influence RGCs, while tissue and genetic engineering can be used to modify the environment within which RGC axons regenerate. These interactions can be assessed at a single location using different combinations of factors, or at multiple locations, for example, testing the effect of factors applied to the cell body and supplying cells/appropriate factors near the growing tips of the axons. Combined pharmacotherapeutic and vector-mediated strategies have resulted in more extensive regrowth of RGCs axons after optic nerve crush injuries beyond the optic chiasm and at least as far as the optic tracts.

17.5 CNS: RETINA

The retina is a thin, layered neural structure consisting of RGCs, bipolar and amacrine cells, photoreceptors (cones and rods), Müller cells, and retinal pigment epithelial (RPE) cells; the latter is attached to a thin membrane called Bruch's membrane. Bruch's membrane is partially synthesized by the RPE cells

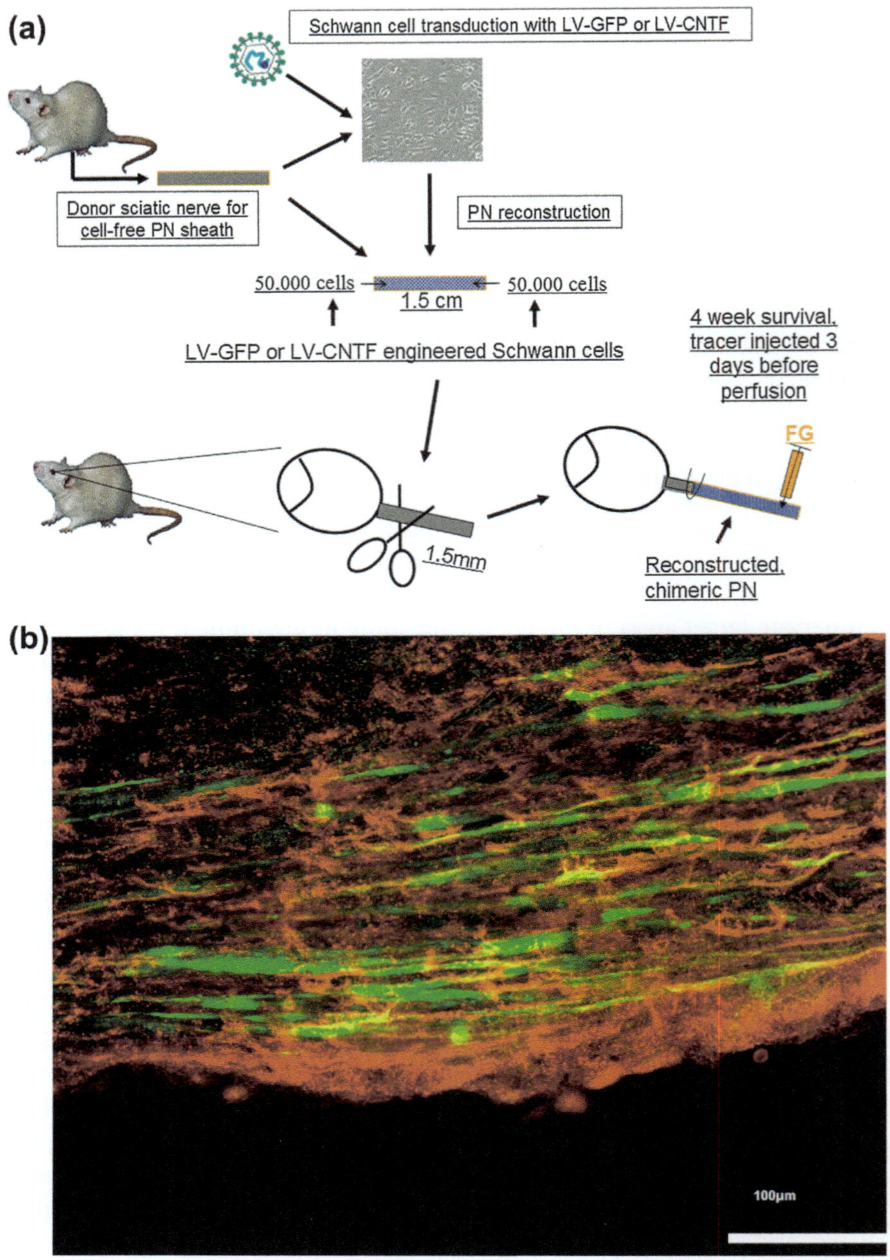

FIGURE 17.11

(a) Chimeric, genetically engineered peripheral nerve grafts act as bridges for axon regeneration. Donor peripheral nerves are freeze-thawed to kill all cells but retain organized extracellular matrix. Sheaths are then reconstituted with host adult Schwann cells that have been expanded in culture and genetically modified using lentiviral (LV) vectors to express appropriate neurotrophic factors. These engineered, chimeric nerves are sutured onto the cut end of adult rat optic nerves. After about 4 weeks, a tracer (fluorogold, FG) is injected into the distal end of the grafts to retrogradely label the retinal ganglion cells (RGCs) in the operated eye that have regenerated axons through the grafts. Animals are perfused 3 days later and the number of surviving and regenerating RGCs is quantified in each eye (see Hu et al., 2005 for further details). (b) Engineered Schwann cells are aligned along the axis of the reconstituted peripheral nerves and express green fluorescent protein (GFP). The red staining shows location of S-100 protein in the tissue, which is located primarily in glial cytoplasm. Reconstituted grafts containing genetically modified Schwann cells have also now been used in peripheral nerve repair (see Godinho et al., 2013). *Figure (a) Courtesy of Ying Hu, The University of Western Australia.*

and contains laminin, collagen type IV, fibronectin, and heparin sulfate proteoglycans. RPE cells maintain the vitality of the photoreceptors. When the RPE and photoreceptors are detached from Bruch's membrane, photoreceptor cell death (and blindness) ensues.

17.5.1 Diseases of the Retina

Diabetic retinopathy, retinitis pigmentosa, and macular degeneration are three major diseases of the eye, while glaucoma—often associated with raised intraocular pressure—results in damage to the optic cup and retina. Age-related macular degeneration (ARMD) is the leading cause of blindness in the United States for people aged >55 years. Current effective treatments for retinal disorders are limited, and TE strategies are currently emerging.

17.5.2 Cell Transplantation

Intraocular cell transplantation is one therapeutic strategy for the treatment of degenerative diseases that primarily affect photoreceptors. Different transplantation methods (retinal sheets, fragments, and dissociated mixed or purified cell suspensions) have been trialled, and different sources of cells have been used, including neural precursors, embryonic and postnatal retinal precursor cells, photoreceptors, neural stem cell lines, and bone marrow stem cells. One experimental strategy for ARMD is the transplantation of RPE cells to the subretinal space. A major challenge is that when RPEs are injected into the subretinal space as a suspension, they are unable to form ordered monolayers and eventually die. Cell transplantation substrates are therefore used as carriers to deliver cells, such as RPEs, iris pigment epithelium cells, and retinal progenitor cells to the retina. The seeded substrate is typically inserted subretinally through a slit in the sclera and choroid. Injected cell suspensions are exposed to high shear forces as they pass through the injecting needle, while therapies involving supporting substrates avoid such stresses. The use of cells on scaffolds for transplantation to the retina thus has advantages over injection of cells as spheres or single cell suspensions, where a large number of cells do not survive the transplantation procedure.

Many different RPE transplantation substrates have been investigated, and Figure 17.12 shows the implantation of a poly(ethylene terephthalate) electrospun sheets. Substrates for subretinal implantation require a relatively thin thickness (Bruch's membrane has an average thickness of 5 μm), but require a certain amount of mechanical integrity to withstand the placement and implantation procedures. Research is ongoing to find an ideal and effective cell transplantation scaffold for optimal delivery of stem cells or precursor cells.

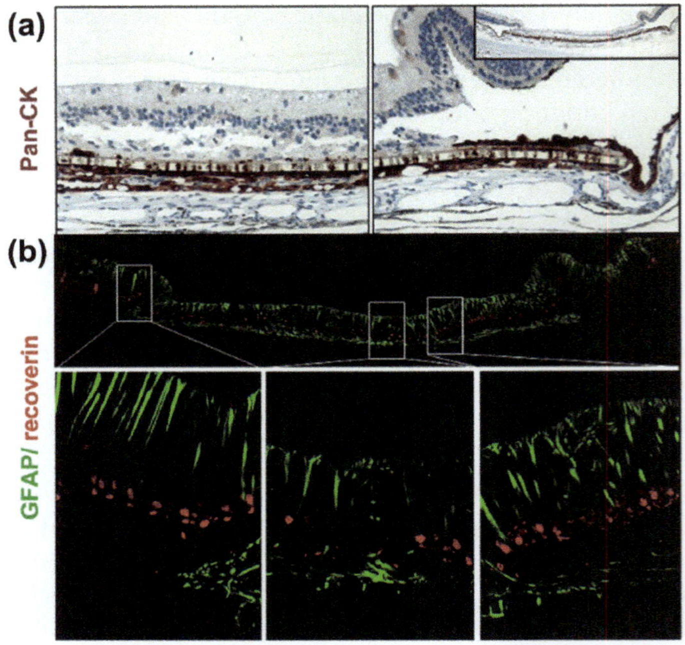

FIGURE 17.12

RPE-seeded electrospun polyethylene terephthalate (PET) scaffolds transplanted sub-retinally in rabbits (a) with Pan-CK showing RPE cells. (b) shows rod photoreceptors (recoverin) and astrocyte (GFAP). *Reproduced from Liu et al. (2014).*

17.6 CNS: BRAIN

17.6.1 Trauma and Stroke

Reconstruction of gray and white matter defects in the brain is especially difficult after large lesions or in the chronic situation. In addition to neuronal death, there is demyelination, axon retraction, and/or aberrant sprouting of injured axons, glial/meningeal scarring, and often progressive tissue cavitation. Under these circumstances, there is a need for cell replacement and some form of neural TE to develop scaffolds that facilitate reconstruction and restore continuity across the traumatized region. The use of supporting scaffolds may allow the formation of a biohybrid tissue that acts as the structural and functional analog of the original tissue.

17.6.2 Neurodegenerative Diseases

Neurodegenerative diseases are often associated with substantial neuronal loss; the pattern of loss varies depending on the type of disease process. Supply of appropriate neurotrophic factors via grafted cells or from slow-release polymers may protect compromised neurons, and alternatively cell replacement strategies may be required.

Some of these neurodegenerative disorders, such as Huntington's disease (HD) and spinocerebellar ataxia, are inheritable and associated with single gene mutations. Other motor neuron disorders and diseases such as Alzheimer's disease (AD) and Parkinson's disease (PD) have more complex etiology, with only a percentage directly associated with dysfunction of a specific gene. It is likely that many of these degenerative conditions are affected by environmental factors as well as some form of genetic susceptibility. AD and PD in particular are diseases associated with the elderly and are becoming increasingly common as life expectancy increases and the population ages.

Early diagnosis and prevention of further deterioration would be the optimal ways of treating these neurodegenerative disorders; however, it is often the case that when symptoms become apparent, there has already been substantial neuronal death. Engraftment of encapsulated cells that secrete the appropriate growth factor(s) have been trialled in experimental models of PD, AD, HD, amyotrophic lateral sclerosis (loss of motor neurons), and chronic pain. Other TE strategies involve attempts to replace the lost neurons with new cells to be incorporated into host circuitries. PD is a multisystem disease, and a major feature is the catastrophic loss of neurons in the midbrain that secrete dopamine. In numerous clinical trials, hundreds of patients with PD have received intracerebral grafts of dopamine-rich fetal human brain tissue; many of these recipients experience some alleviation of their symptoms. Potentially alternative sources of tissue include xenografts and stem cells or neural progenitor cells modified to secrete dopamine. Such approaches, perhaps combined with neuroprotective gene therapy, are potentially also of use in the treatment of motor neuron diseases and HD.

17.6.3 Nanomedicine in the Brain

To reach specific areas of the brain, a noninvasive or minimally invasive delivery of the therapy is key. The use of nanoparticles as delivery systems for difficult-to-access parts of the body is one exciting area of research while the imaging of tumors using nanoparticles is another active area of research. For many drug delivery systems to the CNS, the BBB is a key aspect to consider, and many systemically injected nanoparticles are rapidly removed from circulation. However, there are instances when the vasculature permeability is significantly altered through injury or disease. A cell surface receptor for self-recognition, CD47, has been recombinantly produced and attached onto nanoparticles to significantly increase circulation times. Despite the BBB, a number of papers have demonstrated that nanoparticles can cross from the bloodstream into the brain. With varying degrees of evidence for this, we recommend to source recent reviews on systemic nanoparticle delivery such as by Nagpal et al. and Hwang and Kim. Alternatively, nanoparticles can be delivered to the CNS via intrathecal delivery within an in situ gelling hydrogel to control their release.

Intranasal delivery is also a noninvasive route for delivering nanoparticles to the brain and drug delivery in general.

17.7 NEUROPROSTHESES

TE research on neuroprostheses of the retina, brain, or the ear is increasing, and is driven by the need to improve the device operation. Electrodes either directly stimulate neurons or measure their output and allow control of function or external devices. The convergence of neuroprosthetic devices with TE principles promises to improve neuroprosthesis operation, particularly in the long term. Fattahi et al. have written an excellent review on neuroprostheses that is highly recommended reading.

17.7.1 Cochlear Implants

The cochlear implant is the "success story" of neuroprosthetic devices, developed and gradually improved from the 1970's onwards. A flexible electrode array is inserted into the cochlea where it stimulates spiral ganglion neurons (SGNs), and allows the patient to distinguish noise. As cochlear implant technology evolved, it allowed the profoundly deaf to hear their first sounds including speech discrimination. While the advances are significant, challenges remain in listening to music as well as speech discrimination in places with background noises. This quality of hearing may improve further when combined with TE strategies to better integrate the auditory nerve with the electrode. A review by Wise and Gillespie provides an excellent summary on the challenges and solutions facing cochlea implants.

Most TE approaches focus on drug or growth factor delivery to the SGNs to reduce atrophy, and guide them closer to the electrode surface. "Minimally invasive" intracochlear delivery systems such as via the round window membrane are being also investigated. Maintaining the viability of sensory hair cells has also been attempted using alginate capsules containing BDNF-secreting Schwann cells.

17.7.2 Retinal Prostheses

Retinal prostheses are classified broadly into three types: subretinal, suprachoroidal, and epineural implants (Figure 17.13). Subretinal photodiodes are placed under the retina and contain light-sensitive photodiodes with hundreds of microelectrodes, which stimulate the remaining neurons—there are two notable clinical trials with such devices—Alpha IMS and the Boston Retinal Implant project. Epineural implants have no light-sensitive elements with an image, captured by a camera, transmitted to an implant on top of the retina, where the RGC axons are then stimulated directly. These include the Argus II, Epi-Ret 3, and the Intelligent Medical Implants' device. In this instance, the camera can be

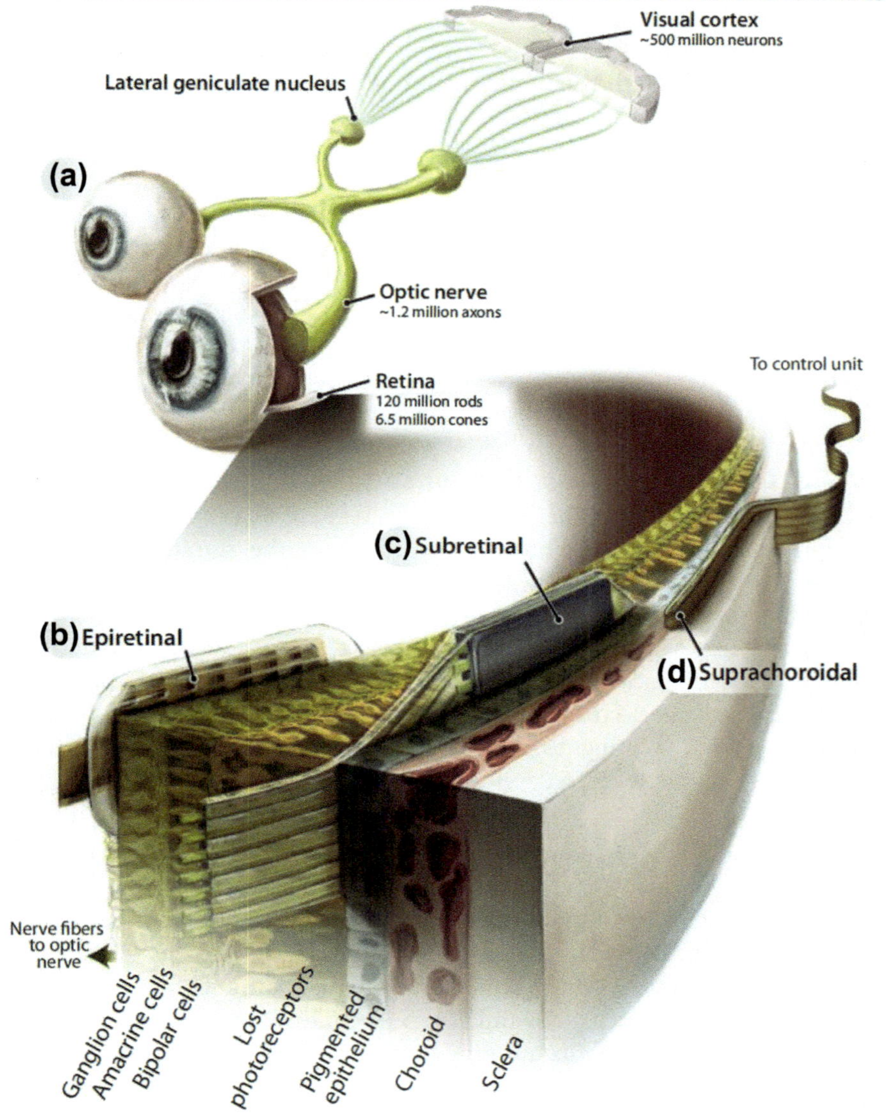

FIGURE 17.13

Illustrations of the visual pathway (a) and placements of three types of retinal implants: epiretinal (b), subretinal (c), and suprachoroidal (d). *Reproduced from Zrenner (2013).*

external to the eye or replaces the natural lens, while fully intraocular devices are now available, which have the benefits of reducing the risk of infection. There remains a need for improved biomaterials for retinal prostheses. Chuang et al. have reviewed various retinal prostheses in the clinic, while Zrenner provides an excellent summary and perspective on retinal prostheses, including the clinic.

17.7.3 Cortical Implants

First developed as single electrodes, then as micromachined electrodes, an ideal recording electrode for the brain should be both small and be able to record many neurons in close proximity. Devices including the Michigan electrode and the Utah array can detect signals in the cerebral cortex and convert this signal into a useful output, with the system termed a brain–machine interface. Alternatively, they can treat neurological disorders, through the stimulation of neurons. Such "deep brain stimulators" have shown benefit to patients with PD and epilepsy. While the literature on cortical implants is extensive, the field has minimal contribution from the TE community so far, and certain limits of cortical implants will likely be solved with the modulation of the surrounding tissue, the development of new electrode materials, and adoption of advanced manufacturing technologies. This electrode/tissue interface between is well reviewed by Fattahi et al. and Lega et al.

17.7.4 Altering the Host Response to Neuroprostheses

A major impediment to the long-term success of many neuroprostheses is the formation of a thick, fibrous capsule around the stimulating electrodes (Figure 17.14(a)). Since electrodes are often hard materials, their placement in nervous tissue and subsequent mechanical mismatch generates significant challenges. Localized increases in GFAP expression indicates the activation of astrocytes, while the recruitment of macrophages to the electrode eventually results in a fibrous capsule that impairs the electrical signal (Figure 17.14(b)). More compliant electrode materials or surface coatings to minimize the extent of the fibrous capsule (Figure 17.14(c)) are key areas for TE development.

17.8 FUTURE APPROACHES

The injury response and healing mechanisms of the PNS are different from those of the CNS. The main issue for PNS regeneration remains the reconnection of severed nerves that are separated from each other above a critical gap length. Fibers, cell transplants, and scaffolds continue to be investigated, and some significant gap lengths are now being bridged with electrospun sheets. Anisotropic scaffolds have morphological similarities to structures within which the bands of Büngner form distal to the PNI, and the next five years should see an increase in their in vivo analysis. Ultimately, the advances in other fabrication processes such as 3D printing may provide the highly resolved scaffold needed for "fascicle-to-fascicle" guided regeneration. Any TE solution must improve over the widely used autograft surgery.

We predict that the use of in situ gelling materials for the treatment of brain and spinal cord trauma will substantially increase over the next decade. Preformed or injectable hydrogels are often considered simple, while in fact they are highly complex, and perform multiple roles required for treating the nervous system. These gels will perform increasing functions, including the survival/differentiation of

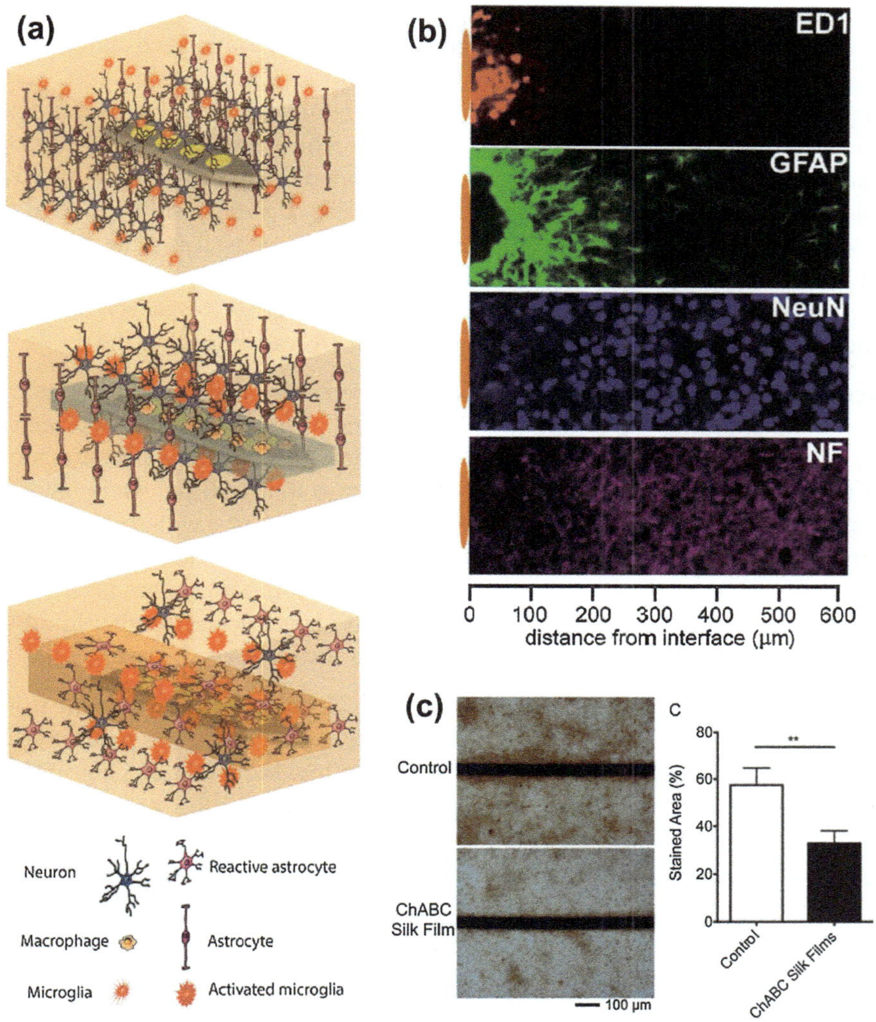

FIGURE 17.14

A major challenge in electrode stability: the fibrotic scar formation. (a) illustrates the environment surrounding an electrode over time, while (b) are fluorescent images of the area surrounding an electrode after 4 weeks. (c) Silk has also been coated onto electrodes to deliver chondroitinase ABC, an enzyme that reduces ECM that is inhibitory to growing neurites. *Figures (a) Reproduced from Fattahi et al. (2014), (b) Reproduced from Biran et al. (2005), (c) Reproduced from Tien et al. (2013).*

transplanted cells and the delivery of nanoparticles for targeted therapeutic delivery. The types of cells available for transplantation are also likely to be greater in the future, including perhaps activation of the patient's own endogenous neural precursor cells. The production of a successful regenerative bridging construct in the CNS requires knowledge of the dynamic vascular, inflammatory and immunological changes that occur over time after an injury, an understanding of the complex molecular and cellular environment in each injured area, and strategies to overcome this complex inhibitory environment. Combining cell transplantation in the contused spinal cord with other repair-promoting interventions, such as pharmacotherapy and gene therapy, allows substantially more "tools" for neurological recovery. Optic nerve injury studies, in which surgery is performed on transgenic animals (particularly mice), has provided important insights into the role of certain genes in axon regeneration. A major challenge will be needed to develop TE protocols that are suitable for use in chronic as well as acute injury situations. Further understanding of the effect of scaffold implantation will continue to provide tissue engineers with new concepts for bridging CNS defects and may one day provide an "off-the-shelf" solution to spinal cord repair.

Neural TE offers a practical role in delivering different therapies such as cell transplantation and drug release and can play a central role in multitherapeutic delivery. The optic nerve is also an excellent CNS model in which to test a combination of therapies, and this could become a model organ to investigate combinatorial approaches. In addition to the optic nerve, the retina is a tissue with surgical accessibility via the vitreous cavity. For this tissue, strategies in delivering polymeric sheets for cell transplantation to the retina will have important ramifications in the delivery of TE strategies to the eye. Improved prosthetic devices that improve activation of the central visual pathways will also advance over the next decade.

Cochlear implants, and other neuroprosthetic devices, will also benefit from advances in TE strategies. Improved neuroprosthesis integration and reduced scarring may lead to better and longer discrete signaling. More diverse neurological injuries and diseases are likely to be tackled with TE approaches. Approaches for the treatment of brain trauma and stroke are likely to involve combinations of therapies and may be distinct from therapies developed to treat degenerative diseases. Diseases such as multiple sclerosis and stroke may also respond to nanoparticle delivered drugs in clinical settings, particularly if systemic or intrathecal delivery can be a route for treating CNS injuries and disorders. The next decade should also see the translation of TE therapies to the clinic for treatment of injuries beyond the PNS. It is even possible that the long sought holy grail of assisting people suffering from SCI will be achieved in 2014–2024 using some aspect of TE principles.

17.9 SUMMARY

- A critical injury gap occurs in peripheral injury, above which the number of successful repairs rapidly declines.
- The Bands of Büngner are Schwann cell columns that form within natural laminin (oriented) scaffolds that are crucial for the regeneration of axons in the PNS.
- Neurotrophic factors play an important, but complicated, role in regenerative therapies.

- The limited regeneration in the CNS is primarily due to the different cellular and molecular environment and also due to the internal properties of the corresponding neurons.
- A contusion injury models the most frequent type of human SCI.
- A range of cell and scaffolds implanted into the injured spinal cord reduce astrocytic scarring, promote tissue sparing, and facilitate axonal regeneration and myelination.
- As in other body systems, cell transplantation and TE in the CNS/PNS is influenced by inflammatory and immune-mediated events. Age and gender differences may also need to be taken into account.
- TE therapies for treatment of traumatic injury (acute/chronic) versus degenerative diseases differ greatly (i.e., what to do and when to do it).
- Genetic modification of cells prior to transplantation into the injured brain and spinal cord can increase neuronal survival and enhance the regeneration response.
- Neural TE strategies are generally multifactorial strategies in terms of both diversity of therapies (i.e., Scaffold *and* cells *and* drugs) and timing of interventions (parallel or sequential).

CLASSICAL EXPERIMENT: SCHWANN CELL AND OEG TRANSPLANTATION IN POLY(ACRYLONITRILE-CO-VINYL CHLORIDE) NERVE GUIDES

The use of Schwann cell transplants within nerve guides was pioneered in the Bunge laboratory at The Miami Project to Cure Paralysis during the 1990s. Schwann cells can be isolated and cultured from peripheral nerves and are therefore autologous. This was demonstrated by Xu and coworkers, who transplanted a poly(acrylonitrile-co-vinyl chloride) nerve guide seeded with a Schwann cells and Matrigel™ into the completely transected adult rat thoracic spinal cord. Several weeks after transplantation, myelinated and unmyelinated axons were present within the Schwann cell cable that connected the opposing cord stumps (Figure shown below). These and other experiments demonstrated the potential of the Schwann cell for spinal cord repair approaches; however, the responding axons grew across the Schwann cell graft but did not exit the graft to grow into the spinal cord tissue beyond; this thereby prevents the formation of new synaptic connections, which are necessary to reestablish axonal circuits involved in controlled motor function. Figure shown below visually depicts the axon penetration and exit of these nerve guides implanted within hemisection models. When Schwann cells are not included within the nerve guides (Matrigel™ only), there is no penetration of axons. While Schwann cell inclusion with the Matrigel™ demonstrates axonal penetration, axons are guided out of the nerve guide via the delivery of BDNF and/or NT-3 into the caudal end of the distal nerve.

OEGs enfold the elongating olfactory axons so that they are prevented from recognizing the growth-inhibitory cues in the CNS environment. This unique feature of OEG was thought to be beneficial for eliciting axonal growth across an injury site such as scar tissue. However, while OEG will ensheath axons, they do not myelinate them; Schwann cells, however, do myelinate axons in the spinal cord. In a pivotal experiment by Ramon-Cueto, Plant and Bunge in the laboratory at The Miami Project to Cure Paralysis in 1998, OEG and Schwann cells were combined. A P(AN-VC) nerve guide was preformed with Schwann cells mixed in Matrigel™ and implanted into the adult thoracic spinal cord. OEGs were then injected into the rostral and caudal stumps.

The results showed that the additional injections of OEG at either end of the Schwann cell channels allowed

Continued...

CLASSICAL EXPERIMENT: SCHWANN CELL AND OEG TRANSPLANTATION IN POLY(ACRYLONITRILE-CO-VINYL CHLORIDE) NERVE GUIDES *Continued*

a greater number of axons to exit the Schwann cell bridging grafts (see Xu et al.). Descending axons such as those of the raphe system (located in the brainstem) grew in areas devoid of many Schwann cells but were seen in areas containing fibroblasts and OEG. Sensory axons also regenerated for long distances—up to 18 mm. Such an experiment demonstrates the utility of multiple treatments for a successful spinal cord repair outcome.

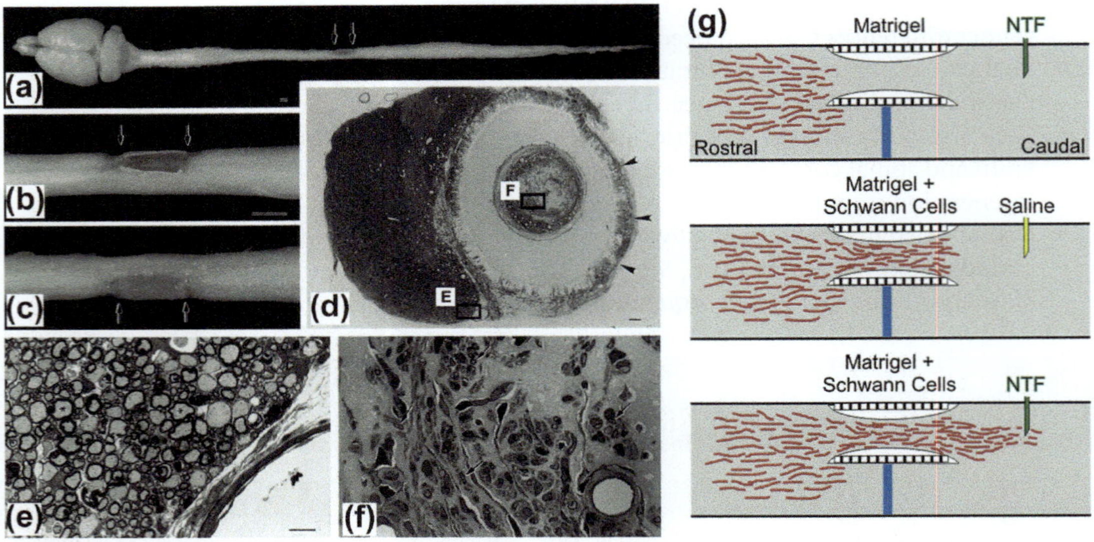

Images of Schwann cell/Matrigel™ transplants in P(AN-VC) nerve guides implanted in hemisection. (a) Photograph of brain and spinal cord, and closer magnifications of the hemisection area and implant (a and c) are shown. (d) A transverse section depicts the nerve guide lumen filled with a tissue bridge and the intact ventral and dorsal horn regions of the spinal cord. Magnifications of d show (e) myelinated axons at the outer interface of the nerve guide and (f) smaller caliber axons present within the lumen. (g) Illustration of the effect of P(AN-VC) nerve guides implanted within hemisections. When Schwann cells are not included within the nerve guides (Matrigel™ only), there is no penetration of axons (red) into the graft. While Schwann cell inclusion with the Matrigel™ demonstrates axonal penetration, the axons can be guided out of the nerve guide via the delivery of BDNF and/or NT-3 into the caudal end of the distal nerve a few millimeters from the implant. Without elevated levels of neurotrophic factor, axons do not exit the implant. The blue line is a lesion made just prior to tracing. *Reproduced and Modified from Xu et al. (1999).*

STATE-OF-THE-ART EXPERIMENT: GUIDING TUMORS TO THEIR DEATH

Glioblastoma multiforme is an aggressive class of brain tumor that can exist deep within the brain, in regions difficult to access by surgery and are often inoperable. Comprising 50% of all brain tumors, this tumor can rapidly migrate along white matter tracts and blood vessels to spread the disease. The Bellamkonda research group hopes that this aggressive growth can also be a solution to remove the glioblastoma. They have engineered guidance tubes that are placed into the brain against the tumor, and guide the tumor cells along the conduit...to their death.

Placing aligned electrospun membranes within these conduits, Jain et al. have encouraged tumor cells to grow toward a hydrogel at the end that contains covalently linked drugs that induce apoptosis. Initially, the in vitro migration of U87MG-eGFP cells on oriented PLL/laminin-coated electrospun PCL membranes was assessed. Topography greatly influenced the migration of the glioblastoma cells, and exhibited a threefold to fivefold higher migration on oriented electrospun sheets. Since these cells will be guided to an apoptosis signal, their survival on an engineered collagen hydrogel was investigated. Cyclopamine was chosen as the toxin to conjugate to the collagen hydrogel as it exerts its effect only in cells that are dependent on the Shh pathway, which excludes neurons. A level of cyclopamine was therefore chosen that induces apoptosis in glioblastoma, but not in cortical neurons.

PCL/polyurethane conduits, filled with the oriented electrospun fibers and end filled with cyclopamine-conjugated collagen, were implanted. Tumor inoculation occurred sevendays prior to conduit implantation and the various regions of the conduit were investigated. Significantly more tumor mass grow within the conduits, while the tumor remaining in the brain reduced in size. One remarkable aspect of this experiment is the lack of efficacy of smooth sheets as guidance structures within the conduits; instead, electrospun membranes provided a significantly improved substrate for these tumor cells to grow upon. While not an ideal solution to glioblastoma removal, applying TE principles to areas such as cancer research provide new concepts that spur on advances in that field.

RECOMMENDED READING

Dalton and Mey, 2008. Material Interactions with PNS/CNS Review.

Sakiyama-Elbert et al., 2012. Scaffolds in the CNS Review.

Deumens et al., 2010. Peripheral Nerve Guide Review.

Kramer et al., 2013. Review on Pluripotent Stem Cell Transplants.

Macaya and Spector, 2012. Injectible Matrices Review.

Chung et al., 2013. The CLARITY Protocol Tissue Clearing Paper.

Kurimoto et al., 2010. Optic Nerve Regeneration.

Zrenner, 2013. Retinal Prosthesis Review.

Fattahi et al., 2014. Neuroprosthesis Review.

Jain et al., 2014. State of the Art Experiment.

REFERENCES

Alilain, W.J., Horn, K.P., Hu, H.M., Dick, T.E., Silver, J., 2011. Nature 475 (7355), 196–201.

Bamber, N.I., Li, H.Y., Aebischer, P., Xu, X.M., 1999. Neural Plast. 6 (4), 103–121.

Bamber, N.I., Li, H.Y., Lu, X.B., Oudega, M., Aebischer, P., Xu, X.M., 2001. Eur. J. Neurosci. 13 (2), 257–268.

Barbour, H.R., Plant, C.D., Harvey, A.R., Plant, G.W., 2013. Bmc Neurosci. 14.

Baumann, M.D., Kang, C.E., Stanwick, J.C., Wang, Y.F., Kim, H., Lapitsky, Y., Shoichet, M.S., 2009. J. Controlled Release 138 (3), 205–213.

Benowitz, L.I., Popovich, P.G., 2011. Curr. Opin. Neurol. 24 (6), 577–583.

Biran, R., Martin, D.C., Tresco, P.A., 2005. Exp. Neurol. 195 (1), 115–126.

Borenstein, J.T., 2011. Expert Opin. Drug. Delivery 8 (9), 1161–1174.

Bozkurt, G., Zahir, T., Mothe, A., Shoichet, M.S., Morshead, C.M., Tator, C.H., 2007. J. Neurotrauma 24 (7), 1285.

Bradbury, E.J., Moon, L.D.F., Popat, R.J., King, V.R., Bennett, G.S., Patel, P.N., Fawcett, J.W., McMahon, S.B., 2002. Nature 416 (6881), 636–640.

Bunge, M.B., Wood, P.M., 2012. Handb. Clin. Neurol. 109, 523–540.

Cheng, Y., Morshed, R.A., Auffinger, B., Tobias, A.L., Lesniak, M.S., 2014. Adv. Drug Delivery Rev. 66, 42–57.

Chuang, A.T., Margo, C.E., Greenberg, P.B., 2014. Br. J. Ophthalmol.

Chung, K., Deisseroth, K., 2013. Nat. Methods 10 (6), 508–513.

Chung, K., Wallace, J., Kim, S.Y., Kalyanasundaram, S., Andalman, A.S., Davidson, T.J., Mirzabekov, J.J., Zalocusky, K.A., Mattis, J., Denisin, A.K., Pak, S., Bernstein, H., Ramakrishnan, C., Grosenick, L., Gradinaru, V., Deisseroth, K., 2013. Nature 497 (7449), 332.

Clements, I.P., Kim, Y.T., English, A.W., Lu, X., Chung, A., Bellamkonda, R.V., 2009. Biomaterials 30 (23–24), 3834–3846.

Dalton, P.D., Mey, J., 2009. Front. Biosci. 14, 769–795.

Deumens, R., Bozkurt, A., Meek, M.F., Marcus, M.A.E., Joosten, E.A.J., Weis, J., Brook, G.A., 2010. Prog. Neurobiol. 92 (3), 245–276.

Dodla, M.C., Bellamkonda, R.V., 2008. Biomaterials 29 (1), 33–46.

Dubovy, P., 2011. Ann. Anat.-Anat. Anz. 193 (4), 267–275.

Eggers, R., de Winter, F., Hoyng, S.A., Roet, K.C.D., Ehlert, E.M., Malessy, M.J.A., Verhaagen, J., Tannemaat, M.R., 2013. Plos One 8 (8).

Fattahi, P., Yang, G., Kim, G., Abidian, M.R., 2014. Adv. Mater. 26 (12), 1846–1885.

Fischer, D., Leibinger, M., 2012. Prog. Retinal Eye Res. 31 (6), 688–701.

Gerardo-Nava, J., Fuhrmann, T., Klinkhammer, K., Seiler, N., Mey, J., Klee, D., Moller, M., Dalton, P.D., Brook, G.A., 2009. Nanomedicine 4 (1), 11–30.

Godinho, M.J., Teh, L., Pollett, M.A., Goodman, D., Hodgetts, S.I., Sweetman, I., Walters, M., Verhaagen, J., Plant, G.W., Harvey, A.R., 2013. Plos One 8 (8).

Gonzalez-Cordero, A., West, E.L., Pearson, R.A., Duran, Y., Carvalho, L.S., Chu, C.J., Naeem, A., Blackford, S.J.I., Georgiadis, A., Lakowski, J., Hubank, M., Smith, A.J., Bainbridge, J.W.B., Sowden, J.C., Ali, R.R., 2013. Nat. Biotechnol. 31 (8), 741.

Gonzalez-Perez, F., Udina, E., Navarro, X., 2013. Int. Rev. Neurobiol. 108, 257–275.

Gupta, D., Tator, C.H., Shoichet, M.S., 2006. Biomaterials 27 (11), 2370–2379.

Haggerty, A.E., Oudega, M., 2013. Neurosci. Bull. 29 (4), 445–459.

Harvey, A.R., Ooi, J.W.W., Rodger, J., 2012. Int. Rev. Neurobiol. 106 (Pt 2), 1–33.

Hellstrom, M., Harvey, A.R., 2011. Curr. Gene Ther. 11 (2), 116–131.

Hodgetts, S.I., Plant, G.W., Harvey, A.R., 2009. Spinal Cord Injury: Experimental Animal Models and Relation to Human Therapy. The Spinal Cord - a Christopher and Dana Reeve Foundation Text and Atlas. Academic Press, UK. pp. 209–237.

Hu, Y., Leaver, S.G., Plant, G.W., Hendriks, W.T.J., Niclou, S.P., Verhaagen, J., Harvey, A.R., Cui, Q., 2005. Mol. Ther. 11 (6), 906–915.

Hutson, T.H., Verhaagen, J., Yanez-Munoz, R.J., Moon, L.D.F., 2012. Gene Ther. 19 (1), 49–60.

Hwang, S.R., Kim, K., 2014. Arch. Pharmacal Res. 37 (1), 24–30.

Hynes, S.R., Lavik, E.B., 2010. Graefes Arch. Clin. Exp. Ophthalmol. 248 (6), 763–778.

Iwasaki, M., Wilcox, J.T., Nishimura, Y., Zweckberger, K., Suzuki, H., Wang, J., Liu, Y., Karadimas, S.K., Fehlings, M.G., 2014. Biomaterials 35 (9), 2617–2629.

Jain, A., Betancur, M., Patel, G.D., Valmikinathan, C.M., Mukhatyar, V.J., Vakharia, A., Pai, S.B., Brahma, B., MacDonald, T.J., Bellamkonda, R.V., 2014. Nat. Mater. 13 (3), 308–316.

Jain, A., Kim, Y.T., McKeon, R.J., Bellamkonda, R.V., 2006. Biomaterials 27 (3), 497–504.

Jain, N.K., Mishra, V., Mehra, N.K., 2013. Expert Opin. Drug. Deliv. 10 (3), 353–367.

Jeffries, E.M., Wang, Y.D., 2013. Biofabrication 5 (3).

Kador, K.E., Goldberg, J.L., 2012. Expert Rev. Ophthalmol. 7 (5), 459–470.

Kang, C.E., Poon, P.C., Tator, C.H., Shoichet, M.S., 2009. Tissue Eng. Part A 15 (3), 595–604.

Kim, Y.T., Caldwell, J.M., Bellamkonda, R.V., 2009. Biomaterials 30 (13), 2582–2590.

King, V.R., Phillips, J.B., Hunt-Grubbe, H., Brown, R., Priestley, J.V., 2006. Biomaterials 27 (3), 485–496.

Kramer, A.S., Harvey, A.R., Plant, G.W., Hodgetts, S.I., 2013. Cell. Transplant. 22 (4), 571–617.

Kriebel, A., Rumman, M., Scheld, M., Hodde, D., Brook, G., Mey, J., 2014. J. Biomed. Mater. Res. Part B-Appl. Biomater. 102 (2), 356–365.

Kurimoto, T., Yin, Y.Q., Omura, K., Gilbert, H.Y., Kim, D., Cen, L.P., Moko, L., Kugler, S., Benowitz, L.I., 2010. J. Neurosci. 30 (46), 15654–15663.

Lega, B.C., Serruya, M.D., Zaghloul, K.A., 2011. Crit. Rev. Biomed. Eng. 39 (1), 5–28.

Li, H.Y., Fuhrmann, T., Zhou, Y., Dalton, P.D., 2013. J. Mater. Sci.-Mater. Med. 24 (8), 2001–2011.

Liu, Z.P., Yu, N., Holz, F.G., Yang, F., Stanzel, B.V., 2014. Biomaterials 35 (9), 2837–2850.

Lundborg, G., 2000. J. Hand Surg.-British Eur. Volume 25B (3), 242–252.

Lynam, D., Bednark, B., Peterson, C., Welker, D., Gao, M.Y., Sakamoto, J.S., 2011. J. Mater. Sci.-Mater. Med. 22 (9), 2119–2130.

Macaya, D., Spector, M., 2012. Biomed. Mater. 7 (1).

Madduri, S., di Summa, P., Papaloizos, M., Kalbermatten, D., Gander, B., 2010. Biomaterials 31 (32), 8402–8409.

McDonald, J.W., Becker, D., Holekamp, T.F., Howard, M., Liu, S., Lu, A.W., Lu, J., Platik, M.M., Qu, Y., Stewart, T., Vadivelu, S., 2004. J. Neurotrauma 21 (4), 383–393.

McGrath, A.M., Novikova, L.N., Novikova, L.N., Wiberg, M., 2010. Brain Res. Bull. 83 (5), 207–213.

Meyer, H., Stover, T., Fouchet, F., Bastiat, G., Saulnier, P., Baumer, W., Lenarz, T., Scheper, V., 2012. Int. J. Nanomed. 7, 2449–2464.

Mittal, D., Ali, A., Md, S., Baboota, S., Sahni, J.K., Ali, J., 2014. Drug. Deliv. 21 (2), 75–86.

Moshayedi, P., Ng, G., Kwok, J.C., Yeo, G.S., Bryant, C.E., Fawcett, J.W., Franze, K., Guck, J., 2014. Biomaterials 35 (13), 3919–3925.

Mothe, A.J., Tam, R.Y., Zahir, T., Tator, C.H., Shoichet, M.S., 2013. Biomaterials 34 (15), 3775–3783.

Nagpal, K., Singh, S.K., Mishra, D.N., 2013. Expert Opin. Drug Delivery 10 (7), 927–955.

Oudega, M., Hagg, T., 1996. Exp. Neurol. 140 (2), 218–229.

Pakulska, M.M., Ballios, B.G., Shoichet, M.S., 2012. Biomed. Mater. 7 (2).

Papa, S., Rossi, F., Ferrari, R., Mariani, A., De Paola, M., Caron, I., Fiordaliso, F., Bisighini, C., Sammali, E., Colombo, C., Gobbi, M., Canovi, M., Lucchetti, J., Peviani, M., Morbidelli, M., Forloni, G., Perale, G., Moscatelli, D., Veglianese, P., 2013. Acs Nano 7 (11), 9881–9895.

Patel, V., Joseph, G., Patel, A., Patel, S., Bustin, D., Mawson, D., Tuesta, L.M., Puentes, R., Ghosh, M., Pearse, D.D., 2010. J. Neurotrauma 27 (5), 789–801.

Pearse, D.D., Pereira, F.C., Marcillo, A.E., Bates, M.L., Berrocal, Y.A., Filbin, M.T., Bunge, M.B., 2004. Nat. Med. 10 (6), 610–616.

Pearson, R.A., Barber, A.C., Rizzi, M., Hippert, C., Xue, T., West, E.L., Duran, Y., Smith, A.J., Chuang, J.Z., Azam, S.A., Luhmann, U.F.O., Benucci, A., Sung, C.H., Bainbridge, J.W., Carandini, M., Yau, K.W., Sowden, J.C., Ali, R.R., 2012. Nature 485 (7396), 99–103.

Pettingill, L.N., Wise, A.K., Geaney, M.S., Shepherd, R.K., 2011. Plos One 6 (4).

Plant, G.W., Bates, M.L., Bunge, M.B., 2001. Mol. Cell. Neurosci. 17 (3), 471–487.

Plant, G.W., Christensen, C.L., Oudega, M., Bunge, M.B., 2003. J. Neurotrauma 20 (1), 1–16.

Ramon-Cueto, A., Plant, G.W., Avila, J., Bunge, M.B., 1998. J. Neurosci. 18 (10), 3803–3815.

Riblett, B.W., Francis, N.L., Wheatley, M.A., Wegst, U.G.K., 2012. Adv. Funct. Mater. 22 (23), 4920–4923.

Ritfeld, G.J., Rauck, B.M., Novosat, T.L., Park, D., Patel, P., Roos, R.A.C., Wang, Y.D., Oudega, M., 2014. Biomaterials 35 (6), 1924–1931.

Ritfeld, G.J., Tewarie, R.D.S.N., Vajn, K., Rahiem, S.T., Hurtado, A., Wendell, D.F., Roos, R.A.C., Oudega, M., 2012. Cell. Transplant. 21 (7), 1561–1575.

Ruitenberg, M.J., Levison, D.B., Lee, S.V., Verhaagen, J., Harvey, A.R., Plant, G.W., 2005. Brain 128, 839–853.

Sakiyama-Elbert, S., Johnson, P.J., Hodgetts, S.I., Plant, G.W., Harvey, A.R., 2012. Handb. Clin. Neurol. 109, 575–594.

Scherer, S.S., Salzer, J.L., 2001. Axon-Schwann cell interactions during peripheral nerve degeneration and regeneration. In: Glial Cell and Development. Oxford Press, New York, pp. 299–330.

Schlosshauer, B., Dreesmann, L., Schaller, H.E., Sinis, N., 2006. Neurosurgery 59 (4), 740–747.

Seiler, M.J., Aramant, R.B., 2012. Prog. Retin. Eye Res. 31 (6), 661–687.

Silver, J., Miller, J.H., 2004. Nat. Rev. Neurosci. 5 (2), 146–156.

Stokols, S., Tuszynski, M.H., 2006. Biomaterials 27 (3), 443–451.

Sun, F., He, Z., 2010. Curr. Opin. Neurobiol. 20 (4), 510–518.

Tan, F., Walshe, P., Viani, L., Al-Rubeai, M., 2013. Trends Biotechnol. 31 (12), 678–687.

Tewarie, R.D.S.N., Hurtado, A., Ritfeld, G.J., Rahiem, S.T., Wendell, D.F., Barroso, M.M.S., Grotenhuis, J.A., Oudega, M., 2009. J. Neurotrauma 26 (12), 2313–2322.

Tien, L.W., Wu, F., Tang-Schomer, M.D., Yoon, E., Omenetto, F.G., Kaplan, D.L., 2013. Adv. Funct. Mater. 23 (25), 3185–3193.

Tresco, P.A., Winslow, B.D., 2011. Crit. Rev. Biomed. Eng. 39 (1), 29–44.

Tsai, E.C., Dalton, P.D., Shoichet, M.S., Tator, C.H., 2006. Biomaterials 27 (3), 519–533.

Tsai, R.K., Discher, D.E., 2008. J. Cell. Biol. 180 (5), 989–1003.

Tysseling-Mattiace, V.M., Sahni, V., Niece, K.L., Birch, D., Czeisler, C., Fehlings, M.G., Stupp, S.I., Kessler, J.A., 2008. J. Neurosci. 28 (14), 3814–3823.

Volpato, F.Z., Fuhrmann, T., Migliaresi, C., Hutmacher, D.W., Dalton, P.D., 2013. Biomaterials 34 (21), 4945–4955.

West, E.L., Pearson, R.A., MacLaren, R.E., Sowden, J.C., Ali, R.R., 2009. Prog. Brain. Res. 175, 3–21.

Wise, A.K., Gillespie, L.N., 2012. J. Neural Eng. 9 (6).

Wood, M.D., Hunter, D., Mackinnon, S.E., Sakiyama-Elbert, S.E., 2010. J. Biomater. Sci.-Polym. Ed. 21 (6–7), 771–787.

Xu, X.M., Chen, A., Guenard, V., Kleitman, N., Bunge, M.B., 1997. J. Neurocytol. 26 (1), 1–16.

Xu, X.M., Zhang, S.X., Li, H.Y., Aebischer, P., Bunge, M.B., 1999. Eur. J. Neurosci. 11 (5), 1723–1740.

Yannas, I.V., Zhang, M., Spilker, M.H., 2007. J. Biomater. Sci. Polym. Ed. 18 (8), 943–966.

Yu, T.T., Shoichet, M.S., 2005. Biomaterials 26 (13), 1507–1514.

Zanin, M.P., Pettingill, L.N., Harvey, A.R., Emerich, D.F., Thanos, C.G., Shepherd, R.K., 2012. J. Control Release 160 (1), 3–13.

Zhang, M., Yannas, I.V., 2005. Adv. Biochem. Eng. Biotechnol. 94, 67–89.

Zrenner, E., 2013. Sci. Transl. Med. 5 (210), 210–216.

CHAPTER 18

Principles of Cardiovascular Tissue Engineering

Emil Ruvinov[1], Yulia Shandalov[2], Shulamit Levenberg[2], Smadar Cohen[1,3,4]

[1]The Avram and Stella Goldstein-Goren Department of Biotechnology Engineering, Ben-Gurion University of the Negev, Beer-Sheva, Israel; [2]Department of Biomedical Engineering, Technion - Israel Institute of Technology, Haifa, Israel; [3]The Center for Regenerative Medicine and Stem Cell (RMSC) Research, Ben-Gurion University of the Negev, Beer-Sheva, Israel; [4]The Ilse Katz Institute for Nanoscale Science and Technology, Ben-Gurion University of the Negev, Beer-Sheva, Israel

LEARNING OBJECTIVES

- Cardiovascular system pathophysiology and the clinical needs addressed by tissue engineering
- Major components of cardiovascular tissue engineering: cells (including clinically relevant cell sources) and biomaterials
- Biomaterial-based strategies for cell or therapeutic molecule delivery
- Major approaches for bioengineering of cardiac patches and strategies to improve patch functionality
- Vascularization strategies of cardiac patches
- Concepts and strategies for the bioengineering of blood vessels
- In situ cardiac reconstruction by acellular biomaterials

18.1 INTRODUCTION

Tissue engineering is of particular interest for the treatment of heart diseases such as myocardial infarction (MI), where large portions of functional tissue are lost with very limited intrinsic regeneration ability. Such loss often leads to significant and progressive deterioration in cardiac contractility and function, eventually resulting in the development of life-threatening conditions and congestive heart failure (CHF). Moreover, the treatment of coronary heart or peripheral vascular diseases often requires vascular reconstruction, such intervention being impossible in a significant proportion of patients, due to the lack of autologous vessel sources. Collectively, the shortage of transplants and donors, the inability of current clinical therapies to restore functional myocardial tissue loss after MI, and the urgent need for functional cardiovascular tissue

Tissue Engineering.
Copyright © 2015 Elsevier Inc. All rights reserved.

substitutes are the main factors contributing to enormous efforts invested in development of cardiovascular tissue engineering and regeneration strategies.

In this chapter, we present an up-to-date review on cardiovascular tissue engineering, providing the readers with essential information regarding advanced and effective strategies for cardiovascular repair and regeneration, to be used either as standalone products or as combinations with current clinical therapies.

18.2 HEART STRUCTURE, DISEASE, AND REGENERATION

18.2.1 The Myocardium

The heart is a vital organ, which is indispensable for maintaining normal homeostasis of the organism. The myocardium, i.e., the cardiac muscle tissue, constitutes about 95% of the heart mass and is responsible for its pumping action. The human myocardium consists of 2 to 3 billion cardiomyocytes (CMs) (75% by volume, ~30% by number); they are striated muscle cells found only in the heart and can be visually distinguished from the skeletal and smooth muscle cells (SMCs).

18.2.2 Myocardial Infarction and Heart Failure

Coronary heart disease (CHD) is now the leading cause of death worldwide. In 2008, CHD was the cause of ~1 out of every 6 deaths in the United States. Nevertheless, recent data show that death rates from CHD have decreased in North America and in many countries in Western Europe (Roger et al., 2012). This decline has been due to improved prevention, diagnosis, and treatment (pharmacological or interventional), in particular reduced rates of cigarette smoking among adults and lower average levels of blood pressure and blood cholesterol. However, the burden of CHD is increasing in developing and transitional countries, partly as a result of increasing longevity, urbanization, and lifestyle changes. It is expected that 82% of the future increase in CHD mortality will occur in developing countries.

Myocardial infarction (MI) is the most common manifestation of CHD, accounting for ~50% of all CHD cases in the United States. MI results from temporary or permanent occlusion of the main coronary arteries (Figure 18.1), causing significant reduction in blood supply to the beating heart muscle (mainly to the left ventricle).

The myocardium (cardiac muscle) tissue that is supplied by the occluded artery becomes ischemic. Because of its high metabolic rate, the myocardium begins to undergo irreversible injury within 20 min of ischemia (Figure 18.1). Five partially overlapping phases can be distinguished in the progression from healthy to infarcted myocardium: (1) cardiomyocyte loss due to an initial ischemic event, (2) acute inflammatory response, (3) extracellular matrix (ECM) remodeling, (4) granulation tissue formation, and (5) scar formation and left ventricular (LV) remodeling. The structural remodeling of the heart (i.e., infarct wall

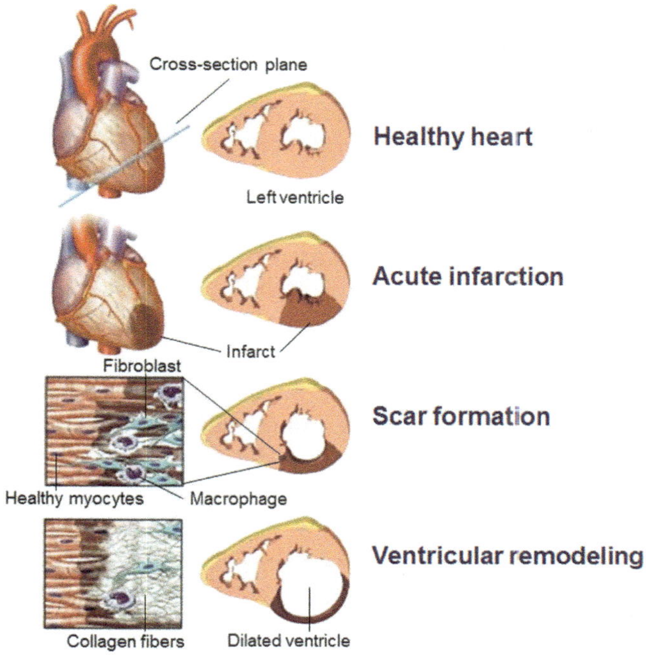

FIGURE 18.1

Heart failure: from acute crisis to chronic disease. In a healthy heart, the LV pumps newly oxygenated blood to the rest of the body, and its walls are normally thick with cardiac muscle fibers. When blood supply to the beating muscle is reduced as a result of coronary artery occlusion, myocytes die from oxygen deprivation, and the infarct develops. Within hours and days, extracellular matrix (ECM) degradation takes place. The infarct is infiltrated by macrophages and collagen-producing myofibroblasts. The infarcted wall of the ventricle becomes thin and rigid. As healthy myocytes die at the border of the scarred area, the infarct expansion continues. Developed pressure overload is initially compensated by hypertrophy of the healthy myocardium. Ultimately, however, this compensating mechanism fails, and infarct wall thinning and ventricle dilatation continues. As a result, the heart is unable to pump effectively, leading to the life-threatening condition of heart failure. *Reprinted with permission from Ruvinov et al. (2012).*

thinning, scar formation, and noninfarcted myocardium hypertrophy) leads to LV functional remodeling, a progressive deterioration in heart muscle function that eventually leads to CHF (Jessup and Brozena, 2003).

Over recent decades, major improvements have been realized in the management of patients with MI (Nabel and Braunwald, 2012). These include in-hospital treatments (e.g., pharmacological lysis, antiplatelet, and antithrombin therapies), interventional therapies, and surgery (e.g., cardiac catheterization, percutaneous coronary intervention, coronary artery bypass surgery, and heart transplantation) as well as drug regimens for prevention and long-term treatment (e.g., aspirin, ACE inhibitors, β-blockers, and statins). The improved management of acute coronary events, however, has led to a significant increase in the number of patients that suffer from chronic conditions, namely CHF.

A major disadvantage of these therapies is their inability to replace, at least partially, cardiac muscle after infarction. Thus, there is a need for alternative approaches that enable to overcome the limitations of standard therapies. The ultimate goal of such novel therapies is the induction of myocardial tissue regeneration (therapeutic, endogenous, or combined) in situ or ex vivo.

18.2.3 Cardiac ECM—Function and Pathological Changes after MI

The architectural complexity of the myocardium and the potential role of the ECM in maintaining the unique myocyte orientations throughout the LV free wall were described by Streeter and Basset (Streeter et al., 1969). Using a structural engineering approach, these authors demonstrated that myocyte orientation and myocardial fiber angles are highly organized and move in a continuous fashion from the endocardium (inner layer) to the epicardium (outer layer of the heart). A network of proteins with highly organized structure and architecture, such as type I and type III collagen, provides structural integrity to adjoining myocytes and contributes to overall LV pump function through the coordination of myocyte shortening (Figure 18.2). The main components of the myocardial ECM and myocardium are listed in Table 18.1.

Significant alterations in the structure and composition of the myocardial ECM occur following MI, which eventually affect the mechanical properties of the LV myocardium. Very early post-MI, the normal ECM within the ischemic region is degraded. As the post-MI period progresses over the next several days, an influx of inflammatory cells into the injured myocardium occurs, which results in further proteolysis of cellular and ECM proteins. In addition, this inflammatory response causes proliferation and differentiation of fibroblasts and other interstitial cells, and the elaboration of bioactive molecules which contribute to a robust synthesis of ECM for the purposes of scar formation (Spinale, 2007). The later phase of post-MI remodeling results in ECM changes within all regions of the LV: the MI region, the viable myocardium within the border zone, and the remote region. This adverse remodeling process has been termed "infarct expansion" and occurs in the absence of additional myocyte injury or alterations in LV loading conditions. It has been postulated that an acceleration of ECM degradation occurs within the myocardium surrounding the MI (border zone) facilitating the infarct expansion process in this later phase of post-MI remodeling (Spinale, 2007). This post-MI remodeling process is a clinically significant problem in that it can lead to LV dilation, systolic and diastolic dysfunction, and progression to heart failure. Indeed, this process, which includes changes in ECM structure and composition, is an independent predictor of morbidity and mortality (Kim et al., 2000).

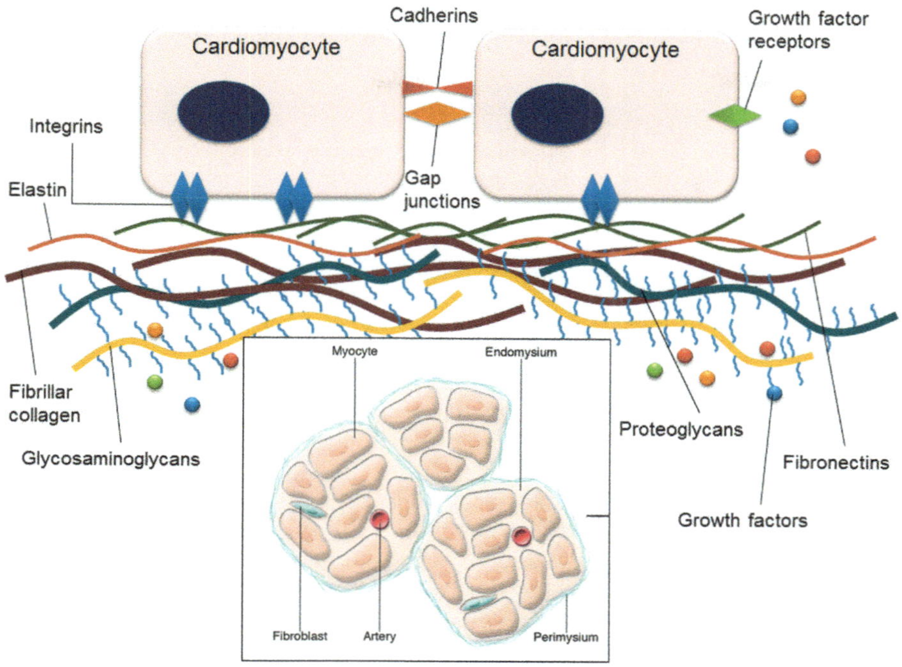

FIGURE 18.2

A schematic presentation of the myocardial environment and major ECM components. Collagen weave surrounding individual myocytes and collagen struts tethering adjacent myocytes comprise the endomysium. Groups of myocytes are bundled within the perimysium. Capillaries and coronary microvessels have free diffusion access to cardiac myocytes throughout the ECM. In the scheme: major ECM components in the myocardium include fibrillar collagen, fibronectin, elastin, proteoglycans, and glycosaminoglycans. The ECM also serves as a controlled reservoir of growth factors. The ECM–cell interactions are mediated by integrins on the cell surface. Cadherins and gap junction proteins comprise the cell–cell interaction complex. *Reprinted with permission from Ruvinov et al. (2012).*

18.2.4 Endogenous Myocardial Regeneration

For several decades, it was assumed that the heart consisted of only terminally differentiated cells and was incapable of intrinsic regeneration. However, recent data have challenged this view, and provided evidence that new CMs can be formed in the adult heart. The field of myocardial regeneration is currently under intensive research, associated with provocative and conflicting reports. Nevertheless, after a major MI the regeneration rates are extremely low, and this can explain the lack of significant restoration of cardiac muscle after severe ischemic injury, the formation of fibrotic scar at the infarct, and progressive deterioration in cardiac function, which eventually lead to CHF.

Table 18.1 Components of the Myocardium and Myocardial ECM (Jugdutt, 2003; Fomovsky et al., 2010)

Component	Main Function
Collagen fibrils (types I and III)	Structural support, maintain shape Force transmission Tensile strength (type I); resilience (type III)
Elastin	Resilience; vessel wall stretch; cardiac wall stretch; and relaxation
Hydrophilic glycosaminoglycans	Diffusion of nutrients, metabolites, growth factors, cytokines, etc.
Proteoglycans	Mechanics, fluid dynamics
Integrins (matrix receptors)	Myocyte–fibroblast/ECM interactions, matrix remodeling
Fibronectin and laminin	Adhesive fibrous proteins
Cells	
Cardiomyocytes	Cardiac muscle cells, responsible for contractile function of the heart
Fibroblasts	Production of fibrillar collagen Differentiate to myofibroblasts after injury
Macrophages	Phagocytosis, inflammatory response
Other cells	Endothelial cells, smooth muscle cells, pericytes, neurons

18.2.5 Potential Therapeutic Targets and Strategies to Induce Myocardial Regeneration

In order to compensate for the low and insufficient intrinsic regeneration ability of the adult heart, strategies for therapeutic regeneration aim to induce myocardial regeneration, improve tissue salvage, facilitate self-repair, reverse or attenuate adverse remodeling, and ultimately achieve long-term functional stabilization and improvement in heart function. Five major processes associated with MI are targeted at present by various experimental regeneration strategies (Ruvinov et al., 2008):

- *Cardioprotection*—the prevention of progressive cardiomyocyte loss following MI by applying various apoptosis-inhibiting reagents or by inducing prosurvival signaling (Garg et al., 2005).
- *Inflammation*—time-adjusted modulation of the post-MI pro/antiinflammatory cytokine/chemokine profile or cellular responses (e.g., granulation tissue formation and macrophage infiltration) in an attempt to induce effective tissue healing and repair and to avoid negative inflammatory effects (e.g., cell death, fibrosis, etc.) (Frantz et al., 2009).
- *ECM remodeling and cardiac fibrosis*—time-adjusted positive modulation of the fibrotic response (i.e., ECM remodeling and scar formation), utilizing recent knowledge on profibrotic signaling, such as modification of the

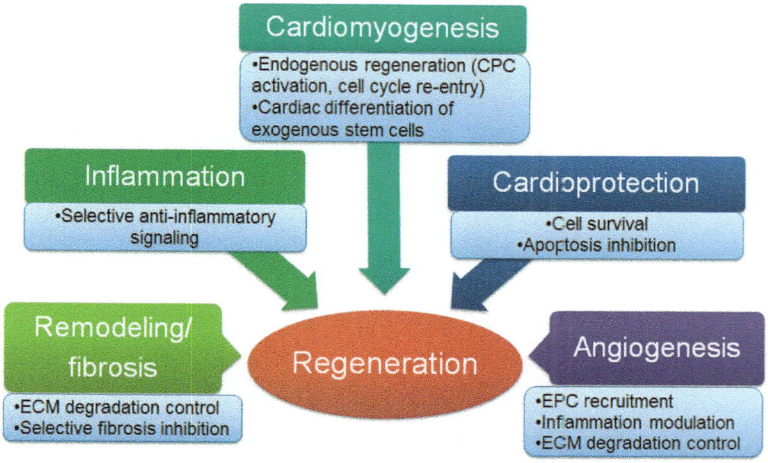

FIGURE 18.3

Therapeutic myocardial regeneration strategies (see text for details). *Reprinted with permission from Ruvinov et al. (2012).*

ratio between metalloproteinases (MMPs) and their inhibitors, which may lead to successful anti-fibrotic therapy (Vanhoutte et al., 2006).

- *Angiogenesis*—increasing the blood supply to ischemic regions is an extensively used approach for effective tissue healing. A variety of proteins, genes, or cells have been tested, aimed at inducing the formation of new vasculature at the infarct site (Renault and Losordo, 2007).
- *Cardiomyogenesis*—myocyte regeneration by activation and/or migration of distinct cell populations with stem- or progenitor-like properties in the adult myocardium which can contribute to de novo myocardium formation after MI. An alternative novel mechanism for endogenous myocyte regeneration is the induction of cardiomyocyte cell cycle reentry by reprogramming of differentiated CMs toward proliferation (Zacchigna and Giacca, 2014).

All these goals can be translated into various therapy strategies aimed at inducing myocardial regeneration (Figure 18.3).

18.3 CELL SOURCES FOR CARDIOVASCULAR TISSUE ENGINEERING AND REGENERATION

Cells represent a critical component in the tissue engineering paradigm. While preclinical research utilizes many readily available cell sources (cell lines, animal-derived isolated cells, etc.), the translation of the developed strategies into the clinic requires human and preferably autologous cell sources of functional cardiovascular cells (Figure 18.4).

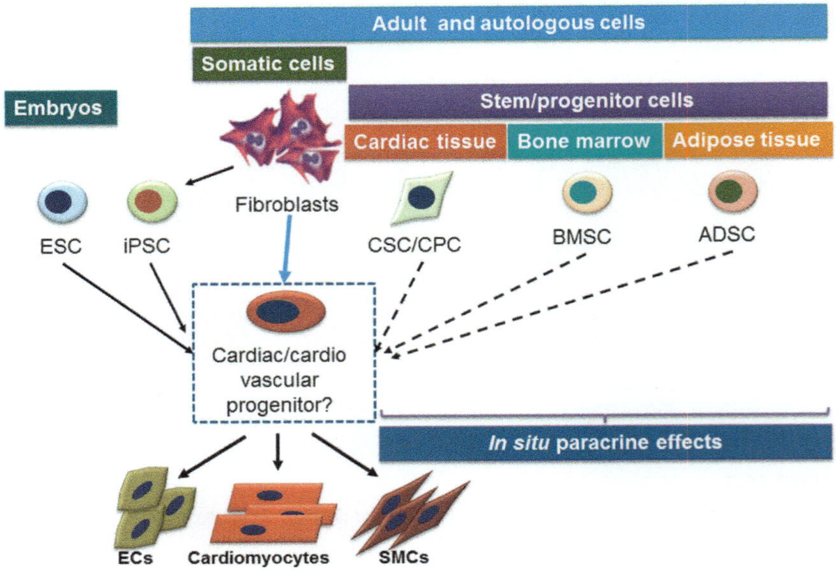

FIGURE 18.4

Clinically relevant cell sources for cardiovascular tissue engineering and regeneration. Blastocyst-derived embryonic stem cells (ESCs) have an established potential to differentiate into cardiomyocytes, ECs and SMCs. An autologous alternative to ESC is represented by introduction of induced pluripotent stem cells (iPSCs) that can be derived from somatic cells (e.g., fibroblasts). Moreover, protocols are being developed for direct reprogramming of somatic cells into cardiovascular cells. The differentiation process goes through several stages with varying efficiency, some leading to formation of cardiac or cardiovascular progenitors. Additional adult and autologous sources for stem/progenitor cells are the bone marrow, adipose, or cardiac tissues. The differentiation potential of these cells is contradictory, and mostly limited. Nevertheless, these cells were proven to exert beneficial paracrine effects in infarcted heart upon transplantation. CSC/CPC, cardiac stem/progenitor cell; BMSC, bone marrow-derived stem cell; ADSC, adipose tissue-derived stem cell (see text for more details). *Reprinted with permission from Ruvinov et al. (2012).*

We briefly list major cell types used for cardiovascular tissue engineering and regeneration in the following section:

- Autologous endothelial cells (ECs) and SMCs—nonimmunogenic autologous ECs and SMCs isolated from patients themselves are the first choice for use in blood vessel engineering.
- Pluripotent stem cell-derived CMs—human CMs produced by directed differentiation of pluripotent stem cells (embryonic stem cells (ESCs) or induced pluripotent stem cells (iPSCs)) (Burridge et al., 2012). More advanced and future clinical applications of these cells will require refinement of differentiation protocols and/or inclusion of maturation steps (Yang et al., 2014).

- Pluripotent stem cell-derived vascular cells—human pluripotent stem cells could also be used for the derivation of vascular cells (ECs and SMCs).
- Mesenchymal stem cells (MSCs)—represent a contemporary alternative in scenarios where human functional cells (e.g., CMs) are unavailable. MSCs can be isolated from bone marrow or adipose tissue.
- Cardiac stem/progenitor cells (CSCs/CPCs)—numerous CSC/CPC pools within the adult heart have been characterized on the basis of stem cell marker expression and cardiomyogenic potential (Matsa et al., 2014).

18.4 BIOMATERIALS—POLYMERS, SCAFFOLDS, AND BASIC DESIGN CRITERIA

Biomaterials and scaffolds are critical components in cardiovascular tissue engineering and regeneration strategies, as standalone or in combination with cells and/or bioactive molecules (Figure 18.5). They have been used as: (1) ECM replacement in acellular strategies; (2) controlled delivery systems for bioactive molecules; (3) vehicles for stem cell delivery and enhanced retention in the damaged tissue; and (4) supporting and guiding matrix for cell organization into a functional tissue, in vitro and in vivo.

18.4.1 Basic Biomaterial Design Criteria

In general, the biomaterials used in cardiovascular tissue engineering and regeneration strategies (Figure 18.5) should comply with the following basic criteria:

- *Biocompatibility*—This term refers to the ability of a material to perform with an appropriate host response in a specific situation (Williams definition) (Williams, 1999). In tissue engineering, biocompatibility refers to the ability of a scaffold to perform as a substrate that will support the appropriate cellular activity, including the facilitation of molecular and mechanical signaling, in order to optimize tissue regeneration, without eliciting any undesirable effects in these cells, or inducing any undesirable local or systemic responses in the host.
- *Mechanical strength*—The strength of a material is its ability to withstand an applied stress without failure. Scaffolds in tissue engineering should have the mechanical properties to contain and protect the seeded or recruited cells and maintain their structure under mechanical perturbations existing during cultivation and at implant site. At the same time, the scaffold mechanical properties should be compatible with the host tissue to allow its integration without interfering with the normal function of the organ. This is especially critical when a biomaterial is used as ECM replacement of damaged myocardium.

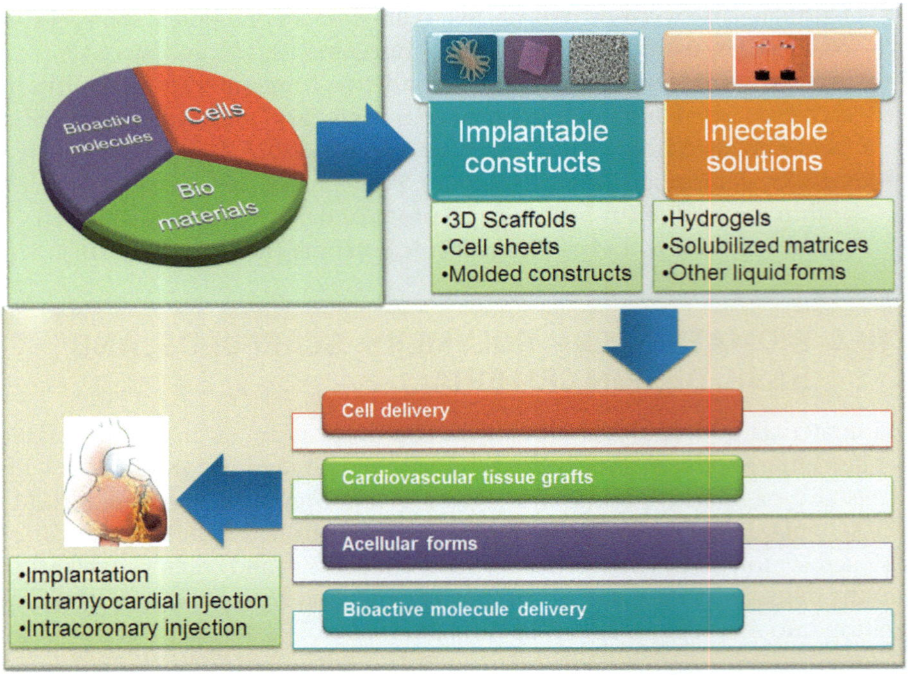

FIGURE 18.5

Paradigms for the use of biomaterials in cardiac tissue engineering and regeneration. Basic components (biomaterials, cells, and bioactive molecules) can be designed in various forms of implantable constructs or injectable solutions. These design strategies can be used for cell delivery, preparation of cardiac patches/grafts, in situ tissue support in acellular forms, or as a platform for the delivery of bioactive molecules. Ultimately, the biomaterial constructs or solutions are delivered to the infarcted heart by implantation, or by using less invasive techniques, such as intramyocardial or intracoronary injections. *Reprinted with permission from Ruvinov et al. (2012).*

- *Biodegradation/bioresorption*—Ideally, the scaffold should disappear from the host when tissue regeneration has been accomplished and normal function is restored. Biodegradable scaffolds can do so via polymer backbone degradation (e.g., hydrolysis, enzymatic cleavage) or by dissolution of the matrix. It is fundamental that the products of this process would be biocompatible and be resorbed by the body or removed from it via excretion from the urine.
- *Scaffold fabrication*—Ideally, this process should use safe reagents which do not affect material properties, such as its cell recognition motifs. For example, cross-linking between polymer chains is often used in hydrogel fabrication from natural materials such as alginate, collagen, hyaluronan, and others. Cross-linking can be physical, where the polymer chains self-assemble due to electrostatic interactions, response to temperature and

irradiation, or chemical, where covalent bonds are introduced between the polymer chains. Chemical cross-linking often changes the material properties (degradability, mechanical strength, and cell recognition) due to the lack of precise control over the position where the cross-link linkages are formed. In addition, chemical cross-linking often involves the use of harsh reagents, thus raising concerns about the material biocompatibility.

- *Scaffold internal morphology*—When used as scaffolding for cells, the matrix should be porous with interconnecting pore structure and pore size larger than 50 μm, to enable cell–cell interactions and support vascularization after implantation.

18.4.2 Biomaterial classification

Polymers commonly used for scaffold fabrication can be categorized by their source origin (natural or synthetic) and by their chemical structure (peptides/proteins, polysaccharides, polyesters, and others).

Natural polymers are usually biocompatible, easy to be chemically and physically modified and to be processed into various structures. Most of them, such as hyaluronan and collagen, possess cell recognition patterns enabling them to stimulate cell response. There are, however, concerns over their use in human therapy because of the risks of pathogen transmission and immune rejection associated with natural polymers that are produced from animal sources.

Synthetic polymers can be designed with versatile properties, such as mechanical strength and biodegradation rate, and can be tailored with functional groups. A disadvantage that can be found in synthetic polymers in comparison with natural polymers is the lack of the biological cues for promoting cell responses.

Table 18.2 lists the main materials used in cardiac tissue engineering strategies, as components in cellular constructs or standalone biomaterial treatments, and presents their main characteristics according to the design criteria listed earlier, namely, biocompatibility, mechanical strength, biodegradability/bioresorption, fabrication/cross-linking, and interactions with cells.

18.4.3 Basic Scaffold Fabrication Forms

Scaffolds can be fabricated in different shapes, sizes, and internal structure (porosity, pore size, and architecture of the pore structure (isotropic or anisotropic)). In tissue engineering of patches, either with seeded cells or without cells, the two main scaffolds in use are hydrogels and macroporous solid scaffolds.

Table 18.2 Main Biomaterials Applied in Cardiovascular Tissue Engineering

Type	Material	Biodegradability/ Control Over Process	Cross-Linking	Mechanical Strength	Cell Adhesion	Biocompatibility/ Immunogenicity
Natural proteins	Collagen	Biodegradable; poor	Chemical	Med-low	Good, cross-linking reduce	Med-immunogenicity after cross-linking
	Gelatin	Biodegradable; poor	Chemical	Low	Good	Low immunogenicity
	Fibrin	Biodegradable; good	Enzymatic	Med-low	Good	Nonimmunogenic
Natural-polysaccharides	Alginate	Bioerodable matrix; good	Physical	Low	Poor	Nonimmunogenic
	Chitosan	Biodegradable; good	Physical, chemical	Med-low	Poor	Low immunogenicity
	Hyaluronan	Biodegradable; poor	Chemical	Low	Good	Nonimmunogenic
Synthetic	PIPAAm	Nonbiodegradable; bioerodable	Chemical	NA	Good (temperature-dependent)	Biocompatible; NA
	Self-assembling peptides	Biodegradable; NA	Physical	NA	Good	Nonimmunogenic
	Polyesters (PCL, PLA, PGA, and copolymers)	Biodegradable; good	Chemical	High	Poor	Med-low-no immunogenicity
Natural	Decellularized ECM	Biodegradable, poor	Chemical	Med-low	Good	Nonimmunogenic, possible pathogen transmission

PIPAAm, poly(N-isopropylacrylamide); PCL, poly(ε-caprolactone); PLA, polylactic acid; PGA, polyglycolic acid; NA, not available.

18.4.3.1 Hydrogels

A hydrogel is a network of polymer chains that are water insoluble, sometimes found as a colloidal gel in which water is the dispersion medium. Hydrogels are superabsorbent (they can contain over 99% water) natural or synthetic polymers. Hydrogels possess a degree of flexibility very similar to natural ECM, due to their significant water content. The hydrogels can be prepared from natural and synthetic polymers by physical/ionic interactions (alginate) or via chemical cross-linking (collagen, hyaluronan (HA), and others). Cells are incorporated/encapsulated in the hydrogel during fabrication. Due to their resemblance to ECM texture, hydrogels are extensively being investigated as ECM replacements for damaged ECM after MI. They are delivered either by intramyocardial injection or by catheter-based techniques via the intracoronary route (Rane and Christman, 2011; Tous et al., 2011).

18.4.3.2 Macroporous Scaffolds

Macroporous scaffolds are characterized by large pore size (50–200 μm in diameter) and matrix porosity (70–90%). The pore size in scaffolds should be at least 50 μm in diameter to enable vascularization (blood vessel penetration) after their implantation. The pore size and architecture as well as the extent of pore interconnectivity are major effectors on cell seeding, cell penetration from the host, and cell organization into a tissue. The most common techniques for preparing macroporous scaffolds are: solvent casting particulate (porogen) leaching, nonsolvent-induced phase separation, thermally induced phase separation, foaming process, microsphere sintering, and electrospinning. Recently proposed fabrication techniques are: rapid prototyping, solid free-form, shape deposition manufacturing, fused deposition modeling, 3D printing, selective laser sintering, stereolithographic technique, and molecular self-assembly (Weigel et al., 2006).

18.5 BIOMATERIALS AS VEHICLES FOR STEM CELLS OR BIOACTIVE MOLECULE DELIVERY

18.5.1 Stem Cell Delivery

The use of biomaterials as a platform or vehicle for cell delivery into the infarcted myocardium can potentially reduce or eliminate the obstacles seen in cell suspension transplantation, such as low engraftment, survival, and unpredictable retention (Segers and Lee, 2011). The biomaterials can protect the implanted cells from the aggressive environment after acute MI. With a proper design, the biomaterials can serve as an artificial ECM, to replace the degraded ECM after MI and provide the required temporal support to allow cell engraftment and retention. Using biomaterial scaffolds, thicker cell constructs could be created in vitro, to achieve high density of cells in future grafts. In addition, biomaterials could provide various biochemical and biophysical cues to enhance cell survival, induce angiogenesis at infarct, and direct and control stem/progenitor cell differentiation, by implementation of controlled delivery mechanisms for growth factors, integration of specific cell–matrix interactions, and more. The successful implementation of biomaterials as cell delivery vehicles promises to yield an improved cell engraftment and long-term functionality.

Several studies have examined the use of biomaterials (natural and synthetic, in hydrogel or scaffold forms) as vehicles for stem cell delivery to the infarct (Segers and Lee, 2011). Due to relatively easy isolation methods, potentially autologous nature, and promising therapeutic potential, most of the studies are focused on biomaterial-assisted delivery of MSCs (Russo et al., 2014). Major mechanisms of improvement reported were increased cell retention and angiogenesis (Segers and Lee, 2011).

18.5.2 Delivery of Bioactive Molecules

Limited success in clinical trials of stem cell transplantation, the lack of true regeneration, together with emerging evidence pointing to paracrine effects as the major mechanism of stem cell-based repair, lead researchers to examine an alternative cell-free strategy aimed at delivery of various bioactive therapeutic molecules to infarcted heart.

The use of bioactive molecules (growth factors, cytokines, and stem cell mobilizing factors) is of a continuous interest in the field of therapeutic myocardial regeneration. The variable effects exerted by these molecules cover almost every target in the regeneration strategies, as discussed in Section 18.2.6.

The systemic delivery of some of the growth factors was found to be beneficial for the restoration of cardiac function in animal models. However, data emerging from clinical studies have been less conclusive (Abdel-Latif et al., 2008). The mixed results obtained with systemic cytokine or growth factor administration are also accompanied by numerous safety concerns and side effects. These include an increased incidence of restenosis, elevated blood pressure and viscosity, thrombolytic events, arrhythmogenesis, and other potential detrimental effects (Vandervelde et al., 2005). In addition, systemic administration requires higher doses of the drug due to unspecific delivery, fast elimination, and extremely low protein stability in the blood. Moreover, most of these molecules have pleiotropic functions, emphasizing the need for careful, local, and time-adjusted interventions. Thus, the combination of bioactive molecules with biomaterials to achieve effective local and temporary delivery seems to be a very attractive option. In such a way, the biomaterial will provide structural temporary matrix support and direct the formation of functional tissue. Simultaneously, it will serve as a temporary depot for sustained delivery of bioactive molecules with spatial and controlled distribution of the desired agent to induce multiple regenerative processes.

18.5.2.1 Injectable Systems

Scaffold- and sheet-like structures have been used for bioactive molecule delivery by local transplantation of the construct (Ruvinov et al., 2012). However, injectable systems composed of biomaterial–bioactive molecule combinations offer an additional advantage to scaffold-based systems, i.e., greater applicability in the clinical settings due to the less invasive route of delivery.

Several polymeric systems have been applied for the delivery of various therapeutic proteins directly into infarcted hearts. One of the best characterized systems is the self-assembling peptide nanofibers developed by the group of Richard Lee (Figure 18.6). This system was shown to induce angiogenesis, myocardial regeneration, and preserve cardiac function in rodent and swine models (Segers and Lee, 2007; Lin et al., 2012). Self-assembling peptides typically are 8–16 amino acids long and are composed of alternating hydrophilic

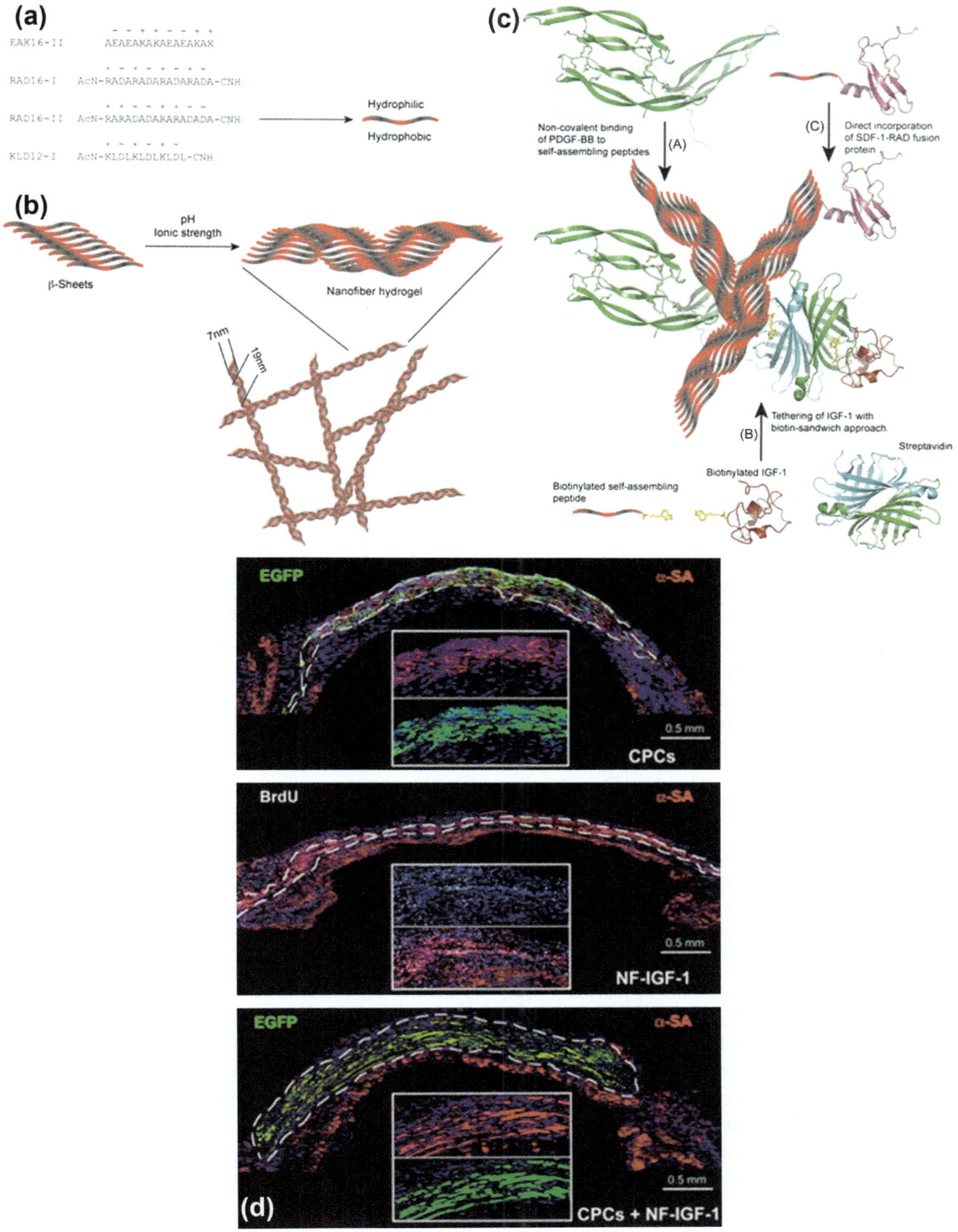

FIGURE 18.6 ***Continued***

and hydrophobic residues. They form stable β-sheets in water, and upon exposure to physiological salt concentration or pH they form a stable hydrogel of flexible nanofibers (7–20 nm in diameter) consisting of more than 99% water.

18.5.2.2 Injectable Affinity-Binding Alginate Biomaterial for Multiple Growth Factor Delivery

To enable a precise control over factor release rate or combinations of growth factors, we engineered the alginate with affinity binding sites for heparin-binding proteins by sulfation of the uronic acid monomers (Freeman et al., 2008). Surface plasmon resonance (SPR) analysis revealed the affinity binding of various heparin-binding proteins to alginate-sulfate (Freeman et al., 2008; Freeman and Cohen, 2009). The binding to alginate-sulfate mimics an example from nature, where many growth factors, chemokines, and cell adhesion molecules, collectively known as heparin-binding proteins, bind heparin and heparan sulfate via high affinity, specific electrostatic interactions with the low- and high-sulfated sequences in these glycosaminoglycans (GAG) (Shriver et al., 2002).

Combination of alginate-sulfate with unmodified alginate represents a unique type of a novel *affinity-binding* alginate biomaterial, which is capable of controlling the presentation and delivery of multiple proteins, while retaining the properties and characteristics of the alginate hydrogel as an ECM-replacing material.

We tested the concept of affinity-binding alginate for the controlled delivery of factors inducing myocardial regeneration in a rat model of MI. The biomaterial system was produced in an injectable form by mixing the alginate-sulfate/growth factor bioconjugates with partially cross-linked alginate solution. This injectable affinity-binding alginate system can be easily delivered into the infarcted heart, where it creates a hydrogel capable of sustained delivery of multiple growth factors (Figure 18.7).

FIGURE 18.6 ***Continued***

Self-assembling peptide nanofiber system for growth factor delivery. (a) Examples of different sequences of self-assembling peptides. Hydrophobic amino acids (alanine (A) or leucine (L)) alternate with positively (lysine (K) and arginine (R)) and negatively charged amino acids (glutamate (E) and aspartate (D)). Hydrophobic amino acids are directed to one side and hydrophilic (positively and negative charged) amino acids to the other side of peptides. (b) Self-assembling peptides are arranged in stable β-sheets at low pH and low ionic strength. Upon exposure to physiological pH and ionic strength, they form stable nanofibers with a diameter of ~10 and 50–200 nm pores. (c) Self-assembling peptides allow delivery and tethering of proteins in different manners. (A) Self-assembling peptides allow long-term delivery in vivo of platelet-derived growth factor-BB (PDGF-BB). PDGF-BB binds to self-assembling peptides noncovalently by electrostatic interactions. (B) Biotinylated variants of self-assembling peptides have been synthesized. They allow binding of every protein that can be biotinylated using streptavidin as a linker. (C) The sequence of self-assembling peptides can be incorporated at the N- or C-terminus of recombinant proteins, allowing incorporation of proteins in the hydrogel. (d) Myocardial repair and regeneration by combined cardiac progenitor cells (CPCs) and nanofiber-Insulin-like growth factor 1 (NF-IGF-1) therapy. Regions of regenerated myocardium (dashed line) in infarcts treated with CPCs (top panel), NF-IGF-1 (middle panel), and CPCs–NF-IGF-1 (bottom panel). In the insert, new myocytes (α-sarcomeric actin [α-SA], red) express enhanced green fluorescent protein (EGFP) (top and bottom panels, green) or are labeled by Bromodeoxyuridine (BrdU) (middle panel, white). *Reprinted with permission from Ruvinov et al. (2012)).*

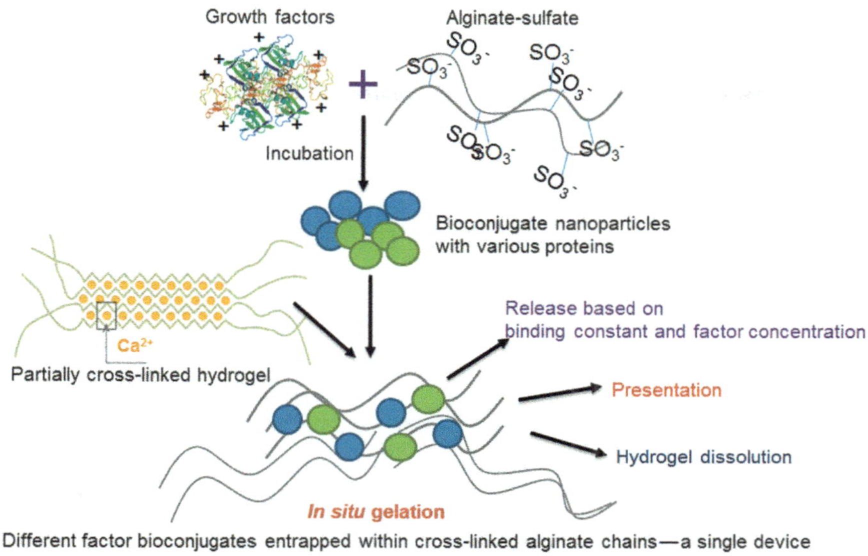

FIGURE 18.7

The concept of injectable affinity-binding alginate biomaterial and mode of protein release and action. *Reprinted with permission from Ruvinov et al. (2012).*

In an acute MI model in rats, we found that the sequential delivery of IGF-1 and hepatocyte growth factor (HGF) from affinity-binding alginate hydrogel reduced scar fibrosis, increased scar thickness, and prevented infarct expansion, 1 and 4 weeks after injection, compared to controls. Furthermore, angiogenesis was induced and cell apoptosis was reduced at the infarct. These therapeutic effects were preserved for 8 weeks after injection (unpublished data). This treatment could possibly be associated with endogenous regeneration of the cardiac muscle (increased incidence of Ki67-postive CMs and GATA-4-positive cell clusters) (Figure 18.8) (Ruvinov et al., 2011).

18.6 BIOENGINEERING OF CARDIAC PATCHES, IN VITRO

The replacement of a large scar tissue after MI and the corrections of congenital heart malformations, such as septal defects, require the transplantation of a fully developed cardiac tissue graft prepared in vitro. The engineered cardiac patch should be thick and has to display the functional and morphological properties of the native cardiac muscle. Once integrated into the heart, the cardiac patch must develop systolic force, withstand diastolic load with appropriate compliance, and form an electrical and functional syncytium with the host myocardium (Zimmermann et al., 2006).

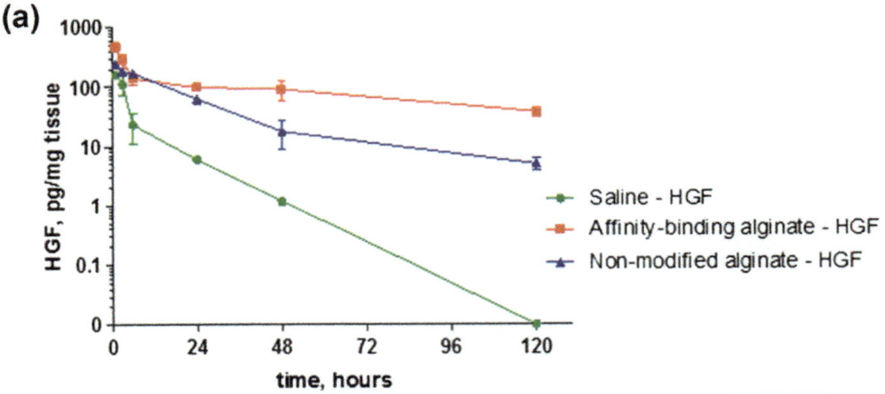

Parameter	Saline - HGF	Affinity-binding alg - HGF	Non-modified alg – HGF
AUC – measure of bioavailability, pg×h/mg tissue	1002	10322	4532

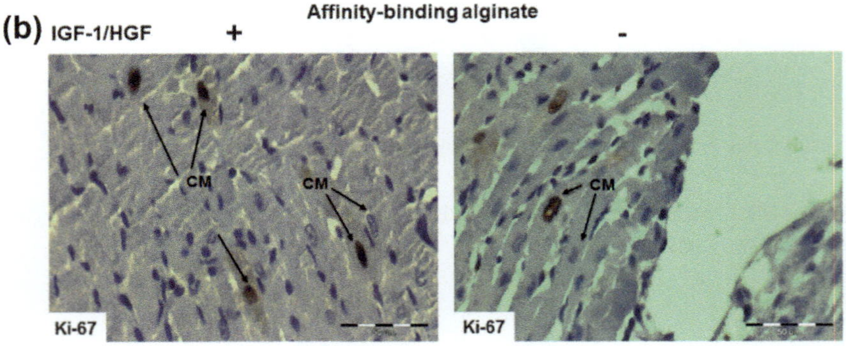

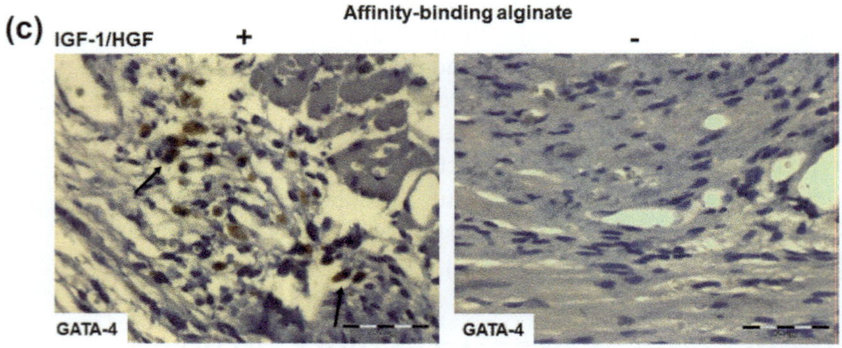

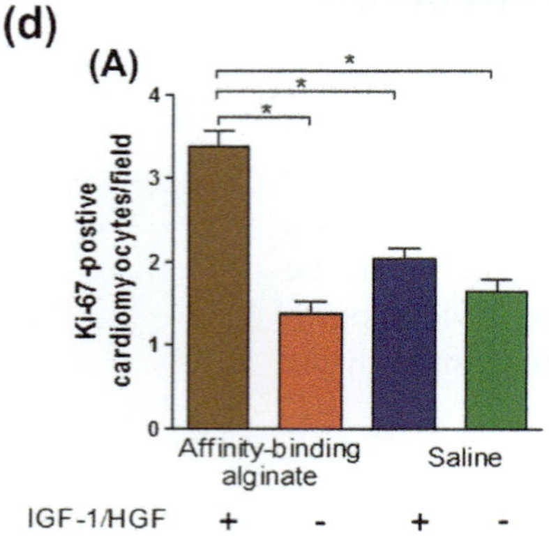

(B)

GATA-4 incidence

	Affinity-binding alg-HGF+IGF-1	Affinity-binding alg	Saline-HGF+IGF-1	Saline
1 week	1/5 (20%)	0/5 (0%)	1/5 (20%)	0/5 (0%)
4 weeks	4/5 (80%)	0/4 (0%)	2/5 (40%)	0/4 (0%)

Three main strategies for in vitro cardiac patch reconstruction have been developed over the last decade or so and they are based on seeding the cardiac cells in scaffolds in the form of hydrogels, polymer layers, and preformed macroporous scaffolds (Figure 18.9) (Hirt et al., 2014).

18.6.1 Cardiac Cell Entrapment in Hydrogels—Engineered Heart Tissue (EHT)

In this strategy, cardiac cells are mixed with a liquid form of the biomaterial, followed by its solidification to create a mixed 3D cell hydrogel. This strategy has been developed by Eschenhagen and collaborators, who cultured embryonic cardiac myocytes in a collagen type I matrix to produce a coherently contracting 3D engineered heart tissue (EHT) (Figure 18.10) (Eschenhagen et al., 1997).

18.6.2 Cell Sheet-Based Cardiac Tissue Engineering

Shimizu and coworkers developed a novel approach in which cardiomyocyte monolayers are assembled to form multilayered heart muscle constructs (Shimizu et al., 2002; Haraguchi et al., 2011). The cardiomyocyte layers were initially formed by cultivating isolated cardiac cells on cell culture surfaces, grafted with the temperature-responsive polymer poly (*N*-isopropylacrylamide) (PIPAAm). Confluent cell monolayers could be detached from the surface as a cell sheet simply by reducing the temperature, without any enzymatic treatments. The released cell layers (2D) were overlaid one on top of the other, forming a multilayered 3D tissue that began to pulsate simultaneously (Figure 18.11) (Shimizu et al., 2002).

18.6.3 Myocardial Tissue Grafts Based on Preformed Implantable Scaffolds

In this approach, cardiac cells are seeded within preformed macroporous scaffolds. The scaffolds are generally prepared by cross-linking (chemical or physical) of biomaterial solution into the desired shape, with a subsequent solidification step and/or drying/freeze-drying. This platform offers important advantages over cell entrapment during hydrogel fabrication, such as the

FIGURE 18.8

Therapeutic efficacy and cardiac tissue regeneration induced by the dual factor delivery from the injectable affinity-binding alginate system in acute MI model. (a) Myocardial tissue retention of HGF in a rat acute MI model. AUC, area under the curve. (b) Mitotic marker Ki67 staining (brown) in healthy/infarct border region, 1 week after injection. Arrows show nuclear localization or unstained nuclei. Bar = 50 μm. CM, cardiomyocytes. (c) Cardiac progenitor cell marker GATA-4 staining (brown) in infarcted hearts (periinfarct region), 4 weeks after injection. Arrows show nuclear localization. Bar = 50 μm. (A) Quantitative analysis of Ki67-positive cardiomyocytes (based on morphological discrimination) in short-term (1 week) experiment. * $p < 0.001$. (B) Incidence of GATA-4-positive cell clusters in different treatment groups. Incidence = (number of animals with positive staining/total number of animals in each group) × 100

Reprinted with permission from Ruvinov et al. (2012) and unpublished data.

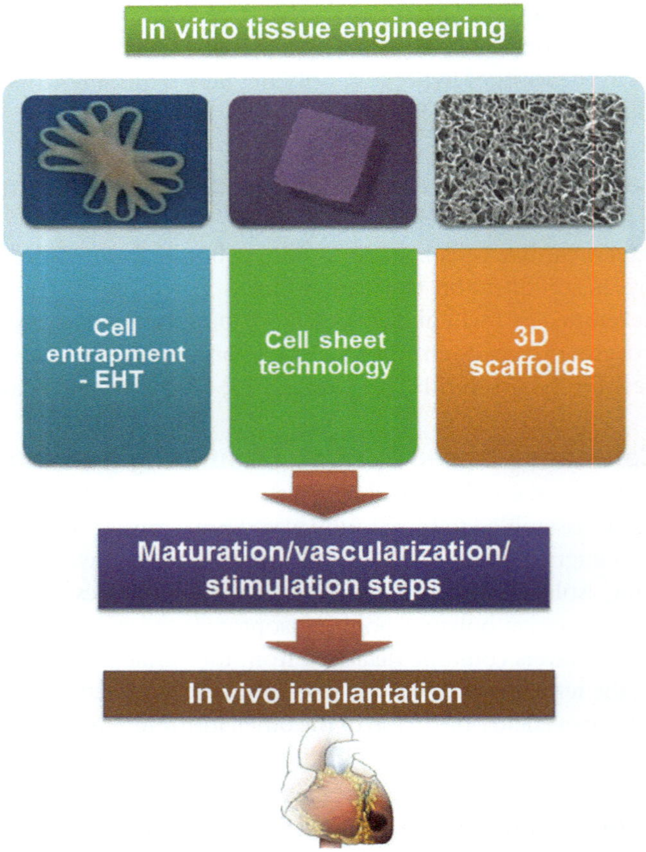

FIGURE 18.9

Strategies for reconstructing the cardiac patch. Three major strategies are used for preparation of cardiac patches: cell entrapment in hydrogels (EHT), cell sheet engineering, and cell seeding in preformed macroporous 3D scaffolds. After initial culture, various stimulation strategies could be applied (i.e., perfusion bioreactors, electrical stimulation, etc.) in order to improve cell organization and tissue maturation. Finally, the constructs can be implanted into infarcted heart (see text for more details). *Reprinted with permission from Ruvinov et al. (2012).*

ability to design and control the porosity of the scaffold prior to cell seeding and reduced exposure of the seeded cells to stress during mixing and molding of the scaffold. Furthermore, the predesigned scaffolds can direct the seeded cells to organize into the gross conformation of native cardiac tissue, for example, by influencing cardiac cell alignment in the porous structure. Morphologically, the scaffold should be porous with an interconnecting pore structure to enable accommodation of a large number of cells and their subsequent organization into a functioning tissue (Shapiro and Cohen, 1997). Pore size should be at least 50 μm to allow high mass transport during in vitro culture and vascularization of the scaffold following implantation, to supply the seeded cells

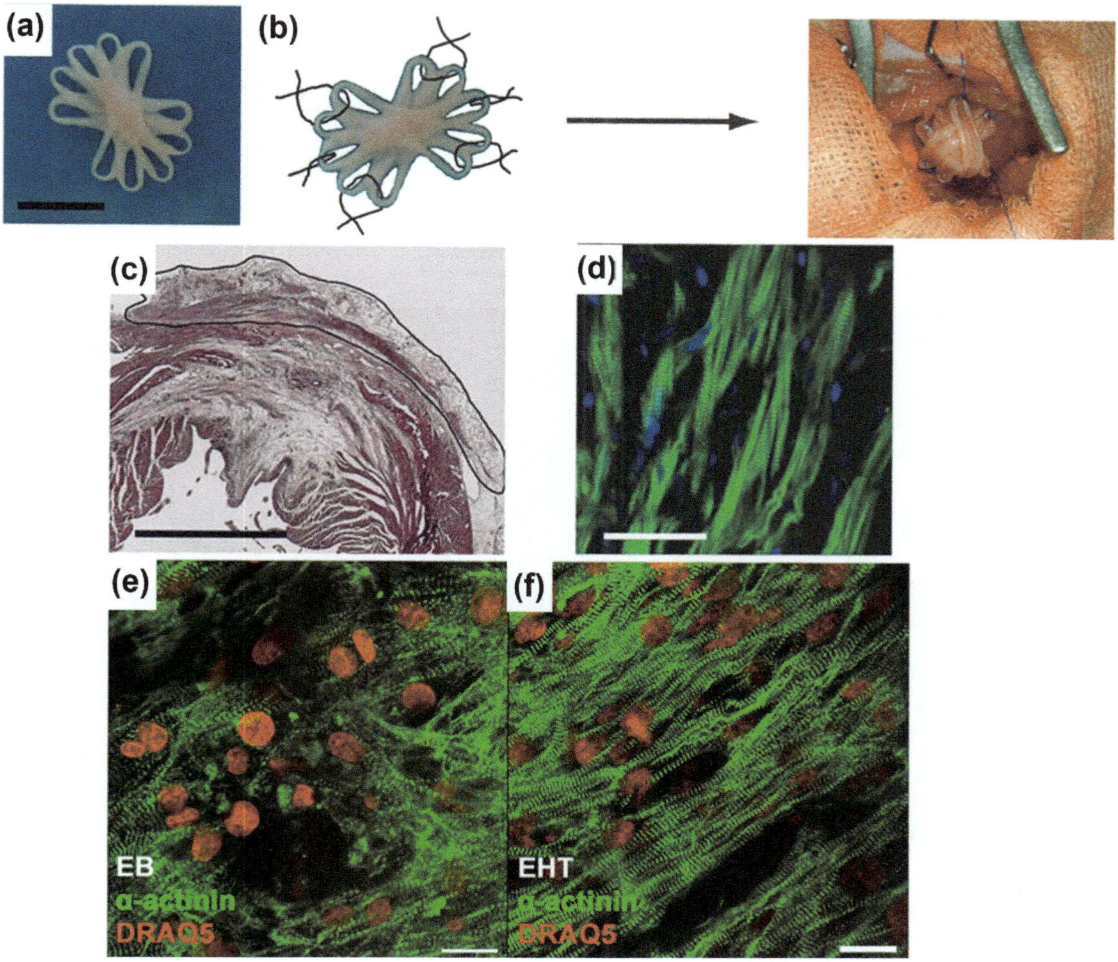

FIGURE 18.10

An engineered heart tissue (EHT) constructed by cell entrapment method. ((a)–(d)) Rat EHT; ((e) and (f)) Human EHT. (a) Multiloop EHT ready for in vivo engraftment. (b) EHTs fixed on the recipient heart. (c) Four weeks after EHT engraftment, the construct formed compact and oriented heart muscle on the top of the infarct scar. (d) Laser confocal microscopy showing the highly differentiated sarcomeric organization of engrafted cardiomyocytes (actin, green; nuclei, blue). (e) Immunofluorescence staining of 2–3 week old embryoid body (EB). (f) Immunofluorescence staining of EHT, showing enhanced alignment and predominant longitudinal orientation along force lines. *Reprinted with permission from Schaaf et al. (2011), Ruvinov et al. (2012).*

with nutrients and to remove secretions (Zmora et al., 2002). At the same time, the scaffold should present the desired mechanical properties and enable its handling in culture and during implantation.

We were among the first groups to report a successful implantation of cardiac cell-seeded porous alginate scaffolds into infarcted rat hearts (Leor et al., 2000; Cohen and Leor, 2004). We found that the seeded fetal rat

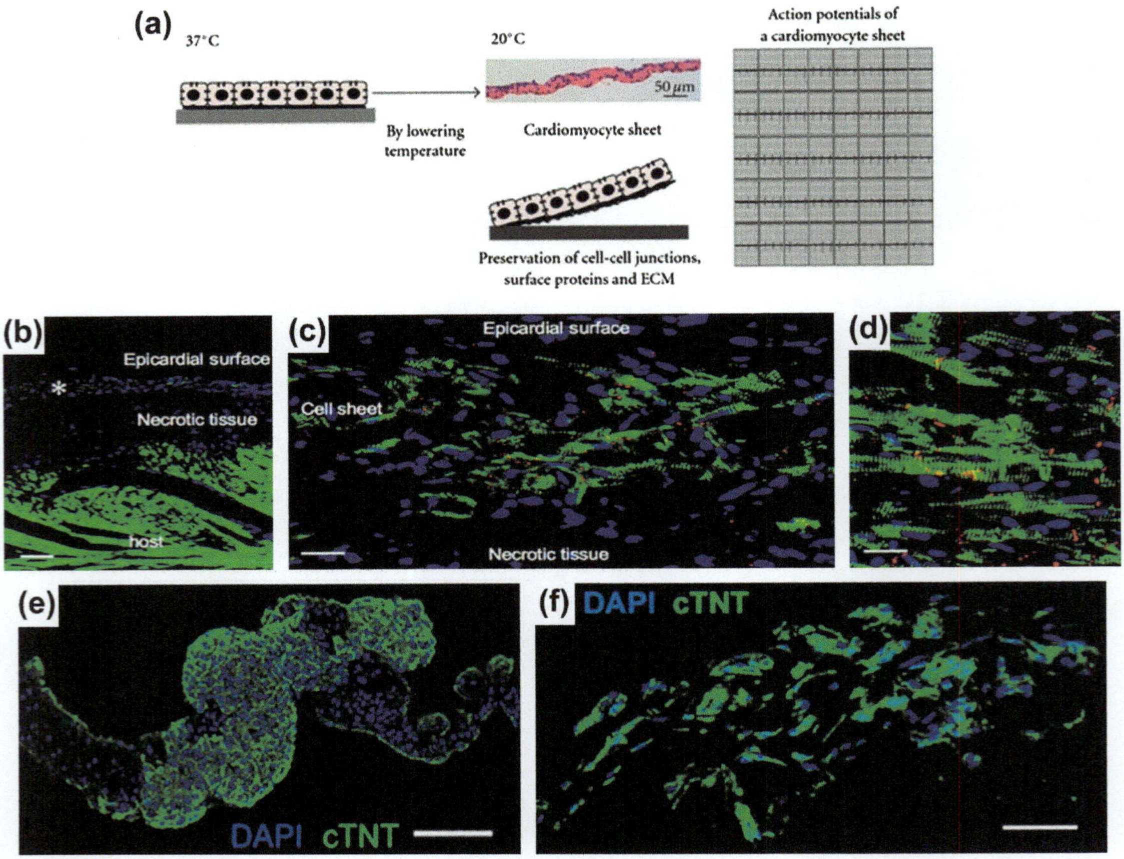

FIGURE 18.11

Cell sheet-based cardiac tissue engineering. (a) Using temperature-responsive culture surfaces, cell structures are preserved and cultured cardiomyocytes (CMs) are released as a contiguous cell sheet. A microphotograph shows the cross-sectional view of a cardiomyocyte sheet. Electrograms show the spontaneous action potentials of the cardiomyocyte sheet. ((b)–(d)) Cardiac tissue sheet grafts using 3D cell sheet manipulation technique integrate with the host heart. Laser confocal microscopic view of the grafted cell sheet and host heart, stained for α-actinin (green), anti-Cx-43 (red), and 4′,6-diamidino-2-phenylindole (DAPI) (blue) to stain the nucleus. (b) An α-actinin-positive cell sheet can be seen at the epicardial surface of the necrotic tissue (*). (c) Clear striation pattern of α-actinin and diffuse Cx-43 staining are seen in the cell sheet. (d) The host image below the necrotic area. Scale bar: 20 μm. ((e)–(f)) Human iPSC-CM-based cell sheets. (e) Immunostaining of the hiPS-CM cell sheet with cardiac Troponin-T (cTNT) antibody (green). The cell nuclei were counterstained with DAPI (blue). (f) In the transplanted regions of mini-pigs with the omentum, cTNT-positive cells are demonstrated by immunohistolabeling (green). The cell nuclei were counterstained with DAPI (blue); scale bar: 50 μm. *Reprinted with permission from Ruvinov et al. (2012).*

cardiac cells retained viability within the scaffolds and within 24 h formed multicellular beating cell clusters (Dar et al., 2002). Following implantation of the cellular constructs into the infarcted myocardium, some of the cells appeared to differentiate into mature myocardial fibers. The graft and surrounding area were populated with a large number of newly formed blood

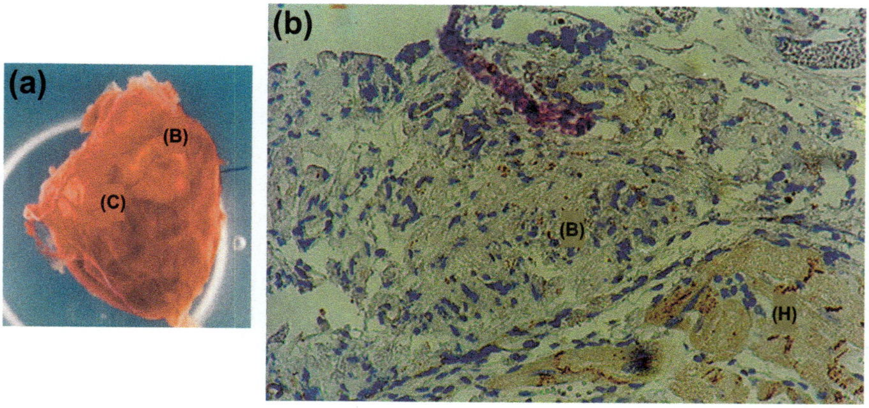

FIGURE 18.12

Bioengineered cardiac patch implantation on infarct. (a) Photograph of heart at week 9 after patch implantation. Visual examination revealed intense neovascularization into implanted patch (B). Note a coronary branch (C) that supplies and covers the patch with extensive network of vessels. (b) Microscopic image of integrated patch, immunostained for connexin-43 (Cx-43) (brown). Cx-43 was localized in normal parallel arrangements in host myocardium (H) and randomly oriented in the patch (B) (original magnification ×200). *Reprinted with permission from Ruvinov et al. (2012).*

vessels, consequently leading to attenuation in LV dilatation and improved heart function (Figure 18.12) (Leor et al., 2000).

The efficacy of this strategy was further proven with patches prepared from other materials, natural or synthetic, emphasizing the potential of this approach for myocardial repair (Ruvinov et al., 2012).

18.6.4 Biomimetic Scaffolds and Integration of Cell–Matrix Interactions

Along with the use of ECM-extracted biological materials as scaffolds, which intrinsically retain signals important for cell–matrix interactions, major efforts have been invested in the intelligent design of novel synthetic materials bio-inspired by the ECM interactions. These materials were created to provide the appropriate interface for cell interactions, including enhanced cell adhesion, growth, migration, and differentiation, while eliminating the potential hazards associated with the use of ECM-extracted materials.

An extensively used strategy applied to achieve such appropriate material surfaces has been by attaching ECM-derived synthetic peptides to the scaffold. The most commonly used peptide is the sequence Arg-Gly-Asp (RGD), the signaling domain derived from fibronectin and laminin (Shin et al., 2003). RGD peptide mediates cell adhesion and signaling via binding of various ECM proteins to integrin receptors on the cell surface (Rosso et al., 2004).

Our group has covalently attached RGD peptide to alginate, in order to explore the possible role of immobilized RGD peptide within the 3D settings of engineering a cardiac tissue in vitro (Shachar et al., 2011). Since alginate is inert and resistant to cell adhesion, it can be used as a "blank canvas" to investigate specific cell–matrix interactions by attaching RGD as well as other biological cues. The RGD peptide decoration of alginate scaffolds accelerated in vitro cardiac tissue regeneration and contributed to a better preservation of the tissue in culture. We also showed that the integration of an additional molecular signal provided by heparin-binding peptides (HBPs) promotes the formation of improved cardiac tissue in vitro (Sapir et al., 2011). We compared cardiac tissue engineering of neonatal rat cardiac cells cultivated within macroporous alginate scaffolds bearing the two peptides, RGD and HBP, to those formed in a single peptide-attached or in unmodified alginate scaffolds (Sapir et al., 2011). Examination of cardiac cell morphology 14 days after seeding in the different scaffolds revealed the characteristic striated fiber organization mainly in the RGD/HBP-attached cell constructs, while in HBP-attached and unmodified alginate cell constructs no such structures were observed (Figure 18.13). Better organization was associated with increase in expression level of Cx-43.

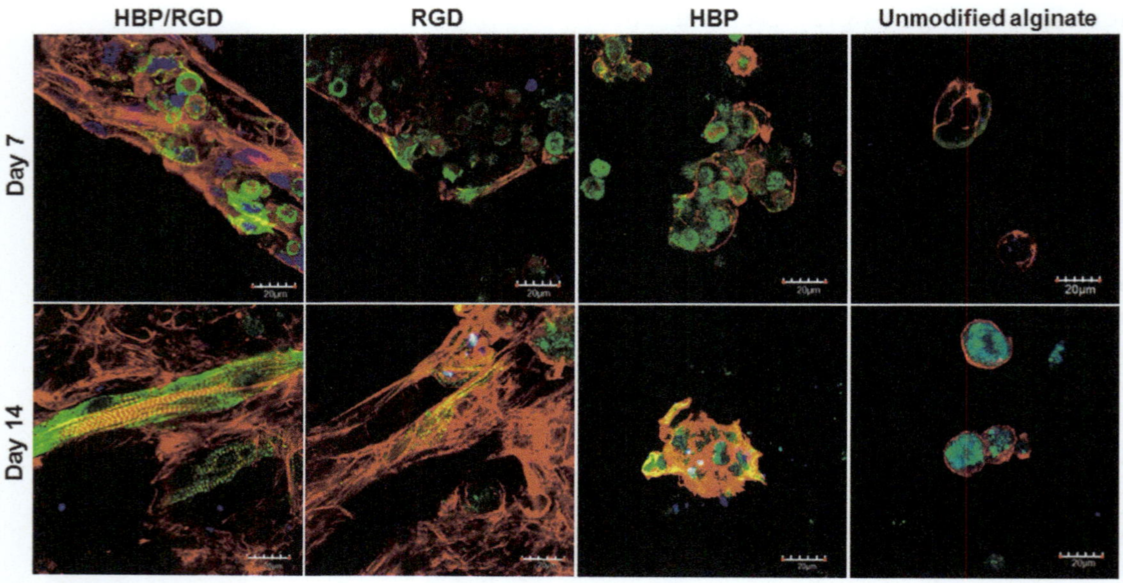

FIGURE 18.13

Integration of multiple cell–matrix interactions into alginate scaffolds promotes cardiac tissue organization. Cardiac cell organization, on days 7 and 14 postcell seeding. Cardiomyocytes are positive for α-actinin (green), while all cells are stained for F-actin (red) and nuclei (blue) (Scale bar: 20 μm). *Reprinted with permission from Ruvinov et al. (2012).*

Collectively, the results of these studies emphasize the importance of integration of multiple and complementary signals into 3D matrices for the recreation of the cardiac microenvironment and successful formation of an engineered cardiac tissue in vitro.

18.6.5 Perfusion Bioreactors and Stimulation Patterns

18.6.5.1 Perfusion Bioreactor Cultivation of Engineered Cardiac Tissue

In order to create relevant cardiac patches in terms of size, diffusional oxygen limitations have to be overcome and this has been accomplished by the use of perfusion bioreactors. The application of perfusion, by forcing the medium into the entire cell-seeded scaffolds, enables efficient mass transfer of oxygen and other soluble factors throughout the entire developing tissue, in a manner similar to the role of the vasculature in tissues.

Perfusion bioreactor operation is based on culture medium convection across a porous cell construct, thus providing a controllable, mechanically active environment for all seeded cells. It was shown that cultivation of engineered tissues in perfusion bioreactors improved the construct size, cellularity, and molecular composition. The bioreactors also made it possible to study effects of (1) mass transport of growth factors and nutrients, (2) hydrodynamic load conditions, and (3) physical stimulation, such as dynamic compression or cyclic stretch, on the developing tissue (Carrier et al., 2002; Radisic et al., 2008).

Our group designed a perfusion bioreactor system able to create homogenous medium flow in cell constructs by installing two cell construct-fixing nets, with an open area of 95.8% (Dvir et al., 2006). Using the unique net architecture, 99.88% of the cell construct volume is perfused by the culture medium. Our perfusion bioreactor was upgraded in another aspect, by insertion of a micromesh, designed to disrupt the media flow for a split second before entering the cell constructs, in order to apply equal shear stress along the bioreactor cross section. Medium reaching the construct core by perfusion enabled the formation of homogenous, thick, and organized (>500 μm) cardiac tissue (Dvir et al., 2006).

18.6.5.2 Inductive Stimulation Patterns

18.6.5.2.1 Mechanotransduction and Physical/Mechanical Stimuli

Mechanotransduction (biochemical response of cells to mechanical stimulation) is a very fundamental process in a living cell. The ability of cells to convert mechanical forces into biochemical regulatory information is crucial for normal physiology and pathological processes in many tissues. CMs employ numerous mechanisms to sense external forces and translate them into biochemical signals (Ruvinov et al., 2012). In such way, CMs can respond to elevated loading conditions, in an attempt to adapt to new mechanical demands and restore wall stress to normal levels, with an increase in cell size, associated with transient and persistent molecular changes, a process generally referred to as cellular hypertrophy. This process and biomechanical dynamics are important in maturation and organization of native cardiac tissue. Thus, the ability of various engineering efforts to recreate these signals in vitro is critical for the formation of functional cardiac patches. As cardiac tissue is an excellent example of a tissue that constantly senses mechanostimulation in vivo, it may respond to a similar stimulation also in vitro.

The effect of mechanical stimulation on cells has been broadly investigated mainly in 2D cultivation systems. In 3D cell cultures, the homogenous distribution of these mechanical stimuli represents another challenge for bioreactor engineers. The most common examples for application of mechanical stimulation are bioreactors, developed to apply mechanical forces via piston/compression systems, substrate bending, hydrodynamic compression, and fluid shear (Ruvinov et al., 2012).

Nowadays, a leading approach in mechanostimulation of the cardiac constructs is via stimulation during construct cultivation within a perfusion bioreactor. The advantages of such bioreactor as a mechanostimulator are obvious; the interstitial medium flow within the porous cell construct also creates a frictional force on the surface of the cells, termed shear stress. Mammalian cells respond to fluid shear stress in different ways, for example, by enhancing proliferation or changing morphology and organization.

Our group has used the perfusion bioreactor described in the previous section to investigate the effect of pulsatile, interstitial fluid flow stimulation on cardiac tissue engineering. The various cellular events taking place in the cell construct, from the cell signaling levels up to the ultrastructural phenotype of the regenerated cardiac muscle tissue, were elucidated in this study (Dvir et al., 2007). When the bioreactor was operated at a shear stress of 0.6 dynes/cm^2, activation of the extracellular signal-regulated kinase (ERK) 1/2 signal transduction pathway was pronounced, followed by enhanced synthesis of the contractile proteins (Troponin-T and sarcomeric α-actinin) and cell–cell interaction proteins (Cx-43 and N-cadherin). More importantly it was shown that long-term cultivation under such pulsatile flow regime promoted organization of thick, highly organized cardiac tissue, resembling the native heart.

18.6.5.2.2 Magnetic Stimulation

Magnetic field has been attracting a great interest as a tool for mechanically stimulating cells, although its signaling pattern within the stimulated cell is still not clear and poorly understood. Magnetic fields can protrude living cells since, unlike the electric field; they are not shielded by membrane potentials, and thus they can influence intracellular organelles. Also, due to their high penetration ability, magnetic fields can reach into deeper tissue layers when applied from a distance.

In stimulation settings, the magnetically mediated actuation "at distance" has a clear advantage compared to other stimulation setups since it enables highly controlled actuation both in vitro and in vivo. Target cells can be stimulated regardless of whether there are intervening structures, as long as these structures do not isolate the magnetic field. In addition, the magnetic field can be coupled with magnetically responsive particles that can be targeted to a specific cell or tissue site both in vitro and in vivo. Such coupling can easily and dynamically control the stress, applied directly by the magnetic particles to a desired area by varying the strength of the applied field.

We explored whether impregnation of alginate scaffolds with magnetically responsive nanoparticles (MNP) and subsequent exposure to an alternating magnetic field at a physiologically relevant frequency (5 Hz) would promote the formation of a functional myocardial tissue. Magnetic stimulation of cell constructs resulted in more mature cardiac culture characterized by anisotropically organized striated cardiac fibers, which preserved its features for longer times than nonstimulated constructs. This was associated with increased Troponin-T expression in the stimulated constructs. A greater activation of AKT protein kinase in cardiac cell constructs after applying a short-term (20 min) external magnetic field indicated the efficacy of magnetic stimulation to actuate at a distance and provided a possible mechanism for its action (Sapir et al., 2014).

It is still not clear what are the exact mechanisms acting on the cells within the nanocomposite scaffold under magnetic stimulation. Based on our studies, one plausible explanation could be the MNP-induced mechanical stimulation. Since the MNPs incorporated within the scaffold wall create an anisotropic surface morphology under application of an alternating magnetic field, they magnetize and may induce local scaffold deformation. Such deformation would apply direct forces of a certain frequency on the cells adhered on the scaffold wall and result in their magneto-mechanical stimulation. Although this hypothesis appears to be feasible, it needs further investigation and confirmation.

18.6.5.2.3 Electrical Stimulation

From the moment it is formed, the heart acts as a sequential contracting syncytium. Although each cardiomyocyte is able to contract spontaneously, the overall heart function is controlled by a group of specialized pacemaker cells. These cells stimulate the generation of cardiac impulse and trigger synchronous contraction. In cardiac tissue engineering, the addition of exogenous electrical stimulation has attributed to the reconstruction of an appropriate environment for tissue regeneration by mimicking the pacemaker activity during heart tissue development.

Radisic et al. (2004) pioneered the application of electrical stimulation in cardiac cell constructs. Electrical signals mimicking those in the native heart (rectangular pulses, 2 ms, 5 V/cm, 1 Hz) were applied on neonatal CMs seeded onto collagen constructs. Already after 8 days of stimulation, the cardiac cells presented higher levels of gap junction proteins (Cx-43) compared to nonstimulated constructs. Electrical stimuli induced functional coupling, amplified contraction amplitude by a factor of 7, and induced a significant level of ultrastructural differentiation (Radisic et al., 2004). Tandon et al. systematically varied the electrode material, amplitude, and frequency of stimulation to determine the conditions that are optimal for cardiac tissue engineering. Carbon electrodes, exhibiting the highest charge-injection capacity and producing cardiac tissues with the best structural and contractile properties, were thus used in tissue engineering studies. Engineered cardiac tissues stimulated at 3 V/cm amplitude and 3 Hz frequency had the highest tissue density, the highest concentrations of cardiac Troponin-1 and Cx-43 and the best-developed contractile behavior (Tandon et al., 2011). Electrical stimulation has been also implemented in tri-cell constructs (CMs, endothelial cells, and fibroblasts) resulting in the formation of vascularized cardiac tissue (Chiu et al., 2011).

The studies mentioned earlier were performed under static conditions, thus the thickness of the engineered cardiac tissue was limited to 100 μm. In order to overcome this drawback, electrical stimulation setups can be integrated into perfusion bioreactors. To this end, a custom-made electrical stimulator was integrated into our previously developed perfusion bioreactor (Figure 18.14). This was achieved by insertion of carbon rod electrodes between the two scaffold mesh holders. Computer models of the electric fields and current density inside the bioreactor and the constructs were created. A successful stimulation of the cell construct in static cultivation mode was achieved at 6 V with a current density of 74.4 mA/cm^2, while in the bioreactor, with the carbon rod electrodes, 5 V was sufficient to achieve the same current density. Already after 4 days under a continuous electrical stimulus in perfusion bioreactor, cell elongation and striation in the cell constructs were clearly detected, as well as enhanced expression level of Cx-43 (Barash et al., 2010).

18.7 VASCULARIZATION OF CARDIAC PATCHES

Integration and survival of thick tissue constructs constitute major engineering challenges, which arise from oxygen diffusion limitations. Insufficient vascularization can result in tissue ischemia and escalate to tissue resorption or

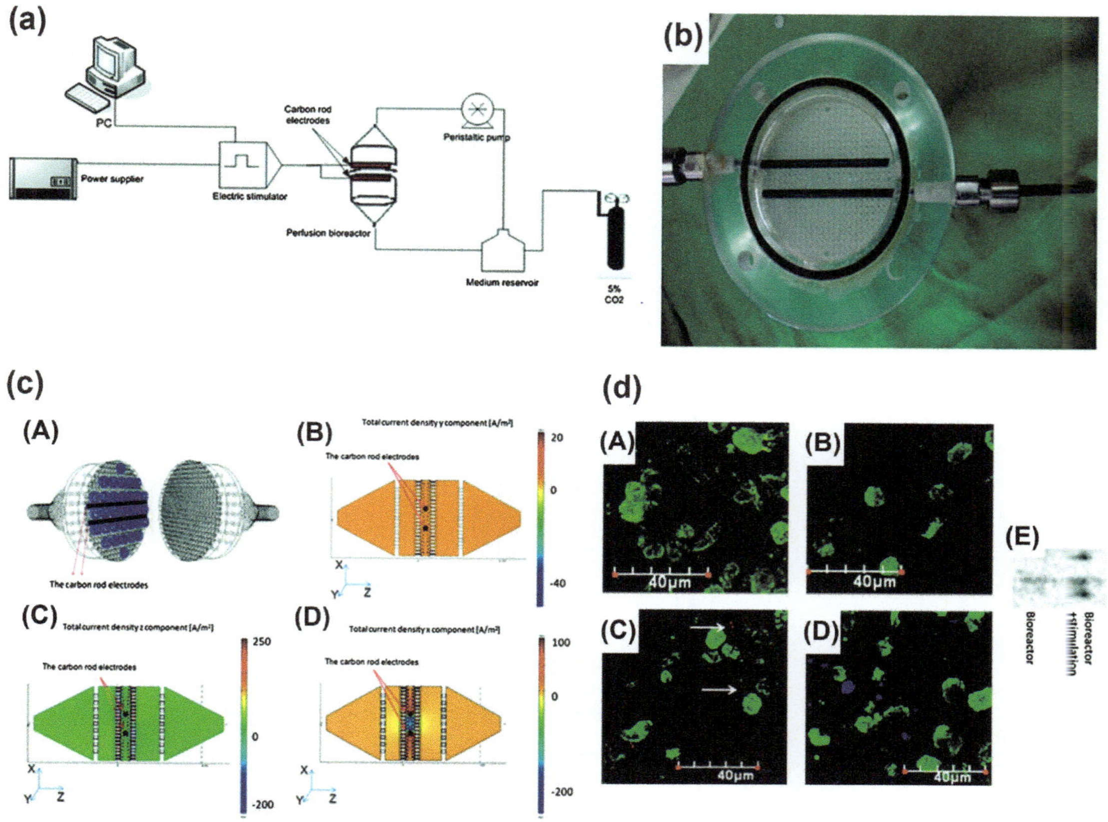

FIGURE 18.14

Electric field stimulation integrated into perfusion bioreactor for cardiac tissue engineering. (a) Bioreactor setup. (b) Configuration of the carbon electrodes integrated into the bioreactor. (c) 3D electric field model in bioreactor. (A) An illustration of the 3D configuration with the carbon electrodes in black and the cell constructs in blue. (B)–(D) The current density in the *y* (B), *z* (C), and *x* (D) directions (1 V between the electrodes). (d) The effect of the electrical stimulation on cell morphology and Cx-43 levels. (A) and (B) Confocal microscopy images of anti-sarcomeric α-actinin immunofluorescence (green) of the cell constructs in the bioreactor with (A) or without (B) electrical stimulation. To-Pro (To-Pro-3 Iodide) was used for nuclear staining (red). The electrical-stimulated cell constructs are those placed in between the carbon electrodes. (C) and (D) Confocal microscopy images of anti-Cx-43 (red), anti-α-sarcomeric actinin (green) immunofluorescence of the cell constructs in a bioreactor with (C) or without (D) electrical stimulation. The white arrows indicate positive staining of Cx-43 between adjacent cells. To-Pro was used for nuclear staining (blue). The electrical-stimulated cell constructs are those placed in between the carbon electrodes. (E) Representative Western blot analysis for Cx-43 after 4 days of cultivation in the bioreactor with or without electrical stimulation. *Reprinted with permission from Ruvinov et al. (2012).*

necrosis (Lesman et al., 2010a). The maximum intercapillary distance within a tissue in vivo is approximately 200 μm, which correlates with the diffusion limit of oxygen; in turn defining the thickness limits of an engineered construct (Radisic et al., 2006). The intercapillary distances in the highly vascularized rat myocardium are approximately 19 μm, thus each myofiber is situated between

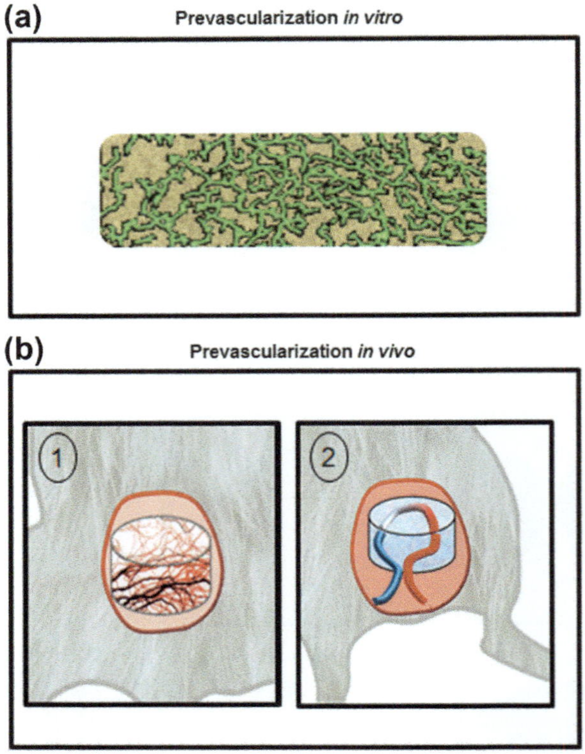

FIGURE 18.15

Different approaches for graft vascularization. (a) Vascularization in vitro by seeding of endothelial cells and supporting cells. (b) Vascularization in vivo by host vessels, for example, (1) graft implantation in the host and vascularization by host vessels, (2) arteriovenous loop in a plastic chamber.

two capillaries, an effective means of addressing the high metabolic demand of this tissue (Korecky et al., 1982; Iyer et al., 2011). Therefore, when designing functional and clinically relevant cardiac constructs, engineering techniques must integrate effective vascularization methods (Tee et al., 2010). Here, we discuss a number of current approaches to overcome diffusion limitations in engineered cardiac tissues (Figure 18.15).

18.7.1 Angiogenesis versus Vasculogenesis

One must first distinguish between two terms: (1) vasculogenesis, which refers to formation of new blood vessels from angioblasts (endothelial cell (EC) precursors), which occurs during the early stages of embryonic development. During this process, the primitive vascular plexus and the primitive heart are formed and (2) angiogenesis, which refers to remodeling and expansion of existing vessels by EC sprouting and subsequent formation of new capillaries (Risau and Flamme, 1995; Patan, 2000; Novosel et al., 2011). Ribatti et al. demonstrate that vasculogenesis is not restricted to the embryonic stages only, but may also occur postnatally, as a consequence of pathological conditions (Ribatti et al., 2001). However, under normal conditions, new vessels in the adult are primarily formed by angiogenesis.

18.7.2 Prevascularization In Vitro

Much effort has been invested into identifying the ideal cell combination for engineering a functional cardiac tissue. A thick graft seeded with CMs only requires support from host vasculature and is likely to undergo necrosis, due to both the extended lag period preceding infiltration of host vessels and the oxygen diffusion limitation (Tee et al., 2010). Incorporation of ECs within the graft enables in vitro generation of vascular-like networks, which can become functional and anastomose with host vasculature postimplantation (Kaully et al., 2009). Various combinations of cells have been described in the literature, including neonatal CMs, hESCs, or iPSCs-derived CMs, and skeletal myoblasts cocultured with ECs, with or without supporting cells (e.g., fibroblasts (FBs), stromal cells, or MSCs). Some of the coculture techniques implemented to obtain early-stage vascular elements are presented in the following section. Levenberg et al. previously demonstrated that coculturing of human umbilical vein endothelial cells (HUVECs) with tissue-specific cells results in enhanced tissue vascularization and graft integration upon implantation of heart, muscle, and pancreatic tissues in comparison with grafts without ECs (Levenberg et al., 2005; Caspi et al., 2007; Kaully et al., 2009; Lesman et al., 2010a,b; Koffler et al., 2011; Kaufman-Francis et al., 2012; Shandalov et al., 2014) (Figure 18.16).

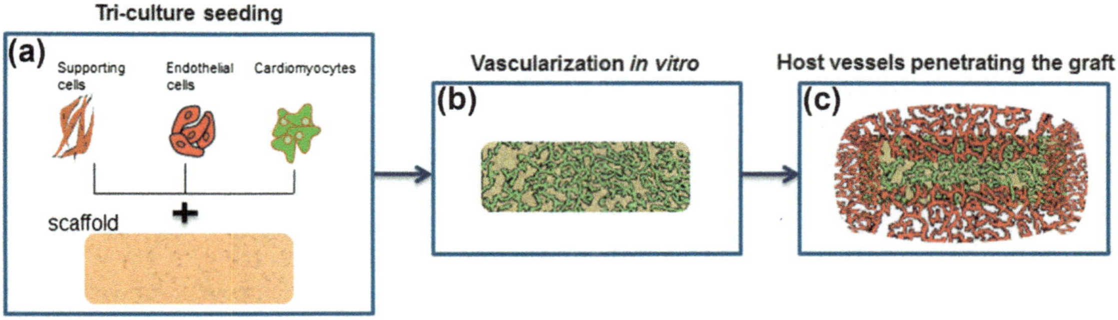

FIGURE 18.16

Vascularization of cardiac patches in vitro by triculture seeding. (a) Triculture seeding of endothelial cells, supporting cells (e.g., fibroblasts, stromal cells), and cardiomyocytes (CMs) on a scaffold. (b) Incubation in vitro to enable formation of vascular networks. (c) Penetration of host vessels into the graft postimplantation.

More specifically, a triculture of human ESC (hESC)-derived CMs, HUVECs, and mouse embryonic fibroblasts (MEFs) seeded on poly-L-lactic-acid (PLLA)/ poly-lactic-co-glycolic-acid (PLGA) scaffolds, resulted in the organization of vessel-like networks and in the enhanced cardiac differentiation in vitro, functionally manifested by spontaneous synchronous contractions (Caspi et al., 2007; Lesman et al., 2010a,b). Upon implantation of the spontaneously contracting patches into the anterior wall of the left ventricle, the vessel networks which originated from the preseeded ECs were perfused and anastomosed with host vasculature and promoted tissue survival and integration (Lesman et al., 2010a,b). Narmoneva et al. (2004) demonstrated that co-seeding of ECs and CMs on peptide gels promoted CM survival, reorganization in interconnected linear networks, and formation of Cx-43-positive gap junctions within 1 day of culture. When the ECs were seeded 1 day earlier to CMs, synchronized spontaneous beating was observed as early as days 2–3 of coculture, while when the two cell types were seeded on the same day, the synchronized beating was only observed on days 6–7 (Narmoneva et al., 2004).

Stevens et al. described a scaffold-free cardiac tissue patch, generated from hESC-derived CMs; however, the patches suffered from a necrotic core, which arose from oxygen and nutrient diffusion limitations and failed to serve as effective grafts upon implantation in rat skeletal muscle (Stevens et al., 2009a,b). These patches were later upgraded by coculturing the CMs with ECs and fibroblasts which increased their mechanical stiffness, without compromising their contraction capacities. Vessel-like structures were observed within the triculture patches in vitro which then anastomosed with host vasculature postimplantation (Stevens et al., 2009b).

Cardiac sheets from rat CMs with rat ECs were layered to form 3D myocardial tissues, and transplanted onto infarcted rat hearts. Increased EC density resulted in improved cardiac function. The vessel networks originating from the preseeded ECs anastomosed with host capillaries in the infarcted myocardium (Sekine et al., 2008).

Iyer et al. compared a triculture of ECs, primary cardiac fibroblasts and rat neonatal CMs seeded simultaneously to sequential seeding of an EC-fibroblast 2-day coculture followed by seeding CMs on matrigel-coated PEG gels (Iyer et al., 2009a). The sequential seeding resulted in the formation of compact contractile cardiac organoids, bearing elongated, Cx-43-expressing CMs, with higher viability in the sequentially cultured gels, when compared to the control (Iyer et al., 2009a). They further demonstrated that the functional properties of the cardiac organoids improved by increasing the CM ratio from 40 to 80% (Iyer et al., 2009b). They next proceeded to sequentially seed each cell type, beginning with the ECs, followed by the fibroblasts 1 day later and the CMs seeded 2 days thereafter. This technique resulted in capillary cord formation and further development into functional cardiac organoids (Iyer et al., 2012). In summary, incorporation of ECs in cultured tissue grafts enhances vessel-like network formation and increases postimplantation, anastamosis potential, and graft perfusion.

18.7.3 Prevascularization In Vivo

Dvir et al. constructed a cardiac patch by seeding rat neonatal CMs onto an affinity-binding alginate scaffold supplemented with IGF-1, stromal cell-derived factor-1 (SDF-1), and vascular endothelial growth factor (VEGF) (Dvir et al., 2009). Patch vascularization was induced in vivo by first implanting the cardiac patch onto the rat omentum for 1 week. In this bioreactor-like environment, host vessels penetrated the cardiac patch. The prevascularized cardiac patches were then implanted in a rat MI model. Four weeks postimplantation, the patches were well integrated with host myocardium, bearing Troponin-T- and Cx-43-expressing cells with a defined cardiac structure. Significant cardiac function improvement and LV remodeling were observed. Moreover, the patches exhibited electrical coupling with host myocardium, as well as a rich vasculature (Dvir et al., 2009).

In vivo prevascularization has also been achieved by culturing the arteriovenous (AV) loop in a chamber with the desired scaffold. The scaffold is then perfused by the AV sprouting vessels, yielding a well vascularized tissue graft, which can then be implanted, to later anastomose with host vasculature. This approach has been applied with rat neonatal CMs seeded on Matrigel in a polycarbonate chamber, which contained a rat femoral AV loop anchored to the inguinal ligament and surrounding tissues with sutures. In this setup, CM survival was ensured and a contractile flap with a defined vascular pedicle was generated (Morritt et al., 2007; Tee et al., 2012). Morritt et al. constructed an AV loop which was placed in a plastic chamber in the rat's groin. Neonatal rat ventricular CMs were seeded on Matrigel within the chamber. Four weeks after implanting the chamber into the rat groin, a viable, vascularized, and spontaneously contracting tissue was observed (Morritt et al., 2007). In another study, after 8 weeks in the groin, Tee et al. transferred the cardiac tissue flaps, now bearing their own patent pedicle, to the neck. Two weeks posttransfer, the tissues preserved their contractile capacity and the cardiac cells exhibited positive staining for desmin, α-smooth muscle actin (SMA), and Cx-43. Additionally, the cardiac tissue flap was transferred to the epicardium and mature striated CMs were observed; however, no direct contact was observed between the myocardium and the transferred tissue (Tee et al., 2012).

Sekine et al. constructed a vascular bed using rat femoral muscle, including the AV vessels and their branches, which was placed in a culture chamber and perfused via the femoral artery (Sekine et al., 2013). A six-layer cardiac tissue was then transplanted to rat neck and the blood vessels were connected by a cuff technique to host vasculature. Two weeks posttransplantation, the vascularized graft anastomosed with host vasculature and maintained its beating capacities and vascularity (Sekine et al., 2013).

Overall, incorporation of functional vessels, whether performed in vitro or in vivo, facilitates scaffold vascularization, which in turn, can form a massive vascular graft that can be applied to replace damaged tissue.

In conclusion, various approaches designed to overcome oxygen diffusion limitations in cardiac tissue engineering have been presented in the literature. However, further enhancement and calibration tasks that remain include characterization of the ideal cell types and combinations, comparison of the effectiveness of vascularization-promoting growth factors, and finally, determination of the ideal perfusion system.

18.8 BIOENGINEERING OF BLOOD VESSELS

18.8.1 Blood Vessels—Structure and Clinical Needs

The cardiovascular system maintains homeostasis of the organism and mediates gas exchange, nutrient and waste transport, and immune defense. These critical functions are fulfilled by the extensive vasculature which consists of a network of blood vessels (arteries, veins, and capillaries). Blood vessels (except capillaries) are structurally built of three layers, which are named from the luminal side outward as the tunica intima (New Latin, "inner coat"), tunica media ("middle coat"), and tunica adventitia ("outer coat") (Figure 18.17). The thickness of each layer varies greatly depending upon the size and type of vessel (arteries vs veins). Moreover, each structural layer consists of distinct cell and ECM composition reflecting its function. The intima contains a monolayer of endothelial cells (ECs), with an underlying basement membrane consisting of loosely oriented type IV collagen and laminin. The EC lining provides a continuous, selectively permeable blood contacting surface that is critical for the maintenance of vessel patency and control of vessel tone. The next layer, the tunica media, is composed of a dense population of concentrically organized SMCs, and is separated from tunica intima by an internal elastic lamina. The elastic structures of the vessels allow the artery to maintain sufficient blood pressure, while sustaining the hemodynamic stress of the cardiac contraction cycle. Finally, the tunica adventitia forms an external layer and contains fibroblasts, collagenous ECM, nerves, and vasa vasorum (a network of small blood vessels that supply blood to the cells of the outer wall part of the vessel). The ECM of the vascular wall is composed of structural proteins (collagen and elastin), adhesion proteins (fibronectin and laminin), GAG, proteoglycans, growth factors, cytokines, and matrix-degrading enzymes and their inhibitors. Overall, the organization, density, spatial geometry, and biochemistry of these ECM components determine mechanical strength, cell response, and final tissue organization of the blood vessel (Zhang et al., 2007; Pankajakshan and Agrawal, 2010).

A broad spectrum of cardiovascular diseases often requires vascular reconstruction. Most prominent examples include blood vessel substitutes for coronary or peripheral bypass surgeries. The use of autologous bypass grafts, including saphenous vein, internal mammary artery, and radial artery bypass grafts, remains an important therapeutic option. However, such autografts are inadequate or unsuitable in approximately one-third of the patients, due to vessel diseases, previous surgery, etc. (Zhang et al., 2007; Seifu et al., 2013). Thus, there is an urgent need for alternative vascular grafts.

Initially, several synthetic materials were found to be possible candidates for vascular replacement, including polyethylene terephthalate (Dacron®), expanded polytetrafluoroethylene (Teflon®), polyurethanes, and others. These materials have been used successfully for large-diameter (>6 mm internal diameter) vessel grafts, where the rate of blood flow is high. However, most blood vessels within the peripheral, cerebral and cardiac (e.g., coronary arteries) vasculature display diameters of less than 6 mm, and such small-sized grafts made from abovementioned synthetic materials showed disappointing clinical results, due to rapid thrombosis and intimal hyperplasia (Zhang et al., 2007; Seifu et al., 2013). The difference in elasticity at the anastomotic site (compliance mismatch) creates turbulence in the blood flow and causes platelet activation, which results in thrombus formation (Seifu et al., 2013).

Motivated by these limitations, the development of tissue-engineered vascular grafts (TEVGs) has progressed significantly over the past two decades. For the first time, engineering of compliant and biologically active healthy like blood vessels was made possible.

18.8.2 Basic Design Principles and Requirements for Blood Vessel Bioengineering

The general approach for vascular tissue engineering is to seed cells onto biodegradable scaffolds first, followed by in vitro culture and/or in vivo implantation. Functional TEVGs should be nonthrombogenic, nonimmunogenic, compliant, vasoactive, matched to native vessels, and fully integrated into the existing network. Also, ideally, the three-layer structure of the native blood vessel (see Figure 18.17) should be recreated on the cellular level (lining of ECs, followed by vascular SMCs, and fibroblasts in the outer layer), with mechanical properties matched to those of native ECM.

Blood vessels belong to a class of soft tissues known as bioviscoelastic solids that demonstrate both viscous and elastic behaviors. At low pressure, elastic fibers and contractile SMCs alone are responsible for vascular mechanical strength and the linear elastic behavior of arteries. At high pressures, recruitment of aligned collagen fibers accounts for the observed nonlinear stress–strain relationship and primarily contributes to the ultimate tensile strength

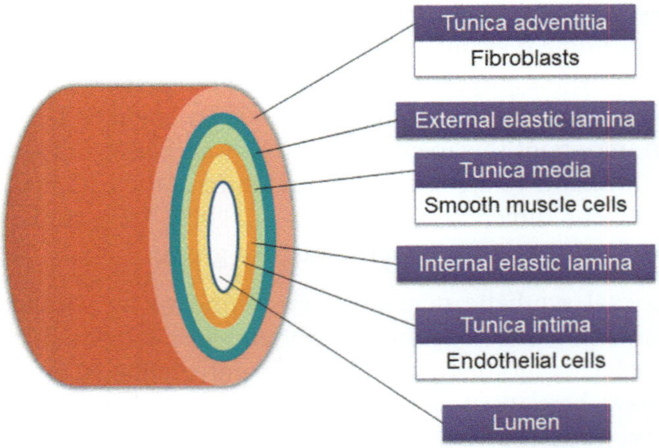

FIGURE 18.17
Schematic representation of various layers in a blood vessel (see text for details).

and stiffness of the native arteries and elastic modulus, respectively. Native arteries thus are compliant and elastic at low pressures and strong at high pressures (Huang and Niklason, 2014). Mechanical considerations in TEVG design should include burst pressure, compliance, fatigue resistance, and other parameters. To mimic native tissue, engineered vascular grafts should have a burst pressure ≥1700 mmHg (equal to that of saphenous vein), fatigue resistance to cyclic physiological loading for 30 days in vitro or during xenogenic implantation, without noticeable dilatation, and compliance of 5–20%/100 mmHg—this parameter is calculated from the percent change in internal radius over a physiologic range of pressure (80–120 mmHg) and is expressed in units of %/100 mmHg (Kumar et al., 2011; Seifu et al., 2013).

The strategies for vascular tissue engineering can be broadly divided into three groups: scaffold-less strategies (self-assembly based on cell sheet technology), in vitro-engineered scaffold-guided grafts (produced from synthetic, natural, or decellularized matrices), and in situ graft endothelialization strategies. The various approaches are summarized in Figure 18.18 and described in the following section.

18.8.3 Scaffold-Free Vascular Grafts Based on Cell Sheet Technology

In a strategy developed in a landmark study by L'Heureux et al., vascular human grafts could be produced from autologous cells without the need for scaffolding material, while relying on the capability of the cells to produce their own ECM (Figure 18.19) (L'Heureux et al., 1998, 2006). Cohesive cellular sheets of human umbilical vein SMCs and adult fibroblasts were cultured in medium enriched

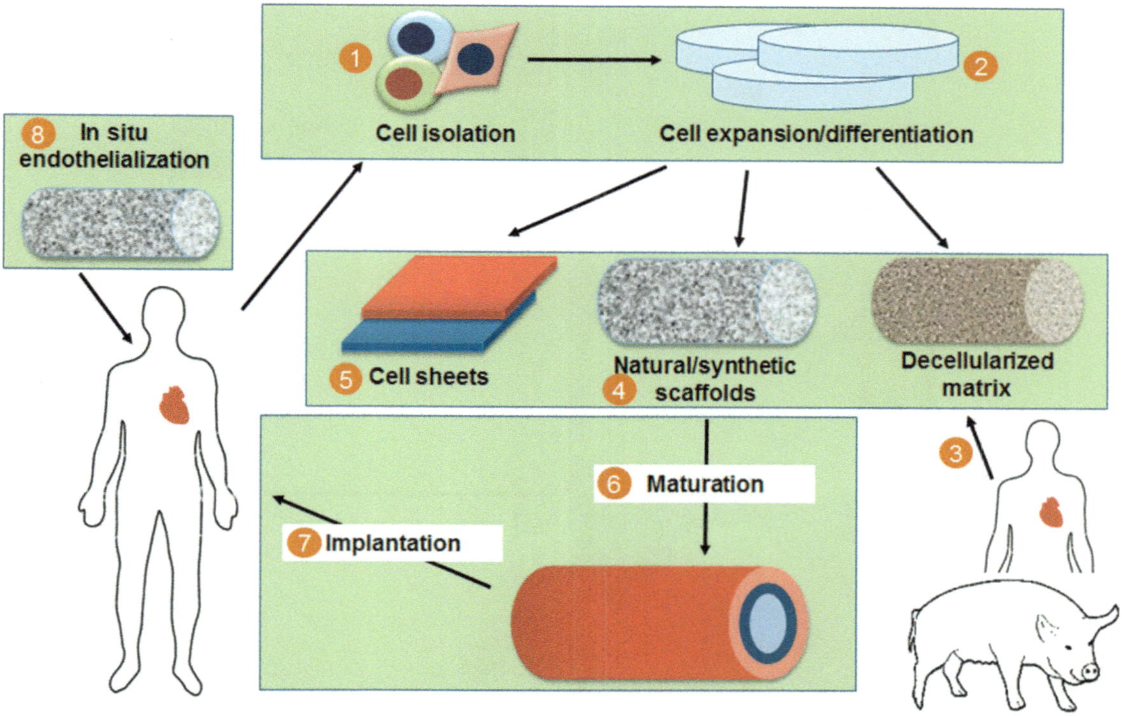

FIGURE 18.18

Vascular tissue engineering. (1) Various cell types (specific differentiated cells (ECs, SMCs, fibroblasts), or stem cells) could be used for TEVGs. The cells are harvested from patient or allogeneic sources. (2) Cells are expanded in culture, and, if required, specific differentiation protocols are applied. Then, cells are seeded for TEVG construction by several strategies. (3) Vessels are harvested from allogeneic, heterogenic, or xenogenic sources, and decellularized to leave a matrix. The matrix is recellularized with the prepared cells. (4) Natural or synthetic polymers are used for construction of scaffolds for subsequent cell seeding. (5) Cell-sheet technology is used to form scaffold-free TEVG. (6) After cell seeding, the constructs can undergo optional maturation steps, including prolonged cultivation, use of bioreactors etc. (7) Fully functional TEVGs are implanted in patients. (8) Alternatively, acellular but biologically functionalized scaffolds could be implanted, to induce in situ endothelialization (see text for details).

with ascorbic acid to induce production of collagen type I. After 30 days of culture, the cell sheets were manually peeled from the culture flask and sequentially rolled on a tubular support to produce media and adventitia equivalents. After a further maturation period of ≥8 weeks, these tubular constructs were endothelialized on the lumen side, to produce vascular grafts with all three layers of native blood vessels. Vascular tissues reconstructed using the cell sheet strategy tend to have extracellular matrix proteins similar to the native tissue, including collagen types I, III, and IV, fibronectin, laminin, and chondroitin sulfates. These tissues also have an adequate burst pressure >2000 mmHg, and functional endothelium. In vivo experiment in dogs for 7 days showed that the engineered vascular tissue could withstand physiological flow conditions (L'Heureux et al., 1998).

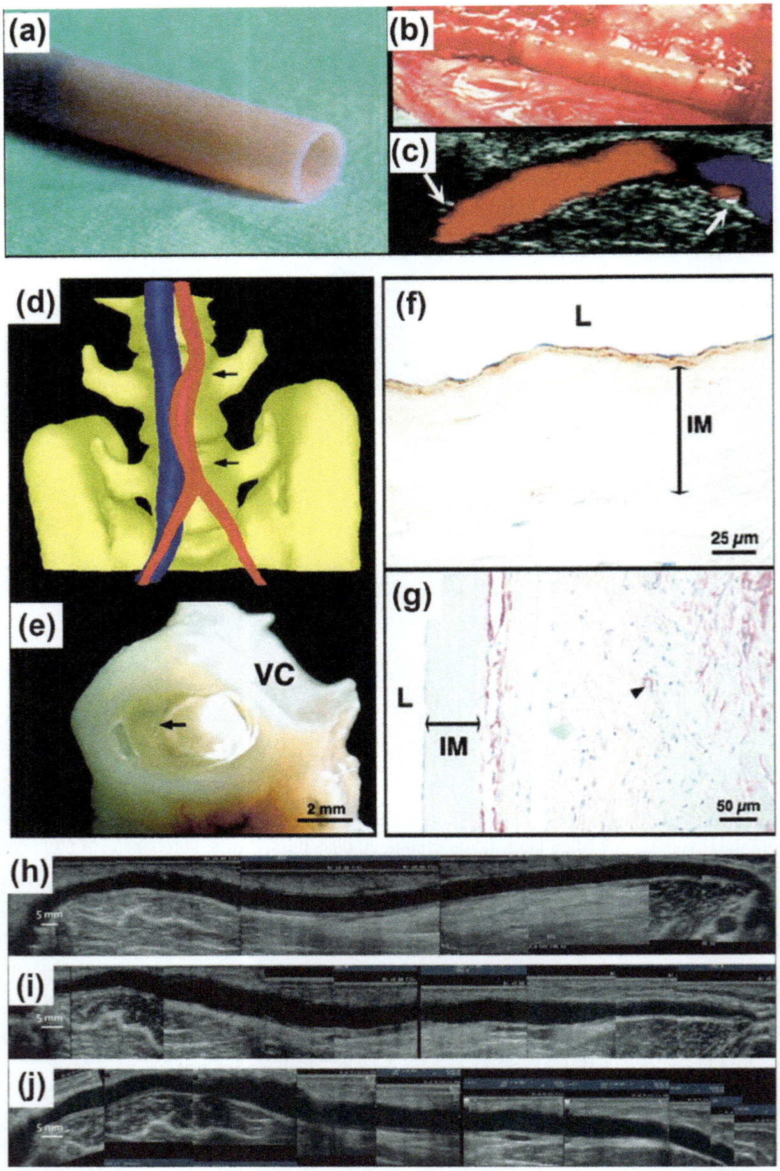

FIGURE 18.19

Scaffold-free cell sheet-based tissue engineering. (a) TEVG before implantation in dogs (4.2 mm internal diameter). (b) TEVG, anastomosed as end-to-end interpositional femoral graft, immediately after removal of the cross-clamps. (c) Doppler-ultrasound imaging at 2 weeks shows a patent vessel with uniform lumen and normal flow (arrows indicate anastomoses). TEVGs were endothelialized with autologous canine endothelial cells. (d)–(g) Implantation of human TEVG in nonhuman primates. (d) CT angiogram after 8 weeks of implantation showing a patent TEBV (red) and the adjacent vena cava (blue). (e) Perfusion-fixed TEVG at 8 weeks shows complete tissue integration and a smooth lumen with no signs of thrombosis, stenosis, or mechanical failure. Arrow indicates suture line. VC: vena cava. (f) Immunolabeling for von Willebrand factor reveals the confluent endothelium and subendothelial matrix (nuclei counterstained in blue). L: lumen, IM: decellularized internal membrane. (g) Immunolabeling for SMC-specific α-actin stains a cell population at the IM/adventitia interface (blue counterstain). Arrow indicates small blood vessel. (h)–(j) Doppler ultrasound images for full graft (25 cm) implanted in patient, in multicenter cohort clinical study. (h) 6 months, (i) 12 months, (j) 18 months. At 20 months, the graft was still patent, and functioned well for hemodialysis access. *Reprinted with permission from L'Heureux et al. (2006), McAllister et al. (2009).*

Later, autologous vascular grafts utilizing the same technology were produced using adult human fibroblasts isolated from skin biopsies from individuals with advanced cardiovascular disease. After a maturation period of 8–10 weeks, the constructs were seeded with endothelial cells. The grafts were tested in rats for up to 180 days and showed higher burst pressures than saphenous vein (3468 mmHg). Moreover, the TEVGs were implanted as interpositional arterial grafts in nonhuman primates. The implanted vessels remained patent at explantation (8 weeks) (L'Heureux et al., 2006). These data are encouraging, and the strategy has already been used in clinical trials. Moreover, this fibroblast sheet-based technology eliminated the need for SMCs, whose early senescence in culture is associated with decreased burst pressures in human models (McKee et al., 2003). However, apparent high cost and extremely long cultivation/maturation times are important drawbacks, limiting the clinical applications of this strategy.

18.8.4 Vascular Grafts Based on Natural Polymers

Natural proteins, such as collagen, elastin, and fibronectin are the major components of ECM in the body. They are the most ideal substrates for cell attachment and cell signaling. Collagen gel was used by Weinberg and Bell (1986) to create the first TEVG. Scaffolds made from naturally occurring biomaterials were, therefore, the first forms of TEVGs. Unfortunately, this construct had poor mechanical strength and could not be used clinically without the use of synthetic sleeves to support the cell-seeded collagen gel. Nearly a decade later, vascular tissues fabricated with purified collagen seeded with vascular cells were implanted in the posterior vena cava of rabbits and dogs (Kanda and Matsuda, 1993; Hirai and Matsuda, 1995). However, these grafts also needed to be strengthened with Dacron®, which caused complications, such as immune rejection and compliance mismatch (Hirai and Matsuda, 1996).

Fibrin gels emerged as an alternative to collagen, in light of the advantage that fibrinogen and thrombin can be readily obtained from a patient's blood (Seifu et al., 2013). The mechanical properties of fibrin gels are poor, but are similar to collagen-based TEVGs. After culturing in vitro for 24 h, cell-seeded fibrin gels were implanted in the jugular vein of chronically heparinized lambs. The grafts showed not only considerable remodeling, with some elastic fiber development after 12 weeks, but also had improved mechanical properties after integrating into the host tissue (Swartz et al., 2005; Yao et al., 2005). Even though substantial dilatation was observed, grafts were patent at the time of explantation, and further application of this technique to human cells is in progress (Peng et al., 2011).

All natural scaffolds have inferior mechanical properties to synthetic scaffolds. Use of the cell-seeding technique and bioreactors for 3D regeneration of vascular tissue can improve the mechanical properties of these constructs (Seifu et al., 2013). The performance and long-term patency of the developed TEVGs still awaits confirmation in animal models.

18.8.5 Vascular Grafts Based on Synthetic Polymers

Various synthetic polymer scaffolds have been, and are being, studied for applications in vascular tissue engineering. Among the most extensively used are degradable polyesters composed of glycolide and lactide, their copolymers (such as polyglycolic acid (PGA), poly-L-lactic acid, and polycaprolactone (PCL)), and polyurethanes (Zhang et al., 2007; Seifu et al., 2013).

Shin'oka et al. developed a pulmonary artery conduit by seeding tubular polyglactin/PGA scaffolds with autologous SMCs followed by ECs. After an in vitro culture period, the TEVGs were implanted into the pulmonary arteries of lambs and evaluated at intervals of 11–24 weeks. Compared to the acellular control, which developed progressive obstruction and thrombosis, all TEVGs were patent and demonstrated a nonaneurysmal increase in diameter, and were histologically similar to adjacent native arteries (Shinoka et al., 1998). A year later, Niklason and colleagues fabricated functional tissue-engineered blood vessels from PGA scaffolds, cultivated under pulsatile conditions. The patency rate was 100% after 4 weeks of implantation in pigs, and the burst strength was >2000 mmHg (Niklason et al., 1999).

In order to characterize the mechanisms behind the transformation of TEVG into a neovessel, Roh et al. (2010) used TEVGs produced by seeding of human bone-marrow mononuclear cells (BM-MNCs) onto 1 mm PGA-poly (DL-caprolactone-co-lactic acid) (P(CL/LA)) tubular scaffolds. The TEVGs were implanted in severe-combined immunodeficient beige (SCID-BG) mice. At the end of a 6-month period, all seeded grafts remained fully patent and functional as venous conduits. Original scaffold material degraded and was replaced by a supportive adventitial layer of collagen fibrils, while a confluent endothelial cell layer developed along the inner lumen. Extensive histological and immunohistochemical analysis revealed that by 1 week postimplantation, human BM-MNCs were no longer detectable within the TEVGs. While these seeded cells did not directly contribute to the architecture of the mature TEVG, a regenerative model was hypothesized in which seeded cells augmented a host inflammatory and healing process via paracrine secretion of various cytokines, rather by differentiation and direct regeneration (Figure 18.20). Due to the observation that seeded grafts demonstrate higher concentrations of monocytes during early graft development compared with unseeded grafts, cytokines involved in monocyte migration were investigated. Notably, TEVGs seeded with alginate microspheres containing monocyte chemoattractant protein-1 (MCP-1), a cytokine known for its role in monocyte migration and collateral vessel development, functioned similar to grafts seeded with human BM-MNCs, revealing mature grafts with an endothelial-lined internal lumen and organized neotissue formation (Roh et al., 2010). A later study in which grafts were seeded with monocytes or with BM-MNC confirmed the critical role that monocytes play in neovessel development. In this study monocyte-seeded scaffolds maintained a significantly

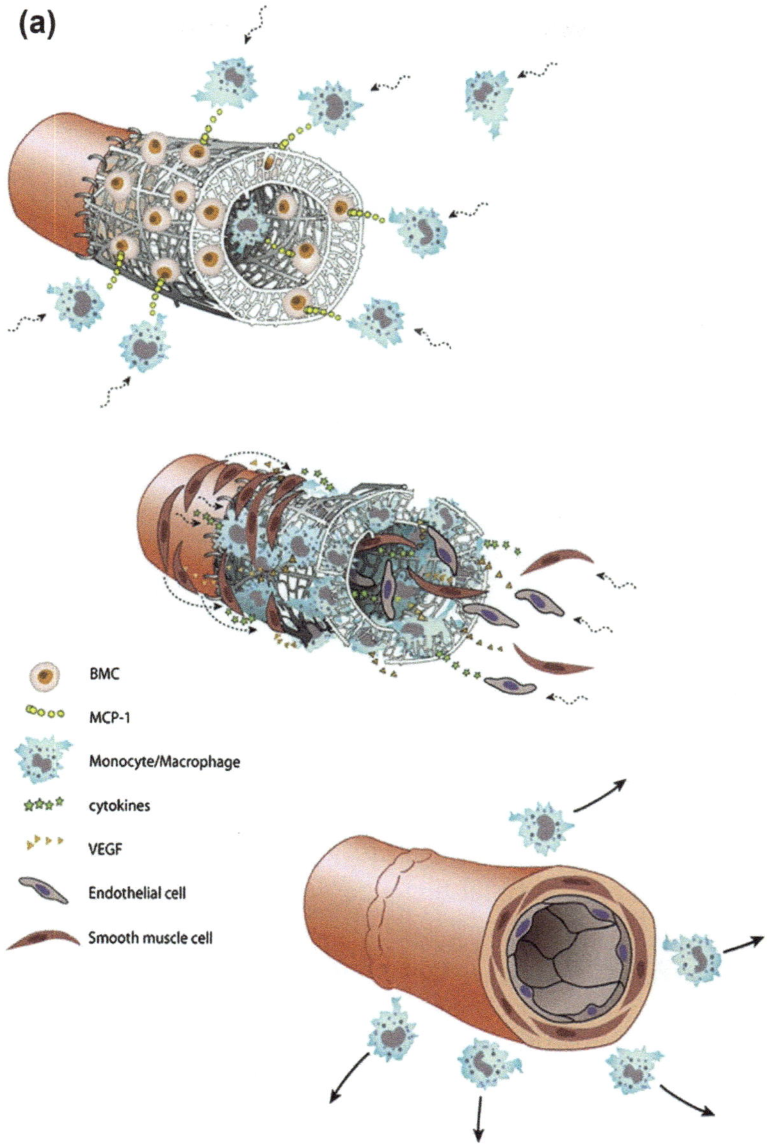

FIGURE 18.20

Vascular tissue engineering based on synthetic biodegradable polymers. (a) Proposed mechanism of vascular transformation of hBMC-seeded biodegradable scaffolds. Early pulse of MCP-1, secreted from seeded hBMCs, enhances early monocyte recruitment to the scaffold. Infiltrating monocytes release multiple angiogenic cytokines and growth factors (i.e., VEGF), which recruit SMCs and ECs to the scaffold. Vascular cells potentially come from circulating progenitors and/or proliferation/migration of mature vascular cells in adjacent vessel segments. ECs and SMCs appropriately organize into a mature blood vessel structure on the luminal surface of the scaffold. As the scaffold degrades, monocytes migrate away, leaving behind a completely autologous neovessel (Roh et al., 2010). (b) Macroscopic finding (left, 18 mm in diameter) and scanning electromicroscopic finding (right) of biodegradable scaffolds used in clinical trials. (c) Pulmonary angiograms of one patient from clinical study. (A) Lateral view of pulmonary angiogram before operation showed peripheral stenosis (white arrowhead) with aneurysmal formation (white arrow) after balloon angioplasty. (B) Same view 6 months after operation. White arrows indicate location of TEVG. *Reprinted with permission from Shin'oka et al. (2005).* (d) Three-dimensional CT 1 year after TEVG implantation. Arrows denote implanted TEVG. The graft is patent without aneurysmal dilation. *Reprinted with permission from Hibino et al. (2010).*

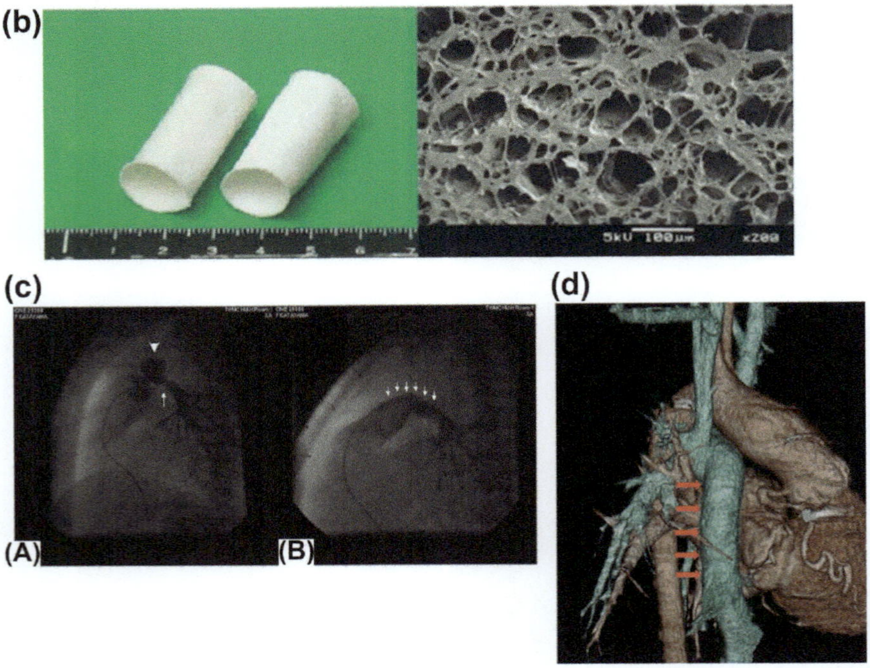

FIGURE 18.20 ***Continued***

larger graft luminal diameter and higher rate of long-term patency compared with scaffolds seeded with BM-MNCs, which were absent of monocytes (Mirensky et al., 2010).

18.8.6 Vascular Grafts Based on Decellularized Matrix

Decellularized allogeneic, xenogenic, or in vitro-engineered tubular tissues that contain an intact and structurally organized ECM of the vessel have been investigated as vascular grafts. Several xenogenic decellularized tubular conduits have been investigated, such as aorta, umbilical arteries, carotid arteries, saphenous veins, and others, isolated from canine, porcine, sheep, and bovine sources (Zhang et al., 2007; Pankajakshan and Agrawal, 2010; Kumar et al., 2011). However, the pathogen transmission from such xenogenic sources is still a major concern.

Niklason and colleagues developed a novel strategy for graft production using in vitro tissue engineering and decellularization, having a potential of significantly reducing the waiting time usually associated with graft preparation (Figure 18.21). Allogeneic banked porcine SMCs were cultured on a biodegradable PGA mesh tubular scaffold in a perfusion bioreactor for 10 weeks. At the conclusion of culture, the fresh tissue-engineered vessel had a robust, dense

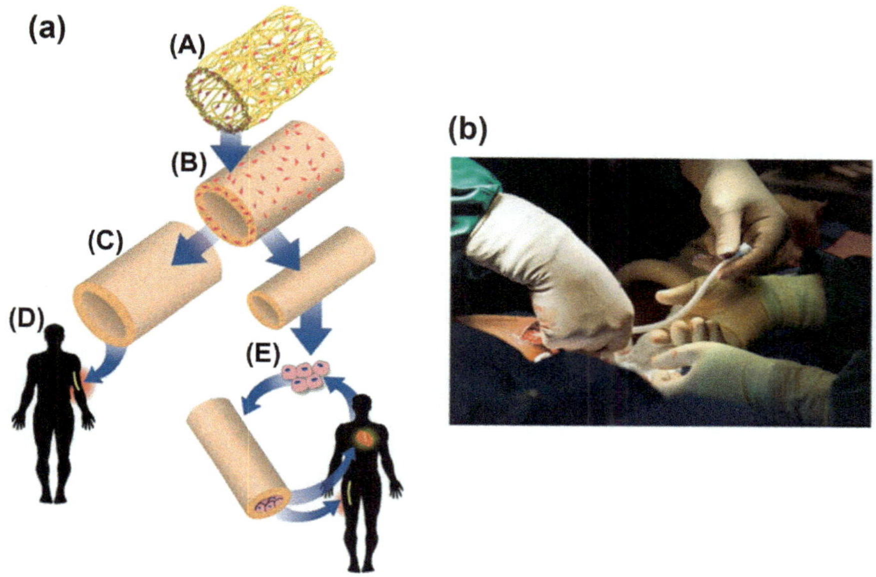

FIGURE 18.21

Decellularized tissue-engineered blood vessels. (a) and (b) Allogeneic human tissue-engineered decellularized vascular grafts. (a) Human SMCs are cultured on the PGA scaffolds (A) to produce the vascular grafts (B) which are then decellularized (C) and stored until the time of patient need. The acellular human-based vascular grafts may be directly implanted without ECs if the diameters are greater than 6 mm (D) or may be seeded with ECs from the patient for small-diameter (3–4 mm) applications (E). (b) Implantation of an engineered human vascular graft into the arm of a patient as an AV shunt. *Reprinted with permission from Quint et al. (2011), Gui and Niklason (2014).*

layer of SMCs with minimal residual scaffold PGA fibers. The constructs were then decellularized to remove cellular content, leaving preformed ECM. The acellular grafts were then seeded with endothelial progenitor cells (EPCs) or ECs on the graft lumen. TEVG were then implanted as end-to-side grafts in the porcine carotid artery. The EPC- and EC-seeded TEVGs all remained patent for 30 days in this study, and demonstrated less neointimal hyperplasia and fewer proliferating cells than did the simple vein grafts (Quint et al., 2011). Later, the same concept was applied for the engineering of human grafts. Banked allogeneic SMCs were used to produce a robust ECM in vitro by culturing the cells for 8–10 weeks on PGA in perfusion bioreactor. Cells were obtained from cadaveric donors, and cellular expansion and banking, thus many grafts can be produced from cells from a single donor. Cells are not harvested from the patient, and therefore no invasive tissue biopsy is required. After decellularization, the graft consists only of human, collagenous ECM, which is nonimmunogenic and thus can be used in any recipient. Importantly, such TEVGs may be stored for months at 4 °C (e.g., in hospitals), and be immediately available for implantation. These 6-mm acellular human grafts have been tested in a baboon arteriovenous model for up to 6 months with 88% patency. Similar 3-mm grafts were cultured from allogeneic canine cells, decellularized, reseeded with autologous canine ECs, and tested in canine models of peripheral and coronary artery bypass with 83% patency up to 1 year. TEVGs integrated well with native vasculature at anastomotic sites and resisted intimal hyperplasia. Infiltration of α-smooth muscle actin-positive cells, ECs on graft lumens, and elastin formation near anastomoses was observed (Dahl et al., 2011). In summary, the developed TEVGs, that could be available without a significant patient wait time, represent a substantial advance over completely autologous tissue engineering approaches, wherein patients must wait for long time periods for grafts to be cultured. These grafts were shown to be functional as arteriovenous conduits and as small-caliber arterial bypasses in the peripheral (carotid) and coronary circulations. Finally, these TEVGs could be stored and be available immediately as "off-the-shelf" products, and serve as a promising alternative for patients who have no available autologous vascular conduit.

18.8.7 In Situ Endothelialization of Vascular Grafts

Many strategies for seeding cells on vascular grafts have been developed to establish a complete endothelium before implantation. Even when these strategies are successful, the hurdles of cell harvesting and seeding may limit clinical applicability, drastically increase graft costs, and/or require a prolonged cultivation time. Thus, instead of relying on autologous or allogeneic cell seeding for the production and implantation of TEVGs, several vascular graft strategies incorporate biofunctional molecules or components to mobilize autologous EPCs or ECs within the vasculature to the graft postimplantation (Melchiorri et al., 2013). These strategies are aimed at inducing cell mobilization, migration, homing, adhesion, and proliferation.

Most widely applied strategies use biofunctionalization of the grafts with growth factors, chemokines, or adhesion molecules, to promote graft endothelialization in vivo.

Zeng et al. (2010) employed nerve growth factors (NGFs), which are known to promote EC migration and proliferation, to improve endothelialization of TEVGs in vivo. NGF was covalently immobilized onto a collagen-cross-linked decellularized matrix from adult rat carotid arteries. The resulting NGF-bound TEVG was transplanted as a carotid artery graft in rats. One month after transplantation, only 1 of 10 control grafts remained open, while others suffered from severe thrombosis. Eight out of ten NGF-bound TEVGs remained patent throughout the experiment. Moreover, the luminal surface of the graft was entirely endothelialized (Zeng et al., 2010).

In attempt to induce cell adhesion to the graft in situ, Zheng et al. fabricated tubular grafts by electrospinning of PCL, after which the surface was functionalized by RGD adhesion/signaling peptide. The healing characteristics of the grafts were evaluated by implanting them in rabbit carotid arteries for 2 and 4 weeks. At both time points, all 10 of the implanted RGD-modified PCL grafts (PCL-RGD) were patent, whereas 4 of the 10 nonmodified PCL grafts were occluded due to thrombus formation. Immunofluorescence staining also showed a 3-fold increase of endothelial coverage of the PCL-RGD grafts in comparison with that of PCL grafts at these two time points. Four weeks after implantation, ~65% of the surface area of the PCL-RGD grafts was covered by a SMC layer, which is almost 23% more than that on the PCL grafts (Zheng et al., 2012).

18.9 IN SITU TISSUE RECONSTRUCTION BY INJECTABLE ACELLULAR BIOMATERIALS

While the search for a suitable cell source for cardiac tissue engineering and regeneration is ongoing, the potential of biomaterials as an acellular therapeutic strategy has been established and continuously expanded over the last decade or so. As mentioned, adverse remodeling after MI has been associated with excessive damage to the cardiac ECM, which plays a critical role in maintenance, integrity, and function of the heart tissue. These dynamic changes and alterations in cardiac ECM composition have a detrimental effect on both systolic and diastolic function (Akhyari et al., 2008; Dobaczewski et al., 2010). Biomaterials can be designed to mimic cardiac ECM properties, such as chemistry, mechanical properties, morphology, hydrogel formation, and degradability, and thus, they are capable of providing temporary tissue support, preserving cardiac function, facilitating self-repair, and inducing cardiac tissue regeneration (Ruvinov et al., 2012).

18.9.1 Injectable Biomaterials

Various natural or synthetic polymers can form hydrogels, namely solid networks of physically or chemically cross-linked polymer chains with varying water content, which can be directly injected into the infarcted heart. Such injectable hydrogels can recapitulate the microenvironment for inducing effective tissue repair, by replacing damaged ECM, providing temporary support for the infarct, and instructing tissue restoration. These effects could prevent the adverse remodeling and progression of heart failure. The application of injectable biopolymers is less invasive than implanting macroporous scaffolds, and is, therefore, more clinically appealing. Various natural or synthetic acellular injectable biomaterials have been found to improve the outcome after MI (Rane and Christman, 2011; Tous et al., 2011).

18.9.2 Injectable Decellularized ECM Matrices

Decellularized ventricular and pericardial ECM can be processed to create a solubilized liquid with the ability to gel via self-assembly at physiological temperature both in vitro and in vivo upon injection into myocardial tissue (Singelyn et al., 2009; Seif-Naraghi et al., 2010). The decellularized ventricular ECM biochemical composition was shown to retain the complexity of proteins, peptides, and GAG of natural ECM. Initial in vivo feasibility testing showed that the solubilized myocardial matrix undergoes gelation in healthy rat myocardial tissue upon direct epicardial injection. The in situ-formed hydrogel had a pore size of ~30 μm and its nanofibrillar structure resembled that of the intact decellularized tissue, prior to processing. Using a similar fabrication procedure, Syngelin et al. prepared an injectable myocardial hydrogel from decellularized porcine ventricular ECM (Singelyn et al., 2012). In a follow-up study, infarcted pigs were treated with percutaneous transendocardial injections of this myocardial matrix hydrogel 2 weeks after MI and evaluated after 3 months. Echocardiography indicated improvement in cardiac function, ventricular volumes, and global wall motion scores. Furthermore, a significantly larger zone of cardiac muscle was found at the endocardium in matrix-injected pigs compared to controls, suggesting that the matrix injection may act to salvage remaining CMs in the infarct. Extensive studies that confirmed biocompatibility and hemocompatibility of the material were performed (Seif-Naraghi et al., 2013). Collectively, these studies demonstrate the feasibility and efficacy of decellularized ventricular ECM-derived in situ-forming hydrogel for the treatment of MI. These promising results pave the way toward translation of this platform to clinical trials in humans.

18.9.3 Alginate Implant: First-in-Man Trial for the Prevention of Heart Failure

We developed an injectable alginate biomaterial which can be delivered by intracoronary injection as a solution. This biomaterial is nonthrombogenic, and when delivered to the infarct-related arteries, it is able to transverse the leaky vessels to the infarct. Then, and only at the acute infarct, due to the high calcium ion concentration and relatively low water content, does the solution undergo additional cross-linking and gelation, forming a hydrogel (Figure 18.22) (Landa et al., 2008; Leor et al., 2009).

These encouraging results have led to a first-in-man clinical trial, proving the safety of intracoronary injection of alginate biomaterial in acute MI patients. Due to the localized physical effect of the biomaterial, it received FDA approval for a regulatory path as a medical device rather than a drug, intended for intracoronary injection to patients following acute MI, for

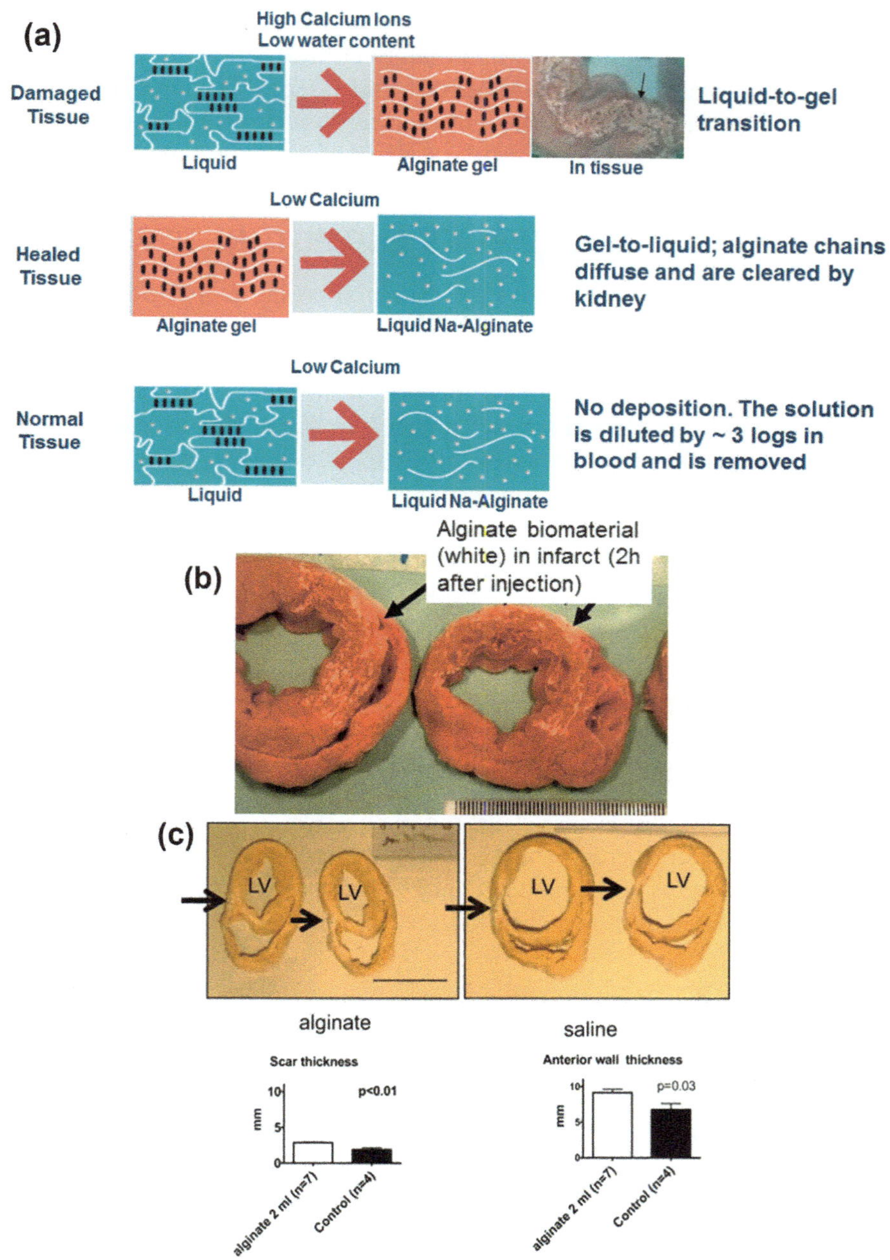

FIGURE 18.22 ***Continued***

prevention of LV remodeling and subsequent CHF. Based on the successful completion of initial safety clinical study, a randomized, multicenter, placebo-controlled, double-blind PRESERVATION I pivotal clinical trial is now underway for the intracoronary injection of alginate implant, now known as Bioabsorbable Cardiac Matrix (BCM) (trial number NCT01226563).

18.10 CONCLUSIONS AND FUTURE PERSPECTIVES

The urgent need for cardiovascular replacement and regeneration therapies has led to a diversity of cardiovascular tissue engineering strategies. Recent advancements in stem cell technology, iPSC derivation, and effective cardiac differentiation and purification techniques offer for a first time previously unavailable cell sources suitable for the engineering of human heart tissues. Continuously evolving biomaterial schemes, novel polymers, bioinspired and "smart" materials, together with micro- and nanotechnological advancements, bioreactor design, and effective maturation/stimulation regimens, offer a list of powerful tools to achieve optimized and long-term functioning grafts.

Biomaterials, when appropriately designed, can be effective vehicles for stem cell delivery, to improve cell retention and function. Moreover, multiple stem cell products responsible for their paracrine beneficial effects can be delivered by biomaterials in a sequential and predicted order to activate the different reparative and regeneration processes in the infarcted heart.

In cardiovascular graft bioengineering, we can already witness significant advancements in terms of graft composition and functionality. Near future progress will include better ECM mimicking, controllable mechanical properties carefully matched to the target myocardial or vascular tissue, introduction

FIGURE 18.22 ***Continued***

Beneficial therapeutic effects of injectable alginate biomaterial on LV remodeling after MI in pigs. (a) A schematic model describing the three possible occurrences after intracoronary injection of the partially cross-linked alginate solution. In a damaged tissue after acute MI, the partially cross-linked alginate solution undergoes gelation at the infarct due to elevated calcium ion concentration. During healing, calcium ion concentration decreases in the tissue, leading to hydrogel dissolution via an exchange reaction of calcium ions with sodium ions. The water soluble alginate molecules can diffuse and be excreted via the urine due to their low molecular weight (less than 50 kDa). Should the alginate reach a normal healthy tissue, no gelation occurs, and the solution is diluted in the systemic circulation and removed through the urine. (b) Macroscopic views of pig heart sections, 2 h after intracoronary injection of calcium cross-linked alginate solution, reveal the in situ deposition of the alginate hydrogel (white areas, arrows), throughout the infarct. (c) Postmortem morphometry at 60 days after MI indicates that the alginate implant increases scar and arterial wall thicknesses as compared to saline treatment. Representative sections of heart treated with alginate (upper left panel) or saline (upper right panel). Morphometry shows that injection of alginate solution (2 ml) increases scar thickness (arrows) as well as anterior wall thickness (lower panels). *Reprinted with permission from Leor et al. (2009), Ruvinov et al. (2012).*

of bioactive, dynamically responsive and bioinspired materials, effective maturation and vascularization techniques, and development of tools for high-throughput and quantitative assessment of graft quality.

We expect that in the following decade extensive and hard work of many dedicated and creative researchers will result in significant technological advancements achieved on the basis of already established principles laid in the chapter, and the concepts for new and maybe even more powerful strategies will be introduced. Such developments will offer true and effective alternatives to currently palliative clinical therapies, and will become a state-of-the-art treatment option for patients with cardiovascular disease.

18.11 SUMMARY

- Low regeneration rate, mainly currently applied palliative therapies, and low availability of donor tissues define the urgent need for the development of new restorative strategies for cardiovascular repair and regeneration.
- Cardiovascular tissue engineering relies on the powerful combination of cells, biomaterials (natural or synthetic), and bioactive molecules (stem cell products or their derivatives).
- Recent advancements in stem biology offer for the first time clinically relevant cell sources for cardiac cells to be used in cardiovascular tissue engineering.
- Biomaterials can be used as a platform or vehicle for cell delivery into the infarcted myocardium, and improve engraftment, survival, retention, and function.
- The "paracrine hypothesis" of stem cell action strengthens the idea behind selective delivery of bioactive molecules; the combination of bioactive molecules with biomaterials allows achieving effective local and sustained delivery with spatial and controlled distribution of the desired agent to induce multiple regenerative processes.
- The replacement of a large damaged tissue requires the transplantation of a fully developed cardiac tissue graft prepared in vitro. Three main strategies for in vitro cardiac patch reconstruction have been developed, based on seeding the cardiac cells in hydrogels, polymer layers, and preformed macroporous scaffolds.
- Generation of human CMs from pluripotent stem cells allows bioengineering of human cardiac grafts for the first time.
- Development of perfusion bioreactors and various stimulation patterns (mechanical, magnetic, and electrical) can promote cell organization and graft maturation.

- Integration and survival of thick tissue constructs rely on effective graft vascularization. Several in vitro (incorporation of endothelial cells) and in vivo (host prevascularization) strategies have been suggested to achieve this goal.
- Bioengineering of small-diameter blood vessels represents a critical challenge in clinical vascular reconstruction; scaffold-less strategies (self-assembly based on cell sheet technology), in vitro-engineered scaffold-guided grafts (produced from synthetic, natural, or decellularized matrices), and in situ graft endothelialization strategies are used for this purpose.
- Along with the research on cell-based cardiac tissue engineering and regeneration, the potential of biomaterials as an acellular therapeutic strategy has been established and continuously expanded. Especially, injectable biomaterials have been found to be effective in temporal ECM replacement, providing mechanical support, and instructing of tissue restoration.

RECOMMENDED READING

Caspi, O., Lesman, A., Basevitch, Y., Gepstein, A., Arbel, G., Habib, I.H., Gepstein, L., Levenberg, S., 2007. Tissue engineering of vascularized cardiac muscle from human embryonic stem cells. Circ. Res. 100 (2), 263–272.

Christman, K.L., et al., 2004. Injectable fibrin scaffold improves cell transplant survival, reduces infarct expansion, and induces neovasculature formation in ischemic myocardium. J. Am. Coll. Cardiol. 44 (3), 654–660.

Dahl, S.L., Kypson, A.P., Lawson, J.H., Blum, J.L., Strader, J.T., Li, Y., Manson, R.J., Tente, W.E., DiBernardo, L., Hensley, M.T., Carter, R., Williams, T.P., Prichard, H.L., Dey, M.S., Begelman, K.G., Niklason, L.E., 2011. Readily available tissue-engineered vascular grafts. Sci. Transl. Med. 3 (68), 68ra69.

Dvir, T., Benishti, N., Shachar, M., Cohen, S., 2006. A novel perfusion bioreactor providing a homogenous milieu for tissue regeneration. Tissue Eng. 12, 2843–2852.

Leor, J., Aboulafia-Etzion, S., Dar, A., Shapiro, L., Barbash, I.M., Battler, A., Granot, Y., Cohen, S., 2000. Bioengineered cardiac grafts. A new approach to repair the infarcted myocardium? Cir. 102 (Suppl. II), 56–61.

Leor, J., Tuvia, S., Guetta, V., Manczur, F., Castel, D., Willenz, U., Petnehazy, O., Landa, N., Feinberg, M.S., Konen, E., Goitein, O., Tsur-Gang, O., Shaul, M., Klapper, L., Cohen, S., 2009. Intracorinjection of in situ forming alginate hydrogel reverses left ventricular remodeling after myocardial infarction in swine. J. Am. Coll. Cardiol. 54 (11), 1014–1023.

Padin-Iruegas, M.E., et al., 2009. Cardiac progenitor cells and biotinylated insulin-like growth factor-1 nanofibers improve endogenous and exogenous myocardial regeneration after infarction. Circulation 120 (10), 876–887.

Roh, J.D., Sawh-Martinez, R., Brennan, M.P., Jay, S.M., Devine, L., Rao, D.A., Yi, T., Mirensky, T.L., Nalbandian, A., Udelsman, B., Hibino, N., Shinoka, T., Saltzman, W.M., Snyder, E., Kyriakides, T.R., Pober, J.S., Breuer, C.K., 2010. Tissue-engineered vascular grafts transform into mature blood vessels via an inflammation-mediated process of vascular remodeling. Proc. Natl. Acad. Sci. USA 107 (10), 4669–4674.

Ruvinov, E., Leor, J., Cohen, S., 2011. The promotion of myocardial repair by the sequential delivery of IGF-1 and HGF from an injectable alginate biomaterial in a model of acute myocardial infarction. Biomaterials 32 (2), 565–578.

Ruvinov, E., Sapir, Y., Cohen, S., 2012. Cardiac Tissue Engineering: Principles, Materials, and Applications. Morgan & Claypool Publishers.

Seif-Naraghi, S.B., Singelyn, J.M., Salvatore, M.A., Osborn, K.G., Wang, J.J., Sampat, U., Kwan, O.L., Strachan, G.M., Wong, J., Schup-Magoffin, P.J., Braden, R.L., Bartels, K., DeQuach, J.A., Preul, M., Kinsey, A.M., DeMaria, A.N., Dib, N., Christman, K.L., 2013. Safety and efficacy of an injectable extracellular matrix hydrogel for treating myocardial infarction. Sci. Transl. Med. 5 (173), 173ra125.

Shimizu, T., Yamato, M., Isoi, Y., Akutsu, T., Setomaru, T., Abe, K., Kikuchi, A., Umezu, M., Okano, T., 2002. Fabrication of pulsatile cardiac tissue grafts using a novel 3-dimensional cell sheet manipulation technique and temperature-responsive cell culture surfaces. Circ. Res. 90 (3), e40–e48.

Zimmermann, W.H., et al., 2006. Engineered heart tissue grafts improve systolic and diastolic function in infarcted rat hearts. Nat. Med. 12 (4), 452–458.

REFERENCES

Abdel-Latif, A., Bolli, R., Zuba-Surma, E.K., Tleyjeh, I.M., Hornung, C.A., Dawn, B., 2008. Granulocyte colony-stimulating factor therapy for cardiac repair after acute myocardial infarction: a systematic review and meta-analysis of randomized controlled trials. Am. Heart J. 156 (2), 216–226.

Akhyari, P., Kamiya, H., Haverich, A., Karck, M., Lichtenberg, A., 2008. Myocardial tissue engineering: the extracellular matrix. Eur. J. Cardiothorac. Surg. 34 (2), 229–241.

Barash, Y., Dvir, T., Tandeitnik, P., Ruvinov, E., Guterman, H., Cohen, S., 2010. Electric field stimulation integrated into perfusion bioreactor for cardiac tissue engineering. Tissue Eng. Part C Methods 16 (6), 1417–1426.

BioLineRx Ltd, 2007. Safety and Feasibility of the Injectable BL-1040 Implant. Study NCT00557531. Available at: http://www.ClinicalTrials.gov (accessed 03.11.13).

Burridge, P.W., Keller, G., Gold, J.D., Wu, J.C., 2012. Production of de novo cardiomyocytes: human pluripotent stem cell differentiation and direct reprogramming. Cell. Stem Cell. 10 (1), 16–28.

Carrier, R.L., Rupnick, M., Langer, R., Schoen, F.J., Freed, L.E., Vunjak-Novakovic, G., 2002. Perfusion improves tissue architecture of engineered cardiac muscle. Tissue Eng. 8 (2), 175–188.

Caspi, O., Lesman, A., Basevitch, Y., Gepstein, A., Arbel, G., Habib, I.H., Gepstein, L., Levenberg, S., 2007. Tissue engineering of vascularized cardiac muscle from human embryonic stem cells. Circ. Res. 100 (2), 263–272.

Chiu, L.L., Iyer, R.K., King, J.P., Radisic, M., 2011. Biphasic electrical field stimulation aids in tissue engineering of multicell-type cardiac organoids. Tissue Eng. Part A 17 (11–12), 1465–1477.

Cohen, S., Leor, J., 2004. Rebuilding broken hearts. Biologists and engineers working together in the fledgling field of tissue engineering are within reach of one of their greatest goals: constructing a living human heart patch. Sci. Am. 291 (5), 44–51.

Dahl, S.L., Kypson, A.P., Lawson, J.H., Blum, J.L., Strader, J.T., Li, Y., Manson, R.J., Tente, W.E., DiBernardo, L., Hensley, M.T., Carter, R., Williams, T.P., Prichard, H.L., Dey, M.S., Begelman, K.G., Niklason, L.E., 2011. Readily available tissue-engineered vascular grafts. Sci. Transl. Med. 3 (68), 68ra69.

Dar, A., Shachar, M., Leor, J., Cohen, S., 2002. Optimization of cardiac cell seeding and distribution in 3D porous alginate scaffolds. Biotechnol. Bioeng. 80, 305–312.

Dobaczewski, M., Gonzalez-Quesada, C., Frangogiannis, N.G., 2010. The extracellular matrix as a modulator of the inflammatory and reparative response following myocardial infarction. J. Mol. Cell. Cardiol. 48 (3), 504–511.

Dvir, T., Benishti, N., Shachar, M., Cohen, S., 2006. A novel perfusion bioreactor providing a homogenous milieu for tissue regeneration. Tissue Eng. 12, 2843–2852.

Dvir, T., Kedem, A., Ruvinov, E., Levy, O., Freeman, I., Landa, N., Holbova, R., Feinberg, M.S., Dror, S., Etzion, Y., Leor, J., Cohen, S., 2009. Prevascularization of cardiac patch on the omentum improves its therapeutic outcome. Proc. Natl. Acad. Sci. USA 106 (35), 14990–14995.

Dvir, T., Levy, O., Shachar, M., Granot, Y., Cohen, S., 2007. Activation of the ERK1/2 cascade via pulsatile interstitial fluid flow promotes cardiac tissue assembly. Tissue Eng. 13 (9), 2185–2193.

Eschenhagen, T., Fink, C., Remmers, U., Scholz, H., Wattchow, J., Weil, J., Zimmermann, W., Dohmen, H.H., Schafer, H., Bishopric, N., Wakatsuki, T., Elson, E.L., 1997. Three-dimensional reconstitution of embryonic cardiomyocytes in a collagen matrix: a new heart muscle model system. FASEB J. 11 (8), 683–694.

Fomovsky, G.M., Thomopoulos, S., Holmes, J.W., 2010. Contribution of extracellular matrix to the mechanical properties of the heart. J. Mol. Cell. Cardiol. 48 (3), 490–496.

Frantz, S., Bauersachs, J., Ertl, G., 2009. Post-infarct remodelling: contribution of wound healing and inflammation. Cardiovasc. Res. 81 (3), 474–481.

Freeman, I., Cohen, S., 2009. The influence of the sequential delivery of angiogenic factors from affinity-binding alginate scaffolds on vascularization. Biomaterials 30 (11), 2122–2131.

Freeman, I., Kedem, A., Cohen, S., 2008. The effect of sulfation of alginate hydrogels on the specific binding and controlled release of heparin-binding proteins. Biomaterials 29 (22), 3260–3268.

Garg, S., Narula, J., Chandrashekhar, Y., 2005. Apoptosis and heart failure: clinical relevance and therapeutic target. J. Mol. Cell. Cardiol. 38 (1), 73–79.

Gui, L., Niklason, L.E., 2014. Vascular tissue engineering: Building perfusable vasculature for implantation. Curr. Opin. Chem. Eng. 3, 68–74.

Haraguchi, Y., Shimizu, T., Yamato, M., Okano, T., 2011. Regenerative therapies using cell sheet-based tissue engineering for cardiac disease. Cardiol. Res. Pract. 2011, 845170.

Hibino, N., McGillicuddy, E., Matsumura, G., Ichihara, Y., Naito, Y., Breuer, C., Shinoka, T., 2010. Late-term results of tissue-engineered vascular grafts in humans. J. Thorac. Cardiovasc. Surg. 139 (2) 431–436, 436 e431–432.

Hirai, J., Matsuda, T., 1995. Self-organized, tubular hybrid vascular tissue composed of vascular cells and collagen for low-pressure-loaded venous system. Cell. Transpl. 4 (6), 597–608.

Hirai, J., Matsuda, T., 1996. Venous reconstruction using hybrid vascular tissue composed of vascular cells and collagen: tissue regeneration process. Cell. Transpl. 5 (1), 93–105.

Hirt, M.N., Hansen, A., Eschenhagen, T., 2014. Cardiac tissue engineering: state of the art. Circ. Res. 114 (2), 354–367.

Huang, A.H., Niklason, L.E., 2014. Engineering of arteries in vitro. Cell. Mol. Life Sci. 71 (11), 2103–2118.

Ikaria Holdings Inc, 2010. IK-5001 for the Prevention of Remodeling of the Ventricle and Congestive Heart Failure after Acute Myocardial Infarction (PRESERVATION 1). Study NCT01226563. Available at: http://www.ClinicalTrials.gov (accessed 03.11.13).

Iyer, R.K., Chiu, L.L., Reis, L.A., Radisic, M., 2011. Engineered cardiac tissues. Curr. Opin. Biotechnol. 22 (5), 706–714.

Iyer, R.K., Chiu, L.L., Vunjak-Novakovic, G., Radisic, M., 2012. Biofabrication enables efficient interrogation and optimization of sequential culture of endothelial cells, fibroblasts and cardiomyocytes for formation of vascular cords in cardiac tissue engineering. Biofabrication 4 (3), 035002.

Iyer, R.K., Chiu, L.L.Y., Radisic, M., 2009a. Microfabricated poly(ethylene glycol) templates enable rapid screening of triculture conditions for cardiac tissue engineering. J. Biomed. Mater. Res. Part A 89A (3), 616–631.

Iyer, R.K., Chui, J., Radisic, M., 2009b. Spatiotemporal tracking of cells in tissue-engineered cardiac organoids. J. Tissue Eng. Regen. Med. 3 (3), 196–207.

Jessup, M., Brozena, S., 2003. Heart failure. N. Engl. J. Med. 348, 2007–2018.

Jugdutt, B.I., 2003. Ventricular remodeling after infarction and the extracellular collagen matrix: when is enough enough? Circulation 108 (11), 1395–1403.

Kanda, K., Matsuda, T., 1993. Behavior of arterial wall cells cultured on periodically stretched substrates. Cell. Transpl. 2 (6), 475–484.

Kaufman-Francis, K., Koffler, J., Weinberg, N., Dor, Y., Levenberg, S., 2012. Engineered vascular beds provide key signals to pancreatic hormone-producing cells. PLoS ONE 7 (7), e40741.

Kaully, T., Kaufman-Francis, K., Lesman, A., Levenberg, S., 2009. Vascularization – the conduit to viable engineered tissues. Tissue Eng. Part B Rev. 15 (2), 159–169.

Kim, H.E., Dalal, S.S., Young, E., Legato, M.J., Weisfeldt, M.L., D'Armiento, J., 2000. Disruption of the myocardial extracellular matrix leads to cardiac dysfunction. J. Clin. Invest. 106 (7), 857–866.

Koffler, J., Kaufman-Francis, K., Shandalov, Y., Egozi, D., Pavlov, D.A., Landesberg, A., Levenberg, S., 2011. Improved vascular organization enhances functional integration of engineered skeletal muscle grafts. Proc. Natl. Acad. Sci. USA 108 (36), 14789–14794.

Korecky, B., Hai, C.M., Rakusan, K., 1982. Functional capillary density in normal and transplanted rat hearts. Can. J. Physiol. Pharmacol. 60 (1), 23–32.

Kumar, V.A., Brewster, L.P., Caves, J.M., Chaikof, E.L., 2011. Tissue engineering of blood vessels: functional Requirements, progress, and future challenges. Cardiovasc. Eng. Technol. 2 (3), 137–148.

L'Heureux, N., Dusserre, N., Konig, G., Victor, B., Keire, P., Wight, T.N., Chronos, N.A., Kyles, A.E., Gregory, C.R., Hoyt, G., Robbins, R.C., McAllister, T.N., 2006. Human tissue-engineered blood vessels for adult arterial revascularization. Nat. Med. 12 (3), 361–365.

L'Heureux, N., Paquet, S., Labbe, R., Germain, L., Auger, F.A., 1998. A completely biological tissue-engineered human blood vessel. FASEB J. 12 (1), 47–56.

Landa, N., Miller, L., Feinberg, M.S., Holbova, R., Shachar, M., Freeman, I., Cohen, S., Leor, J., 2008. Effect of injectable alginate implant on cardiac remodeling and function after recent and old infarcts in rat. Circulation 117 (11), 1388–1396.

Leor, J., Aboulafia-Etzion, S., Dar, A., Shapiro, L., Barbash, I.M., Battler, A., Granot, Y., Cohen, S., 2000. Bioengineered cardiac grafts. A new approach to repair the infarcted myocardium? Circulation 102 (Suppl. II), 56–61.

Leor, J., Tuvia, S., Guetta, V., Manczur, F., Castel, D., Willenz, U., Petnehazy, O., Landa, N., Feinberg, M.S., Konen, E., Goitein, O., Tsur-Gang, O., Shaul, M., Klapper, L., Cohen, S., 2009. Intracoronary injection of in situ forming alginate hydrogel reverses left ventricular remodeling after myocardial infarction in swine. J. Am. Coll. Cardiol. 54 (11), 1014–1023.

Lesman, A., Gepstein, L., Levenberg, S., 2010a. Vascularization shaping the heart. Ann. N.Y. Acad. Sci. 1188, 46–51.

Lesman, A., Habib, M., Caspi, O., Gepstein, A., Arbel, G., Levenberg, S., Gepstein, L., 2010b. Transplantation of a tissue-engineered human vascularized cardiac muscle. Tissue Eng. Part A 16 (1), 115–125.

Levenberg, S., Rouwkema, J., Macdonald, M., Garfein, E.S., Kohane, D.S., Darland, D.C., Marini, R., van Blitterswijk, C.A., Mulligan, R.C., D'Amore, P.A., Langer, R., 2005. Engineering vascularized skeletal muscle tissue. Nat. Biotechnol. 23 (7), 879–884

Lin, Y.D., Luo, C.Y., Hu, Y.N., Yeh, M.L., Hsueh, Y.C., Chang, M.Y., Tsai, D.C., Wang, J.N., Tang, M.J., Wei, E.I., Springer, M.L., Hsieh, P.C., 2012. Instructive nanofiber scaffolds with VEGF create a microenvironment for arteriogenesis and cardiac repair. Sci. Transl. Med. 4 (146), 146ra109.

Matsa, E., Sallam, K., Wu, J.C., 2014. Cardiac stem cell biology: glimpse of the past, present, and future. Circ. Res. 114 (1), 21–27.

McAllister, T.N., Maruszewski, M., Garrido, S.A., Wystrychowski, W., Dusserre, N., Marini, A., Zagalski, K., Fiorillo, A., Avila, H., Manglano, X., Antonelli, J., Kocher, A., Zembala, M., Cierpka, L., de la Fuente, L.M., L'Heureux, N., 2009. Effectiveness of haemodialysis access with an autologous tissue-engineered vascular graft: a multicentre cohort study. Lancet 373 (9673), 1440–1446.

McKee, J.A., Banik, S.S., Boyer, M.J., Hamad, N.M., Lawson, J.H., Niklason, L.E., Counter, C.M., 2003. Human arteries engineered in vitro. EMBO Rep. 4 (6), 633–638.

Melchiorri, A.J., Hibino, N., Fisher, J.P., 2013. Strategies and techniques to enhance the in situ endothelialization of small-diameter biodegradable polymeric vascular grafts. Tissue Eng. Part B Rev. 19 (4), 292–307.

Mirensky, T.L., Hibino, N., Sawh-Martinez, R.F., Yi, T., Villalona, G., Shinoka, T., Breuer, C.K., 2010. Tissue-engineered vascular grafts: does cell seeding matter? J. Pediatr. Surg. 45 (6), 1299–1305.

Morritt, A.N., Bortolotto, S.K., Dilley, R.J., Han, X., Kompa, A.R., McCombe, D., Wright, C.E., Itescu, S., Angus, J.A., Morrison, W.A., 2007. Cardiac tissue engineering in an in vivo vascularized chamber. Circulation 115 (3), 353–360.

Nabel, E.G., Braunwald, E., 2012. A tale of coronary artery disease and myocardial infarction. N. Engl. J. Med. 366 (1), 54–63.

Narmoneva, D.A., Vukmirovic, R., Davis, M.E., Kamm, R.D., Lee, R.T., 2004. Endothelial cells promote cardiac myocyte survival and spatial reorganization: implications for cardiac regeneration. Circulation 110 (8), 962–968.

Niklason, L.E., Gao, J., Abbott, W.M., Hirschi, K.K., Houser, S., Marini, R., Langer, R., 1999. Functional arteries grown in vitro. Science 284 (5413), 489–493.

Novosel, E.C., Kleinhans, C., Kluger, P.J., 2011. Vascularization is the key challenge in tissue engineering. Adv. Drug. Deliv. Rev. 63 (4–5), 300–311.

Pankajakshan, D., Agrawal, D.K., 2010. Scaffolds in tissue engineering of blood vessels. Can. J. Physiol. Pharmacol. 88 (9), 855–873.

Patan, S., 2000. Vasculogenesis and angiogenesis as mechanisms of vascular network formation, growth and remodeling. J. Neurooncol. 50 (1–2), 1–15.

Peng, H.F., Liu, J.Y., Andreadis, S.T., Swartz, D.D., 2011. Hair follicle-derived smooth muscle cells and small intestinal submucosa for engineering mechanically robust and vasoreactive vascular media. Tissue Eng. Part A 17 (7–8), 981–990.

Quint, C., Kondo, Y., Manson, R.J., Lawson, J.H., Dardik, A., Niklason, L.E., 2011. Decellularized tissue-engineered blood vessel as an arterial conduit. Proc. Natl. Acad. Sci. USA 108 (22), 9214–9219.

Radisic, M., Malda, J., Epping, E., Geng, W., Langer, R., Vunjak-Novakovic, G., 2006. Oxygen gradients correlate with cell density and cell viability in engineered cardiac tissue. Biotechnol. Bioeng. 93 (2), 332–343.

Radisic, M., Marsano, A., Maidhof, R., Wang, Y., Vunjak-Novakovic, G., 2008. Cardiac tissue engineering using perfusion bioreactor systems. Nat. Protoc. 3 (4), 719–738.

Radisic, M., Park, H., Shing, H., Consi, T., Schoen, F.J., Langer, R., Freed, L.E., Vunjak-Novakovic, G., 2004. Functional assembly of engineered myocardium by electrical stimulation of cardiac myocytes cultured on scaffolds. Proc. Natl. Acad. Sci. USA 101 (52), 18129–18134.

Rane, A.A., Christman, K.L., 2011. Biomaterials for the treatment of myocardial infarction a 5-year update. J. Am. Coll. Cardiol. 58 (25), 2615–2629.

Renault, M.A., Losordo, D.W., 2007. Therapeutic myocardial angiogenesis. Microvasc. Res. 74 (2–3), 159–171.

Ribatti, D., Vacca, A., Nico, B., Roncali, L., Dammacco, F., 2001. Postnatal vasculogenesis. Mech. Dev. 100 (2), 157–163.

Risau, W., Flamme, I., 1995. Vasculogenesis. Annu. Rev. Cell. Dev. Biol. 11, 73–91.

Roger, V.L., Go, A.S., Lloyd-Jones, D.M., Benjamin, E.J., Berry, J.D., Borden, W.B., Bravata, D.M., Dai, S., Ford, E.S., Fox, C.S., Fullerton, H.J., Gillespie, C., Hailpern, S.M., Heit, J.A., Howard, V.J., Kissela, B.M., Kittner, S.J., Lackland, D.T., Lichtman, J.H., Lisabeth, L.D., Makuc, D.M., Marcus, G.M., Marelli, A., Matchar, D.B., Moy, C.S., Mozaffarian, D., Mussolino, M.E., Nichol, G., Paynter, N.P., Soliman, E.Z., Sorlie, P.D., Sotoodehnia, N., Turan, T.N., Virani, S.S., Wong, N.D., Woo, D., Turner, M.B., 2012. Heart disease and stroke statistics–2012 update: a report from the American heart association. Circulation 125 (1), e2–e220.

Roh, J.D., Sawh-Martinez, R., Brennan, M.P., Jay, S.M., Devine, L., Rao, D.A., Yi, T., Mirensky, T.L., Nalbandian, A., Udelsman, B., Hibino, N., Shinoka, T., Saltzman, W.M., Snyder, E., Kyriakides, T.R., Pober, J.S., Breuer, C.K., 2010. Tissue-engineered vascular grafts transform into mature blood vessels via an inflammation-mediated process of vascular remodeling. Proc. Natl. Acad. Sci. USA 107 (10), 4669–4674.

Rosso, F., Giordano, A., Barbarisi, M., Barbarisi, A., 2004. From cell-ECM interactions to tissue engineering. J. Cell. Physiol. 199 (2), 174–180.

Russo, V., Young, S., Hamilton, A., Amsden, B.G., Flynn, L.E., 2014. Mesenchymal stem cell delivery strategies to promote cardiac regeneration following ischemic injury. Biomaterials 35 (13), 3956–3974.

Ruvinov, E., Dvir, T., Leor, J., Cohen, S., 2008. Myocardial repair: from salvage to tissue reconstruction. Expert Rev. Cardiovasc. Ther. 6 (5), 669–686.

Ruvinov, E., Leor, J., Cohen, S., 2011. The promotion of myocardial repair by the sequential delivery of IGF-1 and HGF from an injectable alginate biomaterial in a model of acute myocardial infarction. Biomaterials 32 (2), 565–578.

Ruvinov, E., Sapir, Y., Cohen, S., 2012. Cardiac Tissue Engineering: Principles, Materials, and Applications. Morgan & Claypool Publishers. Synthesis Lectures on Tissue Engineering, 4 (1), 1–200.

Sapir, Y., Kryukov, O., Cohen, S., 2011. Integration of multiple cell-matrix interactions into alginate scaffolds for promoting cardiac tissue regeneration. Biomaterials 32 (7), 1838–1847.

Sapir, Y., Polyak, B., Cohen, S., 2014. Cardiac tissue engineering in magnetically actuated scaffolds. Nanotechnology 25 (1), 014009.

Schaaf, S., Shibamiya, A., Mewe, M., Eder, A., Stohr, A., Hirt, M.N., Rau, T., Zimmermann, W.H., Conradi, L., Eschenhagen, T., Hansen, A., 2011. Human engineered heart tissue as a versatile tool in basic research and preclinical toxicology. PLoS ONE 6 (10), e26397.

Segers, V.F., Lee, R.T., 2007. Local delivery of proteins and the use of self-assembling peptides. Drug. Discov. Today 12 (13–14), 561–568.

Segers, V.F., Lee, R.T., 2011. Biomaterials to enhance stem cell function in the heart. Circ. Res. 109 (8), 910–922.

Seif-Naraghi, S.B., Salvatore, M.A., Schup-Magoffin, P.J., Hu, D.P., Christman, K.L., 2010. Design and characterization of an injectable pericardial matrix gel: a potentially autologous scaffold for cardiac tissue engineering. Tissue Eng. Part A 16 (6), 2017–2027.

Seif-Naraghi, S.B., Singelyn, J.M., Salvatore, M.A., Osborn, K.G., Wang, J.J., Sampat, U., Kwan, O.L., Strachan, G.M., Wong, J., Schup-Magoffin, P.J., Braden, R.L., Bartels, K., DeQuach, J.A., Preul, M., Kinsey, A.M., DeMaria, A.N., Dib, N., Christman, K.L., 2013. Safety and efficacy of an injectable extracellular matrix hydrogel for treating myocardial infarction. Sci. Transl. Med. 5 (173), 173ra125.

Seifu, D.G., Purnama, A., Mequanint, K., Mantovani, D., 2013. Small-diameter vascular tissue engineering. Nat. Rev. Cardiol. 10 (7), 410–421.

Sekine, H., Shimizu, T., Hobo, K., Sekiya, S., Yang, J., Yamato, M., Kurosawa, H., Kobayashi, E., Okano, T., 2008. Endothelial cell coculture within tissue-engineered cardiomyocyte sheets enhances neovascularization and improves cardiac function of ischemic hearts. Circulation 118 (14 Suppl.), S145–S152.

Sekine, H., Shimizu, T., Sakaguchi, K., Dobashi, I., Wada, M., Yamato, M., Kobayashi, E., Umezu, M., Okano, T., 2013. In vitro fabrication of functional three-dimensional tissues with perfusable blood vessels. Nat. Commun. 4, 1399.

Shachar, M., Tsur-Gang, O., Dvir, T., Leor, J., Cohen, S., 2011. The effect of immobilized RGD peptide in alginate scaffolds on cardiac tissue engineering. Acta Biomater. 7 (1), 152–162.

Shandalov, Y., Egozi, D., Koffler, J., Dado-Rosenfeld, D., Ben-Shimol, D., Freiman, A., Shor, E., Kabala, A., Levenberg, S., 2014. An engineered muscle flap for reconstruction of large soft tissue defects. Proc. Natl. Acad. Sci. USA 111 (16), 6010–6015.

Shapiro, L., Cohen, S., 1997. Novel alginate sponges for cell culture and transplantation. Biomaterials 18 (8), 583–590.

Shimizu, T., Yamato, M., Isoi, Y., Akutsu, T., Setomaru, T., Abe, K., Kikuchi, A., Umezu, M., Okano, T., 2002. Fabrication of pulsatile cardiac tissue grafts using a novel 3-dimensional cell sheet manipulation technique and temperature-responsive cell culture surfaces. Circ. Res. 90 (3), e40–e48.

Shin'oka, T., Matsumura, G., Hibino, N., Naito, Y., Watanabe, M., Konuma, T., Sakamoto, T., Nagatsu, M., Kurosawa, H., 2005. Midterm clinical result of tissue-engineered vascular autografts seeded with autologous bone marrow cells. J. Thorac. Cardiovasc. Surg. 129 (6), 1330–1338.

Shin, H., Jo, S., Mikos, A.G., 2003. Biomimetic materials for tissue engineering. Biomaterials 24 (24), 4353–4364.

Shinoka, T., Shum-Tim, D., Ma, P.X., Tanel, R.E., Isogai, N., Langer, R., Vacanti, J.P., Mayer Jr., J.E., 1998. Creation of viable pulmonary artery autografts through tissue engineering. J. Thorac. Cardiovasc. Surg. 115 (3), 536–545 discussion 545–536.

Shriver, Z., Liu, D., Sasisekharan, R., 2002. Emerging views of heparan sulfate glycosaminoglycan structure/activity relationships modulating dynamic biological functions. Trends Cardiovasc. Med. 12 (2), 71–77.

Singelyn, J.M., DeQuach, J.A., Seif-Naraghi, S.B., Littlefield, R.B., Schup-Magoffin, P.J., Christman, K.L., 2009. Naturally derived myocardial matrix as an injectable scaffold for cardiac tissue engineering. Biomaterials 30 (29), 5409–5416.

Singelyn, J.M., Sundaramurthy, P., Johnson, T.D., Schup-Magoffin, P.J., Hu, D.P., Faulk, D.M., Wang, J., Mayle, K.M., Bartels, K., Salvatore, M., Kinsey, A.M., Demaria, A.N., Dib, N., Christman, K.L., 2012. Catheter-deliverable hydrogel derived from decellularized ventricular extracellular matrix increases endogenous cardiomyocytes and preserves cardiac function post-myocardial infarction. J. Am. Coll. Cardiol. 59 (8), 751–763.

Spinale, F.G., 2007. Myocardial matrix remodeling and the matrix metalloproteinases: influence on cardiac form and function. Physiol. Rev. 87 (4), 1285–1342.

Stevens, K.R., Kreutziger, K.L., Dupras, S.K., Korte, F.S., Regnier, M., Muskheli, V., Nourse, M.B., Bendixen, K., Reinecke, H., Murry, C.E., 2009a. Physiological function and transplantation of scaffold-free and vascularized human cardiac muscle tissue. Proc. Natl. Acad. Sci. USA 106 (39), 16568–16573.

Stevens, K.R., Pabon, L., Muskheli, V., Murry, C.E., 2009b. Scaffold-free human cardiac tissue patch created from embryonic stem cells. Tissue Eng. Part A 15 (6), 1211–1222.

Streeter Jr., D.D., Spotnitz, H.M., Patel, D.P., Ross Jr. J., Sonnenblick, E.H., 1969. Fiber orientation in the canine left ventricle during diastole and systole. Circ. Res. 24 (3), 339–347.

Swartz, D.D., Russell, J.A., Andreadis, S.T., 2005. Engineering of fibrin-based functional and implantable small-diameter blood vessels. Am. J. Physiol. Heart Circ. Physiol. 288 (3), H1451–H1460.

Tandon, N., Marsano, A., Maidhof, R., Wan, L., Park, H., Vunjak-Novakovic, G., 2011. Optimization of electrical stimulation parameters for cardiac tissue engineering. J. Tissue Eng. Regen. Med. 5 (6), e115–125.

Tee, R., Lokmic, Z., Morrison, W.A., Dilley, R.J., 2010. Strategies in cardiac tissue engineering. ANZ J. Surg. 80 (10), 683–693.

Tee, R., Morrison, W.A., Dusting, G.J., Liu, G.S., Choi, Y.S., Hsiao, S.T., Dilley, R.J., 2012. Transplantation of engineered cardiac muscle flaps in syngeneic rats. Tissue Eng. Part A 18 (19–20), 1992–1999.

Tous, E., Purcell, B., Ifkovits, J.L., Burdick, J.A., 2011. Injectable acellular hydrogels for cardiac repair. J. Cardiovasc. Transl. Res. 4 (5), 528–542.

Vandervelde, S., van Luyn, M.J., Tio, R.A., Harmsen, M.C., 2005. Signaling factors in stem cell-mediated repair of infarcted myocardium. J. Mol. Cell. Cardiol. 39 (2), 363–376.

Vanhoutte, D., Schellings, M., Pinto, Y., Heymans, S., 2006. Relevance of matrix metalloproteinases and their inhibitors after myocardial infarction: a temporal and spatial window. Cardiovasc. Res. 69 (3), 604–613.

Weigel, T., Schinkel, G., Lendlein, A., 2006. Design and preparation of polymeric scaffolds for tissue engineering. Expert Rev. Med. Devices 3 (6), 835–851.

Weinberg, C.B., Bell, E., 1986. A blood vessel model constructed from collagen and cultured vascular cells. Science 231 (4736), 397–400.

WHO World Health Organization - the Atlas of Heart Disease and Stroke. http://www.who.int/cardiovascular_diseases/resources/atlas/en/.

Williams, D.F., 1999. The Williams Dictionary of Biomaterials. Liverpool University Press, Liverpool, UK.

Yang, X., Pabon, L., Murry, C.E., 2014. Engineering adolescence: maturation of human pluripotent stem cell-derived cardiomyocytes. Circ. Res. 114 (3), 511–523.

Yao, L., Swartz, D.D., Gugino, S.F., Russell, J.A., Andreadis, S.T., 2005. Fibrin-based tissue-engineered blood vessels: differential effects of biomaterial and culture parameters on mechanical strength and vascular reactivity. Tissue Eng. 11 (7–8), 991–1003.

Zacchigna, S., Giacca, M., 2014. Extracellular and intracellular factors regulating cardiomyocyte proliferation in postnatal life. Cardiovasc. Res. 102 (2), 312–320.

Zeng, W., Yuan, W., Li, L., Mi, J., Xu, S., Wen, C., Zhou, Z., Xiong, J., Sun, J., Ying, D., Yang, M., Li, X., Zhu, C., 2010. The promotion of endothelial progenitor cells recruitment by nerve growth factors in tissue-engineered blood vessels. Biomaterials 31 (7), 1636–1645.

Zhang, W.J., Liu, W., Cui, L., Cao, Y., 2007. Tissue engineering of blood vessel. J. Cell. Mol. Med. 11 (5), 945–957.

Zheng, W., Wang, Z., Song, L., Zhao, Q., Zhang, J., Li, D., Wang, S., Han, J., Zheng, X.L., Yang, Z., Kong, D., 2012. Endothelialization and patency of RGD-functionalized vascular grafts in a rabbit carotid artery model. Biomaterials 33 (10), 2880–2891.

Zimmermann, W.H., Didie, M., Doker, S., Melnychenko, I., Naito, H., Rogge, C., Tiburcy, M., Eschenhagen, T., 2006. Heart muscle engineering: an update on cardiac muscle replacement therapy. Cardiovasc. Res. 71 (3), 419–429.

Zmora, S., Glicklis, R., Cohen, S., 2002. Tailoring the pore architecture in 3-D alginate scaffolds by controlling the freezing regime during fabrication. Biomaterials 23 (20), 4087–4094.

CHAPTER 19

Tissue Engineering of Organ Systems

Jason M. Sandberg, Anthony Atala
Department of Urology, Wake Forest University School of Medicine, Winston-Salem, NC, USA

LEARNING OBJECTIVES

- To understand the basic anatomy and physiology of the organs involved in tissue engineering
- To understand organ failure and the respective diseases causing it
- To understand the importance of cell–cell and cell–matrix interaction in tissue-engineered organs
- To be aware of the importance of adequate mechanical properties of the matrix in organ tissue engineering
- To understand the importance of using the human body as a "bioreactor" to allow final tissue differentiation
- To know the limiting factors for successful in vivo implantation of in vitro engineered organs
- To take note of the most recent biologic constructs in development and their current clinical applications

19.1 INTRODUCTION

Organ replacement, repair, or restoration of organ function is classically treated by organ or tissue transplantation to replace diseased, missing, or malformed organs. In cases where congenital malformation can lead to severe organ agenesis; dysplasia and hypoplasia; or anatomical, structural, and functional deficiencies, it might be necessary to replace the entire or part of the organ. In other cases, cancer, trauma, infection, inflammation, or iatrogenic injuries may lead to organ damage or loss and dictate eventual replacement. Yet three major problems are linked to human organ donation: tissue compatibility, tissue rejection, and donor shortage. Depending on the type of organ, donors and receivers need to be immunologically matched to minimize the risk of rejection, and the donor tissue needs to be of the highest quality to provide the best chance of obtaining long-term function, while for some vital and delicate organs such as pancreas, the number of available donor organs is scarce.

Tissue Engineering.
Copyright © 2015 Elsevier Inc. All rights reserved.

In some organ systems, these problems can be minimized by reconstruction with tissue from other organ systems when native tissue is lacking. For example, in the genitourinary tract, surgeons can use nonurologic tissues, such as intestinal or gastric segments, mucosa from several body sites, and skin with homologous tissues (e.g., cadaver fascia, cadaver, or donor kidney) to reconstruct vital structures. Other options include biopolymers (e.g., bovine collagen) or synthetic polymers (e.g., silicone, polyurethane or polytetrafluoroethylene (PTFE), poly(glycolic acid) (PGA), poly(lactic acid), and poly(lactic-*co*-glycolic acid)), among others. These approaches are, however, often linked to complications, such as metabolic disturbances and even cancer, and rarely replace the entire function of the original tissues because of their inherently different functional parameters.

Although engineering of two-dimensional (2D) and very small three-dimensional (3D) tissue structures has been achieved, construction of true 3D functional tissue structures, acting as organ systems, is still a big challenge. This is even more so if the organ consists of a multitude of different cell types, each having a particular function and influencing each other. Such engineered constructs depend on very complex cell carrier matrices, often combined with biological factors, such as morphogens and extracellular matrix (ECM) components. This supports the respective cell types by promoting vascularization and allowing adequate nutritional support—a key issue in large 3D constructs.

In tissue engineering, biomaterials should act as artificial ECMs, and mimic biologic and mechanical functions of native ECM as much as possible. Biomaterials should also allow controlled delivery of bioactive factors (e.g., cell adhesion peptides, growth factors) and should offer a 3D space for the development of new tissues with appropriate structure and function (Kim and Mooney, 1998). In other words, the optimal biomaterial should be biocompatible, promote cellular interaction and tissue development, and possess proper mechanical and physical properties and act as a delivery vehicle to control the localization of transplanted cells (Brittberg et al., 1994).

Three main types of biomaterials can be used for the engineering of bioartificial organs: naturally derived matrices (e.g., collagen and alginate), acellular tissue matrices (e.g., acellular bladder submucosa and small intestinal submucosa (SIS)), and synthetic polymers. Naturally derived materials and acellular tissue matrices have the advantage of being biologically recognized, but their production needs to be closely controlled and monitored to maintain good and reproducible quality. Synthetic polymers, however, can be reproducibly manufactured in large quantities with good quality and standardized physical properties.

In many cases, a cell source is necessary to fully replace the organ's metabolic, exocrine, or endocrine functions. The source of donor tissue can be xenogeneic (from another species), allogeneic (same species, different individual), or autologous (same individual). Autologous cells are the preferred source as they avoid rejection from the host and preclude the use of immunosuppressive drugs. However, many patients with a specific organ disease no longer posses the specific healthy tissues or cells that should be used for culture expansion and subsequent transplantation. In these instances, donor cells or tissues are required to serve as an appropriate cell source. Another cell type with great potential applicable to many different organs could be the use of stem cells. It is now known that human embryonic stem cells (hESCs), for example, are capable of self-renewal, proliferating in an undifferentiated state while maintaining their pluripotent properties. We also know that we can differentiate these cells into almost any specialized human tissue cell type when providing the correct culture conditions. Researchers are making progress in directing these and other stem cells toward specific cell lineages with the goal of regenerating healthy organs that can replace diseased ones without depending on the availability of donor cells.

According to Folkman and Hochberg (1973), engineered tissues exceeding a volume of 3 mm^3 cannot survive adequately because of impaired nutrition and gas exchange. Vascularization of the regenerating cells is mandatory to engineer large complex tissues.

Progress toward this goal is being achieved by incorporation of angiogenic growth factors into the engineered tissue, by seeding endothelial cells in combination with the other cell types (including stem cells) and by prevascularization prior to cell seeding. There is no doubt that the problem of tissue construct vascularization has to be solved before large entire solid organs can be regenerated.

19.2 UROGENITAL TISSUE ENGINEERING

The urogenital system comprises the urinary tract and the internal and external genital apparatus. The urinary tract is subdivided into the upper tract (comprising the kidney and ureter) and the lower tract (comprising the bladder and urethra). In the female, the ovaries, the uterus, and the vagina belong to the internal genital apparatus. The external apparatus comprises the introitus, in the female the minor and major labia, and in the male the testicles and the penis, and, in particular, the corporal structures (Figure 19.1).

19.2.1 Bladder

Intestinal and, less frequently, gastric segments are normally used as tissues for bladder replacement or repair. However, intestinal tissues are designed to absorb, and when they come into contact with the urinary tract, they also absorb toxic substances normally eliminated with the urine. Therefore,

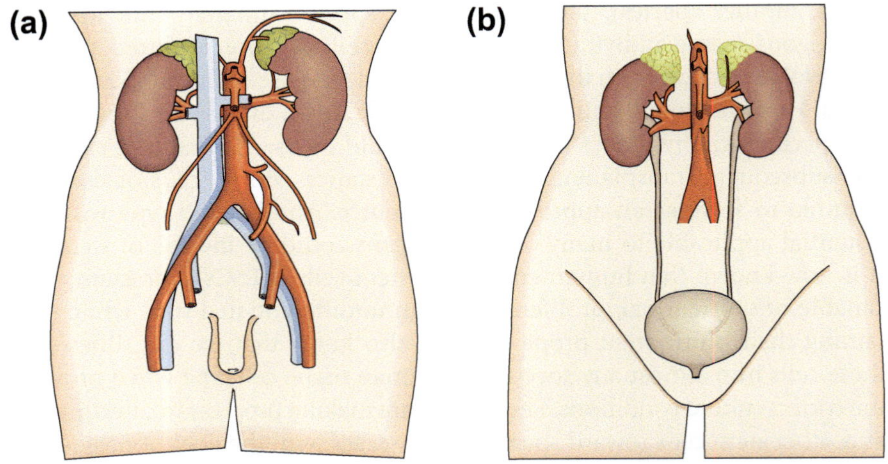

FIGURE 19.1
Schematic representation of the human urogenital system. (a) Vascular anatomy (b) Upper urinary tract.

multiple complications may ensue, such as infection, metabolic disturbances, increased mucus production, stone formation, and even malignancies (Kaefer et al., 1997). The use of gastric tissue may induce metabolic disturbances or even perforation. Owing to the problems encountered with the use of gastrointestinal segments, numerous investigators have attempted alternative reconstructive procedures for bladder replacement or repair, such as the use of seromuscular grafts, matrices for tissue regeneration, and tissue engineering with cell transplantation.

Porcine SIS, a biodegradable, acellular, xenogeneic predominantly collagen-based tissue-matrix graft, was first used in the early 1980s as a matrix for tissue replacement in the vascular field. It has been shown to promote the regeneration of a variety of host tissues, including blood vessels and ligaments (Badylak et al., 1989). Animal experiments have shown that the noncell-seeded SIS matrix used for bladder augmentation is able to regenerate a bladder wall in vivo (Kropp et al., 2004). Allogeneic acellular bladder matrices can also serve as scaffolds for host cellular bladder wall components. The matrices are prepared by mechanically and chemically removing all cellular components from the donor bladder tissue a method called "decellularization" (Yoo et al., 1998). The ECMs serve as vehicles for partial bladder regeneration and relevant graft versus host reaction is not evident.

In several studies using various materials as nonseeded grafts for surgical repair of the bladder, the urothelial layer was able to regenerate normally, but the muscle layer, although present, was not adequately developed (Kropp et al., 2004; Yoo et al., 1998). The grafts contracted to 60–70% of their original size

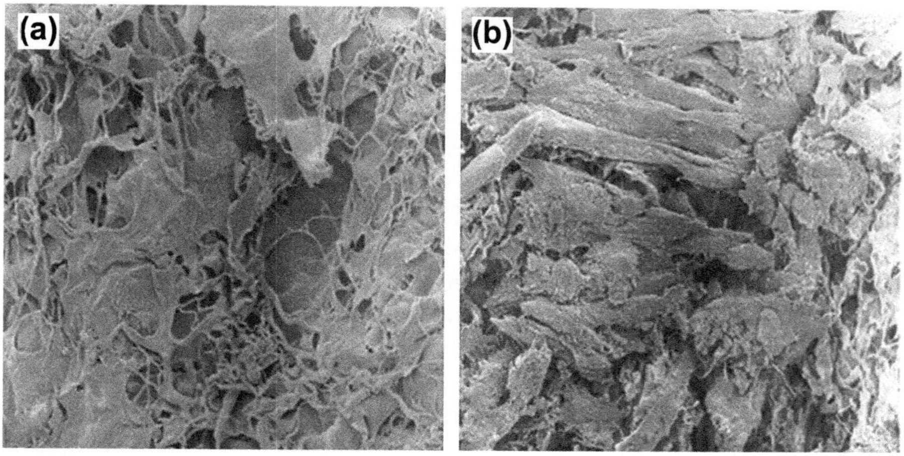

FIGURE 19.2

Scanning electron microscopy of native equine collagen matrix (TissuFleece®) (a) and with human smooth muscle cells cultured for 10 days (b).

(Portis et al., 2000) with little increase in bladder capacity or compliance (Landman et al., 2004). Several studies of acellular matrices indicated that they might provide the necessary environment to promote cell migration, adherence, growth, and differentiation (Hubschmid et al., 2005; Danielsson et al., 2006); with continued research, these matrices may have a clinical role in bladder replacement in the future (Figure 19.2).

In the bladder, the trigone is a smooth triangular region of the inner urinary bladder formed by the two ureteral orifices and the internal urethral orifice. Cell-seeded allogeneic acellular bladder matrices have been used for bladder augmentation in dogs. A group of experimental dogs underwent a trigone-sparing cystectomy and were randomly assigned to one of three groups. One group underwent closure of the trigone without a reconstructive procedure, another underwent reconstruction with a nonseeded bladder-shaped biodegradable scaffold, and the last underwent reconstruction using a bladder-shaped biodegradable scaffold that delivered seeded autologous urothelial cells and smooth muscle cells (Oberpenning et al., 1999).

The cystectomy-only and nonseeded controls maintained average capacities of 22% and 46% of preoperative values, respectively. However, in the cell-seeded tissue-engineered bladder replacements, an average bladder capacity of 95% of the original precystectomy volume was achieved. The compliance of the cell-seeded tissue-engineered bladders was also superior. Histologically, the nonseeded scaffold bladders presented a pattern of normal urothelial cells, however, with a thickened fibrotic submucosa and a thin layer of smooth muscle fibers. The retrieved tissue-engineered bladders showed a normal cellular organization, consisting of a trilayer of urothelium, submucosa, and smooth muscle (Oberpenning et al., 1999). Applying this technology, Atala et al. (2006) performed cystoplasty in seven patients with myelomeningocele with high-pressure, or poorly compliant bladders (State-of-the-Art Experiment). Researchers followed this study with a multicenter phase 2 clinical trial to evaluate bladder capacity, compliance, and safety in 10 pediatric patients with neurogenic bladder secondary to spina bifida (Joseph et al., 2014). Unfortunately, no statistically significant improvement in bladder capacity or compliance was observed, and four patients sustained significant adverse events including bowel obstruction and/or bladder rupture.

19.2.2 Urethra

Bladder-derived acellular collagen matrix has also proven to be a suitable graft for repair of urethral defects, and newly created neourethras have shown normal urothelial lining and organized muscle bundles in a rabbit model (Chen et al., 1999). These results were confirmed clinically in a series of patients with a history of failed hypospadias repair wherein the urethral defects were repaired with human bladder acellular collagen matrices (Atala et al., 1999). The neourethral tissue was created by connecting the matrix to the urethral plate in an onlay fashion. The size of the created neourethra ranged from 5 to 15 cm. After a 3-year follow-up, three of the four patients had a successful outcome in regard to cosmetic appearance and function. One patient who had a 15-cm neourethra created developed a subglandular fistula. Similar results were obtained in pediatric and adult patients with primary urethral stricture disease using the same collagen matrices (El-Kassaby et al., 2003) (Figure 19.3). The described techniques, using nonseeded acellular matrices, have been applied experimentally and clinically with varying success for only urethral repairs by using SIS as a ventral onlay graft for small defects. However, when tubularized urethral repairs were attempted experimentally, adequate functional urethral tissue regeneration was not achieved, and complications, predominantly graft contracture, and stricture formation, were noticed (Le Roux, 2005).

On the contrary, seeded tubularized collagen matrices have performed better in animal studies. In a rabbit model, urethroplasty was performed with tubularized collagen matrices either seeded with cells or without cells. The matrices seeded with autologous cells formed new tissue, which was histologically comparable to native urethra. However, the tubularized collagen matrices without cells led to limited tissue development, fibrosis, and stricture formation. Further, reliable regeneration and repair in subsequent studies could only be achieved at up to 0.5 cm (Dorin et al., 2008).

Still, the ultimate goal remains to provide fully engineered tissues to patients requiring complete substitution urethroplasties, such as those with long strictures or significant penile trauma. Poor success was reported after initial

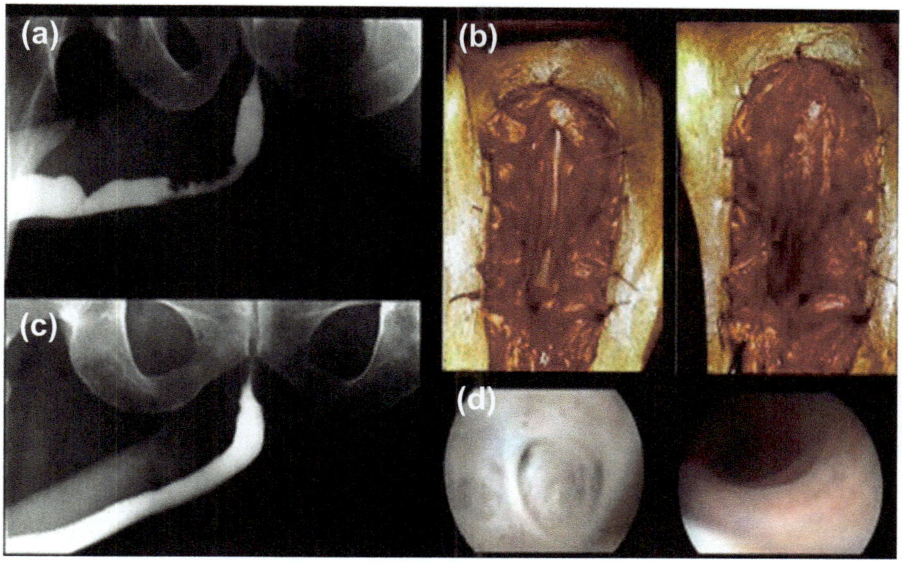

FIGURE 19.3

(a) Representative case of a patient with a bulbar stricture. (b) During the urethral repair surgery, strictured tissue is excised; this preserves the urethral plate on the left side, and the matrix is anastomosed to the urethral plate in an onlay fashion on the right side. Both photographs indicate the area of interest, including the urethra, which appears white in the left photograph. The left photograph, indicates the area of stricture in the urethra. The right photograph indicates the repaired stricture. Note that the engineered tissue now obscures the native white urethral tissue in an onlay fashion in the right photograph. (c) Urethrogram 6 months after repair. (d) Cystoscopic view of the urethra before surgery on the left side, and 4 months after repair on the right side.

attempts in five patients with significant stricture secondary to lichen sclerosis. Keratinocytes and fibroblasts isolated from buccal mucosal biopsies were seeded on acellular dermal grafts used for complete substitution tubularized urethral replacement (Bhargava et al., 2008).

Following this attempt, a study reported the successful implantation of bioengineered urethras into five boys suffering severe strictures (Raya-Rivera et al., 2011). Open bladder biopsies harvested 1-cm^2 full-thickness tissues that were divided into urothelium and smooth muscle, and then cultured separately in the laboratory. These autologous tissues were then seeded onto the intraluminal and outer surfaces of a tubularized PGA acellular matrix for treatment of defects 4–6 cm in length. Of the patients, 100% demonstrated continence with normal flow rates and continence at a median follow-up of nearly 5 years. Further, cellularized implants did not exhibit severe fibrosis, inflammation, or stricture formation, and demonstrated histologically normal tissue on follow-up biopsies. Work in this area continues; recent success demonstrated that lengths of 6 cm were achievable in a canine model by using autologous bladder epithelium on which smooth muscle cells were seeded, and that selecting a combination of the appropriate cell type and biomaterial could lead to a successful clinical outcome (Orabi et al., 2013).

19.2.3 Kidney

Kidney transplantation is superior to dialysis in treating patients with end-stage renal disease (ESRD). However, transplants expose patients to the risk of developing infection, rejection, and on-going costs for medical care. The current demand for kidney transplantation is also substantially greater than that of organ supply; this puts transplantation out of reach for many patients who require it. The alternative is the commonly used dialysis to replace absent renal function, but this is associated with a relatively high morbidity and mortality, and only partially replaces healthy kidney function.

The kidney is surely the most challenging organ in the urinary system to be reconstructed by tissue engineering methods because of its complex structure and function (Figure 19.4). Unlike hollow or viscous organs, the kidney's solid architecture presents the unique challenge of engineering several structural and functional capabilities into a single organ. The challenge is to include the normal kidney's one million highly specialized nephrons working in concert to drain into a single renal pelvis. Two approaches have flourished in creating a bioartificial kidney: some researchers are pursuing the replacement of isolated kidney function parameters with the use of extracorporeal cellular units, while others are aiming at the replacement of total renal function by tissue-engineered bioartificial structures.

Some investigators have sought alternative solutions involving ex vivo cellular systems, in an attempt to assess the viability and physiologic functionality of a cell-seeded device to replace the filtration, transport, and metabolic and endocrinologic functions of the kidney. The initial clinical experience suggests that

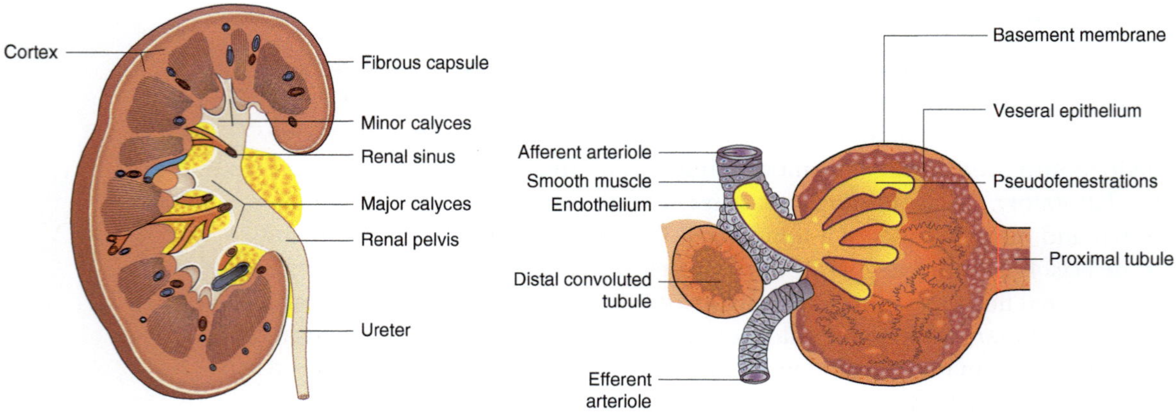

FIGURE 19.4

The complex renal structure. A cross-section of the whole kidney (left) and the glomerular apparatus (right).

renal tubule cell therapy may provide a dynamic and individualized treatment program as assessed by acute physiologic and biochemical indices (Humes et al., 2003).

Two recent studies have demonstrated the possibility of kidney decellularization–recellularization technology for the creation of an implantable bioartificial organ. The first study demonstrated the ability to decellularize intact rat kidneys in a manner that preserved the intricate architecture and seeded them with pluripotent murine embryonic stem cells (ESs) antegrade through the artery or retrograde through the ureter (Ross et al., 2009). In this study, cells seeded onto the scaffolds demonstrated the immunohistochemical markers of differentiation including epithelization, while cells that did not contact the scaffold became apoptotic and formed a potential lumen. The second study in a rhesus monkey model demonstrated the capacity of the scaffold to support $Pax2^+$/vimentin$^+$ cell attachment and migration to recellularize the scaffold after layering with fetal kidney explants (Nakayama et al., 2010).

Follow-up studies using scaffolds and cell explants from different age groups (fetal, juvenile, and adult) showed the extent of cellular repopulation was the greatest with scaffolds from the youngest donors, and with seeding of mixed fetal renal aggregates that formed tubular structures within the kidney scaffold (Nakayama et al., 2011). Even further investigation into both kidney and lung scaffolds using proteomic analysis and seeding with human embryonic stem cells (hESCs) suggests that decellularized scaffolds have an intrinsic spatial ability to influence hESC differentiation by physically shaping cells into tissue-appropriate structures and phenotypes (Nakayama et al., 2013).

Future applications will include the evaluation of bioengineered kidneys in humans with ESRD. In addition, planning for a phase 1 clinical trial to evaluate safety and efficacy of a neokidney composed of autologous renal cells formulated in a gelatin-based hydrogel is underway.

19.3 LIVER TISSUE ENGINEERING

The liver is a very complex organ that has the ability to regrow and regenerate (Figure 19.5). However, liver failure can still occur due to many different causes, including acquired liver disease, inborn errors of metabolism, or drug toxicity. Currently, when a liver fails, the lost function can only be replaced fully by liver transplantation. The options include transplantation of the whole liver or parts of the liver from cadaveric donors, portions of the liver from living donors, or xenotransplantation, typically porcine. Unfortunately, there are far too few livers available for the need worldwide, with approximately five times as many recipients as donors (Pruett, 2002), and xenotransplantation is limited by immune and infectious complications.

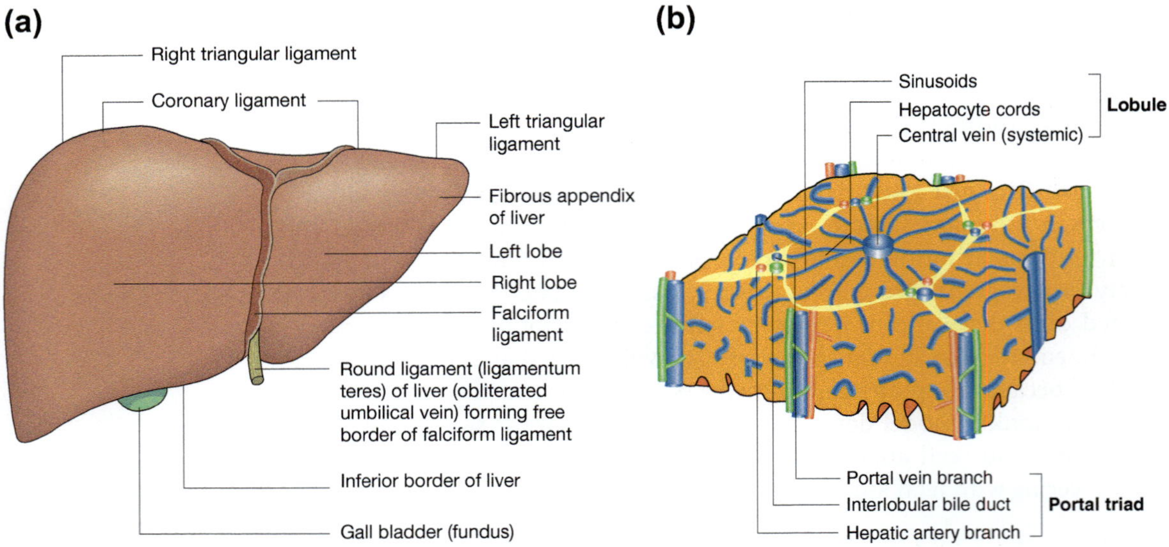

FIGURE 19.5
Liver structure. (a) Surfaces and bed of liver (b) Hepatic architecture.

The simplest method of liver failure treatment using hepatocytes is cell transplantation. The first liver cell transplantation was described in 1979, with the successful transplant of viable rat hepatocytes into the spleen in an animal study (Mito et al., 1979). This and other studies taught us valuable lessons about the importance of the microenvironment to cell survival, as the cells required conditions resembling those of the liver, including a matrix to grow on and a venous blood supply similar physiologically to the hepatic sinusoid to grow and function normally. They also demonstrated the role of nonhepatocyte cells in hepatocyte survival, as coculture with nonparenchymal cells was often required to ensure cell viability (Selden and Hodgson, 2004). Unfortunately, clinical experience shows that many patients treated with hepatocytes for acute liver failure die of sepsis and most trials investigating the treatment of fulminate liver failure with hepatocyte transplantation have given equivocal results (Selden and Hodgson, 2004).

For cells to be useful in transplantation, they must have a high proliferative capacity without loss of function and they must be economically feasible. There are several groups of possible cell sources for liver replacement. These include primary hepatocytes, immortalized cell lines, stem cells, and transdifferentiated cells (Selden and Hodgson, 2004). Primary hepatocytes are the most desirable and commonly used cells. Unfortunately, hepatocyte function is closely tied to the orientation, location, and surrounding environment of the cells. These environmental characteristics are lost in culture, so these cells rapidly and irreversibly lose their function. Therefore, novel methods of cell culture, or alternative sources of cells, must be developed to obtain the necessary hepatocytes by perhaps mimicking the appropriate microenvironment (Selden and Hodgson, 2004).

Human hepatocytes would be the best choice for cell transplantation, but just as liver donors are limited, so are human hepatocytes. Fetal or neonatal hepatocytes may also provide a source of cells. They have finite proliferation capacity, but may be immortalized (Selden and Hodgson, 2004). Another source of hepatocytes is stem cells. Stem cells have an unlimited replication potential and can differentiate into various different cell types. Potential sources of stem cells for hepatocyte generation include ES cells, adult liver progenitor cells, and transdifferentiated nonhepatic cells. ES cells offer great hope, but of course, there are ethical and legal concerns with ES cell use. Also, transdifferentiation of adult nonhepatic cells may also provide a novel cell source for liver replacement therapy (Baccarani et al., 2005). To this end, a small clinical trial investigating autologous transplantation of adult mesenchymal stem cells (MSCs) in patients with cirrhosis has demonstrated an improvement in liver and kidney functions (Kharaziha et al., 2009). As a practical treatment, however, this method has limitations. The small scale of this study (eight patients) and short-term follow-up did not allow for the analysis of a survival benefit. Further, preparation of harvested MSCs required greater than two months in culture, during which time many patients to potentially benefit may die. Another trial using hematopoietic stem cells in patients with viral or autoimmune hepatitis demonstrated similar results (Salama et al., 2010).

Alternatives to human cells include primary porcine hepatocytes. These cells are plentiful and readily available, and have similar functional characteristics as those of human cells (Barshes et al., 2005). This strategy is not without its limitations, as there are possibilities of rejection and infectious complications such as xenozoonosis (e.g., porcine endogenous retroviral infection). To ensure a clinically safe method of transplantation of such cells, immunoprotective strategies are necessary.

Cell encapsulation is one strategy to prevent rejection; it involves the encapsulation of single cells or groups of cells by using specific hydrogels, nanometer-thick coatings, or thin membranes of a nondegradable biomaterial. It does not prevent the development of antibodies to secreted proteins that cross the encapsulation barrier, but does prevent access of host immune cells to the transplanted cells. Systemic immunosuppressants are also useful if needed, but cotransplanting cells with Sertoli cells, or MSCs, can also provide additional immunosuppression locally, or at least dampen the immune response. Sertoli cells are interesting since they are central to the immune privileged environment in the testes. Therefore, coculture of cells with Sertoli cells may provide a level of protection from the host immune response. Preliminary experiments have been performed in which encapsulation of Sertoli cells and hepatocytes in microporous polypropylene, acetonitrile methallyl sulfonate copolymers, and alginate with poly-L-lysine demonstrated improvements in cell survival time and hepatic cellular function (Zheng et al., 2009).

19.3.1 The Bioartificial Liver

Another strategy of liver replacement in the setting of liver failure is bioartificial liver (BAL) replacement. As an evolution of the artificial liver devices, BAL constructs were developed to harness the function of hepatocytes in an extracorporeal system. BAL devices have been under development for >40 years and usually consist of nearly 15 billion isolated hepatocytes with a membrane through which the plasma is circulated (Park and Lee, 2005). There are already several examples of BALs currently undergoing clinical trials, which include ELAD, HepatAssist®, BLSS, AMC-BAL, MELS, RFB, and HBAL/TECA-HALSS. Unfortunately, only two clinical trials that explored the effectiveness of BALs have been reported, and it is clear that many challenges remain before widespread dissemination for clinical use can be achieved (Zhao et al., 2012).

The creation of a complete, implantable bioengineered liver remains the ultimate goal of tissue engineering for liver replacement. Many strategies have been explored as a means to achieve this goal, such as the injection of hepatocytes into vascular beds or the use of implantable seeded scaffolds. One of the main limitations of bioengineering large organs such as the liver is vascularization to provide cells embedded in such constructs with adequate nutrient supply ensuring a proper physiological response. Prior to transplantation, researchers must have a fully functional bioengineered liver, as it must immediately perform all hepatic functions. Vacanti's group used a prevascularized sponge device that was fabricated and seeded with adequate cells in vitro, and then implanted in vivo. Another strategy is to use an implantable scaffold with direct access to the portal blood. One example used SIS grafted between the portal vein and inferior vena cava as a venous conduit, which could be seeded with hepatocytes. Microelectromechanical systems are another strategy, using silicon microfabrication technology to form large sheets of liver tissue, which can then be folded into 3D structures (Kulig and Vacanti, 2004).

Animal studies have shown that the liver architecture remains viable after stripping and can be repopulated with hepatocytes. Hepatic functional data using this strategy was demonstrated for the first time in rats (Uygun et al., 2010). As early as 2011, Soker and Atala were able to generate a fully vascularized liver organoid after decellularization of livers across multiple species (Baptista et al., 2011). After rescue with human fetal liver and endothelial cells, these in vitro constructs were able to withstand fluid flow through the vascular network and generated significantly higher concentrations of urea and albumin than equivalent hepatic cells grown in culture dishes. Badylak's group subsequently endorsed similar results, and provided encouragement that these methods may one day provide real therapeutic benefit to patients (Soto-Gutierrez et al., 2011). Other researchers have been able to successfully generated nonimmunogenic, large-scale scaffolds of clinically significant size from porcine livers (Mirmalek-Sani et al., 2013). As a conglomerate, these studies and others like them are ushering in the next phase of research in regenerating a viable liver construct for therapeutic use in humans by overcoming the challenges of oxygen and nutrient delivery, waste removal, and immunologic rejection. Researchers still must overcome challenges associated with establishing an ideal cell source, maintaining in vivo-like oxygen conditions, achieving bile secretory function, increasing convenience for the patient, and establishing efficacy of the devices overall.

19.4 GASTROINTESTINAL TISSUE ENGINEERING

The current best therapy for replacing diseased or absent intestine is organ transplantation, aimed at restoring intestinal function for adequate nutrient absorption, fluid and electrolyte transport, and restoration of metabolic homeostasis. This involves an extensive operative procedure with its own morbidity and mortality, along with the well-described side effects of chronic systemic immunosuppression. There are also the concerns of inadequate numbers of organ donors and rejection of the transplanted tissue. An excellent alternative to current therapies would be to regenerate intestine using tissue engineering techniques that have been successfully applied to other organs (Chen and Badylak, 2001; Chen and Beierle, 2004).

Early attempts at tissue engineering bowel segments involved patch repairs of intestinal defects. As far back as Kobold and Thal (1963), who used jejunal serosa to repair duodenal defects in dogs. These patches were covered by duodenal epithelium in 2 months. Binnington et al. (1973) repaired jejunal defects in animals with the serosa of the descending colon. The new mucosal covering was noted to grow in from the edges, which covered the serosa. Expanding their work, they used a model of "short bowel syndrome" in pigs to show that the jejunal patch was able to induce increased weight gain (Chen and Badylak, 2001). These "patch" experiments were repeated using different models and strategies, such as colonic serosa or abdominal wall pedicle flap patches in rabbits, with similar results (Chen and Beierle, 2004). The studies concluded that these patches were all able to support the growth of functional neomucosa.

19.4.1 The Intestinal Mucosa

All these studies share the need for intestinal mucosa to grow and spread across the patch surface. Therefore, the regulation of intestinal growth is an area of research that is crucial to the advancement of bowel tissue engineering. Numerous biomolecules have been studied and proven to have an effect on intestinal mucosal growth. Basic fibroblast growth factor (bFGF) delivered to smooth muscle in a mouse model was shown to maintain the viability of small intestinal implanted sphincteric smooth muscle (Orlando et al., 2012). Similar results were reported with the use of platelet-derived growth factor as applied to bioengineered internal anal sphincter constructs (Miyasaka et al., 2011).

Early work involving biomaterials focused on synthetic scaffolds such as Dacron and PTFE to repair intestinal defects in animals. However, these strategies suffered from extrusion into the bowel wall and incomplete neomucosal coverage (Harmon et al., 1979; Watson et al., 1980). The studies demonstrated that selecting the ideal biomaterial is important and that biocompatibility and biodegradability are crucial for a successful in vivo outcome.

19.4.2 Natural Biomaterials for Intestinal Repair

Acellular and natural scaffolds have recently been found to be more appropriate for bowel repair. An example of a widely used acellular biomaterial is SIS. SIS is primarily composed of ECM material, following the removal of the mucosa, external muscle layers, and other cells from the small intestine of pigs. The naturally occurring ECM is rich in components that support angiogenesis (i.e., fibronectin; glycosaminoglycans; collagen types I, III, IV, and V; bFGF; and vascular endothelial growth factor). SIS has also been shown to be nonimmunogenic (Demirbilek et al., 2003). SIS has been extensively investigated as a scaffold for in numerous tissue engineering applications, and is now processed in a standardized fashion in mass quantities; this makes it readily available for tissue engineering studies (Chen and Badylak, 2001).

SIS has been used to cover defects in the esophagus and small bowel in dogs. In these studies, the SIS patch allowed the mucosa to grow over the scaffold, and the histology of the patch was similar to the native intestine (Chen and Badylak, 2001). In another study, tubular noncell containing SIS grafts from Sprague–Dawley rat donors were used to create an ileostomy (Figures 19.6). These grafts were initially covered by inflammatory cells, but by 3 months, they were completely covered by mucosa. Also, the outsides of the tubular grafts were covered with bundles of smooth muscle-like cells and fibrovascular tissue. A similar study performed in the stomach showed similar results. In this study, rats underwent the removal of a full-thickness 1-cm segment of the stomach, with a patch repair using a double-layer patch of porcine-derived SIS. Three weeks postoperatively, the site was covered by granulation tissue and fibrosis, with some growth of normal gastric mucosa at the edges of the graft (de la Fuente et al., 2003). These studies demonstrated that SIS can develop structural features of the normal intestine and may provide an off-the-shelf option for replacement patches for intestinal tissue (Wang et al., 2005). Other studies using various other novel scaffold materials, such as fibrin glue and PGA as a framework for tissue engineering neointestine, have had promising results as well. Hori et al. (2001) used a collagen sponge enzymatically

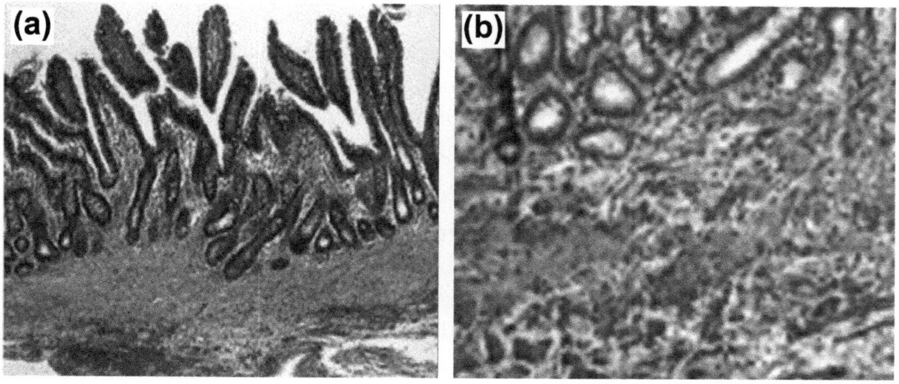

FIGURE 19.6
Histologic photomicrographs of SIS-regenerated rat small bowel tissue at 12 weeks. (a) Well-organized mucosal epithelial layer with smooth muscle cells regeneration. (b) Immunohistochemically stained intestinal tissue suggesting bundles of well-formed smooth muscle fibers with less regularity.

processed from porcine skin as a scaffold for regeneration of stomach and small bowel with good success in a beagle model. Isch et al. (2001) have also performed esophagoplasty by patching defects in the cervical esophagus in a dog with AlloDerm® (human decellularized skin).

19.4.3 Combining Biomaterials with Cells for Intestinal Repair

The key components for this process include a cell source, and an appropriate biocompatible and biodegradable scaffold. Early work in intestinal tissue engineering by Vacanti used just such a strategy. Cells were obtained from an intestinal organoid unit; a scaffold of sheets of collagen-coated, microporous, nonwoven PGA was formed into a tubular shape, and these tubes once seeded with cells were implanted into the omenta of animals (Vacanti, 2003). Vacanti's group harvested cells from intestinal organoids, seeded these on the constructs, and after subsequent implantation, they found it decreased morbidity associated with bowel resections in rats (Grikscheit et al., 2004). Before this study was conducted, Tait et al. (1994) had already described these organoids to be excellent for intestinal tissue engineering because of their unique composition, which includes progenitor cells, epithelium, and mesenchymal stroma. What was very interesting to see is that the implanted devices demonstrated angiogenesis from the omental vessels, crucial for the long-term survival of the constructs. After additional examination in animal studies, the constructs were found to be quite similar to normal small bowel and were anastomosed to the native bowel without causing feeding problems (Chen and Beierle, 2004; Rocha and Whang, 2004). More recently, smooth muscle cells on PGA sheets, a synthetic approach, in combination with fibroblastic keratinocytes have been used to successfully replace portions of the esophagus (Nakase et al., 2008). However, such constructs have not demonstrated physiologic functionality; this shows again that a proper combination of cells and biomaterials is critical for a good outcome. Although these implants have now demonstrated basic metabolic function, one important aspect is still missing, namely, integration with the enteric nervous system that is crucial for a normal bowel motility.

There are however indications that nerve innervation, or at least the stimulation of nerve ingrowth, is possible by using the right cell type. Notable success has been achieved with the implantation and subsequent neovascularization of bioengineered internal anal sphincters in murine models. Implanted constructs have demonstrated maintenance of basal tone and sphincteric relaxation similar to native sphincters (Raghavan et al., 2011). The group described their method to bioengineer functionally innervated gastrointestinal smooth muscle constructs using neuronal progenitor cells and smooth muscle cells that were isolated and cultured from intestinal tissues of adult human donors (Gilmont et al., 2014). This breakthrough in bioengineering functional gastrointestinal tissue in vitro provides not only a useful construct for the examination of the disease processes of the bowel but also a possible "off-the-shelf" source of intestinal tissues with physiological and metabolic functions for clinical use.

19.5 PANCREAS TISSUE ENGINEERING

Diabetes mellitus is a common disease, which affects 4% of the world, and causes significant morbidity and mortality through its effects on vascular/microvascular disease. Diabetes is due to systemic hyperglycemia, resulting from inadequate insulin production by the pancreas and/or resistance to insulin (Kahn, 2004).

The insulin-producing β cells can be found in the pancreatic islets, or islets of Langerhans. The pancreas is composed of acinar cells, ducts, and islets of endocrine cells. Pancreatic cells develop from endodermal progenitor cells. The islets contain a specific set of endocrine cells comprising the β cells, α cells, which produce glucagon, δ cells, which produce somatostatin, ε cells, which produce ghrelin, and PP cells, which produce pancreatic polypeptide. These endocrine cells all work in unison and stimulate, or inhibit, each other through paracrine effects. Although the islets only take up a very small portion (1–2%) of the total pancreas volume, they receive 10–15% of the blood flow that increases when blood glucose levels are elevated. Each islet therefore is densely vascularized, and islets cells are never further away than 100 μm from a blood vessel to provide them with sufficient amounts of nutrients and oxygen and to efficiently release insulin into the bloodstream (Figure 19.7).

There are two main types (a third has been described) of diabetes, types I and II, and they are caused by islet β cell dysfunction, with an absolute or relative

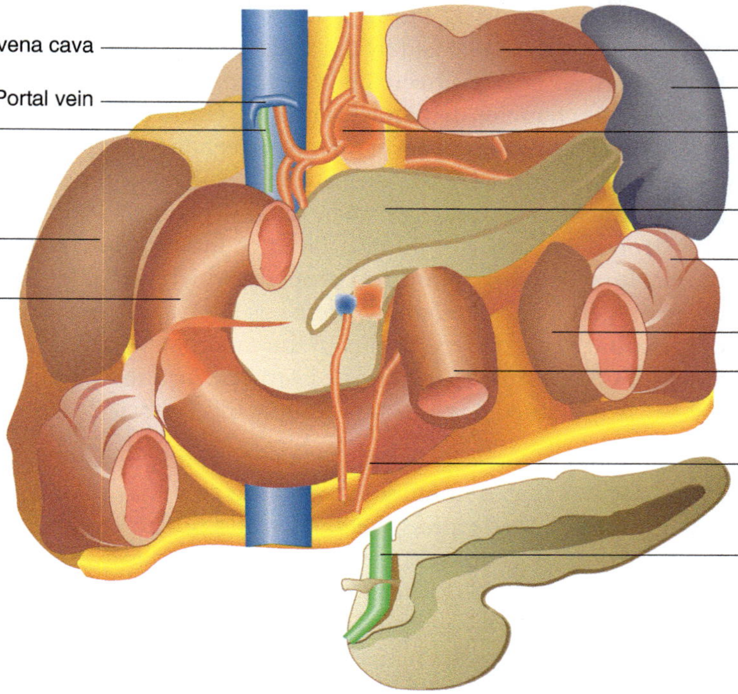

FIGURE 19.7
The pancreas anatomy.

decrease in insulin production. Type I diabetes results from the autoimmune destruction of β islet cells. Type II diabetes is characterized by systemic insulin resistance, an inability of the β cell to rapidly release sufficient insulin in response to oral nutrient stimulation, and to impaired glucose tolerance. Also, β cells may not be able to convert proinsulin, which is the precursor in the cell for forming insulin (Kahn, 2004).

The most effective therapy for replacing lost β cell function, since the discovery of insulin by Banting and Best in 1921, has been insulin replacement therapy (Bliss, 1982). The ability to monitor blood glucose levels and provide patients with insulin on a daily basis multiple times a day works for many patients today. However, it is still very difficult to match insulin levels appropriately to the blood glucose concentration in real time. Scientists therefore continue to search for new therapies to improve the treatment of diabetes mellitus, such as β cell replacement, which seeks to replace the deficient endogenous insulin production especially for type I diabetes patients.

Designing of new β cells involves two approaches: gene therapy and cell replacement strategies. Engineered β cells need to be able to modulate basal insulin release for normal events such as exercise or infection, and respond to continuous changing blood glucose levels. Keep in mind that the islets of Langerhans are innervated by the sympathetic, parasympathetic, and sensory nerve fibers, so that engineered β cells need to respond to neural input. Over life, there are changing demands on insulin release that are the result of changes in insulin sensitivity, such as in puberty, pregnancy, and old age. There is quite a challenging list of criteria an in vitro engineered β cell should obey. The cells should be able to increase their replicative capacity and involute when no longer needed. Cells should also produce insulin and appropriately process proinsulin to insulin to function properly, which is highly dependent on adequate oxygen supply. Preferably, there has to be a way to recreate the influence of gastrointestinal derived factors in the modulation of insulin release (Kahn, 2004). β Cells normally release the hormone insulin in a pulsatile fashion, so the newly generated β cells should mimic this behavior.

A major concern for scientists in β cell engineering is the recurrence of disease. Type I diabetes has an immunological cause, so it may recur with in vitro differentiated cells, if they express the antigens leading to an autoimmune reaction. The causes of type II diabetes are not as clear, so recurrence could be a possibility here as well (Kahn, 2004). Therefore, in vitro generated β cells may require immunosuppression, and these agents (specifically glucocorticoids) can cause insulin resistance, suppress insulin release, and prevent insulin processing. They also have other more severe risks, such as malignancy. Also, proinsulin may pose a cardiovascular risk, so proper processing of the hormones is essential (Kahn, 2004).

19.5.1 Creating New β Cells for Cell Therapy in Type I Diabetes

Based on the aforementioned challenging checklist, the cells developed must have certain physiologic functions, such as insulin gene expression, appropriate posttranslational modification of insulin, glucose-sensitive insulin production, and, ideally, be available in an unlimited supply to be used as cell therapy for diabetes mellitus (Kahn, 2004). All these characteristics are typical of β cells, but this does not mean that one needs β cells to replace them; just cells that act like them would suffice. For example, modification of genes in cells that have similar origins to pancreatic cells may produce β cell-like qualities, such as the transfer of the *Pdx-1* gene into mouse liver cells. These modified liver cells can activate gene expressing of insulin 1 and 2 and the prohormone convertase, but unfortunately are not sensitive to glucose in the environment. It has also been demonstrated that ductal cells also, which comprise the ducts in the pancreas through which digestive enzymes are secreted, can be modified to produce insulin if modified with gene transfection (Kahn, 2004). Zhou et al. have demonstrated the ability to reprogram fully differentiated adult exocrine cells into insulin secreting β-like cells (Zhou et al., 2008). In this study, developmental transcription factors including *Pdx-1*, *Ngn3*, and *MafA* were delivered using a viral vector via injection into the pancreas, and over time, acinar cells began to exhibit an endocrine rather than exocrine phenotype. An alternative approach is to start with cells that already have some β cell characteristics and to modify these as needed to obtain the desired function. For example, Cheung et al. modified intestinal K cells, which normally secrete peptides in response to insulin secretion, using a transgenic model to express insulin in response to glucose sensation and these cells were able to restore normoglycemia to streptozotocin-induced diabetic mice (Cheung et al., 2000) (Figure 19.8).

But what about a stem cell approach for replacement of exogenous insulin therapy? The goal of stem cell therapy, as with all studies of multipotent cells in regenerative medicine, is the directed differentiation of these cells and the

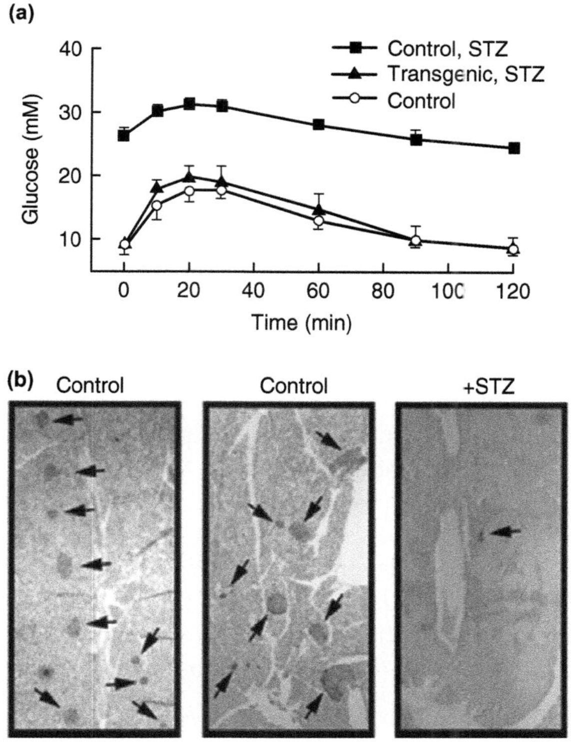

FIGURE 19.8

Production of human insulin from K cells protects transgenic mice from diabetes induced by destruction of pancreatic cells. (a) Oral glucose tolerance tests. Mice were given intraperitoneal injection of streptozotocin (STZ, 200 mg/kg) or an equal volume of saline. On the fifth day after treatment, after overnight food deprivation, glucose (1.5 g/kg body weight) was administered orally by feeding tube at 0 min. Results are means (6SEM) from at least three animals in each group. (b) Immunohistochemical staining for mouse insulin in pancreatic sections from control mice and an STZ-treated transgenic mouse. Arrows indicate mouse islets.

introduction of specific characteristics to these cells via genetic manipulation. There are several different choices of cells with which to work in this field of study, such as ES cells and adult stem cells. However, it is still somewhat unclear if there is a pool of adult stem cells in the pancreas and whether the duct cells can act as islet progenitors, or even if β cells can renew themselves (Kahn, 2004). But perhaps an answer will be found in the near future; a study by Seaberg et al. (2004) described the isolation of pancreas-derived multipotent precursor cells. They isolated these cells by dissociating and plating pancreatic ductal and islet tissue. These tissues formed floating colonies typical of neurospheres. These colonies expressed markers such as PDX-1 and Nestin, characteristic of both pancreatic and neural precursors (Seaberg et al., 2004). It was discovered that these cells could not only develop into neurons, astrocytes, and oligodendrocytes but they also produce insulin-producing β-like pancreatic cells. They were found to be glucose responsive, with increased insulin release in response to insulin. These findings are very encouraging, but require considerably more research to prove clinical utility.

Is there a role for stem cells in diabetes cell therapy? Perhaps, there has been some recent success in exploring strategies using ES cells and other stem cell types that can now be fully differentiated into cells with functional characteristics of native β cells. For example, some researchers have been using precursors, such as adipose-derived stem cells, capable of glucose regulation in mice (Chandra et al., 2009). Yet, it has been considerably challenging to achieve a full β cell phenotype in vitro, and multiple studies have suggested the transplant site milieu to be an important factor necessary for terminal differentiation, including the effect of in situ signals (Orlando et al., 2012). Notable success was achieved when multipotent cells of both adult progenitor (bone marrow) and blastocyst (ES) origin showed β-like potential (Kumar et al., 2013). These rat stem cells were cultured and differentiated ex vivo into β-like cells. On day 21, 20% of progeny expressed Pdx-1 and C-peptide and secreted C-peptide under the stimulus of insulin agonist carbachol. When implanted into diabetic nude mice for 4 weeks, hypoglycemia was reversed in roughly half of the subjects with hyperglycemic recurrence within 24 h of graft removal, which showed that at least to a certain extent differentiated stem cells could be used to treat type 1 diabetes.

Similar to tissue engineering of other organ systems, there is much interest in using scaffolds and/or tissue constructs to regenerate a 3D organ that mimics the natural environment of the native endocrine pancreas compartment. It may even prove necessary for efficient terminal differentiation (Wang and Ye, 2009). Multiple groups have explored this option in rodents with some success, including the constitution of viable vascular networks and insulin secretion for several weeks (Orlando et al., 2012). Although relatively new in the field of tissue engineering, there is clearly an increasing trend toward using regenerative medicine- and bioengineering-based strategies for the creation of a bioartificial organ that mimics the endocrine function of the native pancreas (Forster et al., 2011; Dufour et al., 2005; Buitinga et al., 2013; Yap et al., 2013; Pedraza et al., 2012; Gibly et al., 2011; Blomeier et al., 2006).

19.6 LUNG TISSUE ENGINEERING

As in other fields currently being investigated by tissue engineers, organ transplantation is the preferred treatment and standard of care for organ failure for the lung. The lungs might seem to comprise a relative simple anatomy, but in fact contain an elaborate network of branching bronchi, and ultimately end in very small alveoli that are covered by a thin layer of squamous epithelium that permits diffusion of oxygen into the dense microvascular network lining the epithelium on the inside. The trachea is lined by rings of cartilage to support the tubular structure that otherwise would collapse on itself (Figure 19.9).

The demand for human organs for transplantation is large and growing, and the supply of donor lungs (as with most transplanted organs) does not meet the demand. Developing technologies for lung replacement include stem cell biology, therapeutic cloning, tissue and organ bioengineering, and xenotransplantation (Ogle et al., 2004).

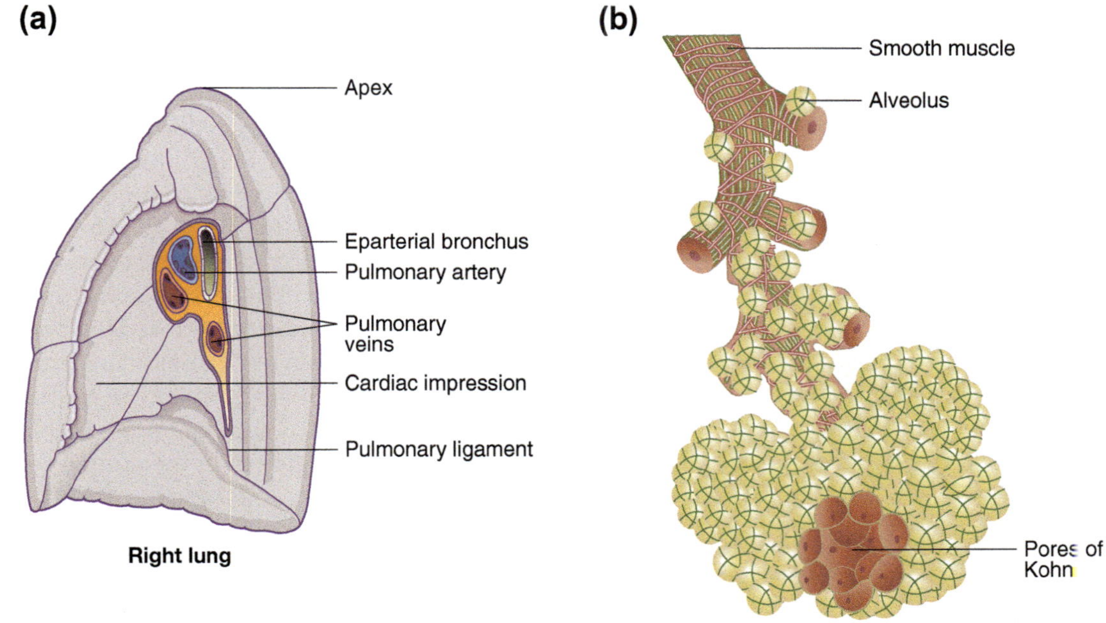

FIGURE 19.9
Anatomy of the lung.

Studies have demonstrated that ES cells can be induced to differentiate in vitro into pulmonary epithelial cells, type II pneumocytes, using a specific medium designed for the maintenance of lung epithelial cells in culture (Rippon et al., 2004). The generation of differentiated airway epithelial tissue had not been reported, however, until Coraux et al. also demonstrated that type II alveolar epithelial cells could be derived from ES cells in a murine model (Coraux et al., 2005). They went on to show that murine ES cells were able to differentiate into nonciliated secretory Clara cells and that type I collagen induced this development. In addition, when these cells were cultured at an air–liquid interface, ES cells were able to form fully differentiated airway epithelium. The cells were composed of basal, ciliated, intermediate, and Clara cells, such as found in the tracheobronchial airway epithelium (Coraux et al., 2005). Another method of inducing the development of ES cells along the pulmonary epithelial lineage is the use of coculture techniques. Coculture of ES cells from a murine model with embryonic mesenchyme from the distal lung promoted the development of pneumocytes. The technique involved differentiating the stem cells first into embryoid bodies (EBs) and then culturing these EBs for one to two weeks with embryonic pulmonary mesenchyme. The EBs demonstrated the development of channels lined with pulmonary epithelial cells including type II pneumocytes. Even more exciting was the observation in some animal studies that ES cells delivered systemically to animals with pulmonary damage seeded at the site of the diseased tissue and developed pulmonary phenotypes. These results encourage the search for an ideal use of ES cells as some form of cell therapy for pulmonary disease (Van Vranken et al., 2005).

The final goal of all this research is the ability to regenerate portions of the lung for replacement of damaged tissue. Clinically, the experience of this scale of bioengineering, while limited, is progressing. In one reported case, a bioengineered airway patch was created for a patient who developed a dehiscence of his tracheobronchial anastomosis following a right completion carinal pneumonectomy. The patch was created by harvesting autologous muscle cells and fibroblasts from the patient, culturing and expanding these cells, and plating them on a collagen matrix derived from decellularized porcine jejunum. The patch was placed over an airway defect, satisfactorily closed the opening, and was immediately airtight. The graft demonstrated neovascularization and excellent coverage with functional and viable ciliated respiratory epithelium. The cellularized patch worked well in this instance, although it is possible that an acellular patch could be used, as there was ingrowth of the normal respiratory epithelium over the patch (Macchiarini et al., 2004) (Figure 19.10).

The above-mentioned procedure has been improved over the last decade, and the investigators have extended their reach to multiple patients. They have shortened the time to obtain tracheal decellularization by using an alternative cell technological approach (Bader and Macchiarini, 2010). The improved approach involves the decellularized human tracheal graft being reseeded

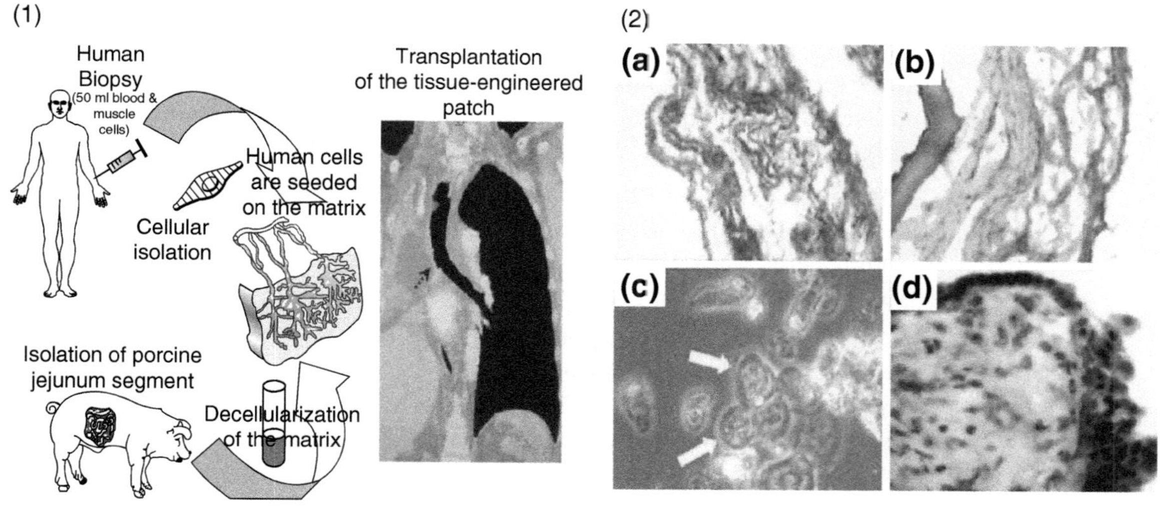

FIGURE 19.10

Engineering process (1) and histologic composition (2) of the bioartificial patch. Muscle cells and fibroblasts are isolated from a biopsy specimen obtained from the patient. These cells are seeded on a biologic matrix representing a collagen network generated from a decellularized porcine jejunal segment. During the incubation period, the cells start to remodel the xenogeneic matrix and replace it with autologous connective tissue. Within 4 weeks, this autologous bioartificial implant can be clinically used. The computed tomographic scan of the chest 6 weeks after graft implantation shows the site (arrow) where the tissue-engineered patch was transplanted. (a) Acellular porcine collagen matrix (hematoxylin and eosin staining). Notice the absence of cellular components and collagen fibers. (b) Reseeded patch before implantation. Antivimentin immunohistology depicting fibroblast and smooth muscle cell seeding. (c) Brush cytology from the patch surface showing the respiratory epithelium (12th week). The arrows indicate epithelial cilia. (d) Immunohistochemical staining for respiratory epithelium (12th week) (Villin antibody).

intraoperatively with autologous cells (bone marrow MSCs and respiratory cells) and conditioned with growing, differentiation, and "boosting" factors. This in vivo tissue-engineered approach was used in five patients with benign tracheal diseases and in two patients with primary tracheal cancers involving the entire trachea (Kalathur et al., 2010). The aforementioned therapy has improved not only lung function for these patients many months after surgery but also the overall quality of life.

19.7 FUTURE DEVELOPMENTS

Today's strategy for tissue engineering predominantly depends upon autologous cells from the diseased organ of the patient. In the case of extensive end-stage organ failure, a tissue biopsy may not yield enough normal cells for culture and ultimate transplantation. Primary autologous human cells cannot always be expanded from a particular organ, as is the case with the pancreas. In these cases, pluripotent human ES cells could become a source of cells from which the desired tissue can be derived. Embryonic stem cells exhibit two remarkable properties: the ability to proliferate in an undifferentiated, but pluripotent state and the ability to differentiate into many specialized cell types. Although these cells have great therapeutic potential, their use is currently limited by biological and ethical factors alike. ES cells are an area of continued fertile research and one day may prove to be an extremely powerful cell source for tissue engineering.

Adult stem cells are ethically more acceptable and have the advantage of not transdifferentiating spontaneously into a malignant phenotype. Because some adult stem cells are notoriously difficult to expand into large quantities, their use for clinical therapy in these cases is limited.

A relatively recent development, induced pluripotent stem cells (iPSCs) represent an exciting alternative to both ES cells and adult stem cells. These cells are derived from adult cells, but can be reprogrammed via genetic engineering to exhibit the pluripotency that was at one time unique to ES cells. For this reason, iPSCs are considered more ethically acceptable than ES cells and hold potential to overcome many of their biological limitations, including those associated with immunogenicity. The development of iPSCs for the use in tissue engineering is an area of fertile research.

Fetal stem cells, harvested from amniotic fluid and placentas, may represent another novel source of stem cells. These cells express markers consistent with human ES cells, such as octamer-binding transcription factor 4 and stage-specific embryonic antigen 4. They are multipotent and are able to differentiate into all three germ layers, and they do not form teratomas, nor do they spontaneously transdifferentiate into a malignant phenotype. A further advantage is that the cells have a high replicative potential and they could be stored for future possible use in the patient, without the risks of rejection and ethical concerns (Siddiqui and Atala, 2004).

Further, growth factors and other signaling molecules alone or in combination with biological or synthetic matrices may be used for cell proliferation and differentiation, tissue guidance, and regeneration, and thus replace cell-based tissue engineering. Once the growth factors to be delivered have been identified, there are several important criteria to take into consideration. First, the delivery must be controlled, and allow sustained and localized delivery of small amounts of growth factors to the tissue repair site, as systemic administration can cause toxicity in other organs. Further, the release profile must be controlled in ensuring frequent exposure to these factors for a relatively long time to obtain the desired effect. Growth factors typically have short half-lives once they are introduced into the body and are eliminated rapidly. Finally, the process of incorporating the factors must not involve harsh solvents or high temperatures that might denature or deactivate the proteins. To meet these criteria, the growth factors could be entrapped within or covalently linked to a polymer matrix and released by matrix degradation.

As the prenatal diagnosis of fetal abnormalities has become extremely accurate, theoretically, a small tissue biopsy could be obtained under ultrasound guidance. These tissues could then be processed and the various cell types could be cultured in vitro. Perhaps an interesting approach to prevent the onset of certain life-long diseases at a very early age would be to apply tissue engineering strategies, to engineer in vitro reconstituted structures of tissues, or organs, which could then be available for reconstruction or transplantation at the time of birth.

19.8 SUMMARY

- The engineering of simple hollow organs (e.g., the bladder) has been achieved, but the construction of more complex hollow organs (e.g., blood vessels) and functional solid organs (e.g., kidney, liver) is the main challenge facing the future of tissue engineering.
- Engineered bladder tissues, created with autologous cells seeded on collagen–PGA scaffolds and wrapped in omentum after implantation, are currently undergoing clinical trials in patients requiring bladder augmentation.
- Today's strategy for tissue engineering predominantly depends upon autologous cells from the diseased organ of the patient. In the case of extensive end-stage organ failure, a tissue biopsy may not yield enough normal cells for culture and ultimate transplantation.
- Primary autologous human cells cannot always be expanded from a particular organ, as is the case with the pancreas. In these cases, pluripotent human ES cells can become a source of cells from which the desired tissue can be derived.
- Nonseeded acellular matrices have functioned well clinically for onlay urethral repairs. However, when tubularized urethral repairs are performed, seeded tubularized collagen matrices are required.
- The kidney is surely the most challenging organ in the urinary system to be reconstructed by tissue engineering methods because of its complex structure and function. To develop a bioartificial kidney, some researchers are pursuing the replacement of isolated kidney function parameters with the use of extracorporeal cellular units, while others are aiming at the replacement of total renal function by tissue-engineered bioartificial structures.
- As in renal replacement strategies, the complexity of liver tissue engineering has driven some effort toward liver replacement therapies with BAL replacement, but so far, human studies have been equivocal. New advances in decellularization technology are moving researchers closer to viable in vitro constructs.
- Cellularized, and possibly acellular, segments of collagen scaffold material offer a possible source of tissue for patch repairs of the tracheobronchial tree.

- Studies into gastrointestinal tissue engineering have demonstrated that SIS, AlloDerm®, and other unseeded scaffolds can develop structural features of the normal intestine when implanted in the bowel, and may provide an off-the-shelf option for replacement patches for intestinal tissue. Additional studies have shown that cells obtained from intestines, seeded on a scaffold, and formed into a tubular shape developed tissues similar to those in the normal small bowel and have been successfully anastomosed to the native bowel in animal studies. More recent studies have demonstrated promising advances in enteric neuronal function, including internal anal sphincter constructs.
- The most effective therapy for diabetes, since the discovery of insulin by Banting and Best in 1921, has been insulin replacement therapy, but this therapy does not match insulin levels appropriately to the blood glucose concentration. The ultimate goal, therefore, is to develop a permanent endogenous replacement for the lack of insulin, and many tissue engineering strategies have been directed toward this goal.

CLASSICAL EXPERIMENT

After several reports in the past about successful culture of animal and human urothelial and bladder smooth muscle cells, it was only in the 1990s that Southgate et al. (1994) characterized in detail cultured human urothelial cells and were able to induce urothelial stratification in vitro. They summarized their research as follows:

The purpose of the work was to establish urothelium as an in vitro model for the study of proliferation, stratification, and differentiation in "complex" epithelia. Normal human urothelial cells were cultured in a serum-free medium. The effects of epidermal growth factor (EGF), cholera toxin (CT), extracellular calcium, and 13-*cis*-retinoic acid on cell growth, morphology, phenotype, and cytodifferentiation were studied using phase-contrast microscopy and indirect immunofluorescence. Stratification-related changes were additionally analyzed by transmission electron microscopy. Under optimized conditions, long-term cultures were successful in 44 (74.5%) out of 59 specimens. Bacterial infection was the most common cause of failure (nine cases).

Primary urothelial cells required an initial plating density of 1×10^4 cells/cm^2 for survival; passaged cells survived much lower plating densities (2.53×10^2 cells/cm^2). CT significantly improved cell attachment, but neither CT nor EGF was essential for growth. By contrast, cells failed to proliferate without bovine pituitary extract. In media containing bovine pituitary extract, CT, and EGF, cultures had a mean population doubling time of 14.7 6 1.8 h, maintained a nonstratified phenotype, and expressed the cytokeratin (CK) profile of basal/intermediate urothelium: CK7, CK8, CK17, CK18, and CK19, with variable expression of CK13. CK20 was not expressed in vitro. CK14 and CK16 were also expressed, suggestive of squamous metaplasia in culture, which could be inhibited with 13-*cis*-retinoic acid. Increasing extracellular calcium from 0.09 to 0.9–4.0 mM slowed down cell proliferation, induced stratification and desmosome formation, and increased expression of E-cadherin. High calcium, EGF, CT, and retinoic acid did not induce markers of late/terminal urothelial cytodifferentiation.

The above-mentioned study describes a simplified technique for the isolation and long-term culture of human urothelial cells. Urothelial cells in vitro are capable of rapid proliferation and can be induced to form integrated stratifying cell layers in high calcium medium.

STATE-OF-THE-ART EXPERIMENT

In this study, transplantable urinary bladder neoorgans were reproducibly created in vitro from urothelial and smooth muscle cells grown in culture from canine native bladder biopsies and seeded onto preformed bladder-shaped polymers. The native bladders were subsequently excised from canine donors and replaced with the tissue-engineered neoorgans. In functional evaluations for up to 11 months, the bladder neoorgans demonstrated a normal capacity to retain urine, normal elastic properties, and histological architecture. This study demonstrated, for the first time, that successful reconstitution of an autonomous hollow organ is possible using tissue engineering methods (Oberpenning et al., 1999). Patients with end-stage bladder disease can be treated with cystoplasty using gastrointestinal segments. The presence of such segments in the urinary tract has been associated with many complications. The authors explored an alternative approach using autologous engineered bladder tissues for reconstruction.

Seven patients with myelomeningocele, a birth defect in which the backbone and spinal canal do not close before birth leading to loss of bladder or bowel control, aged 4–19 years with high-pressure, or poorly compliant bladders, were identified as candidates for bladder repair. A bladder biopsy was obtained from each patient. Urothelial and muscle cells were grown in culture, and seeded on a biodegradable bladder-shaped scaffold made of collagen, or a composite of collagen and polyglycolic acid. About 7 weeks after the biopsy, the autologous engineered bladder constructs were used for reconstruction and implanted either with, or without an omental wrap. Serial urodynamics, cystograms, ultrasounds, bladder biopsies, and serum analyses were done. Follow-up range was 22–61 months with a mean of 46 months. Postoperatively, the mean bladder leak point pressure decrease at capacity, the volume, and compliance increase were the greatest in the composite engineered bladders with an omental wrap. Bowel function returned promptly after surgery. No metabolic consequences were noted, urinary calculi did not form, mucus production was normal, and renal function was preserved. The engineered bladder biopsies showed an adequate structural architecture and phenotype This study showed that engineered bladder tissues, createc with autologous cells seeded on collagen–polyglycolic acid scaffolds, and wrapped in omentum after implantation, can be used in patients who need cystoplasty (Atala et al., 2006) (Figures 19.11 and 19.12).

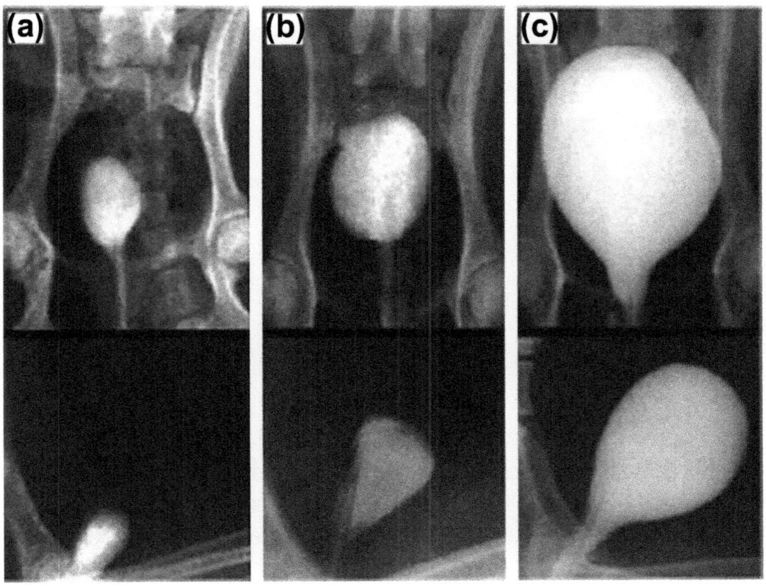

FIGURE 19.11

Radiographic cystograms 11 months after subtotal cystectomy. (a) Subtotal cystectomy without reconstruction (group A), (b) polymer-only implant (group B), and (c) tissue-engineered neoorgan (group C).

Continued...

STATE-OF-THE-ART EXPERIMENT *Continued*

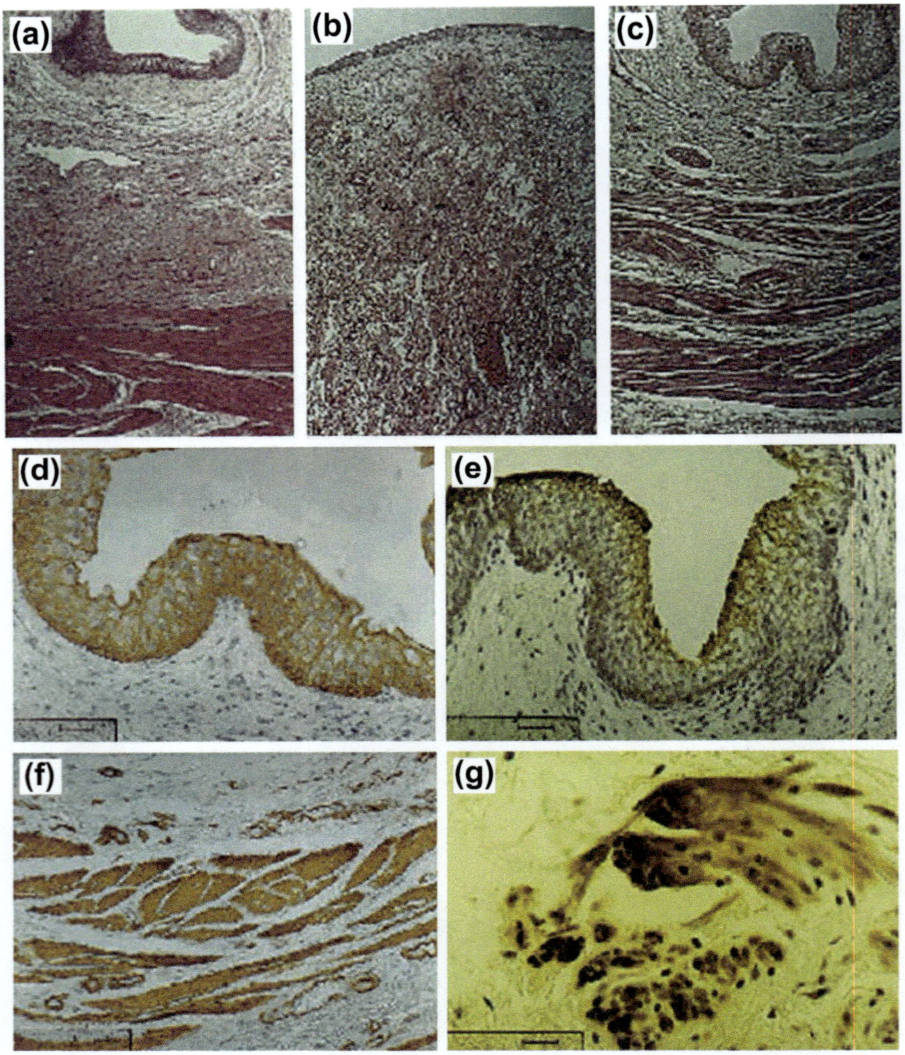

FIGURE 19.12

Histological and immunochemical analyses of implants 6 months after surgery. Hematoxylin and eosin histologic results are shown for (a) normal canine bladder (group A), (b) the bladder dome of the cell-free polymer reconstructed bladder (group B), and (c) the tissue-engineered neoorgan (group C). Immunocytochemical staining of tissue-engineered neoorgan revealed; (d) positive staining of the epithelial cell layers with pancytokeratin AE1/AE3 antibodies; (e) urothelial differentiation-related membrane proteins, which constitute the apical plaques of the asymmetric unit membrane in normal urothelium; (f) anti-α smooth muscle actin staining of phenotypically normal smooth muscle; and (g) positive staining with S-100 antibodies binding to neural tissue and nerve sheaths.

REFERENCES

Atala, A., et al., 1999. A novel inert collagen matrix for hypospadias repair. J. Urol. 162, 1148–1151.

Atala, A., et al., 2006. Tissue-engineered autologous bladders for patients needing cystoplasty. Lancet 367, 1241–1246.

Baccarani, U., et al., 2005. Human hepatocyte transplantation for acute liver failure: state of the art and analysis of cell sources. Transpl. Proc. 37, 2702–2704.

Bader, A., Macchiarini, P., 2010. Moving towards in situ tracheal regeneration: the bionic tissue engineered transplantation approach. J. Cell. Mol. Med. 14, 1877–1889.

Badylak, S.F., et al., 1989. Small intestinal submucosa as a large diameter vascular graft in the dog. J. Surg. Res. 47, 74–80.

Baptista, P.M., et al., 2011. The use of whole organ decellularization for the generation of a vascularized liver organoid. Hepatology 53, 604–617.

Barshes, N.R., et al., 2005. Support for the acutely failing liver: a comprehensive review of historic and contemporary strategies. J. Am. Coll. Surg. 201, 458–476.

Bhargava, S., et al., 2008. Tissue-engineered buccal mucosa urethroplasty-clinical outcomes. Eur. Urol. 53, 1263–1269.

Binnington, H.B., et al., 1973. A technique to increase jejunal mucosa surface area. J. Pediatr. Surg. 8, 765–769.

Bliss, M., 1982. Banting's, Best's, and Collip's accounts of the discovery of insulin. Bull. Hist. Med. 56, 554–568.

Blomeier, H., Zhang, X., Rives, C., Brissova, M., Hughes, E., Baker, M., et al., 2006. Polymer scaffolds as synthetic microenvironments for extrahepatic islet transplantation. Transplantation 82 (4), 452–459.

Brittberg, M., et al., 1994. Treatment of deep cartilage defects in the knee with autologous chondrocyte transplantation. N. Engl. J. Med. 331, 889–895.

Buitinga, M., Truckenmüller, R., Engelse, M.A., Moroni, L., Hoopen Ten, H.W.M., van Blitterswijk, C.A., et al., 2013. Microwell scaffolds for the extrahepatic transplantation of islets of Langerhans. PLoS ONE 8 (5), e64772.

Chandra, V., et al., 2009. Generation of pancreatic hormone-expressing islet-like cell aggregates from murine adipose tissue-derived stem cells. Stem Cells 27, 1941–1953.

Chen, F., et al., 1999. Acellular collagen matrix as a possible "off the shelf" biomaterial for urethral repair. Urology 54, 407–410.

Chen, M.K., Badylak, S.F., 2001. Small bowel tissue engineering using small intestinal submucosa as a scaffold. J. Surg. Res. 99, 352–358.

Chen, M.K., Beierle, E.A., 2004. Animal models for intestinal tissue engineering. Biomaterials 25, 1675–1681.

Cheung, A.T., et al., 2000. Glucose-dependent insulin release from genetically engineered K cells. Science 290, 1959–1962.

Coraux, C., et al., 2005. Embryonic stem cells generate airway epithelial tissue. Am. J. Respir. Cell Mol. Biol. 32, 87–92.

Danielsson, C., et al., 2006. Polyesterurethane foam scaffold for smooth muscle cell tissue engineering. Biomaterials 27, 1410–1415.

De La Fuente, S.G., et al., 2003. Evaluation of porcine-derived small intestine submucosa as a biodegradable graft for gastrointestinal healing. J. Gastrointest. Surg. 7, 96–101.

Demirbilek, S., et al., 2003. Using porcine small intestinal submucosa in intestinal regeneration. Pediatr. Surg. Int. 19, 588–592.

Dorin, R.P., et al., 2008. Tubularized urethral replacement with unseeded matrices: what is the maximum distance for normal tissue regeneration? World J. Urol. 26, 323–326.

Dufour, J., Rajotte, R., Zimmerman, M., Rezania, A., Kin, T., Dixon, D., et al., 2005. Development of an ectopic site for islet transplantation, using biodegradable scaffolds. Tissue Eng. 11, 1323–1331.

El-Kassaby, A.W., et al., 2003. Urethral stricture repair with an off-the-shelf collagen matrix. J. Urol. 169, 170–173. Discussion 173.

Folkman, J., Hochberg, M., 1973. Self-regulation of growth in three dimensions. J. Exp. Med. 138, 745–753.

Forster, N.A., Penington, A.J., Hardikar, A.A., Palmer, J.A., Hussey, A., Tai, J., et al., 2011. A prevascularized tissue engineering chamber supports growth and function of islets and progenitor cells in diabetic mice. Islets 3 (5), 271–283.

Gibly, R.F., Zhang, X., Graham, M.L., Hering, B.J., Kaufman, D.B., Lowe, W.L., et al., 2011. Extrahepatic islet transplantation with microporous polymer scaffolds in syngeneic mouse and allogeneic porcine models. Biomaterials 32 (36), 9677–9684.

Gilmont, R.R., et al., 2014. Bioengineering of physiologically functional intrinsically innervated human internal anal sphincter constructs. Tissue Eng. Part A. 20 (11–12), 1603–1611.

Grikscheit, T.C., et al., 2004. Tissue-engineered small intestine improves recovery after massive small bowel resection. Ann. Surg. 240, 748–754.

Harmon, J.W., et al., 1979. Fate of dacron prostheses in the small bowel of rabbits. Surg. Forum 30, 365–366.

Hori, Y., et al., 2001. Tissue engineering of the small intestine by acellular collagen sponge scaffold grafting. Int. J. Artif. Organs 24, 50–54.

Hubschmid, U., et al., 2005. In vitro growth of human urinary tract smooth muscle cells on laminin and collagen type I-coated membranes under static and dynamic conditions. Tissue Eng. 11, 161–171.

Humes, H.D., et al., 2003. Renal cell therapy is associated with dynamic and individualized responses in patients with acute renal failure. Blood Purif. 21, 64–71.

Isch, J.A., et al., 2001. Patch esophagoplasty using AlloDerm as a tissue scaffold. J. Pediatr. Surg. 36, 266–268.

Joseph, D.B., et al., 2014. Autologous cell-seeded biodegradable scaffold for augmentation cystoplasty: phase 2 study in children and adolescents with spina bifida. J. Urol 191 (5), 1389–1395.

Kaefer, M., et al., 1997. Continent urinary diversion: the children's hospital experience. J. Urol. 157, 1394–1399.

Kahn, S.E., 2004. Engineering a new beta-cell: a critical venture requiring special attention to constantly changing physiological needs. Semin. Cell. Dev. Biol. 15, 359–370.

Kalathur, M., et al., 2010. Translating tissue-engineered tracheal replacement from bench to bedside. Cell. Mol. Life Sci. 67, 4185–4196.

Kharaziha, P., et al., 2009. Improvement of liver function in liver cirrhosis patients after autologous mesenchymal stem cell injection: a phase I-II clinical trial. Eur. J. Gastroenterol. Hepatol. 21, 1199–1205.

Kim, B.S., Mooney, D.J., 1998. Development of biocompatible synthetic extracellular matrices for tissue engineering. Trends Biotechnol. 16, 224–230.

Kobold, E.E., Thal, A.P., 1963. A simple method for the management of experimental wounds of the duodenum. Surg. Gynecol. Obstet. 116, 340–344.

Kropp, B.P., et al., 2004. Reliable and reproducible bladder regeneration using unseeded distal small intestinal submucosa. J. Urol. 172, 1710–1713.

Kulig, K.M., Vacanti, J.P., 2004. Hepatic tissue engineering. Transpl. Immunol. 12, 303–310.

Kumar, A., et al., 2013. Reversal of hyperglycemia by insulin-secreting rat bone marrow- and blastocyst-derived hypoblast stem cell-like cells. PLoS ONE 8, e63491.

Landman, J., et al., 2004. Laparoscopic mid sagittal hemicystectomy and bladder reconstruction with small intestinal submucosa and reimplantation of ureter into small intestinal submucosa: 1-year follow up. J. Urol. 171, 2450–2455.

Le Roux, P.J., 2005. Endoscopic urethroplasty with unseeded small intestinal submucosa collagen matrix grafts: a pilot study. J. Urol. 173, 140–143.

Macchiarini, P., et al., 2004. First human transplantation of a bioengineered airway tissue. J. Thorac. Cardiovasc. Surg. 128, 638–641.

Mirmalek-Sani, S.H., et al., 2013. Immunogenicity of decellularized porcine liver for bioengineered hepatic tissue. Am. J. Pathol. 183, 558–565.

Mito, M., et al., 1979. Studies on ectopic liver utilizing hepatocyte transplantation into the rat spleen. Transpl. Proc. 11, 585–591.

Miyasaka, E.A., et al., 2011. In vivo growth of a bioengineered internal anal sphincter: comparison of growth factors for optimization of growth and survival. Pediatr. Surg. Int. 27, 137–143.

Nakase, Y., et al., 2008. Intrathoracic esophageal replacement by in situ tissue-engineered esophagus. J. Thorac. Cardiovasc. Surg. 136, 850–859.

Nakayama, K.H., et al., 2010. Decellularized rhesus monkey kidney as a three-dimensional scaffold for renal tissue engineering. Tissue Eng. Part A 16, 2207–2216.

Nakayama, K.H., et al., 2011. Renal tissue engineering with decellularized rhesus monkey kidneys: age-related differences. Tissue Eng. Part A 17, 2891–2901.

Nakayama, K.H., et al., 2013. Tissue specificity of decellularized rhesus monkey kidney and lung scaffolds. PLoS ONE 8, e64134.

Oberpenning, F., et al., 1999. De novo reconstitution of a functional mammalian urinary bladder by tissue engineering. Nat. Biotechnol. 17, 149–155.

Ogle, B., et al., 2004. Fusion of approaches to the treatment of organ failure. Am. J. Transpl. 4 (Suppl. 6), 74–77.

Orabi, H., et al., 2013. Cell-seeded tubularized scaffolds for reconstruction of long urethral defects: a preclinical study. Eur. Urol. 63, 531–538.

Orlando, G., et al., 2012. Regenerative medicine as applied to general surgery. Ann. Surg. 255, 867–880.

Park, J.K., Lee, D.H., 2005. Bioartificial liver systems: current status and future perspective. J. Biosci. Bioeng. 99, 311–319.

Pedraza, E., Coronel, M.M., Fraker, C.A., Ricordi, C., Stabler, C.L., 2012. Preventing hypoxia-induced cell death in beta cells and islets via hydrolytically activated, oxygen-generating biomaterials. Proc. Natl. Acad. Sci. USA 109 (11), 4245–4250.

Portis, A.J., et al., 2000. Laparoscopic augmentation cystoplasty with different biodegradable grafts in an animal model. J. Urol. 164, 1405–1411.

Pruett, T.L., 2002. The allocation of livers for transplantation: a problem of Titanic consideration. Hepatology 35, 960–963.

Raghavan, S., et al., 2011. Successful implantation of bioengineered, intrinsically innervated, human internal anal sphincter. Gastroenterology 141, 310–319.

Raya-Rivera, A., et al., 2011. Tissue-engineered autologous urethras for patients who need reconstruction: an observational study. Lancet 377, 1175–1182.

Rippon, H.J., et al., 2004. Initial observations on the effect of medium composition on the differentiation of murine embryonic stem cells to alveolar type II cells. Cloning Stem Cells 6, 49–56.

Rocha, F.G., Whang, E.E., 2004. Intestinal tissue engineering: from regenerative medicine to model systems. J. Surg. Res. 120, 320–325.

Ross, E.A., et al., 2009. Embryonic stem cells proliferate and differentiate when seeded into kidney scaffolds. J. Am. Soc. Nephrol. 20, 2338–2347.

Salama, H., et al., 2010. Autologous hematopoietic stem cell transplantation in 48 patients with end-stage chronic liver diseases. Cell. Transpl. 19, 1475–1486.

Seaberg, R.M., et al., 2004. Clonal identification of multipotent precursors from adult mouse pancreas that generate neural and pancreatic lineages. Nat. Biotechnol. 22, 1115–1124.

Selden, C., Hodgson, H., 2004. Cellular therapies for liver replacement. Transpl. Immunol. 12, 273–288.

Siddiqui, A., Atala, A., 2004. In: Lanza, R.A. (Ed.), Amniotic Fluid-derived Pluripotential Cells. Academic Press, Philadelphia, PA.

Soto-Gutierrez, A., et al., 2011. A whole-organ regenerative medicine approach for liver replacement. Tissue Eng. Part C Methods 17, 677–686.

Southgate, J., et al., 1994. Normal human urothelial cells in vitro: proliferation and induction of stratification. Lab. Invest. 71, 583–594.

Tait, I.S., et al., 1994. Progressive morphogenesis in vivo after transplantation of cultured small bowel epithelium. Cell. Transpl. 3, 33–40.

Uygun, B.E., et al., 2010. Organ reengineering through development of a transplantable recellularized liver graft using decellularized liver matrix. Nat. Med. 16, 814–820.

Vacanti, J.P., 2003. Tissue and organ engineering: can we build intestine and vital organs? J. Gastrointest. Surg. 7, 831–835.

Van Vranken, B.E., et al., 2005. Coculture of embryonic stem cells with pulmonary mesenchyme: a microenvironment that promotes differentiation of pulmonary epithelium. Tissue Eng. 11, 1177–1187.

Wang, X., Ye, K., 2009. Three-dimensional differentiation of embryonic stem cells into islet-like insulin-producing clusters. Tissue Eng. Part A 15, 1941–1952.

Wang, Z.Q., et al., 2005. Morphologic evaluation of regenerated small bowel by small intestinal submucosa. J. Pediatr. Surg. 40, 1898–1902.

Watson, L.C., et al., 1980. Small bowel neomucosa. J. Surg. Res. 28, 280–291.

Yap, W.T., Salvay, D.M., Silliman, M.A., Zhang, X., Bannon, Z.G., Kaufman, D.B., et al., 2013. Collagen IV-modified scaffolds improve islet survival and function and reduce time to Euglycemia. Tissue Eng. Part A 19 (21–22), 2361–2372.

Yoo, J.J., et al., 1998. Bladder augmentation using allogenic bladder submucosa seeded with cells. Urology 51, 221–225.

Zhao, L.-F., et al., 2012. Key challenges to the development of extracorporeal bioartificial liver support systems. Hepatobiliary Pancreat. Dis. Int. 11, 243–249.

Zheng, M.H., et al., 2009. Mixed microencapsulation of rat primary hepatocytes and sertoli cells improves the metabolic function in a D-galactosamine and lipopolysaccharide-induced rat model of acute liver failure. Cytotherapy 11, 326–329.

Zhou, Q., et al., 2008. In vivo reprogramming of adult pancreatic exocrine cells to beta-cells. Nature 455, 627–632.

CHAPTER 20

Organs-on-a-Chip

Chungmin Han[1], Joseph M. Labuz[2], Shuichi Takayama[2,3], Jaesung Park[1,4]

[1]Department of Mechanical Engineering, Pohang University of Science and Technology, Pohang, Gyeongbuk, Republic of Korea; [2]Department of Biomedical Engineering, College of Engineering, University of Michigan, Ann Arbor, MI, USA; Biointerfaces Institute, University of Michigan, NCRC, MI, USA; [3]Macromolecular Science and Engineering Center, College of Engineering, University of Michigan, Ann Arbor, MI, USA; [4]School of Interdisciplinary Bioscience and Bioengineering, Pohang University of Science and Technology, Pohang, Gyeongbuk, Republic of Korea

LEARNING OBJECTIVES

- Learn why one would want to construct organs on a chip
- Learn challenges associated with constructing organs on a chip
- Learn basic concepts of organs on a chip
- Learn about microfluidics
- Learn about specific organs on a chip

20.1 INTRODUCTION

Organs-on-a-chip are microfluidic cell and tissue culture systems that recapitulate key physiologies and functions of specific organs for further biological and medical purposes. Here, as in most disciplines of biology, the relationship between form and function remains paramount. Drawing from microengineering/nanoengineering (structure) and cell biology (function), major advances have been generated by understanding and harnessing this nuanced interplay (Park and Shuler, 2003). Microfluidic devices are commonly made of polydimethylsiloxane (PDMS) using soft lithography techniques (Xia and Whitesides, 1998). Due to conventional advantages (e.g., good cell culture compatibility, easy fabrication, and easy flow control) (Whitesides, 2006), these experimental platforms are capable of generating unique structural *topology* to reproduce key organ *functions* on a microscale chip. In turn, advances in life sciences provide a wealth of information on how the *function* of various biological units (e.g., an alveolus or a nephron) demands a specific *topology* to carry out its task. Such efforts are further supported by cell culture studies that enable us to isolate almost all types of cells and to characterize and culture them in vitro (Freshney, 2011). This

Tissue Engineering.
Copyright © 2015 Elsevier Inc. All rights reserved.

marriage of topology and function—which fundamentally mirrors our modern understanding of life—makes organ-on-a-chip studies remarkably compelling and powerful endeavors. This chapter will further examine the roots of organ-on-a-chip efforts, review recent advances in the area, and assess how these developments may impact the broader fields of science and medicine.

20.2 CONCEPT OF ORGAN-ON-A-CHIP

20.2.1 Motivation for Organ-on-a-Chip

The concept of organ-on-a-chip evolved from conventional in vitro cell culture technique that was developed to overcome the limitations of in vivo systems. Although in vivo systems, such as animal models, help us understand many biological phenomena, they are usually too complicated for mechanistic studies. For example, we might find a metastatic region in the lung a few weeks after a cancer cell injection, but a rat model cannot fully explain how cancer cells migrate to the lung, how they escape from the circulatory system, and how they invade into lung tissue. On the other hand, in vitro systems commonly have well-defined and well-controllable environments. In vitro culture environments allow us to control the types and numbers of cells as well as apply well-defined stimuli. In a dish, for instance, scientists can apply a growth factor and observe its effect on cells while tightly controlling the cell number, cell type, dose profile, and environmental conditions.

However, this simplicity can also mislead. Confounding factors, ancillary pathways, and other important elements may not be preserved from in vivo to in vitro studies (Abbott, 2003). Discrepancies between in vitro and in vivo drug test results provide an excellent illustration of this problem (Hartung and Daston, 2009). To overcome these limitations, modified in vitro cell culture techniques, such as trans-well insert culture and 3D extracellular matrix (ECM) culture, have been developed. These systems, however, are still too simple to mimic the complexity of in vivo systems and as such can only solve a small portion of these problems. Both the trans-well and 3D ECM cultures provide only limited numbers of compartmented spaces and can only be applied to static cell cultures. Conversely, cells in the body inhabit and interact across a plethora of diverse, dynamic niches. In short, in vivo systems are too complicated, and in vitro systems are too simple to study the details of biological phenomena. Therefore, well-defined, mechanistic studies of these processes demand model systems that bridge this divide (van der Meer and van den Berg, 2012).

Organ-on-a-chip technology is meant to fill this gap (Baker, 2011). Many kinds of organ-on-a-chip devices have been developed including blood vessels, lung, liver, and others (van der Meer and van den Berg, 2012). These devices capably reproduce select functions of specific organs and have been applied to answer important biological and medical questions (Roxanne, 2005). In the following section, more details about the unique advantages of organ-on-a-chip devices will be explained and, in the later sections, real organ-on-a-chip device examples and their characteristics will be further introduced.

20.2.2 The Unique Advantages of Organ-on-a-Chip

The two most well-recognized advantages of organ-on-a-chip devices are their ability to provide (1) physiologically reasonable and (2) spatiotemporally controllable microenvironments (El-Ali et al., 2006). Together, these benefits make the organ-on-a-chip concept a powerful tool for biological studies. In particular, the first advantage is greatly favorable when applied as a drug-testing platform. Conventional drug-testing techniques, such as in vitro culture and animal tests, often do not provide environments that are physiologically relevant to humans, whereas organ-on-a-chip does (Esch et al., 2011). The second advantage is greatly useful when it is applied to in situ mechanistic studies such as real-time cell migration research (Seker et al., 2011).

20.2.2.1 Drug-Testing Platform

During the twentieth century, substantial resources have been devoted to developing new drugs. The conventional way to develop a new drug is to, first, find as many drug candidates as possible based on the principles of biology and medicine and, second, screen out inappropriate candidates with cell and animal tests. The final, human tests are performed in clinical trials and are the most difficult steps to pass throughout the entire drug approval process. Only 11 percent of test-registered drugs, which are already tested with cells and animals, managed to pass clinical trials during 1991–2000. Inappropriate or insufficient preclinical testing platforms have been considered as the major cause of this low clinical success rate (Kola and Landis, 2004; Hartung and Daston, 2009).

In vitro cell-based drug tests are a good example of insufficient/inappropriate preclinical testing tools. These assays are usually performed using immortalized cell lines with characteristics different from the cells in our body and tissues consisting of very few types of cells (Pan et al., 2009). Results from drug tests with only limited types of cells lack heterotypic interaction—interactions between cells of different types—and therefore cannot fully represent the response of human body. Along with a dearth of cell types, the simple microenvironments of in vitro 2D cell culture might also contribute to this discrepancy since many studies have shown that cells in a 2D environment act differently compared to in a 3D environment with mechanical or chemical stimuli (Hartung and Daston, 2009). This simple 2D environment lacks cell–cell interactions, one of the biggest and most widely acknowledged limitations of in vitro testing (Esch et al., 2011).

Shuler and his colleagues have developed the concept of "body-on-a-chip" to overcome this lack of cell–cell communication in conventional in vitro drug-testing platforms (Shuler and Esch, 2010). These devices, also called microcell culture analogs (μCCAs), consist of multiple, systemically connected tissue culture compartments (Esch et al., 2011). These compartments provide cells with isolated spaces that serve as discrete, distinct tissue mimicking units (Moraes et al., 2012). At the same time, these connections allow cells to communicate with other compartments. Thus, many important pharmacological concepts, such as pharmacokinetics, can be studied (Sung et al., 2010). The positions of the tissue culture compartments in the system can be ordered physiologically so that the tissue interactions and drug-response directionality in the human body can be reproduced (Figure 20.1(a)) (Shuler and Esch, 2010). Moreover, the flow rates of each tissue culture compartment can be adjusted to mimic the flow of the corresponding tissues or organs in vivo, so that drug distributions and shear stress profiles of each organ can also be reproduced within the system (Sin et al., 2004). One outstanding example of this concept is Shuler and colleagues' animal-on-a-chip device that reproduced Naphthalene's liver-mediated lung toxicity (Viravaidya et al., 2004). In this study, a four-chamber μCCA device (lung, liver, fat, and other tissues compartments) was designed to reproduce the pharmacokinetics of Naphthalene in the human body. Naphthalene was first metabolized by the liver compartment, and then, the resulting metabolites circulated to the lung compartment where cytotoxicity occurred. Since the drug toxicity occurs in organs other than the organ of interest, these side effects could never have been detected in vitro if there were no multiple compartments or no network connecting the discrete organ analogs.

Another way in which organ-on-a-chip approaches support advanced drug testing is by mimicking the organs' unique structural, chemical, and mechanical characteristics. While Shuler's works have ably reproduced the communication of tissues, they typically cannot mimic important aspects of tissue function, such as barrier functions, because the compartments are not designed to mimic tissue microstructure (Esch et al., 2010). Well-designed microchip features can direct cell organization toward tissue-like forms (Figure 20.1(b)), and it is evident that such structures can enhance cell survival and function (Martin, 1996). Moreover, these chip microstructures can also contribute to building well-defined chemical gradients that could not be generated in conventional in vitro dish cultures. One of the most recognized products of this approach is a lung-on-a-chip device developed by Huh et al. (2010). This approach mimicked the unique structural and mechanical characteristics of the lung alveolus. In particular, periodic mechanical stretching and the microporous membrane structure of the alveolar–capillary interface were reproduced using a flexible PDMS chip. This mimicking of specific mechanical and structural characteristics significantly contributed to reproducing better organ function such as tighter alveolar–capillary barrier junctions and a more sensitive cytokine response. Another good example of a microstructure-mimicking organ-on-a-chip device is the blood–brain barrier (BBB) chip developed by Booth and Kim (2012). In this study, they mimicked

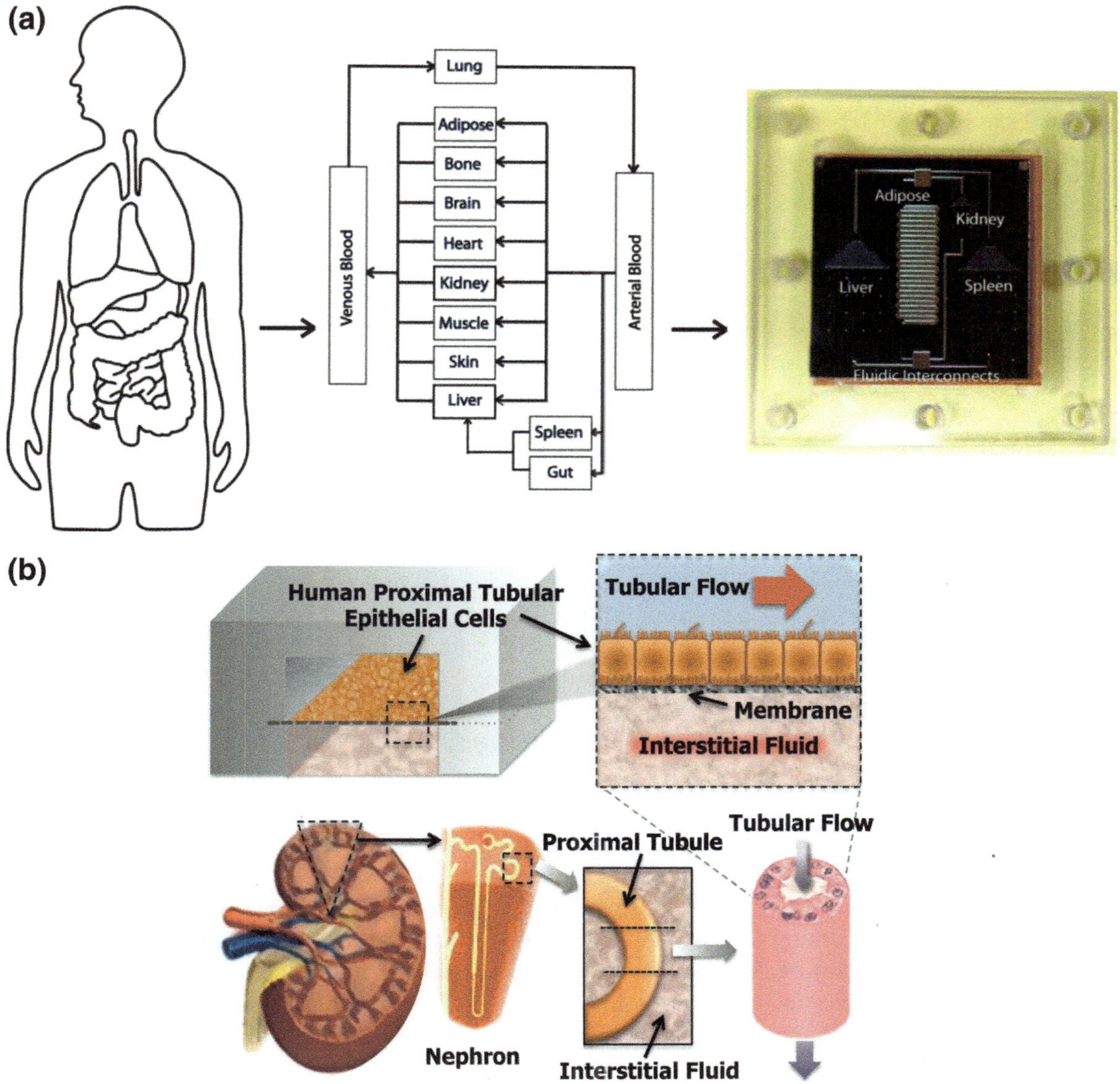

FIGURE 20.1 Two types of organ-on-a-chip devices.
(a) Schematic of a body-on-a-chip device consisting of interconnected organ compartments. The flow directions and rates were designed to mimic physiologic communication between organs. *(Figure reproduced from Shuler and Esch (2010).)* (b) Schematic of proximal tubule-on-a-chip device that mimics a unique organ microenvironment. The ECM-coated porous membrane and flow-induced shear stress provide cells with a more in vivo-like environment that contributes to better cell function. *(Figure reproduced from Jang et al. (2013).)*

not only the dynamic fluidic environment of BBB but also the very thin dual cell layer structure, supported here by a porous membrane. Both of these structurally well-reconstructed biological barrier tissues provide us physiologically relevant drug-testing platforms because almost all drugs must pass through biological barriers to reach target sites. Therefore, these tissue microstructure mimicking organs-on-a-chip present a more relevant platform for permeability or accessibility screening than previously discussed multicompartmental devices.

Besides the previously mentioned merits over conventional in vitro drug-testing platforms, organs-on-a-chip also enjoy advantages over in vivo paradigms such as animal models. First of all, the use of microfluidic chips and cultured cells for testing has fewer associated ethical issues than does the direct use of animals. Although a certain amount of animals will still be sacrificed for primary cell sources, organ-on-a-chip devices are a clearly more efficient way of using animals for research. For example, in the case of conventional in vivo animal test, one animal life must be sacrificed for only one test, while, in the case of an organ-on-a-chip device, over a 100 drug-testing organs-on-a-chip devices can be prepared from only one animal life because such chips are usually highly miniaturized. Additionally, different types of organ cells from one animal can also be utilized for different organ-on-a-chip devices at the same time. In many cases, chip-based studies are also quicker than in vivo experiments. For example, a mouse tumor graft model usually takes at least a month, and those tumor-grafted mice survive with pain for the entire test period. In the case of organ-on-a-chip devices, this ethical issue can be circumvented.

20.2.2.2 In Situ Biological Study Platforms

Most biological research has been carried out by applying a given stimulus and observing the response. These "cause and effect" studies have given us a tremendous amount of information regarding which factors might govern which phenomena. This crude information, however, usually fails to provide the mechanistic insight vital to deducing fundamental principles of biology. This shortcoming can be attributed to the limitations of current techniques, which often require that biological samples be prepared in particular ways such as by using plastic dishes, endpoint analysis, and static culture. Therefore, if we are able to observe biological phenomena at the site of origin, we could better understand and apply them to further purposes such as drug discovery and diagnostics.

Most conventional in vivo imaging techniques have spatial and temporal resolution limitations as well as penetration limitations. Magnetic resonance imaging, for example, has a relatively good spatial resolution compared to other techniques but is still insufficient for cell level studies. This problem is further exacerbated by poor temporal resolution and high cost. In contrast, in vitro imaging techniques such as conventional microscopy have enough spatial and temporal resolution for cell level studies and are also relatively cheap.

However, these microscopy techniques usually require static conditions (e.g., fixation) and certain preparation steps that may disturb the original state of samples (especially in vivo ones), so that the results from the studies do not faithfully represent the biological phenomena under investigation. Recently, more advanced microscopy techniques, such as intravital microscopy and multiphoton microscopy, have been developed to overcome those limitations of conventional in vivo and in vitro imaging systems (Sumen et al., 2004; So and Dong, 2000). However, the observation area of intravital microscopy is only limited to a surgically prepared window, which has a narrow observation area and requires a complicated surgical procedure. Multiphoton microscopies usually do not require such an invasive preparation but necessitate highly complex and expensive laser equipment and only offer penetration depths on the order of millimeters. Thus, such techniques can only be applied for the observation of peripheral tissues.

Since organ-on-a-chip devices serve as a bridge between too simple in vitro and too complicated in vivo environments, these technologies represent one possible solution to such imaging issues. By inducing processes such as intravasation (Zervantonakis et al., 2012) or vasculogenesis (Kim et al., 2013) to occur on-chip, researchers can reliably interrogate these phenomena in real time. Organs-on-a-chip are usually made of a thin glass plate and transparent PDMS materials and can integrate with existing microfluidic components. Therefore, these chips can be easily observed with conventional imaging techniques such as light, epifluorescence, and confocal microscopies. Although these chips are not exactly identical to tissues and organs of in vivo environments, this approach can still generate critical information on processes that are difficult to study with conventional in vitro or in vivo techniques.

20.3 EXAMPLES OF ORGAN-ON-A-CHIP

20.3.1 Blood Vessels-on-a-Chip

20.3.1.1 Introduction

Since cells cannot survive without the nutrients and oxygen supplied by circulating blood, the blood vessel network is important and the most widely spread structure in the body. Blood vessels not only play a role as the transportation of nutrients and oxygen but they also provide a route for leukocyte migration, which is closely related to important immune responses such as inflammation. The importance of blood vessels in medical research is increasing as new studies highlight interactions between blood vessels and immune or cancer cells. For example, cancer cell circulation and cancer cell escape from the network (extravasation) have been recognized among the most important phenomena in cancer biology (Chaffer and Weinberg, 2011; Jeon et al., 2013).

The basic building blocks of the vascular system are endothelial cells, which are closely related to the characteristics of the system. The morphological and functional characteristics of the vessel network vary significantly according to its locations in the body. In addition to changes in endothelial cell behavior, these differences may also be attributed to surrounding supportive components, such as muscle cells and ECM at corresponding body sites. The surrounding muscle cell layer on arteries and veins contributes to the flexibility of the vessel system and the comparatively thin ECM layers on capillaries contribute to high vessel permeability at tissue–capillary interfaces. These differences can also be affected by disease. For example, cholesterol deposition inside of arteries induces vessel wall stiffening that leads to atherosclerosis while the secretion of cytokines and growth factors by cancer cells permeabilizes the capillary–tissue interface; this results in severe tumor progressions such as metastasis.

Despite the physiological and pathological importance of blood vessel networks, basic mechanisms of those biological phenomena have been poorly understood because of the absence of proper model systems. Under this circumstance, the development of organ-on-a-chip techniques for blood vessel research has substantially contributed to a better understanding of vascular behavior (Franco and Gerhardt, 2012). In the following section, a few examples of blood vessels-on-a-chip will be introduced with their characteristics and following contributions.

20.3.1.2 Examples of Blood Vessels-on-a-Chip

A basic blood vessel-on-a-chip approach to study the vessel–tissue interface is a microfluidic channel divided by an endothelial cell monolayer on a supporting polymer membrane. In this configuration, the compartment with the endothelial cell monolayer side represents the inside of blood vessel (apical side), and the polymer membrane side represents a certain tissue environment (basal side). A recent example of this design strategy is a microfluidic endothelium device for studying adhesion of metastatic cancer cell developed by Takayama et al. (Song et al., 2009). To reproduce metastatic tropism of breast cancer cells with this device, MDA-MB-231 breast cancer cells were made to flow through the apical side with the physiologically relevant flow rate and metastasis-related cytokines were introduced into the basal side (Figure 20.2(a)). As a result, a significant number of breast cancer cells were attached only on the specific region of the endothelial cell layer where the cytokines were applied from the basal side (Figure 20.2(b)). Further, they suggested a prospective cytokine-receptor mechanism of cancer cell adhesion, which could further be applied to cancer drug discoveries.

Another outstanding example of a blood vessel-on-a-chip device is the functional, perfusable 3D microvascular network developed by Jeon et al.

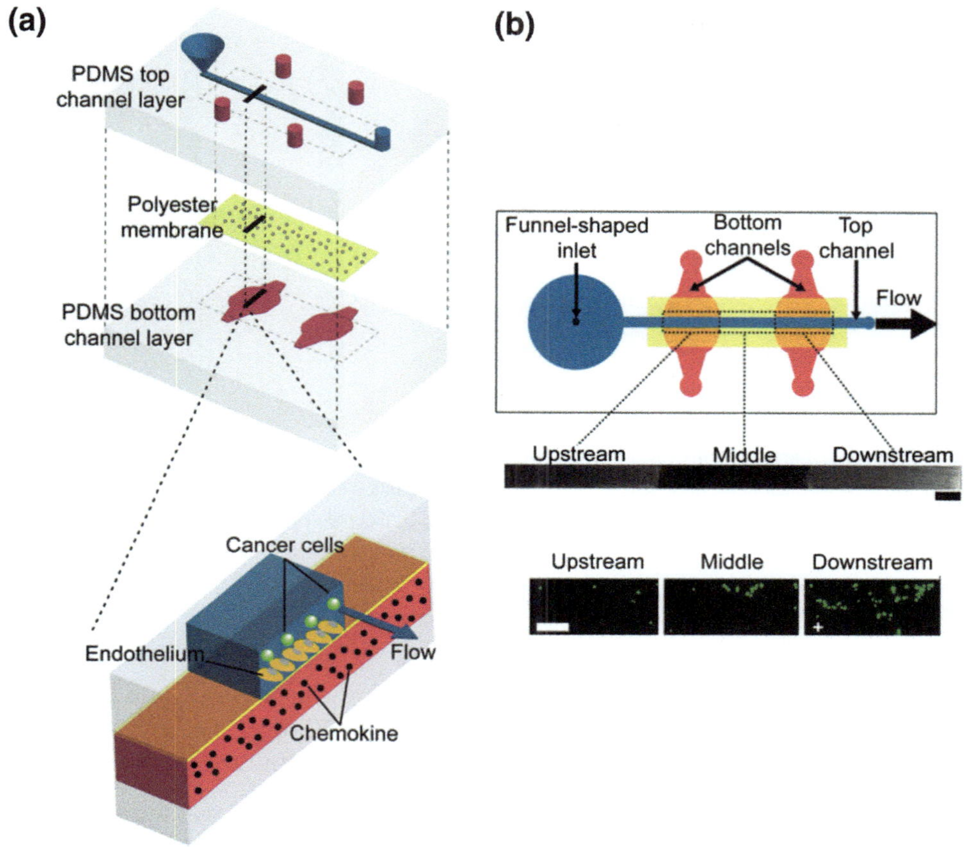

FIGURE 20.2 Microfluidic endothelium device for cancer cell adhesion study.
(a) Schematic of microfluidic endothelium device. (b) Significantly more cancer cells were attached downstream of the device when a metastasis-inducing chemokine was selectively applied from the bottom channel. This tissue–vessel interface device can be further utilized for interaction mechanism studies. Figure reproduced from Song et al. (2009).

(Kim et al., 2013). In this study, the researchers successfully observed spontaneous vasculogenesis using fibrin ECM gels and supporting lung fibroblasts. Because no artificial structural guides (e.g., polymer membranes) were necessary, the reconstituted vessel structures were more physiologically relevant (Figure 20.3(a)). These reproduced vessel structures were visualized with immunofluorescence techniques to confirm their morphological characteristics (Figure 20.3(b)) and were also perfused with fluorescein isothiocyanate (FITC)-dextran and polystyrene beads to check the permeability of the whole vessel structure and tightness of vessel junctions (Figure 20.3(c)). Moreover, angiogenesis was also reproduced on the same device by using a different cell-seeding configuration so that the angiogenic sprouting and vessel growth were able to be observed daily until connection between the branches was

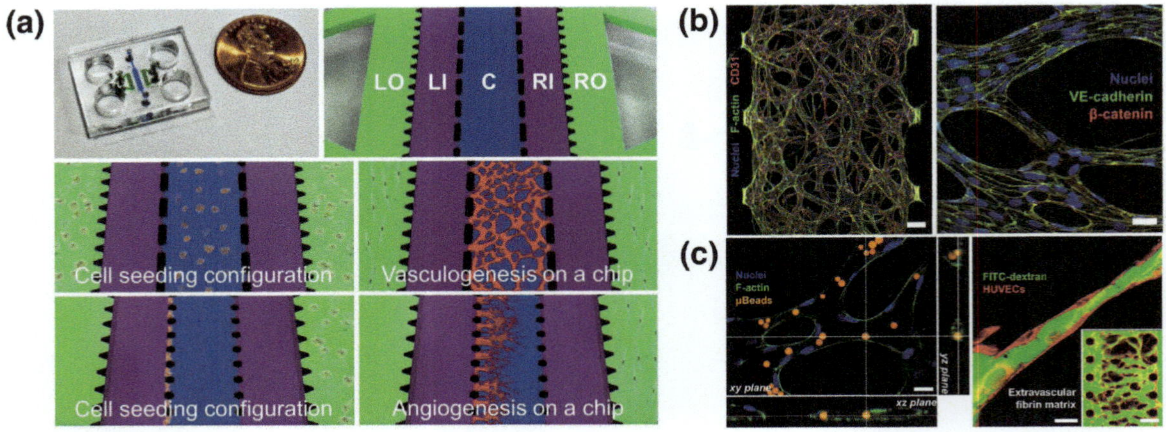

FIGURE 20.3 **Functional, perfusable microvascular network-on-a-chip device.**
(a) Schematics of the microvascular network-on-a-chip device. LO and RO regions are for stromal cells, LI and RI are for media, and C is for endothelial cells. Using different stromal cell and endothelial cell-seeding configurations, both vasculogenesis and angiogenesis were reproduced. (b) Overall structure of vessel network (left) and junction-related immunostained vessel structure (right). (c) Fluorescent bead perfusion test to confirm lumen space (left) and FITC perfusion test to confirm junction intactness (right). *Figure reproduced from Kim et al. (2013).*

complete. In addition to these endothelial cell-only studies, they incorporated vessel-supporting pericytes to the ECM gel and, as a result, endothelial–pericyte interactions were successfully reconstituted using the same device.

More recently, Kamm et al. observed cancer cell extravasation in an in vitro microvascular network (Chen et al., 2013). Notably, this study was built on the principles of the two previously introduced works: first, a focus on studying cell–vessel interactions and second, physiologically relevant vascular networks. In this study, they built a microvascular network via spontaneous vasculogenesis (Figure 20.4(a)). Metastatic cancer cells were then introduced into the preformed microvascular network to reproduce the phenomenon of extravasation on-chip. With this experimental platform, they observed the entire process of cancer cell extravasation using high-resolution time-lapse microscopy (Figure 20.4(b)) and assayed extravasation efficiency of different kinds of cell lines and different subpopulations of the cells. As a result, significant differences in extravasation efficiency of three different cell lines (MDA-MB-231, HT-1080, and MCF-10A) were measured; this correlated well with their previously known metastatic potential of each. More interestingly, the researchers observed differences in extravasation efficiency depending on whether single cells were trapped within or adhered to the microvasculature. Data such as these could provide important clues regarding whether circulating cancer cells escape the vasculature through an adherence-mediated or entrapment-mediated mechanism (Figure 20.4(c)).

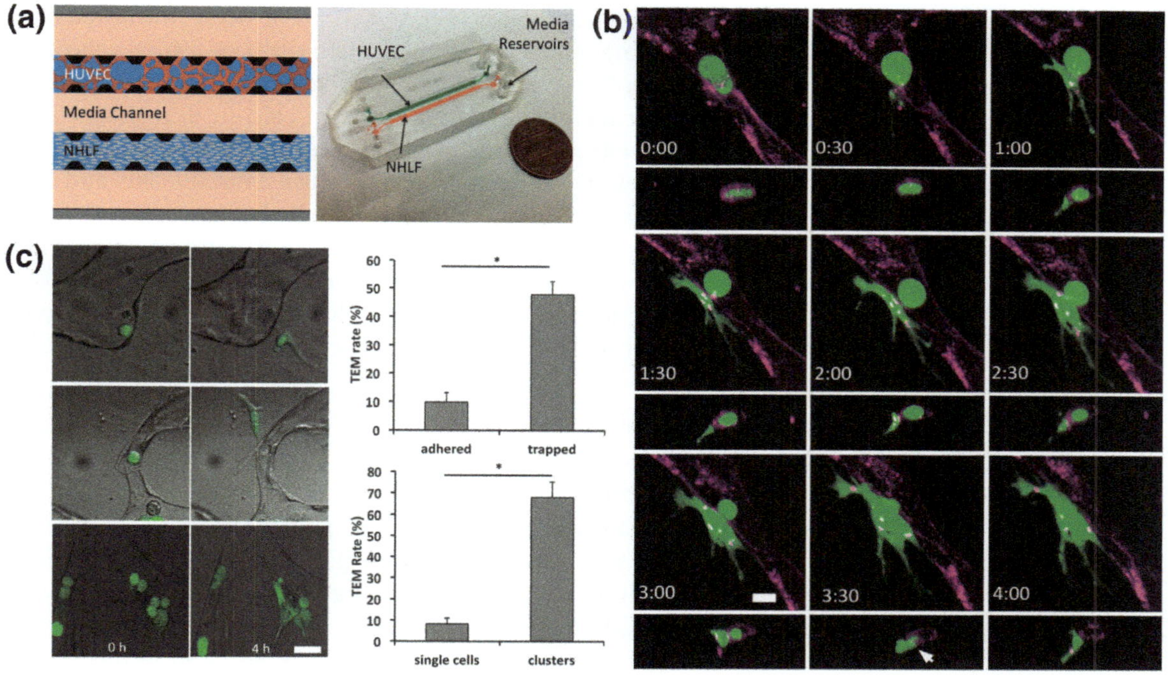

FIGURE 20.4 Microvascular network platform for tumor cell extravasation.
(a) Schematic design and picture of microvascular network platform. (b) Time-lapse confocal images of MDA-MB-231 cell (green) extravasation from preformed vessel network (purple). (c) Comparison of trans-endothelial migration rate between adhered and entrapped tumor cells (right top) and between single and cluster cells (right bottom). This result suggests that an adhesion-based mechanism alone may not be able to fully explain extravasation. *Figure reproduced from Chen et al. (2013).*

20.3.1.3 Limitations and Further Directions

One major problem of existing blood vessel-on-a-chip devices is that they have been too focused on cancer or immune cell transmigration studies. Even though cancer is one of the most urgent problems in medicine and biology, the most fundamental purpose of the vascular system remains the transportation and exchange of nutrients and oxygen, not cancer cells. This research bias results in a dearth of studies on nutrient transportation and exchange, which in turn hinders further integration of blood vessels into human-on-a-chip devices. The absence of studies incorporating critical transportation-related blood components, such as red blood cells, serum albumins, and lipid particles, clearly show this bias. Therefore, blood vessel-on-a-chip devices for reconstituting transportation (connection between other organ chips) and exchange (organ–vessel interfaces) functions should be more actively investigated.

Some functions of vessel system in current devices have generally been imitated by introducing artificial microfluidic components. Connections between organs in chip, for example, have been mimicked using fluidic channels, and interface functions between organ and vessels have been mimicked using porous polymer membranes. However, these attempts have limitations caused by the imperfections of artificial materials. Polymer membranes, for example, act as a basement membrane for endothelial cells in devices but at the same time interrupt the transmigration of cells due to the polymer barrier. An alternative blood-vessel-on-a-chip approach is to use natural ECMs and spontaneous vasculogenesis, which are more physiologic but also present further obstacles. Because vessels in these chips grow randomly within a defined ECM area, the resulting networks are difficult to embed into other devices to organize vessel–tissue interfaces. Therefore, new blood vessels-on-a-chip devices using more physiologic materials that can be integrated into other chip designs should be developed.

20.3.2 Lung-on-a-Chip

20.3.2.1 Introduction

The lung is the gas exchange organ of the human body, and therefore, has very unique structural characteristics. Because gas exchange mainly occurs by passive diffusion, the lung has an extremely large surface area-to-volume ratio achieved by continuous branching. The biggest airway of lung is called primary bronchus, and the groups of smaller airways after subsequent branchings are called bronchioles. At the end of these bronchioles, grape-like microchambers called alveoli are clustered and serve as functional units where the actual gas exchange takes place. To increase gas exchange efficiency, alveoli consist of a thin, single-cell layer epithelium surrounded by ECM and capillary vessels. Some are even known to have small pores in their walls. Unlike other cells in the human body, alveolar epithelial cells are subject to gas and liquid phases. Accordingly, a thin mucus layer protects them from drying out and also from infectious particles. Since the lung does not have any muscles, inspiration and expiration are controlled by pressure changes in the thoracic cavity generated by the diaphragm. This periodic up-and-down motion also causes cyclical stretching of lung tissues (Huh et al., 2008).

Due to the special microenvironment of the lung, cells originating from lung tissues usually require special in vitro culture conditions such as growth factors, ECM and serum (Yamaya and Finkbeiner, 1992; Kitamura et al., 1990). In addition, some studies have shown that the lung environment-like conditions such as an air–liquid interface and periodic stretching favorably contribute to in vitro culture of lung cells regarding long-term maintenance, protein secretion, and stimuli response (de Jong et al., 1994; Tschumperlin et al., 2003). These results, however, were only tested in separate in vitro cultures and have not been observed simultaneously because reconstituting all such characteristics on a 2D surface is almost impossible. Recently, microfluidics techniques have been recruited to reconstitute lung characteristics on a chip and have achieved breakthroughs in the field.

20.3.2.2 Examples of Lung-on-a-Chip

One early lung-on-a-chip study using compartmentalized microfluidic channels is the microfluidic airway system developed by Takayama et al. (Huh et al., 2007). This chip comprises upper and lower PDMS chambers, which are separated by a porous polyester membrane. Epithelial cells were seeded on the upper side of the membrane with culture media in both channels. After the confluence was achieved, air and media were made to flow through the upper and lower channels, respectively, to reproduce an air–liquid interfacial environment (Figure 20.5(a)). With this 3D in vitro airway system, the researchers recreated lung injury caused by liquid plug generation and rupture (Figure 20.5(b)). Those liquid plugs are known to be caused by malfunction of the surfactant mucus layer, a common symptom of many lung diseases such as asthma, pneumonia, etc. As a result, cells exposed to a higher number of liquid plug ruptures showed a significantly lower viability. In particular, cells near the location of plug rupture exhibited more death. Based on this observation, numerical simulations regarding pressure near the plug were carried out, and

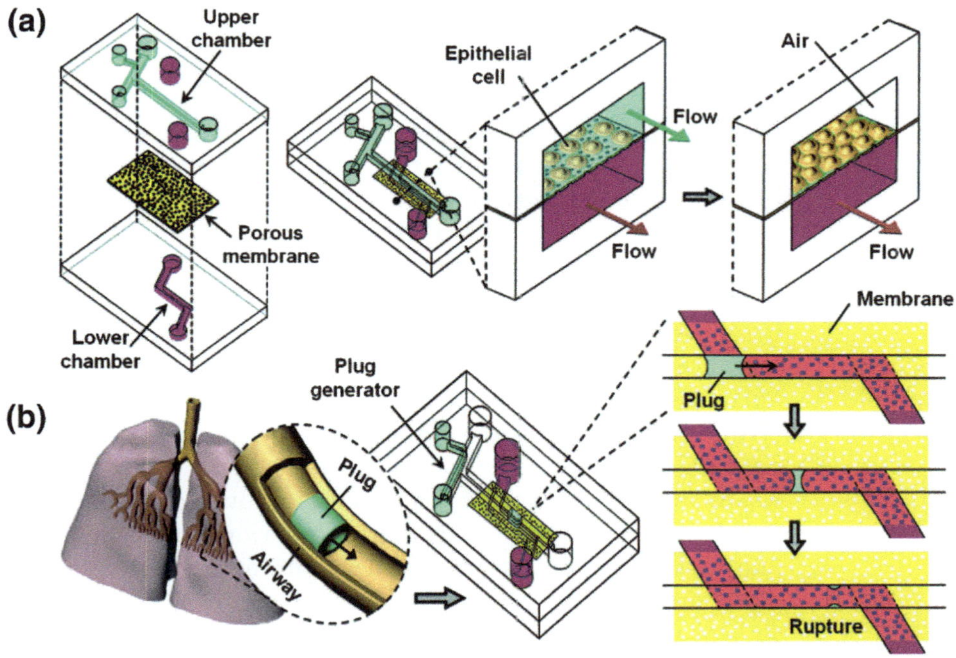

FIGURE 20.5 Microfluidic airway system for the study of liquid plug rupture effect in lung.
(a) Schematic of a microfluidic airway system to mimic the thin alveolar membrane. After epithelial cells reach confluence, the medium is removed from the top channel and an air–liquid interface is formed. (b) Liquid plugs are generated, propagated along the channel, and finally ruptured downstream. Rupture of the plug causes substantial mechanical stress and, as a result, adversely affects cell viability. *Figure reproduced from Huh et al. (2007).*

the result showed dramatic change of wall pressure and wall shear pressure, which may be responsible for injury of the cells.

While previous studies reproduced some basic characteristics of the lung such as a single-cell layer and air–liquid interfaces, the lung-on-a-chip device developed by Huh et al. (2010) reproduced almost all important characteristics of the lung. The configuration of the chip is similar to the one described by Takayama et al. except that this device includes two side chambers that are used for the generation of periodic stretching motions. The two side chambers were connected to a computer-controlled vacuum device that can apply a cyclic pressure to the chambers. As a result, the main chamber and dividing membrane were stretched and shrunk along the longitudinal direction. In addition, endothelial cells were seeded on the underside of the polymer membrane to reproduce the epithelium, basement membrane, and endothelium structures of the lung alveolus, so that lung–blood vessel interaction studies could be conducted in this chip (Figure 20.6(a)). With this chip configuration, Huh et al. first recapitulated pulmonary inflammation and immune cell transmigration by incorporating cytokines or bacteria into the epithelial side and neutrophils into the endothelial side. Moreover, the effects of the lung's stretching motion were demonstrated using nanosized silica particles, which proved that the many immune responses of the lung (intercellular adhesion molecule-1 (ICAM-1) expression, reactive oxygen species generation and particle translocation) were closely related to its mechanical stretching motions. In a following study with the same device, the researchers reproduced pulmonary edema induced by drug toxicity, which has been clinically reported (Huh et al., 2012). A clinically relevant dose of interleukin-2 (IL-2) was introduced through the endothelium side of the device to induce pulmonary edema, and a few days after the treatment, liquid leakage across the membrane was observed under a phase-contrast microscope. This compromised barrier function due to IL-2 treatment was maximized by periodic stretching, which correlates well with the pulmonary inflammation result obtained in the previous study. After the researchers successfully established edema in the lung-on-a-chip device, they identified new potential therapeutics, which alleviate IL-2 toxicity. Moreover, to validate their results and demonstrate the potential for the lung-on-a-chip device to be used as a drug test platform, the results from an in vivo whole lung barrier permeability test using mice were compared with the results from their lung-on-a-chip device and, surprisingly, these in vivo animal results and in vitro chip results showed almost no differences (Figure 20.6(b)).

20.3.2.3 Limitations and Further Directions

In spite of these outstanding achievements, current lung-on-a-chip devices still have problems in reproducing the basic gas exchange function of the lung. Although Huh's study measured oxygen saturation percentage from

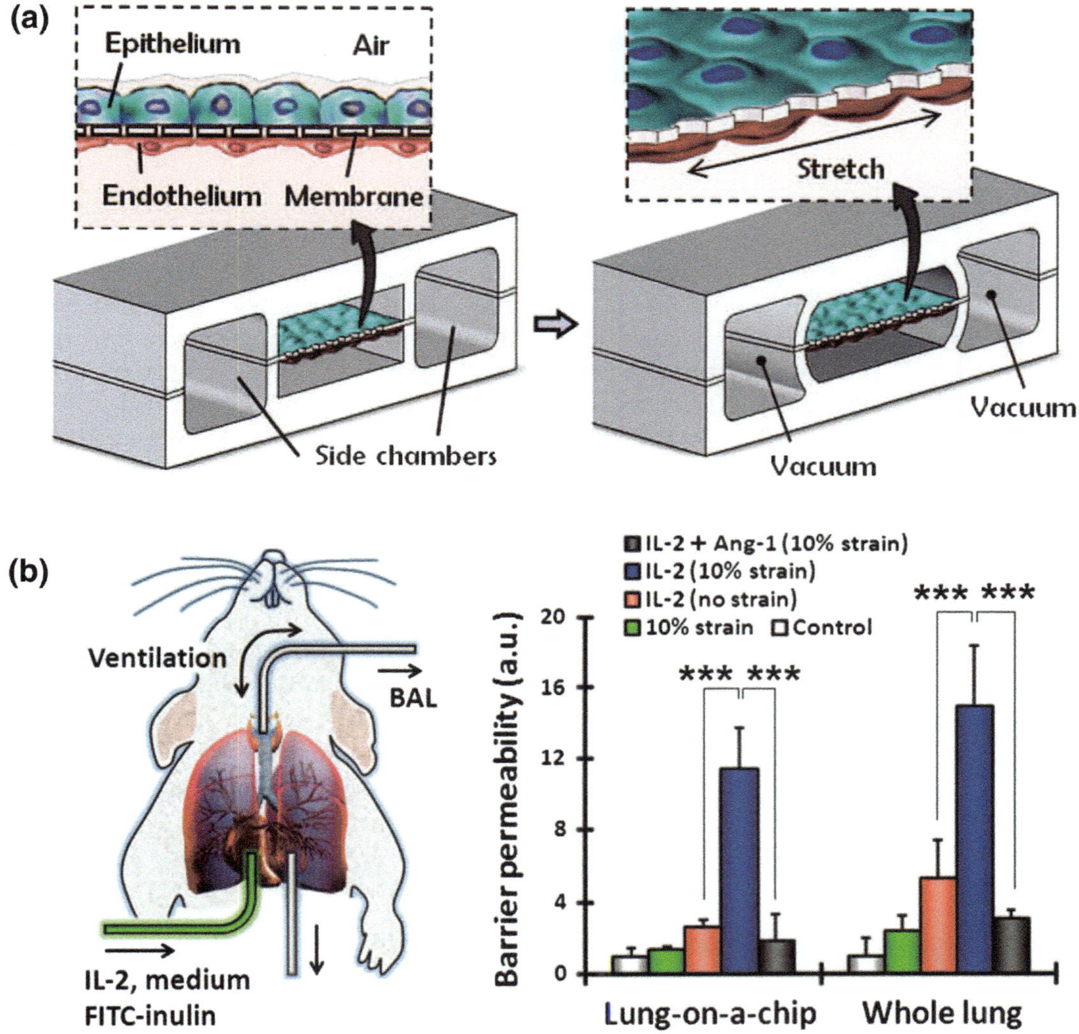

FIGURE 20.6

Lung-on-a-chip device: (a) Schematic of lung-on-a-chip device that has epithelium and endothelium dual cell layers. The epithelial side is subjected to air–liquid interface culture while the endothelial side is subjected to fluid flow culture. Vacuum-connected side chambers induce mechanical stretching of the membrane. *(Figure reproduced from Huh et al. (2010).)* (b) The potential for this lung-on-a-chip device to serve as drug test platform is validated by comparing in vivo and in vitro results in response to the same stimuli. *(Figure reproduced from Huh et al. (2012).)*

deoxygenized culture medium (Huh et al., 2012), this was carried out only as a simple comparison between normal and diseased environments in their chips and not as a fundamental study of gas exchange. Potkay and colleagues conducted a more careful investigation of on-chip gas exchange for an artificial lung. Notably, their device offered comparable performance to a commercial extracorporeal membrane oxygenation device using air—rather than pure oxygen—as the ventilating gas (Potkay et al., 2011). Burgess et al. (2009) developed another gas exchange device that was shown to support endothelial cell growth over a period of 10 days. Both efforts, however, were motivated to build organ augmentation systems rather than miniaturized models for drug discovery or mechanistic studies. Accordingly, there remains a substantial lack of lung-on-a-chip devices capable of examining the processes of gas exchange. Since the lung is the only organ for gas exchange in the human body, reproduction of proper gas exchange in lung-on-a-chip devices is a critical step for further organ-on-a-chip development. However, in most current lung-on-a-chip devices, the physical effects of this gas exchange have been neglected. Computational studies have shown that either symmetric or asymmetric bifurcations may induce stress gradients along an airway, and that this stress variation may cause different physiological responses (Luo and Liu, 2008; Liu et al., 2003). Unfortunately, current lung-on-a-chip devices ignore these mechanical cues or even actively mitigate them (Long et al., 2012).

However, as respiratory drug inhalation becomes recognized as one of the most efficient drug delivery routes, the need for proper drug test platforms involving this route is also growing. Although the possibility of lung-on-a-chip devices as respiratory drug inhalation test platforms seemed to be shown in the nanoparticle translocation study from Huh et al. (2010), a few limitations such as unrealistic drug (or particle) treatment methods and nonphysiologic interfacial area between epithelium and endothelium still remain.

20.3.3 Liver-on-a-Chip

20.3.3.1 Introduction

Liver, the biggest organ in the human body, is known to play a major role in processes such as metabolism, detoxification, and biochemical production. Because of its wide range of activity, the liver requires very high oxygen supply and, therefore, tremendous amounts of blood from related organs are directed to the liver through the hepatic portal vein and hepatic arteries not only for delivering oxygen but also for delivering other metabolites and nutrients. Almost all drugs are expected to interact with the liver because of its metabolism and detoxification functions. Therefore, liver toxicity is one of the most important checkpoints for newly developed drugs. Most liver functions are mediated by hepatocytes, which occupy about 80% of the liver's total volume. Interestingly, the liver comprises relatively simple functional subunits that belie its complex function. These hexagonal-shaped structures are known as hepatic lobules and contain millions of hepatocytes and complex vasculature, including blood vessels and secretory ducts, which are spread throughout the lobules. In particular, the capillary vessels branched into the lobules have a special structure known as a sinusoid (open pore capillary), so that the exchange of metabolites and nutrients between hepatocytes and blood is more efficient (Jindal et al., 2008).

Since the liver is responsible for many important roles, most liver diseases are very crucial. In addition, there are no fundamental cures for severe liver diseases except organ transplantation or natural recovery. For these reasons, many researchers have tried to develop liver supporting devices, which can reproduce the functions of liver inside or outside of body. At first glance, reproducing liver function in an in vitro environment seems very simple because hepatocytes are responsible for almost all functions of the liver. Unfortunately, hepatocytes are known as one of the most difficult cells to be maintained in vitro and considerable effort has been devoted to identifying important supporting factors for their in vitro *culture* such as oxygen supply, proper ECMs, and soluble factors (Dunn et al., 1991; Allen et al., 2001).

Recently, armed with this basic knowledge regarding in vitro hepatocyte culture, people started to use hepatocytes to build a liver-on-a-chip device for drug test purposes. In the following section, examples of liver devices from bioartificial liver (BAL) supports to liver-on-a-chip device for drug testing will be introduced.

20.3.3.2 Examples of Liver-on-a-Chip

The earliest liver-on-a-chip efforts were BAL devices—centimeter-scale bioreactors that incorporate isolated hepatocytes to achieve liver function outside of body. These devices have been developed in various configurations such as hollow fiber, flat-plate, perfused beds/scaffolds, and encapsulation (Allen et al., 2001). Another important factor for improved BAL function is the condition of incorporated hepatocytes. Because hepatocytes usually lose their functions soon after isolation, specific culture conditions for maintaining phenotype, such as ECM configuration, coculture with supporting cells and 3D spheroid formation, have been investigated. One widely recognized example is extracorporeal flat-plate BAL device developed by Toner et al. (Shito et al., 2003). In this study, they adapted a basic flat-plate BAL configuration to include a flat glass substrate, polycarbonate casing, and highly gas permeable polyurethane membrane. Porcine primary hepatocytes were isolated and maintained on the glass substrate with ECM (rat tail collagen) coating, which is known to support hepatocyte function in vitro. With this device configuration, they tested the life-supporting ability of the device on rats whose liver function had been compromised by D-galactosamine injections. Animal survival rates five times higher than controls were observed with the BAL device, and important indicators of liver function such as histology, ammonia concentration, and prothrombin time also showed significant differences.

After the concept of BAL was established, a new approach for better hepatocyte maintenance in BAL devices enabled by microfabricated structures was reported by Park et al. (2005). In this study, they utilized a similar flat-plate device but with integrated microgroove structures on the glass substrate, which is fabricated using photosensitive SU-8 lithography. Rat primary hepatocytes were seeded between the microgrooves, and supporting 3T3-J2 cells were then seeded on the same substrate the following day. With this device configuration, Park and colleagues achieved a higher hepatocyte survival under high flow rate conditions due to the shear shielding effect of microfabricated groove structures. This better hepatocyte maintenance under high shear conditions results in high oxygen supply to the seeded hepatocytes that further contributes to better overall function.

Meanwhile, Lee et al. suggested a more chip-like approach for maintaining hepatocytes by using microfluidic configuration that mimics the structure of the liver sinusoid (Figure 20.7(a)) (Lee et al., 2007). Distinct from previous hepatocyte-maintaining techniques that focused on ECM configuration and coculture, this microfluidic approach focused on reproducing key structural features and physiologic nutrient supply rate of the liver without ECM or supporting cells. With this unique design, they achieved a similar level of primary hepatocyte viability compared to previous ECM and/or supporting-cell supplemented cultures. In addition, due to the well-designed vessel–hepatocyte interface in this device, they successfully reproduce delivery of hepatotoxic diclofenac through sinusoid-mimicking gaps and demonstrated potential of the device as a drug test platform. This liver-on-a-chip approach for drug test platform was further developed in the perfused multiwell device for 3D liver developed by Griffith et al. (Figure 20.7(b)) (Domansky et al., 2010). The most recognized feature of this work is that it incorporates multiple tissue engineered liver components into multiwell platforms with pneumatic pumps. The liver components are prepared by seeding primary hepatocytes and primary liver sinusoidal endothelial cells to the ECM-coated scaffold in reactor wells. Fresh, oxygen-rich medium is continuously circulated from reservoir wells by programmed pneumatic pumps to build physiologic oxygen concentration gradients. With this configuration, they expect the device to be utilized for high-throughput liver toxicology and metabolism tests.

20.3.3.3 Summaries and Further Directions

Unlike other organs-on-a-chip, liver devices have generally focused on better supporting hepatocyte growth and function, instead of reproducing of liver structure—largely because most of the liver functions are carried out by hepatocytes. Considering that the liver recruits a large portion of blood flow and metabolizes almost all drugs in our body, previously developed liver devices mainly aimed for drug toxicology test platforms, and only a few

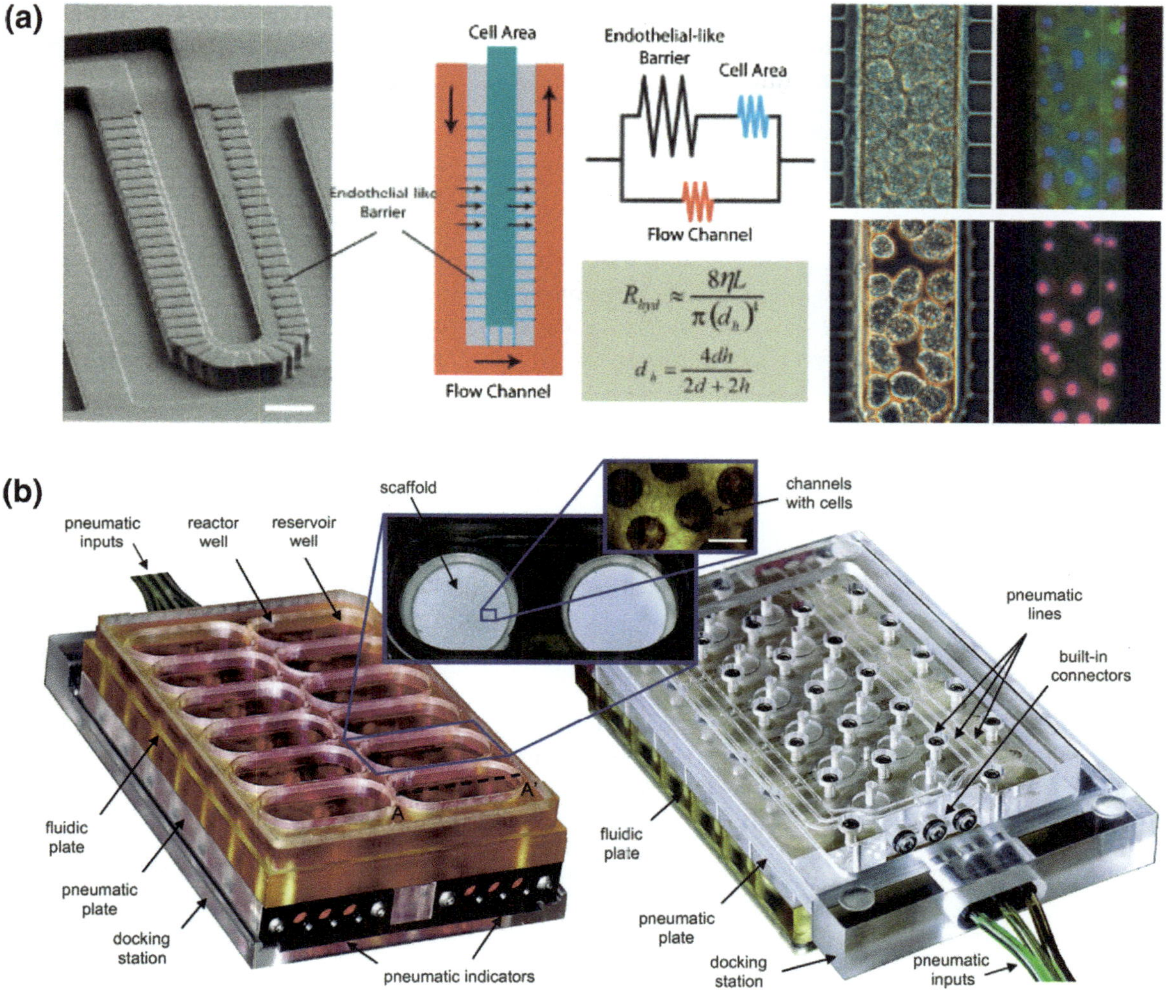

FIGURE 20.7

Liver-on-a-chip devices: (a) Schematic of a sinusoid-mimicking liver-on-a-chip device. (Left) The endothelial-like barrier acts to reduce fluid shear stress. (Mid) The initial seeding density of hepatocytes affects cell survival. Densely seeded cells showed a higher survival (right top) and sparsely seeded cells showed a lower survival (Right bottom). *(Figure reproduced from Lee et al. (2007).)* (b) Schematic of a 3D liver multiwell plate device. One well comprises one liver cell-containing bioreactor and one media-containing reservoir. Computer-controlled pneumatic pumps generate mixing flow between the bioreactor and reservoir to enhance oxygen supply to the hepatocytes. *(Figure reproduced from Domansky et al. (2010).)*

devices were further integrated with other organ components to study pharmacokinetics and liver-mediated effects of drugs (Viravaidya et al., 2004). However, liver devices with primary hepatocytes are challenging to integrate into broader networks because of their high oxygen consumption and environmental dependency. Therefore, the development of more physiologically relevant drug-testing platforms requires further advances that render primary hepatocyte-derived liver-on-a-chip devices compatible with other organ-on-a-chip modules.

However, such drug-oriented development of in vitro liver analogs has resulted in a lack of microenvironment-mimicking liver chips that in turn restrains mechanistic studies of liver function and disease. For example, liver–vessel interactions, including immune cell migration, cannot currently be performed because there is no liver device reproducing liver-blood vessel interfaces using biological materials. Development of these environment-mimicking liver chips could act as a biological research platform for many unexplained behaviors of the liver such as immune tolerance and regeneration.

20.3.4 Other Organs-on-a-Chip

20.3.4.1 Brain-on-a-Chip

Although many aspects of the central nervous system remain a mystery, researchers have been able to reproduce a few key characteristics of the brain-on-a-chip. One of the most recognized works in neuron function study on a chip is a high-density nanowire transistor array developed by Lieber et al. (Figure 20.8(a)) (Patolsky et al., 2006). In this study, the research team built a hybrid structure that has junctions between nanowire transistors and live mammalian neurons. This structure was prepared by deposition of poly-lysine on a predesigned nanowire array so that neurons could easily adhere. After these nanowire–neuron junctions were generated, electrical stimuli were applied at different junction sites, and the resulting shapes and amplitudes were measured to study signal propagation characteristics.

Another good example of a brain-on-a-chip device is the microfluidic in vitro BBB model for barrier function study and drug development platform, developed by Kim et al. (Figure 20.8(b)) (Booth and Kim, 2012). This device consists of four different layers of PDMS for two flow channels, a porous membrane to serve as a cell attachment substrate and two electrodes for *trans*-endothelial electrical resistance (TEER) measurement. In the completed device, porous membranes were seeded with b.End3 (brain endothelial cell line) cells and C8D1A astrocyte cells to mimic the structure of in vivo BBB membrane. With this configuration, they showed significantly higher TEER of reproduced BBB than static trans-well study and also reproduced important functional hallmarks such as histamine-induced TEER drop and recovery.

20.3.4.2 Heart-on-a-Chip

The heart is the core organ of our circulatory system, continuously pumping blood throughout the whole body. Due to the importance of heart function, newly developed drugs are often tested with cardiomyocytes to check their

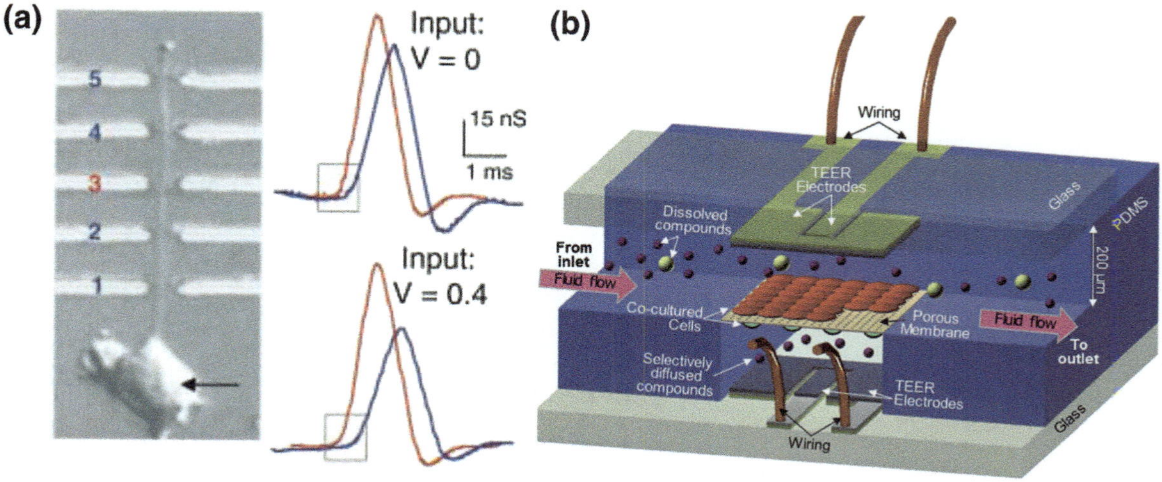

FIGURE 20.8

Brain-on-a-chip devices: (a) Microscopic view of neuron and nanowire junctions. Stimulus was applied at the cell body (black arrow) and nanowire 3 was used as an additional input source. Red lines are signals from nanowire 1 and blue lines are signals from nanowire 5 as results of stimulus on the cell body. The presence of an additional 0.4 V input stimulus at nanowire 3 delays signal propagation and reduces the signal amplitude, which indicates the inhibition of signal transfer. *(Figure reproduced from Patolsky et al. (2006).)* (b) Schematic of BBB-on-a-chip device. A channel with endothelial cells mimics the abluminal region and another channel with astrocyte cells mimics the luminal side of the BBB. Each channel has embedded electrodes to measure TEER of the reconstituted BBB membrane. *(Figure reproduced from Booth and Kim (2012).)*

cardiotoxicity. As a result, various kinds of heart-on-a-chip device for cardiotoxicity tests have also been developed. The on-chip cardiomyocyte cell network system developed by Yasuda et al. is a good example (Kaneko et al., 2007). In this study, they made four-cell or nine-cell cardiomyocytes networks on a thin agarose layer using temperature responsive sol–gels (Figure 20.9(a)). These array-patterned cardiomyocytes demonstrated stable and uniform beating intervals so that unstable beating (arrhythmia) caused by cardiotoxic drugs could easily be observed. To validate the stability of the device as a drug-screening platform, they induced arrhythmia on-chip by treating the cells with a well-known cardiotoxin (haloperidol) followed by rescue of regular beating after a washout step (Figure 20.9(b)).

Other kinds of heart-on-a-chip device for cardiac muscle studies have also been suggested (Grosberg et al., 2011). In one study, cardiomyocytes were seeded as an anisotropic monolayer on an arrayed elastomeric thin film. Contraction by the cardiomyocytes bent the thin film and measuring the resulting curvature of the structure allowed the researchers to quantitatively estimate the frequency and magnitude of the contractile forces generated by the cells. In this way, the effect of epinephrine on both force and frequency was also observed.

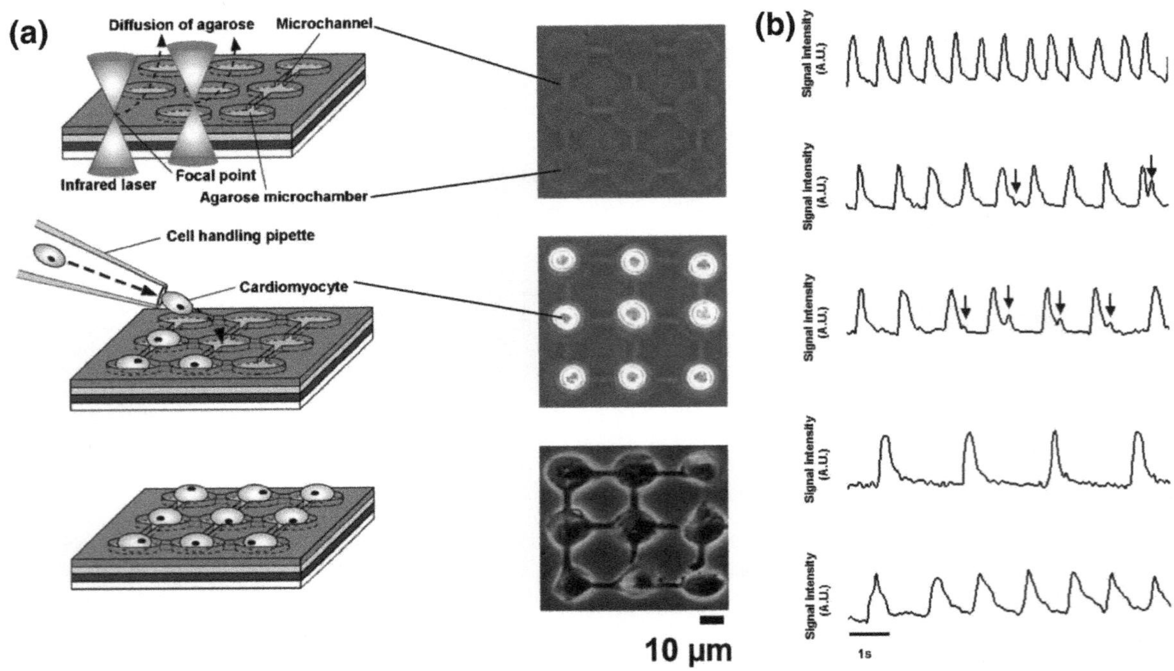

FIGURE 20.9
Cardiomyocyte cell network-on-a-chip device *(figure reproduced from Kaneko et al. (2007))*: (a) Schematics (left) and microscopic views (right) of a cardiomyocyte cell network-on-a-chip device. The pattern was fabricated using temperature-sensitive agarose layers and an infrared laser. Cardiomyocytes were then seeded on the network patterned agarose layer. (b) Cardiotoxic haloperidol treatment induces irregular beating. Upon buffer exchange, the original stable rhythm is completely restored.

20.3.4.3 Kidney-on-a-Chip

The kidney is a key homeostasis-regulating organ that governs blood pressure and electrolyte concentration by controlling water and ion absorption. This basic function of the kidney was reproduced with a simple multilayer microfluidic device containing renal tubular cells (Jang and Suh, 2010). The device consisted of two layers of PDMSs for fluid channels and a porous membrane acting as a cell substrate. Rat inner medullary collecting duct (IMCD) cells were seeded on the porous membrane, and continuous fluidic shear was applied to the cells through the upper channel of the device to generate an in vivo like microenvironment. As a result of applying shear stress, IMCD cells showed enhanced polarization and reorganized cytoskeletal structure. This reproduced IMCD monolayer in chip device showed important hormone-dependent osmotic pressure regulation and sodium transportation response. Moreover, cytotoxicity and recovery were also observed with and without fluid shear in subsequent work (Jang et al., 2013).

20.4 CONCLUSION

Mimicking the unique microstructure of organs has enabled many organ-on-a-chip devices to successfully reconstitute the functions of many different organs in a miniaturized environment. Previous studies on organ physiology and tissue microenvironments, such as fluid shear, chemical gradient and cell density, have also greatly contributed to developing more physiologically relevant organ-on-a-chip devices.

The development of organ-on-a-chip devices has helped widen our understanding of biological phenomena. Blood vessel-on-a-chip devices give us valuable information about cancer cell adhesion mechanisms and also enabled us to observe tumor trans-endothelial migration phenomena with high spatial and temporal resolution. The importance of mechanical stretching on physiologic functioning of the lung was suggested by several lung-on-a-chip devices, and the importance of flow rate and shear stress has been discussed in liver-on-a-chip and kidney-on-a-chip studies. Moreover, various kinds of primary cell culture techniques have been suggested from the process of developing liver-on-a-chip devices and the importance of tissue-specific oxygen concentration on in vitro cell maintenance was also emphasized in such studies.

Beyond basic science, organ-on-a-chip devices have also contributed to pharmacological studies. Shuler's concepts of animal-on-a-chip and body-on-a-chip have reconstituted organ–organ and tissue–tissue interactions on a chip, thereby enabling pharmacokinetic studies and side effect screening in vitro. Drug-testing platforms for individual organs have also been improved by organ-on-a-chip devices. Blood vessel-on-a-chip devices provided an angiogenesis assay platform, which can further be utilized as drug test platform for antiangiogenic drugs for cancer therapy. Liver-on-a-chip device also provided a conventional multiwell-like drug test platform, which is very convenient to manipulate while, at the same time, very physiologically relevant.

Despite the many outstanding outcomes from previous organ-on-a-chip studies, there is still room for further improvement. One goal for further development would be integration of individual organ modules into systems-on-a-chip (e.g., digestive system-on-a-chip) and eventually a human-on-a-chip. Therefore, if we could integrate sufficient organ-on-a-chip devices to mimic a system in our body, plenty of pathological and physiological studies could be carried out using the platform. Of course, integration of these on-chip systems carries a new, different set of questions and problems that have only recently begun to be considered. Designing this next generation of physiologically relevant models to faithfully capture inter-organ interactions for both known *and unknown* pathways will be a major challenge for future studies and will likely require substantial innovations

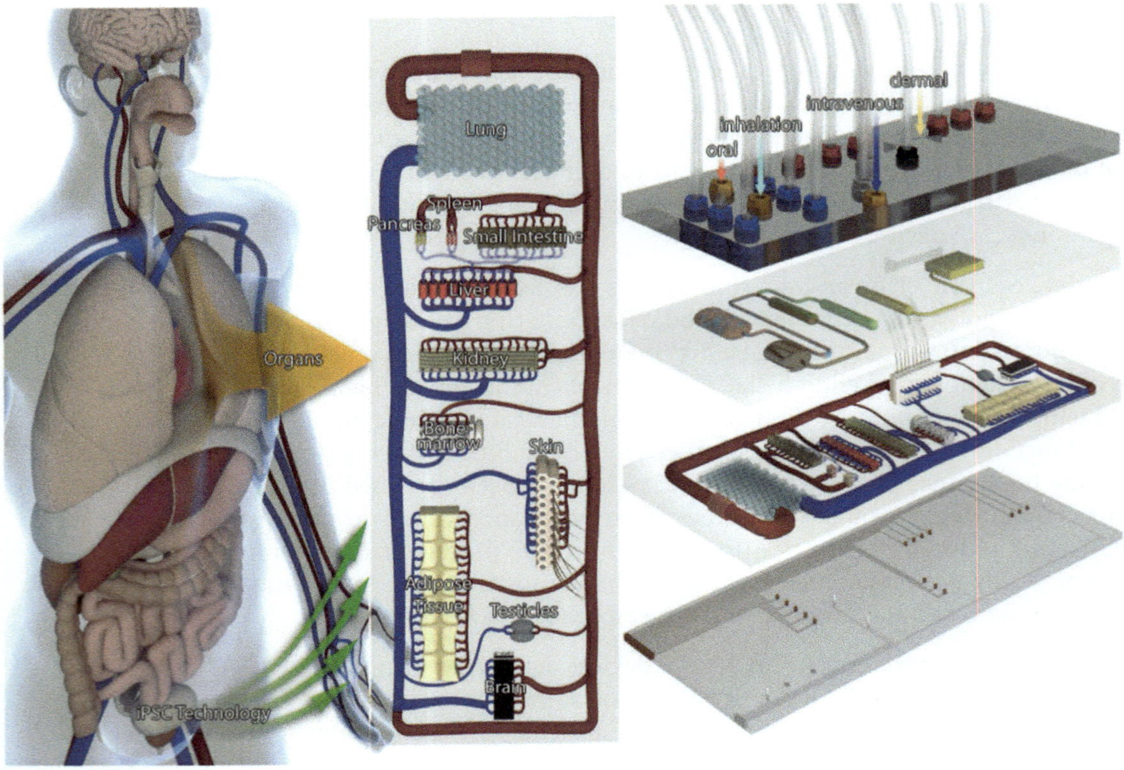

FIGURE 20.10
Prospective design of a "human-on-a-chip". *Figure reproduced from Materne et al. (2013).*

in both engineering and design (Wikswo et al., 2013; Moraes et al., 2013; Materne et al., 2013) (Figure 20.10).

This proposed human-on-a-chip system consists of 11 organ equivalents connected by one close loop circulation system. This device has multiple routes of administration and sensing as well as mechanical actuators to mimic organ motions such as heartbeat, intestine movements, and lung breathing motions.

Nevertheless, these future goals remain dependent on the further development of individual organ-on-a-chip systems. Recreating fundamental organ functions, such as transportation in blood vessels and gas exchange in the lung, should be a major point of emphasis going forward. Progress in this field would not only serve as a basis for more effective integrative studies but also advance the cardinal goal of organ-on-a-chip studies—the high-fidelity recapitulation in vivo phenomena within a controllable, miniaturized in vitro platform.

20.5 SUMMARY

- Organs-on-a-chip are microfluidic cell and tissue culture systems capable of recapitulating key physiologies and functions of specific organs for further biological and medical purposes.
- In many cases, in vivo studies (e.g., animal models) are too complicated while in vitro platforms (e.g., petri dishes) are too limited for detailed studies of biological phenomena.
- Organ-on-a-chip technology can bridge this gap—providing unique structural and biological cues along with precise spatial and temporal controls.
- High attrition rates during drug development are likely due to insufficient/inappropriate preclinical testing tools.
- Blood vessels-on-a-chip may be generated by on-chip vasculogenesis or by using thin, porous polymer films to mimic the basement membrane.
- Recent advances in lung-on-a-chip technology have demonstrated the importance of integrating mechanical cues (e.g., cyclic stretching) in organ model systems.
- Key features of other organs, such as liver, brain, heart, and kidney, have been recapitulated in engineered microsystems.
- Future efforts should focus on reproducing fundamental organ functions (e.g., gas exchange in the lung or O_2/nutrient transport in blood vessels) to better study these critically important processes as well as create organ modules that can be more practically integrated to form organ networks.
- Systems- or humans-on-a-chip would be a powerful extension of the organ-on-a-chip concept; however, critical questions regarding system design remain unanswered.

CLASSICAL EXPERIMENT

Adsorption, distribution, metabolism, elimination, and potential toxicity are all important measurements for any drug compound. However, until work by the Shuler group at Cornell, these tests were run almost exclusively on animal models. Rather than following the norm or turning to conventional in vitro culture methods, the researchers developed a new paradigm—the μCCA or "animal-on-a-chip."

This simple microfluidic device consisted of four chambers that—importantly—were able to fluidically communicate with each other. To study the toxic effects of naphthalene, Shuler's group built a four-compartment system consisting of liver, lung, fat, and "other tissues" chambers.

When applied to the system, naphthalene is metabolized by the liver compartment, to generate compounds

Continued...

CLASSICAL EXPERIMENT *Continued*

that can either give rise to reactive oxygen species (which cause oxidative stress and eventually cell death) or bind to glutathione, among several other possibilities. After tuning residence times and flow distributions to match in vivo conditions among the four compartments, naphthalene was administered to the μCCA and flow was recirculated. Over a span of 6h, the lung compartment showed substantial glutathione depletion—a marked difference from control experiments with naphthalene but lacking a functioning liver compartment (Viravaidya et al., 2004).

The paper lays the foundation for the goal of systems- and bodies-on-a-chip. By creating a network of interacting culture spaces, Shuler et al. were able to recapitulate a key metabolic toxicity pathway that would have been undetectable in practically any other in vitro system. This goal of an in vitro, human analog for drug screening (as well as mechanistic studies of disease and other applications) has only grown more popular with time. By combining Shuler's concept of a μCCA with advanced organ-on-a-chip models, we may be able to build models that cannot only recreate, but also predict, various toxicity pathways.

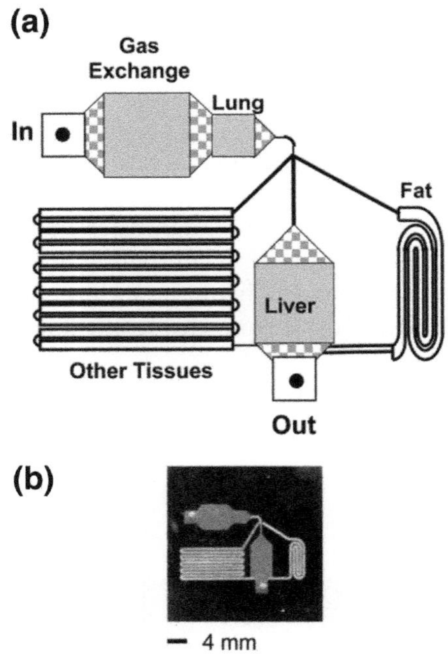

(a) A schematic diagram of a 1″ × 1″ μCCA design. Dimension of each chamber are as followed (w × l × d): lung (2 mm × 2 mm × 20 μm), liver (3.5 mm × 4.6 mm × 20 μm), fat (0.42 mm × 50.6 mm × 100 μm), and other tissues (0.4 mm × 109 mm × 100 μm). Note that, according to the physiologically based pharmacokinetic model, 25%, 9%, and 66% of flow from the lung chamber go to liver, fat, and other tissues chambers, respectively. The fat chamber represents a slowly perfused organ and contains no cells. (b) The actual size of μCCA; the scale bar indicates 4 mm.

STATE-OF-THE-ART EXPERIMENT

As the gas exchange unit of the human body, the lung is subject to a variety chemical and mechanical cues that are applied across length scales on the order of microns. Microfluidic techniques, therefore, are particularly well suited to model this complex, yet fundamentally critical, environment in vitro. Early efforts leveraged a compartmentalized microfluidic design to build a model system consisting of a single epithelial cell layer cultured at an air-liquid interface. Importantly, Takayama et al. introduced physical cues in the form of liquid plugs that exposed cells to interfacial tension during propagation and even greater stresses at rupture (Huh et al., 2007). This marriage of environmental, structural, and mechanical cues highlighted the potential for microfluidic systems to capture the nuances of the organ microenvironment in vitro.

In a landmark article, Ingber and colleagues built off previous lung-on-a-chip efforts and introduced their own version of the device (Huh et al., 2010). In addition to the epithelia layer, the researchers included an endothelial layer, more effectively mimicking the alveolar-capillary interface where gas exchange takes place. Also of note, a thin silicone membrane replaced the rigid membrane previously used by researchers. This allowed for the entire device to be stretched—recreating the cyclic tension and relaxation of the breathing cycle. These two key innovations enabled on-chip observation of neutrophil response during infection and inflammation. Further studies indicated that cyclic mechanical strain was a key mediator of nanoparticle toxicity. In a subsequent study, the group compared the response of their lung-on-a-chip device to an excised mouse model, finding that both responded to IL-2 and mechanical stimulation in the same way (Huh et al., 2012).

For many, this work finally provided tangible proof that a sufficiently realistic microfluidic organ model system could be practically realized. The Ingber group's integration of structural, mechanical, and chemical cues provides a blueprint for future organ-on-a-chip efforts.

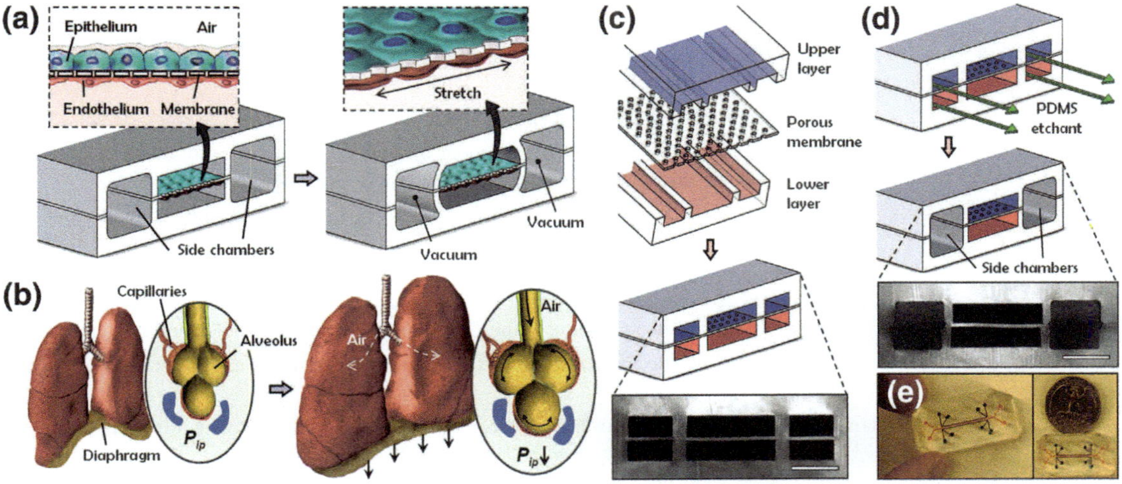

Biologically inspired design of a human breathing lung-on-a-chip microdevice. (a) The lung mimic device uses compartmentalized microchannels to form an alveolar–capillary barrier on a thin. (b) During inhalation in the living lung, contraction of the diaphragm causes a reduction in intrapleural pressure (P_{ip}), leading to physical stretching of the alveolar–capillary interface. (c) PDMS layers are aligned and irreversibly bonded to form sets of parallel microchannels separated by a 10-μm-thick PDMS membrane. Scale bar, 200 μm. (d) Selective etching of the membrane layers in the side channels produces two large chambers where applied vacuum causes mechanical stretching. Scale bar, 200 μm. (e) Image of an actual lung-on-a-chip microfluidic device viewed from above.

REFERENCES

Abbott, A., August 2003. Biology's new dimension. Nature 424, 870–872.

Allen, J.W., Hassanein, T., Bhatia, S.N., 2001. Advances in bioartificial liver devices. Hepatology 34 (3), 447–455.

Baker, M., 2011. A living system on a chip. Nature, 4–7.

Booth, R., Kim, H., 2012. Characterization of a microfluidic in vitro model of the blood–brain barrier (μBBB). Lab Chip 12 (10), 1784–1792.

Burgess, K.A., et al., 2009. Towards microfabricated biohybrid artificial lung modules for chronic respiratory support. Biomed. Microdevices 11 (1), 117–127.

Chaffer, C.L., Weinberg, R.A., 2011. A perspective on cancer cell metastasis. Science 331 (6024), 1559–1564.

Chen, M.B., et al., 2013. Mechanisms of tumor cell extravasation in an in vitro microvascular network platform. Integr. Biol. 5 (10), 1262–1271.

Domansky, K., et al., 2010. Perfused multiwell plate for 3D liver tissue engineering. Lab Chip 10 (1), 51–58.

Dunn, J.C., Tompkins, R.G., Yarmush, M.L., 1991. Long-term in vitro function of adult hepatocytes in a collagen sandwich configuration. Biotechnol. Prog. 7 (3), 237–245.

El-Ali, J., Sorger, P.K., Jensen, K.F., 2006. Cells on chips. Nature 442 (7101), 403–411.

Esch, M.B., King, T.L., Shuler, M.L., 2011. The role of body-on-a-chip devices in drug and toxicity studies. Annu. Rev. Biomed. Eng. 13, 55–72.

Esch, M.B., Sung, J.H., Shuler, M.L., 2010. Promises, challenges and future directions of microCCAs. J. Biotechnol. 148 (1), 64–69.

Franco, C., Gerhardt, H., August 23, 2012. Blood vessels on a chip. Nature 488, 465–466.

Freshney, R.I., 2011. Culture of Animal Cells, sixth ed. John Wiley & Sons, Inc.

Grosberg, A., et al., 2011. Ensembles of engineered cardiac tissues for physiological and pharmacological study: heart on a chip. Lab Chip 11 (24), 4165–4173.

Hartung, T., Daston, G., 2009. Are in vitro tests suitable for regulatory use? Tox. Sci. 111 (2), 233–237.

Huh, D., et al., 2012. A human disease model of drug toxicity-induced pulmonary edema in a lung-on-a-chip microdevice. Sci. Transl. Med. 4 (159), 159ra147.

Huh, D., et al., 2007. Acoustically detectable cellular-level lung injury induced by fluid mechanical stresses in microfluidic airway systems. Proc. Natl. Acad. Sci. U.S.A. 104 (48), 18886–18891.

Huh, D., et al., 2008. Engineering pulmonary epithelia and their mechanical microenvironments. In: Micro and Nanoengineering of the Cell Microenvironment, pp. 503–534.

Huh, D., et al., 2010. Reconstituting organ-level lung functions on a chip. Science 328 (5986), 1662–1668.

Jang, K.-J., et al., 2013. Human kidney proximal tubule-on-a-chip for drug transport and nephrotoxicity assessment. Integr. Biol. 5 (9), 1119–1129.

Jang, K.-J., Suh, K.-Y., 2010. A multi-layer microfluidic device for efficient culture and analysis of renal tubular cells. Lab Chip 10 (1), 36–42.

Jeon, J.S., et al., 2013. In vitro model of tumor cell extravasation. PLoS One 8 (2), e56910.

Jindal, R., et al., 2008. Microscale hepatic tissue engineering. In: Micro and Nanoengineering of the Cell Microenvironment, pp. 557–578.

de Jong, P.M., et al., 1994. Ciliogenesis in human bronchial epithelial cells cultured at the air-liquid interface. Am. J. Respir. Cell Mol. Biol. 10 (3), 271–277.

Kaneko, T., Kojima, K., Yasuda, K., 2007. An on-chip cardiomyocyte cell network assay for stable drug screening regarding community effect of cell network size. Analyst 132 (9), 892–898.

Kim, S., et al., 2013. Engineering of functional, perfusable 3D microvascular networks on a chip. Lab Chip 13 (8), 1489–1500.

Kitamura, H., et al., 1990. Growth and differentiation of human distal airway epithelial cells in culture. Effects of small amounts of serum in defined medium. Lab. Invest. 63 (3), 420–428.

Kola, I., Landis, J., 2004. Can the pharmaceutical industry reduce attrition rates? Nat. Rev. Drug Discov. 3 (8), 711–715.

Lee, P.J., Hung, P.J., Lee, L.P., 2007. An artificial liver sinusoid with a microfluidic endothelial-like barrier for primary hepatocyte culture. Biotechnol. Bioeng. 97 (5), 1340–1346.

Liu, Y., So, R.M.C., Zhang, C.H., 2003. Modeling the bifurcating flow in an asymmetric human lung airway. J. Biomech. 36 (7), 951–959.

Long, C., et al., 2012. Design optimization of liquid-phase flow patterns for microfabricated lung on a chip. Ann. Biomed. Eng. 40 (6), 1255–1267.

Luo, H.Y., Liu, Y., 2008. Modeling the bifurcating flow in a CT-scanned human lung airway. J. Biomech. 41 (12), 2681–2688.

Martin, L., November 1996. Effect of extracellular function, matrix hepatocytes topology cultured on cell in a sandwich configuration and physiological responsiveness. FASEB 10, 1471–1484.

Materne, E.-M., Tonevitsky, A.G., Marx, U., 2013. Chip-based liver equivalents for toxicity testing–organotypicalness versus cost-efficient high throughput. Lab Chip 13 (18), 3481–3495.

Van der Meer, A.D., van den Berg, A., 2012. Organs-on-chips: breaking the in vitro impasse. Integr. Biol. 4 (5), 461–470.

Moraes, C., et al., 2013. On being the right size: scaling effects in designing a human-on-a-chip. Integr. Biol. 5 (9), 1149–1161.

Moraes, C., et al., 2012. Organs-on-a-chip: a focus on compartmentalized microdevices. Ann. Biomed. Eng. 40 (6), 1211–1227.

Pan, C., et al., 2009. Comparative proteomic phenotyping of cell lines and primary cells to assess preservation of cell type-specific functions. Mol. Cell. Proteomics 8 (3), 443–450.

Park, J., et al., 2005. Microfabricated grooved substrates as platforms for bioartificial liver reactors. Biotechnol. Bioeng. 90 (5), 632–644.

Park, T.H., Shuler, M.L., 2003. Integration of cell culture and microfabrication technology. Biotechnol. Prog. 19 (2), 243–253.

Patolsky, F., et al., 2006. Detection, stimulation, and inhibition of neuronal signals with high-density nanowire transistor arrays. Science 313 (5790), 1100–1104.

Potkay, J.A., et al., 2011. Bio-inspired, efficient, artificial lung employing air as the ventilating gas. Lab Chip 11 (17), 2901–2909.

Roxanne, K., May 2005. Meet the stripped down rat. Nature 435, 12–13.

Seker, E., et al., December 2011. Solving medical problems with BioMEMS. IEEE Pulse, 51–59.

Shito, M., et al., 2003. Efficacy of an extracorporeal flat-plate bioartificial liver in treating fulminant hepatic failure. J. Surg. Res. 111 (1), 53–62.

Shuler, M.L., Esch, M.B., 2010. Body-on-a chip: using microfluidic systems to predict human responses to drugs. Pure Appl. Chem. 82 (8), 1635–1645.

Sin, A., et al., 2004. The design and fabrication of three-chamber microscale cell culture analog devices with integrated dissolved oxygen sensors. Biotechnol. Progress 20 (1), 338–345. Available at: http://www.ncbi.nlm.nih.gov/pubmed/14763861.

So, P., Dong, C., 2000. Two-Photon Excitation Fluorescence Microscopy, vol. 2Annual Review of Nonprofit Scientific Publisher. 399–429.

Song, J.W., et al., 2009. Microfluidic endothelium for studying the intravascular adhesion of metastatic breast cancer cells. PLoS One 4 (6), e5756.

Sumen, C., et al., 2004. Intravital microscopy: visualizing immunity in context. Immunity 21 (3), 315–329.

Sung, J.H., Kam, C., Shuler, M.L., 2010. A microfluidic device for a pharmacokinetic-pharmacodynamic (PK-PD) model on a chip. Lab Chip 10 (4), 446–455.

Tschumperlin, D.J., et al., 2003. Mechanical stress triggers selective release of fibrotic mediators from bronchial epithelium. Am. J. Res. Cell Mol. Biol. 28 (2), 142–149.

Viravaidya, K., Sin, A., Shuler, M.L., 2004. Development of a microscale cell culture analog to probe naphthalene toxicity. Biotechnol. Prog. 20 (1), 316–323.

Whitesides, G.M., 2006. The origins and the future of microfluidics. Nature 442 (7101), 368–373.

Wikswo, J.P., et al., 2013. Scaling and systems biology for integrating multiple organs-on-a-chip. Lab Chip 13 (18), 3496–3511.

Xia, Y., Whitesides, G.M., 1998. Soft lithography. Annu. Rev. Mater. Sci. 28, 153–184.

Yamaya, M., Finkbeiner, W., 1992. Differentiated structure and function of cultures from human tracheal epithelium. Am. J. Physiol. 262, 713–724.

Zervantonakis, I.K., et al., 2012. Three-dimensional microfluidic model for tumor cell intravasation and endothelial barrier function. Proc. Natl. Acad. Sci. U.S.A. 109 (34), 13515–13520.

CHAPTER 21

Product and Process Design: Toward Industrial TE Manufacturing

Ioannis Papantoniou[1,2], Athanasios Mantalaris[3], Maarten Sonnaert[1,4], Toon Lambrechts[1,5], Jean-Marie Aerts[1,5], Liesbet Geris[1,6], Jan Schrooten[1,4]

[1]Prometheus, Division of Skeletal Tissue Engineering, KU Leuven, Leuven, Belgium; [2]Skeletal Biology and Engineering Research Center, KU Leuven, Leuven, Belgium; [3]Biological Systems Engineering Laboratory, Centre for Process Systems Engineering, Department of Chemical Engineering, Imperial College London, London, UK; [4]Department of Metallurgy and Materials Engineering, KU Leuven, Leuven, Belgium; [5]Division M3-BIORES: Measure, Model & Manage Bioresponses, KU Leuven, Heverlee, Belgium; [6]Biomechanics Research Unit, Universite de Liège, Liège, Belgium

LEARNING OBJECTIVES

- An understanding of the challenges that tissue-engineered products will face at the translational phase
- Overview of the sequence of process steps involved in tissue engineering manufacturing
- Possible strategies to manufacture tissue-engineered products
- How current technologies fit and can be best used for autologous and allogeneic tissue-engineered products
- How to determine quality characteristics and link them to the production process environment
- Using monitoring and control for real-time follow-up of TE product maturation
- Understand the importance of models for the design of novel tissue-engineered products and bioprocess optimization
- What is downstream process in the case of TE manufacturing
- Integration of process steps in larger schemes, "whole bioprocessing" approach

21.1 INTRODUCTION

The steady increase of early- and late-stage clinical trials involving cell therapy applications, as well as the presence of approved commercial tissue-engineered (TE) products in the market, strongly indicates that the cell therapy industry is

Tissue Engineering.
Copyright © 2015 Elsevier Inc. All rights reserved.

Table 21.1 Abbreviations for Commonly Used Terms Related to Tissue Engineering Manufacturing

TE	Tissue engineering
ATMP	Advanced therapeutic medicinal product
EMA	European Medicines Agency
FDA	Food and Drug Administration
cGMP	Current good manufacturing practice
QbD	Quality by design
DoE	Design of experiments
CQA	Critical quality attribute
CPP	Critical process parameter

on its way to evolve into a novel healthcare sector (Mason et al., 2013; Mason and Manzotti, 2010). Apart from clinical efficacy, TE products will need to be manufactured in cost-effective, scalable, and robust bioprocesses that at the same time also meet the requirements of regulatory bodies in terms of quality control (QC) and good manufacturing practice (GMP) (dos Santos et al., 2011; Ratcliffe et al., 2011) (Table 21.1). TE product manufacturing will require a series of process steps from biopsy harvest and stem cell isolation for autologous strategies, or starting from master cell bank frozen vials for allogeneic strategies, to end product formulation and implantation. Currently, stem and progenitor cell isolation, expansion, differentiation, and three-dimensional (3D) tissue construct formation consist of a series of conventional manual and static techniques (Figure 21.1), heavily depending on operator expertise. Hence these process steps are still suboptimal and uncontrolled, with high risk of contamination and processing inconsistency (Mason and Hoare, 2007; Placzek et al., 2009).

In addition, manual operations are an additional source of complexity to TE construct production, where the fusion of numerous biological "raw" materials is required to manufacture complex final products. Furthermore, end product quality attribute profile (such as cell number, phenotype, extracellular matrix content/morphology) will be constantly affected by the bioprocess environment. Hence the translation of laboratory-scale produced 3D engineered constructs to clinically effective and economically viable products requires the development of efficient and robust manufacturing processes that will ensure consistent product quality according to regulatory requirements addressing key barriers to translation (French et al., 2014). An increasing number of recent review papers is highlighting the need for a concerted effort, between academic, clinical, regulatory, and industrial partners in order to outline the landscape for research and translation suggesting a global standardization and harmonization

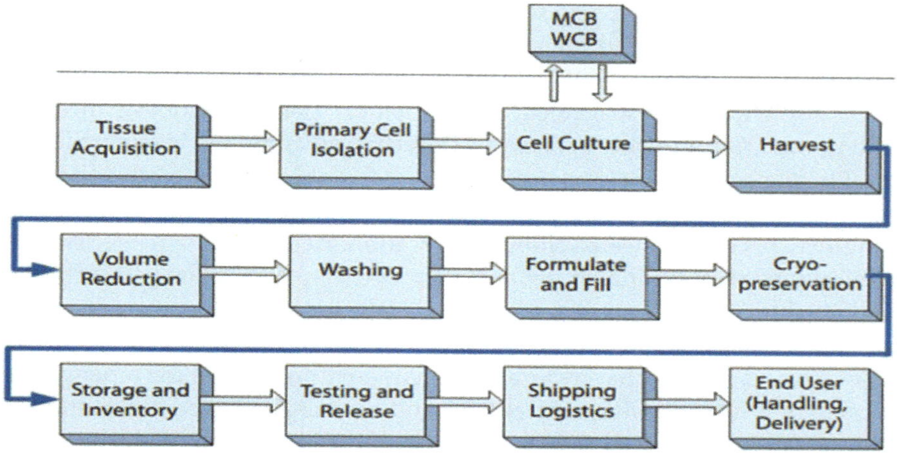

FIGURE 21.1

Sequence of units of operations involved in tissue-engineered (TE) manufacturing. Processes from biopsy acquisition (or master cell bank (MCB)) through cell expansion, cell harvest condensation, and formulation are all components of the pipeline that will be required for cell and TE construct production.

(European Science Foundation how to advance cell-based advanced therapies in Europe: http://www.esf.org/index.php?eID=tx_nawsecuredl&u=0&file=fileadmin/be_user/research_areas/emrc/RNPs/Remedic/2013/White_paper_REMEDIC_22_July_2013_final_v1.pdf&t=1396269348&hash=e5e0dbdbaa744fed8a1123d88adbe12571469553).

This global attempt should follow rational and orchestrated approaches rather than trial and error ones, from stem cell niche to process scale (Kirouac and Zandstra, 2008), taking into account the development of commercially viable and robust bioprocesses that will ultimately allow for a cost-effective production of TE products (http://scec.gatech.edu/sites/default/files/StemCellEngineering-CompleteReport.pdf) (Nerem, 2012).

21.1.1 Satisfying cGMP Requirements

Current good manufacturing practice (cGMP) regulations were first introduced by the US Food and Drug Administration and are also implemented in Europe (Commission Directive 2003/94/EC). cGMP regulations aim to ensure the identity, quality, safety, purity, and potency of TE products (FDA cGMP guidelines, http://www.fda.gov/Drugs/GuidanceComplianceRegulatoryInformation/Guidances/ucm064971.htm). All clinical trials currently in the United States are obliged to produce all cell products following FDA cGMP guidelines (Code 21 CFR 210–211, 312, 600, and 1271). Equivalent regulatory rules have also been established in Europe (EU Directive 2003/94/EC) and many other countries and regions internationally (ICHQ7).

To achieve this, all processing stages should adhere to the guidelines by using cGMP-grade reagents and materials as well as cGMP-compliant systems that operate under current cGMP conditions. Besides the manufacturing process itself, also a quality management system that can precisely monitor, measure, and record critical quality parameters in each processing step is required. This quality management systems need to ensure that the final TE product is manufactured according to the desired safety, quality, quantity, purity, and potency requirements of the targeted clinical application. Hence to obtain effective and clinically relevant cell-based therapeutic products, manufacturing in stringent and reliable context is needed. More specifically for TE product manufacturing all "raw" materials, like biomaterials, substrates, culture media, or biomolecules (growth factors) that create an adequate environment for the propagation of stem cells in an undifferentiated state and/or their controlled differentiation into more mature cells, should (at the start of the process) individually satisfy cGMP requirements.

Current good manufacturing practice (cGMP) requirements were established to be flexible in order to allow each manufacturer to decide individually how to best implement the necessary controls by using scientifically sound design, processing methods, and testing procedures. The "c" in cGMP stands for "current," meaning that technologies and systems that are up-to-date should be employed in order to comply with the existing regulations. However, given the complexity and range of the expected TE products a variety of customized "rules" would be required at least at the early phases to conduct a product-based characterization and therefore it should be noted that cGMPs are only minimum requirements. Approved cell-based advanced therapeutic medicinal products (ATMPs) such as MACI (matrix-induced autologous chondrocyte implantation), aiming to repair cartilage defects (Genzyme/Sanofi, France), and ChondroCelect® (TiGenix, Belgium) are the only licensed cell-based ATMPs on the market in Europe and could provide influential blueprints for their cGMP procedures (Warren, 2013) (Table 21.1).

21.1.2 Quality by Design: Pathway for Optimized Translational Strategies

Quality by design (QbD) is a systematic approach to drive development that begins with predefined objectives and emphasizes product and process understanding and process control, based on sound science and quality risk management. The application of the QbD principles in TE bioprocess development has been brought forward as a strategy to ensure that manufactured TE products will possess predefined quality characteristics addressing emerging regulatory requirements and ensuring a functional in vivo behavior, facilitating thus the road to clinical use and commercialization (Rathore and Winkle, 2009). Primary QbD considerations that need to be taken into account while designing bioprocesses should focus on (1) the materials to

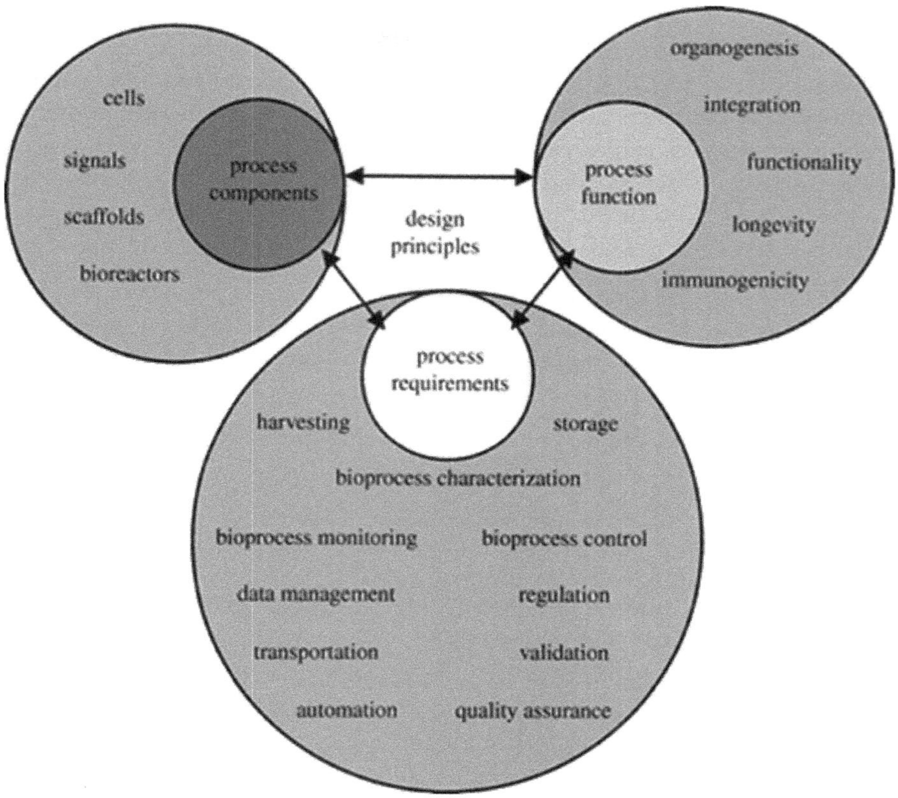

FIGURE 21.2
Design principles for stem cell and tissue engineering bioprocessing. *Figure from Placzek et al. (2009).*

be used (process components), (2) the various manipulations and assessments of the materials (process requirements), and (3) the performance of the output product(s) (process function). In the case of TE products, their performance assessment would consist on the their regenerative potential upon implantation (Figure 21.2).

Understanding and refining of the manufacturing process includes:

- To explore and identify the material attributes and process parameters that can have an effect on final TE product critical quality attributes (CQAs)
- To determine functional relationships that link CQAs to critical process parameters (CPP)

This means that out of a range of final quality characteristics of a TE product (quality target profile) only a subset may be linked to its efficacy and therapeutic potential, and should be therefore controlled and built in during its manufacturing stage. This should be carried out for each process step involved in the manufacturing of TE products. For instance, homogeneity, during the scaffold seeding process step, may be a more important quality attribute than the seeded cell number for the therapeutic value of the TE construct, and therefore should be termed a CQA. This means that the manufacturer should guarantee homogeneity prior to maximizing cell presence. It should be noted that the quality attributes of upstream processes will not only affect the performance of subsequent processes but also end-product quality profile and its performance in vivo. Those TE product quality attributes that will result in the desirable in vivo behavior should be identified and included in the optimization loop.

One of the most important concepts in QbD is termed as "design space" and its definition was given by the international conference on harmonization of technical requirements for Registration of Pharmaceuticals for Human Use ((ICH) Q8 guidance document), as "the multidimensional combination and interaction of input variables that have been demonstrated to provide an assurance of quality."

Multifactorial design of experiments (DoE) is a component of QbD strategies and has been suggested as an important tool to develop efficient procedures that enable multiple factor experiments. As TE process and products intrinsically consist of a multitude of variables, DoE allows data to be obtained in such a way that it yields consistent and objective conclusions on the involvement and statistical significance of all investigated parameters but also of their interactions. Hence process parameters can be linked with TE product quality attributes via statistical correlations named response surfaces (Figure 21.3). Furthermore, DoE requires a smaller number of experimental runs, while, at the same time, covering a broader knowledge space than a one factor at a time

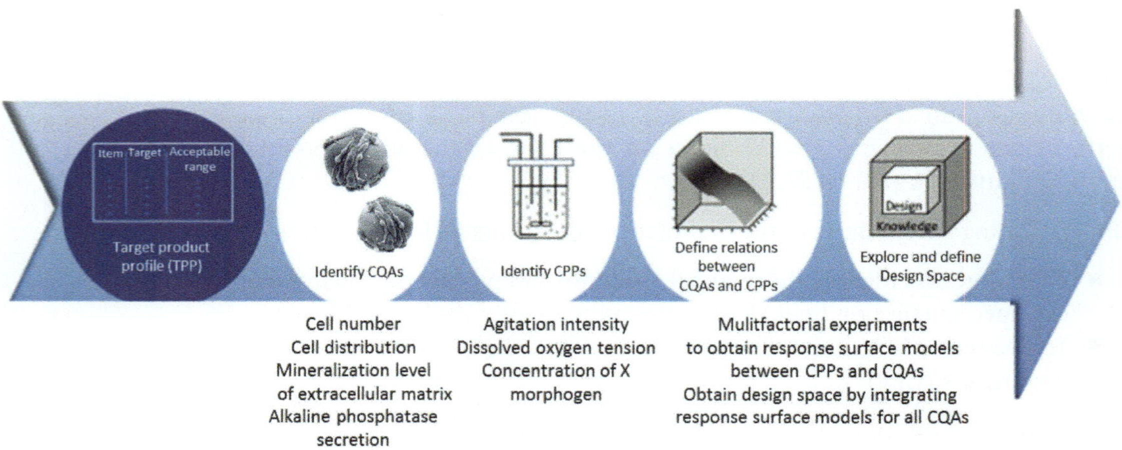

FIGURE 21.3

Typical approach for QbD implementation in bioprocess development and the role of multifactorial design of experiments and analysis. A case study for bone tissue engineering is shown. *Adapted from Mercier et al. (2014).*

approach. DoE has been used for optimizing complex medium compositions for stem cell culture (Lim et al., 2012). Multifactorial DoE was employed to investigate the effect of inoculation (seeding) density and agitation rate for stirred suspension bioreactors while expanding human embryonic stem cell aggregates. Significant interaction effects between inoculation density and agitation rate specifically in the case of exponential growth rates were observed. This study showed that a stepwise optimization may result in missing out on the real optimal operation regime (Hunt et al., 2014).

21.1.3 Allogeneic versus Autologous Strategies

Tissue-engineered therapies are often divided into two main categories: "patient-specific"/autologous and "off-the-shelf"/allogeneic (Wang et al., 2013a). Pioneering cell therapies (e.g., blood cell transfusions, bone marrow transplantation) were "patient-specific," due to the need for histocompatibility matching. More recent therapies, however, are based on cell types that are less immunogenic (e.g., mesenchymal stem cell (MSCs)) and could therefore be more amenable to the development of "off-the-shelf" products, more similar to traditional biopharmaceutical products (see Section 23.1.2). An "off-the-shelf" cell therapy product is likely to be mass produced in large-scale processes, taking advantage of bioprocess technology and also efficient scale-up strategies. The manufacturing process usually involves the establishment of master and working cell banks (facilities that store *cells* of specific genome for the purpose of future use in a product or medicinal needs) that will then be used for the production of large numbers of cells for subsequent processing. Cells will need to go through a method of isolation and banking, followed by bioprocessing (expansion, controlled differentiation, and sorting) in a centralized manufacturing facility to produce the cell and neotissue type of choice for clinical use (Figure 21.4). In contrast, "patient-specific" therapies are based in autologous strategies or in tight histocompatibility matching and thus production is likely to be performed on an individual basis (Figure 21.4).

Each of these approaches will likely require different solutions in terms of bioprocessing. Scalability can be implemented either in volume-based metrics (scale-up) or in unit-based metric (scale-out). "Off-the-shelf" therapeutic products will most likely be manufactured based on a "scale-up" strategy, which consists in the development and optimization of cell processes that begin at a small scale while subsequently, progressively increase the volume of the system, while maintaining the efficiency of the process. This approach, however, may not be appropriate for "patient-specific" therapies, which will have to be "scaled-out" in tightly controlled bioprocesses. This means that the process will have to be extensively replicated to increase the final cell number due to the large number of strictly parallel culture events. Also cell performance unpredictability across this parallel processing will have to be taken into account due to "donor

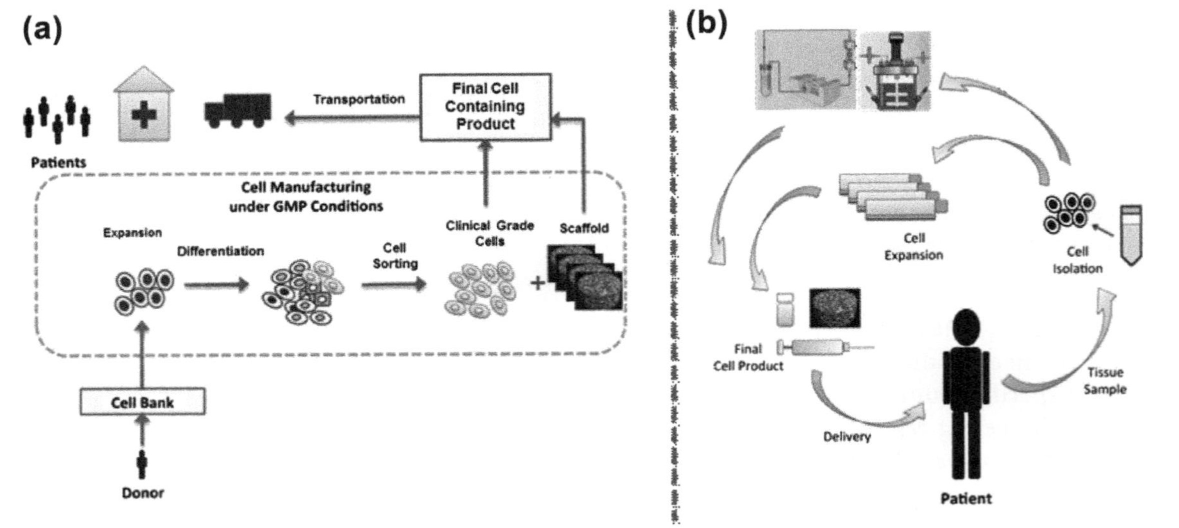

FIGURE 21.4
(a) Flow chart for the production of an "off-the-shelf" cell therapy product. Allogeneic treatments are likely to be mass-produced using large-scale bioprocessing. (b) Flow chart for the production of a "patient-specific" cell therapy product. Such therapies are based on autologous strategies or in tight histocompatibility matching, and thus production is likely to be performed on an individual basis. *Adapted from Fernandes et al. (2014).*

to donor" variability. However, in the context of "allogeneic" tissue engineering a combined approach composed of an initial step aiming at scale-up for cell expansion, followed by a "scaled-out" perfusion culture for subsequent differentiation stages or "neotissue" formation, could be followed.

21.2 BIOREACTOR SYSTEMS FOR TE PRODUCT MANUFACTURING

Bioreactors as already discussed will play a crucial role for TE product development. They provide the means to overcome limitations of conventional manual methods by (1) delivering sufficient cell numbers for multiple doses per batch and hence provide scale-up potential (Yeatts and Fisher, 2011; Zweigerdt et al., 2011; Wang et al., 2013b) and (2) helping to develop structurally defined and functionally effective complex 3D engineered constructs at the patient-scale using scale-out strategies (Wendt et al., 2009; Salter et al., 2012) (Table 21.2). The ability to develop automated bioreactor systems with steady-state control of bioprocess parameters such as flow rate/shear stress, dissolved oxygen tension, pressure drop (for perfusion bioreactor systems), and nutrient supply minimizes complexity and variability of the environment that the cells will experience (Martin et al., 2009; dos Santos et al., 2013). At present commercial platforms that succeeded in automating 2D cell culture such as Fraunhofer's—"The Tissue Factory" (http://www.tissue-factory.com/en/The_Tissue_Factory.html)—are available, showing the maturation of the field. However, automation will ultimately rely on information

Table 21.2 Bioreactor Systems for Cell Culture Advantages Disadvantages Role in the TE Manufacturing Pipeline

Type	Advantages	Disadvantages	Scale-up/Scale-out
Stirred suspension	Homogeneity in culture environment due to mixing Allows cell–cell interactions	Excessive agglomeration between aggregates and microcarriers Sensor readouts may prove inaccurate due to diffusion within 3D aggregates	Scale-up
Wave bag	Disposable minimizes contamination risks Easy to scale-up	Cost Control of process parameters	Scale-up
Fixed bed/perfusion	Allows cell–cell and cell–ECM interactions Incorporation of bioactive scaffolds Direct control over shear stress development Accurate sensor readouts at the outlet of the bioreactors	Low volume Spatial gradient build-up No sampling of cells is possible during culture Difficulties in cell harvest	Scale-out
Hollow fiber	Easy to monitor and control Low shear stress	Gradient build up at the hollow fiber interface 3D cell growth may alter flow patterns within the hollow fibers	Scale-out
Multiplate	Allows image follow up Close to the low risk commonly used paradigm of the T-flask	Nonphysiological cell culture environment Difficulties in cell harvest	Scale-up

provided by the use of closed bioreactor systems that incorporate noninvasive online monitoring modalities. These modalities will allow the follow up of the quality characteristics of manufactured ATMPs during stem cell culture. Reproducible bioprocesses that will deliver the quantity and quality of the required cellular and tissue product should be the cornerstone for successful translation strategies (Yeo et al., 2013).

21.2.1 Bioreactors for Allogeneic Strategies

Allogeneic products will most likely be produced in large quantities and will therefore require suspension culture systems that will allow for scale-up in order to meet the demand. Bioreactor platforms available for producing adherent cells—including planar technologies, packed-bed systems, and suspension platforms such as microcarriers and aggregate cultures—have been employed for their potential to satisfy batch requirements at different scales. Microcarrier-based bioreactor systems such as stirred tanks (Chen et al., 2013; dos Santos et al., 2011) and spinner flasks (stirred suspension bioreactors) (dos Santos et al., 2011) and also wave bags (Timmins et al., 2012) have been successfully used for human MSC expansion. Recently, a number of studies employed xeno-free components for the expansion of stromal MSCs as well as for human pluripotent stem(hPS) cells (Fan et al., 2014) and induced pluripotent stem (iPS) cells (Wang et al., 2013b) showing alignment with GMP requirements. Cell expansion using 3D aggregates has also been achieved in shake flasks for MSCs (Frith et al., 2010) and hPS cells (Abbasalizadeh et al., 2012) and in hollow fiber systems (Nold et al., 2013) with the retention of their differentiation potential (Figure 21.3). Furthermore, commercial stem cell expansion systems already exist. These platforms, like the Terumo Quantum® hollow fiber and the single-use Integrity Xpansion™ multiplate bioreactor system (see Figure 21.5), provide a GMP platform with enhanced reproducibility and scalability.

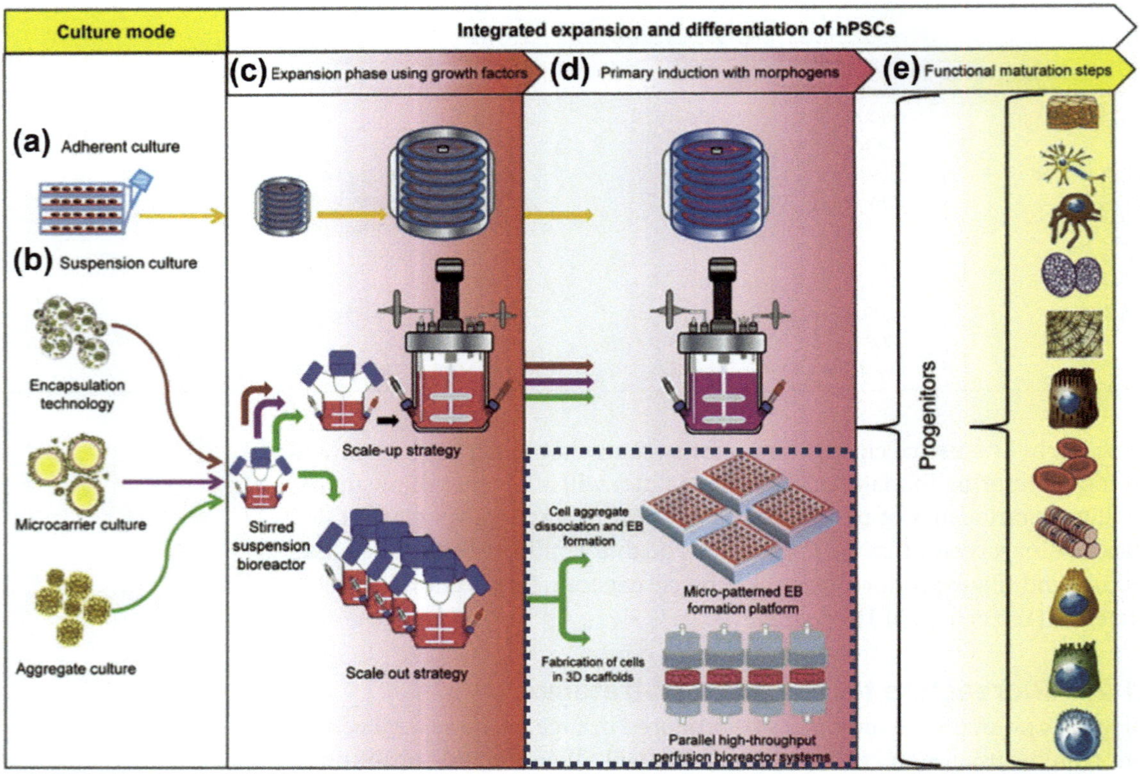

FIGURE 21.5

Scalable platforms for integrated expansion and differentiation of hPS cells including integrated expansion and differentiation of hPS cells in dynamic adherent culture systems (a) and suspension culture systems (b) comprising encapsulation technology, microcarrier-based cultures, and aggregate culture in stirred suspension bioreactors. Differentiation could be initiated after reaching desired undifferentiated cell numbers (c) by replacing the expansion media with differentiation media for early differentiation in the same expansion culture system or dissociating the cell aggregates in aggregate culture systems and transferring the single cells to high-throughput perfusion systems or micropatterned cell aggregates based on targeted differentiation (d) and then starting functional maturation steps using different morphogens to produce target progenitor or terminally differentiated cells (e). *From Saeed and Baharvand (2013).*

21.2.2 Bioreactors for Autologous Strategies

As the reintroduction of cells to the patient will require a rapid in vitro stage prior to implantation, tightly controlled systems providing reliable monitoring potential should be used. For the cells-seeded-on-3D-scaffold paradigm, the flow-through perfusion bioreactor is the reactor of choice (Marolt et al., 2012). Upon cell attachment to the carrier/scaffold, packed-bed perfusion systems provide continuous control options that overcome the limitations of traditional, standard culture methodologies. A broad range of variations on these bioreactor systems has been described to date in literature for the production of 3D neo-tissue constructs (Frohlich et al., 2010; Grayson et al., 2010; Papantoniou et al., 2012; Kim and Ma, 2012). For example, bone repair applications via the intramembranous (Yeatts et al., 2014; Janssen et al., 2010) as well as endochondral bone forming (Scotti et al., 2013) pathway have been reported upon implantation of human MSCs expanded in perfusion bioreactors in animal models.

21.3 QUALITY CONTROL FOR TE PRODUCTS—A MULTISCALE APPROACH

Tissue-engineered product characterization and analysis will be of higher complexity than for other biochemically derived biological products. Moreover, QC is of high importance due to potential manufacturing process variability and the living nature of the product. The development of therapeutic stem cell containing products such as 3D TE constructs will require insightful QC that will guarantee identity, quality, purity, safety, and potency. In most cases, a QC pipeline will also help to further optimize the manufacturing processes. Cell identity is normally measured by marker gene and protein expression analysis using quantitative real-time polymerase chain reaction (qPCR), enzyme-linked immunosorbent assay (ELISA), and flow cytometry. For cells that satisfy this standard set of identity criteria, multilineage differentiation assays that are typically associated with stem cell properties will be needed to further identify stem cell populations. For example, MSCs should be able to differentiate to the osteogenic, adipogenic, and chondrogenic lineage. However, these established analysis methods are not the only cell-related QC options, noninvasive measurements could also provide information regarding the properties of the cultured cells and tissue. Upon thorough validation, they could be sufficient for the determination of TE product quality attributes. For example, such methods have been employed for the determination of cell content during perfusion bioreactor operation via metabolic activity assays (Zhou et al., 2013b) oxygen consumption (Santoro et al., 2012; Lambrechts et al., 2014), heat production due to cell growth (Santoro et al., 2011), and metabolic imaging (Ward et al., 2013). However, in order to obtain more insightful and accurate information, where volume averaged sensor readouts might prove insufficient, fluorescent-based imaging of oxygen sensitive fluorescent microbeads (Lambrechts et al., 2013) could fill in the gap.

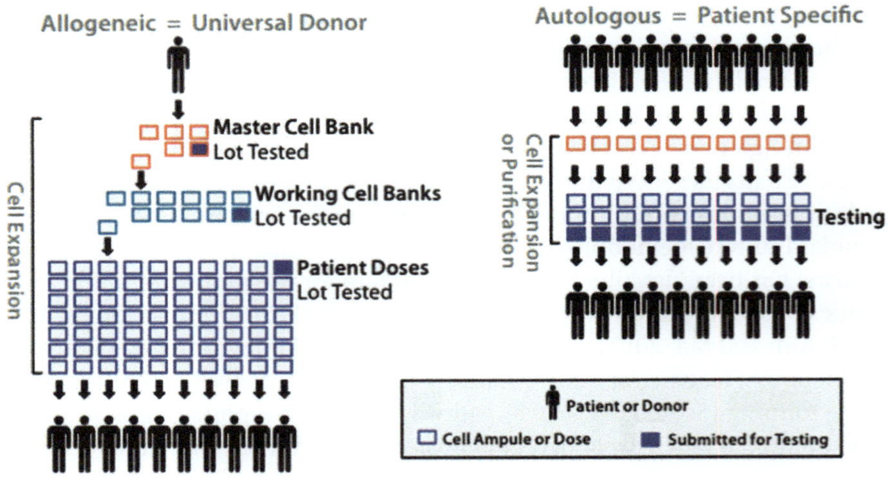

FIGURE 21.6

Quality control (QC) strategies for allogeneic and autologous TE products. For the first strategy, representative lots could be tested due to the homogeneous starting cell population. In the second case, a more stringent QC strategy should be employed since the variability introduced in cell populations due to the variety of donors will impact on process performance and TE product quality attributes. *Figure from Brandenberger et al. (2011).*

A crucial aspect toward noninvasive QC for TE products would be the development of novel methods or the incorporation of existing technologies (ELISA, mass spectrometry) in bioreactor set-ups that would allow online biochemical readouts (i.e., gene expression, protein secretion, cell surface antigen expression). It should be noted that when dealing with clinical delivery of TE products, quality properties regarding spatial properties should also be taken into account. For example, extracellular matrix quantity and its distribution throughout the produced TE construct require a tailored characterization approach. Imaging techniques such as computed tomography could provide possibilities to obtain this information for TE constructs in a nondestructive way (Papantoniou et al., 2013). Overall, the QC strategy followed will be determined by the manufacturing context that will be required for the clinical delivery of the TE product (Figure 21.6).

21.4 ONLINE DATA-BASED MONITORING–CROSS-TALK BETWEEN PROCESS PARAMETERS AND TE CONSTRUCT QUALITY ATTRIBUTES

Online noninvasive bioprocess monitoring has gained attention since the FDA launched the process analytical technology (PAT) initiative in which biopharmaceutical companies are encouraged to adopt monitoring tools to ensure predefined final product quality (Teixeira et al., 2009). However, monitoring is only the first step toward the controlled production of tissue constructs with a robust clinical success rate. Ultimately an improved control of the TE construct quality is envisioned through the measuring, modeling, and managing of critical cell culture parameters (M^3C) (Carrondo et al., 2012). The complexity and inherent variability of biological systems makes their control a challenging task. This becomes even more challenging when in vitro processes aim to mimic the developmental stages of embryo or organ maturation requiring spatial and temporal manipulation of the bioprocess (Lenas et al., 2009; Ingber et al., 2006). However, interesting opportunities arise within GMP environments for TE production as large amounts of high-quality data related to stem cell and TE construct properties will be continuously gathered. Examples are, among others, measurements on environmental culture conditions (dO_2, pH, CO_2), image analysis (i.e., confluency, morphological features, collagen presence), and omics datasets. With the rise of automated cell culture processes the amount of datasets will only increase, both sample-wise due to higher sampling frequency and also parameter-wise due to the development of new modalities. Monitoring of critical culture parameters on its own is the key for process control and fault detection, but the real challenge will be to translate common process readouts into interpretable TE construct quality attributes. At the final stage this could provide the potential to link end TE product characteristics to in vivo performance and ultimately predict in vivo outcome upon implantation (Figure 21.7).

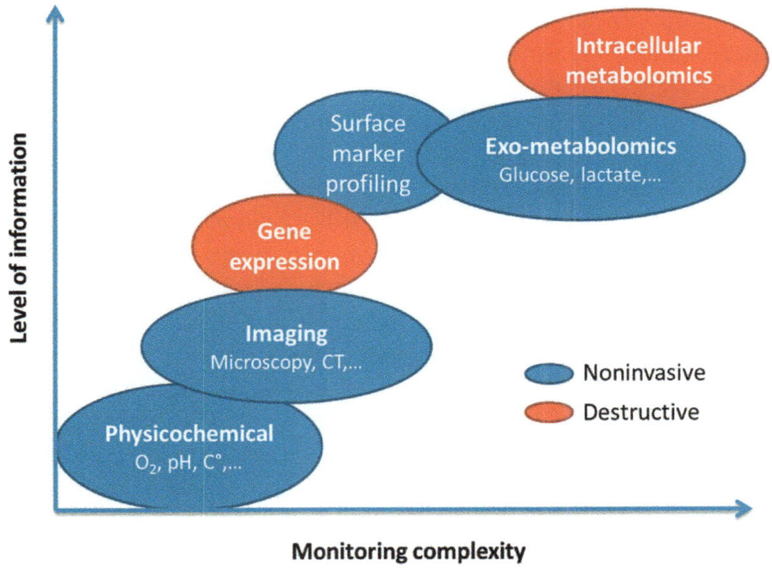

FIGURE 21.7

Schematic ranking of the trade-off between the complexity of analysis techniques and the level of information they can provide. The selection of the most appropriate analysis technique used is application specific. The challenge will be to gain the right level of information with (a combination of) the most accessible technologies.

21.4.1 Image Data Readouts

Visual inspection, either manual or automated, is common practice to assess cellular characteristics and can often be linked to, for example, stem cell differentiation. Certain process development challenges, such as the reprogramming success rate of iPS cells (Smith et al., 2010) or the likelihood of differentiation commitment (Roccio et al., 2013), can be nondestructively assessed in early culture phases with imaging-based algorithms. The next step is to leverage this information from monitoring algorithms to predict quality attributes of the construct. For example, Matsuoka et al. (Matsuoka et al., 2013) used phase-contrast microcopy to obtain time-series of morphological changes to subsequently develop a prediction model of osteogenic differentiation. In this model computational machine learning (ridge regression) was applied to make morphology-based predictions of cellular quality (alkaline phosphatase activity and mineral deposition at day 14 and 21, respectively, after osteogenic induction) taking into account patient-specific variance (Figure 21.8). However, this method is only suitable for 2D cell culture systems. The extra dimension when culturing in 3D TE constructs would require for more powerful imaging modalities such as computed tomography systems to be incorporated in bioreactor operation (Hagenmuller et al., 2010) to obtain spatial information regarding the whole tissue under development.

21.4.2 Online Sensor Data Readouts

The incorporation of multiple sensors to frequently monitor environmental culture conditions during culture has been suggested by PAT strategies. Although effective for process monitoring and fault detection, dO_2 and pH readouts are no direct measures of cell or tissue construct quality. Several strategies exist to indirectly extract more quantitative information on TE quality attributes such as cell number and cellular activity based on oxygen measurements in bioreactors (Santoro et al., 2012). Volkmer et al. (Volkmer et al., 2012) not only monitored the oxygen concentration within a TE scaffold, but also controlled the cellular microenvironment based on an oxygen concentration triggered feedback loop to steer the medium feed rate. However, provided the biological complexity of cells or a TE construct, it is unlikely that environmental physicochemical measurements alone will deliver a detailed signature of the construct quality and be able to robustly predict the clinical effectiveness of the construct. Meanwhile, dynamics in metabolite accumulation or nutrient consumption have the potential to yield more in-depth assessment of the cellular status (Patti et al., 2012). Glucose/lactate measurements (Schop et al., 2009), soluble signaling factors (Csaszar et al., 2013), and metabolic assays in bioreactors are currently being investigated (Zhou et al., 2013b). By measuring cellular metabolites that are secreted into the medium, the so-called exometabolomics, noninvasive information on the intracellular state up to the

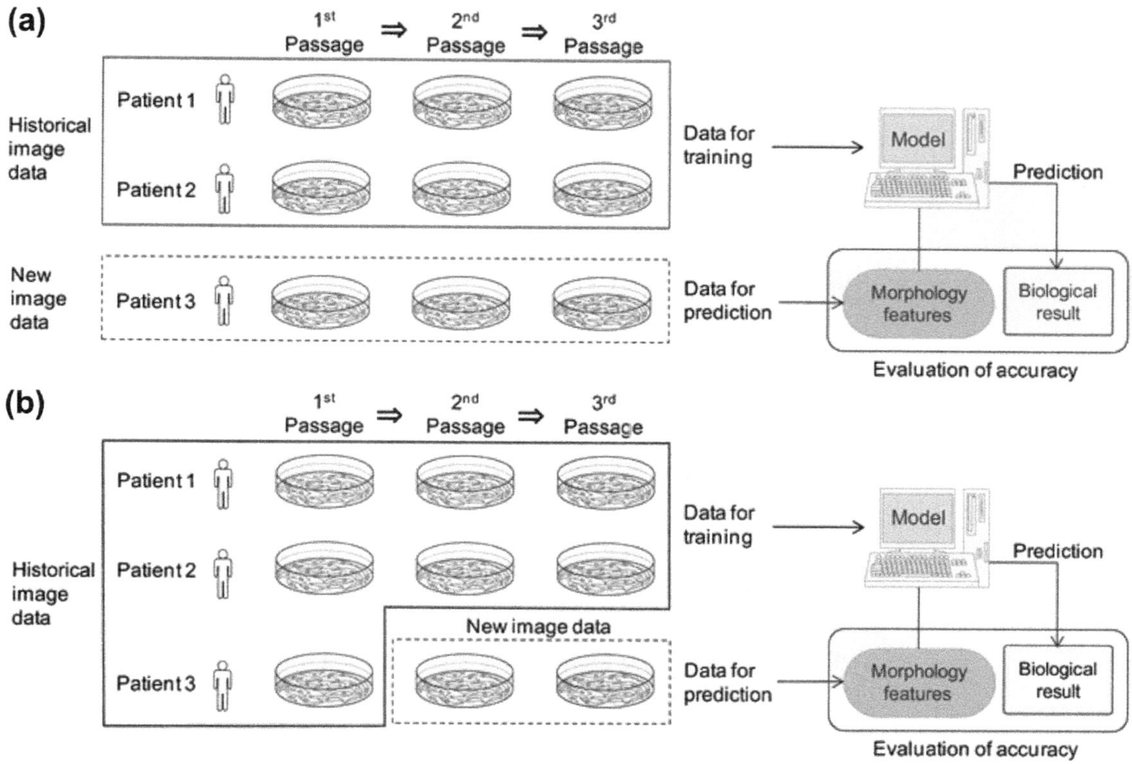

FIGURE 21.8

Historical image data collected from a range of patients were used to train algorithms (Scenario I). (a) Images from all passages of patient 3 were used for prediction. (b) Ongoing prediction scheme: Trained by historical patient datasets and a partial dataset from the new patient. For example, for the prediction of cell potential of patient 3, Scheme I uses images of patients 1 and 2 only. Scheme II uses images of patients 1 and 2, together with some images of patient 3. *Figure from Matsuoka et al. (2013).*

proteomic and genomic level can be obtained. Metabolomics and especially exometabolomics have great potential to become a noninvasive holistic physicochemical characterization of TE cultures.

21.4.3 Online Data Treatment

Ultimately the development of data-based algorithms that will be able to process data online/in real time are essential for the automatic adjustment of bioprocess parameters based on variations of cell behavior (Carrondo et al., 2012). This evolution will be crucial for the final optimization of ATMP manufacturing protocols and for the delivery of reproducible high-quality ATMP end products. Such an approach, in synergy with already existing mechanistic approaches, will enhance our understanding of the highly complex TE bioprocesses and allow the development of predictive models for in vitro cell behavior, and hopefully in vivo outcome.

Table 21.3 Advantages and Disadvantages of Various Modeling Approaches

	Mechanistic Models	Data-Based Models	Data-Based Mechanistic
Providing insight into system	+++	−−−	+
Cost of development	−−−	++	−
Online interrogation	−−	++	++
Controller friendly	+	+++	+++
References	Aehle (2012), Kirouac (2009)	Roberts (2011), Volkmer et al. (2012)	Lambrechts et al. (2014), Ashoori (2009), Bennett et al. (2008)

In order to increase the TE construct quality, there is a need for mathematical approaches that are able to systematically integrate multiple data sources and provide us with predictive models for optimization and control of the TE processes. Generally, models can be divided into mechanistic models (deductive approach based on a priori knowledge) and data-driven models (inductive approach based on observed data). From the former, many examples can be found within the TE field under the form of conservation equations, stoichiometric models, reaction kinetics, etc. Data-based models are found within the biological process technology under the form of regression models, principal component analysis, and neural networks. Mechanistic models may be superior in providing insight into the system under consideration; however, they require a significant cost of development, more difficult parameterization, and are generally harder to solve. Data-based models provide less insight into the system, but in situations where the data logging surpasses the speed of analysis they provide an ideal basis for online prediction and control. An interesting direction regarding the GMP production of TE constructs, where both increased automation and process understanding are pursued, is the data-based mechanistic approach in which the initial "black-box" model identification stage is followed by a mechanistic interpretation of the model parameters and model structure (Table 21.3). This approach generally results in physically/biologically meaningful low-order models that are able to describe the dominant behavior of a biological system and can be used in real-time controllers.

21.5 ENHANCING IN VIVO PERFORMANCE: AN IN SILICO MEDIATED APPROACH FOR TE PRODUCT DESIGN

The challenge the TE field now faces is to translate (in vivo) regenerative processes to an in vitro environment on an engineering scale, thus making the process measurable and controllable with a limited number of parameters. Nature uses a very complex system of regulatory mechanisms compounded by a huge amount of redundancy to deal with the panoply of internal and external influences. The intricate interplay between all these factors is too complex to be interpreted without the help of computational, in silico, modeling (Jukes et al., 2008; Ingber et al., 2006).

21.5.1 Model Variety and Classification

In silico models exist in many shapes and forms, allowing for very different questions to be tackled. Typically, models are categorized based on the information they use to build the model (Janes et al., 2004). When using only experimental data (such as the large datasets generated by high-throughput techniques), models are called empirical or data-driven. When they are based on researcher interpretation of the data, and the proposed hypotheses on the mechanisms of action, we talk about mechanistic or hypothesis-driven models. Another classification often used for in silico models is based on the spatiotemporal scales that are encompassed. Intracellular or subcellular models most often focus on the regulatory networks defining cellular behavior (Kinney and McDevitt, 2013). Cellular models consider cells as individual agents that can interact with each other and their environment. A level higher, the tissue/organ level, cells are represented in a more continuous way by means of their density. This level is typically suited to look at the overall effect of cellular actions and interactions in the form of neotissue formation in TE scaffolds, i.e., the influence of external influences such as mechanical modulators, nutrient supply, and chemicals. Also the influence of the in vivo environment on the TE product is described at this level. Finally, at the population level, in silico clinical trials are increasingly carried out, especially in the field of critical care and cancer treatments (Roelofs et al., 2012), to optimize the trial design and patient stratification as to limit the expenses and avoid negative side-effects as much as possible.

Given the current underlying difficulty in estimating parameter values, many models may remain limited in the mechanistic insight they provide and in their capability to predict system dynamics in unforeseen conditions. To investigate the impact of the chosen parameter values on the simulation results, sensitivity analyses should be carried out. Sensitivity analyses appear under many different forms, ranging from the most frequently used one-at-the-time (OAT) analysis where only one parameter is altered in each simulation to multifactorial DoE Run approaches providing information on the interaction between experimental process parameters. Finally, in order to explicitly account for the noisy character of the data and processes, stochasticity can be introduced into the in silico models. Explicitly taking into account the uncertainty in the experimental data, can result in the retrieval of a wide range of parameter sets, all capable of fitting the data. Perhaps surprisingly, some predictions will still be very well constrained even in the face of this enormous parameter uncertainty, a property which was coined "sloppy" (Gutenkunst et al., 2007). Even though sloppiness is not unique to biological systems, it is particularly relevant to biology because the collective behavior of most biological systems is much easier to measure in vivo than the values of individual parameters. Using sloppy parameter analysis (SPA), concrete predictions can be extracted from models long before their parameters are even roughly known (Brown et al., 2004), and when a system is not already well-understood, it can be more profitable to design experiments to directly improve predictions of interesting system behavior rather than to improve estimates of parameters. In order for TE to benefit maximally from the advantages in silico modeling can bring, an integrative approach is indispensable in which the development of the in vitro, in vivo, and in silico components of research go hand in hand from the start.

21.5.2 In Silico Implantation of TE Constructs

The host environment is a crucial component of the TE design strategy and the ability to predict the interaction between the host and the TE product is pivotal for many of the processes discussed in this chapter. Silico models allow to combine knowledge on basic biology and TE product behavior to study the effect of, e.g., in vivo scaffold dissolution on local in vivo cell biology (Carlier et al., 2011) and/or blood vessel formation (Checa and Prendergast, 2010). Patients presenting with structurally and/or genetically challenged healing environments pose additional challenges to the TE strategy, but it is most often in those patients that normal healing is impaired and thus TE solutions are required. Silico models are applied to both study the etiology of impaired healing (Geris et al., 2010) and to design novel therapeutic strategies that are able to overcome the additional patient-specific hurdles. These models provide an additional level of (mechanistic) understanding to the data-driven empirical models which use multiparametric techniques to link in vitro characteristics to observed in vivo behavior. Taking into account that parameter quantification of scale-specific models is already difficult, the problem is even worse for multiscale models where parameters have to be defined at different scales. It is widely accepted that as more quantitative experimental data become available to build and constrain the parameter values, the models will be more likely to describe observed behaviors accurately.

21.5.3 Whole Bioprocess Virtual TE

One important application of mathematical models is the study of individual components of TE products such as the cells and the scaffold. Biological systems (TE product behavior being one of them) are naturally multiscale and to understand their behavior fully we must understand the interaction of a number of processes that may occur on diverse temporal and spatial scales. Models focusing on the cell compartment are used to identify the biological state of the cell using statistical methods applied to large datasets (Mentink et al., 2013) or more mechanistic methods built on knowledge of specific relevant pathways. Combining both approaches allows to identify not only the precise state of the cells of the TE product but also of optimized culture conditions providing the potential to push or keep cells in the desired state (see State-of-the-art experiment (Csaszar et al., 2012)). Models regarding the scaffold focus on the mechanical (Song et al., 2013), biochemical (Demol et al., 2011; Bjork and Tranquillo, 2009), and morphological aspects (Saito et al., 2012) of its design with the aim of understanding its influence on the behavior of the seeded cells. More and more these in silico models are combined with a description of the physical environment that bioreactors present to the TE product during the in vitro culture process (in terms of, e.g., fluid flow, mechanical stimulation, and mass transport)(Causin et al., 2013). Optimization of initial cell seeding and initial cellular differentiation is predicted through the adaptation of the bioreactor protocol and scaffold morphology (Spencer et al., 2013). Additionally, neotissue growth during extended culture of the TE products can be captured by multiphysics models combining a description of the physical bioreactor environment with a mechanistic description of cellular behavior and matrix production (Guyot et al., 2014) (Figure 21.9). Once validated, these multiphysics models can become an inherent part of bioprocess development

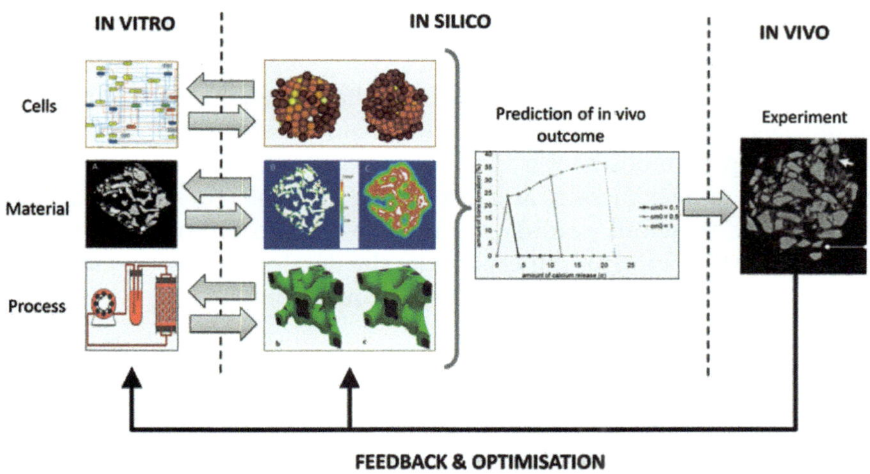

FIGURE 21.9

In vitro–in silico–in vivo optimization loop. The in vitro part (left) shows the raw materials of TE product manufacturing: cells (showing a gene regulatory network, Kerkhofs et al., 2012), biomaterials (i.e., CaP-collagen carrier visualized with nano-computed tomography, courtesy G. Kerkchofs) and process environment (schematic representation of perfusion bioreactor set-up). The properties of these raw materials can be optimized using in silico models for all the constituents of a TE product (middle): cells (aggregating on a microbead, Smeets et al., 2013), biomaterials (degradation and calcium release from calcium containing scaffolds, courtesy V. Manhas), and process environment (neotissue growth in scaffold, Guyot et al., 2014). Additionally, in silico models can combine the raw materials and make predictions on optimal combinations leading to maximal bone formation in vivo (influence of seeding density on bone formation in calcium containing scaffolds, Carlier et al., 2011). In vivo experiments (bone formation in CaP scaffolds, Roberts et al., 2012) close the in vitro–in silico–in vivo optimization loop and confirm the in silico predictions or suggest model adaptations and in more focused vitro experiments.

loop providing an insight view into the TE product in culture by allowing to liaise the bioreactor sensor readouts (e.g., pressure drop) with the biological interpretation in terms of local neotissue growth in the TE product.

21.6 DOWNSTREAM PROCESSING IN TE MANUFACTURING

As already discussed, automation of current lab-scale processes for regenerative medicine and tissue engineering is widely considered to be essential for the successful clinical implementation and industrial translation of these emerging technologies (Ratcliffe et al., 2011; Abbasalizadeh and Baharvand, 2013). As already discussed in this chapter, the use of bioreactor systems is considered to be an indispensable step for cell culture and expansion as part of TE strategies. Although bioreactor systems already enable automated cell expansion in a GMP environment, a series of manual, operator-dependent downstream postprocessing steps are still required to deliver a finalized product. For example, cell harvest steps for the recovery of cells from microcarriers or 3D cell aggregates and also subsequent seeding steps will be needed to reach the end product. Process steps such as cell harvest and upconcentration can be managed on a lab scale for a limited amount of systems. However, this lab-scale approach poses severe limitations when scale-up and scale-out challenges need to be addressed, as it impacts on the quality of the manufactured TE products. Therefore, next to continuing to develop and optimize strategies enabling GMP-compliant cell expansion, the development of suitable downstream processing methods should receive equal attention.

21.6.1 Cell Harvest/Recovery

Cell harvest is proving to be one of the most challenging and critical process steps limiting the industrial implementation of current processes (Cierpka et al., 2013; Liu et al., 2013). Despite the development and intensive research activities in bioreactor technology development, subsequent processes are currently lacking equal attention. Enzymatic retrieval of the cells from the culture surface is the gold standard up till now despite the associated disadvantages such as (1) potential damage to the cells, (2) cost of the procedure, and (3) availability of the GMP grade proteolytic dissociation enzymes (Abbasalizadeh and Baharvand, 2013; Cierpka et al., 2013). The long standing use of these reagents for cell recovery has, however, resulted in the development of highly efficient, minimally invasive methodologies, which have already been successfully implemented in commercial available systems (Roberts et al., 2012; Jones et al., 2013). Similar methodologies have been developed for microcarrier-based cell expansion (Chen et al., 2013; Goh et al., 2013; Zhou et al., 2013a), although the versatility of this culture platform also enables to develop novel, enzyme-independent, cell recovery methods based on, for example, temperature-dependent behavior of the surface (Tamura et al., 2012). Recently, the use of carrier-free culture methods for anchorage-dependent cells based on cell aggregate formation has emerged as a novel, promising concept for cell culture. Although the classical enzymatic methods can be used to recover cells expanded in these culture systems (Abbasalizadeh et al., 2012), the potential to develop enzyme-free methodologies has significant advantages such as the reduced consumable cost, reduced final volumes, and the removal of a potential harmful reagent. Although enzyme-free methods for aggregate dissociation are currently being developed, they are often associated with a significant decrease in viability and other adverse effects ranging from spontaneous differentiation to apoptosis (Abbasalizadeh and Baharvand, 2013; Ratcliffe et al., 2011; Cierpka et al., 2013). Novel methods are, however, being developed which might enable enzyme-free cell recovery from microcarrier and aggregate culture systems (Wallman et al., 2011). However, for 2D-based systems such as multiplate stacks and hollow fiber systems enzymatic cell recovery will probably remain the gold standard. It should be noted that enzymatic dissociation media will require further optimization both in composition and concentrations and a validation in a 3D context should also carried be out. Recently, Nienow et al. (2014) introduced a dynamic harvest methodology to retrieve expanded cells from microcarriers. However, it is evident that only minimal information is currently available and there is ample space for cell harvest process optimization research efforts (Chen et al., 2013).

21.6.2 Volume Reduction

Volume reduction and washing process requirements will be directly linked to the harvest volume, which will be a consequence of the bioreactor platform employed for cell culture and also the volumes used during the harvest process. Independent of the culture and harvest methodology used, the recovered cells will in most cases be suspended in large volumes of medium requiring subsequent upconcentration prior to further processing. Dependent on the final application of the process, different strategies might be of interest although mostly the physical differences between the cells and the fluid in which they are suspended are exploited. When the process targets cell expansion for allogeneic products, a closed production pipeline capable of continuously processing large volumes of cells should be aimed at such as continuous centrifugation (used for the collection of blood plasma (Burd and Schembri, 1993)) or aqueous two-phase separation systems (based on solubility differences (Gonzalez-Gonzalez and Rito-Palomares, 2013)). Other, more recent, developments for constant cell separation aim at the use of microfluidic-based devices such as shown in Figure 21.10, although significant progress still needs to be made in scale-up and scale-out prior to industrial implementation (Autebert et al., 2012; Gossett et al., 2010). For autologous products the use of disposable systems is a more viable approach as it allows for the use of simple, single-use

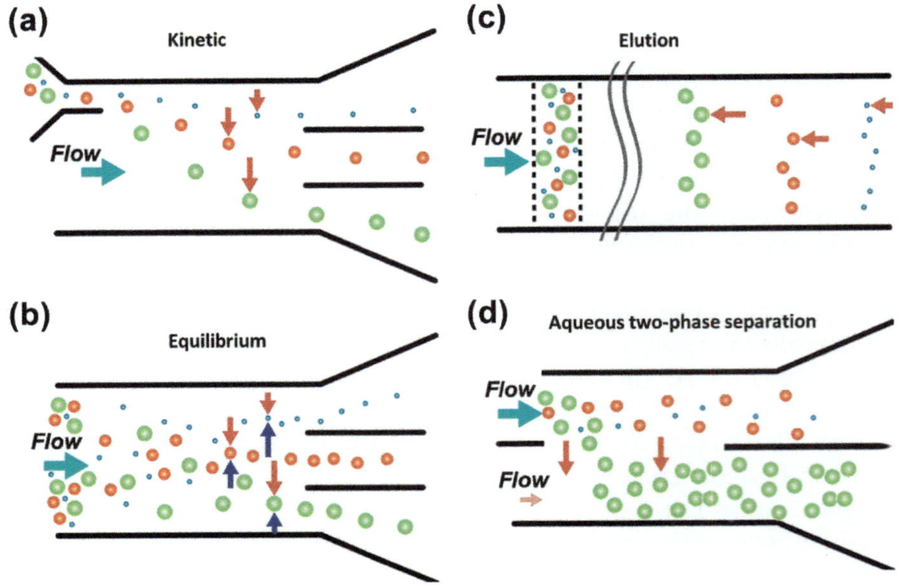

FIGURE 21.10
Generalized separation methods employed in microfluidic devices. (a) Continuous kinetic methods depend on the rate of cell deflection perpendicular to primary channel flow. (b) Continuous equilibrium methods involve migration to property-dependent equilibrium positions. (c) Elution methods depend on forces antiparallel to primary flow to create differential retention. (d) Subcategory of equilibrium-based separation methods based on differences in solubility in two different aqueous phases. *Adapted from Gossett et al. (2010).*

separation methods. Currently, different clinical grade expansion systems such as the Quantum bioreactor (Terumo BCT, USA) and VueLife culture bags (CellGenix, USA) use batch centrifugation for cell concentration (Rojewski et al., 2013; Spanholtz et al., 2011). The scale-out potential of this methodology is, however, limited as it still requires a significant amount of manual handlings – and the integration of alternative strategies such as cross-flow filtration might be required for the successful industrial implementation.

21.6.3 Cell Selection

Cell selection will be of predominant importance before the cell expansion but the use of mixed cell populations could require additional purification steps subsequent to the harvest. Although various methods are available for this purpose, ranging from separation based on physical properties to the use of aptamers and advanced tag-free methods. Significant advances in process development are still required to enable their online implementation (Diogo et al., 2012). While separation based on physical properties such as centrifugation, aqueous two-phase separation, and filtration are reasonably high throughput, these methods offer a low specificity, making them more suited for volume reduction than for high-resolution cell selection (Abbasalizadeh and Baharvand, 2013; Diogo et al., 2012). Monoclonal antibody-based techniques such as fluorescent activated cell sorting (FACS) and magnetic activated cell sorting (MACS) offer a significantly higher resolution, although the labeling of cells with monoclonal antibodies might affect cell function and could induce unwanted responses (Abbasalizadeh and Baharvand, 2013). Aptamer-based cell recognition is a more novel technique which might address these issues, although obtaining a higher throughput becomes once more a limiting factor. The use of affinity chromatography-based methods could enable the required high throughput and could also be integrated in both disposable systems for autologous therapy as well as in a production line for allogeneic products. In combination with additional recognition elements such as aptamers high specificity for the selection of a specific cell population could be obtained. The use of tag-free methodologies would, however, be more beneficial still. Parallelization of novel microfluidic-based developments such as field flow fractionation and dielectrophoresis could enable high-throughput separation of cell suspension; however, these developments are still in their infancy and will require further development especially when scale-up challenges will need to be faced (Abbasalizadeh and Baharvand, 2013; Fernandes et al., 2014; Diogo et al., 2012).

21.6.4 Integrating Units of Operation

Dependent on the culture system used, multiple challenges still need to be met to develop integrated, controlled, and automated systems. Octane Biotech (Kingston, Canada) already delivered proof of concept with the development of the Cocoon system (Martin et al., 2009) which allows the automated processing of a tissue biopsy into a TE construct using a closed and controlled environment. As this system was developed to facilitate the entire process from biopsy to final TE construct, it includes all the required up- and downstream processing modules. Once commercialized, it could offer great potential for bringing lab-scale tissue engineering processes to the clinic. However, in order to efficiently operate and optimize interdependable sequences of units of operation, a "whole-bioprocess" optimization strategy should be followed from early on during development. An understanding of how upstream processes will impact on the performance of subsequent process steps and also on final TE product quality attributes should be kept in mind during bioprocess design and development.

21.7 TOWARD EFFICIENT TE PRODUCT TRANSLATION

The term "advanced" in the acronym ATMPs signifies the novelty and lack of prior history in dealing with living products. Hence the establishment of novel strategies that will have to emerge from a rapidly evolving research landscape is required. Major scientific breakthroughs, such as the introduction of iPS technology, can have major impact on the way "current" manufacturing strategies may be structured up to that point. However, the translation of scientific discoveries to clinic trials and subsequently to the market is a long procedure that will require input from various directions. Recent initiatives at both continental and global scale (European interdisciplinary summit on ATMPs, white paper REMEDIC, May 2–3, 2013, Vienna; Global assessment of stem cell engineering, 2014) have managed to conceptually streamline research developments in stem cell and tissue engineering into a more concrete direction toward translation not only to the clinic but also to the market. The joint "forces" between academia industry and regulatory bodies will be key for the translation of TE products in a hospitable market/environment. Minimum requirements should be clearly defined and baselines given based on a constant discussion regarding international harmonization and standardization. Also some "pioneer" products could set some initial guidelines toward this direction, although enhanced communication and knowledge dissemination will be needed (see Section 23.1.2). Hospital exemption, minimizing national and subnational differences, reimbursement policies, predictive preclinical efficacy and safety testing, need for innovative systems for preclinical testing, product characterization, and product potency will have to be addressed. A recently published 2-year global survey on stem cell engineering funded by the National Science Foundation (USA) concluded that there needs to be an increased role for interdisciplinary research platforms where biologist and engineers and the engineering approach will provide a foundation for the generation of new markets and future economic growth (Loring et al., 2014). The recent establishment of institutes such as the Cell Therapy Catapult, London, UK (https://ct.catapult.org.uk/), and the Centre for Commercialization of Regenerative Medicine, Toronto, Canada (http://www.ccrm.ca/), shows a trend of integration that is required for the realization of the hopes and promises of regenerative medicine.

There is need for technologies that will allow extract information from the performance and fate of implanted TE constructs in order to provide minimal quality requirements that should be met by the TE product quality profile. This will directly provide information for process engineers to target and customize accordingly the production pipeline to meet minimal criteria as obtained from in vivo monitoring. Regarding in vitro TE manufacturing platforms and practices, it may be stated that currently the production of sufficient quantities of stem cells (embryonic, iPS, and mesenchymal) is feasible even in xenogeneic-free platforms, as recently demonstrated by a number of publications. However, further improvements in expansion and differentiation efficiencies based on the manipulation of process parameters are still achievable. Furthermore, the incorporation of developmental biology principles during in vitro 3D TE production will require more elaborate strategies for efficient spatiotemporal delivery of the appropriate stimuli.

Future challenges in bioprocessing and manufacturing will include advanced and sophisticated monitoring platforms that allow monitoring at the cellular level. Developments in monitoring and control strategies that will be tailored for not only specific stem cell types but also for more complex 3D TE construct culture should be pursued by the incorporation of multiple sensors in bioreactor systems. This would provide information- rich processes for the manufacturing of TE products that could meet regulatory demands. Furthermore, monitoring robust bioprocesses (hence minimized variability) can provide high-quality, low-noise data for modelers in order to make more rapid progress in the years to come. Modeling approaches that will be able to capture the dynamic and transient nature of stem cell behavior by predicting characteristics not only of the final cell populations but also of the 3D neotissue will be required. High-throughput platforms will still be needed not only to determine optimal combinations between stem cell populations and biomaterial composition and texture but also for optimizing multicomponent differentiation media formulations.

Recently, the development of decision tools that may screen and identify optimal manufacturing strategies, incorporating cost of goods (CoG), was demonstrated recently for an allogeneic therapy production scenario where various cell expansion strategies were compared computationally in terms of cost-efficiency (Simaria et al., 2014) (Figure 21.11). Similar studies should

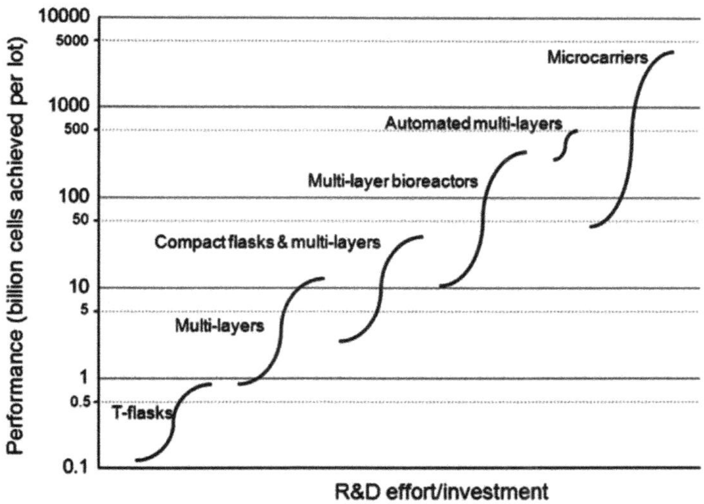

FIGURE 21.11

Conceptual illustration of a technology S-curve showing the evolution of expansion technologies used in cell therapy manufacture. The limits of each S-curve correspond to the amount of cells achieved by the smallest and largest size of each technology type when using the maximum number of units. Automated multilayers refer to L-40 and cL-120. The *x*-axis represents qualitatively the R&D effort required for a company currently using T-flasks to change to other cell expansion technologies. *From Simaria et al. (2014).*

include more units of operation taking into account risk assessment as introduced by donor to donor variability and its implications, especially for autologous production scenaria. The integration of bioprocess economics with bioprocess optimization to assess the economic competitiveness of available systems (i.e., planar, microcarrier-based cell expansion technologies) will be an asset to map optimum TE product translation strategies.

21.8 SNAPSHOT SUMMARY

- The translation of tissue-engineered products to the market will require a new set of technological challenges to be addressed. In order to achieve this, there is a need for engineering principles to be applied to bring biologic discoveries to the market.
- There are two main categories of tissue-engineered products based on the origin of the cells in the TE: (1) allogeneic therapies may result in "off-the-shelf" products ready to be administered to patients and (2) autologous "patient-specific" therapies based on the use of the patient's own cells for the production of the TE product.
- To produce TE products without risk of contamination, GMP grade raw materials and facilities should be the base for TE manufacturing.
- The production of TE products will consist on a series of process steps which need to be designed, understood, and optimized. Currently, there is lack of process characterization while manual operations are still dominant.
- Quality control for TE products should take into account the complexity of the product. A mutliscale approach characterizing cells but also tissue properties (3D) will be needed; hence, apart from standard techniques for cell characterization novel ones should be developed.
- For TE products the end product is linked directly to the process due to the responsiveness of the cultured cells to their environment. Monitoring and controlling the culture environment may be indirectly linked to the quality properties of the TE construct.
- Data derived from sensors should be analyzed by algorithms, if possible, online to extract maximum information and aid in decision making during manufacturing.
- Bioreactor systems can delivery sufficient volumes for industrial production of allogeneic TE products while providing control options and automation potential which is crucial for autologous treatments.

- Downstream processes will be required for the harvest and recovery of cells, their purification, and potentially reseeding to scaffolds. Currently there is lack of information on this aspect of TE manufacturing; however, in a commercial setting this will become necessary.
- Mathematical modeling may be applied for the design of TE constructs by running cost-effective in silico experiments. This can be applied both for in vitro as well as in vivo case studies. Models characterizing cell–scaffold interactions, cell signaling, tissue growth, and in vivo behavior should eventually be linked across scales.

STATE-OF-THE-ART EXPERIMENT

Current expansion strategies expose hematopoietic stem and progenitor cell (HSC) populations to uncontrolled culture environments with limited dynamics due to the unregulated accumulation of inhibitory endogenous factors. This results in compromised expansion levels that could ideally be obtained. In vitro (and in vivo) hematopoiesis is a dynamic and regulated process, through nonlinear feedback control. It should be noted that the rate of factor secretion varies widely among individual factors and does not always correlate with the exponential growth rates of total cell expansion. Overall, a highly complex culture environment is developed with factor concentration dynamics following the dynamics of specific lineage subpopulations or resulting from multiple interacting feedback networks.

In a very interesting work, Csaszar et al. used a simplified model of HSC differentiation in which cell fates were regulated by feedback from secreted molecule cell interaction networks among various cells from different points in the developmental hierarchy. The model was employed to develop a culture strategy by which the controlled and specific inhibition of negative regulators of HSC differentiation allowed the global control of the cell population dynamics. As a preliminary step, they identified numerous factors present in cultures with the ability to inhibit the expansion of HSC and HSC progenitor cells. In order to predict culture strategies that would maximize the abundance of HSC and progenitors, these factors were used as feedback candidates in the computational model. Using two commonly utilized bioreactor setups, namely fed-batch and perfusion systems, their calculations predicted that the fed-batch dilution approach would outperform other methods. When tested experimentally, the predicted protocol gave significantly higher average expansion of HSC and CD34+ progenitor cells after 12 days (figure as shown below).

The ability to modulate secreted factor concentrations and measure corresponding functional outputs of cell expansions will allow for a more precise study of links between specific endogenous protein secretion and lineage subpopulations and their associated cell–cell interactions. This strategy serves as a robust clinically relevant system for rapid and automated in vitro cell expansion as well as a platform for further study of the regulation of cell–cell interactions in vitro and in vivo.

Continued...

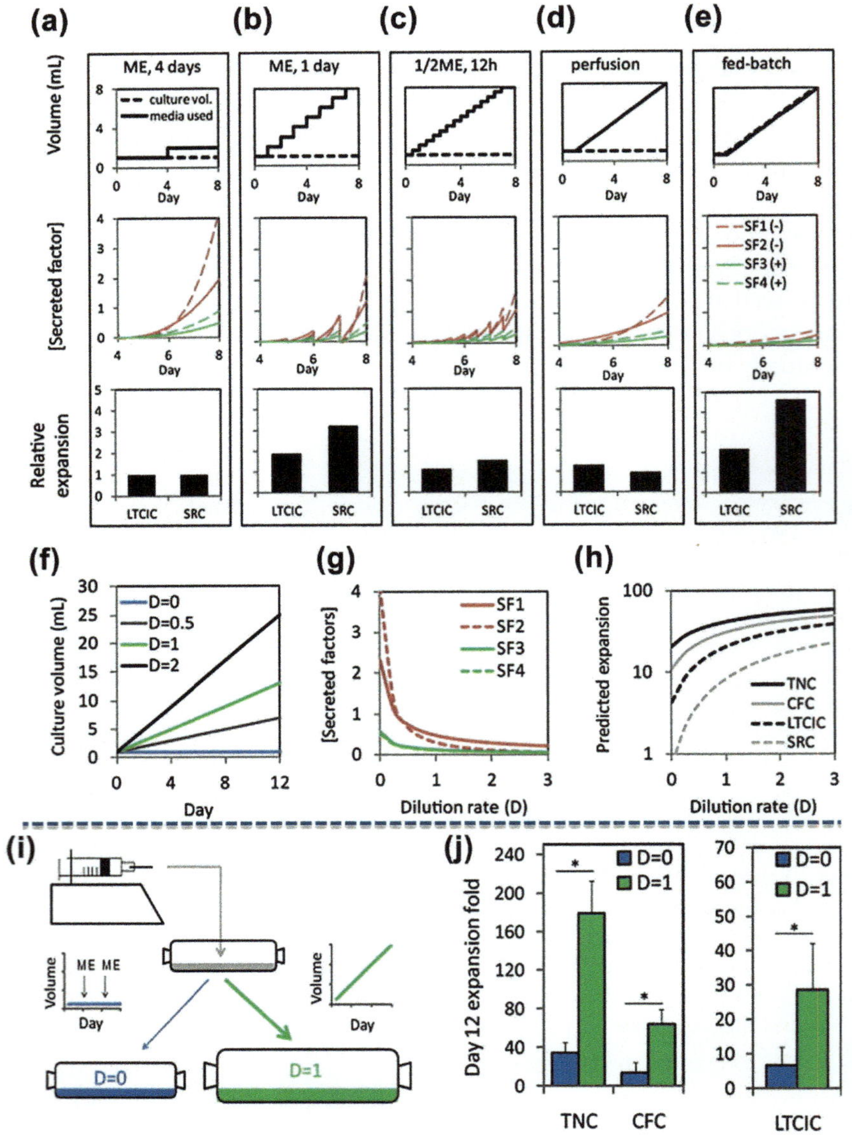

Computational simulations predict a fed-batch strategy, at moderate dilution rates, to greatly reduce secreted factor concentrations enhancing cell expansion. (a–e) Simulated volume, secreted factor concentrations, and relative expansions under different media manipulation strategies; (f) "control" culture with complete media exchange (ME) every 4 days; (g) culture with complete media exchange every 24h; (h) culture with 50% media exchange every 12h; (f) perfusion culture with one unit of media perfused every 24h; (g) fed-batch culture with one unit of media added every 24h. Conditions (f–h) are normalized to same media and cytokine requirements (one additional unit of media every 24h). (f) Media volume requirements for a fed-batch culture at different constant dilution rates, assuming a 1ml initial volume; (g) Predicted effect of increasing constant dilution rate of fed-batch strategy on secreted factor concentrations. (h) Predicted effect of increasing constant dilution rate of fed-batch strategy on population expansions. The fed-batch strategy was experimentally validated in vitro to give significantly improved expansion of progenitor cells (i) Schematic of experimental set-up comparing control (D = 0) strategy with 100% media exchange (ME) every 4 days to fed-batch (D = 1) strategy. (j) Expansions of TNC, CFC, and LTC-IC, after 12 days of culture. $n>5$. Total nucleated cells (TNCs), colony-forming cells (CFCs), long-term culture-initiating cells (LTC-ICs).

REFERENCES

Abbasalizadeh, S., Baharvand, H., 2013. Technological progress and challenges towards cGMP manufacturing of human pluripotent stem cells based therapeutic products for allogeneic and autologous cell therapies. Biotechnol. Adv. 31, 1600–1623.

Abbasalizadeh, S., Larijani, M.R., Samadian, A., Baharvand, H., 2012. Bioprocess development for mass production of size-controlled human pluripotent stem cell aggregates in stirred suspension bioreactor. Tissue Eng. Part C. Methods 18, 831–851.

Aehle, M., Bork, K., Schaepe, S., Kuprijanov, A., Horstkorte, R., Simutis, R., Lübbert, A., 2012. Increasing batch-to-batch reproducibility of CHO-cell cultures using a model predictive control approach. Cytotechnology 64, 623–634.

Ashoori, A., Moshiri, B., Khaki-sedigh, A., Reza, M., 2009. Optimal control of a nonlinear fed-batch fermentation process using model predictive approach. J. Process Control 19, 1162–1173.

Autebert, J., Coudert, B., Bidard, F.C., Pierga, J.Y., Descroix, S., Malaquin, L., Viovy, J.L., 2012. Microfluidic: an innovative tool for efficient cell sorting. Methods 57, 297–307.

Bennett, M.R., Pang, W.L., Ostroff, N.A., Baumgartner, B.L., Nayak, S., Tsimring, L.S., Hasty, J., 2008. Metabolic gene regulation in a dynamically changing environment. Nature 454, 1119–1122.

Bjork, J.W., Tranquillo, R.T., 2009. Transmural flow bioreactor for vascular tissue engineering. Biotechnol. Bioeng. 104, 1197–1206.

Brandenberger, R., Burger, S., Campbell, A., Fong, T., Lapinskas, E., Rowley, J.A., 2011. Cell therapy bioprocessing. BioProcess Int. 9 (S1), 30–37.

Brown, K.S., Hill, C.C., Calero, G.A., Myers, C.R., Lee, K.H., Sethna, J.P., Cerione, R.A., 2004. The statistical mechanics of complex signaling networks: nerve growth factor signaling. Phys. Biol. 1, 184–195.

Burd, T.L., Schembri, C.T., 1993. Separation Chamber Located Radially Outward from Sample Chamber and Connected by Flow Restrictive Channel. Google Patents.

Carlier, A., Chai, Y.C., Moesen, M., Theys, T., Schrooten, J., Van Oosterwyck, H., Geris, L., 2011. Designing optimal calcium phosphate scaffold-cell combinations using an integrative model-based approach. Acta Biomater. 7, 3573–3585.

Carrondo, M.J.T., Alves, P.M., Carinhas, N., Glassey, J., Hesse, F., Merten, O.W., Micheletti, M., Noll, T., Oliveira, R., Reichl, U., Staby, A., Teixeira, A.P., Weichert, H., Mandenius, C.F., 2012. How can measurement, monitoring, modeling and control Advance cell culture in industrial biotechnology? Biotechnol. J. 7, 1522–1529.

Causin, P., Sacco, R., Verri, M., 2013. A multiscale approach in the computational modeling of the biophysical environment in artificial cartilage tissue regeneration. Biomech. Model Mechanobiol. 12, 763–780.

Checa, S., Prendergast, P.J., 2010. Effect of cell seeding and mechanical loading on vascularization and tissue formation inside a scaffold: a mechano-biological model using a lattice approach to simulate cell activity. J. Biomech. 43, 961–968.

Chen, A.K.L., Reuveny, S., Oh, S.K.W., 2013. Application of human mesenchymal and pluripotent stem cell microcarrier cultures in cellular therapy: achievements and future direction. Biotechnol. Adv. 31, 1032–1046.

Cierpka, K., Elseberg, C.L., Niss, K., Kassem, M., Salzig, D., Czermak, P., 2013. hMSC production in disposable bioreactors with regards to GMP and PAT. Chem. Ing. Tech. 85, 67–75.

Csaszar, E., Chen, K., Caldwell, J., Chan, W., Zandstra, P.W., 2013. Real-time monitoring and control of soluble signaling factors enables enhanced progenitor cell outputs from human cord blood stem cell cultures. Biotechnol. Bioeng. 111, 1258–1264. http://dx doi.org/10.1002/bit.25163.

Csaszar, E., Kirouac, D.C., Yu, M., Wang, W.J., Qiao, W.L., Cooke, M.P., Boitano, A.E., Ito, C., Zandstra, P.W., 2012. Rapid expansion of human hematopoietic stem cells by automated control of inhibitory feedback signaling. Cell Stem Cell 10, 218–229.

Demol, J., Lambrechts, D., Geris, L., Schrooten, J., Van Oosterwyck, H., 2011. Towards a quantitative understanding of oxygen tension and cell density evolution in fibrin hydrogels. Biomaterials 32, 107–118.

Diogo, M.M., Da Silva, C.L., Cabral, J.M., 2012. Separation technologies for stem cell bioprocessing. Biotechnol. Bioeng. 109, 2699–2709.

Dos Santos, F., Andrade, P.Z., Abecasis, M.M., Gimble, J.M., Chase, L.G., Campbell, A.M., Boucher, S., Vemuri, M.C., Da Silva, C.L., Cabral, J.M.S., 2011. Toward a clinical-grade expansion of mesenchymal stem cells from human sources: a microcarrier-based culture system under xeno-free conditions. Tissue Eng. Part C-Methods 17, 1201–1210.

Dos Santos, F.F., Andrade, P.Z., Da Silva, C.L., Cabral, J.M.S., 2013. Bioreactor design for clinical-grade expansion of stem cells. Biotechnol. J. 8, 644–654.

Fan, Y., Hsiung, M., Cheng, C., Tzanakakis, E.S., 2014. Facile engineering of xeno-free microcarriers for the scalable cultivation of human pluripotent stem cells in stirred suspension. Tissue Eng. Part A 20, 588–599.

Fernandes, T.G., Rodrigues, C.A.V., Diogo, M.M., Cabral, J.M.S., 2014. Stem cell bioprocessing for regenerative medicine. J. Chem. Technol. Biotechnol. 89, 34–47.

French, A., Bure, K., Brindley, D.A., 2014. CASMI TSCC launch event, Paris, France, july 2013: an assessment of the key barriers to the commercialization and clinical adoption of pluripotent stem cell therapies. Rejuvenation Res. 17, 84–88.

Frith, J.E., Thomson, B., Genever, P.G., 2010. Dynamic three-dimensional culture methods enhance mesenchymal stem cell properties and increase therapeutic potential. Tissue Eng. Part C-Methods 16, 735–749.

Frohlich, M., Grayson, W.L., Marolt, D., Gimble, J.M., Kregar-Velikonja, N., Vunjak-Novakovic, G., 2010. Bone grafts engineered from human adipose-derived stem cells in perfusion bioreactor culture. Tissue Eng. Part A 16, 179–189.

Geris, L., Reed, A.A.C., Vander Sloten, J., Simpson, A.H.R.W., Van Oosterwyck, H., 2010. Occurrence and treatment of bone atrophic non-unions investigated by an integrative approach. Plos Comput. Biol 6.

Goh, T.K., Zhang, Z.Y., Chen, A.K., Reuveny, S., Choolani, M., Chan, J.K., Oh, S.K., 2013. Microcarrier culture for efficient expansion and osteogenic differentiation of human fetal mesenchymal stem cells. Biores. Open. Access 2, 84–97.

Gonzalez-Gonzalez, M., Rito-Palomares, M., 2013. Aqueous two-phase systems strategies to establish novel bioprocesses for stem cells recovery. Crit. Rev. Biotechnol. Epub ahead of print, http://dx.doi.org/10.3109/07388551.2013.794125.

Gossett, D.R., Weaver, W.M., Mach, A.J., Hur, S.C., Tse, H.T.K., Lee, W., Amini, H., Di Carlo, D., 2010. Label-free cell separation and sorting in microfluidic systems. Anal. Bioanal. Chem. 397, 3249–3267.

Grayson, W.L., Frohlich, M., Yeager, K., Bhumiratana, S., Chan, M.E., Cannizzaro, C., Wan, L.Q., Liu, X.S., Guo, X.E., Vunjak-Novakovic, G., 2010. Engineering anatomically shaped human bone grafts. Proc. Natl. Acad. Sci. USA 107, 3299–3304.

Gutenkunst, R.N., Waterfall, J.J., Casey, F.P., Brown, K.S., Myers, C.R., Sethna, J.P., 2007. Universally sloppy parameter sensitivities in systems biology models. Plos Comput. Biol. 3, 1871–1878.

Guyot, Y., Papantoniou, I., Chai, Y.C., Van Bael, S., Schrooten, J., Geris, L., 2014. A computational model for Cell/ECM growth on 3D surfaces using the level set method: a bone tissue engineering case study. Biomech. Model Mechanobiol. Epub ahead of print, http://dx.doi.org/10.1007/s10237-014-0577-5.

Hagenmuller, H., Hitz, M., Merkle, H.P., Meinel, L., Muller, R., 2010. Design and validation of a novel bioreactor principle to combine online micro-computed tomography monitoring and mechanical loading in. Bone Tissue Eng. Rev. Sci. Instrum. 81, 014303.

Hunt, M.M., Meng, G.L., Rancourt, D.E., Gates, I.D., Kallos, M.S., 2014. Factorial experimental design for the culture of human embryonic stem cells as aggregates in stirred suspension bioreactors reveals the potential for interaction effects between bioprocess parameters. Tissue Eng. Part C-Methods 20, 76–89.

Ingber, D.E., Mow, V.C., Butler, D., Niklason, L., Huard, J., Mao, J., Yannas, I., Kaplan, D., Vunjak-Novakovic, G., 2006. Tissue engineering and developmental biology: going Biomimetic. Tissue Eng. 12, 3265–3283.

Janes, K.A., Kelly, J.R., Gaudet, S., Albeck, J.G., Sorger, P.K., Lauffenburger, D.A., 2004. Cue-signal-response analysis of TNF-induced apoptosis by partial least squares regression of dynamic multivariate data. J. Comput. Biol. 11, 544–561.

Janssen, F.W., Van Dijkhuizen-Radersma, R., Van Oorschot, A., Oostra, J., De Bruijn, J.D., Van Blitterswijk, C.A., 2010. Human tissue-engineered bone produced in clinically relevant amounts using a semi-automated perfusion bioreactor system: a preliminary study. J. Tissue Eng. Regen. Med. 4, 12–24.

Jones, M., Varella-Garcia, M., Skokan, M., Bryce, S., Schowinsky, J., Peters, R., Vang, B., Brecheisen, M., Startz, T., Frank, N., Nankervis, B., 2013. Genetic stability of bone marrow-derived human mesenchymal stromal cells in the quantum system. Cytotherapy 15, 1323–1339.

Jukes, J.M., Both, S.K., Leusink, A., Sterk, L.M.T., Van Blitterswijk, C.A., De Boer, J., 2008. Endochondral bone tissue engineering using embryonic stem cells. Proc. Natl. Acad. Sci. USA 105, 6840–6845.

Kerkhofs, J., Roberts, S., Luyten, F., Van Oosterwyck, H., Geris, L., 2012. Relating the chondrocyte gene network to growth plate morphology: from genes to phenotype. PLoS One 7. http://dx.doi.org/10.1371/journal.pone.0034729.

Kim, J., Ma, T., 2012. Perfusion regulation of hMSC microenvironment and osteogenic differentiation in 3D scaffold. Biotechnol. Bioeng. 109, 252–261.

Kinney, M.A., Mcdevitt, T.C., 2013. Emerging strategies for spatiotemporal control of stem cell fate and morphogenesis. Trends Biotechnol. 31, 78–84.

Kirouac, D.C., Zandstra, P.W., 2008. The systematic production of cells for cell therapies. Cell Stem Cell 3, 369–381.

Kirouac, D.C., Madlambayan, G.J., Yu, M., Sykes, E.A., Ito, C., Zandstra, P.W., 2009. Cell-cell interaction networks regulate blood stem and progenitor cell fate. Mol. Syst. Biol. 5, 293.

Lambrechts, D., Roeffaers, M., Kerckhofs, G., Roberts, S.J., Hofkens, J., Van De Putte, T., Van Oosterwyck, H., Schrooten, J., 2013. Fluorescent oxygen sensitive microbead incorporation for measuring oxygen tension in cell aggregates. Biomaterials 34, 922–929.

Lambrechts, T., Papantoniou, I., Sonnaert, M., Schrooten, J., Aerts, J.M., 2014. Model-based cell number quantification using online single-oxygen sensor data for tissue engineering perfusion bioreactors. Biotechnol. Bioeng. 111, 1982–1992. Epub ahead of print, http://dx.doi.org/10.1002/bit.25274.

Lenas, P., Moos, M., Luyten, F.P., 2009. Developmental engineering: a new paradigm for the design and manufacturing of cell-based products. Part I: from three-dimensional cell growth to biomimetics of in vivo development. Tissue Eng. Part B-Rev. 15, 381–394.

Lim, M., Shunjie, C., Panoskaltsis, N., Mantalaris, A., 2012. Systematic experimental design for bioprocess characterization: elucidating transient effects of multi-cytokine contributions on erythroid differentiation. Biotechnol. Bioprocess Eng. 17, 218–226.

Liu, N., Zang, R., Yang, S.T., Li, Y., 2013. Stem cell engineering in bioreactors for large-scale bioprocessing. Eng. Life Sci. 14, 4–15.

Loring, J.F., McDevitt, T.C., Palecek, S.P., Schaffer, D.V., Zandstra, P.W., Nerem, R.M., 2014. A global assessment of stem cell engineering. Tissue Eng. Part A. Epub ahead of print, http://dx.doi.org/10.1089/ten.tea.2013.0468.

Marolt, D., Campos, I.M., Bhumiratana, S., Koren, A., Petridis, P., Zhang, G., Spitalnik, P.F., Grayson, W.L., Vunjak-Novakovic, G., 2012. Engineering bone tissue from human embryonic stem cells. Proc. Natl. Acad. Sci. USA 109, 8705–8709.

Martin, I., Smith, T., Wendt, D., 2009. Bioreactor-based roadmap for the translation of tissue engineering strategies into clinical products. Trends Biotechnol. 27, 495–502.

Mason, C., Hoare, M., 2007. Regenerative medicine bioprocessing: building a conceptual framework based on early studies. Tissue Eng. 13, 301–311.

Mason, C., Manzotti, E., 2010. Regenerative medicine cell therapies: numbers of units manufactured and patients treated between 1988 and 2010. Regen. Med. 5, 307–313.

Mason, C., Mason, J., Culme-Seymour, E.J., Bonfiglio, G.A., Reeve, B.C., 2013. Cell therapy companies make Strong progress from October 2012 to March 2013 amid mixed stock market sentiment. Cell Stem Cell 12, 644–647.

Matsuoka, F., Takeuchi, I., Agata, H., Kagami, H., Shiono, H., Kiyota, Y., Honda, H., Kato, R., 2013. Morphology-based prediction of osteogenic differentiation potential of human mesenchymal stem cells. Plos One 8. http://dx.doi.org/10.1371/journal.pone.0055082.

Mercier, S.M., Diepenbroek, B., Wijffels, R.H., Streefland, M., 2014. Multivariate PAT solutions for biopharmaceutical cultivation: current progress and limitations. Trends Biotechnol. 32, 329–336.

Mentink, A., Hulsman, M., Groen, N., Licht, R., Dechering, K.J., Van Der Stok, J., Alves, H.A., Dhert, W.J., Van Someren, E.P., Reinders, M.J.T., Van Blitterswijk, C.A., De Boer, J., 2013. Predicting the therapeutic efficacy of MSC in bone tissue engineering using the molecular marker CADM1. Biomaterials 34, 4592–4601.

Nerem, R., 2012. A global assessment of stem cell engineering. J. Tissue Eng. Regen. Med. 6, 406.

Nienow, A.W., Rafiq, Q.A., Coopman, K., Hewitt, C.J., 2014. A potentially scalable method for the harvesting of hMSCs from microcarriers. Biochem. Eng. J. 85, 79–88.

Nold, P., Brendel, C., Neubauer, A., Bein, G., Hackstein, H., 2013. Good manufacturing practice-compliant animal-free expansion of human bone marrow derived mesenchymal stroma cells in a closed hollow-fiber-based bioreactor. Biochem. Biophys. Res. Commun. 430, 325–330.

Papantoniou, I., Chai, Y.C., Luyten, F.P., Schrooten, J., 2012. Process quality engineering for bioreactor-driven manufacturing of tissue engineered constructs for bone regeneration. Tissue Eng. Part C. Methods. 19, 596–609.

Papantoniou, I., Sonnaert, M., Geris, L., Luyten, F.P., Schrooten, J., Kerckhofs, G., 2013. Three dimensional characterization of tissue-engineered constructs by contrast enhanced nanofocus computed tomography. Tissue Eng. Part C. Methods. 20, 177–187.

Patti, G.J., Yanes, O., Siuzdak, G., 2012. Metabolomics: the apogee of the omics trilogy. Nat. Rev. Mol. Cell Biol. 13, 263–269.

Placzek, M.R., Chung, I.M., Macedo, H.M., Ismail, S., Blanco, T.M., Lim, M., Cha, J.M., Fauzi, I., Kang, Y., Yeo, D.C.L., Ma, C.Y.J., Polak, J.M., Panoskaltsis, N., Mantalaris, A., 2009. Stem cell bioprocessing: fundamentals and principles. J. R. Soc. Interface 6, 209–232.

Ratcliffe, E., Thomas, R.J., Williams, D.J., 2011. Current understanding and challenges in bioprocessing of stem cell-based therapies for regenerative medicine. Br. Med. Bull. 100, 137–155.

Rathore, A.S., Winkle, H., 2009. Quality by design for biopharmaceuticals. Nat. Biotechnol. 27, 26–34.

Roberts, I., Baila, S., Rice, R.B., Janssens, M.E., Nguyen, K., Moens, N., Ruban, L., Hernandez, D., Coffey, P., Mason, C., 2012. Scale-up of human embryonic stem cell culture using a hollow fibre bioreactor. Biotechnol. Lett. 34, 2307–2315.

Roberts, S.J., Geris, L., Kerckhofs, G., Desmet, E., Schrooten, J., Luyten, F.P., 2011. The combined bone forming capacity of human periosteal derived cells and calcium phosphates. Biomaterials 32, 4393–4405.

Roccio, M., Schmitter, D., Knobloch, M., Okawa, Y., Sage, D., Lutolf, M.P., 2013. Predicting stem cell fate changes by differential cell cycle progression patterns. Development 140, 459–470.

Roelofs, E., Engelsman, M., Rasch, C., Persoon, L., Qamhiyeh, S., De Ruysscher, D., Verhaegen, F., Pijls-Johannesma, M., Lambin, P., Consortium, R., 2012. Results of a multicentric in silico clinical trial (ROCOCO) comparing radiotherapy with photons and protons for non-small cell lung cancer. J. Thorac. Oncol. 7, 165–176.

Rojewski, M.T., Fekete, N., Baila, S., Nguyen, K., Furst, D., Antwiler, D., Dausend, J., Kreja, L., Ignatius, A., Sensebe, L., Schrezenmeier, H., 2013. GMP-compliant isolation and expansion of bone marrow-derived MSCs in the closed, automated device quantum cell expansion system. Cell Transplant. 22, 1981–2000.

Saeed, A., Baharvand, H., 2013. Biotechnol. Adv. 31, 1600–1623.

Saito, E., Liu, Y.F., Migneco, F., Hollister, S.J., 2012. Strut size and surface area effects on long-term in vivo degradation in computer designed poly(L-lactic acid) three-dimensional porous scaffolds. Acta Biomater. 8, 2568–2577.

Salter, E., Goh, B., Hung, B., Hutton, D., Ghone, N., Grayson, W.L., 2012. Bone tissue engineering bioreactors: a role in the clinic? Tissue Eng. Part B Rev. 18, 62–75.

Santoro, R., Braissant, O., Muller, B., Wirz, D., Daniels, A.U., Martin, I., Wendt, D., 2011. Real-time measurements of human chondrocyte heat production during in vitro proliferation. Biotechnol. Bioeng. 108, 3019–3024.

Santoro, R., Krause, C., Martin, I., Wendt, D., 2012. On-line monitoring of oxygen as a non-destructive method to quantify cells in engineered 3D tissue constructs. J. Tissue Eng. Regen. Med. 6, 696–701.

Schop, D., Janssen, F.W., Van Rijn, L.D.S., Fernandes, H., Bloem, R.M., De Bruijn, J.D., Van Dijkhuizen-Radersma, R., 2009. Growth, metabolism, and growth inhibitors of mesenchymal stem cells. Tissue Eng. Part A 15, 1877–1886.

Scotti, C., Piccinini, E., Takizawa, H., Todorov, A., Bourgine, P., Papadimitropoulos, A., Barbero, A., Manz, M.G., Martin, I., 2013. Engineering of a functional bone organ through endochondral ossification. Proc. Natl. Acad. Sci. USA 110, 3997–4002.

Simaria, A.S., Hassan, S., Varadaraju, H., Rowley, J., Warren, K., Vanek, P., Farid, S.S., 2014. Allogeneic cell therapy bioprocess economics and optimization: single-use cell expansion technologies. Biotechnol. Bioeng. 111, 69–83.

Smeets, B., Odenthal, T., Tijskens, E., Ramon, H., Van Oosterwyck, H., 2013. Quantifying the mechanical micro-environment during three-dimensional cell expansion on microbeads by means of individual cell-based modeling. Comp. Meth. Biomech. Biomed. Eng. 16, 1071–1084.

Smith, Z.D., Nachman, I., Regev, A., Meissner, A., 2010. Dynamic single-cell imaging of direct reprogramming reveals an early specifying event. Nat. Biotechnol. 28, 521–526.

Song, M.J., Dean, D., Tate, M.L.K., 2013. Mechanical modulation of nascent stem cell lineage commitment in tissue engineering scaffolds. Biomaterials 34, 5766–5775.

Spanholtz, J., Preijers, F., Tordoir, M., Trilsbeek, C., Paardekooper, J., De Witte, T., Schaap, N., Dolstra, H., 2011. Clinical-grade generation of active NK cells from cord blood hematopoietic progenitor cells for immunotherapy using a closed-system culture process. Plos One 6.

Spencer, T.J., Hidalgo-Bastida, L.A., Cartmell, S.H., Halliday, I., Care, C.M., 2013. In silico multiscale model of transport and dynamic seeding in a bone tissue engineering perfusion bioreactor. Biotechnol. Bioeng. 110, 1221–1230.

Tamura, A., Kobayashi, J., Yamato, M., Okano, T., 2012. Temperature-responsive poly (N-isopropylacrylamide)-grafted microcarriers for large-scale non-invasive harvest of anchorage-dependent cells. Biomaterials 33, 3803–3812.

Teixeira, A.P., Oliveira, R., Alves, P.M., Carrondo, M.J.T., 2009. Advances in on-line monitoring and control of mammalian cell cultures: supporting the PAT initiative. Biotechnol. Adv. 27, 726–732.

Timmins, N.E., Kiel, M., Gunther, M., Heazlewood, C., Doran, M.R., Brooke, G., Atkinson, K., 2012. Closed system isolation and scalable expansion of human placental mesenchymal stem cells. Biotechnol. Bioeng. 109, 1817–1826.

Volkmer, E., Otto, S., Polzer, H., Saller, M., Trappendreher, D., Zagar, D., Hamisch, S., Ziegler, G., Wilhelmi, A., Mutschler, W., Schieker, M., 2012. Overcoming hypoxia in 3D culture systems for tissue engineering of bone in vitro using an automated, oxygen-triggered feedback loop. J. Mater. Sci. Mater. Med. 23, 2793–2801.

Wallman, L., Akesson, E., Ceric, D., Andersson, P.H., Day, K., Hovatta, O., Falci, S., Laurell, T., Sundstrom, E., 2011. Biogrid–a microfluidic device for large-scale enzyme-free dissociation of stem cell aggregates. Lab. Chip 11, 3241–3248.

Wang, Y., Cheng, L., Gerecht, S., 2013a. Efficient and scalable expansion of human pluripotent stem cells under clinically compliant settings: a view in 2013. Ann. Biomed. Eng. 42, 1357–1372.

Wang, Y., Chou, B.K., Dowey, S., He, C.X., Gerecht, S., Cheng, L.Z., 2013b. Scalable expansion of human induced pluripotent stem cells in the defined xeno-free E8 medium under adherent and suspension culture conditions. Stem Cell Res. 11, 1103–1116.

Ward, A., Quinn, K.P., Bellas, E., Georgakoudi, I., Kaplan, D.L., 2013. Noninvasive metabolic imaging of engineered 3D human adipose tissue in a perfusion bioreactor. Plos One 8.

Warren, V., 2013. Understanding regenerative medicine: a commissioner's viewpoint. Regen. Med. 8 (2), 227–232.

Wendt, D., Riboldi, S.A., Cioffi, M., Martin, I., 2009. Bioreactors in tissue engineering: scientific challenges and clinical perspectives. Adv. Biochem. Eng. Biotechnol. 112, 1–27.

Yeatts, A.B., Both, S.K., Yang, W., Alghamdi, H.S., Yang, F., Fisher, J.P., Jansen, J.A., 2014. In vivo bone regeneration using tubular perfusion system bioreactor cultured nanofibrous scaffolds. Tissue Eng. Part A 20, 139–146.

Yeatts, A.B., Fisher, J.P., 2011. Bone tissue engineering bioreactors: dynamic culture and the influence of shear stress. Bone 48, 171–181.

Yeo, D., Kiparissides, A., Cha, J.M., Aguilar-Gallardo, C., Polak, J.M., Tsiridis, E., Pistikopoulos, E.N., Mantalaris, A., 2013. Improving embryonic stem cell expansion through the combination of perfusion and bioprocess model design. Plos One 8.

Zhou, L., Kong, J., Zhuang, Y., Chu, J., Zhang, S., Guo, M., 2013a. Ex vivo expansion of bone marrow mesenchymal stem cells using microcarrier beads in a stirred bioreactor. Biotechnol. Bioprocess Eng. 18, 173–184.

Zhou, X.H., Holsbeeks, I., Impens, S., Sonnaert, M., Bloemen, V., Luyten, F., Schrooten, J., 2013b. Noninvasive real-time monitoring by alamarblue (R) during in vitro culture of three-dimensional tissue-engineered bone constructs. Tissue Eng. Part C-Methods 19, 720–729.

Zweigerdt, R., Olmer, R., Singh, H., Haverich, A., Martin, U., 2011. Scalable expansion of human pluripotent stem cells in suspension culture. Nat. Protoc. 6, 689–700.

SUGGESTED PAPERS

Abbasalizadeh, S., Baharvand, H., 2013. Technological progress and challenges towards cGMP manufacturing of human pluripotent stem cells based therapeutic products for allogeneic and autologous cell therapies. Biotechnol. Adv. 31, 1600–1623.

Chang, T., Zimmerley, M.S., Quinn, K.P., Lamarre-Jouenne, I., Kaplan, D.L., Beaurepaire, E., Georgakoudi, I., 2013. Non-invasive monitoring of cell metabolism and lipid production in 3D engineered human adipose tissues using label-free multiphoton microscopy. Biomaterials 34, 8607–8616.

Csaszar, E., Kirouac, D.C., Yu, M., Wang, W.J., Qiao, W.L., Cooke, M.P., Boitano, A.E., Ito, C., Zandstra, P.W., 2012. Rapid expansion of human hematopoietic stem cells by automated control of inhibitory feedback signaling. Cell Stem Cell 10, 218–229.

Diogo, M.M., da Silva, C.L., Cabral, J.M., 2012. Separation technologies for stem cell bioprocessing. Biotechnol. Bioeng. 109, 2699–2709.

Hunt, M.M., Meng, G., Rancourt, D.E., Gates, I.D., Kallos, M.S., 2014. Factorial experimental design for the culture of human embryonic stem cells as aggregates in stirred suspension bioreactors reveals the potential for interaction effects between bioprocess parameters. Tissue Eng. Part C Meth. 20, 76–89.

Kinney, M.A., McDevitt, T.C., 2013. Emerging strategies for spatiotemporal control of stem cell fate and morphogenesis. Trends biotechnol. 31, 78–84.

Lambrechts, D., Roeffaers, M., Kerckhofs, G., Roberts, S.J., Hofkens, J., Van de Putte, T., Van Oosterwyck, H., Schrooten, J., 2013. Fluorescent oxygen sensitive microbead incorporation for measuring oxygen tension in cell aggregates. Biomaterials 34, 922–929.

Simaria, A.S., Hassan, S., Varadaraju, H., Rowley, J., Warren, K., Vanek, P., Farid, S.S., 2014. Allogeneic cell therapy bioprocess economics and optimization: single-use cell expansion technologies. Biotechnol. Bioeng. 111, 69–83.

Wu, J., Tzanakakis, S., 2013. Deconstructing stem cell population heterogeneity: single cell analysis and modeling approaches. Biotechnol. Adv. 31, 1047–1062.

Yeo, D., Kiparissides, A., Cha, J.M., Aguilar-Gallardo, C., Polak, J.M., Tsiridis, E., Pistikopoulos, E.N., Mantalaris, A., 2013. Improving embryonic stem cell expansion through the combination of perfusion and bioprocess model design. PloS One.

CHAPTER 22

Clinical Translation

Johan Joly[1], Johan Vanlauwe[2], Dieter Van Assche[1,3], Ilona Reischl[4], Frank P. Luyten[1,5]

[1]Division of Rheumatology, UZ Leuven, Leuven, Belgium; [2]Faculty of Physical Education and Physical Therapy, Faculty of Medicine and Pharmacy, VU Brussels and Department of Orthopedics & Traumatology, UZ Brussel, Brussels, Belgium; [3]Department of Rehabilitation Sciences, Faculty of Kinesiology and Rehabilitation Sciences, KU Leuven, Leuven, Belgium; [4]Division clinical trials, Institute Surveillance, Federal Office for Safety in Health Care/Austrian Agency for Health and Food Safety, Vienna, Austria; [5]Skeletal Biology & Engineering Research Center, KU Leuven, Leuven, Belgium

LEARNING OBJECTIVES

- Get acquainted with the basic principles of the clinical development of conventional drugs
- Learn about the documents required for implementation of a clinical trial
- Understand some typical challenges encountered with the clinical translation of tissue-engineered products
- Identify special points to consider in the clinical translation of tissue-engineered products
- Get familiar with regulatory aspects in the clinical development of tissue-engineered products

22.1 INTRODUCTION

Since the beginning of this century, the field of tissue engineering (TE) has been changing rapidly, with new scientific and technology developments, leading to paradigm shifts in this field (Lenas et al., 2009a, 2009b). Concomitantly, the path of clinical development and related regulation and legislation has also evolved, resulting in the perception that regulatory requirements and statutes have become uncontrollable "moving targets" that may hamper progress in the field. The use of new approaches and technology in humans implies providing answers to new and unexpected questions and clearing challenging regulatory and ethical hurdles. Historically, devices and cell and tissue products had fewer regulatory requirements than the established class of drugs/medicinal

Tissue Engineering.
Copyright © 2015 Elsevier Inc. All rights reserved.

products, and these requirements mostly focused on safety. Since the birth of a new class of products, it was recognized that many tissue-engineered products, in particular cell-based combination products, should meet the requirements of a demonstration of effectiveness that apply for conventional drugs and new regulation as well as legislation was put in place. These products are categorized as Advanced Therapeutic Medicinal Products (ATMPs) under Regulation (EC) N. 1394/2007 in Europe (Official Journal of the European Union (OJL), 2007) and as Human Cells, Tissues, and Cellular and Tissue-Based Products (HCTPs) in the US (Code of Federal Regulations (CFR), 2005).

In order to be classified as an ATMP in Europe, a given biological product must fulfill the legal definition of a medicinal product and in addition the definition for somatic cell therapy, a gene therapy, or a tissue-engineered product. In the US, HCTPs are regulated under Section 351 of the Public Health Service Act and/or the Food Drug & Cosmetic (FD&C) Act because they do not meet all of the criteria in 21 Code of Federal Regulation (CFR) 1271.10 as determined by the Food and Drug Administration (FDA), and are regulated as drugs and/or biological products.

We will present an overview of issues and hurdles that may be encountered and will try to provide a framework on how to translate these innovative therapeutic products and approaches in a practical and safe way into daily clinical practice with respect to the perspective of the patient and worldwide standards of good clinical practice (GCP) as outlined in the guideline of the International Conference on Harmonization (ICH) of Technical Requirements for Registration of Pharmaceuticals for Human Use, ICH E6 (ICH E6, 1996).

22.1.1 Historical Perspective

To better understand the development of regulation and legislation in the field of TE and the ATMPs or HCTPs, it is important to understand how the execution of clinical trials changed over time.

Historically, the rules of performing clinical trials were developed and adapted as a result of incidents occurring with new treatments. As an example, an important event in this context was the 'Elixir Sulfanilamide' incident (1937) (Ballentine, 1981), which was called "taste of raspberries, taste of death." 'Elixir Sulfanilamide' was a drug used to treat Streptococcal infection. Because the infection did not exclude children, a syrup was developed that was more pleasing to children. Therefore, the 'Sulfanilamide' was dissolved in diethylene glycol and the juice of raspberries was added. Testing was done on the solution to evaluate taste and smell as was required at that time. But 107 people, most of them children, died because of the deadly ingredient diethylene glycol, commonly known as antifreeze. As a consequence of this event, the FD&C Act (1938) was adopted and stated that beside demonstration of efficacy, data on safety had to be evaluated before approval by the FDA.

During the Second World War (1940–1945), unethical medical experiments were performed by German doctors on concentration camp inmates and civilians of occupied countries. This horror resulted in the Nuremberg Code (1946), the first ethical code for experiments on humans with voluntary consent as, most important principle (Annas and Grodin, 1992; Markman and Markman, 2007).

The Softenon (thalidomide) incident (1956) (Akhurst, 2010) confronted scientists with the fact that for new synthetic medicines, not only safety and short-term effects were important but also the long-term effects had to be studied. Softenon was used as effective treatment against morning sickness in pregnant woman but many birth defects were seen in babies of mothers who took the medicine. This event led to an amendment of the FD&C, better known as the Kefauver–Harris Drug Amendment (1962) (FDA Consumer Health Information, 2012). This amendment expanded the requirements of the regulatory bodies (FDA) to manufacturers to not only provide evidence of safety in the short and longer term but also to demonstrate efficacy before approval. A direct consequence of this amendment was the development of the foundation for the current model of performing clinical trials in the development of medicinal products (Davidoff, 2011).

In 1964, the World Medical Association (WMA) adopted the "Declaration of Helsinki" (WMA, 2013) which stipulated the ethical principles for medical research involving human subjects (Markman and Markman, 2007). The basis of this declaration was the Nuremberg Code and the declaration led to the most important principle still used in clinical trials: good clinical practice. GCP is the international ethical and scientific standard used in the design, execution, verification, analysis, and reporting of clinical research in the "human model." This concept is to ensure that results generated by clinical trials are scientifically rigorous, reliable, and collected with respect to the integrity and privacy of the research subjects in such a way as to maximize generalizable knowledge beneficial to the human community.

In the beginning, several GCP guidelines were developed resulting in country-specific interpretations of this universal concept. Fortunately, in 1997 an international conference was organized to harmonize the guidelines between the USA, Europe, and Japan. As a consequence, clinical trials conducted according to the ICH E6 GCP guidance can be submitted in support of marketing authorizations (MAs) in all regions of the ICH (ICH E6, 1996).

Since then, the ICH guidelines, including ICH-GCP, were referred to in legislations worldwide. For Europe, these guidelines are part of the European clinical directive (2001/20/EC) (L, 2001). A directive sets a legislative goal for European member states that needs to be translated into National law, whereas a regulation is binding in its exact wording for the entire EU. The implementation of the clinical trials directive into national laws gave rise to a degree of disharmonization between distinct European member states and resulting difficulties for applicants of multinational trials. For example, in Belgium, no distinction is made in clinical research with registered medicines compared to development of new medicines, while this difference is made in other member states as, for example, in the Netherlands. As a consequence, investigator initiated research should be registered centrally as EudraCT (European Union Drug Regulating Authorities Clinical Trials) in Belgium, but not in the Netherlands. To ensure harmonization and to remove obstacles for clinical researchers, the legislative framework for clinical trials in preparation in the EU will take the form of a regulation (European Commission, 2014a, 2014b).

For the US, ICH-GCP guidelines were published as a guidance document in the Federal Register in May 1997 and are not legally binding. Parts of the guidelines were adapted as regulations and are legally binding, i.e., ICH E2 on Clinical Safety Data Management. Regarding other items, differences remain between the ICH guidelines and the FDA requirements; for example, on IRB (Institutional Review Board) responsibilities (ICH 3.1) and investigator requirements (ICH 4), ICH is more demanding, whereas on IRB Composition (ICH 3.2), FDA has supplementary obligations (Below, 2008).

22.1.2 Framework of Clinical Development for Conventional Medicinal Products

As mentioned earlier, clinical research involving drug development is strictly regulated. It is divided into different phases; typically, over a decade of research and development is needed to get a central MA.

Conventional small molecule drug discovery programs typically start with large-scale screening using appropriate models, in contrast to TE products. Nonclinical research further involves testing in cell and tissue cultures, and in vivo experiments in several relevant animal models.

Clinical drug development in humans is divided into different phases, each with its specific focus as outlined in ICH E8 (ICH E8, 1997).

In conventional models of drug development, the first phase (Phase I, human pharmacology), where feasible involves healthy volunteers and its objectives include safety, the assessment of tolerance, the definition of pharmacokinetics (PK), and pharmacodynamics (PD), an exploration of drug metabolism and interactions with a first assessment of activity. Over the last years, there has been a shift of using patients instead of healthy volunteers in this phase of drug development. This suggests that the physician responsible for the treatment of the patient needs to be closely involved, as standard care and disease management needs to be provided at all times. Early on, it is important to determine the highest safe dose of the product with carefully chosen dose escalation studies in small cohorts, typically starting with a single dose step-up design followed by a multiple dose scheme to determine the disposition of the drug.

In the second phase (Phase II, therapeutic exploratory), the use for the targeted indication might be explored and more detailed investigations into the dosage for subsequent studies conducted in small groups of patients. The phase II study should provide the proof of concept, evaluate the first signs of efficacy and safety and thus form a solid basis for most optimal confirmatory study design, endpoints, and methodologies.

The last phase (Phase III, therapeutic confirmatory) before an MA is called pivotal, and aims to provide data for assessing the benefit/risk in a sufficiently large group of patients to support licensing. Therefore, efficacy needs to be demonstrated in this phase and the safety profile should be established. This involves, in general, at least two independent studies, in which the second study is anticipated to confirm the results of the first pivotal trial. In some cases, a single trial may suffice, if adequately designed and controlled.

In the context of tissue-engineered products, deviations from this "classic" phasic model are seen with adaptive protocol designs and shifts from healthy volunteers to patients. New medicinal products are required not only to prove efficacy in the short term but also in the long run. Under the prerequisite that efficacy has been demonstrated, European Medicines Agency (EMA) and FDA provide a path for a conditional MA with the requirement of providing data on the efficacy and safety within a specified time with postmarketing commitments where appropriate.

Commercial developers conduct the majority of the clinical trials in the first three phases of clinical research. Phase IV studies, on the contrary, are conducted on approved medicinal products. The distinction between phase IV studies and noninterventional studies is that in the latter no interventions beyond routine practice are permitted, that treatment needs to be done according to the conditions of the MA and that patients can only be included after the treatment decision has been taken. Randomization and additional diagnostic tests, for example, therefore belong to a phase IV setting. Noninterventional studies serve to study a drug in everyday practice and are therefore suited to address questions on its use or relevance and to detect rare safety signals in daily clinical practice.

Considerations other than MA, although outside the scope of this chapter, may be important in health policy decision making. Health Technology Assessment (HTA) is intended to support policy makers with information on how to provide safe, effective, patient-focused, and cost-effective treatments. In this context, effectiveness in comparison with existing standard of care might be informative. In a recent EU initiative, Eunethta has been created to facilitate harmonization between national assessment bodies (http://www.eunethta.eu) and EMA provides the option of joint EMA/HTA scientific advice (http://www.ema.europa.eu). Moreover, it may be in certain cases useful to design a single trial to provide information sufficient not only for MA but to guide or inform intelligent decisions regarding reimbursement.

22.2 CLINICAL TRANSLATION OF TISSUE-ENGINEERED PRODUCTS

Translating the above regulations from classical medicinal products to cell and combination products is challenging in this young and quickly developing field. In the following, the main hurdles will be discussed to present a rough framework for the development of products that can be categorized within the ATMPs/HCTPs with particular relevance to the field of TE. Most of these products are legally categorized as ATMPs in Europe (OJL, 2007) and HCTPs in the US (CFR, 2005), the active substances being engineered/manipulated cells and/or tissues. Additional substances (scaffolds, matrices, devices, etc.) can be combined with the ATMPs/HCTPs. The definition of a "Combined advanced therapy medicinal product" requires that the ATMP/HCTP incorporate, as an integral part, one or more medical devices that act in their primary action on the human body. Most of the ATMPs/HCTPs used in TE contain cells of human origin. Even so, these products are heterogeneous with regard to the origin (autologous or allogeneic), type of cells (e.g., adult mature cells, adult progenitors, stem cells), and in the complexity of the product. Standardization of the process therefore needs to include the procurement step to ensure that the starting material for manufacture is consistent. Further, the development of this type of ATMPs/HCTPs requires in-depth characterization of the product/cells intended for the treatment and to ensure that the final product can be manufactured robustly and reproducibly (Table 22.1). A facility with documented proof of adherence to GMP (good manufacturing practice) is essential for a standardized manufacturing process. Consistent manufacture is vital, but

Table 22.1 Specifications of the Product: Development of This Type of ATMPs/HCTPs Requires In-Depth Characterization of the Product/Cells Intended for the Treatment and to Ensure That the Final Product Can Be Manufactured Robustly and Reproducibly

Test	Method	GMP Goal
Identity	Morphology Biochemical markers	Quality Safety
Dose	Viable cell number	Quality Efficacy
Potency	Relevant bioassay potentially replaced by marker profile	Efficacy
Purity	% Cell viability Residual impurities	Safety
Safety	Sterility Mycoplasma Endotoxins Adventitious viruses	Safety

particularly challenging for tissue-engineered ATMPs/HCTPs, some of which are patient-tailored. In the case of Autologous Chondrocyte Implantation (ACI), for example, chondrocytes obtained from an articular cartilage biopsy are used to repair joint surface defects. The amount of chondrocytes needed depends on the size of the lesion (Saris et al., 2009). Thus, it is important to have a certified GMP facility that can provide a product containing characterized chondrocytes in the required amount.

22.2.1 Preclinical Phase

Before entering the human model, preclinical data should be obtained. Characterization of the product involves testing the widest variety of potential quality attributes that is feasible. Sufficient information must be submitted to ensure proper identification of the product, purity, quality, and strength. Although challenging, the development of a suitable potency assay will be needed prior to pivotal trials and should therefore be considered early on in development. Potency is defined as the quantitative measure of biological activity based on the attributes of the product, which is linked to the relevant biological properties (ICH Q6B, 1999; FDA, 2014). The assay should be based on the intended biological effect and ideally be related to the clinical response. As in the development of medicinal products, in vitro assessments including cell and tissue cultures are used most commonly. Alternatively, in vivo models can be proposed, but are more challenging. As pointed out earlier, this is necessary to characterize the ATMP and to assess if the product can perform its intended functions with an acceptable safety profile. This last aspect is frequently quite challenging, as in many cases the knowledge on the mode of action, though extremely useful, is limited. Indeed, predicting the cause of the therapeutic effect remains one of the most challenging aspects of cell-based therapies. Gaining in-depth understanding during this phase of research is desirable for further development and optimization of the product, and in extenso the manufacturing process of the ATMP/HCTP. Following in vitro assessments, in vivo studies are required. Clinically relevant animal models have to be identified and applied. Mice and rats can be used to generate data on both "behavior" and safety of the product but this needs to be confirmed in a relevant in vivo environment. Proof of concept/feasibility testing is also needed to provide evidence that the proposed therapy offers sufficient promise to justify exposing human subjects to the risks inherent in any cellular therapy (OJL, 2009). Depending on the clinical problem and/or patient context and considerations of allometric scaling, relevant physiology, or biomechanics, this is often done in a larger animal model, ideally one that resembles the intended clinical situation closely. In the example of ACI, mice and rats cannot mimic the condition of interest. Mimicking the human situation as close as possible should be the goal. Biological findings that are conserved across species provide confidence in the behavior of the product. Gathering of nonclinical data is a time-consuming undertaking, not only due to the need for different in vitro and in vivo models, but also for follow-up that is sufficiently long to ensure the safety of the product and durability of the therapeutic effect. Beside its predominant intent of use such as, for instance, regeneration of a joint surface defect, special interest should be focused on safety concerns including the biodistribution, i.e., whether and in what quantities cells will migrate through the body to unintended sites, tissues, and organs including the gonads. The ATMP/HCTP used in a preclinical setting does not always equal the final cell product used in humans. For example, human cells used in animal models can trigger immunological reactions with the rejection of the donor cells. Thus, in some cases testing may better be done with homologous cells to clearly identify the mode of action than to investigate the human cells in an immunologically impaired animal model. Ideally, one seeks to obtain similar data with both the homologous and the "real" product, or at least identify the similarities and differences and their potential impact in the human model.

22.2.2 Clinical Phase

The clinical phase in TE is not simply copying the expertise and experience of the different phases from the development of medicinal products. To set up a clinical program for ATMPs/HCTPs, it is highly recommended to contact the regulatory agencies (EMA and/or FDA) early in the process to discuss the program. Both agencies have specific committees to aid in the classification of a product as ATMP/HCTP. This classification can be made early in the development process and is not legally binding. Depending on the stage of the development and the availability of new data, this classification can be adapted. The committees mentioned earlier are also involved in providing recommendations in the development of ATMPs/HCTPs. Within EMA, the Committee for Advanced Therapies (CAT) (CAT, 2010) provides a tailored certification procedure for ATMPs. In the US, Request for Designation is considered by the Tissue Reference Group (FDA, 2013a), consisting of representatives from the Center for Biologics Evaluation and Research (CBER) and the Center for Devices and Radiological Health's (CDRH), and, where applicable, the Office of Combination Products. The review center and office identified by this process will provide the primary contact for advice, though inter-center consultation.

Both EMA and FDA are involved in centralized licensing procedures. At the end of the route, in Europe the CAT will present a report and an opinion to the Committee for Medicinal Products for Human use (CHMP). The CHMP will finalize the procedure and send the final opinion to the Commission, to the Member States, and to the applicant. The Commission will then take a final decision on the procedure (Van Wilder, 2012). In the US, if classified as a Biological License Application, the Office of Cellular, Tissue, and Gene Therapies (OCTGT) of CBER will evaluate the license application and either approve it or provide a Complete Response to the sponsor. The basis of the evaluation to grant an MA by the regulatory agencies for an ATMP/HCTP is done using a risk-based approach based on the totality of data on the product and the indication (Arcidiacono et al., 2012).

Since 2004, FDA and EMA have started a dialogue on the development of ATMPs/HCTPs known as the Parallel Scientific Advice (PSA). This forum facilitates harmonization of the views of the different stakeholders, allowing them to avoid duplication of efforts in the development of ATMPs/HSTPs and provide consistent advice to regulated industry. The idea of the PSA became more formalized in 2008 when FDA and EMA established the US FDA-EMA ATMP cluster, which was joined by Health Canada in 2012 (Arcidiacono et al., 2012).

Thus, first steps toward harmonization are taken. However, FDA and EMA currently can provide different advices or have different requirements. Therefore, the ICH has organized a Regulatory Forum on cellular therapy intended to create an environment and roadmap to enhance the availability of safe and efficacious ATMPs/HCTPs worldwide. More recently, regulation in Japan has also undergone substantial changes but a more detailed comparison and discussion is outside the scope of this chapter.

22.3 TYPICAL CHALLENGES FOR TISSUE ENGINEERING ENCOUNTERED IN THE CLINICAL PHASE

As described earlier, it is not possible to copy the process of the development of conventional medicinal products to develop an ATMP/HCTP. These challenges have been recognized and illustrated by a recent guidance of an Advisory Committee in the US discussing on the topic of trial design for cell therapies (FDA, 2013b). Unfortunately, randomized controlled trials (RCTs) are not common, and are quite challenging to execute in the field of TE, typically seeking to repair or regenerate damaged tissues such as skin, cartilage, bone, intervertebral discs, myocardial tissue, and peripheral nerves and tendons. Thus, only few examples exist. However, RCTs remain the only accepted standard of proving effectiveness and safety. An example of such RCT is given in the State-of-the-Art/Classical Experiment.

In the next paragraphs, an overview of a number of, *but not all*, challenging issues will be covered that need attention in setting up such trials (Figure 22.1).

22.3.1 Exploratory Trial

For starters, due to the type of products covered, the concept of phase 1 research on healthy volunteers is not appropriate and will not add relevant information. Most guidelines from regulatory agencies will allow an exploratory trial establishing safety and proof of concept as a starting point. This involves a small group of patients, sequentially escalating doses, and looking for biological effects that reflect to some extent the results in the animal models. It further

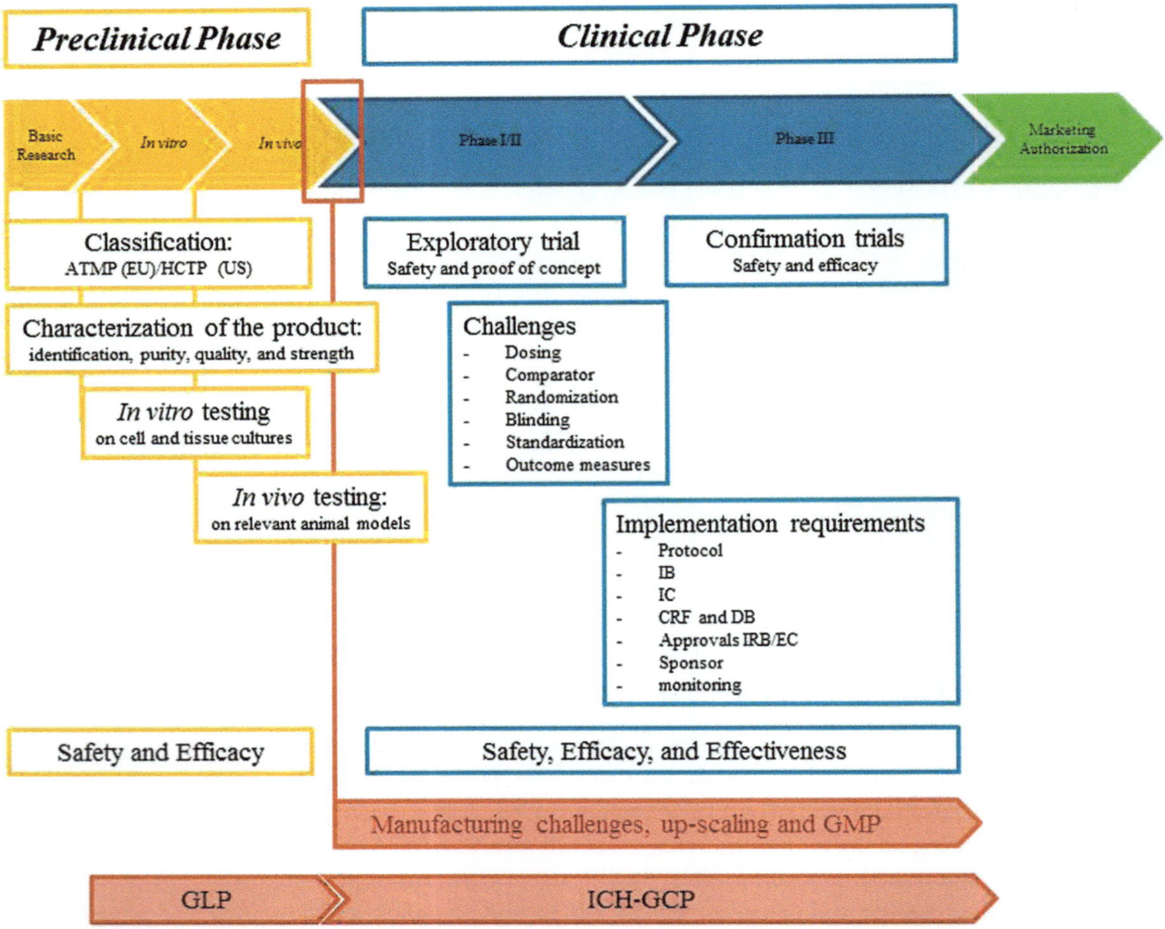

FIGURE 22.1

Schematic overview on the preclinical and clinical development path of tissue-engineered products. (Used abbreviations: ATMP: Advanced Therapeutic Medicinal Product; EU: European Union; HCTP: Human Cells, Tissues, and Cellular and Tissue-Based product; US: United States; GMP: good manufacturing practice; GLP: good laboratory practice; ICH-GCP: International Conference on Harmonization-Good Clinical Practice; IB: Investigator Brochure; IC: informed consent; CRF: Case Report Form; DB: Database; IRB: Institutional Review Board; EC: Ethical Committee).

evaluates the feasibility of the treatment in the human model and aims to detect catastrophic safety issues that could not be predicted in animal studies or unexpected findings.

22.3.2 Dosing

For TE products discussed here, conventional notions of dose do not always apply. Initial doses should be chosen so as to offer a realistic possibility of therapeutic benefit, for this reason, a major departure from the logic of conventional small-molecule trials. In many cases, there appears to be a threshold dose, and the maximum dose is limited not by potential toxicity, but by other determinants including the size of the site of administration, the number of cells that can be created, or other factors defining a "maximum feasible" as opposed to "maximum tolerated" dose.

22.3.3 Defining the Comparator

As the trials in TE frequently involve surgical procedures, one of the first hurdles to clear is to define the comparator, i.e., the currently accepted standard treatment for this specific clinical problem. It is not unusual to be confronted with the lack of a generally accepted comparator or control group, and differences in treatment approaches across the countries and continents become evident. Choosing the standard of care in daily practice as a comparator can raise the demand for supplementary research by the regulatory agencies, because the standard of care is not necessarily evidence based. The standard of care should be proven superior to placebo, but in a number of cases this was/has not been done. The widespread use of coronary artery bypass prior to a careful analysis of outcomes was a particularly compelling case in point (Varnauskas, 1988). Another example relates to articular cartilage repair, where the technique of microfracturing was selected as comparator (Saris et al., 2008). Subsequently, it turned out no one had proven that this therapy/technique was superior to placebo or a conservative approach without surgical intervention, and the regulatory bodies demanded proof of the superiority. Therefore, the effectiveness of microfracture had to be evaluated in addition to the proposed experimental therapy. An extra burden on the control groups in ATMP/HCTP RCTs is that in most cases the techniques to apply the ATMP/HCTP are more demanding than in classical medicinal products (pills, infusions, or subcutaneous injections), and bear risks on their own (Table 22.1).

By analogy, a given device is not always the best on the market, but in the hands of an experienced surgeon the results can be excellent. Therefore, the best device cannot guarantee the best efficacy when applied by someone who is not trained for and experienced with the technique. This brings us to a next point of attention; the complexity of the procedure to apply the ATMP/HCTP can influence the efficacy of the ATMP/HCTP and certainly change the relevant timelines for assessment (Figure 22.2). Before starting a clinical trial, it may be important to train the investigators in all procedures when applying the ATMP/HCTP. One must be aware that most techniques require a learning curve, and so a standardization program should be in place to make sure the registered efficacy is a result of the difference in treatment between groups and not of technical skill with which the ATMP/HCTP and control intervention are applied. In order to standardize this within a trial, few options are available

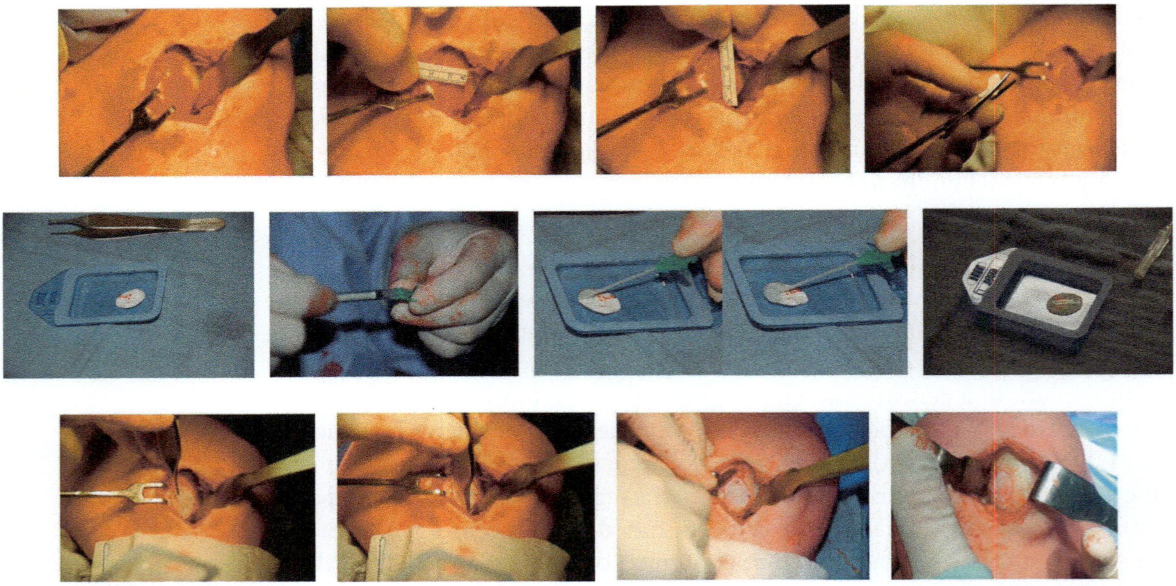

FIGURE 22.2
Example of the complexity of an ATMP/HCTP administration: Autologous Chondrocyte Implantation in a joint surface lesion of the femoral condyle using a collagen membrane. This procedure starts with debriding and measuring the lesion, followed by cutting the membrane to the appropriate size and seeding the chondrocytes. The final step in this procedure is the application and suturing of the cell-seeded membrane to the lesion.

ranging from training the procedures on animal models to appointing a single surgeon who will be applying the ATMP/HCTP in all patients.

In view of this, it is critical to consult all partners involved including the key opinion leaders and regulators to come to a broadly agreed comparator group and the appropriate technique to "deliver" the ATMP/HCTP to the patient.

22.3.4 Randomization

Randomization in clinical trials is a tool to make sure the situation at the start of the trial is identical in the different groups studied, typically the treatment group and the comparator. Thus, every group should have a similar distribution of demographic data and disease parameters. Moreover, randomization also implies that every participant has the same chance of receiving the treatment studied. Randomization procedures are standardized and independent of the investigator. Most of the time, a computerized system is used for the assignment of patients to groups. For ATMPs/HCTPs this already implies a hurdle: how can you organize this randomization procedure when decisions are made in the operating theatre. Solutions can be proposed and include, for instance, calling in a central phone number that blindly assigns the patient to a treatment arm.

22.3.5 Blinding a Trial

The idea of double-blind use in RCTs is to guarantee that the situation between groups is kept identical throughout the trial. When both the patient and treating physician (investigator) are unaware of the treatment assignment, all decisions are based on the clinical presentation and are independent of the treatment/procedure given.

In the field of TE this is not an easy task. Most of these ATMPs/HCTPs are applied by specific (surgical) procedures, not allowing for a full double-blind process. In addition, any surgical intervention has a substantial placebo effect that can persist for years. However, sham controls are not always acceptable, and therefore this point needs to be discussed with the key opinion leaders in the field and the regulatory bodies.

Furthermore, the idea of double blind is to ensure that the evaluation of the patients is done on clinical presentation and not on whether a patient received the experimental ATMP/HCTP or the control treatment. Therefore, strategies of blinded evaluation have to be implemented for patients in an RCT with these types of ATMPs/HCTPs. The investigator who has to make the assessments essential for the evaluation of the ATMP/HCTP has to be independent. In practice, the evaluation should be done by a clinician not involved in the initial treatment. Some of these techniques of independent evaluation can be adopted from experience in other fields, as for example, from trials in rheumatoid arthritis.

22.3.6 Standardization of Patient Care and Follow-Up

The standardization of the postintervention patient care and rehabilitation is a crucial point, not only to determine efficacy and safety but also to improve the recovery of sometimes complex procedures such as orthopedic interventions.

In case a surgical technique is applied, an appropriate standardized wound care and rehabilitation plan is required. Again, the option should be taken to delegate one person to follow up on all patients treated within the RCT or make sure all different care givers within the follow up are appropriately trained. Take appropriate measures to educate and train physical therapists, provide a consensus on standardized identical rehabilitation programs, stipulate an implementation plan, define criteria to measure progress in rehabilitation over a wide postoperative time, monitor patients' activities both in rehabilitation and during activities of daily living, and if possible organize supervised education and training (Figure 22.3).

In the case of ACI, the ATMP/HCTP is typically applied with a two-step surgical procedure: harvesting chondrocytes arthroscopically and weeks later implanting the cell product with an open knee surgery. In contrast, in the control group

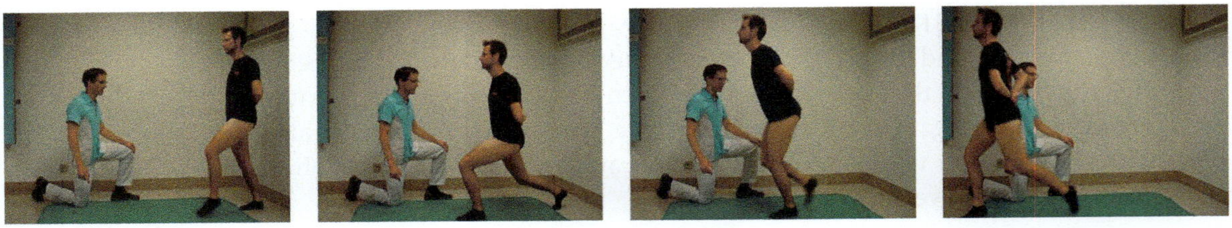

FIGURE 22.3
Physiotherapy as part of a standardized postoperative rehabilitation program is required to optimize wound care and rehabilitation after surgery. The design of such standardized program is to facilitate comparison between groups by defining, measuring, and monitoring progress.

the microfracturing is done in a single procedure. The postoperative rehabilitation following an ACI procedure may result in another approach than the one used in the control group. Accordingly, the trial should be designed to minimize the influence of the procedure on postoperative management (Van Assche et al., 2011).

22.3.7 Outcome Measures

The same caution should be applied for the primary and secondary outcome measures or endpoints. What seems to be a "hard" outcome in a clinical trial with medicinal products may not be so evident in TE. For example, pain is an outcome typically used in trials with medicinal products in musculoskeletal diseases such as rheumatoid arthritis, and it is validated for these diseases. For TE, this outcome would not necessarily be measuring what is intended. When a patient is treated with ACI and the appropriate open knee procedure is required, pain will be less relevant immediately after the procedure for evaluation of the efficacy, and only become of relevance months after surgery (Saris et al., 2009). Other "hard" structural outcomes, validated for development of medicinal products, are requested by the regulatory agencies in RCTs for ATMPs/HCTPs. However, most of the time they are not validated for tissue-engineered products and no standards or validated tools are available. This may mean that simultaneously with the trial, or even before starting the trial, suitable, clinical, or surrogate endpoints will have to be defined, evaluated, and validated. These endpoints should reflect the long-term efficacy within a reasonable timeframe.

22.4 IMPLEMENTATION OF A CLINICAL TRIAL

22.4.1 Protocol

The process of developing a clinical trial starts with the development of the protocol. Content requirements for the protocol are outlined at the highest level in ICH E6 (ICH E6, 1996), with further information on the level of clinical trial guidance documents and guidelines. As described earlier, it is recommended to contact the regulatory agencies to discuss the protocol as soon as possible in the process of development.

The protocol is the key document that contains all information on the conduct of the specific trial. It is a blue print of the complete process of what is going to be done, why it is to be done, how it has to be done, and how it will be evaluated. The protocol starts with a general part on the rationale and introduction, the study design, the objectives, and the flow chart. In the next part of the protocol, all items are discussed in more detail. The definition of the study population with inclusion and exclusion criteria, the primary and secondary outcome measures, the study duration, the enrolment period, the number of investigative sites, the clinical evaluations required per visit, stopping rules, informed consent procedure, randomization, statistical evaluation, laboratory tests, and technical investigations will need to be described. In the final part of the protocol more general issues will be addressed in more detail, including items such as how to handle safety and how the safety committee is organized, the obligations of the investigator, compliance with the protocol and GCP, the Declaration of Helsinki, IRB/Ethics committee (EC), informed consent process, data acquisition, source documents, retention of documents, confidentiality, deviations from the protocol and how to handle them, monitoring plan, final report and publication policy, proposed analysis plan, and study material. The final version of the protocol has to be signed on the protocol signature page by the principal investigator (PI) confirming adherence to the protocol. The protocol is also referred to within the clinical trial agreement.

22.4.2 Investigator Brochure (IB)

The IB is the document where information on the ATMP/HCTP, relevant to the research program or on further products which are relevant to the study, can be found. It is intended to facilitate investigators' understanding of the product and contains information to support the rationale for dosing and scheduling schemes. The available data start with the nonclinical studies and pharmacology in vitro, followed by the pharmacokinetics and studies on product distribution and behavior in animals. The research on toxicology, safety, and genotoxicity is also handled in the IB. Following the nonclinical data, all potential clinical studies in humans on PD and PK are less relevant for cell-based products. On the contrary, dosage and rationale for dosage chosen, safety, and therapeutic rationale have to be outlined. As a consequence of the above, the IB is a dynamic document that needs to be updated during development as new information on the ATMP/HCTP becomes available. These updates need to be discussed with all participants in the clinical program and evaluated for their potential implications with regard to safety and efficacy.

22.4.3 Investigational Medicinal Product Dossier (IMPD) and Investigational New Drug (IND) Application

The Investigational Medicinal Product Dossier (IMPD) in Europe is the document that captures the information on the manufacture and control of a given investigational medicinal product (IMP) during the clinical trial stage and also includes the relevant information on reference product and placebo for a given clinical trial. The IMPD is continuously updated as experience is gained with manufacturing and reflects the summary of the current and all historic manufacturing processes of the ATMP/HCTP.

In the US, a medicinal product in development is called an Investigational New Drug (IND) and the documentation that needs to be submitted in support of clinical trials equally needs to capture extended information on the source, characterization, production, formulation, stability, storage, and handling of the product. The explicit requirements are presented in the US Code of Federal Regulations, Title 21, Section 312.23(a)(7) (21CFR312.(a)(7)). This section is called the Chemistry, Manufacturing and Controls (CMC) part of the IND (FDA, 2008).

On both continents, detailed information is required on how the quality of the ATMP is guaranteed through the use of GMP approved facilities.

The information contained in the IMPD/IND CMC forms the core of module 3 (Quality) of the Common Technical Document format for an MA.

22.4.4 Informed Consent

Patients need to consent to the study in general and to all related procedures, as well as understand their rights as participants before any study-related procedure is performed. To comply, patients and investigator have to sign the informed consent form in duplicate, one copy for the patient the other for the investigator file to be kept at site. The procedure has to be clearly documented in the patient medical file.

The informed consent contains an explanation of the disease studied, the aim of the study, the study procedures, possible side effects of the procedures/ATMP/HCTP, insurance, patient rights, privacy, IRB/EC review, possible costs, data handling, and alternative procedures including the right to withdraw or refuse without prejudice to their medical care. The content of the consent form can be found in ICH-GCP guidelines (www.ich.org).

22.4.5 Case Report Form and Database

Before starting an RCT, a case report form (CRF) and database (DB) have to be constructed to collect and store data for further analysis. This can be done on paper but there is a shift to using digital systems. This provides the advantage of storing data immediately into the underlying DB without the need for tedious manual operations. Collecting data in a clinical trial is done within the medical file of the patient before entering the data into the CRF; this implies the CRF can be verified and checked for inconsistencies. Important in all systems, paper or digital, is to have track records available. Every change in the system has to be captured and should be traceable and accounted for by a registered user. Registered users should be listed on the delegation of authority (DoA) document specifying the appointed tasks. Notwithstanding these precautions, it is important to emphasize that the PI is the one with the overall responsibility.

Collection of data on safety, such as adverse events (AEs) and concomitant medication (conmed) should be outlined in the protocol. Due to the nature of the ATMP/HCTP and the procedure required to apply the ATMP/HCTP, the protocol needs to be very clear on criteria used to define an adverse event, as opposed to what is considered an expected consequence of the procedure. Not only for the AEs but also for the conmed, when an operative technique is used to administer the ATMP/HCTP it is required to have all anesthetics recorded or entered in the CRF (local or general).

An important point to consider when developing a CRF and DB is the privacy of the human subject. In several countries, a privacy commission is installed by the government to follow up on measures taken to protect the integrity of the participants and to approve the way data collection will be done.

22.4.6 Institutional Review Board (IRB)/Ethics Committee (EC) and Competent Authority (CA)

When all logistics are in place, the next important step to take before entering patients in a clinical protocol is to obtain approval from an IRB (USA) or EC (EU) as well as the national Competent Authority (CA). The priority for both the CAs and the IRB/EC is to protect participants in a clinical trial. While it is impossible to completely split the tasks, the focus of CAs will naturally lie with regulatory compliance, adherence to guidelines, and manufacturing details, whereas the ECs will focus on ethical issues and clinical practice. For ECs, a file has to be submitted by the coordinating investigator to the central IRB/EC as well as the local IRBs/ECs, in case of a multicenter study. This file should contain the final protocol, the final informed consent form in all languages needed, an insurance certificate, all documents that are patient related, such as diary cards, IB, suitability of the study center, curriculum vitae of the investigator(s), contract with the investigator, and registration form to the CA (EudraCT in Europe).

For Europe, the process of handling a file by the EC and the composition of the EC is consolidated in the European directive (EC/2001/20, to be replaced by the future clinical trials Regulation) (OJL, 2001). The EC has 3 calendar days to validate a submission. The timeline for reasoned opinion is maximum 60 days. In case of ATMPs/HCTPs, the period to decide for an EC can be extended by 30 days, and in case extra consultations are required by a further 90 days. A consequence of the directive, and clinical trial approval being member state responsibility, is that the approval has to be granted by only one EC per member state. This Leading EC has to give a single opinion after consulting the local ECs. In case of amendments to an approved protocol, the central EC is the only one to provide an opinion. A local EC can make comments on the complete file. However, its primary responsibility is to evaluate the ability of the local investigator to perform the protocol and determine whether the required resources are available. In case of ATMPs, this is important because frequently multiple disciplines will be involved in the execution of this type of trials.

When the EC has collected all comments from the local ECs and expert consults, questions/deficiencies will be passed through to the coordinating investigator and sponsor for reply. After adequate reply, the final approval can be granted.

For the US, this process is regulated by the CFR (45 CFR 46) (CFR, 2009) where the policy on IRBs is described. The approval of an IRB may be subject to further review by the local institution to authorize the clinical research to be performed within the institution.

22.4.7 Monitoring, Audits, and Inspections

Measures have to be taken to assure quality and reliability of the registered data and to comply with GCP.

As an internal quality assurance, standard operating procedures (SOPs) are highly recommended to describe how all procedures, ranging from obtaining informed consent to taking a blood pressure, are done on site and for the protocol in specific. When SOPs are in place, everyone at site has to adhere to them. Beside the description of how to handle certain procedures, SOPs also describe what actions have to be taken when a procedure is done differently.

On top of this, external quality assurance should be in place. The first step in this external quality assurance is a monitoring plan to check for consistency between the data entered in the CRF/DB and the source data in the patient file. This quality control is important to make sure the outcome of an RCT is reliable and can be trusted. On top of this monitoring plan, audits need to be organized by an independent organization appointed by the sponsor.

Regulatory agencies will organize inspections to rigorously check the processes of data collection and the validity of the data collected. These inspections can occur in the EU during the clinical trial stage and during an MA. In the US, inspections are usually done at the prelicensing stage (preapproval inspection).

22.4.8 Sponsor

The sponsor of the CT can be a pharmaceutical company, a governmental agency, an academic institution, or an individual. According to GCP, the sponsor organizes a clinical trial and takes ultimate responsibility for the quality and integrity of the trial data. The sponsor is not the PI, and cannot be disqualified in contrast to the PI. Detailed sponsor responsibilities on the conduct of a trial can be found in the GCP guidelines.

For very innovative products, such as ATMPs/HCTPs, the sponsor will often be a university or an academic research unit who also take on the role of manufacturer of these products (Lincker et al., 2014).

22.5 SPECIAL POINTS TO CONSIDER

22.5.1 Define the Patient

As the therapeutic effect and potential toxicity of living cell-based products are particularly sensitive to the microenvironment, defining the patient and thus the environmental effects are of importance. Therefore, testing the products in the preclinical phase in models that reflect the patient, and thus the proper environment, appears to be highly desirable. In addition, the handling of the product will be guided by the surgical procedures required and thus needs to be mimicked in the preclinical studies. Furthermore, other factors that may be of importance include, but are not limited to, the age and gender of the patient and the presence of comorbidities such as diabetes or other metabolic diseases.

22.5.2 Manufacturing Challenges and Upscaling

Entering the human model implies a treatment scale largely exceeding the scale of small animal models. Therefore, allometric scaling needs attention, and it is imperative that the preclinical studies in larger animal models address this appropriately. The scale-up also includes scaling up the manufacturing process (e.g., producing 4×10^8 cells, instead of 4×10^6 cells) and the adaptation of the surgical procedures and instrumentation. The complexity of upscaling is frequently underestimated and leads potentially to nonrobust products of inferior quality whose survival after implantation may be negatively affected by its increased size merely as a result of mass transport limitations.

It is worth mentioning that, if the manufacturing process has been established using cells of healthy donors, verification is essential to ensure patient cells will behave comparably, during manufacture, shelf life, and transport.

22.5.3 Exploratory Trials in the Academic Environment

Academic research is of invaluable importance in this quickly evolving field. Typically, exploratory clinical trials in an academic setting can move the field forward. However, this academic research often cannot provide efficacy and safety data collected in rigorous multicenter pivotal clinical trials. The cost for the development of treatments to live up to standards set out by the regulators and to provide a well-defined quality product, produced within a GMP approved facility, is high and independent academic research is mostly not capable to provide this full clinical development track, with some exceptions for orphan indications. For this, the hospital exemption was introduced in Europe.

22.5.4 Hospital Exemption (EU Only)

When ATMPs are used in clinical trials to obtain an MA the pathway of the central filing has to be followed. Only two exemptions are foreseen in regulation 1394/2007, firstly concerning products that were legally on the market before the regulation became applicable and secondly the "hospital exemption". ATMPs already on the market had to be compliant with the regulation no later than December 30, 2012, or they were withdrawn. The "hospital exemption" to the centralized licensing procedure can be used to apply patient-tailored ATMPs on a nonroutine basis by a medical practitioner upon prescription. Since the hospital exemption is considered a purely national framework, the rules implemented in member states differ. This might be one of the reasons why the topic of the hospital exemption triggered the most responses and conflicting views in the public consultation on the 5-year review of the ATMP regulation. A change to the legislation is therefore likely, as outlined in the recent report by the European Commission (European Commission, Health and Consumers Directorate-General, 2013).

22.5.5 Voluntary Harmonization Procedure (VHP) (EU Only)

The European Voluntary Harmonization Procedure was introduced in 2009 as a pilot for the future joint assessment of clinical trials by concerned member states (HMA, 2013). It was established by the Clinical Trial Facilitation Group, a working group of the Heads of Medicines Agencies (HMA) and consists of a joint assessment phase for multinational clinical trials by the CA of concerned member states followed by an accelerated national authorization procedure. It is open to all trials involving two or more member states willing to participate and involves a validation phase and a joint assessment phase (voluntary) followed by the legally required national submission. The benefit for the applicant lies in the provision of joint comments, which facilitates dossier harmonization across member states.

22.5.6 Combination Products

Combination products typically consist of cells, scaffolds or biomaterials, and also growth factors. This implies not only a strict characterization of the starting materials such as the cells, the biomaterials, and growth factors, but also in-depth studies of the combination of the above. For instance, cell behavior may be strongly affected when deposited on a biomaterial, and potency assays for cells delivered on a scaffold may thus be different when the same cells are injected as cell suspensions. Also, distinct lots of biomaterials may be quite different when looking at the material characteristics such as pore size, and that may affect the biological behavior of the combination product. Importantly, the manufacturing process to make the combination product is quite challenging and needs special attention and validation, with the need to demonstrate that data obtained in later stages of the development are comparable to those obtained in the early phases.

22.6 CONCLUSION AND FUTURE PERSPECTIVES

As described, the process of developing an ATMP/HCTP in TE is challenging and evolving. In recent years, much effort has been directed at creating a framework for development of these products, but during this process new hurdles have arisen. From the regulatory point of view, FDA as well as EMA, pathways have been created to facilitate the development of TE ATMPs/HCTPs. Nevertheless, due to the unique nature of each ATMP/HCTP no single solution can be often applied generically to all situations. The most important message that can be given is to reach out to the agencies very early in the development of scientific and regulatory advice.

We further believe it is critical to respect the basic rules and regulations for drug development, also based on the historical incidents and background of the regulation and clinical development that evolved over the last century. This includes good clinical practice, the "first do not harm" principle, good manufacturing practices, and proper assessment of efficacy. Indeed, providing a treatment that does not work is also a safety issue.

Finally, special attention in this field needs to go to the patient consent. Indeed, as for many of these approaches, there is a poor understanding of the underlying mechanisms of action, and such aspects as drug (cells) clearance are not yet fully appreciated; the patient needs to be a "full" partner in these novel treatment approaches. Once the cells are in, we cannot remove them, and some of the cell products may persist for the lifespan of the patient, thereby allowing the possibility for unforeseeable late safety events. Because of dedicated research and development, some treatments have ultimately reached the clinic. This has brought important progress for patients in the past, and will continue to do so, even with a limited understanding of the new treatment approach. Let us hope this will also be true for the novel class of advanced therapeutic products. The development of new enabling technologies and a stimulating regulatory environment will be key factors in the success of these approaches.

22.7 SNAPSHOT SUMMARY

- Clinical translation involves a preclinical as well as a clinical dossier.
- Preclinical studies include small and large animal models relevant to the patient.
- Large animal models provide essential information on feasibility and safety aspects.
- Exploratory trials in the human model may identify unexpected behavior of tissue-engineered products.
- Exploratory trials are providing information on potential efficacy and proof of principle.
- Regulatory bodies should be involved as early as possible, not only for classification but also for guidance in nonclinical and clinical development.
- A regulatory framework for TE products is evolving around the world.
- Pitfalls for the clinical development of TE products include dosing, defining the comparator, randomization, blinding, outcome measures, and standardization of patient care and follow-up.
- Upscaling not only implies allometric upscaling but also scale up of manufacturing and surgical approaches.
- A patient-centered approach and patient consent are essential.

STATE-OF-THE-ART/CLASSICAL EXPERIMENT

Autologous Chondrocyte Implantation (ACI) as example for clinical translation in Advanced Therapy Medicinal Products.

Autologous Chondrocyte Implantation (ACI) and more specifically the Characterized Chondrocyte Implantation is an example for clinical translation and implementation of a new ATMP/HCTP in the field of tissue engineering. This example shows the different steps to take and pitfalls to be avoided to get a marketing authorization (MA) for this type of products.

Despite the fact that the technique of ACI is in use since 1994, Knutsen et al. (2004) were the first to perform a randomized controlled trial (RCT) to challenge the technique versus microfracture (MF). This well-designed RCT aimed to compare the outcome of the techniques by looking at the ability of hyaline cartilage formation. Within this trial attention was paid to the classical design of a clinical trial with randomization, defining and scoring of outcome parameters, training the surgeons, and attempting to harmonize the rehabilitation program.

Saris et al. (2008) took this a step further and introduced a characterization of the cell product. Indeed, in this trial special attention was taken to define and characterize the final cell product to make sure it consisted of a specific dose of viable chondrocytes able to produce hyaline cartilage. There was an advanced preclinical training of the surgeons and as per protocol imposed rehabilitation and follow-up. There were fixed audits to verify compliance with the protocol.

Despite the attention paid to the design of the RCT, pitfalls arose during the execution of the trial. The most striking example hereof was the comparator selected. MF was selected as it was the standard of care in daily practice. But the existence/acceptance of MF as standard of care in daily practice did not imply that it was proven to be superior to placebo or to a nonsurgical, conservative approach.

In the course of subsequent discussions with the regulatory bodies, it turned out no RCTs had convincingly demonstrated superiority of MF. Indeed, the natural

STATE-OF-THE-ART/CLASSICAL EXPERIMENT *Continued*

course of symptomatic joint surface defects in the knee appeared to be insufficiently documented. Therefore, the effectiveness of MF had to be evaluated in addition to the proposed experimental therapy.

In conclusion, clinical translation of ATMPs/HCTPs is challenging but as for all new conventional drugs, necessary to obtain MA. When new treatment approaches are investigated, an in-depth analysis is needed not only with respect to the commonly accepted clinical practice, but also looking critically into the evidence supporting the existing clinical practice.

REFERENCES

Akhurst, R.J., 2010. Taking thalidomide out of rehab. Nat. Med. 16 (4), 370–372.

Annas, G.J., Grodin, M.A. (Eds.), 1992. The Nazi Doctors and the Nuremberg Code: Human Rights in Human Experimentation. Oxford University Press, New York.

Arcidiacono, J.A., Blair, J.W., Benton, K.A., 2012. US Food and Drug Administration international collaborations for cellular therapy product regulation. Stem Cell. Res. Ther. 3, 38.

Ballentine, C., 1981. Taste of sulfanilamide disaster. FDA Consum. Mag. [June Issue Online] Available from: http://www.fda.gov/AboutFDA/WhatWeDo/History/ProductRegulation/SulfanilamideDisaster/default.htm (accessed 15.04.14).

Below, P., 2008. Where the FDA regulations end and ICH GCP begins: the differences and similarities between the two. SoCRA Source, 48–53.

Clinical Trials Facilitation Groups, 2013. Guidance Document for Sponsors for a Voluntary Harmonisation Procedure (VHP) for the Assessment of Multinational Clinical Trial Applications. [Online] Available from: http://www.hma.eu/fileadmin/dateien/Human_Medicines/01-About_HMA/Working_Groups/CTFG/2013_06_CTFG_VHP.pdf (accessed 15.04.14).

Code of Federal Regulations, 2005. Part 1271 Human Cells, Tissues, and Cellular and Tissue-Based Products. [Online] Available from: http://www.gpo.gov/fdsys/pkg/CFR-2005-title21-vol8/xml/CFR-2005-title21-vol8-part1271.xml (accessed 15.04.14).

Code of Federal Regulations, 2009. Title 45, Public Welfare, Department of Health and Human Services, Part 46, Protection of Human Subjects. [Online] Available from: http://www.hhs.gov/ohrp/policy/ohrpregulations.pdf (accessed 15.04.14).

Committee for Advanced Therapies (CAT), 2010. Challenges with advanced therapy medicinal products and how to meet them. Nat. Rev. 9, 195–201.

Committee for Advanced Therapies (CAT), 2013. Guideline on the Risk-Based Approach According to Annex I, Part IV of Directive 2001/83/EC Applied to Advanced Therapy Medicinal Products 11 February 2013. EMA/CAT/CPWP/686637/2011 [Online] Available from: http://www.ema.europa.eu/docs/en_GB/document_library/Scientific_guideline/2013/03/WC500139748.pdf (accessed 15.04.14).

Davidoff, F., 2011. Heterogeneity: we can't live with it, and we can't live without it. BMJ Qual. Saf. 20 (Suppl. 1), i11–i12.

European Commission, 2014a. Medicinal Products for Human Use, Clinical Trials. [Online] Available from: http://ec.europa.eu/health/human-use/clinical-trials/index_en.htm (accessed 15.04.14).

European Commission, 2014b. Report from the Commission to the European Parliament and the Council in Accordance with Article 25 of Regulation (EC) No 1394/2007 of the European Parliament and of the Council on Advanced Therapy Medicinal Products and Amending Directive 2001/83/EC and Regulation (EC) No 726/2004. [Online] Available from: http://ec.europa.eu/health/files/advtherapies/2014-03_atmp_report.pdf (accessed 15.04.14).

European Commission, Health and Consumers Directorate-General, 2013. Regulation (EC) No. 1394/2007 on Advanced Therapy Medicinal Products: Summary of the Respons to the Public Consultation. [Online] Available from: http://ec.europa.eu/health/files/advtherapies/2013_05_pc_atmp/2013_04_03_pc_summary.pdf (accessed 15.04.14).

FDA, 2008. Guidance for FDA Reviewers and Sponsors: Content and Review of Chemistry, Manufacturing, and Control (CMC) Information for Human Somatic Cell Therapy Investigational New Drug Applications (INDs). [Online] Available from: http://www.fda.gov/downloads/BiologicsBloodVaccines/GuidanceComplianceRegulatoryInformation/Guidances/Xenotransplantation/ucm092705.pdf (accessed 15.04.14).

FDA Consumer Health Information, 2012. Kefauver-Harris Amendments Revolutionized Drug Development. [Online] Available from: http://www.fda.gov/downloads/ForConsumers/ConsumerUpdates/UCM322886.pdf (accessed 15.04.14).

FDA, 2013a. Tissue Reference Group. [Online] Available from: http://www.fda.gov/BiologicsBloodVaccines/TissueTissueProducts/RegulationofTissues/ucm152857.htm (accessed 15.04.14).

FDA, 2013b. Draft Guidance for Industry: Considerations for the Design of Early-phase Clinical Trials of Cellular and Gene Therapy Products. [Online] Available from: http://www.fda.gov/downloads/BiologicsBloodVaccines/GuidanceComplianceRegulatoryInformation/Guidances/CellularandGeneTherapy/UCM359073.pdf (accessed 15.04.14).

FDA, 2014. Cellular and Gene Therapy Guidance Documents: Guidance for Industry Potency Tests for Cellular and Gene Therapy Products. [Online] Available from: http://www.fda.gov/BiologicsBloodVaccines/GuidanceComplianceRegulatoryInformation/Guidances/CellularandGeneTherapy/default.htm (accessed 15.04.14).

Heads of Medicines Agencies (HMA), 2013. Guidance Document for Sponsors for a Voluntary Harmonisation Procedure (VHP) for the Assessment of Multinational Clinical Trial Applications. [Online] Available from: http://www.hma.eu/fileadmin/dateien/Human_Medicines/01-About_HMA/Working_Groups/CTFG/2013_06_CTFG_VHP.pdf (accessed 15.04.14).

International Conference on Harmonisation of Technical Requirements for Registration of Pharmaceuticals for Human Use, 1996. ICH Harmonised Tripartite Guideline Specifications: Guideline for Good Clinical Practice E6. [Online] Available from: http://www.ich.org/fileadmin/Public_Web_Site/ICH_Products/Guidelines/Efficacy/E6_R1/Step4/E6_R1__Guideline.pdf (accessed 15.04.14).

International Conference on Harmonisation of Technical Requirements for Registration of Pharmaceuticals for Human Use, 1997. ICH Harmonised Tripartite Guideline Specifications: General Considerations for Clinical Trials E8. [Online] Available from: http://www.ich.org/fileadmin/Public_Web_Site/ICH_Products/Guidelines/Efficacy/E8/Step4/E8_Guideline.pdf (accessed 15.04.14).

International Conference on Harmonisation of Technical Requirements for Registration of Pharmaceuticals for Human Use, 1999. ICH Harmonised Tripartite Guideline Specifications: Test Procedure and Acceptance Criteria for Biotechnological/Biological Products Q6B. [Online] Available from: http://www.ich.org/fileadmin/Public_Web_Site/ICH_Products/Guidelines/Quality/Q6B/Step4/Q6B_Guideline.pdf (accessed 15.04.14).

Knutsen, G., Engebretsen, L., Ludvigsen, T.C., Drogset, J.O., GrØntvedt, T., Solheim, E., Strand, T., Roberts, S., Isaksen, V., Johansen, O., 2004. Autologous chondrocyte implantation compared with microfracture in the knee. A randomized trial. J. Bone Jt. Surg. Am. 86, 455–464.

Lenas, P., Moos Jr., M., Luyten, F.P., 2009a. Developmental engineering: a new paradigm for the design and manufacturing of cell-based products. Part I: from three-dimensional cell growth to biomimetics of in vivo development. Tissue Eng. Part B Rev. 15 (4), 381–394.

Lenas, P., Moos Jr., M., Luyten, F.P., 2009b. Developmental engineering: a new paradigm for the design and manufacturing of cell-based products. Part II: from genes to networks: tissue engineering from the viewpoint of systems biology and network science. Tissue Eng. Part B Rev. 15 (4), 395–422.

Lincker, H., Ziogas, C., Carr, M., Porta, N., Eichler, H.G., 2014. Where do new medicines originate from in the EU? Nat. Rev. Drug. Discov. 13, 92.

Markman, J.R., Markman, M., 2007. Running an ethical trial 60 years after the Nuremberg code. Lancet Oncol. 8, 1139–1146.

Official Journal of the European Communities, 2001. Directive 2001/20/EC of the European parliament and of the council of 4 April 2001 on the approximation of the laws, regulations and administrative provisions of the member states relating to the implementation of good clinical practice in the conduct of clinical trials on medicinal products for human use. Off. J. Eur. Comm. L121, 34–44. [Online] Available from: http://eur-lex.europa.eu/LexUriServ/LexUriServ.do?uri=OJ: L:2001:121:0034:0044:en:PDF (accessed 15.04.14).

Official Journal of the European Union, 2009. Directives, commission directive 2009/120/EC of 14 September 2009 amending directive 2001/83/EC of the European parliament and of the council on the community code relating to medicinal products for human use as regards advanced therapy medicinal products (Text with EEA relevance). Off. J. Eur. Union L242, 3–12. [Online] Available from: http://eur-lex.europa.eu/LexUriServ/LexUriServ.do?uri=OJ: L:2009:242:0003:0012:EN: PDF (accessed 15.04.14).

Official Journal of the European Union, 2007. Regulation (EC) No 1394/2007 on advanced therapy medicinal products, of the European parliament and the council of 13 November 2007. Off. J. Eur. Union L324, 121–137. [Online] Available from: http://eur-lex.europa.eu/legal-content/EN/TXT/PDF/?uri=CELEX:32007R1394&from=EN (accessed 15.04.14).

Saris, D.B.F., Vanlauwe, J., Victor, J., Almqvist, K.F., Verdonk, R., Bellemans, J., Luyten, F.P., 2009. Characterized chondrocyte implantation results in better clinical outcome at 36 months in a randomized trial compared to microfracture. Am. J. Sports Med. 37 (Suppl. 1), 10S–19S.

Saris, D.B., Vanlauwe, J., Victor, J., Haspl, M., Bohnsack, M., Fortems, Y., Vandekerckhove, B., Almqvist, K.F., Claes, T., Handelberg, F., Lagae, K., van der Bauwhede, J., Vandenneucker, H., Yang, K.G., Jelic, M., Verdonk, R., Veulemans, N., Bellemans, J., Luyten, F.P., 2008. Characterized chondrocyte implantation results in better structural repair when treating symptomatic cartilage defects of the knee in a randomized controlled trial versus microfracture. Am. J. Sports Med. 36 (2), 235–246.

Van Assche, D., Van Caspel, D., Staes, F., Saris, D.B., Bellemans, J., Vanlauwe, J., Luyten, F.P., 2011. Implementing one standardized rehabilitation protocol following autologous chondrocyte implantation or microfracture in the knee results in comparable physical therapy management. Physiother. Theory Pract. 27 (2), 125–136.

Van Wilder, P., 2012. Advanced therapy medicinal products and exemptions to the regulation 1394/2007: how confident can we be? an exploratory analysis. Front. Pharmacol. 3, 1–9. http://dx.doi.org/10.3389/fphar.2012.00012.

Varnauskas, E., The European Coronary Surgery Study Group, 1988. Twelve-year follow-up of survival in the randomized European coronary surgery study. N. Engl. J. Med. 319, 332–337.

World Medical Association, 2013. World medical association declaration of Helsinki: ethical principles for medical research involving human subjects. JAMA 310 (20), 2191–2194.

CHAPTER 23

Ethical Issues in Tissue Engineering

Stellan Welin
Department of Culture and Communication, Linköping University, Linköping, Sweden

LEARNING OBJECTIVES

- To understand the distinction between values and facts
- To acquire a basic knowledge of consequentialism and deontology
- To distinguish between ethics and law
- To understand the main views of the human biological materials
- To acquire basic knowledge of the Helsinki principles for research on donated human biological material
- To understand the ethical issues involved in commercialization of biological material
- To understand how new technological developments may create ethical problems but also may solve ethical problems

> Science can discuss the causes of desires, and the means for realizing them, but it cannot contain any genuinely ethical sentences ...
>
> **Russell (1998, p. 25)**

> The man who shuns and fears everything and stands up to nothing becomes a coward; the man who is afraid of nothing at all, but marches up to every danger, becomes foolhardy ... Thus temperance and courage are destroyed by excess and deficiency and preserved by the means.
>
> **Aristotle (1955, p. 94)**

23.1 INTRODUCTION

Tissue engineering will hopefully be able to make many wishes come true. There may be biocompatible implants for repair of the body; there may be tissue and organs grown from human cells; there may be new possibilities of connecting the body and its nervous system to artificial limbs. Is not all of this good? Who can be against it?

There are many controversial areas, however. One was related to the use of human cells, in particular, the use of embryos to produce human embryonic

Tissue Engineering.
Copyright © 2015 Elsevier Inc. All rights reserved.

stem cells. Some of these problems have been overcome with new development. There are also questions about the commercialization of human biological material—after all, there are biotech firms who want to make a profit—and there are issues of donation and control of human biological material. Human cells contain DNA and hence much information of the donor. How should that be safeguarded and who should be in control?

Completely different issues deal with the idea of "naturalness" and the traditional medico-ethical concept that the right thing is to cure and help, not to improve and enhance human beings. Where is the limit when we stop being human? Who is to judge? Applications of tissue engineering in the animal sphere opens up the possibility to produce meat to eat by culturing animal cells in bioreactors. Is that "natural?" How will such an application of tissue engineering change our relation to animals?

In the future lurks even more dramatic possibilities. Synthetic biology could become a more radical version of tissue engineering by allowing scientists and engineers to work not just with existing biological material but to actually construct new organisms. What benefits and threats pose synthetic biology? How should they be regulated?

Questions like these cannot be answered by science alone. They deal with values and normative principles. Before turning to a discussion of the issues mentioned earlier, I will give a short overview of ethics, the "theory" of good and evil, right and wrong. In this chapter, I arrive at certain, sometimes controversial (but hopefully convincing), positions on various issues. Suffice it to say that other writers have arrived at other conclusions. In the end everyone has to think for themselves.

23.2 MORALITY, ETHICS, AND VALUES

With the help of language we can state facts, propose hypotheses, approve or disapprove, give promises, etc. Sometimes we describe the world. "Today it will rain" expresses what is often called a *proposition*, which is either true or false. Some propositions are more complicated, e.g., a statement describing the whole human genome, but, nevertheless, they are either true or false. *A common idea is that science (including biomedicine) and social science are concerned with finding true propositions about the world and reliable predictions.* This is, however, not the whole story. Science and social science also try to give explanations, to make the natural and social world intelligible.

Statements about values (what is good or bad, right or wrong) and norms (what one ought to do) are different. It is a common idea that there is a gap between statements of facts and statements of values and norms. The Scottish enlightenment philosopher David Hume stated that one can never derive or infer value or moral conclusions from statements of pure facts. There is an unbridgeable gap between "ought" and "is." For example, from the fact that Sweden is a monarchy it does not follow that having a hereditary king as head of state is morally superior to having an elected president. (Of course, the opposite does not follow either.) From the (contested) fact that males are biologically determined for certain aggressive behavior, it does not follow that this behavior is morally right.

However, in many situations we use statements and concepts that have both a *factual* part (describing the world) and an *evaluative* or *normative* part (saying what is good or bad, making recommendations). For example, *democracy* is such a concept, *intelligence* is another. To make things even worse, very often people agree just that the evaluative part that democracy or intelligence is good, but disagree strongly about the factual parts. Unfortunately, this is the case with many of the key concepts in social science. In natural science, the concepts are usually less value-laden. In medicine, the concept of health obviously has this dual character as well as the even more contested notion of "being normal" in a psychiatric or psychological sense. It is presumably a good thing to be normal, but what is really normal?

Science cannot make recommendations based *only* on facts. There must always be a normative or value component added, according to the view of the division of the realm of facts and values. Medicine can never—based solely on medical facts—show that a long life is better than a short one. At most, it may give recommendations on how to live and behave, if you *wish to have a long life*.

We use language not just for describing the world, but also for inducing behavior in others. "Shut the door" is such an imperative, obviously neither true nor false. If we think that the other person should not just shut the door once, but every time, we may express this by saying "always shut the door" or "you ought to shut the door." This is a rather trivial example, but most of the *moral norms* regulating (hopefully) our behavior can be expressed in similar ways. "Always tell the truth" and "one ought to keep one's promises" are typical examples of the everyday morality many people subscribe to, or at least believe they subscribe to, and follow.

There are many norms as explained earlier. Some are *conventions* like "always drive on the right side of the road." We have agreed to that norm; it is very dangerous if people differ locally on the issue on what side to drive. Such a convention can be changed (like in Sweden in 1967) and it can be different (like in the UK). A norm like "always tell the truth" is different. It is not an agreement in the sense of a convention.

The simplest way to characterize a *moral norm for a person P* is that it is an *imperative or "ought" statement* ("always tell the truth" or "one ought to tell the truth") that:

1. P tries to follow in his/her life—even if no one can observe P.
2. P is ashamed if he/she breaks the norm *even if no one finds out.*

If someone follows a particular norm ("always tells the truth," for instance) for fear of punishment, then it is not a moral norm *for that particular person* in the sense just stated. A moral norm is followed from "inner pressure," not because of threats from others. The legal laws are other kinds of norms. The say, like the moral norms, what you should and should not do. However, in contradistinction to moral norms, legal norms are associated with sanctions like "you will be punished if you do not follow the law." In happy circumstances, the law incorporates the most important moral norms together with other prescriptions relating to conventions. Even if it is not a moral norm that we should drive on the right side of the road, the law, nevertheless, prescribes that we should all drive on the same side and will punish transgressors.

The reason for having laws and a legal system can be stated very easily: not everybody will follow the (most important) moral norms without the threat of punishment. In a society with perfectly moral citizens, one would not need a legal system. If all scientists were highly moral, one would not need legislation on matters related to research and technological development.

Not every moral norm finds its way into the legal system. You are not punished by the law if you lie to your partner or are unfaithful even if the behavior is morally blameworthy. That would be very intrusive and, apart from that, we do not agree on all moral norms. In fact, in modern western societies there has been a diminishing regulation of what we may call personal behavior. This is obvious not only in the area of sexual behavior, but also in matters of what you may say and think.

Different persons in the same society may adhere to different norms and have different values. Usually the difference is even bigger if we look to different cultures and other times. For example, the equal rights of men and women are not very old. One hundred years ago a woman could not decide herself if she wanted to marry a man (we assume here that the man wanted to marry the woman). The father of the bride to be—if he was alive—had that particular right. The woman must consent, but that was only a necessity, not a sufficient condition for marriage. The development toward an equal standing for women to men is now considered a moral progress.

The different moral and value systems of individuals, social groups, and culture can be investigated empirically. This is done by social science methods of interviews, observations of behavior, and questionnaires. Over a period of 100 years there have been substantial shifts in the value system of the western world. Very few today would regard one human race as superior to another, few believe in different rights and behavioral norms for men and women, and animals have entered the frame as something to be included in moral considerations.

To find out about the existing moral systems and values in a society is very interesting. However, it is *not* an answer to the *normative* questions:

What should I do?
What values should I have?

Or phrased alternatively: *What moral should I have*? This is a question dealt with in the philosophical inquiries on ethics. Unfortunately, there is no consensus on this important question. Philosophers disagree among themselves on the ultimate foundation of ethics as well on the content. However, it has been possible to reach some consensus in some areas, even if that consensus has not reached the seminars of philosophers. For example, the Universal Declaration of Human Rights (http://www.un.org/Overview/rights.html) and the Helsinki Declaration on Biomedical Research on Humans (http://www.wma.net/e/policy/b3.htm) can be seen as kinds of principles that can be defended and understood from many different, more fundamental philosophical starting points. An influential attempt to present such a common framework for the biomedical area was the attempt to develop the "Belmont principles," where respect for persons (autonomy), the principle of beneficence (do good, do not harm), and justice play a central role into a coherent whole (Beauchamp and Childress, 2001).

Following the Helsinki Declaration and various scandals in biomedical research, there are now ethics committees in place who review biomedical research before it begins. There is also extensive regulation of pharmaceutical drugs and growing regulation on medical technologies and materials used for implantation in the human body. Much of this regulation can be viewed as an attempt to translate the ethical principles and values in the Helsinki Declaration into legal regulation. There are similar regulations on animal experimentation.

There is a growing consensus that the area of moral concern covers not just humans, but all *sentient beings*, i.e., beings that experience and feel pain and pleasure. The philosopher Peter Singer (2011) has famously written that we should show "equal respect for interests" regardless of who has the interest (Singer, 2011, p. 49). It does not matter if the interest belongs to a white or black person, to a man or a woman, nor if it is an animal or a human being. Not showing equal respect in this way amounts, according to Singer, to *racism*, *sexism*, or *speciesism*, respectively. Singer advocates a certain kind of equality between all sentient beings.

However, there are still distinctions to be made. A human being normally has a larger number of and/or more complex interests than, say, a mouse. If a human being is killed, more is lost than if a mouse dies. However, enough mice may outweigh a human being. This position is controversial and quite contrary to many ethical systems stressing the *dignity of humans*, where humans are seen as special and there is a clear dividing line between nonhuman animals and humans. Such a special position can be argued for from a religious perspective (God created man special) or from some property (like reason), which distinguish humans from animals.

23.3 MORAL PROBLEMS RELATING TO THE SOURCE OF MATERIAL FOR TISSUE ENGINEERING

In tissue engineering one usually needs to start from some biological material. It may be cells from an animal or a human being that are removed and, for example, placed on some matrix to grow into a structure. Most tissue-engineered structures are meant to eventually find their way into humans to be used as a treatment or replacement. This means that it is preferable to use human biological material as a starting point.

The use of nonhuman biological material will pose the same technical and medical problems as xenotransplantation, i.e., the idea of transplanting organs from animals into humans. One particular problem facing the use of xenogenic biological material is the risk of infection from retrovirus in the source material. In the transplantation area there is a serious shortage of human organs, which may be an argument to try xenotransplantation in a controlled form anyhow—when and if the source animal is suitable and the technology for treating rejections is more developed. No such shortage exists for human cells. From a moral point of view (for the good of the patient to have the material implanted), human material should be used to avoid a variety of technical problems.

There are *general moral problems* relating to the use of human material. They have to do with *donation, informed consent, ownership*, and *control* of the resulting material. There are also *special moral problems* dealing with cells derived from human embryos, like embryonic stem cells, or from aborted human fetuses. The general problems also apply in the case of material from human embryos and aborted fetuses. Some of the problems involving human embryos have been overcome due to development of technology, in particular the possibility to reverse the development of cells and project them back to a more "primitive" stage.

23.3.1 Biological Material Derived from Human Embryos

In 1998, the first human embryonic stem cell line was derived from an embryo (Thomson et al., 1998). Since then there have been many more such human embryonic stem cell lines derived. In principle, such a human embryonic stem cell line may stay in the same condition forever. Other types of stem cells and cells can be derived from the embryonic stem cells by putting them in a suitable environment. This could be a very good source of human cells. There is still more to be learned about the development of the embryonic stem cell into an adult cell, but in due time we can expect such cell lines to be very good sources.

23.3.1.1 Two Views of the Human Embryo

To establish an embryonic stem cell line, "surplus" in vitro fertilization (IVF) embryos are usually used, after they have been cultured for some days. At the blastocyst stage (around 5 days), the inner cell mass containing the embryonic stem cells is extracted and transferred to another medium for further culturing. In this process the embryo is destroyed. One argument is that destroying the embryo in this way is morally wrong; the embryo has the same moral status as a new-born child, because a full human life begins at conception (this may be called *conceptualism*). The Catholic Church holds a version of conceptualism, as did the American President George Bush, when the work with human embryonic stem cells started [2]. The Catholic Church is still opposed to embryo destruction while the position has changed in American politics.

Some conceptualists use an argument from potentiality to defend the full moral status of the newly formed embryo. An embryo, if left on its own in a suitable environment, i.e., a uterus, will develop by itself into a child unless interfered with. The embryo has an *active potentiality* to become a child. One may then argue "backwards." The new-born child is a full human being from a moral point of view. The fetus at an advanced stage, say just before birth, is very similar to the new-born child. And so it goes on. There is no specific breaking point when the fetus or embryo becomes human. To put it simply, the development from conception to a new-born child is the development *of* a human being—not *to* a human being. This position is relatively new and has been strengthened by the modern theory of genetics and DNA [3]. The DNA of the grown-up is produced at conception and it starts working as a whole shortly afterward.

The most obvious problem of this argument for active potentiality is, however, that the IVF embryo does not have active potentiality. It needs help to be implanted; if we just let "nature to proceed" it would never develop into a child on its own.

Another conception of the embryo is related to the sentient being approach and may be called *gradualism*. The simple idea is that the early embryo—as just a single cell—cannot feel or experience anything. It is not yet a sentient being and we do not owe the same kind of moral respect toward it as we do toward sentient beings. Somewhere along its development it becomes sentient. The belief is that this sentience is tightly coupled to the development of the brain. When is the fetus sentient enough to count as a human? It seems highly unlikely that there is a sudden jump just at birth. It occurs either before birth or after. Many people believe that it occurs before birth, while others do not [4]. Unfortunately, while it may eventually be possible to measure the degree of sentience (there are of course great problems in characterizing such a measure) this will not decide the morally important question: *What degree of sentience is enough to count as human sentience?*

Gradualism does not solve all problems. First, it is not clear from gradualism what moral respect we owe to the embryo. Is it different from what we owe to other human cells? Most people would think so. There is something special about an embryo. It has to do with the possibility to develop into a child. Do not all cells have this possibility? Cloning of animals actually proceeds from somatic cells.

There is, however, a difference. The embryo—when inside a female uterus—is capable of this development on its own in a way that the unfertilized egg or the somatic cell is not. A somatic cell—even if its cell nucleus after fusion with a denucleated unfertilized egg may develop into a child—is analogous to the sperm or unfertilized egg. Here, I am aware that this discussion is quite reminiscent of the Catholic one. However, the difference from conceptualism is that gradualism denies that the embryo has the same value as the born child. Adherents to gradualism can hold, as do adherents of conceptualism, that the moral value of the embryo is higher than that of other cells, including a sperm and egg pair not yet fused. However, conceptualism and gradualism differ on the moral status of the newly conceived embryo.

But how high is the moral value of the embryo? Is it so high as to rule out destroying the embryo in order to derive human embryonic stem cells? While some people may not agree with this, there may be an amount of justification, e.g., a possible good outcome, which may outweigh the harm to the embryo. The good outcome that most easily comes to mind is that human embryonic stem cells may, in the future, be used for treating human illnesses. There are other uses that may outweigh the harm, e.g., substituting research human embryonic stem cells for research on sentient animals.

23.3.1.2 Using Cells from Aborted Fetuses

In the gradualistic view, it is also easier to accommodate abortion from a moral point of view than from the conceptualistic view. However, all arguments on abortion are based on the basic fact that the fetus is inside the body of the woman. These arguments do not apply to IVF embryos in petri dishes, as they are not inside the body of a woman (they can be protected and safeguarded without infringement to the autonomy of the woman and her right to bodily self-determination). It is a consistent—although perhaps rare—position to hold that embryos in a petri dish should be more protected than fetuses in a woman.

Suppose you subscribe to the gradualist theory. You accept (early) abortion because you believe in a woman's right to bodily self-determination, and in the stage of pregnancy when the fetus is not sentient (in a human way) the right of bodily self-determination takes precedence. On the other hand, you believe that the value of a human embryo is too high to warrant its use for derivation of stem cells. Attempts have been made to defend its use anyhow, if it is a surplus embryo in an IVF context [5]. If not used for implantation or stored for later infertility treatment, it will be discarded, wasted. Surely it is better to use it for something beneficial. In the gradualist view most will regard a more developed fetus as more "valuable" than the early embryo.

Still, it is possible to make an argument that it is better to use fetal material from aborted fetuses than using embryonic material. The surplus argument can be applied. In a country where abortion is legal up to a certain gestational time, women decide for themselves if they want an abortion. After abortion, the aborted fetus is "surplus" in different sense to the IVF embryo in the petri dish—it is not the physicians or the scientists, about to derive stem cells, who produced the surplus material. In the case of the aborted fetus one may argue that it is better to use it than to simply waste it. The morally relevant distinction is that the physician and scientist in the IVF case bring about the surplus character, while this is not the case in abortion. Of course, as in the IVF case, the use of biological material should depend on the informed consent of the woman (in the IVF case consent is usually sought from the couple).

23.3.1.3 The Risk of Instrumentalization

Some critics of the use of embryonic and fetal material have warned about the risk of *instrumentalization* of human life: human beings may no longer be seen as goals in themselves, but as means to produce something else. In particular, human embryos may be seen in this way. Such critics claim that this will undermine our moral community. Such a view has been forcefully argued by the German philosopher and social scientist, Jurgen Habermas. His arguments apply to the harvesting and use of human embryos, even if they are designed primarily against reproductive cloning and genetic modification of the child to be. Such a cloned or modified person would not, according to Habermas (2003), be "the author of his or her own life" and this will undermine any possibility to participate in the moral community. Unfortunately, Habermas' argument also seems to apply to many other parental decisions like what school the child would attend or what kind of religious or secular upbringing the child would get. He needs to assume that genetics is much more important than other factors in determining the future of a child, which is hard to believe. This is not a question of ethics. It is an empirical question—hard to research in the detail needed—of the causal force of the genetic set up compared to various social and environmental factors.

23.3.1.4 The Importance of Words

There is an interesting earlier example of how change of words changes a debate. In 1984, both Houses of the British Parliament rejected a proposal in the Warnock report to allow research on human embryos for the first 14 days of their existence. The members of parliament (MPs) were very opposed to research on human embryos. However, after the research community came up with the clever idea that the oozyte in its early stages was not an embryo at all, but a "pre-embryo," the pendulum swung. The British parliament approved research on pre-embryos and the MPs could continue to maintain their opposition to embryo research (Mulkay, 1994). This paved the way to MPs approving research on human pseudo-embryos while still rejecting research on human embryos.

23.3.2 Human Stem Cells without Embryo Destruction

The big ethical hurdle in dealing with human embryonic stem cells was its relation to the destruction of human embryos. Recent developments have done away with this particular problem. Sometimes new technological developments can—not solve—but dissolve ethical problems.

23.3.2.1 Induced Pluripotent Stem (iPS) Cells

In 2006, Shinya Yamanaka and Kazutoshi Takahashi reported that they had managed to produce pluripotent stem cells from mouse fibroblast cells (Takahashi and Yamanaka, 2006). By introducing four genes—selected from a list of originally 24—they managed to produce pluripotent stem cells, called induced pluripotent stem (iPS) cells. Since then such iPS cells have been produced in many ways.

The iPS cells are in many ways similar to embryonic stem cells. They are pluripotent like the embryonic stem cells and have the same risk of producing tumors. The original method to produce the iPS cells was by viral transfection and some of the genes used were oncogenes. Since then nonviral and nonintegrating methods have been developed, where the factors are only transient (no new genetic material remains in the cells), which decrease the tumor risk. For use in humans, such nonviral production of the iPS cells is to be preferred. Unfortunately, it seems at present that these DNA-free methods are not so efficient (Martins, Vunjak-Novakovic and Reis 2014).

The main advantages of iPS cells in the human case are twofold when compared with embryonic stem cells. One, they bypass completely all ethical discussions of the moral status of the embryo and the moral problems of destroying embryos. This is an illustration of feature often overlooked: some ethical problems can completely disappear due to technological advances. Second, iPS cells in the human case can be derived from the cells of the patient to be treated. This has the great advantage of overcoming immune reactions, which is always present for embryonic stem cells as they are derived from embryos.

There are many interesting applications. One is to test the efficacy of a drug intended for a patient. The drug can be tested in vitro on iPS cells derived from that particular patient. There have also been attempts to program iPS cells to be used in cardiac tissue engineering (Eschenhagen, 2011). However, the efficiency of the production is still not very good for the non-DNA methods. There are also still concerns for the tumor inducing effects of iPS cell, which seem to depend on how the iPS cells are derived (Miura et al., 2009).

In contrast to the use of human embryonic stem cells the remaining ethical issue is one of risk. How large risk is acceptable for treatment of patients with iPS cells? This question has no clear answer as it depends heavily on the severity of the condition to be treated.

23.3.2.2 Adult Stem Cells

One may hope that a better understanding of the role of the niches of the various adult stem cells will allow for better culturing and expansion. The interesting point from an ethical point of view is that none of these sources entails the destruction of embryos. Such sources are thus acceptable to those who oppose the use of human embryonic stem cells. A clear illustration is the talk the late Pope John Paul II gave to the International Transplantation Society Congress in Rome in 2000. In that talk, Pope John Paul II declared that donation of organs after death is an act of Christian love and endorsed the brain-related death criteria. However, while being somewhat flexible on organ donation at the end of life, Pope John Paul II condemned, in strong words, the destruction of human embryos to produce human embryonic stem cells. Instead, he urged scientists to use adult stem cells.

As with the previously mentioned iPS cells, adult stem cells have an advantage that there is no immune reactions, if cells from the patient to be treated are used. One interesting line of research is to find ways to stimulate the growth and proliferation of adult stem cells in situ, thus bypassing the need for iPS cells or embryonic stem cells. The use of adult stem cells of the body itself is, of course, the safest method compared to iPS cells. But as usual there are also questions of efficiency and cost as well as security.

It is now time to turn to general ethical problems associated with using human cells for tissue engineering.

23.3.3 Donation and Informed Consent

Even if, in most legal systems, you are not strictly the "owner" of your body or its part, you have the ultimate authority over its use. This is the simple principle of *autonomy* enshrined in most medicolegal regulations and a very important point in the Helsinki Declaration—the principles guiding biomedical research on humans. You have to give consent to any invasive procedure involving your body and to all medical treatment in general, otherwise the procedure is illegal. There are some exceptions related to cases of emergency. If arriving unconsciously at an emergency ward, the physicians do not, of course, have to wait until you wake up. They are allowed to act on *presumed consent*. For minors the parents will have a greater say.

When biological material from aborted fetuses is donated, it is usually the woman who has to give consent. This can be interpreted in two ways. The most straightforward interpretation is to view donation from an aborted fetus as similar to the donation of biological material that has been removed from the woman. The fetus is regarded as part of her body; the abortion is similar to an amputation. Consent has to be sought if consent is needed for the use of material from an amputated leg. In Sweden, for example, discarded biological material (that would otherwise be wasted) could formerly be used without consent (no one is asked), but is now included in the category where permission from the former "owner" must be sought.

The second interpretation may be that a fetus is not simply a part of a woman's body but a separate individual. The material from an aborted fetus could be viewed as being similar to a donation from a dead child. Then it is clear that consent is needed.

In the case of embryo donation for stem cell research, in most cases it is the couple, wherever possible, who together have to consent to donation. The case of embryo donation for research, or derivation of stem cells, is best seen as being different from the interpretations of donation from an aborted fetus. Firstly, at the time of donation the IVF embryos are alive and, hence, the donation cannot be compared to donation from a deceased. Secondly, the IVF embryo is not a part of the woman's body so cannot be seen as an amputated limp. Thirdly, there are many alternative uses for an IVF embryo: it can be implanted in a uterus, frozen for later use in infertility treatment, donated to another couple (at present not legally allowed in many countries), or simply destroyed and discarded.

23.3.4 Ownership, Control, and Commercialization of Human Biological Material

As stated earlier, even if it is your own body, you do not strictly own it according to most legal systems. You are not allowed to sell the whole body or part of it. In some countries commerce is allowed with sperm, human eggs, and tissues as well as cells but buying and selling of human organs for transplantation is forbidden in nearly all countries.

23.3.4.1 The Transplantation Paradigm

There are many views to support this nonownership with regard to the human body. One common argument is that it is *degrading* to do commerce with body parts; it is contrary to *human dignity*, etc. However, everyone agrees that an individual should decide on what will happen to his or her body and body parts. In transplantation it is the will of the deceased (before death) which is the most important (although in some cases the relatives might have a greater say). In all cases of living donations (whether of organs, tissues, or cells), it is the individual who decides, unless they are, for example, a minor, or unconscious, etc.

There are, however, severe limitations in the area of autonomous choice in the donation of organ transplantation. Suppose someone, while still living, signs up for organ transplantation after death. There are many things that we do not let this person decide. The donor-to-be is not allowed to specify the race, political opinion, gender, etc., of the future recipient. Some may question why; Would it be unethical to donate organs only to persons with, for example, similar political views? Would it be morally worse not to donate at all?

In organ donation from a dead person, the only decision for the donor is whether to donate or not; the rest is up to the hospital (or in the case of cells for tissue engineering, the scientists) to decide. In the *transplantation paradigm* there is no influence or control of the future of your former cells (you may choose which organs to donate and if your organs may be used for other medical purposes if not suitable for transplantation).

23.3.4.2 The Research Paradigm

If you donate cells or tissues for use in tissue engineering, what kind of influence or control are you entitled to regarding the future fate of your (former) cells? It seems that there are two conflicting ways of thinking at present. One has been discussed earlier. The other may be called the *research paradigm*.

The idea propounded in the Helsinki Declaration is that donation should be done freely after full informed consent. It is, of course, far from clear what "full and adequate information" is. The usual understanding is that it should include information about what kind of research is to be done, what the aim of the research is, and so on. Hence, the first difference from the transplantation paradigm is that you can choose the kind of research you donate to; in the transplantation paradigm the only choice is to donate or not to donate.

Sometimes there is a second difference as well. The donor of material may have a say in the future of the donated material. The cells or the tissue may not be used for research other than the one covered by the donation consent. The donor may revoke their consent and claim the material "back" [6]. This idea has actually been incorporated in the Swedish Biobank Act, although it is somewhat vaguely phrased. The donor may at any time request that his or her donated material be destroyed. The head of the Biobank may then decide whether to do this or to make the material nonidentifiable (how this can be done if the DNA is still there is indeed a mystery).

To summarize then, in the case of the transplantation paradigm, at organ (or tissue or cell) donation for transplantation, the control over the material is transferred to the recipient, and the donor has no more say over his or her former cells. The research paradigm and some of the ideas underlying it in the Helsinki Declaration rest on the idea of the *extended body*. The materials donated for research (e.g., cells and tissues) are still part of the donor when they are cultured and experimented upon in the laboratory (it is part of your autonomy and bodily self-determination to have a continued say in what happens to these (former) cells and/or tissues). Many might think that this analogy is misleading and false, and it can also complicate life for the scientists who work with the donated material. While there should be regulations in place, they should not give the donor a veto in the use of the material.

23.3.4.3 Commercialization of Cells and Tissues

Many companies have been set up for commercial exploitation of tissue engineering based on donated human cells and tissues. Can this tissue-engineered product be sold for profit? Should the original donors be compensated in financial terms? When, if ever, are the donors' interests in the products engineered from "their" cells to be considered void?

In both the transplantation and the research paradigm, the altruistic and nonfinancial character of the donation is often stressed. Donors should not have a part of the future profit or any financial compensation. Whether right or wrong, this is often deemed unethical. A better argument for noncompensation may be mustered from the *authorship criteria for scientific articles*. Usually there are many contributions made by various actors in the production of a scientific article. Some contributions are considered so small that it is enough to get a "thank you" in the article. Some are considered so important that the person should be on the authorship list. Basically, an author should have made a *significant intellectual contribution* in the production of the article [7]. We can translate this to the donation process: *only those who make a significant contribution to the production of the resulting tissue-engineering products have a right to be compensated*. The contribution of cellular material is not significant enough.

One school of thought is that the transplantation paradigm is preferable to the research paradigm. Hence, the donor gives away his or her material and then *the control is transferred* to the recipient (a research group, a university, a company). There is no ownership transferred if we do not believe in the ownership of human biological material. In Europe, such ownership is not granted. Obviously, if a company does not own the human biological material it cannot be sold [8]. Indeed, just receiving the donated material for free and then selling it would be unfair to the donors.

There are, however, other ethically acceptable ways that a profit may be made. One is to put in intellectual effort, knowledge, and other resources to develop a new product. It is a common idea, going back at least to the British philosopher John Locke, that ownership can be based on the labor put into the process of extracting something from nature. It is thus fair that researchers and companies can make a profit from products *produced* from donated cells. Making profits from services provided is, of course, also acceptable.

There may be a problem if the donation was to a public institution, say a university, and the cells (or the products) are then transferred to a company and a profit is made. This is a problem relating to the public/private divide and there is no general answer. In Sweden, for example, the government and parliament, who basically pay for research at the universities, allow scientists to create companies and commercialize their products, where the companies get the profits. The basic idea is that this arrangement is good for society. In other countries, there may be different rules.

23.3.4.4 The Morality of Intellectual Property Rights in Human Biological Material

One may believe that there are no *special* moral problems related to intellectual property rights in biological material in general and human biological material (removed from the body) in particular. (As long as we legally recognize ownership over animals, we should not get upset if there are patents in animals.) Intellectual property rights, like patents, are smaller bundles of rights and entitlement, and do not give the holder the same right as the owner. Patents are furthermore limited in time. Patents in human biological material permanently removed from the body pose no *special* problems [9].

There are, however, *general* problems relating to intellectual property rights. The first is related to who should be the owner of the patent—Is it the researcher, or the company, or university that the researcher is working in? Is it morally appropriate that it is part of the employment contract that an employee signs away his or her intellectual property rights to the company? The rules for being an inventor/researcher are similar to *the authorship criteria for scientific articles.* The intellectual property rights themselves are, however, more analogous to copyrights. A copyright can be sold and bought; authorships hopefully are never sold and bought. It seems hard to find a unique position based on ethical arguments on the question where the ownership of the patent should be placed.

Another general problem is what you are entitled to do given the patent. A patent is a temporary monopoly—only the patent holder has the right to exploit the invention commercially under the period when the patent is valid. A monopoly means that there is only one provider of the product or service. Can any price be charged? It is possible to develop an argument for an upper limit of that price. The argument runs as follows: First, there is a kind of consequentialist reason for the patent system. Without the patent system, certain useful products will not be developed [10]. Second, the monopoly price should only be enough to encourage investment and research resulting in the new product. Excess pricing is bad from the point of view of the public.

There should, therefore, be a limit to the price for the product and service provided. This is obviously more important in the health sector than it is in the beauty sector. The problem is similar to the pricing of pharmaceutical drugs.

If we look at the legal situation things get very complicated. There are different patent systems across the world. In the United States "anything under the sun" can be patented according to a Supreme Court ruling. This is not the case in Europe. Furthermore, an American patent can be challenged at any time in an ordinary court; European patents are more protected once they have been decided. In the biotech industry patents are very important. The Court of Justice of the European Union recently decided that human embryonic stem cells cannot be patented, because they are too close to the human embryo [11]. It may prove difficult to develop therapies without such patent protection but on the other hand there seems to be possibilities to patent other steps in the procedure. In the United States there are patents on embryonic stem cells. iPS cells have been granted patents both in the United States and Europe.

23.4 NEW TECHNOLOGIES: NEW POSSIBILITIES AND NEW DANGERS

New technologies introduce new risks and new responsibilities. For example, before the advent of fast and reliable transportation and the storage of food, it was not the responsibility of Europeans or North Americans to relieve starving people in Africa; indeed, it was not possible to do this. Today it is possible, and hence it is our responsibility to do what we can.

New technologies may thus enable us to help where it was previously impossible. It may also relieve us of some direct personal responsibility, and it may help us to avoid morally bad acts. In the following text, I will give two different examples where successful tissue engineering may produce a moral gain and dissolve moral issues.

23.4.1 Organs and Tissues for Transplantation

One objective of tissue engineering is to be able to produce human organs originating from human cells. The ability to produce organs and tissues will drastically change our lives. One obvious area is the possibility to remove the scarcity of organs for transplantation. Replacements can be made in time, without having to wait for a medical crisis; it would do away with the ugly issues of distributing scarce organs to waiting patients in desperate situations; and it would do away with the ethical problems of organ donation. Some people prefer to be buried with all their organs still in their body and this can then be realized without hurting someone else. (At present, if a potential organ suitable for transplantation is buried along with the person, at least one potential recipient of that organ is harmed.)

Such a regime of producing organs and tissues for transplantation would be a good thing. It will save lives and improve quality of living. It would also do away with an even more serious moral problem that we seldom discuss—that of living organ donation. Some organs can be donated while we are still alive (e.g., a kidney). As we have two kidneys, the risk of removing one is not very high, while a life without any functioning kidneys at all is very difficult—and usually shorter. The dilemma, then, is should I join a donor scheme? Should you? However, if there were other sources available for kidneys we would not have to face this moral dilemma. Technological development produces ethical problems, but may also do away with others.

New problems may however be ahead. The cost of transplantation is at present regulated by the scarcity of organs. If organs can be grown, much more transplantations can be done. This will be particularly difficult for welfare states, where transplantation is paid for by tax money. Should we have an age limit, giving priority to the young ones? Should rich people be allowed to pay for grown organs, thus buying themselves a longer life?

23.4.2 Cultured Meat

If it will be possible to grow human organs, obviously animal organs may be grown as well. Why kill a pig in order to eat some pig muscles when the muscles can be grown by tissue engineering? Killing sentient beings, i.e., beings that can experience pain and suffering, have wishes, etc., can be seen as a morally serious issue (Singer, 2011).

Some might say that it will be cheaper to feed the whole pig and then slaughter it than to tissue engineer the various edible parts. A simple guess is that when this technology develops it should prove cheaper to supply these nutrients and energy than to slaughter a complete pig (with a lot of inedible parts). Furthermore, even if tissue-engineered meat turned out to be more expensive, you might argue that it can be worth paying extra for ethical reasons.

There are animal ethics reasons for cultured meat and there are also environmental issues. Recently, there has been an increased awareness that conventional meat production, in particular beef, makes a big contribution to greenhouse gases. A preliminary life cycle study on cultured meat shows quite favorable results compared with conventional meat (Tuomisto and Teixeira de Mattos, 2011). Furthermore, there is at present quite a growing demand for meat, especially beef, which it can be difficult to meet.

The technology of cultured meat is in many ways a conservative technology. You can still eat your beef; it is only the case that the beef had been produced in another way than growing inside an animal who is slaughtered to get the beef. Inducing people to eat cultured meat instead of going vegetarian—which is another alternative—is probably much easier (Welin, 2013).

23.4.3 CRISPR/Cas9, Genome Editing, and Gene Therapy

There is much interest in engineering tissues and organs for use in medicine, but there is also need for possibilities to repair and correct genomic mutations that have bad effects. Clustered Regularly Interspaced Short Palindromic Repeats (CRISPR) is a newly discovered part of the immune system in bacteria and archaea. It serves as a defense against horizontal gene transfer. A CRISPR consists of a series of conserved repeated sequence with nonrepetitive "spacers" placed in between (Zhang et al., 2014). The spacers are incorporated from invading foreign DNA and serve as templates to produce very specific RNA, which in turn guide the associated CRISPR Associated Sequences protein that will cleave the foreign DNA in a specific place.

The CRISPR/Cas9 system can be used for genome editing by making it possible to precisely manipulate very specific genomic element. It has proven to be very effective in inducing mutations. Zhang et al. (2014) reports, for example, that a very specific mutation was created by injection of CRISPR/Cas9 in the embryo of zebra fish. Similar trials have been conducted in drosophila and various bacteria as well as in plants. The claim is that CRISPR/Cas9 systems are much more effective in inducing specific genetic changes in embryos than other ways of inserting genes. There are, however, risks that there will be off-target mutations too and various ways to reduce that risk have been tried (Sander and Joung, 2014).

The system has also been used to correct genetic deficiencies in human cells. As reported in Lokody (2014), CRISPR/Cas9 systems corrected a mutation in the cystic fibrosis transmembrane conductance regulator (CFTR) gene. The defect gene could be deleted and a functioning one inserted. This opens up for interesting therapeutical applications to humans.

One such application would be gene therapy targeted against cystic fibrosis. However, there is probably a long way to go to manage to get an effective somatic gene therapy to work. The general problem is to be able to transfect all the relevant cells in the body with a defect gene. There are also risks associated with massive transfections. A more promising target is probably some blood disease, where the cells can be targeted and transfected outside the body ex vivo. After manipulation by a suitable CRISPR/Cas9 system, the blood cells could be transferred back to the body. Similar procedures could be envisaged for bone marrow, etc. From an ethical point of view this could be beneficial, and given that the risks are reasonable and informed consent is obtained such procedures could be tried in clinical trials.

CRISPR/Cas9 technology opens up for a more troubling application to humans, namely, germ-line gene therapy. Somatic gene therapy, as the case mentioned earlier, does not pose any special ethical concerns. In somatic gene therapy only somatic cells are targeted and such a procedure is from an ethics point of view completely similar to other somatic treatments. There can, of course, be a risk that the specific gene therapy will affect the germ-line but at least the ex vivo procedures seem safe from this perspective. Germ-line gene therapy on the other hand targets the germ-line, say injected into a human embryo. The genetic change brought about in Zebra fish discussed earlier was such a germ-line gene therapy. It is much more effective and that is essentially what is used to produce transgenic animals. But there are special problems with germ-line on humans. One way to phrase it is that changes in the human germ-line and hence in the human genome infringes on the "integrity of the human genome."

In animal research, if the transgenic animal produced in the procedure has a serious (not intended) defect it is killed. This we do not do with humans. It would be morally wrong to kill a born child for this reason—or for any reason. What about doing the germ-line gene therapy in a human embryo—most probably ex vivo in conjunction with IVF—and then implanting the embryo in a woman, let it develop, and do prenatal diagnostic to see if the development is satisfactory.

Or perhaps we can do it even better. Think of cystic fibrosis. A one-cell human embryo with two defect genes is transfected with a suitable CRISPR/Cas9 system and the genetic defects are (hopefully) corrected. The embryo could develop up to at most eight-cell state. Then one cell can be removed and investigated for the status of the CFTR gene. If correct, the embryo will be implanted and hopefully a healthy child will be born.

Many ethicists including me would be in favor of this germ-line procedure for cystic fibrosis. Those who oppose must point to the violation of some principle, for example, that it is morally wrong to change the human germ-line. The position of most ethicists can be stated as that it is admissible to change the human germ-line in order to correct pathologies.

What about other changes? Suppose we know the genetic background of male baldness. This is not considered a disease but have some drawbacks nevertheless. No American president has been bald since Ike Eisenhower. So if you have plans for your child to become an American president, you better try a CRISPR/Cas9 procedure if it is a male embryo.

In general, the CRISPR/Cas9 technology opens up for many directed genetic changes of an embryo. I feel disturbed by that prospect. One way to limit the changes is to stick to the idea of curing an illness but unfortunately the concepts of disease and illness are not quite clear. First of all, conditions formally regarded as disease such as homosexuality are now in the Western world simply regarded as a normal variation. And some conditions, like strong shyness, were formerly just a personality trait but is now a pathology. So where do we draw the line? Who should draw the line?

One thing is clear. It is not easy to rule out such embryonic changes by taking the children to be born into considerations. Once the child is born and is old enough to complain about its constitution, we can simply answer that if these things had not been done he or she would not exist. Most people believe very strongly in the idea that one descends from just a particular fertilized egg with a specific genetic set up. Had the embryo been different it had not been the same person that had been born. This is, for example, the intuition shared by many people with disabilities, where the condition is genetic. They may say: "I am happy that my mother did not do preimplantation genetic diagnosis and had chosen another healthy embryo. Then I would not have existed."

These questions bring us deep into metaphysics but those who want an alternative to the common idea of our embryonic origin can consult Derek Parfit for ideas (Parfit 1984).

23.4.4 Synthetic Biology

Application of engineering methods to biological systems offers possibilities of standardization and more control. There is strong interest in applying ideas from computer science to produce "genetic switches" and to control the development of cells and elementary organisms. More specifically, the hope is to develop a kind of tool kit by using biological parts in new ways. Another line of work has been the search for minimal genomes. The most far reaching possibility is to be able to create new organism, not from cells, but from assembled nonorganic material combined with say some DNA strands.

This last possibility raises many biosecurity issues. Today, dangerous biological materials like some highly pathogenic viruses can be kept under control in high security laboratories. We can restrict who can work with such material and how it can be distributed. This is, of course, not absolutely secure—nothing is!—but there is a fair chance of keeping such material out of the hands of terrorists. If similar dangerous pathogens can be bioengineered from other material not under such strict control, then we simply cannot control the production of the pathogens. In a report from the ethics group of the European Union on synthetic biology, the group stresses the precaution of only working inside laboratories and to ensure that the new organism should not be able to survive on its own outside (European Group on Ethics in Science and New Technologies, 2009).

The more futuristic possibility of engineering completely new organisms raises many concerns. From the ethical point of view as outlined earlier, there is no general ethics of life. There have been attempts to create such an ethics, but there is absolutely no consensus on that. That said, it seems that the fact that an engineered device is alive does not matter from an ethical point of view. If that organism is sentient, then there is really an ethical issue. Even more so, if that sentient organism turns out to be a person. A "person organism" can for ethical reasons not be destroyed (killed) and if it is self-replicating, one may argue that our moral concerns stretch out to all its descendants. It is the nightmare of Mary Shelly's Frankenstein (see Box "Science that Shock the World") coming true.

A good advice is to stay clear of all sentient organisms in the future and at all costs void creating person-like synthetic beings. Once such person-like organisms are created, all the moral obligations we owe to persons come into force. We should not kill them, but we should respect their autonomy and if they want we should allow them to reproduce. Creating such person-like organisms that are able to reproduce is an irreversible act, if we adhere to our ethics.

Is this a real risk? Can we ever produce such person-like organism with abilities to think, reason, and having emotions? My view is that nature has already provided us with a proof of principle. It is possible to create person-like organisms (namely, humans and higher apes) from "simpler" biological building blocks.

23.5 SOME QUESTIONS FOR THE FUTURE

- Tissue engineering and generative medicine will prolong life, how will our societies cope with that? Will the Earth be overpopulated, will we be forced to have restrictions (permission) on conceiving a new child? What world will it be if there are many simultaneous generations living?
- Can we cope with a situation where the west will get healthier and live longer and the poor of the world will be denied that?
- Should those who pursue tissue engineering, generative medicine, and synthetic biology take responsibility for the future?

23.6 NOTES

1. Unfortunately there are many such scandals. The Tuskagee syphilis study ending in 1972, where poor black men were studied, not informed, and not treated, and the Human Radiation Experiments, where the effect of radiation was tested on unaware subjects, are among the most well-known from the postwar era (e.g., see Faden et al., 1996).

2. President Bush allowed American researchers to use federal funds when they worked with human embryonic stem cell lines established before August 2001, the time of his decision, but not use federal funds to develop new ones. In spring of 2005, American researchers in the private sector were allowed to develop new human embryonic stem cell lines. To make things more complex (at least for non-Americans), the Federal Drug Administration is prepared to review all kinds of stem cells—whether fulfilling Bush's criteria of acceptability or not—on an equal basis.
3. In the middle ages the Catholic Church, following Aristotle, regarded the quickening of the fetus as the significant point, not the conception.
4. There have been arguments that it is only some time after birth that the newborn child is sentient enough to count as a *human person*. Killing a child before that time would not be the same evil as killing an older person (see Kuhse and Singer, 1985).
5. Actually, in most of the early regulations on research on human embryonic stem cells this "surplus" character of the extra IVF embryos was regarded as ethically more acceptable than, for example, creating an embryo for research. This was the case in the Swedish guidelines in 2001. This has been changed to allow for therapeutic cloning.
6. This was the situation for Swedish human embryonic stem cell lines before the Biobank Act. Permission was granted by the Institutional Review Board for culturing for a limited time and then new permission had to be sought from the donors, who could refuse.
7. The rules given by the International Committee of Medical Journals Editors can be found at http://www.icmje.org/#author. Another important point is that all authors should read and approve the final version.
8. An in-depth treatment of the commercialization issue from an American point of view is given in Resnic (2002).
9. Patents in IVF embryos, donated organs, etc., are problematic, however.
10. A stronger consequentialist argument is perhaps that abolishing the patent system today may give rise to adverse effects. It is hard to know what would have happened if the patent system had never been introduced.
11. Decision by the Court of Justice of the European Union October 18, 2011. Available at: http://curia.europa.eu/jcms/upload/docs/application/pdf/2011-10/cp110112en.pdf (accessed 24.02.14.).

SOME ETHICS THEORIES

There is a wide variety of theories of ethics with both normative statements and values. On the foundation of ethics one distinguishes between *deontological* and *consequentialist* theories. There are also ethical theories that regard ethics as being based on *agreements* under certain conditions (Figure 23.1).

Deontological theories are often characterized by "right" precedes "good." One can do the right action (the duty) without looking at the consequences (the "good"). Some deontological theories seek the foundation for ethics by rationally deriving norms for action. An example is given by the various versions of Kantian ethics. By pure reasoning every rational being can (in principle) find out what their duty is in various circumstances. Such an approach means that our emotions and feelings are not considered important in ethics: it is our rational capacity that gives us a moral.

An example of a consequentialist theory is *utilitarianism*: "Always act in such a way that your action produces as much good in the world as possible." This is a version of "the Greatest Happiness Principle" originally pronounced by the British philosopher Jeremy Bentham. There are no absolute duties; you always have to consider the consequences before acting.

FIGURE 23.1
Ethics by agreement.

Continued...

SOME ETHICS THEORIES *Continued*

FIGURE 23.2
Dividing the cake: distributive justice.

Furthermore, there are various concepts of the "good." Some say the good is pleasure (hedonists), others think fulfillment of preferences is good, still others would pronounce self-fulfillment, etc. A well-known representative of a kind of preference-utilitarianism today is Peter Singer. Recently the British philosopher Derk Parfit has made an ambitious attempt to combine some of the Kantian ideas with a consequentialist theory by placing *reasons* instead of preferences in the forefront but still relating to the outcome of actions (Parfit, 2013) (Figure 23.2).

In the consequentialist tradition sentient beings are included in the moral deliberations. Already Bentham stated that the important thing was not if they can think or reason but if they can suffer. A new development in the broad concequentalist tradition is the idea that perhaps humans are special—they are *persons* with certain mental and emotional capacities. A somewhat different ethics applies to persons than to "merely" sentient beings. While the moral considerations in relation to merely sentient beings are basically "do not cause suffering," the consideration toward persons includes respecting autonomy and do not kill. There are animals that obviously are persons in this sense, like the great apes. How should we treat them?

Basing ethics on *agreements* can be done in two ways. One is hypothetical agreement in a hypothetical situation. A well-known example is John Rawls' *Theory of Justice* as a result of rational and prudent agents agreeing on basic principles for society in a situation where they are aware that they have interests, but do not know the details nor basic facts about themselves. A prudent person, not sure whether they are a man or woman, will not endorse principles that discriminate against either gender.

Another idea is to base ethics on actual agreements reached in a free exchange of ideas without threats or power involved. This is basically the idea of the discursive ethics, developed by Jurgen Habermas. It is based upon agreements between actual persons, but in a very special situation.

Some prefer to view the agreement as something that has evolved as mutually beneficial rational agreements between persons pursuing their own interests. One example of this view is held by David Gauthier, a Canadian–American philosopher.

It is noteworthy that all the views presented in this fact box so far put special emphasis on consciousness (consequentialists often emphasize on all kinds of consciousness; i.e., sentient beings), deontologists and agreement proponents often emphasize on human consciousness. The nonconscious world is a matter of ethics only in so far as it has effects on sentient beings

SOME ETHICS THEORIES *Continued*

and/or humans. There have been various attempts to develop ethics encompassing either all living organisms or maybe everything. Arne Naess and "deep ecology" is an example of the first. Some examples of environmental ethics may exemplify the "ethics of everything."

Which view of all the ones presented here might be the correct one? There is no agreement, nor will there ever be. More and more work in bioethics, applied ethics, and political philosophy might be characterized as developing a *political conception of ethics*, producing principles and values acceptable from a wide variety of more basic ethical theories.

SCIENCE THAT SHOOK THE WORLD

The First "Test Tube Baby (1978)"

Louise Brown was born in July 1978. She was the first child to be born from IVF treatment. Her mother had blocked Fallopian tubes, which means that the eggs cannot pass down from the ovaries and be fertilized. Dr Patrick Steptoe and Dr Robert Evans, in the UK, had been working on methods for IVF for many years. Some years earlier they had managed to fertilize eggs outside the female body, but they had not succeeded in having a pregnancy sustained for more than a few weeks. They had made about 80 such attempts (Figure 23.3).

In the unsuccessful attempts, the physicians had cultured the fertilized eggs until they had evolved into 64 cells. This time they implanted a younger fertilized egg after 2.5 days. The pregnancy continued normally, but some days before the expected birth the mother's blood pressure increased and Louise Brown was delivered by a caesarean section.

This first instance has been followed by many more trials, and IVF is now an established standard treatment. The first apprehended risk of health problems for the IVF babies can now be disregarded; the remaining problems seem to be due to a higher prevalence of twin births in older mothers. However, it is worth noting that there were no regulated clinical trials—this development took place completely as clinical treatment.

Remaining ethical issues have to do with what happens to the fertilized eggs which are not implanted. They can be thrown away, frozen for later use, or donated for research, including derivation of human embryonic stem cells. In many countries IVF is heavily commercialized.

Baby Fae: Receiving a Baboon Heart (1984)

In October 1984, a baby girl was born, known in the literature as Baby Fae. She had a lethal underdeveloped left side of the heart. Most babies with this problem die shortly after birth. Dr Leonard Bailey at Loma Linda University Center, USA, decided to try to transplant a baboon heart in Baby Fae. Although heart transplants were successful in older people, there had been no successful infant-to-infant heart transplant (although this did happen 1 year later).

Baby Fae lived for a few days with the baboon heart. She died from massive organ failure when her blood started to clump together. The Baby Fae transplantation had been controversial from the start. First, the common scientific opinion was that the possibility of success was very low. Second, it had been questioned if and how sufficient was the information given to the mother of the baby girl. Third, a healthy baboon had been sacrificed even though the possibility of success was extremely low.

Xenotransplantation (cross-species transplantation) still remains an active field, although with less optimism than just before the year 2000, even if it has not yet hit the wider clinic community. Some cell transplantation protocols have been approved and executed, but no whole-organ transplantation is in sight. The field is still struggling to overcome rejection. It has also been discovered that there is a risk that the endogenous retrovirus in the pigs (the preferred "donor") may infect humans. To avoid the worst rejections the pigs must be transgenic. It is fair to say that transgenic pigs at present are not good enough to avoid rejections. (Rejection is studied in baboons with transplanted transgenic pig organs.) The most hotly discussed ethical issues are

Continued...

SCIENCE THAT SHOOK THE WORLD *Continued*

FIGURE 23.3
A test tube baby.

related to animal rights and to the risk of spreading a disease by transplanting pig organs. Some are worried that xenotransplants will also affect our "humanness."

Xenotransplantation was conceived as one way of overcoming the shortage of human organs for transplantation; to tissue engineer organs from human cells—preferably cells from the recipient. This will avoid most of the problems related to xenotransplantation.

The Cloning of Dolly (1996)

Few scientific reports have affected the public more than the news of the cloning of a sheep. Making a genetic copy of an adult mammalian was believed to be impossible, but, nevertheless, Ian Wilmut and his colleagues at Roslin Institute, UK, managed to do it. They took a mammary cell from a Finn Dorset sheep and transferred its nucleus to an enucleated

SCIENCE THAT SHOOK THE WORLD *Continued*

egg from a Scottish Blackface sheep. A tiny pulse of electricity made it start to develop. The egg was transferred to a surrogate mother, the egg donor, and Dolly was born after a normal pregnancy. Dolly is a genetic copy of the Finn Dorset sheep where the cell nucleus came from (Figure 23.4). (Her mitochondrial DNA came from the donor of the egg, however.)

Wilmut and coworkers performed 276 attempts before they succeeded. Since Dolly, many other mammalian species have been cloned. The birth of Dolly immediately started a big public discussion on cloning. In the United States, the President asked the National Bioethics Advisory Board to prepare a report on cloning. They claimed that cloning should not be ruled out from basic ethical principles, and should not be undertaken at this present time. The risk to the child is too high. Various health problems have been seen in the other cloned mammalians and Dolly developed arthritis at a surprisingly young age. She is now dead.

Those who favor human cloning tend to see it as a prolongation of IVF. It started with "conventional IVF," as in the case of Louise Brown. The next step was the technology of injecting a single sperm into the egg in the case when the male has very little or nonmobile sperms. Cloning—according to this line of thought—would just be a case of IVF with no sperm at all. (The idea is to use an adult cell from the man and inject its nucleus into an enucleated egg from the woman, although other combinations do exist.) Most critics focus on health risks for the child and on the "unnaturalness" of the procedure. It is less frequently discussed that the cloning technology may give a woman the opportunity of having children completely on her own. A woman has both cells and eggs.

FIGURE 23.4
A dilemma of clones.

SCIENCE OUT OF CONTROL

A recurrent theme in many novels about science is that science becomes out of control—humans create something that cannot be controlled. Here are some examples.

Mary Shelley: *Frankenstein* (1831)

Victor Frankenstein, a scientist, creates a living creature in his laboratory from cadaveric human material from the newly buried dead. The new creature has many human traits. He is alone and demands that Frankenstein creates a woman for him. This is done, but then Frankenstein becomes afraid—he has created a new race, capable of procreating and multiplying, and perhaps even destroying humankind itself. He kills the female creature. The original monster promises revenge and threatens to kill Victor Frankenstein's future bride. This is done on the eve of the wedding. Frankenstein vows to kill the monster and the case takes him to the North Pole.

This is a remarkable—and still readable—book, written by the young Mary Shelley to entertain her poet-husband Percy Shelley, the poet Lord Byron, and their friends. The metaphor of "Frankenstein's monster" is still very alive, e.g., in the term "Frankenfood."

Aldous Huxley: *Brave New World* (1932)

The novel was inspired partly by a book by the biologist and friend—J.B.S. Haldane's *Daedalus or Science and the Future*. Huxley's novel is set 600 years after Ford—or Freud as he was mysteriously called when he pronounced on psychology.

In Huxley's tale we are taken to the London central hatchery, where human embryos are "grown" on an industrial scale. Each fertilized egg gives rise to up to 96 identical individuals. The embryos are treated differently to create different people. Some are heavily radiated (the latest thing in the 1930s!) to lower intelligence. Society consists of different castes of humans and each has learned from earliest childhood to love their position. The world functions without friction. Nobody has to be unhappy—there is a drug "Soma" which keeps them happy.

However, there is a reservation in the wilderness where people live "natural lives." The hero of the book, a member of the highest caste who is groomed for world leadership, is permitted to go into the wilderness. There he meets people who live differently to him, and have strange feelings and aspirations. He finds a woman at the reservation who was originally from the civilized world. She and her son are taken back to London but neither can adjust. The best part of the novel is when "his Fordship" himself, one of the world governors, discusses the rationale behind the smooth scientific running of the world.

Huxley's dystopia does not feature individual crazy scientists—as in Shelley's *Frankenstein*. Instead he paints a picture of a—at least in principle—benevolent society based completely on science. In that world, the individual is nothing and everything is for the collective happiness of everyone. It is a *utilitarian nightmare*.

Margaret Atwood: *Oryx and Crake* (2003)

This is a novel that starts after a huge catastrophe in an indeterminate future. As the story unfolds we learn that it may not be such a distant future. The cities of today are still there; however, some are abandoned after the rise of the oceans. It is a world of colossal social breakdown, where the rich live in specially guarded places and the areas outside are worse than the worst criminal downtown areas of big American cities. There is Crake, a brilliant young boy who turns to biosciences, the young girl Oryx from a poor village somewhere in Asia, and the only survivor from the catastrophe, known from the start as Snowman.

The story is told backwards, but the following summary is the "forward" version of it:

Crake advances quickly through the scientific world and is soon head of the "Paradise project" of one of the big Pharmas. Here he is working on two things. One is a "super pill" that cures almost everything, including sexual dysfunctioning. (It also makes people sterile, but that is not advertised.) The other project is the creation by gene technology of new—and better—animals and humans. In particular, humans are engineered to avoid all destructive passions, to be out of grasp of religion, etc. All these are things that Crake and his colleagues think are very dangerous and destructive. Oryx is brought in to teach the new humans, and Jim—Snowman—is an old friend of Crake and responsible for security in "Paradise."

SCIENCE OUT OF CONTROL *Continued*

The super pills cause an intended (by Crake) pandemic that wipes out most of humanity. In the end only Snowman and the new humans are left together with the new animals. It all starts over again. Maybe it is in the end rather similar to what happened to our "old" humanity. A scary book about what greed and hubris can do if combined with enough scientific talent and money.

REFERENCES

Aristotle, 1955. The Ethics of Aristotle: The Nicomachean Ethics. Penguin Classics, London.

Beauchamp, T., Childress, J., 2001. Principles of Biomedical Ethics, fifth ed. Oxford University Press, New York.

Eschenhagen, T., 2011. The beat goes on: human heart muscle from pluripotent stem cells. Circ. Res. 109, 2–4.

European Group on Ethics in Science and New Technologies, 2009. Ethics of synthetic biology. Opinion 25. Available at: http://ec.europa.eu/bepa/european-group-ethics/docs/opinion25_en.pdf (accessed 26.02.14.).

Faden, R.R., Lederer, S.E., Moreno, J.D., 1996. US medical researchers, the Nuremberg doctors trial, and the Nuremberg code. A review of findings of the advisory committee on human radiation experiments. J. Am. Med. Assoc. 276, 1667–1671.

Habermas, J., 2003. The Future of Human Nature. Polity Press, Cambridge.

Kuhse, H., Singer, P., 1985. Should the Baby Live? Oxford University Press, Oxford.

Lokody, I., 2014. Cerrecting genetic defects with CRISPR-Cas9. Nat. Rev. Genet. 15, 63.

Martins, A.M., Vunjak-Novakovic, M., Reis, R.L., 2014. The current status of IPS cells in cardiac research and their potential for tissue engineering and regenerative medicine. Stem Cell. Rev. Rep. 10, 177–190.

Miura, K., Okada, Y., Tajashi, A., et al., 2009. Variation in the safety of induced pluripotent stem cell lines. Nat. Biotechnol. 27 (8), 743–745.

Mulkay, M., 1994. The Triumph of the pre-embryo: interpretation of the human embryo in parliamentary debate over embryo research. Soc. Stud. Sci. 24, 611–639.

Parfit, D., 1984. Reasons and Persons. Oxford University Press.

Parfit, D., 2013. What Matters. Oxford University Press.

Resnic, D.B., 2002. The commercialization of human stem cells: ethical and policy issues. Health Care Anal. 10, 127–154.

Russell, B., 1998. Science and ethics. In: Rachel, J. (Ed.), Ethical Theory 1: The Question of Objectivity. Oxford University Press, Oxford.

Sander, J.D., Joung, K., 2014. CRIPSPR/Cas systems for editing, regulating and targeting genomes. Nat. Biotechnol. 32 (4), 347–355.

Singer, P., 2011. Practical Ethics, third ed. Cambridge University Press, Cambridge.

Takahashi, K., Yamanaka, S., 2006. Induction of pluripotent stem cells from mouse embryonic and adult fibroblast cultures by defined factors. Cell 126 (4), 663–667.

Thomson, J.A., Itskovitz-Eldor, J., Shapiro, S., et al., 1998. Embryonic stem cell lines derived from human blastocysts. Science 282, 1145–1147.

Tuomisto, H.L., Teixeira de Mattos, M.J., 2011. Environmental impacts of cultured meat production. Environ. Sci. Technol. 45 (14), 6117–6123.

Welin, S., 2013. Introducing the new meat. Problems and prospects. Nordic J. Appl. Ethics 7 (1), 24–37.

Zhang, F., Wen, Y., Guo, X., April 7, 2014. CRISPR/CAS9 for genome editing: progress, implications and challenges. Hum. Mol. Genet. 23 (R1), R40–R46. http://dx.doi.org/10.1093/hmg/ddu125 advance publication.

Index

Note: Page numbers followed by "b", "f" and "t" indicate boxes, figures and tables respectively.

H

I